The structure, function and development of brain systems utilizing a major class of neurotransmitters (namely the catecholamines) are described comprehensively for members of all living vertebrate classes. Much of the information presented is new. The book provides an excellent overview of the current knowledge on organization and evolution of the catecholaminergic systems of the brain and is the first to give a comprehensive, comparative overview for any single class of neurotransmitter. The final chapter synthesizes and integrates foregoing chapters emphasizing that brain catecholamines are very ancient and appear to be one of the key systems of the brain. The book will be of interest to researchers and postgraduates of neuroscience, neurobiology, zoology and physiology.

Phylogeny and Development of Catecholamine Systems in the CNS of Vertebrates

Phylogeny and Development of Catecholamine Systems in the CNS of Vertebrates

Edited by
WILHELMUS J. A. J. SMEETS
Vrije Universiteit, Amsterdam, The Netherlands

and ANTON REINER
University of Tennessee, Memphis, USA

Published by the Press Syndicate of the University of Cambridge
The Pitt Building, Trumpington Street, Cambridge CB2 1RP
40 West 20th Street, New York, NY 10011-4211, USA
10 Stamford Road, Oakleigh, Melbourne 3166, Australia

© Cambridge University Press 1994

First published 1994

Printed in Great Britain at the University Press, Cambridge

A catalogue record for this book is available from the British Library

Library of Congress cataloguing in publication data

Phylogeny and development of catecholamine systems in the CNS
  of vertebrates / edited by Wilhelmus J.A.J. Smeets and Anton Reiner.
     p. cm.
  Includes bibliographical references and index.
  ISBN 0-521-44251-6 (hc)
  1. Sympathetic nervous system—Evolution. 2. Dopaminergic
mechanisms. 3. Noradrenergic mechanisms. 4. Adrenergic mechanisms.
5. Developmental neurophysiology. I. Reiner, Anton.
QP368.5.P48 1995
596'.0188–dc20   94–8285
                                                                     CIP

ISBN 0 521 44251 6 hardback

Contents

List of contributors — xiii

Preface — xv

Part I Phylogenetic aspects of catecholamine systems in the CNS of vertebrates

1 The study of catecholaminergic perikarya and fibers in the nervous system: methodological considerations and technical limitations
A. Reiner — 1
- Historical overview — 1
- Catecholamine synthesis and metabolism — 3
- Induced fluorescence for catecholamine localization — 3
- Immunohistochemical localization — 4
- References — 5

2 Catecholamine systems in the brain of cyclostomes, the Lamprey, *Lampetra fluviatilis*
J. Pierre, J. P. Rio, M. Mahouche & J. Repérant
- Introduction — 7
- Distributional maps of catecholamine cell bodies and fibers — 7
  - *Tyrosine hydroxylase* — 7
  - *Dopamine* — 12
  - *Comparison with previous studies of catecholamines in lampreys* — 13
- Relationships of catecholamines with other neurotransmitter systems — 13
  - *Interactions with other monoamines: serotonin and histamine* — 13
  - *Interactions with peptides* — 13
- Relationships of catecholamines and known pathways — 15
  - *Sensory systems* — 15
  - *THi somata contacting the CSF* — 15
  - *Motor systems* — 15
- Comparison with Myxiniforms — 16
  - *Maps of dopamine* — 16
  - *Relationships between dopamine and other neurotransmitters* — 16
- Concluding remarks — 16
- Abbreviations — 17
- References — 18

3 Localization of catecholamines in the brains of Chondrichthytes (cartilaginous fishes)
S. L. Stuesse, W. L. R. Cruce & R. G. Northcutt — 21
- Introduction — 21
  - *Phylogenetic considerations* — 21
  - *Previous catecholaminergic research* — 21
- Descriptions of catecholaminergic cell bodies and fibers — 22
  - *Spinal cord* — 23
  - *Myelencephalon* — 30
  - *Metencephalon* — 32
  - *Mesencephalon* — 33
  - *Diencephalon* — 35
  - *Telencephalon* — 35
- Relationships between catecholamines and functional systems — 37
  - *Olfactory system* — 37
  - *Visual system* — 37
  - *Octaval and lateral line systems* — 39
  - *Motor systems* — 39
  - *Other* — 39
- Relationships between catecholamines and other neurotransmitter systems — 40
  - *Other monoamines* — 40
  - *Acetylcholine* — 40
  - *Peptides and others* — 41
- Concluding remarks — 41
- Abbreviations — 42
- References — 43

vii

4 Catecholamines in the brains of Osteichthyes (bony fishes)

J. Meek — 49

Introduction — 49
- The phylogenetic relations of bony fishes — 49
- General organization of the brain of Osteichthyes — 53
- Organization of the present survey — 54

Catecholamines in teleostean brains — 54
- Tyrosine hydroxylase (TH) — 54
- Dopamine (DA) — 60
- Noradrenaline (NA) — 69
- Adrenaline — 70
- Formaldehyde-induced fluorescence (FIF) studies — 71

Catecholamines in the brains of non-teleostean actinopterygians — 71
Catecholamines in sarcopterygian brains — 72
Concluding remarks — 72
Abbreviations — 73
References — 74

5 Catecholamine systems in the CNS of amphibians

A. González & W. J. A. J. Smeets — 77

Introduction — 77
- Phylogenetic considerations — 77
- Amphibian catecholamine research in historical perspective — 77

Distribution of catecholamines in the brain of amphibians: theme and variations — 78
- Dopamine — 78
- Noradrenaline — 91
- Adrenaline — 96

Comparison of the DA and NA innervation of brain structures — 97
Sites of origin of DA and NA fibers — 98
Concluding remarks — 99
Abbreviations — 100
References — 101

6 Catecholamine systems in the CNS of reptiles: structure and functional correlations

W. J. A. J. Smeets — 103

Introduction — 103
Distribution of catecholamines in the brain of reptiles: theme and variations — 104
- Dopamine — 104
- Noradrenaline — 116
- Adrenaline — 118
- Catecholamine containing cells with deviating immunohistochemical properties — 119

Catecholamines and functional systems — 122
- Sensory systems — 122
- Motor systems — 124

Relationships of catecholamines with other neurotransmitter systems — 124
- Interactions with other monoamines: serotonin and histamine — 125
- Interactions with acetylcholine — 125
- Interactions with peptides — 126

Concluding remarks — 127
Abbreviations — 128
References — 129

7 Catecholaminergic perikarya and fibers in the avian nervous system

A. Reiner, E. J. Karle, K. D. Anderson & L. Medina — 135

Introduction — 135
Rationale — 135
Distribution of catecholamines: perikarya — 136
- Retina – A17 — 136
- Olfactory bulb – A16 — 136
- Diencephalon – the hypothalamic A11–A15 — 136
- Pretectum – nucleus pretectalis pars medialis — 147
- Mesencephalon – A8–A10 — 147
- Rhombencephalon – the locus coeruleus complex: A4, A6 and A7 — 148
- Rhombencephalon – the noradrenergic A5, A3, A2 and A1 — 149
- Rhombencephalon – the adrenergic C3, C2 and C1 — 149
- Spinal cord — 150

Distribution of catecholamines: fibers and terminals — 150
- Retina — 150
- Olfactory bulb — 150
- Telencephalon — 150
- Diencephalon — 157
- Pretectum and mesencephalon — 159
- Rhombencephalon — 160
- Spinal cord — 162

Connectivity and function — 162
- Retina – A17 — 162
- Olfactory bulb – A16 — 164
- Hypothalamic A14 and A15 groups — 164
- Diencephalon – A13 — 165
- Diencephalon – A12 — 165
- Diencephalon – A11 — 165
- Pretectum – PTM — 165
- Mesencephalon – A8–A10 — 165
- Rhombencephalon – A6 and A4 — 170
- Rhombencephalon – A5 — 171
- Rhombencephalon – C2 and A2 — 171
- Rhombencephalon – C1 and A1 — 172

Abbreviations — 172
References — 175

8 Catecholamine systems in mammalian midbrain and hindbrain: theme and variations

K. Kitahama, I. Nagatsu & J. Pearson — 183

Introduction — 183

Midbrain CA cell groups	183
A10 cell group	183
A9 cell group	188
A8 cell group	188
DA and L-DOPA	189
Monoamine oxidase in the midbrain	189
CA cell groups in the pons	189
A6 cell group	190
A4, A5 and A7 cell groups	193
CA cell groups in the medulla oblongata	193
Dorsomedial medulla oblongata	195
Ventrolateral medulla oblongata	196
AADC-ir neurons in the subependymal layer of the spino-medullary junction and spinal cord	197
Summary and conclusion	197
Abbreviations	198
References	199

9 Catecholaminergic neuronal systems in the diencephalon of mammals

Y. Tillet	207
Introduction	207
Catecholaminergic neuronal groups	207
Caudal diencephalon	207
Intermediate diencephalon	210
Rostral diencephalon	211
PNMT-like immunoreactive neurons	221
Distribution of catecholamine-containing fibers in the diencephalon	223
Caudal diencephalon	223
Intermediate hypothalamus	225
Anterior hypothalamus and preoptic area	225
Thalamus	225
Innervation of catecholamine cell groups of the diencephalon	226
Interaction with other neuronal systems	227
Monoaminergic systems (histamine, serotonin)	227
Other classical neurotransmitters (GABA, Acetylcholine)	227
Peptidergic systems	227
Catecholamines and steroids	230
Afferents to catecholaminergic neurons	230
Functional implications	230
Catecholamines and reproduction	230
Interactions between catecholamines and opiate peptide-containing neurons	232
Interactions between catecholamine neurons	232
Interactions between catecholamines and serotonin	233
Interactions between catecholamine and GABA-containing neurons	233
Interactions between catecholamines and neurotensin-containing neurons	233
Interactions between catecholamines and NPY-containing neurons	233
Interactions between catecholamines and peptide-containing neurons of the PVN and SON	233
Interactions between catecholamines and GHRH- and SRIF-containing neurons	234
Conclusion	235
Abbreviations	236
References	236

10 Catecholaminergic innervation of the basal ganglia in mammals: anatomy and function

A. Reiner	247
Introduction	247
General basal ganglia anatomy	247
General distribution of dopamine input to basal ganglia	249
Intrinsic organization of basal ganglia	251
Organization of connections between striatum and dopaminergic midbrain neurons	254
Topography of DA input to striatum	254
Topography of striatal input to A8–A10	254
Organization of basal ganglia cell types	255
Connections between midbrain dopamine neurons and striatal neurons at the ultrastructural and single cell level	255
Dopaminergic input to the striatum	255
Striatal input to midbrain dopaminergic neurons	256
Receptors on striatal neurons and tegmental dopamine neurons	256
Dopamine uptake mechanisms in the striatum	258
Functional considerations	259
Function of dopaminergic input to the striatum: cellular actions of dopamine	259
Function of striatal input to midbrain dopamine neurons: cellular and behavioral actions of GABA and neuropeptides	260
General scheme of basal ganglia functional organization	261
Role of dopamine in overall basal ganglia functions	262
Human diseases involving dopamine and the basal ganglia	263
Noradrenergic and adrenergic input to basal ganglia	263
Noradrenergic and adrenergic input topography and LM features	264
Receptors, uptake mechanisms and functional role of noradrenergic and adrenergic input	264
References	264

11 Telencephalic dopamine cells in monkeys, humans, and rats

M. Dubach	273
Introduction	273
Telencephalic dopamine cell distribution	273
Monkeys	273

Humans	278
Rats	280
Striatal dopamine cells: anatomical context	280
Neostriatum	281
Ventral striatum	281
Telencephalic dopamine cells: ontogeny in rats	282
Marginal zone ontogeny in rats	282
Prenatal marginal zone ontogeny	283
Postnatal marginal zone ontogeny	283
Telencephalic dopamine cells: developmental significance	283
Distribution	283
Plasticity	284
Species differences in biochemistry	284
Functional comments	285
Summary and conclusions	286
Abbreviations	287
References	287

12 Comparative anatomy of the catecholaminergic innervation of rat and primate cerebral cortex

B. Berger & P. Gaspar	293
Introduction	293
General methodological remarks	293
The cortical dopamine system	294
The cortical dopamine system of rats	294
The cortical DA system of primates	299
The dopaminoceptive population	304
The cortical NA system	306
Distributional maps in rodents	306
Distributional maps in primates	307
Origin in the brainstem and collateralization pattern	307
Heterogeneity within the NA coeruleo-cortical pathway?	308
The noradrenoceptive population	309
Comparative developmental sequence of the DA and NA innervations in rats and primates	310
Interactions between the catecholaminergic innervation and other neurotransmitters	310
Functions of the cortical DA and NA system	311
Cortical DA system	311
Cortical NA system	311
Phylogenetic trends and genetic differences of DA and NA neurons	313
Abbreviations	314
References	315

Part II Developmental aspects of catecholamine systems in the CNS of vertebrates 325

13 Development of central catecholamine neurons in teleosts

P. Ekström, T. Honkanen & B. Borg	325
Introduction	325
Development of catecholaminergic neuronal cell groups and pathways	326
Relationship with the development of sensory systems	338
Relationship with the development of other neurotransmitter systems	339
Functional aspects of the early catecholamine systems	339
Abbreviations	340
References	341

14 Developmental aspects of catecholamine systems in the brain of anuran amphibians

A. González, O. Marin, R. Tuinhof & W. J. A. J. Smeets	343
Introduction	343
Notes on the development of anuran amphibians	344
Development of catecholamine systems in Xenopus	344
Late embryonic stages	344
Premetamorphic stages	349
Prometamorphic stages	350
Metamorphic climax	352
Development of catecholamine systems of Rana ridibunda	353
Concluding remarks	355
Comparative aspects of development of CA systems in the brain of amphibians	355
Temporal correlation between the development of CA systems and other developmental aspects of the CNS of anurans	357
Abbreviations	359
References	359

15 Ontogenesis of catecholamine systems in the brain of the lizard Gallotia galloti

L. Medina, L. Puelles & W. J. A. J. Smeets	361
Introduction	361
Material and methods	361
Development of catecholamine systems in Gallotia galloti	362
Spatio-temporal sequence of appearance of CA cell bodies in the embryonic brain of Gallotia	372
Transient expression of CA cell bodies in the brain of Gallotia	375
Development of catecholamine fiber tracts and innervation	375
Abbreviations	377
References	377

16	**Development of neurons expressing tyrosine hydroxylase and dopamine in the chicken brain: a comparative segmental analysis**	
	L. Puelles & L. Medina	381
	Introduction	381
	Material and methods	381
	Morphological background	382
	Development of TH/DA cells and fiber tracts	383
	Early stages (5–7 days of incubation)	383
	Intermediate stages (8–10 days of incubation)	383
	Late embryonic stages – mature pattern	388
	Segmental localization pattern	391
	Temporal patterns and the problem of migrations	395
	Comparison with other tetrapods	396
	Amphibians	396
	Reptiles	396
	Mammals	397
	Abbreviations	400
	References	401
17	**Ontogeny of catecholaminergic neurons in the CNS of mammalian species: general aspects**	
	G. A. Foster	405
	Introduction	405
	Distributional maps of catecholaminergic neurones in the developing rat brain	406
	Dopamine	406
	Noradrenaline	425
	Adrenaline	426
	Disparate appearance of catecholamine parameters	426
	Abbreviations	429
	References	430
18	**Hypothalamic catecholaminergic systems in ontogenesis: development and functional significance**	
	M. V. Ugrumov	435
	Introduction	435
	Architectonics of the hypothalamic catecholamine system	435
	Cell bodies	435
	Nerve fibers	440
	Morpho-functional characteristics of hypothalamic catecholamine neurons	441
	Genesis and morphology	441
	Expression of biochemical phenotype	442
	Genetic sexual dimorphism	443
	Afferent innervation	444
	Efferent innervation	444
	Neurohormonal control of the hypothalamic catecholamine system	446
	Hypothalamo-hypophysial (neuro) hormonal factors	446
	Hormones of peripheric endocrine glands	446
	Maternal hormones	447
	Functional significance of catecholamines	447
	Neuropeptide gene expression	447
	Sexual differentiation of the hypothalamus	447
	Adenohypophysial hormone secretion	449
	Conclusions	449
	Abbreviations	449
	References	450
19	**Sexual differentiation of central catecholamine systems**	
	I. Reisert, E. Küppers & C. Pilgrim	453
	Introduction	453
	Sex-specific development of catecholamine systems *in vivo*	453
	Sex-specific development of catecholamine neurons *in vitro*	455
	Sex-specific vulnerability of catecholamine systems during development	458
	References	460

Part III Catecholamines in the CNS of vertebrates: current concepts of evolution and functional significance 463

20	**Catecholamines in the CNS of vertebrates: current concepts of evolution and functional significance**	
	W. J. A. J. Smeets & A. Reiner	463
	Introduction	463
	Evolutionary considerations	463
	Evaluation of the techniques used in catecholamine research	464
	Comparative analysis of catecholamine systems of vertebrates using the A1–A17/C1–C3 nomenclature	465
	Caudal rhombencephalon: A1–A3/C1–C3	465
	Rostral rhombencephalon: A4–A7	467
	Midbrain: A8–A10	468
	Diencephalon: A11–A15	468
	Olfactory bulb: A16	469
	Retina: A17	470
	Non-classified catecholamine cell groups in the CNS of vertebrates	470
	Spinal cord	470
	Pretectum and habenular region	471
	Hypothalamic periventricular organ	470
	Cortex and basal forebrain	472
	Comparative analysis of catecholamine systems of vertebrates using a segmental approach	473
	Current concepts of evolution of catecholamine systems in the CNS of vertebrates	473

Functional significance of catecholamine 476
  systems in the CNS of vertebrates
Concluding remarks 477
References 478

Index 482

List of contributors

K. D. Anderson
 Regeneron Pharmaceuticals, Inc.
 Tarrytown, NY 10591
 USA
B. Berger
 INSERM U 106
 Batiment de Pediatrie
 Hôpital de la Salpêtrière
 47 Bld de l'Hopital
 75651 Paris Cedex 13
 France
B. Borg
 Department of Zoology
 University of Stockholm
 Stockholm
 Sweden (see Ekström)
W. L. R. Cruce
 Neurobiology Department
 Northeastern Ohio Universities College of Medicine
 PO Box 95
 Rootstown, OH 44272–0095,
 USA
M. Dubach
 Department of Psychiatry & Behavioral Sciences
 Regional Primate Research Center
 University of Washington SJ-50
 Seattle, WA 98195
 USA
P. Ekström
 Department of Zoology
 University of Lund
 Helgonavägen 3, Lund
 Sweden
G. A. Foster
 Department of Physiology
 University of Wales College of Cardiff
 PO Box 902
 Cardiff CF1 1SS
 UK
P. Gaspar (see Berger)

A. González
 Departamento de Biologia Celular
 Facultad de Biologia
 Universidad Complutense
 28040 Madrid
 Spain
T. Honkanen (see Ekström)
E. J. Karle (see Reiner)
K. Kitahama
 Département de Médecine Expérimentale
 CNRS URA 1195
 INSERM U 52
 Faculté de Médecine
 Université Claude Bernard
 8 avenue Rockefeller
 69373 Lyon, Cedex 08
 France
E. Küppers (see Reisert)
M. Mahouche (see Pierre)
O. Marin (see González)
L. Medina
 Department of Anatomy & Neurobiology
 University of Tennessee-Memphis
 875 Monroe Avenue
 Memphis, TN 38163
 USA
J. Meek
 Department of Anatomy & Embryology
 Katholieke Universiteit Nijmegen
 PO Box 9101
 6500 HB Nijmegen
 The Netherlands
I. Nagatsu
 Department of Anatomy
 Fujita Health University
 School of Medicine
 Aichi
 Japan

R. G. Northcutt
  Neurobiology Unit
  Scripps Institution of Oceanography &
  Department of Neurosciences
  University of California, San Diego
  La Jolla, CA 92093
  USA

J. Pearson
  Department of Pathology
  New York University Medical Center
  550 First Avenue
  New York, NY 10016
  USA

J. Pierre
  Laboratoire d'Anatomie Comparée
  Muséum National d'Histoire Naturelle
  55 rue Buffon
  75005 Paris, France

C. Pilgrim (see Reisert)

L. Puelles
  Department of Morphological Sciences
  Faculty of Medicine
  University of Murcia
  Murcia, Spain

A. Reiner
  Department of Anatomy & Neurobiology
  University of Tennessee-Memphis
  875 Monroe Avenue
  Memphis, TN 38163
  USA

I. Reisert
  Abteilung Anatomie und Zellbiologie
  Universität Ulm
  D-89069 Ulm
  Germany

J. Repérant
  INSERM U 106
  Hôpital de la Salpêtrière
  75013 Paris
  France

J. P. Rio
  INSERM U 106
  Hôpital de la Salpêtrière
  75013 Paris
  France

W. J. A. J. Smeets
  Graduate School Neurosciences Amsterdam
  Research Institute Neurosciences Vrije Universiteit
  Faculty of Medicine
  Department of Anatomy & Embryology
  van der Boechorststraat 7
  1081 BT Amsterdam
  The Netherlands

S. L. Stuesse
  Neurobiology Department
  Northeastern Ohio Universities College of Medicine
  PO Box 95
  Rootstown, OH 44272–0095
  USA

Y. Tillet
  Laboratoire de Neuroendocrinologie Sexuelle
  Station de Physiologie de la reproduction des Mammifères Domestiques
  INRA
  37380 Nouzilly
  France

R. Tuinhof
  Department of Animal Physiology
  University of Nijmegen
  Toernooiveld 1
  6525 ED Nijmegen
  The Netherlands

M. V. Ugrumov
  Institute of Developmental Biology
  Russian Academy of Sciences
  26 Vavilov Str.
  Moscow 117808
  Russia

Preface

During the last decade, a large body of data about the structure and organization of catecholamine systems in the central nervous systems of vertebrates has been accumulated by means of immunohistochemical methods, and the point has now been reached where some degree of integration and overview is desirable. Although several excellent reviews of catecholamine systems in vertebrates exist, they deal exclusively with mammals. Moreover, the available reviews often consider only one species, e.g. rats or cats, with implicit assumptions about the generality of the conclusions presented, to the extent of maintaining the view (almost as an act of faith) that such works are unequivocally of direct relevance to the study of man. Since catecholamine systems are known to be involved in human diseases such as Parkinson's disease, schizophrenia and depression, it is understandable why catecholamine research has been primarily focussed on mammals. The notion is growing, however, that there might be significant differences between the brains of rats and humans.

The main goal of the present book is to treat the topic of structure and function of catecholamine systems in the CNS of vertebrates in a wide range of vertebrate species, including representatives of all the extant classes of vertebrates from cyclostomes (e.g. lampreys) to mammals including human. Data have also been accumulated on the development of catecholamine systems in a wide range of vertebrate species. Such information is also presented in the present book. As can be seen from the list of contents, there exists a large amount of information about many non-mammalian species, as well as about at least six mammalian species from four different orders. By bringing these data together within a phylogenetic and ontogenetic framework, it may be possible to produce new insights into some of the fundamental principles governing the distribution, anatomical interrelations and integrative functioning of the diverse components of the catecholamine systems of the CNS. A better understanding of these basic principles of organization and function may, in turn, lead to new directions or hypotheses for neuroanatomical or experimental analysis, with the further possible outcome of more effective pharmacological intervention in human neurological disorders.

The initiative for preparing the present book was taken when the first editor was invited to organize and chair a session on phylogenetic and developmental aspects of catecholamine systems during the 7th International Catecholamine Symposium, held in Amsterdam, The Netherlands, 22–26 June 1992. The response to his first letter of invitation was so overwhelming that a two-day session was required. The invited neuroscientists presented a review of the organization and function of catecholamine systems of the vertebrate class in which they are specialized. All presenters agreed to prepare a chapter for this book, based to a large extent on their presentation at the Symposium.

The book is divided into three parts. In the first part (i.e. phylogenetic aspects of catecholamine systems in the CNS of vertebrates), a brief introductory chapter on methodology and limitations of catecholamine research is followed by chapters surveying the anatomy and function of the CNS catecholamine systems in each of the seven extant classes of vertebrates: cyclostomes, cartilaginous fish, bony fish, amphibians, reptiles, birds and mammals (including humans). In each chapter, the following information is given (if available): distribution of catecholamines and their synthesizing enzymes (with an emphasis on within-class variability), relationships of catecholamines with functional systems, and relationships of catecholamines with other neurotransmitter systems. To avoid redundancy, the chapter contributors were asked to minimize comparisons among vertebrate classes, and to focus on general and derived traits within the class under consideration. In the second part (i.e. developmental aspects of catecholamine systems in the CNS of vertebrates), the chapters deal with the ontogeny of these systems in bony fish, amphibians, reptiles, birds and mammals. Special attention is paid to the appearance of the first immunodetectable amounts of enzyme or catecholamine in the various catecholamine cell groups, to the maturation of cells and fibers, and to the relationship between the development of catecholamine systems and other developmental events. Again, the authors were asked to minimize between-class comparisons. In the final two chapters of this part of the book, specialized topics such as the involvement of catecholamines in the regulation of neurohormones and the sexual differentiation of catecholamine systems are discussed. In the third and final part of the book, the information about catecholamine systems in both adult and developing brains provided for each vertebrate class in the preceding chapters is synthesized and an effort is made to point out the major evolutionary implications and the basic principles of organization and function of CNS catecholamine systems.

We are well aware that this book is the product of a collaboration of many persons. First, we are grateful to the chapter contributors who have expended considerable time and effort to provide an up-to-date review of data on catecholamine systems in the various vertebrate classes. We also appreciate their willingness to accept our guidelines in order to get a book that is well balanced in text as well as in illustrations. Our work has been generously supported by the Department of Anatomy and Embryology of the Vrije Universiteit of Amsterdam and the Department of Anatomy and Neurobiology of the University of Tennessee, Memphis. Finally, the authors thank Cambridge University Press, in particular Dr Robert M. Harrington, for their support and help during the publication of this book.

Amsterdam/Memphis, Wilhelmus J. A. J. Smeets
Spring 1994   Anton Reiner

Part I: Phylogenetic aspects of catecholamine systems in the CNS of vertebrates

1

The study of catecholaminergic perikarya and fibers in the nervous system: methodological considerations and technical limitations

A. Reiner

Historical overview

Catecholamines are a group of organic compounds consisting of a benzene ring with two adjacent hydroxyl groups (the catechol portion of the catecholamine) and an opposing amine side chain (Fig. 1.1). The major catecholamines used by the nervous system are dopamine, noradrenaline and adrenaline. The latter two were first discovered for their role in peripheral cardiovascular functions and their abundance in sympathetic nerve terminals and adrenal medulla, from which they are released to produce their biological actions (Barger & Dale, 1910; Loewi, 1933; von Euler, 1956; McGeer, Eccles & McGeer, 1978). Subsequently, dopamine was also shown to be vasoactive and produce strong pressor responses (Holtz, 1939), although at that time it was uncertain whether dopamine and noradrenaline themselves had biological potency or served only as adrenaline precursors. Catecholamines were much later discovered to also be abundant in the CNS by Vogt (1954). Carlsson and his coworkers (1959) then showed that dopamine, noradrenaline and adrenaline must all play individual roles in the CNS since the levels of each did not covary in a consistent fashion from brain region to brain region. In particular, high levels of dopamine were observed in the striatum. The then recently discovered ability of reserpine to deplete catecholaminergic terminals of their catecholamine stores, and the discovery that reserpine treatment in humans led to Parkinsonian symptoms, led Carlsson to propose that the catecholamines in general played neuroactive roles in the CNS. More specifically, he proposed that striatal dopamine might be essential for extrapyramidal motor system function and its depletion might be involved in the genesis of the symptoms of Parkinson's disease. Subsequent pathological studies and investigations into the effectiveness of dopamine precursors in ameliorating Parkinson's disease confirmed Carlsson's insights (McGeer et al., 1978).

It was thus clear by the late 1950s that catecholamines were involved in neurotransmission in the central and peripheral nervous systems. During the course of these various studies on catecholamines, much was learned about the enzymes involved in the synthesis and degradation of catecholamines (McGeer et al., 1978). Central to the further elucidation of the role and importance of catecholamines in neural function was the development during the 1950s (serendipitously at first) and early 1960s of methods for localizing catecholamines relying on the conversion of catecholamines in tissue into fluorescent products by exposure of the tissue to formaldehyde (Eränkö, 1955; Falck et al., 1962). The ability to map catecholamines and to distinguish biochemically among them made it possible, for example, to con-

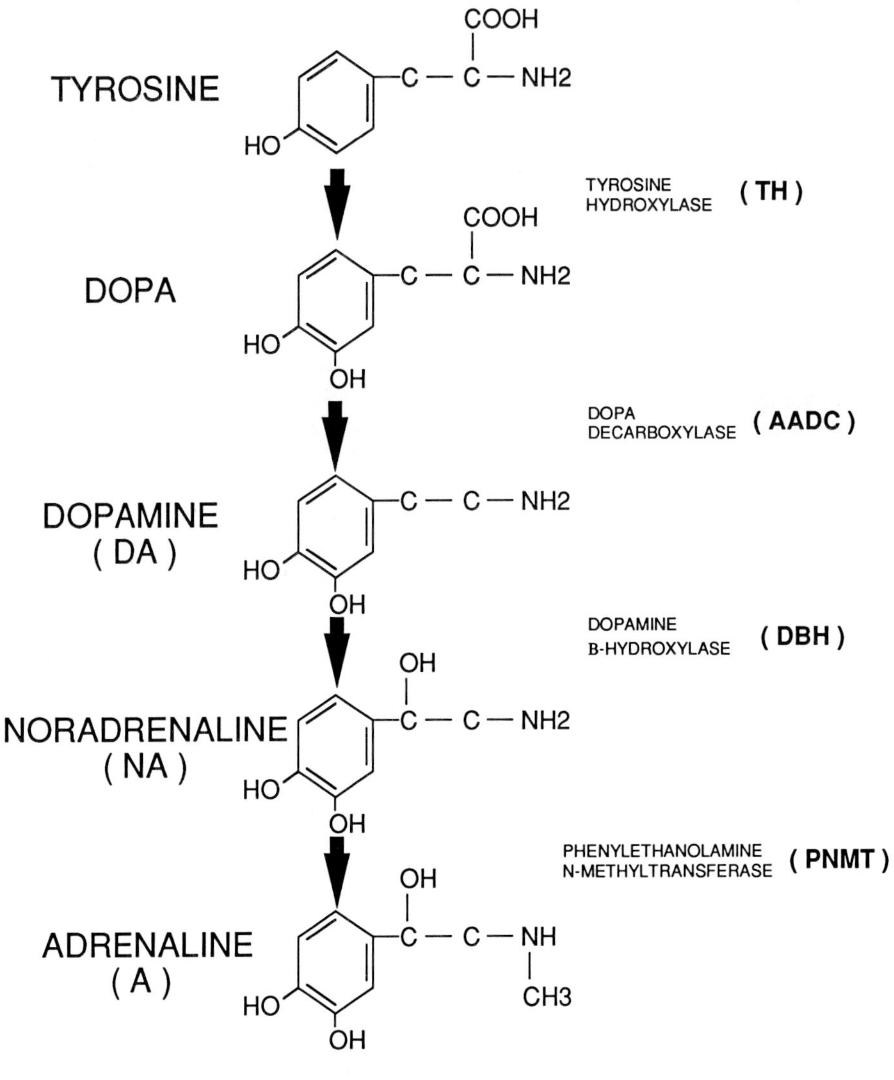

Fig. 1.1. Schematic indicating the molecular structure of the different catecholamines found in nervous tissue, and the identity of the enzymes involved in the serial transformation of one catecholamine into another, beginning with the non-catecholamine precursor tyrosine.

firm the importance of dopamine in basal ganglia function and to reveal the loss of dopaminergic substantia nigra neurons as the basis of the impaired control of movement in Parkinson's disease. The earliest methods for CA localization, however, had their limitations in the sensitivity and selectivity of what they could reveal, which hindered progress in understanding the diverse roles of the catecholamines in nervous system function. With the recent advent of immunohistochemical strategies for catecholamine localization, such shortcomings have been overcome and many new details of the organization of catecholamine systems in diverse species have been revealed. The purpose of the present chapter is to provide an overview of these different methods and evaluate current immunohistochemical strategies for catecholamine localization. We believe the superiority of the immunohistochemical labeling strategies justifies the focus of this book on data derived from the application of such strategies.

Catecholamine synthesis and metabolism

Catecholamines are synthesized in the CNS by the serial conversion of one catecholamine into another, beginning with the dietary amino acid tyrosine as the precursor (McGeer et al., 1978; Cooper, Bloom & Roth, 1986; Siegel et al., 1989). Tyrosine (which is not a catecholamine) is converted into the catecholamine dihydroxyphenylalanine (the acronym DOPA is used for this catecholamine) by addition of a hydroxyl group to the catechol ring by the enzyme tyrosine hydroxylase (TH). The enzyme TH is the rate limiting enzyme in catecholamine synthesis (McGeer et al., 1978; Cooper et al., 1986). DOPA is then converted into dopamine (DA) by decarboxylation of the amine group by the enzyme DOPA decarboxylase. Because this enzyme can also perform a similar action on other aromatic amino acid substrates, it is also termed aromatic amino acid decarboxylase (AADC). Dopamine has long been regarded as one of the major biologically active end products of CA synthesis, while DOPA has generally been regarded as merely a precursor for dopamine. Recent immunohistochemical data, however, raise the possibility that DOPA may be the terminal biologically active catecholamine produced by some neurons (Smeets & Steinbusch, 1990; Vincent & Hope, 1990). In addition to its biologically active role, dopamine also serves as precursor for the synthesis of noradrenaline (NA). This is carried out by the addition of a hydroxyl group to the carbon atom nearest the catechol ring by the enzyme dopamine B-hydroxylase (DBH). Noradrenaline is also a major catecholamine end product in the CNS, although to some extent in the CNS and greatly in the periphery, noradrenaline is converted to adrenaline by methylation of the terminal amide group by the enzyme phenylethanolamine N-methyltransferase (PNMT) (McGeer et al., 1978; Siegel et al., 1989).

Catecholamines are broken down by several degradative enzymes, including monoamine oxidase (MAO) and catechol-*o*-methyl transferase (COMT). There are two isoforms of MAO in the nervous system, termed MAOA and MAOB (McGeer et al., 1978; Cooper et al., 1986; Siegel et al., 1989). Both degrade catecholamines into their corresponding aldehyde, and MAO can also degrade noncatecholaminergic monoamines (such as indoleamines). MAO is thought to mainly be intracellular and located within presynaptic terminals and in glia, and play a major role in degrading dopamine after its uptake from the synaptic cleft (McGeer et al., 1978; Cooper et al., 1986; Siegel et al., 1989). In contrast, COMT is thought to be postsynaptic and membrane bound (McGeer et al., 1978; Cooper et al., 1986; Siegel et al., 1989). The two different forms of MAO differ somewhat in their tissue distribution (McGeer et al., 1978). Further, the two forms differ among species in their relative distributions in comparable tissues. For example, MAOB is abundant in the striatum and in the substantia nigra of cats and primates (in whom the MAOB is localized to glial cells in the nigra), while MAOA is abundant in the striatum and in the substantia nigra of rats (Fowler & Strolin-Benedetti, 1983; Westlund et al., 1985, 1988; Snyder & D'Amato, 1986; Schneider & Markham, 1987). Although MAO appears to be the major CNS degradative enzyme for catecholamines, the uptake of catecholamines into presynaptic terminals or glia that is necessary for MAO to degrade catecholamines also plays a role in terminating the synaptic action of catecholamines. This uptake is carried out by membrane bound carrier proteins in an energy dependent fashion (McGeer et al., 1978; Cooper et al., 1986; Siegel et al., 1989). The protein structure, localization and binding properties of some catecholamine carrier proteins have been elucidated in recent years (Siegel et al., 1989; Richfield, 1991; Kilty, Lorang & Amara, 1991; Shimada et al., 1991; Cline et al., 1992).

Strategies for localizing catecholamines in tissue can utilize methods that detect the catecholamines themselves, the synthetic enzymes, the degradative enzymes or the uptake complexes. Localization of receptors for catecholamines also provides insights into the tissues using catecholamines and the types of catecholamines involved in those tissues. For specifically detecting the neurons synthesizing catecholamines, however, localization of the catecholamines and/or the enzymes is required, since degradative enzymes, uptake proteins and receptor proteins may all be postsynaptic in localization. Two major anatomical methodologies have been used for catecholamine localization, formaldehyde induced fluorescence (FIF) and immunohistochemistry. The features and relative merits of these methodologies are considered in the next sections.

Induced fluorescence for catecholamine localization

Other than biochemical assays of specific tissues, the first strategies for catecholamine localization involved histochemical detection of MAO (Koelle & Valk, 1954; Glenner, Burtner & Brown, 1957; Shimizu, Morikawa & Okada, 1959; Pearse, 1972). As noted above, although this revealed the CNS regions in which catecholaminergic transmission was prominent, it did not specifically reveal the location of catecholamine neurons, fibers or pathways. During the early 1950s, the Finnish investigator Eränkö (1955) discovered that sections of formaldehyde fixed adrenal medulla were fluorescent. Similar observations were made of so-called enterochromaffin cells in the GI tract (Pearse, 1972). Although formaldehyde fixation did produce fluorescence in cells of various tissues, the method was of low sensitivity and revealed few labeled cells. Subsequently, the chemical changes induced in catecholamines (and indoleamines) that rendered them fluorescent following exposure to formaldehyde were determined (Corrodi, Hillarp & Jonsson, 1964; Pearse, 1972). At the same time, this research group also demonstrated that freeze drying tissue and then exposing it to hot formaldehyde vapors produced fluorescent labeling of many cells and fibers in diverse tissues (Falck et al., 1962). This methodological advance for catecholamine localization was followed by development of a somewhat more sensitive FIF method called the glyoxylic acid technique (Björklund et al., 1972). The development of these sensitive FIF methods then ushered in the era during which the catecholaminergic and indoleaminergic neuron and fiber systems of the brain, spinal cord and periphery were extensively mapped (Dahlstrom & Fuxe, 1964; McGeer et al., 1978; Björklund & Lindvall, 1984; Hökfelt et al., 1984; Cooper et al., 1986). These methods made it possible to distinguish catecholaminergic cells from indoleaminergic cells, since they fluoresced at slightly different wavelengths. Similar distinctions could be made for fibers, but owing to the smaller size of fibers such distinctions were at times equivocal. By combining lesions with FIF labeling, it was possible to elucidate many of the brain catecholaminergic pathways.

The FIF method possessed several limitations, however. For example, catecholamines and indoleamines were always both labeled in the tissue, unless some kind of toxic pretreatment was employed to eliminate one system. More significantly, the FIF method did not make it possible to distinguish different catecholamines by their labeling. Finally, as became evident when more sensitive methods became available, the FIF methods were not sufficiently sensitive to reveal catecholaminergic systems in all their complexity and abundance. Nonetheless, FIF methods were valuable and provided the basis for recognizing diverse specific populations of catecholaminergic and indoleaminergic cell groups in the brain (Dahlstrom & Fuxe, 1964; Björklund & Lindvall, 1984; Hökfelt et al., 1984). Twelve different catecholaminergic cell groups were recognized in rat brain and they were designated A1–12, with A1 the most caudal. A similar terminology was introduced for the indoleaminergic (specifically 5HT+, i.e. serotonergic) cell groups of brain, with individual 5HT+ groups being assigned the letter 'B' as a prefix followed by a number (with B1 being the most caudal). At the time, the FIF methods were thought to be very sensitive since they were an improvement over prior methods and because they revealed brain pathways that had not been revealed by pathway tracing methods based on silver staining of degenerating fibers (McGeer et al., 1978; Björklund & Lindvall, 1984; Hökfelt et al., 1984; Cooper et al., 1986).

Immunohistochemical localization

Immunohistochemical labeling methods became widely used to map catecholamines beginning in the 1980s. This approach became available as: 1) the enzymes involved in catecholamine synthesis were purified and used to make antisera against the specific enzymes; and 2) methods were developed for making conjugates of the catecholamines that were immunogenic and could be used to make antisera directed against specific catecholamines (Björklund & Lindvall, 1984; Hökfelt et al., 1984; Smeets & Steinbusch, 1990). It thus became possible to use immunohistochemical methods to determine which neurons contained specific catecholamine enzymes or specific catecholamines themselves. This provided a strategy for determining more precisely than possible with FIF methods the types of catecholamines produced as a biologically active end product by the various catecholaminergic neurons of the brain (Björklund & Lindvall, 1984; Hökfelt et al., 1984; Smeets & Steinbusch, 1990). For example, an investigator could determine that a neuron was dopaminergic if it contained TH and AADC, but did not contain DBH or PNMT. If this cell was also only labeled for dopamine, but not for noradrenaline or adrenaline, then it would be doubly clear that the cell in question was dopaminergic. Similarly, a noradrenergic neuron would be one that contains TH, AADC and DBH, as well as noradrenaline. Such a neuron might also label for dopamine, since in this case it would serve as a precursor for noradrenaline. Although in many cases, it has been possible to neatly categorize catecholaminergic neurons in the brain as dopaminergic or noradrenergic on this basis, classification as to the type of catecholaminergic neuron has not been straightforward in all cases in which immunolabeling is observed for catecholamines or their enzymes. For example, neurons have been encountered in brain that label for catecholamines but not for the enzymes needed for the synthesis of such catecholamines. One instance of this is the paraventricular organ in nonmammals, whose cells label for dopamine and noradrenaline, but not for TH or DBH (Smeets & Gonzalez, 1990; Smeets & Steinbusch, 1990). Further, neurons containing TH are present in many regions of the hypothalamus in birds and mammals that have not been observed to contain prominent numbers of catecholaminergic neurons by either FIF methods or direct immunolabeling for catecholamines (see chapters by Reiner and Tillet in this book). The interpretation of such a finding is as yet uncertain. It is possible that dopamine is too low in abundance to be detected in these neurons or that these neurons make DOPA as their terminal catecholamine and not dopamine. The evidence that neurons at these sites do not label immunohistochemically for AADC suggests that the latter is a likely correct interpretation (Smeets & Steinbusch, 1990; Vincent & Hope, 1990). It is possible, however, that these hypothalamic neurons actually contain DOPA decarboxylase in too low a level to be detected immunochemically and that these neurons are after all dopaminergic. Thus, in applying an immunolabeling strategy for determining the distribution of specific catecholamines in the brain, the possibilities that antisera sensitivity could vary or that enzyme or catecholamine levels could be variable, thereby leading to false negatives, need to be considered. Thus, it is not always a straightforward exercise to use immunolabeling strategies for elucidating the identities of the catecholamines synthesized and released by specific neurons of the brain. Nonetheless, immunolabeling is the best anatomical approach available and clear conclusions have been reached for many brainstem regions. Further, by combining data on receptor localization, on uptake mechanism characterization and on neuronal function, it is possible to overcome some of the ambiguities arising from the immunolabeling data alone.

As immunolabeling methods became available, it quickly became clear that they were more sensitive than FIF methods for catecholamine detection and mapping. Many more catecholaminergic cells were revealed throughout the brain and the fiber abundance was shown to be greater than realized (Björklund & Lindvall, 1984; Hökfelt et al., 1984). The difference between FIF and immunolabeling in catecholamine cell abundance was particularly striking in hypothalamus. The A1–12 system was consequently revised to A1–17, including three new hypothalamic regions, the olfactory bulb and the retina, based on tyrosine hydroxylase immunolabeling (Björklund & Lindvall, 1984; Hökfelt et al., 1984). Further improvements in immunolabeling methods led to recognition of additional diencephalic and midbrain cell groups that were then awkwardly fit in with the existing A1–17 nomenclature. Immunolabeling for PNMT revealed the presence of adrenergic cell groups overlapping some of the previously recognized noradrenergic cell groups (Hökfelt et al., 1984). These came to be known as the C1–3 groups. Further, the distinctness and separateness of some previously recognized catecholaminergic cell groups came to be seen as dubious (e.g. A3 and A7) based on the detailed mappings made possible by the more sensitive immunolabeling methods (Hökfelt et al., 1984; see Kitahama chapter in this book).

Thus, immunolabeling appears the currently best strategy for localizing catecholaminergic cells and fibers (although *in situ* hybridization histochemistry for the mRNA for the catecholamine enzymes can also be used for perikaryal localization). The immunolabeling approach is consequently the main approach used for generating the data presented in the subsequent chapters. Data from FIF studies are noted where relevant, either for historical reasons or because of the absence of other data. The following chapters will generally not note the advantages of immunolabeling for catecholamine localization, since that has been the purpose of this chapter and we wish to avoid undue repetition of this technical matter.

References

Barger, G. & Dale, H. H. (1910). The presence in ergot and physiological activity of B-imidozolylethylamine. *Journal of Physiology*, **40**, 38–40.

Björklund, A. & Lindvall, O. (1984). Dopamine-containing systems in the CNS. In *Handbook of Chemical Neuroanatomy. Vol. 2: Classical Transmitters in the CNS. Part 1*, ed. A. Björklund & T. Hökfelt, Amsterdam: Elsevier, pp. 55–122.

Björklund, A., Lindvall, O. & Svensson, L. A. (1972). Mechanisms of fluorophore formation in the histochemical glyoxylic acid method for monoamines. *Histochemie*, **32**, 113–31.

Carlsson, A. (1959). Occurrence, distribution and physiological role of catecholamines in the nervous system. *Pharmacological Reviews*, **11**, 490–93.

Cline, E. J., Scheffel, U., Boja, J. W., Mitchell, W. M., Carroll, F. I., Abraham, P., Lewin, A. H. & Kuhar, M. J. (1992). *In vivo* binding of [^{125}I]RTI-55 to dopamine transporters: Pharmacology and regional distribution with autoradiography. *Synapse*, **12**, 37–46.

Cooper, J. R., Bloom, F. E. & Roth, R. H. (1986). *The Biochemical Basis of Neuropharmacology*. Oxford University Press, New York, NY.

Corrodi, H., Hillarp, N. A. & Jonsson, G. (1964). Fluorescence methods for the histochemical demonstration of monoamines. 3. Sodium borohydride reduction of the fluorescent compounds as a specificity test. *Journal of Histochemistry and Cytochemistry*, **12**, 582–6.

Dahlström, A. & Fuxe, K. (1964). Evidence for the existence of monoamine containing neurons in the central nervous system. *Acta Physiologica Scandinavica*, **62**, Suppl. 232, 1–55.

Eränkö, O. (1955). Histochemistry of the adrenal medulla in rats and mice. *Endocrinology*, **57**, 363–8.

Falck, B., Hillarp, N. A., Thieme, G. & Thorp, A. (1962). Fluorescence of catecholamines and related compounds with formaldehyde. *Journal of Histochemistry & Cytochemistry*, **10**, 348–54.

Fowler, C. J. & Strolin-Benedetti, M. (1983). The metabolism of dopamine by both forms of monoamine oxidase in the rat brain and its inhibition by cimoxatone. *Journal of Neurochemistry*, **40**, 1534–41.

Glenner, G. G., Burtner, H. J. & Brown, G. W. (1957). The histochemical demonstration of monoamine oxidase activity in tetrazolium salts. *Journal of Histochemistry & Cytochemistry*, **5**, 591–600.

Hökfelt, T., Martensson, R., Björklund, A., Kleinau, S. & Goldstein, M. (1984). Distributional maps of tyrosine hydroxylase-immunoreactive neurons in the rat brain. In *Handbook of Chemical Neuroanatomy. Volume 2 Classical Transmitters in the CNS. Part 1*, ed. A. Björklund & T. Hökfelt, Amsterdam: Elsevier, pp. 277–379.

Holtz, P. (1939). Dopadecarboxylase. *Naturwissenschaften*, **27**, 72–5.

Kilty, J. E., Lorang, D. & Amara, S. G. (1991). Cloning and expression of a cocaine-sensitive rat dopamine transported. *Science*, **154**, 578–9.

Koelle, G. B. & Valk, A. T. (1954). Physiological implications of the histochemical localization of monoamine oxidase. *Journal of Physiology*, **126**, 434–47.

Loewi, O. (1933). Problems connected with the principle of humoral transmission of nervous impulses. *Proceedings of the Royal Society of London (Biology)*, **118**, 299–316.

McGeer, P. L., Eccles, J. C. & McGeer, E. G. (1978). *Molecular Neurobiology of the Mammalian Brain*. Plenum Press, New York, NY.

Pearse, A. G. E. (1972). *Histochemistry. Theoretical & Applied*. Volume 2. Third Edition. Churchill Livingstone, New York, NY.

Richfield, E. K. (1991). Quantitative autoradiography of the dopamine uptake complex in rat brain using [³H] GBR 1235: binding characteristics. *Brain Research*, **540**, 1–13.

Schneider, J. S. & Markham, C. H. (1987). Immunohistochemical localization of monoamine oxidase-B in the cat brain: clues to understanding N-methyl-4-phenyl-1,2,3,6-tetrahydropyridine (MPTP) toxicity. *Experimental Neurology*, **97**, 465–81.

Shimada, S., Kitayama, S., Lin, C. L., Patel, A., Nanthakumar, E., Gregor, P., Kuhar, M. & Uhl, G (1991). Cloning and expression of a cocaine-sensitive dopamine transporter complementary DNA. *Science*, **254**, 576–8.

Shimizu, N., Morikawa, M. & Okada, Z. (1959). Histochemical studies of monoamine oxidase of the brain of rodents. *Zeitschrift für Zellforschung und Mikroskopische Anatomie*, **49**, 389–400.

Siegel, G., Agranoff, B., Albers, R. W. & Molinoff, P. (1989). *Basic Neurochemistry*. Raven Press, New York, NY.

Smeets, W. J. A. J. & González, A. (1990). Are putative dopamine-accumulating cell bodies in the hypothalamic periventricular organ a primitive brain character of non-mammalian vertebrates? *Neuroscience Letters*, **114**, 248–52.

Smeets, W. J. A. J. & Steinbusch, H. W. M. (1990). New insights into reptilian catecholaminergic systems as revealed by antibodies against the neurotransmitters and their synthetic enzymes. *Journal of Chemical Neuroanatomy*, **3**, 25–43.

Snyder, S. H. & D'Amato, R. J. (1986). MPTP: a neurotoxin relevant to the pathophysiology of Parkinson's disease. *Neurology*, **36**, 250–8.

Vincent, S. R. & Hope, B. T. (1990). Tyrosine hydroxylase containing neurons lacking aromatic amino acid decarboxylase in the hamster brain. *Journal of Comparative Neurology*, **295**, 290–8.

Vogt, M. (1954). The concentration of sympathin in different parts of the central nervous system under normal conditions and after the administration of drugs. *Journal of Physiology*, **123**, 451–81.

von Euler, U. S. (1956). *Noradrenalin*. Charles C. Thomas, Springfield, IL.

Westlund, K. N., Denney, R. M., Kochersperger, L. M., Rose, R. M. & Abell, C. W. (1985). Distinct monoamine oxidase A and B populations in primate brain. *Science*, **230**, 181–3.

Westlund, K. N., Denney, R. M., Rose, R. M. & Abell, C. W. (1988). Localization of distinct monoamine oxidase A and monoamine oxidase B cell populations in human brainstem. *Neuroscience*, **25**, 439–56.

2
Catecholamine systems in the brain of cyclostomes, the lamprey, *Lampetra fluviatilis*

J. Pierre, J. P. Rio, M. Mahouche, and J. Repérant

Introduction

Lampreys (Petromyzontiforms) and hagfishes (Myxiniforms), which constitute the cyclostomes, are the only living vertebrates which emerged more than 400 million years ago. Recent discoveries of new kinds of extinct agnathans (Forey & Janvier, 1993; Wilson & Caldwell, 1993), and advances in systematic methods (molecular systematics) suggest that lampreys are now generally considered to be the sister group of gnathostomes, and hagfishes that of myopterygians (lampreys and gnathostomes). Data on catecholamine (CA) distribution in the brain of the hagfish is limited to a few histochemical studies on monoamine oxidase (MAO) distribution (Tsuneki, Urano & Kobayashi, 1974; Kusunoki, Kadota & Kishida, 1981, 1982), and one immunohistochemical study on dopamine (DA) distribution (Kadota, Goris & Kusunoki, 1993).

The lamprey central nervous system (CNS) is believed to have retained many primitive traits (Nieuwenhuys, 1977) which make it, therefore, of particular interest for comparative analysis of the organization of the catecholamine systems. In the lamprey, *Lampetra japonica*, a histochemical study of the distribution of MAO in the hypothalamo-hypophyseal system has been carried out (Tsuneki et al., 1975), whereas several studies using the formaldehyde-induced histofluorescence technique (FIF, Falck et al., 1962) have demonstrated the distribution of catecholamine cell bodies and fibers in the brain of lampreys. Most of these studies concerned the involvement of catecholamines in the hypothalamo-hypophyseal axis (Honma, 1969; Honma & Honma, 1970; Ochi & Hosoya, 1974; Tsuneki et al., 1975; Konstantinova, 1973). The overall catecholamine distribution throughout the brainstem and spinal cord of *Lampetra fluviatilis* was studied with the same technique by Baumgarten (1972).

An important step was made by the development of antibodies against enzymes involved in the catecholamine biosynthesis. Antibodies raised against tyrosine hydroxylase (TH) have now been successfully applied in the brain of *Lampetra fluviatilis* (Brodin et al., 1990a; present study). The most recent development of antisera against dopamine (DA) and noradrenaline enabled a further differentiation between dopamine and noradrenaline systems.

The main goal of the present study is to provide a detailed description of the location of catecholamine containing cell bodies and fibers in *Lampetra fluviatilis* as demonstrated by means of TH-immunohistochemistry (THi). A further differentiation into dopamine and noradrenaline systems will be discussed on the basis of our preliminary results with DA antibodies and those with a noradrenaline antiserum in a previous study (Steinbusch et al., 1981).

Distributional maps of catecholamine cell bodies and fibers

Tyrosine hydroxylase
TH-immunohistochemistry was chosen for localization of catecholamines throughout the brain and spinal cord of *Lampetra fluviatilis*. Brains were fixed with 4 % paraformaldehyde in phosphate buffer 0.12 M pH 7.4, embedded in polyethylene glycol and serially sectioned (18 µm) either in the transverse or sagittal plane. The sections were incubated first in a rabbit anti-TH (Eugene Tech. International, Inc [TH antiserum against purified bovine adrenal TH] or Chem-

icon, International, Inc [denatured TH from rat pheochromocytoma]). Washed sections were incubated in the second antibody (biotinylated goat anti-rabbit immunoglobulin Vector), washed again and transferred to a solution of avidin biotin horseradish peroxidase (Vector). Peroxidase was visualized by the diaminobenzidine reaction and sections were finally collected onto gelatinized slides, dehydrated and mounted with Eukitt. The specificity of the procedure was verified either by replacing the primary antibody with normal rabbit serum or by omitting the primary antibody. In neither case did we observe any labeling. The macroscopic analysis reveals that the *Lampetra fluviatilis* brain is extremely small, its total length is approximately 8 mm, whereas its maximal width and height measure about 2.5 mm. The principal parts of the vertebrate neuraxis, namely the telencephalon, diencephalon, mesencephalon and rhombencephalon, can be readily identified (Fig. 2.1(*a*), (*b*)). The anatomical nomenclatures of Heier (1948), Schober (1964) and Nieuwenhuys (1977) were adopted.

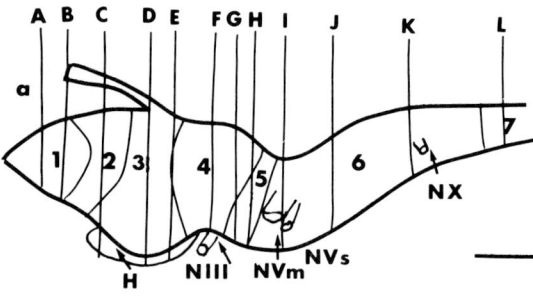

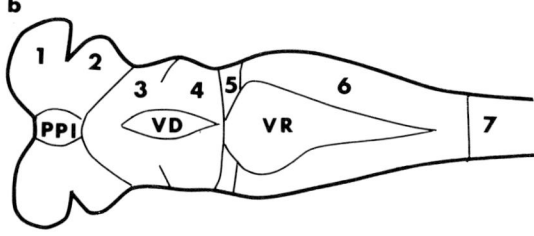

Fig. 2.1. Schematic representations of a lateral (*a*) and dorsal (*b*) view of the lamprey brain. The lines A to L represent the different planes of sections of Fig. 2.2. The major brain subdivisions are: (1) olfactory bulb; (2) telencephalon; (3) diencephalon; (4) mesencephalon; (5) isthmus rhombencephali; (6) rhombencephalon proper; (7) spinal cord. Scale bar, 1 mm.

Telencephalon In the lamprey, the telencephalon is usually subdivided into three parts: 1) the olfactory bulbs, 2) the cerebral hemispheres and 3) the telencephalon medium.

The olfactory bulbs are clearly evaginated, whereas the telencephalon proper is partly evaginated (the cerebral hemispheres) and unevaginated (the telencephalon medium or impar). The same general morphology is evident in the telencephalon of other groups (elasmobranches, dipnoans, amphibians, and amniotes). However, in the lamprey, the telencephalon impar is more extensive than in any of these groups. The olfactory bulb glomeruli (Fig. 2.2(*a*)), which are of variable size and localized at the periphery, are organized in a distinct monolayer (GO). The inner granular layer (IGL), located beneath the mitral cell layer (MCL), is a cell-rich layer. The granular layer of the olfactory bulb is separated from the cerebral hemisphere by a transitional zone, the nucleus olfactorius anterior (NOa, Fig. 2.2(*b*)). The telencephalon proper is constituted by a dorsal pallial region comprising the primordium hippocampi (PH), the lobus subhippocampalis (LS), the primordium pallii dorsalis (PPD) and the primordium piriformis (PP), and by a subpallial ventral region composed of the nucleus septi lateralis (NSL) and the corpus striatum (CS, Fig. 2.2(*c*)). The NSL, the PH, and a part of the CS are situated in the telencephalon impar.

THi cell bodies. Most TH immunoreactive (THi) cell bodies of the olfactory bulb (Fig. 2.2(*a*), (*b*)) and the nucleus olfactorius anterior (NOa, Fig. 2.2(*b*)) are bipolar and elongated or ovoid-shaped. Two cell processes leave the perikaryon in opposite directions: one running toward the glomerulus, the other toward more caudal structures. THi cell bodies of the olfactory bulb are mainly observed in the IGL and, less frequently, around the glomeruli.

THi fibers. The olfactory bulb glomeruli (Fig. 2.2(*a*), 2.2(*b*)) are enriched in varicosities whereas the remaining olfactory bulb displays a dense THi fiber plexus. The cerebral hemispheres and the telencephalon impar are more heterogeneously immunoreactive. A moderately dense plexus of THi fibers was observed in the dorsal pallial region (Fig. 2.2*c*). The ventral part of the telencephalon, in particular the NSL and the CS, are characterized by dense plexuses of THi fibers.

Diencephalon According to the nomenclature of Schober (1964), the following subregions were easily

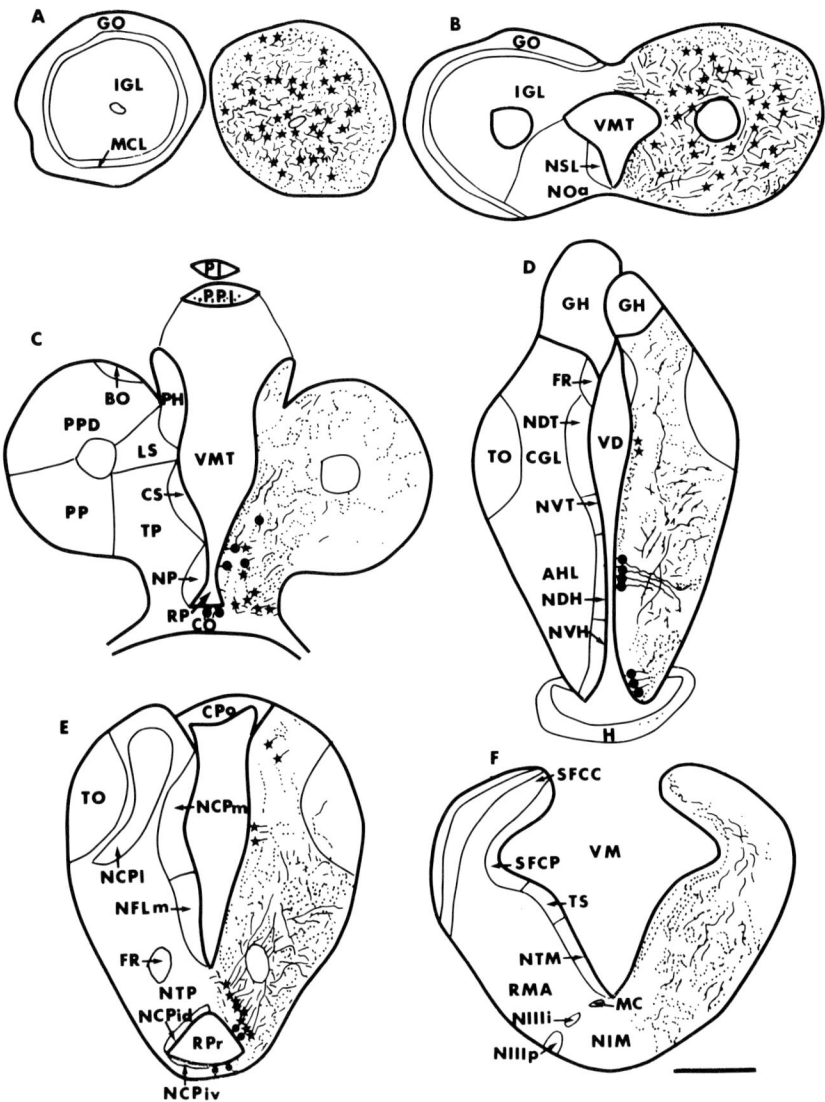

Fig. 2.2. (a)–(l) Camera lucida drawings of transverse sections of the lamprey brain and spinal cord, showing the distribution of tyrosine hydroxylase (TH) immunoreactive cell bodies [(●) in contact with the cerebrospinal fluid (CSF), (★) without contact with the CSF] and fibers and terminals. Scale bar, 0.5 mm.

identified (Fig. 2.2(d), 2.2(e)); 1) the epithalamus composed of the so-called parietal organs (pineal and parapineal) and the asymmetrical habenular ganglia (GH), 2) the thalamus constituted by the nn. dorsalis thalami (NDT), ventralis thalami (NVT), corpus geniculatum laterale (CGL), 3) the pretectum containing dorsally, the lateral and medial subnuclei of the posterior commissure (NCPL and NCPm), 4) the hypothalamus, the anterior part of which comprises the n. preopticus (NP) and the n. commissurae postopticae (NCP), in the intermediate part the n. dorsalis hypothalami (NDH) and the n. ventralis hypothalami (NVH) and in the posterior part, the n. commissurae postinfundibularis composed of a dorsal (NCPid) and a ventral (NCPiv) part, and the nucleus tuberculi posterioris (NTP) and 5) the hypophysis (H).

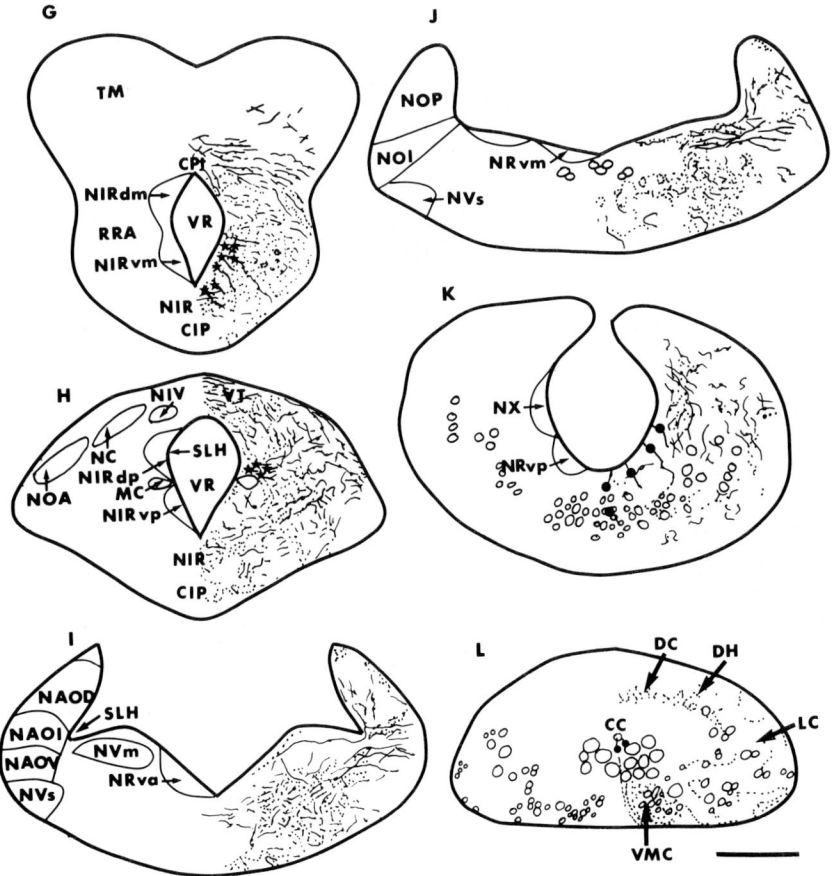

Fig. 2.2. (a)–(l) (cont.)

THi cell bodies. Most THi somata in the diencephalon are observed in the hypothalamus. The rostralmost immunopositive neurons (Fig. 2.2(c)) lie within the NP. The floor of the preoptic recess contains two THi cell types: ovoid-shaped and large diameter cell bodies (18–20 μm) and elongated and small diameter cell bodies (10–12 μm). More caudally, THi fusiform somata are observed in the NCP (Fig. 2.3(a)), one of the cell processes runs toward the ventricle whereas the other crosses the commissura postoptica. Medially, THi somata (diameter 10–15 μm) are identified in the NDH and NVH (Fig. 2.2(d)). The THi cell bodies of the NDH (Fig. 2.3(b)) give off two processes oriented perpendicularly to the ventricular wall: one reaches the ventricle and terminates as a bulb and the other runs in an opposite direction and constitutes a fine-caliber fiber plexus. Caudally, two groups of THi somata are present, one situated in the NCPid and NCPiv, the other in the NTP (Fig. 2.2(e)), Fig. 2.3(c)) in which densely immunoreactive large diameter (20–25 μm) ovoid-shaped cell bodies are organized in a fan-like manner (Baumgarten, 1972, named this cell group the nucleus paratubercularis posterior). The THi somata of the NDH and NVH correspond to the paraventricular organ pars anterior of gnathostomes while those of the commissural postinfundibular positive neurons correspond to the paraventricular organ pars posterior of gnathostomes. In the thalamus, the NDT contains a few THi cell bodies (Fig. 2.2(d)), and in the pretectum, THi somata are observed in the NCPl and NCPm (Fig. 2.2(e)).

THi fibers. With the exception of the habenular ganglia, the diencephalon contains numerous THi fibers that are heterogeneously distributed (Fig. 2.2(c)–2(e)). Most of them are observed in a neuropil zone, lateral

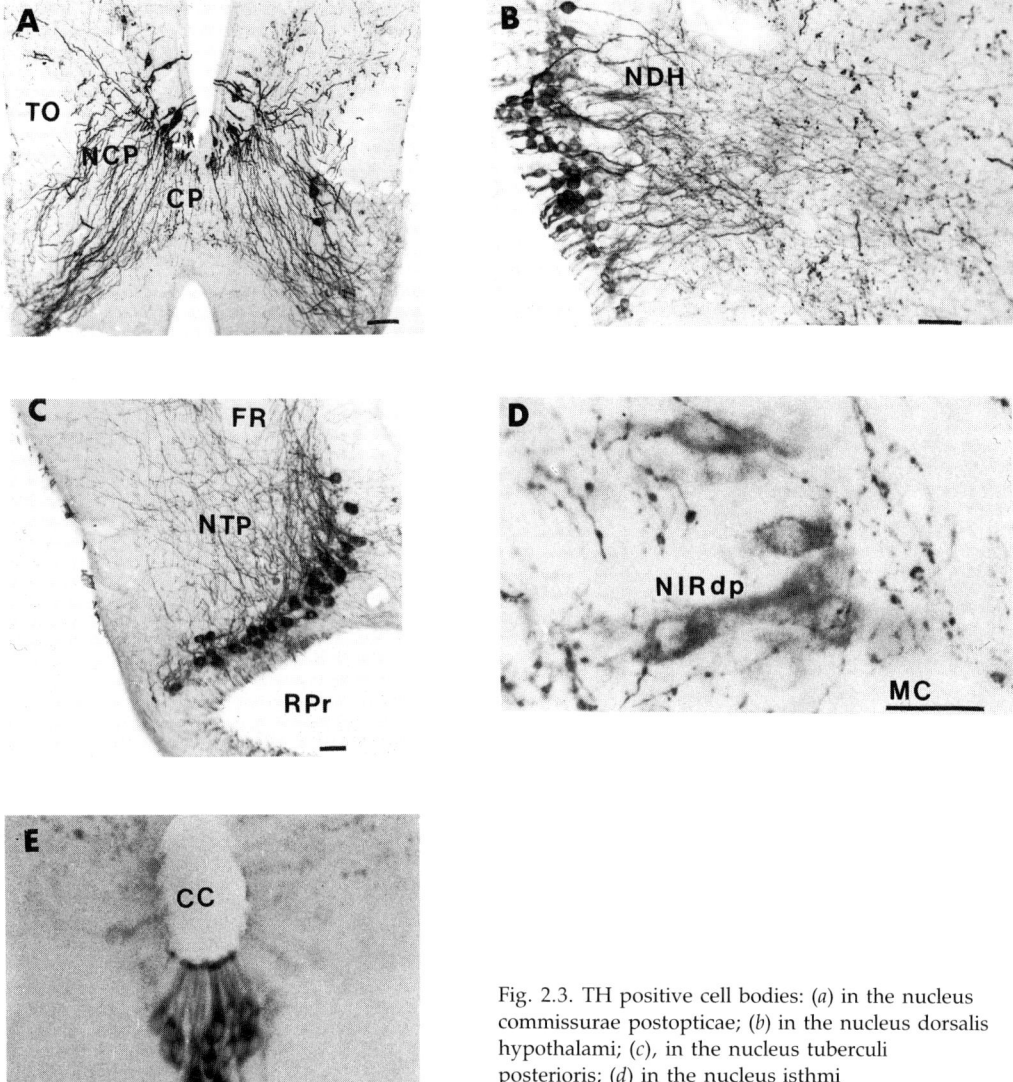

Fig. 2.3. TH positive cell bodies: (a) in the nucleus commissurae postopticae; (b) in the nucleus dorsalis hypothalami; (c), in the nucleus tuberculi posterioris; (d) in the nucleus isthmi rhombencephali dorsalis; (e) around the central canal of spinal cord. Scale bars: (a) 100 μm; (b)–(e), 40 μm.

to the nn. NP, NCP, and NDH as well as in an area skirting the nucleus fasciculi longitudinalis medialis (NFLm). A few large varicose fibers are located in an area lateral to the NVT and NDT and these fibers could be traced to their somata in the NTP. The neurohypophysis contains only a few THi fibers.

Mesencephalon The mesencephalon of the lamprey can be divided in two regions viz. a dorsal part composed of the tectum mesencephali (TM) and the torus semi-circularis (TS), and a ventral part, the tegmentum mesencephali composed of the oculomotor nuclei (NIIIi, NIIIp), the nucleus tegmenti mesencephali (NTM) and the rostral part of the nucleus interpeduncularis mesencephali (NIM). Although THi somata have never been observed in the mesencephalon of lamprey, numerous THi fibers are found in the reticular mesencephalic area (RMA) and in the NIM. Varicosities are also dense in the TS. In the tectum, only the deep layers (stratum fibrosum et cellulare centrale, SFCC and stratum fibrosum et cellulare periventriculare, SFCP) contain THi positive fibers (Fig. 2.2(*f*)).

Rhombencephalon The rhombencephalon or hindbrain, localized between the mesencephalon and spinal cord, comprises two components of unequal size: a short, rostrally located isthmus rhombencephali and a caudal more extensive region, the rhombencephalon proper. A ventricular groove, the sulcus limitans of His (SLH), which extends from the central part of the isthmus to the caudal tip of the rhombencephalon, delineates the dorsal alar plate from the ventral basal plate (Nieuwenhuys, 1977). The alar plate contains, in addition to a very small cerebellum, several small cell masses placed either medially (such as the nn. isthmi rhombencephali dorsalis posterior (NIRdp) and motorius nervi trochlearis (NIV), or laterally (such as nn. cerebelli (NC), octavomotorius (NOA, NOI, NOP) and areae octavolateralis (NAOD, NAOI, NAOV). The basal plate can be divided into medial and lateral zones which extend throughout the rhombencephalon. The medial zone situated against the ventricle and extending throughout the rhombencephalon, contains the nn. isthmi rhombencephali dorsalis medialis (NIRdm), ventralis medialis (NIRvm) and ventralis posterior (NIRvp), the nn. rhombencephali ventralis anterior (NRva), medialis (NRvm) and posterior (NRvp). These nuclei correspond to the reticular nuclei. This zone of the basal plate has often been described as somato-motor (Nieuwenhuys, 1977) or motor (Baumgarten, 1972). The lateral zone of the basal plate, which Nieuwenhuys describes as viscero-motor contains a number of small cellular agglomerates (the motor nuclei of the cranial nerves V, VII, IX and X), and, in addition, a large Mauthner cell.

THi cell bodies. In the medial part of the isthmic region, THi somata are found at the level of the NIRvm (Fig. 2(*g*)). Caudally, immunopositive fusiform cell bodies are localized in a periventricular zone, laterally to the IVth ventricle surface, corresponding to the NIRdp (Fig. 2.2(*h*), Fig. 2.3(*d*)). Moreover, numerous THi varicosities are observed in close vicinity of these cell bodies. Within the rhombencephalon proper, THi somata lie ventrally to the n. motorius nervi vagus and emit processes directed toward the ventricle (Fig. 2.2(*k*)). These cells extend rostrocaudally from mid-medullary levels to medullospinal levels.

THi fibers. In the isthmic region, numerous varicose fiber plexuses are observed especially in the commissura post-tectalis (CPt, Fig. 2.2(*g*)), and in the tractus vestibulocerebellaris (VT, Fig. 2.2(*h*)). In the rhombencephalon proper, varicose fibers and varicosities are extremely dense in its ventral portion throughout its rostro-caudal extent. Another plexus is present in the nn. octavomotorius and the area octavolateralis (Fig. 2.2(*i*), 2.2(*j*)).

Spinal cord Caudal to the obex, transversally-cut sections reveal a spinal cord whose shape is ellipsoid rostrally (Fig. 2.2(*l*)) and more flattened caudally. The dorsal root entry zone is referred to as to the dorsal horn (DH), and the region located medially to this as the dorsal column (DC). In addition, the axon tracts are subdivided into lateral (LC) and ventromedial (VMC) columns. Motoneurons and interneurons are located in the lateral gray cell column on each side of the central canal (CC). THi somata are observed ventral to the central canal and may contact the cerebrospinal fluid (Fig. 2.2(*l*)), Fig. 2.3(*e*)). This intrinsic system of THi cells gives rise to a fiber plexus in the ventromedial column, and numerous THi fibers and varicosities are present in the DH and DC.

Dopamine

The chemical concentration of dopamine and noradrenaline in several *Lampetra fluviatilis* brains has been measured by Baumgarten (1972). The dopamine content is much higher (0.57 μg/g) than that of noradrenaline (0.04 μg/g), suggesting that the TH maps largely represent the distribution of dopamine. In the present study, we have also tested an antidopamine antibody in *Lampetra fluviatilis*. Our preliminary results show that the dopamine antiserum reveals an almost identical pattern of immunoreactivity as the TH antiserum. Minor differences are that TH antibodies stained cell bodies in internal granular layer (IGL) of the olfactory bulb, and in the nucleus isthmi rhombencephali dorsalis posterior (NIRdp), while anti-DA antibodies did not. At the hypothalamic level (NDH, NVH, NCPid, NCPiv), dopamine positive cell bodies seem to be more numerous than THi neurons. In several brainstem regions (NIRVm, NRvp), there are more cells immunoreactive to TH than to dopamine. Some of these THi cells are very likely noradrenergic, for example those located in the isthmic and rhombencephalon proper (cf. Smeets & Steinbusch, 1990). However, it is unlikely that the cell bodies in the forebrain (cells bodies in IGL of the olfactory bulb), which are TH positive and dopamine negative, could be noradrenergic, raising the possibility that L-DOPA may be the end product in these cells.

Comparison with previous studies of catecholamines in lampreys

Distribution of catecholamine cell bodies In general, the distribution of catecholamine (CA) containing cell bodies as determined using the FIF method in the whole brain (Baumgarten, 1972) and the hypothalamus (from the level of the preoptic recess to the posterior recess (Konstantinova, 1973) of the lamprey (*Lampetra fluviatilis*) resembles the pattern observed in the present immunohistochemical study. However, it should be noted that Baumgarten (1972) never observed CA somata within the olfactory bulb, the n. praeopticus, the rhombencephalon and spinal cord. By means of immunohistochemistry Brodin *et al.* (1990a) have identified in the hypothalamus the same cell groups that we have described here: 'the postinfundibular commissural nucleus, the dorsal and ventral hypothalamic nuclei and the postoptic commissural nucleus'. They also reported a weak TH immunolabeling in somata located ventrally to the central canal of the spinal cord, whereas our results and those of Ochi and Hosoya (1974) show a strong immunolabeling. It is noteworthy that Steinbusch *et al.* (1981) have observed noradrenaline immunoreactive cell bodies in two regions: 1) the ventral parts of the thalamus and hypothalamus and 2) at the junction between the medulla oblongata and the spinal cord.

Distribution of catecholamine fibers The localization of catecholaminergic fibers and terminals has been only studied by Baumgarten (1972) in the whole brain using the histofluorescent methods. The use of immunohistochemistry has allowed us to corroborate his results, except for the olfactory bulb in which he did not observe any catecholaminergic positive fibers. In contrast to Honma (1969) and Honma and Honma (1970), Baumgarten (1972), Tsuneki *et al.* (1975) and our group have revealed catecholamine fibers in the posterior neurohypophysis.

Relationships of catecholamines with other neurotransmitter systems

Catecholamine systems are well developed in the lamprey CNS and likely interact with other neurotransmitter systems. In this section, special attention will be paid to the distribution maps of CA containing cell somata, fibers and terminals with other monoamines and/or peptides to study the possibilities of interactions.

Interactions with other monoamines: serotonin and histamine

Several immunohistochemical studies have supplied maps of the distribution of serotonin-immunoreactive (5HTi) perikarya and fibers in the river lamprey brain (*Lampetra fluviatilis*: Steinbusch & Nieuwenhuys, 1979; Steinbusch *et al.*, 1981; Brodin *et al.*, 1988, 1990a; Pierre *et al.*, 1992). On the contrary, only one study has been devoted to the distribution of histaminergic neurons (Hi) in this species (Brodin *et al.*, 1990b). Cell somata in NDH, NVH, NCPid, NCPiv of the mid- and posterior hypothalamus are THi, 5HTi, and Hi. In none of these regions was any colocalization of TH, 5HT and histamine observed in the same cell body. THi, 5HTi and Hi fibers from positive NDH and NVH somata mainly innervate the telencephalon (NOa, NSL and CS: Steinbusch & Nieuwenhuys, 1979; Brodin *et al.*, 1990a,b). Brain structures that are densely innervated with catecholaminergic as well as with histaminergic and serotoninergic fibers are the nucleus commissurae postopticae and area hypothalamica lateralis. It is worth noting that the THi somata of the NTP are innervated by 5HTi and Hi fibers arising from NCPid positive somata and that the THi somata of the IGL of the olfactory bulb and NOa are innervated by 5HTi fibers arising from neurons of the hypothalamus as well as neurons of the rhombencephalon (Pierre *et al.*, 1992).

In the NCPl and NCPm of the pretectum and in the NIRvm and NIRvp of the rhombencephalon, the distribution of THi somata shows a considerable overlap with that of 5HTi cell bodies (Brodin *et al.*, 1988; Pierre *et al.*, 1992). The 5HTi fibers of these somata (NIRvm and NIRvp) project to the spinal cord level (Brodin *et al.*, 1986, 1988, 1989a), where a considerable overlap of THi and 5HTi fibers occurs. In the mesencephalon, the structures which contain dense THi and 5HTi fibers and terminals with a partial overlap are the SFCC and SFCP of the optic tectum, the torus semi-circularis and the reticular mesencephalic area. Rhombencephalic structures densely innervated by THi and 5HTi fibers are the rhombencephalic reticular area, the nucleus interpeduncularis rhombencephali and the nucleus octavomotorius. In the cord, the THi cell bodies which are localized ventrally to the central canal with one process contacting the CSF, contain also 5HT (Brodin *et al.*, 1988). The most obvious overlap of THi/5HTi fibers is observed in VMC, LV, of the spinal cord.

Interactions with peptides

Several studies in the CNS have clearly shown that neuropeptides do not only act as neurohormonal or neuromodulatory agents, but that they also fulfill several neurotransmitter criteria, e.g. their presence in axon terminals, a calcium-dependent release and effects on neuronal excitability.

Neurotensin (NT). NT is a tridecapeptide which has been isolated from bovine hypothalamus. The distribution of NT cell bodies and fibers has been studied in *Lampetra fluviatilis* CNS by Brodin et al., 1990a. Furthermore, the anatomical relationship between NTi and THi perikarya and fibers has been examined with immunofluorescence in certain brain areas. In double stained sections, THi and NTi cell bodies are codistributed in NP, NCPiv and NCPïd; the THi somata in the commissurae posterioris nuclei have been observed to be located ventrally to the NTi cell bodies. Moreover, THi somata in NDH and NVH appeared somewhat smaller and more often located closer to the ventricle surface when compared to NTi somata. In consecutive sections, some NTi fibers have been traced rostrally and dorsally from the midhypothalamus (NDH, NVH) toward the CS and NSL and overlapped THi fibers originating from these same cell groups. In the spinal cord, the THi somata localized ventrally to the central canal also contain NTi as well as other substances, e.g. tachykinin (Van Dongen et al., 1986), somatostatine (SS), (Buchanan et al., 1987). Among the somata, Brodin et al., 1989b, 1990c, have shown the colocalization of GABA and SS and GABA and NT, respectively.

Cholecystokinin (CCK). It is a 33 amino-acid peptide and the octapeptide form (CCK-8) is thought to predominate in the CNS. Brodin et al. (1988) have described the codistribution of CCK-8 immunoreactive neuronal elements with that of 5HT in *Lampetra fluviatilis*. Both have been observed in the NRvp without any colocalization, these immunopositive somata areas projecting to the spinal cord level. Groups of CCK-8 cell bodies have been observed in the NDH, NVH and in the NCPid and NCPiv of the hypothalamus. 5HT and CCK-8 have been reported to be colocalized in NDH, NVH somata. Dense CCK-8i fiber plexuses are located in the hypothalamus and striatum in which THi fibers overlap. In comparison with the density of THi fibers in the rhombencephalon, that of CCK-8i fibers was low.

Pancreatic polypeptide (PP). Brodin et al., 1989a have demonstrated by radioimmunoassay that PP has different molecular forms, the peptide YY (PYY) and the neuropeptide Y (NPY), which are stored in separate neuronal populations. The lamprey brain NPY shares about 80% homology with the porcine neuropeptide (Rawitch, Pollock & Brodin, 1992). The distribution maps of these two peptides have been only achieved at the brainstem and spinal cord level. PYYi cell somata have been observed in or close to the mesencephalic and rhombencephalic reticular formation. In NIRvm, PYYi somata codistribute with other substances such as TH, CCK and 5HT and, moreover, the projection of PYY somata is like that of THi reticulo-spinal cell bodies. NPYi perikarya have been reported in the rostral and caudal rhombencephalic alar plate and a codistribution of NPY and TH has been observed in NIRdp cell somata. In the spinal cord, the THi somata located ventrally to the central canal also contain PYY.

Substance P (SP). Its distribution has been studied in sea lampreys (*Petromyzon marinus*) and pacific lampreys (*Entosphenus tridentatus*) by Nozaki and Gorbman, 1986. Codistribution of SPi and THi somata was observed in several nuclei, i.e. in NP and NCP of the anterior hypothalamus, and in the NDH, NVH of the middle hypothalamus. A codistribution of these two substances has been only reported in *Entosphenus* in the NIRvm at the isthmorhombencephalic level. Additionally, a partial overlapping of THi and SPi fibers occurs in the CS, NP, NCP and in the area hypothalamica lateralis.

Other neuropeptides. Somatostatin (SST-14, SST-34), Cheung, Plisetskaya & Youson, 1990 in *Petromyzon marinus*; luteinizing hormone releasing hormone (LH-RH), Crim, Urano & Gorbman, 1979 in *Entosphenus tridentatus*; Met-enkephalin, β endorphine, Nozaki and Gorbman, 1984 in *Entosphenus tridentatus*; FMRF amide, Ohtomi, Fujii & Kobayashi, 1989 in *Lampetra japonica*, a river lamprey. At the telencephalic level, we have demonstrated the presence of THi somata in the olfactory bulb IGi, an area that contains also SST-34i cell somata (Cheung et al., 1990). THi, LH-RHi and Met-enkephalin somata have been reported to codistribute in the nuclei praeopticus and commissura postoptica of the anterior hypothalamus. Usually, most of these immunoreactive somata do not colocalize with any neurosecretory cell bodies that project to the neurohypophysis. The only neuropeptide which mimics the preopticohypophyseal neurosecretory system is vasotocin (Goossens, Dierickx & Vandesande, 1977 in *Lampetra fluviatilis*. More caudally, NDH and NVH cell somata contain TH, SST-14, SST-34, β endorphin and FMRF amide and mainly project to CS and NSL. In the posterior hypothalamus, and overlapping, of THi somata with SST-14i and SST 34i cells bodies has been observed in NCPid and NCPiv.

Relationships of catecholamines and known pathways

Sensory system

Olfactory system The fibers of the olfactory nerve spread over the surface of the olfactory bulb and branch into the underlying glomeruli. In the glomeruli olfactori, among the DAi cells some of them can be compared to horizontal cells and the others to periglomerular cells. In this layer, the dendrites of some IGL cell bodies are either DAi, or L-DOPA, but eventual synaptic contacts between these dendrites and those of mitral cells have not been demonstrated yet. We can conclude that L-DOPA and dopamine are the catecholamines involved in olfactory information processing and that DA innervation of the olfactory bulb is most likely and exclusively provided by intrinsic DA neurons. The targets of the secondary olfactory fibers, i.e. the striatum, the nucleus praeopticus, the hypothalamus and the nucleus tuberculi posterioris (Northcutt & Puzdrowski, 1988, Polenova & Vesselkin, 1993) are innervated by THi somata and fibers, suggesting that catecholamines may modulate the olfactory input at different levels.

Visual system In the retina, De Miguel and Wagner, 1990, have observed that THi somata were located mainly in the innermost part of the inner nuclear layer with their processes preferentially running in the inner plexiform. In addition, a dense plexus of THi processes have been noted in the outer plexiform and nuclear layers. These data suggest that most THi cells in the lamprey retina represent interplexiform cells. In contrast, no THi fibers have been observed in the optic tract. The nucleus commissura postoptica is the unique visual center which contains THi somata. Tracing studies have demonstrated that the tectum receives information from the nucleus dorsalis thalami and the nucleus tuberculi posterioris (Baumgarten, 1972). These two structures contain THi somata and may represent the afferent catecholaminergic source to the tectum.

THi somata contacting the CSF
Some TH positive cells of the nn. praeopticus and commissurae postopticae and all those in nn. dorsalis and ventralis hypothalami and commissurae postinfundibularis are in contact with the CSF. They give rise to short projections which arborize in the hypothalamus and/or the hypophysis as well as to long extrahypothalamic pathways reaching telencephalic structures and the brainstem. These projections suggest that the CSF-contacting TH positive cells may modulate the activity of some telencephalic and brainstem nuclei, according to the functional state of the hypothalamus or even according to the rate of diverse hormones or other chemical substances in the CSF.

Motor systems

Basal ganglia At the caudal hypothalamic level, we have demonstrated TH positive cell bodies of the n. tuberculi posterioris, which give rise to ascending processes that border the thalamus before projecting to the telencephalon (particularly at the level of the corpus striatum) and to descending processes which project to the deep tectal layers and to the brainstem. Similar observations were reported by Baumgarten (1972) who additionally observed ascending projections from the NTP toward the nucleus praeopticus. These observations are in favor of the hypothesis of Baumgarten (1972) that the nucleus tuberculi posterioris represents a cytoarchitectonic entity equivalent to the substantia nigra of reptiles and mammals. Until now, no data are available in lampreys regarding a striatal projection from the midbrain cell groups.

Brainstem and spinal cord Our immunohistochemical studies did not reveal a dense catecholamine innervation of the motor nuclei of cranial nerves. The brainstem contains distinct plexuses of THi fibers, in the alar plate, in the nuclei octavomotorius (NOA, NOI, NOP) or vestibular nuclei. No tracing studies has reported the origin of these fibers.

The descending THi fibers from the posterior rhombencephalic nuclei (NRvp) have been observed sometimes in close apposition to spinal neurons in lateral column as well as the THi varicosities of the intraspinal neurons in the ventromedial column. During steady swimming, the lamprey maintains a horizontal orientation of the body, but it can also swim at a definite angle in relation to the horizon ('uphill and down hill') and in all these cases the body is oriented with the dorsal side up (Deliagina et al., 1992). The reticulospinal system plays an important role in mediating vestibular reflexes to the spinal cord which are of critical importance for maintaining the body orientation during locomotion. The role of CA in these systems is still unknown.

Rovainen, 1979, has shown that L-DOPA elicits

swimming activities in lamprey spinal cord, but its excitatory effects resemble those of acidic amino-acids and are not mimicked by application of dopamine. The possible effects of catecholamines on single neurons and synapses of lamprey are also unknown.

Comparison with myxiniforms

The brain of the hagfish differs to some extent from that of the lamprey. The brain of adult pacific hagfish, *Eptatretus burgeri*, has reduced ventricles and optic nerves, subcutaneous eyes without lenses, a laminated pallium and no pineal gland, all of which are well developed in the lamprey. In a recent anatomical and hodological study in the pacific hagfish, Wicht and Northcutt (1992) have shown that the forebrain ventricular system consists of two major subdivisions, the third ventricle, which is partially vestigial, and lateral ventricles, which are entirely vestigial. The hypothalamic area shows little differentiation. A densely-packed row of cells surround the ventricular lumen (infundibular ventricle). Laterally, cells represent the nucleus infundibularis. Three other ventricular structures are connected with the infundibular ventricle: the posterior and lateral infundibular recess, both parts of which are characterized by a small, but open ventricular cavity and a dense population of ependymal cells, and rostral to both of them, the neurohypophyseal recess. It is worth noting that *Eptatretus burgeri* which is considered nearly blind due to its particular eye anatomy, has been reported to have an elaborate primary visual system and displays many secondary projections (retinopetal projections originates from two cells groups in the rostral mesencephalic tegmentum and bilaterality of retinofugal projections which reach preoptic, thalamic and pretectal, as well as tectal targets) and a distinct set of central targets which are common to macrophthalmic species as lampreys (Wicht & Northcutt, 1990).

Maps of dopamine

The distribution of dopamine-containing cell bodies and fibers has been histochemically studied by Kadota *et al.* (1993) in the hagfish, *Eptatretus burgeri*. Dopamine-containing cell bodies are present only in the hypothalamus. One cell type, with a club-like process and an ependymofugal process, is in contact with the CSF. These cells are abundant along the ependymal wall of the ventral part of the posterior hypothalamic ventricle (infundibular ventricle) and its lateral recess. A second type is bipolar or multipolar, not in contact with CSF and is located in the nucleus hypothalamicus (a group dorsal to the lateral recess. According to Kadota *et al.* (1993); dopamine-containing fibers are widely located thoughout the brain and spinal cord and the immunoreactive distribution is low. If these results are compared with our data on the distribution of dopamine positive neurons and fibers in the lamprey, it appears that the distribution of dopamine in hagfishes is more limited.

Relationships between dopamine and other neurotransmitters

An immunohistochemical study of the distribution of serotonin in the brain of the hagfish (*Eptatretus burgeri*) carried out by Kadota (1991) has revealed that serotoninergic neurons are located in the same hypothalamic region as dopaminergic neurons. On the other hand, dopamine positive fibers were very rare or absent in the densely serotonin innervated dorsal pallial area, the corpus interpeduncularis and in the rhombencephalon at the level of the nucleus motorius nervi trigemini.

FMRF-amide positive cell bodies (Chiba & Homna, 1992), somatostatin positive perikarya (Nozaki & Gorbman, 1983) have been reported in the same hypothalamic region which contains dopamine positive somata. In a comparative study of the distribution of LH-RH immunoreactivity in the pacific hagfish (*Eptatretus stouti*) and the lamprey (*Entosphenus tridentata*), Crim *et al.* (1979) reported the presence of LH-RH positive neurons at the level of the nucleus praeopticus only in the lamprey.

The distribution of acetylcholinesterase (AChE) has also been studied in the brains of several cyclostomes (Wächtler, 1974, 1975, 1983). A decrease in AChE activity from caudal to rostral brain has been reported in the lamprey as well as in several low vertebrates. The myxine shows some deviations mainly due to the high level of AChE activity in the diencephalon and telencephalon including the bulbus olfactorius.

Concluding remarks

Catecholamine systems or at least the dopamine system, are well developed in the brain of lampreys, a member of the most primitive group of living vertebrates. The TH-and often DA-immunoreactive neurons are numerous at the diencephalic and mesodiencephalic levels compared to those in the mesencephalon and rhombencephalon. It should be noted that the neurons of the nuclei dorsalis and ventralis hypothalami which probably correspond to the paraventricular organ pars anterior of gnathostomes contain CSF contacting cells that are THi/DAi.

In higher vertebrates, it seems that these cell groups are TH negative and that a lateral TH positive group exists, which is absent in the lamprey. The noradrenergic system in the lamprey, on the contrary, is poorly developed according to Steinbusch *et al.* (1981).

Acknowledgements

This work was supported by INSERM and MNHN (J. E. DRED). Thanks are due to S. Arnold, D. Le Cren and to G. Sanchez for their excellent technical assistance.

Abbreviations

AHL	area hypothalamica lateralis	NIRdp	nucleus isthmi rhombencephali dorsalis posterior
CC	central canal	NIRvm	nucleus isthmi rhombencephali ventralis medialis
CGL	corpus geniculatum laterale		
CIP	commissura interpeduncularis	NIRvp	nucleus isthmi rhombencephali ventralis posterior
CO	chiasma opticum		
CP	commissura postoptica	NOa	nucleus olfactorius anterior
CPo	commissura posterior	NOA	nucleus octavomotorius anterior
CPt	commissura post-tectalis	NOI	nucleus octavomotorius intermedius
CS	corpus striatum	NOP	nucleus octavomotorius posterior
DC	dorsal column	NP	nucleus praeopticus
DH	dorsal horn	NRva	nucleus rhombencephali ventralis anterior
FR	fasciculus retroflexus		
GH	ganglion habenula	NRvm	nucleus rhombencephali ventralis medialis
GO	glomeruli olfactorii		
H	hypophysis	NRvp	nucleus rhombencephali ventralis posterior
IGL	internal granular layer		
LC	lateral column	NSL	nucleus septi lateralis
LS	lobus subhippocampalis	NTM	nucleus tegmenti mesencephali
MC	Müller cell	NTP	nucleus tuberculi posterioris
MCL	mitral cell layer	NVH	nucleus ventralis hypothalami
NAOD	nucleus areae octavolateralis dorsalis	NVT	nucleus ventralis thalami
NAOI	nucleus areae octavolateralis intermedius	N IIIi	nucleus motorius nervi oculomotorii intermedius
NAOV	nucleus areae octavolateralis ventralis	N IIIp	nucleus motorius nervi oculomotorii profundus
NC	nucleus cerebelli		
NCP	nucleus commissurae postopticae	NIV	nucleus motorius nervi trochlearis
NCPid	nucleus commissurae postinfundibularis pars dorsalis	NVm	nucleus motorius nervi trigemini
		NVs	nucleus sensibilis nervi trigemini
NCPiv	nucleus commissurae postinfundibularis pars ventralis	NX	nucleus motorius nervi vagus
		PH	primordium hippocampi
		PI	pineal organ
NCPL	nucleus commissurae posterioris lateralis	PP	primordium piriformis
		PPD	primordium pallii dorsalis
NCPm	nucleus commissurae posterioris medialis	PPI	parapineal organ
		RMA	reticular mesencephalic area
NDH	nucleus dorsalis hypothalami	RP	recessus praeopticus
NDT	nucleus dorsalis thalami	RPr	recessus posterior
NFLm	nucleus fasciculi longitudinalis medialis	RRA	rhombencephali reticular area
NIM	nucleus interpeduncularis mesencephali	SFCC	stratum fibrosum et cellulare centrale
NIR	nucleus interpeduncularis rhombencephali	SFCP	stratum fibrosum et cellulare periventriculare
NIRdm	nucleus isthmi rhombencephali dorsalis medialis	SLH	sulcus limitans of His
		TM	tectum mesencephali

TO	tractus opticus	VMC	ventromedial column
TP	telencephalic peduncle	VMT	ventriculus medius telencephali
TS	torus semi-circularis	VR	ventriculus rhombencephali
VD	ventriculus diencephali	VT	vestibulocerebellar tract
VM	ventriculus mesencephali		

References

Baumgarten, H. G. (1972). Biogenic amines in the cyclostome and lower vertebrate brain. *Progress in Histochemistry and Cytochemistry*, **4**, 1–90.

Brodin, L., Buchanan, J. T., Hökfelt, T., Grillner, S. & Verhofstad, A. A. J. (1986). A spinal projection of 5-hydroxytryptamine neurons in the lamprey brainstem; evidence from combined retrograde tracing and immunohistochemistry. *Neuroscience Letters*, **67**, 53–7.

Brodin, L., Buchanan, J. T., Hökfelt, T., Grillner, S., Rehfeld, J. F., Frey, P., Verhofstad, A. A. J., Dockray, G. J. & Walsh, J. H. (1988) Immunohistochemical studies of cholecystokinin-like peptides and their relation to 5 HT, CGRP, and bombesin immunoreactivities in the brainstem and spinal cord of lampreys. *Journal of Comparative Neurology*, **271**, 1–18.

Brodin, L., Rawitch, A., Taylor, T., Ohta, Y., Ring, H., Hökfelt, T., Grillner, S. & Terenius, L. (1989a). Multiple forms of pancreatic polypeptide related compounds in the lamprey CNS: partial characterization and immunohistochemical localization in the brainstem and spinal cord. *Journal of Neuroscience*, **9**, 3428–42.

Brodin, L., Ring, H., Grillner, P., Persson, H., Hökfelt, T. & Grillner, S. (1989b). Comparative aspects of some neurotransmitter systems in the lamprey CNS. *European Journal of Neuroscience* (suppl. 2), **12**, 1.

Brodin, L., Theodorsson, E., Christenson, J., Cullheim S., Hökfelt, T., Brown, J. C., Buchan, A., Panula P., Verhofstad, A. A. J. & Goldstein, M. (1990a). Neurotensine-like peptides in the CNS of lampreys: chromatographic characterization and immunohistochemical localization with reference to aminergic markers. *European Journal of Neuroscience*, **2**, 1095–109.

Brodin, L., Hökfelt, T., Grillner, S. & Panula, P. (1990b). Distribution of histaminergic neurons in the brain of the lamprey *Lampetra fluviatilis* revealed by histamine-immunohistochemistry. *Journal of Comparative Neurology*, **292**, 435–42.

Brodin, L., Dale, N., Christenson, J., Storm-Mathisen, J., Hökfelt, T. & Grillner, S. (1990c) Three types of GABA-immunoreactive neurons in the lamprey spinal cord. *Brain Research*, **508**, 172–5.

Buchanan, J. T., Brodin, L., Hökfelt, T., Van Dongen, P. A. M. & Grillner, S. (1987). Survey of neuropeptide-like immunoreactivity in the lamprey spinal cord. *Brain Research*, **408**, 299–302.

Cheung, R., Plisetskaya, E. M. & Youson, J. H. (1990). Distribution of two forms of somatostatin in the brain, anterior intestine, and pancreas of adult lampreys (*Petromyzon marinus*). *Cell and Tissue Research*, **262**, 283–92.

Chiba, A. & Honma, Y. (1992). FMRF amide-immunoreactive structures in the brain of the brown hagfish, *Paramyxine atami*: relationship with neuropeptide Y-immunoreactive structures. *Histochemistry*, **98**, 33–8.

Crim, J. W., Urano, A. & Gorbman, A. (1979). Immunocytochemical studies of luteinizing hormone-releasing hormone in brains of agnathan fishes. 1. Comparisons of adult pacific lamprey (*Entosphenus tridentata*) and the pacific hagfish (*Eptatretus stouti*). *General and Comparative Endocrinology*, **37**, 294–305.

Deliagina, T. G., Orlovsky, G. N., Grillner, S. R. & Wallen, P. (1992). Vestibular control of swimming in lamprey. II Characteristics of spatial sensitivity of reticulospinal neurons. *Experimental Brain Research*, **90**, 489–98.

De Miguel, E. & Wagner, H. J. (1990). Tyrosine hydroxylase immunoreactive interplexiform cells in the lamprey retina. *Neuroscience Letters*, **113**, 151–5.

Falck, B., Hillarp, N. A., Thieme, G. & Torp, A. (1962). Fluorescence of catecholamines and related compounds condensed with paraformaldehyde. *Journal of Histochemistry and Cytochemistry*, **10**, 348–54.

Forey, P. & Janvier, P. (1993). Agnathans and the origin of jawed vertebrates. *Nature*, **361**, 129–34.

Goossens, N., Dierickx, K. & Vandesande, F. (1977). Immunocytochemical demonstration of the hypothalamo-hypophysial vasotocinergic system of *Lampetra fluviatilis*. *Cell and Tissue Research*, **177**, 317–23.

Heier, P. (1948). Fundamental principles in the structure of the brain. A study of the brain of *Petromyzon fluviatilis*. *Acta Anatomica* (suppl.), **8**, 1–123.

Honma, S. (1969). Fluorescence microscopic observations on the brain of the lamprey, *Lampetra japonica*. *Archives of Histologica Japonica*, **31**, 167–78.

Honma, S. & Honma, Y. (1970). Histochemical demonstration of monoamines in the hypothalamus of the lamprey and ice-goby. *Bulletin of the Japanese Society of Scientific Fisheries*, **36**, 125–34.

Kadota, T. (1991). Distribution of 5HT (serotonin) immunoreactivity in the central nervous system of the inshore hagfish, *Eptatretus burgeri*. *Cell and Tissue Research*, **266**, 107–16.

Kadota, T., Goris, R. C. & Kusunoki, T. (1993). Dopamine neurons in the hagfish brain. *The Anatomical Record* (suppl. 1), in press.

Konstantinova, M. (1973). Monoamines in the liquor contacting nerve cells in the hypothalamus of the lamprey, *Lampetra fluviatilis* L., *Zeitschrify für Zellforschung*, **144**, 549–57.

Kusunoki, T., Kadota, T. & Kishida, R. (1981). Chemoarchitectonics of the forebrain of the hagfish, *Eptatretus burgeri*. *Journal für Hirnforschung*, **22**, 285–98.

Kusunoki, T., Kadota, T. & Kishida, R. (1982). Chemoarchitectonics of the brainstem of the hagfish, Eptatretus burgeri. *Journal für Hirnforschung*, **23**, 109–19.

Nieuwenhuys, R. (1977). The brain of the lamprey in a comparative perspective. *Annual New York Academy Science*, **299**, 97–145.

Northcutt, R. G. & Puzdrowski, R. L. (1988). Projections of the olfactory bulb and nervus terminalis in the silver lamprey. *Brain Behavior and Evolution*, **32**, 96–107.

Nozaki, M. & Gorbman, A. (1983). Immunocytochemical localization of somatostatin and vasotocin in the brain of the Pacific hagfish, *Eptatretus stouti*. *Cell and Tissue Research*, **229**, 541–50.

Nozaki, M. & Gorbman, A. (1984). Distribution of immunoreactive sites for several components of pro-opiocortin in the pituitary and brain of adult lampreys, *Petromyzon marinus* and *Entosphenus tridentatus*. *General and Comparative Endocrinology*, **53**, 335–52.

Nozaki, M. & Gorbman, A. (1986). Occurrence and distribution of substance P-related immunoreactivity in the brain of adult lampreys, *Petromyzon marinus* and *Entosphenus tridentatus*. *General and Comparative Endocrinology*, **62**, 217–29.

Ochi, J. & Hosoya, Y. (1974). Fluorescence microscopic differentiation of monoamines in the hypothalamus and spinal cord of the lamprey using a new filter system. *Histochemistry*, **40**, 263–6.

Ohtomi, M., Fujii, K & Kobayashi, H. (1989). Distribution of FMRF amide-like immunoreactivity in the brain and neurohypophysis of the lamprey, *Lampetra japonica*. *Cell and Tissue Research*, **256**, 581–4.

Pierre, J., Repérant, J., Ward, R., Vesselkin, N. P., Rio, J.-P., Miceli, D. & Kratskin, I. (1992). The serotoninergic system of the brain of the lamprey, *Lampetra fluviatilis*: an evolutionary perspective. *Journal of Chemical Neuroanatomy*, **5**, 195–219.

Polenova, O. A. & Vesselkin, N. P. (1993). Olfactory and Nonolfactory projections in the river lamprey (*Lampetra fluviatilis*) telencephalon. *Journal für Hirnforschung* (in press).

Rawitch, A. B., Pollock, H. G. & Brodin, L. (1992). A neuropeptide Y (NPY)-related peptide is present in the river lamprey CNS. *Neuroscience Letters*, **140**, 165–8.

Rovainen, C. M. (1979). Neurobiology of Lampreys. *Physiological Reviews*, **59** (4), 1008–77.

Schober, W. (1964). Vergleichend-anatomische Untersuchungen am Gehirn der Larven und adulten Tiere von *Lampetra fluviatilis* (Linné 1758) und *Lampetra planeri* (Bloch 1784). *Journal für Hirnforschung*, **7**, 107–209.

Smeets, W. J. A. J. & Steinbusch, H. W. M. (1990). New insights into the reptilian catecholaminergic systems as revealed by antibodies against the neurotransmitters and their synthetic enzymes. *Journal of Chemical Neuroanatomy*, **3**, 25–43.

Steinbusch, H. W. M. & Nieuwenhuys, R. (1979). Serotoninergic neuron systems in the brain of the lamprey, *Lampetra fluviatilis*. *The Anatomical Record*, **193**, 693.

Steinbusch, H. W. M., Verhofstad, A. A. J., Penke, B., Varga, J. & Joosten, H. W. J. (1981). Immunohistochemical characterization of monoamine containing neurons in the central nervous system by antibodies to serotonin and noradrenalin. A study in the rat and the lamprey *Lampetra fluviatilis*. *Acta Histochemica* (suppl.), **24**, 107–22.

Tsuneki, K., Urano, A. & Kobayashi, H. (1974). Monoamine oxidase and acetylcholinesterase in the neurohypophysis of the hagfish, *Eptatretus burgeri*. *General and Comparative Endocrinology*, **24**, 249–56.

Tsuneki, K., Kobayashi, H., Yanagisawa, M. & Bando, T. (1975) Histochemical distribution of monoamines in the hypothalamo-hypophysial region of the lamprey, *Lampetra japonica*. *Cell and Tissue Research*, **161**, 25–32.

Van Dongen, P. A. M., Hökfelt, T., Grillner, S., Verhofstad, A. A. J., Steinbusch, H. W. M., Cuello, A. C. & Terenius, L. (1985). Immunohistochemical demonstration of some putative neurotransmitters in the lamprey spinal cord and spinal ganglia: 5-hydroxytryptamine, tachykinin, and neuropeptide-Y-immunoreactive neurons and fibers. *Journal of Comparative Neurology*, **234**, 501–22.

Van Dongen, P. A. M., Theodorsson-Norheim, E., Brodin, L., Hökfelt, T., Grillner, S., Peters, A., Cuello, A. C., Forssmann, W. G., Reinecke, M., Singer, E. A. & Lazarus, L. H. (1986). Immunohistochemical and chromatographic studies of peptides with tachykinin-like immunoreactivity in the central nervous system of the lamprey. *Peptides*, **7**, 297–13.

Vigh, B. & Vigh-Teichmann, I. (1973). Comparative ultrastructure of the cerebrospinal fluid contacting neurons. *International Revue of Cytology*, **35**, 189–51.

Wächtler, K. (1974). The distribution of acetylcholinesterase in the cyclostome brain. I. *Lampetra planeri*. *Cell and Tissue Research*, **152**, 159–270.

Wächtler, K. (1975). The distribution of acetylcholinesterase in the cyclostome brain. II. *Myxine glutinosa*. *Cell and Tissue Research*, **159**, 109–20.

Wächtler, K. (1983). The acetylcholine-system in the brain of cyclostomes with special references to the telencephalon. *Journal für Hirnforschung*, **24**, 63–70.

Wicht, H. & Northcutt, R. G. (1990). Retinofugal and retinopetal projections in the pacific hagfish, *Eptatretus stouti* (Myxinoidea). *Brain, Behavior and Evolution*, **36**, 315–28.

Wicht, H. & Northcutt, R. G. (1992). The forebrain of the pacific hagfish: a cladistic reconstruction of the ancestral craniate forebrain. *Brain, Behavior and Evolution*, **40**, 25–64.

Wilson, M. V. H. & Caldwell, M. W. (1993) New silurian and devonian forktailed 'thelodonts' are jawless vertebrates with stomachs and deep bodies. *Nature*, **361**, 442–4.

3
Localization of catecholamines in the brains of Chondrichthyes (cartilaginous fishes)

S. L. Stuesse, W. L. R. Cruce and R. G. Northcutt

Introduction

Phylogenetic considerations

Cartilaginous fishes are represented by two major radiations which diverged over 350 million years ago, the elasmobranchs and the holocephalians (Carroll, 1988). There have been several hundred million years for the two groups to develop independently (Lovejoy et al., 1991). Fossil evidence suggests that living holocephalians are a much older form than modern elasmobranchs which arose approximately 200 million years ago (Carroll, 1988). Within the elasmobranchs, several major groups evolved, including squalomorph sharks, galeomorph sharks, and batoids (skates, rays and sawfish). The elasmobranchs are widely distributed in the oceans. There are about 350 species of sharks, and an even greater number of skates and rays (McEachran, 1990). Among sharks, galeomorphs have the largest brains relative to their body weight, in fact, the brain weight body weight ratio of some cartilaginous fishes overlaps that of mammals (Bauchot, Platel & Ridet, 1976; Northcutt, 1978, 1989). Squalomorph sharks have smaller brain to body weights and retain more primitive characters than do galeomorph sharks (Compagno, 1973, 1977; Carroll, 1988; Northcutt, 1989). The heterodontids, horn sharks, which may have retained many primitive characteristics, are a key species to examine. Heterodontids alternately have been grouped with extinct, ancestral sharks (Maisey, 1984) or with galeomorph sharks (Compagno, 1973).

The sister-group to all Elasmobranchii is the Holocephali (Schaeffer & Williams, 1977; Carroll, 1988; Northcutt, 1989). The living Holocephali, consisting of one order composed of six genera, have many specializations that distinguish them from elasmobranchs (Nelson, 1984; Carroll, 1988). Compared to elasmobranchs, the brain of a representative holocephalian, *Hydrolagus colliei*, is unusual in that the telencephalon is small relative to the brainstem, the diencephalon is narrow and elongated, and the olfactory bulbs are small (compare top to bottom in Fig. 3.1; Northcutt, 1978; Smeets, Nieuwenhuys & Roberts, 1983). Because holocephalians are very delicate, do not survive well in captivity, and live in deep waters, few studies have been done on their brains (for examples, see Kuhlenbeck & Niimi, 1969; Smeets et al., 1983; Lovejoy et al., 1991). Among batoids, there is considerable brain variation, for example, the thornback guitarfish, *Platyrhinoidis triseriata*, has a relatively smaller cerebellum and telencephalon than either rajoid skates to which it is closely related, or its sister group of rays (Compagno, 1977; Northcutt, 1978, 1989).

Previous catecholaminergic research

Early studies describing the distribution of catecholamines in Chondricthyans relied on either the histochemical fluorescent techniques of Falck and Hillarp (Wilson & Dodd, 1973; Meurling & Björklund, 1970) or the distribution of an enzyme present in monoamine-containing pathways (serotonin and catecholamines), monoamine oxidase (Urano, 1971; Kusunoki, Tsuda & Takashima, 1973). These older techniques are not as sensitive as the immunohistochemical techniques that have been used recently (Meredith & Smeets, 1987; Roberts & Meredith, 1987; Northcutt, Reiner & Karten, 1988; Stuesse, Cruce & Northcutt, 1990, 1991a,b; Stuesse & Cruce 1991, 1992, 1993; Stuesse, Stuesse, & Cruce, 1992; Meade,

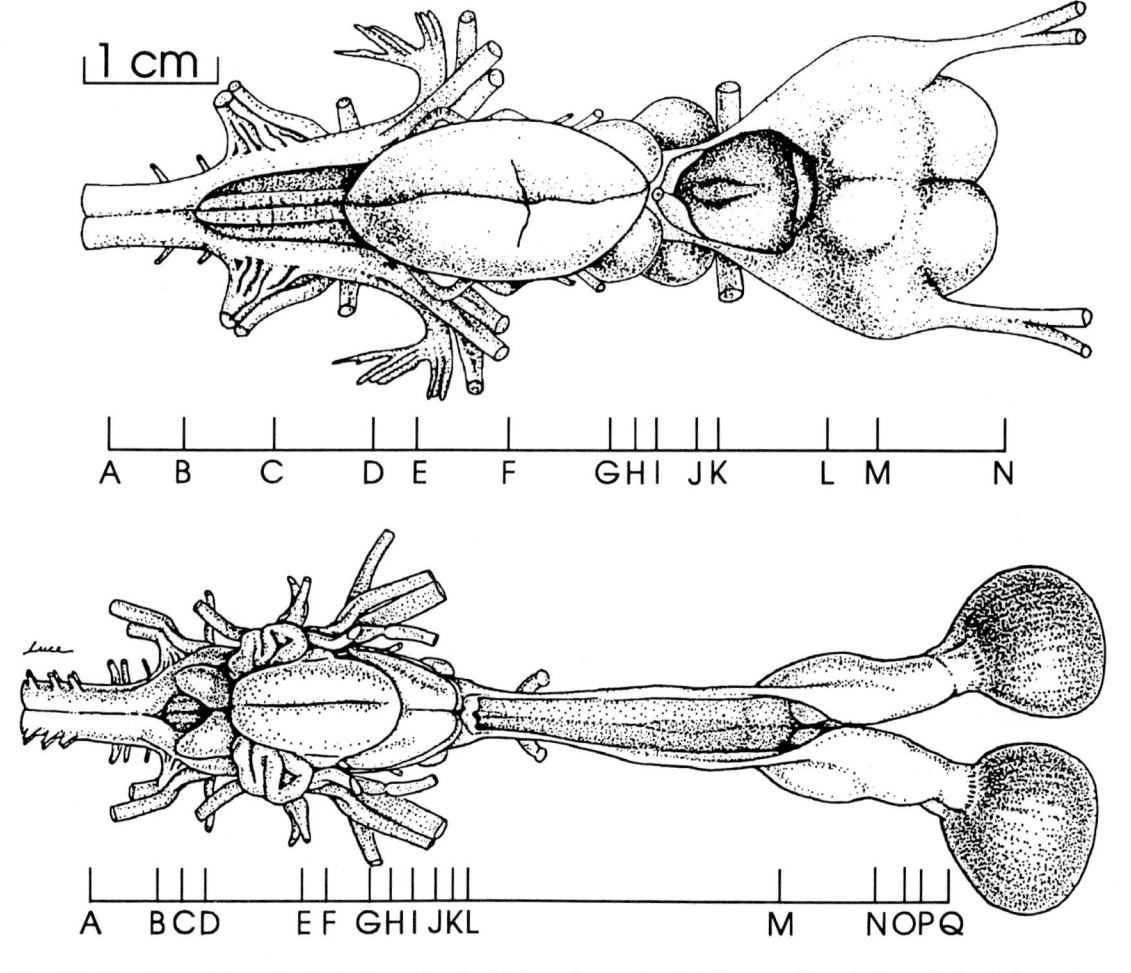

Fig. 3.1. Top: Dorsal view of a brain from the shark *Heterodontus francisci*. Bottom: Dorsal view of a brain from the holocephalian, *Hydrolagus colliei*. Letters below the brain indicate the levels of the transverse sections shown in Figs. 3.2 and 3.3, respectively. Modified from Northcutt, 1978.

Stuesse, & Cruce, 1993), nor do the older techniques permit a clear distinction between the presence of indoleamines and catecholamines (Dahlström & Füxe, 1964). However, the general distribution patterns of monoamines could be discerned and subsequent work has built on this foundation. For immunohistochemical identification of catecholamines in the brains of cartilaginous fishes, the most commonly used antibodies have been those against tyrosine hydroxylase. In studies by Meredith and her coworkers (Meredith & Smeets, 1987; Roberts & Meredith, 1987) on the forebrain, midbrain, and spinal cord of a skate, *Raja radiata*, antibodies were directed against dopamine, thus clarifying which of the cells and fibers contain dopamine.

Descriptions of catecholaminergic cells and fibers

The results described in this paper are based on the study of a variety of cartilaginous fishes. The fishes examined and their taxonomic relationships are shown in Table 3.1. There are no reports yet on the distribution of catecholamines in the superorder Squatinomorphii, angel sharks, and the literature on the order Carcharinoformes in the galeomorph sharks is still sparse. The one study of Carcharinoformes published since 1984 used histofluorescence and described the distribution of aminergic neurons in the hypothalamus (Rodriguez-Moldes & Anadón, 1987). Many of the detailed descriptions of cell loca-

Table 3.1. *Catecholamine localization in the central nervous system of the class Chondrichthyes*

Subclass Holocephali: Hydrolagus colliei, Stuesse & Cruce 1991; Stuesse et al., 1992
Subclass Elasmobranchii
   Superorder Squalomorphii, Order Squaliformes
      Squalus acanthias, Northcutt et al., 1988; Stuesse & Cruce 1992; Stuesse et al., 1992
   Superorder Batoidea
      Order Rajiformes
         Raja radiata, Meurling & Björklund; Meredith & Smeets 1987; Roberts & Meredith 1987
         Raja binoculata and *Raja rhina,* Stuesse et al., 1992
         Playrhinoidis triseriata, Stuesse et al., 1990, 1992
      Order Myliobatiformes
         Myliobatis californica, Stuesse et al., 1992
Superorder Squatinomorphii. No reports in literature
Superorder Galeomorphii
   Order Heterodontiformes
      Heterodontus japonicus, Kusunoki et al., 1973
      Heterodontus francisci, Stuesse et al., 1991, 1992
   Order Carchariniformes
      Mustelus manazo, Kusunoki et al., 1973
      Scyliorhinus canicula, Wilson & Dodd 1973; Rodriguez-Moldes & Anadón 1987
      Triakis scyllia, Urano, 1971; Kusunoki et al., 1973

Compagno's taxonomic scheme is used (1973, 1977).

tion and sizes described in this chapter were obtained from studies in our laboratories on the sharks *Heterodontus francisci* (Stuesse et al., 1991b, 1992) and *Squalus acanthias* (Stuesse & Cruce 1992; Stuesse et al., 1992), the batoids *Myliobatis, Raja* (Stuesse et al., 1991a, 1992), and *Platyrhinoidis* (Stuesse et al., 1990, 1992), and the holocephalian, *Hydrolagus* (Stuesse & Cruce 1991; Stuesse et al., 1992). When the localization and the shapes of tyrosine hydroxylase immunoreactive (THi) cells are similar for the various species studied, only one description is given. Differences between species are noted. The detailed fiber description is based mostly on analysis of *Heterodontus* and *Hydrolagus* brains in the authors' laboratories and has not been published previously. We chose one fish, *Heterodontus,* as an example of the results and illustrated fibers and cells in representative transverse sections from this fish (Fig. 3.2). For comparison purposes, we have also presented cell distribution in the holocephalian, *Hydrolagus colliei* (Fig. 3.3). The levels at which these sections were obtained are shown in Fig. 3.1. We used the atlas of Smeets and his colleagues for general chondrichthyan terminology (Smeets et al., 1983), but we used mammalian terminology for the reticular formation. For telencephalic divisions in *Hydrlagus,* we followed the nomenclature in the Smeets atlas, but for the other fish, we used the nomenclature for *Squalus* (Northcutt et al., 1988). We used two telencephalic nomenclatures because the telencephalon of *Hydrolagus* is so different from the other fishes that homologies between areas are very hypothetical. A unified nomenclature for the telencephala of cartilaginous fishes has not yet been published. The reticular nomenclature has been used in our previous publications on cartilaginous fishes and is based on the reticular descriptions in mammals by Newman (1985a,b). The results are described in a caudal to rostral order.

Spinal cord

Kusunoki and his coworkers (1973) reported that the central grey and substantia gelatinosa of the spinal cord of three types of sharks contained monoamine oxidase activity. Roberts and Meredith (1987) later described dopaminergic cells in the entire rostrocaudal extent of the spinal cord in a ray, *Raja radiata.* The cells are located in the subependymal layer, just ventral to the central canal. Typically, two or three dopaminergic cells were found per 40 µm thick section. These oval, 10–15 µm diameter cells extend one short process into the central canal and a longer process ventrally. By use of an antibody directed against tyrosine hydroxylase, we found that THi cells ventral to the central canal are present in every cartilaginous fish that we examined. We only inspected the rostral spinal cord, but, like Roberts' and Meredith's description (1987), most of the cells are oval (8 × 10 µm to 8 × 12 µm) and are clustered just ventral to the central canal (Figs. 3.2(a), 3.3(a)). These cells have a short process that extends to the central canal and a longer process that extends along the midline to the ventral surface of the spinal cord (Fig. 3.4(a)). Unlike the cells described by Roberts and Meredith (1987), we found from two to 10 cells per section, that some of the cells have round somata (8 × 8 µm), and that these cells continue rostrally into the myelencephalon to the level of the obex. An occasional cell with an oval soma (10 × 24 µm) was seen in the dorsolateral portion of the dorsal horn (Roberts & Meredith, 1987; Stuesse et al., 1990, 1991b; Stuesse & Cruce, 1991, 1992).

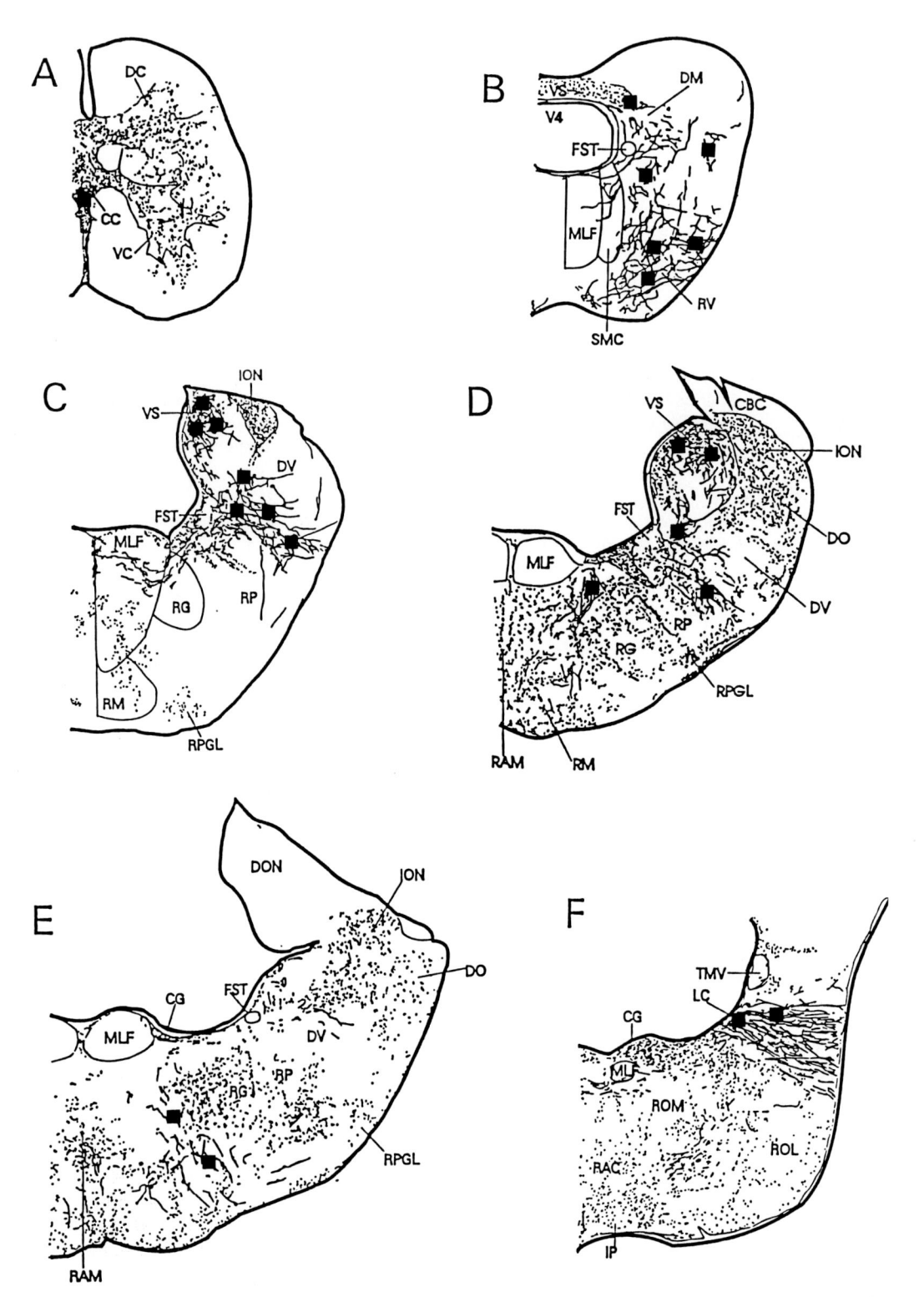

Fig. 3.2. Drawings of transverse sections from the brain of a heterodontid shark, *Heterodontus francisci*, showing the locations of tyrosine hydroxylase immunoreactivity (non-outline lines). The general location of THi cells is indicated by the squares. The sections are arranged from caudal to rostral, and the level of each section is shown in Fig. 3.1, top. For abbreviations, see list.

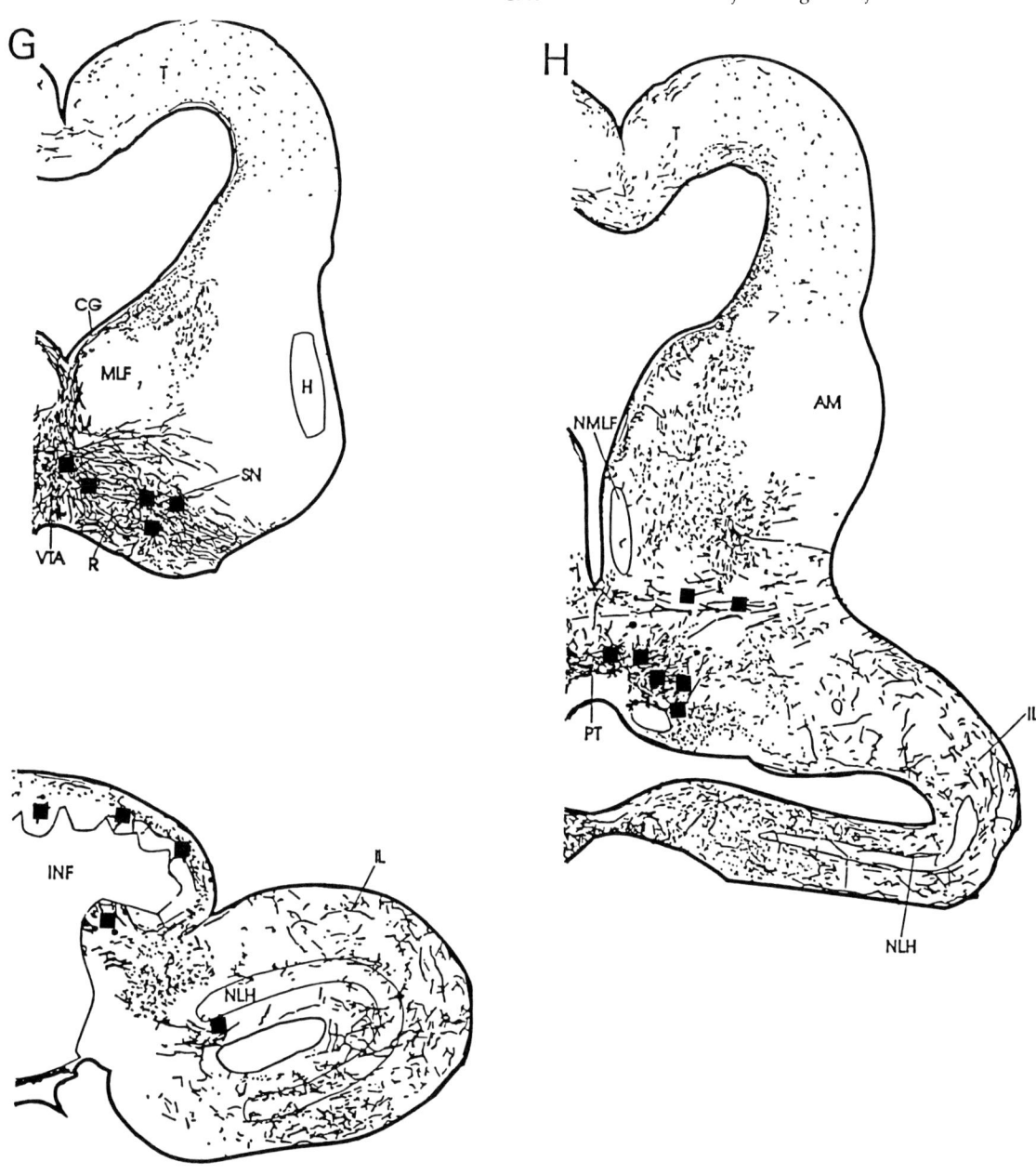

Catecholamines in brain of cartilaginous fishes 25

Fig. 3.2. (*cont.*)

Fig. 3.2. (cont.)

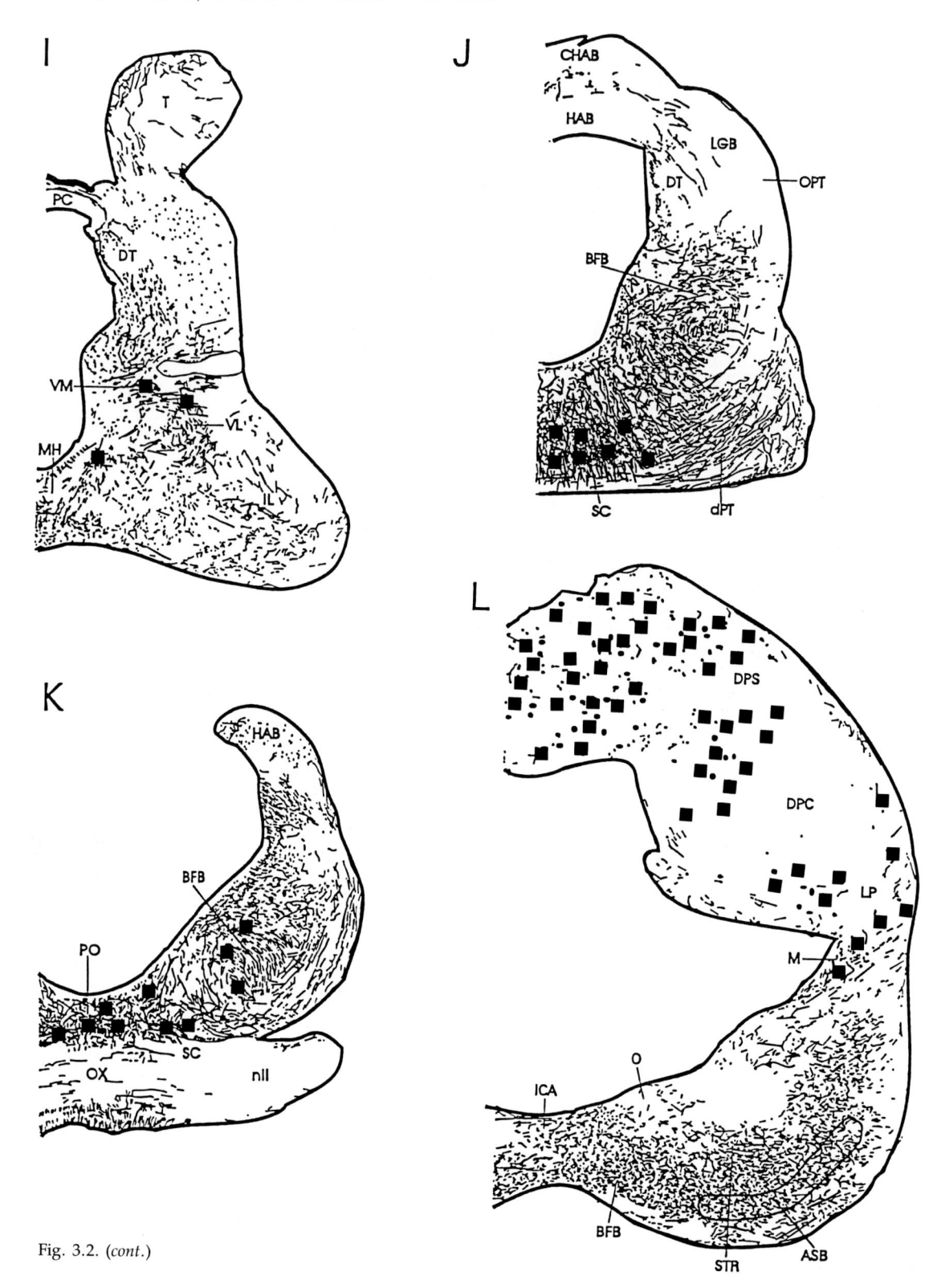

Fig. 3.2. (cont.)

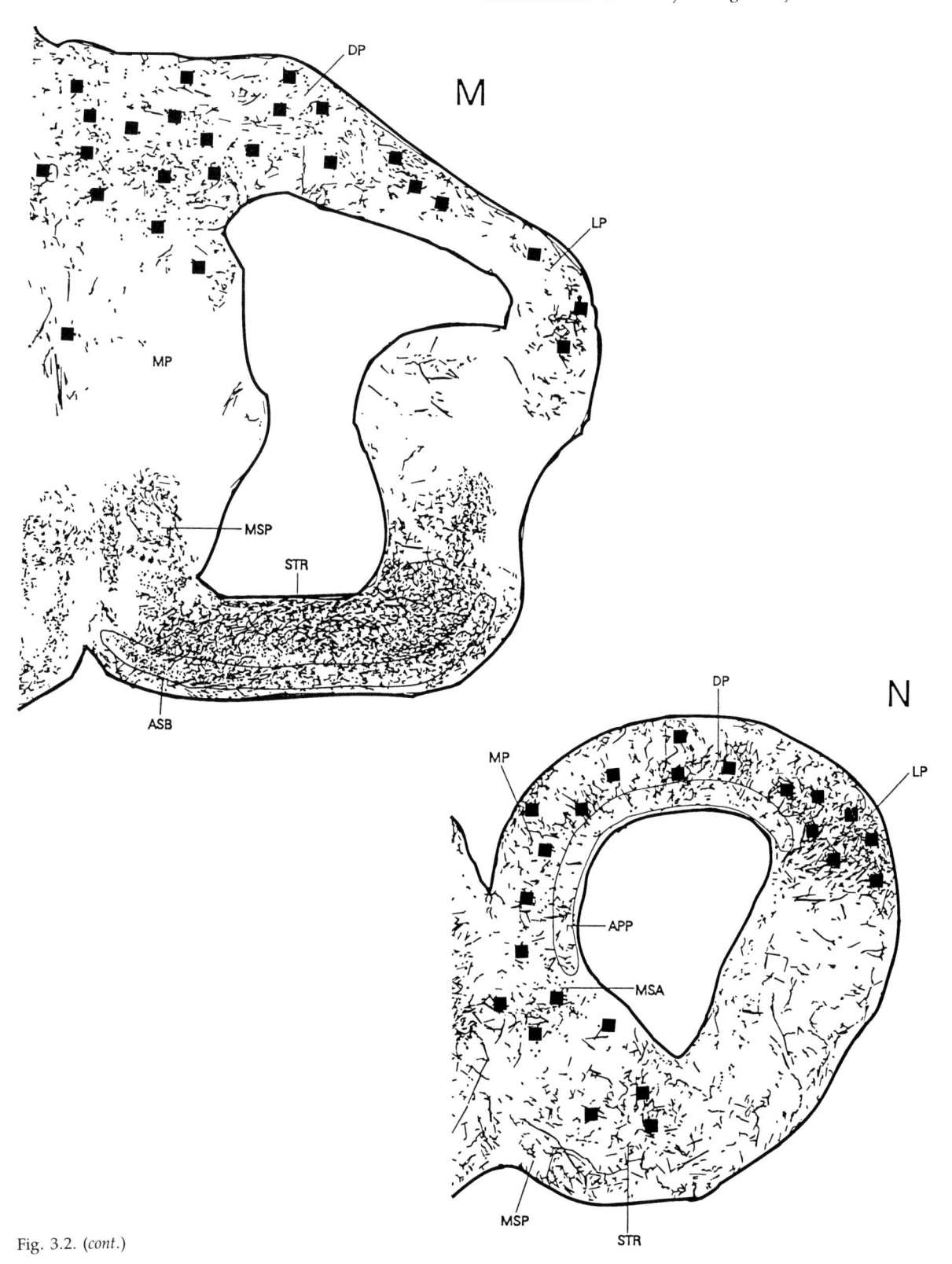

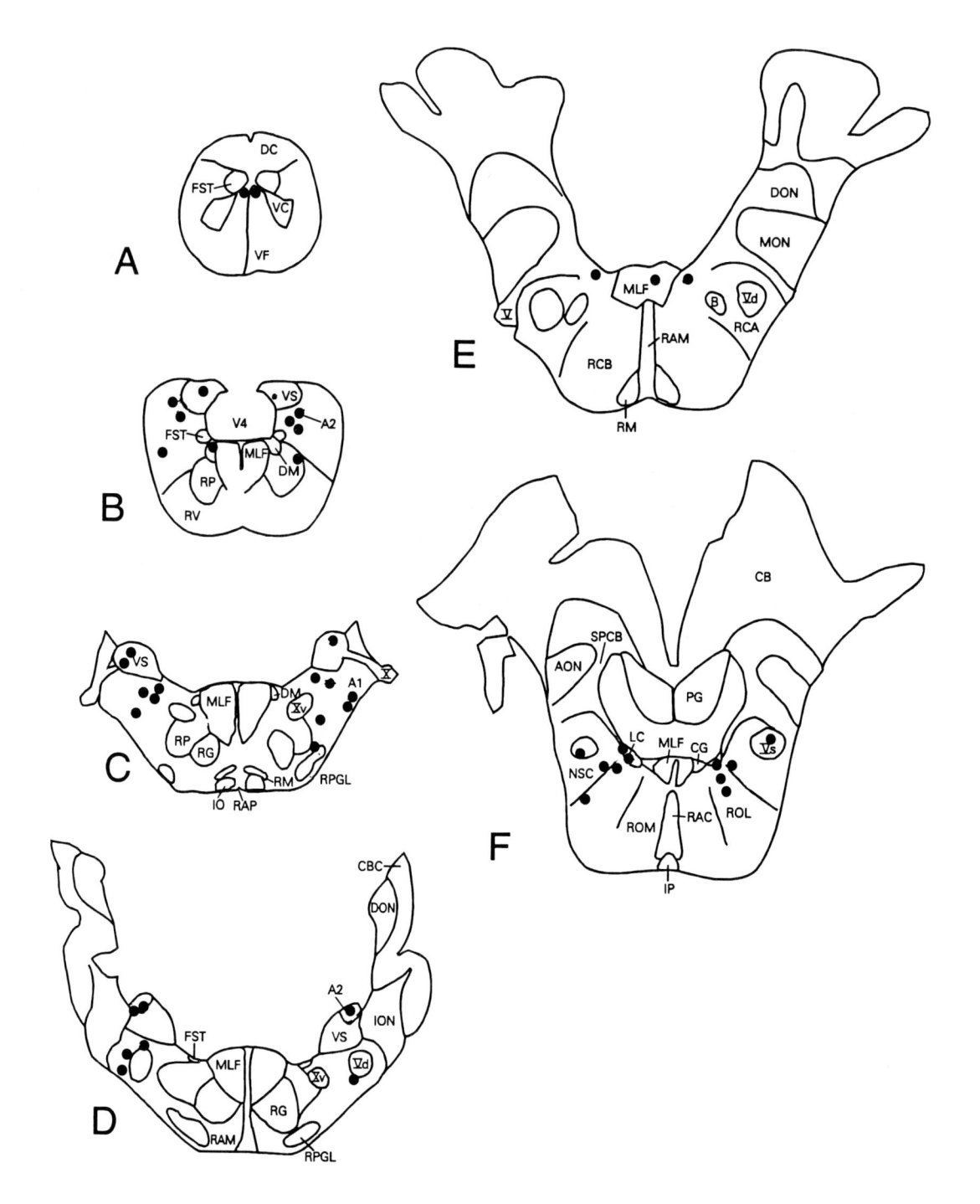

Fig. 3.3. Drawings of transverse sections from the brain of a holocephalian, *Hydrolagus colliei*, showing the locations of tyrosine hydroxylase positive cells (dots). The sections are arranged from caudal to rostral, and the level from which each section was taken is shown in Fig. 3.1, bottom. For abbreviations, see list.

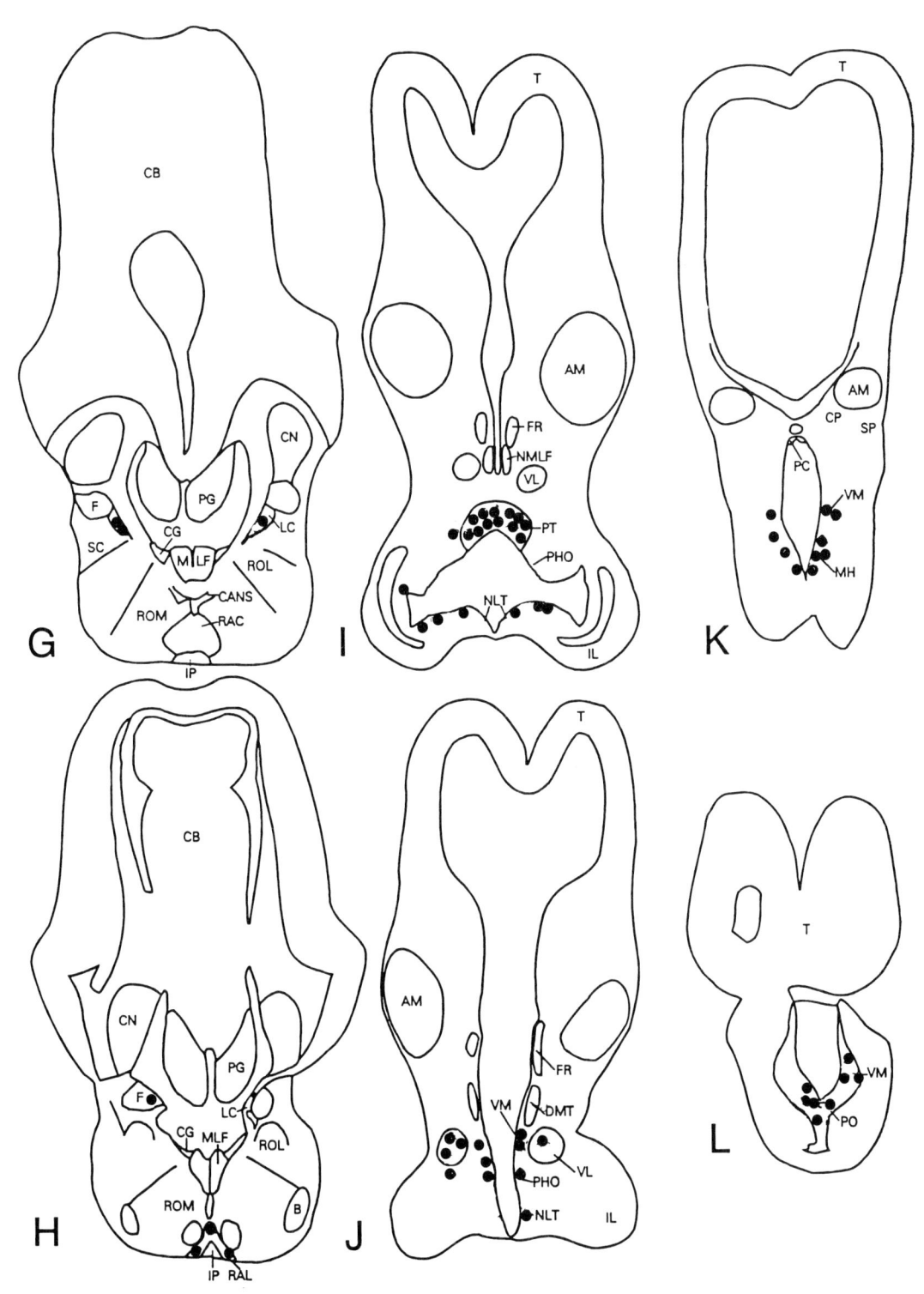

Fig. 3.3. (*cont.*)

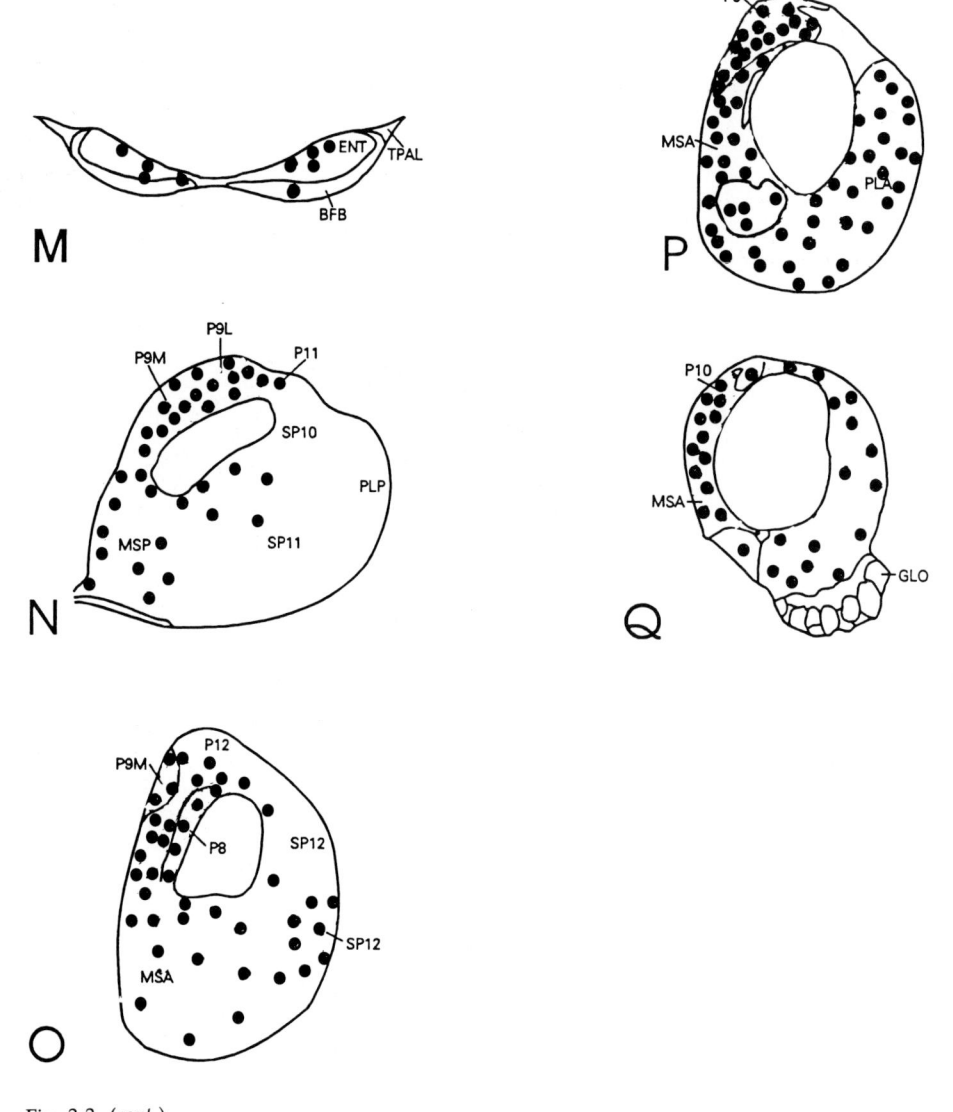

Fig. 3.3. (cont.)

Dopaminergic-positive fibers are confined to fine, varicosity-bearing processes in gray matter and are densest in the ventral horn (Roberts & Meredith, 1987). Tyrosine hydroxylase-positive fibers are sparsely scattered throughout the dorsal and ventral horns, including the substantia gelatinosa, and extend into the lateral and ventrolateral funiculi (Fig. 3.2(a)).

Myelencephalon
The ventral sensory columns of sharks contain moderate to strong monoamine oxidase activity (Kusunoki et al., 1973). At least some of this activity is due to catecholamines as tyrosine hydroxylase-positive cells are found in the visceral sensory complex of sharks, skates, rays, and the holocephalian (Figs. 3.2(c),(d), 3.3(b)–(d), 3.4(b)). From caudal to the obex to the level of nerve X (Fig. 3.4), the THi cells in the visceral sensory complex are located dorsally and are somewhat scattered, while at the level of nerve IX, a cluster of THi cells is found in a dorsolateral subnucleus of the visceral sensory complex.

More cells are found in the caudal portions of the complex than in the rostral portions. Most of the cells are small with round (8 × 8 μm) or oval (6 × 10 μm to 8 × 12 μm) somata, but in the rostral portion of the vagal lobe, occasionally rounded, larger multipolar cells (15 × 20 μm) are located very close to the lip of the fourth ventricle.

Multipolar THi cells with medium (8 × 12 μm to 12 × 12 μm) or large (12 × 20 μm to 20 × 25 μm) somata are loosely scattered throughout the dorsal and ventrolateral reticular formation from the caudal myelencephalon to the level of entry of cranial nerve X (Figs. 3.2(b), 3(c), 5(c)). Their somata are rounded, oval, or triangular. At caudal levels, the cells are mainly in the ventral portion of the brain in nucleus reticularis ventralis (RV), and their processes have no particular orientation. At more rostral levels, many of the THi cells are in nucleus reticularis parvocellularis or lateral to it. In the mid-myelencephalon, some of their processes have a ventrolateral to dorsomedial

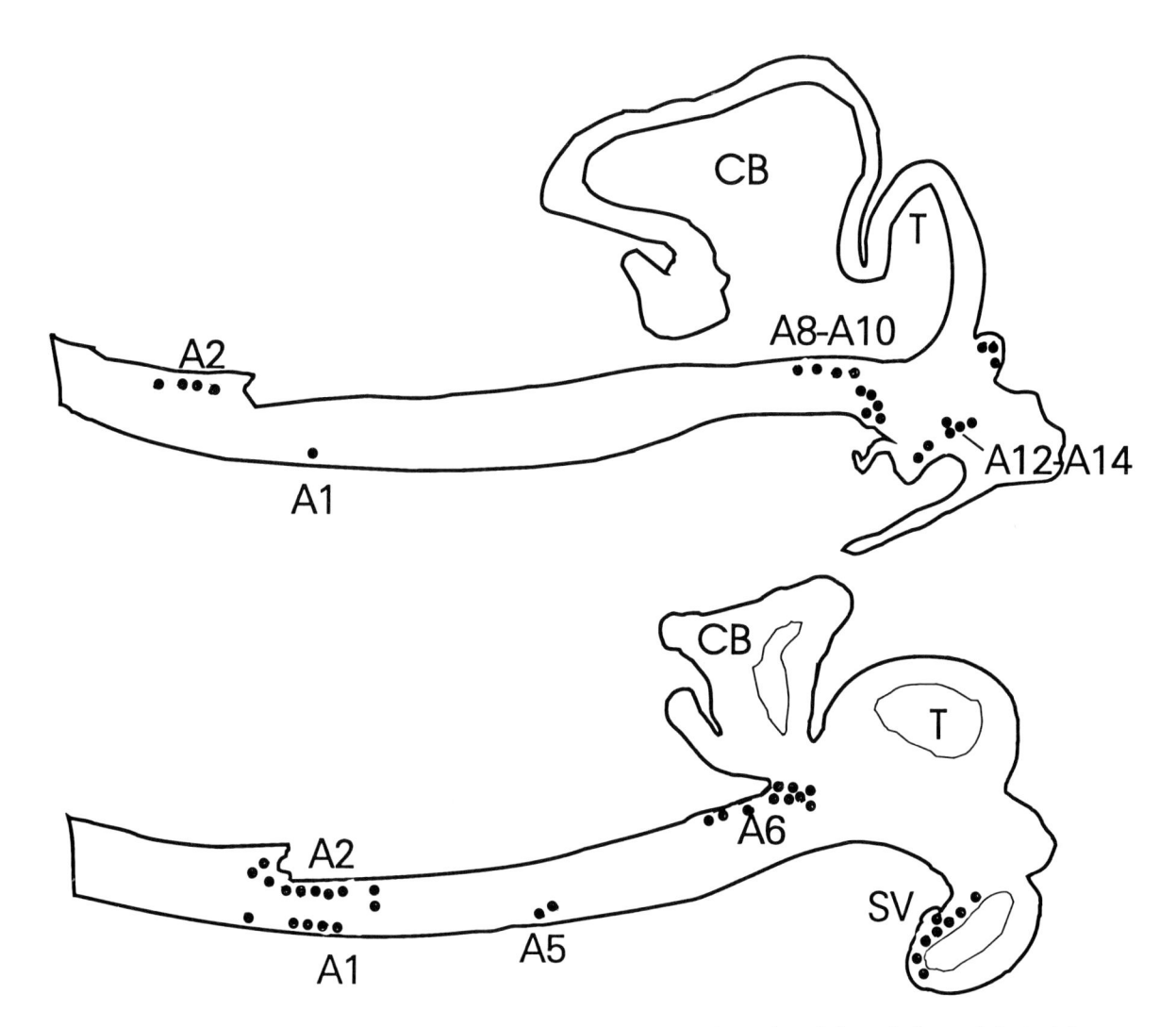

Fig. 3.4. Drawings of parasagittal sections from the brainstem (rostral spinal cord through diencephalon) of the spiny dogfish, *Squalus acanthias*, showing the locations of tyrosine hydroxylase positive cell groups that may be homologous to mammalian cell groups (the 'A', catecholaminergic, series). Other abbreviations are SV: saccus vasculosus, T: tectum, CB: cerebellum. The top drawing is 0.5 mm from the midline while the bottom drawing is 1.0 mm from the midline. The relationship between these cells groups and others described in the literature is summarized in Table 3.2.

orientation (Fig. 3.2(c)), but others have no particular orientation. The dorsal extension of this cell group is ventral to the visceral sensory lobe near the level of entry of cranial nerve X. In the batoids, these cells tend to be much closer to the ventrolateral surface of the brainstem than is the case in the sharks and in a holocephalian. These THi myelencephalic cell groups extend through the myelencephalon to the level of entry of cranial nerve VIII. Additionally, in *Hydrolagus*, the caudal portion of the motor nucleus of the vagus nerve also contains THi cells. The rhombencephalon lacks THi cells from the level of the entry of cranial nerve VII to past the level of entry of nerve V.

The caudal myelencephalon, at the level of the obex, contains very few THi fiber varicosities (Fig. 3.2(b)). The visceral sensory complex, rostral to the obex has lightly scattered fibers, as do the remnants of the ventral horns. At slightly more rostral myelencephalic levels (levels of entry of cranial nerves IX and X), fine THi varicosities of moderate intensity are present in the intermediate nucleus of the octavolateralis area and in or near the dorsal motor nucleus of cranial nerve X (Fig. 3.2(c)). Light THi varicosities are found along the ventral edge of the fourth ventricle in the central gray, along the midline of the brain (in raphe nuclei), and in several reticular nuclei, reticularis (r.) magnocellularis, r. gigantocellularis, r. parvocellularis, and r. paragigantocellularis lateralis. At this level of the brain, fibers from the THi cells in the brainstem are seen sweeping medially through the medial longitudinal fasciculus to the midline where they ascend in the brainstem (Fig. 3.2(c)). At midmyelencephalic levels, light THi fiber varicosities cover most of the reticular formation (Fig. 3.2(d)). The descending nucleus of cranial nerve V is the only major nucleus devoid of THi fiber varicosities. Moderately dense varicosities continue in the intermediate nucleus of the octavolateralis area, and very dense varicosities are in the motor nucleus of cranial nerve VII. These varicosities spread in a narrow band from the motor nucleus cranial of nerve VII in the dorsal brainstem to the pial surface where they are concentrated just dorsal to the ventrolateral sulcus.

Metencephalon

In *Squalus* and in *Heterodontus*, between the levels of cranial nerves V and VII, a few THi cells are located in the reticular formation near the rostrocaudal transition from r. gigantocellularis to r. pontis caudalis pars beta, more rostrally. These multipolar cells with oval somata (10 × 16 µm to 12 × 16 µm) have processes oriented in a dorsal to ventral direction. In contrast, *Platyrhinoidis* has many THi cells (15 × 20 µm) located in the ventral metencephalon. Some of the cells are in r. magnocellularis (RM) and have processes oriented parallel to the ventral surface of the brain. Others are in the more ventral portions of nucleus r. pontis caudalis (RCA and RCB) and continue rostrally in the lateral portions of reticularis pontis oralis (ROL). Tyrosine hydroxylase positive cells in this location are also found in *Squalus*, but not in *Hydrolagus* (Fig. 3.3(e)) or in *Heterodontus*. In addition, a few small cells with oval somata (8 × 10 µm), each with one prominent process, are in the central gray near the nucleus of cranial nerve VII. These cells are present in all the fishes examined except *Hydrolagus*.

In the caudal metencephalon, fine THi terminals are spread over most of the reticular formation, with the exception of r. magnocellularis, where they are very sparse (Fig. 3.2(d),(e)). Terminals of moderate density are present in the central gray, in the intermediate octavolateralis nucleus, and in raphe magnus. The descending nucleus and tract of V and the dorsal octavolateralis nucleus are devoid of THi terminals (Fig. 3.2e)).

In the rostral metencephalon, a cluster of THi cell are in locus coeruleus near the ventrolateral lip of the fourth ventricle and in the subcoeruleus area ventrolateral to the locus coeruleus (Figs. 3.2(f), 3.3(F),(g), 3.5(d)). The somata of locus coeruleus cells are triangular (16 × 20 µm) or oval (8 × 16 µm) with processes that extend ventrolaterally towards the pial surface of the brainstem. The cell group ventrolateral to the locus coeruleus, the nucleus reticularis subcoeruleus (Fig. 3.3(g)) contained multipolar cells which are larger (25 × 13 µm or 20 × 25 µm) and more elongated than those in the first group. The dendrites of cells in the second group have a dorsomedial to ventrolateral orientation. Moderately dense THi fibers terminate in the central gray, in another area ventromedial to the locus coeruleus, in the nucleus raphe centralis superior, and in the interpeduncular nucleus. Tyrosine hydroxylase positive fibers cross in the dorsal metencephalon in the midline just ventral to the medial longitudinal fasciculus, and in more ventral stria that traverse the reticular formation. However, the nuclei reticularis pontis oralis pars medialis and pars lateralis have very few THi terminals.

There are no THi cells in the cerebellum. Most of the cerebellum had very fine, very light THi fiber terminals that are distributed uniformly in the granular layers. Thus at least some of the faint monoamine oxidase activity described in sharks by

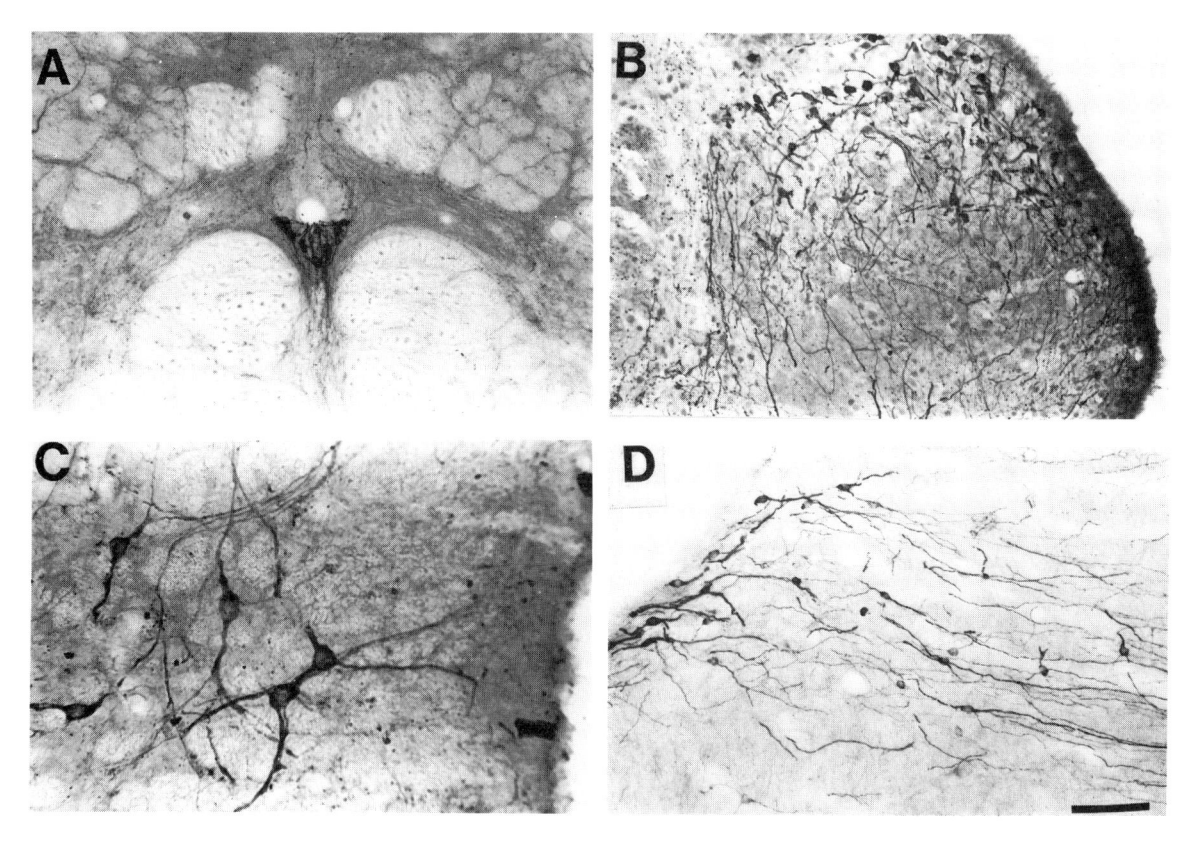

Fig. 3.5. Photomicrographs of examples of tyrosine hydroxylase positive cells in brains of cartilaginous fishes. (a) cells ventral to the central canal of the spinal cord (*Heterodontus*), (b) A1 cell group in the myelencephalon (*Platyrhinoidis*), (c) A2 cell group within the visceral sensory lobe (*Squalus*), (d) A6 cell group in the locus coeruleus (*Heterodontus*). Bar = 100 μm for panels (a), (b) and (d) and 60 μm for (c).

Kusunoki and his colleagues (1973) could be due to catecholamines.

Mesencephalon

Two groups of dopaminergic cells have been identified in the mesencephalon in a ray, *Raja radiata* (Meredith & Smeets, 1987). These groups are both present at the level of cranial nerve III, and have been identified as a ventral tegmental area and a putative substantia nigra. These two mesencephalic DAi cell groups extend into the hypothalamus where they merge with the DAi cell group in the posterior tubercle. Tyrosine-hydroxylase positive cells were subsequently described in a similar position in a shark, *Squalus acanthias*, by Northcutt and his collaborators (1988). All subsequent elasmobranchs examined contain THi cells in these two area (Figs. 3.2 g, 3.6 (a), (b)). The cells in the ventral tegmental area have oval (10 × 12 μm to 13 × 20 μm) somata that are elongated in the dorsoventral direction. The processes of many of these cells extend dorsally up the midline to the third ventricle and ventrally to the pial surface of the brain where they spread out laterally. Tyrosine hydroxylase positive cells on the lateral edges of the ventral tegmental nucleus have processes that extend laterally. Tyrosine hydroxylase positive cells within the substantia nigra tend to be scattered around the red nucleus. The somata of many of these cells are medium (12 × 20 μm to 14 × 18 μm) or large (20 × 32 μm). The somata of cells within the substantia nigra have a heterogeneous morphology (fusiform, polygonal, or triangular) and most have processes with no predominant dendritic orientation, however, some cells have processes that extend laterally. Unlike the elasmobranchs, the holocephalian, *Hydrolagus*, lacks THi cells in the substantia nigra and ventral tegmental area. However, at the level of the oculomotor nerve, THi cells with tiny

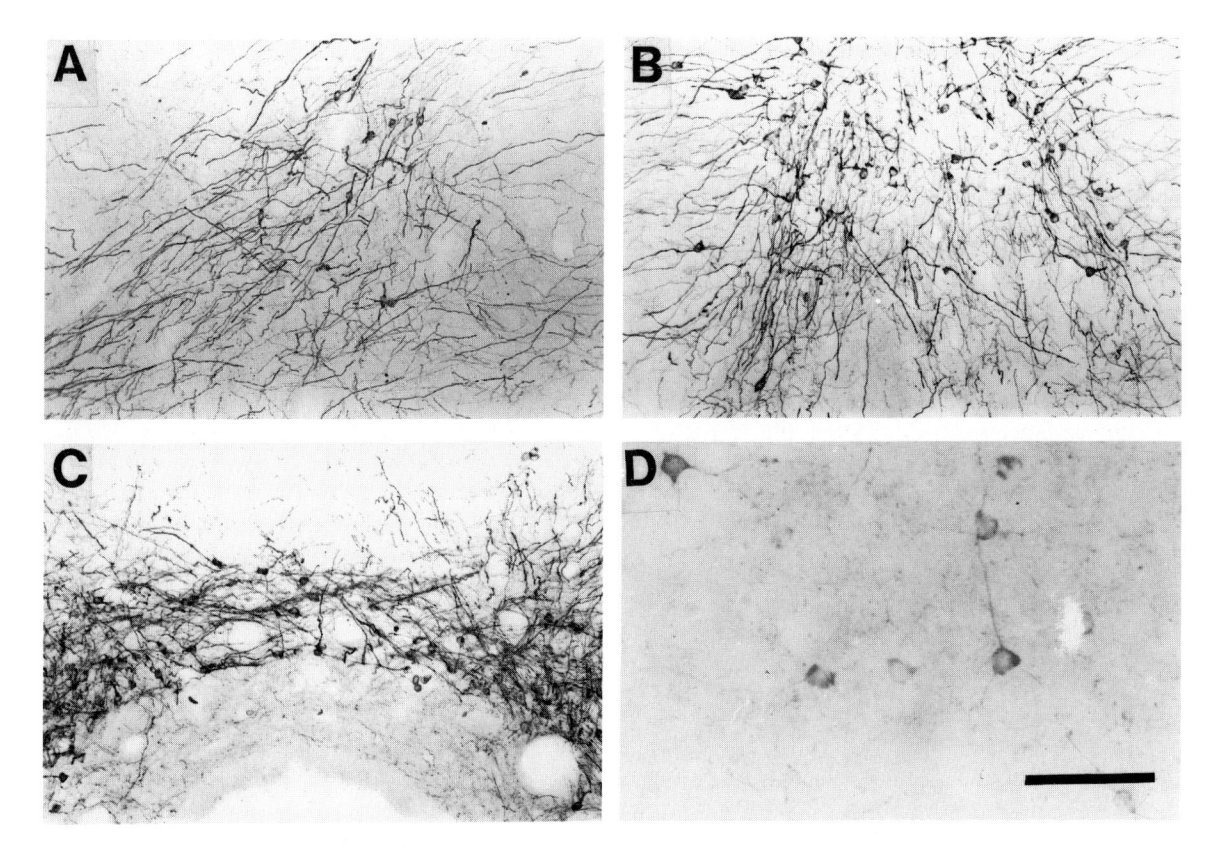

Fig. 3.6. Photomicrographs showing examples of tyrosine hydroxylase positive cells in brains of cartilaginous fishes. (a) substantia nigra (*Heterodontus*), (b) ventral tegmental area (*Heterodontus*), (c) posterior tuberculum (*Hydrolagus*), (d) dorsal telencephalon (*Hydrolagus*). Bar = 200 µm for panels (a), (b), and (c) and 50 µm for (d).

fusiform (2 × 12 µm) or triangular (4 × 6 µm) somata are found just under the dorsomedial sulcus in *Hydrolagus* but not in the other fish examined.

In *Heterodontus*, a few THi cells with rounded somata (8 × 8 µm) are found in the most superficial layer of the optic tectum. In *Squalus*, similarly shaped THi cells are found in intermediate layers of the rostral optic tectum. THi cells were not seen in the tectum in the other fishes examined.

Tyrosine hydroxylase positive fibers, many of which are dopaminergic (Meredith & Smeets, 1987) are scattered over most of the midbrain tegmentum except the nucleus ruber (Fig. 3.2g; Northcutt et al., 1988). They are of light intensity in the lateral parts of the tegmentum and slightly denser in the other parts. Moderately dense THi fibers are in the area surrounding the red nucleus, in the central gray, and in the deep layers of the tectum. A bundle of THi fibers ascends in the midbrain tegmentum to the tectum (Fig. 3.2 (g); Northcutt et al., 1988). Within the tectum, the distribution of monoaminergic fibers varies in the different species of fishes (Kusunoki et al., 1973: Stuesse & Cruce, unpublished observations). In *Heterodontus francisci*, THi fibers are fairly evenly distributed throughout all the tectal layers, but are slightly denser in deeper layers. In deeper layers, THi fibers tend to run transversely through the layers and to cross in the tectal commissure. In *Squalus* (Northcutt et al., 1988) and *Hydrolagus*, THi fibers are almost totally absent in intermediate and superficial layers, but otherwise, the pattern is similar to that in seen in *Heterodontus*. In *Myliobatus* and *Raja*, THi fibers in the tectum are very light and are concentrated in the outermost layers with very little label in intermediate and deep layers. THi fibers could be seen ascending through deep and intermediate layers to reach the superficial tectal layers in these latter two fish. However, DAi fibers in Raja are concentrated in the deep layers and the more lateral parts of the tectum (Meredith & Smeets, 1987).

Diencephalon

Monoaminergic cells are present in the medial hypothalamic nucleus and the nucleus of the inferior lobe of the hypothalamus (Wilson & Dodd, 1973; Meredith & Smeets, 1987: Rodriguez-Moldes & Anadon, 1987; Northcutt et al., 1988; Stuesse et al., 1990, 1991b; Stuesse & Cruce, 1991, 1992) as well as in the pituitary (Meurling & Björklund, 1970; Wilson & Dodd, 1973). Most of these cells are probably dopaminergic (Meredith & Smeets, 1987). Within the hypothalamus, THi cells with oval somata (8 × 12 µm) are found in the nucleus periventricularis hypothalami (PT, Figs. 3.2(h), 3.3(i), 3.6(c) and the medial hypothalamic nucleus (MH, Figs. 3.2(i), 3.3(k)). The processes of these cells have no particular orientation. THi labeled cells in the posterior tuberculum (the nucleus tuberculi posterioris of Meredith & Smeets, 1987) are located just rostral to the saccus vasculosus and continue into the midbrain THi/DAi cell groups. The processes of these multipolar cells (12 × 12 µm to 15 × 20 µm) have no particular orientation in the dorsal portion of this nucleus, but in the caudoventral portion, the cells have horizontally oriented processes that extend laterally. Many THi cells with rounded soma (12 × 12µm) are also scattered in the saccus vasculosus. The nucleus lobi lateralis of the hypothalamus (NLH, Fig. 3.2(g)), ventral to the third ventricle, also contains scattered THi cells, but few if any of these cells are dopaminergic (Meredith & Smeets, 1987). In addition, THi cells with small somata (8 × 8µm) are found in the infundibulum.

Tyrosine hydroxylase positive fiber varicosities and dopaminergic fibers of moderate density are present in the inferior lobe of the hypothalamus (Fig. 3.2(g)), Meredith & Smeets, 1987; Northcutt et al., 1988; and unpublished observations of the authors). These THi fibers are absent within the dorsal and ventral nuclei in the inferior lobes. Many of the THi fibers traverse the inferior lobes. There are dense THi fibers in the preoptic-hypophyseal tract and in the posterior tuberculum. Again, some or all of these fibers may be dopaminergic (Meredith & Smeets, 1987). The medial hypothalamic nucleus contains few THi fibers.

The ventromedial (VM, Figs. 3.23.(i), 3(j)–(l)) and ventrolateral (VL, Fig. 3.3(j)) nuclei of the thalamus contains THi cells. Somata in the ventrolateral thalamus are oval (8 × 12µm to 15 × 18µm) with processes oriented dorsomedially to ventrolaterally. The ventromedial thalamic cells are of similar size, but their processes have no particular orientation.

Light THi fiber varicosities, which may be dopaminergic (Meredith & Smeets, 1987), cover all major thalamic nuclei (Fig. 3.2(i)). A tract containing THi fibers traverses the diencephalon just lateral to the thalamic nuclei and ventromedial to the anterior mesencephalic nucleus (Fig. 3.2(i)). This tract appears to arise in the mesencephalic tegmentum and projects to the tectum (Northcutt et al., 1988; unpublished observations of the authors).

THi (Stuesse et al., 1990, 1991b; Stuesse & Cruce, 1991, 1992) and DAi cells (Meredith & Smeets, 1987) are scattered rostral to, caudal to, and within the optic chiasm in the suprachiasmatic nucleus (SC, Fig. 3.2(j)). This cell group continues rostrally into the preoptic nucleus of the rostral diencephalon dorsal to the pallial decussation (PO, Figs. 3.2(k), 3.3(l)). Cells in the caudal preoptic area have bipolar somata (12 × 18µm) with processes oriented in the dorsoventral axis. Most of the more rostral cells have oval somata (12 × 16µm to 15 × 18µm), although some are fusiform (8 × 20µm). Their processes had no particular orientation. THi fibers are dense in the preoptic and suprachiasmatic areas. THi fibers of moderate density are found in the anterior commissure and in the basal telencephalic fasciculi (BFB, Fig. 3.2(j)–(l)), unpublished observations of the authors). Dopaminergic fibers are found in these same locations (Meredith & Smeets, 1987).

Telencephalon

Telencephala of cartilaginous fishes consist of two hemispheres with a ventricle in each. In batoids and some galeomorph sharks, the telencephalon is so expanded that the ventricles are almost obliterated in the rostral portions. In addition, in these latter two types of fishes, a large midline mass has developed. In *Hydrolagus*, the ventricle remains large and midline masses are minimal. The telencephalon can be further divided into a dorsal pallium (P) and a ventral subpallium (SP). The ventral subpallium contains among its structures the striatum, an area superficialis basalis, and a large telencephalic tract, the fasciculus basalis telencephala (basal forebrain bundle) which serves as the major fiber tract between the telencephalon and the rest of the brainstem (Smeets et al., 1983). Cell masses within the telencephalon vary within the cartilaginous fishes and, because of our sparse knowledge about these fishes, are difficult to compare. In batoids and in *Hydrolagus*, many histologically identified cell masses have been given numbers that run from a caudal to a rostral direction. Thus P3 is pallial mass number three. In the discussion that follows, reference is made to this nomenclature.

All of the chondrichthyans contain many THi cells scattered in portions of the telencephalon (Northcutt et al., 1988; Stuesse et al., 1990). The heaviest label is in the dorsal and dorsolateral pallium. In elasmobranchs, the lateral and dorsal pallium (LP, DP, DPC, DPS, Fig. 3.2(*l*)–(*n*)) contain many scattered THi cells with rounded, multipolar somata (12 × 12μm, Fig. 3.6(*d*)). Many fewer cells are scattered in the medial pallium (MP, Fig. 3.2(*m*)). A few scattered THi fibers are in the dorsal and the lateral pallium, but these fibers seem to be mostly the processes of the THi cells that are found in this area and not terminals that originate from cells in other brain areas. In the dorsal telencephalon, THi fibers are especially dense in the most lateral part of the pallium (LP, Fig. 3.2(*n*)), in area periventricularis pallialis (APP, Fig. 3.2(*n*)) and in the septal nuclei (MSP, MSA, Fig. 3.2(*m*), (*n*)).

Dopaminergic cells in the dorsal pallium were described by Meredith & Smeets (1987) in *Raja*, so at least some of the tyrosine-positive cells should contain dopamine. However, Meredith and Smeets did not describe immunopositive cells in the lateral pallium, and thus these cells may be noradrenergic or adrenergic. In addition to cartilaginous fishes (Northcutt et al., 1988; Stuesse et al., 1990) tyrosine-hydroxylase positive neurons in the telencephalon have been described in mammals, in a lungfish, and in a ray-finned fish (Specht et al., 1981; Gaspar et al., 1987; Reiner & Northcutt, 1987, 1992; Hornung, Tork & De Tribolet, 1989) although histofluorescent cells have not been found there (see Reiner & Northcutt, 1992). At least some of the telencephalic cells are DAi (Meredith & Smeets, 1987). The DAi/THi cells in chondrichthyan telencephala may be interneurons because their cell bodies are small, and their processes never seem to project to other areas of the brain.

The area superficialis basalis (ASB, Fig. 3.2(*l*), (*m*)), in the ventral pallium, contains many fewer of these small, round cells than does the dorsal pallium (PDS, Fig. 3.2(*l*)). In the ventral telencephalon, THi fibers are heavy in the basal telencephalic fasciculi (BFB, Fig. 3.2(*l*) and in the striatum (ASB, Fig. 3.2(*l*), (*m*); Northcutt et al., 1988). At least some of the fibers in the basal telencephalic fasciculi are dopaminergic (Meredith & Smeets, 1987). A basal nuclear group, nucleus O, was free from THi fibers (O, Fig. 3.2(*l*)).

In *Hydrolagus*, the basal telencephalic fasciculi, area superficialis basalis, and the subpallial region (SP5) contain moderately dense fibers, with fibers being heavier in ASB than in the SP5. The SP5 area contains a few THi cells. The SP6 area is almost free of THi fibers and cells. The pallial region 8 (P8, Fig. 3.3(*o*)) contains moderately dense fibers and THi cells with small somata (4 × 6 to 4 × 8 μm). Tyrosine hydroxylase positive label is denser in the P8 region than in the rest of the pallium. This THi label extends rostrally throughout the dorsal pallium (P9M, P9L, P11, Fig. 3.3(*n*)–(*q*) medially into the medial pallium, and ventrally into the septal nucleus (MSA, Fig. 3.3(*n*)–(*q*)). THi fibers of light density traverse the SP7 and SP8 areas. There are very few THi terminals in SP7 or SP8. In the dorsal telencephalon, THi cells and fibers are present in P9, both lateral and medial portions, and extend rostrally into the P12 area (Fig. 3.3(*n*)–(*q*)). In the rostral telencephalon, very heavy THi cell and fiber label extends in the pallium along the lateral border of the lateral ventricle and forms a swath of cells that travels ventromedially (Fig. 3.3(*p*)). The nucleus that contains these dense THi fibers and cells is not named in the Smeets atlas (1983), but the nucleus is in the medial part of the lateral anterior pallium. Area 10 of the pallium (P10) has fewer fibers and cells than do the P8, P9 and P12 areas. A mesh of very light THi fibers is present in the other areas of the telencephalon that are not specifically mentioned above. THi fibers are heavy in the granule cell layers of the olfactory bulb and can be traced entering the olfactory bulb. The granular layer of the olfactory bulb also contains THi cells (Northcutt et al., 1988).

In a skate, *Raja radiata* (Meredith & Smeets, 1987), many dopaminergic cells are scattered throughout the telencephalon and the olfactory bulbs. The dorsal pallium (P2, P3, P7, and Pdc) contains especially dense fibers and cells. Cells in the rostral portion of the dorsal pallium tend to be rounded and to have one or two short processes while in more caudal portions of the dorsal pallium the cells look more triangular and have three processes. The telencephalon ventral and lateral to the P2 in the mid- to caudal telencephalon also contains many dopaminergic cells, however, the lateral pallium itself does not. The ventral telencephalon is devoid of dopaminergic cell bodies but rich in dopaminergic fibers. Like the other cartilaginous fishes (Northcutt et al., 1988, and unpublished results of the authors) these fibers are concentrated in the striatum and in the area superficialis basalis, while nucleus O contains no dopaminergic fibers.

The retina of cartilaginous fishes contains THi cells and fibers within the amacrine cells and in inner plexiform layers (Bruun, Ehinger, & Sytsma, 1984; Brunken, Witkovsky, & Karten, 1986). The distribution varies somewhat in the species that were examined, but the cells in the nuclear layer send their axons into the inner plexiform layers. In addition, some THi immunoreactivity was present in the optic fiber layer, but no THi immunoreactive ganglion cells were present (Brunken et al., 1986). At least some of these cells are dopaminergic (Bruun et al., 1984).

Relationships between catecholamines and functional systems

Chondrichthyans contain most of the major catecholaminergic cell groups described in other vertebrates (Table 3.2 and see other chapters). Based on position, immunohistochemical localizations, cytoarchitectonics, and the distribution of these cell groups in other vertebrates, we hypothesize that many catecholaminergic nuclei in cartilaginous fishes are homologous to those first described and named in rats (Dahlström & Füxe, 1964). In vertebrates, the THi cells in the metencephalon and myelencephalon mostly contain either noradrenaline or adrenaline (Armstrong et al., 1982; Moore & Card, 1984; Hökfelt et al., 1984a,b; Füxe et al., 1985; Kalia, Füxe & Goldstein, 1985; Ekström et al., 1986; Reiner & Vincent, 1986; Vincent, 1988; Steeves et al., 1987; Sas, Maler & Tinner, 1990) while those in the mesencephalon and diencephalon probably contain dopamine (Björklund & Lindvall, 1984; Meredith & Smeets, 1987; Vincent, 1988; Hornby & Piekut, 1990; Meek et al., 1989; Sas et al., 1990). There are several exceptions to this distribution. The medulla contains some dopamine immunoreactive cells (Hökfelt et al., 1984b; Reiner & Vincent, 1986; Sas et al., 1990; Smeets & Steinbusch, 1990). Also, cells in the solitary complex and area postrema that are immunopositive for tyrosine hydroxylase but not for catecholamines may use L-dopa as a transmitter (Smeets & Steinbusch, 1990).

The brains of cartilaginous fishes are blanketed by a diffuse network of THi fibers. This network varies from very sparse varicosities to relatively dense fiber plexus, but a few areas are spared. For example, THi fibers are almost totally absent in the extreme caudal portion of the myelencephalon, but most of the myelencephalon has light to very dense catecholaminergic fibers. Unfortunately, almost nothing is known about the function of catecholaminergic cells and fibers in cartilaginous fishes.

Olfactory system

In general, the pallium of all vertebrates can be divided into dorsal, medial, and lateral subdivisions (Northcutt & Kicliter, 1980; Northcutt et al., 1988; Clairambault & Timmel, 1990; Reiner & Northcutt, 1992). Secondary olfactory fibers terminate in the lateral pallium (Bruckmoser & Dieringer, 1973; Smeets, 1983) which, in elasmobranchs, contains dense THi fibers and cells. Moreover, the localization of DAi/THi fibers and cells within the granular cell layer of the olfactory bulb (Meredith & Smeets, 1987; Northcutt et al., 1988) would suggest that catecholamines play a role in the processing of olfactory information.

Visual system

THi fibers and cells in the retina in rays, skates, and sharks, indicate that catecholamines play a role in vision in cartilaginous fishes (Bruun et al., 1984; Brunken et al., 1986). The retina projects to the suprachiasmatic/preoptic area of the hypothalamus (Northcutt, 1979, 1991; Smeets, 1981a; Repérant et al., 1986) which contains many THi cells. The tectum, which is rich in THi fibers, receives projections from the retina, the pretectum, and the thalamus (Northcutt & Wathey, 1980; Smeets, 1982; Ebbesson, 1984; Northcutt, 1991). However, the retinal projections are not the source of THi in the tectum because retinal ganglion cells are not THi positive (Brunken et al., 1986). From the tectum, information is sent to the brainstem, to rhombencephalic reticular formation, to the spinal cord, to the thalamus, and back to the pretectal area (Smeets, 1981b; Smeets & Timerick, 1981). The tectum of elasmobranchs is an area of multimodal integration (Bodznick, 1990) that responds to auditory, mechanical and cutaneous stimulation, and to weak electrical currents in the water (Bullock, 1984) and may be involved in visuomotor integration in cartilaginous fishes (Ebbesson, 1984). Visually evoked potentials can be recorded from the tectum of elasmobranchs (Bullock et al., 1990). Although ablation of the tectum in the more primitive squaloid sharks (which includes *Squalus*) causes blindness, tectal ablation in the nurse shark (which belongs to the more advanced galeomorph group) does not abolish the ability to discriminate between black and white (Graeber, 1984). Thus, it has been suggested that there are two patterns of visual organization in sharks. If this is so, the pathways and the distribution of neurochemicals within them have yet to be elucidated. The distribution of THi fibers in the tecta varied among cartilaginous fishes, however, in his review article of visual pathways in elasmobranchs, Northcutt (1991) emphasizes the commonalities in all elasmobranch visual pathways.

The posterior tuberal nucleus of the hypothalamus is another area that contains many THi cells and may be involved in visual processing. In *Raja*

Table 3.2. Major Brainstem catecholaminergic cell groups

	Elasmobranchs	Teleost fishes	Amphibians	Reptiles	Birds	Mammals
	Sas et al., 1990	Ekström et al., 1990	Yoshida et al., 1983 (Y) González & Smeets, 1991 (G)	Hornby & Piekut, 1990 Hornby et al., 1987	Smeets & Steinbusch, 1990 Wolters et al., 1984	Dubé & Parent, 1981 (D) Kiss & Péczely, 1987 (U)	Hökfelt et al., 1984 a, b
Diencephalon						
PO	Nucleus preopticus periventricularis	Preoptic area	Nucleus preopticus parvocellularis	Preoptic area (G)		Preoptic (K)	A14
SC	Suprachiasmatic nucleus		Nucleus suprachiasmaticus	Suprachiasmatic region		Supraoptic (K)	A15
VL	Ventrolateral thalamus	Ventral lateral thalamus					
VM	Ventromedial thalamus	Ventral medial thalamus					A14
MH						Medial hypothalamus (K)	A14
PT	Periventricular nucleus of posterior tuberculum	Posterior tuberal nucleus	Posterior tubercle (G), nucleus infundibularis dorsalis (Y)	Periventricular hypothalamic nucleus		Medial hypothalamic region (K)	
Mesencephalon						
SN			Rostral midbrain (Y)	Substantia nigra	Midbrain tegmentum (D)		A9 & A8
VTA			mesencephalic tegmentum (G)	Ventral tegmental area			A10
Metencephalon						
LC	Locus coeruleus	Locus coeruleus	Isthmic cells (G & Y)	Locus coeruleus	Locus coeruleus (D)	A6
NSC	Subcoerules	Isthmal cells	Cells ventral to nucleus isthmii (Y)	Subcoeruleus	Subcoeruleus (D)	A6v
Myelencephalon						
VS	Dorsal and rostral medullary tegmentum, dorsomedian postopercular region	Medullospinal (A2)	Nucleus tractus solitarius	Nucleus tractus solitarius, area postrema	Nucleus solitarius (D)	A2
A1	Reticular formation	Medullary tegmentum	Ventral medial medulla and other medullary areas		Reticular formation	Reticular formation (D)	A1

erinacea, scattered cells in the posterior tuberal nucleus project to an area of the medial pallium that responds to electrosensory and visual input (Bodznick & Northcutt, 1984). Whether any of these THi cells project to the medial pallium is not known, and the medial pallium contains very few THi cells and fibers.

Octaval and lateral line systems
The lateral line system of cartilaginous fishes consists of electrosensory and mechanoreceptive organs (Bodznick & Northcutt, 1984). These two systems form separate fiber pathways within the brainstem (Boord & Northcutt, 1982, 1988). Based on correlating the pathways with THi localizations, there is no evidence that catecholamines are involved in the processing of electrosensory information. For example, the electrosensory system projects to the dorsal octavolateralis nucleus, to nucleus B, to the lateral mesencephalic nucleus, to two nuclei in the diencephalon, and to the medial third of the telencephalon, all of which are almost devoid of THi fibers and somata (Bodznick & Northcutt, 1984; Schweitzer & Lowe, 1984; Schweitzer, 1986; Boord & Northcutt, 1988; Puzdrowski & Leonard, 1993). The mechanoreceptive intermediate octavolateralis nucleus (Boord & Northcutt, 1982, 1988; Puzdrowski & Leonard, 1993) contains THi fibers, but most other nuclei important for the processing of mechanoreceptive information (Boord & Northcutt, 1982; 1988), such as the lateral mesencephalic nuclear complex, do not. In *Squalus*, the medial pallium is dominated by input from the lateral line (Northcutt *et al.*, 1988) but is low in THi fibers and cells. However, the central portion of the tectum, which receives lateral line input (Bodznick, 1990), has many THi fibers. The ocataval columns, where primary sensory information from the vestibular system and inner ear terminates (Puzdrowski & Leonard, 1993), contain THi fibers but no cells.

Motor systems
Several studies identifying spinal projecting neurons have been done in elasmobranchs (Cruce & Northcutt, 1985, 1986; Timerick, Roberts, & Paul, 1992). As no colocalization studies been done, it is not possible to determine whether any of the THi cells in the brainstem project to the cord. However, in all chondrichthyans, THi cells are present in the lateral tegmental area within the myelencephalic reticular formation (Stuesse *et al.*, 1990, 1991b; Stuesse & Cruce, 1991, 1992), and some of these cells might project to the cord. The myelencephalic THi cells are separated into a dorsal (A2) and a ventral (A1) group in the rays (Stuesse *et al.*, 1990), but in the holocephalian and the sharks, they are distributed in dorsal and ventral areas with bridging cells between them (Stuesse *et al.*, 1991b; Stuesse & Cruce, 1991, 1992). The lateral tegmentum at the level of entry of cranial nerve VII contains another clearly identifiable cell group in most of the cartilaginous fishes. This group may be the homologue of the A5 group in rats (Hökfelt *et al.*, 1984b) and is present in all cartilaginous fishes except *Hydrolagus* (Stuesse *et al.*, 1990, 1991b; Stuesse & Cruce, 1991, 1992). In mammals, both A1 and A2 cell groups are involved in cardiovascular, gastrointestinal, and other aspects of autonomic regulation (Moore & Guyenet, 1985; Blessing & Willoughby, 1987). The A5 group projects to the intermediolateral cell column of the spinal cord and has been implicated in autonomic control (Loewy, McKeller & Saper, 1979). The other major spinal projecting group which has THi neurons is the ventromedial nucleus of the thalamus. The function of these THi cells in fishes is still unknown.

The degree of development of the midbrain tegmental groups varies across cartilaginous fishes. The ventral mesencephalon of elasmobranchs contains many THi cells, but these cells are lacking in the holocephalian (Stuesse *et al.*, 1990, 1991b; Stuesse & Cruce, 1991, 1992). In rays, the THi cells tend to separate into at least two groups, those in the ventral tegmental area and those surrounding the red nucleus (Stuesse *et al.*, 1990; Meade *et al.*, 1992; Stuesse & Cruce, 1993). These two nuclei may be homologous to the A10 and A8/A9 cell groups described in rats, respectively (Meredith & Smeets, 1987; Hökfelt *et al.*, 1984b). In sharks, the THi cells in the ventral tegmental area and substantia nigra merge into a single cell group with *Heterodontus* having more cells in the ventral mesencephalon than does *Squalus* (Stuesse & Cruce, 1992). The pattern of a single merged cell group is consistent with the embryonic origin of the substantia nigra and the ventral tegmental nuclei in mammals, where they derive from one cell mass (see Fallon & Loughlin, 1985). Thus this pattern in *Squalus* may represent the primitive (ancestral) condition.

Other
In cartilaginous fishes and teleosts, the greatest concentration of THi cells in the brain are the fluid-contacting cells in the hypothalamus, hypothalamic recess, periventricular organ, and pituitary (Rodriguez-Moldes & Anadon, 1988; Chevins, 1972). These fluid-contacting hypothalamic cells are found in all vertebrate groups except mammals and have

been described by many investigators (see final chapter in this book for a discussion). Cells with a very similar structure are found ventral to the central canal of the spinal cord. These catecholamine cells may play a role in either sensing levels of chemicals within the cerebrospinal fluid or in secreting substances into the cerebrospinal fluid.

A small, well-circumscribed nucleus is located on the lip of the fourth ventricle and projects to spinal cord, cerebellum, and telencephalon (Cruce et al., 1988; Fiebig, 1988). The cells and processes within this group spread ventrolaterally to the pial surface of the ventral tegmentum (Stuesse et al., 1990, 1991b; Stuesse & Cruce, 1991, 1992). Based on connectivity and immunohistochemical studies in the thornback guitarfish (Cruce et al., 1988; Fiebig, 1988; Fiebig & Bleckmann, 1989), this group is likely homologous to the noradrenergic cells found in the locus coeruleus and subcoeruleus (A6) of mammals (Dalström & Füxe, 1964, Hökfelt et al., 1984b).

Several THi cell groups are in the hypothalamus of cartilaginous fishes, but because rostral regions in the brain vary more than caudal regions among vertebrates, we are unsure of the homologies of these THi groups. The homology, if any, of the cells in the posterior tuberculum to THi cells in other vertebrates is especially problematical. One suggestion is that these catecholaminergic cells may correspond to the tuberal hypothalamic cell group (A12–A14) that projects to the pituitary in mammals (Björklund & Lindvall, 1984; Meredith & Smeets, 1987; Stuesse et al., 1991). These diencephalic dopaminergic cells in rat also project to the spinal cord, to the hypothalamus, and to the thalamus (Björklund & Lindvall, 1984). Alternatively, this group may correspond to the ventral tegmental area/substantia nigra of amniotes or it might be a nucleus that is unique to anamniotes but which corresponds to the subthalamic region of amniotes (See Reiner & Northcutt, 1992, for a discussion of these possibilities).

Thalamic inputs terminate in the pars centralis of the dorsal pallium in a galeomorph shark (Schroeder & Ebbesson, 1974; Luiten, 1981), thus this region may be homeoplaseous to isocortex in mammals (Northcutt et al., 1988). The dorsal pallium is rich in THi fibers and cells.

Relationships between catecholamines and other neurotransmitter systems

The distribution of neurochemicals in the telencephalon has been recognized as a very powerful tool for determining homologous structures across classes of vertebrates (Contestabile, Bissoli & Niso, 1990). In the section that follows, we will point out nuclei and general areas in which catecholamines and other neuromodulators or neurotransmitters have been localized. To date, no colocalization studies, in which more than one neurochemical has been localized to a single cell, have been done in elasmobranchs. However, if neurochemicals are found in the same vicinity, they may well interact in the control of behavior and internal homeostasis.

Other monoamines

The distribution of serotonin has been described in the central nervous system of cartilaginous fishes (Ritchie & Leonard, 1982; Ritchie et al., 1983, 1984; Cameron, Penderleith, & Snow, 1990; Stuesse et al., 1991a,b, 1992; Stuesse & Cruce, 1991, 1992; Meade, Stuesse & Cruce et al., 1992, 1993). For the most part, THi fibers and cells are concentrated in different parts of the central nervous system than are serotoninergic fibers and cells. However, there are areas in which there are dense accumulations of both serotonin and catecholaminergic fibers. For example, the spinal cord of sharks and rays has a rich serotoninergic network of fibers (Ritchie & Leonard, 1982; Ritchie et al., 1984; Cameron et al., 1990), the location of which overlaps that of THi fibers. The location of the THi/DAi cells does not correspond to the location of serotoninergic cells that are found more ventrally. The visceral sensory lobe has a rich plexus of serotoninergic fibers in addition to the catecholaminergic cells and fibers (Stuesse et al., 1992). Also, the hypothalamic periventricular organ contains serotonin (Watson, 1980; Ritchie et al., 1983; Yoshida et al., 1983; Muerling & Rodriguez, 1990; Stuesse & Cruce, 1991, 1992; Stuesse et al., 1990, 1991a,b; Yamanaka et al., 1990). All areas of the telencephalon contain serotoninergic fibers but not cells (Northcutt et al., 1988; Yamanaka et al., 1990). Serotoninergic cells are found in inner nuclear cells of the retina, however, the serotoninergic cells are either bipolar or are in a different class of amacrine cells than the THi cells (Bruun et al., 1984; Brunken et al., 1986). The presence of serotoninergic and THi fibers in some of the inner plexiform layers suggests that there might be an interaction between the two neurochemically distinguishable cells in the retina.

Acetylcholine

In dogfish shark, the concentration of choline acetyltransferase, a marker for acetylcholine, is about equal in basal and pallial telencephalic areas (Contestabile et al., 1990), many of which also contain THi cells and fibers. A detailed localization of acetylcholine in the brains of cartilaginous fishes has not yet been published.

Peptides and others

Many of the peptides found in amniotes have been localized to the central nervous system of cartilaginous fishes, and their locations are often similar in the two groups of vertebrates. Within the spinal cord, substance P, somatostatin, bombesin, neuropeptide Y, and calcitonin gene-related peptide immunoreactive fibers are concentrated in the substantia gelatinosa, but some of the substances are also distributed in the nucleus proprius and in all portions of the dorsal horn (Cameron et al., 1990), areas in which THi fibers are also found. The visceral sensory lobe has a rich plexus of substance P and enkephalinergic fibers that overlaps the distribution of THi fibers and cells (Stuesse et al., 1992). The lobe also contains cells and fibers that are immunopositive for the calcium binding protein, calbindin-D_{28k} (Rodriguez-Moldes et al., 1990). In addition to catecholaminergic cells, the posterior tubercular region in the galeomorph shark, Scyliorhinus canicula, is immunoreactive for the tetrapeptide, FMRF, and atrial natriuretic factor-related peptide, leading to the speculation that this area may play a role in control of pituitary secretions (Vallarino et al., 1990, 1991). The hypothalamic periventricular organ contains leu-enkephalin (Reiner, 1987; Stuesse & Cruce, 1991, 1992; Stuesse et al., 1990, 1991a,b) and somatostatin (Meurling & Rodriguez, 1990). These are all areas with THi cells.

The locations of gonadotropin hormone-releasing hormone positive (GHRHi) cells in elasmobranchs (Wright & Demski, 1991) do not overlap those of THi cells, however, THi cells in the substantia nigra/ventral tegmental area appear to be located just ventral to a very large GHRHi cell group, and the cell groups could well interact. These GHRHi cells are probably involved in reproductive control. Enkephalin positive amacrine cells are present in the retina of elasmobranchs, but they are found in cells that are distinct from those that are THi (Brunken et al., 1986).

Within the telencephalon, the lateral pallium contains less substance P and enkephalin immunoreactivity than the more medial, pars superficialis of the dorsal pallium, an area that may be homologous with the medial dorsal pallium in other vertebrates, i.e. the cingulate cortex of rats (Northcutt et al., 1988). The pars centralis of the dorsal pallium in Squalus is low in THi, substance P positive, and enkephalin positive fibers and cells (Northcutt et al., 1988).

Within the subpallium, Northcutt and his colleagues have hypothesized that the area superficialis basalis and the area periventricularis ventrolateralis in Squalus correspond to the amniote globus pallidus and striatum (caudate-putamen), respectively. These are areas that are rich in THi, substance P, and enkephalin fibers (Northcutt et al., 1988). In a galeomorph shark, Scyliorhinus canicula, both these nuclei are immunopositive for atrial natriuretic factor-related proteins (Vallarino et al., 1990).

Concluding remarks

The distribution of catecholamine cells in the brain of cartilaginous fishes bears some similarity to the distribution in amniotes, but there are differences not only between cartilaginous fishes and amniotes, but also within the cartilaginous fishes themselves. For example, no THi cells were found in the ventral tegmental area/substantia nigra of the holocephalian, Hydrolagus. However, based on neuropeptide and tyrosine-hydroxylase distribution, the telencephalon of Hydrolagus bears many similarities to those of the sharks, Squalus (Northcutt et al., 1988) and Heterodontus. The lateral pallium was exceptionally dense with THi fibers and cells. Moreover, nuclei in the subpallium that correspond to basal ganglia and the striatum are probably present in Hydrolagus even though a putative substantia nigra could not be identified. Although we now have a good start on determining the location of catecholaminergic fibers and cells in many representatives of cartilaginous fishes, our lack of knowledge about the function of these brains presently makes comparisons difficult within the fishes and between cartilaginous fishes and other vertebrate groups. This paucity of functional and behavioral studies on these fishes makes any statements about the role of these catecholaminergic groups in these fishes extremely speculative.

Acknowledgements

Thanks are due to Dr Ted Bullock for his support and helpfulness during two very fruitful summers at UCSD. We thank the staff of the University of Washington's Friday Harbor Laboratories and the director, Dr Dennis Willows, for their cordiality, helpfulness, and support while a portion of our research was done. Supported by grant NS25895 (WLRC and SS) and NS24669 (RGN) from the National Institutes of Health and a Research Challenge grant from the Ohio Board of Regents.

Abbreviations

A1–A14	various catecholaminergic cell groups
AM	anterior mesencephalic nucleus
AON	anterior octaval nucleus
APP	area periventricularis pallialis
ASB	area superficialis basalis
B	nucleus B
BFB	basal forebrain bundle
CANS	commissura ansulata
CB	cerebellum
CBC	cerebellar crest
CC	central canal
CG	central gray
CHAB	habenular commissure
CN	cerebellar nucleus
CP	central pretectal nucleus
DC	dorsal column
DM	dorsal motor nucleus of cranial nerve X
DMT	dorsomedial thalamic nucleus
DO	descending octaval nucleus
DON	dorsal octavolateralis nucleus
DP	dorsal pallium
DPC	dorsal pallium, central portion
DPS	dorsal pallium, superficial portion
dPT	decussation of the pallial tract
DT	dorsal thalamus
DV	descending nucleus of cranial nerve V
ENT	nucleus entopeduncularis
F	nucleus F
FR	fasciculus retroflexus
FST	fasciculus medianus of Stieda
GLO	glomeruli olfactorii
H	nucleus H
HAB	habenula
ICA	nucleus of the anterior commissure
IL	inferior lobe of the hypothalamus
INF	infundibulum
IO	inferior olivary nucleus
ION	intermediate octavolateralis nucleus
IP	interpeduncular nucleus
LC	locus coeruleus
LGB	lateral geniculate body
LP	lateral pallium
M	nucleus M
MH	medial hypothalamic nucleus
MLF	medial longitudinal fasciculus
MP	medial pallium
MSA	medial septal nucleus, anterior part
MSP	medial septal nucleus, posterior part
nII	cranial nerve II
NLH	nucleus of the inferior hypothalamic lobe
NLT	lateral tuberal nucleus
NMLF	nucleus of the medial longitudinal fasciculus
NSC	nucleus reticularis subcoeruleus
P8–12	areas pallialis 8–12
P9L	area pallialis pars lateralis
P9M	area pallialis pars medialis
PC	posterior commissure
PG	granular prominence
PHO	periventricular hypothalamic organ
PLA	area pallialis, pars anterioris
PLP	area pallialis, pars posterioris
PO	preoptic nucleus
PT	posterior tuberculum
O	nucleus O
ob	olfactory bulb
oe	olfactory epithelium
OPT	optic tract
OX	optic chiasm
R	red nucleus
RAC	n. raphe centralis superior
RAL	n. raphe linearis
RAM	n. raphe magnus
RAP	n. raphe pallidus
RCA	n. reticularis pontis caudalis, pars alpha
RCB	n. reticularis pontis caudalis, pars beta
RG	n. reticularis gigantocellularis
RM	n. reticularis magnocellularis
ROL	n. reticularis pontis oralis, pars lateralis
ROM	n. reticularis pontis oralis, pars medialis
RP	n. reticularis parvocellularis
RPGL	n. reticularis paragigantocellularis lateralis
RV	n. reticularis ventralis
SC	suprachiasmatic nucleus
SMC	spinal motor column
SN	substantia nigra
SP	superficial pretectal nucleus
SP10–12	areas subpallialis 10–12
SPCB	spinocerebellar tract
STR	striatum (APVL of Northcutt et al., 1988)
SV	saccus vasculosus
T	tectum
TMV	tract of the mesencephalic nucleus
TPAL	tractus pallii
V	cranial nerve V
V4	fourth ventricle
VC	ventral column
Vd	descending nucleus of V
VF	ventral fasciculus

VL	ventrolateral thalamic nucleus
VM	ventromedial thalamic nucleus
Vs	sensory trigeminal nucleus
VS	visceral sensory complex
VTA	ventral tegmental area
X	cranial nerve X
Xv	rostroventral nucleus of the vagus nerve

References

Armstrong, D. M., Ross, C. M., Pickel, V. M., Joh, T. H. & Reis, D. J. (1982). Distribution of dopamine-, noradrenaline-, and adrenaline-containing cell bodies in the rat medulla oblongata: demonstrated by the immunocytochemical localization of catecholamine biosynthetic enzymes. *Journal of Comparative Neurology*, **212**, 173–87.

Bauchot, D. M., Platel, R. & Ridet, J. M. (1976). Brain–body weight relationships in Selachii. *Copeia*, 1976, 305–10.

Björklund, A. & Lindvall, O. (1984). Dopamine containing systems in the CNS. In *Handbook of Chemical Neuroanatomy*, vol. 2, Part 1A, ed. A. Björklund & T. Hökfelt, pp. 55–122. Amsterdam: Elsevier.

Blessing, W. W. & Willoughby, J. O. (1987). Depressor neurons in rabbit caudal medulla do not transmit the baroreceptor-vasomotor reflex. *American Journal of Physiology*, **253**, H777–86.

Bodznick, D. (1990). Elasmobranch vision: multimodal integration in the brain. *Journal of Experimental Zoology Suppl.*, **5**, 108–16.

Bodznick, D. & Northcutt, R. G. (1984). An electrosensory area in the telencephalon of the little skate, *Raja erinacea*. *Brain Research*, **298**, 117–224.

Boord, R. L. & Northcutt, R. G. (1982). Ascending lateral line pathways to the midbrain of the clearnose skate, *Raja eglanteria*. *Journal of Comparative Neurology*, **207**, 174–82.

Boord, R. L. & Northcutt, R. G. (1988), Medullary and mesencephalic pathways and connections of lateral line neurons of the spiny dogfish *Squalus acanthias*. *Brain, Behavior and Evolution*, **32**, 67–88.

Bruckmoser, P. & Dieringer, N. (1973). Evoked potentials in the primary and secondary olfactory projection areas of the forebrain in Elasmobranchia. *Journal of Comparative Physiology*, **87**, 65–74.

Brunken, W. J., Witkovsky, P. & Karten, H. J. (1986). Retinal neurochemistry of three elasmobranch species: an immunohistochemical approach. *Journal of Comparative Neurology*, **243**, 1–12.

Bruun, A., Ehinger, B. & Sytsma, V. M. (1984). Neurotransmitter localization in the skate retina. *Brain Research*, **295**, 233–48.

Bullock, T. T. (1984). Physiology of the tectum mesencephali in elasmobranchs. In *Comparative Neurology of the Optic Tectum*, ed. H. Vanegas, pp. 47–68, New York: Plenum Press.

Bullock, T. H., M. H. Hoffman, J. G. New & F. K. Nahm (1990). Dynamic properties of visual evoked potentials in the tectum of cartilaginous and bony fishes, with neuroethological implications. *Journal of Experimental Zoology Supplement*, **5**, 142–55.

Cameron, A. A., Penderleith, M. B. & Snow, P. J. (1990). Organization of the spinal cord in four species of elasmobranch fish: cytoarchitecture and distribution of serotonin and selected neuropeptides. *The Journal of Comparative Neurology*, **297**, 210–18.

Carroll, R. L. (1988). Sharks and other cartilaginous fish. In *Vertebrate Paleontology and Evolution*, pp. 1–698. New York: W. H. Freeman.

Chevins, P. F. D. (1972). Ultrastructure of the pituitary complex in the genus *Raia* (Elasmobranchii). *Zeitschrift für Zellforschung*, **130**, 193–204.

Clairambault, P. & Timmel, J. F. (1990). Developmental organization of the amphibian pallium. In *The Forebrain in Nonmammals. New Aspects of Structure and Development*. ed. W. K. Schwerdtfeger & P. Germroth, pp. 29–41, New York: Springer-Verlag.

Compagno, L. J. V. (1973). Interrelationships of living elasmobranchs. In *Interrelationships of Fishes*, ed. P. H. Greenwod, Miles, R. S. & Patterson, C., pp. 15–61. New York: Academic Press.

Compagno, L. J. V. (1977). Phyletic relationships of living sharks and rays. *American Zoologist*, **17**, 303–22.

Contestabile, A., R. Bissoli & R. Niso (1990). Regional distribution of neurotransmitter-related markers: a quantitative microchemical approach to the study of telencephalic evolution. In *The Forebrain in Non-mammals: New Aspects of Structure and Development*. ed. W. K. Schwerdtfeger & P. Germroth, *Experimental Brain Research Series*, vol. 19, pp. 183–96. Berlin: Springer-Verlag.

Cruce, W. L. R., & Northcutt, R. G. (1985). Spinal projecting nuclei in the brainstem of an elasmobranch, the thornback guitarfish, *Platyrhinoidis triseriata*. *Neuroscience Abstract*, **11**, 1313.

Cruce, W. L. R., & Northcutt, R. G. (1986). Spinal projecting nuclei in the brainstem of a galeomorph shark, *Heterodontus francisci*. *Society for Neuroscience Abstract*, **12**, 1546.

Cruce, W. L. R., Fiebig, E., Stuesse, S. L. & Bleckmann, H. (1988). Locus coeruleus and subcoeruleus in an elasmobranch: telencephalic, cerebellar, and spinal cord connectivities and immunohistochemistry. *Society for Neuroscience Abstract*, **14**, 53.

Dahlström, A. & Füxe, K. (1964). Evidence for the existence of monoamine-containing neurons in the central nervous system. 1. Demonstration of monoamines in cell bodies of brain neurons. *Acta Physiologica Scandinavia*, **62**, Suppl. 232, 1–55.

Dubé, L. & Parent, A. (1981). The monoamine-containing neurons in avian brain. I. A study of the brainstem of the chicken (*Gallus domesticus*) by means of fluorescence and acetylcholinesterase histochemistry. *Journal of Comparative Neurology*, **196**, 695–708.

Ebbesson, S. O. E. (1984). Structure and connections of the optic tectum in elasmobranchs. In *Comparative Neurology of the Optic Tectum*, ed. H. Vanagas, pp. 33–46. New York: Plenum.

Ekström, P., Reschke, M., Steinbusch, H. & van Veen, T. (1986). Distribution of noradrenaline in the brain of the teleost *Gasterosteus aculeatus* L.: an immunohistochemical analysis. *Journal of Comparative Neurology*, **254**, 297–313.

Ekström, P., Honkanen, T. & Steinbusch, H. W. M. (1990). Distribution of dopamine-immunoreactive neuronal perikarya and fibers in the brain of a teleost, *Gasterosteus aculeatus* L. Comparison with tyrosine-hydroxylase, and dopamine-B-hydroxylase-immunoreactive neurons. *Journal of Chemical Neuroanatomy*, **3**, 233–60.

Fallon, J. H. & Loughlin, S. E. (1985). Substantia nigra. In *The Rat Nervous System*, Vol. 1. Forebrain and Midbrain, ed. G. Paxinos, pp. 353–74. New York: Academic Press.

Fiebig, E. (1988). Connections of the corpus cerebelli in the thornback guitarfish, *Platyrhinoidis triseriata* (Elasmobranchii). A study with WGA-HRP and extracellular granule cell recording. *Journal of Comparative Neurology*, **268**, 567–83.

Fiebig, E. & Bleckmann, H. (1989). Cell groups afferent to the telencephalon in a cartilaginous fish (*Platyrhinoidis triseriata*). (Elasmobranchii). A WGA-HRP study. *Neuroscience Letters*, **105**, 57–62.

Füxe, K., Agnati, L. F., Kalia, M., Goldstein, M., Andersson, K. & Harfstrand, A. (1985). Dopaminergic systems in the brain and pituitary. In *Basic and Clinical Aspects of Neuroscience*, ed. E. Flückiger, E. E. Müller & M. O. Thorner, pp. 1–25. Berlin: Springer-Verlag.

Gaspar, P., Berger, A., Febvret, A., Vigny, A., Krieger-Poulet, M. & Borri-Voltattorni, C. (1987). Tyrosine hydroxylase-immunoreactive neurons in the human cerebral cortex: a novel catecholaminergic group? *Neuroscience Letters*, **80**, 257–62.

González, A. & Smeets, W. J. A. J. (1991). Comparative analysis of dopamine and tyrosine hydroxylase immunoreactivities in the brain of two amphibians, the anuran *Rana ridibunda* and the urodele *Pleurodeles waltlii*. *Journal of Comparative Neurology*, **303**, 457–77.

Graeber, R. C. (1984). Behavioral correlates of tectal function in elasmobranchs. In *Comparative Neurology of the Optic Tectum*, ed. H. Vanagas, pp. 69–92, New York: Plenum Press.

Hökfelt, T., Johansson, O. & Goldstein, M. (1984a). Central catecholamine neurons as revealed by immunohistochemistry with special reference to adrenaline neurons. In *Chemical Neuroanatomy*, Vol. 2: *Classical Transmitters in the CNS*, Part 1, ed. A. Björklund & T. Hökfelt, pp. 157–276. Amsterdam: Elsevier.

Hökfelt, T., Martensson, R., Björklund, A., Kleinau, S. & Goldstein, M. (1984b). Distributional maps of tyrosine-hydroxylase-immunoreactive neurons in the rat brain. In *Handbook of Chemical Neuroanatomy*, Vol. 2: *Classical Transmitters in the CNS*, Part 1, ed. Björklund, A. & T. Hökfelt, pp. 277–379. Amsterdam: Elsevier.

Hornby, P. J., Piekut, D. T. & Demski, L. S. (1987). Localization of immunoreactive tyrosine hydroxylase in the goldfish brain. *Journal of Comparative Neurology*, **26**, 1–14.

Hornby, P. J. & Piekut, D. T. (1990). Distribution of catecholamine-synthesizing enzymes in goldfish brains: presumptive dopamine and norepinephrine neuronal organization. *Brain, Behavior and Evolution*, **35**, 49–64.

Hornung, J. P., Tork, I. & De Tribolet, N. (1989). Morphology of tyrosine hydroxylase-immunoreactive neurons in the cerebral cortex. *Experimental Brain Research*, **76**, 12–20.

Kalia, M., Füxe, K. & Goldstein, M. (1985). Rat medulla oblongata. II. Dopaminergic, noradrenergic (A1 and A2) and adrenergic neurons, nerve fibers, and presumptive terminal processes. *Journal of Comparative Neurology*, **233**, 308–32.

Kiss, J. Z. & Péczely (1987). Distribution of tyrosine-hydroxylase (TH)-immunoreactive neurons in the diencephalon of the pigeon (*Columba livia domestica*). *Journal of Comparative Neurology*, **257**, 333–46.

Kuhlenbeck, H. & Niimi, K. (1969). Further observations on the morphology of the brain in the Holocephalian elasmobranchs *Chimaera* and *Callorhynchus*. *Journal für Hirnforschung*, **11**, 267–314.

Kusunoki, T., Tsuda, Y. & Takashima, F. (1973). The chemoarchitectonics of the shark brain. *Journal für Hirnforschung*, **14**, 13–26.

Loewy, A. D., McKeller, S. & Saper, C. B. (1979). Direct projections from the A5 catecholaminergic cell group to the intermediolateral cell column. *Brain Research*, **174**, 309–14.

Lovejoy, D. A., Sherwood, N. M., Fischer, W. H., Jackson, B. C., Rivier, J. E., & Lee, T. (1991). Primary structure of gonadotropin-releasing hormone from the brain of a holocephalan (Ratfish: *Hydrolagus colliei*) *General and Comparative Endocrinology*, **82**, 152–61.

Luiten, P. G. M. (1981). Two visual pathways to the tectum in the nurse shark (*Ginglymostoma cirratum*). II. Ascending thalamo-telencephalic connections. *Journal of Comparative Neurology*, **196**, 539–48.

McEachran, J. D. (1990). Diversity in rays: why are there so many species? *Chondros*, **2**, 1–6.

Maisey, J. G. (1984). Higher elasmobranch phylogeny and biostratigraphy. *Zoology Journal of the Linnean Society*, **84**, 33–54.

Meade, C. A., Stuesse, S. L. & Cruce, W. L. R. (1992). Enkephalinergic and tyrosine hydroxylase immunoreactivity in the pretectal area and ventral mesencephalon of cartilaginous fishes. *Society of Neurosciences Abstracts*, **18**, 326.

Meade, C. A., Stuesse, S. L., & Cruce, W. L. R. (1993). Distribution of neurochemicals in the tecta of cartilaginous fishes. *Society of Neurosciences Abstracts*, **19**, 159.

Meek, J., Joosten, H. W. J. & Steinbusch, H. W. M. (1989). Distribution of dopamine immunoreactivity in the brain of the mormyrid teleost *Gnathonemus petersii*. *Journal of Comparative Neurology*, 281, 362–83.

Meredith, G. E. & Smeets, W. J. A. J. (1987). Immunocytochemical analysis of the dopamine system in the forebrain and midbrain of *Raja radiata*: evidence for a substantia nigra and ventral tegmental area in cartilaginous fish. *Journal of Comparative Neurology*, 265, 530–48.

Meurling, P. & Björklund, A. (1970). The arrangement of neurosecretory and catecholamine fibers in relation to the pituitary intermedia cells of the skate, *Raja radiata*. *Zeitschrift für Zellforschung*, 108, 81–92.

Meurling, P. & Rodriguez, E. M. (1990). The paraventricular and posterior recess organs of elasmobranchs: a system of cerebrospinal fluid-contacting neurons containing immunoreactive serotonin and somatostatin. *Cell and Tissue Research*, 259, 463–73.

Moore, R. Y. & Card, J. P. (1984). Noradrenaline-containing neuron systems. In *Handbook of Chemical Neuroanatomy*, Vol. 2: Classical Transmitters in the CNS, Part 1, ed. A. Björklund & T. Hökfelt, pp. 123–56. Amsterdam: Elsevier.

Moore, S. D. & Guyenet, P. G. (1985). Effect of blood pressure on A2 noradrenergic neurons. *Brain Research*, 338, 169–72.

Nelson, J. S. (1984). *Fishes of the World*. pp. 1–523. New York: John Wiley.

Newman, D. B. (1985a). Distinguishing rat brainstem reticulospinal nuclei by their neuronal morphology. I. Medullary nuclei. *Journal für Hirnforschung*, 26, 187–226.

Newman, D. B. (1985b). Distinguishing rat brainstem reticulospinal nuclei by their neuronal morphology. II. Pontine and mesencephalic nuclei. *Journal für Hirnforschung*, 26, 385–418.

Northcutt, R. G. (1979). Retinofugal pathways in fetal and adult spiny dogfish, *Squalus acanthias*. *Brain Research*, 162, 219–30.

Northcutt, R. G. (1978). Brain organization in the cartilaginous fishes. In *Sensory Biology of Sharks, Skates and Rays*, ed. E. S. Hodgson & R. F. Mathewson, pp. 117–93. Arlington, VA: Office of Naval Research.

Northcutt, R. G. (1989). Brain variation and phylogenetic trends in elasmobranch fishes. *Journal of Experimental Zoology Supp.*, 2, 83–100.

Northcutt, R. G. (1991). Visual pathways in elasmobranchs: organization and phylogenetic implications. *Journal of Experimental Zoology Suppl.*, 5, 97–107.

Northcutt, R. G. & Kicliter, E. E. (1980). Organization of the amniote telencephalon. In *Comparative Neurology of the Telencephalon*, ed. Ebbesson, S.O.E., pp. 203–55, New York: Plenum.

Northcutt, R. G. & Wathey, J. C. (1980) Guitarfish possess ipsilateral as well as contralateral retinofugal projections. *Neuroscience Letters*, 20, 237–42.

Northcutt, R. G., Reiner, A. & Karten, H. J. (1988). An immunohistochemical study of the telencephalon of the spiny dogfish, *Squalus acanthias*. *Journal of Comparative Neurology*, 277, 250–67.

Parent, A., Poitras, D. & Dubé, L. (1984). Comparative anatomy of central monoaminergic systems. In *Handbook of Chemical Neuroanatomy, Classical Transmitters in the CNS*, vol. 2, Part I, ed. A. Björklund & T. Hökfelt, pp. 409–39. Amsterdam: Elsevier.

Puzdrowski, R. L. & Leonard, R. B. (1993). The octavolateral systems in the stingray, *Dasyatis sabina*. I. Primary projections of the octaval and lateral line nerves. *Journal of Comparative Neurology*, 332, 21–37.

Reiner, A. (1987). The distribution of proenkephalin-derived peptides in the central nervous system of turtles. *Journal of Comparative Neurology*, 259, 65–91.

Reiner, A. & Northcutt, R. G. (1987). An immunohistochemical study of the telencephalon of the African lungfish. *Journal of Comparative Neurology*, 256, 463–81.

Reiner, A. & Northcutt, R. G. (1992). An immunohistochemical study of the telencephalon of the Senegal bichir (*Polypterus senegalus*). *Journal of Comparative Neurology*, 319, 359–86.

Reiner, P. B. & Vincent, S. R. (1986). The distribution of tyrosine hydroxylase, dopamine-β-hydroxylase, and phenylethanolamine-N-methyltransferase immunoreactive neurons in the feline medulla oblongata. *Journal of Comparative Neurology*, 248, 518–31.

Repérant, J., Miceli, D., Rio, J. P., Peyrichoux. J., Pierre, J., & Kirpitchnikova, E. (1986). The anatomical organization of retinal projections in the shark *Scyliorhinus canicula* with special reference to the evolution of the selachian primary visual system. *Brain Research Reviews*, 11, 227–48.

Ritchie, T. C. & Leonard, R. B. (1982). Immunocytochemical demonstration of serotoninergic cells, terminal and axons in the spinal cord of the stingray, *Dasyatis sabina*. *Brain Research*, 240, 334–7.

Ritchie, T. C., Livingston, C. A., Hughes, M. G., McAdoo, D. J. & Leonard, R. B. (1983). The distribution of serotonin in the CNS of an elasmobranch fish: immunocytochemical and biochemical studies in the Atlantic stingray, *Dasyatis sabina*. *Journal of Comparative Neurology*, 221, 429–43.

Ritchie, T. C., Roos, L. J., Williams, B. J. & Leonard, R. B. (1984). The descending and intrinsic serotoninergic innervation of the elasmobranch spinal cord. *Journal of Comparative Neurology*, 224, 395–406.

Roberts, B. L. & Meredith, G. E. (1987). Immunohistochemical study of a dopaminergic system in the spinal cord of the ray, *Raja radiata*. *Brain Research*, 437, 171–5.

Rodriguez-Moldes, I. & Anadón, R. (1987). Aminergic neurons in the hypothalamus of the dogfish, *Scyliorhinus canicula* L. (Elasmobranch). A histofluorescence study. *Journal für Hirnforschung*, 6, 685–93.

Rodriguez-Moldes, M. I. & Anadón, R. (1988). Ultrastructural study of the evolution of globules in coronet cells of the saccus vasculosus of an elasmobranch (*Scyliorhinus canicula* L.), with some observations on cerebrospinal fluid-contacting neurons. *Acta Zoologica* (Stockholm), 4, 217–24.

Rodriguez-Moldes, I., Timmermans, J. P. Adriaesen, D., De Groodt-Lasseel, M. H. A., Scheuermann, D. W., &

Anadon, R. (1990). Immunohistochemical localization of calbindin-D$_{28k}$ in the brain of a cartilaginous fish, the dogfish (*Scyliorhinus canicula* L.). *Acta Anatomica*, 137, 293–302.

Sas, E., Maler, L. & Tinner, B. (1990). Catecholaminergic systems in the brain of a gymnotiform teleost fish: an immunohistochemical study. *Journal of Comparative Neurology*, 292, 127–62.

Schaeffer, B. & Williams, M. (1977). Relationships of fossil and living elasmobranchs. *American Zoologist*, 17, 293–302.

Schroeder, D. M. & Ebbesson, S. O. E. (1974). Nonolfactory telencephalic afferents in the nurse shark (*Ginglymostoma cirratum*). *Brain, Behavior and Evolution*, 9, 121–55.

Schweitzer, J. (1986). Functional organization of the electroreceptive midbrain in an elasmobranch (*Platyrhinoidis triseriata*). A single unit study. *Journal of Comparative Physiology. A, Sensory, Neural, and Behavioral Physiology*. 158, 43–58.

Schweitzer, J. & Lowe, D. (1984). Mesencephalic and diencephalic cobalt-lysine injections in an elasmobranch: evidence for two parallel electrosensory pathways. *Neuroscience Letters*, 44, 317–22.

Smeets, W. J. A. J. (1981a). Retinofugal pathways in two chondrichthyans, the shark *Scyliorhinus canicula* and the rat *Raja clavata*. *Journal of Comparative Neurology*, 195, 1–11.

Smeets, W. J. A. J. (1981b). Efferent tectal pathways in two chondrichthyans, the shark *Scyliorhinus canicula* and the ray *Raja clavata*. *Journal of Comparative Neurology*, 195, 13–23.

Smeets, W. J. A. J. (1982). The afferent connections of the tectum mesencephali in two chondrichthyans, the shark *Scyliorhinus canicula* and the ray *Raja clavata*. *Journal of Comparative Neurology*, 205, 139–52.

Smeets, W. J. A. J. (1983). The secondary olfactory connections in two chondrichthians, the shark *Scyliorhinus canicula* and the ray *Raja clavata*. *The Journal of Comparative Neurology*, 218, 334–44.

Smeets, W. J. A. J., Nieuwenhuys, R. & Roberts, B. L. (1983). *The Central Nervous System of Cartilaginous Fishes: Structure and Functional Correlations*. New York: Springer-Verlag, pp. 1–266.

Smeets, W. J. A. J. & Steinbusch, H. W. M. (1990). New insights into the reptilian catecholaminergic systems as revealed by antibodies against the neurotransmitters and their synthetic enzymes. *Journal of Chemical Neuroanatomy*, 3, 25–43.

Smeets, W. J. A. J. & Timerick, S. J. B. (1981). Cells of origin of pathways descending to the spinal cord in two chondrichthyans, the shark *Scyliorhinus canicula* and the ray *Raja clavata*. *Journal of Comparative Neurology*, 202, 473–91.

Specht, L. A., Picket, V. M., Joh, T. H., & Reis, D. J. (1981). Light-microscopic immunocytochemical localization of tyrosine hydroxylase in prenatal rat brain. II. Late ontogeny. *Journal of Comparative Neurology*, 199, 255–76.

Steeves, J. D., Taccogna, C. A., Bell, K. A. & Vincent, S. R. (1987). Distribution of phenylethanolamine-N-methyltransferase (PNMT)-immunoreactive neurons in the avian brain. *Neuroscience Letters*, 76, 7–12.

Stuesse, S. L. & Cruce, W. L. R. (1991). Immunohistochemical localization of serotoninergic, enkephalinergic, and catecholaminergic cells in the brainstem of a cartilaginous fish, *Hydrolagus colliei*. *Journal of Comparative Neurology*, 309, 535–48.

Stuesse, S. L. & Cruce, W. L. R. (1992). Distribution of tyrosine hydroxylase, serotonin, and leu-enkephalin immunoreactive cells in the brainstem of a shark, *Squalus acanthias*. *Brain, Behavior and Evolution*, 39, 77–92.

Stuesse, S. L. & Cruce, W. L. R. (1993). Immunohistochemical localization of tyrosine hydroxylase in the brains of cartilaginous fishes. *Anatomical Record Suppl.*, 1, 108.

Stuesse, S. L., Cruce, W. L. R. & Northcutt, R. G. (1990). Distribution of tyrosine hydroxylase and serotonin immunoreactive cells in the central nervous system of the thornback guitarfish, *Platyrhinoidis triseriata*. *Journal of Chemical Neuroanatomy*, 3, 45–58.

Stuesse, S. L., Cruce, W. L. R. & Northcutt, R. G. (1991a). Serotoninergic and enkephalinergic cell groups in the brainstem of the bat ray and two skates. *Brain, Behavior and Evolution*, 38, 39–52.

Stuesse, S. L., Cruce, W. L. R. & Northcutt, R. G. (1991b). Localization of serotonin, tyrosine hydroxylase, and leu-enkephalin immunoreactive cells in the brainstem of the horn shark, *Heterodontus francisci*. *Journal of Comparative Neurology*, 308, 277–92.

Stuesse, S. L., Stuesse, D. C. & Cruce, W. L. R. (1992). Immunohistochemical localization of serotonin, leu-enkephalin, tyrosine hydroxylase, and substance P within the visceral sensory area of cartilaginous fish. *Cell and Tissue Research*, 268, 305–16.

Timerick, S. J. B., Roberts, B. L., & Paul, D. H. (1992) Brainstem neurons projecting to different levels of the spinal cord of the dogfish *Scyliorhinus canicula*. *Brain, Behavior and Evolution*, 39, 93–100.

Urano, A. (1971). Monoamine oxidase in the hypothalamo-hypophysial region in the brown smooth dogfish, *Triakis scyllia*. *Endocrinology Japonica*, 18, 27–36.

Vallarino, M., Feuilloley, M., Gutkowska, J., Cantin, M. & Vaudry, H. (1990). Localization of atrial natriuretic factor (ANF)-related peptides in the central nervous system of the elasmobranch fish *Scyliorhinus canicula*. *Peptides*, 11, 1175–81.

Vallarino, M., Salsotto-Cattaneo, M. T., M. Feuilloley, M. & Vaudry, H. (1991). Distribution of FMRFamide-like immunoreactivity in the brain of the elasmobranch fish *Scyliorhinus canicula*. *Peptides*, 12, 1321–8.

Vincent, S. R. (1988). Distributions of tyrosine hydroxylase-, dopamine-B-hydroxylase-, and phenylethanolamine-N-methyltransferase-immunoreactive neurons in the brain of the hamster (*Mesocricetus auratus*). *The Journal of Comparative Neurology*, 268, 584–99.

Watson, A. D. H. (1980). The distribution of aminergic neurons and their projections in the brain of the teleost

Myoxocephalus scorpius. Cell and Tissue Research, **208,** 299–312.

Wilson, J. F. & Dodd, J. M. (1973). Distribution of monoamines in the diencephalon and pituitary of the dogfish, *Scyliorhinus canicula* L. *Zeitschrift für Zellforschung,* **137,** 451–69.

Wolters, J. G., ten Donkelaar, H. J. & Verhofstad, A. A. J. (1984). Distribution of catecholamines in the brain stem and spinal cord of the lizard *Varanus exanthematicus*: an immunohistochemical study based on the use of antibodies to tyrosine hydroxylase. *Neuroscience,* **13,** 469–93.

Wright, D. E. & Demski, L. S. (1991). Gonadotropin hormone-releasing hormone (GnRH) immunoreactivity in the mesencephalon of sharks and rays. *The Journal of Comparative Neurology,* **307,** 49–56.

Yamanaka, S., Honma, Y., Ueda, S., & Sano, Y. (1990). Immunohistochemical demonstration of serotonin neuron system in the central nervous system of the Japanese dogfish, *Scyliorhinus torazame* (Chondrichthyes). *Journal of Hirnforschung,* **31,** 385–97.

Yoshida, M., Nagatsu, I., Kondo, Y., Karasawa, Ohno, T., Spatz, M. & Nagatsu, T. (1983). Immunohistochemical localization of the neurons containing catecholamine-synthesizing enzymes and serotonin in the brain of bullfrog *(Rana catesbeiana). Acta Histochemica Cytochemica,* **16,** 245–58.

4
Catecholamines in the brains of Osteichthyes (bony fishes)

J. Meek

Introduction

Bony fishes are by far the most abundant group of vertebrates, with about 25 000–30 000 living species (Nelson, 1969). In only about 20 species out of this large number, aspects of the distribution of catecholamines in the central nervous system have been studied, to be reviewed in the present chapter. Most studies devoted to this subject have used the formaldehyde-induced fluorescence (FIF) technique, and in only about ten species also the recently developed and more sensitive immunohistochemical techniques have been applied, using specific antibodies to different catecholamines or their synthesizing enzymes. To be able to evaluate the significance of the findings in only a few species of bony fishes out of the about 30 000 available, some introductory remarks on the phylogenetic relations of the species studied and their overall brain organization seem to be appropriate.

The phylogenetic relations of bony fishes

Fishes are generally subdivided into two major groups: the cartilaginous fishes or Chondrichthyes and the bony fishes or Osteichthyes. The latter are usually subdivided into the ray-finned fishes or Actinopterygii and the lobe- or fleshy-finned fishes, the Sarcopterygii. Actinopterygii comprise the Polypteri, Chondrostei, Holostei and Teleostei (Table 4.1). Interestingly, these groups represent subsequent stages or gradations of actinopterygian evolution, the polypteriforms being the most ancient and primitive group. Chondrostei are considered as descendants of the Palaeoniscoidei, abundant in paleozoic times, but supplanted by Holostean groups in mesozoic times. In turn, the Holostei were largely replaced in the cenozoic area by the expanding and now abundant Teleostei. However, several species of Polypteriform, Chondrostean and Holostean fishes are still extant (Table 4.1). So, the study of organizational aspects of the brains of these fishes, including the distribution of catecholamines, might well reveal important phylogenetic trends, assuming that the brains of the presently living species are similar and unchanged compared with the ancestral species, just as is their skeleton on which the establishment of phylogenetic relationships is based (e.g. Lauder & Liem, 1983a, b).

Teleosts are presently the most abundant group of fishes, with over 25 000 species (Nelson, 1969). They have complex phylogenetic relationships, open for different interpretations (cf. Greenwood et al., 1966; Nelson, 1969; Grzimek, 1973; Lauder & Liem, 1983a, b). For simplicity, I will confine this survey to some remarks on the teleostean species used for immunohistochemical studies of the central distribution of catecholamines, i.e. the European eel, a mormyrid fish, the goldfish, a catfish, a gymnotid fish and the stickleback (Fig. 4.1; Table 4.2). These appear to represent a phylogenetically diverse sample of the teleostean fishes. The European eel (*Anguilla anguilla*) and the mormyrid fish *Gnathonemus petersii* belong to the phylogenetically primitive groups of the Osteoglossomorpha and the Elepomorpha, respectively. In contrast, the three–spined stickleback *Gasterosteus aculeatus* belongs to the most advanced neoteleostean group of Acanthopterygii. The remaining species, i.e. the goldfish *Carassius auratus*; the Gymnotid fish *Apteronotus leptorhynchus* and the African catfish *Clarias gariepinus* belong to the euteleostean Ostaryophysi (Table 4.2).

The sarcopterygian or lobe-finned bony fishes comprise only seven presently living species. To

Table 4.1 *The systematic relations and subdivisions of bony fishes (osteichthyes)* (see e.g. Greenwood et al., 1966; Grzimek, 1973; Lauder & Liem, 1983a, b)

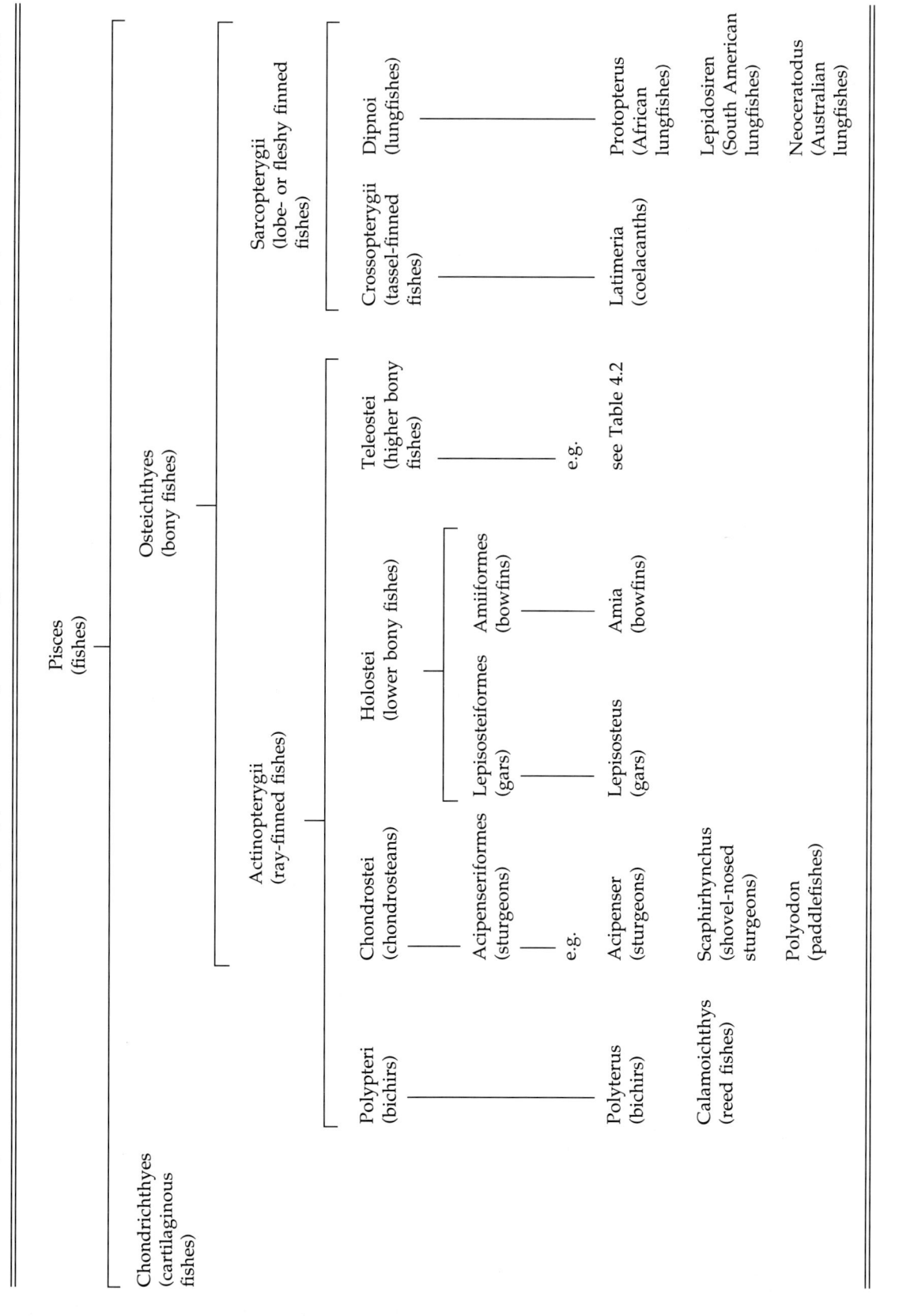

Pisces (fishes)					
Chondrichthyes (cartilaginous fishes)	**Osteichthyes** (bony fishes)				
	Actinopterygii (ray-finned fishes)			Sarcopterygii (lobe- or fleshy finned fishes)	
	Polypteri (bichirs)	Chondrostei (chondrosteans)	Holostei (lower bony fishes)	Teleostei (higher bony fishes)	Crossopterygii (tassel-finned fishes)	Dipnoi (lungfishes)
		Acipenseriformes (sturgeons)	Lepisosteiformes (gars)	Amiiformes (bowfins)			
		e.g.			e.g.		
	Polyterus (bichirs)	Acipenser (sturgeons)	Lepisosteus (gars)	Amia (bowfins)	see Table 4.2	Latimeria (coelacanths)	Protopterus (African lungfishes)
	Calamoichthys (reed fishes)	Scaphirhynchus (shovel-nosed sturgeons)					Lepidosiren (South American lungfishes)
		Polyodon (paddlefishes)					Neoceratodus (Australian lungfishes)

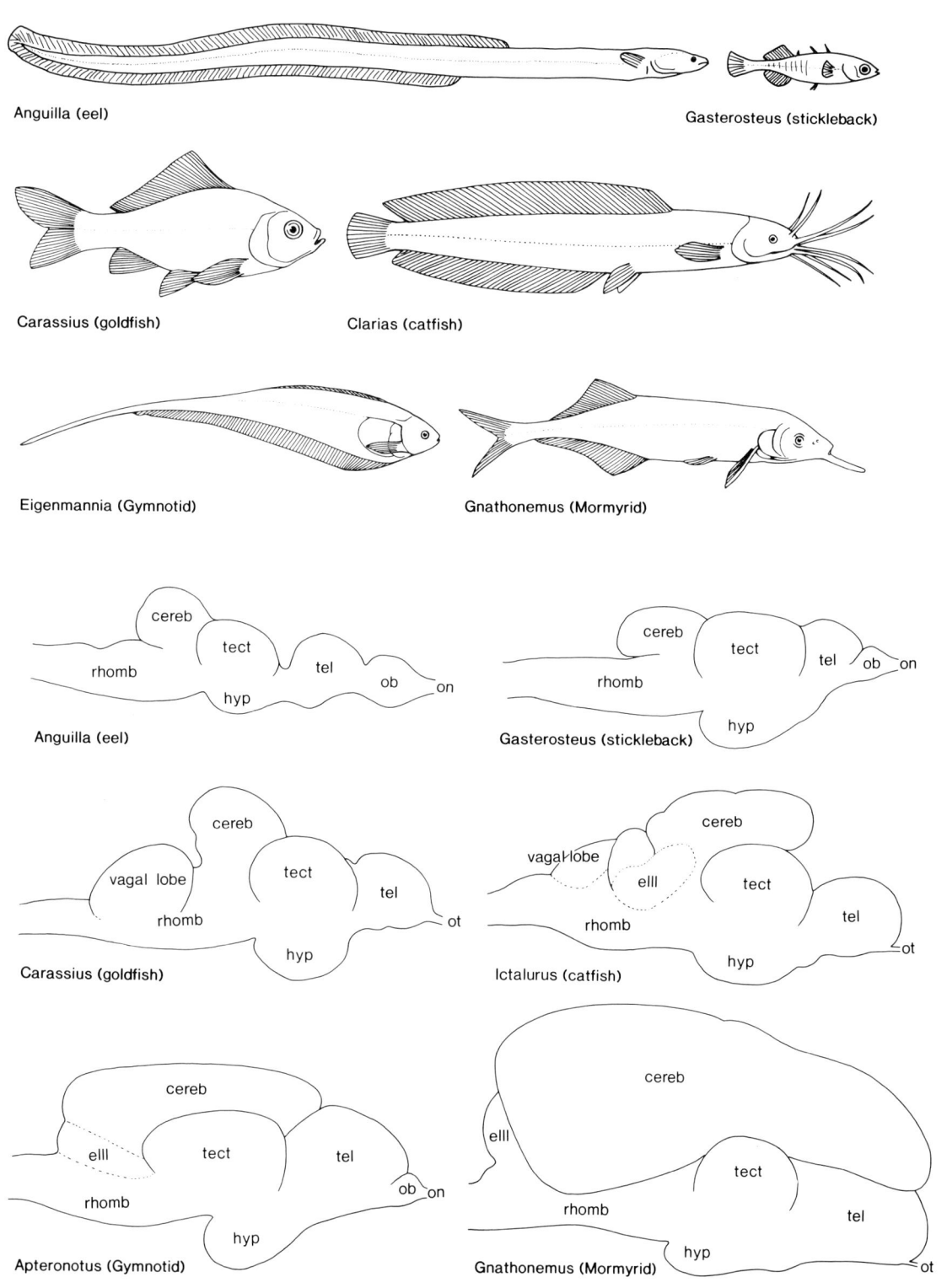

Fig. 4.1. Lateral views of a number of teleostean fishes (above) and brains (below) used to study the distribution of catecholamines. The Figure shows the variability in morphology as explained on pp. 53–54. Notice that the brains have not been drawn at the same magnification, but at the same size, thus emphasizing the *relative* size of different brain regions, neglecting their absolute size!

Table 4.2 *The systematic place of the teleostean species in which the distribution of catecholamines in the brain has been studied with immunohistochemical techniques (see Lauder & Liem, 1983a, b)*

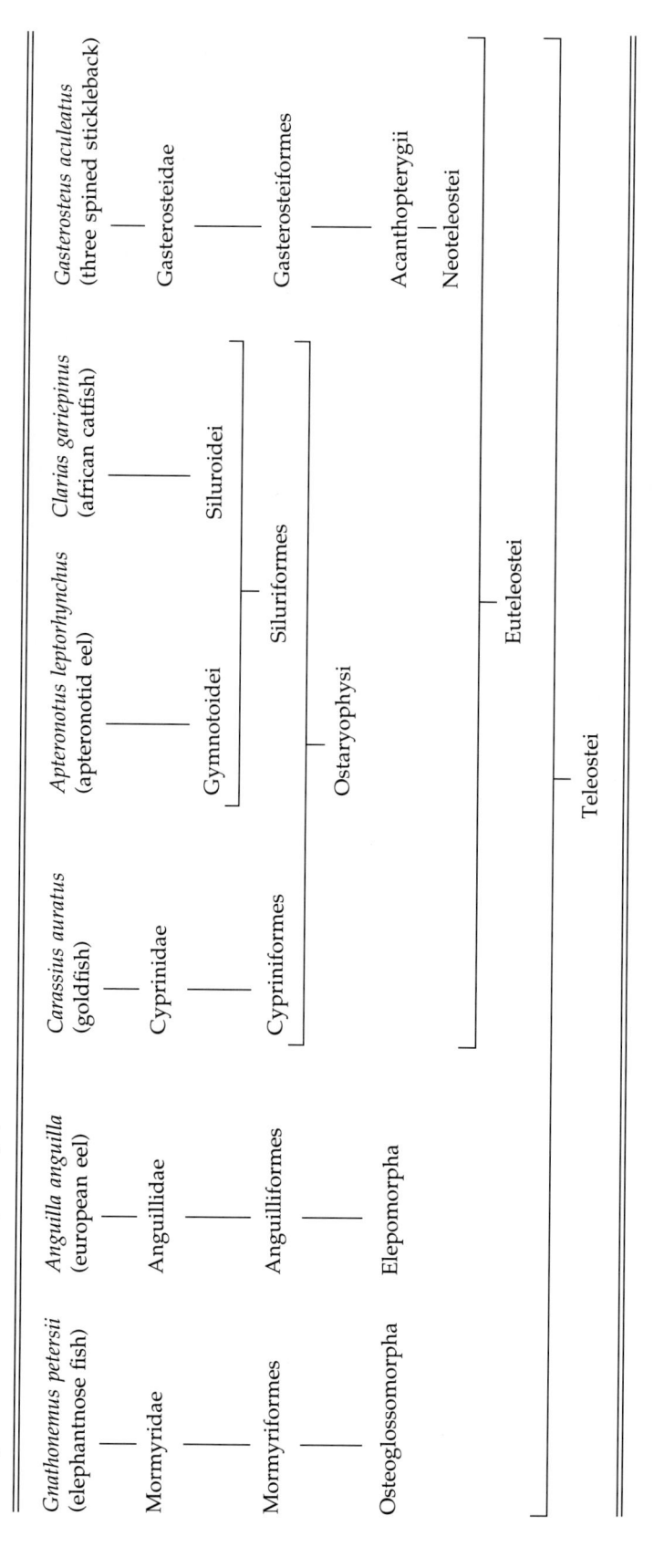

Gnathonemus petersii (elephantnose fish)	*Anguilla anguilla* (european eel)	*Carassius auratus* (goldfish)	*Apteronotus leptorhynchus* (apteronotid eel)	*Clarias gariepinus* (african catfish)	*Gasterosteus aculeatus* (three spined stickleback)
Mormyridae	Anguillidae	Cyprinidae	Gymnotoidei	Siluroidei	Gasterosteidae
Mormyriformes	Anguilliformes	Cypriniformes	Siluriformes		Gasterosteiformes
			Ostaryophysi		Acanthopterygii
Osteoglossomorpha	Elepomorpha				Neoteleostei
			Euteleostei		
		Teleostei			

these belongs to the famous coelacanth *Latimeria chalumnae*, discovered only in 1938 (Forey, 1988). The other six species are lungfishes or Dipnoi, which comprise four species of African lungfishes (*Protopterus*), one species of the South American genus *Lepidosiren*, and one species of the Australian genus *Neoceratodus* (Table 4.1).

General organization of the brain of Osteichthyes

Teleostean brains Since teleosts constitute by far the majority of bony fishes living to date, their brains have been studied extensively. They show a variety of sensory specializations, embedded in a general framework of teleostean brain organization (for review see Nieuwenhuys, 1982; Nieuwenhuys & Pouwels, 1983; Nieuwenhuys & Meek, 1985). The latter can be observed in teleosts without marked sensory specializations, such as the eel and the stickleback (Fig. 4.1). Their brains comprise a rostrally located olfactory bulb, a small telencephalon, a marked tectum and cerebellum, and an elongated rhombencephalon (Fig. 4.1). The olfactory bulbs are sessile in these species, i.e. directly attached to the telencephalon, but they may be displaced towards the olfactory organ in other species. These include cyprinids, catfishes and mormyrids, where the olfactory bulbs are connected with the telencephalon by long olfactory tracts (Fig. 4.1). The telencephalon is formed, during ontogeny, by a process of eversion instead of evagination, which is the principal morphogenetic event in the forebrain of other vertebrates. (Nieuwenhuys & Meek, 1990a; Fig. 4.2). Consequently, the telencephalon of teleosts and other actinopterygians does not consist of hollow hemispheres around a lateral ventricle, but of massive hemipheres with a ventricular surface located superficially (Fig. 4.2).

Sensory specializations observed in teleosts used to study the distribution of catecholamines include internal gustatory, external gustatory, and electrosensory lateral line specializations (see Nieuwenhuys, 1982; Nieuwenhuys & Meek, 1985). Internal gustatory specializations are especially well developed in cyprinids, which have a palatal organ in the roof of the mouth, densely occupied with taste buds (Sibbing, 1984). These project to a dorsal rhombencephalic specialization: the vagal lobe (Fig. 4.1; Morita, Murakami & Ito, 1983). External gustatory specializations, i.e. large amounts of taste buds in the skin of the animal, are present in cyprinids, but especially well developed in silurids. Consequently,

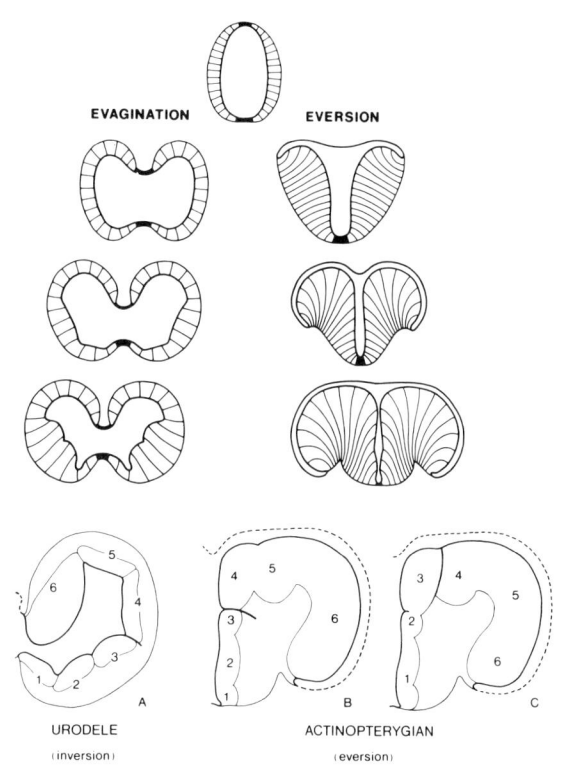

Fig. 4.2. Schematic transverse sections through the developing telencephalon of a urodele (left) and an actinopterygian fish (right), showing the process of evagination and inversion as present in most vertebrates, compared with the process of eversion as displayed by the actinopterygian telencephalon (after Nieuwenhuys and Meek 1990a; see p. 53). The lower Figures show two possible ways for a comparative topological homology, in which 1 indicates: the medial subpallium or septum; 2: the central subpallium or olfactory tubercle; 3: the lateral subpallium or striatal primordium; 4: the lateral or piriform pallium; 5: the dorsal or general pallium; and 6: the medial or hippocampal pallium. A further discussion on possible amniote and actinopterygian telencephalic homologies is presented on pp. 65–69.

these teleosts have a well-developed facial lobe (Finger, 1983), located between their vagal lobe and the electrosensory lateral line lobe (elll) (Fig. 4.1). The latter receives input from electroreceptive lateral line organs, which are specialized derivatives of the generally present teleostean mechanosensory lateral line system (Finger 1986). Electrosensory lateral line lobes are particularly well developed in active electroreceptive fishes, which have an electric organ and specialized tuberous electroreceptors involved in electrocommunication and electrolocation. Both gymnotids and mormyrids are electric fishes and have large ellls (Carr & Maler, 1986; Bell & Szabo, 1986), only partly visible at the outer surface of the brain (Fig. 4.1). Mormyrids have, in addition, an enormous cerebellum, covering the other parts of the brain almost completely (Fig. 4.1; Meek & Nieuwenhuys, 1991).

Primitive actinopterygian brains The brains of the primitive actinopterygian fishes, i.e. the Polypteri, Chondrostei and Holostei, are rather similar to the brains of teleosts without marked sensory specializations. Typical examples are shown in Fig. 4.3. They all have an everted telencephalon (Nieuwenhuys & Meek, 1990a). Their midbrain tectum and cerebellum are rather small compared with teleosts.

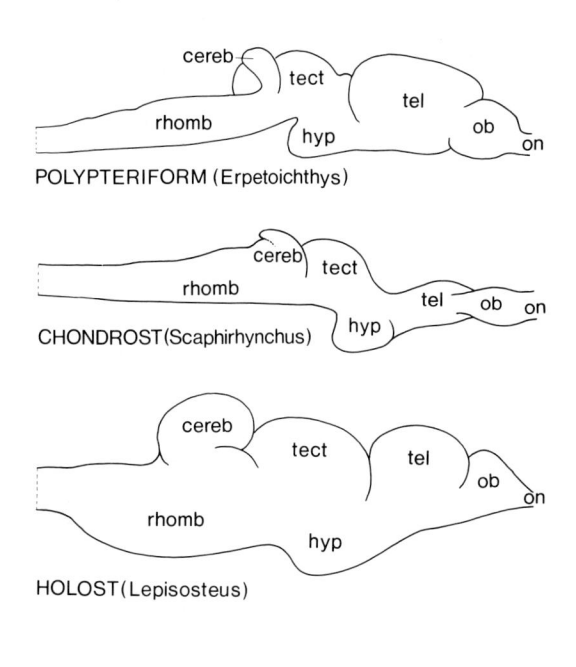

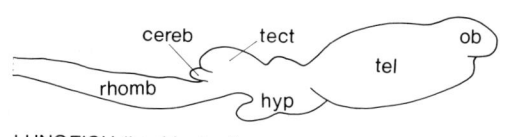

POLYPTERIFORM (Erpetoichthys)

CHONDROST (Scaphirhynchus)

HOLOST (Lepisosteus)

LUNGFISH (Lepidosiren)

Fig. 4.3. Lateral views of the brains of representatives of three groups of non-teleostean (more primitive) actinopterygian fishes, and one group of sarcopterygian fishes (see p. 54). Similar to Fig. 4.1, the brains have been drawn at the same size, emphasizing the relative size of different subdivisions, but neglecting their absolute size.

The brains of sarcopterygian fishes The brains of sarcopterygian fishes differ substantially from those of actinopterygian fishes. They have larger telencephala, not (Dipnoi) or not entirely (Crossopterygians) formed by eversion but by evagination (Nieuwenhuys & Meek, 1990b). They have a very small midbrain tectum and corpus cerebelli, but a marked caudolateral cerebellar extension or auricle, and a very elongated rhombencephalon. An example is shown in Fig. 4.3.

Organization of the present survey

The present survey of the distribution of catecholamines in the brain of bony fishes will primarily concentrate on immunohistochemical studies performed in teleosts. The distribution of tyrosine hydroxylase-, dopamine-, noradrenaline-, dopamine-β-hydroxylase and phenyl-n-methyl transferase immunoreactivity will be described, discussed, and compared with studies using formaldehyde-induced fluorescence techniques. This will yield a good insight in the general pattern and variability of the catecholamine systems of teleosts, since the species studied form quite a representative sample of both the phylogenetic and functional diversity encountered in teleosts, as argued on p. 49 and pp. 53, 54, respectively.

After the survey of the distribution of catecholamines in teleosts, the chapter will be continued with some notes on the distribution of these monoamines in other actinopterygian fishes and in sarcopterygians. This part can be short, since only a few studies have been devoted to this subject.

Catecholamines in teleostean brains

Tyrosine Hydroxylase (TH)

The distribution of TH gives an overview of the overall population of catecholamine synthesizing neurons, without information, however, about which of the catecholamines, dopamine, noradrenaline, adrenaline or possibly L-DOPA, is the endproduct of TH enzymatic activity, and thus used for neurotransmission or neuromodulation (see Chapter 1). The overall distribution of TH immunoreactivity has been studied in the goldfish (Yoshida et al., 1983; Hornby, Piekut & Demski, 1987; Hornby & Piekut, 1990), the stickleback (Ekström, Honkanen & Steinbusch, 1990), the gymnotid fish Apteronotus (Sas, Maler & Tinner, 1990) and the mormyrid fish Gnathonemus (Meek & Joosten, 1993). The results will be reviewed in the next paragraphs, without further reference to these studies. Reference will only be made to some other studies, devoted to details or subpopulations of TH-immunoreactive (THir) neurons in teleosts, and not yet cited above.

Cell bodies The results of the studies enumerated above on the distribution of TH-immunoreactivity are summarized in Figs. 4.4 and 4.5 as far they concern cell bodies. In general, THir neurons occur in the olfactory bulb, the ventral telencephalon, the preoptic region, the ventral thalamus, the hypothalamus, the pretectal region, the isthmic region, the caudal reticular formation and the area postrema of the brains of teleosts. However, a number of species differences and the use of different names by different authors deserve some further comments. Going from rostral to caudal the following may be noticed in this respect.

THir neurons have been observed in the *olfactory bulbs* of the stickleback and the gymnotid fish *Apteronotus* (Fig. 4.3) as well as of the cyprinid fishes *Tinca tinca*, the tench, and *Barbus meridionalis*, a barbel, and the rainbow trout *Salmo gairdneri* (Alonso et al., 1989). In these teleosts a subpopulation of granule cells in the internal cell layer displays TH immunoreactivity. In goldfish and *Gnathonemus*, the olfactory bulbs were not included in the histochemical studies, but there is no reason to assume that the situation is different in the olfactory bulbs of these species.

In the *telencephalon*, THir neurons have been observed in ventral telencephalic (V) subregions indicated as the dorsal (Vd) and supracommissural (Vs) subdivision in *Gasterosteus*, as the ventral (Vv) and dorsal (Vd) subdivision in *Carassius*, as Vv, Vd and Vs in *Gnathonemus* and as the ventral (Vv), central (Vc) and supracommissural (Vs) subdivision in *Apteronotus* (Fig. 4.3). The latter has a caudal extension indicated as the entopeduncular nucleus by Sas et al. (1990). All regions indicated as Vv, Vd, Vc and/or Vs are located rostral and dorsal to the anterior commissure, and the THir neurons described in these regions of different teleosts might well be comparable. More caudolateral to the anterior commissure and its caudal continuation, i.e. the lateral forebrain bundle, another group of THir neurons may be present, located at the transition zone of the dorsal and ventral telencephalon. Hence it was indicated as area ventralis intermedius (Vi) by Sas et al. (1990) and Meek & Joosten (1993) (Fig. 4.4). However, Hornby et al. (1987) named this group in the goldfish area dorsalis pars centralis (Dc), suggesting that it is not a subpopulation of ventral telencephalic neurons, but of dorsal, or pallial, telencephalic neurons. Remarkably, similar THir neurons seem to be absent in the stickleback (Fig. 4.4).

In the *preoptic region*, most studies agree in the distinction of an anterior preoptic THir cell group, located ventrocaudal to the anterior commissure, a suprachiasmatic population of THir cells, located dorsocaudal to the optic chiasm, and a posterior preoptic cell group, located dorsal to the suprachiasmatic nucleus along the preoptic ventricular recess. (Figs. 4.4, 4.5 and 4.11). In *Gasterosteus*, anterior preoptic THir neurons seems to be absent. In *Gnathonemus*, a small population of partly cerebrospinal fluid contacting neurons along the preoptic recess, just above the anterior preoptic cell group, displays THir as well (Fig. 4.5(b)).

In the *hypothalamus*, the caudal continuation of the preoptic region, large THir neurons occur lateral to the paraventricular organ (PVO). The latter is a population of cerebrospinal fluid (CSF)-contacting neurons along the infundibulum of the third ventricle, to be described in detail below (pp. 60–61). The large THir neurons lateral to the PVO are quite numerous in *Apteronotus* and *Gnathonemus*, and have been designated as the periventricular nucleus of the posterior tuberculum (tpp.; Sas et al., 1990) or as the magnocellular hypothalamic nucleus (nmh.; Meek et al., 1989, see Fig. 4.11(d)). These neurons can not only be identified on the basis of their position but also on the basis of their dorsally extending processes (Fig. 4.5 (e)). Similar neurons, although less numerous, have been indicated as PVO accompanying cells in the stickleback and the goldfish (Fig. 4.4).

A second group of hypothalamic THir cells is located in the nucleus posterior tuberis (npt) of all teleosts studied. This nucleus is located caudal to the infundibulum. Some neurons located lateral to the infundibulum, more ventral than the large PVO accompanying cells, may display THir as well, including neurons indicated as lateral hypothalamic (lh) neurons in *Apteronotus*, and as nucleus lateralis tuberis (nlt) and nucleus inferior tuberis (nit) in *Gnathonemus*. The PVO, i.e. CSF-contacting neurons along the infundibulum or infundibular recessi, is not THir in *Gasterosteus*, only very weakly at some places in *Gnathonemus*, and only in a few subdivisions in *Carassius* and *Apteronotus*, indicated as the anterior and posterior PVO (pvoa and pvop) in *Carassius*, and as the first part of nucleus recessus lateralis pars medialis (nrlm$_1$), nucleus recessus lateralis pars inferior (nrli), and nucleus of the infundibular recess (nir) in *Apteronotus*. It is presently uncertain whether differences in THir of parts of the PVO indicate true

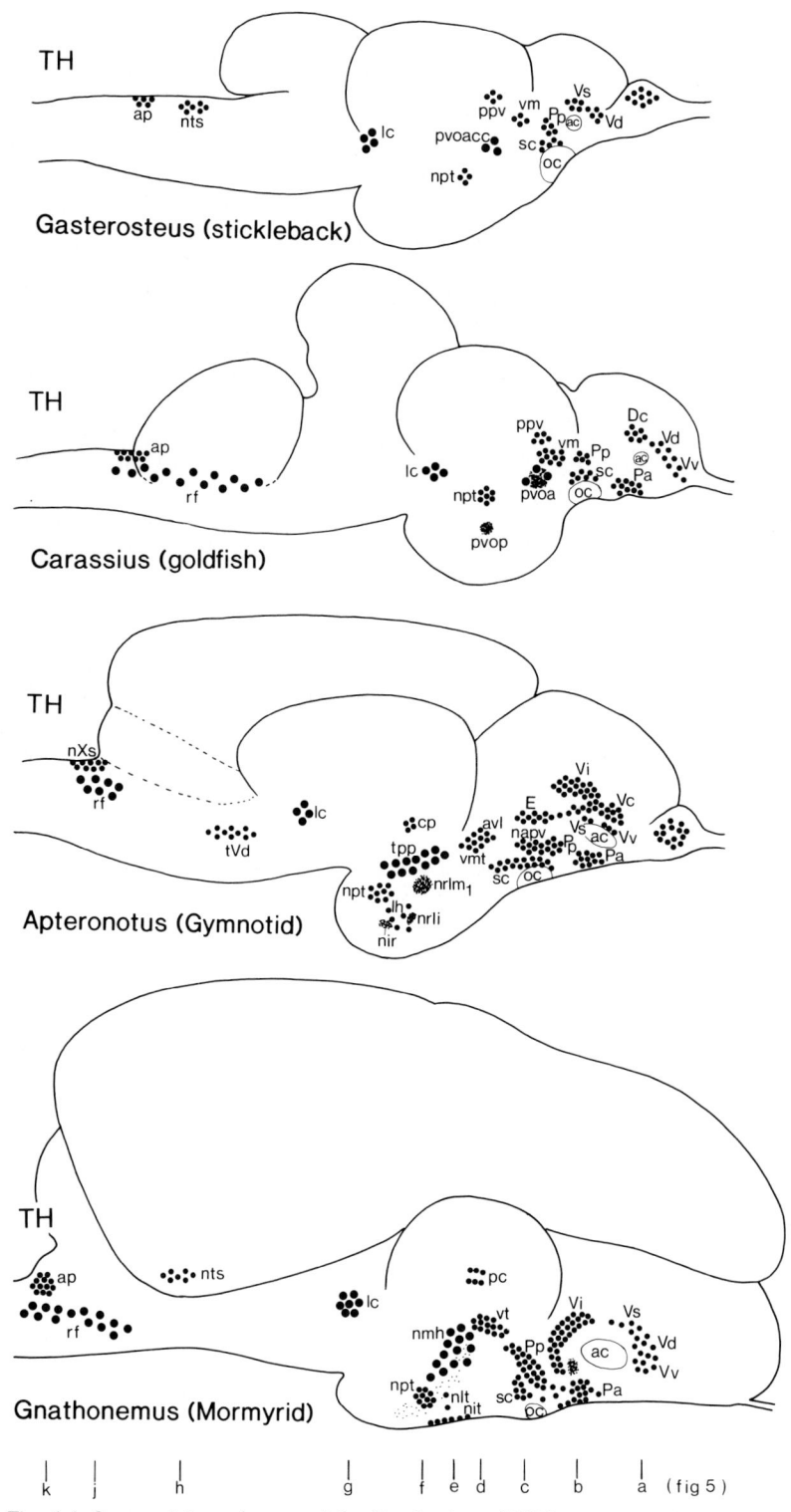

Fig. 4.4. Summarizing schemes of the distribution of TH immunoreactive neurons in the brains of four teleosts, projected on lateral views of these brains. The size of the dots encodes the relative size of the neurons in three groups (small, medium and large), while stipples encode small CSF-contacting neurons. For details and references, see pp. 55–58. Transverse sections through the brain of *Gnathonemus* are presented in Fig. 4.5 at the levels indicated.

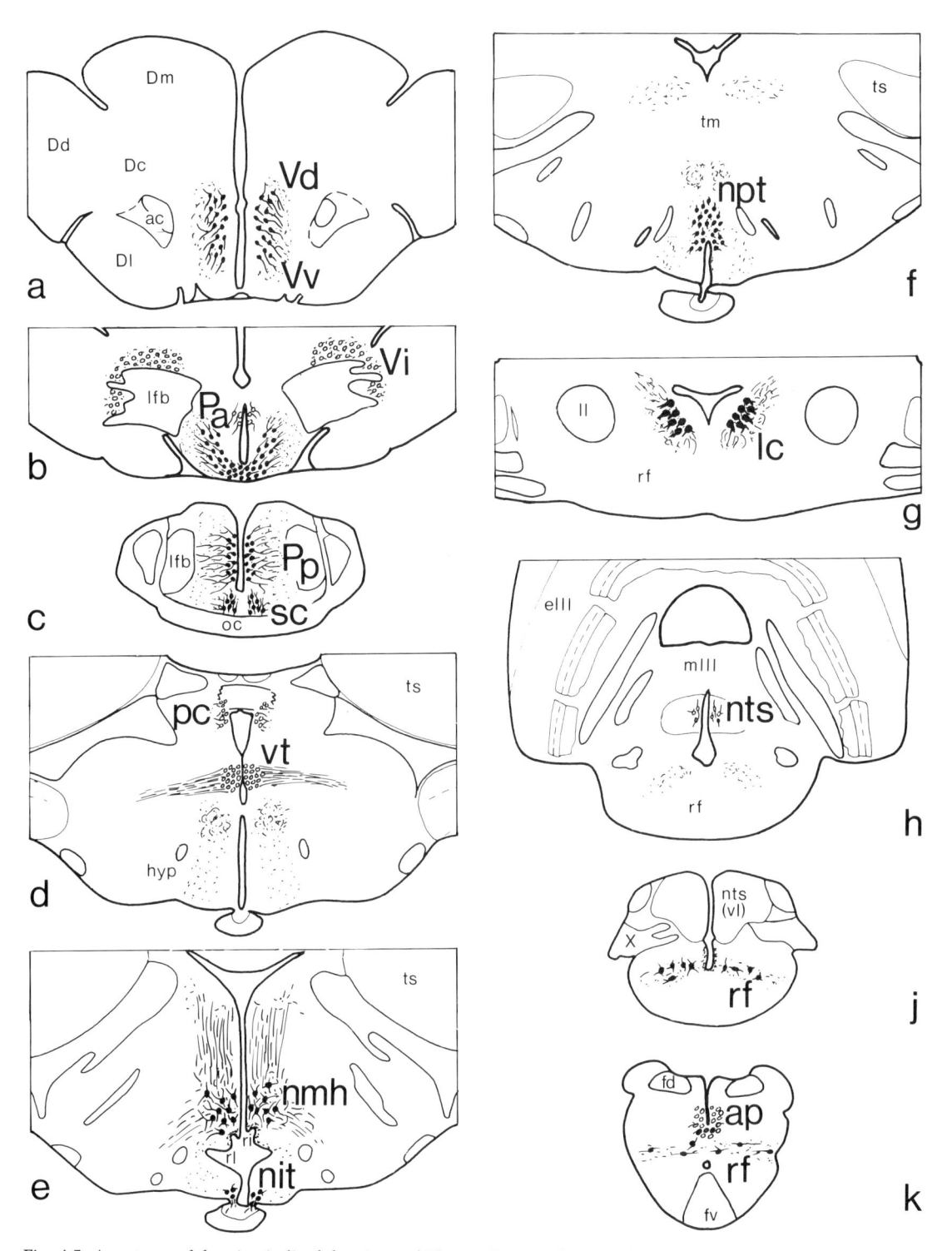

Fig. 4.5. A rostrocaudal series (*a–k*) of drawings of (the medioventral parts of) transverse sections through the brain of *Gnathonemus* at the levels indicated in Fig. 4.4, showing the distribution and shape of TH immunoreactive neurons and fibers. Lightly immunoreactive neurons have been indicated by empty circles, and more heavily immunoreactive neurons have been drawn in black.

species differences or may have technical reasons. Using different antibodies against TH in *Gnathonemus*, we observed different degrees of labeling of the PVO, which were, however, correlated with different degrees of labeling of raphe neurons, suggesting some cross-reactivity with serotonin or serotonin-related components (Meek & Joosten, 1993). Since the PVO contains large amounts of serotonin (e.g. Kah & Chambolle, 1983; Yoshida *et al.*, 1983, goldfish; Ekström & van Veen, 1984, *Gasterosteus*; Meek & Joosten, 1989 and Grant *et al.* 1989, *Gnathonemus*; and Johnston, Maler & Tinner 1990, *Apteronotus*), the degree of involvement of such a cross-reactivity should be carefully investigated before definite conclusions about possible species differences in PVO THir can be drawn in teleosts.

In the *thalamus* a relatively weakly THir cell group occurs in ventral subdivisions indicated as the ventromedial thalamic nucleus (vm) in *Gasterosteus* and *Carassius*, as the ventromedial nucleus (vmt) and ventrolateral area (avl) in *Apteronotus* and as the ventral thalamus (vt) in *Gnathonemus* (Fig. 4.4). All these groups seem to be comparable, since they all have long, laterally extending THir processes (Fig. 4.5(*d*)). In the transition zone between the thalamus and the mesencephalon, i.e. the *pretectal region*, located around the posterior commissure, a few weakly THir neurons are present as well in all teleosts studied, in a nucleus designated as the nucleus periventricularis pretectalis pars ventralis (ppv) in *Gasterosteus* and *Carassius*, as the central posterior thalamic nucleus (cp) in *Apteronotus*, or as neurons around the posterior commissure (pc) in *Gnathonemus* (Figs. 4.4, 4.5(*d*)).

In the *mesenphalon* proper, no THir neurons occur. This contrasts sharply with the situation in most other vertebrate groups, where a strongly THir dopaminergic substantia nigra and ventral tegmental area, or comparable regions, are present. These are absent, or at least not catecholaminergic, in teleosts. Mesencephalic THir neurons are restricted to the *isthmus* region, the transition zone between the mesencephalon and rhombencephalon, a location which is characteristic for the locus coeruleus (lc) of vertebrates (Fig. 4.4, 4.5(*g*)).

In the *rhombencephalon*, THir neurons occur in the caudal reticular formation (rf) of *Carassius*, *Apteronotus* and *Gnathonemus* (Fig. 4.4). Their location is just in the boundary zone between the dorsal and ventral rhombencephalon, in proximity of the descending trigeminal tract and the motor nucleus of the vagal nerve. Hence, they were indicated as such by Roberts, Meredith & Maslam (1989) in the eel (Fig. 4.6). Remarkably, such neurons were not observed in the stickleback (Fig. 4.4). In this fish only small THir neurons were observed in the nucleus of the solitary tract (nts, Fig. 4.4). Similar neurons are present in *Gnathonemus*, but not in goldfish and *Apteronotus*. In the latter species still another population of THir neurons, located in the nucleus of the descending trigeminal tract, has been described. In the *obex region*, the transition zone between the rhombencephalon and the spinal cord, THir neurons occur in all teleosts studied in the area postrema (Fig. 4.4, 4.5(*k*)). A detailed description of the area postrema of goldfish and its THir neurons has been presented by Morita & Finger (1987).

Fibers In general, not too much attention has been paid to the distribution of THir fibers in teleosts, since antibodies directed against dopamine or noradrenaline yield better pictures of the distribution of catecholaminergic fibers than TH immunoreactivity. The synthesis of catecholamines seems to be restricted largely to the cell bodies and proximal parts of dendrites, and thus TH immunoreactivity shows only a subpopulation of catecholamine containing fibers.

Comments and pictures on the distribution of THir fibers in teleosts have been presented by Hornby *et al.* (1987) and Hornby & Piekut (1990) for the goldfish; by Sas *et al.* (1990), for *Apteronotus* and by Meek & Joosten (1993) for *Gnathonemus*. Most details have been presented by Sas *et al.* (1990), who showed the presence of a light to moderate density of THir fibers in almost all subdivisions of the gymnotid brain. A basically similar widespread distribution is observed in *goldfish* and *Gnathonemus*.

L-DOPA producing cells? All neurons that synthesize catecholamines have to contain TH as well as DA, since the latter is the metabolic precursor for noradrenaline and adrenaline, and may also by itself be used for catecholaminergic neurotransmission or neuromodulation (see also Section 2.2.). However, particularly in *Gnathonemus* a number of THir cell groups in the ventral telencephalon, the preoptic region and the thalamus do not, or only partly, display DA immunoreactivity (cf Figs. 4.4, 4.6 and 4.11; see pp. 60–61). Similar, although less distinct observations have been made by Ekström *et al.* (1990), who state that in the tel-and diencephalon of *Gasterosteus* in general more THir neurons than DAir neurons occur.

The observations mentioned above raise the question whether THir but not DAir neurons, indicated with open circles in the summarizing Fig. 4.10, might be considered as L-DOPA-producing cells, sim-

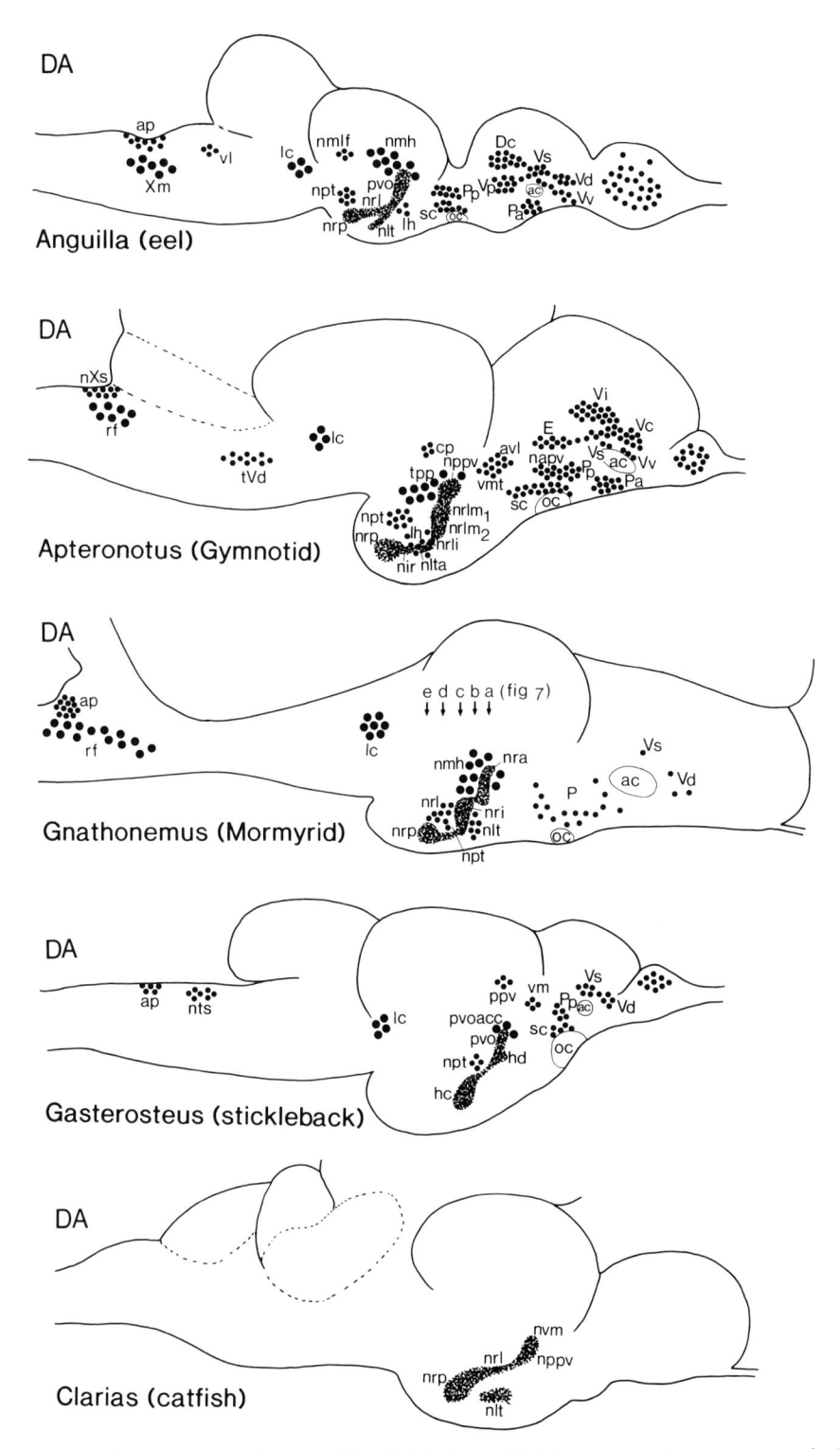

Fig. 4.6. Summarizing schemes of the distribution of DA immunoreactive neurons in the brains of five teleosts, as described by a number of studies discussed on pp. 60–61. Transverse sections through the brain of *Gnathonemus* are presented in Fig. 4.7 at the levels indicated.

ilar to those described in the mammalian hypothalamus (Okamura et al., 1988; Mons, Dubrovsky & Tramu, 1991). To make this plausible, it should be excluded that THir but not DAir neurons are depleted dopaminergic cells. This is not possible at present. It should be noted that Sas et al. (1990) do not mention quantitative differences in TH-or DA-immunoreactivity in the cell groups under discussion, and that the DAir neuronal populations found in the eel by Roberts et al. (1989) cover the complete spectrum of THir neurons observed in other teleosts (cf Fig. 4.4 and 4.6). So, the presence of L-DOPA producing cells in teleosts can only be evaluated when more knowledge becomes available about the physiology and functional significance of THir cells in different behavioral conditions.

Dopamine (DA)
The distribution of dopamine immunoreactivity has been studied in three of the four species also used for TH immunoreactivity, i.e. the stickleback *Gasterosterus* (Ekström, Honkanen & Steinbusch, 1990); the gymnotid fish *Apteronotus* (Sas et al. 1990) and the mormyrid fish *Gnathonemus* (Meek, Joosten & Steinbusch, 1989) as well as in the eel *Anguilla* (Roberts et al. 1989) and the catfish *Clarias* (Corio, Peute & Steinbusch, 1991). In addition, Kah et al. (1984) studied DA-immunoreactivity in the preoptic region of the goldfish. The results of these studies will be summarized in the next sections, without further reference to them.

Cell bodies In the teleostean species used for both TH-and DA immunohistochemistry, i.e. *Gasterosteus*, *Gnathonemus* and *Apteronotus*, DA immunoreactivity occurs in the same nuclei, and thus probably in the same neurons, as TH immunoreactivity, except for the possibly L-DOPA producing neurons discussed on pp. 58–60, plus the hypothalamic paraventricular organ (PVO, cf Figs. 4.4 and 4.6). In the eel, a similar pattern is observed, whereas for *Clarias* only DAir neurons in the PVO have been reported. Since most DAir cell groups of *Gasterosteus*, *Gnathonemus* and *Apteronotus* have already been dealt with in the previus sections, the present section will focus on DA immunoreactivity in the eel and on the PVO.

In the eel, telencephalic DAir cell groups are located in the olfactory bulb, the ventral, dorsal, supracommissural and postcommissural subregions of the ventral telencephalon (Vv, Vd, Vs and Vp, respectively) as well as in the transition region between the dorsal and ventral telencephalon, indicated as area dorsalis pars centralis by Roberts et al. (1989; Dc) but as area ventralis pars intermedialis (Vi) by others (see section 2.1.; cf Figs. 4.4 and 4.6). In the preoptic region, an anterior, posterior and suprachiasmatic DAir cell group is present (Pa, Pp and sc, respectively). In the diencephalon, DAir neurons are found in the PVO, to be discussed below, in the magnocellular hypothalamic nucleus (nmh), which accompanies part of the PVO in the lateral hypothalamic area (lha) and in the nucleus posterior tuberis (npt). The few DAir neurons observed in the mesencephalic nucleus of the medial longitudinal fascicle (nmlf) of the eel might be comparable to the neurons indicated as pretectal or central posterior thalamic in other species (cf p. 58 and Figs. 4.4 and 4.6). DAir neurons in the isthmic region represent the teleostean locus coeruleus (lc, Fig. 4.6). Rhombencephalic DAir neurons occur in the eel in the nucleus of the solitary tract or vagal lobe (vl), in the caudal reticular formation at the boundary with, or within, the vagal motor nucleus (Xm) and in the area postrema (ap). A basically similar pattern of DAir neurons is found in *Apteronotus, Gnathonemus* and *Gasterosteus*, sometimes indicated with a different nomenclature (cf pp. 55–58; see Fig. 4.6), and with the exception of several tel- and diencephalic cell groups in *Gnathonemus* that display TH- but not DA immunoreactivity, as discussed on pp. 58–60.

As shown in Fig. 4.6, the paraventricular organ (PVO) is consistently labelled in all teleosts studied thus far, including also *Clarias*. The PVO is a population of cerebro-spinal fluid (CSF)-contacting neurons located along the infundibulum and its recessi of the third ventricle. In gross features, the location and arrangement of DAir CSF-contacting PVO neurons is similar in all teleosts studied, starting rostrodorsally along the rostrodorsal tip of the infundibulum, a region also indicated as the organum vasculosum hypothalami (Gomez-Segade, Anadón & Gomez-Segade, 1989), and connected via more caudoventrally located regions around the proximal part of the lateral recess with the most caudoventral part of the population: the nucleus of the posterior recess of the infundibulum (Figs. 4.6, 4.7). However, different subregions have been indicated with different names by different authors in different fishes, as summarized in Fig. 4.6. In spite of the different names used, the distribution of DAir neurons in the PVO of different teleosts seems to be basically similar. Whether true and significant species differences exist can only be evaluated when more data become available concerning the connections, functional significance and further chemical differentiation of the subdivisions distinguished. In this respect, it should be noted that

Catecholamines in brain of Osteichthyes 61

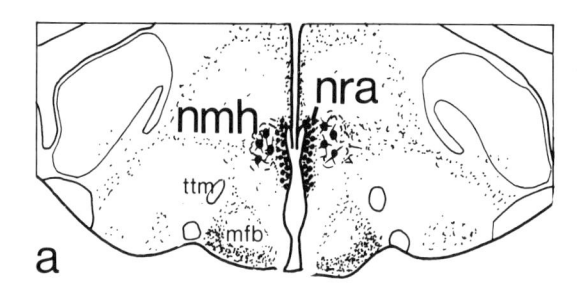

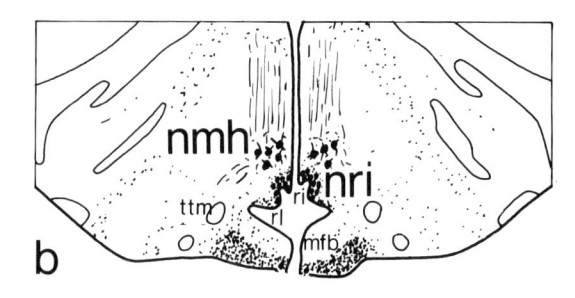

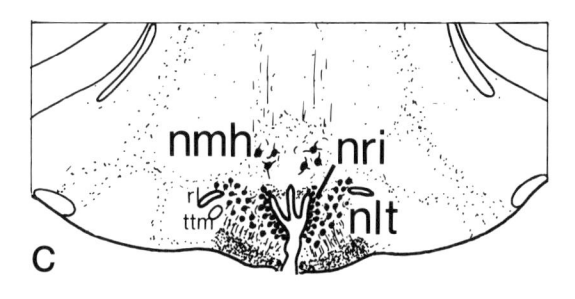

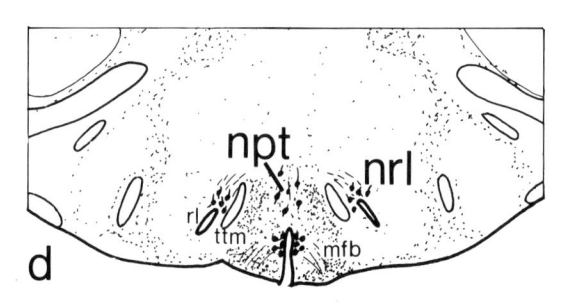

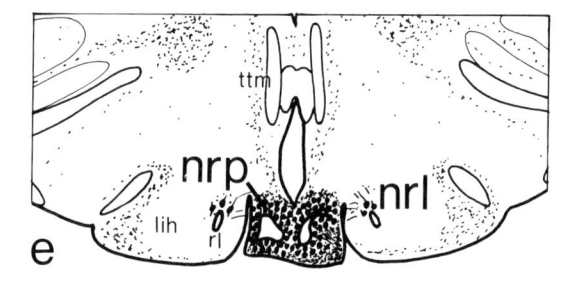

in almost all subdivisions of the teleostean PVO that contain DAir CSF-contacting neurons, serotonin (5Ht)-immunoreactive neurons occur as well (e.g. Frankenhuis-van den Heuvel & Nieuwenhuys, 1984, trout; Ekström & van Veen, 1984, stickleback; Grant et al., 1989; Meek & Joosten, 1989, *Gnathonemus*; Ekström & Ebbesson, 1989, salmon; Johnston, Maler & Tinner, 1990, *Apteronotus*; and Corio et al., 1991, catfish).

Evaluation of true species differences in teleostean DA immunoreactivity is not only hampered by the use of different nomenclatures in different fishes and lack of data about physiological and connectional properties of the DAir neurons, but also by the use of different antibodies against DA, which apparently induces differences as well. The results compared in Fig. 4.6 are based on three different antibodies: An antibody against a glutaraldehyde-coupled DA-conjugate raised in rabbits by Dr Buijs was used in the eel (Roberts et al., 1989); a similar antibody, raised by Dr Steinbusch, was used in *Apteronotus*, *Gasterosteus* and *Gnathonemus* (Sas et al., 1990; Ekström et al., 1990 and Meek, Joosten & Steinbusch, 1989, respectively), but an antibody against a formaldehyde-coupled DA-conjugate raised in rabbits by Dr Steinbusch was used in the catfish *Clarias* (Corio et al., 1991) The lack of DAir neurons located outside the PVO in *Clarias* is probably due to a lower sensitivity of this formaldehyde-fixation requiring antibody compared with the glutaraldehyde-related antibodies used in the other studies summarized in Fig. 4.6, or to lack of fixation of dopamine in formaldehyde-fixed material. However, the same problem does not seem to be involved in the different DA immunoreactivity pattern in *Gnathonemus*, as compared with other teleosts and TH immunoreactivity (see also pp. 58–60), since in *Gnathonemus* the same antibody and fixation procedure was used as in *Apteronotus* and *Gasterosteus*.

Fibers Descriptions and pictures of the distribution of DAir fibers in teleostean brains have been presented by Meek et al. (1989) for *Gnathonemus*; by Roberts et al. (1989) for *Anguilla*; by Ekström et al. (1990)

Fig. 4.7. A rostrocaudal (a–e) series of drawings of (the medioventral parts of) transverse sections through the PVO and surrounding hypothalamic regions of *Gnathonemus* at the levels indicated in Fig. 4.6, showing the distribution of DA immunoreactive neurons and fibers (see also pp. 60–63). (modified after Meek et al., 1989).

for *Gasterosteus* and by Corio *et al.* (1991) for *Clarias*. It has already been discussed on p. 61 that the antibody used by the latter authors seems to be less sensitive than those used by others, which means that the picture for *Clarias* is probably less complete than that for other teleosts. Sas *et al.* (1990) give an extensive description of the distribution of DAir fibers in *Apteronotus*, however, without pictures. Their pictures are restricted to TH immunoreactivity with the statement that DA immunoreactivity is basically similar.

The studies just mentioned show that DAir fibers are widely distributed in teleostean brains and occur in a variety of tel-, di-, mes- and rhombencephalic regions, with the highest density in the medial forebrain bundle and the surrounding ventral telencephalic, preoptic and hypothalamic regions (Fig. 4.7). A detailed evaluation of the distribution of DAir fibers in the teleostean brains studied is faced with several technical differences that cannot always be distinguished from species differences. They include not only the use of different antisera, but also of different dilutions and different ways to encode fiber densities in drawings. An example of the latter is shown in Fig. 4.8, where the distribution of DAir fiber density in the telencephalon of different teleosts

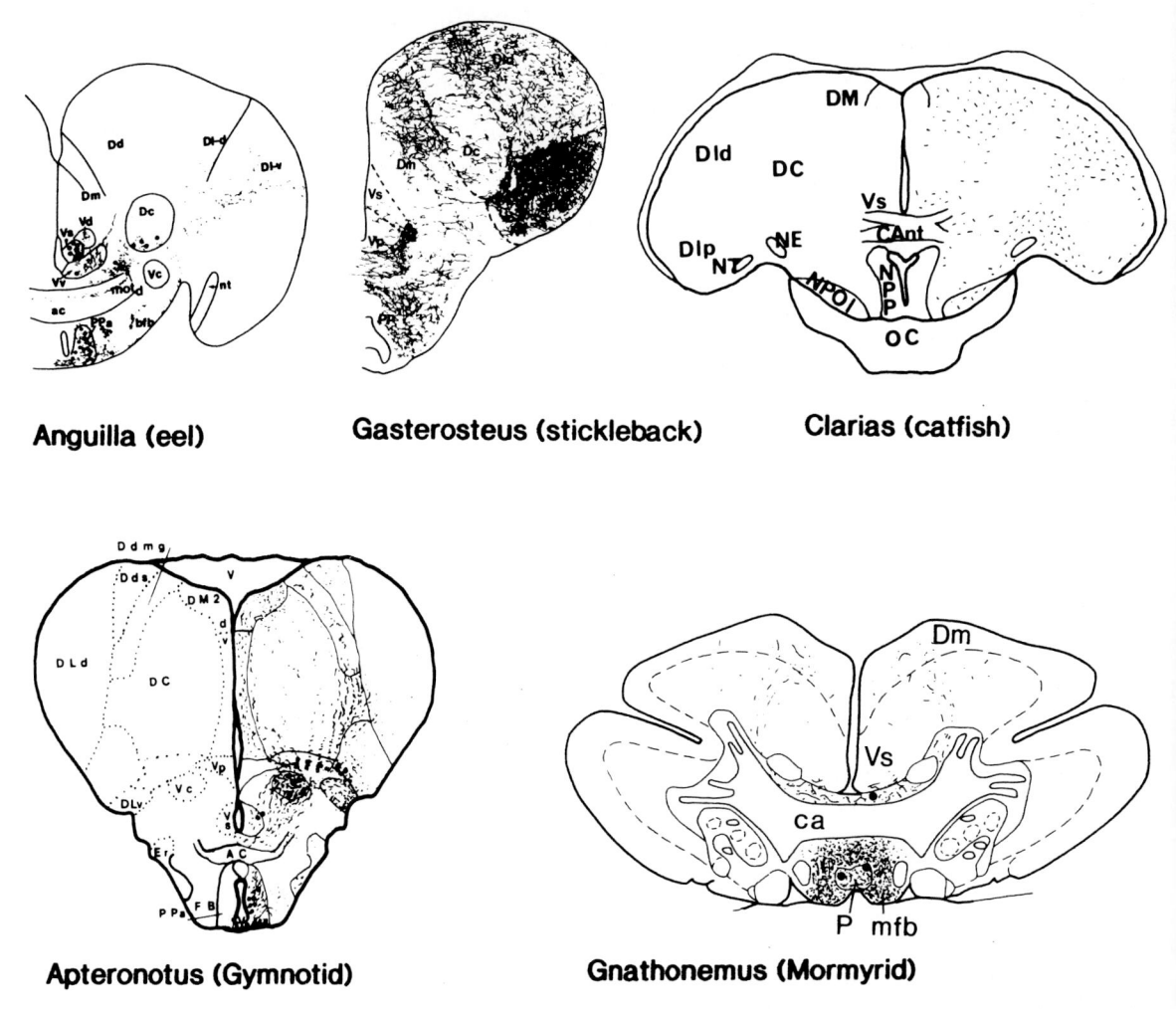

Fig. 4.8. Comparison of different distributions and ways of presentation of the DA immunoreactivity in the everted telencephalon of five teleosts at the level of the anterior commissure, reproduced with permission from Roberts *et al.* (1989) for the eel, Ekström *et al.* (1990) for the stickleback; Corio *et al.* (1991) for catfish: Sas *et al.* (1990) for *Apteronotus* and Meek *et al.* (1989) for *Gnathonemus*. For details and abbreviations used in this Figure, the reader is reffered to the original papers. A discussion of the results summarized in this Figure is presented on pp. 62–63.

as presented by different authors, is compared. Comparison of drawings and photographs presented for different teleosts shows that the huge differences in overall density of telencephalic DAir fibers, as suggested by these drawings, partly reflect different criteria used to make drawings rather than true species differences in fiber density. Thus, only relative differences in DAir fiber density can be evaluated, whereas data on the absolute levels of DA innervation as well as on the origin of DAir fiber populations are presently lacking.

The studies enumerated above agree in a moderate to dense DA innervation of the olfactory bulbs, but present substantial differences with respect to the DA innervation of other regions of the everted telencephalon. In the area dorsalis (D), the densest innervation occurs in pars lateralis ventralis (Dlv) and pars posterior (Dp) in both the stickleback and the eel, where other regions receive a moderate (stickleback) or very low (eel) DA innervation. However, in *Gnathonemus* only the medial part (Dm) receives a (moderate) DA innervation, but the lateral part (Dl) not at all, while in *Apteronotus* Dm, Dlv and Dp receive a similar DA innervation (Fig. 4.8). In the ventral telencephalic region (V) the region lateroventral to the dorsal part (Vd), indicated in the stickleback as the lateral part (Vl) and in *Apteronotus* as the central part (Vc), receives the most dense DA innervation, with a low to moderate density in other ventral telencephalic subdivisions.

The preoptic region receives a moderate to dense DA innervation in all teleosts studied, just as the tuberal hypothalamic region around the paraventricular organ. The inferior hypothalamic lobe and dorsal diencephalic regions, including the nucleus glomerulosus and preglomerulosus, the thalamus and the pretectal region, are generally sparsely provided with DAir fibers, except for a more densely innervated pretectal nucleus in the stickleback and the eel.

In the mesencephalon, a dense DA innervation is observed in the interpeduncular nucleus of the stickleback and the eel, but not in that of *Gnathonemus* and *Apteronotus*. Likewise, in the eel a particularly dense DA innervation occurs in the oculomotor and trochlear nucleus, which is absent in other species. The laminar distribution described for tectal DAir fibers differs substantially in different teleosts: in the stickleback, they occur particularly in the stratum opticum (SO) and the stratum griseum centrale (SGC), but in the eel they seem to avoid the stratum opticum (SO), as well as the stratum marginale (SM) and the stratum periventriculare (SPV). However, in *Gnathonemus* and *Apteronotus*, the SPV is the most densely innervated layer, followed by the stratum griseum centrale (SGS) and stratum album centrale (SAC). In the laminated torus semicircularis of *Apteronotus*, DAir fibers particularly occur in the deeper layers 8b, 8c and 9, while, of the six toral nuclei of *Gnathonemus* particularly the mediodorsal and ventroposterior nuclei are densely innervated. The anterior exterolateral toral nucleus of *Gnathonemus* is completely devoid of DAir fibers. Remarkably, the precerebellar nucleus lateralis valvulae of the stickleback receives a dense DAir innervation, whereas that of *Gnathonemus* is devoid of DAir fibers, except for the most laterocaudal subdivision bordering the locus coeruleus.

In the rhombencephalon, ventrally located regions generally receive a denser DAir innervation than dorsally located centers. Moderate densely innervated ventral regions include the reticular formation of all teleosts, the raphe nuclei of the stickleback, the cranial nerve motor nuclei of the eel, the descending trigeminal nucleus of *Apteronotus* and the subependymal ventricular region of the vagal motor nucleus of *Gnathonemus*. The electrosensory lobes of the electric fishes *Apteronotus* and *Gnathonemus* as well as the cerebella of all teleosts are almost completely devoid of DAir fibers. However, some DA innervation of Purkinje cells has been reported for the stickleback.

DA as neurotransmitter or precursor After the description of DA immunoreactivity in teleostean brains, it seems appropriate to devote some comments on its functional significance. Dopamine may be used as a neuroactive substance, i.e. a neurotransmitter or neuromodulator involved in interneuronal communication, and neurons using dopamine for these purposes are generally indicated as dopaminergic. However, dopamine may also be used as a precursor for noradrenaline or adrenaline, to be converted by the enzymes DBH and PNMT, respectively (see pp. 69–71). In the latter situation, it may be expected that the neurons do not only display TH and/or DA immunoreactivity, but also NA and DBH immunoreactivity. This appears to be the case in the locus coeruleus and the caudal reticular formation (see p. 69), which thus probably are not dopaminergic but noradrenergic or adrenergic. Neurons in these regions probably use dopamine only to synthesize other catecholamines, although it cannot definitely be excluded that a subpopulation uses dopamine as a neurotransmitter as well, all or not in combination with noradrenaline or adrenaline. In

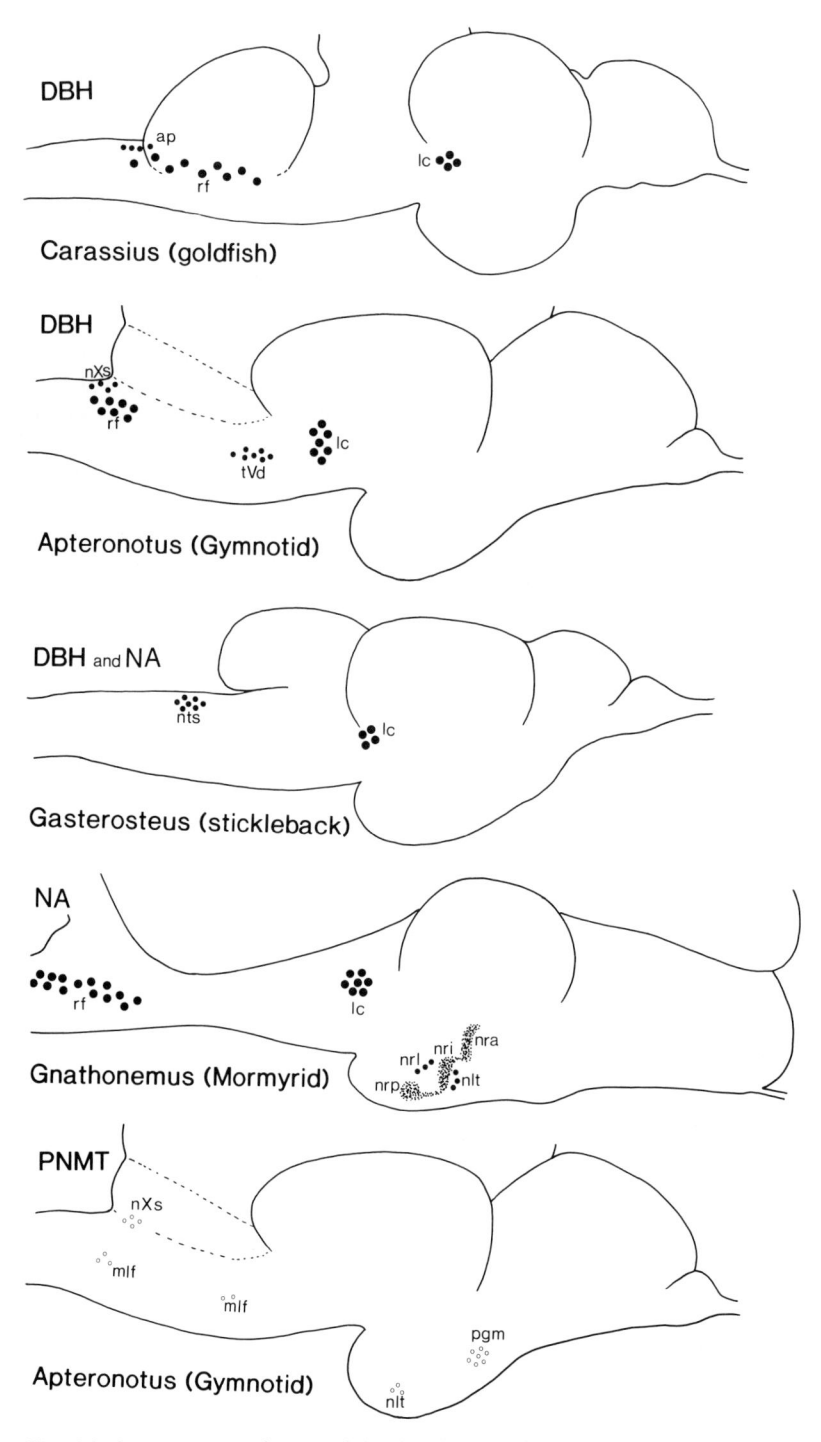

Fig. 4.9. Summarizing schemes of the distribution of DBH-, NA- and PNMT-immunoractive neurons in the brains of four teleosts, as described by a number of studies discussed on pp. 69–71.

spite of these uncertainties, such DAir neurons have been indicated as noradrenergic in the summarizing Fig. 4.10, and not as dopaminergic.

Dopaminergic neurons, defined as neurons that use dopamine for signal transduction, may be expected to display both TH and DA immunoreactivity, but not NA or DBH immunoreactivity. Such neurons, indicated as black dots in the summarizing Fig. 4.10, appear to comprise the majority of the catecholaminergic neuronal population. They include a number of parvocellular forebrain nuclei, located in the ventral telencephalon, preoptic region, hypothalamus, thalamus and pretectum; the magnocellular hypothalamic PVO accompanying cells; a few parvocellular neurons in or around the nucleus of the solitary tract and the area postrema. However, it should be stressed again that the mesencephalic tegmentum does not contain dopaminergic neurons, and that there is consequently no teleostean substantia nigra or ventral tegmental area. Unfortunately, the precise contribution of the distinct dopaminergic cell groups to the dopaminergic innervation of different teleostean brain regions is uncertain. The same holds for their functional significance and relation with the PVO and its efferent fibers. The latter appears to be a different type of DAir cell group, to be discussed below, since it is strongly DA immunoreactive (as well as serotonin-immunoreactive), but does not, or hardly, contain TH (Fig. 4.9 and 4.11).

DA acquiring cells? The majority of CSF-contacting PVO neurons along the infundibulum of the third ventricle in teleosts show the remarkable phenomenon that they are DA immunopositive but TH immunonegative. This suggests that they cannot synthesize dopamine themselves, but have to acquire it from external sources. Consequently, they may be indicated as DA acquiring cells. However, it should be noted that it cannot definitely be excluded that they synthesize DA. A subpopulation displays well-detectable levels of TH immunoreactivity (see pp. 55–58) and thus, excluding possible cross reactivities as discussed on p. 58, might well be able to synthesize dopamine. Other cells might still contain very small amounts of TH, not detected, or only faintly, by the current immunohistochemical procedures (see Fig. 4.11(a)–(f)), but enough to synthesize well-detectable amounts of dopamine over long times, when it would not be released, but would be stored within the neurons.

When we still consider large numbers of CSF-contacting DAir PVO cells as dopamine-acquiring, and not synthesizing, neurons, the most intriguing question concerns the source of the dopamine acquired. The CSF-contacting nature of these neurons immediately points to the CSF (e.g. Meek et al., 1989; Sas et al., 1990; Ekström et al., 1990), the more since Vigh-Teichmann & Vigh (1974) have demonstrated that CSFcontacting processes are primarily receptive, i.e. picking up substances from the CSF and transporting it to the cell body. Intraventricular dopamine might be released into the CSF by subependymal axons, as, for example, found in a DAir subependymal plexus along the preoptic recess of *Gnathonemus* (Meek et al., 1989; Fig. 4.11(h)) and a similar intraventricular release and PVO-uptake mechanism might be present for serotonin (Meek & Joosten, 1989) and noradrenaline (Meek et al., 1993). However, there is presently no definite proof of uptake of monoamines in the CSF by PVO cells, and several ultrastructural characteristics, including the presence of dopamine in dense-cored vesicles within the CSF contacting processes, suggest release rather than uptake (Meek et al., 1992). Moreover, the location of the PVO along all kinds of recesses of the blind ending infundibulum seems to be more efficient for a gradual and constant release, than for uptake from the CSF.

Apart from the CSF, DAir PVO cells might acquire dopamine also from other sources, including possible DA synthesizing (THir) parts of the PVO, and the magnocellular PVO accompanying cells. In particular, the close relation of these cells with large parts of the PVO, as well as the fact that during ontogeny their TH and DA immunoreactivity precedes that of the PVO, support this suggestion (Ekström, Chapter 13). The functional significance of an intra or extraventricular DA acquiring mechanism is presently unknown. The only thing we know is that parts of the PVO are involved in the innervation of the pituitary (Fryer et al., 1985), and that intraventricular dopamine is very effective in eliciting various aggressive as well as non-aggressive behavioral responses in gymnotid fish (Maler & Ellis, 1987).

Is there a teleostean striatum? An intriguing observation in teleosts concerns the absence of dopaminergic midbrain neurons comparable to those of the substantia nigra and/or ventral tegmental area. In other vertebrate groups, such neurons are the most conspicuous population of dopaminergic neurons, and project massively to the striatum by way of a mesencephalotelencephalic projection. Thus, the question arises whether a teleostean striatum is absent as well, or is present but receives a dopaminergic projection from extra-mesencephalic sources,

66 *J. Meek*

○ : TH⁺,DA⁻,NA⁻   : DOPA-ergic or depleted CA-cells

● : TH⁺,DA⁺,NA⁻ or DBH⁻ : Dopaminergic cells

✱ : TH⁺,DA⁺,NA⁺ or DBH⁺ : Noradrenergic cells

□ : TH⁻,DA⁻,DBH⁻,PNMT⁺ : Adrenergic cells ????

★ : TH⁻,DA⁺   : Dopamine acquiring cells?

✪ : TH⁻,NA⁺   : Noradrenaline acquiring cells?

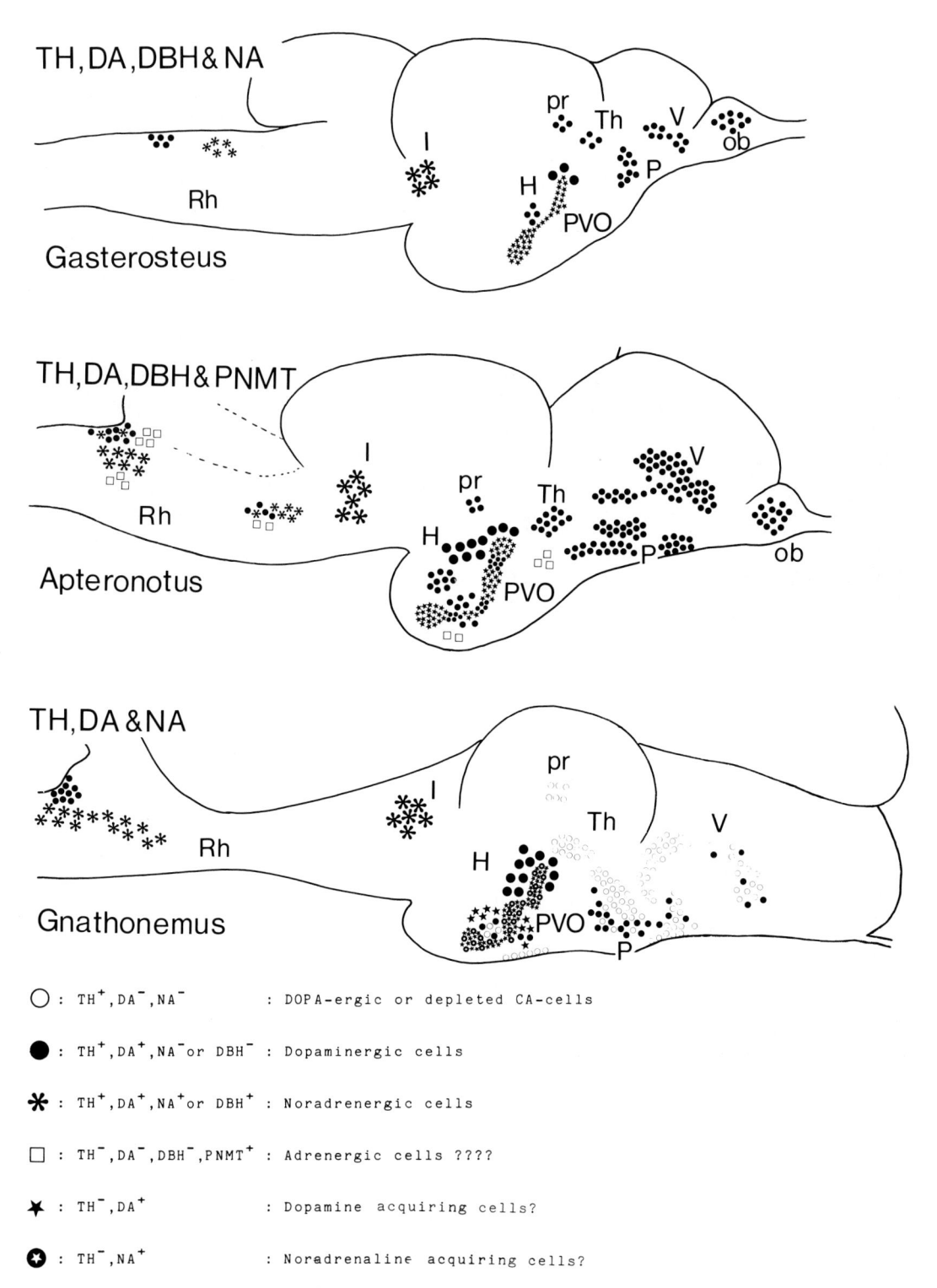

Fig. 4.10. Summarizing scheme of possibly DOPA-ergic and presumed dopaminergic, noradrenergic, adrenergic, dopamine acquiring and noradrenaline acquiring cell groups in the brains of three teleosts, projected on lateral views of these brains.

Catecholamines in brain of Osteichthyes 67

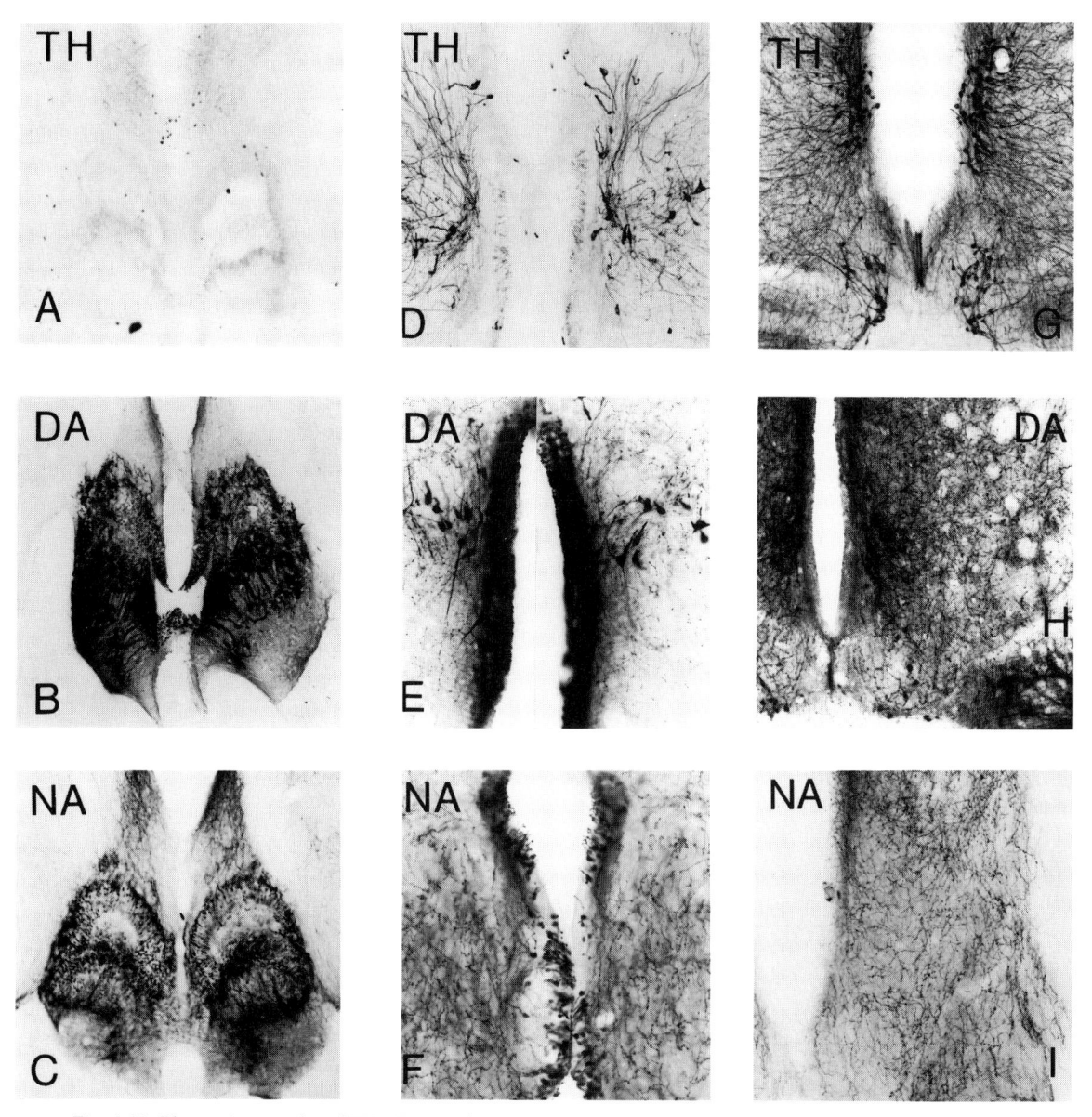

Fig. 4.11. Photomicrographs of TH-, DA- and NA-immunoreactive neurons and fibers in the brain of *Gnathonemus petersii* at the level of the posterior infundibular recess (a), (b), (c); cf Fig. 4.7(e), the anterior infundibular recess (d), (e), (f); cf Fig. 4.7(a) and the preoptic recess (g), (h), (i); cf Fig. 4.5(c).

or is present without a dopaminergic innervation. This concerns an interesting aspect of comparative neuroanatomy and the role of chemical neuroanatomical criteria in the establishment of homologies between different vertebrate groups, complicated by the everted structure of the teleostean telencephalon (p. 53, Fig. 4.2). Therefore, some comments on this subject seem to be appropriate at this place.

A variety of structural aspects of nuclei may be used to establish homologies, starting with the relative position of these nuclei, supplemented with criteria based on afferent and efferent connections, internal structural organization and chemoarchitecture. The mammalian, and in general, the amniote striatum, is characterized by a position in the lateral subpallium at the boundary with the pallium (Fig.

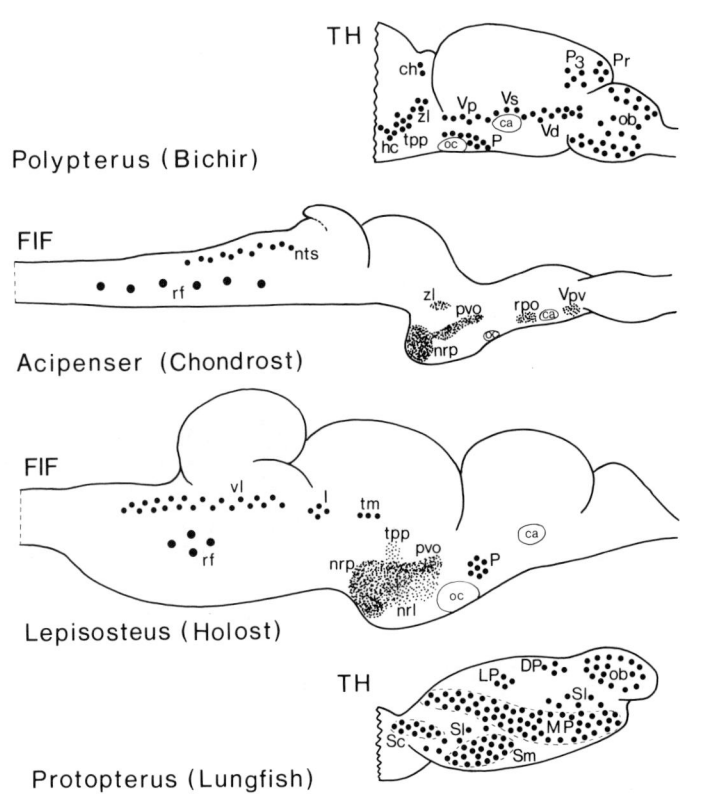

Fig. 4.12. Summarizing schemes of the data that are presently available on the distribution of catecholamines in the brains of non-teleostean bony fishes (see pp. 71–72 for details).

4.2), and by afferent and efferent connections with cortical areas, descending projections to premotor regions, the presence of many multipolar spiny neurons, and a dopaminergic mesencephalotelencephalic innervation. The corresponding topolgial position in teleosts is area ventralis telencephalis pars dorsalis (Vd), or, depending on the criteria used to distinguish a teleostean pallium and subpallium, area dorsalis telencephalic pars medialis (Dm; Fig. 4.2; Nieuwenhuys & Meek, 1990a). However, striatum-like reciprocal connections with dorsal (pallial) telencephalic (cortical) regions and descending output are particularly found for area dorsalis pars centralis (Dc; Murakami, Morita & Ito, 1983; Nieuwenhuys & Meek, 1990a), where also many spiny multipolar neurons are present (Meek & Nieuwenhuys, personal observations). The dopaminergic innervation of the telencephalon of teleosts makes the picture even more confusing: there is a large interspecific variability (see p. 63 and Fig. 4.8), with in general the densest innervation in Dlv, Dp, Vc and Vl (see above). However, neither Dm, nor Vd or Dc belongs to the most heavily DA innervated regions.

Thus, the study of DA immunoreactivity supports the conclusion of Nieuwenhuys & Meek (1990a) that a teleostean telecephalic region homologous to the striatum of amniotes cannot be unequivocally identified presently and even may be absent, just as a dopaminergic substantia nigra or ventral tegmental area. They suggest that the everted actinopterygian telencephalon has a basically different functional organization than the evaginated telencephalon of other vertebrate groups, which would mean that the establishment of homologies between subdivisions of evaginated and everted telencephala is not possible. Reiner & Northcutt 1992, however, propose another possibility after their study of the more simply everted brain of *Polypterus* (see p. 71). They suggest that the periventricular nucleus of the posterior tubercle of this fish (see Fig. 4.12), which is homologuous with the teleostean nucleus of the lateral recess, is comparable with the amniote substantia nigra and ventral tegmental area, on the basis of its neurochemical differentiation. Since this region probably projects to the rather densely TH-innervated dorsal part of the ventral telencephalon (Vd), they suggest that this region represents the actinopterygian striatum. This would be in agreement with topological considerations (see Fig. 4.12).

Noradrenaline (NA)
The characteristics of noradrenaline containing neurons and terminals have been investigated using antibodies against the noradrenaline synthesizing enzyme dopamine β-hydroxylase (DBH), as well as against a glutaraldehyde coupled noradrenaline–bovine serum albumin conjugate. As will be discussed in the next sections, both methods give similar results, except for the PVO.

DBH immunoreactive cell bodies The properties and distribution of DBH immunoreactivity have been investigated in the goldfish (Hornby & Piekut 1988, 1990), the stickleback (Ekström et al., 1990) and the gymnotid fish Apteronotus (Sas et al., 1990). All studies demonstrated a population of large DBH-ir neurons in the isthmus region, corresponding with the TH-and DAir cell group in that region, and thus interpreted as the teleostean noradrenergic locus coeruleus (Fig. 4.9). In goldfish and gymnotid fish, the DA- and THir neurons of the caudal rhombencephalon display DBH immunoreactivity as well. However, in Gasterosteus such neurons are absent, confirming similar observations with TH- and DA immunoreactivity (cf Figs. 4.4 and 4.6). DBHir neurons in the area postrema, as observed in goldfish and gymnotid fish, are absent as well in stickleback. In turn, the latter has some DBHir neurons in the nucleus of the solitary tract, not observed in Carassius or Apteronotus. Finally, the TH- and DAir neurons in the nucleus of the descending trigeminal tract of Apteronotus appear to the DBHir as well (Fig. 4.9).

DBH immunoreactive fibers The distribution of DBHir fibers has been described and presented in an atlas for goldfish (Hornby & Piekut, 1988, 1990) and Apteronotus (Sas et al., 1990). As discussed for THir fibers, it is not certain whether DBH is present in all parts of NA-synthesizing neurons, or only in proximal regions. Nevertheless, DBHir fibers have a very widespread distribution in Apteronotus, and occur in all brain regions, including the cerebellum, which is almost devoid of TH- and DA-immunoreactivity in this species. In Carassius, the distribution of DBH immunoreactivity seems to be less widespread, and e.g. absent in the cerebellum, lateral diencephalic and dorsal rhombencephalic regions. However, it has already been discussed (pp. 61–63) that it is difficult to make quantitative comparisons between species, since different technical procedures and criteria for drawings are used by different authors in different fishes (cf. Hornby & Piekut, 1988, 1990 and Sas et al., 1990). In both Carassius and Apteronotus, the distribution of DBH immunoreactivity seems to be less dense and widespread than that of NA immunoreactivity as observed in Gasterosteus and Gnathonemus (see pp. 69–70).

NA immunoreactive cell bodies The distribution of immunoreactivity against an antibody directed against noradrenaline itself, coupled by glutaraldehyde to bovine serum albumin, has been investigated in Gasterosteus (Ekström et al., 1986) and Gnathonemus (Meek, Joosten & Hafmans, 1993). In Gasterosteus, the results agree with the results on DBH-immunoreactivity as described above, showing a NAir locus coeruleus or coeruleus-like complex (see Ekström et al., 1986 for a detailed discussion on the teleostean locus coeruleus), but not a NAir rhombencephalic reticular cell group. A few neurons in the nucleus of the solitary tract, and some CSF-contacting postobecular cells, possibly belonging to the area postrema, are weakly NAir as well in Gasterosteus (Ekström et al., 1986).

In Gnathonemus, the distribution of NA immunoreactivity differs substantially from that observed in Gasterosteus: in addition to a distinct NAir locus coeruleus, most, if not all, DA- and THir caudal reticular cells are NA immunoreactive. A substantial number of CSF-contacting PVO cells, and a few non-CSF-contacting neurons along the lateral recess of the infundibulum, are NAir as well (Figs. 4.9 and 4.11).

NA immunoreactive fibers The distribution of NA ir fibers, studied in Gasterosteus (Ekström et al., 1986) and Gnathonemus (Meek et al., 1993) is very widespread. All regions of the brains of these teleosts appear to receive at least some noradrenergic innervation, including the cerebellum, which is not, or is hardly innervated by dopaminergic fibers. An interesting observation in Gnathonemus is the presence of a number of subependymal plexuses along the forebrain ventricular surface, i.e. a dorsal and ventral telencephalic, and a preoptic one. Conceivably, these plexuses might be involved in intraventricular release of noradrenaline. In Gasterosteus, such subependymal NAir fiber plexuses seem to be absent. Instead, noradrenergic fibers are rather evenly distributed over distinct telencephalic regions, including the olfactory bulbs, area dorsalis and area ventralis telencephali. The same holds for the remaining subdivisions of the brain of *Gasterosteus*, where in almost any region a low to moderate NAir fiber density has been observed.

Interesting features of the distribution of NAir fibers in the brain of *Gnathonemus* concern, apart from the subependymal plexuses, the involvement of a noradrenergic innervation in electrosensory specializations and the huge cerebellum. In the electrosensory rhombencephalic lateral line lobe, a distinct NAir fiber lamination is present, with the highest density in the ganglionic layer. In the electrosensory torus semicircularis, NAir fiber density is high in the ventroposterior nucleus, moderate in others, but completely absent in the anterior exterolateral nucleus. In different parts of the mormyrid gigantocerebellum, NAir fibers terminate predominantly in the granular layer, but may also form parallel fibers in the molecular layer. In some subdivisions of the well differentiated precerebellar nucleus lateralis valvulae of *Gnathonemus*, NAir fibers establish large club endings around adendritic cells. All these features suggest that the NAir fiber system is phylogenetically rather plastic, and much more involved in and adapted to a number of brain variations and specializations of *Gnathonemus* than the DAir fiber system (Meek *et al.*, 1993).

NA as neurotransmitter or precursor Cells that use noradrenaline as a neuroactive substance in interneuronal communication processes, thus as a neurotransmitter or neuromodulator, are indicated as noradrenergic. They may be expected to display TH-, DA-, DBH- as well as NA-immunoreactivity, which is the case for the teleostean locus coeruleus, and the caudal reticular cell group (Fig. 4.10). In addition, a few noradrenergic cells are probably present in the area postrema of *Carassius* and *Apteronotus*, in the nucleus of the solitary tract of *Gasterostus*, and in the descending trigeminal nucleus of *Apteronotus* (Figs. 4.9 and 4.10). Whether the fact that these groups occur only in some species is a true species difference, or due to technical differences or behavioural conditions before animal perfusion, is presently unknown.

Apart from its function as a neurotransmitter or -modulator, NA may also be used as a precursor for adrenaline. To be sure that NAir or DBHir neurons are noradrenergic, it should be certain that they do not contain PNMT or adrenaline. This actually has only been investigated in *Apteronotus* (see pp. 70–71 and Fig. 4.9). In this teleost, PNMT-immunoreactivity suggests that the cell groups enumerated above are indeed noradrenergic and not adrenergic. To be sure in other species, further experiments using PNMT- or adrenaline immunoreactivity are necessary, however.

NA acquiring cells? In *Gnathonemus*, CSF-contacting PVO cells do not only seem to acquire dopamine from external sources, as discussed above (p. 65), but also noradrenaline. Although DBH immunoreactivity has not been investigated in *Gnathonemus*, the PVO of other teleosts does not show any DBHir neurons. In other vertebrates it has been demonstrated convincingly that NAir PVO cells are DBH negative (e.g. Smeets & Steinbusch, 1989; 1990, using a lizard; see also other chapters of this book), and the same probably holds for *Gnathonemus*. Consequently, a subpopulation of CSF-contacting PVO cells in *Gnathonemus* has been indicated as noradrenaline-acquiring in the summarizing Fig. 4.10. It is presently unknown whether these cells acquire both dopamine and noradrenaline, or only one of them. Likewise, the source of the acquired noradrenaline is uncertain as well, although the subependymal telencephalic plexuses are good candidates (Meek *et al.*, 1993).

Remarkably, NAir PVO cells were not observed in *Gasterosteus* (Ekström *et al.*, 1986). It is possible that this reflects a true species difference, and that the PVO of the quite small and simply organized brain of *Gasterosteus* is less differentiated and more restricted with respect to its functional organization, than the much larger PVO along a number of specialized recessi of *Gnathonemus*. However, several technical differences or physiological conditions before animal perfusion might also underlie the differences observed. More data on the distribution of NA immunoreactivity in different teleostean species are necessary to investigate whether noradrenaline-acquiring PVO cells are a general teleostean phenomenon, or restricted to a smaller or larger subset of the total teleostean population.

Adrenaline

Indications about the presence of adrenaline in the brain of teleosts has been obtained in only one species, i.e. the gymnotid fish *Apteronotus* (Sas *et al.*, 1990), by means of PNMT-immunohistochemistry. Studies using antibodies against adrenaline itself in teleosts are presently not available.

PNMT-immunoreactive cells Sas *et al.* (1990) have restricted their study of PNMT immunoreactivity in

Apteronotus to cell bodies. PNMTir cell bodies occur in the diencephalic nucleus preglomerulosus pars medialis (pgm), nucleus lateralis tuberis (nlt), around the longitudinal medial fascicle (mlf) and in the sensory vagal nucleus (nXs) (Fig. 4.9). Remarkably, none of these neuronal populations displays other catecholamine-related immunoreactivity, i.e. neither TH- or DBH-, nor DA immunoreactivity. Thus, it is uncertain whether they really synthesize adrenaline, and use it in interneuronal communication. However, their position corresponds largely with several groups described in mammals (Sas et al. 1990), and it might be possible that they still contain low levels of synthesizing enzymes and precursors, or that they synthesize adrenaline from externally acquired noradrenaline. Presumptive adrenergic cells have been indicated with open circles instead of dots in Fig. 4.9, because of the uncertainties just discussed.

Formaldhyde-induced fluorescence (FIF) studies

The presence and distribution of catecholamine containing neurons and fibers can not only be studied with immunohistochemical techniques, but also with the older formaldehyde-induced fluorescence (FIF) techniques. These have been applied in a large number of studies preceding the immunohistochemical investigations on teleosts as decribed above, but also in some recent studies. Studies published before 1982, dealing with *Anguilla, Carassius, Lepomis, Leuciscus, Misgurnus, Myoxocephalus, Perca* and *Salmo* have been reviewed by Parent (1983) and Parent, Poitras & Dubé (1984). Later FIF studies dealt with *Blennius* (Kotrschal & Adam, 1983), *Carassius* (Fryer, Boudreault Chateauvert & Kirby, 1985; Bonn 1987), *Chelon* (Gomez-Segade, Anadon & Gomez-Segade, 1989), *Eigenmannia* (Bonn & Kramer 1987) and *Ictalurus* (Ekstöm & van Veen 1982). The data presented by the FIF studies just enumerated fit well in the picture presented in Fig. 4.10. However, they did not show telencephalic or preoptic catecholaminergic cell groups, owing to the limitations discussed in Chapter 1.

Catecholamines in the brains of non-teleostean actinopterygians

Although there are only a few studies dealing with the distribution of catecholamines in non-teleostean actinopterygian fishes, they interestingly cover all three major groups that are presently still living, i.e. the Polypteri, Chondrostei and Holostei (see p. 49 and Table 4.1).

In *Polypterus senegalus*, the Senegal bichir, Reiner & Northcutt (1992) studied the distribution of THir cells and fibers in the forebrain, in combination with the distribution of a number of other neurochemical substances. THir neurons were present in (see Fig. 4.12) the external cell layer of the olfactory bulb; in the rostral part of the third- or most distal-dorsal telencephalic or pallial zone (Pr and P3); in the dorsal (Vd) supracommissural (Vs) and postcommissural (Vp) part of the ventral telencephalic area; in the periventricular preoptic nucleus (P); in the proximity of the habenular commissure (ch), in the zona limitans diencephali, i.e. the boundary region between the thalamus and hypothalamus (zl) and in the periventricular nucleus of the posterior tubercle (tpp), which is comparable with the teleostean nuclei around the lateral recess of the infundibulum. THir fibers occur particularly in the glomerular layer of the olfactory bulb, the ventral telencephalon and a variety of diencephalic nuclei, but not, or very sparsely, in the dorsal telencephalon. On the basis of the density of TH- as well as substance P-immunoreactive fibers, Reiner & Northcutt (1992) suggest that dorsal part of the ventral telencephalon (Vd) may be homologuous to the basal ganglia, i.e. striatum and pallidum, of amniotes. As already discussed on pp. 68–69 they suggest that the diencephalic periventricular nucleus of the posterior tubercle (tpp) is, on the basis of its chemical differentiation, comparable with the amniote substantia nigra and ventral tegmental area, and might well project to Vd.

In the chondrostean bony fish *Acipencer ruthenus*, the sterlet, the distribution of aminergic neurons has been studied by means of FIF techniques by Kotrschal, Krautgartner & Adam (1985). Catecholamine related (green) fluorescent cell groups were present (see Fig. 4.12) in CSF-contacting cells along the telencephalic ventricle, rostral to the anterior commissure (Vpv); along the preoptic recess (rpo); and in the PVO, which extends from the rostrodorsal tip of the infundibulum via the region around the lateral recess, to the nucleus of the posterior recess, just as in teleosts (pp. 60–61). In the brainstem, green fluorescent neurons were observed in the reticular formation of the rhombencephalon (rf) and in the dorsal aspect of the viscerosensory zone of *Acipenser* (Fig. 4.12), which corresponds with the teleostean nucleus of the solitary tract (nts) or vagal lobe.

In the holostean gar *Lepisosteus osseus*, Parent & Northcutt (1982) found green fluorescent CSF-contacting catecholaminergic neurons in large density along the epenymal wall of the lateral and posterior infundibular recess, i.e. in the PVO, while non-CSF-contacting catecholaminergic neurons were observed in the preoptic region (p), the isthmus region (I), the basal midbrain tegmentum (tm), the vagal lobe (vl) and the lateral reticular formation (rf) (Fig. 4.12).

It is difficult to compare the results for the non-teleostean actinopterygians just described with the results reviewed for teleostean species, since different techniques were used, and immunohistochemical studies for chondrosteans and holosteans are lacking. Nevertheless, similarities between the different actinopterygian groups seem to predominate. In all groups, the catecholaminergic (dopaminergic and/or noradrenergic) PVO dominates the picture, with additional groups in the ventral telencephalon, preoptic region, isthmus region and caudal rhombencephalon. Mesencephalic tegmental dopaminergic cells have until now not been found in actinopterygian fishes, except perhaps in *Lepisosteus*, where a few weakly fluorescent cells have been found in the mesencephalic tegmentum (Parent & Northcutt, 1982). However, it is presently uncertain whether these might represent a primordial dopaminergic substantia nigra. The fact that not all teleostean cathecholaminergic cell groups have been demonstrated in chondrosteans and holosteans may well be due to the relative insensitivity of the FIF technique, which equally did not demonstrate in teleosts all cell groups that were later demonstrated by immunohistochemistry.

Catecholamines in sarcopterygian brains

Data on the distribution of catecholamines in the brains of sarcopterygian bony fishes are restricted to dipnoi or lungfishes. To my knowledge, there are presently no data on catecholamines in the brain of the only living species of crossopterygian fishes, *Latimeria chalumnae*, because of its very limited availability (Forey, 1988).

Data on catecholamines in the brains of dipnoi are restricted to the African genus *Protopterus*, and only deal with its forebrain. It should be mentioned that the forebrain of lungfishes is evaginated, just as in amphibians and amniotes, and not everted as in actinopterygians (pp. 53–54, Fig. 4.12). Reiner & Northcutt (1987) studied the distribution of TH-immunoreactivity in the forebrain of *Protopterus annectens*, and found THir neurons predominantly (see Fig. 4.12) in the internal granular layer of the olfactory bulb, in several subdivisions of the medial pallium (MP), in the medial and central subpallium (Sm and Sc respectively) and in the preoptic region (P). A few THir neurons occur in addition (Fig. 4.12) in the lateral subpallium (Sl) and the lateral and dorsal pallium (LP and DP). A marked difference with actinopterygian species concerns the presence of THir neurons in a mesencephalic tegmental region, comparable to the substantia nigra of land vertebrates (Reiner & Northcutt, 1987). THir fibers are predominantly located in the lateral subpallium, but also in the medial subpallium. It should be noted that the distinction of pallial and subpallial (sub)-fields is not unequivocal in lungfishes, and that e.g. Nieuwenhuys & Meek (1990b) and von Bartheld, Collin & Meyer (1990) have presented a different interpretation of the telencephalon of lungfishes than Reiner & Northcutt (1987). However, these differences mainly concern the medial (subpallial versus pallial) telencephalic wall, but do not interfere with an unequivocal distinction of lateral structures in lungfishes, including a lateral subpallial (primordial) striatum or basal ganglia.

It is presently uncertain whether the PVO of lungfishes, extensively described and characterized by von Bartheld & Meyer (1990) and von Bartheld *et al.* (1990), contains catecholamines. It does not seem to contain TH (Reiner & Northcutt 1987), but antibodies against DA or NA have not been applied in lungfishes. Therefore, more data are necessary for a full comparison with, on the one hand, their fellow bony fishes, the actinopterygians, and on the other hand amphibians, to which the lungfishes are generally supposed to be ancestral (e.g. Reiner & Northcutt, 1987; Nieuwenhuys & Meek, 1990b and von Bartheld *et al.*, 1990).

Concluding remarks

The present chapter shows that immunohistochemical and formaldehyde induced fluorescence techniques have yielded detailed information about the distribution of catecholaminergic neurons and fibers in the brains of teleostean bony fishes. The teleostean catecholamine system is dominated by the hypothalamic paraventricular organ, where dopamine and noradrenaline acquiring cerebrospinal fluid contacting neurons are intermingled with serotonin-containing ones. A similar system seems to be present in more primitive actinopterygians. Dopaminergic neurons occur, in addition, in the teleostean telencephalon, diencephalon and rhombencephalon, but not in the midbrain tegmentum. This means that a mesencephalic substantia-nigra/ventral tegmental-like region and a telencephalic striatum-like region is probably lacking (see also Chapter 20). Noradrenergic neurons occur in the locus coeruleus and caudal rhombencephalon. Data on the distribution of adrenaline are too limited for general conclusions.

Data on the functional significance of the catecholamine systems in bony fishes are scarce. The precise projections and functional involvements of the distinct catecholamine containing cell groups are unknown. Equally, studies on receptor binding, receptor characterization and interactions of catecholamines with other neurotransmitter systems are lacking or too scarce for any general conclusion. So, it may be concluded that we presently have a rather good insight in the constraints and variability of the distribution of catecholamines in the central nervous system of bony fishes, in particular teleosts, but that a variety of further studies is necessary to assess their functional significance in this group of verterbrates.

Acknowledgements

I would like to thank Mr T. G. M. Hafmans for photographical assistence, and Ms S. Rohde for typing the manuscript.

Abbreviations

ac	anterior commissure
ap	area postrema
avl	area ventrolateralis thalami
cereb	cerebellum
ch	neurons around the habenular commissure
cp	central posterior thalamic nucleus
Dc	area dorsalis telencephali pars centralis
Dd	area dorsalis telencephali pars dorsalis
Dl	area dorsalis telencephali pars lateralis
Dm	area dorsalis telencephali pars medialis
DP	dorsal pallium
E	entopeduncular nucleus
elll	electrosensory lateral line lobe
fd	funiculus dorsalis
fv	funiculus ventralis
H(yp)	hypothalamus
hc	hypothalamus caudalis
hd	hypothalamus dorsalis
I	isthmic region
lc	locus coeruleus
lfb	lateral forebrain bundle
lh	lateral hypothalamus
ll	lateral lemniscus
LP	lateral pallium
mfb	medial forebrain bundle
mlf	medial longitudinal fascicle
mlll	mechanosensory lateral line lobe
MP	medial pallium
napv	nucleus anterior periventricularis
nir	nucleus of the infundibular recess
nit	nucleus inferior tuberis
nlt	nucleus lateralis tuberis
nlta	anterior part of nucleus lateralis tuberis
nmh	nucleus magnocellularis hypothalami
nmlf	nucleus of the medial longitudinal fascicle
nppv	nucleus posterior periventricularis
npt	nucleus posterior tuberis
nra	nucleus recessi anterioris
nri	nucleus recessi intermedii
nrl	nucleus recessi lateralis
nrli	nucleus recessi lateralis pars inferior
nrlm	nucleus recessi lateralis pars medialis
nrp	nucleus recessi posterioris
nts	nucleus tractus solitarii
nvm	nucleus ventromedialis thalami
nXs	sensory vagal nucleus
ob	olfactory bulb
oc	optic chiasm
on	olfactory nerve
ot	olfactory tract
P	preoptic region
Pa	anterior preoptic cell group
pc	nucleus of the posterior commissure
pgm	nucleus preglomerulosus pars medialis
Pp	posterior preoptic cell group
ppv	nucleus periventricularis pretectalis, pars ventralis
Pr	rostral cap of the pallium
PVO	paraventricular organ
pvoa	anterior part of PVO
pvoacc	PVO accompanying cells
pvop	posterior part of PVO
P3	third (or lateral) pallial subfield
rf	reticular formation
rh(omb)	rhombencephalon
ri	recessus intermedius (infundibuli)
rl	recessus lateralis (infundibuli)
rpo	recessus preopticus organ
sc	suprachiasmatic nucleus
Sc	central subpallium
Sl	lateral subpallium
Sm	medial subpallium
tect	tectum mesencephali
tel	telencephalon

Th	thalamus	vl	vagal lobe
tm	tegmentum mesencephali	vm(t)	ventromedial thalamus
tpp	periventricular nucleus of posterior tuberculum	Vp	area ventralis telencephali pars postcommissuralis
ts	torus semicircularis	Vpv	area ventralis telencephali pars periventricularis
ttm	tractus telencephalo-mesencephalicus (part of lfb)	Vs	area ventralis telencephali pars supracommissuralis
tVd	nucleus of the descending trigeminal tract	vt	ventral thalamus
V	area ventralis telencephali	Vv	area ventralis telencephali pars ventralis
Vc	area ventralis telencephali pars centralis	zl	zona limitans diencephali
Vd	area ventralis telencephali pars dorsalis	X	vagal nerve
Vi	area ventralis telencephali pars intermedius	Xm	vagal motornucleus

References

Alonso, J. R., Coveñas, R., Lara, J., Arévalo, R., de Léon, M. & Aijon, J. (1989). Tyrosine Hydroxylase immunoreactivity in a subpopulation of granule cells in the olfactory bulb of teleost fish. *Brain, Behavior and Evolution*, **34**, 318–24.

Bartheld, C. S. von, Collin, S. P. & Meyer, D. L. (1990). Dorsomedial telencephalon of bony fishes: A pallial or subpallial structure? Criteria based on histology, connectivity and histochemistry. *Journal of Comparative Neurology*, **294**, 14–29.

Bartheld, C. S. von & Meyer, D. L. (1990). Paraventricular organ of the bony fish *Protopterus dolloi*: morphology and projections of CSF-contacting neurons. *Journal of Comparative Neurology*, **297**, 410–34.

Bell, C. C. & Szabo, T. (1986). Electroreception in mormyrid fish: central anatomy. In *Electroreception*, eds. Bullock, T. H. and Heiligenberg, W. pp.375–421. New York: Wiley.

Bonn, U. (1987). Distribution of monoamine containing neurons in the brain of a teleost, *Carassius auratus* (Cyprinidae). *Journal für Hirnforschung*, **28**, 529–44.

Bonn, U. & Kramer, B. (1987). Distribution of monoamine-containing neurons in the brain of the weakly electric teleost, *Eigenmannia lineata*. *Zeitschrift fur Mikroskopisch-Anatomische Forschung, Leipzig*, **101**, 339–62.

Carr, C. E. & Maler, L. (1986). Electroreception in gymnotiform fish. Central anatomy and physiology. In *Electroreception*. ed. Bullock, T. H. and Heiligenberg, W. pp.319–74. New york: Wiley.

Corio, M., Peute, J. & Steinbusch, H. W. M. (1991). Distribution of serotonin- and dopamine immunoreactivity in the brain of the teleost *Clarias gariepinus*. *Journal of Chemical Neuroanatomy*, **4**, 79–95.

Ekström, P. & Ebbesson, S. O. E. (1989). Distribution of serotonin-immunoreactive neurons in the brain of Sockeye Salmon Fry (*Oncorhynchus nerka*). *Journal of Chemical Neuroanatomy*. **2**, 201–13.

Ekström, P., Honkanen, T. & Steinbusch, H. W. M. (1990). Distribution of dopamine-immunoreactive neuronal perikarya and fibers in the brain of a teleost, *Gasterosteus aculeatus*. Comparison with TH- and DBH-IR neurons. *Journal of Chemical Neuroanatomy*, **3**, 233–60.

Ekström, P., Reschke, M., Steinbusch, H. W. M. & van Veen, T. (1986). Distribution of noradrenaline in the brain of the teleost *Gasterosteus aculeatus* L: an immunohistochemical analysis. *Journal of Comparative Neurology*, **254**, 297–13.

Ekström, P. & van Veen, T. (1982). The monoaminergic paraventricular organ in the teleost *Ictalurus nebulosus* Le Sueur, with special reference to its vascularization. *Acta Zoologica* (Stockholm), **63**, 45–54.

Ekström, P. & van Veen, T. (1984). Distribution of 5-hydroxytryptamine (serotonin) in the brain of the teleost *Gasterosteus aculeatus* L. *Journal of Comparative Neurology*, **226**, 307–20.

Finger, T. E. (1983). The gustatory system in teleost fish. In *Fish Neurobiology* Vol 1: *Brainstem and Sense Organs*. ed. Northcutt, R. G. and Davis, R. E. pp.285–310. Ann Arbor: University of Michigan Press.

Finger, T. E. (1986) Electroreception in catfish: behaviour, anatomy and electrophysiology. In *Electroreception* ed. Bullock, T. H. and Helligenberg, W. pp.287–318. New York: Wiley.

Forey, P. L. (1988). Golden jubilee for the coelacanth *Latimeria chalumnae*. *Nature*, **336**, 727–32.

Frankenhuis-van den Heuvel, T. H. M. & Nieuwenhuys, R. (1984) Distribution of serotonin-immunoreactivity in the diencephalon and mesencephalon of the trout, *Salmo gairdneri*; cell bodies, fibers and terminals. *Anatomy and Embryology*, **169**, 193–204.

Fryer, J. N., Boudreault-Chateauvert, C. & Kirby, R. P. (1985). Pituitary afferents originating in the paraventricular organ (PVO) of the goldfish hypothalamus. *Journal of Comparative Neurology*, **242**, 474–84.

Gomez-Segade, P., Anadón, R. & Gomez-Segade, L. (1989). Monoaminergic systems in the hypothalamus of the acanthopterygian *Chelon labrosis*, with special reference to the organum vasculosum hypothalami. *Acta Zoologica* (Stockholm), **70**, 1–11.

Grant, K., Clausse, S., Libouban, S. & Szabo, T. (1989) Serotoninergic neurons in the mormyrid brain and their projections to the pre-electromotor and primary electrosensory centers: immunohistochemical study. *Journal of Comparative Neurology*, **281**, 114–28.

Greenwood, P. H., Rosen, D. E., Weitzman, S. H. & Myers, G. S. (1966). Phyletic studies of teleostean fishes, with a provisional classification of living forms. *Bulletin of the American Museum of Natural History*, **131**, 339–456.

Grzimek, B. (ed.) (1973). *Animal life Encyclopedia. Vol 4, Fishes I.* New York: Van Nostrand-Reinhold.

Hornby, P. J. & Piekut, D. T. (1988). Immunoreactive dopamine-β-hydroxylase in neuronal groups in the goldfish brain. *Brain Behavior and Evolution*, **32**, 252–56.

Hornby, P. J. & Piekut, D. T. (1990). Distribution of catecholamine-synthesizing enzymes in goldfish brains: presumptive dopamine and norepinephrine neuronal organization. *Brain Behaviour and Evolution*, **35**, 49–64.

Hornby, P. J., Piekut, D. T. & Demski, L. S. (1987). Localization of immunoreactive tyrosine hydroxylase in the goldfish brain. *Journal of Comparative Neurology*, **261**, 1–14.

Johnston, S. A., Maler, L. & Tinner, B. (1990). The distribution of serotonin in the brain of *Apteronotus leptorhynchus* – an immunohistochemical study. *Journal of Chemical Neuroanatomy*, **3**, 429–66.

Kah, O. & Chambolle, P. (1983). Serotonin in the brain of the goldfish, *Carassius auratus*. An immunocytochemical study. *Cell and Tissue Research*, **234**, 319–33.

Kah, O., Chambolle, P., Thibault, J. & Geffard, M. (1984). Existence of dopaminergic neurons in the preoptic region of the goldfish. *Neuroscience Letters*, **48**, 293–8.

Kotrschal, K. & Adam, H. (1983). The aminergic system in the brain of *Blennius incognitus* (Bath 1968) (Teleostei, Perciformes). *Cell and Tissue Research*, **229**, 403–9.

Kotrschal, K., Krautgartner, W. D. & Adam, H. (1985). Distribution of aminergic neurons in the brain of the sterlet *Acipenser ruthenus* (Chondrostei, Actinopterygii). *Journal für Hirnforschung*, **26**, 65–72.

Lauder, G. V. & Liem, K. F. (1983a). Patterns of diversity and evolution in ray-finned fishes. In *Fish Neurobiology Vol 1: Brainstem and Sense Organs.* ed. Northcutt, R. G. and Davis, R. E. pp. 1–24. Ann Arbor: University of Michigan Press.

Lauder, G. V. & Liem, K. F. (1983b). The evolution and interrelationships of the actinopterygian fishes. *Bulletin of the Museum of Comparative Zoology*, **150**, 95–197.

Maler, L. & Ellis, W. G. (1987). Inter-male agressive signals in weakly electric fish are modulated by monoamines. *Behavioral Brain Research*, **25**, 75–81.

Meek, J. & Joosten, H. W. J. (1989). The distribution of serotonin in the brain of the mormyrid teleost *Gnathonemus petersii*. *Journal of Comparative Neurology*, **281**, 206–24.

Meek, J. & Joosten, H. W. J. (1993). Tyrosine hydroxylase immunoreactive cell groups in the brain of the teleost fish *Gnathonemus petersii*. *Journal of Chemical Neuroanatomy*, **6**, 431–46.

Meek, J., Joosten, H. W. J. & Hafmans, T. G. M. (1992). Catecholamines in the paraventricular organ of the mormyrid teleost *Gnathonemus petersii*. Abstracts of the 7th Catecholamine Symposium Amsterdam p. 205.

Meek, J., Joosten, H. W. J. & Hafmans, T. G. M. (1993). Distribution of noradrenaline-immunoreactivity in the brain of the mormyrid teleost *Gnathonemus petersii*. *Journal of Comparative Neurology*, **328**, 145–60.

Meek, J., Joosten, H. W. J. & Steinbusch, H. W. M. (1989) The distribution of dopamine-immunoreactivity in the brain of the mormyrid teleost *Gnathonemus petersii*. *Journal of Comparative Neurology*, **281**, 362–83.

Meek, J. & Nieuwenhuys, R. (1991). The palisade pattern of mormyrid Purkinje cells. A correlated light and electron microscopic study. *Journal of Comparative Neurology* **306**, 156–92.

Mons, N., Dubourg, P. & Tramu, G. (1991). Preparation and characterization of a specific antibody for the immunohistochemical detection of L-Dopa in paraformaldehyde-fixed rodent brains. *Brain Research*, **554**, 122–9.

Morita, Y. & Finger, T. E. (1987). Area postrema of the goldfish, *Carassius auratus*: ultrastructure, fiber connections, and immunocytochemistry. *Journal of Comparative Neurology*, **256**, 104–16.

Morita, Y., Murakami, T. & Ito, H. (1983). Cytoarchitecture and topographic projections of the gustatory centers in a teleost, *Carassius carassius*. *Journal of Comparative Neurology*, **218**, 378–94.

Murakami, T., Morita, Y. & Ito, H. (1983). Extrinsic and intrinsic fiber connections of the telencephalon in a teleost, *Sebastiscus marmoratus*. *Journal of Comparative Neurology*, **216**, 115–31.

Nelson, G. J. (1969). Origin and diversification of teleostean fishes. *Annals of the New York Academy of Science*, **167**, 18–30.

Nieuwenhuys, R. (1982). An overview of the organization of the brain of actinopterygian fishes. *American Zoologist*, **22**, 287–310.

Nieuwenhuys, R. & Meek, J. (1985). Constructional principles of the brainstem in anamniotes, with emphasis on actinopterygian fishes. In *Fortschritte der Zoologie Band 30: Vertebrate Morphology.* ed. Ducker and Fleisher. pp.515–28. Stuttgart–New York: Gustav Fischer Verlag.

Nieuwenhuys, R. & Meek, J. (1990a). The telencephalon of actinopterygian fishes. In *Cerebral Cortex. Vol.8A: Comparative Aspects of Cortical Structure.* ed. Jones, E. G., & Peters, A. pp.31–73. New York: Plenum Press.

Nieuwenhuys, R. & Meek, J. (1990b). The telencephalon of sarcopterygian fishes. In *Cerebral Cortex. Vol.8A: Comparative Aspects of Cortical Structure.* ed. Jones, E. G. & Peters, A. pp.75–106. New York: Plenum Press.

Nieuwenhuys, R. & Pouwels, E. (1983). The brain stem of actinopterygian fishes. In *Fish Neurobiology, Vol.1: Brain Stem and Sense Organs.* ed. Northcutt, R. G. & Davis, R. E. pp.25–87. Ann Arbor: University of Michigan Press.

Okamura, H., Nikahara, K., Mons, N., Ibata, I., Jouvet, M. & Geffard, M. (1988). L-DOPA immunoreactive neurons in the rat hypothalamic tuberal region. *Neuroscience Letters* **95**, 42–6.

Parent, A. (1983) The monoamine-containing neuronal systems in the teleostean brain. In *Fish Neurobiology, Vol.2: Higher Brain Areas and Functions*. eds. Davis, R. E. & Northcutt, R. G. pp.285–315. Ann Arbor: University of Michigan Press.

Parent, A., Poitras, D. & Dubé, L. (1984). Comparative anatomy of central monoaminergic systems. In *Handbook of Chemical Neuroanatomy Vol. 2: Classical transmitters in the CNS Part 1*. eds. A. Björklund & T. Hökfelt. pp 409–39. Amsterdam: Elsevier Science Publishers B.V.

Parent, A. & Northcutt, R. G. (1982). The monoamine-containing neurons in the brain of the garfish, *Lepisosteus osseus*. *Brain Research Bulletin*, **9**, 189–204.

Reiner, A. & Northcutt, R. G. (1987). An immunohistochemical study of the telencephalon of the African lungfish *Protopterus annectens*. *Journal of Comparative Neurology*, **256**, 463–81.

Reiner, A. & Northcutt, R. G. (1992). An immunohistochemical study of the telencephalon of the Senegal Bichir (*Polypterus senegalis*). *Journal of Comparative Neurology*, **319**, 359–86.

Roberts, B. L., Meredith, G. E. & Maslam, S. (1989). Immunocytochemical analysis of the dopamine system in the brain and spinal cord of the European eel, *Anguilla anguilla*. *Anatomy and Embryology*, **180**, 401–12.

Sas, E., Maler, L. & Tinner, B. (1990). Catecholaminergic systems in the brain of a gymnotiform teleost fish: an immunohistochemical study. *Journal of Comparative Neurology*, **292**, 127–62.

Sibbing, F. A. (1984). Food handling and mastication in the carp. Thesis, University of Wageningen, The Netherlands.

Smeets, W. J. A. J. & Steinbusch, H. W. M. (1989). Distribution of noradrenaline immunoreactivity in the forebrain and midbrain of the lizard *Gekko gecko*. *Journal of Comparative Neurology*, **285**, 453–66.

Smeets, W. J. A. J. & Steinbusch, H. W. M. (1990). New insights into the reptilian catecholaminergic systems as revealed by antibodies against the neurotransmitters and their synthetic enzymes. *Journal of Chemical Neuroanatomy*, **3**, 25–43.

Terlou, M., Ekengren, B. & Hiemstra, K. (1978). Localization of monoamines in the forebrain of two salmonid species, with special reference to the hypothalamohypophysial system. *Cell and Tissue Research*, **190**, 417–34.

Vigh-Teichmann, I. & Vigh, B. (1974). The infundibular cerebrospinal-fluid contacting neurons. *Advances in Anatomy, Embryology and Cell Biology*, **50/2**, 1–90.

Yoshida, M., Nagatsu, I., Kawakami-Kondo, Y., Karasawa, N., Spatz, M. & Nagatsu, T. (1983). Monoaminergic neurons in the brain of goldfish as observed by immunohistochemical techniques. *Experientia*, **39**, 1171–4.

5
Catecholamine systems in the CNS of amphibians

A. González and W. J. A. J. Smeets

Introduction

Phylogenetic considerations

Living ampibians comprise three different groups or orders: Anura or Salientia (frogs and toads, approximately 3400 species), Urodela or Caudata (newts and salamanders, approximately 350 species), and Gymnophiona or Apoda (caecilians, approximately 160 species) (Duellman & Trueb, 1986). The orders are separated from each other by extensive structural variation and distinct lifestyle. From their earliest appearance in fossil record, the three orders are easily distinguished. Yet, there are several characters, e.g. structure of teeth, middle ear bones and vertebral–skull articulation, which have led to the conclusion that extent amphibians represent a monophyletic group, the Lissamphibia (Parsons & Williams, 1963; Carroll, 1988). The monophyletic hypothesis of Parsons and Williams is widely accepted, although a diphyletic origin of modern amphibians cannot be completely ruled out (see e.g. Jarvic, 1960).

Most paleontologists agree that amphibians emerged from crossopterygian fishes, the Ripidistia, some 350–400 million years ago, and that they rapidly split into the three extant orders. The living amphibians constitute a remnant of a once dominant vertebrate group that characterizes the evolutionary transition from an aquatic to a terrestrial lifestyle. However, the modern amphibians clearly differ from Paleozoic amphibians, including the ancestral labyrinthodonts that gave rise to the first really successful terrestrial vertebrates, the reptiles (Romer & Parsons, 1977). Therefore, despite their rather primitive level of organization relative to other tetrapods, living amphibians cannot be considered ancestral to reptiles.

Amphibian catecholamine research in historical perspective

The basis for our current knowledge of the catecholamine systems in amphibians was laid by studies using the original formaldehyde induced fluorescence (FIF) technique, developed by Falck et al. (1962), or its subsequent modifications (see, e.g. Parent, Poitras & Dubé, 1984: Sims, 1986; Corio & Doerr-Schott, 1988; Lamas et al., 1988). During the last decade, the development of antibodies against the enzymes that are involved in the biosynthesis of catecholamines has given new impetus to catecholamine research in amphibians. Particularly, the antibodies against tyrosine hydroxylase (TH) were successfully applied to the brains of anurans (Yoshida et al., 1983; Smeets & González, 1990; González & Smeets, 1991; González, Tuinhof & Smeets, 1993) and urodeles (Franzoni et al., 1986; Corio, Thibault & Peute, 1990, 1992; Smeets & González, 1990; González & Smeets, 1991). Unfortunately, antibodies against dopamine-β-hydroxylase (DBH) and phenylethanolamine-N-methyltransferase (PNMT), enzymes that are involved in the conversions of dopamine into noradrenaline, and noradrenaline into adrenaline, respectively, have been proven less successful in brains of amphibians. Hitherto, DBH and PNMT immunoreactive neuronal structures have been described only in two anurans, i.e. *Rana catesbeiana* (Yoshida et al., 1983) and *Xenopus laevis* (González & Smeets, 1993).

With the development of antibodies against

dopamine (Geffard et al., 1984) and noradrenaline (Steinbusch et al., 1981; Steinbusch & Tilders, 1987) it became possible to demonstrate specifically the distribution of these catecholamines in the the CNS of amphibians (Smeets & González, 1990; González & Smeets, 1991; Corio et al., 1992; González et al., 1993; González & Smeets, 1993). Moreover, the antibodies against DA and NA also enabled us to investigate the possibility of mutual influences of the two catecholamine systems.

In this chapter, we provide detailed information on the catecholamine systems in the adult brains of amphibians. In our studies, we have been investigating in detail the organization of catecholamine systems in the anurans, Xenopus laevis and Rana ridibunda, and the urodeles, Pleurodeles waltlii and Siren lacertina. By comparing the results obtained in the various species, general and derived conditions of the catecholamine systems in amphibians are inferred, then we have determined more specifically the contribution of DA and/or NA fibers to the catecholaminergic innervation of brain structures. Finally, by comparing hodological data with the distribution of DA and NA fibers, the sites of their origin and putative role of each catecholamine in the functioning of the brain of amphibians is discussed.

Distribution of catecholamines in the brain of amphibians: theme and variations

In this section, we provide first a detailed description of the distribution of catecholaminergic cell bodies and fibers in the brains of Xenopus, as core species of anurans, and of Pleurodeles, as representative of urodeles. Subsequently, these distribution patterns are compared to those observed in other amphibians in order to determine general and derived conditions for each catecholamine system in amphibians.

Dopamine

Since the development of antibodies raised against dopamine (DA), several studies have been carried out to demonstrate selectively the DA system in amphibians (Smeets & González, 1990; González & Smeets, 1991; González et al., 1993). In the species studied, the DA antibodies revealed an excellent staining of fibers. Cell bodies are also well stained with the DA antiserum, but comparison with adjacent sections stained with the TH antiserum revealed a generally better staining of dendrites with the latter antiserum. As can be expected on the basis of the catecholamine biosynthetic pathway, more cell bodies are immunopositive to TH than to DA antibodies. This suggests that these cells are either noradrenergic or adrenergic. Comparison of adjacent sections leads to the conclusion that DAi perikarya are also THi. One obvious exception is the nucleus of the periventricular organ, where cells are immunopositive for DA but immunonegative for TH antibodies suggesting that these cells accumulate rather than metabolize dopamine.

Cell bodies

Anura. The most rostral DAi and THi cell bodies in Xenopus are located in the olfactory bulb around the glomeruli and both in the external plexiform layer and mitral cell layer (Figs. 5.1 (*a*), 5.2, 5.3). The cells in the glomerular layer are considerably larger than those in the latter two layers. Rostrally, where the two bulbs are fused, cell bodies are present in the midline and some of their processes cross to the contralateral side. With the TH antiserum additional cell bodies are found in the internal granular layer. No DAi or THi cell bodies are found in the telencephalon proper (Fig. 5.1 (*b*)–(*d*)).

The diencephalon harbors the majority of the DAi and THi cell bodies. On the basis of their morphology, the DAi and THi cell bodies in the diencephalon can be subdivided into two categories. One category comprises cell bodies that are in direct contact with the cerebrospinal fluid (CSF) of the third ventricle by means of short processes with club-like endings. The other category is constituted by cells that do not contact the ventricle. Most of the DAi and THi cells in the anterior preoptic area and the DAi cells of the periventricular organ (Fig. 5.1 (*d*), (*f*)) are CSF-contacting. The cell bodies in the remaining cell groups belong to the second category, although some cells of the suprachiasmatic nucleus may be liquor-contacting (Fig. 5.1 (*e*)).

Numerous perikarya, immunopositive for both DA and TH, are located in the preoptic area (Fig. 5.4). They are widely distributed around the preoptic recess bordering the ventricular wall. At rostral chiasmatic levels, another group of DAi and THi cell bodies is found in the suprachiasmatic nucleus (Fig. 5.5). This group can be subdivided into a medial portion that consists of cell bodies which lie adjacent to the ventricle and probably contact the CSF, and a lateral portion of which the cells lie further away from the ventricle and have processes that are directed laterally or ventrally. In the caudal half of the diencephalon, four groups of DAi cell bodies can be recognized, viz. in the posterior thalamic nucleus, the

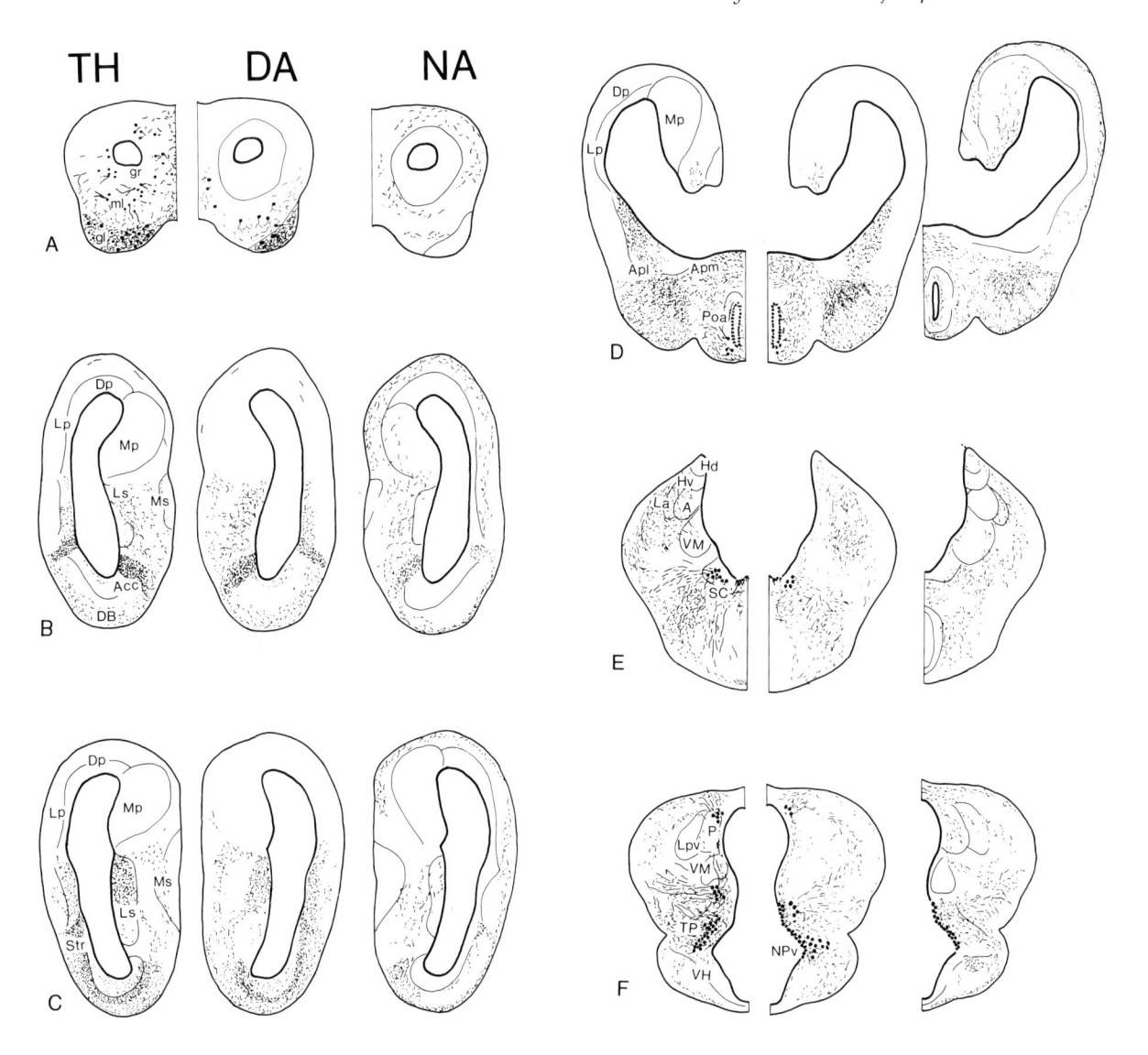

Fig. 5.1. Diagrams of transverse sections through the brain of *Xenopus laevis* from rostral (A) to caudal (N) showing the position of immunoreactive cell bodies (large dots) and fibers (small dots, wavy lines). Tyrosine hydroxylase immunoreactive (THi) cells and fibers are shown on the left, dopamine immunoreactive (DAi) cells and fibers in the middle and noradrenaline immunoreactive (NAi) cells and fibers on the right.

periventricular organ, an area lateral to the periventricular organ, and the nucleus of the posterior tubercle. THi cells are observed in similar locations, except for the periventricular organ. The cells in the posterior thalamic nucleus stain only weakly with DA antibodies but particularly well with the TH antiserum. They lie in the deep layers of the nucleus and possess long processes that course through the superficial cell layers to the neuropil where they branch profusely and reach the midbrain tectum. Caudally, the cell group extends underneath the posterior commissure (Fig. 5.1 (f), (g)). The DAi cell bodies in the periventricular organ lie in several rows in the subependymal layer and have apical processes that protrude into the ventricle thus contacting the CSF (Fig. 5.6). The cells lying lateral to the periventricular organ and here referred to as the accompanying cells of the periventricular organ, do not have

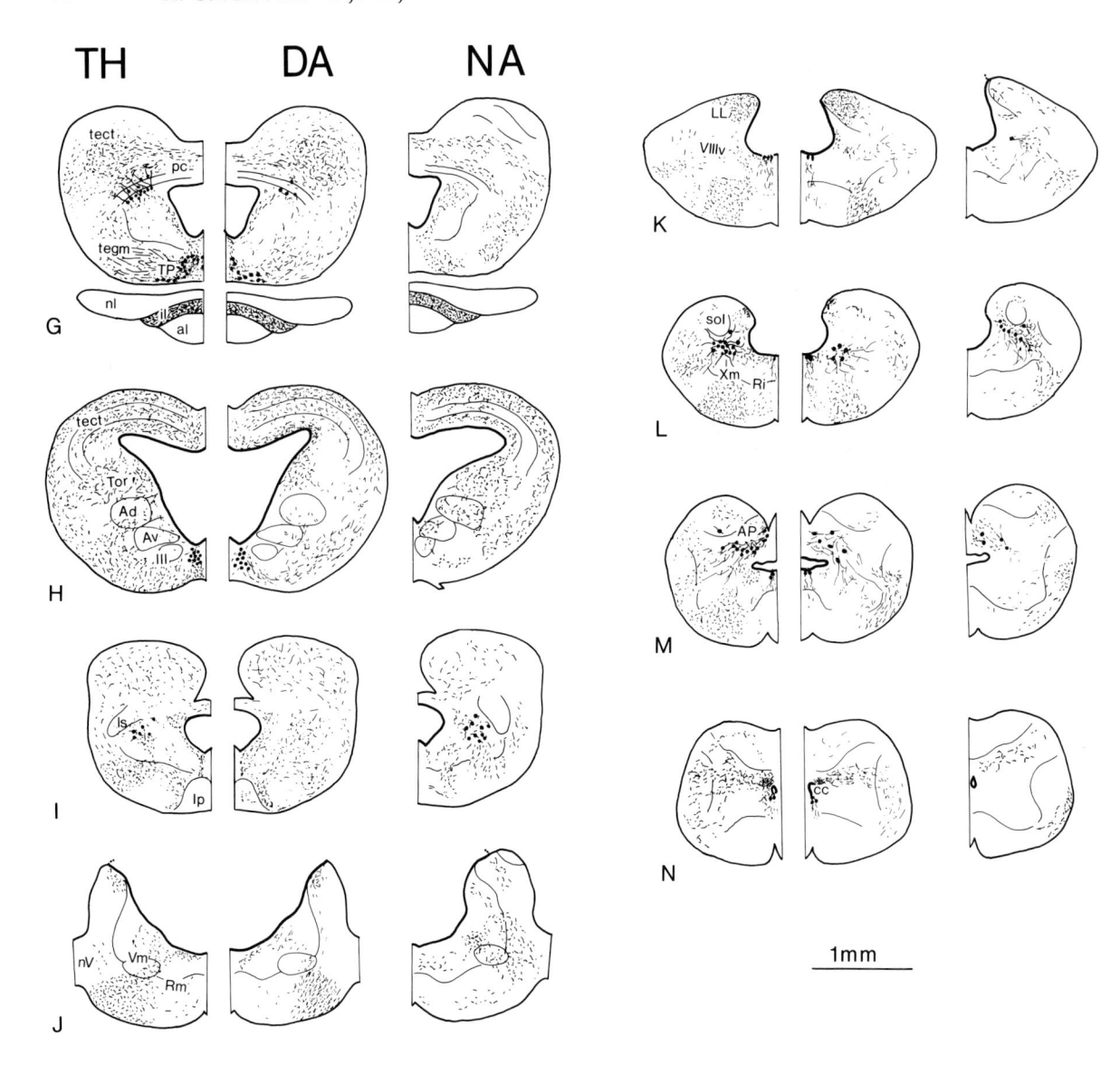

Fig. 5.1. (cont.)

processes that contact the CSF. Within the nucleus of the posterior tubercle, two separate populations of DAi and THi cell bodies can be distinguished, a dorsomedial and a ventrolateral group (Fig. 5.7).

In the brainstem, DAi and THi cell bodies are present in the medial part of the midbrain tegmentum, around the solitary tract, and in the area postrema at the obex level (Figs. 5.1 (h), (l), (m)). The cells in the midbrain tegmentum constitute, rostrally, separate groups on both sides of the midline which appear to be caudal continuations of the dorsomedial cell groups in the posterior tubercle. More caudally, a few cells lie in the midline. The majority of the DAi and THi cells extend their dendrites mediolaterally, but the few cell bodies lying in the midline have processes that are directed mainly dorsoventrally. At caudal rhombencephalic levels, DAi and THi cell bodies lie around the solitary tract. The cells are large and multipolar and are mainly located medial and ventral to the tract. At the level of the obex, the cells

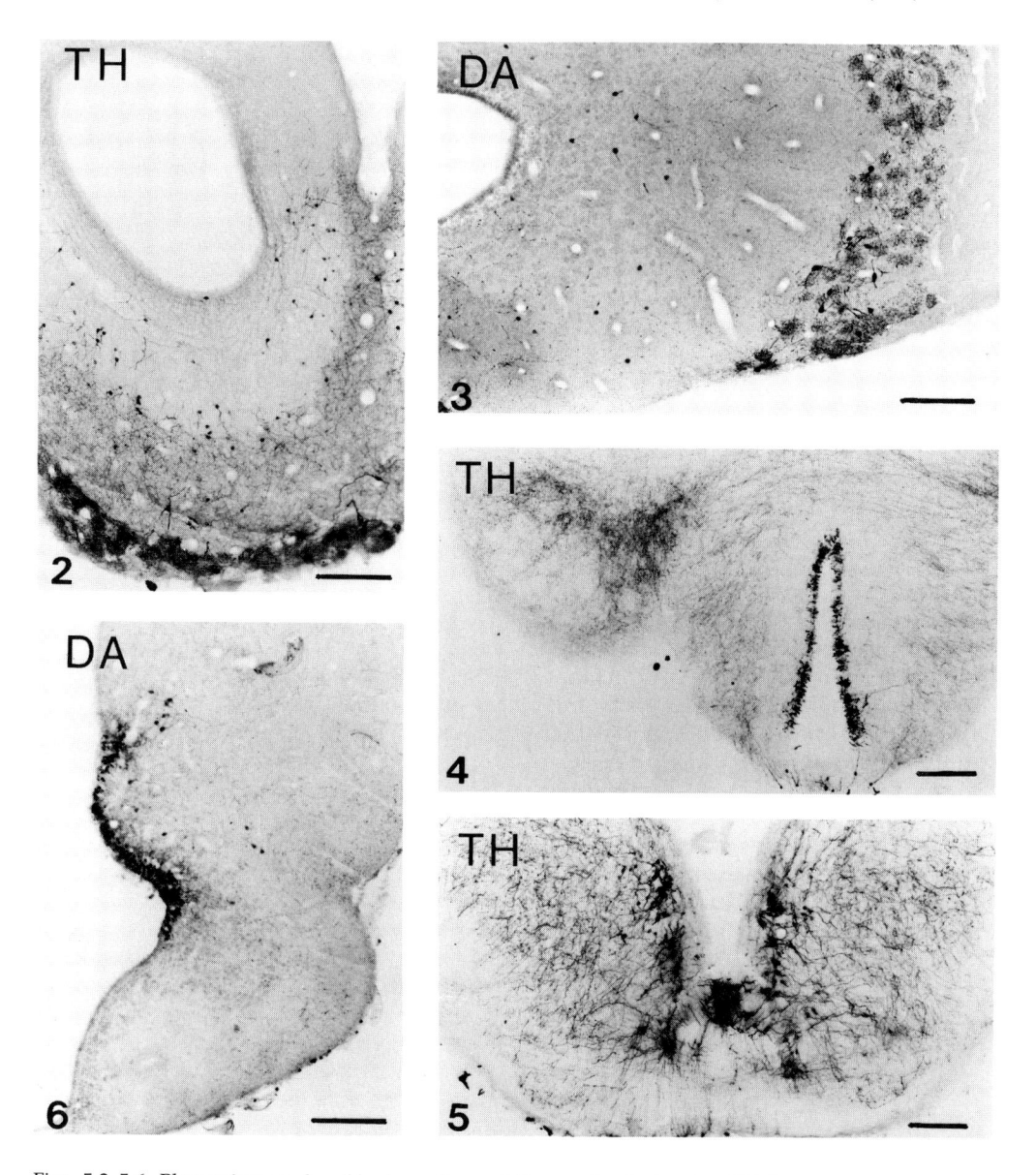

Figs. 5.2–5.6. Photomicrographs of brain sections of *Xenopus laevis* showing THi/DAi cell bodies and fibers in the olfatory bulb, as seen in a transverse (Fig. 5.2) and horizontal (Fig. 5.3) section; THi/DAi cell bodies and fibers in transverse sections through the anterior preoptic area (Fig. 5.4), suprachiasmatic nucleus (Fig. 5.5) and the periventricular organ (Fig. 5.6). Bars = 200 μm.

of both sides fuse above the ventricle, in the area postrema (Fig. 5.8). Additional cell bodies occur in the ependymal layer along the midline of the caudal rhombencephalon and along the central canal of the spinal cord where they contact the CSF (Fig. 5.1 (*k*)–(*n*)). They form, at least in the cervical spinal cord, a continuous column lying ventral to the central canal.

Urodela. The distribution of the DAi and THi cell bodies in the newt, *Pleurodeles waltlii*, largely resembles that of *Xenopus* (Fig. 5.9). As in the clawed frog, DAi cell bodies are found in the glomerular and mitral cell layers, whereas the TH antiserum labels additional cells in the internal granular layer (Fig. 5.9(*a*)). In the telencephalon proper, neither DAi nor

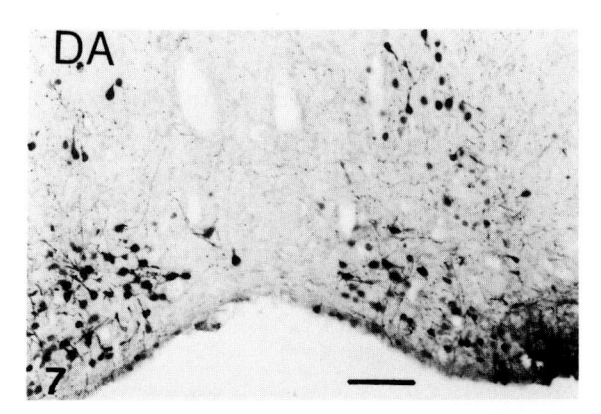

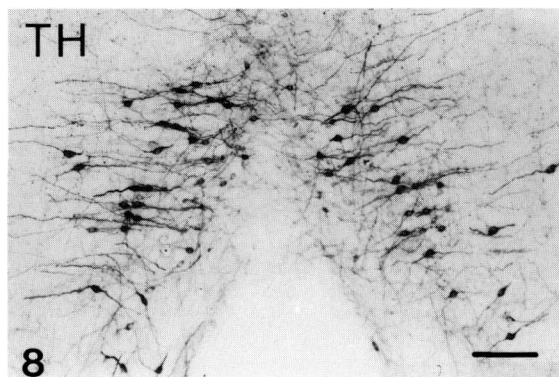

Figs. 5.7–5.8. Photomicrographs of a transverse and a horizontal section through the brain of *Xenopus laevis* showing DAi cell bodies in the nucleus of the posterior tubercle (Fig. 5.7), slightly rostral to the level shown in Figure 5.1H, and THi neurons in the area postrema (Fig. 5.8). Bars = 100 μm.

THi cell bodies are observed (Fig. 5.9(*b*), (*c*)). In the diencephalon, DAi cell bodies are present in the preoptic area (Fig. 5.9(*d*), 5.10), the suprachiasmatic nucleus (Figs. 5.9(*f*), 5.11), the periventricular organ, the dorsomedial and ventrolateral portions of the nucleus of the posterior tubercle (Fig. 5.9(*g*)), and the posterior thalamic nucleus. Except for the periventricular organ, THi cell bodies are found in the same loci. In the brainstem, DAi and THi perikarya lie in the midbrain tegmentum, along the medial margin of the solitary tract, and in the ependymal and subependymal layers along the midline of the caudal rhombencephalon (Figs. 5.9(*h*)–(*k*), 5.13). Cells that are immunopositive for both antisera are also found in the ventral wall of the central canal of the spinal cord (Fig. 5.9(*l*)).

Fibers

Anura. The DA and TH antisera reveal an almost identical distribution of fibers suggesting that the TH antibodies label primarily dopaminergic fibers (Fig. 5.1). The olfactory bulb contains many immunoreactive fibers in the glomerular layer. In the external plexiform and the mitral cell layers, a considerable number of DAi/THi fibers, probably dendrites, are present. Only a few fibers are observed in the internal granular layer. In the telencephalon proper, the DAi and THi fibers are almost exclusively confined to the subcortical areas (Fig. 5.1(*b*), (*c*)). The most obvious plexus of immunoreactive fibers occurs in the nucleus accumbens. At very rostral levels, a single plexus is visible, but more caudally, this plexus is subdivided into ventromedial and ventrolateral portions separated by a region that is poor in immunoreactive fibers (Fig. 5.14). Another distinct plexus of DAi and THi fibers is found in the striatum (Fig. 5.15). The immunoreactive fibers course primarily in the intermediate zone, probably contacting the proximal parts of the dendrites of the striatal neurons. In the lateral septum, the DAi and THi fibers are inhomogeneously distributed and occur predominantly in its dorsolateral part (Figs. 5.1(*c*), 5.15)). The plexus of the amygdaloid complex is rostrally continuous with that of the striatum. Numerous DAi and THi fibers cross the midline in the anterior commissure and interconnect the amygdalar immunoreactive plexuses. A distinct plexus is also found in an area that encompasses the lateral forebrain bundle (Fig. 5.4).

At rostral diencephalic levels, a rather dense plexus of immunoreactive fibers is observed in the lateral preoptic area, which is primarily formed by the coarse fibers of immunoreactive cell bodies in the suprachiasmatic nucleus. Other areas in the diencephalon that contain a moderately dense plexus of DAi and THi fibers are the ventral habenular nucleus, the anterior and central thalamic nuclei and the hypothalamus (Fig. 5.1(*e*), (*f*)). Numerous immunoreactive fibers are present in the lateral hypothalamus coursing toward the median eminence where they distribute to the intermediate lobe of the hypophysis (Fig. 5.1(*e*)).

Both antisera reveal an almost identical staining pattern of fibers in the brainstem. A rather dense network of immunoreactive fibers is found in the midbrain tectum (Figs. 5.1(*h*), 5.16). Although all

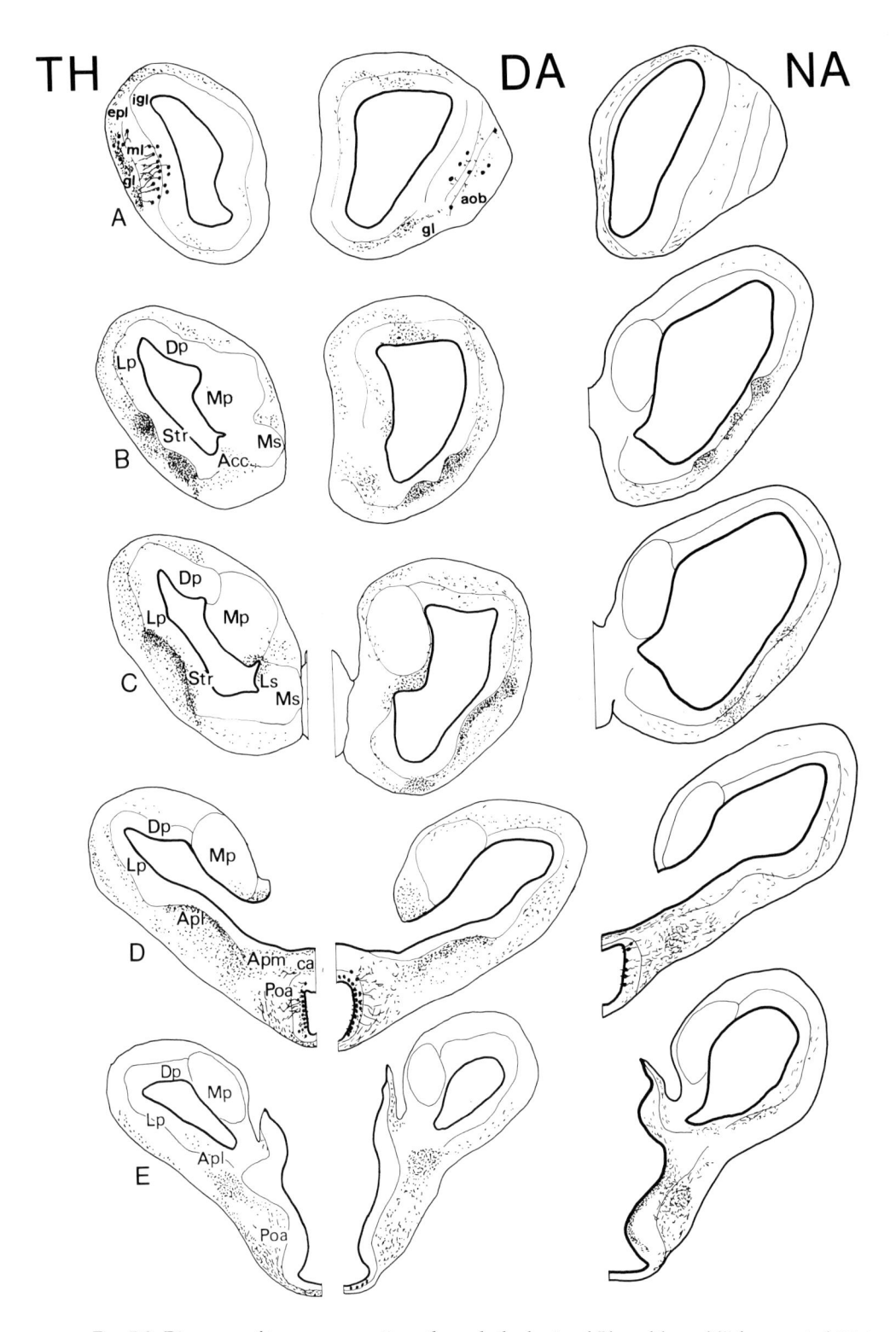

Fig. 5.9. Diagrams of transverse sections through the brain of *Pleurodeles waltlii* from rostral (A) to caudal (L) showing the distribution of immunoreactive cell bodies (large dots) and fibers (small dots, wavy lines). THi cells and fibers are drawn on the left, DAi cells and fibers in the middle, and NAi cells and fibers on the right.

Fig. 5.9. (cont.)

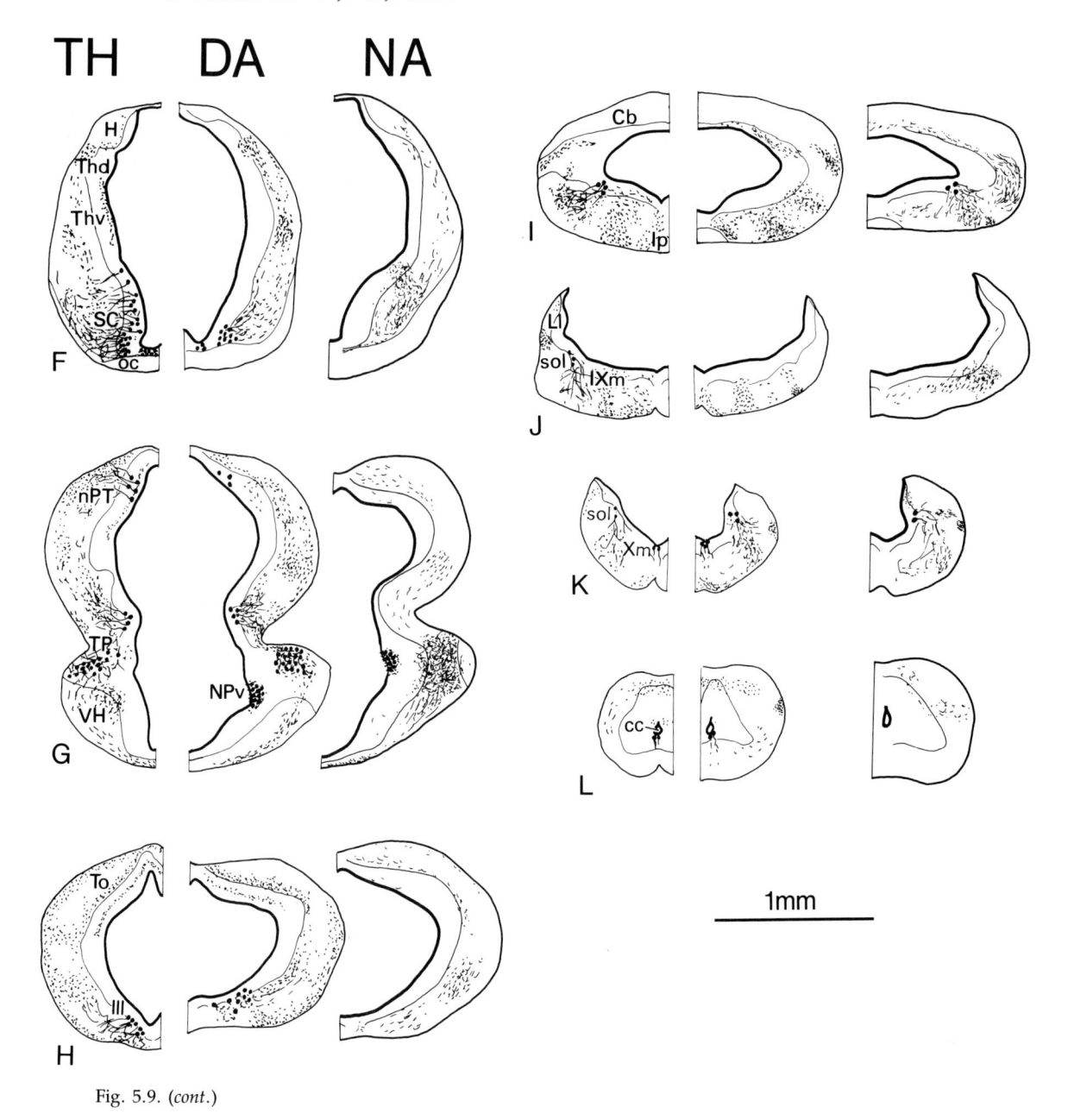

tectal layers contain DAi and THi fibers, their density in the superficial and deep tectal zones is higher than that in the intermediate zone. Ventrally, in the midbrain tegmentum, moderately dense plexuses of immunoreactive fibers are present in the periventricular gray, the raphe and the ventrolateral tegmentum. These fibers can be traced caudally, in a similar position, to cervical spinal cord levels (Fig. 5.1(h)–(n)). In the dorsal alar plate of the rhombencephalon numerous DAi and THi fibers are found in the lateral line area (Fig. 5.1(k)).

Urodela. In *Pleurodeles*, as in *Xenopus*, the majority of the DAi and THi fibers in the forebrain are confined to subpallial areas (Fig. 5.9(b)–(d)). Nevertheless, more or less developed networks of DAi and

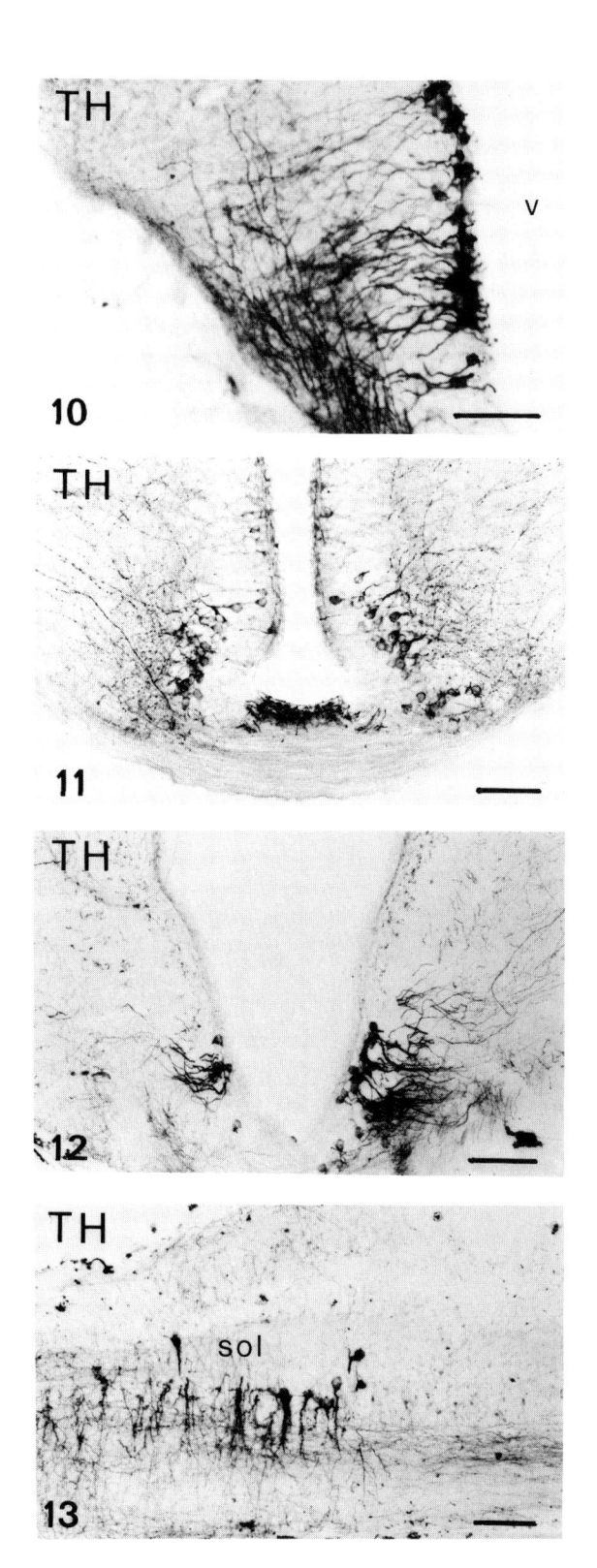

THi fibers are also observed in all pallial areas. The most dense network is found in the dorsal pallium. At rostral levels, the DAi fibers lie mainly within the cell layer. More caudally, they are primarily confined to the neuropil. A moderate dense plexus of DAi and THi fibers is found in the neuropil of the lateral pallium. The medial pallium contains immunoreactive fibers primarily in its superficial zone. In *Pleurodeles*, the striatum contains numerous DAi fibers which show a patchy appearance at rostral telencephalic levels (Figs. 5.9(b), 5.17). Caudally, the fibers are more uniformly distributed and form a band adjacent to the cell layer (Figs. 5.9(g), 5.18). Rather dense plexuses of DAi/THi fibers are further observed in the ventral thalamus, the deep and superficial layers of the midbrain tectum, and in the ventrolateral portion of rhombencephalic tegmentum (Figs. 5.9(f)–(l), 5.19).

General and derived conditions of the amphibian dopamine system

Cell bodies. As mentioned before, apart from brains of *Xenopus* and *Pleurodeles* that are stained with antisera to TH and DA, additional material was available of R. ridibunda (TH, DA) and S. lacertina (TH). When the distribution of DAi/THi cell bodies in the brains of the various amphibians are compared with each other, it is obvious that there exists a basic pattern. For example, all amphibians studied so far have in common that, in the olfactory bulbs, DAi cell bodies in the glomerular layer are substantially larger in size than those in the mitral cell layer. Further, in none of the amphibians studied are DAi cell bodies observed in the telencephalon proper. To the basic features of the diencephalic portion of the DA system belongs the presence of DAi cells in the anterior preoptic area, the medial and lateral parts of the suprachiasmatic nucleus, the periventricular organ (including its accompanying cells), the dorsomedial and ventrolateral parts of the posterior tubercle, and the posterior thalamic or pretectal nucleus. However,

Figs. 5.10–5.13. Photomicrographs of transverse (Figs. 5.10–5.12) and sagittal (Fig. 5.13) sections through the brain of *Pleurodeles waltlii* showing THi cell bodies and fibers at the level of the anterior preoptic area (Fig. 5.10), suprachiasmatic nucleus (Fig. 5.11) and rostral brainstem tegmentum (Fig. 5.12). Fig. 5.13 shows THi cell bodies adjacent to the solitary tract in a sagittal plane. Bars = 100 μm.

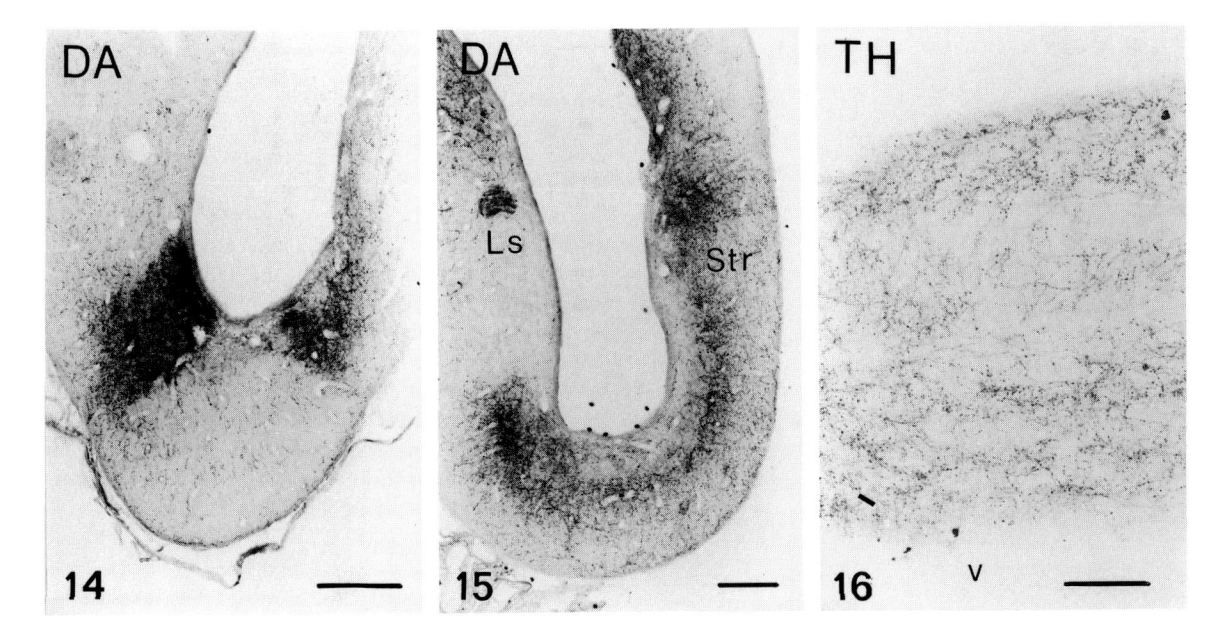

Figs. 5.14–5.16. Photomicrographs of transverse sections through the brain of *Xenopus laevis* showing DAi/THi fibers in the nucleus accumbens (Fig. 5.14), caudal striatum (Fig. 5.15), and mesencephalic tectum (Fig. 5.16). Bars = 200 μm.

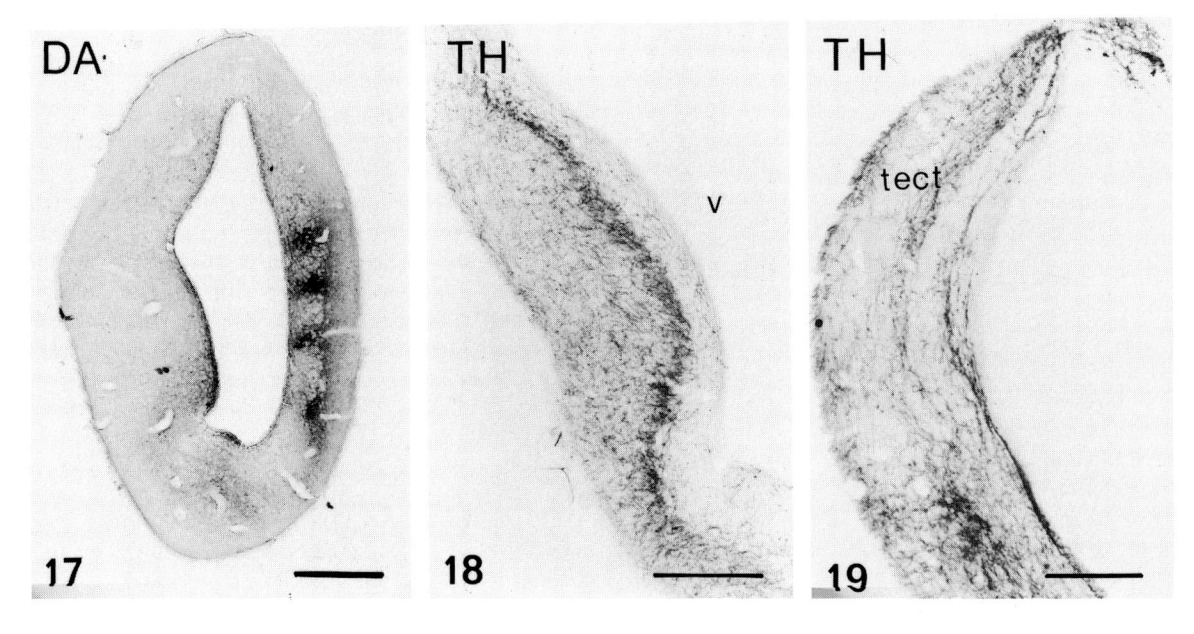

Figs. 5.17–5.19. Photomicrographs of transverse sections through the brain of *Pleurodeles waltlii* showing DAi/THi fibers in the rostral striatum (Fig. 5.17), caudal striatum (Fig. 5.18) and mesencephalic tectum (Fig. 5.19). Bars = 250 μm.

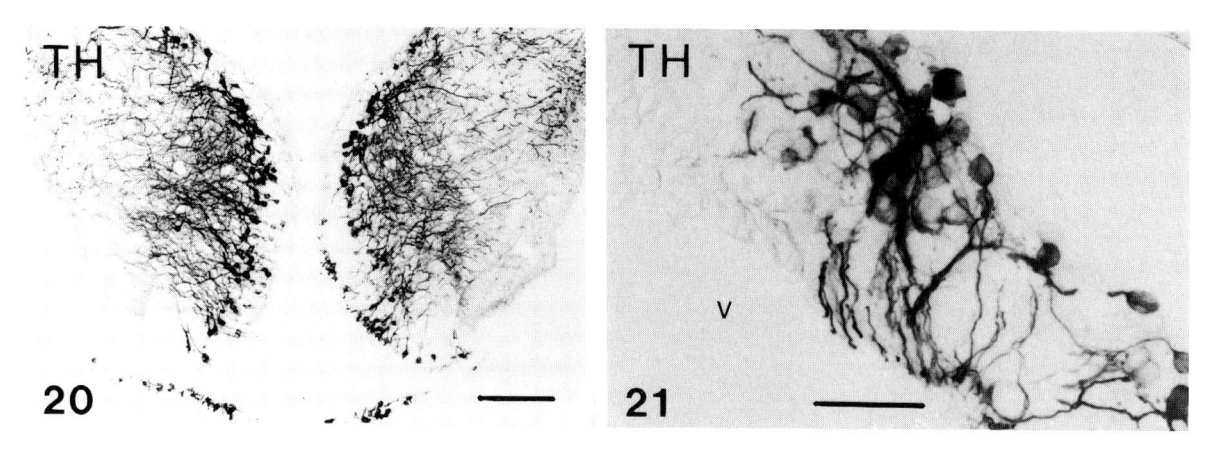

Figs. 5.20–5.21. Photomicrographs of transverse sections through the brain of *Siren lacertina* showing THi cell bodies and fibers in the suprachiasmatic nucleus (Fig. 5.20) and the lateral extent of the posterior tubercle (Fig. 5.21). Bars = 50 µm (Fig. 5.20) and 25 µm (Fig. 5.21).

some features of the dopaminergic system of *Siren*, as observed in TH stained material, deserve comments. First, it should be noted that in the, generally well stained, material of *Siren* no THi cell bodies are observed in the posterior thalamic nucleus. Another remarkable feature of this species is that only a few THi cell bodies are found in the anterior preoptic area. These cells are detached from the ventricular surface. Contrary to the preoptic DAi/THi cell group in the other amphibian species studied, in *Siren* no THi CSF-contacting cells are observed in the ependymal lining of the diencephalic ventricle. On the other hand, the number of CSF-contacting THi perikarya in the suprachiasmatic nucleus is markedly larger in *Siren* than in *Pleurodeles* (cf. Figs. 5.11, 5.20). Compared to other urodeles, in *Siren* the ventrolateral part of the posterior tubercle, sometimes called the dorsal infundibular nucleus, also contains considerably more THi cells, some of which extend processes that clearly reach the ventricle (Fig. 5.21).

As regards the brainstem, in all four species studied DAi/THi cell bodies are found in the midbrain tegmentum and, at caudal rhombencephalic levels, around the solitary tract, and in the ependymal and subependymal layers along the midline. Also the presence of DAi/THi CSF-contacting cells, predominantly ventral to the central canal of the spinal cord, appears to be a basic feature of the DA system of amphibians. Although the distributions of DAi/THi neurons in the brainstems of the amphibians studied so far largely resemble each other, some differences should be noted. A major difference is observed in the midbrain tegmentum (Fig. 5.22). Whereas in *R. ridibunda*, the dopaminergic cell group in this brain subdivision is unpaired (Fig. 5.23), in *P. waltlii* the corresponding DAi/THi cells do not constitute a single cell group but lie on both sides of the midline (Fig. 5.12). In *X. laevis* and *S. lacertina*, an intermediate condition is found. The major portion of the midbrain DAi/THi cell group is paired, but in its caudal part the cells on both sides of the midline fuse together constituting a single group (Fig. 5.24). In all amphibian species studied, however, the midbrain dopaminergic cell groups merge rostrally with the DAi/THi cell groups of the dorsomedial posterior tubercle. The different conditions observed in the adult brains of the four species studied probably reflect merely differences in degree of fusion of the initially bilateral cell groups during the development (see Chapter 14).

Another, more general difference is that the DAi/THi cells in the urodeles studied are smaller in number but larger in size than those in the anurans studied. Finally, a remarkable feature observed only in the brain of *Siren* is the presence of THi cell bodies in the caudal brainstem, at the level of the glossopharyngeal motor nucleus, with processes that extend into the glossopharyngeal nerve (Fig. 5.25).

Fibers. In the brains of the four species studied, the majority of the DAi/THi fibers in the forebrain are confined to subpallial areas. However, several remarkable differences are noted. For example, whereas in *Rana* and *Xenopus* immunoreactive fibers are absent in cortical areas, in *Pleurodeles* more or less developed plexuses of DAi and THi fibers are

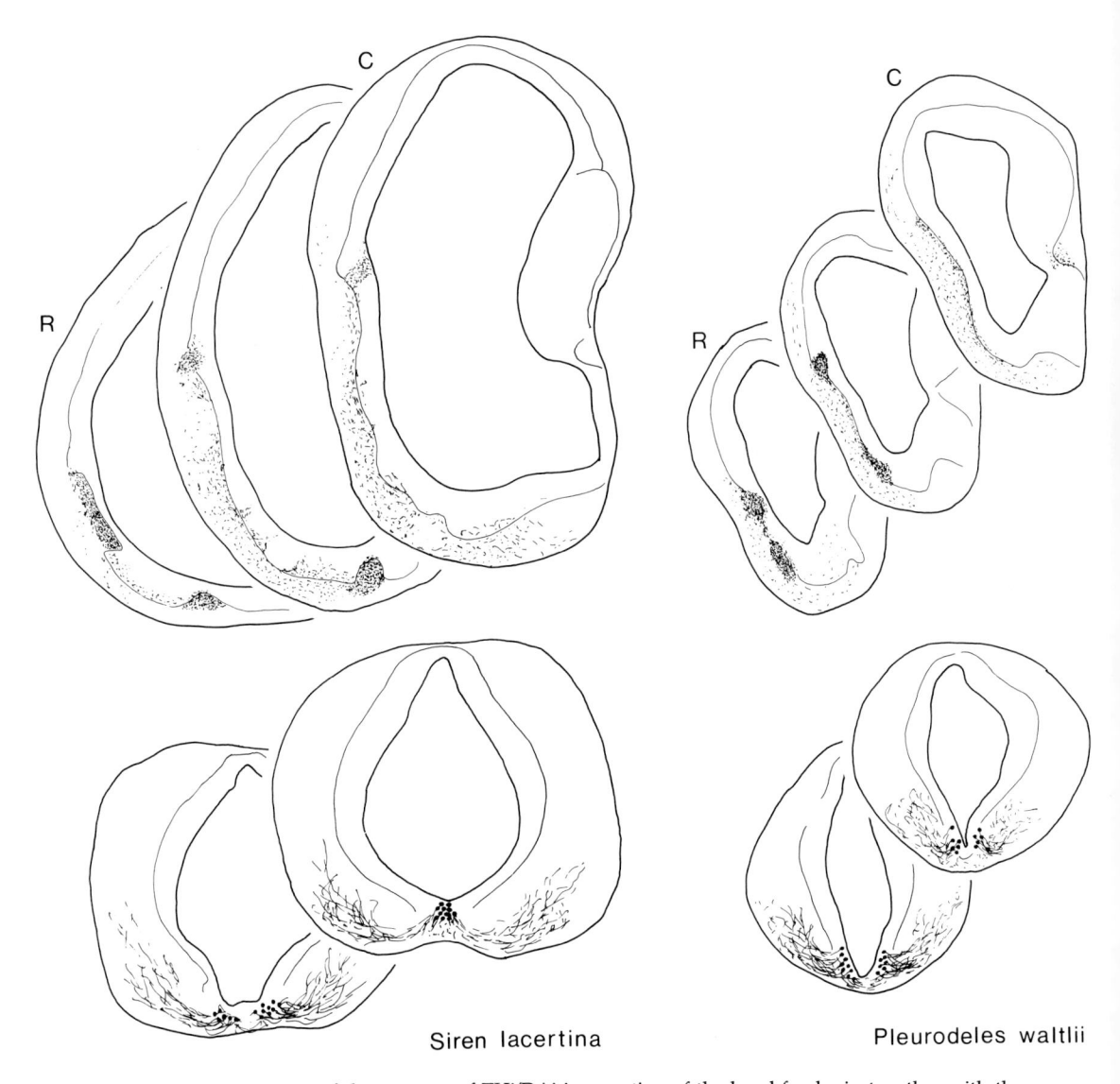

Fig. 5.22. Schematic drawings of the patterns of THi/DAi innervation of the basal forebrain together with the corresponding localization of the mesencephalic dopaminergic cell bodies in the brains of *Siren lacertina*, *Pleurodeles waltlii*, *Rana ridibunda* and *Xenopus laevis*.

observed in all pallial areas. In *Siren*, of which only TH-stained material was available, no immunoreactive fibers were observed in pallial areas. However, we consider these results as tentative, since the immunostaining in the forebrain of this species was certainly not optimal. Another difference concerns the dopaminergic innervation of the basal forebrain (Fig. 5.22). Whereas in *Pleurodeles* the striatum contains the densest DAi and THi plexus, in *Rana* the nucleus accumbens is the most dense innervated structure (Fig. 5.26). On the basis of a paired or unpaired midbrain dopaminergic cell group and the observed predominance of DA/TH immunoreactivity in the nucleus accumbens or striatum, it has been suggested that the mesolimbic system is particularly well developed in frogs, whereas the mesostriatal system prevails in the newt (González & Smeets, 1991). Our observations in *Siren* favor this notion, since the midbrain cell group is largely unpaired and THi fibers are mainly confined to the striatum where

Rana ridibunda   Xenopus laevis

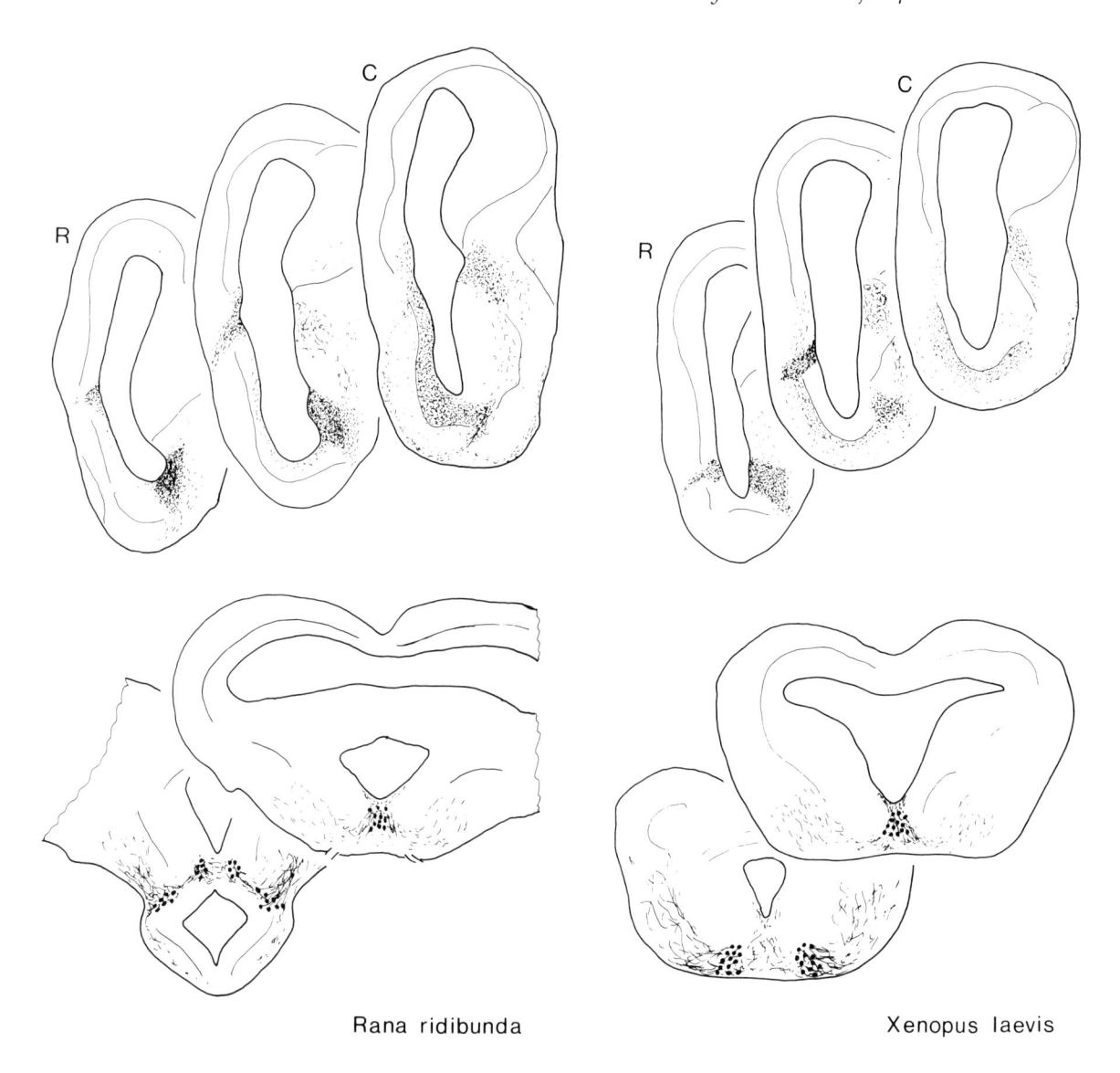

Fig. 5.22. (cont.)

they form distinct plexuses both in its dorsolateral and ventromedial portions. However, our findings in *Xenopus* do not unequivocally support this notion. Although the midbrain DA cell group in the latter species is largely unpaired, the nucleus accumbens and the striatum are almost equally densely innervated. Also differences in the site of termination of the mesostriatal fibers are recognized. A feature that *Xenopus* shares with *Pleurodeles* is that the DAi and THi fibers probably contact the proximal dendrites of striatal cells. In *Rana*, on the contrary, the majority of the immunoreactive fibers lie within the cellular layer (Figs. 5.22, 5.27). Currently, studies combining immunohistochemical and tract tracing techniques are in progress to find possible fundamental differences in mesostriatal connections.

Major species differences in the catecholaminergic innervation of brainstem structures are found in the tectum and the octavolateral area. In *Rana*, the distinct laminar organization of the tectum is

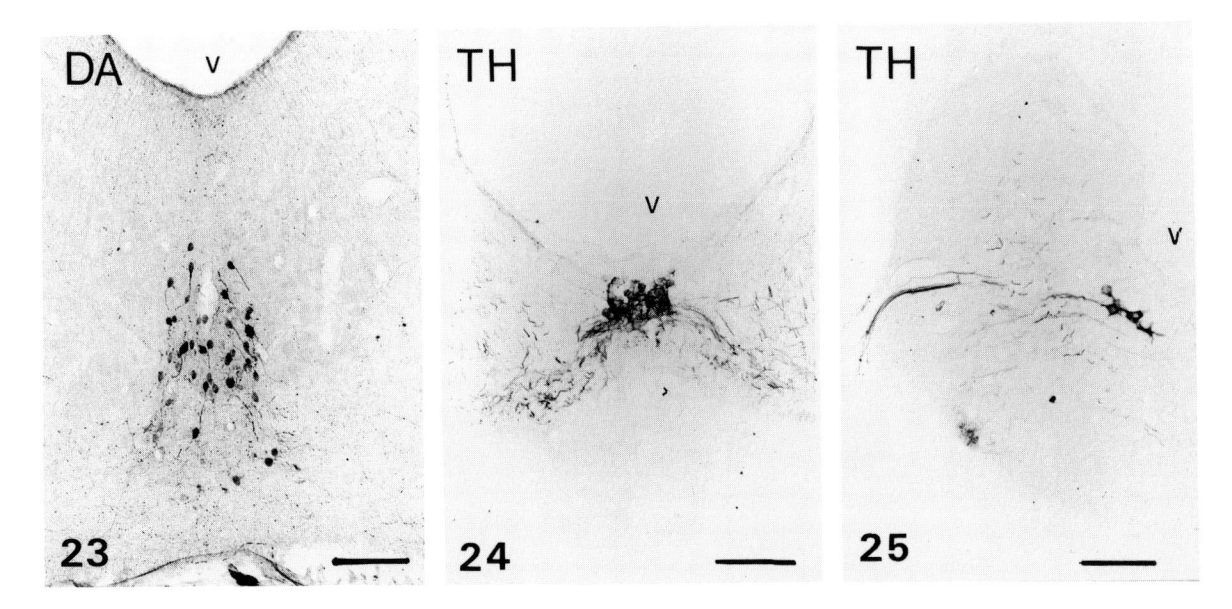

Figs. 5.23–5.25. Photomicrographs of transverse sections showing DAi cell bodies in the midbrain of *Rana ridibunda* (Fig. 5.23), THi cells in a similar position in *Siren lacertina* (Fig. 5.24) and THi cell bodies and fibers at caudal rhombencephalic levels of *Siren* (Fig. 5.25). Bars = 100 μm.

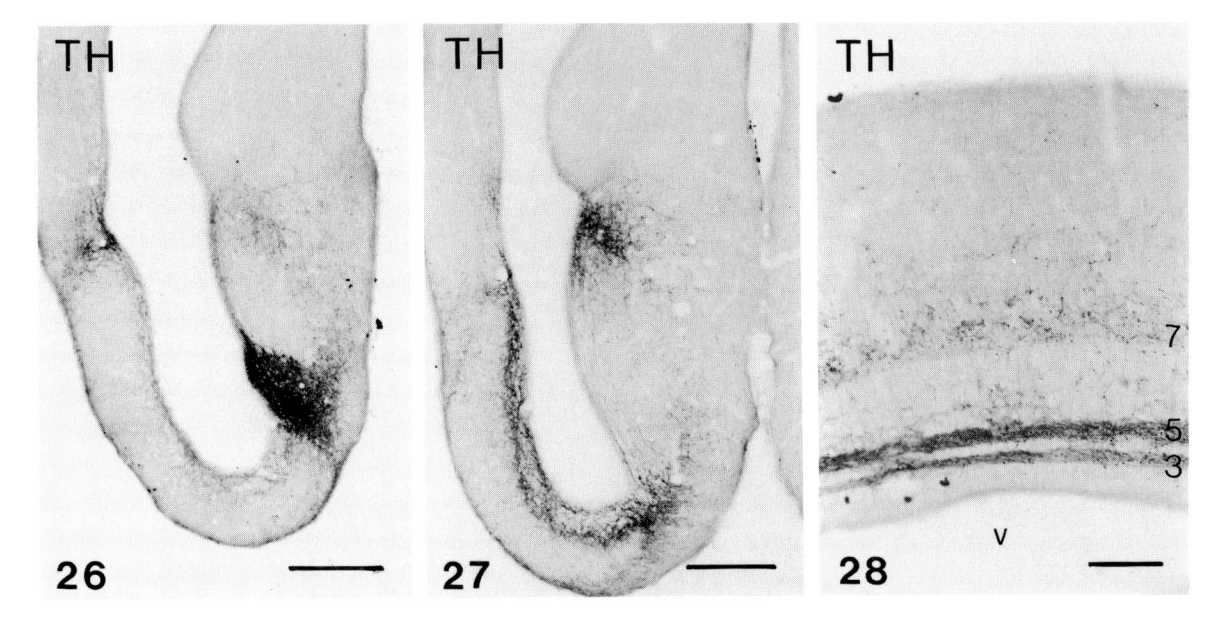

Figs. 5.26–5.28. Photomicrographs of transverse sections through the brain of *Rana ridibunda* showing THi fibers in the nucleus accumbens (Fig. 5.26), caudal striatum (Fig. 5.27) and midbrain tectum (Fig. 5.28). Bars = 500 μm (Figs. 5.26 and 5.27) and 100 μm (Fig. 5.28).

reflected in the distribution of DAi/THi fibers which occupy almost exclusively the deep fiber layers 3, 5 and 7 (Fig. 5.28). The differences observed in the catecholaminergic innervation of the octavolateral area depends on the presence or absence of the lateral line system. For example, *Rana* which lose its lateral line system during metamorphosis, lacks most of the rich DA innervation in the dorsal alar plate as observed in *Xenopus*.

Noradrenaline

Data on the distribution of noradrenaline (NA) in the brain of amphibians are sparse. The location of DBHi cell bodies has been described for the bullfrog *Rana catesbeiana* (Yoshida et al., 1983). Our own studies by means of antibodies against DBH as well as noradrenaline made it possible to determine precisely the noradrenergic contribution to the catecholaminergic innervation of the brain of *Xenopus* (González & Smeets, 1993) and *Pleurodeles* (unpublished observations). For this review, also DBH stained material of *R. ridibunda* was available.

Cell bodies

Anura. The antibodies against DBH and NA applied to the brain sections of *Xenopus* revealed patterns of immunostaining that are constant from animal to animal. As was expected on the basis of their position in the biosynthetic pathway of catecholamines, the DBH antiserum stains more cell bodies than the NA antiserum suggesting that some cells in the CNS of *Xenopus* metabolize adrenaline. Immunostaining of noradrenergic fibers is most distinct in material processed with the NA antiserum. On the contrary, noradrenergic cell bodies stain generally more darkly with the DBH antiserum.

In the brain of *Xenopus*, noradrenaline immunoreactive (NAi) cell bodies are clustered in three distinct cell groups. From rostral to caudal, cells are found in the nucleus of the hypothalamic periventricular organ, the isthmic region, and the caudal brainstem at the level of the obex (Fig. 5.1(*f*), (*l*), (*m*)). The most conspicuous NAi cell group is observed in the nucleus of the periventricular organ extending rostrocaudally from the dorsal hypothalamus at mid-diencephalic level to the lateral extension of the infundibular recess (Fig. 5.29). The cell bodies show a laminar organization, 4–5 cell layers thick, within the ependymal and subependymal zones. All cells seem to be in direct contact with the cerebrospinal fluid (CSF) by means of processes with club-like endings protruding into the ventricle. Other processes

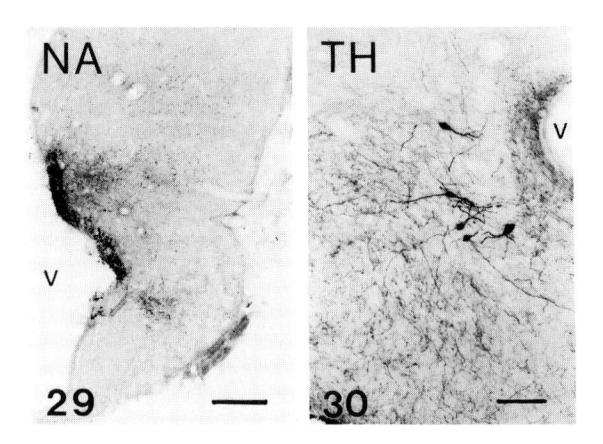

Figs. 5.29–5.30. Photomicrographs of transverse sections through the brain of *Xenopus laevis* showing NAi cell bodies and fibers in the caudal part of the periventricular organ (Fig. 5.29) and THi cell bodies in the locus coeruleus (Fig. 5.30). Bars = 250 μm (Fig. 5.29) and 100 μm (Fig. 5.30).

are directed laterally but their exact course is difficult to trace. Remarkably, these NAi cell bodies are immunonegative for the TH- as well as the DBH- antisera which suggests that they lack the enzymes to synthesize catecholamines.

The second group of NAi cell bodies lies in the isthmic region. The cells are located medial and ventral to the isthmic nucleus and extend caudally just to the rostral pole of the trigeminal nuclear complex. In this area, the DBH and NA antibodies do not reveal differences in position or in number of putative noradrenergic cell bodies. The morphology of the cells, however, is best appreciated in sections stained with TH antibodies (Fig. 5.30). The cells are multipolar with long processes that are directed ventrally, and short processes that extend dorsally into the area between the ventricle and the prominent isthmic nucleus.

The third group of NAi cellbodies, located in the caudal brainstem, consists of cells that are scattered along the ventral and medial aspects of the solitary tract. The morphology of these cells resembles that of the isthmic cell group in having several long processes directed ventrally into the tegmentum. Caudally, the cells continue into the area postrema where they lie dorsolateral to the central canal. The latter cells differ in morphology from the more rostral cells. They are mainly bipolar and their processes are oriented in a medio-lateral direction. It should be noted that the immunoreactivity of the NAi cell bodies in the caudal brainstem shows considerable variation in intensity. On the whole, the cells stain more intensely with the TH and DBH antisera.

Urodela In the brain of *Pleurodeles*, four groups of NAi cell bodies are present. In contrast to *Xenopus*, where immunoreactive cells are only found in the periventricular organ, the locus coeruleus, and an area adjacent to the solitary tract, in the newt additonal NAi perikarya are observed in the anterior preoptic area (Fig. 5.31). The number of the CSF-contacting cells in the preoptic area that stain with the NA antiserum is much smaller than that of DAi cells. Moreover, whereas there appears to be a good match of THi and DAi cells in this area, the DBH antiserum fails to demonstrate immunoreactive cell bodies in a corresponding position. This finding resembles the situation observed in the periventricular organ, where the CSF-contacting cells are immunopositive for the NA antiserum, but immunonegative for the DBH antiserum.

The cells of the locus coeruleus of *Pleurodeles* form a compact group at isthmic levels, immediately ventral to the cerebellum (Figs. 5.32, 5.33). Their processes extend ventrally where they merge with longitudinally coursing THi fibers. Contrary to the locus coeruleus where no apparent differences are observed in the number and distribution of DBHi and NAi perikarya, the number of DBHi cell bodies adjacent to the solitary tract notably exceeds that stained with NA antibodies. No DBHi or NAi cells are observed in the ependymal layer along the midline of the caudal rhombencephalon or the central canal of the spinal cord.

Fibers

Anura The distribution of noradrenergic fibers in the brain of *Xenopus*, as demonstrated with the NA antiserum, is essentially the same as that obtained with the DBH antibodies. Nevertheless, the visualization of thin noradrenergic fibers in certain brain areas, viz. pallium, lateral line area, and cerebellum, could only be achieved in series stained with NA antibodies.

The most rostral NAi fibers are found in the olfactory bulbs, mainly in the mitral cell layer (Fig. 5.1(*a*)). In the ventral part of each bulb, a few, coarse fibers are present in the external plexiform layer adjacent to the glomerular layer. The glomerular layer and the internal granular layer are devoid of NAi fibers. The NAi fibers in the dorsal parts of the olfactory bulbs are caudally continuous with the NAi fibers in the dorsal pallium.

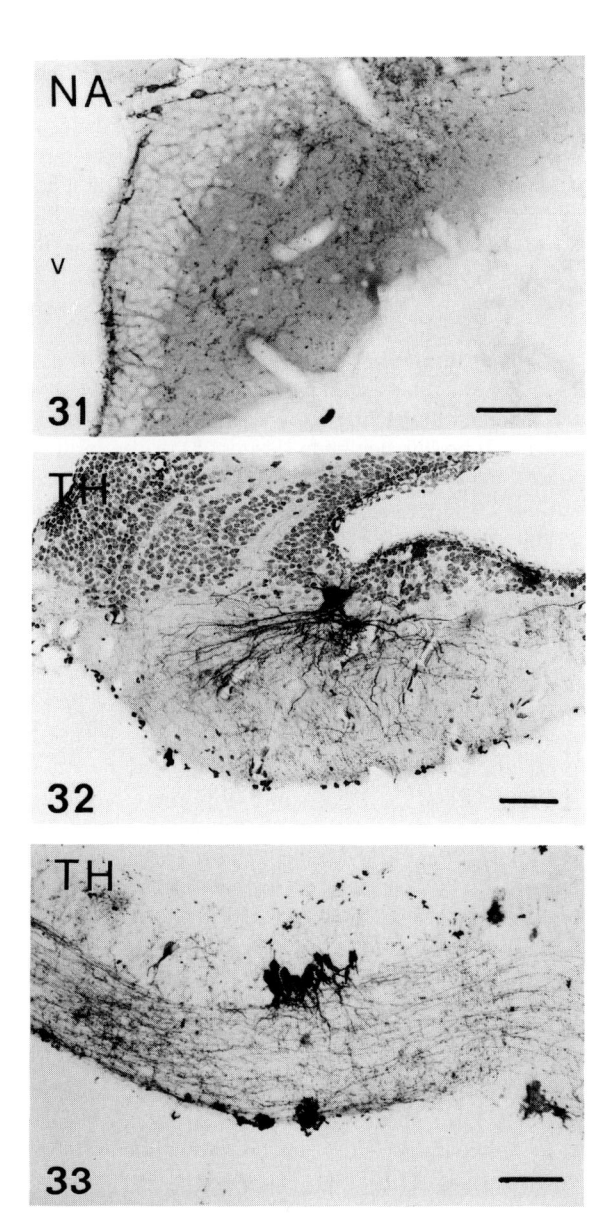

Figs. 5.31–5.33. Photomicrographs of brain sections of *Pleurodeles waltlii* showing NAi cell bodies and fibers in the anterior preoptic area (Fig. 5.31) and THi cell bodies in the locus coeruleus in a transverse (Fig. 5.32) or sagittal (Fig. 5.33) plane. Bars = 100 μm.

In the telencephalon proper, NAi fibers and terminals are present in both pallial and subpallial areas, although those in the latter areas prevail (Fig. 5.1(*b*)–(*d*)). Throughout the rostrocaudal extent of the hemispheres, thin NAi fibers were observed in the superficial zones of all pallial areas. In the subpallium, distinct NAi fiber plexuses are present in the nucleus accumbens, the nucleus of the diagonal band, the dorsolateral part of the striatum, the medial amygdala, and in an area that encompasses the lateral forebrain bundle (Fig. 5.34). Moderate plexuses are found in the septum and the submeningeal zones of both the striatum and the nucleus accumbens. The NAi fibers in the septum are mainly confined to its intermediate portion; only a few fibers can be traced into the lateral septum. Numerous NAi fibers cross the midline in the anterior commissure and interconnect the amygdalar immunoreactive plexuses.

The densest NAi fiber plexus is observed at mid-diencephalic levels in the hypothalamus where the fibers lie in a band ventrolateral to the nucleus of the periventricular organ. This plexus extends caudally into the infundibulum as a round-shaped immunoreactive neuropile in the dorsolateral portion of the ventral hypothalamus (Fig. 5.35). Another dense plexus is present in the posterior tubercle where numerous NAi fibers seem to cross the midline. Other moderate to dense hypothalamic plexuses of NAi fibers are located in the area lateral to the nucleus of the periventricular organ and in the submeningeal zone of the infundibular region. The fibers in the latter region can be traced via the median eminence to the intermediate lobe of the pituitary (Fig. 5.36). In the dorsal part of the diencephalon, moderate plexuses of NAi fibers are present in the subependymal zone and the posterior thalamic nucleus.

In the midbrain, numerous NAi fibers were observed in both the deep and superficial tectal layers, but only a few immunoreactive fibers are present in the central zone. The fibers in the superficial and deep tectal zones are continuous ventrally with those in the superficial and periventricular zones of the tegmentum. Moderate to dense plexuses of NAi

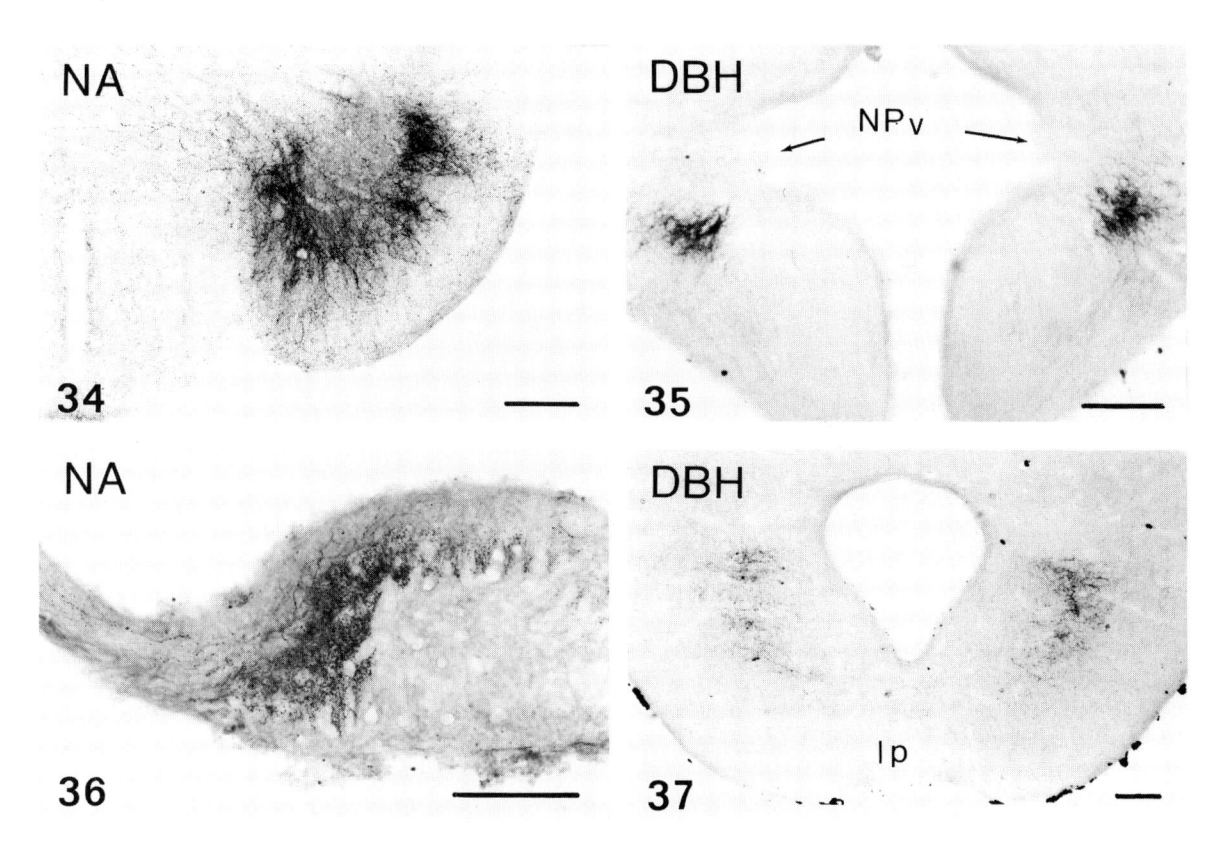

Figs. 5.34–5.37. Photomicrographs of transverse sections through the brain of *Xenopus laevis* showing NAi fibers in an area adjacent to the lateral forebrain bundle (Fig. 5.34), DBHi fibers in the infundibular hypothalamus (Fig. 5.35), NAi fibers that innervate the intermediate lobe of the hypophysis (Fig. 5.36) and DBHi fibers in the mesencephalic tegmentum (Fig. 5.37). Bars = 200 μm (Figs. 5.34–5.36) and 100 μm (Fig. 5.37).

fibers in the periventricular zone comprise the tegmental nuclei (anterodorsal, anteroventral, posterodorsal, posteroventral) and the torus semicircularis (Figs. 5.1(*h*), 5.37). A considerable number of immunoreactive fibers lie dorsal and lateral to the interpeduncular nucleus whereas only a few fibers traverse the nucleus.

The rhombencephalon and the spinal cord are, compared to the forebrain and midbrain, less densely innervated by NAi fibers (Fig. 5.1(*j*)–(*l*)). In the rostral rhombencephalon, moderate to dense plexuses of NAi fibers are found in the superior raphe nucleus, the locus coeruleus, the secondary visceral nucleus, the trigeminal motor and principal sensory nuclei, and the cerebellar nucleus. Some fibers form a bundle and cross the midline via the cerebellar commissure, whereas others distribute to the ipsilateral cerebellum. Throughout the rhombencephalon, the majority of the NAi fibers are confined to the ventrolateral tegmentum where they run in a longitudinal direction. From this bundle, fibers are issued to the superior olive and the octavolateral area. Near the obex, a conspicuous plexus of NAi fibers lies adjacent to the solitary tract and the vagal motor nucleus. Caudal to the obex, some fibers course to the area postrema, whereas others run further caudally and can be traced at least to thoracic spinal cord levels (Fig. 5.1(*m*), (*n*)).

Urodela. Whereas only a few fibers are present in the olfactory bulbs of *Pleurodeles*, the telencephalon proper is well innervated by NAi fibers (Fig. 5.9(*b*)–(*d*)). The most rostrally located, distinct plexuses are observed in the striatum where they show a patchy appearance (Fig. 5.38). Except for the lateral pallium, the pallial areas are almost completely devoid of NAi fibers. At caudal telencephalic levels, moderately dense plexuses are found in the medial amygdala and the lateral preoptic area. A dense plexus of NAi fibers is observed immediately caudal to the NAi CSF-contacting cells of the preoptic area within the subependymal layer (Fig. 5.39). This plexus has only a limited rostrocaudal extent and is considered to be constituted by the processes of the latter cells, since such a plexus is not seen in adjacent sections stained with DBH antibodies. In the diencephalon, the regions lateral to the suprachiasmatic nucleus and the periventricular organ are densely innervated by NAi fibers (Fig. 5.40). Other areas, such as the dorsal and ventral thalamus, and the median eminence contain moderate to dense networks of immunoreactive fibers. In the brainstem, dense plexuses of NAi fibers are present in an area ventrolateral to the locus coeruleus, in the parabrachial region, and in the motor nuclei of the cranial nerves. At obex levels, immunoreactive fibers are observed in the solitary tract area and area postrema, and can be traced further caudalward into the spinal cord.

General and derived conditions of the amphibian noradrenaline system The finding that the cells in the periventricular organ of *Xenopus* and *Pleurodeles* are immunopositive for the NA antiserum but immunonegative for the TH- and DBH-antisera suggests that these cells accumulate not only dopamine but also noradrenaline. Similar findings have been reported for reptiles (Smeets & Steinbusch, 1989, 1990).

In the brain of *R. catesbeiana*, Yoshida *et al.* (1983) found two groups of cells that were immunoreactive with antibodies against DBH but not with PNMT. Both groups lie in the caudal brainstem: one in the nucleus of the solitary tract and the adjacent reticular formation, and the other along the midline of the caudal one-third of the rhombencephalic tegmentum. The latter group consists of CSF-contacting cells. Yoshida *et al.* (1983) mentioned also the presence of THi cell bodies in the isthmic tegmentum but did not report DBH immunoreactivity in these cells. As shown by our studies, in *Xenopus* as well as *Pleurodeles*, there are three groups of NAi cell bodies, i.e. in an area adjacent to the solitary tract, in the presumed amphibian homologue of the locus coeruleus, and in the hypothalamic periventricular organ. For clarity, it should be noted that, in *Xenopus* and *Pleurodeles*, only the cells adjacent to the solitary tract and in the locus coeruleus are immunopositive for the DBH antiserum. The lack of immunostaining with the DBH antiserum in CSF-contacting cells of the periventricular organ of all amphibian species studied so far (including *R. ridibunda*, own observations) is most likely due to the accumulating nature of those cells.

The discrepancies between *R. catesbeiana* and the three species studied by us (*X. laevis*, *R. ridibunda*, *P. waltlii*) in the isthmic region and caudal brainstem are less easy to explain. In the species studied by us, the cell bodies in the locus coeruleus are immunoreactive with the TH-, DBH- and/or NA antisera but not with the DA antiserum. The lack of DBH immunoreactivity in the isthmic cell group of *R.catesbeiana* is most likely due to a staining failure, since in amniotes as well as anamniotes the THi cell bodies in a corresponding position always stain with DBH- or NA antibodies. A similar explanation could be given for the failure of the NA antibodies to stain the CSF-contacting cells along the midline in the caudal brain-

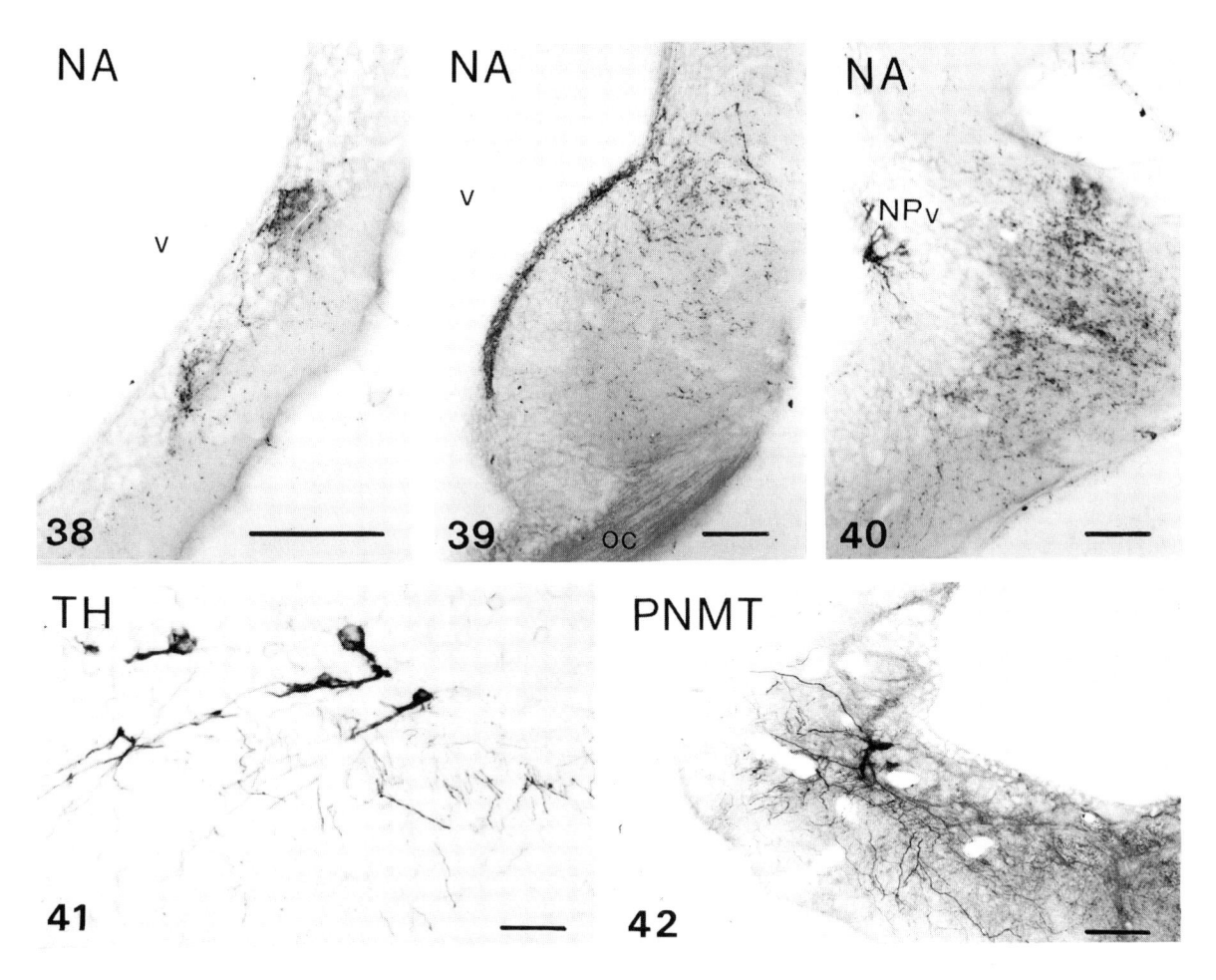

Figs. 5.38–5.42. Photomicrographs of transverse sections showing NAi fibers in the striatum (Fig. 5.38), preoptic area (Fig. 5.39) and hypothalamus (Fig. 5.40) of *Pleurodeles waltlii* and THi cell bodies in the locus coeruleus of *Siren lacertina* (Fig. 5.41). Fig. 5.42 illustrates PNMTi cell bodies in the caudal rhombencephalon of *Pleurodeles waltlii*. Bars = 100 µm (Figs. 5.38–5.40 and 5.42) and 25 µm (Fig. 5.41).

stem of the amphibians studied by us. However, there are two arguments that make such an assumption unlikely: (1) not only the NA antibodies but also the DBH-antibodies fail to stain these cells, and (2) CSF-contacting cells located in a corresponding position are strongly immunoreactive with TH- and DA-antisera in the anurans, *R. ridibunda* and *X. laevis*, and the urodele, *P. waltlii*. This suggests that the cells are dopaminergic (González & Smeets, 1991; González et al., 1993). It seems, therefore, more likely that we are dealing either with a false-positive staining in *R. catesbeiana* or with a true species difference. The latter possibility implies that in some amphibians the CSF-contacting cells in the caudal brainstem have lost the capacity to metabolize noradrenaline, whereas others have retained it. A study by means of NA antibodies of the brain of *R. catesbeiana* could probably solve this question.

With respect to the distribution of NAi/DBHi fibers it appears that the patterns in the various species studied are quite similar. Only two features deserve some attention. The first feature concerns the noradrenergic innervation of basal forebrain structures. In *Xenopus*, the densest plexus of NAi/DBHi fibers is found in the ventromedial part of the hemisphere, largely overlapping the area that receives a strong dopaminergic input. In *Pleurodeles*, on the contrary, the striatum contains several patches of NAi fibers but also in the latter species it is obvious that a considerable overlap of NAi and DAi fibers occurs.

Our preliminary observations in R. ridibunda show that a similar codistribution of DAi and DBHi fibers is present in the basal forebrain. Thus, although there are differences in the distribution of noradrenergic fibers in the basal forebrain, there is a strong resemblance in the relationship between noradrenergic and dopaminergic fibers in striatal structures. The second feature that has to be mentioned concerns the morphology of the cells in the locus coeruleus. Whereas in the anurans R. ridibunda and X. laevis, as well as in the newt, P. waltlii, the cell processes are directed mainly ventrally or ventrolaterally covering almost the entire tegmental wall, in S. lacertina the processes of the corresponding cells constitute only a relatively thin lamina of coarse fibers which can be traced caudally to the level of the glossopharyngeal motor nucleus (Figs. 5.41, 5.43). During their course, the latter fibers stay adjacent to the boundary between the gray and white matter.

Adrenaline

Data on the distribution of PNMTi and, therefore, putative adrenergic neuronal structures in the brains of amphibians are fragmentary. Yoshida et al. (1983) were the first to report PNMTi cell bodies in the brain of an amphibian, i.e. the bullfrog R. catesbeiana. They found immunoreactive cell bodies in and around the nucleus of the solitary tract. For the present study, we have applied the PNMT antibodies to the brains of R. ridibunda, X. laevis, and P. waltlii. The PNMT antiserum stained satisfactorily only in *Pleurodeles* and yielded a Golgi-like staining of cell bodies. On each side of the brain, about 70 PNMTi cell bodies are found at caudal brainstem levels (Figs. 5.42, 5.44). They lie primarily ventral to the solitary tract, rostral to the obex. The PNMTi cells are characterized by numerous processes which arborize profusely in the lateral and ventrolateral parts of the rhombencephalic tegmentum. In addition, a number of processes courses ventromedially. Since the processes of the PNMTi cell bodies occupy almost the entire tegmentum in the transverse plane, these cell bodies seems to have a strategic position with respect to the long ascending and descending fiber tracts. Although the catecholamine cell group in the caudal brainstem appears to contain DAi, NAi as well as PNMTi cells, its major part is constituted, at least in *Pleurodeles*, by PNMTi perikarya.

Whereas the putative dendritic cell processes stain well with the PNMT antiserum, visualization of axonal processes is very arduous. Nevertheless, a few, weakly stained PNMTi fibers could be traced through the brainstem tegmentum to the basal forebrain where they probably also terminate within the

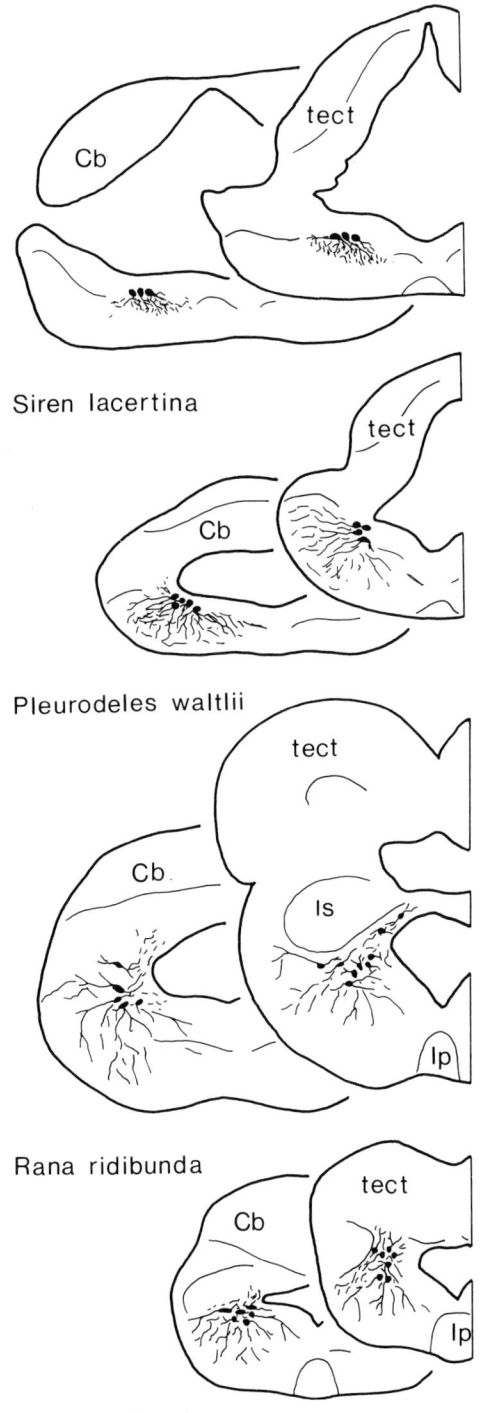

Fig. 5.43. Schematic drawings comparing the localization and morphology of the cells in the locus coeruleus of *Siren lacertina, Pleurodeles waltlii, Rana ridibunda* and *Xenopus laevis*.

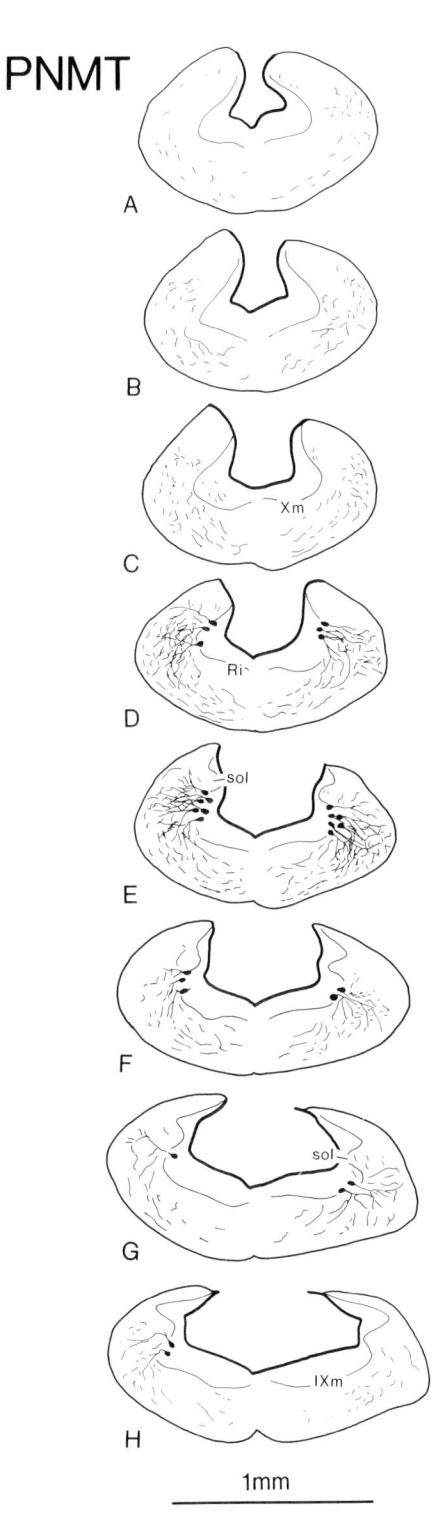

Fig. 5.44. Diagrams of transverse sections through the caudal rhombencephalon of *Pleurodeles waltlii* from caudal (A) to rostral (H) showing the distribution and morphology of PNMTi neurons.

nucleus accumbens. It is obvious that we are still in the beginning of unraveling the adrenergic system of amphibians and that this system deserves special attention in the near future.

Comparison of the DA and NA innervation of brain structures

The studies on the distribution of dopaminergic and noradrenergic cell bodies and fibers in the brains of *Xenopus* and *Pleurodeles* by means of antibodies that are specifically directed against DA and NA, respectively, offer the opportunity to differentiate between the two catecholamines in the two amphibian species.

In the olfactory bulb, there is some overlap of DAi and NAi fibers in the plexiform layers and the mitral cell layer. On the contrary, in the glomerular layer, the catecholaminergic fibers are exclusively dopaminergic. In the telencephalon proper, extensive overlap between DAi and NAi fibers was noted in the nucleus accumbens, the striatum, the medial amygdala, and in the area encompassing the lateral forebrain bundle. Although the ventromedial portion of the striatum of *Xenopus* is reached by DAi- as well as by NAi-fibers, the two catecholaminergic fiber systems are located in different laminae: the DAi fibers course primarily in the intermediate zone, probably contacting the proximal parts of the dendrites of the striatal neurons, whereas the NAi fibers run in the superficial zone and most likely contact the distal parts of the dendrites.

In the diencephalon, the most obvious overlap of the two systems is found in the nucleus of the periventricular organ. On the basis of DA- and NA-immunostained material, the possibility cannot be excluded that DA and NA are contained within the same cell bodies. Double-labeling procedures are needed to confirm this hypothesis. The distribution patterns of DAi and NAi fibers in the other parts of the diencephalon match each other quite well. The most striking overlap is observed in the intermediate lobe of the pituitary. There are no diencephalic cell masses that are exclusively innervated by either DA or NA fibers.

In the brainstem, the distribution patterns of DAi and NAi fibers are largely the same, although the innervation of the tectum and the lateral line area of *Xenopus* by the two catecholamines deserves some comment. DAi fibers are mainly confined to the cell-free fiber layers in the deep zone of the tectum. More superficial layers contain a much smaller number of DAi fibers. The NAi fibers, on the contrary, appear to be more numerous in the superficial zone than in the deep tectal zone. *Xenopus* which is strictly aquatic, differs from other anuran amphibians in having a persisting lateral line system which is represented in the brain by a well organized lateral line area in the rhombencephalic alar plate (Will et al., 1985a, b). By means of antibodies against DA and NA, it is now demonstrated that the DAi fibers outnumber the NAi fibers in the lateral line area. This finding suggests that DA plays a major role in processing or modulating lateral line information.

Sites of origin of DA and NA fibers

Information on the presumed sites of origin of the DA- and NA innervation of neuronal structures in the brains of amphibians is obtained by comparing the location of DAi and NAi cells with data on their connectivity. The dense DA innervation of the glomerular layer of the olfactory bulbs is most likely supplied by the DAi periglomerular cells. This notion is corroborated by the results of a previous study (Tohyama et al., 1977) by means of the FIF technique which revealed no changes in histofluorescence in the olfactory bulbs after total sections immediately caudal to the bulbs. The origin of the NAi fibers in the olfactory bulb has not been demonstrated experimentally, yet. Since the bulbs do not have intrinsic NAi perikarya, the source of the immunoreactive fibers must lie outside the olfactory bulbs. Two possible candidates are the nucleus of the periventricular organ and the locus coeruleus. The possibility that the CSF-contacting cells of the hypothalamic periventricular organ are the origin of these fibers seems to be unlikely, since 1) the fibers in the olfactory bulb are also DBHi, whereas the cells in the organ are DBH immunonegative, and 2) a study by von Bartheld and Meyer (1990) has shown that the cells of this organ do not project to the olfactory bulb in lungfishes. Therefore, the locus coeruleus is considered by us as the most likely source of noradrenergic input to the olfactory bulb, even though Kemali & Guglielmotti (1987) were unable to label retrogradely cells in this nucleus after HRP application to the main and accessory olfactory bulbs of *R. esculenta*.

Hodological studies in amphibians have revealed that the major projections to the striatum come from thalamic nuclei (Wilczynski & Northcutt, 1983; Wicht & Himstedt, 1988). These fibers terminate primarily in the central neuropil of the striatum which, in *R. ridibunda* and *P. waltlii*, lacks a dopaminergic innervation (González & Smeets, 1991). Other projections to the striatum originate from the supra-chiasmatic nucleus and the nucleus of the posterior tubercle (Wilczynski & Northcutt, 1983; Dubé, Clairambault & Malacarne, 1990). The terminal fields of these projections match with the plexus of DAi fibers in the striatum. This finding suggests that the latter two nuclei are the sources of the dopaminergic innervation of the striatum in the two species. A projection to the nucleus accumbens, arising from the midbrain tegmentum, has been suggested by Kicliter (1979) in the frog, *R. pipiens*. In *Xenopus*, contrary to *R. ridibunda* and *P. waltlii*, the DAi fibers are located primarily in the central neuropil of the striatum (González et al., 1992). Unfortunately, no information is available on whether this difference is reflected in a correspondingly different site of termination of the striatal afferent connections.

Several studies have dealt with the afferent connections of the tectum mesencephali of amphibians (e.g. Wilczynski & Northcutt, 1977; Finkenstädt, Ebbesson & Ewert, 1983; Rettig, 1988; Zittlau, Claas & Munz, 1988). In these studies, retrogradely labeled neurons were observed in the posterior thalamic nucleus following injections of horseradish peroxidase (HRP) in the tectum. Our previous studies of the distribution of DAi cell bodies in the brain of amphibians (González & Smeets, 1991; González et al., 1993) have revealed the presence of immunoreactive cell bodies in a corresponding position. This suggests that the posterior thalamic nucleus contributes to the dopaminergic innervation of the amphibian midbrain tectum. Hitherto, no tectal inputs have been reported from cells that match, in position, the dopaminergic cell bodies in the midbrain tegmentum. On the basis of the location of retrogradely labeled cells in the isthmic region after tectal HRP injections (Wilczynski & Northcutt, 1977; Finkenstädt et al., 1983), it seems very likely that the noradrenergic innervation of the midbrain tectum originates from the amphibian homologue of the mammalian locus coeruleus. Although we were first reluctant to label the isthmic NA cell group as locus coeruleus (González & Smeets, 1991), the recent demonstration of a telencephalic projection originating in the isthmic region of amphibians (Dubé, Clairambault & Malacarne, 1990; Lázár & Kozicz, 1990) led us now to use this label.

The DA innervation of the medulla probably arises from both hypothalamic and rhombencephalic DAi cell groups (Tóth, Csank & Lázáar, 1985; Naujoks-Manteuffel & Manteuffel, 1988). In addition, the rather dense plexus of DAi fibers in the superior olivary nucleus may have its origin in the posterior thalamic nucleus (cf. Feng, 1986).

A comparison of the distribution of DAi and NAi cell bodies with data obtained by retrograde tracing studies yields several candidates for supraspinal sources of catecholaminergic input to the spinal cord. Following injections of HRP or cobalt in the spinal cord of several species of amphibians, retrogradely labeled cells were observed, among others, in the midbrain tegmentum, the isthmic region, and adjacent to the solitary tract (ten Donkelaar et al., 1981; Tóth et al., 1985; Naujoks-Manteuffel & Manteuffel, 1988). The DAi fibers of the spinal cord are, therefore, likely to originate from the DAi cell groups in the midbrain and solitary tract area, whereas the NAi fibers may arise from the locus coeruleus and, possibly, also the solitary tract area. An additional DA input to the spinal cord may be provided by the DAi cells in the posterior tubercle (ten Donkelaar et al., 1981; Naujoks-Manteuffel & Manteuffel, 1988). Furthermore, the DAi CSF-contacting cells adjacent to the central canal certainly contribute to the DA innervation of the spinal cord of amphibians.

Dense plexuses of DAi and NAi fibers are present in the intermediate lobe of the hypophysis of *Xenopus* (González et al., 1993; González & Smeets, 1993), and, probably, also of other amphibians (cf. Prasada Rao & Hartwig, 1974). Previous experimental studies have provided evidence for a role of the amphibian hypothalamic periventricular organ in color change mechanisms (van Oordt et al., 1972). This notion was further supported by the demonstration of monoaminergic fibers linking the periventricular organ with the intermediate lobe (Prasada Rao & Hartwig, 1974). However, the present survey shows that the cells of the hypothalamic periventricular organ of amphibians indeed contain DA and NA but lack the enzymes TH and DBH. Since the intermediate lobe also contains dense THi and DBHi plexuses, it seems unlikely that the TH- and DBH-immunonegative cells of the organ are the source of the catecholaminergic innervation of the hypophysis.

Recently, electron microsopical studies have provided evidence for the coexistence of DA, GABA and NPY (de Rijk, van Strien & Roubos, 1992) or TH, GABA and NPY (Tonon et al., 1992) in varicosities of fibers that innervate the intermediate lobe of anurans. The presence of cell bodies containing TH/DA, GABA and NPY in the suprachiasmatic nucleus of anurans makes the latter nucleus a good candidate to be the origin of these fibers (Tuinhof et al., 1992). The immunohistochemical demonstration of a noradrenergic innervation of the intermediate lobe is in agreement with results of previous receptor studies which suggest that not only dopamine but also noradrenaline and adrenaline are involved in the control of background adaptation (Verburg-van Kemenade et al., 1986; de Koning, Jenks & Roubos 1992). The present finding of a dense plexus of NAi fibers in the intermediate lobe of the hypophysis provides anatomical support of such an involvement. Most likely, the DBHi/NAi fibers to the intermediate lobe originate from the locus coeruleus. This notion is corroborated by the presence of NPY immunoreactive cell bodies in the isthmic region of anurans, in a position similar to that of the NAi perikarya (Danger et al., 1985; Cailliez et al., 1987). However, the possible coexistence of NA and NPY in the locus coeruleus as well as in the nerve terminals of the intermediate lobe of amphibians awaits confirmation by double labeling techniques.

Concluding remarks

From this survey it is clear that, in contrast to what has been suggested previously (Parent et al., 1984), both anurans and urodeles possess distinct dopaminergic cell groups in the midbrain tegmentum. There seems to be no direct relationship between the appearance of a midbrain DA cell group and the transition to a terrestrial lifestyle. However, remarkable differences exist in the organization of the midbrain cell group (paired or unpaired) as well as in the DA innervation of striatal structures. As mentioned before, only detailed studies with combined immunohistochemical and tract-tracing techniques may shed more light upon the significance of these differences in mesotelencephalic pathways.

A common feature of all amphibians studied so far is that THi cells occur in the olfactory bulb which do not stain with antisera against DA, DBH, NA or PNMT. Similar observations have been made in the olfactory bulb of reptiles (Smeets & Steinbusch, 1990). Since these cells are immunoreactive for TH but not for DA or DBH, they contain, most likely, L-DOPA as end product (see also Chapter 6). Another general feature of catecholamine systems in amphibians is the presence of CSF-contacting cells in the hypothalamic periventricular organ that are immunopositive for DA and NA antisera, but immunonegative for TH and DBH antisera. The notion that these cells accumulate rather than metabolize the catecholamines is supported by the results of an earlier study by Nakai, Ochiai & Shioda (1977). The latter authors found that, after intraventricular injections, tritiated dopamine was taken up by cells with processes that contact the CSF (for further details, see Chapter 6).

In most previous studies of catecholamine systems in the CNS of vertebrates, the phylogenetic constancy of these systems has been emphasized. However, this survey of catecholamine systems in the brain of amphibians has demonstrated that, within a single class, remarkable differences are observed. It is obvious that, in the near future, the immunohistochemical studies have to be extended to other amphibian species in order to draw more solid conclusions. In particular, studies of the catecholamine systems in the brain of representatives of the third amphibian radiation, i.e. caecilians, are needed. Since they differ in many aspects from anurans and urodeles (being almost blind, burrowing, lacking legs, and displaying external segmentation of the body), caecilians may be of special interest for studying primitive and derived features of the amphibian catecholamine systems.

Finally, it has become also evident that our knowledge of the noradrenaline and adrenaline system is limited. In this regard, it is worth mentioning that the concentration of adrenaline has been reported to be considerably higher in amphibians than in most other vertebrate species (Cooney, Conaway & Mefford, 1985). Yet, by means of PNMT immunohistochemistry we found only a limited distribution of immunoreactive cell bodies and fibers. The latter finding seems to be contradictory to the biochemically observed predominance of adrenaline, but administration of an inhibitor of PNMT (dichloro-a-methyl-benzylamine) revealed only a limited effect on the adrenaline concentration suggesting that most of the adrenaline is not synthesized within the brain (Cooney et al., 1985).

Acknowledgements

The authors are much indebted to Dr R. Tuinhof for providing the *Xenopus* DA material, Mrs B. Jorritsma-Byham and Ms R. G. P. Wismans for technical assistance, and Mr D. de Jong for preparing the photomicrographs. The study was financially supported by Spanish Research Grant DGICYT PB900628 and by NATO Collaborative Grant CRG 910970.

Abbreviations

A	anterior thalamic nucleus
Acc	nucleus accumbens
Ad	nucleus anterodorsalis tegmenti
al	anterior lobe of the pituitary
AP	area postrema
Apl	amygdala, pars lateralis
Apm	amygdala, pars medialis
Av	nucleus anteroventralis tegmenti
C	central thalamic nucleus
cc	central canal
DB	diagonal band of Broca
Dp	dorsal pallium
gl	glomerular layer of the olfactory bulb
gr	granule cell layer of the olfactory bulb
Hd	dorsal habenula
Hv	ventral habenula
il	intermediate lobe of the pituitary
Ip	nucleus interpeduncularis
Is	nucleus isthmi
La	lateral thalamic nucleus, pars anterior
Lc	locus coeruleus
LH	lateral hypothalamic nucleus
LL	lateral line nucleus
Lp	lateral pallium
Lpv	lateral thalamic nucleus, pars posteroventralis
Ls	lateral septum
ml	mitral cell layer of the olfactory bulb
Mp	medial pallium
Ms	medial septum
nl	neural lobe of the pituitary
NPv	nucleus of the periventricular organ
nV	nervus trigeminus
P	posterior thalamic nucleus
pc	posterior commissure
Pd	nucleus posterodorsalis tegmenti
Poa	anterior preoptic area
Pv	nucleus posteroventralis tegmenti
Ri	nucleus reticularis inferior
Rm	nucleus reticularis medius
SC	nucleus suprachiasmaticus
sol	solitary tract
Str	striatum
tect	tectum mesencephali
tegm	tegmentum mesencephali
Tor	torus semicircularis
TP	tuberculum posterius
v	ventricle
VH	ventral hypothalamic nucleus
VM	ventromedial thalamic nucleus
III	nucleus nervi oculomotorii
Vm	nucleus motorius nervi trigemini
VIIIv	ventral nucleus of octaval nerve
Xm	nucleus motorius nervi vagi

References

Cailliez, D., Danger, J.-M., Andersen, A. C., Polak, J. M., Pelletier, G., Kawamura, K., Kikuyama, S. & Vaudry, H. (1987). Neuropeptide Y (NPY)-like immunoreactive neurons in the brain and pituitary of the amphibian *Rana catesbeiana*. *Zoological Science*, **4**, 123–34.

Carroll, R. L. (1988) *Vertebrate Paleontology and Evolution*. New York: Freeman.

Cooney, M. M., Conaway, C. H. & Mefford, I. N. (1985). Epinephrine, norepinephrine and dopamine concentrations in amphibian brain. *Comparative Biochemistry and Physiology*, **82C**, 395–7.

Corio, M. & Doerr-Schott, J. (1988). The monoaminergic system in the diencephalon of the newt tadpole. *Triturus alpestris*. A histofluorescence study. *Journal für Hirnforschung*, **29**, 377–84.

Corio, M., Thibault, J. & Peute, J. (1990). Topographical relationships between catecholamine- and neuropeptide-containing fibers in the median eminence of the newt, *Triturus alpestris*. An ultrastructural immunocytochemical study. *Cell and Tissue Research*, **259**, 561–6.

Corio, M. Thibault, J. & Peute, J. (1992). Distribution of catecholaminergic and serotoninergic systems in forebrain and midbrain of the newt. *Triturus alpestris* (Urodela). *Cell and Tissue Research*, **268**, 377–87.

Danger, J. M., Guy, J., Benyaminia, M., Jegou, S., Leboulenger, F., Cote, J., Tonon, M. C., Pelletier, G. & Vaudry, H. (1985). Localization and identification of neuropeptide Y (NPY)-like immunoreactivity in the frog brain. *Peptides*, **6**, 1225–36.

de Koning, H. P., Jenks, B. G. & Roubos, E. W. (1992). Evidence for three different catecholamine receptors on pituitary melanotrope cells of *Xenopus laevis*. *Abstracts 7th International Catecholamine Symposium*, Amsterdam, p.68.

de Rijk, E. P. C. T., van Strien, F. J. C. & Roubos, E. W. (1992). Demonstration of coexisting catecholamine (dopamine), amino acid (GABA), and peptide (NPY) involved in inhibition of melanotrope cell activity in *Xenopus laevis*: a quantitative ultrastructural, freeze-substitution immunocytochemical study. *Journal of Neuroscience*, **12**, 864–71.

Dubé, L., Clairambault, P. & Malacarne, G. (1990). Striatal afferents in the newt *Triturus cristatus*. *Brain, Behavior and Evolution*, **35**, 212–26.

Duellman, W. E. & Trueb, L. (1986). *Biology of Amphibians*. New York: McGraw-Hill.

Falck, B., Hillarp, N. A., Thieme, G. & Thorp, A. (1962). Fluorescence of catecholamines and related compounds with formaldehyde. *Journal of Histochemistry and Cytochemistry*, **10**, 348–54.

Feng, A. S. (1986). Afferent and efferent innervation patterns of the superior olivary nucleus of the leopard frog. *Brain Research*, **364**, 167–71.

Finkenstädt, Th., Ebbesson, S. O. E. & Ewert, J.-P. (1983). Projections to the midbrain tectum in *Salamandra salamandra* L. *Cell and Tissue Research*, **234**, 39–55.

Franzoni, M. F., Thibault, J., Fasolo, A., Martinoli, M. G., Scanari, F. & Calas, A. (1986). Organization of tyrosine-hydroxylase immunopositive neurons in the brain of the crested newt, *Triturus cristatus* carnifex. *Journal of Comparative Neurology*, **251**, 121–34.

Geffard, M., Buijs, R. M., Seguela, P., Pool, C. W. & Le Moal, M. (1984). First demonstration of highly specific and sensitive antibodies against dopamine. *Brain Research*, **294**, 161–5.

González, A. & Smeets, W. J. A. J. (1991). Comparative analysis of dopamine and tyrosine hydroxylase immunoreactivities in the brain of two amphibians, the anuran *Rana ridibunda* and the urodele *Pleurodeles waltlii*. *Journal of Comparative Neurology*, **303**, 457–77.

González, A. & Smeets, W. J. A. J. (1993). Noradrenaline in the brain of the South African clawed frog *Xenopus laevis*. A study with antibodies against noradrenaline and dopamine-β-hydroxylase. *Journal of Comparative Neurology*, **331**, 363–74.

González, A., Tuinhof, R. & Smeets, W. J. A. J. (1993). Distribution of tyrosine hydroxylase- and dopamine-immunoreactivities in the brain of the South African clawed frog *Xenopus laevis*. *Anatomy and Embryology*, **187**, 193–201.

Jarvik, E. (1960). *Théories de l'Evolution des Vertébrés, Reconsiderées à la Lumière des Récentes Découvertes sur les Vertébrés Inferieurs*. Masson, Paris.

Kemali, M. & Guglielmotti, V. (1987). A horseradish peroxidase study of the olfactory system of the frog, *Rana esculenta*. *Journal of Comparative Neurology*, **263**, 400–17.

Kicliter, E. (1979). Some telencephalic connections in the frog, *Rana pipiens*. *Journal of Comparative Neurology*, **185**, 75–86.

Lamas, J., Rodicio, C., Caruncho, H. & Anadon, R. (1988). Monoaminergic systems of the hypothalamus of ten amphibian species: a histofluorescence study. *Journal für Hirnforschung*, **3**, 289–97.

Lázár, G. & Kozicz, T. (1990). Morphology of neurons and axon terminals associated with descending and ascending pathways of the lateral forebrain bundle in *Rana esculenta*. *Cell and Tissue Research*, **260**, 535–48.

Nakai, Y., Ochiai, H. & Shioda, S. (1977). Cytological evidence for different types of cerebrospinal fluid-contacting subependymal cells in the preoptic and infundibular recesses of the frog. *Cell and Tissue Research*, **176**, 317–34.

Naujoks-Manteuffel, C. & Manteuffel, G. (1988). Origins of descending projections to the medulla oblongata and rostral medulla spinalis in the urodele *Salamandra salamandra* (Amphibia). *Journal of Comparative Neurology*, **273**, 187–206.

Parent, A., Poitras, D. & Dubé, L. (1984). Comparative anatomy of central monoaminergic systems. In *Handbook of Chemical Neuroanatomy, Vol. 2. Classic Transmitters in the CNS, Part I*, ed. A. Björklund & T. Hökfelt, pp.409–39. Amsterdam: Elsevier.

Parsons, T. S. & Williams, E. E. (1963). The relationships of the modern amphibia. *Quarterly Review of Biology*, **38**, 26–53.

Prasada Rao, P. D. & Hartwig, H. G. (1974). Monoaminergic tracts of the diencephalon and innervation of the pars intermedia in *Rana temporaria*. A fluorescence and microspectrofluorimetric study. *Cell and Tissue Research*, **151**, 1–26.

Rettig, G. (1988). Connections of the tectum opticum in two urodeles, *Salamandra salamandra* and *Bolitoglossa subpalmata*, with special reference to the nucleus isthmi. *Journal für Hirnforschung*, **29**, 5–16.

Romer, A. S. & Parsons, T. S. (1977). *The Vertebrate Body*. Philadelphia, Saunders.

Sims, J. (1986). Identification of a second type of catecholaminergic neuron in the spinal cord of the axolotl salamander. *Experimental Neurology*, **93**, 428–33.

Smeets, W. J. A. J. & González, A. (1990). Are putative dopamine-accumulating cell bodies in the hypothalamic periventricular organ a primitive brain character of nonmammalian vertebrates? *Neuroscience Letters*, **114**, 248–52.

Smeets, W. J. A. J. & Steinbusch, H. W. M. (1989). Distribution of noradrenaline immunoreactivity in the forebrain and midbrain of the lizard *Gekko gecko*. *Journal of Comparative Neurology*, **285**, 453–66.

Smeets, W. J. A. J. & Steinbusch, H. W. M. (1990). New insights into the reptilian catecholaminergic systems as revealed by antibodies against the neurotransmitters and their synthetic enzymes. *Journal of Chemical Neuroanatomy*, **3**, 25–43.

Steinbusch, H. W. M., Verhofstad, A. A. J., Penke, B., Varga, J. & Joosten, H. W. J. (1981). Immunohistochemical characterization of monoamine-containing neurons in the central nervous system by antibodies to serotonin and noradrenaline. A study in the rat and the lamprey. *Acta Histochemica*, suppl. **24**, 107–22.

Steinbusch, H. W. M. & Tilders, F. (1987). Immunohistochemical techniques for light-microscopical localization of dopamine, noradrenaline, adrenaline, serotonin and histamine in the central nervous system. In *Methods in Neurosciences, Vol. 10, Monoaminergic Neurons: Light Microscopy and Ultrastructure*, ed. Steinbusch, H. W. M. pp. 125–66. Chichester: Wiley.

ten Donkelaar, H. J., de Boer-van Huizen, R., Schouten, F. T. M. & Eggen, S. J. H. (1981). Cells of origin of descending pathways to the spinal cord in the clawed toad (*Xenopus laevis*). *Neuroscience*, **6**, 2297–312.

Tohyama, M., Yamamoto, K., Satoh, K., Sakumoto, T. & Shimizu, T. (1977). Catecholamine innervation of the forebrain in the bullfrog, *Rana catesbiana*. *Journal für Hirnforschung*, **18**, 223–8.

Tonon, M. C., Bosler, O., Stoeckel, M. E., Pelletier, G., Tappaz, M. & Vaudry, H. (1992). Co-localization of tyrosine hydroxylase, GABA and neuropeptide Y within axon terminals innervating the intermediate lobe of the frog *Rana ridibunda*. *Journal of Comparative Neurology*, **319**, 599–605.

Tóth, P., Csank, G. & Lázár, G. (1985). Morphology of the cells of origin of descending pathways to the spinal cord in *Rana esculenta*. A tracing study using cobaltic-lysine complex. *Journal für Hirnforschung*, **26**, 365–83.

Tuinhof, R., de Rijk, E. P. C. T., van Strien, F. J. C., Wismans, R. G. P., Smeets, W. J. A. J. & Roubos, E. W. (1992). Role of coexisting dopamine, GABA and NPY in the control of background adaptation of the amphibian *Xenopus laevis*. *Abstracts 7th International Catecholamine Symposium*, Amsterdam, p.319.

van Oordt, P. G. W. J., Goos, H. J. T., Peute, J. & Terlou, M. (1972). Hypothalamo-hypophysial relations in amphibian larvae. *General Comparative Endocrinology*, Suppl. **3**, 41–50.

Verburg-van Kemenade, B. M. L., Tonon, M. C., Jenks, B. G. & Vaudry, H. (1986). Characteristics of receptors for dopamine in the pars intermedia of the amphibian *Xenopus laevis*. *Neuroendocrinology*, **44**, 446–56.

von Bartheld, C. S. & Meyer, D. L. (1990). Paraventricular organ of the lungfish *Protopterus dolloi*: morphology and projections of CSF-contacting neurons. *Journal of Comparative Neurology*, **297**, 410–34.

Wicht, H. & Himstedt, W. (1988). Topologic and connectional analysis of the dorsal thalamus of *Triturus alpestris* (Amphibia, Urodela, Salamandridae). *Journal of Comparative Neurology*, **267**, 545–61.

Wilczynski, W. & Northcutt, R. G. (1977). Afferents to the optic tectum of the leopard frog: an HRP study. *Journal of Comparative Neurology*, **173**, 219–30.

Wilczynski, W. & Northcutt, R. G. (1983). Connections of the bullfrog striatum: afferent organization. *Journal of Comparative Neurology*, **214**, 321–32.

Will, U., Luhede, G. & Görner, P. (1985a). The area octavolateralis in *Xenopus laevis*. I. the primary afferent projections. *Cell and Tissue Research*, **239**, 147–61.

Will, U., Luhede, G. & Görner, P. (1985b). The area octavolateralis in *Xenopus laevis*. II. Second order projections and cytoarchitecture. *Cell and Tissue Research*, **239**, 163–75.

Yoshida, M., Nagatsu, I., Kondo, Y., Karasawa, N., Ohno, T, Spatz, M. & Nagatsu, T. (1983). Immunohistochemical localization of the neurons containing catecholamine-synthesizing enzymes and serotonin in the brain of the bullfrog (*Rana catesbeiana*). *Acta Histochemica Cytochemica*, **16**, 245–58.

Zittlau, K. E., Claas, B. & Munz, H. (1988). Horseradish peroxidase study of tectal afferents in *Xenopus laevis* with special emphasis on their relationship to the lateral-line system. *Brain, Behavior and Evolution*, **32**, 208–19.

6

Catecholamine systems in the CNS of reptiles: structure and functional correlations

W. J. A. J. Smeets

Introduction

From a comparative point of view, reptiles are of particular interest for studying brain evolution, since they are believed to be ancestral to both birds and mammals. The first fossil records of reptiles are dated from the Upper Carboniferous period, about 260 million years ago. Several distinct branches were already present at that time suggesting that reptilian evolution was well under way. As the first terrestrial vertebrates largely independent of water, reptiles conquered all habitats on earth and flourished principally in the Mesozoic era. Before the beginning of the Tertiary Period, most branches of the complex reptilian evolutionary tree died out. Of the ancient reptiles lacking temporal openings in the skull, the subclass Anapsida, only about 200 species of turtles and tortoises (Order Chelonia or Testudinata) remain. About 20 species of crocodilians (Order Crocodilia) are all that is left of the subclass Archosauria that once contained the dinosaurs. A third extant subclass, the Lepidosauria, is today represented by two orders, viz. Rhynchocephalia and Squamata. Of the Rhynchocephalia, one single species has survived, the tuatara (Sphenodon) of New Zealand, but members of the Order Squamata (lizards, snakes, amphisbaenians) have found numerous niches in which, largely protected from the competition of the mammals, they could flourish again with about 5000 species at present.

Crocodilians and turtles have received special attention because the former were believed to be most closely related to the reptilian stem from which birds evolved, whereas turtles were thought to have the closest relation to the stem reptiles that gave rise to mammals. However, as has recently been pointed out again by Ulinski (1990), the idea of a close relationship between turtles and mammals is based upon a misunderstanding of evolutionary relationships since the lines leading to mammals had separated off before the first turtles appeared in the fossil record. Therefore, as has been stressed time after time (see e.g. Ebbesson, 1984; Northcutt, 1984; Smeets, 1988a, b; Brauth, 1990; Northcutt, 1990; Ulinski, 1990), a better strategy to study the evolution of the brain is by carefully analyzing several selected characters in a number of properly chosen representatives of each vertebrate class. In this chapter, such an approach has been taken to study the catecholamine systems in the brain of reptiles.

A great number of studies using the formaldehyde-induced fluorescence (FIF) technique as developed by Falck et al. (1962), has laid the basis of our understanding of the catecholaminergic systems of reptiles (for reviews, see Parent, 1979; Parent, Poitras & Dubé, 1984). With this technique, however, it is difficult to discriminate between the various catecholamines, i.e. dopamine, noradrenaline and adrenaline. With the introduction of immunohistochemical techniques by means of antibodies raised against the catecholamine-synthesizing enzymes of mammals, about 20 years ago (see e.g. Goldstein et al., 1971; Hökfelt et al., 1973), new impetus was given to research on catecholamines in vertebrates. Whereas such studies have extended our knowledge of the catecholamine systems in mammals, their usefulness in nonmammalian vertebrates varies considerably. In reptiles, for example, antibodies against tyrosine hydroxylase (TH) have been successfully applied to the brains of lizards (Wolters, ten Donkelaar & Verhofstad, 1984; Smeets & Steinbusch, 1990; Lopez et al., 1992; Medina & Smeets, 1992), turtles

(Halasz et al., 1982; Brauth et al., 1983; Smeets & Steinbusch, 1990; Kiehn, Rostrup & Moller, 1992), and crocodiles (Brauth et al., 1983; Brauth, 1988). In contrast, the interspecies cross-reactivities of dopamine-β-hydroxylase (DBH) and phenylethanolamine-N-methyltransferase (PNMT) antibodies are poor and studies dealing with the distribution of these enzymes in brains of reptiles limited (Smeets & Steinbusch, 1989; Smeets & Jonker, 1990). However, the development of antibodies against dopamine (DA) conjugated to a carrier protein (Geffard et al., 1984; Steinbusch, de Vente & Schipper, 1986; Steinbusch & Tilders, 1987) and against noradrenaline (NA) conjugated likewise (Steinbusch et al., 1981; Steinbusch & Tilders, 1987) now enable the specific demonstration of dopamine and noradrenaline in the brains of mammalian as well as non-mammalian vertebrates.

In this chapter, a survey is given of the distribution of the various catecholamines in the brain of reptiles. Most of the studies have been carried out in the Tokay gekko, *Gekko gecko*, a lizard species that has been, and still is, used in our department as a model to study brain organization in reptiles. Immunohistochemical (Smeets, Hoogland & Voorn, 1986b; Russchen, Smeets & Hoogland, 1987a; Smeets & Steinbusch, 1988, 1989, 1990; Hoogland & Vermeulen-van der Zee, 1990), hodological (Russchen, Smeets & Lohman, 1987b; Russchen & Jonker, 1988; Hoogland & Vermeulen-van der Zee, 1989; Smeets, 1992), as well as pharmacological studies (Stoof et al., 1987; Henselmans, Hoogland & Stoof, 1991) have contributed to a better understanding of brain organization in this lizard species. Parallel immunohistochemical studies performed in other lizard species, turtles and snakes, have further contributed to our understanding of general and derived features of the catecholaminergic systems in reptiles.

Distribution of catecholamines in the brain of reptiles: theme and variations

For clarity, first a detailed description of the distribution of catecholaminergic perikarya and fibers in the lizard *Gekko gecko* is given. Subsequently, the distribution patterns as observed in other reptiles will be compared with the pattern observed in *Gekko* and an attempt will be made to determine general and derived conditions for each catecholamine system in reptiles.

The nomenclature used in this survey for the forebrain and midbrain is the same as that described by Smeets, Hoogland & Lohman (1986a) and Smeets, Jonker & Hoogland, 1987). The nomenclature for the brainstem is taken from ten Donkelaar et al. (1987).

Dopamine
Since the development of antibodies raised against dopamine (DA), several studies have been carried out to demonstrate selectively the DA system in reptiles (Smeets et al., 1986b, 1987; Smeets, 1988a, b; Smeets & Steinbusch, 1990). In all reptiles studied, the DA antibodies revealed an excellent staining of fibers. Cell bodies are also well stained with the DA antiserum, but comparison with adjacent sections stained with a TH antiserum revealed a generally better staining of dendrites with the latter antiserum.

Cell bodies In the lizard *G. gecko*, the most rostral DAi/THi cells are found in the olfactory bulbs. They have a round to oval shape and show short processes. Most of the cells surround the glomeruli, but a few cells lie at some distance from the glomeruli in the external plexiform layer. In the telencephalon proper, DAi cell bodies were never observed.

Numerous DAi cells are present in the diencephalon, where they constitute six cell groups (Fig. 6.1). The most rostral DAi/THi group lies immediately caudodorsal to the optic chiasm (Fig. 6.2). Some of its cells lie in the most caudal portion of the periventricular preoptic area as delineated by Smeets et al. (1986a), but the majority is found in the ventral part of the periventricular hypothalamic nucleus arching around the ventral tip of the third ventricle. Slightly further caudally, the bi- and tripolar neurons of this DAi/THi cell group are arranged in a row parallel to the ventricular surface (Fig. 6.3) giving off fibers that course ventrolaterally toward the median eminence. At more caudal levels, these cells extend to a second group of DAi/THi cell bodies that are found in the dorsal part of the periventricular hypothalamic nucleus (Figs. 6.1(*e*), 6.3). The caudal portion of the latter DAi/THi cell group lies lateral to the periventricular organ and consists of cells that have predominantly mediolaterally directed processes (Figs. 6.1(*f*), 6.22). The third group of DAi (but TH immunonegative) cells is located in the periventricular organ (Figs. 6.1(*f*), 6.22). The cells are arranged in several cell layers and seem to be in direct contact with the cerebrospinal fluid (CSF) of the third ventricle by means of short, clublike processes. Another group of DAi (but TH immunonegative) cell bodies is found in the ependymal layer of the infundibular recess. The cells closely resemble the DAi cells of the periventricular organ in having short, clublike processes that contact

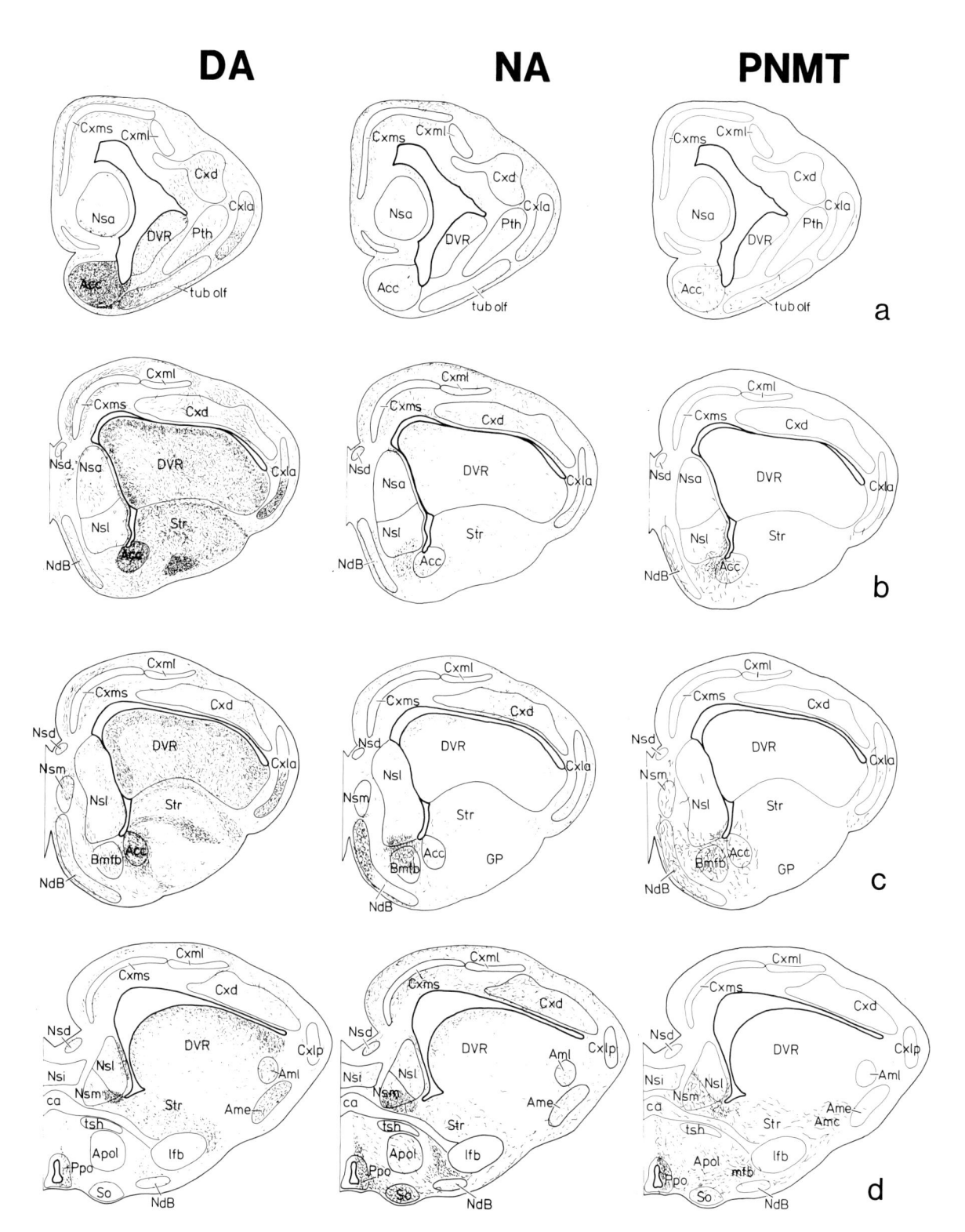

Fig. 6.1. Drawings of a series of transverse hemisections through the brain of the lizard *Gekko gecko* showing the distribution of catecholaminergic cell bodies (large dots) and fibers (small dots, wavy lines) in three panels: immunoreactivity for dopamine to the left, noradrenaline in the middle, and PNMT to the right (modified after Smeets *et al.*, 1986b; Smeets & Steinbusch, 1989; Smeets & Jonker, 1990).

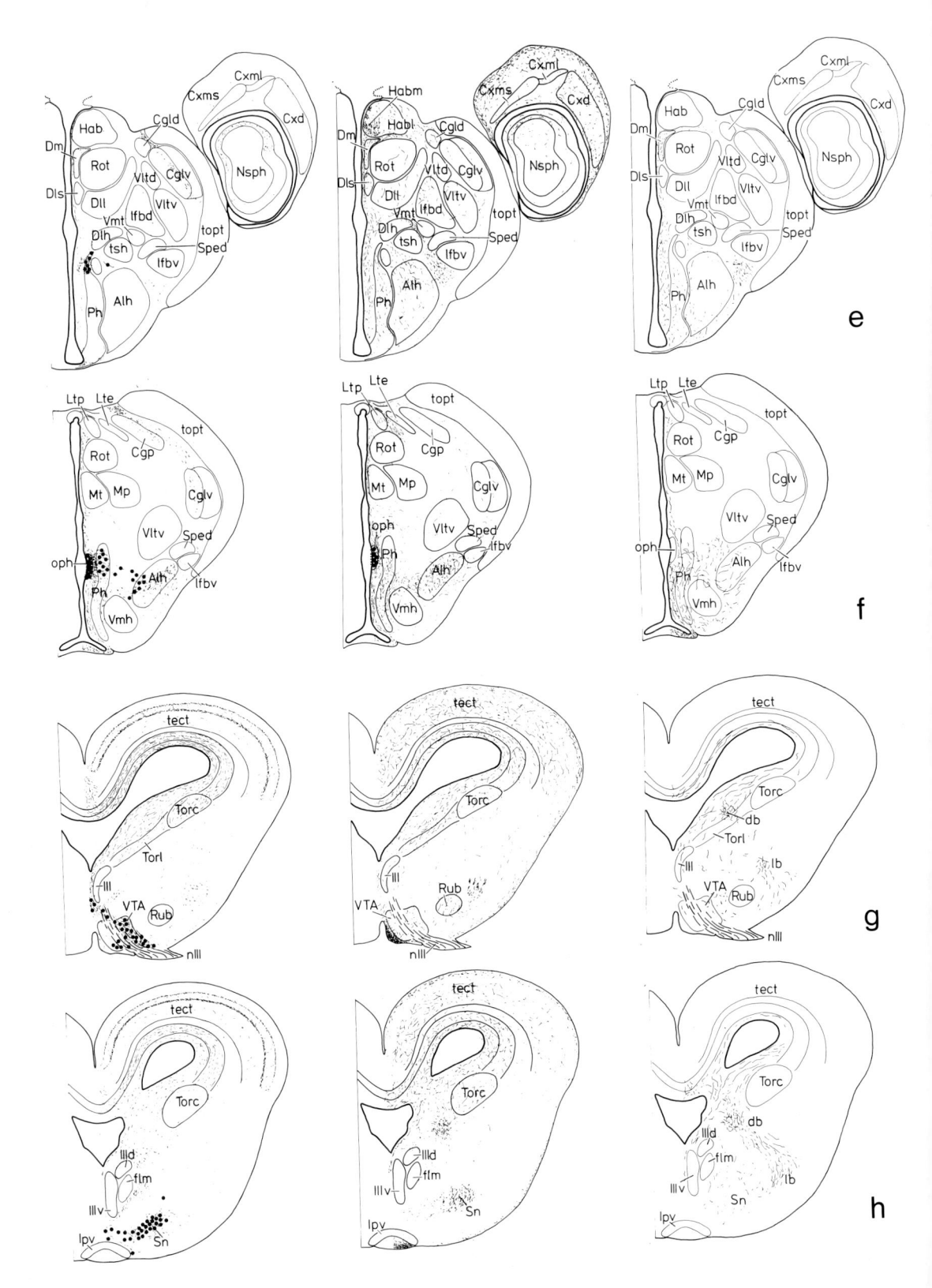

Fig. 6.1. (*cont.*)

Catecholamine systems in CNS of reptiles 107

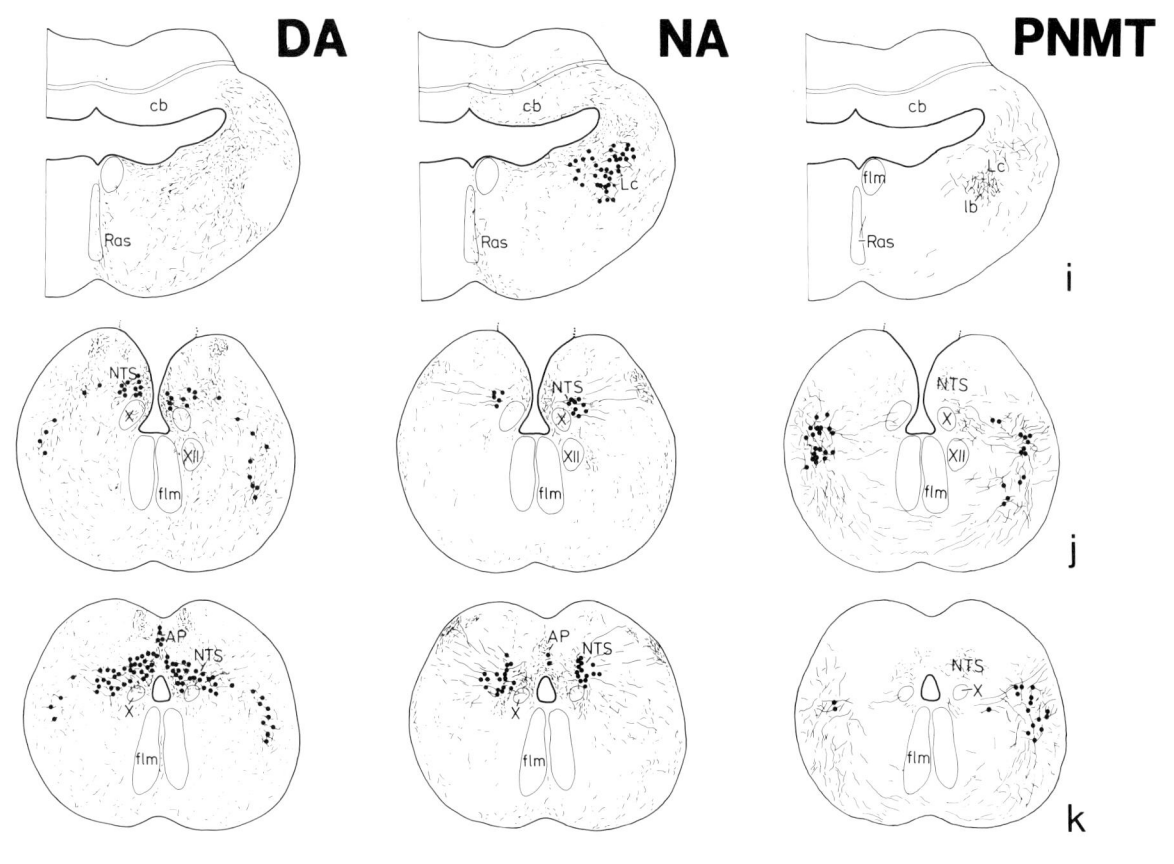

Fig. 6.1. (*cont.*)

the CSF. Apart from these four periventricular DAi cell groups, another group of DAi/THi cell group is found in the lateral hypothalamic area at the level of the periventricular organ (Figs. 6.1(*f*), 6.22). With regard to cell shape and orientation of cell processes these DAi/THi cells resemble those of the caudal portion of the dorsal periventricular hypothalamic nucleus. A sixth group of DAi/THi cells is found in the pretectal posterodorsal nucleus (Fig. 6.4). Compared with the DAi/THi cells in the hypothalamic areas, these cells are considerably smaller in size, stain less intensely, and show only short processes.

The majority of the DAi/THi cells in the brain of *Gekko* are located in in the mesencephalic tegmentum (Fig. 6.1(*g*), (*h*)). On the basis of their position and the direction of their processes, the mesencephalic DAi/THi cells have been divided into three groups: ventral tegmental area (VTA), substantia nigra, pars compacta (SNc), and the RA8 group (the presumed reptilian homologue of the retrorubral area of mammals). The most rostral group, the VTA, consists of cells that generally do not show long processes. Rostrally they have a ventromedial position, whereas caudally they intermingle with the oculomotor nerve (Fig. 6.1(*g*)). Further caudally the DAi/THi cells of the VTA are found adjacent to the interpeduncular nucleus medially, whereas caudolaterally the cells are more or less continuous with those of the SNc (Figs. 6.1(*h*), 6.6). The latter group contains DAi/THi cells that can be distinguished from those in the VTA primarily by their long and predominantly medio-laterally oriented processes. At caudal midbrain levels the DAi/THi cells of the SNc merge with those of the RA8 group. The cells of the latter DAi/THi cell group extend caudally to the isthmic region, occupying a dorsal position, lateral to the trochlear nucleus. Contrary to the cells in the SNc, the cells of the RA8 group have only short processes which have a random orientation.

At caudal brainstem levels, DAi/THi cells are found in the lateral part of the reticular formation, the nucleus of the solitary tract and the area postrema

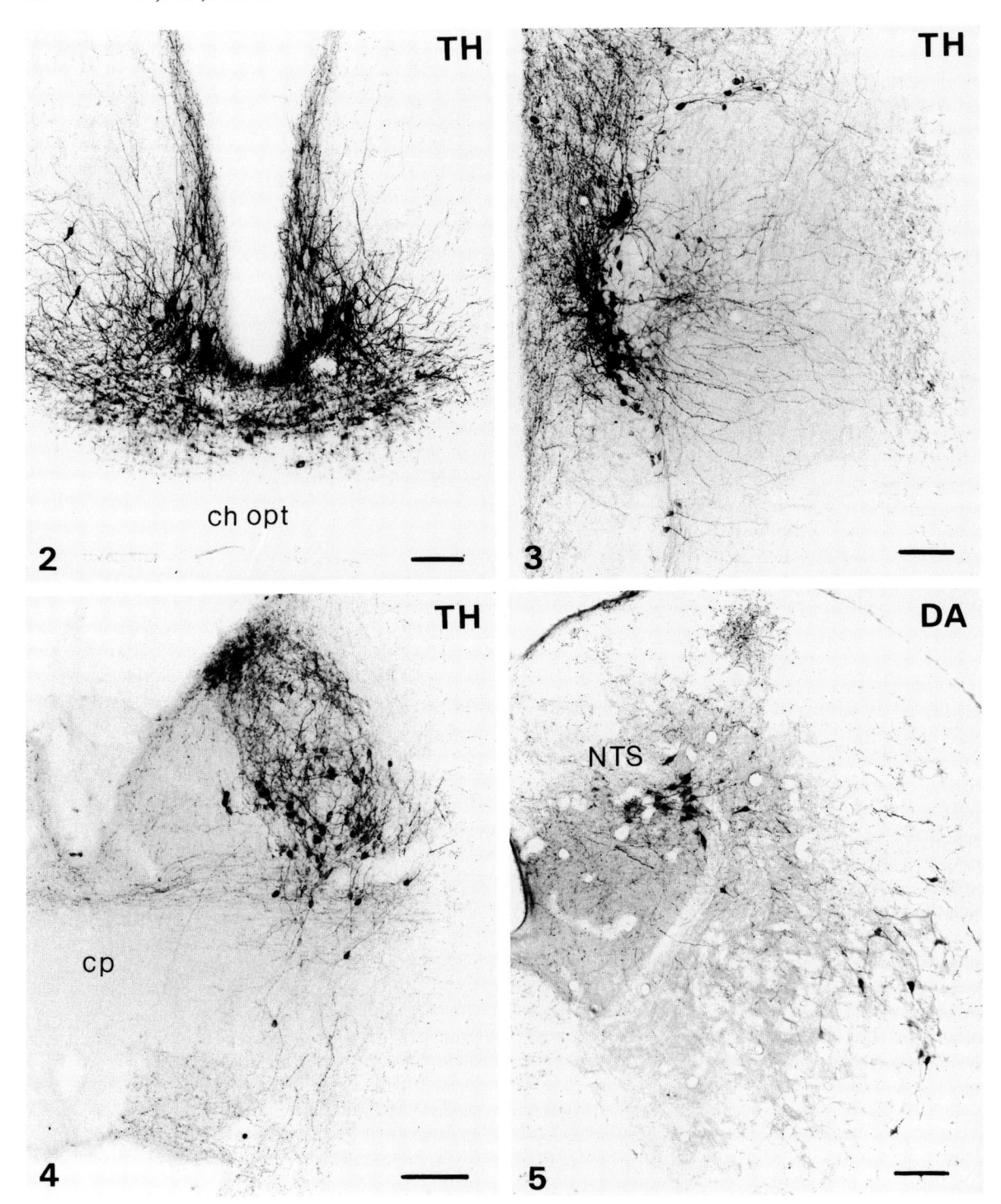

Figs. 6.2–6.5. Photomicrographs of transverse sections through the brain of the lizard *Gekko gecko* showing DAi/THi cell bodies at the transition of the preoptic area to the ventral part of the periventricular hypothalamic nucleus (2), in the periventricular hypothalamic nucleus at slightly more caudal levels (3), in the pretectal posterodorsal nucleus (4), and at the level of the solitary tract nucleus (5). Bars = 0.1 mm.

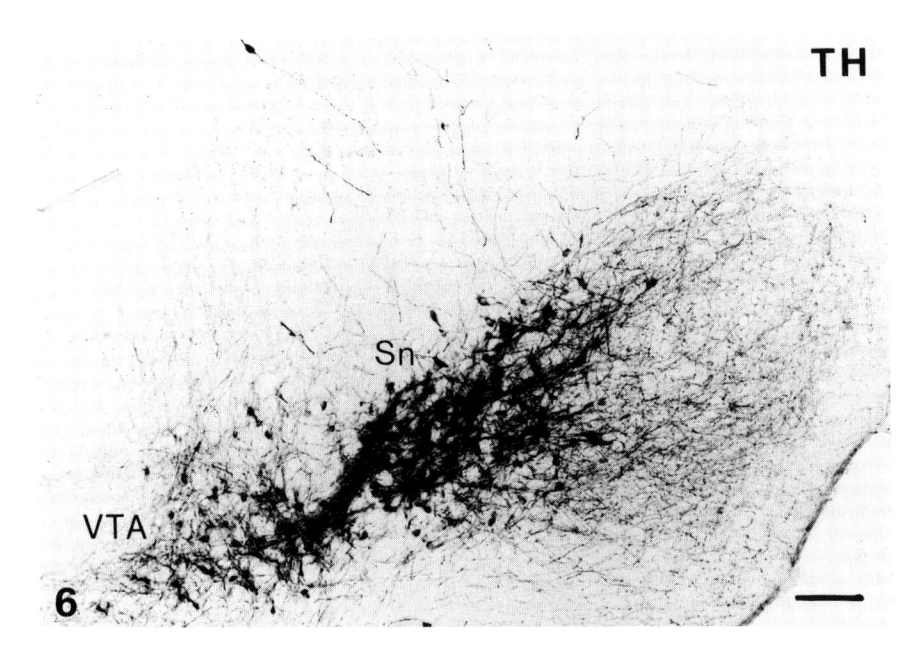

Fig. 6.6. Photomicrograph of a transverse section through the midbrain tegmentum of the lizard *G.gecko* showing the THi cell bodies in the ventral tegmental area (VTA) and substantia nigra (Sn). Note the medial to lateral orientation of the processes of the cells in the Sn. Bar = 0.1 mm.

(Figs. 6.1(*j*), (*k*), 6.5). The cells in the lateral reticular formation have processes that run predominantly parallel to the meningeal surface, whereas the DAi/THi cells ventromedial to the solitary tract have dendrites with a mainly mediolateral orientation. At the level of the obex, the DAi/THi cells of the latter group of both sides merge into the area postrema which lies along the midline dorsal to the ventricle. The morphology of the DAi/THi cells in the area postrema differs from that of the solitary tract group in having mainly dorsoventrally oriented cell processes.

Throughout the spinal cord, DAi/THi cells are found lying ventral to the central canal. The cells are bipolar and are characterized by a fairly thick, straight process of variable length, that contacts the CSF and another thin and often branching process that is directed ventralwards. The number of these cells varies from 1–3 in 40 µm thick, transverse sections.

Fibers In the olfactory bulbs of *Gekko gecko*, DAi/THi fibers are predominantly confined to the internal granular layer, where they course parallel to the ventricular surface. A few DAi/THi fibers are found in the internal plexiform layer. Although the glomerular layer shows a high immunoreactivity, no distinct fibers or terminals can be recognized, except for the short processes of the periglomerular cells.

Dopaminergic fibers and varicosities are observed throughout the telencephalon proper, with the highest density in its ventral part (Fig 6.1(*a*)–(*d*). The olfactory tubercle contains numerous DAi/THi fibers in its medial portion, whereas its lateral portion is only sparsely innervated. The most densely innervated area in the basal forebrain is the nucleus accumbens (Fig. 6.1(*a*), (*b*)) which is characterized by a more dense DAi/THi plexus in its rostral than in its caudal part. Two other structures in the ventral portion of the telencephalon received a strong DA innervation. These are the striatum and the dorsal ventricular ridge (DVR). In the most rostral portion of the striatum the distribution of DAi/THi fibers has a patchy appearance (Fig. 6.1(*b*)). More caudally the striatum shows a zonal rather than a patchy pattern of DAi/THi fibers (Fig. 6.1(*c*)). The large-celled portion of the striatum, i.e. the globus pallidus, is almost devoid of fibers, although baskets of DAi/THi fibers surrounding the soma and proximal dendrites of some large cells can be recognized (Fig. 6.1(*c*)). The entire caudal pole of the striatum is densely innervated by DAi/THi fibers without a zonal or patchy pattern (Fig. 6.1.(*d*). The DA innervation of the DVR shows the following features (Fig. 6.1(*a*)–(*d*)): (1) the superficial zone contains considerably more DAi/THi fibers than the central core; (2) the lateral half of the

DVR shows a more dense innervation than its medial half; and (3) there is a decrease in the density of DAi/THi fibers in rostrocaudal direction, in particular with respect to the central core.

In the amygdaloid complex a DAi/THi plexus is found in the lateral and external amygdaloid nuclei and in the nucleus sphericus (Fig. 6.1(d),(e)). In the latter nucleus many DAi/THi fibers are found surrounding the cells in the dorsal half of the nucleus, whereas the ventral half receives only a sparse DA innervation. Only a few fibers are present in the central core of the nucleus.

The anterior and lateral septal nuclei of *Gekko* are characterized by baskets of DAi/THi fibers surrounding unstained somata and proximal processes (Figs. 6.1(a)–(c), 6.7). Many DAi/THi fibers were further observed in the medial septal nucleus, whereas the unpaired (impar) and dorsal septal nuclei contain only a small number of fibers. In the ventromedial wall of the hemisphere a moderate number of DAi/THi fibers is located in the nucleus of the diagonal band of Broca and in the bed nucleus of the medial forebrain bundle.

Compared to basal telencephalic areas, the cortical areas are only weakly innervated by DAi/THi fibers, except for the lateral cortex. The ventral part of the anterior subdivision of the lateral cortex contains a dense plexus of DAi/THi varicosities in its cell layer (Fig. 6.1(a)–(c)). The dorsal part of the anterior lateral cortex and the entire posterior lateral cortex, on the contrary, are sparsely innervated by DAi/THi fibers. A moderate number of DAi/THi fibers course to the dorsal cortex, where they terminate mainly in the lateral and medial subdivisions (Fig. 6.1(b)–(d)). In the small-celled medial cortex, DAi/THi fibers are located on both sides of the cellular layer. The large-celled part of the medial cortex does not receive DA fibers, although fibers pass through its molecular layer.

In the telencephalon impar, the majority of the DAi fibers are found within the periventricular preoptic area and in an area ventromedial to the lateral forebrain bundle that partly matches the area occupied by the medial forebrain bundle.

In the diencephalon, the habenular nuclei receive a moderate number of fibers, whereas the dorsomedial nucleus and the cell layer of the lateral geniculate body of the thalamus receive a relatively strong DA innervation. The other thalamic nuclei contain only a few DAi/THi fibers. In the hypothalamus, numerous immunoreactive fibers are found in an area ventromedial to the lateral forebrain bundle, in the median eminence, and in the periventricular fiber layer.

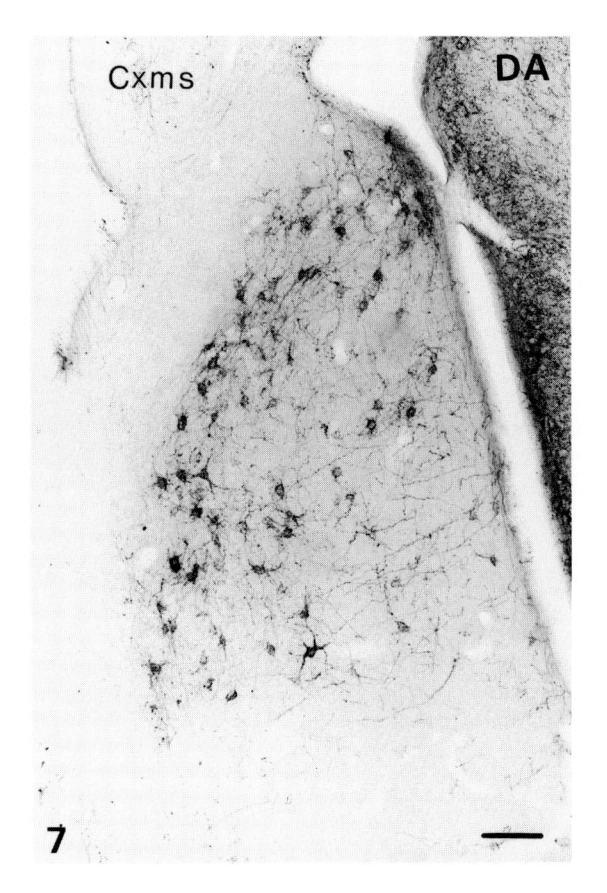

Fig. 6.7. Photomicrograph of a transverse section through the anterior septal nucleus of *G.gecko* showing the baskets of DAi fibers surrounding the somata and proximal dendrites. Bar = 0.1 mm.

In the tectum mesencephali of *Gekko* three distinct laminae of DAi/THi fibers are present: two in the superficial tectal zone and one in the periventricular zone (Fig. 6.1(g), (h)). The most superficial tectal lamina is very thin and matches tectal layer 11 as defined by Northcutt (1978). The other DAi/THi fiber layer in the superficial tectal zone coincides with tectal layer 9, whereas the broad periventricular DAi/THi fiber lamina matches tectal layers 2–5. Laterally the tectal periventricular DA fiber zone is continuous with the periventricular zone of the tegmentum. The rostral part of the central nucleus of the torus semicircularis contains a moderate plexus of DAi/THi fibers (Fig. 6.1(g)), but its caudal part is devoid of immunoreactive fibers (Fig. 6.1(h)).

In the brainstem, DAi/THi fibers are widely, although not uniformly, distributed (Fig. 6.1(i)–(k)). Moderate to dense plexuses of immunoreactive fibers are present in the parabrachial region, locus coeruleus, raphe nuclei, periventricular zone, trigeminal motor and princeps nuclei, reticular formation, and superior olive. Other regions, e.g. cerebellar nuclei, vestibular nuclei and cochlear nuclei receive a less dense DA innervation.

Rather dense plexuses of DAi/THi fibers in the spinal cord of *Gekko* are predominantly located in the dorsal horn of the grey matter. The fibers are not clearly confined to particular areas as defined by Kusuma, ten Donkelaar & Nieuwenhuys (1979), although they show a preference for the areas I and II, the medial part of the dorsal horn and the dorsal part of area X. Only a few DAi/THi fibers were observed in the ventral horn.

Reptilian dopamine system: general and derived conditions

Dopamine containing cell bodies When the distribution of DAi cell bodies in the brain of *Gekko gecko* is compared with that in the brains of other lizards, turtles and snakes, as determined with the same antibodies (Smeets et al., 1987; Smeets, 1988a, b), it is obvious that a basic pattern exists (cf. Figs. 6.1, 6.8, 6.9). For instance, all reptiles studied so far have in common that, in the olfactory bulbs, the majority of DAi cells surround the glomeruli, whereas a smaller number are located in the external plexiform layer (Fig. 6.21). Further, in none of the reptiles studied are DAi cell bodies observed in the telencephalon proper.

The basic features of the diencephalic portion of the DA system include the presence of DA containing cell bodies in both the rostroventral and caudodorsal parts of the periventricular hypothalamic nucleus, the hypothalamic periventricular organ, the dorsal wall of the infundibular recess, the lateral hypothalamic area, and the pretectal posterodorsal nucleus. However, two additional groups of DAi/THi cell bodies have been identified in the diencephalon of turtles and snakes that apparently do not fit into this basic pattern. One group, demonstrated in the turtle *Pseudemys scripta elegans*, consists of cells that lie in the preoptic recess, immediately dorsal to the optic chiasm (Fig. 6.8(*d*)). In lizards and snakes, such a DAi/THi cell group could not be delineated. However, comparable cells might be located dorsal to the caudal pole of the optic chiasm, in the rostral part of the ventral periventricular hypothalamic DAi/THi cell group of Squamata. The second group of diencephalic DAi/THi cells that does not belong to the basic reptilian pattern is a group of large cells that are found only in snakes (Figs. 6.9(*e*), 6.10). The cells are located in an area that, for *Python*, has been previously labeled the magnocellular ventrolateral thalamic nucleus by Molenaar and Fizaan-Oostveen (1980). The intriguing question arises whether this DAi/THi cell group is involved in behavior that is particular to snakes. Finally, it should be noted that, between the various reptiles studied, considerable variation exists in the number of CSF-contacting cells. For example, the number of CSF-contacting DAi cells in the hypothalamic periventricular organ of *Pseudemys* is much larger than that in the corresponding organ of lizards and snakes (cf. Figs. 6.1(*f*), 6.8(*g*), 6.9(*e*), 6(*f*)).

As regards the brainstem, all reptiles studied so far show DAi/THi cell bodies in the ventral tegmental area, the substantia nigra, the reptilian A8 group of the midbrain, and, at caudal brainstem levels, in the lateral part of the reticular formation, the nucleus of the solitary tract and the area postrema. Three groups of DAi/THi cell bodies are not consistently found in the reptiles studied. The first group concerns cell bodies that lie within the periventricular zone of the midbrain. Such cells are found in some lizards, i.e. *Varanus exanthematicus* and *Anolis carolinensis* (Wolters et al., 1984; Lopez et al., 1992; own, unpublished observations). Particularly in *Anolis*, several DAi/THi cell bodies are easily recognized dorsal to the decussating trochlear fibers (Fig. 6.11). In the other lizard species and snakes, such cells were never observed, but in the turtle *Pseudemys*, a few scattered cells were found in the nucleus laminaris of the torus semicircularis. Another group of DAi/THi cells is found only in the isthmic region of snakes and consists of rather large cell bodies rostral and ventrolateral to the locus coeruleus, from which they can easily be distinguished on the basis of cell size, morphology and position (Figs. 6.9(*j*), 6.12). A third group of DAi/THi cells located at the level of the eighth cranial nerve and consisting of small cells that lie close to the ependymal lining of the fourth ventricle, is easily recognized in turtles (Fig. 6.8(*m*)) and snakes (Fig. 6.9(*k*)), but absent in most lizards. Only in *Varanus*, a few DAi/THi cell bodies were observed in a corresponding position. Following the nomenclature of Molenaar (1977), this group has been labeled the prevagal part of the solitary tract nucleus.

Although the other brainstem DAi/THi cell groups are similar in overall distribution, considerable differences in number and appearance of the cells were noted between the various species studied. For example, the number of DAi/THi cell bodies in the midbrain tegmentum of both *Pseudemys* and *Python* is substantially larger than that in the lizards studied (cf. Figs. 6.1, 6.8, 6.9). Moreover, the mesencephalic dopaminergic cells in turtles and lizards have been divided into three groups (VTA, SNc,

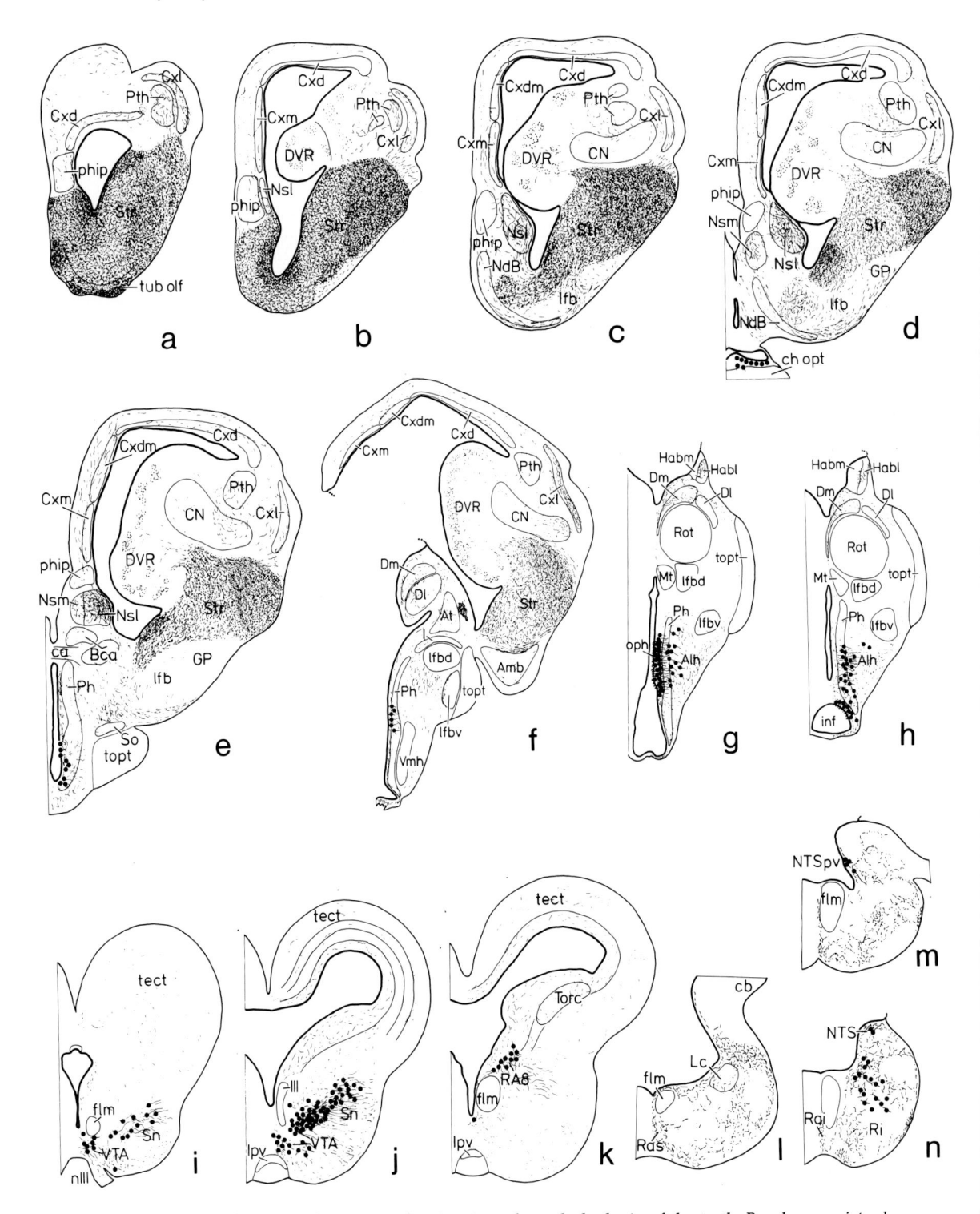

Fig. 6.8. Drawings of a series of transverse hemisections through the brain of the turtle *Pseudemys scripta elegans* showing the distribution of DAi cell bodies (large dots) and fibers (small dots, wavy lines). Modified after Smeets et al. (1987).

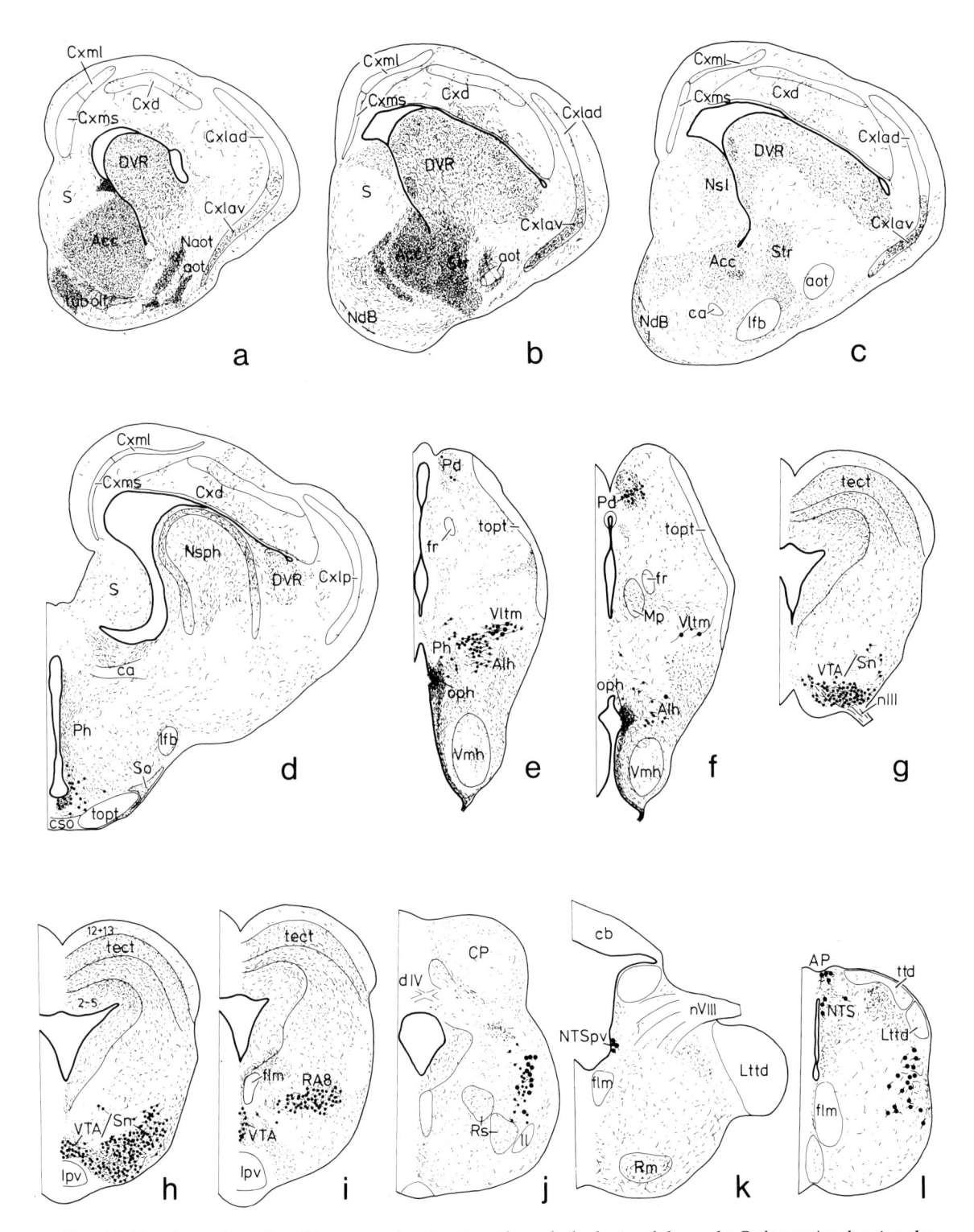

Fig. 6.9. Drawings of a series of transverse hemisections through the brain of the snake *Python regius* showing the distribution of DAi cell bodies (large dots) and fibers (small dots, wavy lines). Modified after Smeets (1988b).

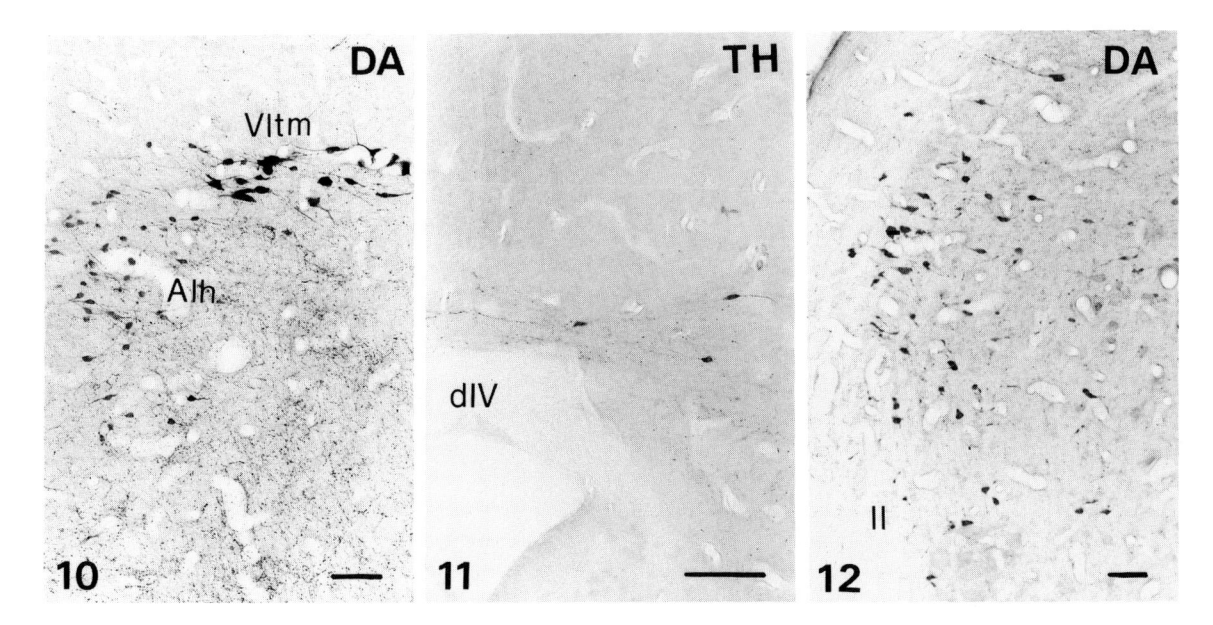

Figs. 6.10–6.12. Photomicrographs showing immunoreactive cell bodies in the lateral hypothalamic area and the magnocellular ventrolateral thalamic nucleus of *Python regius* (10), in an area dorsal to the trochlear nerve decussation (dIV) of the lizard *Anolis carolinensis* (11), and in the isthmic region adjacent to the lateral lemniscus (ll) of *Python regius* (12). Bars = 0.1 mm.

RA8) on the basis of the position of the cells and the direction of their processes. Such a tripartition of the mesencephalic DAi/THi cell bodies could not be made in the snakes *Python* and *Thamnophis*, since, in these species, the cells of the presumed SNc lack long, laterally directed processes. The absence of these processes in snakes may be related to the position of the pars reticulata of the substantia nigra, which can be defined by its substance P plexus (see also p. 127). Species differences are also observed in the arrangment of the DAi/THi cell bodies in the caudal brainstem (cf. Figs. 6.1(*j*), (*k*), 6.8(*n*), 6.9(*l*)).

Dopamine containing fibers The distribution of DAi/THi fibers and varicosities shows considerably more variations than that of the DAi/THi cell bodies in the different species. Yet, the reptiles studied have a number of features in common (cf. Figs. 6.1, 6.8, 6.9, 6.13). These include: 1) the DA innervation of the olfactory bulbs, where the DAi/THi fibers and terminals are mainly confined to the glomerular and internal granular layer; 2) the much more dense DA innervation of ventral telencephalic areas when compared to cortical areas; 3) the inhomogeneous distribution within the striatum, nucleus accumbens, and olfactory tubercle; 4) the DA innervation of the septal area where characteristic pericellular baskets of DAi/THi fibers are observed and the densest innervation is found in the lateral septal nucleus; 5) the DA innervation of the dorsal portion of the nucleus sphericus of the amygdaloid complex; and 6) the overall distribution of DAi fibers in the diencephalon and the brainstem tegmentum.

Although all reptiles studied show these general features, several differences are noted. First, it is evident that the striatum in *Pseudemys* and *Python* receives a much stronger innervation than the corresponding area in *Gekko*. However, substantial differences in the DA innervation of the striatum occur also within the reptilian radiation of lizards. In *Varanus*, *Podarcis* and *Gallotia* which belong to the group of intensive foragers (Regal, 1978), the striatum is characterized by a more dense DAi/THi plexus than the corresponding region in *Gekko* and *Eublepharis*, which are considered as sit-and-wait predators. The fact that the density of DAi/THi fibers in the striatum of different species is not clearly correlated with the number of mesencephalic DAi/THi cells, suggests that a high density of striatal DAi/THi fibers is due to a high degree of arborization of the mesotelencephalic fibers. It should be further noted that the DA innervation of the striatum in all reptilian species studied

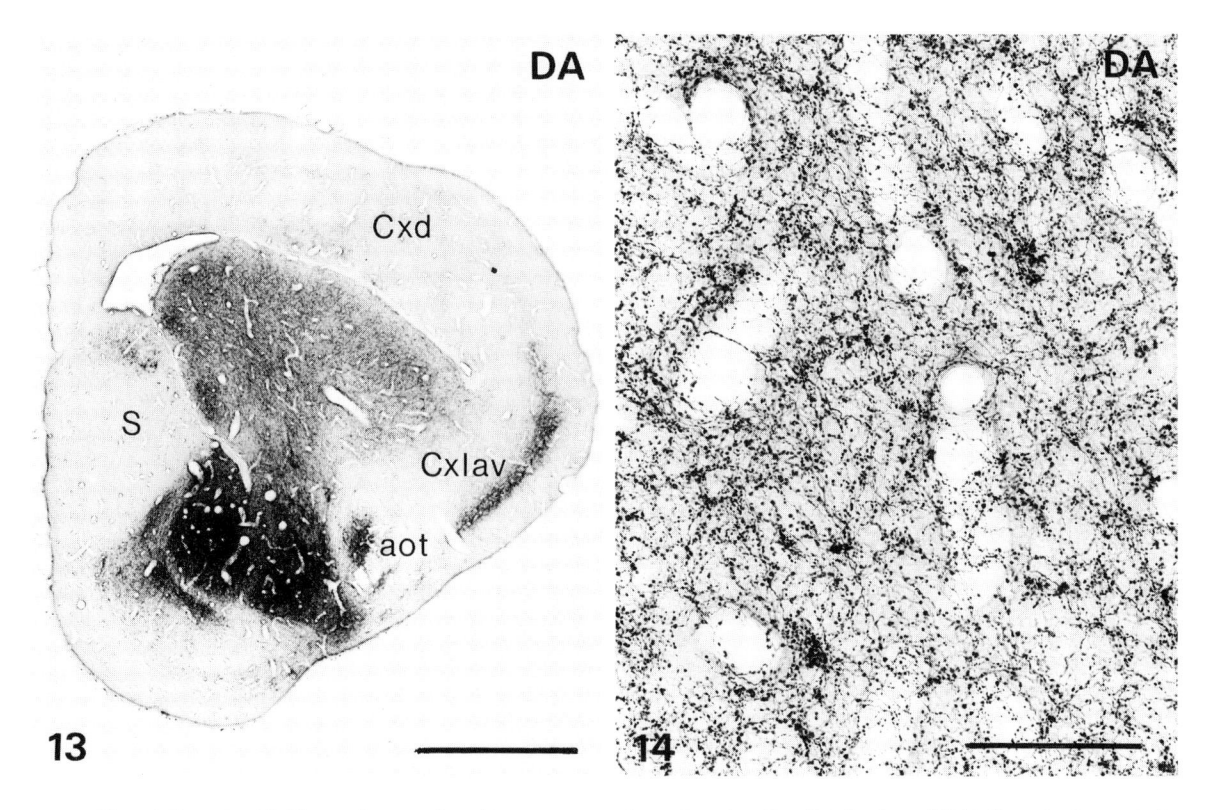

Figs. 6.13 and 6.14. Photomicrographs of transverse sections showing the distribution of DAi fibers in the rostral part of the telencephalon of *Python regius*. The level of Fig. 6.13 is comparable to that of Fig. 6.9(*b*), whereas Fig. 6.14 is a higher magnification of the dense DA innervation of the dorsal ventricular ridge. Bar = 1 mm (13), 0.1 mm (14).

shows a nonuniform compartmentalization, which is most distinct in the gekkonid species (Russchen *et al.*, 1987*b*).

Another difference in the distribution of DAi fibers between the various reptiles studied is found in the lateral cortex. The lizards and the snakes studied have in common that the ventral part of the anterior lateral cortex is characterized by a dense DA innervation, whereas the dorsal part of this cortical region as well as the posterior lateral cortex are almost completely devoid of DAi/THi fibers (Fig. 6.1, 6.9, 6.13). In the turtle *Pseudemys*, on the contrary, the lateral cortex receives fibers throughout its rostrocaudal extent without differences in density between dorsal and ventral portions (Fig. 6.8).

The Squamata also contrast with turtles in having a considerably more dense DA innervation of the midbrain tectum. In *Pseudemys*, a moderate number of fibers is found in the periventricular zone with only a few fibers distributed to the other tectal layers, but in lizards and snakes numerous fibers are observed in both periventricular and superficial tectal layers. However, within the Squamata the tectal DA innervation varies from extremely laminated (*Gekko*) to almost uniform (*Python, Thamnophis*).

A striking difference is also observed in the DA innervation of the dorsal ventricular ridge (DVR). Whereas in turtles (Smeets *et al.*, 1987) and, probably also in crocodiles (Brauth & Kitt, 1980), the DVR is only weakly innervated by DAi/THi fibers, a generally dense innervation is present in lizards and snakes (cf. Figs. 6.1, 6.8, 6.9, 6.14). An exception to the latter might be the DVR of the lizard *Anolis carolinensis*, where only a weak DAi/THi fiber plexus was observed (Lopez *et al.*, 1992; own, unpublished observations).

The small-celled medial cortex of lizards and snakes shows the greatest variation in DA innervation. Whereas in *Gekko* distinct bundles of DAi/THi fibers course on both sides of the cellular layer, in

Podarcis the fibers are mainly observed internal to the cellular layer of this cortical region. In *Varanus*, DAi/THi fibers and varicosities are almost exclusively confined to the cellular layer, but in *Eublepharis* and in the two snake species studied only a few DAi/THi fibers are observed in the small-celled medial cortex. Although a small-celled medial cortex is not recognized in turtles, it is noteworthy that, in *Pseudemys*, DAi/THi fibers occupy almost the entire neuronal wall of the medial cortex coursing parallel to the meningeal and ventricular surfaces.

Noradrenaline

The distribution of noradrenaline in the CNS of reptiles has been studied by means of antibodies against NA and the enzyme DBH (Smeets, 1988*b*; Smeets & Steinbusch, 1989, 1990). The mapping of noradrenergic cells and fibers is based on sections of the lizard *Gekko gecko* stained with antibodies against NA, since this antiserum revealed the best staining of fibers. An obvious example is the noradrenergic, sympathetic innervation of the walls of cerebral blood vessels. This is easily observed in material stained with NA antibodies, but could not be recognized in DBH-stained material. Similarly, throughout the brain numerous NAi fibers, but no DBHi fibers, were observed piercing through or coursing parallel to the meningeal surface. It is not clear whether these are projection fibers of NAi cell bodies in the CNS or processes of pial cells.

Cell bodies The distribution of NA containing neurons is more restricted than that of DA cells. In the lizard *Gekko gecko*, the most rostral group of NAi cells is found in the hypothalamic periventricular organ (Fig. 6.1(*f*)). The cells are arranged in several layers and appear to contact the CSF of the third ventricle. Slightly more caudally, another small group of NAi cells is observed in the ependymal layer of the infundibular recess. These cells seem also to contact directly the CSF. Remarkably, the cells in both the periventricular organ and the ependymal layer of the infundibular recess do not stain with the DBH antiserum.

At isthmic levels, NAi/DBHi cells are present in the locus coeruleus, and more caudally, at obex levels, in the nucleus of the solitary tract and area postrema (Fig. 6.1(*i*)–(*k*)). The NAi/DBHi cell bodies of the locus coeruleus occupy a rather large area and possess generally three or four processes that do not show a preferential orientation (Fig. 6.15). The NAi/DBHi cells at the level of the obex are, in general, bipolar and their processes are oriented in a mediolateral direction (Fig. 6.16). The DBH antiserum additionally stains cells in the ventrolateral tegmentum of the caudal rhombencephalon, which are putative adrenergic cell bodies (see p. 118).

Fibers In the olfactory bulb of *G. gecko*, NAi/DBHi fibers are predominantly confined to the internal granular and plexiform layer. Some fibers are observed in the mitral cell layer, whereas others surround the glomeruli. Numerous NAi/DBHi fibers with almost no varicosities are found in the olfactory peduncles.

In the telencephalon proper, NAi/DBHi varicose fibers are present in cortical as well as in subcortical areas (Fig. 6.1(*a*)–(*e*)). On both sides of the cellular layer of the small-celled medial cortex, a plexus of NAi/DBHi fibers is present. The cells of the large-celled medial cortex are not in close contact with NAi fibers, except probably for the most distal part of their apical and basal dendrites. As in the small-celled medial cortex, the NAi/DBHi fibers in the dorsal cortex form a plexus superficial and deep to the cellular layer with the highest density in the superficial layer (Fig. 6.17). However, compared to the small-celled medial cortex, NAi/DBHi fibers in the dorsal cortex are less uniformly distributed terminating predominantly in its medial and intermediate portions with an increasing density in caudal direction (Fig. 6.1(*a*)–(*e*)). The lateral cortex receives a relatively weak noradrenergic input. Throughout its rostrocaudal extent, NAi/DBHi fibers are found in the molecular layer. In addition, particularly in the ventral part of the anterior lateral cortex, fibers are also observed in the celular layer.

In the ventral telencephalon, dense NAi/DBHi plexuses are found in the bed nucleus of the medial forebrain bundle, the vertical limb of the nucleus of the diagonal band of Broca, and the caudoventral part of the septal area. A very weak noradrenergic input reaches the remaining parts of the septal area, the nucleus accumbens, the striatum, and the olfactory tubercle. The external, lateral, and spherical nuclei of the amygdaloid complex receive a weak to moderate NA innervation. The DVR is almost devoid of NAi fibers, except for its dorsomedial portion, where numerous varicose fibers are present in the superficial cell plate (Fig. 6.1(*b*)–(*d*)).

Diencephalic cell masses that show a dense NA innervation are the supraoptic nucleus and the medial habenular nucleus (Fig. 6.1(*d*)–(*f*)). In addition, numerous NAi/DBHi fibers are found in an area ventromedial to the lateral forebrain bundle that partly matches the region occupied by the medial

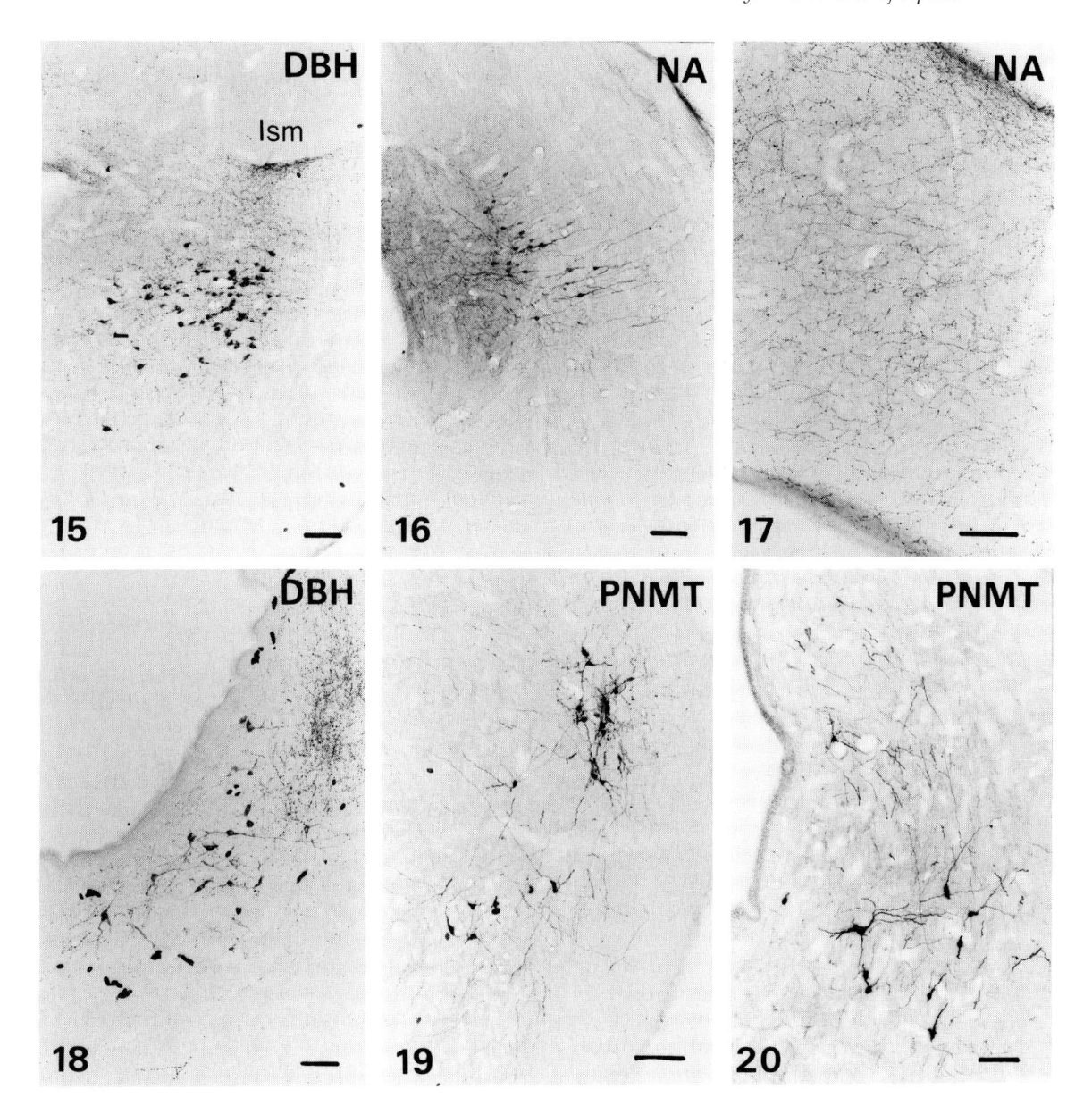

Figs. 6.15–6.20. Photomicrographs of transverse sections showing DBHi/NAi cells and fibers in the locus coeruleus (15), the solitary tract nucleus (16) and the dorsal cortex (17) of *Gekko*, the locus coeruleus of *Pseudemys* (18), and PNMTi cell bodies in the caudal brainstem of *Gekko* (19) and *Pseudemys* (20). Bars = 0.1 mm.

forebrain bundle. A moderate to dense plexus of NAi fibers is present in the subependymal layer of both the thalamus and the hypothalamus, in the periventricular hypothalamic nucleus, and in the lateral hypothalamic area. The remaining diencephalic areas receive only a sparse noradrenergic innervation.

In the midbrain, a very dense NAi/DBHi plexus is found in the ventromedial part of the VTA (Fig. 6.1(g)). Caudally, the plexus continues into an area ventral to the interpeduncular nucleus (Fig. 6.1(h)). Another conspicuous NAi/DBHi plexus is present in an area that matches the region containing the dopaminergic cell bodies of the substantia nigra, pars compacta (SNc). A moderate to dense plexus is further observed in the retrorubral (RA8) dopaminergic cell group at caudal midbrain levels. The remaining parts of the midbrain tegmentum contain only a few diffusely arranged NAi/DBHi fibers, except for the periventricular layers including the torus semicircularis, where numerous fibers run parallel to the ventricular surface. The tectum shows a moderate to dense plexus of NAi/DBHi fibers in all tectal layers with the exception of layer 6, which contains only a few fibers. The NAi/DBHi fibers in the periventricular zone are continuous with those in the periventricular layers of the tegmentum.

NAi/DBHi fibers are widely distributed throughout the rhombencephalon (Fig. 6.1(i)–(k)). A very dense plexus borders ventrally the magnocellular isthmic nucleus (Fig. 6.15). Other moderate to dense plexuses of NAi/DBHi fibers are found in the parabrachial region, the griseum centrale, the motor and sensory trigeminal nuclei, the raphe nuclei and the periventricular region medial to the vagal motor nucleus and the solitary tract nucleus. The cerebellum receives a substantial NA innervation. The pattern seems to be quite homogeneous throughout the cerebellum. Numerous NAi/DBHi fibers are found in the granule cell layer where they form a loose network that is evenly distributed throughout the layer. A small number of NAi fibers pierce through the Purkinje cell layer and enter the molecular cell layer.

In the spinal cord, the distribution of NAi/DBHi fibers resembles that of the DAi fibers except for the numerous NAi fibers and varicosities in or close to the meningeal surface which are not immunoreactive with the DA antiserum.

Reptilian noradrenergic system: general and derived conditions In contrast to dopamine, information on noradrenaline in the brains of other reptilian species is limited. In our laboratory, only one brain of *Pseudemys* has been stained with NA antibodies, whereas two other brains were immunoreacted with the DBH antiserum. Brains of other reptiles stained with the latter antiserum include *Varanus* ($n = 2$) and *Anolis* ($n = 2$). Although the staining with the NA antiserum in the brain of the turtle, *Pseudemys*, was not optimal, the distribution of NAi cell bodies appeared to be basically similar to the pattern observed in *Gekko*. The material immunoreacted with the DBH antiserum yielded comparable results in *Pseudemys* with two exceptions: 1) the NAi CSF-contacting cells in the hypothalamic periventricular organ do not react with the DBH antiserum; 2) the

DBH antiserum additionally stains cells in the ventrolateral tegmentum of the caudal rhombencephalon which are putative adrenergic cell bodies (see p. 119). A notable difference between turtles on the one hand, and the squamate reptiles studied so far on the other hand, is that in turtles the NA cells of the locus coeruleus lie closer to the ventricle and are more compactly organized (cf. Figs. 6.15, 6.18).

With respect to the distribution of NAi and DBHi fibers, the limited material available shows obvious similarities in the nucleus accumbens, the bed nucleus of the medial forebrain bundle, the preoptic area, and the subependymal layer of both the thalamus and hypothalamus, where dense immunoreactive plexuses are present. In the brainstem, the most remarkable common feature is the presence of rather dense NAi or DBHi plexuses in areas that contain dopaminergic cell bodies. Notable species differences in distribution of noradrenergic fibers are probably present in the medial cortex and the superficial tectal zone in the midbrain. In contrast to *Gekko*, in *Pseudemys* and in the lizards, *Varanus* and *Anolis*, these areas are almost devoid of immunoreactive fibers. However, further studies with antibodies against NA and DBH are necessary for drawing definitive conclusions on general and derived conditions of the noradrenergic systems in reptiles.

Adrenaline

Adrenaline is the final product of the catecholamine biosynthetic pathway. Whereas adrenergic cells in the adrenal medulla of mammals have been demonstrated with antibodies against adrenaline (Verhofstad et al., 1980, 1982), these antibodies fail to stain putative adrenergic structures in the central nervous system (Steinbusch & Tilders, 1987). Therefore, our knowledge of the location of adrenergic cell bodies and fibers in the brain is exclusively based on PNMT immunohistochemistry. Hitherto, only one study has dealt in detail with the distribution of PNMT in the brain of a reptile, the lizard *G. gecko* (Smeets & Jonker, 1990).

Cell bodies In the brain of *G. gecko*, PNMTi cell bodies are found only caudally in the brainstem at the level of the obex (Figs. 6.1(j), (k), 6.19). The cells form a continuous, loose aggregation in the ventrolateral tegmentum of the rhombencephalon. The rostrocaudal extent of this cell group is approximately 1 mm. Two-thirds of the group lie rostral and one-third caudal to the obex. Approximately 400 cells are present on each side. At rostral levels, the group of PNMTi cells can be subdivided into a dorsolateral and ventromedial group (Fig. 6.19). The cells in the dorsolateral group are more tightly clustered than those in the ventromedial one. In both subgroups, the cells have long processes, some of which extend to the meningeal surface, while others run parallel to it and sometimes cross the midline. A few, isolated PNMTi cells are often seen in the area between the dorsal motor nucleus of the vagus and the PNMTi cell group in the ventrolateral tegmentum. No immunoreactive perikarya were found in the nucleus of the solitary tract, the medial longitudinal fascicle or the hypothalamus.

Fibers PNMTi fibers are present throughout the brain of *G. gecko*, extending rostrally as far as the olfactory peduncle. In the telencephalon the PNMTi fibers are almost completely confined to subcortical regions (Fig. 6.1(a)–(d). Only a few fibers are found in the lateral cortex. Moderate to dense immunoreactive plexuses are observed in the caudoventral septal region, the medial septal nucleus, the nucleus of the diagonal band of Broca, the bed nucleus of the medial forebrain bundle and the nucleus accumbens. Also the central amygdaloid nucleus contains a considerable number of PNMTi fibers.

In the preoptic area and the periventricular zone of the hypothalamus, dense plexuses of PNMTi fibers are found (Fig. 6.1(d)–(f)). Also the medial forebrain bundle contains numerous immunoreactive fibers coursing rostrally, parallel to each other. A moderate to dense innervation by PNMTi fibers is observed in the lateral portions of the hypothalamus and the preoptic area. Thalamic and epithalamic structures are devoid of PNMTi fibers, with the exception of the dorsomedial thalamic nucleus which contains a rather dense plexus.

At caudal brainstem levels, PNMTi fibers are present in the nucleus of the solitary tract. Further rostrally, PNMTi fibers collect into a distinct bundle that courses in the ventrolateral rhombencephalic tegmentum, indicated as the longitudinal bundle (Fig. 6.1(g)–(i)). The bundle traverses the area that contains the locus coeruleus. Within the confines of the locus coeruleus, dense plexuses of varicose PNMTi fibers are found. A number of fibers leave the longitudinal bundle and course dorsolaterally to the parabrachial region (Fig. 6.1(i)). The main bundle continues rostrally into the tegmentum of the midbrain where it gives off a dorsal fiber bundle (Fig. 6.1(g), (h)). The varicose fibers of the latter bundle form a dense plexus in the periventricular zone of the tegmentum and, to a lesser extent, in the tectum. The cerebellum is devoid of PNMTi fibers, although occasionally an immureactive fiber can be observed in the granule cell layer. Only a few PNMTi fibers are present in the cervical spinal cord.

Reptilian adrenergic system: general and derived conditions Very little is known about the adrenergic system of other reptilian species. The major problem is that the PNMT antiserum raised in mammals does not crossreact well with the corresponding enzyme in reptiles. The crossreactivity varies between species. As reported above, in *Gekko* the PNMT antiserum revealed well stained perikarya and fibers. The same antibodies are also suited to demonstrate putative adrenergic cell bodies and fibers in the turtle, *Pseudemys scripta elegans* (Fig. 6.20). However, preliminary studies with PNMT antibodies in two other lizards (*Varanus, Anolis*) and a snake (*Thamnophis*) resulted in weakly immunostained cell bodies only in *Varanus*, whereas PNMTi fibers could not be recognized in any of these species.

In spite of the limited data, several features of the distribution of PNMTi structures in the brains of reptiles deserve comments. As in *Gekko*, PNMTi cell bodies in the brain of *Pseudemys* are located exclusively at caudal brainstem levels (Fig. 6.20). The majority of the cells lie dispersed throughout the ventrolateral rhombencephalic tegmentum and are, with respect to their position and cell morphology, readily comparable to the PNMTi cell group in the gekkonid brain. A striking difference between the two species, however, is observed in the dorsomedial portion of the rhombencephalic tegmental area. In *Gekko* no cell bodies are found in the nucleus of the solitary tract, but in *Pseudemys* several PNMTi cell bodies are observed in the latter area. Thus, whereas in *Gekko* (and probably also *Varanus*) only one group of PNMTi cells is present which, on the basis of its position, may be compared to the C1 cell group of rats, *Pseudemys* possess two cell groups, which are putatively homologous to the C1 and C2 cell groups of rats.

The distribution of PNMTi fibers in the brain of *Pseudemys* largely resembles the pattern observed in *Gekko*. Thus it appears that the innervation of the single group of PNMTi cells in *Gekko* matches that of the two cell groups in *Pseudemys*.

Catecholamine containing cells with deviating immunohistochemical properties

Based on the biosynthesis of catecholamines, it might be expected that dopaminergic neuronal structures stain with TH and DA antibodies, but not with DBH, NA or PNMT antisera. Noradrenergic structures should stain with TH, DBH, and NA antibodies, whereas adrenergic elements should be immunoreactive for TH, DBH and PNMT antibodies. In general, the results obtained with the five antibodies of the present report are in agreement with the expectations (Smeets & Steinbusch, 1990). However, two remarkable discrepancies are noted: 1) some cell bodies stain with TH antibodies, but are immunonegative for DA, DBH, NA or PNMT; 2) some cell bodies exhibit DA- or NA-immunoreactivity, but no TH- or DBH-immunoreactivity.

Cells that stain with antibodies against TH, but not with any of the other antibodies related to the catecholamine biosynthetic pathway are found in several places in the brain of both *Gekko* and *Pseudemys*. The most obvious example is found in the olfactory bulb (Fig. 6.21). Whereas antibodies against DA stain cell bodies around the glomeruli and, to a lesser degree, in the external plexiform layer, TH antibodies stain additional cells in the superficial portion of the internal granular layer. The latter group of cells do not stain with DA, DBH, NA or PNMT antisera. Halasz et al. (1982) also using antibodies against TH in *Pseudemys*, reported identical results, but they considered the cells in the internal granular layer to be dopaminergic based on their immunonegative staining with DBH or PNMT antisera. However, in all reptilian species studied so far, these cells do not exhibit DA-immunoreactivity, nor DBH-, NA-, or PNMT-immunoreactivity. Similar observations, i.e. staining with TH antibodies but not with one of the other antisera used, have been made in the rostral pole of the dorsal cortex and in the pretectal nucleus of *Gekko*, and in the primordium hippocampi of *Pseudemys*.

There are several possibilities to explain why some cells stain only with the TH antiserum such as

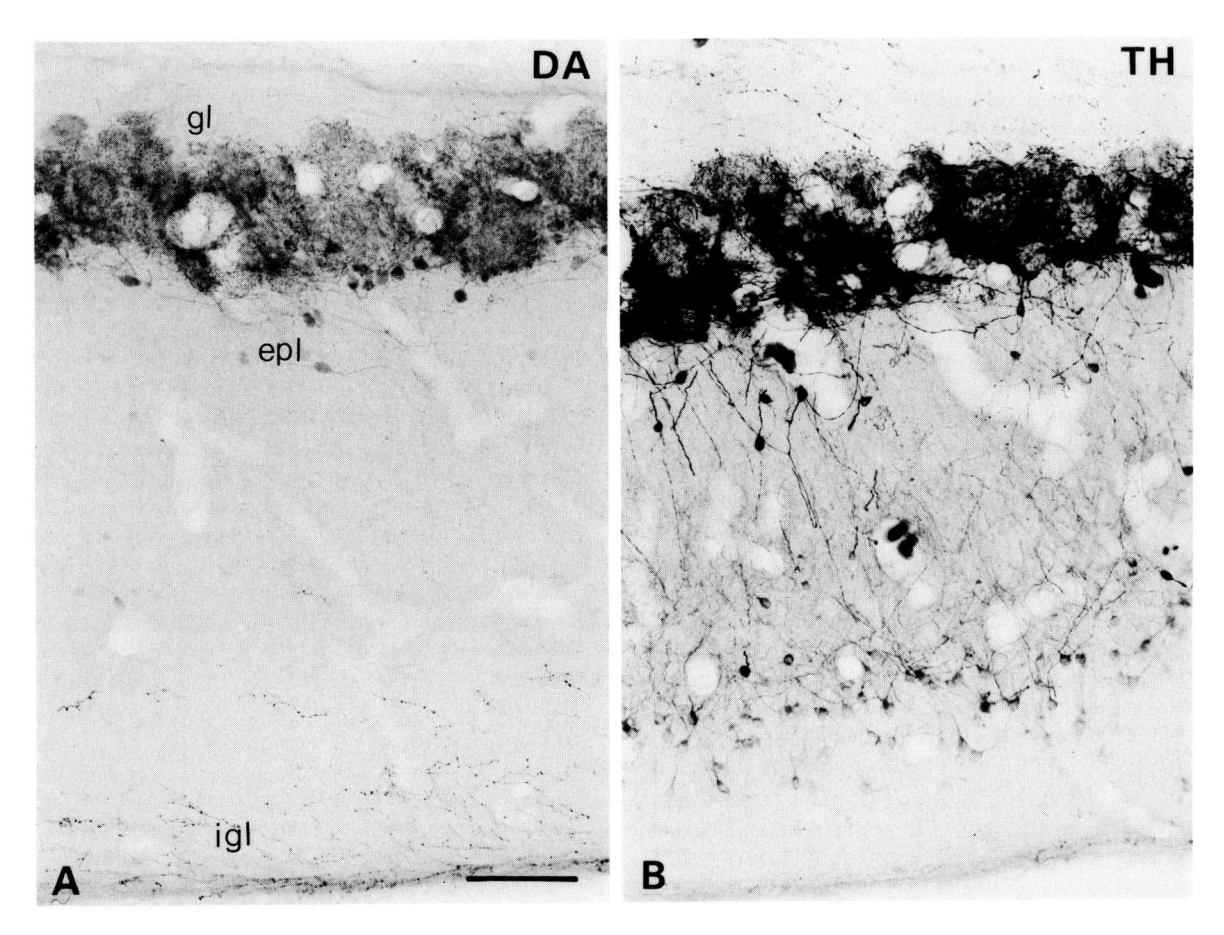

Fig. 6.21. Photomicrographs showing DA (A) and TH (B) immunoreactive cell bodies and fibers in the olfactory bulb of *Pseudemys*. Note the absence of DA cell bodies in the internal granular layer. Bar = 0.1 mm.

a low, immunohistochemically non-detectable level of DA content, a crossreactivity with an unknown tissue antigen, or an expression of TH in cells that do not produce catecholamines at all (for a more detailed discussion, see Smeets & Steinbusch, 1990). However, the most straightforward possibility is the presence of DOPA as the endproduct of enzymatic activity in these cells. This suggestion is supported by recent studies of the arcuate nucleus in mammals (Meister et al., 1988; Okamura et al., 1988a, b, c). Studies with antibodies against AADC and/or L-DOPA are needed to clarify the true nature of these THi cells in reptiles.

The second discrepancy, i.e. the presence of cells that are immunopositive for DA and NA antisera, but immunonegative for TH and/or DBH antibodies, is observed in the hypothalamic periventricular organ (Fig. 6.22). DA/NA immunopositive, but TH/DBH immunonegative neurons have in common that they are in direct contact with the CSF of the third ventricle. It should be noted that not all CSF-contacting cells in the diencephalon share this feature since DAi CSF-contacting cells in the preoptic recess and in the ependymal wall of the infundibular recess of *Pseudemys* stain intensely with TH antibodies.

Several possible explanations have been discussed in a previous paper (Smeets & Steinbusch, 1990), of which the most straightforward one is that the cells in the periventricular organ lack the enzymes involved in the biosynthesis of the catecholamines. This implies that the DA- and NA-immunoreactivities in the cells are the result of a local uptake mechanism. An obvious source of extracellular DA and NA could be the CSF of the third ventricle. Uptake of catecholamines from the CSF by these neurons would be in line with the proposed receptive nature of the CSF-contacting processes of the cells in the periventricular organ (Vigh-Teighman & Vigh, 1974). A recent study in *Gekko* by means of injections of the dopamine synthesis inhibitor alpha-methylparatyrosine (α-MPT)

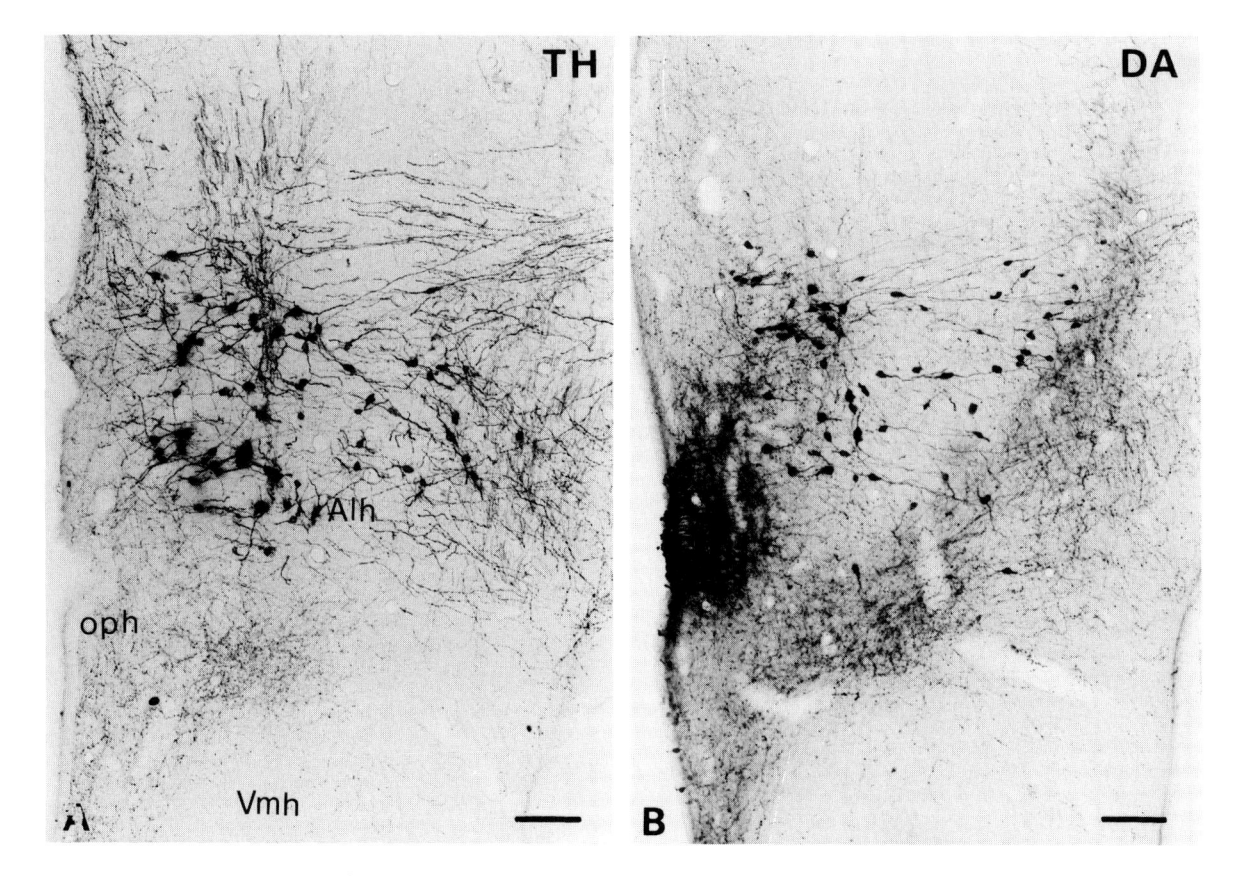

Fig. 6.22. Photomicrographs showing TH (A) and DA (B) immunoreactive cells and fibers in the caudal hypothalamus of *Gekko*. Note the absence of THi cells in the periventricular organ (oph). Bars = 0.1 mm.

revealed that DA cell groups are differently affected by this drug (Smeets, Kidjan & Jonker, 1991). While α-MPT does not alter the intensity of the DA immunoreactivity in the cells of the periventricular organ, it dramatically decreases the intensity of immunostaining in the other DA cell groups. This finding supports the notion that the cells in the reptilian periventricular organ do not synthesize but accumulate catecholamines and that the CSF may play an important role in catecholamine neurotransmission.

It should be noted that a periventricular organ can not be recognized in all reptiles. In Nissl stained sections of the agamid lizards, *Calotes versicolor* (Prasado Rao & Subhedar, 1977) and *Agama agama* (Smeets, unpublished observations), no specializations of the ependymal layer are found in the hypothalamus at levels that correspond to those of the periventricular organs in other reptiles. The notion that this organ may be lacking in, at least, some agamid lizards has been corroborated by the failure to stain DAi CSF-contacting cells in the correponding area of *C. versicolor* (Subhedar & Rama Krishna, personal communication).

Catecholamines and functional systems

In this section, it is attempted to relate the distribution of dopamine, noradrenaline and adrenaline to functional systems in the brain of reptiles. First, the catecholaminergic involvement in sensory systems, particularly olfaction, vision, and dorsal ventricular ridge, will be discussed. Subsequently, the distribution of the various catecholamines will be discussed with regard to motor systems, viz. the basal ganglia, the brainstem and the spinal cord.

Sensory Systems
Except for the nuclei that are involved in olfactory and visual information processing, sensory systems are not densely innervated by catecholaminergic fibers. Of the thalamic nuclei, only the dorsomedial thalamic nucleus receives a rather strong DA, NA as well as adrenergic input, which are probably derived from the periventricular hypothalamic nucleus, the locus coeruleus, and the ventromedial portion of the caudal brainstem PNMTi cell group, respectively (Hoogland, 1982).

Olfactory system Apart from DA and putative DOPA cells, the olfactory bulbs of reptiles contain numerous DAi and NAi fibers. Since no PNMTi fibers are found, it is concluded that dopamine and noradrenaline are the catecholamines that are involved in olfactory information processing. Of the two catecholamines, dopamine probably plays a major role in view of the strong immunoreactivity observed in the glomerular layer where the olfactory nerve fibers are known to contact the dendrites of the mitral cells (Nieuwenhuys, 1967; Halasz & Shepherd, 1983). The DA innervation of the olfactory bulb is most likely exclusively provided by intrinsic DA neurons for two reasons: 1) DAi fibers are not observed in the olfactory peduncle; 2) retrograde tracing studies do not label cells in areas that are known to contain DA cell bodies (Martinez-Garcia et al., 1991). Physiological and pharmacological studies have demonstrated that DA may affect indirectly the activity of mitral cells, possibly by acting through interneurons (Nowycky, Halasz & Shepherd, 1983; Halasz & Shepherd, 1983).

The NA fibers are most likely derived from the locus coeruleus (Martinez-Garcia et al., 1991). Although some studies have dealt with the role of the NA innervation of the olfactory bulb (for refs., see Halasz & Shepherd, 1983), its functional significance is as yet unclear.

The primary olfactory formation comprises in most reptiles not only a main bulb, but also an accessory bulb, receiving the vomeronasal nerve (for review, see Halpern & Kubie, 1984). No clear differences are observed by us in the organization of the catecholamines in these two subdivisions of the olfactory bulb formation. However, it is now generally accepted that the vomeronasal and main olfactory systems in reptiles are segregated with respect to the course and termination sites of their respective fibers (for review, see Lohman & Smeets, 1993). The main bulb projects primarily to the anterior olfactory nucleus, the olfactory tubercle, and the lateral cortex, with an additional projection to the rostral parts of the external and central amygdaloid nuclei. The accessory bulb, on the contrary, projects primarily to the nucleus sphericus, the interstitial cells of the accessory olfactory tract, and, at least in some reptiles, to the caudal parts of the central amygdaloid nucleus and the bed nucleus of the stria terminalis.

The present survey reveals that the targets of the secondary olfactory fibers, i.e. the lateral cortex, the olfactory tubercle and some parts of the amygdaloid complex (Reiner & Karten, 1985; Lohman & Smeets, 1993), are innervated by DAi and, although less densely, by NAi and PNMTi fibers suggesting that catecholamines, in particular dopamine, may modulate olfactory input at different levels.

Visual system There is abundant evidence that catecholamines are involved in visual information processing. This involvement may take place in the retina as well as in primary visual centers. Since the discovery that there exists a subpopulation of dopaminergic amacrine cells in the retina of rats (Malmfors, 1963), several reviews on their distribution in other vertebrate classes, including reptiles, have appeared (Ehinger, 1982; Brecha, 1983; Witkovsky, Eldred & Karten, 1984; Wulle & Wagner, 1990). The perikarya of DA amacrine cells invariably are present at the junction of the inner nuclear and inner plexiform layers. A remarkable feature of the retina of turtles is the apparent absence of dopaminergic interplexiform cells (Witkovsky et al., 1984; Nguyen-Legros et al., 1985), despite the potent action of dopamine on the physiological response of the large-field horizontal cell of the turtle retina (Gerschenfeld et al., 1983).

In reptiles, retinal ganglion cells that are known to give rise to the optic tract fibers, are never found to be immunopositive for TH and DA antibodies. Still, a perfect overlap exists between the retinofugal fibers and DAi/THi fibers in the thalamic visual relay centers and in the tectum of the midbrain, particularly in *Gekko gecko* (Smeets et al., 1986b; Medina & Smeets, 1992). The absence of DAi/THi fibers in the optic tract suggests that dopamine from extraretinal sources modulates visual information processing at these sites (Medina & Smeets, 1992).

Hodological studies have established that the tectum in reptiles receives afferents from, among others, the hypothalamic periventricular nucleus, the pretectal posterodorsal nucleus, and the substantia nigra (Welker, Hoogland & Lohman, 1983; ten Donkelaar et al., 1987; Medina & Smeets, 1991). Since all three nuclei contain dopaminergic cell bodies, they are good candidates for the source of the dopaminergic innervation of the midbrain tectum. The noradrenergic input to the tectum is most likely derived from cells in the locus coeruleus (ten Donkelaar et al., 1987). No information is available about the origin of DAi/THi afferents to thalamic visual centers in reptiles.

Dorsal ventricular ridge and dorsal cortex Other brain structures that are known to be involved in sensory information processing and deserve some comment, are the dorsal ventricular ridge (DVR) and the dorsal cortex. From a comparative point of view, these two telencephalic structures are of particular interest, since both have been considered as putative reptilian homologue of the mammalian neocortex (Nauta & Karten, 1970; Kirsche, 1972; see also Lohman & Smeets, 1990).

Many investigations have demonstrated that at least three major sensory modalities reach the DVR: visual, auditory, and somatosensory (for review see Ulinski, 1983; Lohman & Smeets, 1990). In brief, visual information reaches the rostrolateral part of the DVR, auditory input is relayed to the medial portion, whereas somatosensory input reaches, depending on the species studied, the intermediate part or the caudolateral part of the ridge. Apart from these segregated thalamic inputs, the DVR receives a diffuse projection from the dorsomedial thalamic nucleus. On the basis of these thalamic inputs, the low reactivity for catecholamines and acetylcholine esterase in turtles, as well as embryological data, it has been suggested that the DVR in reptiles is comparable to the mammalian isocortex (Parent & Olivier, 1970; Parent, 1973; see also Reiner, 1993). However, Lohman and Smeets (1990) have critically reviewed the current data on the DVR of reptiles and questioned this notion again. For example, the sensory information, that reaches the neocortex of mammals, is processed through successive stages of unimodal and polymodal elaboration. There is no indication of such an integration between sensory modalities within the DVR of reptiles. On the contrary, the connections within the DVR are predominantly organized radial to the telencephalic ventricle and also within the projections of the DVR to the striatum; this segregation of sensory modalities is maintained (González et al., 1990). Another difference between the reptilian DVR and the neocortex of mammals is, that the DVR efferent projections remain, for the most part, within the telencephalon (Hoogland, 1977; Voneida & Sligar, 1979) and do not project to the thalamus, brainstem motor nuclei and reticular formation, nor to the spinal cord. Therefore, it seems to be more appropriate to deal with the DVR as a separate entity, unique to reptiles and birds, than to relate it to cortical or striatal structures.

Since cortical and subcortical telencephalic areas receive a dopaminergic input, our studies of catecholamines in the CNS of reptiles can not solve this problem. Nevertheless, it is now evident that two patterns of DA innervation are present within the class of Reptilia: a weak DA innervation of the DVR is found in turtles and, probably, also in crocodilians, but a dense DA innervation of the DVR is present in squamate reptiles. The dopaminergic input to the DVR most likely originates from the midbrain DA cell group (ten Donkelaar & de Boer-van Huizen, 1988). Taking the observed weak DA innervation of the corresponding structure in birds (Chapter 7) into consideration, the high DA content of the DVR of squamates seems to be the derived condition.

Although a considerable overlap of DAi and NAi fibers occurs in the dorsal cortex of *Gekko gecko*, there is some indication that DAi fibers reach primarily the lateral portion of this cortical region, whereas NAi fibers terminate predominantly in its medial and intermediate portions. This immunohistochemically identified topography is in agreement with a previous hodological study (Bruce & Butler, 1984) which revealed retrogradely labeled cells in the locus coeruleus only after HRP injections in the medial part of the dorsal cortex.

Motor systems

As was first noted by the British neurologist Hughlings Jackson, at the end of the nineteenth century, motor systems consist of separate neural circuits which are hierarchically organized. These neural circuits are located in four distinct areas: (1) the spinal cord; (2) the brainstem including the motor nuclei of the cranial nerves and the reticular formation; (3) the motor cortex; and (4) the premotor cortex. In addition to the four levels of hierarchy, two other parts of the brain are important for motor function, i.e. the cerebellum and the basal ganglia. In the following sections, we will first deal with the telencephalic basal ganglia structures, i.e. the nucleus accumbens and the striatum, in relation to dopamine. Subsequently, the catecholaminergic inputs to the motor centers in the brainstem and spinal cord will be discussed. The two other levels of motor hierarchy, i.e. the motor and premotor cortex are apparently missing in the brains of reptiles.

Basal ganglia During the last decade, the use of new and powerful methods, such as immunohistochemistry for the localization of neurotransmitters and various axonal transport techniques for the tracing of neuronal connections, has considerably enlarged our knowledge of the basal ganglia organization in the brain of reptiles. For example, striatal structures, i.e. the nucleus accumbens and the striatum, are easily recognized in all reptiles studied so far because of their dense plexuses of DAi fibers (Smeets et al., 1986b, 1987; Russchen et al., 1987a; Smeets, 1988a). Other features, such as the presence of cholinergic, enkephalinergic and substance P containing neurons (see pp. 125–127), further support the notion that the nucleus accumbens and the striatum of reptiles are homologous to the structures with the same name in the forebrain of mammals.

Early hodological studies already revealed that a major projection to the basal forebrain arises from the catecholaminergic cell group located in the midbrain tegmentum (Parent, 1976; Brauth & Kitt, 1980). In these studies no distinction was made between the afferents to the nucleus accumbens and to the striatum. Recently, González et al. (1990) found in *Gekko gecko* a clear medial to lateral topography of the mesostriatal projections: the nucleus accumbens receives input from the ventral tegmental area, whereas the major input to the striatum originates in the substantia nigra. Only the rostromedial part of the striatum receives a sparse input from the ventral tegmental area.

Studies of the efferent connections of striatal structures in the forebrain of *G. gecko* have revealed that the nucleus accumbens and the striatum project back to the midbrain dopaminergic cell groups. According to Russchen and Jonker (1988), the striatum projects to the substantia nigra, whereas the efferents of the nucleus accumbens reach predominantly the ventral tegmental area. A re-examination of the striatal outputs in *Gekko* revealed additional, substantial projections from the nucleus accumbens to the substantia nigra and retrorubral area which suggests that the major striatal output systems in reptiles and mammals are basically the same (Smeets, 1992).

Brainstem and spinal cord Our immunohistochemical studies did not reveal a dense dopaminergic innervation of the motor nuclei of cranial nerves, except for the motor trigeminal nucleus and the dorsal motor nucleus of the vagus. The latter nuclei also receive noradrenergic and adrenergic input. In general, the brainstem reticular formation does not contain distinct plexuses of catecholaminergic fibers. Only the raphe nuclei, in particular the nucleus raphes superior, are well innervated by DAi fibers and, although less densely, by NAi and PNMTi fibers. The origin of these catecholaminergic fibers is, as yet, unknown.

Rather dense plexuses of DAi and NAi/DBHi fibers are present in the spinal cord of all reptiles studied. A comparison of the location of DAi and NAi/DBHi cells with data on connectivity (ten Donkelaar, Kusuma & de Boer-van Huizen, 1980; ten Donkelaar et al., 1987) yields several putative sources of origin of the DAi fibers projecting to the spinal cord. Candidates are the periventricular hypothalamic nucleus and the nucleus of the solitary tract (ten Donkelaar et al., 1980; Woodson & Kunzle, 1982).

Relationships of catecholamines with other neurotransmitter systems

During the last decade, a considerable number of immunohistochemical studies have revealed the location of perikarya and fibers that belong to a certain neurotransmitter system on the basis of their chemical compound. Since catecholamines are widely distributed throughout the CNS of reptiles, it is very likely that they interact, at many places, with those other neurotransmitter systems. There are several possibilities of interactions: (1) catecholamines are colocalized with other transmitters in the same cells; (2) catecholaminergic fibers contact neurons that contain a certain transmitter; (3) catecholamine cell bodies are contacted by fibers of other transmitter systems; and (4) catecholaminergic fibers and fibers containing another transmitter impinge together on certain cell bodies. With these possibilities in mind, the following paragraphs present a survey of interactions of catecholamines with other monoamines, acetylcholine, and peptides.

Interactions with other monoamines: serotonin and histamine

Several immunohistochemical studies have provided detailed maps of the distribution of serotonin-immunoreactive (5-HTi) perikarya and fibers in the brains of lizards (Wolters et al., 1985; Smeets & Steinbusch, 1988; Guirado et al., 1989; Bennis et al., 1990), snakes (Challet et al., 1991), and turtles (Ueda, Takeuchi & Sano, 1983; Kiehn et al., 1992). However, only one study (Inagaki et al., 1990) has dealt with the distribution of histaminergic neuronal elements in the brain of a reptile, the turtle *Chinemys reevesii*.

The presence of histaminergic cell bodies exclusively in the posterior part of the ventral hypothalamus makes colocalization with any catecholamine highly unlikely. In contrast, the distribution of 5-HTi cell bodies shows at several places a considerable overlap with that of catecholaminergic cell bodies. An obvious match occurs in the hypothalamic periventricular organ. Other brain structures which contain both DAi and 5-HTi cell bodies are the ventral tegmental area and the retrorubral RA8 area (Smeets, 1988b; Kiehn et al., 1992). In none of these regions, however, colocalization of dopamine and serotonin in the same cell body has been demonstrated.

DAi fibers may contact the histaminergic cell bodies in the posterior hypothalamus. In turn, histaminergic fibers heavily innervate the medial part of the midbrain tegmentum and thus may directly influence the DAi/THi cell bodies in the ventral tegmental area. Brain structures that are densely innervated by catecholaminergic as well as histaminergic fibers are the lateral habenular nucleus and the lateral hypothalamic area.

In general, serotoninergic cell groups are not particularly well innervated by catecholaminergic fibers. Equally, catecholaminergic cell groups generally lack dense plexuses of 5-HTi fibers. An obvious exception is the pretectal posterodorsal DA cell group in snakes, where a very dense plexus of 5-HTi fibers was found in the brain of the viper, *Vipera aspis* (Challet et al., 1991). Also the ventral tegmental area and the substantia nigra contain rather dense plexuses of 5-HTi fibers (Brauth et al., 1983; Smeets & Steinbusch, 1988), as is the case in avian and mammalian species. In various brain regions of *G. gecko*, a considerable overlap of 5-HTi and DAi fibers occurs (cf. Smeets et al., 1986b; Smeets & Steinbusch, 1988). In the olfactory bulb, the internal granular layer appears to be the major area of overlap. In the telencephalon proper there is a considerable overlap in the inner half of the nucleus accumbens, some parts of the striatum, the lateral amygdaloid nucleus, the small-celled medial cortex, the ventral portion of the anterior lateral cortex, and the medial forebrain bundle. The septal area of *Gekko* shows characteristic pericellular baskets formed by immunoreactive axons of serotoninergic and dopaminergic cells. Considering their distribution, the 5-HTi and DAi/THi baskets surround different populations of septal cells. The most obvious overlap of DAi/THi and 5-HTi fibers is observed in the tectum of the midbrain where tectal layer 9 receives a very strong input from both monoaminergic systems. Also in tectal layers 2–5 and layer 11 5HTi and DAi fibers and terminals overlap, but here their density is lower than in layer 9.

Interactions with acetylcholine

Recent immunohistochemical studies have confirmed the existence of cholinergic neurons in the CNS of reptiles by means of antibodies against choline acetyltransferase (ChAT), the enzyme involved in the conversion of choline to acetylcholine (Mufson et al., 1984; Brauth et al., 1985; Hoogland & Vermeulen-van der Zee, 1990; Medina et al., 1993). In particular, the ChAT immunoreactive cell bodies in striatal structures have received much attention. In contrast to the situation in mammals, the ChAT immunoreactive neurons in *Caiman* (Brauth et al., 1985) and *Gekko* (Hoogland & Vermeulen-van der Zee, 1990) are not dispersed throughout the striatum, but are concentrated in its caudoventral portion. In the striatum of *Gekko*, the zones containing dense ChAT immunoreactive neuropil are virtually devoid of ChAT immunoreactive cell bodies. In rats, on the contrary, the majority of the cholinergic cell bodies are located within zones of dense ChAT immunoreactive neuropil (Phelps, Houser & Vaughn, 1985; Meredith, Blank & Groenewegen, 1989). Remarkably, in another lizard, *Gallotia galloti*, both patterns were observed (Medina et al., 1993). At rostral and intermediate striatal levels, the ChAT immunoreactive cell bodies lie within plexuses of immunoreactive varicose fibers, whereas at caudal striatal levels the majority of the cell bodies are located around the immunoreactive plexuses. The possible functional significance of the observed differences in organization of striatal cholinergic elements has still to be elucidated.

From clinical and biochemical studies it is well known that a dopamine-acetylcholine interaction in the mammalian striatum exists (for review, see Lehmann & Langer, 1983). For example, parkinsonian patients can be treated with dopamine receptor agonists as well as with anticholinergic drugs. In vivo, the release of acetylcholine can be inhibited by dopamine and drugs that stimulate dopamine receptors. The hypothesis is that the activity of the cholinergic interneurons in the mammalian striatum is regulated by an excitatory glutamatergic, predominantly cortical input and an inhibitory dopaminergic input from the midbrain tegmentum. The dopamine receptor that mediates an inhibition of the acetylcholine release, displays the characteristics of a D2 receptor. The presence of D1 and D2 receptors in the basal ganglia of reptiles has recently been demonstrated (Richfield, Young & Penney, 1987; Stoof et al., 1987).

The dopamine-acetylcholine interaction in the striatum and the nucleus accumbens of rats and the lizard *Gekko gecko* was studied at the level of receptor-mediated inhibition of the depolarization-induced release of radiolabeled acetylcholine in an in vitro superfusion system (Stoof et al., 1987). It appeared that in the gekkonid striatum the selective D2 receptor agonist quinpirole (LY 171555) was unable to inhibit the release of acetylcholine. This is in sharp contrast with the striatum of rats, where an almost complete inhibition was obtained. Interestingly, under similar conditions, only a 50% inhibition of the release of acetylcholine was observed in the nucleus accumbens of rats. In subsequent studies (Henselmans & Stoof, 1991; Henselmans, Hoogland & Stoof, 1991), it was shown that the condition found in *Gekko* may hold for reptiles in general, and that the nucleus accumbens of rats is not homogeneous with respect to the regulation of acetylcholine release upon D2 dopamine and N-methyl-D-aspartate (NMDA) receptor activation. In the rostrolateral part of the nucleus accumbens, D2 receptor activation results in a similar inhibition as found in the striatum, but in the caudomedial part no significant inhibition could be detected. The NMDA induced release of acetylcholine was smaller in the caudomedial part as compared to the rostrolateral part of the nucleus accumbens. In this respect, the striatum of reptiles resembles the caudomedial part of the nucleus accumbens of rats.

The dorsal cortex of *Gekko* has, on the basis of its projections to the striatum, been compared to the ventral subiculum of mammals (Hoogland & Vermeulen-van der Zee, 1989). Remarkably, the caudomedial part of the nucleus accumbens of rats receives its most dense cortical input from the ventral part of the hippocampus.

Interactions with peptides

During the 1970s, it became evident that neuropeptides, not only act as hormonal agents, but also play a major role in interneuronal communication. The literature dealing with neuropeptides in the CNS of reptiles has recently been extensively reviewed by Reiner (1992). In the following paragraphs we will confine ourselves, therefore, to those peptides that show an intimate relationship with catecholamines.

Interactions with gut–brain peptides: substance P, cholecystokinin, neuropeptide Y During the last decade, it was discovered that many polypeptides that were initially isolated from the gut are also present in the CNS of vertebrates. The precise role of these so-called gut–brain peptides in the brain is uncertain, but they display many of the characteristics of a neurotransmitter in terms of localization, release and receptor-mediated actions on postsynaptic neurons. Among the many gut–brain peptides studied in the CNS of reptiles, substance P (SP), cholecystokinin (CCK) and neuropeptide Y (NPY) have received relatively much attention, either because of their co-occurrence with monoamines or because of their involvement in basal ganglia organization.

Substance P. The distribution of SP cell bodies and fibers has been studied in several reptilian species (Brauth et al., 1983; Reiner et al., 1984; Wolters, ten Donkelaar & Verhofstad, 1986; Terashima, 1987; Kadota et al., 1988; Anderson & Reiner, 1990; Reiner & Anderson, 1990). When the distribution of SP fibers is compared with that of catecholaminergic neurons, an obvious match of these neurotransmitter systems is observed in the periventricular hypothalamic nucleus, the pretectal posterodorsal nucleus, the substantia nigra and the nucleus of the solitary tract (Reiner et al., 1984; Wolters et al., 1986; Medina &

Smeets, 1992). It is also noted that extant amniotes have inherited many features of basal ganglia organization from a common ancestor. For example, all amniotes studied so far are characterized by extensive numbers of striatonigral and striatopallidal projection neurons in which SP and dynorphin are colocalized (for review, see Anderson & Reiner, 1990; Reiner & Anderson, 1990). Nevertheless, some remarkable differences exist within the class of Reptilia (Brauth et al., 1983; Smeets, 1991). With respect to the relationship between dopamine cells and SP fibers in the substantia nigra and the retrorubral area, two different patterns are observed: one pattern with little overlap between the DA cells and SP fibers (lizards, turtles) and another with an extensive overlap (crocodiles, snakes). The two patterns have also been found in mammals (Reiner, Karten & Solina, 1983; Haber & Groenewegen, 1989). On the basis of an out-group comparison with anamniotes, it is suggested that the pattern with little overlap represents the primitive condition, whereas that with considerable overlap has to be considered as a case of parallel homoplasy (Smeets, 1991).

Cholecystokinin. CCK is a 33 amino acid peptide that, as its name indicates, stimulates gall bladder contraction. It is also found in the CNS of vertebrates, including reptiles, where shorter peptide fragments of CCK, in particular the C-terminal octapeptide (CCK8), predominate (Reiner & Beinfeld, 1985). A comparison of the distribution of CCK8 immunoreactive neuronal elements with that of catecholaminergic elements in the brain of turtles reveals several places where interactions between these transmitter systems may occur. Codistribution of CCK8 and catecholamine cell bodies is observed in the periventricular hypothalamic nucleus, ventral tegmental area, substantia nigra, locus coeruleus and the nucleus of the solitary tract, including its prevagal portion. Moreover, catecholamine fibers may impinge upon CCK8 cell bodies in several other places, e.g. the nucleus accumbens and the septum. Other areas, like the medial and dorsal cortices, the periventricular zone of the midbrain and the parabrachial region, are innervated by both catecholamine and CCK8 fibers.

Neuropeptide Y. Although only one study has provided a complete overview of the distribution of NPY in the brain of a reptile, the lizard *Gallotia galloti* (Medina et al., 1992), it is obvious that a close relationship exists between this neuropeptide and catecholamine systems at several places in the brain. For example, the distribution of NPY immunoreactive cell bodies matches that of DAi/THi cell bodies in the periventricular hypothalamic nucleus, ventral tegmental area, substantia nigra and retrorubral area. Moreover, at many places codistribution of NPY and CA immunoreactive fibers occurs, particularly in the nucleus accumbens, striatum, DVR, amygdaloid complex, hypothalamus and the periventricular zone of the midbrain. Coexistence of dopamine with NPY has been demonstrated for the intermediate lobe of the pituitary of amphibians (see Chapter 5), but data on such relationships in reptiles are still lacking.

Interactions with neurohypophyseal peptides: vasotocin and mesotocin Biochemical studies have shown that in the CNS of reptiles arginine-vasotocin (AVT) and mesotocin (MST) are present, two peptides that are considered to function similarly to the mammalian vasopressin and oxytocin, respectively (Acher, 1981). Immunohistochemical studies of the distribution of AVT in the brain of lizards (Stoll & Voorn, 1985; Thepen et al., 1987), turtles and snakes (Smeets, Sevensma & Jonker, 1990) have revealed that AVTi cell bodies are present in the supraoptic nucleus, the paraventricular nucleus and the bed nucleus of the stria terminalis. The MSTi cell bodies are found in the same nuclei, except for the bed nucleus of the stria terminalis. DAi and AVTi cell bodies lie close to each other, but the two neurotransmitters are most likely not colocalized. Nevertheless, intimate relationships seem to exist between the two neuropeptidergic systems and catecholamines. For example, all three catecholamines innervate profusely the paraventricular AVT/MST cell group, whereas the reverse holds for the catecholaminergic cell groups in the ventral tegmental area, substantia nigra, locus coeruleus, and the nucleus of the solitary tract where elaborate networks of AVTi fibers are found. Co-distribution of AVT and catecholaminergic fibers occurs in limbic structures, such as the nucleus accumbens, the septum, the nucleus of the diagonal band of Broca, and the amygdaloid complex.

Concluding remarks

A survey of our present knowledge of catecholamine systems in the central nervous system of reptiles has been presented in the preceeding sections. Within the class of Reptilia, many general conditions exist with regard to the distribution of catecholamine cell bodies and fibers. However, remarkable differences are also observed between the various species studied, such as the dopaminergic innervation of striatal structures and the dorsal ventricular ridge, the presence of large, DAi/THi cell bodies in the diencephalon and isthmic region of snakes, and the relative abundance of CSF-contacting DAi cells in the diencephalon of turtles. Unfortunately, the significance of these differences are not understood yet.

A notable gap in our knowledge of catecholamines in the CNS of reptiles concerns the distribution of noradrenaline and adrenaline. Only for the lizard *Gekko gecko*, detailed information on these systems is available. Other shortcomings are the lack of information on the distribution of receptors and on the functional significance of the relationships between catecholamines and sensory-, motor- or other neurotransmitter systems. The present survey intends to provide some common ground upon which experimental studies can be launched to solve the problems left.

Acknowledgements

I wish to thank Drs J. Meek and T. Reiner for their helpful comments on the manuscript. I am also grateful to my colleagues Drs H. J. Groenewegen, P. V. Hoogland, A. H. M. Lohman, F. T. Russchen, P. Voorn and to my Spanish collaborators Drs A. Gonzalez and L. Medina for their continuous interest in comparative aspects of catecholamine research. I am also much indebted to Mr P. Goede, Mr A. J. Jonker, Mrs B. Jorritsma-Byham and Mrs E. Vermeulen-van der Zee for their excellent technical assistance and to Mr D. de Jong for preparing high quality photomicrographs during all those years.

Abbreviations

Acc	nucleus accumbens
Alh	area lateralis hypothalami
Amc	nucleus centralis amygdalae
Ame	nucleus externus amygdalae
Aml	nucleus lateralis amygdalae
Apol	area preoptica lateralis
Bmfb	bed nucleus of the medial forebrain bundle
ca	commissura anterior
cb	cerebellum
Cgld	corpus geniculatum laterale, pars dorsalis
Cglv	corpus geniculatum laterale, pars ventralis
Cgp	corpus geniculatum pretectale
Cxd	cortex dorsalis
Cxla	cortex lateralis, pars anterior
Cxlp	cortex lateralis, pars posterior
Cxml	cortex medialis, large-celled part
Cxms	cortex medialis, small-celled part
db	dorsal bundle
Dlh	nucleus dorsolateralis hypothalami
Dll	nucleus dorsolateralis thalami, large-celled part
Dls	nucleus dorsolateralis thalami, small-celled part
Dm	nucleus dorsomedialis thalami
DVR	dorsal ventricular ridge
flm	fasciculus longitudinalis medialis
GP	globus pallidus
Hab	ganglion habenulae
Habl	nucleus lateralis habenulae
Habm	nucleus medialis habenulae
Ipv	nucleus interpeduncularis, pars ventralis
lb	longitudinal bundle
Lc	locus coeruleus
lfb	lateral forebrain bundle
lfbd	lateral forebrain bundle, dorsal peduncle
lfbv	lateral forebrain bundle, ventral peduncle
Lte	nucleus lentiformis thalami, pars extensa
Ltp	nucleus lentiformis thalami, pars plicata
Mp	nucleus medialis posterior
Mt	nucleus medialis thalami
NdB	nucleus of the diagonal band of Broca
Nsa	nucleus septalis anterior
Nsd	nucleus septalis dorsalis
Nsi	nucleus septalis impar
Nsl	nucleus septalis lateralis
Nsm	nucleus septalis medialis
Nsph	nucleus sphericus
NTS	nucleus tractus solitarii
nIII	nervus oculomotorius
oph	organon periventriculare hypothalami
Ph	nucleus periventricularis hypothalami
Ppo	periventricular preoptic area
Pth	pallial thickening
Ras	nucleus raphes superior
RA8	reptilian equivalent of mammalian A8
Rot	nucleus rotundus
Rub	nucleus ruber
SNc	substantia nigra, pars compacta
SNr	substantia nigra, pars reticulata
So	nucleus supraopticus
Sped	nucleus suprapeduncularis
Str	striatum
tect	tectum mesencephali

topt	tractus opticus
Torc	nucleus centralis of the torus semicircularis
Torl	nucleus laminaris of the torus semicircularis
tsh	tractus septohypothalamicus
tub olf	tuberculum olfactorium
Vltd	nucleus ventrolateralis thalami, pars dorsalis
Vltv	nucleus ventrolateralis thalami, pars ventralis
Vmh	nucleus ventromedialis hypothalami
Vmt	nucleus ventromedialis thalami
VTA	ventral tegmental area
III	nucleus nervi oculomotorii
IIId	nucleus dorsalis nervi oculomotorii
IIIv	nucleus ventralis nervi oculomotorii
IV	nucleus nervi trochlearis
X	nucleus motorius dorsalis nervi vagi
XII	nucleus nervi hypoglossi

References

Acher, R. (1981). Evolution of neuropeptides. *Trends in Neurosciences*, **4**, 225–9.

Anderson, K. D. & Reiner, A. (1990). Extensive co-occurrence of substance P and dynorphin in striatal projection neurons: An evolutionary conserved feature of basal ganglia organization. *Journal of Comparative Neurology*, **295**, 339–369.

Bennis, M., Gamrani, H., Geffard, M., Calas, A. & Kah, O. (1990). The distribution of 5-HT immunoreactive systems in the brain of a saurian, the chameleon. *Journal für Hirnforschung*, **31**, 563–74.

Brauth, S. E. (1988). Catecholamine neurons in the brainstem of the reptile *Caiman crocodilus*. *Journal of Comparative Neurology*, **270**, 313–26.

Brauth, S. E. (1990). Histochemical strategies in the study of neural evolution. *Brain, Behavior and Evolution*, **36**, 100–15.

Brauth, S. E. & Kitt, C. A. (1980). The paleostriatal system of *Caiman crocodilus*. *Journal of Comparative Neurology*, **189**, 437–65.

Brauth, S. E., Kitt, C. A., Price, D. L. & Wainer, B. H. (1985). Cholinergic neurons in the telencephalon of the reptile *Caiman crocodilus*. *Neuroscience Letters*, **58**, 235–40.

Brauth, S. E., Reiner, A., Kitt, C. A. & Karten, H. J. (1983). The substance P-containing striatotegmental path in reptiles: an immunohistochemical study. *Journal of Comparative Neurology*, **219**, 305–27.

Brecha, N. C. (1983). Retinal neurotransmitters: histochemical and biochemical studies. In *Chemical Neuroanatomy*, ed. Emson, P. C. pp.85–129. New York: Raven Press.

Bruce, L. L. & Butler, A. B. (1984). Telencephalic connections in lizards. I. Projections to cortical areas. *Journal of Comparative Neurology*, **229**, 585–601.

Challet, E., Pierre, J., Reperant, J., Ward, R. & Miceli, D. (1991). The seotoninergic sytem of the brain of the viper, *Vipera aspis*. An immunohistochemical study. *Journal of Chemical Neuroanatomy*, **4**, 233–48.

Ebbesson, S. O. E. (1984). Evolution and ontogeny of neural circuits. *The Behavioral and Brain Sciences*, **7**, 321–66.

Ehinger, B. (1982). Neurotransmitter systems in the retina. *Retina*, **2**, 305–21.

Falck, B., Hillarp, N. A., Thieme, G. & Torp, A. (1962). Fluorescence of catecholamines and related compounds with formaldehyde. *Journal of Histochemistry and Cytochemistry*, **10**, 348–54.

Geffard, M., Buijs, R. M., Seguela, P., Pool, C. W. & Le Moal, M. (1984). First demonstration of highly specific and sensitive antibodies against dopamine. *Brain Research*, **294**, 161–5.

Gerschenfeld, H. M., Neyton, J., Piccolino, M. & Witkovsky, P. (1983). L-horizontal cells of the turtle: network organization and coupling modulation. *Biomedical Research*, **3**, 21–34.

Goldstein, M., Fuxe, K., Hökfelt, T. & Joh, T. H. (1971). Immunohistochemical studies on phenylethanolamine-N-methyltransferase, DOPA-decarboxylase and dopamine-β-hydroxylase. *Experientia*, **27**, 951–2.

González, A., Russchen, F. T. & Lohman, A. H. M. (1990). Afferent connections of the striatum and the nucleus accumbens in the lizard *Gekko gecko*. *Brain Behavior and Evolution*, **36**, 39–58.

Guirado, S., de la Calle, A., Gutierrez, A. & Davila, J. C. (1989). Serotin innervation of the cerebral cortex in lizards. *Brain Research*, **488**, 213–20.

Haber, S. & Groenewegen, H. J. (1989). Interrelationship of the distribution of neuropeptides and tyrosine hydroxylase immunoreactivity in the human substantia nigra. *Journal of Comparative Neurology*, **290**, 53–68.

Halasz, N. & Shepherd, G. M. (1983). Neurochemistry of the vertebrate olfactory bulb. *Neuroscience*, **10**, 579–619.

Halasz, N., Nowycky, M., Hökfelt, T., Shepherd, G. M., Markey, K. & Goldstein, M. (1982). Dopaminergic periglomerular cells in the turtle olfactory bulb. *Brain Research Bulletin*, **9**, 383–9.

Halpern, M. & Kubie, J. L. (1984). The role of the ophidian vomeronasal system in species-typical behavior. *Trends in Neurosciences*, **7**, 472–7.

Henselmans, J. M. L., Hoogland, P. V. & Stoof, J. C. (1991). Differences in the regulation of acetylcholine release upon D2 dopamine and N-methyl-D-aspartate receptor activation between the striatal complex of reptiles and the neostriatum of rats. *Brain Research-r*, **566**, 8–12.

Henselmans, J. M. L. & Stoof, J. C. (1991). Regional differences in the regulation of acetylcholine release upon D2 dopamine and N-methyl-D-aspartate receptor activation

in rat nucleus accumbens and neostriatum. *Brain Research*, **566**, 1–7.
Hökfelt, T., Fuxe, K., Goldstein, M. & Joh, T. (1973). Immunohistochemical studies of three catecholamine-synthesizing enzymes: aspects and methodology. *Histochemie*, **29**, 325–39.
Hoogland, P. V. (1977). Efferent connections of the striatum in *Tupinambis nigropunctatus*. *Journal of Morphology*, **152**, 229–46.
Hoogland, P. V. (1982). Brainstem afferents to the thalamus in a lizard, *Varanus exanthematicus*. *Journal of Comparative Neurology*, **210**, 152–62.
Hoogland, P. V. & Vermeulen-van der Zee, E. (1989). Efferent connections of the dorsal cortex of the lizard *Gekko gecko* studied with *Phaseolus vulgaris*-leucoagglutinin. *Journal of Comparative Neurology*, **285**, 289–303.
Hoogland, P. V. & Vermeulen-van der Zee, E. (1990). Distribution of choline acetyltransferase immunoreactivity in the telencephalon of the lizard *Gekko gecko*. *Brain, Behavior and Evolution*, **36**, 378–90.
Inagaki, N., Panula, P., Yamatodani, A. & Wada, H. (1990). Organization of the histaminergic system in the brain of the turtle *Chinemys reevesii*. *Journal of Comparative Neurology*, **297**, 132–144.
Kadota, T., Kishida, R., Goris, R. C. & Kusunoki, T. (1988). Substance P-like immunoreactivity in the trigeminal sensory nuclei of an infrared-sensitive snake. *Agkistrodon blomhoffi*. *Cell Tissue Research*, **253**, 311–17.
Kiehn, O., Rostrup, E. & Moller, M. (1992). Monoaminergic systems in the brainstem and spinal cord of the turtle *Pseudemys scripta elegans* as revealed by antibodies against serotonin and tyrosine hydroxylase. *Journal of Comparative Neurology*, **325**, 527–47.
Kirsche, W. (1972). Die Entwicklung des Telencephalons der Reptilien und deren Beziehung zur Hirn-Bauplanlehre. *Nova Acta Leopoldina*, **204**, 1–78.
Kusuma, A., ten Donkelaar, H. J. & Nieuwenhuys, R. (1979). Intrinsic organization of the spinal cord. In *Biology of the Reptilia, Vol. 10: Neurology B*, ed. Gans, C., Northcutt, R. G. & Ulinski, P. pp. 59–109. London: Academic Press.
Lehmann, J. & Langer, S. Z. (1983). The striatal cholinergic interneuron: Synaptic target of dopaminergic terminals. *Neuroscience*, **10**, 1105–20.
Lohman, A. H. M. & Smeets, W. J. A. J. (1990). The dorsal ventricular ridge and cortex of reptiles in historical and phylogenetic perspective. In *The Neocortex. Ontogeny and Phylogeny*, NATO ASI Series, Vol. 200, ed. Finlay, B. L., Innocenti, G. & Scheich, H. pp. 59–74. New York: Plenum.
Lohman, A. H. M. & Smeets, W. J. A. J. (1993). Overview of the main and accessory olfactory bulb projections in reptiles. *Brain Behavior and Evolution*, **41**, 147–55.
Lopez, K. H., Jones, R. E., Seufert, D. W., Rand, M. S. & Dores, R. M. (1992). Catecholaminergic cells and fibers in the brain of the lizard *Anolis carolinensis* identified by traditional as well as whole-mount immunohistochemistry. *Cell Tissue Research*, **270**, 319–37.

Malmfors, T. (1963). Evidence of adrenergic neurons with synaptic terminals in the retina of rats demonstrated with fluorescence and electron microscopy. *Acta Physiologica Scandinavica*, **58**, 99–100.
Martinez-Garcia, F., Olucha, F. E., Teruel, V., Lorente, M. J. & Schwerdtfeger, W. K. (1991). Afferent and efferent connections of the olfactory bulbs in the lizard *Podarcis hispanica*. *Journal of Comparative Neurology*, **305**, 337–47.
Medina, L., Marti, E., Artero, C., Fasolo, A. & Puelles, L. (1992). Distribution of neuropeptide-like immunoreactivity in the brain of the lizard *Gallotia galloti*. *Journal of Comparative Neurology*, **319**, 387–405.
Medina, L. & Smeets, W. J. A. J. (1991). Comparative aspects of the basal ganglia-tectal pathways in reptiles. *Journal of Comparative Neurology*, **308**, 614–29.
Medina, L. & Smeets, W. J. A. J. (1992). Cholinergic, monoaminergic and peptidergic innervation of the primary visual centers in the brain of the lizards *Gekko gecko* and *Gallotia galloti*. *Brain Behavior and Evolution*, **40**, 157–81.
Medina, L., Smeets, W. J. A. J., Hoogland, P. V. & Puelles, L. (1993). Distribution of choline acetyltransferase immunoreactivity in the brain of the lizard *Gallotia galloti*. *Journal of Comparative Neurology* **331**, 261–85.
Meister, B., Hökfelt, T., Steinbusch, H. W. M., Skagerberg, G., Lindvall, O., Geffard, M., Joh, T. H., Cuello, A. C. & Goldstein, M. (1988). Do tyrosine hydroxylase-immunoreactive neurons in the ventrolateral arcuate nucleus produce dopamine or only L-Dopa? *Journal of Chemical Neuroanatomy*, **1**, 59–64.
Meredith, G. E., Blank, B. & Groenewegen, H. J. (1989). The distribution and compartmental organization of the cholinergic neurons in nucleus accumbens of the rat. *Neuroscience*, **31**, 327–45.
Molenaar, G. J. (1977). The rhombencephalon of *Python reticulatus*, a snake possessing infrared receptors. *Netherlands Journal of Zoology*, **27**, 133–80.
Molenaar, G. J. & Fizaan-Oostveen, J. L. F. P. (1980). Ascending projections from the lateral descending and common sensory trigeminal nuclei in *Python*. *Journal of Comparative Neurology*, **189**, 555–72.
Mufson, E. J., Desan, P. H., Mesulam, M. M., Wainer, B. H. & Levey, A. I. (1984). Choline acetyltransferase-like immunoreactivity in the forebrain of the red-eared pond turtle (*Pseudemys scripta elegans*). *Brain Research*, **323**, 103–8.
Nauta, W. J. H. & Karten, H. J. (1970). A general profile of the vertebrate brain, with side lights on the ancestry of cerebral cortex. In *The Neurosciences Second Study Program*, ed. Schmitt, F. O. pp. 1–26. New York: Rockefeller Univ.
Nguyen-Legros, J., Versaux-Botteri, C., Vigny, A. & Raoux, N. (1985). Tyrosine hydroxylase immunohistochemistry fails to demonstrate dopaminergic interplexiform cells in the turtle retina. *Brain Research*, **339**, 323–8.
Nieuwenhuys, R. (1967). Comparative anatomy of olfactory centres and tracts. In *Sensory Mechanisms, Progress in*

Brain Research, Vol. 23, ed. Y. Zotterman, pp. 1–64. Amsterdam: Elsevier.

Northcutt, R. G. (1978). Forebrain and midbrain organization in lizards and its phylogenetic significance. In *Behavior and Neurology of Lizards*, ed. Greenberg, N. & MacLean, P. D. pp. 11–64. Rockville, National Institute of Mental Health.

Northcutt, R. G. (1984). Evolution of the vertebrate nervous system: Patterns and processes. *American Zoologist*, **24**, 701–16.

Northcutt, R. G. (1990). Ontogeny and phylogeny: a re-evaluation of conceptual relationships and some applications. *Brain, Behavior and Evolution*, **36**, 116–40.

Nowycky, M. C., Halasz, N. & Shepherd, G. M. (1983). Evoked field potential analysis of dopaminergic mechanisms in the isolated turtle olfactory bulb. *Neuroscience*, **8**, 717–22.

Okamura, H., Kitahama, K., Raynaud, B., Nagatsu, I., Borri-Volttatorni, C. & Weber, M. (1988a). Aromatic L-amino acid decarboxylase (AADC)-immunoreactive cells in the tuberal region of the rat hypothalamus. *Biomedical Research*, **9**, 261–7.

Okamura, H., Kitahama, K., Nagatsu, I. & Geffard, M. (1988b). Comparative topography of dopamine- and tyrosine hydroxylase-immunoreactive neurons in the rat arcuate nucleus. *Neuroscience Letters*, **95**, 347–53.

Okamura, H., Kitahama, K., Mons, N., Ibata, Y., Jouvet, M. & Geffard, M. (1988c). L-DOPA-immunoreactive neurons in the rat hypothalamic tuberal region. *Neuroscience Letters*, **95**, 42–6.

Parent, A. (1973). Distribution of monoamine-containing nerve terminals in the brain of the painted turtle, *Chrysemys picta*. *Journal of Comparative Neurology*, **148**, 153–66.

Parent, A. (1976). Striatal afferent connections in the turtle (*Chrysemys picta*) as revealed by retrograde axonal transport of horseradish peroxidase. *Brain Research*, **108**, 25–36.

Parent, A. (1979). Monoaminergic systems of the brain. In *Biology of the Reptilia*, Vol. 10, ed. C. Gans, R. G. Northcutt & P. S. Ulinski, pp. 247–85. London; Academic Press.

Parent, A. & Olivier, A. (1970). Comparative histochemical study of the corpus striatum. *Journal fur Hirnforschung*, **12**, 73–81.

Parent, A., Poitras, D. & Dubé, L. (1984). Comparative anatomy of central monoaminergic systems. In *Handbook of Chemical Neuroanatomy, Vol. 2. Classical transmitters in the CNS, Part I*, ed. A. Björklund & T. Hökfelt, pp. 409–39. Amsterdam: Elsevier.

Phelps, P. E., Houser, C. R. & Vaughn, J. E. (1985). Immunocytochemical localization of choline acetyltransferase within the rat neostriatum: A correlated light and electron microscopic study of cholinergic neurons and synapses. *Journal of Comparative Neurology*, **238**, 286–307.

Prasado Rao, P. D. & Subhedar, N. (1977). A cytoarchitectonic study of the hypothalamus of the lizard, *Calotes versicolor*. *Cell Tissue Research*, **180**, 63–85.

Regal, P. J. (1978). Behavioral differences between reptiles and mammals: an analysis of activity and mental capabilities. In *Behavior and Neurology of Lizards*, ed. N. Greenberg & P. D. MacLean, pp. 183–202. Rockville: National Institute of Mental Health.

Reiner, A. (1992). Neuropeptides in the nervous system. In *Biology of the Reptilia, Vol. 17, Neurology C, Sensory Motor Integration*, ed. Gans C. & Ulinski, P. S., pp. 587–739. London: University Chicago Press.

Reiner, A. (1993). Neurotransmitter organization and connections of turtle cortex: implications for the evolution of mammalian isocortex. *Comparative Biochemistry and Physiology*, **104A**, 735–48.

Reiner, A. & Anderson, K. D. (1990). The patterns of neurotransmitter and neuropeptide co-occurrence among striatal projection neurons: conclusions based on recent findings. *Brain Research Reviews Bulletin*, **15**, 251–65.

Reiner, A. & Beinfeld, M. C. (1985). The distribution of cholecystokinin-8 in the central nervous system of turtles: an immunohistochemical and biochemical study. *Brain Research Bulletin*, **15**, 167–81.

Reiner, A. & Karten, H. J. (1985). Comparison of olfactory bulb projections in pigeons and turtles. *Brain, Behavior and Evolution*, **27**, 11–27.

Reiner, A., Karten, H. J. & Solina, A. R. (1983). Substance P: localization within paleostriatal-tegmental pathways in the pigeon. *Neuroscience*, **9**, 61–85.

Reiner, A., Krause, J. E., Keyser, K. T., Eldred, W. D. & McKelvy, J. F. (1984). The distribution of substance P in turtle nervous system: A radioimmunoassay and immunohistochemical study. *Journal of Comparative Neurology*, **226**, 50–75.

Richfield, E. K., Young, A. B. & Penney, J. B. (1987). Comparative distribution of dopamine D–1 and D–2 receptors in the basal ganglia of turtles, pigeons, rats, cats, and monkeys. *Journal of Comparative Neurology*, **262**, 446–63.

Russchen, F. T. & Jonker, A. J. (1988). Efferent connections of the striatum and the nucleus accumbens in the lizard *Gekko gecko*. *Journal of Comparative Neurology*, **276**, 61–80.

Russchen, F. T., Smeets, W. J. A. J. & Hoogland, P. V. (1987a). Histochemical identification of pallidal and striatal structures in the lizard *Gekko gecko*. Evidence for compartmentalization. *Journal of Comparative Neurology*, **256**, 239–41.

Russchen, F. T., Smeets, W. J. A. J. & Lohman, A. H. M. (1987b). On the basal ganglia of a reptile: The lizard *Gekko gecko*. In *Basal Ganglia: Structure and Function*, ed. Carpenter M. B. & Jayaraman A. pp. 261–81. New York: Plenum Press.

Smeets, W. J. A. J. (1988a). Distribution of dopamine immunoreactivity in the forebrain and midbrain of the snake *Python regius*: a study with antibodies against dopamine. *Journal of Comparative Neurology*, **271**, 115–29.

Smeets, W. J. A. J. (1988b). The monoaminergic systems of reptiles investigated with specific antibodies against seotonin, dopamine, and noradrenaline. In *The Forebrain of Reptiles: Current Concepts of Structure and Func-

tion, ed. Schwerdtfeger W. K. & Smeets, W. J. A. J., pp. 97–109. Basel: Karger.

Smeets, W. J. A. J. (1991). Comparative aspects of the distribution of substance P and dopamine immunoreactivity in the substantia nigra of amniotes. *Brain, Behavior and Evolution,* **37**, 179–88.

Smeets, W. J. A. J. (1992). Comparative aspects of basal forebrain organization in vertebrates. *European Journal of Morphology,* **30**, 23–36.

Smeets, W. J. A. J, Hoogland, P. V. & Lohman, A. H. M. (1986a). A forebrain atlas of the lizard *Gekko gecko. Journal of Comparative Neurology,* **254**, 1–19.

Smeets, W. J. A. J., Hoogland, P. V. & Voorn, P. (1986b). The distribution of dopamine immunoreactivity in the forebrain and midbrain of the lizard *Gekko gecko*: an immunohistochemical study with antibodies against dopamine. *Journal of Comparative Neurology,* **253**, 46–60.

Smeets, W. J. A. J. & Jonker, A. J. (1990). Distribution of phenylethanolamine-N-methyltransferase-immunoreactive perikarya and fibers in the brain of the lizard *Gekko gecko. Brain, Behavior and Evolution,* **36**, 59–72.

Smeets, W. J. A. J., Jonker, A. J. & Hoogland, P. V. (1987). Distribution of dopamine in the forebrain and midbrain of the red-eared turtle, *Pseudemys scripta elegans,* reinvestigated using antibodies against dopamine. *Brain, Behavior and Evolution,* **30**, 121–42.

Smeets, W. J. A. J., Kidjan, M. N. & Jonker, A. J. (1991). α-MPT does not affect dopamine levels in the periventricular organ of lizards. *NeuroReport,* **2**, 369–72.

Smeets, W. J. A. J., Sevensma, J. J. & Jonker, A. J. (1990). Comparative analysis of vasotocin-like immunoreactivity in the brain of the turtle *Pseudemys scripta elegans* and the snake *Python regius. Brain, Behavior and Evolution,* **35**, 65–84.

Smeets, W. J. A. J. & Steinbusch, H. W. M. (1988). Distribution of serotonin immunoreactivity in the forebrain and midbrain of the lizard *Gekko gecko. Journal of Comparative Neurology,* **271**, 419–34.

Smeets, W. J. A. J. & Steinbusch, H. W. M. (1989). Distribution of noradrenaline immunoreactivity in the forebrain and midbrain of the lizard *Gekko gecko. Journal of Comparative Neurology,* **285**, 453–66.

Smeets, W. J. A. J. & Steinbusch, H. W. M. (1990). New insights into the reptilian catecholaminergic systems as revealed by antibodies against the neurotransmitters and their synthetic enzymes. *Journal of Chemical Neuroanatomy,* **3**, 25–43.

Steinbusch, H. W. M. & Tilders, F. J. H. (1987). Immunohistochemical techniques for light-microscopical localization of dopamine, noradrenaline, adrenaline, serotonin and histamine in the central nervous system. In *Monoaminergic Neurons: Light Microscopy and Ultrastructure. IBRO Handbook Series: Methods in the Neurosciences,* Vol.10, ed. Steinbusch, H. W. M. pp. 125–66. Chichester: Wiley.

Steinbusch, H. W. M., Verhofstad, A. A. J., Penke, B., Varga, J., & Joosten, H. W. J. (1981). Immunohistochemical characterization of monoamine-containing neurons in the central nervous system by antibodies to serotonin and noradrenaline. A study in the rat and the lamprey. *Acta Histochemica,* Supp. **24**, 107–22.

Steinbusch, H. W. M., de Vente, J. & Schipper, J. (1986). Immunohistochemistry of monoamines in the central nervous system. In *Neurochemistry Today, Neurology and Neurobiology,* Vol.20, ed. Panula, P., Paivarinta, H., Soinila, S. pp. 75–105. New York: Alan Liss.

Stoll, C. J. & Voorn, P. (1985). The distribution of hypothalamic and extrahypothalamic vasotocinergic cells and fibers in the brain of a lizard, *Gekko gecko*: presence of a sex difference. *Journal of Comparative Neurology,* **239**, 193–204.

Stoof, J. C., Russchen, F. T., Verheijden, P. F. H. M. & Hoogland, P. V. J. M. (1987). A comparative study of the dopamine–acetylcholine interaction in telencephalic structures of the rat and of a reptile, the lizard *Gekko gecko. Brain Research,* **404**, 273–81.

ten Donkelaar, H. J., Bangma, G. C., Barbas-Henry, H. A., de Boer-van Huizen, R. & Wolters, J. G. (1987). The brain stem in a lizard, *Varanus exanthematicus. Advances in Anatomy Embryology and Cell Biology,* **107**, 1–168.

ten Donkelaar, H. J. & de Boer-van Huizen, R. (1988). Brainstem afferents to the anterior dorsal ventricular ridge in a lizard (*Varanus exanthematicus*). *Anatomy and Embryology,* **177**, 465–75.

ten Donkelaar, H.J ., Kusuma, A. & de Boer-van Huizen, R. (1980). Cells of origin of pathways descending to the spinal cord in some quadrupedal reptiles. *Journal of Comparative Neurology,* **192**, 827–51.

Terashima, S. (1987). Substance P-like immunoreactive fibers in the trigeminal sensory nuclei of the pit viper, *Trimeresurus flavoviridis. Neuroscience,* **23**, 685–91.

Thepen, Th., Voorn, P., Stoll, C. J., Sluiter, A. A., Pool, C. W. & Lohman, A. H. M. (1987). Mesotocin and vasotocin in the brain of the lizard *Gekko gecko.* An immunocytochemical study. *Cell and Tissue Research,* **250**, 649–56.

Ueda, S., Takeuchi, Y. & Sano, Y. (1983). Immunohistochemical demonstration of serotonin neurons in the central nervous system of a turtle (*Clemnys japonica*). *Anatomy and Embryology,* **168**, 1–19.

Ulinski, P. S. (1983). *The Dorsal Ventricular Ridge. A Treatise on Forebrain Organization in Reptiles and Birds.* New York: Wiley.

Ulinski, P. S. (1990). The cerebral cortex of reptiles. In *Cerebral Cortex, Vol. 8A, Comparative Structure and Evolution of Cerebral Cortex,* Part I, ed. E. G. Jones & A. Peters, pp. 139–215. New York: Plenum.

Verhofstad, A. A. J., Steinbusch, H. W. M., Penke, B., Varga, J. & Joosten, H. W. J. (1980). Use of antibodies to norepinephrine and epinephrine in immunohistochemistry. In *Histochemistry and Cell Biology of Autonomic Neurons, SIF Cells and Paraneurons,* ed. Eranko, O., Soinila, S. & Paivarinta H., pp. 185–93. New York: Raven Press.

Verhofstad, A. A. J., Steinbusch, H. W. M., Joosten,

H. W. J., Penke, B., Varga, J. & Goldstein, M. (1982). Immunocytochemical localization of noradrenaline, adrenaline and serotonin. In *Immunocytochemistry: Practical Applications in Pathology and Biology*, ed. Polak, J. M. & Van Noorden, S. pp. 143–68. Bristol: John Wright & Sons.

Vigh-Teichmann, I. & Vigh, B. (1974). The infundibular cerebrospinal-fluid contacting neurons. *Advances in Anatomy Embryology and Cell Biology*, **50**, 1–90.

Voneida, T. J. & Sligar, C. M. (1979). Efferent projections of the dorsal ventricular ridge and the striatum in the Tegu lizard, *Tupinambis nigropunctatus*. *Journal of Comparative Neurology*, **186**, 43–64.

Welker, E., Hoogland, P. V. & Lohman, A. H. M. (1983). Tectal connections in *Python reticulatus*. *Journal of Comparative Neurology*, **220**, 347–54.

Witkovsky, P., Eldred, W. & Karten, H. J. (1984). Catecholamine- and indolamine-containing neurons in the turtle retina. *Journal of Comparative Neurology*, **228**, 217–25.

Wolters, J. G., ten Donkelaar, H. J., Steinbusch, H. W. M. & Verhofstad, A. A. J. (1985). Distribution of serotonin in the brain stem and spinal cord of the lizard *Varanus exanthematicus*: an immunohistochemical study. *Neuroscience*, **14**, 169–73.

Wolters, J. G., ten Donkelaar, H. J. & Verhofstad, A. A. J. (1984). Distribution of catecholamines in the brain stem and spinal cord of the lizard *Varanus exanthematicus*: an immunohistochemical study based on the use of antibodies to tyrosine hydroxylase. *Neuroscience*, **13**, 469–93.

Wolters, J. G., ten Donkelaar, H. J. & Verhofstad, A. A. J. (1986). Distribution of some peptides (substance P, [leu]enkephalin, [met]enkephalin) in the brain stem and spinal cord of a lizard, *Varanus exanthematicus*. *Neuroscience*, **18**, 917–46.

Woodson, W. & Kunzle, H. (1982). Distribution and structural characterization of neurons giving rise to descending spinal projections in the turtle, *Pseudemys scripta elegans*. *Journal of Comparative Neurology*, **212**, 336–348.

Wulle, I. & Wagner, H.-J. (1990). GABA and tyrosine hydroxylase immunocytochemistry reveal different patterns of colocalization in retinal neurons of various vertebrates. *Journal of Comparative Neurology*, **296**, 173–8.

7

Catecholaminergic perikarya and fibers in the avian nervous system

A. Reiner, E. J. Karle, K. D. Anderson and L. Medina

Introduction

Various aspects of the distribution of catecholaminergic (CA) perikarya and fibers in the avian brain have been described in previous studies. The earliest studies on this topic used catecholamine histofluorescence methods (for a review see Parent, 1979; Parent, Poitras & Dubé, 1984). Since these methods cannot readily distinguish between neurons containing different catecholamines (e.g. dopamine versus adrenalin or noradrenalin), these studies were unable to characterize definitively the distribution of the different types of catecholaminergic neurons and fibers. Immunohistochemical studies using antisera against each of the CA synthetic enzymes or against specific catecholamines have recently made it possible to make such distinctions. Further, by their greater sensitivity, immunohistochemical methods have made it possible to more fully determine the distribution of catecholaminergic perikarya and fibers than possible with catecholamine histofluorescence methods. Several recent studies have, in fact, been carried out in a few avian groups (chickens, pigeons and song birds) using immunohistochemical staining for catecholamine synthetic enzymes or for specific catecholamines. The published information on the distribution of catecholamines in birds is, however, yet incomplete (many studies having focussed only on specific brain regions) and data are available for only a few species. We have been investigating in detail the organization of the catecholaminergic systems in pigeons. In one line of studies, we have used light microscopic single-label immunohistochemical methods to discriminate among and map the perikaryal and fiber distributions of the different catecholamines (dopamine, noradrenalin and adrenalin), using antisera against dopamine (DA), and against the synthetic enzymes tyrosine hydroxylase (TH), dopamine β-hydroxylase (DBH) and phenyl N-methyltransferase (PNMT). In a second set of studies, we have used a variety of approaches to investigate the connectivity of midbrain dopaminergic cell groups with the forebrain. The various available data help clarify a number of features of the distribution and function of catecholamines in the avian brain. The purpose of the present chapter is to summarize these data.

Rationale

The following presentation will be organized into several sections. In the first section we will focus on our own data on the distribution of TH, DBH, PNMT and DA in the central nervous system in pigeons. These studies allow us to reach some conclusions about the distribution of DA+, NA+ and adrenergic perikarya and fibers in the avian brain. In the course of this description, we will also discuss relevant previously published information on pigeons and other avian species. Where possible we will point out differences among birds in CA distribution, but it should be noted that comparisons among avian species are hindered by the paucity of data and by possible technical differences among studies. The data on the distribution of CA+ perikarya are presented in the next section of this chapter on a region by region basis, and the information on CA+ fiber distribution is presented afterwards. In the final section of this chapter, information on the connectivity of the CA cell groups with other brain regions and the functions of these cell groups or regions is presented. In this final section, we summarize our own data on the reciprocal circuitry interconnecting basal ganglia and the midbrain dopaminergic cell groups.

Distribution of catecholamines: perikarya

The nomenclature we will use for CA+ perikarya in birds is adapted from that used for mammals (Hökfelt et al., 1984), as recently utilized for chickens by Kuenzel, Kirtinitis & Saidel (1992). This has been done because the overall distribution of catecholaminergic perikarya in birds closely resembles that in mammals, as well as in other amniotes (Parent et al., 1984; Smeets, 1988; Smeets & Jonker, 1990; Smeets & Steinbusch, 1990). As in other vertebrate groups, the CA+ cell groups of the avian retina, forebrain and midbrain appear to be predominantly dopaminergic (with some hypothalamic neurons possibly being dopa-ergic), while the CA+ cell groups of the avian hindbrain are typically noradrenergic or adrenergic.

Retina: A17

Numerous TH+ cells are found in the pigeon retina at the border between the inner nuclear (INL) and inner plexiform layers (IPL). These cells are spaced at approximately 200–500 μm intervals and give rise to fibers ramifying mainly in the outer IPL, with a few in the inner IPL. These cells fall into two subgroups, an intensely labeled group and a more lightly labeled group (Britto et al., 1988). The former of these appear to be amacrine cells, while retrograde labeling studies show that as many as 60% of the more lightly labeled TH+ cells are displaced ganglion cells projecting to the contralateral nucleus of the basal optic root (nBOR) (Britto et al., 1988). Finally, we also observed some TH+ cells in the outer INL, with processes ramifying in the INL. Amacrine cells that contain TH similar to those described in pigeons have also been reported in chicken retina (Kagami et al., 1991). Catecholamine histofluorescence studies indicate the TH+ amacrine cells in birds to be dopaminergic (Araki, Maeda & Kimura, 1983). None of the various studies has reported the presence of dopaminergic interplexiform cells in birds.

Olfactory bulb: A16

We observed numerous TH+ (but no DA+) perikarya in a periglomerular position and in the outer plexiform layer of the olfactory bulb in pigeons (Fig. 7.1(a)). Kuenzel et al. (1992) have observed similar results in chickens. In pigeons, the processes of the TH+ neurons of the olfactory bulb enter the glomeruli. Labeling with anti-DA in pigeons reveals a dense band of DA+ fibers in the neuropil in the outer plexiform layer and in the glomerular zone of the olfactory bulb (Fig. 7.1(b)), thus suggesting that the TH+ olfactory bulb neurons are DA+ and ramify widely in the outermost olfactory bulb.

Diencephalon: the hypothalamic A11–A15

Five different hypothalamic CA+ cell groups have been identified in mammals based on TH immunolabeling: 1) an A15 in the lateral hypothalamus throughout its extent; 2) an A14 in the medial hypothalamus throughout most of its extent; 3) an A13 in the zone between thalamus and hypothalamus at midhypothalamic levels; 4) an A12 in the tuberal hypothalamus; and 5) an A11 along the midline in the caudalmost fused portion of the hypothalamus. Using the criteria applied in mammals, we could recognize all five of these in pigeons using TH immunolabeling, with some possessing subdivisions similar to those in mammals. Kiss & Peczely (1987) have observed results similar to ours on the distribution of TH+ neurons in pigeon hypothalamus. These five hypothalamic groups do not, however, label for DBH or PNMT. Further, although these groups can be identified by immunolabeling for TH, they do not all label throughout their entire extents for DA. It is uncertain whether the inability to label all of the hypothalamic CA groups for dopamine reflects a technical problem (antisera sensitivity) or is indicative that some hypothalamic CA neurons do not synthesize DA as their final catecholamine.

Each of these cell groups is described in the paragraphs below. It should be noted that each of these groups is defined as a distinct entity largely by its position in the hypothalamus. Several of the groups, however, span diverse levels of the hypothalamus and there is no evidence by connections or by functional data that each of these hypothalamic CA cell groups in birds is a singular entity. Finally, it is interesting to note that catecholamine histofluorescence studies have typically not detected all of these hypothalamic CA cell groups in birds. For example, only caudal parts of A14 and A15 (Parent, 1979; Takatsuki et al., 1981; Dubé & Parent, 1981) and the rostral part of A11 (Parent, 1979; Shiosaka et al., 1981; Takatsuki et al., 1981) were previously detected with CA histofluorescence, while no parts of A12 and A13 were detected (Parent, 1979). In contrast, such studies had found the hypothalamic paraventricular organ (PVO) in birds to be rich in CA neurons (Fuxe & Ljunggren, 1965; Soest, Farner & Oksche, 1973; Parent, 1979). This cell group cannot be classified into the A11-A15 nomenclature, it is not found in mammals and its neurons do not label for TH.

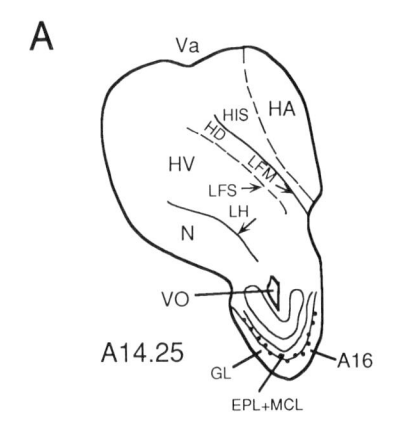

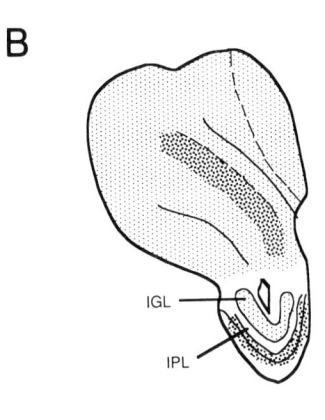

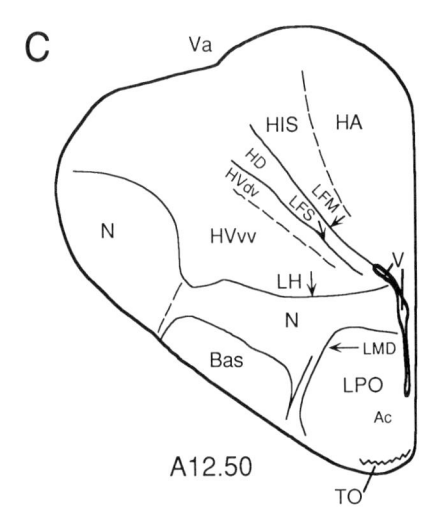

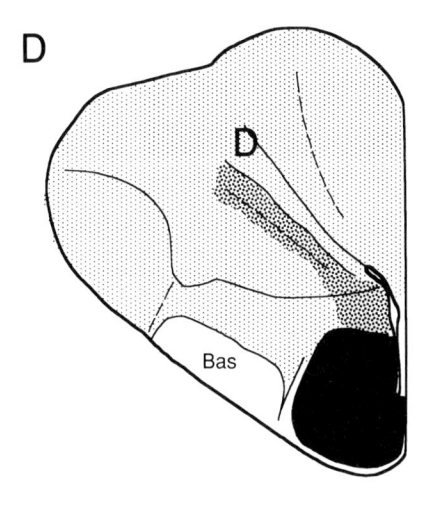

Figs. 7.1–7.9. Line drawing schematics of transverse sections from rostral telencephalic (1) to spinal cord (9) levels. The sections in the left hand column in each panel (A, C, E and G) show the distribution of THi/DBH− (solid) dots or TH+/DBH+ (solid rhomboids) cell bodies and the major cell groups and immunonegative fiber bundles, while the sections in the right hand column of each panel (b), (d), (f) and (h) show the distribution of TH+ fibers and terminals. The density of shading in (b), (d), (f) and (h) is scaled to the relative abundance of TH+ fibers. The numbers to the lower left of sections (a), (c), (e) and (g) show the anterior-posterior level of the sections according to the Karten and Hodos (1967) atlas.

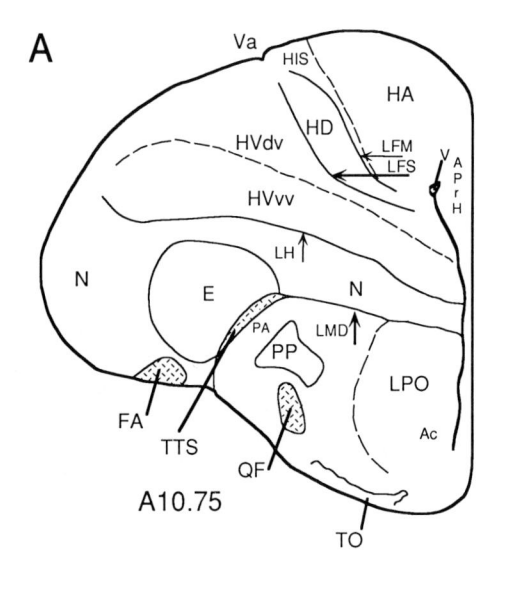

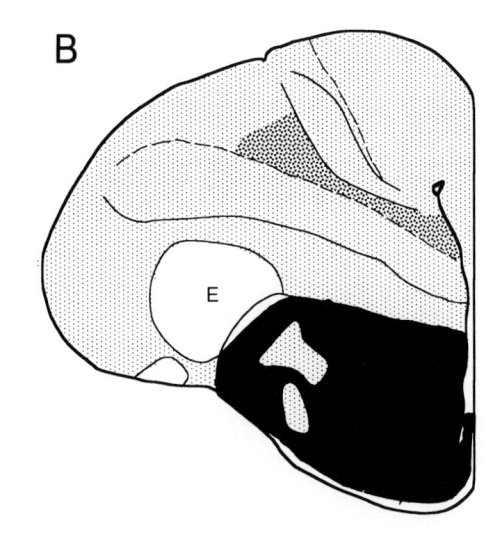

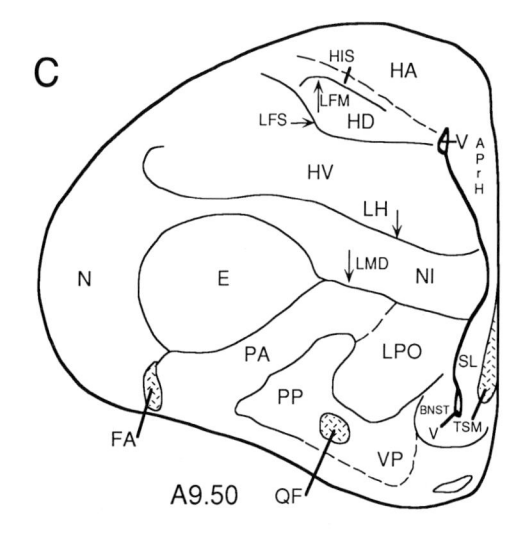

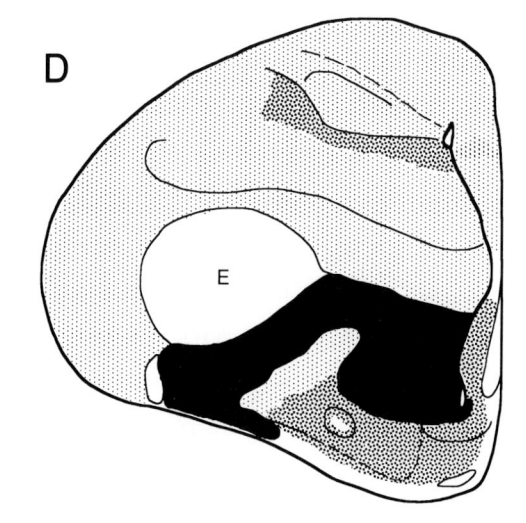

Fig. 7.2.

Catecholamines in avian brain 139

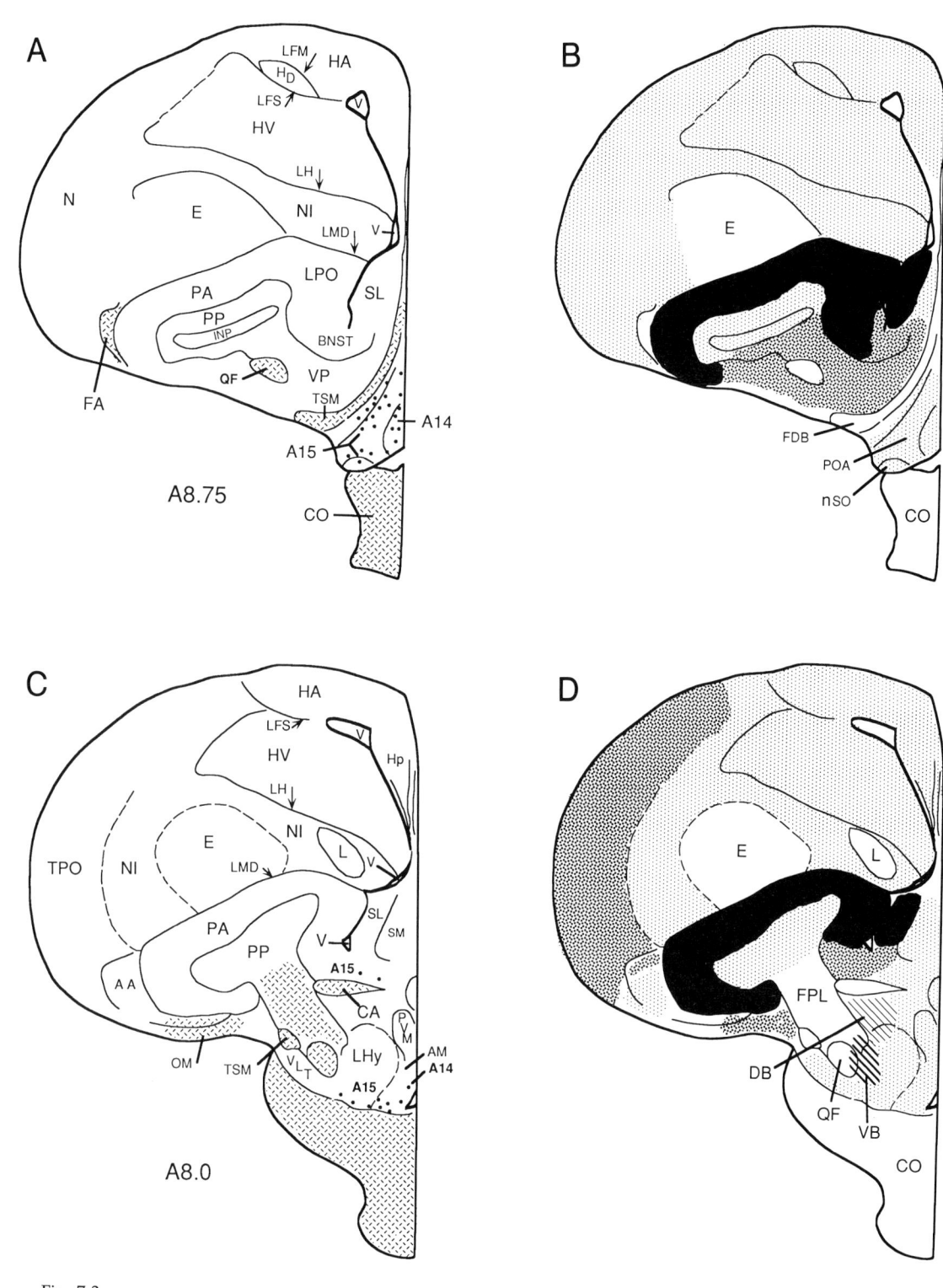

Fig. 7.3.

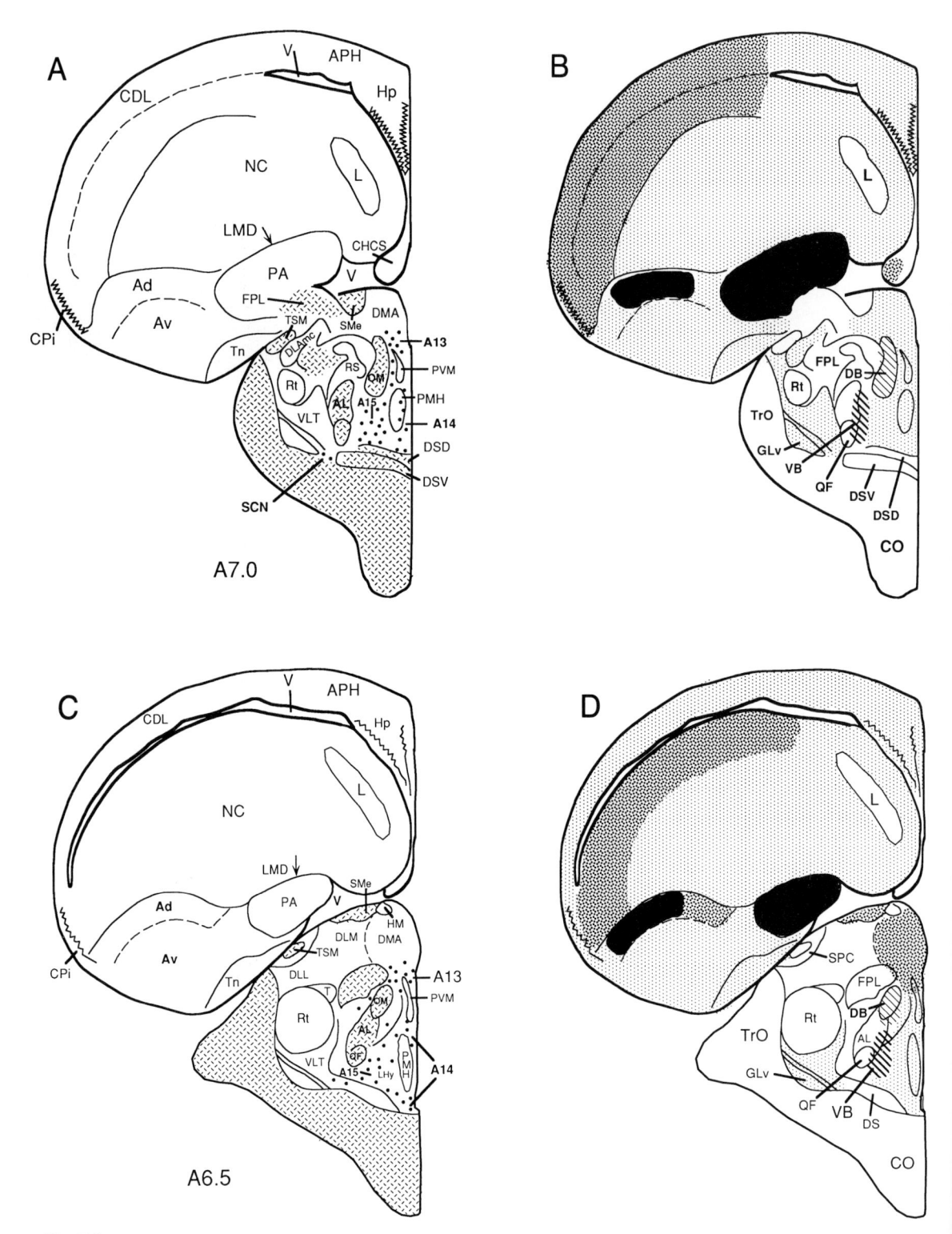

Fig. 7.4.

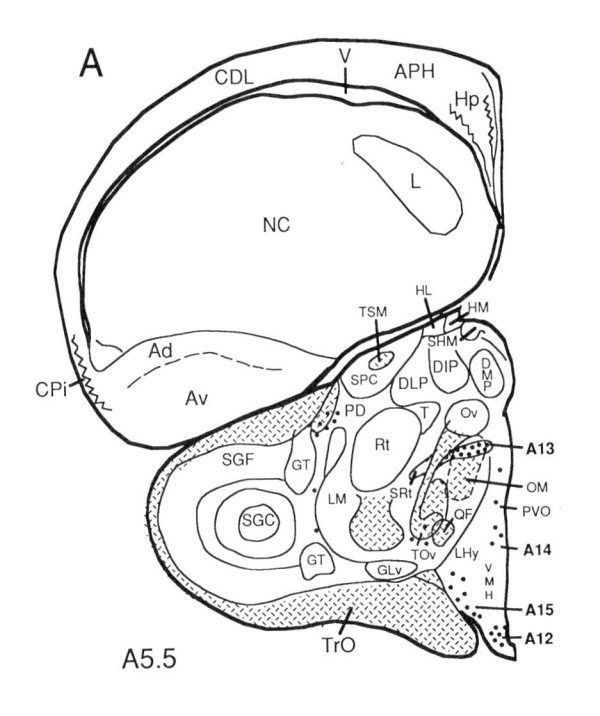

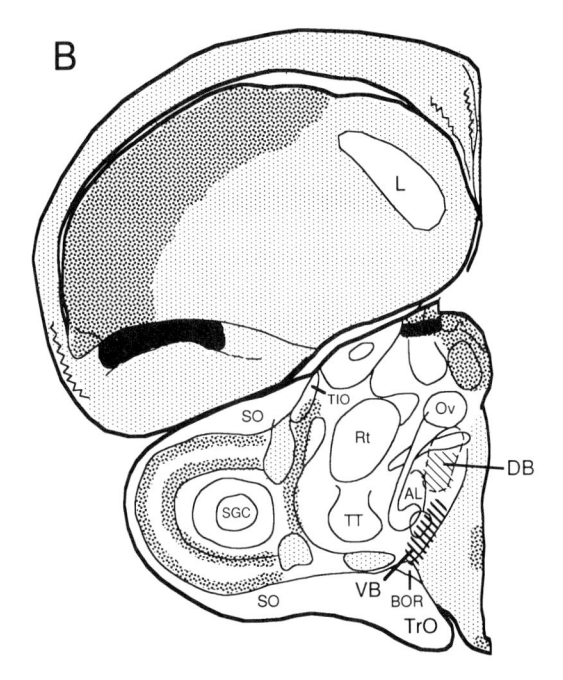

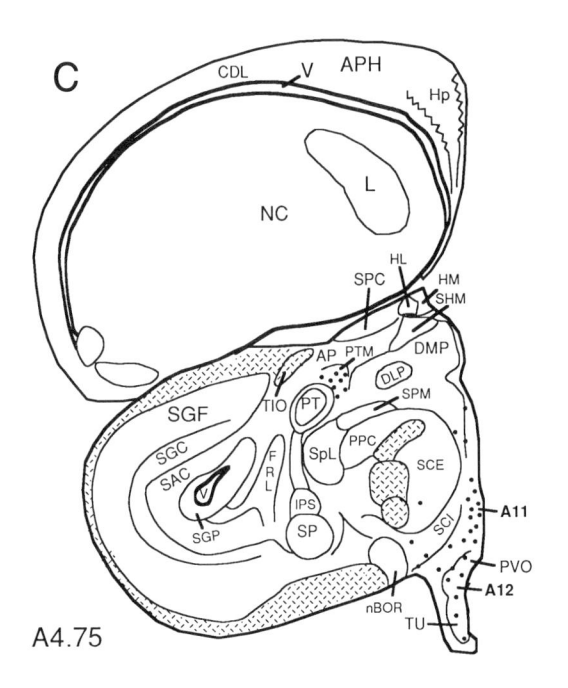

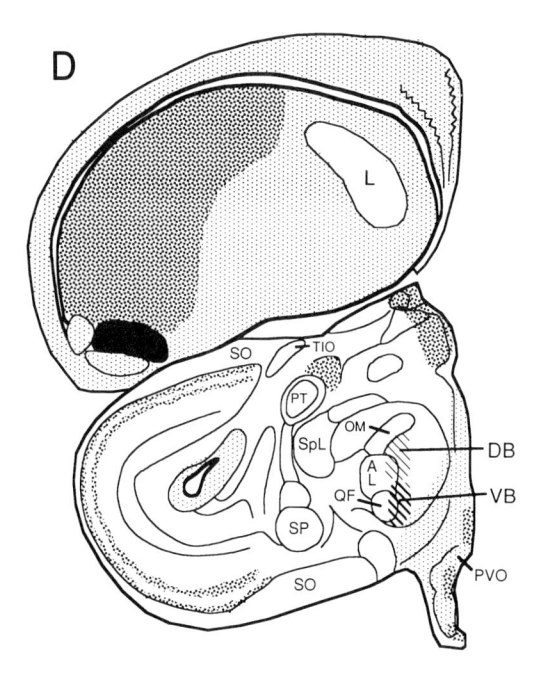

Fig. 7.5.

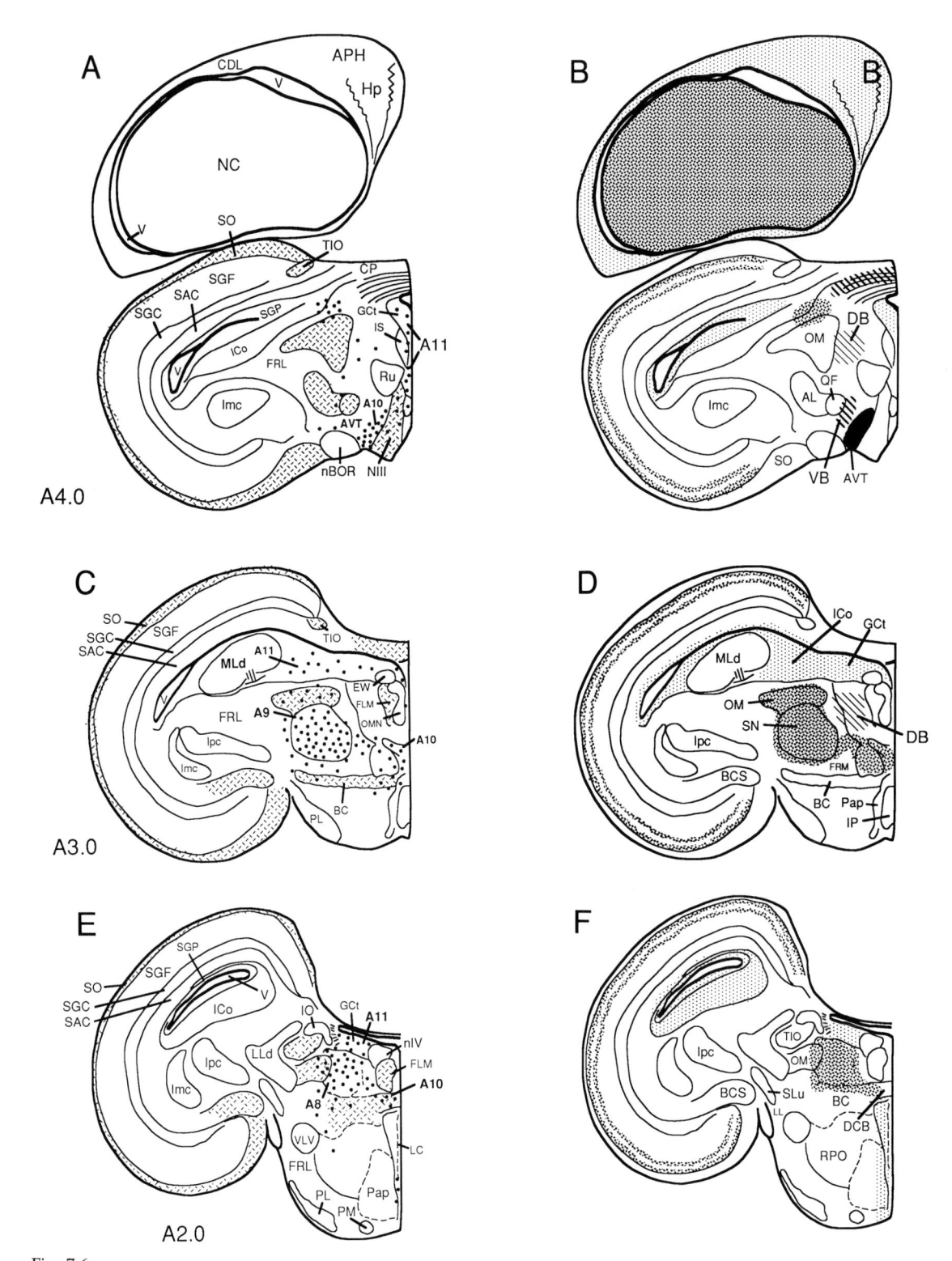

Fig. 7.6.

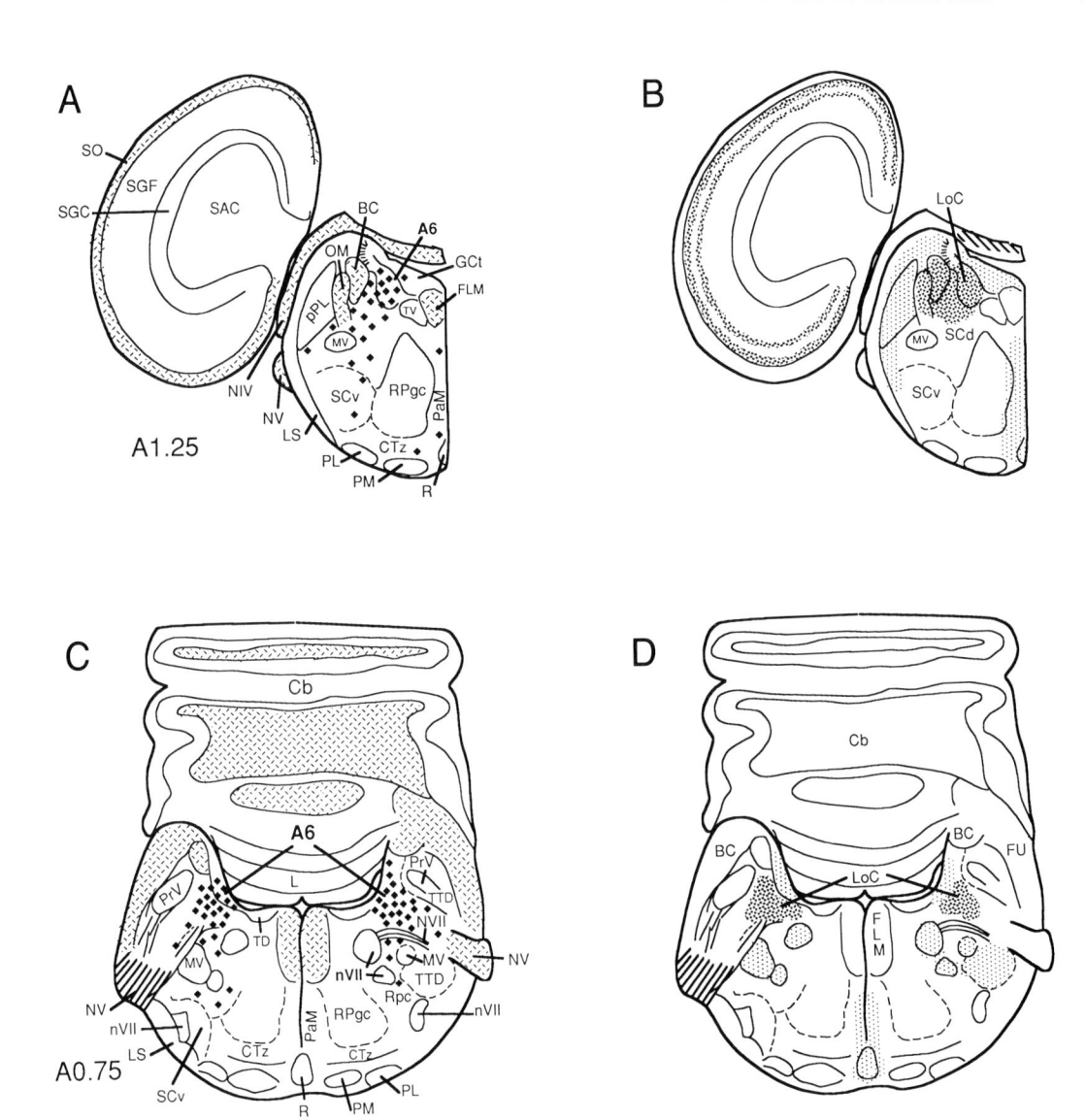

Fig. 7.7.

144   A. Reiner et al.

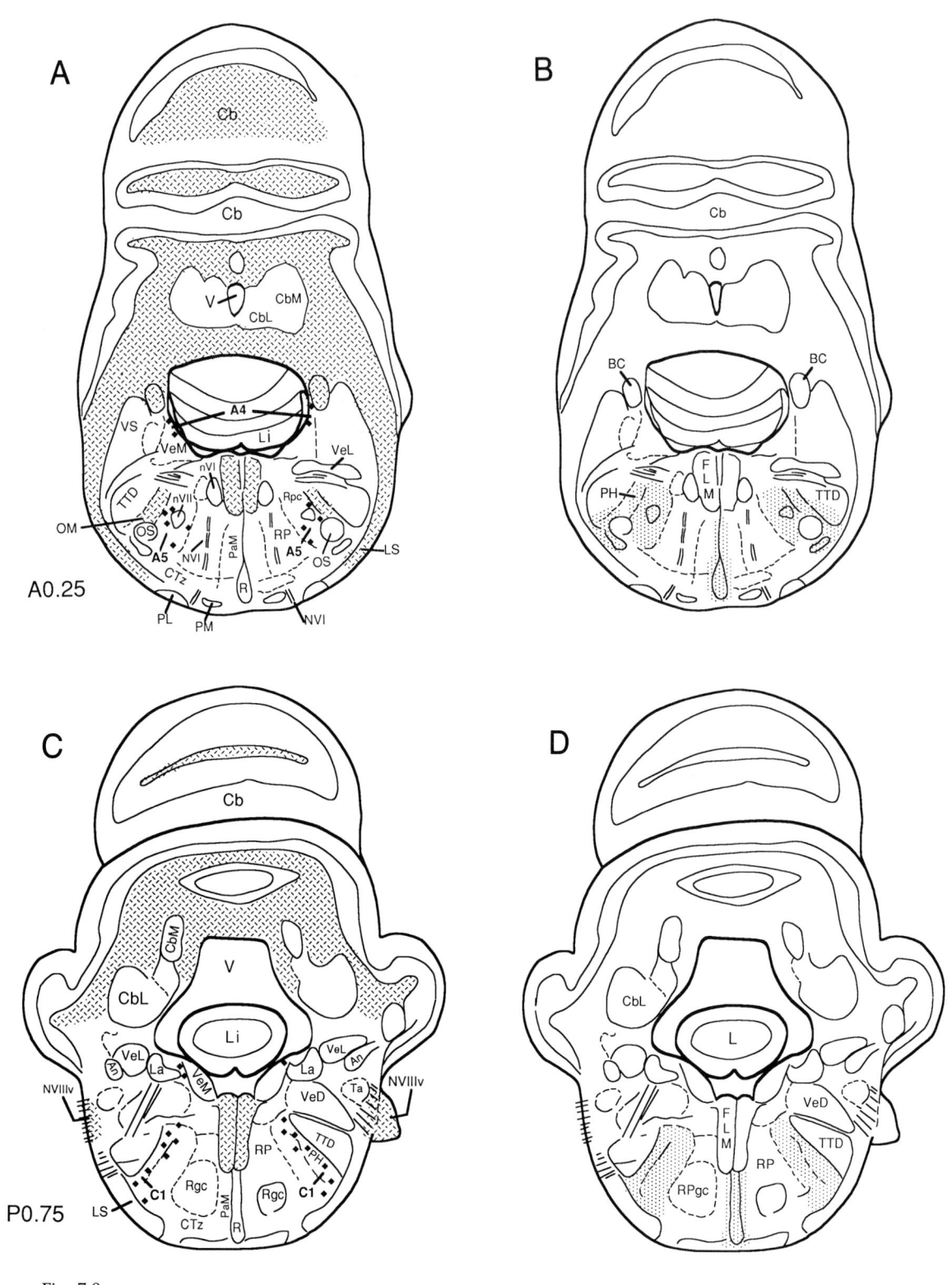

Fig. 7.8.

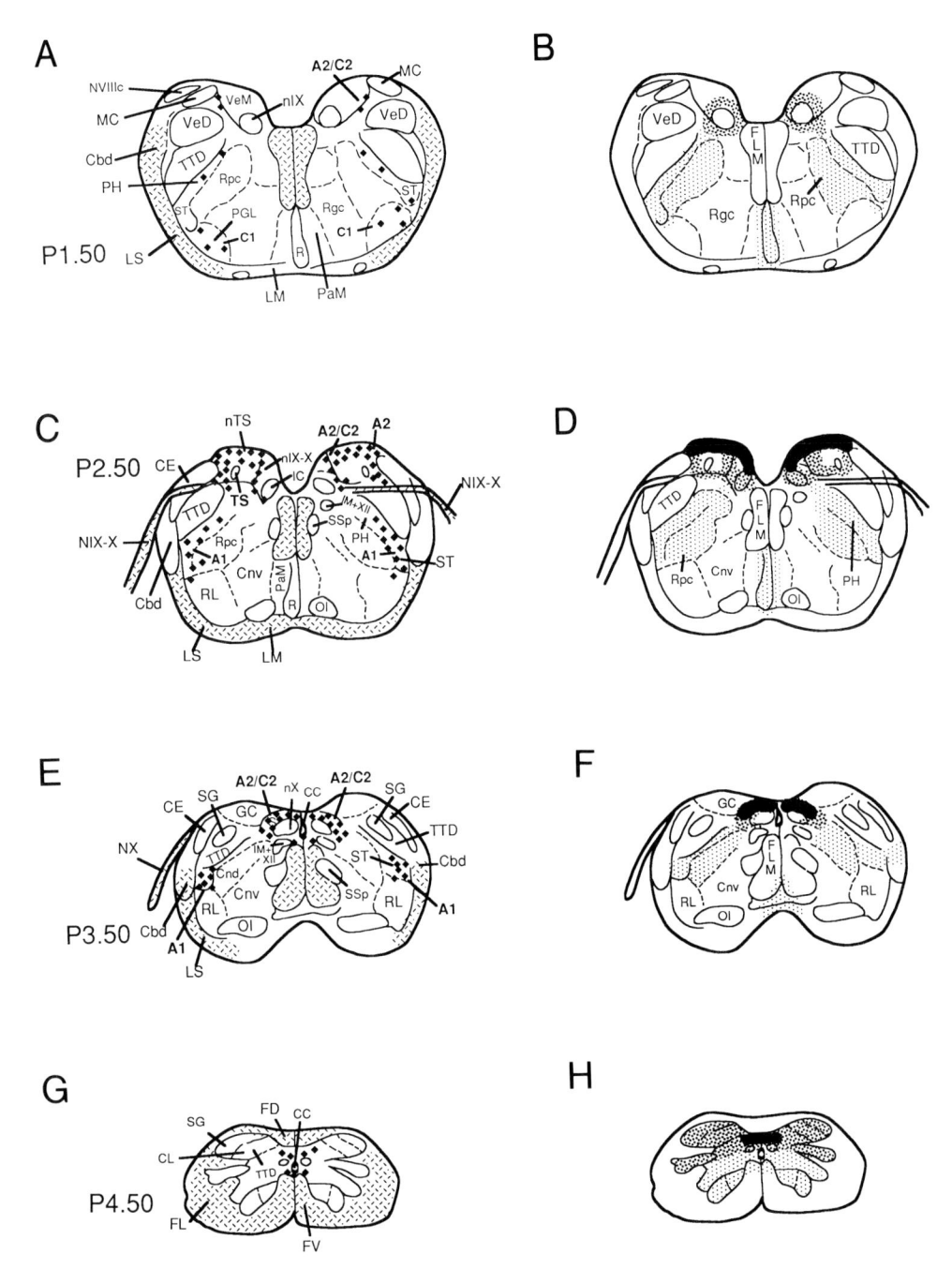

Fig. 7.9.

A15. The avian A15 has both dorsal and ventral parts. The perikarya of dorsal A15 (A15d) immunolabel for TH (but not DA) and lie above and medial to the fibers of the anterior commissure (AC) within the bed nucleus of the stria terminalis (BNST) (Fig. 7.3(*c*)). The A15d cell group is limited in its extent and consists of only a few neurons. The ventral part of the A15, by contrast, is an extensive cell group and some of its neurons at its intermediate levels label for DA (Figs. 7.3(*a*), 7.3(*c*), 7.4(*a*), 7.4(*c*), 7.5(*a*), 7.5(*c*), 7.12(*d*), 7.15(*a*)–(*c*)). Throughout its extent, the A15v lies lateral to the periventricular hypothalamic cell groups. The A15v begins at the level of preoptic area, where its neurons are found within the hypothalamus above the optic chiasm, including within the supraoptic nucleus (nSO) and the suprachiasmatic nucleus (SCN) (Gamlin, Reiner & Karten, 1982) (Fig. 7.4(*a*)). The A15v extends as far caudally as the tuberal hypothalamus. Some TH+ neurons that appear to be the most lateral and caudal part of A15v lie medial and dorsal to the nucleus of the basal optic root (nBOR) (Reiner, Karten & Solina, 1983*b*). At some levels, the A15v extends dorsally as far as the occipitomesencephalic tract (OM) (Fig. 7.4(*a*), (*c*). Since more dorsal parts of the A14 have a lateral extension that meets this dorsal part of A15v, it can be difficult to define a boundary between these two cell groups at such dorsal levels. Among other avian species, Bailhache & Balthazart (1993) observed a TH+/DBH– A15 cell group in quail with the same distribution as we observed in pigeons and Kuenzel *et al.* (1992) have observed an A15v but not an A15d in chickens using anti-TH and anti-DA immunolabeling. Finally, Bottjer (1993) saw the mid-hypothalamic and caudal hypothalamic portions of A15v in zebra finch using anti-TH immunohistochemistry. These various results indicate that an A15 is typically present in birds and that this cell group is largely dopaminergic.

A14. The A14 cell group is periventricular in position in pigeons and parallels A15v in its rostral–caudal extent (Fig. 7.3(*a*)–5(*c*), 15(*a*)–(*c*)). Like the A15v, the neurons of this cell group label for TH, with perikarya in its central part also labeling for DA. At the levels of the median eminence (ME), A14 extends dorsally into the region of the paraventricular nucleus and the dorsomedial hypothalamic nucleus. At these levels, a dorsolateral extension of A14 (which can be called A14d) joins the dorsalmost part of A15d (Fig. 7.4(*a*). Kuenzel *et al.* (1992) observed an A14 in chickens and noted that its neurons label immunohistochemically for both TH and DA. Bottjer (1993) observed the A14 with anti-TH immunolabeling in zebra finch, while Bailhache & Balthazart (1993) observed the neurons of A14 in quail to be TH+ and DBH– and possess a distribution similar to that in pigeons. These various results indicate that an A14 is typically present in birds and that this cell group is largely dopaminergic.

A13. An A13 cell group consisting of two parts (a medial and a lateral) is present in pigeons (Figs. 7.4(*a*), 4(*c*), 5(*a*)). Both parts are located in the border region between hypothalamus and thalamus, i.e. the subthalamus. The more medial part consists of neurons that label for TH and DA and are located near the ventricle (Figs. 7.4(*a*), (*c*). The region containing these TH+ neurons includes nucleus paramedianus internus (PMI) caudally, but it extends considerably anteriorly as well (Kiss & Peczely, 1987). This region, which is largely unnamed and is co-extensive with the dorsal thalamus, corresponds topographically to the medial part of the mammalian zona incerta (Hökfelt *et al.*, 1984). The more lateral part of A13 extends below nucleus ovoidalis (the primary auditory cell group of the thalamus) (Karten, 1968) within the nucleus subrotundus (SRt) (Fig. 7.5(*a*)). These TH+ cells in SRt label only lightly for TH and do not label for DA. Kuenzel *et al.* (1992) have observed a TH+ and DA+ population of neurons in PMI and they identified PMI as A13 in chickens, while Bailhache & Balthazart (1993) observed the medial part of A13 in quail and noted that its neurons immunolabel for TH but not DBH.

A12. A small group of TH+ cells is located within the periventricular zone of the tuberal hypothalamus just above the median eminence (ME) (Figs. 7.5(*a*), (*c*), 7.16(*a*)–(*c*)). A few of these label immunohistochemically for DA in pigeons. This cell group appears to represent the A12 group. It contains very few TH+ or DA+ perikarya, however, compared to the A12 group of mammals (Hökfelt *et al.*, 1984). Other authors using immunolabeling methods have not observed TH+ cells in the tuberoinfundibular region of the avian hypothalamus and have thus concluded that there is no A12 in birds (Kiss & Peczely, 1987; Kuenzel *et al.*, 1992; Bailhache & Balthazart, 1993). Nonetheless, we did observe a small scattering of TH+ cells in the tuberoinfundibular region in pigeons. Based on their position, we have called them the A12. We recognize, however, that studies looking at the connectivity and function of this cell group are required to definitively establish its identity.

A11. An A11 cell group whose neurons are rich in TH can be observed along the midline at caudal levels of the hypothalamus where it is fused across the midline (Reiner et al., 1983b; Kiss & Peczely, 1987) (Figs. 7.5(c), 16(a)–(e)). This region lies just rostral to the midbrain and the A11 cell group extends caudally and dorsally into the midbrain (Figs. 7.6(a), (c), (e)). At rostral midbrain levels, neurons of A11 extend dorsally into the periaqueductal midbrain gray (Reiner et al., 1983b) (PAMG: identified as the central gray in Fig. 7.6(a)). More caudally as the PAMG becomes continuous with the central gray (GCt) and intercollicular region (ICo) (which line the lower part of the tectal ventricle), TH+ neurons of the A11 cell group extend into ICo and GCt (Figs. 7.6(c), (e)). TH+ neurons in A11 extend to the caudalmost levels of the midbrain. Many neurons in all parts of A11 also label for DA. Since the A11 cell group largely is located along the hypothalamic midline and within the central gray, it is possible that following the mammalian convention by combining the CA+ neurons of these two regions into a single entity is inappropriate. Little is known, however, about the connections or functions of these neurons whereby to evaluate this issue. Kuenzel et al. (1992) have identified TH+ and DA+ neurons of PMI, PAMG and GCt as representing A11 in chickens, while Bailhache & Balthazart (1993) observed a TH+/DBH- A11 in quail with the same distribution as in pigeons. The latter authors also reported a group of TH+/DBH- cells in the rostral medial habenular region.

Paraventricular organ (PVO). A specialized peripendymal structure called the paraventricular organ (PVO) is located at caudal levels of the hypothalamus. The PVO is found at the dorsal edge of the third ventricle where it is overlain by the fused portion of the hypothalamus (Fig. 7.5(c), 16(b), (c)). The PVO consists of a tight clump of neurons that send their processes through the ependyma into the ventricle, where they can be in contact with the CSF (Parent, 1979). Some A14 TH+ cells are found lateral to PVO, some A11 TH+ cells are found caudal and dorsolateral to PVO, and some A12 TH+ cells are found ventrolateral to PVO. No TH+ cells, however, are found in PVO (Fig. 7.16(c)). Similar observations have been made in previous studies in pigeons (Reiner et al., 1983b; Kiss & Peczely, 1987), quail (Bailhache & Balthazart, 1993) and zebra finch (Bottjer, 1993). PVO, however, is extremely rich in DA+ cells that extend processes into the ventricle (Fig. 7.16(b)). Kuenzel et al (1992) and Smeets & Gonzalez (1990) also reported cells of the PVO in chicken to be TH- and DA+. The PVO is the only hypothalamic cell groups in birds in which catecholamine histofluorescence studies revealed a substantial number of CA+ perikarya (Fuxe & Ljunggren, 1965; Soest et al., 1973; Parent, 1979). We have observed that cells of PVO in pigeons also contain serotonin (5HT). Cells of PVO have also been observed to contain 5HT in chickens (Parent, 1979) and quail (Cozzi et al., 1991). Thus, while PVO cells contain biogenic amines (DA and 5HT), there is no current evidence that they contain the enzymes for their synthesis.

Pretectum: nucleus pretectalis pars medialis (PTM)

The PTM in pigeons is rich in TH+ perikarya and their processes (Figs. 7.5(c), 6(a), 16(d), (e)). This cell group is found medial to the pretectal nucleus (PT) at midpretectal levels (Fig. 7.5(e)), but PTM also extends rostral (Fig. 7.5(c)) and caudal to PT (Fig. 7.6(a)). The neurons of PTM also label for DA. At the level of PTM, a handful of TH+ perikarya were also observed in the pretectal region termed the nucleus lentiformis mesencephali. Smeets & Gonzalez (1990) have observed a similar TH+/DA+ cell group as PTM in chickens. Similarly, Bottjer (1993) observed TH+ perikarya in PTM in zebra finch and Bailhache & Balthazart (1993) observed TH+/DBH- perikarya in PTM in quail. The available data thus indicate the presence of DA+ cells in the PTM in a variety of avian species. CA+ neurons were not, however, observed in PTM in CA histofluorescence studies (Parent, 1979; Dubé & Parent, 1981).

Mesencephalon: A8–A10

The A8–A10 cell groups represent one large continuous field of dopaminergic neurons in the midbrain tegmentum. The A10 is the most rostral and medial component, the A9 more intermediate and lateral and the A8 the most caudal. The neurons of this cell group label intensely for TH and DA, but not for DBH or PNMT. The components of this field are discussed separately below.

A10. A distinct A10 cell group is present in pigeons in the ventral tegmental area (AVT) lateral to the oculomotor nerve at rostral midbrain levels (Figs. 7.6(a), 7.17(a), (b)), around the interpeduncular nucleus at levels just caudal to the oculomotor nerve, and below the medial longitudinal fasciculus at the level of the trochlear nucleus (Figs. 7.6(c), (e), 7.17(c)–(e)). Some TH+ and DA+ neurons are also present within and medial to the oculomotor nerve. A small group of TH+ perikarya lateral to the PAMG region may represent a dorsal part of A10 (Fig. 7.6(a)). We have observed that more caudal and dorsal portions of the A10 overlap the 5HT+ neurons of the dorsal raphe in pigeons. Parent et al. (1984) have reported a similar observation in chickens using histofluorescence methods.

A9. The A9 cell group in pigeons consists of a broad field of TH+/DA+ perikarya occupying much of the dorsal and lateral parts of the tegmentum in pigeons (Figs. 7.6(c), 7.17(c), (e)). Medially, this field is continuous with the A10 cell group. In the sagittal plane, the A9 can be seen to be oriented from rostroventrally to caudodorsally. As a consequence, in the transverse plane the rostral A9 appear is in the ventral tegmentum while the caudal A9 is situated in the dorsal tegmentum. The A9 is very densely packed with TH+/DA+ perikarya and their processes. Some TH+/DA+ cells are present in the occipitomesencephalic tract (OM) where it is contiguous with the A9. Caudal levels of A9 show some mixing of dopaminergic neurons with 5HT+ perikarya of the laterodorsal raphe, as observed by us immunohistochemically and as observed in histofluorescence studies in other avian species (Parent & Dubé, 1981; Parent et al., 1984; Yamada, Takeuchi & Sano, 1984). It should be noted that when the A9 cell group of birds was originally described by Karten & Dubbeldam (1973) using histofluorescence methods, they termed it the nucleus tegmentipedunculopontinus pars compacta (TPc). This was done because the tegmental distribution of the avian A9 and its position within the tegmentum did not appear to resemble that of the mammalian substantia nigra pars compacta, but did resemble that of the A8, which at that time was termed TPc (Karten & Dubbeldam, 1973). Since this term in mammals is now used to refer to a cholinergic cell group (Armstrong et al., 1983), continued use of the term TPc for the avian A9 only promotes confusion. Hence, we use the term SN to refer to the avian A9, since we believe the two to be homologous (Anderson, Karle & Reiner, 1991).

A8. A cell group whose neurons are TH+ and DA+ occupies the region below the GCt that Karten & Hodos (1967) termed the rostral part of locus coeruleus in their atlas. Since these neurons do not label for DBH (Reiner et al., 1983b), this cell group is clearly not part of locus coeruleus, whose neurons are noradrenergic (i.e. TH+/DBH+). We consider this caudal tegmental field of TH+/DA+ neurons to represent the A8 cell group and we have named the A8 the caudal part of SN (SNc). The A8 is located at the levels of the lateral lemniscus and it lies dorsomedial to this fiber bundle (Figs. 7.6(e), 7.17(d)). The A8 shows some mixing of dopaminergic neurons with 5HT+ perikarya of the laterodorsal raphe in pigeons and other avian species (Parent & Dubé, 1981; Parent et al., 1984; Yamada et al., 1984; Cozzi et al., 1991).

Other studies. Kuenzel et al. (1992) have recognized a dopaminergic A8–A10 in chickens, as has Smeets (1991). Kuenzel et al. (1992) termed A9 the TPc, based on his atlas (Kuenzel & Masson, 1988), and recognized the A8 as a caudal continuation of A9. Bottjer (1993) reported the presence of many TH+ cells in the A8–A10 in zebra finches and Bailhache & Balthazart (1993) reported TH+/DBH- (i.e. dopaminergic) perikarya in A8–A10 in quail. Histofluorescence studies typically reported the presence of CA+ cells in the A8–A10 field in a variety of avian species, including chickens (Ikeda & Gotoh, 1971; Dubé & Parent, 1981; Guglielmone & Panzica, 1984), zebra finch (Lewis et al., 1981), warbling grass parakeet (Shiosaka et al., 1981; Takatsuki et al., 1981), and pigeons (Bertler et al., 1964; Fuxe & Ljunggren, 1965). Bertler et al. (1964) and Fuxe & Ljunggren (1965) both referred to their tegmental CA+ cell group as the substantia nigra. In chicken, Knigge & Piekut (1985) showed that corticotropin releasing factor containing (CRF+ cells) are present in abundance in the TH+ fields of A8–A10 and that CRF is possibly present in DA neurons.

Rhombencephalon: the locus coeruleus complex: A4, A6, A7

The locus coeruleus is A6, a noradrenergic cell group located at the lateral edge of the isthmic central gray. Two smaller noradrenergic cell groups that appear to be continuations of the A6 have been identified in mammals: a more rostrally situated A7 and a more caudally situated A4 (Hökfelt et al., 1984). An A6 and A4 are clearly present in birds, while the presence of an A7 is more uncertain. The neurons of LoC are DBH+/TH+/PNMT − (and hence noradrenergic) in pigeons and occupy the region just below the central gray at isthmic and rostral pontine levels (Figs. 7.7(a), (c), 7.14(e), 7.18(a), (b)). The TH+/DBH+ cells of LoC extend into the dorsal and ventral subcoeruleus regions (SCd and SCv) and caudally extend to the rostral level of the cerebellar brachium (i.e. the A4) (Fig. 7.7(c)). At the level of LoC, some TH+ are also found along the midline (Fig. 7.7(a)). The A6 cell groups in chickens consists of TH+ neurons scattered within LoC and SCv, according to Kuenzel et al. (1992). CA histofluorescence studies have also noted LoC to be rich in CA+ neurons in chickens (Dubé &

Parent, 1981; Chikasawa, Fujioka & Watanabe, 1982; Guglielmone & Panzica, 1982, 1984), and in warbling grass parakeet (Shiosaka et al., 1981). Immunolabeling has been used to demonstrate TH+ neurons in LoC in zebra finch (Bottjer, 1993) and TH+/DBH+ neurons in LoC in quail (Bailhache & Balthazart, 1993). The number of noradrenergic neurons in SCv is greater in some avian species than in others (Bailhache & Balthazart, 1993). Finally, Knigge & Piekut (1985) showed that A4 and A6 neurons in chickens may also contain CRF.

Although a clearly distinct A7 is not evident in pigeons, the group of TH+/DBH+ cells extending ventrolaterally from locus coeruleus into the lateral pontine region dorsal to SCv may represent A7 (Fig. 7.7(a)). Other investigators, however, have not identified a distinct A7 in birds. The A4 cell group in pigeons consists of a few TH+/DBH+/PNMT − perikarya scattered along the ventricle within the medial vestibular nucleus and in the brachium of the cerebellum. This cell group appears to be a caudal continuation of LoC. Kuenzel et al. (1992) observed an A4 cell group in chickens.

Rhombencephalon: the noradrenergic A5, A3, A2 and A1

The A1–A3 and A5 cell groups in pigeons consist of groups of TH+/DBH+ (i.e noradrenergic) perikarya that are associated with the autonomic nervous system by their connections. They are located in the caudal pons and medulla, with the A1, A3 and A5 located ventrolaterally and the A2 located dorsomedially. Knigge & Piekut (1985) showed that these neurons in chickens may also contain CRF.

The A5 is located dorsomedial to the superior olive and around the facial motor nucleus (Fig. 7.8(a)). Kuenzel et al. (1992) identified a similar TH+ cell group in chickens and Bailhache & Balthazart (1993) noted this cell group in quail and reported that its neurons are TH+/DBH+. Histofluorescence studies in chickens (Dubé & Parent, 1981) and in warbling grass parakeet (Shiosaka et al., 1981) noted CA+ neurons in this same region. The A3 cell group in mammals is identified as a TH+/DBH+/PNMT − population of neurons located medial and above the inferior olive (Hökfelt et al., 1992). A comparable cell group has not been observed with TH or DBH labeling in pigeons (Fig. 9C,E), chickens (Kuenzel et al., 1992), or quail (Bailhache & Balthazart, 1993). The A2 cell group in pigeons consists of TH+/DBH+/PNMT − perikarya that completely surround the solitary tract, except ventrally (Figs. 7.9(c), (e)). The A2 also has a ventrally directed arm that is situated between DMN and nTS (Figs. 7.9(a), (c), (e), 7.19(b), (c)). Some of the A2 cells are located in the area postrema. The A2 cell group extends to spinal cord levels (Fig. 7.9(e)). An A2 has also been identified as a TH+ cell group in chickens (Kuenzel et al., 1992) and zebra finch (Bottjer, 1993), and as a TH+/DBH+ cell group in quail (Bailhache & Balthazart, 1993). Catecholamine histofluorescence has demonstrated this cell group in pigeons (Katz & Karten, 1979), chickens (Dubé & Parent, 1981; Guglielmone & Panzica, 1984) and warbling grass parakeet (Shiosaka et al., 1981). The neurons of A1 in pigeons are found as a strand of neurons within the PH and as a cluster within the nucleus subtrigeminalis (ST) (Figs. 7.9(c), (e), 7.14(f), 7.19(b), (d)). Caudally, the A1 neurons continue into spinal cord. The A1 in chickens has been identified as residing within the nucleus reticularis subtrigeminalis by Kuenzel et al. (1992). A similar localization has been reported with CA histofluorescence in chickens (Dubé & Parent, 1981; Guglielmone & Panzica, 1984) and warbling grass parakeet (Shiosaka et al., 1981). Bottjer (1993) reported TH+ neurons in A1 in zebra finch and Bailhache & Balthazart (1993) noted TH+/DBH+ neurons in A1 in quail.

Rhombencephalon: the adrenergic C3, C2 and C1

The C1–C3 cell groups in birds consist of groups of TH+/DBH+/PNMT+ (i.e adrenergic) perikarya that are associated with the autonomic nervous system by their connections. A TH+/DBH+/PNMT+ group of neurons along the midline within and dorsal to the medial longitudinal fasciculus has been identified as the C3 cell group in mammals (Hökfelt et al., 1984). This cell group in mammals is found at caudal pontine and rostral medullary levels. No TH+, DBH+ or PNMT+ neurons, however, have been observed in this region in pigeons (Fig. 7.8(c), 7.9(a)), chickens (Kuenzel et al., 1992), quail (Bailhache & Balthazart, 1993) or ducks (Steeves et al., 1987). The C2 cell group in pigeons is made up of TH+/DBH+/PNMT+ cells surrounding the solitary tract on its dorsal, lateral and medial borders (DMN) (Figs. 7.9(a), (c), (e), 7.14(b)–(d)). These adrenergic neurons are intermingled with the noradrenergic neurons of the A2 cell group. Although we have observed PNMT+ neurons in nTS in pigeons, we are yet uncertain of the relative numbers of adrenergic C2 and noradrenergic A2 neurons in nTS. Steeves et al. (1987) used anti-PNMT to show plentiful adrenergic neurons throughout the nTS/DMN region, in area postrema (i.e. the taenia choroidea) along the surface of the fourth ventricle, and along the IX-X cranial nerve as it traverses the medulla in Pekin ducks. Thus, adrenergic neurons are abundant in C2 in some if not all avian species. The C1 group consists of perikarya individually labeled for TH, DBH and PNMT present in the plexus of Horsley (PH) and the lateral paragigantocellular nucleus (PGL) at caudal pontine levels, beginning just caudal to the facial motor nucleus (Figs. 7.8(c), 7.9(a),(b),(c), 7.16(a)). The C1 cell group extends to the level of area postrema (Fig. 7.9(c)). The C1 cell group in pigeons consists of a dorsomedially to ventrolaterally oriented band of cells, with a dorsomedial group (largely in PH) and a ventrolateral group (largely in PGL). The noradrenergic neurons of the A1 cell group intermingle with the adrenergic neurons of the C1. Steeves et al. (1987) used anti-PNMT to demonstrate that there are numerous adrenergic neurons in the PH and PGL region in Pekin ducks, as well as in the region dorsolateral to the inferior olive (termed the subtrigeminal region in pigeons).

Spinal cord
Within the spinal cord of pigeons, TH+/DA+ cells are present at the inferior edge of the central canal (Fig. 7.9(g)). These neurons send their processes through the ependymal lining of the central canal and appear to be in contact with the ventricular cavity. In addition, TH+ perikarya are present dorsal to the central canal of pigeons. In chicks, TH+ cells are also found below the central canal and along the lateral and inferior parts of the dorsal horn (Wallace et al., 1987). Okado et al. (1991) found that TH+ cells in chicken are located mainly at cervical levels and are most abundant immediately below and above the central canal. The cells below the canal send processes that penetrate into the central canal.

Distribution of catecholamines: fibers and terminals

Immunohistochemical labeling for TH, DBH, PNMT and DA revealed many features of the distributions of specific catecholamines in fibers and terminals of the pigeon brain. Note that the distributions of TH+ and DA+ fibers were in very close concordance with each other, while the distribution of DBH+ fibers was different from that for TH or DA. These findings and biochemical data discussed below suggest that the distribution of TH+ fibers in avian brain largely reflects the distribution of dopaminergic fibers, while the distribution of DBH+ fibers largely reflects the distribution of noradrenergic or adrenergic fibers. The schematics accompanying the text therefore provide separate maps of the TH+ (Figs. 7.1–7.9) and DBH+ fibers (Figs 7.10–7.11). Details of the distributions of dopaminergic and noradrenergic fibers and terminals are described below.

Retina
The TH+ cells found in the avian retina give rise to a dense fiber arborization in layer 1 of the (outer) IPL, and a sparser arborization in the inner IPL (Britto et al., 1988). The TH+ cells in the outer INL ramify mainly in the INL. The TH+ amacrine cells in the inner INL of chickens show similar ramification patterns to those in pigeons (Kagami et al., 1991). Examination of sections through the avian eye reveals CA+ fibers and terminals in nonretinal portions of the eye also. For example, CA+ fibers arising from the superior cervical ganglion innervate choroidal blood vessels (Guglielmone & Cantino, 1982) and iris dilator muscles (Kirby, Diab & Mattio, 1978). The innervation of choroidal vessels is likely to play a role in regulating choroidal blood flow (Fitzgerald, Vana & Reiner, 1990), while the innervation of the iris dilator muscle is involved in reflexive pupil dilation (Reiner et al., 1983a).

Olfactory bulb
The TH+ cells in the olfactory bulb (Fig. 7.1(b)), which we assume to be dopaminergic, have processes that appear to form a dense band of DA+ fibers within the neuropil in the OPL and in the glomerular zone. In addition, DBH+ fibers are sparsely scattered in all olfactory bulb layers.

Telencephalon
The telencephalon is extremely rich in TH+ and DA+ fibers and relatively sparser in DBH+ fibers. We will describe the TH+ fiber distribution first and describe our basis for believing that the bulk of this is dopaminergic. Subsequently we will describe the DBH+ fiber distribution in the telencephalon.

TH+/DA+ fibers. The basal telencephalon in birds is very rich in CA+ fibers and terminals and the bulk of these are dopaminergic. For example, biochemical studies have shown that the levels of DA are much higher in the striatal part of the basal ganglia than in other parts of the telencephalon (Bertler et al., 1964; Juorio & Vogt, 1967; Karten & Dubbeldam, 1973). Further, Divac, Mogenson & Bjorklund (1985) showed that DA is approximately five times as abundant as NA in the pigeon basal telencephalon. Divac, Mogenson & Bjorklund (1988) also showed that DA levels are higher in the telencephalon of mixed breed pigeons than in White Carneaux pigeons

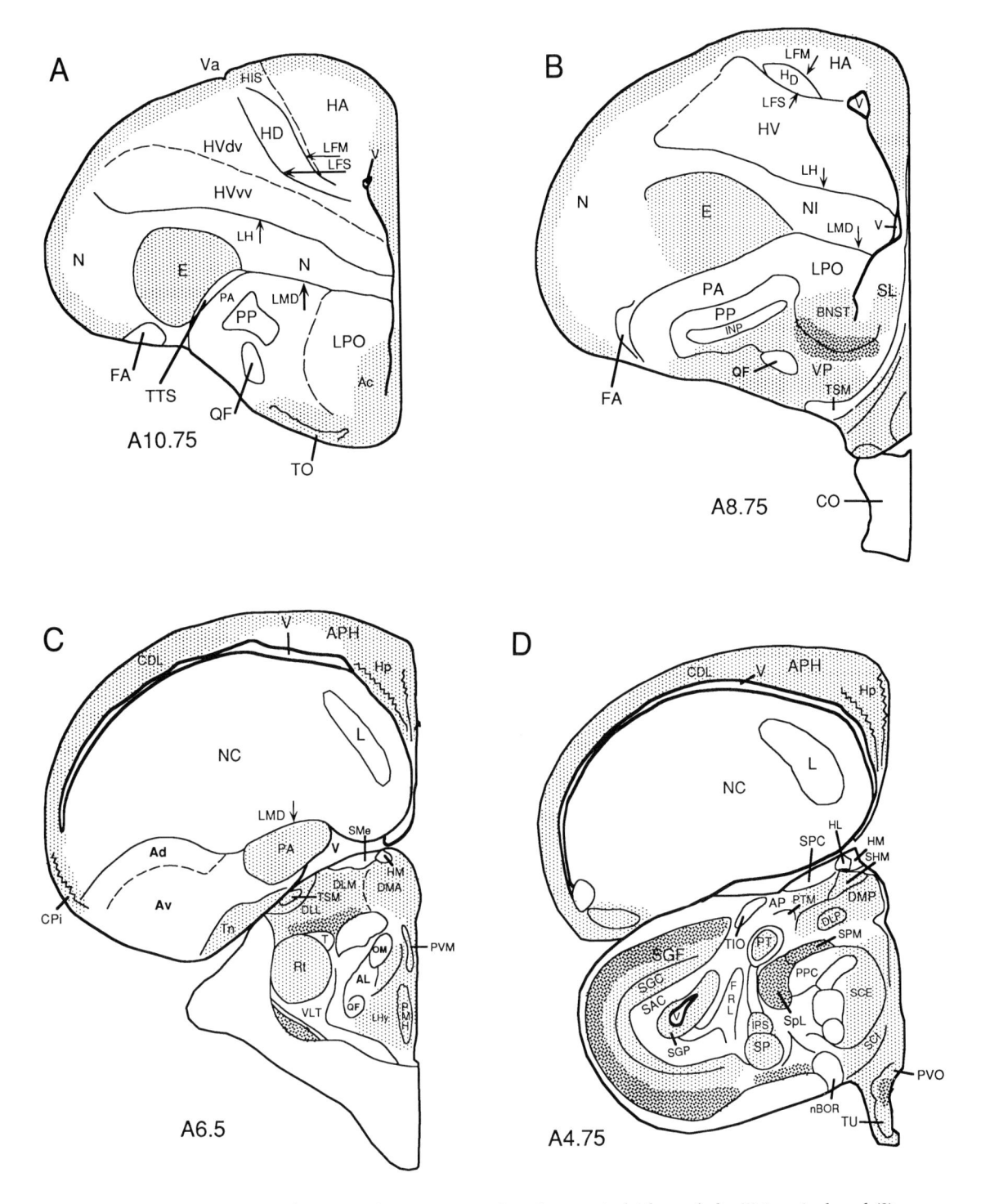

Figs. 7.10–7.11. Line drawing schematics of transverse sections from rostral telencephalic (1) to spinal cord (9) levels. The sections show the distribution of DBH+ fibers and terminals. The density of shading is scaled to the relative abundance of DBH+ fibers. The numbers to the lower left of each section show the anterior-posterior level of the sections according to the Karten and Hodos (1967) atlas.

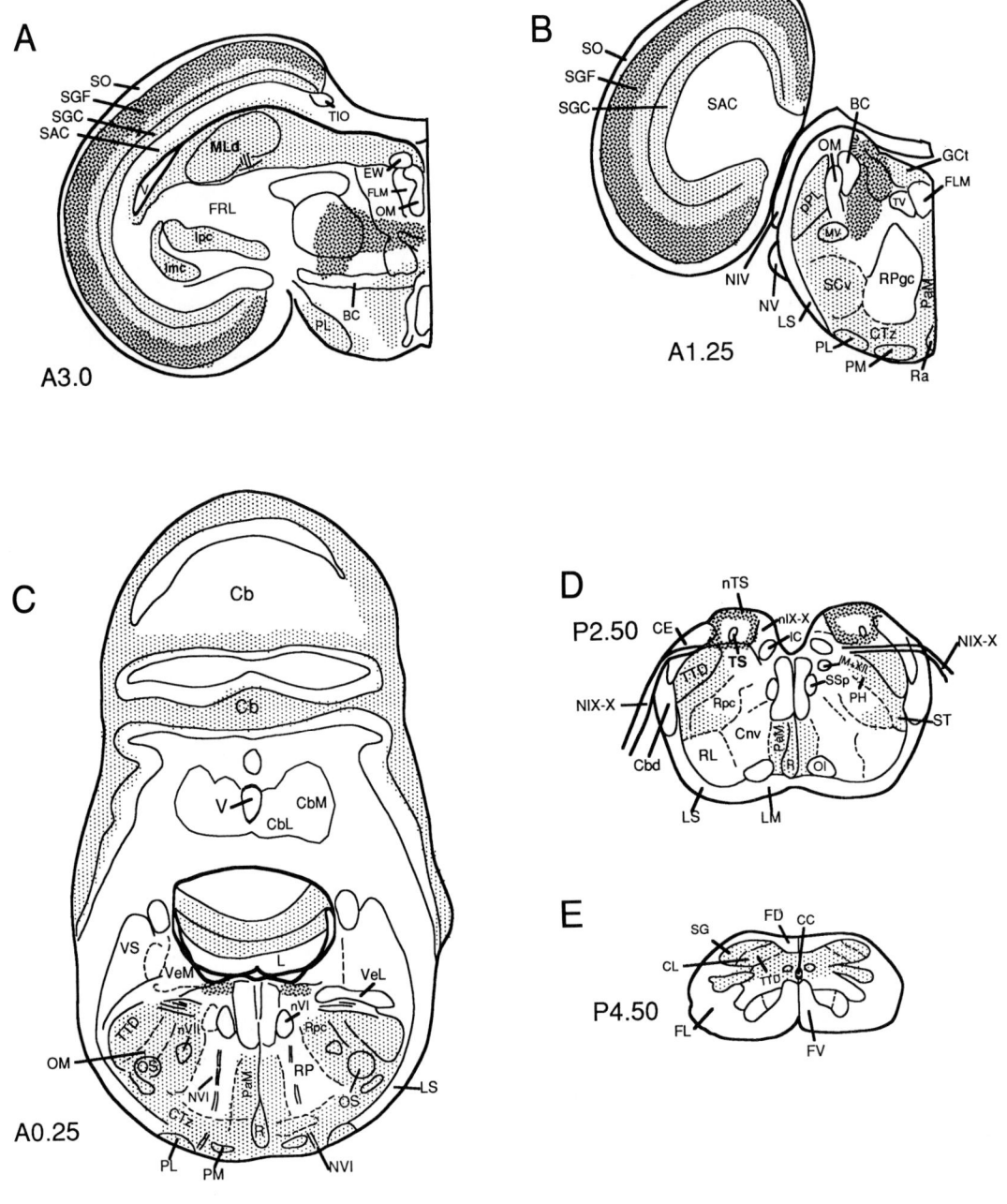

Fig. 7.11.

(which are commonly used in research). As in mammals, the avian basal telencephalon is a large complex field that can be divided into two subdivisions, a somatic and a visceral. Other authors have used the terms somatic and limbic or dorsal and ventral to refer to these two same subdivisions. We find these terms ambiguous because 'limbic' is an ill-defined term and because all so-called dorsal components of the dorsal basal forebrain are not dorsal to all so-called ventral components. We prefer the terms somatic and visceral because these terms reflect the functional relations of these parts of the basal forebrain.

Both the somatic and visceral striatum are made up of striatal and pallidal components. The somatic striatum consists of the medially situated lobus parolfactorius (LPO) and the laterally situated paleostriatum augmentatum (PA). The LPO and PA are extremely rich in TH+ and DA+ fibers and terminals (Figs. 7.1 (d), 7.2(b), (d), 7.4(b), (d), 7.12(a)–(d), 7.13(a), (b). The visceral striatal cell groups include the nucleus accumbens (Ac), the bed nucleus of the stria terminalis (BNST) and the olfactory tubercle (TuOl). The BNST region surrounds the tip of the inferior horn of the lateral ventricle, the Ac occupies the region lateral and dorsal to BNST (note this region includes part of what the Karten & Hodos atlas has previously included in LPO), and the TuOl occupies the base of the rostral telencephalon. These visceral regions are also extremely rich in TH+/DA+ terminals, although

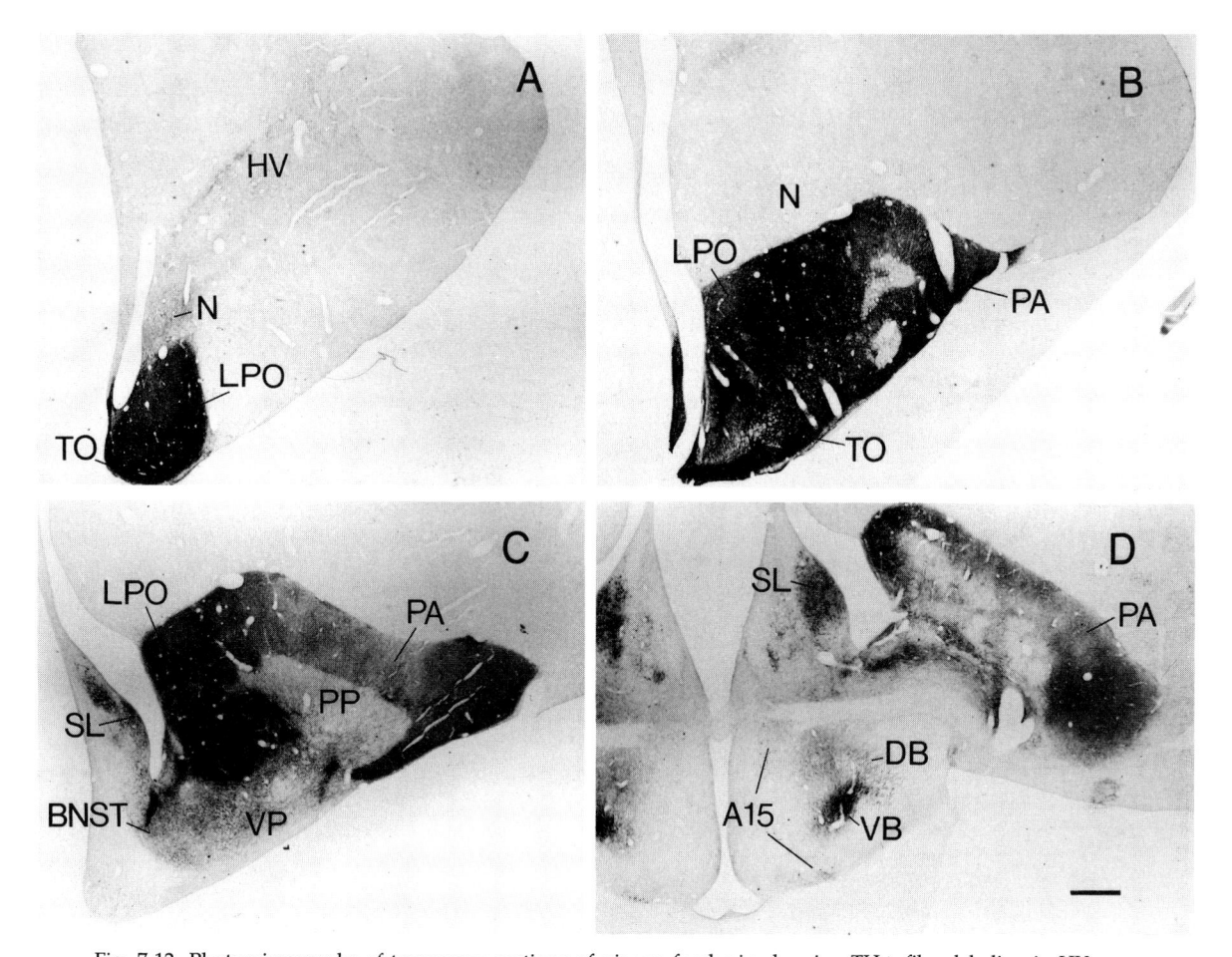

Fig. 7.12. Photomicrographs of transverse sections of pigeon forebrain showing TH+ fiber labeling in HV, neostriatum and LPO of the rostral telencephalon (a), in LPO and PA somewhat more caudally (b), in LPO, PA, PP and VP at midtelencephalic levels (c) and in the septum, basal ganglia and preoptic area at the level of the anterior commissure (d). The scale bar is equal to 200 μm and the magnification is the same for all four photomicrographs. The midline is to the left and dorsal to the top in all photomicrographs.

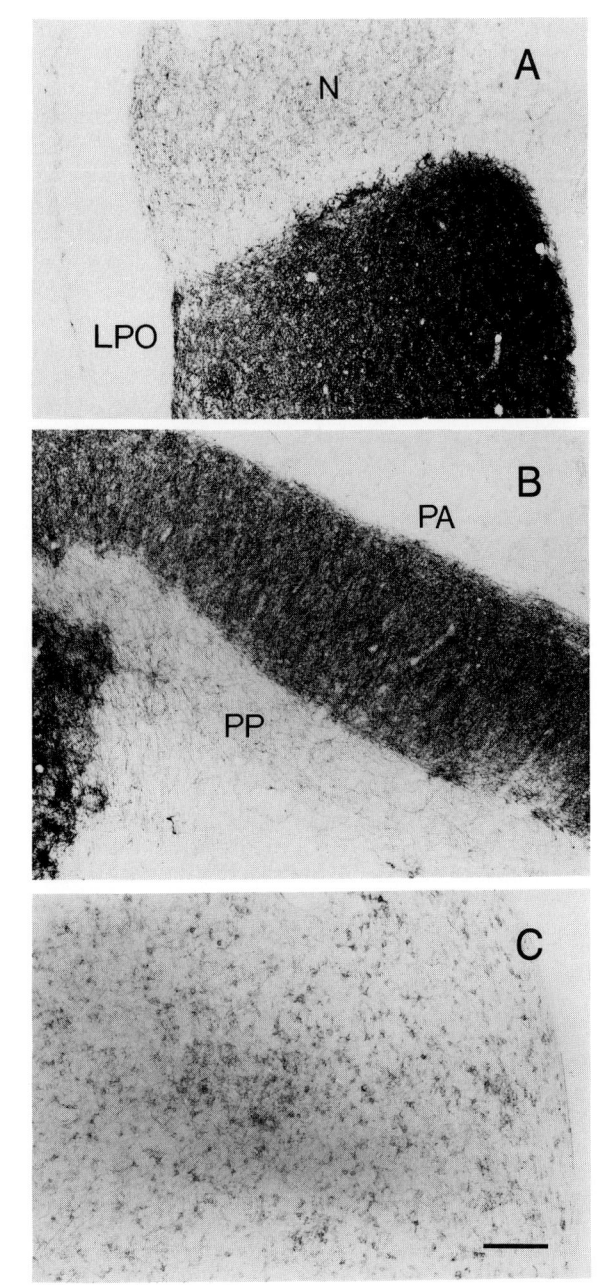

Fig. 7.13. High power photomicrographs of transverse sections illustrating the TH+ fiber distribution in pigeon in rostral LPO and neostriatum (a), in PA and PP (b) and in the TPO (c). The scale bar is equal to 100 μm and the magnification is the same for all three photomicrographs. The midline is to the left and dorsal to the top in all photomicrographs.

the BNST is less rich than the Ac and olfactory tubercle. The somatic and visceral striatum have also been shown to be richly innervated by CA+ fibers in diverse avian species by immunohistochemical methods (Bottjer, 1993; Bailhache & Balthazart, 1993) and histofluorescence methods (Bertler et al., 1964; Parent & Olivier, 1970; Karten & Dubbeldam, 1973; Lewis et al., 1981; Takatsuki et al., 1981; Parent et al., 1984; Divac & Mogenson, 1985). One prominent species and sex difference is that the dorsal portion of LPO associated with song control in male zebra finches (i.e. area X, which has no counterpart in non-passerine birds) is richer in TH+ and CA+ fibers than other parts of the surrounding LPO (Bottjer, 1993; Lewis et al., 1981).

TH+ fibers are also present in considerable abundance in such pallidal areas as the paleostriatum primitivum (the somatic pallidum) (Figs. 7.3(b), (d), 7.12(c), 7.13(b) and the ventral pallidum (the visceral pallidum) (Figs. 7.3(b), 7.12(c)). It seems likely that these pallidal fibers are DA+, although we have yet to confirm this immunohistochemically. The presence of such CA+ fibers in pallidal regions has not been reported in CA histofluorescence studies of avian brain, but they have been observed in other immunolabeling studies (Bailhache & Balthazart, 1993). Finally, the septal nuclei are also very rich in TH+/DA+ fibers and terminals in birds (Figs. 7.2(d), 7.3(b), (d), 7.12(c), (d) (Bertler et al., 1964; Parent & Olivier, 1970; Karten & Dubbeldam, 1973; Lewis et al., 1981; Takatsuki et al., 1981; Parent et al., 1984; Divac & Mogenson, 1985; Bottjer, 1993; Bailhache & Balthazart, 1993). Within the septum, such fibers and terminals completely outline the perikarya and proximal dendrites of many septal neurons.

The pallial parts of the avian telencephalon consist of the dorsal ventricular ridge (DVR), the Wulst and the hippocampal complex. TH+ fibers and terminals are relatively uniformly distributed throughout the various subdivisions of these pallial regions (Figs. 7.1(b), (d), 7.2(b), (d), 7.3(b), (d), 7.5(b), (d), 7.6(b). In all cases, the fibers and terminals appear as pericellular nests surrounding individual neurons (Fig. 7.13(c)), with a high percentage of the neurons in each region possessing such nests (25–50%). Within the DVR (which includes the regions termed the neostriatum, hyperstriatum ventrale, the ectostriatum, the nucleus basalis and the archistriatum), four regions are particularly abundant in TH+ fibers or rich in fibers that label intensely: 1) the dorsal part of the hyperstriatum ventrale (HVdv) (Figs. 7.1(b), (d), 7.2(b), (d), 7.12(a), (b)); 2) the part of the DVR we have previously termed the pallium externum (comprising the temporo-parietal-occipital area and the rostral part of the dorsolateral corticoid area) and the dorsolateral neostriatum caudale (NCL) (Veenman & Reiner, 1994; Veenman et al., 1994) (Figs. 7.3(d), 7.4(b), (d), 7.5(b), (d), 7.6(b)); 3) the dorsal archistriatum (Figs. 7.4(b), (d) 7.3(b)); and 4) the medial frontal neostriatum (Figs. 7.1(d), 7.12(a), 7.13(a)). The TH+ fiber distribution in the archistriatum did not distinguish between the somatic (dorsolateral) and visceral (ventromedial, including nucleus taenia) archistriatum (Figs. 7.4(b), (d), 7.5(b)) (Zeier & Karten, 1971). One further noteworthy feature of the TH+ fiber distribution in the avian DVR is that the DVR targets of specific sensory input from the thalamus (i.e. the nucleus basalis, ectostriatum, and field L) are relatively poor in TH+ fibers (Figs. 7.1(d): nu. basalis, Figs. 7.2(b), (d), 7.3(b), (d): ectostriatum, and Figs. 7.3(d), 7.4(b), (d): field L). Finally, we did not observe major regional variation in TH+ fiber distribution in either the Wulst (which includes the hyperstriatum accessorium, hyperstriatum intercalatum supremum and hyperstriatum dorsale) or hippocampal complex (which includes the prehippocampal area, hippocampus and parahippocampal area).

Histofluorescence studies have revealed CA+ fibers to also be highly abundant in the PE and NCL in chickens (Parent et al., 1984), and immunolabeling studies have revealed that among DVR regions the PE and Ad in quail are particularly rich in TH+ fibers and terminals (Bailhache & Balthazart, 1983). Bottjer (1993) has shown that the song nuclei of the DVR in male zebra finches (namely the lateral magnocellular anterior neostriatum, the so-called higher vocal center part of the neostriatum and the archistriatum robustus) are richer in TH+ fibers than surrounding parts of the DVR. Finally, some authors have described subtle regional variations in CA+ or TH+ fiber distributions in the Wulst and hippocampal complex of various avian species (Takatsuki et al., 1981; Parent et al., 84; Shimizu & Karten, 1990; Krebs, Erichsen & Bingman, 1991).

The extent to which the TH+ terminals in the Wulst, DVR and hippocampal complex contain DA or NA is incompletely known, although it seems likely that the TH+ terminals in HVdv, the pallium externum, the dorsal archistriatum and the frontal medial neostriatum contain DA since we have found that terminals in these regions label intensely and in abundance for DA. Further, DBH+ terminals (as discussed further below) are sparse in these four regions in pigeons. In addition, Waldmann & Güntürkün (1993) have shown that DA+ fibers in the DVR of pigeons form pericellular nests similar in appearance to those formed by TH+ fibers. Finally, Divac and his coworkers have shown using histofluorescence and biochemical methods that the predominant catecholamine in the pigeon DVR is dopamine (Divac & Mogenson, 1985; Divac et al., 1985, 1988). DA levels are higher in the DVR of mixed breed than pure White Carneaux pigeons (Divac et al., 1988). The facts that the PE and NCL portion of the DVR is a pallial structure and is particularly rich in dopaminergic fibers (together with some behavioral evidence) have been used by Divac and his coworkers to argue that the PE and NCL is the avian correspondent of the mammalian prefrontal cortex (Mogenson & Divac, 1982; Divac et al., 1985, 1988; Reiner, 1986). The merits of the interesting idea that the PE and NCL is the avian analogue (perhaps homologue) of mammalian prefrontal cortex have been discussed by Reiner (1986a).

DBH+ fibers. Our immunohistochemical studies revealed DBH+ fibers to be generally much sparser and more uniformly distributed in the pigeon telencephalon than TH+ fibers (Fig. 7.10). DBH+ fibers were most abundant in the visceral part of the striatum and the VP (Figs 7.10(b), 7.14(a)). Fibers and terminals labeled for DBH in pigeons were also abundant in the APH and the thin rim of pallium above the lateral ventricle termed the CDL. The DBH+ fibers were, however, no greater in abundance in the TH+ rich regions of the DVR (i.e. PE, Ad, medial neostriatum and HVdv) than elsewhere in the telencephalon. DBH+ fibers were in fact absent from Ad in pigeons. The DBH+ fibers were different in appearance from the TH+ fibers. Whereas the TH+ fibers formed a dense mat in the striatal regions of the basal telencephalon and pericellular nests in the DVR, Wulst and hippocampal complex, the DBH+ fibers consisted of scattered varicosity-laden axonal processes that did not make pericellular nests. This morphological appearance and the clear association of the TH+ fibers with the distribution of DA suggests that anti-TH in the avian telencephalon mainly labels DA+ fibers and terminals and anti-DBH mainly labels NA+ fibers and terminals. The observation of Divac et al. (1985) that NA is uniformly distributed and present in generally low levels in pigeon telencephalon is consistent with this conclusion. Bailhache & Balthazart (1993) in quail reported a distribution of DBH+ fibers in quail very similar to that described here in pigeons (for example, most abundant in visceral striatum, but sparse in PE and Ad). DBH+ fibers in abundance were observed by these authors in the nucleus taenia, thereby suggesting that the presence of DBH+ fibers might dis-

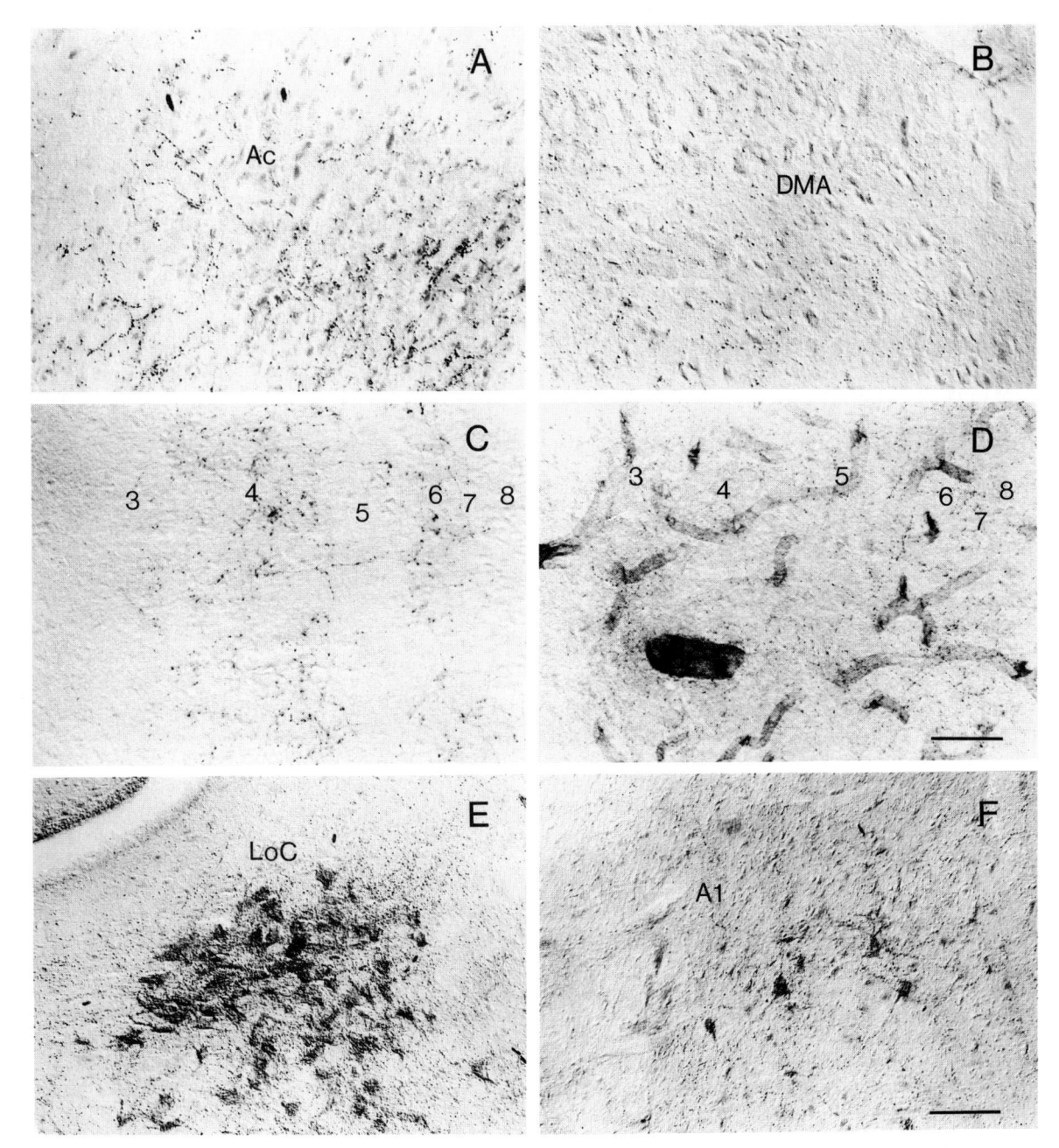

Fig. 7.14. Photomicrographs of transverse sections of pigeon forebrain showing DBH+ fibers in the nucleus accumbens – bed nucleus of the stria terminalis region of the telencephalon (a), DBH+ fibers in the dorsomedial anterior thalamic nucleus (b), TH+ fibers in the upper tectal layers of the lateral tectum (c); DBH+ fibers in the same tectal layers of lateral tectum (d); DBH+ neurons in the LoC (e); and DBH+ neurons in A1 of the ventrolateral medulla (f). The scale bar is equal to 50 μm in (d) and the magnification is the same for (a)–(d), while the scale bar is equal to 100 μm in (f) and the magnification is the same for (e) and (f). The midline is to the right in (b), (c), (d) and (f), and to the left in (a) and (e). The tectal layers are indicated by number in (c) and (d). Dorsal is to the top in all photomicrographs.

tinguish between visceral and somatic archistriatum. We have also observed that DBH+ fibers in pigeons are relatively more abundant in the nucleus taenia than in other parts of the archistriatum.

Diencephalon

In contrast to the telencephalon, TH+, DA+ and DBH+ fiber labeling in the diencephalon is light. In general, our results for TH+ fiber distribution in pigeon diencephalon are similar to those of Kiss & Peczely (1987).

TH+/DA+ fibers. Several features distinguish the TH+ fiber labeling pattern in the diencephalon. First, the hypothalamus in pigeons contains a distribution of TH+ fibers that is relatively uniform and moderate in abundance (Figs. 7.3(b), (d), 7.4(b), (d), 7.5(b), (d), 7.12(d), 7.15(a)–(c), 7.16(a)–(c)). Histofluorescence

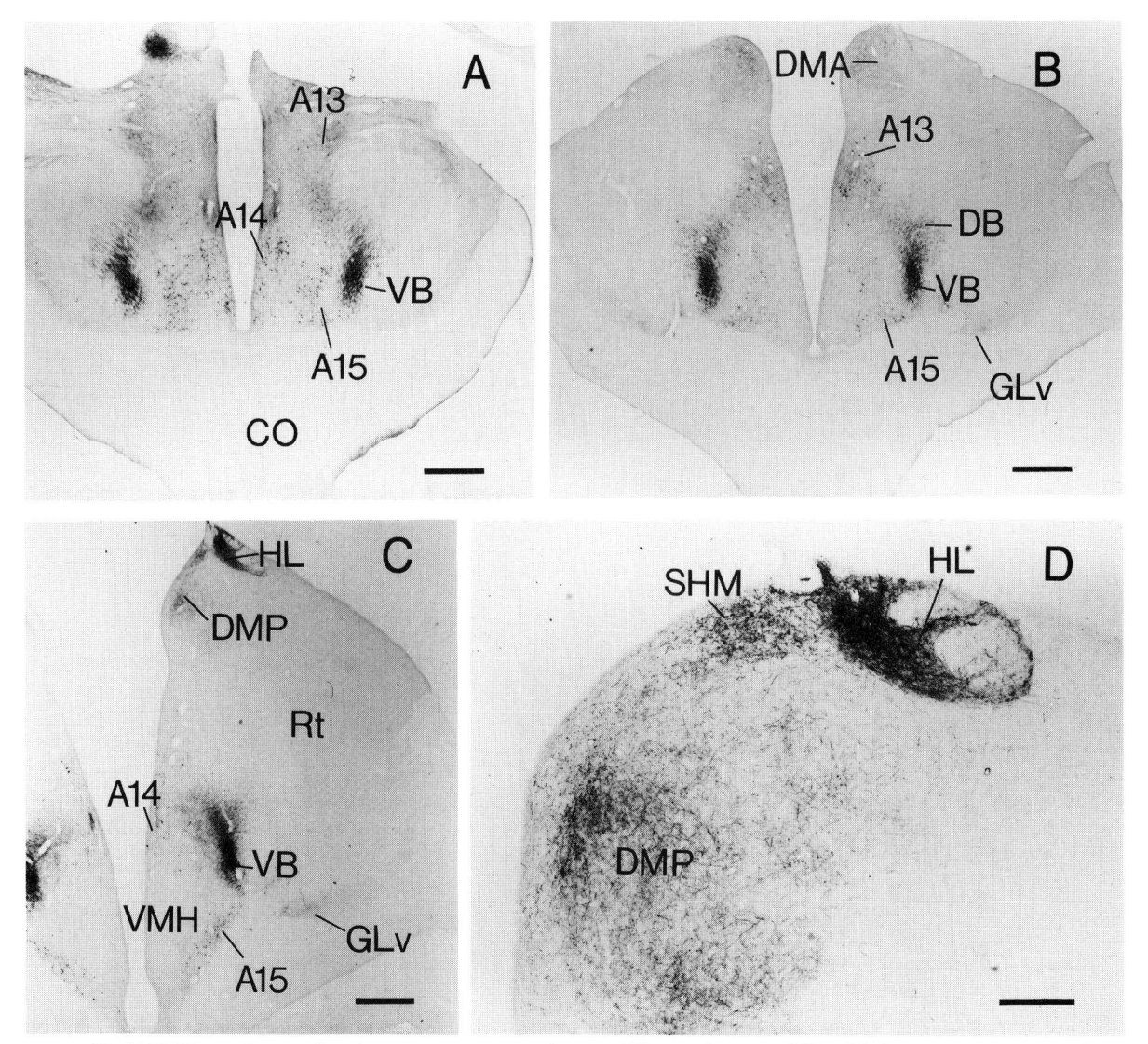

Fig. 7.15. Photomicrographs of transverse sections showing TH+ perikarya and fiber labeling in pigeon at rostral diencephalic (a), middiencephalic (b) and caudal diencephalic (c) levels. Photomicrograph (d) shows a higher magnification view of TH+ fiber labeling in the dorsomedial thalamus and habenular region at caudal diencephalic levels. The scale bar is equal to 200 μm in (a), (b) and (c), while in (d) it equals 100 μm. Both sides of the diencephalon are shown in (a) and (b), while in (c) and (d) the midline is to the left. Dorsal is to the top in all photomicrographs.

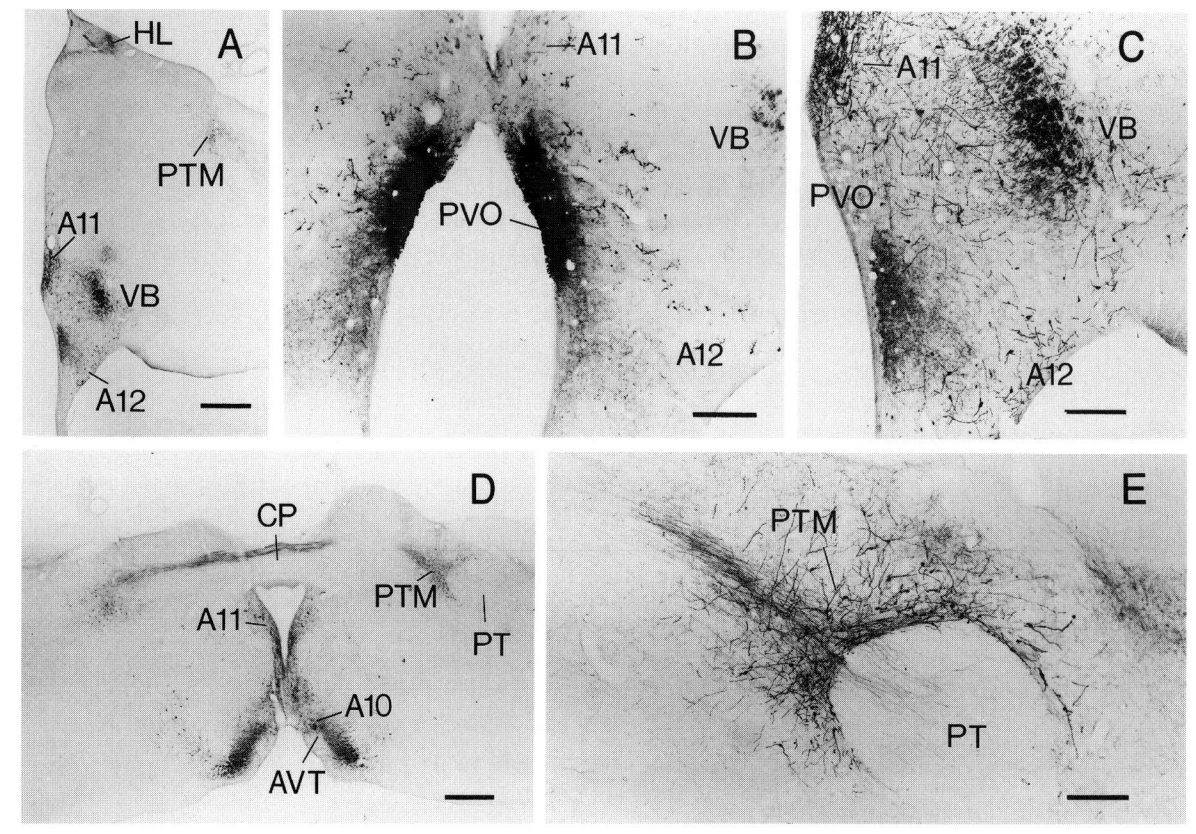

Fig. 7.16. Transverse sections through the pigeon diencephalon and pretectum immunolabeled for TH or DA. Photomicrograph (a) shows TH+ fiber labeling in pigeon at caudal diencephalic/pretectal levels. Photomicrograph (b) shows a high power view of DA+ perikaryal labeling in PVO. Photomicrograph (c) shows a high power view of TH+ labeling at the level of PVO. Photomicrograph (d) shows TH+ labeling at the level of the posterior commissure. Photomicrograph (e) shows a high power view of TH+ cell and fiber labeling in PTM and PT. The scale bars in (a) and (d) equal 200 μm, while those in (b), (c) and (e) equal 100 μm. Both sides of the brain are shown in (b) and (d), while in (a), (b) and (e) the midline is to the left. Dorsal is to the top in all photomicrographs.

studies also had shown that the hypothalamus in birds contains a moderate abundance of CA+ fibers (Parent, 1979; Takatsuki et al., 1981; Parent et al., 1984). We did, however, observe several regions within the hypothalamus that were somewhat richer in TH+ fibers (for example the periventricular zones superior and inferior to the PVO) (Figs. 7.5(b), (d), 7.16(a)–(c)). Although the DA+ fiber labeling pattern in the pigeon hypothalamus did not differ dramatically from the TH+ fiber labeling pattern, we did note that the magnocellular paraventricular nucleus (PVM) stands out from the surrounding hypothalamus as very rich in DA+ fibers. The TH+ fiber abundance in the PVM was, however, similar to surrounding hypothalamus. Finally, the inner zone of the median eminence is rich in TH+ fibers, but the outer zone is poor. The outer zone of the ME has also been reported to be poor in CA+ fibers in histofluorescence studies (Parent, 1979; Takatsuki et al., 1981). Kiss, Jurani & Kvaltinova (1983) in a study on Japanese quail using radioenzymatic methods found that NE is about 3–4 times as abundant as DA in the avian hypothalamus and DA is twice as abundant as adrenalin. They found that the medial hypothalamus generally contained higher levels of catecholamines than the lateral and that the highest levels (by far) of catecholamines were present in the posterior hypothalamus (which contains PVO).

The second major feature of the distribution of TH+ fibers in the avian diencephalon is that the dorsomedial thalamus (including the dorsomedial anterior and posterior nuclei) (Figs. 7.4(*b*), (*d*), 7.5(*b*), (*d*), 7.15(*b*)–(*d*)) and the habenular region are rich in TH+ fibers (7.5(*b*), 7.15(*d*)). Finally, the specific sensory nuclei of the thalamus are devoid or nearly devoid of evident TH+ fibers (7.4(*b*), (*d*), 7.5(*b*), 7.15(*b*), (*c*)). These regions include nucleus rotundus, nucleus ovoidalis, and the dorsal lateral geniculate nucleus (which consists of the dorsolateral lateral and anterior nuclei). Güntürkün & Karten (1991) have noted some differential distribution of TH+ fibers in the dorsal lateral geniculate of birds, namely low levels are present centrally and caudally, but higher levels are located in more rostral and dorsal parts of this cell group. These observations are consistent with our own. TH+ fiber labeling was also observed in two prominent fiber tracts, a dorsal one within the OM that labeled for TH but not DA (which we term the dorsal bundle) and a ventral one medial to the ansa lenticularis that labeled for TH and DA (which we term the ventral bundle) (Figs. 7.3(*d*), 7.4(*b*), (*d*), 7.5(*b*), (*d*), 7.12(*d*), 7.15(*a*)–(*c*). Finally, within the more ventral parts of the thalamus, we observed moderate numbers of TH+ fibers in the ventral lateral geniculate nucleus (GLv) (Figs. 7.4(*b*), (*d*), 7.15(*b*), (*c*)).

DBH+ fibers. The distribution of DBH+ fibers in the diencephalon is of interest with respect to the possibility that the anti-DBH we used preferentially labels NA+ fibers while the anti-TH preferentially labels DA+ fibers. We observed that DBH+ fibers in pigeons were very abundant along the periventricular portion of the hypothalamus, with a somewhat lesser amount of DBH+ fibers in the more lateral hypothalamus (Fig. 7.10(*c*), (*d*)). DBH+ fibers were abundant in the PVN region and in the so-called stratum cellulare internum (SCI) region, but low in the tuberoinfundibular and ventromedial nucleus (VMN) regions (Fig. 7.10(*d*)). In general, the abundance of DBH+ fibers in the hypothalamus was comparable to that in the visceral striatum. Interestingly, the PVO was very rich in DBH+ fibers, as were the zones below and lateral to PVO. This may account for the high levels of NA in PVO reported by Kiss *et al.* (1983). We also found many DBH+ fibers and terminals in the DMA/DMP cell groups of the dorsal thalamus (Figs. 7.10(*c*), (*d*), 7.14(*b*)), and nearly as many in more lateral dorsal thalamic cell groups (i.e. dorsointermediate dorsal thalamic nuclei) and in the dorsal lateral geniculate (i.e. the dorsolateral lateral and anterior thalamic cell groups). Only scattered fibers were present in nucleus rotundus and nucleus ovoidalis. Finally, the GLv is very rich in DBH+ fibers (more so than for TH+ fibers), mainly in its outer retinorecipient zone.

Bailhache & Balthazart (1993) found a distribution of TH+ fibers and DBH+ fibers in the diencephalon of quail very similar to that observed in pigeons. They noted that TH+ fibers were abundant in the preoptic area and throughout the hypothalamus, particularly medially. They noted that DBH+ and TH+ fibers were very similar in their distribution in hypothalamus, with the exception that the dopaminergic ventral bundle did not label for DBH. They noted that both TH+ fibers and DBH+ fibers were very rich in the PVN region in quail and sparse in the ventromedial nucleus (VMN) and ME. Bailhache & Balthazart (1993) also found that TH+ fibers and DBH+ fibers were present in the dorsal thalamic cell groups, especially along the midline and in the dorsal lateral geniculate. TH+ fibers appeared to be more abundant than DBH+ fibers in these cell groups.

Pretectum and mesencephalon

Note that we include the pretectum with our description of the mesencephalon, as is the conventional practice, although some authors have argued that the pretectum should be considered part of the diencephalon for embryological reasons (Puelles, Amat & Martinez-de-la-Torre, 1987; Medina *et al.*, 1993).

TH+/DA+ fibers. The pretectum and midbrain show heterogeneous labeling patterns for TH+ fibers. Within the pretectum, the PTM is rich in processes associated with the TH+ neurons of PTM. Some of these processes may be dendrites, while others appear to be axons entering and leaving the PTM (Figs. 7.5(*d*), 7.6(*b*), 7.16(*d*,*e*)). The axons leaving PTM can be followed to the posterior commissure, where they cross in the superficial part of the posterior commissure and course toward the PTM of the opposite side of the brain (Figs. 7.6(*d*), 7.16(*d*),(*e*). Fibers and terminals labeled for TH also appear to be present in PTM. Thus, PTM appears to have a homotypic projection to the PTM of the opposite side of the brain. Some TH+ fibers (seemingly coursing to or from PTM) travel through PT (7.16(*d*),(*e*). Bailhache & Balthazart (1993) also observed TH+ fibers in PTM in quail. Also at pretectal levels, TH+ fibers were observed in the tectal gray (GT) and the lentiformis mesencephali nucleus (LM) (Fig. 7.5(*d*)).

Within the optic tectum, two separate bands of TH+ fibers were observed, one in layers 3–4 and one in layers 6–8 (Figs 7.5(*d*), 7.6(*b*),(*d*),(*f*), 7.7(*b*)). In addition, TH+ fibers were observed in layer 15 of the tectum (also termed the stratum griseum periventriculare). The TH+ fibers in layer 15 were continuous with TH+ fibers in ICo and GCt, with which layer 15 is confluent (Fig. 7.6(d)). Rodman & Karten (1991) have studied the distribution and origin of TH+ fibers in the pigeon tectum. They report that the upper band of TH+ fibers is centered in layer 4 and upper layer 5, while the lower TH+ band is centered in layer 7. They also noted a dense plexus of TH+ fibers in layer 10–11 and in 15. The origins of the TH+ fibers in the tectum are discussed in the connectivity section. Note that our anti-DA antisera did not label tectal fibers. Bottjer (1993) reported the presence of TH+ fibers in the outer tectum in zebra finches, while Bailhache & Balthazart (1993) reported the presence of TH+ fibers in the superficial tectal layers in quail. Histofluorescence studies have reported CA+ fibers in superficial tectal layers in warbling grass parakeet (Shiosaka et al., 1981; Takatsuki et al., 1981).

Within the tegmentum, rich plexuses of TH+/DA+ processes representing the dendrites of dopaminergic neurons were observed in A10 (Figs. 7.6(b), 7.16(d), 7.17(a),(b),(c),(e)), A9 (Figs. 7.6(d), 7.17(c,e)) and A8 (Figs. 7.6(f), 7.17(d,e)). The TH+/DA+ fibers of the ventral bundle could be observed to merge with the AVT/SN region at rostral tegmental levels (Fig. 7.6(b)). The ventral bundle, whose axons are DA+, contains the efferent axons of the dopaminergic tegmental neurons that course to the striatum of the telencephalon, as well as to other telencephalic sites (Bertler et al., 1964; Kitt & Brauth, 1986b). The dorsal bundle was observed to course as a loose collection of TH+/DBH+ fibers within the OM in the dorsal tegmentum throughout the midbrain (Figs. 7.6(b),(d),(f)). At caudal tegmental and rostral pontine levels, TH+ fibers were observed along the raphe region at the midline (Figs. 7.6(f), 7.7(b)).

DBH+ fibers. The distribution of DBH+ fibers in the midbrain that we observed was different from the distribution of TH+ fibers (Fig. 7.11). At pretectal levels, DBH+ fibers and terminals were abundant in caudal GLv and in GT. In contrast to TH+ labeling, PT and PTM contained only scattered DBH+ fibers, while the medial spiriform nucleus (SpM) and the lateral spiriform nucleus (SpL) were rich in DBH+ fibers and terminals. In contrast, TH+ fibers were virtually undetectable in SpM and SpL (Fig. 7.5(d)). In addition, the subpretectal nucleus (SP), which was poor in TH+ fibers (Fig. 7.5(d)), was also quite rich in DBH+ fibers. Periaqueductal regions such as ICo, GCt and PAMG and the inferior colliculus (i.e. the dorsal lateral mesencephalic nucleus, MLd) were also in general richer in DBH+ fibers than they were in TH+ fibers. Bailhache & Balthazart (1993) also reported the presence of TH+ and DBH+ fibers in GCt and ICo in quail.

TH+ and DBH+ fiber patterns in the tectum also differed (Fig. 7.14(c),(d)). While TH+ fibers were localized to some tectal layers, DBH+ fibers were present in all layers, with a heightened abundance in superficial retinorecipient layers, particularly layer 5. Thus, anti-TH and anti-DBH do not label the identical populations of CA+ fibers in the tectum. In addition, DBH+ fibers were observed in layer 15 of the tectum. The DBH+ fibers in layer 15 were continuous with DBH+ fibers in ICo and GCt. The origins of the tectal DBH+ fibers in the tectum are discussed in the connectivity section. Bailhache & Balthazart (1993) also reported the presence of DBH+ fibers in the superficial tectal layers in quail.

Within the optic lobe, the lateral reticular formation in pigeons contains many clear DBH+ fibers, but the magnocellular (Imc) and parvocellular (Ipc) isthmic nuclei (which are isthmic nuclei located within the subventricular optic lobe) are poor in DBH+ fibers. The tegmental reticular formation also contains DBH+ fibers. Many DBH+ fibers and terminals are present in the A8–A10 DA cell fields, with many being pericellular, and DBH+ fibers and terminals are conspicuous in the raphe region. Thus, in the case of the midbrain also, the DBH+ fibers may represent primarily NA terminals while the TH+ fibers may represent primarily DA+ terminals. Bailhache & Balthazart (1993) reported the presence of TH+ and DBH+ fibers in the A8-10 region in quail, while histofluorescence studies have reported CA+ fibers in the A8-A10 cell fields in warbling grass parakeet (Shiosaka et al., 1981; Takatsuki et al., 1981).

Rhombencephalon

At isthmic levels in pigeons, TH+/DBH+ processes (presumably dendrites of A6 neurons) were observed in LoC and the region surrounding it (Figs. 7.7(b),(d), 7.18(a),(b)). The TH+/DBH+ fibers of the noradrenergic dorsal bundle were observed to merge with LoC rostromedially (Fig. 7.6(f)). Bailhache & Balthazart (1993) reported that TH+ and DBH+ fibers are present in LoC and SCv in quail also. Throughout the pigeon hindbrain, TH+ fibers were observed along the midline within the raphe region (Figs. 7.7(b),(d), 7.8(b),(d), 7.9(b),(d),(e)). Sparse and scat-

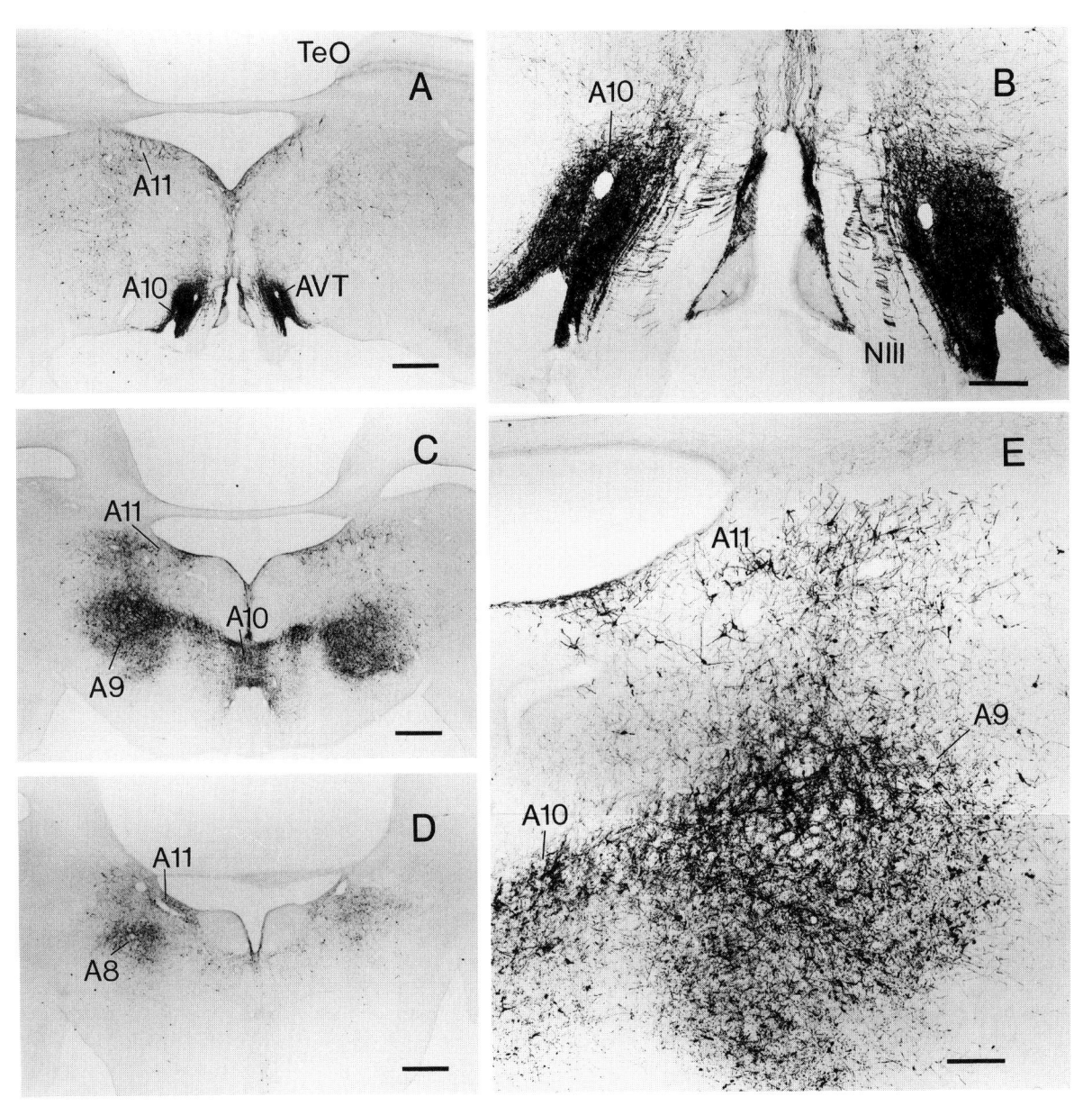

Fig. 7.17. Transverse sections through the pigeon midbrain immunolabeled for TH. Photomicrograph (a) shows TH+ cell and fiber labeling in pigeon in AVT at low magnification, while photomicrograph (b) shows the TH+ cell and fiber labeling in AVT at a higher magnification. Photomicrographs (c) and (d) show the TH+ cell and fiber labeling in the midbrain tegmentum at the level of A9 (c) and A8 (d). Photomicrograph E shows a high power view of TH+ cell and fiber labeling in the A9 at a level slightly caudal to that shown in (c). The scale bars in (a), (c) and (d) equal 200 μm, while those in (b) and (e) equal 100 μm. Both sides of the brain are shown in (a–d), while in (e) the midline is to the left. Dorsal is to the top in all photomicrographs.

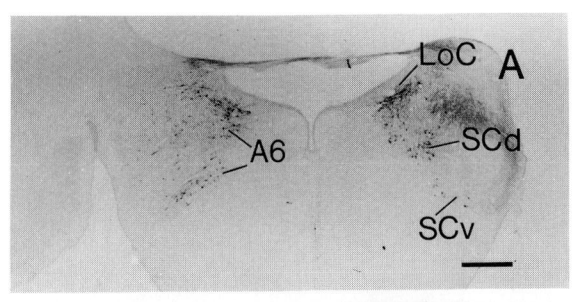

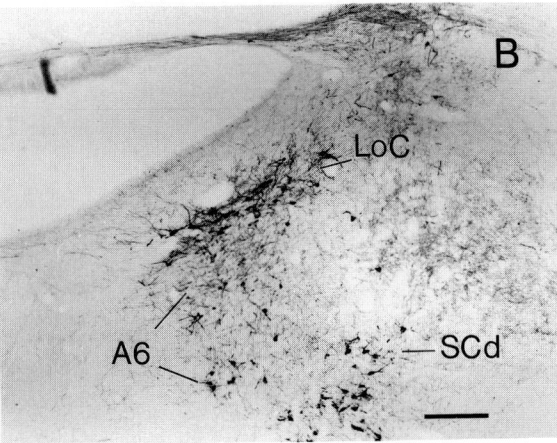

Fig. 7.18. Transverse sections through the pigeon isthmic region immunolabeled for TH. Low power (a) and high power (b) view of TH+ cell and fiber labeling in locus coeruleus in pigeon. The scale bar in (a) equals 200 μm, while that in (b) equals 100 μm. Both sides of the brain are shown in (a), while in (b) the midline is to the left. Dorsal is to the top in both photomicrographs.

tered TH+ fibers were also present within and around the pontine nuclei and the trapezoid body region. Fibers and terminals labeled for TH were also observed in the trigeminal and facial motor nuclei in the pons, and in the plexus of Horsley (PH) and the parvocellular reticular nucleus (Rpc) throughout the hindbrain (Figs. 7.7(d), 7.8(b),(d), 7.9(b),(d), 7.19(a),(b)). Caudal to the PH, TH+ fibers were present in the subtrigeminal nucleus, while caudal to the Rpc TH+ fibers were present in the central dorsal nucleus of the medulla (Cnd) (Figs. 7.9(d),(f)). Although TH+ fibers were largely absent from the pigeon cerebellum, we did observe a large bundle of TH+ fibers to enter and course medially within the superior medullary velum from the region of LoC (Fig. 7.18(b)). Finally, the upper part of the nTS/DMN complex was very rich in TH+ fibers (Figs. 7.9(c),(e), 7.19(b)–(d)).

DBH+ fibers. In general, we found that the distribution of DBH+ fibers in the pons and medulla in pigeon was very similar to that for TH+ fibers (Figs. 7.11(c),(d), 7.14(f)), while PNMT+ fibers were limited largely to the dorsal part of the nTS/DMN complex. DBH+ fibers, however, were more abundant than TH+ fibers among the vestibular nuclei and DBH+ fibers were conspicuously absent from the deep cerebellar nuclei and the cochlear nuclei. In contrast, TH+ fibers were rarely entirely absent from any hindbrain cell group. Bailhache & Balthazart (1993) reported that TH+ fibers but not DBH+ fibers are present within the paramedian reticular formation throughout the quail hindbrain. In addition, they reported that TH+ and DBH+ fibers are present in the A1 and A2 cell fields. Histofluorescence methods have revealed a distribution of CA+ fibers in warbling grass parakeet similar to that for TH+/DBH+ fibers in pigeons (Shiosaka et al., 1981).

Although TH+ fibers and terminals were largely absent from the pigeon cerebellum, we observed DBH+ fibers and terminals to be abundant in the cerebellum of pigeon. The DBH+ fibers and terminals were prominent around Purkinje cells and in the molecular layer at the level of the basal dendrites of the Purkinje cells. In addition, DBH+ fibers and terminals were scattered in the granule cell layer and in the outermost molecular layer. Using histofluorescence methods, Mugnaini & Dahl (1975) have also shown an abundance of presumptive NA+ fibers in the granule cell layer, in the molecular layer and around Purkinje cells in the cerebellum of chickens.

Spinal cord
The dorsal and lateral horns of the spinal cord in pigeons are rich in TH+ fibers (Fig. 7.9(h)). The ventral horns are poorer in TH+ fibers. The region above the central canal is the region of sympathetic preganglionic neurons, called the column of Terni in birds (Cohen & Karten, 1974). This region is rich in TH+ fibers (Fig. 7.9(h). The distribution of DBH+ fibers is similar to that of TH+ fibers (Fig. 7.11(e)). A similar distribution of CA+ fibers has been reported in warbling grass parakeet (Shiosaka et al., 1981) and chickens (Okado et al., 1991).

Connectivity and function

Retina: A17
The function of dopaminergic amacrine cells in birds is not known. Amacrine cells in general are thought to play a role in shaping the center-surround organ-

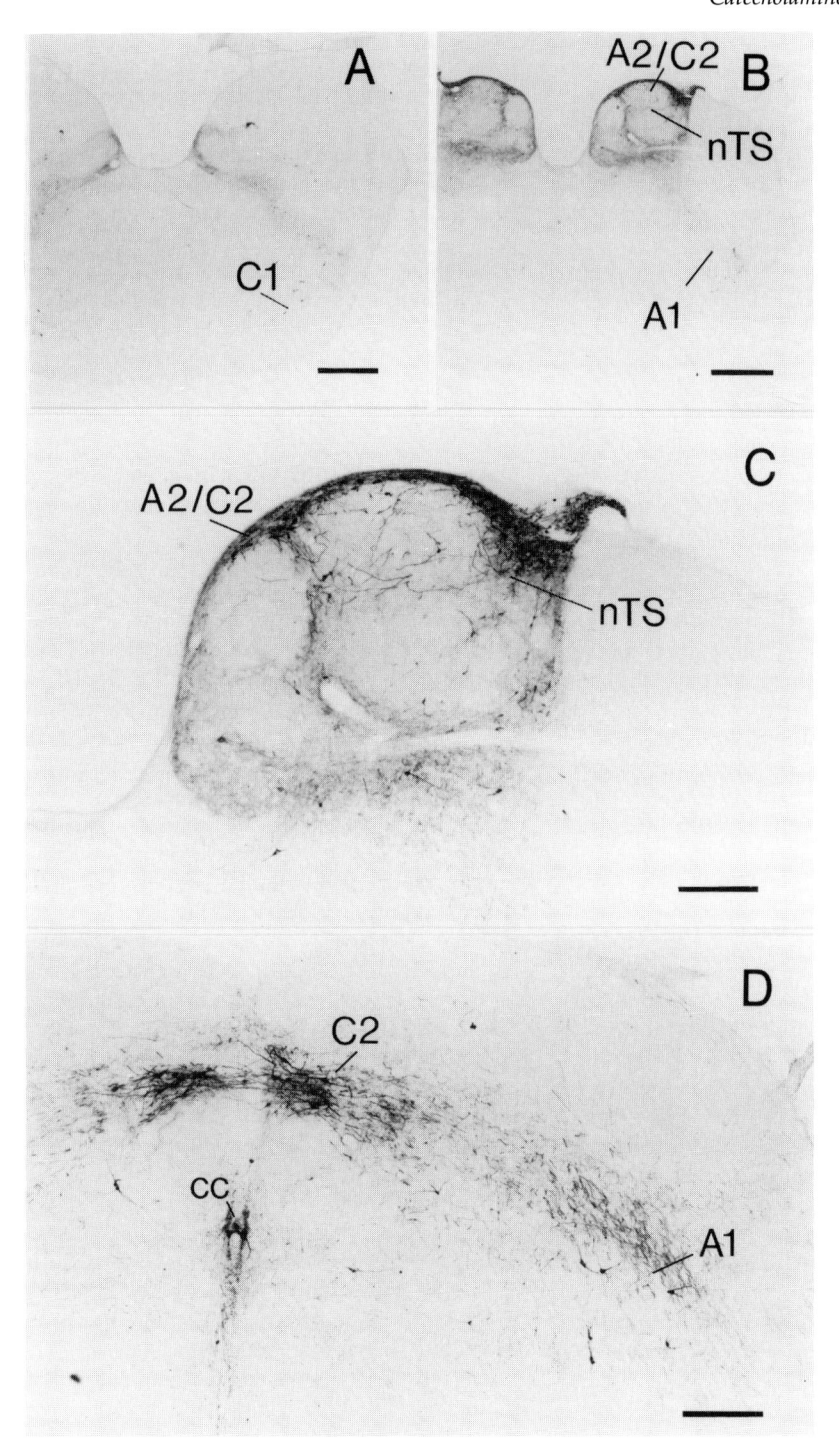

Fig. 7.19. Transverse sections through the pigeon medulla immunolabeled for TH. Photomicrographs (a) and (b) show TH+ cell and fiber labeling in the rostral (a) and caudal (b) parts of the A1/C1 and A2/C2 fields. Photomicrograph (c) shows a high power view of the nTS region shown in (b). Photomicrograph (d) shows a high power view of the TH+ fiber and cell labeling at the level of the obex. The scale bars in (a) and (b) equal 200 μm, while those in (c) and (d) equal 100 μm. Both sides of the brain are shown in (a), (b) and (d), while in (c) the midline is to the left. Dorsal is to the top in all photomicrographs.

ization of ganglion cell receptive fields based on inputs from bipolar cells (Miller, 1989). The dendrites of the TH+ amacrine cells in the IPL presumably interact with ganglion cell dendrites in the IPL to this end. In diverse nonavian species, dopaminergic amacrine cells have been found to diminish electrotonic coupling between neighboring neurons and thereby alter the paths of retinal information flow from photoreceptors to ganglion cells (Miller, 1989). Several lines of evidence also show that some of the TH+ retinal cells in pigeons are displaced ganglion cells (DGCs) that project to the contralateral nBOR (Britto et al., 1988, 1989). The TH+ DGCs appear to be one of several neurochemically specific types of DGCs projecting to nBOR (Britto & Hamassaki, 1991). The putative functional differences associated with such neurochemical differences are uncertain.

Olfactory bulb: A16

The role of the dopaminergic periglomerular cells has not been explored in birds. Based on studies in other vertebrate groups, in whom these cells are also present, it seems likely that the olfactory bulb DA+ cells in birds are involved in inhibitory interactions with the distal dendrites of mitral cells within the glomeruli (Halasz et al., 1977; Halasz, Nowycky & Shepherd, 1983; Nowycky, Halasz & Shepherd, 1983).

The hypothalamic A14 and A15 groups

In general, catecholamines within the hypothalamus of birds have been implicated in reproductive behavior. Considerably more information is available on the connectivity and function of the medial portions of the hypothalamus than of the lateral. Kiss & Peczely (1987) have identified the general region of the hypothalamus occupied by the A14 as the homologue of the mammalian PVN and they suggest that the TH+ cells of PVN may project to the cord, among other places (as is the case in mammals). In fact, Cabot, Reiner & Bogan (1982) have shown that many neurons in the more dorsal parts of the A14 region project to the spinal cord in pigeons. The medial hypothalamic region inferior to the anterior commissure contains many vasotocin+ cells in pigeons, chickens and canaries (Berk et al., 1982; Voorhuis, De Kloet & De Wied, 1991), which is reminiscent of the mammalian PVN and therefore further supports the notion that the A14 cells at least partly lie within PVN. Since the PVN is involved in the control of bodily homeostasis, this suggests that the A14 neurons in birds may play a role in such visceral functions (Kiss & Peczely, 1987). The A14 region in general is also known to give rise to ascending projections to telencephalic and diencephalic limbic structures, as well as descending projections to autonomic and visceral areas of the brainstem and cord (Berk & Butler, 1981; Korf, 1984). The studies on the projections of the medial hypothalamus, however, have not specifically examined whether TH+ neurons are among the neurons giving rise to the extra-hypothalamic projections. The medial hypothalamus also receives inputs from such diverse regions as the septum, LoC, the lateral hypothalamic area and the nTS (Korf, 1984), which is consistent with a role of this region in homeostatic control of visceral functions. As discussed further below, the input from LoC, the abundance of NA and the presence of DBH+ fibers in the hypothalamus are consistent with the possibility that LoC is the source of a major CA projection to the hypothalamus.

Hypothalamic catecholamines play a role in reproductive behavior in birds and this role involves interactions between gonadal steroids and CAs at the level of individual hypothalamic nuclei. One particular part of the A14 field, termed the medial preoptic area (POM), appears to play a major role in such behavior (although the specific role of the TH+ neurons in this field is not established). The POM is located inferior to the anterior commissure in birds and it is larger in males than females (Adkins-Regan & Watson, 1990). The POM is essential for male copulatory behavior in quail (Grant & Stumpf, 1975; Adkins-Regan & Watson, 1990). The POM contains gonadal steroid concentrating cells (Watson & Adkins-Regan, 1989), and these neurons also contain high levels of the enzyme aromatase (Schumacher & Balthazart, 1987), which is necessary for conversion of testosterone to estrogen. Conversion of testosterone to estrogen is, in turn, necessary for testosterone to stimulate male copulatory behavior in quail (Adkins-Regan et al., 1980; Balthazart & Schumacher, 1985). Catecholamines (particularly NE) appear to act on POM to inhibit expression of male sexual behavior (Balthazart et al., 1992). The inhibitory effects of NE on male quail reproductive behavior are via α1 and α2 adrenergic receptors (Barclay, 1989). Further, NA concentrations in female quails have been found to be higher in hypothalamic sites such as POA and POM (as well as in other brain regions) than in males (Ottinger et al., 1986). This could explain why testosterone fails to elicit male mating behavior in female quail, since the high NE may inhibit male reproductive behavior and thereby block the effects of testosterone (Balthazart et al., 1992). Blocking TH function with alpha-methyl-tyrosine in quail also reveals that NA function in the hypothalamus is more significant in females than males, while DA function in POM appears more important in males than females (Balthazart et al., 1992). Catecholamines also influence reproductive behavior in quail by influencing luteinizing hormone (LH) secretion from the pituitary. For example, NE stimulates LH release in birds (Scanes, Rabii & Buonomo, 1982; Buonomo, Rabii & Scanes, 1983; Balthazart & Ball, 1989), acting via α1 and α2 adrenergic receptors (Scanes et al., 1982). DA also stimulates LH release in birds (Buonomo, Rabii & Scanes, 1981; Scanes et al., 1982). Finally, NA is involved in vocal displays related to sexual behavior, such as song production in male songbirds (DeLanerolle & Youngren, 1978; Barclay, Johnson & Cheng, 1985; Barclay & Harding, 1988, 1990). The precise neuronal sites, however, at which catecholamines influence LH release and male vocal behavior are uncertain. It is possible that some of the sites may be non-hypothalamic or that some of the CA+ fibers involved may not be of hypothalamic origin.

The hypothalamus is also involved in other vegetative functions that are affected by catecholamines. Extensive data reviewed by Marley (1983) indicates that NA agonists injected intravenously or subcutaneously produce hypothermia, akinesia and blood pressure increases in chickens. Hypothalamic sites of action may be involved in these effects. Hypothalamic catecholamines may also modulate food and water intake in birds (Ravazio & Paschoalini, 1992).

Diencephalon – A13
The connectivity and function of the A13 cells in birds is uncertain. The A13 cell group is located within the zona incerta, which in mammals receives diverse inputs (e.g. from cortex, the ventral lateral geniculate nucleus, the deep cerebellar nuclei, the trigeminal complex and the spinal cord) and has diverse projections (e.g. to spinal cord, superior colliculus, hypothalamus and neurohypophysis) (Carpenter & Sutin, 1983; Jones, 1985). In rats, the A13 neurons within the zona incerta specifically give rise to the so-called incerto-hypothalamic DA+ projection system (Bjorklund, Lindvall & Nobin, 1975). The targets of this include various hypothalamic sites such as the dorsal and anterior hypothalamus and the medial preoptic region (Bjorklund & Lindvall, 1984). Thus, the specific functions of the A13 may be complex and involve visceral and autonomic functions.

Diencephalon: A12
In mammals, the A12 provides a heavy DA+ innervation of the outer ME. No such innervation of the ME has been observed in birds and it thus seems appropriate that the avian A12 cell group is meager by mammalian standards (Oksche & Farner, 1974; Parent, 1979; Kiss & Peczely, 1987). Since dopamine inhibits prolactin release in mammals (Leong, Frawley & Neill, 1983), this finding in birds is consistent with the prevailing view that DA is not a primary prolactin release-inhibiting factor in birds (Juorio & Vogt, 1967; El Halawani et al., 1984).

Diencephalon: A11
The specific connectivity and function of the A11 neurons in birds is unknown. Since TH+ A11 neurons in mammals project to the cord (Skagerberg & Lindvall, 1985), it is likely that some of the rostral A11 neurons in birds do project to the spinal cord. Consistent with this interpretation, Cabot et al. (1982) have shown that cells in this region resembling TH+ A11 neurons project to the spinal cord in pigeons. Based on data in mammals, the avian A11 could also have hypothalamic and telencephalic projections (Takada, Li & Hattori, 1988; Takada, 1990). Finally, the mesencephalic A11 cells may belong functionally to the A8–A10 tegmental DA+ neurons.

Pretectum: PTM
The TH+ neurons of PTM give rise to several projections. One evident projection from the PTM crosses in the dorsal portion of the posterior commissure. Although the various targets of this crossed PTM projection are uncertain, it seems likely that one target is the opposite PTM. Recently, Rodman & Karten (1991) used retrograde labeling methods to show that TH+ neurons in PTM project bilaterally to the tectum. Thus, PTM is a likely source of many of the TH+ fibers in the tectum. Since PTM neurons are DA+, this would imply that many of the tectal TH+ fibers are DA+.

Mesencephalon: A8–A10
Considerable information is available on the connectivity and function of the dopaminergic tegmental neurons in birds. To organize this information, we will separately present the data on the outputs of these neurons, on their inputs and on the functions of the nigrostriatal system.

Efferent projections to basal ganglia. The A8–10 cell groups give rise to ascending projections to the ipsilateral telencephalon (Reiner, Brauth & Karten, 1984a). Bertler et al. (1964) first demonstrated this by showing that substantia nigra lesions in birds virtually eliminated DA from the ipsilateral telencephalon, particularly the basal telencephalon. The projection of these dopaminergic neurons to the striatum has since been demonstrated using anterograde and retrograde pathway tracing methods (Brauth, Ferguson & Kitt, 1978; Reiner et al., 1984a; Bons & Oliver, 1986; Kitt & Brauth, 1986a, b) and by studying the pattern of loss of CA+ or TH+ terminals following destruction of the A8–A10 neurons or the ventral bundle (Lewis et al., 1981; A. Reiner & K. D. Anderson, unpublished observations). The targets of the A8–A10 projection include the somatic striatum, the visceral striatum, the somatic pallidum (PP), the ventral pallidum, and the INP, as well as such pallial structures as the dorsal archistriatum, the medial neostriatum, the dorsal hyperstriatum ventrale and the PE (Bertler et al., 1964; Karten & Dubbeldam, 1973; Brauth et al., 1978; Kitt & Brauth, 1986a, b). The projection to the striatum (including INP) appears topographically organized, with more rostromedial parts of the tegmental DA neuron field projecting to rostromedial striatum, and more caudolateral parts of the tegmental DA neuron field projecting to more caudolateral striatum (Kitt & Brauth, 1986a, b). EM immunohistochemical studies show that within the striatum the dopaminergic terminals are small (less than one micron in diameter) and often flattened (Karle, Anderson & Reiner, 1989, 1991, 1992) (Fig. 7.20(a), (b)). The results are similar regardless of whether anti-TH or anti-DA is used, further reinforcing the notion that anti-TH primarily labels DA+ terminals in the telencephalon. The DA+ terminals typically make symmetric synapses, with dendritic spines and shafts of striatal neurons representing the most common postsynaptic targets (Fig. 7.20(a), (b)) (Karle et al., 1989, 1991, 1992). Striatal perikarya themselves also receive input from DA+ terminals to some extent (Karle et al., 1989, 1991, 1992). The synaptic morphology and targets of these dopaminergic terminals in medial striatum (LPO) is indistinguishable from that in lateral striatum (PA).

The neuronal identity of the targets of the DA+ input has been addressed in EM immunohistochemical double-label studies. The two most abundant types of striatal neurons are those containing the neuropeptides substance P (SP), dynorphin (DYN) and the neurotransmitter GABA and those containing the enkephalin (ENK) neuropeptides and the neurotransmitter GABA (Reiner et al., 1983b, 1984b; Reiner & Anderson, 1990; Anderson & Reiner, 1990a; Reiner, 1986a, b, c). Both types of neurons are found throughout the mediolateral extent of the striatum, with the SP/DYN/GABA neurons and the ENK/GABA neurons of the medial striatum projecting to the A8–A10 field (Kitt & Brauth, 1981; Reiner et al., 1983b; Anderson & Reiner, 1990a, 1991a; Reiner & Anderson, 1990) and the SP/DYN/GABA neurons and the ENK/GABA neurons of the lateral striatum projecting to PP (Reiner et al., 1984b; Reiner & Anderson, 1990). In addition, small populations of interneurons co-containing parvalbumin, LANT6 and GABA (Reiner & Carraway, 1985, 1987; Reiner & Anderson, 1993), containing acetylcholine (Medina & Reiner, 1994), and co-containing somatostatin and neuropeptide Y are found in the avian striatum (Anderson & Reiner, 1990b). EM immunohistochemical double-labeling studies in pigeons have shown that both of the major types of striatal neurons in LPO (i.e. SP+ projection neurons and ENK+ projection neurons) receive direct input from TH+ terminals (Karle et al., 1991, 1992). In both cases, this input primarily targets dendritic shafts and dendritic spines (Fig. 7.20(b)), with SP+ and ENK+ neurons being similar in the ratio of TH+ terminals on dendritic spines compared to dendritic shafts (Karle et al., 1991, 1992). The possibility cannot yet be excluded, however, that despite their similarity in the ratio of axospinous vs. axodendritic TH+ input one of these two types of striatal neurons receives a greater abundance of TH+ terminals. Possible differences between these two types of neurons in their dopaminergic input is of significance because SP+ striatal neurons appear to respond differently to DA than do ENK+ striatal neurons (Anderson & Reiner, 1990a; Reiner & Anderson, 1990). It will be of interest in future studies to examine the ultrastructural features

Fig. 7.20. Electron photomicrographs of immunolabeling in the pigeon striatum (a) and (b) and substantia nigra (c) and (d). Photomicrograph (a) shows a DA-immunolabeled terminal (t*) in the medial striatum (LPO) making a symmetric contact with a dendritic shaft (D) of an unidentified neuron. Photomicrograph (b) shows a DAB-labeled dendrite (d) with attached spine of an SP+ LPO neuron. Note that the dendrite is contacted (thick arrow) by a silver-intensified immunogold-labeled TH+ terminal (t*2), while the spine is also contacted (longer, thinner arrow) by a silver-intensified immunogold-labeled TH+ terminal (t*1). Photomicrograph (c) shows a longitudinally sectioned silver-intensified immunogold-labeled TH+ dendrite (d) in the substantia nigra contacted (arrows) by numerous DAB-labeled SP+ terminals. Photomicrograph (d) shows a transversely sectioned silver-intensified immunogold-labeled TH+ dendrite (d) in the substantia nigra contacted (arrows) by a large DAB-labeled SP+ terminal (t*) and two unlabeled terminals (t1 and t2). All scale bars are equal to 500 nm.

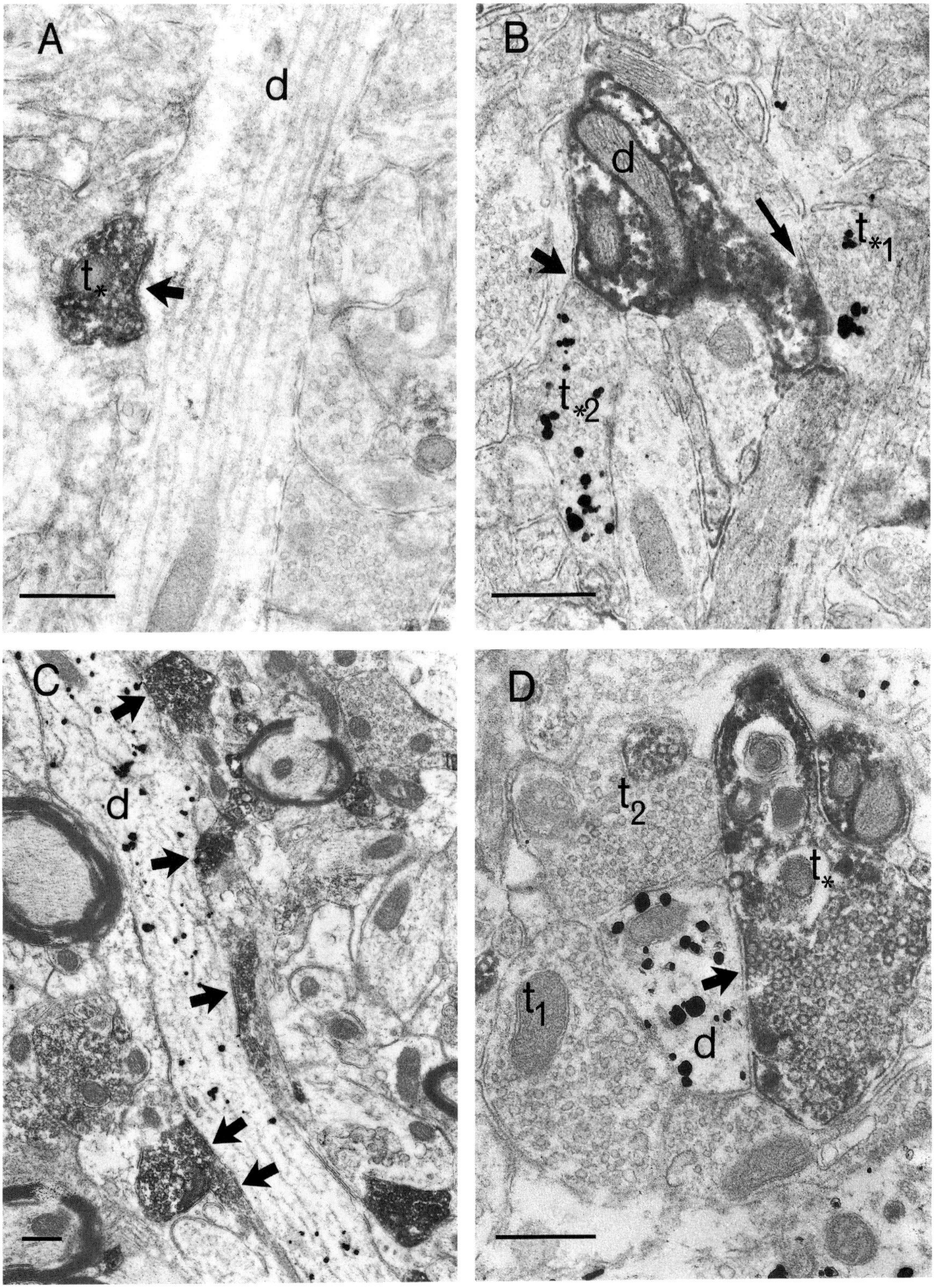
A
d
t*
B
d
t*1
t*2
C
d
D
t2
t*
t1
d

of the DA+ input to the three major types of striatal interneurons.

Actions of DA on basal ganglia function. Dopamine acts on striatal neurons in birds via D1 and D2 dopamine receptors, which are present in conspicuously high levels in the avian striatum, with both being present in greater abundance in the lateral striatum (Richfield, Young & Penney, 1987; Richfield et al., 1988; Dietl & Palacios, 1988). The concentration of D1 receptors in the avian striatum is only 10% that in mammalian striatum, while the concentration of D1 in the avian pallidum is 5% that in mammals (Richfield et al., 1987; Dietl & Palacios, 1988). In contrast, D2 receptor concentrations in pigeon striatum were 50% of those in mammalian striatum, while pallidal levels of D2 were 5% of those in mammals (Richfield et al., 1987; Dietl & Palacios, 1988). It is thus interesting to note that D1 levels exceed D2 levels in mammalian striatum, but in pigeons it is the opposite (Richfield et al., 1987). The DA receptors in birds show the same binding characteristics to several ligands as in mammals, suggesting that they are structurally similar to those in mammals (Covelli et al., 1981; Richfield et al., 1987). The D1 and D2 receptors in the striatum are now primarily thought to reside postsynaptically on striatal neurons (Anderson & Reiner, 1991b). Dopamine receptors in the pallidum may reside on either the terminals of striatopallidal projection neurons or on pallidal neurons themselves (Richfield et al., 1987). In addition, Richfield et al. (1987) found moderately high levels of D1 and D2 in the avian AVT-SN regions. Similar results have been observed in mammals, in whom the nigral D1 receptors are thought to be localized on striatonigral terminals and the D2 nigral receptors are thought to be autoreceptors on the DA neurons themselves (Richfield et al., 1987; Anderson & Reiner, 1991b).

Excessive stimulation of dopamine receptors (e.g. by systemic injection of dopamine agonists such as apomorphine) produces stereotypic pecking and/or vocalization in birds (Nistico & Stephenson, 1979; Gargiulo et al., 1981; Goodman et al., 1983; Nistico, Rotiroti & Stephenson, 1983; Akbas et al., 1984; Reiner et al., 1984a). This effect is pharmacologically specific to DA receptors since it can be blocked by dopamine receptor antagonists (Nistico & Stephenson, 1979; Gargiulo et al., 1981; Akbas et al., 1984). Conversely, DA antagonists promote tonic immobility and catalepsy in chicks (Sanberg & Mark, 1983). Similarly, Goodman et al. (1982) showed that depleting the pigeon striatum of DA by lesions of SN resulted in retardation of voluntary movements (i.e. bradykinesia and akinesia). Antipsychotic drugs (i.e. DA antagonists) have a similar effect in pigeons (Barrett, 1983). Since in mammals systemic DA agonists also produce pharmacologically specific stereotypic behavior and DA antagonists (or loss of DA neurons) yield bradykinesia/akinesia (Reiner et al., 1988; Albin, Young & Penney, 1989), the results in birds show that the DA system plays a similar role in promoting movement in both birds and mammals (Sanberg & Mark, 1983; Reiner et al., 1984a). Further, the effect of DA on motor behavior in birds and mammals is primarily mediated via the striatum (Goodman & Stitzel, 1977; Nistico et al., 1983; Sanberg & Mark, 1983; Reiner et al., 1984a; Albin et al., 1989). Not surprisingly therefore, unilateral lesions of the avian basal ganglia itself (Rieke, 1980) or of the A8-A10 cell group (Rieke, 1981) result in unilateral disturbances in posture and movement control. Thus, all available evidence favors the view that the nigrostriatal system and the basal ganglia in general play similar roles in the initiation and control of movement in birds and mammals (Reiner et al., 1984a; Reiner & Anderson, 1990).

Some data are available on the neuronal mechanisms by which dopamine affects basal ganglia-mediated movement control. Many lines of evidence in both mammals and birds show that the SP+ striatal neurons are differently affected by the nigrostriatal DA input than are ENK+ neurons (Reiner & Anderson, 1990; Anderson & Reiner, 1990a, 1991b). Activation of dopamine receptors by non-specific agonists such as apomorphine increases the levels of SP in SP+ striatal perikarya and in the terminals of these neurons in striatal target areas, while it decreases the levels of ENK in ENK+ striatal perikarya and in terminals in their targets areas. Conversely, blockade of dopamine receptors with a non-specific antagonist such as haloperidol or removal of the dopamine input to the striatum by 6-hydroxydopamine lesions of the tegmental dopamine neurons results in the opposite effects – decreased SP in SP+ perikarya and their terminals and increased ENK in ENK+ perikarya and their terminals. Thus, DA appears to stimulate SP+ striatal neurons and inhibit ENK+ striatal neurons. The cellular basis of this difference in response is yet uncertain and it is uncertain if the difference between SP+ and ENK+ neurons stems from a difference in the types of DA receptors present on these two neurons, the abundance of the DA input or other aspects of their striatal circuitry (Anderson & Reiner, 1991b; Karle et al., 1992; Surmeier et al., 1993). Nonetheless, the results show that

DA agonists promote stereotypy and increase the activity of SP+ striatal neurons, while DA antagonists yield bradykinesia/akinesia and increased ENK levels in striatal neurons. These outcomes are consistent with models of basal ganglia function that implicate the SP+ striatal projection neurons in promoting intentional voluntary movements and implicate the ENK+ striatal neurons in inhibiting unwanted movements (Reiner et al., 1988; Albin et al., 1989, 1990; Reiner & Anderson, 1990;). Since blocking GABA transmission in the basal ganglia by local antagonists in birds produces stereotyped movements, while local inhibitors of GABA-transaminase (which degrades GABA) block the stereotyped behaviors produced by apomorphine (Nistico et al., 1983), it seems likely that the stereotyped movements are normally suppressed by the ENK+/GABA+ striatopallidal neurons, but released from this inhibition by the inhibitory effects of DA on the ENK+ neurons. The possible interactions of DA with other types of striatal neurons in influencing basal ganglia function and movement (for example striatal cholinergic neurons) requires further exploration. For example, Nistico et al. (1983) have shown that the effects of DA agonists in birds on stereotyped behavior are blocked by drugs that enhance cholinergic transmission.

Striatal input to A8-A10. The A8–A10 field of tegmental dopaminergic neurons receives input from striatal and pallidal neurons (Reiner et al., 1984a). The somatic striatal input arises exclusively from neurons of the medial striatum (LPO), while neurons in the lateral striatum (PA) project to the paleostriatum primitivum (Karten & Dubbeldam, 1973; Reiner et al., 1984a; Kitt & Brauth, 1981). The striatal input to the nigra appears to be topographically organized, with rostromedial LPO projecting to A10 and more caudolateral LPO projecting to A8-A9 (Kitt & Brauth, 1981; Reiner et al., 1984a). The vast majority of the striatonigral projection neurons in pigeons (95% or more) have been found to be SP/DYN/GABA neurons of medial striatum (Anderson & Reiner, 1990a, 1991a; Reiner & Anderson, 1990). The remainder of the striatonigral projection neurons have been found to cocontain ENK and GABA (Anderson & Reiner, 1990a, 1991a; Reiner & Anderson, 1990). Terminals of striatal origin cocontaining either SP, DYN and GABA or ENK and GABA are present in the A8-A10 field (Anderson & Reiner, 1990a,b; Reiner & Anderson, 1990). There is some evidence that the ENK+ striatal input to the tegmentum may also contain the neuropeptide neurotensin (Reiner & Carraway, 1987; Reiner & Anderson, 1990). In birds, the striatal terminal field largely overlies the entire A8-A10 field (note that in some mammalian species the input to A9 is largely ventrolateral to it) (Reiner et al., 1983b; Smeets, 1991). EM immunohistochemical double-label studies show that both SP+ terminals and ENK+ terminals end on DA+ tegmental neurons, as well as non-DA+ tegmental neurons (Fig. 7.20(c),(d) (Anderson, Karle & Reiner, 1990, 1991). The striatal input to both types of tegmental target neurons makes predominantly symmetric synapses. In the tegmental fields examined, about half of the terminals contacting target structures contacted DA+ structures and half contacted non-DA+ structures (presumably GABAergic neurons within the tegmental dopaminergic field, Veenman & Reiner, 1994). Both SP+ terminals and ENK+ terminals end mainly on dendritic shafts, and infrequently on perikarya, of DA neurons. SP+ terminals, however, tended to terminate on thinner DA+ dendrites than did ENK+ terminals. Because of these various findings, we further sought to determine the extent to which SP+ terminals and ENK+ terminals ended on the same postsynaptic tegmental profiles. Using triple-label EM immunohistochemical labeling, we found that SP+ terminals and ENK+ terminals frequently both contacted the same postsynaptic unlabeled tegmental dendrites (Reiner, Anderson & Karle, 1992). In contrast, SP+ and ENK+ terminals were very rarely observed to both contact the same dopaminergic dendrites. These results indicate that while SP+ and ENK+ striatal terminals do appear to synapse upon the same tegmental non-dopaminergic (presumably GABAergic) neurons, they either do not synapse upon the same dopaminergic neurons or they synapse upon disparate parts of the same dopaminergic neurons. In either case, the results of our triple label studies show that SP+ and ENK+ striatal neurons may differ in their influence on tegmental DA neurons.

Pallidal Input to A8–A10. As noted, pallidal neurons also project to the A8–A10 field of DA neurons (Karten & Dubbeldam, 1973; Kitt & Brauth, 1981; Reiner et al., 1984a). The neurons giving rise to this input contain the neurotensin-related hexapeptide LANT6 and the neurotransmitter GABA (Reiner & Carraway, 1985, 1987; Reiner & Anderson, 1993). It seems likely that this input also influences the activity of tegmental DA+ neurons, since GABA and neurotensin binding sites (which could bind LANT6) are present among (presumably to some extent on) these neurons (Brauth et al., 1986; Dietl, Cortes & Palacios, 1988; Veenman et al., 1994). EM immunohistochemical studies on the pallidal input to the tegmentum have not yet been carried out. Thus, it is uncertain if this input actually ends on tegmental DA+ neurons. Nonetheless, tegmental DA neurons in mammals are rich in neurotensin receptors (Palacios & Kuhar, 1981). The tegmental dopaminergic neurons also receive a prominent 5HT input+ from a currently undetermined portion of the raphe (Yamada et al., 1984; Yamada & Sano, 1985; A. Reiner, unpublished observations). In addition, the A8–A10 field receives an enkephalinergic input that is not of striatal origin (Reiner, 1986c). The exact origin of the nonstriatal ENK+ fibers and terminals in the A8–A10 field is currently uncertain.

Function of striatal and pallidal inputs to A8–A10. Although the function of the various neuropeptides and GABA found in striatal terminals is uncertain, it is assumed that they act to influence the activity of both DA+ and non-DA+ tegmental neurons. Consistent with these conclusions, GABA/benzodiazepine, opiate and neurotensin receptors are found within the field of tegmental dopaminergic neurons (Brauth et al., 1986; Dietl et al., 1988; Reiner et al., 1989; Veenman et al., 1994). Rieke (1982) has reported that injection of GABA agonists into SN in birds yields contralateral turning and rotation, as such injections do in mammals. Further, systemic DA agonists and antagonists modulate this effect, thereby showing that it is mediated by an action of the GABA agonist on the DA neurons. Rieke (1981) has shown that SN lesions in birds also result in turning and rotation to the opposite side. Thus, GABA appears to have an inhibitory effect on tegmental DA neurons, as in mammals (Reiner & Anderson, 1990). Based on work in nonavians, it seems likely that neurotensin, LANT6, and opioid peptides also all have inhibitory effects on tegmental DA neurons in birds, while SP and related tachykinins are likely to have excitatory effects (Brauth et al., 1986; Reiner & Carraway, 1987; Dietl et al., 1988; Reiner et al., 1989; Reiner & Anderson, 1990; Veenman et al., 1994).

Other projections of A8–A10. The A8–A10 DA neurons also project heavily to medial neostriatum, HVdv, Ad, PE, and caudolateral neostriatum (NCL) and more lightly to the rest of the DVR, Wulst and hippocampal complex (Kitt & Brauth, 1986b). As noted above, Divac and his coworkers (Mogenson & Divac 1982; Divac & Mogenson, 1985; Divac et al., 1985) have argued that the PE/NCL is analogous to the mammalian prefrontal cortex (Reiner, 1986a). All evidence indicates that the A8–A10 DA neurons also project to epithalamic, thalamic, and hypothalamic sites, as well as to the central gray (including PAMG, GCt and ICo) (Fuxe & Ljunggren, 1965; Kitt & Brauth, 1986b; Nozaki & Kobayashi, 1975; Parent, 1979). The DA projections to the forebrain from the tegmentum are via the VB (or medial forebrain bundle, as it could also be called) (Parent et al., 1984). Consistent with the input of DA+ terminals to the entire telencephalic pallium, with some regions of higher innervation, D1 receptors have been found to be high throughout the avian telencephalon (except for the ectostriatum), with somewhat greater abundance in the Ad and PE/NCL, as well as parts of Wulst (Richfield et al., 1987; Dietl & Palacios, 1988). The levels of D1 in Ad and PE are, in fact, nearly as high as those in the striatum (Dietl & Palacios, 1988). In general, however, D2 levels are low in the Wulst, DVR and hippocampus (Richfield et al., 1987; Dietl & Palacios, 1988). In mammals also, D2 levels in the pallium are much lower than D1 levels (Dietl & Palacios, 1988). Moderate levels of D1 receptors and much lower levels of D2 receptors were also observed in the cerebellum and tectum, as is also the case in mammals (Dietl & Palacios, 1988). Finally, D2 receptors were relatively more abundant in such thalamic nuclei as DLA and SPC and in the pretectal nucleus PT (Dietl & Palacios, 1988). Note that the PT is, however, poor in TH+ and DA+ fibers. The role of dopamine at these various levels of the forebrain and midbrain is uncertain, but it is possible that DA exerts some of its effects on pecking and vocalization (Nistico et al., 1983) via some of these regions.

Rhombencephalon: A6, A4

The noradrenergic neurons of locus coeruleus have widespread projections to all levels of the avian CNS and account for most of the DBH+ fibers and perhaps some of the TH+ fibers in the brain and spinal cord (Kitt & Brauth, 1986a). While LoC is the likely source of DBH+ fibers in the tectum, LoC does not appear to be the source of many of the TH+ fibers in the avian tectum, since many arise from PTM and neurotoxic lesions of LoC only moderately reduce TH+ fibers in the tectum (Rodman & Karten, 1991). LoC does, however, project extensively to the cerebellum (Mugnaini & Dahl 1975; Tohyama, 1976; Dube & Parent, 1981; Yurkewicz et al., 1981). Some data suggest that LoC is the major source of DBH+/NA+ fibers in pallial regions of the telencephalon, while the SCd region projects to septal and hypothalamic sites (Tohyama et al., 1974; Kitt & Brauth, 1986a). LoC and to some extent SCd and SCv also project to the spinal cord (mainly thoracic levels) (Smolen, Glazer &

Ross, 1979; Cabot et al., 1982; Chikasawa, Fujioka & Watanabe, 1983).

The distribution and characteristics of adrenergic receptors have been studied in several avian species. Based on work in chickens, distinct α1-, α2- and β-adrenergic receptors all appear to be present in birds and display binding characteristics generally similar to those in mammals (Dermon & Kouvelas, 1988). For example, adrenalin has higher affinity for α2-adrenergic receptors in birds than does NA, as also true in mammals (Ball et al., 1989). Further, the binding affinity of α1-adrenergic receptors has been found to be similar among chickens (Dermon & Kouvelas, 1988), starling, song sparrow (Nock, Ball & Wingfield, 1986) and quail (Balthazart, Ball & McEwen, 1989). The affinity of the α1-adrenergic receptor is slightly higher than in mammals, however, but the number of sites is lower than in mammals, which is consistent with the fact that the amount of NA and adrenalin is higher in avians than in mammals (Kiss et al., 1983; Balthazart & Ball, 1989). Note that α2-adrenergic receptors are thought to be mainly presynaptic, while α1- and β-adrenergic receptors are thought to be postsynaptic (Dermon & Kouvelas, 1988). The three receptor types show some individual differences in their distribution, but all tend to be found in regions that contain DBH+ fibers. For example, α2-adrenergic receptors are high in MLd, Ipc and Imc, while α1-adrenergic receptors are high in MLd but low in Ipc and Imc. All three receptor types are abundant in the superficial tectum in adult chickens. Within the cerebellum, α1-and β-adrenergic receptors are both abundant in the molecular layer of the cerebellum, while only α1-adrenergic receptors are abundant in the granule cell layer. Balthazart et al. (1989) have obtained similar results for α1 adrenergic receptors in the brain of the Japanese quail. They additionally found high levels of α1, adrenergic receptors present in the ventromedial archistriatum, in the hippocampal complex, in CDL and outer PE, in dorsal thalamus and the dorsal geniculate and in PT, all regions that tend to be richer in DBH+ fibers. Somewhat lower levels of α1-adrenergic receptors were observed in the superficial tectum and in the hypothalamus. Ball et al. (1989) also found β2-adrenergic receptors in the brain of the Japanese quail to be greatest in DBH+ fiber-rich regions. For example, they found very high levels in septum, hypothalamus (especially POM and POA), and tuberal/mammillary hypothalamus. In addition, β2-adrenergic receptors were also abundant in DMA, and lower levels were observed in hyperstriatum, LPO, ICo, and superficial tectum.

Thus, the features of the distribution of adrenergic receptors are consistent with the apparent distribution of NA+ fibers and terminals in birds and many aspects of their distribution are similar to those in rats (e.g. high levels in the tectum and cerebellum and high levels of α1-adrenergic receptors in the telencephalon) (Dermon & Kouvelas, 1988; Balthazart et al., 1989). Anterior hypothalamic levels of α1-adrenergic receptors tend to be higher in mammals than birds, however (Balthazart et al., 1989). As discussed above, the NA system exerts an inhibitory influence on reproductive behavior in ring dove and Japanese quail (Balthazart, Liboulle & Sante, 1988; Barclay, 1989) and affects LH release (Scanes et al., 1982; Balthazart & Ball, 1989; Barclay, 1989). Presumably the NA system involved is the LoC projection to the hypothalamus. Nonetheless, the sites at which NA exerts its inhibitory influences on reproductive behavior is uncertain and the specific involvement of the coerulean NA system in such behavior needs to be established.

Rhombencephalon: A5

The projections and specific function of this cell group in birds have not been studied. This cell group, however, is likely to be similar to the C1/A1 group in connections and function (Cohen & Cabot, 1979; Loewy, Wallach & McKellar, 1981).

Rhombencephalon: C2, A2

Since the nTS is divided into different functional regions based on the visceral structure each region receives input from, different fields of CA cells in this region may receive different gustatory and visceral inputs and play a role in different functions (Dubbeldam, Karten & Menken, 1976; Katz & Karten, 1979, 1983). For example, Katz & Karten (1979) have shown that vagal afferents from the aortic arch in pigeons terminate in the lateralmost nTS. These afferents may carry information from baroreceptors and chemoreceptors in the aorta wall and play a role in control of blood pressure and heart rate (Nonidez, 1935; Jones, 1973; Katz & Karten, 1979). It is uncertain, however, if the CA+ neurons in this nTS region are involved in these functions. Caudal nTS receives vagal sensory information from esophagus, proventriculus and lungs (Katz & Karten, 1983), while rostral nTS receives taste input from the tongue (Berk, 1991). These inputs tend to end in central nTS regions (Berk, 1991). Since the C2/A2 neurons surround these central regions, it seems that the majority of the C2/A2 neurons do not receive direct axosomatic visceral sensory input, although inputs onto the dendrites of the C2/A2 neurons extending centrally seems likely (Berk, 1991). Arends, Wild & Zeigler (1988) reported projections from nTS in pigeons to ventrolateral pontine RF, to the parabrachial region, to the C1/A1 region, to the vagal motor nucleus, to the PVN region of the hypothalamus, to the medial thalamic wall (including DMA) and to the BNST, but did not examine which of these projections arises from CA+ neurons. Berk (1991) specifically studied the projections of A2 TH+ neurons by combining immunolabeling with retrograde labeling in pigeons. TH+ nTS neurons (particularly of caudal nTS) project to the periventricular hypothalamus. Within the A2 region, 60–80% of the neurons projecting to medial hypothalamus are TH+ (Berk, 1991). Central nTS neurons (which are typically not TH+) also project to the medial hypothalamus (Berk, 1991). Finally, the TH+ neuron rich parts of nTS also project to the BNST, the SCE and SCI of the caudal hypothalamus, the AVT, LoC and the parabrachial nuclei (Berk, 1987; Arends et al., 1988). Thus, these regions may derive some of their adrenergic and noradrenergic fibers from nTS.

The region of the taenia choroidea of the avian hindbrain seems comparable to mammalian area postrema since it contains a dense cluster of CA+ neurons, has an open bloodbrain barrier (for detection of blood borne substances) and does not project to telencephalon (Berk, 1991). Arends et al. (1988) report that this cell group projects to the parabrachial region and ventrolateral medulla in pigeons, and that lateral anterior nTS also is a source of the nTS projection to the parabrachial region (Arends et al., 1988).

Rhombencephalon:- C1, A1

The C1/A1 cell group projects to the spinal cord in birds (Cabot et al., 1982; Chikasawa et al., 1983). It seems likely that the C1/A1 projection terminates in the column of Terni, based on data in mammals (in whom C1/A1 projects to the sympathetic preganglionic neurons of the spinal cord) (Cohen & Cabot, 1979; Loewy et al., 1981; Goldstein & Kopin, 1990) and based on the abundance of TH+ fibers in the column of Terni. The column of Terni in birds contains the sympathetic preganglionic neurons of the avian spinal cord (Cohen & Karten, 1974; Cabot et al., 1982). The C1/A1 region is involved in control of blood flow and other sympathetic functions by means of the input to the column of Terni (Cohen & Cabot, 1979; Loewy et al., 1981; Caverson, Ciriello & Calarescu, 1983; Goldstein & Koplin, 1990). Consistent with this interpretation, pressor responses have been obtained in pigeons by stimulation of the C1/A1 region (Macdonald & Cohen, 1973).

Acknowledgements

We thank Chris Laverack and Pat Lindaman for their excellent technical assistance during the course of this research. We would like to thank Dr Harvey J. Karten for allowing us to use digitized versions of the images from the Karten and Hodos (1967) pigeon brain atlas for constructing our schematics, and we also thank him for support and encouragement during early phases of this research. We also gratefully acknowledge receiving unpublished information and data from W. Kuenzel (University of Maryland) on his own work on TH and DA localization in chicken brain. Finally, we thank R. N. Buijs (Netherlands Institute for Brain Research) for providing us with his anti-dopamine antiserum for some of our studies and we thank T. Joh (Cornell Medical Center) for providing us with his anti-tyrosine hydroxylase and anti-dopamine-β-hydroxylase for our early studies on TH and DBH localization. Other studies on DA, TH, DBH and PNMT localization were carried out using antisera from INCSTAR. This research has been supported by NIH grants NS-19620 and EY-05298 (A.R.), a postdoctoral fellowship from the Neuroscience Center for Excellence of the University of Tennessee at Memphis (K.D.A.) and a postdoctoral fellowship from the Ministry of Education and Science in Spain (L.M.).

Abbreviations

AA	nucleus archistriatalis anterior
Ac	nucleus accumbens
Ad	archistriatum pars dorsalis
AL	ansa lenticularis
AM	nucleus anterior medialis hypothalami
An	nucleus angularis
AP	area pretectalis
APH	area parahippocampalis
APrH	area prehippocampalis
Av	archistriatum pars ventralis
AVT	area ventralis (Tsai)
A1–A15	A1–A15 catecholaminergic cell groups
Bas	nucleus basalis
BC	brachium conjunctivum

BCS	brachium colliculi superioris	GCt	substantia grisea centralis
BNST	bed nucleus of the stria terminalis	GL	glomerular layer
BOR	basal optic root	GLv	nucleus geniculatus lateralis, pars ventralis
CA	anterior commissure		
Cb	cerebellum	GT	nucleus griseum tectalis
Cbd	tractus spinocerebellaris dorsalis	HA	hyperstriatum accessorium
CbL	nucleus cerebellaris lateralis	HD	hyperstriatum dorsale
CbM	nucleus cerebellaris intermedius	HIS	hyperstriatum intercalatum superior
CC	central canal	HL	nucleus habenularis lateralis
CDL	dorsolateral corticoid area	HM	nucleus habenularis medialis
CE	nucleus cuneatus externus	Hp	hippocampus
CHCS	tractus cortico-habenularis et cortico-septalis	HV	hyperstriatum ventrale
		HVdv	hyperstriatum ventrale dorso-ventrale
CL	nucleus cervicalis lateralis	HVvv	hyperstriatum ventrale ventro-ventrale
Cnd	nucleus centralis medullae oblongatae, pars dorsalis	Ico	nucleus intercollicularis
		IGL	internal glomerular layer
Cnv	nucleus centralis medullae oblongatae, pars ventralis	IM	nucleus intermedius
		Imc	nucleus isthmi, pars magnocellularis
CO	chiasma opticum	INP	nucleus intrapeduncularis
CP	commissura posterior	IO	nucleus isthmo-opticus
CPi	cortex piriformis	IP	interpenduncular nucleus
CTz	corpus trapezoideum (Papez)	Ipc	nucleus isthmi, pars parvocellularis
C1, C2	C1 and C2 catecholaminergic cell bodies	IPL	internal plexiform layer
DB	dorsal bundle of TH+ fibers	IPS	nucleus interstitio-pretecto-subpretectalis
DCB	decussatio brachiorum conjunctivorum		
DIP	nucleus dorsointermedius posterior thalami	IS	nucleus interstitialis (Cajal)
		L	L field
DLAmc	nucleus dorsolateralis anterior thalami, pars magnocellularis	La	nucleus laminaris
		LC	nucleus linearis caudalis
DLL	nucleus dorsolateralis anterior thalami, pars lateralis	LFM	lamina frontalis suprema
		LFS	lamina frontalis superior
DLM	nucleus dorsolateralis anterior thalami, pars medialis	LH	lamina hyperstriatica
		LHy	nucleus lateralis hypothalami
DLP	nucleus dorsolateralis posterior thalami	Li	lingula
DMA	nucleus dorsomedialis anterior	LL	lemniscus lateralis
DSD	decussatio supraoptica dorsalis	LLd	nucleus lemnisci lateralis, pars dorsalis (Groebbels)
DSV	decussatio supraoptica ventralis		
E	ectostriatum	LM	nucleus lentiformis
EPL	external plexiform layer	LMD	lamina medullaris dorsalis
EW	nucleus of Edinger-Westphal	LoC	locus coeruleus
FA	tractus fronto-archistriatalis	LPO	lobus parolfactorius
FD	funiculus dorsalis	LS	lemniscus spinalis
FDB	fasciculus diagonalis Brocae	MC	nucleus magnocellularis
FL	funiculus lateralis	MCL	mitral cell layer
FLM	fasciculus longitudinalis medialis	MLd	nucleus mesencephali lateralis, pars dorsalis (inferior colliculus)
FPL	fasciculus prosencephali lateralis		
FRL	formatio reticularis lateralis mesencephali	MV	nucleus motorius nervi trigemini
		N	neostriatum
FRM	formatio reticularis medialis mesencephali	nBOR	nucleus of the basal optic root
		NCL	caudolateral neostriatum
FU	fasciculus uncinatus (Russell)	NI	neostriatum intermedium
FV	funiculus ventralis	N III–XII	III–XII cranial nerves
GC	nuclei gracilis et cuneatus	n IV	nucleus nervi trochlearis

n IX	nucleus nervi glossopharyngei	SG	substantia gelatinosa Rolandi (trigemini)
nSO	nucleus supraopticus	SGC	stratum griseum centrale of the tectum
nTS	nucleus of the tractus solitarius	SGF	stratum griseum et fibrosum superficiale of the tectum
n VI	nucleus nervi abducentis		
n VII	nucleus nervi facialis	SGP	stratum griseum periventriculare of the tectum
N VIIIc	ramus caudalis N VIII		
N VIIIv	ramus ventralis N VIII	SHM	nucleus subhabenularis medialis
nX	nucleus motorius dorsalis nervi vagi	SL	nucleus septalis lateralis
OI	nucleus olivaris inferior	SLu	nucleus semilunaris
OM	tractus occipitomesencephalicus	SM	nucleus septalis medialis
OMN	nucleus nervi oculomotorii	SN	substantia nigra
OS	nucleus olivaris superior	SMe	stria medullaris
Ov	nucleus ovoidalis	SO	stratum opticum of the tectum
PA	paleostriatum augmentatum	SP	nucleus subpretectalis
PaM	nucleus paramedianus	SPC	nucleus superficialis parvocellularis (Nucleus tractus septomesencephalici)
Pap	nucleus papillioformis		
PD	nucleus pretectalis diffusus	SpL	nucleus spiriformis lateralis
PGL	nucleus paragigantocellularis lateralis	SPM	nucleus spiriformis medialis
PH	plexus of Horsley	SRt	nucleus subrotundus
PL	nucleus pontis lateralis	SSp	nucleus supraspinalis
PM	nucleus pontis medialis	ST	nucleus subtrigeminalis
PMH	nucleus medialis hypothalami posterioris	T	nucleus triangularis
		Ta	nucleus tangentialis (Cajal)
pPL	nucleus paraprincipalis nervi trigemini	TD	nucleus tegmenti dorsalis (Gudden)
POA	nucleus preopticus anterior	TeO	optic tectum
PP	paleostriatum primitivum	TIO	tractus isthmo-opticus
PPC	nucleus principalis precommissuralis	Tn	nucleus taeniae
PrV	nucleus sensorius principalis nervi trigemini	TO	tuberculum olfactorium
		TOv	tractus nuclei ovoidalis
PT	nucleus pretectalis	TPO	area temporo-parieto-occipitais
PTM	nucleus pretectalis medialis	TrO	optic tract
PVM	nucleus periventricularis magnocellularis	TS	tractus solitarius
PVO	periventricular organ	TSM	tractus septomesencephalicus
QF	tractus quintofrontalis	TT	tratus tectothalamicus
R	nucleus raphes	TTD	tractus descendens nervi trigemini
Rgc	nucleus reticularis gigantocellularis	TTS	Tractus thalamostriaticus
RL	nucleus reticularis lateralis	TU	nucleus tuberis
Rt	nucleus rotundus	TV	nucleus tegmenti ventralis (Gudden)
RP	nucleus reticularis pontis caudalis	V	ventricle
RPO	nucleus reticularis pontis oralis	Va	vallecula
Rpc	nucleus reticularis parvocellularis	VB	ventral bundle of TH+ fibers
RPgc	nucleus reticularis pontis caudalis, pars gigantocellularis	VeD	nucleus vestibularis descendens
		VeL	nucleus vestibularis lateralis
RS	nucleus reticularis superior	VeM	nucleus vestibularis medialis
Ru	nucleus ruber	VLT	nucleus ventrolateralis thalami
SAC	stratum album centrale of the tectum	VLV	nucleus ventralis lemnisci lateralis
SCd	nucleus subcoeruleus dorsalis	VMH	nucleus ventromedialis hypothalami
SCE	stratum cellulare externum	VO	olfactory ventricle
SCI	stratum cellulare internum	VP	ventral paleostriatum
SCN	nucleus suprachiasmaticus	VS	nucleus vestibularis superior
SCv	nucleus subcoeruleus ventralis	XII	nucleus motorius nervi hypoglossi

References

Adkins-Regan, E. & Watson, J. T. (1990). Sexual dimorphism in the avian brain is not limited to the song system of songbirds: a morphometric analysis of the brain of the quail (*Coturnix japonica*). *Brain Research*, **514**; 320–26.

Adkins-Regan, E., Boop, J. J., Koutnik, D. L., Morris, J. B. & Pniewski, E. E. (1980). Further evidence that aromitization is essential for the activation of copulation in male quail. *Physiology and Behavior*, **24**; 441–6.

Akbas, O., Verimer, T., Onur, R. & Kayaalp S. O. (1984). The effects of yohimbine and neuroleptics on apomorphine-induced pecking behavior in the pigeon. *Neuropharmacology*, **23**, 1261–4.

Albin, R. L., Young, A. B. & Penney, J. B. (1989). The functional anatomy of basal ganglia disorders, *Trends in Neurosciences*, **12**, 366–75.

Albin, R. L., Reiner, A. Anderson, K. D. Penney, J. B. & Young, A. B. (1990). Striatal and nigral neuron subpopulations in rigid Huntington's disease: implications for the functional anatomy of chorea and rigidity-akinesia. *Annals of Neurology*, **27**, 357–65.

Anderson, K. D. & Reiner, A. (1990a). The extensive co-occurrence of substance P and dynorphin in striatal projection neurons: an evolutionarily conserved feature of basal ganglia organization. *Journal of Comparative Neurology*, **295**, 339–69.

Anderson, K. D. & Reiner, A. (1990b). The distribution and relative abundance of neurons in the pigeon forebrain containing somatostatin, neuropeptide Y or both. *Journal of Comparative Neurology*, **299**, 261–82.

Anderson, K. D. & Reiner, A. (1991a). Striatonigral projection neurons: a retrograde labeling study of the percentages that contain substance P or enkephalin. *Journal of Comparative Neurology*, **303**, 658–73.

Anderson, K. D. & Reiner, A. (1991b). Immunohistochemical localization of DARPP-32 in striatal projection neurons and striatal interneurons: implications for the localization of D1 dopamine receptors on different types of striatal neurons. *Brain Research*, **568**, 235–43.

Anderson, K. D., Karle, E. J. & Reiner, A. (1990). An EM study of the relationship between ENK+ boutons and DA+ nigral neurons in pigeons. *Society for Neuroscience Abstracts*, 16: 236.

Anderson, K. D., Karle, E. J. & Reiner, A. (1991). Ultrastructural single- and double-label immunohistochemical studies of substance P-containing terminals and dopaminergic neurons in the substantia nigra in pigeons. *Journal of Comparative Neurology*, **309**, 341–62.

Araki, M., Maeda, T. & Kimura, H. (1983). Dopaminergic cell differentiation in the developing chick retina. *Brain Research Bulletin*, **10**, 97–102.

Arends, J.J.A., Wild, J. M. & Zeigler, H. P. (1988). Projections of the nucleus of the tractus solitarius in the pigeon (*Columba livia*). *Journal of Comparative Neurology*, **278**, 405–29.

Armstrong, D. M., Saper, C. B., Levey, A. I., Wainer, B. H. & Terry, R. D. (1983). Distribution of cholinergic neurons in rat brain: demonstrated by immunocytochemical localization of choline acetyltransferase. *Journal of Comparative Neurology*, **216**, 53–68.

Bailhache, T. & Balthazart, J. (1993). The catecholaminergic system of the quail brain: immunocytochemical studies of dopamine β-hydroxylase and tyrosine hydroxylase. *Journal of Comparative Neurology*, **329**, 230–56.

Ball, G. F., Nock, B., McEwen, B. S. & Balthazart, J. (1989). Distribution of α2-adrenergic receptors in the brain of the Japanese quail as determined by quantitative autoradiography: implications for the control of sexually dimorphic reproductive processes. *Brain Research*, **491**, 68–79.

Balthazart, J. & Schumacher, M. (1985). Role of testosterone metabolism in the activation of sexual behavior in birds. In *Neurobiology*, ed. Gilles, R. & Balthazart, J. Berlin: Springer-Verlag.

Balthazart, J., Ball, G. F. & McEwen, B. S. (1989). An autoradiographic study of α1-adrenergic receptors in the brain of the Japanese quail (*Coturnix coturnix japonica*). *Cell and Tissue Research*, **258**, 563–8.

Balthazart, J. & Ball, G. F. (1989). Effects of the noradrenergic neurotoxin DS4-P on luteinizing hormone levels, catecholamine concentrations, α2-adrenergic receptor binding and aromatase activity in the brain of the Japanese quail. *Brain Research*, **492**, 163–75.

Balthazart, J., Foidart, A. Sante, P. & Hendrick, J. C. (1992). Effects of α-methyl-para-tyrosine on monoamine levels in the Japanese quail: Sex differences and testosterone effects. *Brain Research Bulletin*, **28**, 275–88.

Balthazart, J., Liboulle, J. M. & Sante, P. (1988). Stimulatory effects of the noradrenergic neurotoxin DS4-P on sexual behavior in the male quail. *Behavior Proceedings*, **17**, 27–44.

Barclay, S. R. (1989). The role of the noradrenergic system in male courtship behavior in the ring dove. Doctoral Dissertation, Rutgers University, Newark, NJ.

Barclay, S. R., Johnson, A. & Cheng, M. F. (1985). Male courtship vocalization and the noradrenergic system. *Society for Neuroscience Abstracts*, **11**, 736.

Barclay, S. R. & Harding, C. F. (1988). Androstenedione modulation of monoamine levels and turnover in hypothalamic and vocal control nuclei in the male zebra finch. *Brain Research*, **459**, 333–43.

Barclay, S. R. & Harding, C. F. (1990). Differential modulation of monoamine levels and turnover rates by estrogen and/or androgen in hypothalamic and vocal control nuclei of male zebra finches. *Brain Research*, **523**, 251–62.

Barrett, J. E. (1983). Comparison of the effects of antipsychotic drugs on the schedule-controlled behavior of squirrel monkeys and pigeons. *Neuropharmacology*, **22**, 519–24.

Berk, M. L. (1987). Ascending axonal projections of the nucleus tractus solitarius in the pigeon. *Society for Neuroscience Abstracts*, **13**, 730.

Berk, M. L. (1991). Distribution and hypothalamic projection of tyrosine hydroxylase containing neurons of the nucleus of the solitary tract in the pigeon. *Journal of Comparative Neurology*, 312, 391–403.

Berk, M. L. & Butler, A. B, (1981). Efferents of the medial preoptic nucleus and medial hypothalamus in the pigeon. *Journal of Comparative Neurology*, 203, 379–99.

Berk, M. L., Reaves, T. A. Jr., Hayward, J. N., & Finkelstein, J. A. (1982). The localization of vasotocin and neurophysin neurons in the diencephalon of the pigeon, Columba livia. *Journal of Comparative Neurology*, 204, 392–406.

Bertler, A., Falck, B., Gottfreiss, C. G., Ljunggren, L. & Rosengren, E. (1964). Some observations on adrenergic connections between the mesencephalon and the cerebral hemispheres. *Acta Pharmacologica et Toxicologica*, 21, 283–9.

Bjorklund, A. & Lindvall, O. (1984). Dopamine-containing systems in the CNS. In *Handbook of Chemical Neuroanatomy. Vol. 2: Classical Transmitters in the CNS. Part 1*, ed. Bjorklund, A. & Hokfelt, T. pp 55–122. Amsterdam: Elsevier.

Bjorklund, A., Lindval, O. & Nobin, A. (1975) Evidence of an incerto-hypothalamic dopamine neuron system in the rat. *Brain Research*, 89, 29–42.

Bons, N. & Oliver, J. (1986). Origin of the afferent connections to the parolfactory lobe in quail shown by retrograde labeling with a fluorescent neuron tracer. *Experimental Brain Research*, 63, 125–34.

Bottjer, S. W. (1993). The distribution of tyrosine hydroxylase immunoreactivity in the brains of male and female zebra finch. *Journal of Neurobiology*, 24, 51–69.

Brauth, S. E., Ferguson, J. L. & Kitt, C. A. (1978). Prosencephalic pathways related to the paleostriatum of the pigeon (Columba livia). *Brain Research*, 147, 205–21.

Brauth, S. E., Kitt, C. A., Reiner, A. & Quirion, R. (1986). Neurotensin receptors in the forebrain and midbrain of the pigeon. *Journal of Comparative Neurology*, 253, 358–73.

Britto, L. R. G. & Hamassaki, D. E. (1991). A subpopulation of displaced ganglion cells of the pigeon retina exhibits substance P-like immunoreactivity. *Brain Research*, 546, 61–8.

Britto, L. R. G. & Hamassaki, D. E. (1993). Different subsets of displaced ganglion cells in the pigeon retina exhibit cholecystokinin-like and enkephalin-like immunoreactivities. *Neuroscience*, 52, 403–13.

Britto, L. R. G., Keyser, K. T. Hamassaki, D. E. & Karten, H. J. (1988). Catecholaminergic subpopulation of retinal displaced ganglion cells project to the accessory optic nucleus in the pigeon (Columba livia). *Journal of Comparative Neurology*, 269, 109–17.

Britto, L. R. G., Hamassaki, D. E., Keyser, K. T. & Karten, H. J. (1989). Neurotransmitters, receptors, and neuropeptides in the accessory optic system: An immunohistochemical survey in the pigeon (Columba livia). *Visual Neutoscience*, 3, 463–75.

Buonomo, F. C., Rabii, H. J. & Scanes, C. G. (1981). Aminergic involvement in the control of luteinizing hormone secretion in the domestic fowl. *General and Comparative Endocrinology*, 45, 162–6.

Buonomo, F. C., Rabii, J. & Scanes, C. G. (1983). Pharmacological studies on the noradrenergic control of luteinizing hormone secretion in the domestic fowl. *General and Comparative Endocrinology*, 49, 358–63.

Cabot, J. B., Reiner, A. & Bogan, N. (1982). Avian bulbospinal pathways: Anterograde and retrograde studies of cells of origin, funicular trajectories and laminar terminations. In *Progress in Brain Research*, ed. H. G. J. M. Kuypers & G. F. Martin, vol. 57, pp.79–108. Amsterdam, Elsevier.

Carpenter, M. B. & Sutin, J. (1983). *Human Neuroanatomy*. Baltimore, MD: Williams & Wilkins.

Caverson, M. M. & Ciriello, J. & Calarescu, F. R. (1983). Direct pathway from cardiovascular neurons in the ventrolateral medulla to the region of the upper thoracic cord: an anatomical and electrophysiological investigation in the cat. *Journal of the Autonomic Nervous System*, 9, 451–75.

Chikasawa, H., Fujioka, T. & Watanabe, T. (1982). Catecholamine-containing neurons in the mesencephalic tegmentum of the chicken. Light, fluorescence and election microscopic studies. *Anatomy and Embryology*, 164, 303–13.

Chikasawa, H., Fujioka, T. & Watanabe, T. (1983). Bulbar catecholaminergic neurons projecting to the thoracic spinal cord of the chicken. *Anatomy and Embryology*, 167, 411–23.

Cohen, D. H. & Cabot, J. B. (1979). Toward a cardiovascular neurobiology. *Trends in Neuroscience* Nov., 1–3.

Cohen, D. H. & Karten, H. J. (1974). The structural organization of the avian brain: an overview. In *Birds, Brain and Behavior*. pp.29–73. San Francisco: Academic Press.

Covelli, V., Memo, M., Spano, P. F. & Trabucchi, M. (1981). Characterization of dopamine receptors in various species of invertebrates and vertebrates. *Neuroscience*, 6, 2077–9.

Cozzi, B., Viglietti-Panzica, C., Aste, N., Panzica, G. C. (1991). The serotonergic system in the brain of the Japanese quail. *Cell and Tissue Research*, 263, 271–84.

DeLanerolle, N. C. & Youngren, O. M. (1978). Chick vocalization and emotional behavior influenced by apomorphine. *Journal of Comparative and Physiological Psychology*, 92, 416–30.

Dermon, C. R. & Kouvelas, E. D. (1988). Binding properties, regional ontogeny and localization of adrenergic receptors in chick brain. *International Journal of Developmental Neuroscience*, 6, 471–82.

Dietl, M. M., & Palacios, J. M. (1988). Neurotransmitter receptors in the avian brain. I. Dopamine receptors. *Brain Research*, 439, 354–9.

Dietl, M. M., Cortes, R. & Palacios, J. M. (1988). Neurotransmitter receptors in the avian brain. I. GABA-benzodiazepine receptors. *Brain Research*, 439, 366–71.

Divac, I. & Mogenson, J. (1985). The prefrontal 'cortex' in the pigeon. Catecholamine fluorescence. *Neuroscience*, 151, 677–82.

Divac, I., Mogenson, J. & Bjorklund, A. (1985). The prefrontal 'cortex' in the pigeon. Biochemical evidence. *Brain Research*, **332**, 365–8.

Divac, I., Mogenson, J. & Bjorklund, A. (1988). Strain differences in catecholamine content of pigeon brains. *Brain Research*, **444**, 371–3.

Dubbeldam, J. L., Karten, H. J. & Menken, S. B. (1976). Central projections of the chorda tympani nerve in the mallard, *Anas platyrhynchos* L. *Journal of Comparative Neurology*, **170**, 415–20.

Dubé, L & Parent, A. (1981). The monoamine-containing neurons in the avian brain: I. A study of the brainstem of the chicken (*Gallus domesticus*) by means of fluorescence and acetylcholinesterase histochemistry. *Journal of Comparative Neurology*, **196**, 695–708.

El Halawani, M. E., Burke, W. H., Millam, J. R., Fehrer, S. C. & Hargis, B. M. (1984). Regulation of prolactin and its role in Gallinaceous bird reproduction. *Journal of Experimental Zoology*, **232**, 521–9.

Fitzgerald, M. E. C., Vana, B. & Reiner, A. (1990). Control of choroidal blood flow by the nucleus of Edinger – Westphal: a laser-Doppler study. *Investigative Ophthalmology and Visual Sciences*, **31**, 2483–92.

Fuxe, K. & Ljunggren, L. (1965). Cellular localization of monamamines in the upper brain stem of the pigeon. *Journal of Comparative Neurology*, **125**, 355–92.

Gamlin, P. D. R., Reiner, A. & Karten, H. J. (1982). Substance P-containing neurons of the avian suprachiasmatic nucleus project directly to the nucleus of Edinger – Westphal. *Proceedings of the National Academy of Sciences, USA*, **79**, 3891–5.

Gargiulo, Nistico, G., Rotiroti, D. Silvestri, R. & Stephenson, J. D. (1981). Stereotyped behavior in fowls elicited by apomorphine given into the optic ventricle and into the nucleus spiriformis lateralis. *British Journal of Pharmacology*, **72**, 124P.

Goldstein, D. S. & Koplin, I. J. (1990). The autonomic nervous system and catecholamines in normal blood pressure control and in hypertension. In *Hypertension, Diagnosis and Management*, ed. Laragh, J. H. & Brenner, B. M. pp 711–47. New York: Raven Press.

Goodman, I. J. & Stitzel, R. E. (1977). Corpus striatal (paleostriatal complex) interventions and stereotyped behavior in pigeons. *Society for Neuroscience Abstract*, **3**; 37.

Goodman, I. J., Zacny, A., Osman, A., Azzaro, A. & Donovan, C. (1982). Lesion-produced telencephalic catecholamine imbalances and altered operant pecking rates in pigeons. *Physiology and Behavior*, **29**, 1045–50.

Goodman, I. J., Zacny, A., Osman, A., Azzaro, A. & Donovan, C. (1983). Dopaminergic nature of feeding-induced behavioral stereotypies in stressed pigeons. *Pharmacology, Biochemistry and Behavior*, **18**, 153–18.

Grant, L. D. & Stumpf, W. E. (1975). Hormone uptake sites in relation to CNS biogenic amine systems. In *Anatomical Neuroendocrinology* ed. Stumpf, W. E. & Grant, L. D. pp. 212–27. Basel: Karger.

Guglielmone, R. & Cantino, D. (1982). Autonomic innervation of the ocular choroid membrane in the chicken. A fluorescence-histochemical and electron-microscopic study. *Cell and Tissue Research*, **222**, 417–31.

Guglielmone, R. & Panzica, G. C. (1982). Topographic, morphological and developmental characterization of the nucleus loci coerulei in the chicken. A Golgi and fluorescence-histochemical study. *Cell and Tissue Research*, **225**, 95–110.

Guglielmone, R. & Panzica, G. C. (1984). Typology, distribution and development of the catecholamine-containing neurons in the chicken brain. *Cell and Tissue Research*, **237**, 67–79.

Gunturkun, O. & Karten, H. J. (1991). An immunohistochemical analysis of the lateral geniculate complex in the pigeon (*Columba livia*). *Journal of Comparative Neurology*, **314**, 721–49.

Halasz, N., Ljungdahl, A., Hokfelt, T., Johansson, O., Goldstein, M., Park, D. & Biberfeld, P. (1977). Transmitter histochemistry of the rat olfactory bulb. I. Immunohistochemical localization of monomaine synthesizing enzymes. Support for intrabulbar, periglomerular dopamine neurons. *Brain Research*, **126**, 455–74.

Halasz, N., Norwycky, M. C. & Shepherd, G. M. (1983). Autoradiographic analysis of [^{3}H] dopamine and [^{3}H]dopa uptake in the turtle olfactory bulb. *Neuroscience*, **8**, 705–15.

Hökfelt, T., Martensson, R., Björklund A., Kleinau, S. & Goldstein, M. (1984). Distributional maps of tyrosine hydroxylase-immunoreactive neurons in the rat brain. In *Handbook of Chemical Neuroanatomy, Volume 2 Classical Transmitters in the CNS, Part I*, ed. A. Björklund & T. Hökfelt, pp. 277–379. Amsterdam: Elsevier.

Ikeda, H. & Gotoh, J. (1971). Distribution of monoamine-containing cells in the central nervous system of the chicken. *Japanese Journal of Pharmacology*, **21**, 577–80.

Jones, E. J. (1985). *The Thalamus*. New York: Plenum Press.

Jones, D. R. (1973). Systemic arterial baroreceptors in ducks and the consequences of their denervation on some cardiovascular responses to diving. *Journal of Physiology (London)*, **234**, 499–518.

Juorio, A. V. & Vogt, M. (1967). Monoamines and their metabolites in the avian brain. *Journal of Physiology (London)*, **189**, 489–518.

Kagami, H., Sakai, H., Uryu, K., Kaneda, T. & Sakanaka, M. (1991). Development of tyrosine hydroxylase-like immunoreactivity in the chick retina: three dimensional analysis. *Journal of Comparative Neurology*, 308: 356–70.

Karle, E., Anderson, K. D. & Reiner, A. (1989). The nigrostriatal projection system in pigeons: an LM and EM immunohistochemical study using antisera against tyrosine hydroxylase. *Society for Neuroscience Abstracts*, **15**, 902.

Karle, E. J., Anderson, K. D. & Reiner, A. (1991). ENK+ striatal neurons and their input from nigral DA+ neurons at the EM level. *Society for Neuroscience Abstracts*, **17**, 457.

Karle, E. J., Anderson, K. D. & Reiner, A. (1992). Ultrastructural double-labeling directly reveals synaptic contact between dopaminegic terminals and substance P-containing striatal neurons in pigeons. *Brain Research*, **572**, 303–9.

Karten, H. J. (1968). The organization of the ascending auditory pathway in the pigeon (*Columba livia*). II Telencephalic projections of the nucleus ovoidalis thalami. *Brain Research*, **11**: 134–53.

Karten, H. J. & Dubbeldam, J. L. (1973). The organization and projections of the paleostriatal complex in the pigeon (*Columbia livia*). *Journal of Comparative Neurology*, 148: 61–89.

Karten, H. J. & Hodos, W. (1967). *A stereotaxic atlas of the brain of the pigeon (Columba livia)*. Baltimore: Johns Hopkins University Press.

Katz, D. M. & Karten, H. J. (1979). The discrete anatomical localization of vagal aortic afferents within a catecholamine-containing cell group in the nucleus solitarius. *Brain Research*, **171**, 187–95.

Katz, D. M. & Karten, H. J. (1983). Visceral representation within the nucleus of the tractus solitarius in the pigeon. *Journal of Comparative Neurology*, **218**, 142–73.

Kirby, M. L., Diab, I. M. & Mattio, T. G. (1978) Development of adrenergic innervation of the iris and fluorescent ganglion cells in the choroid of the chick eye. *Anatomical Record*, **191**, 311–20.

Kiss, J. Z. & Peczely, P. (1987). Distribution of tyrosine-hydroxylase (TH-immunoreactive neurons in the diencephalon of the pigeon (*Columba livia domestica*). *Journal of Comparative Neurology*, **257**, 333–46.

Kiss, A., Jurani, M. & Kvaltinova, Z. (1983). Distribution of catecholamines in individual hypothalamic nuclei of Japanese quail. *Journal of Endocrinology*, **8**, 125–70.

Kitt, C. A. & Brauth, S. E. (1981). Projections of the paleostriatum upon the midbrain tegmentum in the pigeon. *Neuroscience*, **6**, 1551–66.

Kitt, C. A. & Brauth, S. E. (1986a). Telencephalic projections from midbrain and isthmal cell groups in the pigeon. II. Locus coeruleus and subcoeruleus. *Journal of Comparative Neurology*, **247**, 69–91.

Kitt, C. A. & Brauth, S. E. (1986b). Telencephalic projections from midbrain and isthmal cell groups in the pigeon. II. The nigral complex. *Journal of Comparative Neurology*, 247: 92–110.

Knigge, K. M. & Piekut, D. T. (1985). Distribution of CRF- and tyrosine hydroxylase- immunoreactive neurons in the brainstem of the domestic fowl (*Gallus domesticus*). *Peptides*, **6**: 97–101.

Korf, H. W. (1984). Neuronal organization of the avian paraventricular nucleus: Intrinsic, afferent and efferent connections. *Journal of Experimental Zoology*, **232**, 387–95.

Krebs, J. R., Erichsen, J.T. & Bingman, V. P. (1991). The distribution of neurotransmitters and neurotransmitter-related enzymes in the dorsomedial telencephalon of the pigeon (*Columbia livia*). *Journal of Comparative Neurology*, **314**, 467–77.

Kuenzel, W. J. & M. Masson (1988). *Stereotaxic Atlas of the Brain of the Chick (Gallus domesticus)*. Baltimore: Johns Hopkins University Press.

Kuenzel, W. J., Kirtinitis, J. & Saidel, W. (1992). Comparison of tyrosine hydroxylase (TH) vs. dopamine (DA) specific antibody procedures for mapping DA-containing perikarya throughout the chick brain. *Society for Neuroscience Abstracts*, **18**, 329.

Leong, D. A., Frawley, L. S. & Neill, J. D. (1983). Neuroendocrine control of prolactin secretion. *Annual Review of Physiology*, **45**, 73–81.

Lewis, J. W. Ryan, S. M. Arnold, A. P. & Butcher, L. L. (1981) Evidence for a catecholaminergic projection to area X in the zebra finch. *Journal of Comparative Neurology*, **196**, 347–54.

Loewy, A. D., Wallach, J. H. & McKellar, S. (1981). Efferent connections of the ventral medulla oblongata in the rat. *Brain Research Reviews*, **3**, 63–80.

Macdonald, R. L. & Cohen, D. M. (1973) Heart rate and blood pressure responses to electrical stimulation of the central nervous system in the pigeon (*Columba livia*). *Journal of Comparative Neurology*, **150**, 109–36.

Marley, E. (1983). Behavioral and electrochemical effects of catecholamines and allied substances in avian species. In *Progress in Nonmammalian Brain Research. Volume II*, ed. G. Nistico & L. Bolis, pp. 79–88. Boca Raton: CRC Press.

Medina, L., Smeets, W. J .A. J., Hoogland, P. V. & Puelles, L. (1993). Distribution of choline acetyltransferase immunoreactivity in the brain of the lizard *Gallotia galloti*. *Journal of Comparative Neurology*, **331**, 261–85.

Medina, L. & Reiner, A. (1994). Distribution of CHAT immunoreactive perikarya and fibers in the pigeon brain. *Journal of Comparative Neurology* (in press).

Miller, R. F. (1989) The physiology and morphology of the vertebrate retina. In *Retina, Volume One. Basic Science & Inherited Retinal Disease*, ed. Ryan, S. J., Ogden, T. E. & Schachat, A. P. pp. 83–106. St Louis: The C. V. Mosby Co.

Mogenson, J. & Divac, I. (1982). The prefrontal 'cortex' in the pigeon. Behavioral evidence. *Brain, Behavior and Evolution*, **21**, 60–6.

Mugnaini, E. & Dahl, A. L. (1975). Mode of distribution of aminergic fibers in the cerebellar cortex of the chicken. *Journal of Comparative Neurology*, **162**, 417–32.

Nistico, G. & Stephenson, J. D. (1979). Dopaminergic mechanisms in birds. *Pharmacological Research Communications*, **11**, 555–70.

Nistico, G., Rotiroti, D. & Stephenson, J. D. (1983). Neurotransmitters and stereotyped behavior in birds. In *Progress in Nonmammalian Brain Research. Volume II*, ed. Nistico, G. & Bolis, L. pp. 89–105. Boca Raton: CRC Press.

Nock, B., Ball, G. F. & Wingfield, J. C. (1986). Regional localization of α1 and α2 adrenergic binding in two species of wild songbird using tritium sensitive film autoradiography. *Society for Neuroscience Abstracts*, **12**, 143.

Nonidez, J. R. (1935). The presence of depressor nerves in the aorta and carotid of birds. *Anatomical Record*, **62**, 47–73.

Nowycky, M. C., Halasz, N. & Shepherd, G. M. (1983). Evoked field potential analysis of dopaminergic mech-

anisms in the isolated turtle olfactory bulb. *Neuroscience* **8**, 717–22.
Nozaki, M. & Kobayashi, H. (1975). Monoamine fluorescence in the median eminence of the Japanese quail, *Coturnix coturnix japonica*, following medial basal hypothalamic deafferentation. *Cell and Tissue Research*, **164**, 425–34.
Okado, N., Ishihara, R., Ito, R., Homma, S. & Kohno, K. (1991). Immunohistochemical study of tyrosine-hydroxylase-positive cells and fibers in the chicken spinal cord. *Neuroscience Research*, **11**, 108–18.
Oksche, A. & Farner, D. S. (1974). Neurohistological studies of the hypothalamo-hypophyseal system of Zonotrichia leucophrys gambelli (Aves, Passeriformes) with special attention to its role in the control of reproduction. In *Advances in Anatomy, Embryology and Cell Biology*. New York: Springer.
Ottinger, M. A., Schumacher, M., Clarke, R. N., Duchala, C. S. & Balthazart, J. (1986). Comparison of monoamine concentrations in the brains of adult male and female Japanes quail. *Poultry Sciences*, **65**, 1413–20.
Ravazio, M. R. & Paschoalini, M. A. (1992). Modulation of food and water intake by catecholamines injected into the lateral ventricle of the pigeon brain. *Brazilian Journal of Medical and Biological Research*, **25**, 841–44.
Palacios, J. M. & Kuhar, M. J. (1981). Neurotensin receptors are located on dopamine-containing neurons in rat midbrain. *Nature*, **294**, 587–9.
Parent, A. (1979). Anatomical organization of monoamine- and aceylcholinesterase-containing neuronal systems in the vertebrate hypothalamus. In *Handbook of the Hypothalamus, Vol. I: Anatomy of the Hypothalamus*, ed. Morgane, P. J. & Panksepp, J. pp. 511–54. New York: Marcel Dekker Inc.
Parent, A. & Olivier, A. (1970). Comparative histochemical study of corpus striatum. *Journal für Hirnforschung*, **12**, 73–81.
Parent, A., Poitras, D. & Dubé, L. (1984). Comparative anatomy of central monoaminergic systems. In *Handbook of Chemical Neuroanatomy. Vol. 2: Classical Transmitters in the CNS. Part I*, ed. Bjorklund, A. & Hokfelt, T. pp. 409–439. Amsterdam: Elsevier.
Puelles, L., Amat, J. A. & Martinez-de-la-Torre, M. (1987). Segment-related, mosaic neurogenetic pattern in the forebrain and mesencephalon of early chick embryos: I. Topography of AChE-positive neuroblasts up to stage HH18. *Journal of Comparative Neurology*, **266**, 247–68.
Ravazio, M. R. & Paschoalini, M. A. (1992). Modulation of food and water intake by catecholamines injected into the lateral ventricle of the pigeon brain. *Brazilian Journal of Medical & Biological Research*, **25**, 841–4.
Reiner, A. (1986a). Is prefrontal cortex found only in mammals? *Trends in Neurosciences*, **9**, 298–300.
Reiner, A. (1986b). The co-occurrence of substance P-like immunoreactivity and dynorphin-like immunoreactivity in striatopallidal and striatonigral projection neurons in birds and reptiles. *Brain Research*, **371**, 155–61.
Reiner, A. (1986c). Transmitter-specific projections from the basal ganglia to the tegmentum in pigeons. *Society for Neuroscience Abstracts*, **12**, 873.
Reiner, A., Albin, R. L., Anderson, K. D., D'Amato, C. J., Penney, J. B. & Young, A. B. (1988). Differential loss of substance P-containing striatopallidal and enkephalin-containing striatopallidal projections in Huntington's Disease. *Proceedings of the National Academy of Sciences, USA*, **85**, 5733–7.
Reiner, A. & Anderson, K. D. (1990). The patterns of neurotransmitter and neuropeptide co-occurrence among striatal projection neurons: conclusions based on recent findings. *Brain Research Reviews*, **15**, 251–65.
Reiner, A. & Anderson, K. D. (1993). Co-occurrence of GABA, parvalbumin and the neurotensin-related hexapeptide LANT6 in pallidal, nigral and striatal neurons in pigeons and monkeys. *Brain Research*, **624**, 317–35.
Reiner, A., Anderson, K. D. & Karle, E. J. (1992). A pre-embedding triple-label EM immunohistochemical method as applied to the study of multiple inputs to defined nigral neurons. *Society for Neuroscience Abstracts*, **18**, 1127.
Reiner, A., Brauth, S. E. & Karten, H. J. (1984a). Evolution of the amniote basal ganglia. *Trends in Neurosciences*, **7**, 320–5.
Reiner, A., Brauth, S. E., Kitt, C. A. & Quirion, R. (1989). The distribution of mu, delta and kappa opiate receptors in the pigeon forebrain and midbrain. *Journal of Comparative Neurology*, **280**, 359–82.
Reiner, A. & Carraway, R. E. (1985). The presence and phylogenetic conservation of a neurotensin-related hexapeptide in neurons of globus pallidus. *Brain Research*, **341**, 365–71.
Reiner, A. & Carraway, R. E. (1987). Immunohistochemical and biochemical studies on Lys^8–Asn^9–Neurotensin$^{8-13}$ (LANT6)-related peptides in the basal ganglia of pigeons, turtles and hamsters. *Journal of Comparative Neurology*, **257**, 453–76.
Reiner, A., Davis, B. M., Brecha, N. C. & Karten, H. J. (1984b). The distribution of enkephalin-like immunoreactivity in the telencephalon of the adult and developing domestic chicken. *Journal of Comparative Neurology*, **228**, 245–62.
Reiner, A., Karten, H. J., Gamlin, P. D. R. & Erichsen, J. T. (1983a) Parasympathetic control of ocular function: functional subdivisions and connections of the avian nucleus of Edinger–Westphal. *Trends in Neurosciences*, **6**, 140–5.
Reiner, A., Karten, H. J. & Solina, A. R. (1983b). Substance P: Localization within paleostriatal-tegmental pathways in pigeons. *Neuroscience*, **9**, 61–85.
Richfield, E. K., Albin, R. L., Reiner, A., Young, A. B. & Penney, J. B. Jr. (1988). Receptor binding patterns in the basal ganglia of pigeons. *Society for Neuroscience Abstracts*, **14**, 1022.
Rieke, G. L. (1980). Kainic acid lesions of pigeon paleostriatum: a model for study of movement disorders. *Physiology and Behavior*, **24**, 683–7.
Richfield, E. K., Young A. B. & Penney J. B. (1987) Compar-

ative distribution of D-1 and D-2 receptors in the basal ganglia of turtles, pigeons, rats, cats and monkeys. *Journal of Comparative Neurology*, 262: 446–463.

Rieke, G. L. (1981). Movement disorders and lesions of pigeon brainstem analogues of basal ganglia. *Physiology and Behavior*, **26**, 379–84.

Rieke, G. L. (1982). The TPc, the avian substantia nigra: pharmacology and behavior. *Physiology and Behavior*, **28**, 755–63.

Rodman, H. R. & Karten, H. J. (1991). Laminar distribution and origins of the catecholaminergic innervation of the optic tectum in the pigon (*Columba livia*). *Society for Neuroscience Abstracts*, **17**, 112.

Sanberg, P. R. & Mark, R. F (1983). The effect of striatal lesions in the chick on haloperidol-potentiated tonic immobility. *Neuropharmacology*, **22**, 253–7.

Scanes, C. G., Rabii, J., & Buonomo, F. (1982). Brain amines and the regulation of anterior pituitary secretion in the domestic fowl. In *Aspects of avian endocrinology: practical and theoreticals implications*, ed. C. G. Scanes, M. A. Ottinger, A. Kenny, J. Balthazart, J. Crenshaw & J. Chester-Jones, pp. 13–32. Lubbock: Texas Tech. Press.

Schumacher, M. & Balthazart, J. (1987). Neuroanatomical distribution of testosterone-metabolizing enzymes in the Japanese quail. *Brain Research*, **422**, 137–48.

Shimizu, T. & Karten, H. J. (1990). Immunohistochemical analysis of the visual Wulst of the pigon (*Columba livia*). *Journal of Comparative Neurology*, **300**, 346–69.

Shiosaka, S., Takatsuki, K., Inagaki, S., Sakanaka, M., Takagi, H., Senba, E., Matsuzaki, T. & Tohyama, M. (1981). Topographic atlas of somatostatin-containing neuron system in the avian brain in relation to catecholamine-containing neuron system. II. Mesencephalon, rhombencephalon and spinal cord. *Journal of Comparative Neurology*, **202**, 115–24.

Skagerberg, G. & Lindvall, O. (1985). Organization of diencephalic dopamine neurons projecting to the spinal cord in the rat. *Brain Research*, **342**, 340–51.

Smeets, W. J. A. J. (1988). The monoaminergic systems of reptiles investigated with specific antibodies against serotonin, dopamine and noradrenalin. In *The Forebrain of Reptiles. Presented at the International Symposium on Recent Advances in Understanding the Structure & Function of the Forebrain in Reptiles. Frankfurt/M. 1987*, ed. Schwerdtfeger, W. & Smeets W. J. A. J., pp. 97–109. Basel: Karger.

Smeets, W. J. A. J. (1991). Comparative aspects of the distribution of substance P and dopamine immunoreactivity in the substantia nigra of amniotes. *Brain, Behavior and Evolution*, **37**, 179–88.

Smeets, W. J. A. J. & González, A. (1990). Are putative dopamine-accumulating cell bodies in the hypothalamic periventricular organ a primitive brain character of non-mammalian vertebrates? *Neuroscience Letters*, **114**, 248–52.

Smeets, W. J. A. J. & Jonker, A. J. (1990). Distribution of phenylethanolamine-N-methyltransferase-immunoreactive perikarya and fibers in the brain of the lizard *Gekko gecko*. *Brain, Behavior and Evolution*, **36**, 59–72.

Smeets, W. J. A. J. & Steinbusch, H. W. M. (1990). New insights into reptilian catecholaminergic systems as revealed by antibodies against the neurotransmitters and their synthetic enzymes. *Journal of Chemical Neuroanatomy*, **3**, 25–43.

Smolen, A. J., Glazer, E. J. & Ross, L. L. (1979). Horseradish peroxidase histochemistry combined with glyoxylic acid-induced fluorescence used to identify brain stem catecholaminergic neurons which project to the chick thoracic spinal cord. *Brain Research*, **191**, 417–28.

Soest, S. W., Farner, D. S & Oksche, A (1973). Fluorescence microscopy of neurons containing primary catecholamine in the ventral hypothalamus of the white-crowned sparrow, *Zonotricha leucophrys gambelli*. *Zeitschrift für Zellforschung und Mikroskopische Anatomie*, **141**, 1–17.

Steeves, J. D., Taccogna, C. A. Bell, K. A. & Vincent, S. R. (1987). Distribution of phenylethanolamine-N-methyltransferase (PNMT)-immunoreactive neurons in the avian brain. *Neuroscience Letters*, **76**, 7–12.

Surmeier, D. J., Reiner, A., Levine, M. S. & Ariano, M. A. (1993). Are neostriatal dopamine receptors co-localized? *Trends in Neurosciences*, **16**, 299–305.

Takada, M. (1990). The A11 catecholamine cell group: another origin of the dopaminergic innervation of the amygdala. *Neuroscience Letters*, **118**, 271–85.

Takada, M., Li, Z. K. & Hattori, T. (1988). Single thalamic dopaminergic neurons project to both the neocortex and spinal cord. *Brain Research*, **455**, 346–52.

Takatsuki, K., Shiosaka, S., Inagaki, S., Sakanaka, M., Takagi, H., Senba, E., Matsuzaki, T. & Tohyama M. (1981). Topographic atlas of somatostatin-containing neuron system in the avian brain in relation to catecholamine-containing neuron system. II. Telencephalon and diencephalon. *Journal of Comparative Neurology*, **202**, 103–13.

Tohyama, M. (1976). Comparative anatomy of cerebellar catecholamine systems from teleosts to mammals. *Journal für Hirnforschung*, **17**, 43–60.

Tohyama, M., Maeda, T. Hashimoto, J., Shreshta, G. R. & Tamura, O., Shimizu, N. (1974). Comparative anatomy of the locus coreuleus. I. Organization and acsending projections of the catecholamine containing neurons in the pontine region of the bird, *Melopsittacus undulatus*. *Journal für Hirnforschung*, **15**, 319–30.

Veenman, C. L. & Reiner, A. (1990). Telencephalic and thalamic inputs to the striatal portion of the pigeon basal ganglia. *Society for Neuroscience Abstracts*, **16**, 246.

Veenman, C. L. & Reiner, A. (1991). An anterograde study of corticostriatal projections to the basal ganglia in pigeons. *Society for Neuroscience Abstracts*, **17**, 652.

Veenman, C. L. & Reiner, A. (1994). The distribution of GABA-containing neurons and fibers in the forebrain and midbrain of pigeons, with particular reference to the basal ganglia and its projection targets. *Journal of Comparative Neurology*, **339**, 209–50.

Veenman, C. L., Albin, R. L., Richfield, E. C. & Reiner,

A. (1994). The distribution of GABA receptors in the forebrain and midbrain of pigeons, with particular reference to the basal ganglia and its projection targets. *Journal of Comparative Neurology*, **344**, 161–90.

Voorhuis, T. A. M., De Kloet, E. R. & De Wied, D. (1991). Ontogenetic and seasonal changes in immunoreactive vasotocin in the canary brain. *Developmental Brain Research*, **61**, 23–31.

Waldman, C. & Güntürkün, O. (1993). The dopaminergic innervation of the pigeon caudolateral forebrain: immunocytochemical evidence for a 'prefrontal cortex' in birds? *Brain Research*, **600** 225–34.

Wallace, J. A., Mondragon, R. M., Allgood, P. C., Hoffman, T. J. & Maez, R. R. (1987) Two populations of tyrosine hydroxylase-positive cells occur in the spinal cord of chick embryo and hatchling. *Neuroscience Letters*, **83**, 253–8.

Watson, J. A. & Adkins-Regan, E. (1989) Neuroanatomical localization of sex steroid concentrating cells in the Japanese quail (*Coturnix japonica*): autoradiography with [3H]- testosterone, [^{3}H]-estradiol and [^{3}H]- dihydrotestosterone. *Neuroendocrinology*, **49**; 51–64.

Yamada, H. & Sano, Y. (1985). Immunohistochemical studies on the serotonin neuron system in the brain of the chicken (*Gallus domesticus*) – II. The distribution of nerve fibers. *Biogenic Amines*, **2**, 21–36.

Yamada, H., Takeuchi, Y. & Sano, Y. (1984). Immunohistochemical studies on the serotonin neuron system in the brain of the chicken (*Gallus domesticus*) – II. The distribution of the neurons somata. *Biogenic Amines*, **1**, 283–94.

Yurkewicz, L., Marchi, M., Lauder, J. M. & Giacobini, E. (1981). Development and aging of noradrenergic cell bodies and axon terminals in the chicken. *Journal of Neurosciences Research*, **6**, 621–41.

Zeier, H., & Karten, H. J. (1971). The archistriatum of the pigeon: organization of afferent and efferent connections. *Brain Research*, **31**, 313–26.

8
Catecholamine systems in mammalian midbrain and hindbrain: theme and variations

K. Kitahama, I. Nagatsu and J. Pearson

Introduction

Numerous studies by means of formaldehyde-induced fluorescence and immunohistochemical techniques have revealed the distribution of the different catecholamines (CA) in the brain of mammals, in particular of rats (for review, see Björklund & Lindvall, 1984; Hökfelt, Johansson & Goldstein, 1984a; Hökfelt et al., 1984b; Moore & Card, 1984). It is now well known that dopamine (DA) cells are concentrated in the hypothalamus and midbrain, noradrenaline (NA) cells in the pons and medulla, and adrenaline cells in the rostral medulla oblongata. The CA cells were originally classified numerically into 14 anatomically discrete groups in the rat brain (Dahlström & Fuxe, 1964). This classification is generally applicable to other species, e.g. the cat (Fig. 8.1), although notable differences between species are observed. For example, CA cell bodies are more widely dispersed in the human midbrain and cat pontine tegmentum than in the rat. The comparative anatomy of central CA systems of mammals other than the rat has been briefly reviewed by Parent, Poitras & Dubé (1984). The present chapter seeks to describe in more detail novel aspects of CA structures in the midbrain and lower brainstem of mammals other than the rat. We have studied therefore, CA systems in a variety of mammals including human, monkey, rabbit, guinea pig, hamster, mouse, cat, dog, sheep, monkey, domestic pig and cow.

Midbrain CA cell groups

As mentioned above, the rat mesencephalic DA histofluorescent cells were originally classified into three groups, i.e. A8, A9 and A10, according to their localization (Dahlström & Fuxe, 1964). Later, based on TH-immunohistochemical data, these CA cells were regrouped by Hökfelt et al. (1984a, b) into subgroups: A9v (ventral), A9l (lateral, dorsal), A10dr (dorsal, rostral), A10dc (dorsal, caudal), A10vr (ventral, rostal) and A10c (caudal). The latter classification is adaptable to the many rodent species studied and is essentially applicable to cats (Fig. 8.1(a), (b)), dogs and monkeys, but hardly to the human midbrain. Fig. 8.2 presents schematic illustration of frontal sections of the midbrain of several mammalian species. It is obvious that there are species differences in brain size, cell number and distribution patterns. Fig. 8.3 and 8.4 illustrate examples of TH-ir cell morphology and distribution in the rat, cat, pig, monkey and human midbrain.

A10 cell group

The A10 group is composed of clusters of DA neurons lying in the midline of the mesencephalon, i.e. the ventral tegmental area (VTA), central gray, Edinger – Westphal, interfascicular, central and rostral linear, dorsal raphe and supramammillary nuclei.

The ventral tegmental area is situated in the reticular formation adjacent to the interpeduncular nucleus, the red nucleus and substantia nigra. On the basis of cytological criteria, the comparative anatomy of this area has been documented in the rat, cat, monkey and human (Halliday & Törk, 1986; also see Oades & Halliday, 1987). The VTA is subdivided into two cell groups: paranigral and parabrachial pigmented nuclei. In the human brain, at the level of the rootlet of the oculomotor nerve, TH-ir neurons are closely packed ventrally in the medially located paranigral nucleus but are more dispersed in the dorsally adjacent parabrachial pigmented nuclei (lower

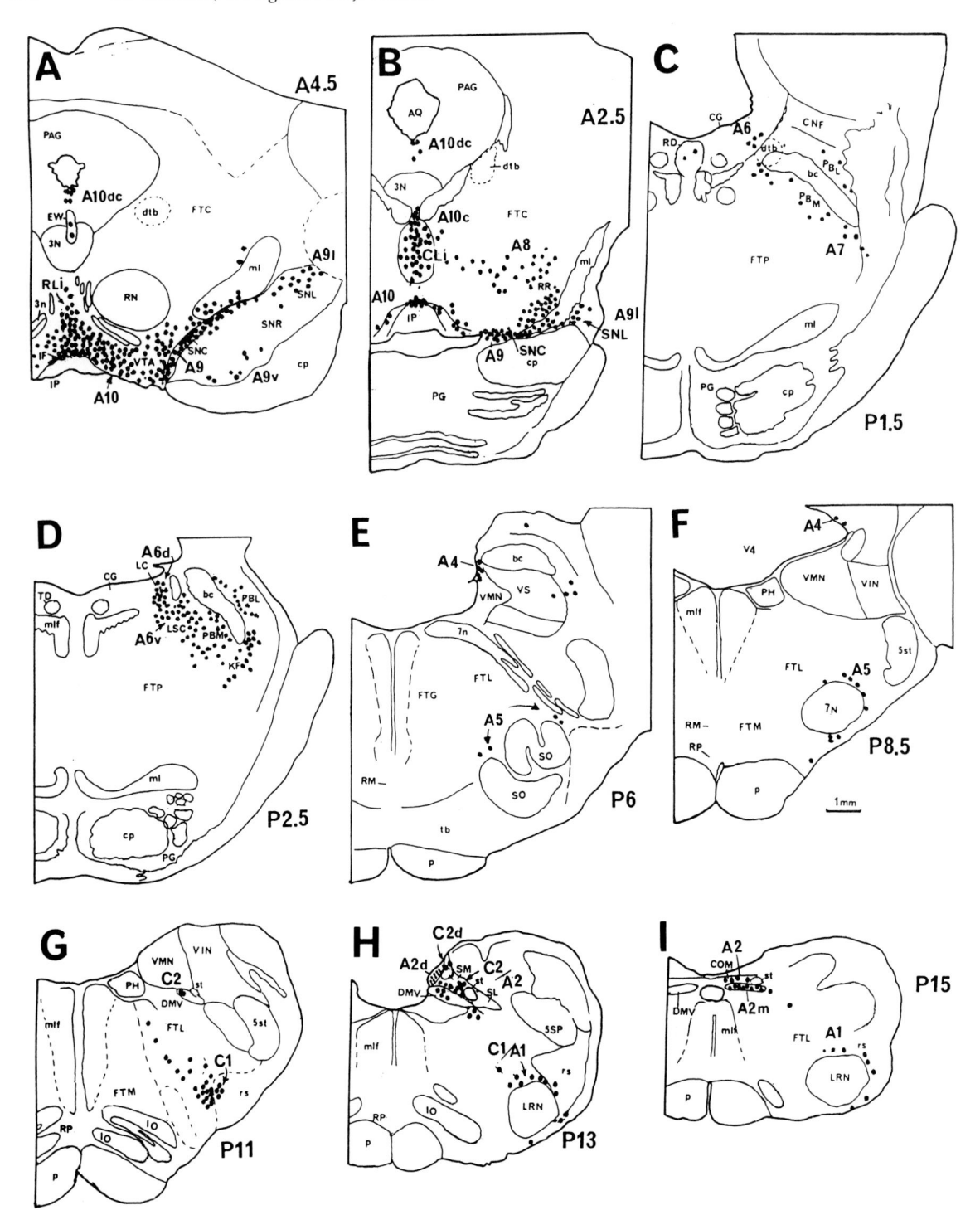

Fig. 8.1. Semischematic drawings of the distribution of TH-ir neurons in the cat midbrain, pons and medulla oblongata displayed in rostrocaudal order from A4.5 to P15 levels of Horsley and Clark stereotaxic plane.

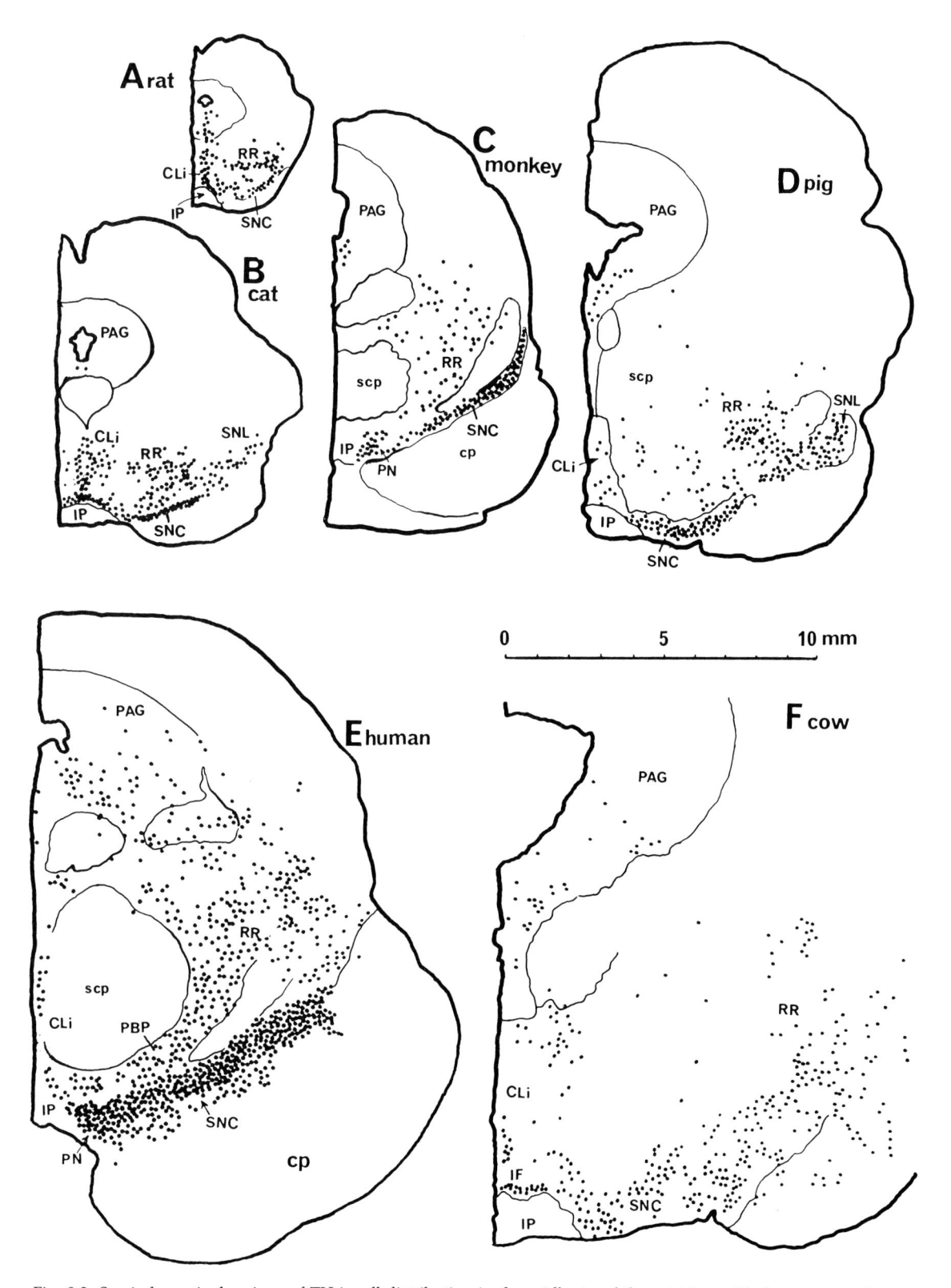

Fig. 8.2. Semischematic drawings of TH-ir cell distribution in the midbrain of the rat (*a*), cat (*b*), Japanese monkey (*c*), domestic pig (*d*), human (*e*) and cow (*f*). Note differences in brain size and cell distribution pattern.

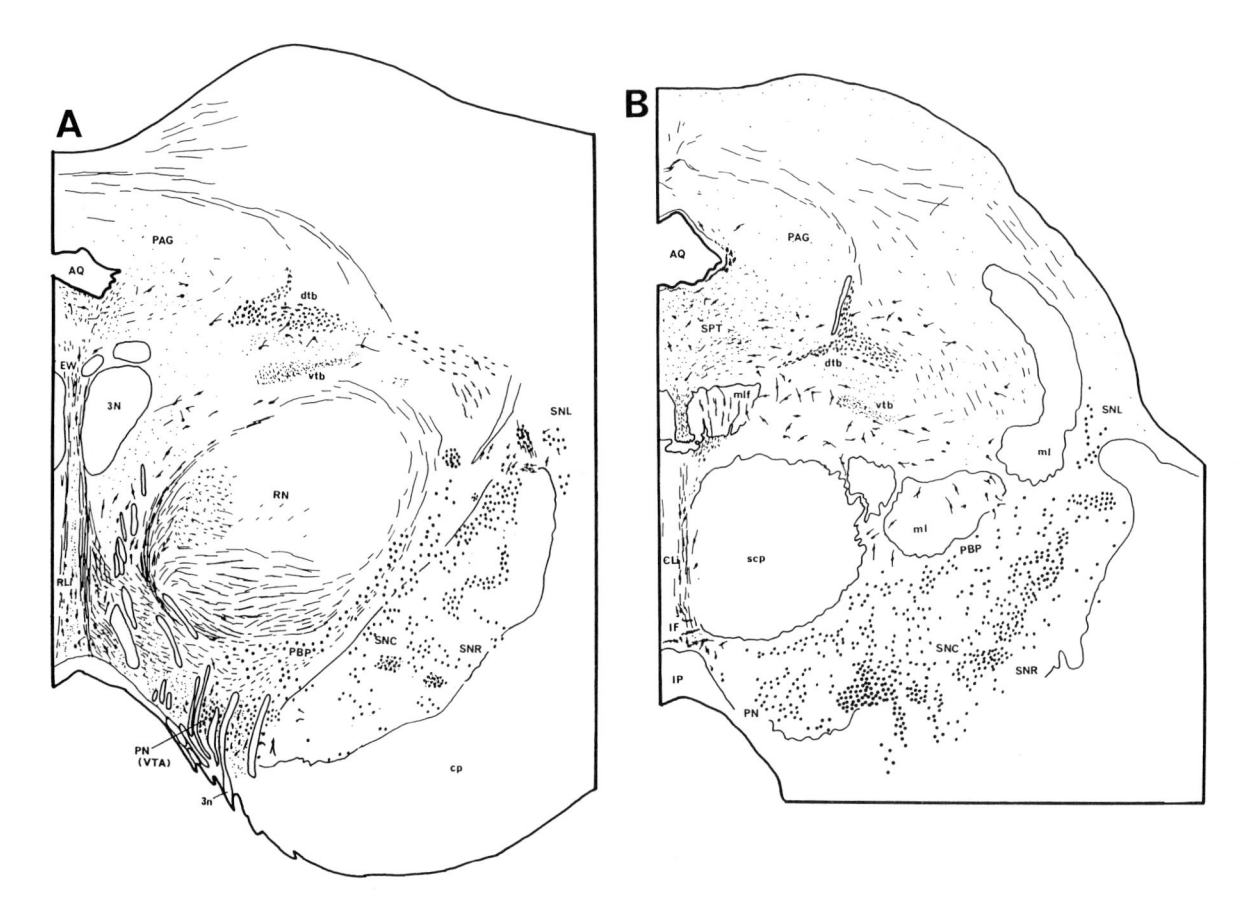

Fig. 8.3. Semischematic drawings of half of the human midbrain stained by TH-immunohistochemistry. (a) at the level of the red nucleus. (b) at the level of the decussation of the superior cerebellar peduncle.

part of Fig. 8.3(a) (Pearson et al., 1983, 1990). The latter group extends widely ventral to the red nucleus. In the cat, the most prominent CA cell group is situated in the paranigral nucleus at the level of the oculomotor nerve exit (Fig. 8.4(a)). Slightly caudally, many CA cells are packed ventrally situated in the paranigral nucleus and dispersed dorsally in the parabrachial pigmented nucleus which join laterally with A9 cells in its lateral part. No distinct boundaries are found between A10 and A9 cell groups as in the rat. TH-/DA-ir cell distribution has been well described in the rat (Swanson, 1982) and guinea pig (Smits, Steinbusch & Mulder, 1990). VTA CA cells give rise to the mesolimbic system that projects to the nucleus accumbens, olfactory tubercle and amygdaloid complex (Fallon & Moore, 1978a, b). Cholecystokinin (CCK) is present in DA cells of the ventral mesencephalon of the rat and is transported to the target areas mentioned above (Seroogy et al., 1989). This coexistence also occurs in the human and monkey, but not in the guinea pig nor hamster (Schalling et al., 1990).

In the midline of the rat mesencephalon (Fig. 8.4(b)), numerous CA cells are seen in the Edinger-Westphal nucleus, the central and rostral linear nuclei and the interfascicular nucleus (Swanson, 1982; Hökfelt et al., 1976, 1984b). This group is defined as A10c. Substantial numbers of TH-ir cells in the central linear and interfascicular nuclei are found to project to the nucleus accumbens and lateral septum (Swanson, 1982). The cat intermediate (central) linear nucleus (Fig. 8.4(c)), contains many medium-sized CA cells, and continues in the rostral linear nucleus (Fig. 8.4(c)), which consists of two diverging dorsoventral laminae of densely packed CA cells on each side of the midline, medial to the exiting fibers of the oculomotor nerve (Wiklund, Léger & Persson 1981). In the human homologous

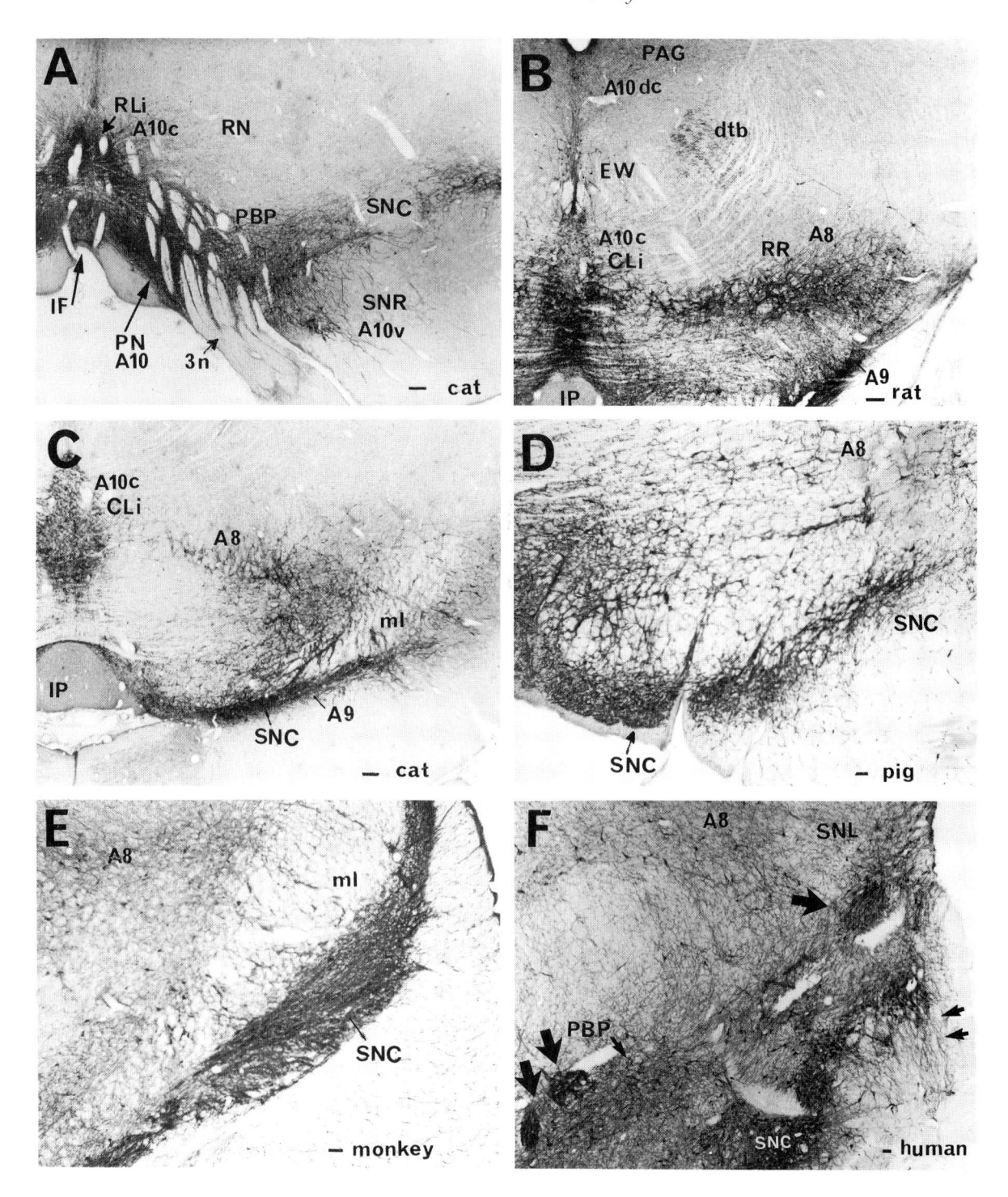

Fig. 8.4. Photomicrographs illustrating TH-ir neuronal structures in the ventral midbrain of the cat (*a*), rostral part), rat (*b*), cat (*c*) caudal part), pig (*d*), monkey (*e*), human (*f*). Bars=100 μm.

region (Fig. 8.3(a)), dorsal to the interfascicular nucleus are bilaterally symmetrical plate-like aggregates of TH-ir cells; these cells increase in number in the rostral linear nucleus (Pearson et al., 1983, 1990).

Rostrally the A10 CA cell group extends to the supramammillary nucleus which lies dorsal to the medial mammillary nucleus (not shown). The lateral mammillary nucleus receives a weak dopaminergic projection from the medial, and stronger projections from the lateral, caudal supramammillary nucleus (Gonzalo-Ruiz et al., 1992). This group named A10vr group is clearly distinguishable in the rat, cat and Japanese monkey, but difficult to identify in the human corresponding region. An A10 dr cell group seen within or near the rat habenular complex is undetectable in the guinea pig (Smits, Steinbusch & Mulder, 1990) and cat.

Very small TH-ir cells are seen in the central gray along the floor of the aqueduct of all the species studied. In the rat (Fig. 8.4(b)), some extend dorsally along the lateral aspect of the aqueduct, but they do not invade the ependymal layer. This pattern is also observable in the human (Fig. 8.4(b)) and monkey, but not in the cat and domestic pig. CA fluorescent cells have been demonstrated in the dorsal raphe nucleus (A10dc) (Lindvall & Björklund, 1974; Ochi & Shimizu, 1978). The nucleus accumbens receives CA axons from this group (Stratford & Wirtshafter, 1990). The number of the dorsal raphe TH-ir cells varies according to rostrocaudal levels. In the rat, hamster, guinea pig, domestic pig, monkey and cat, numerous DA cells form a flower bouquet pattern in the rostral dorsal raphe nucleus (Hökfelt et al., 1984b; Unguez & Schneider, 1988; Vincent, 1988; Smits, Steinbusch & Mulder, 1990). In the sheep and cow, however, TH-ir cells are dispersed throughout the nucleus. Similarly, in the human (upper part of Fig. 8.4(b)), pyramidal darkly stained TH-ir cells extend in low density in the supratrochlear nucleus ventrolateral to the aqueduct (Olszewski & Baxter, 1954).

A9 cell group

In general, the A9 cell group extends in the substantia nigra. CA neurons are closely packed especially in the medial part of the substantia nigra pars compacta (SNC) at the level of the interpeduncular nucleus of the rat (Fig. 8.3(a)) and cat (Fig. 8.3(b)). This part in the cat is named densocellular zone (Jiménez-Castellanos & Graybiel, 1987). In many species, the SNC A9 neurons form a horizontally elongated band. They project mainly to the dorsal striatum (caudate nucleus and putamen) (Beckstead, Domesick & Nauta, 1979; Fallon & Moore, 1978a,b) (see Chapter 10). Such packed nigral DA cells extend further laterally up to the lateral surface of the caudal midbrain of the human (Figs. 8.2(e), 8.3(b)) and monkey (Figs. 8.2(c), 8.3(d)). In the cow, SNC TH-ir cells are distributed in a slightly more dispersed manner (Fig. 8.2(f)). In the north American opossum, this part of the SNC is absent (Hazlett, Ho & Martin, 1991). In rats and cats, TH-ir cells sometimes form compact clusters. In the human (Fig. 8.3(a), and as indicated by large arrows in Fig. 8.4(f)), there are many cell dense islands, some of which send long dendritic processes ventrally to the cerebral peduncle (indicated by double arrows in he right part of Fig. 8.4(f)) (Pearson et al., 1983). Numerous cell clusters in the human SNC have finger-like extensions in the direction of the pars reticulata which are also well developed in the squirrel monkey (Arsenault, Parent & Descarries, 1988).

CA cells of A9v group are situated in the pars reticulata ventral to the SNC. They are not large in number in the cat and sometimes form small clusters. Only a few isolated CA cells are found in the sheep homologous region. In contrast, human SNR contained a substantial number of TH-ir cells forming many finger-like extensions. The zona reticulata DA neurons are found to project exclusively to striatal targets in a topographically defined fashions (Deutch, Goldstein & Roth, 1986). In the substantia nigra pars lateralis (SNL), CA cells are loosely dispersed in many species. This cell group termed A91 group is well developed in the human (Fig. 8.4(b)) and domestic pig (Østergaard, Holm & Zimmer, 1992).

A8 cell group

In the rat, as shown in Fig. 8.4(b), CA cells form an horizontally elongated band in the reticular formation (retrorubral region) caudal to the red nucleus, and extend ventrally to merge with A9 cells at more rostral levels (Köhler & Goldstein, 1984; Hökfelt et al., 1984b). Originally, A8 CA fluorescent cells related to the lateral lemniscus were defined as the lateral lemniscal group in the rat brain (Dahlström & Fuxe, 1964). In cats, the A8 group is represented by a population of CA fluorescent or TH-ir cells restricted to an area medial to the lateral lemniscus (Poitras & Parent, 1978; Miachon et al., 1984), but TH-ir cells extend more medially up to the central linear nucleus (Figs. 8.1(b), 8.4(c)) as in the rat. With the aid of DA-immunohistochemistry (Geffard et al., 1984), DA-ir neurons are visible in this region of the cat (unpublished). TH-ir cells of this group in the cat, domestic pig (Figs. 8.2(d), 8.4(d)) and cow (Fig. 8.2(f)) are more loosely clustered than those in rats. In the human (Figs. 8.2(e), 8.3(b)), such cells are more extensively dispersed than in these species; throughout the entire extent of the central tegmental field caudal and dorsal to the red nucleus (Pearson et al., 1983, 1990). In the macaque, CA cells analogous to A8 are present within and dorsomedial to the medial lemniscus and dorsolateral to the red nucleus (Garver & Sladek, 1975). We confirmed that many TH-ir cells are widely dispersed in the central reticular formation dorsomedial to the lateral lemniscus of the Japanese monkey (Figs. 8.2(c), 8.4(d)). CA cells of A8 group send their axons to the ventral entorhinal cortex, striatum, amygdala, the bed nucleus of the stria terminalis and ventral pallidum (Jiménez-Castellanos & Graybiel, 1987; and see Björklund & Lindvall, 1984)).

DA and L-DOPA

L-DOPA, the natural precursor of the neurotransmitter DA, is formed endogenously by TH, the first enzyme for CA synthesis, but rapidly converted to DA by the second enzyme AADC. In fact, cerebral L-DOPA has been hardly detectable by means of fluorometric methods. Using a newly developed L-DOPA immunohistochemical technique, immunostaining for L-DOPA is clearly demonstrable in midbrain cell groups although less intense than that observed for DA (Kitahama et al., 1988a). L-DOPA evidently accumulates in stainable amounts in neurons before being decarboxylated by the enzyme aromatic L-aminoacid decarboxylase (AADC). It should be noted that the intensity of L-DOPA-immunoreactivity shows regional variation, being weak to moderate in the lateral SNC, and strong elsewhere. L-DOPA-ir axons, although small in number, are detectable in the central nucleus of the amygdala and entorhinal cortex whose cells of origin are located in the A10 DA cell group as mentioned above. By contrast, TH- and AADC-immunostainings are not visually heterogeneous in intensity. This discrepancy might indicate that cells with approximately the same levels of enzyme protein either produce their neurotransmitters at different rate, have quantitatively variable modes of storage, or are heterogeneous with regard to transmitter release.

Monoamine oxidase in the midbrain:

DA is degraded to 3,4-dihydroxyphenyl acetic acid (DOPAC) by monoamine oxidase (MAO), especially by type A MAO (MAO-A) in the rat (Neff, Yang & Fuentes, 1974). Intrastriatal microdialysis has revealed that inhibition of MAO-A induces a marked increase of DA concentration by blocking breakdown and concomitant decrease of that of DOPAC and HVA in the striatum (Colzi et al., 1990). When studied by *in vitro* voltammetry analysis, MAO-A inhibition leads to a disappearance of DOPAC signal in the striatum after nearly 3 hours (Clement, Grote & Wesemann, 1990).

A loss of striatal MAO-A activity was reported after lesions of the substantia nigra (Agid, Javoy & Youdim, 1976; Damarest, Smith & Azzaro, 1980). Therefore, it has been supposed that MAO-A is transported from nigral cells to the striatum, and immunohistochemical studies using antibodies against MAO-A have been applied to visualize DA cell bodies in the primate brain (Westlund et al., 1985, 1988; Konradi et al., 1988). Unexpectedly, the great majority of the nigral pigmented perikarya were negative for MAO-A; approximately only 10% of melanin-containing neurons of the substantia nigra presented a positive immunoreactivity for MAO-A (Moll et al., 1990). MAO-A-immunoreactivity is low in the human striatum (Westlund et al., 1988). MAO enzyme histochemical study shows that the feline nigrostriatal pathway and caudate nucleus present only weak MAO-A activity (Kitahama et al., 1991). MAO-A activity is weaker in the striatum than in other parts of the brain (Saavedra, Brownstein & Palkovits, 1976). MAO-A inhibitor binding in these regions is weak in comparison with the LC (Saura Marti et al., 1990). This array of findings raises questions concerning mechanism of catabolism of DA in cell bodies and their terminals where it is highly concentrated.

On the other hand, in the ventral midbrain, densely packed astrocytes which contain type B MAO (MAO-B) are demonstrated (Westlund et al., 1985, 1988; Schneider & Markham, 1987). Systemic injection of 1-methyl-4-phenyl-1,2,3,6-tetrahydropyridine (MPTP) provokes a syndrome like Parkinson's disease, associated with destruction of nigral DA cells (Langston et al., 1983; Schneider, Yuwiler & Markham, 1986; Williams & Schneider, 1989). MPTP is a good substrate for MAO-B, which transforms it to 1-methyl-4-phenyl-pyridium ion (MPP$^+$), a specific neurotoxin for nigral cells. Treatment with MAO inhibitor prevents the MPTP-induced syndromes (Langston et al., 1984). This suggests the possibility that astrocytes densely packed in the basal midbrain produce a high concentration of MPP$^+$ to be taken up by nigral DA cells and that the primary active process is the production of MPP$^+$ in astrocytes and that neuronal death is a 'bystander' effect.

CA cell groups in the pons

In the rat pons, NA cells are classified into four groups, A4, 5, 6 and 7 (Dahlström & Fuxe, 1964). In earlier studies, several research groups determined the chemical cytoarchitecture of CA fluorescent neurons in the pons of the cat, dog and rabbit (Pin, Jones & Jouvet, 1968; Maeda et al., 1973; Chu & Bloom, 1974; Jones & Moore, 1974; Ishikawa, Shimada & Tanaka, 1975; Shimada, Ishikawa & Tanaka, 1976; Blessing, Chalmers & Howe, 1978; Poitras & Parent, 1978; Wiklund, Léger & Persson, 1981; Jones & Friedman, 1983). In the human brain, Olson, Nyström & Seiger (1973) reported weak CA fluorescence caudally in some perikarya of the locus coeruleus (LC). The morphology and topography of histofluorescent cells and fibers have been described in the brain of the human fetus (Nobin & Björklund, 1973). Most early primate studies were effected in monkey brains (Battista et al., 1972; DiCarlo, Hubbard & Pate, 1973; Hubbard & DiCarlo, 1973; Felten, Laties & Carpenter, 1974, Felten & Sladek, 1983; Garver and Sladek, 1975; Jacobowitz and MacLean, 1977; Murray, Dominguez & Martinez, 1982; Tanaka, Ishikawa & Shimada, 1982). Pontine CA neurons contain TH and the NA synthesizing enzyme dopamine-β-hydroxylase (DBH) (Swanson & Hartman, 1975; Grzanna & Molliver, 1980; Bérod et al., 1982). They also contain its degrading enzyme MAO (Shimizu, Morikawa & Okada, 1959), especially MAO-A (Westlund et al., 1985, 1988; Kitahama et al., 1986a, 1991, 1993).

A6 cell group

The most prominent pontine NA cell cluster is that of the locus coeruleus (LC) which is termed A6. Fig. 8.5 presents photomicrographs of the dorsolateral pontine region stained by DBH-immunohistochemistry in six different species (rat, cat, pig, cow, human and monkey). In the rat, the A6 NA cells are aggregated in the LC (Fig. 8.5(a)), and classified into two subgroups; dorsal (LCd) and ventral (LCv) (Swanson, 1976). A similar distribution is observed in the mouse pontine structure (Touret, Valatx & Jouvet, 1982). On the other hand, LC cells are dispersed in carnivores and ruminants (for review, see Russel, 1955). The cat LC complex (Fig. 8.1(c), (d)) is of the dispersed type in having a more prominent parabrachial group than rodents and primates. As presented in Fig. 8.6(b), feline LC cells are aggregated in the dorsal part of the LC (LCd) embedded in the lateral central gray and its immediate ventral area LC alpha, but extensively distributed in the more ventrally located locus subcoeruleus (LSC), Kölliker-Fuse (KF) nucleus ventral to the brachium conjunctivum (bc) and the medial and lateral parabrachial nuclei. Most of them contain enkephalin-immunoreactivity (Léger et al., 1983).

The general pontine CA cell distribution pattern in the dog is similar to that described in the cat (Ishikawa, Shimada & Tanaka, 1975; Barnes et al., 1988). However, the LCd and dorsolateral parabrachial region contain fewer CA cell bodies than feline LC complex. The dorsal (or lateral) parabrachial nucleus is generally poor in or devoid of CA cell bodies in many species, but it should be noted that it contains numerous TH-ir cells in the guinea pig brain (Vincent, 1988). In the sheep, CA cells are more extensively dispersed than those in the cat and dog (Tillet & Thibault, 1989). Recently, we have found that pig DBH-ir cells are also widely distributed in the pontine tegmentum. Its LCd contains only a few labeled cells, instead, a small cluster is seen in a region ventrolateral to the LCd (Fig. 8.5(c)). Similar results are obtained in the bovine brain (Fig. 8.5(d)), but the ventrolateral cluster is not evident.

In the human brain (upper part of Fig. 8.5(e)), the LCd contains many closely packed DBH-ir cell bodies as have been reported for TH-ir ones (Pickel et al., 1980; Pearson et al., 1983; Robert et al., 1984; Chan-Palay & Asan, 1989a,b). Numerous DBH-ir cells are dispersed ventrolateral to it in the LSC (lower part of Fig. 8.6(e)) (Kemper, O'Conner & Westlund, 1987), but they are scant in the peribrachial region. In the monkey, a distribution pattern is similar to, but more compact than that of humans (Felten & Sladek, 1983; Tanaka et al., 1982; Westlund & Coulter, 1980). Fig. 8.5(f) shows DBH-ir cell bodies aggregated in the dorsolateral pontine tegmentum of the Japanese monkey.

It should be noted that the number of human LC cells decreases with age; there is 25–60% loss by 90 years (Vijayshankar & Brody, 1979). In patients suffering from Parkinson's disease and Alzheimer's disease, there is a marked loss of LC cells associated with diminished NA and DBH in the cerebral cortex (Iversen et al., 1983; Hornykiewicz & Kish, 1984; Palmer et al., 1987; Chan-Palay & Asan, 1989a). The LC NA cells send their axons to many brain regions; the telencephalon (see Chapters 10, 12), diencephalon, midbrain, pons, cerebellum and medulla oblongata. In the hindbrain, dense plexuses of NA axons originated within the LC are found in the pontine, cochlear, principal and spinal sensory trigeminal nuclei (Levitt & Moore, 1979). This was evidenced by bilateral LC electrolytic lesions. Treatment with N-(2-chloroethyl)-N-ethyl-2-bromobenzylamine (DSP-4) specifically destroys cortical NA terminals arising from the LC in the rat (Fritschy & Grzanna, 1989). Absence or marked decrease in NA terminals is also observed in the hippocampus, thalamus, tectum, spinal trigeminal nuclei, and dorsal horn. DSP-4 decreases plasma luteinizing

CA systems in mammalian brainstem 191

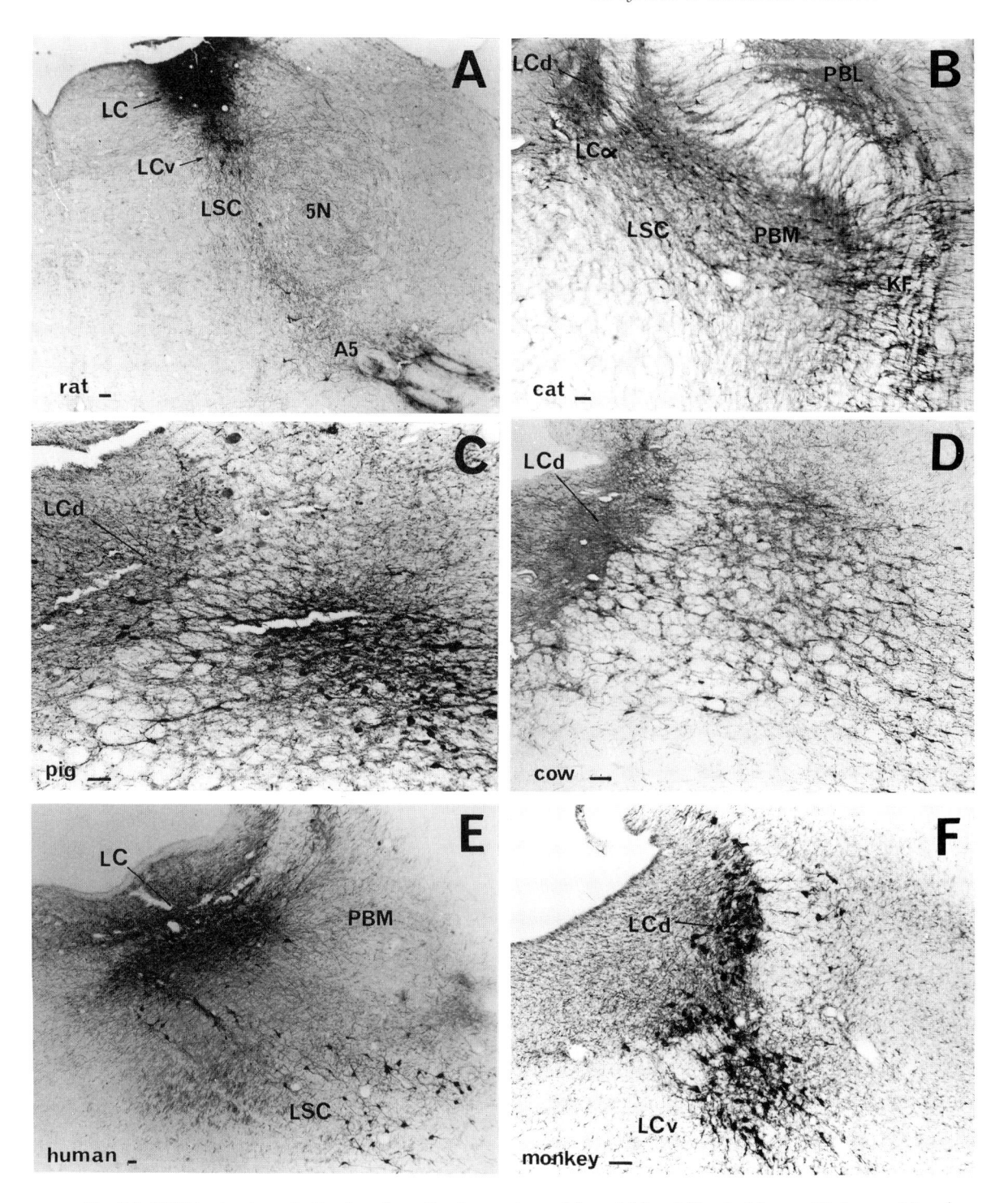

Fig. 8.5. DBH-ir structures in the dorsal pontine tegmentum of the rat (*a*), cat (*b*), pig (*c*), cow (*d*), human (*e*) and Japanese monkey (*f*). Bars=100 μm.

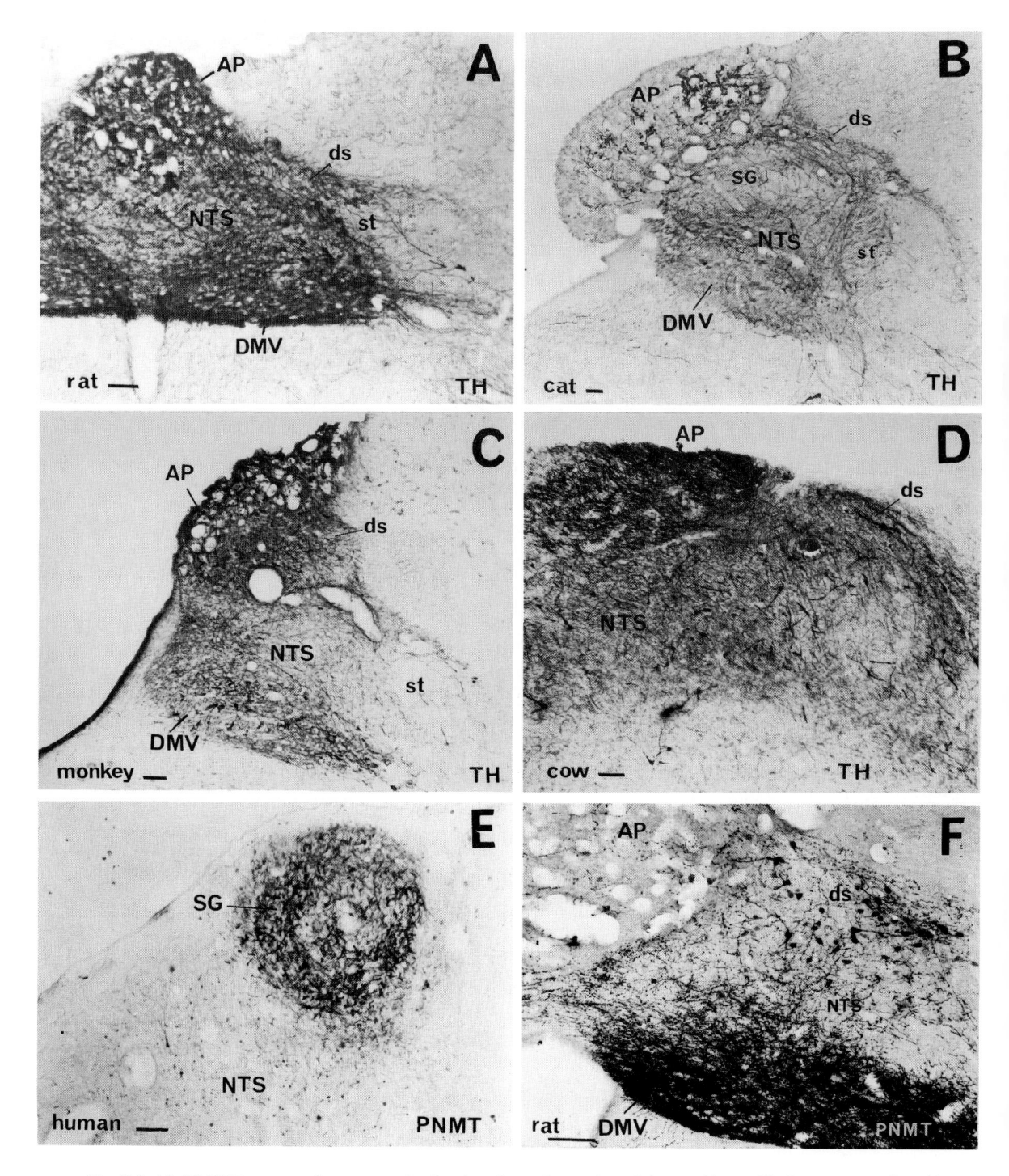

Fig. 8.6. (a)–(d) TH-ir neuronal structures in the dorsal vagal complex of the rat (*a*), cat (*b*), Japanese monkey (*c*) and cow (*d*). Bars=100 μm. (e)–(f):, PNMT-immunostaining in the dorsal vagal complex of the human (*d*) and rat (*f*). Bars=100 μm.

hormone levels and enhances the activating effects of testosterone on copulatory behavior in adult male quails (Balthazart & Ball, 1989), but does not impair inhibitory avoidance learning in the rat (Cornwell-Jones et al., 1989). Currently, we have not yet sufficient experiments to interpret the involvement of noradrenergic system in sleep-waking mechanism of the cat.

A4, A5 and A7 cell groups

A4 cell group. The A4 cell group of the cat is present in an area along the surface and the roof of the fourth ventricle (Fig. 8.1(*e*), (*f*)) as it is in the rat, and is continuous with the posterior portion of the A6 LCd group. Cells of this group project to the cerebellum (Olson & Fuxe, 1971). In the human, equivalent cells are found in the pigmented nucleus of the cerebellar tegmentum (Pearson et al., 1983). Felten & Sladek (1983) found them at the edge of the lateral recess of the fourth ventricle of the monkey. They have proposed to consider this group as a caudal extension of the LC. This group cannot be found in the homologous region of the sheep (Tillet & Thibault, 1989).

A5 cell group. A5 NA cell group in the cat is represented, in part, by a small number of DBH-ir cells situated dorsal to the superior olive and ventromedial to the root of the facial nerve (Fig. 8.1(*e*)), but does not extend further forward. A caudal extension of A5 of the cat is seen dorsal and ventral to the facial nucleus (Fig. 8.1(*f*)), and appears to continue in the line of medullary PNMT-ir C1 cell group.

In the rat, the A5 group is positioned ventromedial to the root of the facial nerve and dorsal to the superior olive complex (A5-dorsal) and extends more rostrally than in the cat, reaching up to a narrow area dorsolateral to the medial trapezoid body (A5-rostral) (Grzanna, Chee & Akeyson, 1987). In rats, but not in cats, the A5 group is rostrally continuous with A7 group. In the human brain, DBH-ir cells are dispersed in the caudal ventral pontine reticular formation (Robert et al., 1984), but it is somewhat difficult to define A5 cell group, because the cells that presumably constitute such groups are continuous with the caudal extension of LSC group. In the sheep, the A5 group is isolated from other groups and no DBH-ir cells are found at the level of the mid portion of the motor trigeminal nucleus (Tillet & Thibault, 1989). Neurons situated in A5 area have significant projections to the central nucleus of the amygdala, periformical area of the hypothalamus, midbrain aqueductal gray, parabrachial area and the nucleus of the solitary tract (Byrum & Guyenet, 1987), and ventral horn of the spinal cord (Lyons, Fritschy & Grzanna, 1989).

A7 cell group. Feline CA fluorescent cell bodies, equivalent to the rodent A7 group are found in the rostrolateral pontine reticular formation related to the lateral lemniscus (Maeda et al., 1973; Wiklund, Léger & Persson, 1981). Chu & Bloom (1974) suggested that NA cells situated close to the tip of the brachium conjunctivum in the anterior pons of the cat correspond to rat A7 cells. In the monkey, this group is found in the rostral pons independently from A6 cells (Felten & Sladek, 1983) but, in humans, the discrimination is difficult (Pearson et al., 1983), since it is restricted in the LSC, but not more rostrally (Robert et al., 1984). In the sheep, A7 extends from the level of the mid portion of the pons up to the level of the inferior colliculus.

Feline A7 cells project to the facial nucleus, magnocellular tegmental field, and motor trigeminal nucleus (Luppi et al., 1988; Fort et al., 1989). In the rat, cells of this group send axons to the motor trigeminal nucleus and spinal cord (Lyons & Grzanna, 1988; Clark & Proudfit, 1991). Many authors have proposed that the numerical classification of A4, A5, A6 and A7 groups should be consolidated under the single term locus coeruleus (Felten et al. 1974; Hubbard & DiCarlo, 1973), but the heterogeneity of cell characters such as the projection and pharmacological sensitivity (i.e. DSP-4) should be considered before this consolidation.

CA cell groups in the medulla oblongata

CA cells are distributed in the ventrolateral and dorsomedial medulla oblongata. In the rat, ventrolaterally located CA fluorescent cells were originally termed A1 and dorsomedial cells A2. Since the discovery of PNMT-ir cells, additional ventrolateral and dorsomedial subgroups have been classified as C1 and C2 (Hökfelt, Johansson & Goldstein, 1974; Hökfelt et al., 1984a,b). A1 and A2 groups are now redefined as caudally situated PNMT-negative TH-/DBH-positive cells. In the rat and hamster, many TH-, DBH- and PNMT-ir cells are seen in the dorsal midline of the rostral medulla oblongata (see Fig. 8.7(*a*)). These cells are classified as C3 cell group (Howe et al., 1980), but in most other species, the C3 group is not detectable.

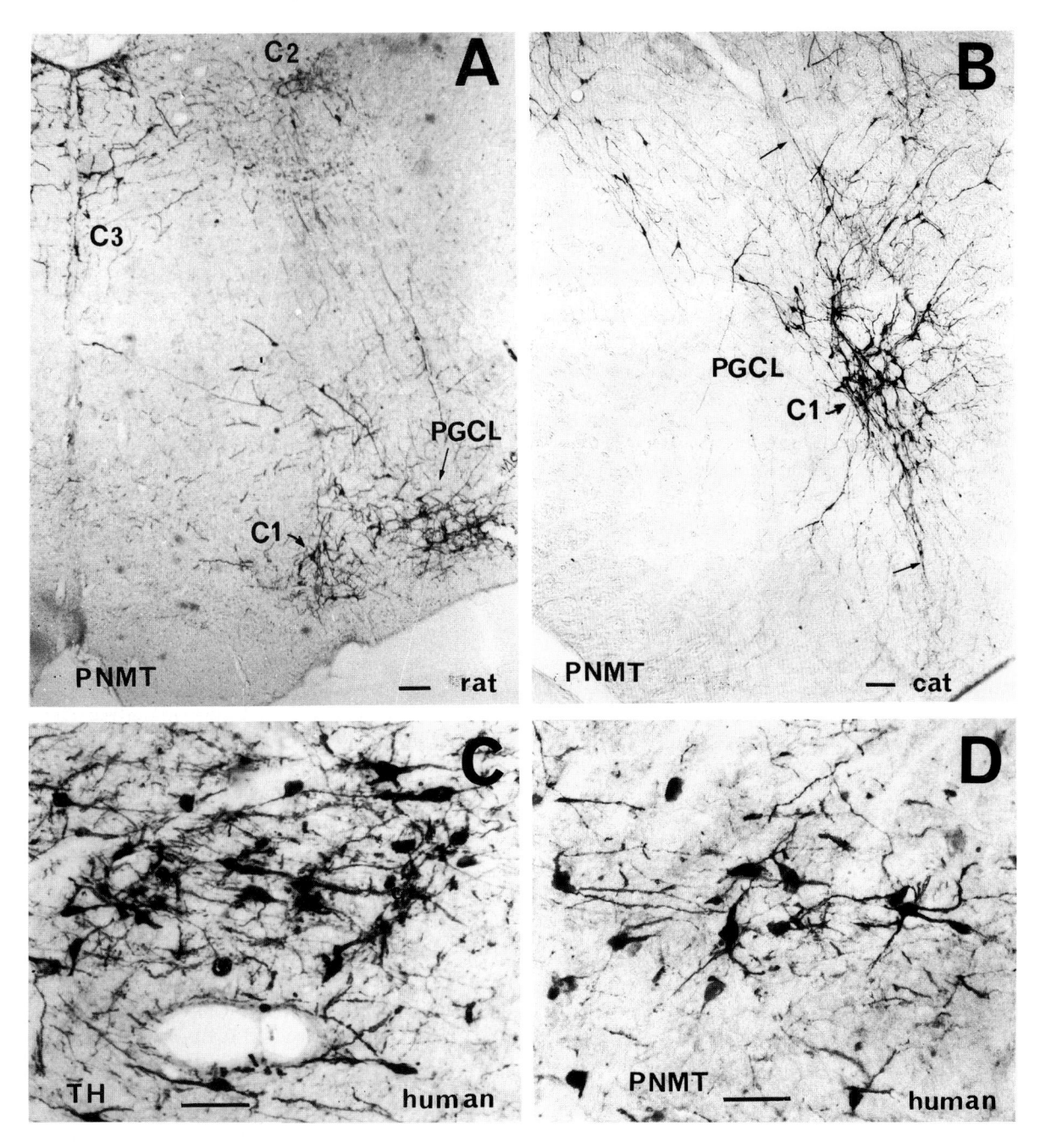

Fig. 8.7. (a),(b) PNMT-ir cell bodies in the ventrolateral medulla oblongata of the rat (*a*) and cat (*b*). Bars=100 μm. (*c*),(*d*): TH-ir (*c*) and PNMT-ir (*d*) neurons in the human rostral ventrolateral medulla oblongata (adjacent sections). PNMT-ir cells are less in number than TH-ir ones. Bars=100 μm.

Dorsomedial medulla oblongata

Nucleus of the solitary tract (NTS). In the rat rostral dorsal medulla, at the level just caudal to the facial nucleus, TH-ir cells are seen in the medial portion of the dorsal motor nucleus of the vagus (DMV), but only a few in the NTS. Slightly caudally, they increase in number and are distributed in the dorsal part of the NTS, including the dorsal strip at the level immediately rostral to the area postrema (AP) (Fig. 8.6(*a*)), and extend further caudally to the commissural subnucleus of the NTS.

In the cat, intensely TH-ir cells are distributed in the ventral and dorsal subnuclei of the NTS at the level of the area postrema (Fig. 8.6(*b*), also Fig. 8.1(*g*)). There are some small labeled cells in the dorsal strip dorsolateral to the substantia gelatinosus (subnucleus of the NTS) which is devoid of them. More caudally at the level of the obex (Fig. 8.1(*i*)), they are distributed in the medial and commissural subnuclei, and some are found in the DMV. Slightly caudally at the level of the spinomedullary junction. TH-ir cells decrease in number and are restricted to the commissural subnucleus. A similar distribution of TH-ir cells is observed in the Japanese monkey: in this species, numerous TH-ir cells are concentrated in the dorsal NTS (Fig. 8.6(*c*)). In the bovine homologous region, a large number of TH-ir cell bodies form an horizontally elongated cluster in the dorsal portion of the nucleus (Fig. 8.6(*d*)).

In the human, TH-ir cells are distributed within the NTS, ventrally along its ventromedial periphery and dorsally in the upper part of the NTS (Pearson *et al.*, 1983,1990; Arango *et al.*, 1988). They are also seen in the substantia gelatinosus (Takahashi *et al.*, 1986). On the other hand, PNMT-ir cells are very few in the human NTS but many small ones are tightly packed in the substantia gelatinous subnuleus (Fig. 8.6(*e*)) (Kitahama *et al.*, 1985,1988*b*). These cells are round or oval, and have one or two processes. It should be noted that sometimes we have not been able to locate this small PNMT-ir cell aggregate in all human specimens. This may be related to its small size, to variability of antigen content, or to differences between individuals. The group is also found in the neonatal swine, but is DBH-negative (Ruggiero, Anwar & Gootman, 1992). We obtained similar results in adult pigs (not shown). Similar results have been also reported in the monkey (Mittendorf, Denoroy & Flügge, 1988; Carlton, Honda & Denoroy, 1989), but the PNMT-ir cells are localized in an area surrounding the substantia gelatinosus subnucleus as we observe to be the case in the cat (Kitahama *et al.*, 1986*b*). In the rat, as shown in Fig. 8.6(*f*), a distinct aggregation of small PNMT-ir cells is seen in the dorsal strip and termed the C2d group (Hökfelt *et al.*, 1984*a*). These cells also show TH- and DBH-immunoreactivities (Kalia, Fuxe & Goldstein, 1985*a,b*). In the sheep, similar cells extend horizontally and are TH-/DBH-/PNMT-positive, but PNMT-ir cells are fewer than TH-/DBH-ir ones (Tillet, 1988). In the cow, several small PNMT-ir cells are observed in a homologous region (not shown), where TH-ir cells are not abundant.

Dorsal motor nucleus of the vagus (DMV). In the rat, TH-ir cells are seen throughout the DMV from levels of the pyramidal decussation up to levels rostral to the nucleus prepositus hypoglossi. The rostral cells contain all the enzymes necessary to synthesize adrenaline (Kalia *et al.*, 1985*b*). The caudal group (Fig. 8.6(*a*)) lacks PNMT and has been named A2m (Hökfelt *et al.*, 1984*b*). As AADC immunoreactivity has been undetectable in A2m cells in the rat (Jaeger *et al.*, 1984) and house-shrew (Karasawa *et al.*, 1991), they have been supposed to contain L-DOPA. Manier *et al.* (1990) demonstrated L-DOPA-immunoreactivity in rat A2m cells but, also found DA in those at the obex level. Presumably, therefore, AADC is present below the level of immunocytochemical detectability.

In the cat, only a small or no TH-ir cells are observed in the DMV (Fig. 8.6(*b*)). A small number of CA fluorescent cells have been reported only in the caudal half of the DMV, but neurons of the rostral part do not contain detectable CA fluorescence even where many perikarya contain strong MAO enzymatic activity (Blessing, Frost & Furness, 1980; Wiklund, Léger & Persson, 1981; Jones & Friedman, 1983; Kitahama *et al.*, 1987*b*). Nevertheless, according to a series of experiments in our laboratory (Kitahama *et al.*, 1987*a,b*; 1990*a*;1992), we were able to visualize TH-immunoreactivity in cat rostral DMV neurons after treatment with parachlorophenylalanine (PCPA), which decreases concentrations of DA and 5-HT, by inhibiting activity of tryptophan and phenylalanine hydroxylases. Similarly located perikarya show distinct immunoreactivity to AADC without any treatment. TH- and DA-immunoreactivities are demonstrable after treatment with colchicine. Since inhibition of monoamine oxidase fails to reveal DA in these cells, its absence in non-colchicine-treated animals cannot be due to rapid deamination. It is likely that DA is synthesized by TH and AADC in DMV cells and is very rapidly removed from the perikarya. In the monkey DMV, several TH-ir cell bodies are detectable without any treatment (Fig. 8.6(c)).

Dorsal motor vagal cells send their axons to the heart, lung and stomach (Kalia & Mesulam, 1980; Norgren & Smith, 1988) and are considered important for parasympathetic autonomic regulation. CA cells in the rat DMV have been labeled by retrograde tracers (horseradish peroxidase, or true blue) injected within the preganglionic terminals (Gwyn, Ritchie & Coulter, 1985; Kalia et al., 1984). This nucleus contains many cholinergic cells (Vincent & Reiner, 1987). Cells containing both TH and choline acetyltransferase are present in the mid-portion of the medial part of the DMV (Armstrong et al., 1990; Manier, Mouchet & Feuerstein, 1987). Thus, vagal action might be influenced by DA, released with acetylcholine at nerve terminals.

In the AP of the cat (Fig. 8.6(b)) many TH-, AADC- and DBH-ir cells are packed together, but PNMT-ir cells are limited in number and confined to its ventral periphery. The human AP, which contains numerous TH- and DBH-ir cells, is relatively small in size in comparison with its brain dimension. Similar results are obtained in the Japanese monkey (Fig. 8.6(c)). In contrast, the bovine AP is large and horizontally elongated and contains numerous TH- and DBH-ir cells (Fig. 8.6(d)). In the rat AP (Fig. 8.6(a)) there are numerous TH-ir and DBH-ir cells intermingled with TH-positive but DBH-negative cells (Hökfelt et al., 1984b) which suggests the presence of DA neurons intermingled with many NA ones.

However, it is puzzling that the feline AP cells display no CA histofluorescence (Blessing, Frost & Furness, 1980; Wiklund, Léger & Persson, 1981), although they do contain immunoreactive TH and AADC. In the cat AP, at present, immunoreactivities to DA and L-DOPA are undetectable (unpublished observations). Small rounded, weakly CA fluorescent cells that are closely packed and lack processes are present in the human developing AP (Nobin & Björklund, 1973). In the rat, CA histofluorescence can be seen only after treatment with alpha-methyl-DOPA (Fuxe, Hamberger & Malmfors, 1966), which is converted to alpha-methyl-DA by AADC, indicating that AADC in these cells is active, but that the immunoreactive TH present does not normally produce L-DOPA. This suggests that all the large number of variously produced antibodies used for staining crossreact with an antigen having sequence homology with TH but lacking TH activity or that the activity of the TH is suppressed. Similar observations have been made in the anterior hypothalamus and paraventricular thalamus of the cat (Kitahama et al., 1989, 1990b), as well as in the forebrain of the hamster (Vincent & Hope, 1990).

Ventrolateral medulla oblongata

In most mammals, CA-fluorescent neurons form a pair of longitudinal columns through the ventrolateral part of the medulla oblongata, from the pyramidal decussation up to the facial nucleus. The basic distribution of TH- and DBH-ir neurons is, in general, similar to that reported by CA histofluorescence methods (Levitt & Moore, 1979; Howe et al., 1980). Rostrally situated TH- and DBH-ir cells are also immunopositive for PNMT and considered to be adrenergic. Figure 8.7(a),(b) presents coronal sections stained by PNMT-immunohistochemistry at the level rostral to the AP of the rat and cat, respectively.

In the cat, several authors have reported that the ventrolateral cells contain the CA synthesizing enzymes TH, AADC, DBH and PNMT (Ciriello, Caverson & Polosa, 1986; Reiner & Vincent, 1986; Ruggiero et al., 1986; Kitahama et al., 1986b, 1990a). Caudally, TH-/AADC-/DBH-ir but PNMT-negative cells are restricted to the area dorsal to the lateral reticular nucleus, and rostrally at levels between the AP and caudal tip of the facial nucleus, TH-/AADC-/DBH-ir and PNMT-positive cells are loosely and sometimes tightly aggregated in the ventrolateral part of the reticular formation including the lateral paragigantocellular nucleus (PGCL) (Fig. 8.7(b)). They send their axons to the spinal cord (Carlton et al., 1991), LC, central gray, hypothalamus 1987a,b; Van Bockstaele et al., 1989; Van Bockstaele & Aston-Jones, 1992; Palkovits et al., 1992), but not to the cerebral cortex.

There are some species differences in ventrolateral C1 cell distribution (see Halliday & McLachlan, 1991a,b). In rodents (rat, hamster), as shown in Fig. 8.7(a), C1 is a compact aggregate situated mainly in the PGCL (Hökfelt et al., 1974, 1984a; Ruggiero et al., 1985; Vincent, 1988). In other animals such as primates and carnivores, numerous CA cells extend dorsally into the reticular formation. Particularly, in the human medulla, CA cells in the reticular formation form a band connecting the dorsomedial and ventrolateral parts (Pearson et al., 1983, 1990; Kitahama et al., 1985, 1988b; Kemper et al., 1987; Arango et al., 1988; Halliday et al., 1988). In the sheep, TH-/DBH-ir cells are distributed in a limited region; a few labeled cells are seen caudal to the facial nucleus and restricted to the area ventrolateral to the nucleus ambiguus and do not extend to more dorsal part of the reticular formation (Tillet & Thibault, 1989). We find a similar distribution in the cow and swine medulla.

The number of PNMT-ir cells in the rostral medulla of adult animals is, in general, equal to, or in many cases smaller than, TH-ir ones. Fig. 8.7(c),(d) presents an example in the human rostral ventrolateral medulla (Kitahama et al., 1988b). Halliday et al. (1988) described this evidence in more detail by counting the number of TH- and PNMT-ir cell bodies. The dog medulla contains more TH-ir cells than PNMT-ir ones (Iwamoto et al., 1989). In the cow and pig, TH-ir cells are dispersed in the ventrolateral medulla, but PNMT-ir neurons are distinctly fewer in number and restricted mainly to the ventrolateral surface of the medulla (in preparation). Tillet (1988) reported the absence of PNMT cells in C1 of the sheep brain. This is not due to the lack of crossreactivity, since the PNMT antibody can recognize C2 cells in the same sections. PNMT-immunoreactivity is absent from TH-ir cells of the guinea pig medulla and adrenaline is not detectable biochemically (Cumming et al., 1986; McLachlan et al., 1989).

Adrenaline is present in the human medulla although in much lower concentration than NA (Mefford et al., 1978). In the rat, there is a small amount of adrenaline in the C1 area (Lambás-Señas et al., 1985; Saavedra, Kvetnansky & Kopin, 1979; Mefford, 1987). Nevertheless, in the intermediolateral cell columns (IML) of the spinal cord which are densely innervated by PNMT terminals, adrenaline levels are below the limit of sensitivity of HPLC assay (less than 0.05% of NA content) (Sved. 1989, 1990). Thus, it may be postulated that these terminals are either not adrenergic or, if they are adrenergic then they either produce and store very small quantities of the neurotransmitter or use it at a very high rate in a region with an extremely efficient degradation machinery.

AADC-ir neurons in the subependymal layer of the spinomedullary junction and spinal cord

At the level of medullo-spinal junction and extending caudally to the spinal cord of the cat, oval or round AADC-ir cells are seen beneath the ependymal layer in the area surrounding the central canal (the rostral extension of the tenth area of Rexed). Labeled medium-sized cell bodies in the pyramidal decussation and small ones in the subependymal layer project to the ependyma of the central canal. Some have processes projecting into the lumen of the central canal where they form bulbous enlargements, contain vesicles of various sizes and granularity (Jaeger et al., 1983). As is also the case in the rat, they are TH-, DBH- and PNMT-negative (Jaeger et al., 1983; Nagatsu et al., 1988). TH-ir cells are present in the gray matter but not in the subependymal layer of the central canal.

Neurons showing CA histofluorescence were reported in the subependymal layer of the central canal of the lamprey spinal cord (Ochi & Hosoya, 1974). Similar cells were found to contain TH- and DA-immunoreactivity in the reptiles (Smeets & Steinbusch, 1990) and are now considered to be a general feature of vertebrate (see Chapter 20).

In the subependymal layer of the cat central canal, we find small MAO- active cells. Some labeled processes at the surface of the ependymal layer are in contact with the CSF. Their morphology and distribution pattern of such neurons is similar to the rat and cat AADC-ir cells described above. In a preliminary study 5-HT-immunoreactivity is visible in cells along the central canal after intraperitoneal injection of 5-HTP without MAO inhibitor. This may be due to the decarboxylation of 5-HTP to 5-HT by the enzyme AADC. Presumably the degradation of the 5-HT by MAO is slow. It is possible that the MAO in processes which contact CSF regulates the concentration of monoamines produced by the action of AADC on precursors taken up from the liquid. Another possibility is that MAO plays a role in deaminating of free monoamines in the fluid. Harmful trace amines synthesized by AADC (Jaeger et al., 1983) might be degraded by MAO.

Summary and conclusion

We described above species differences in distribution of TH-ir cells in the midbrain and lower brainstem of several mammals. Although their overall distribution patterns are fundamentally similar, these differences are sometimes very important according to species and regions. Therefore, results obtained in the rat are not always directly applicable to the other species including the human. Comparative studies are always indispensable and helpful in order to understand human brain organization and function. Furthermore, we should emphasize that neurons containing the other synthesizing enzymes, AADC, DBH and PNMT as well as their products, DA, NA and adrenaline are differently distributed according to the species studied.

TH-immunohistochemistry is very sensitive and a large number of studies have confirmed and extended the results obtained by histofluorescence technique. Neurons immunoreactive to the initiating enzyme TH overlap the position of CA-fluorescent cells in animal brains, but are more numerous than the latter, indicating that some neurons containing CA synthesizing enzymes do not necessarily contain detectable CA. For example, feline area postrema neurons which possess TH-, AADC- and DBH-immunoreactivities do not contain detectable L-DOPA, DA nor NA under normal condition. Some arcuate hypothalamic TH-ir cells do not show immunoreactivity to AADC, the next enzyme required for DA synthesis in the rat (Meister et al., 1988; Okamura et al., 1988a,b,c). In the human hypothalamus, the current authors have observed that a large number of TH-ir neurons present in the supraoptic and paraventricular nuclei do not possess AADC-immunoreactivity. While DA blockers have profound effects on human neurosecretory functions, there is currently no direct information regarding the DA content of these human hypothalamic nuclei. It is evident, however, that the presence of a neurotransmitter in a nerve cell cannot be directly and reliably assumed on the basis of the fact that the cell contains one or more of the necessary synthetic enzymes. In the midbrain and lower brainstem, further investigation will be required to confirm this supposition.

Acknowledgements

This work was supported by CNRS (URA1195), INSERM (U52), DRET (Grant 91–130), Subvention de Rhône-Alpes pour le programme régional de la Recherche en Neuroscience, Université Claude Bernard, a grant-in-aid for Scientific Research on Priority Areas, Japanese Ministry of Education, Science and Culture (0162300, 01570035, 01570031, 01480112) and grant-in-aid from Fujita Health University. We are very grateful to Drs Bérod, A., Buda, C., Debilly, G., Denoroy, L., Geffard, M., Goldstein, M., Jouvet, M., Maeda, T., Mons, N., Okamura, H., Raynaud, B., Roussel, B., Sakamoto, N., Sastre, J. P., Satoh, K., Tillet, Y. and Weber, M. for their generous gift of antisera and their many helpful suggestions.

Abbreviations

A1-A10	A1-A10 CA cell groups
3N	oculomotor nucleus
3n	oculomotor nerve
5-HT	5-hydroxytryptamine (serotonin)
5-HTP	5-hydroxytryptophan
5N	motor trigeminal nucleus
5SP	spinal trigeminal nucleus
5st	spinal trigeminal tract
7N	facial nucleus
AADC	aromatic L-aminoacid-decarboxylase
AP	area postrema
AQ	aqueduct
CG	central gray
CLi	central linear nucleus
CNF	cuneiform nucleus
COM	commissural nucleus, subnucleus of the NTS
cp	cerebral peduncle
DA	dopamine
DBH	dopamine-β-hydroxylase
DMV	dorsal motor nucleus of the vagus
DOPA	3, 4-dihydroxyphenylalanine
DOPAC	3, 4-dihydroxyphenylacetic acid
dtb	dorsal tegmental bundle
EW	Edinger–Westphal nucleus
FTC	central tegmental field
FTG	gigantocellular tegmental field
FTL	lateral tegmental field
FTP	pontine tegmental field
HVA	homovanillic acid
IF	interfascicular nucleus
IO	inferior olive
IP	interpeduncular nucleus
KF	Kölliker–Fuse nucleus
LC	locus coeruleus
LCα	locus coeruleus α
LCd	dorsal part of the locus coeruleus
LCv	ventral part of the locus coeruleus
LRN	lateral reticular nucleus
LSC	locus subcoeruleus
MAO	monoamine oxidase
MAO-A	type A monoamine oxidase
MAO-B	type B monoamine oxidase
ml	medial lemniscus
mlf	medial longitudinal fasciculus
NA	noradrenaline
NTS	nucleus of the solitary tract
p	pyramidal tract
PAG	periaqueductal gray
PBL	lateral (dorsal) parabrachial nucleus
PBM	medial parabrachial nucleus.
PBP	parabrachial pigmented nucleus
PCPA	parachlorophenylalanine
PG	pontine gray
PGCL	lateral paragigantocellular nucleus
PH	nucleus praepositus hypoglossi

PN	paranigral nucleus	SNC	substantia nigra pars compacta
PNMT	phenylethanolamine N-methyltransferase	SNL	substantia nigra pars lateralis
		SNR	substantia nigra pars reticularis
RD	dorsal raphe nucleus	SO	superior olive
RLi	rostral linear nucleus	SPT	supratrochlear nucleus
RM	raphe magnus nucleus	st	solitary tract
RN	red nucleus	tb	trapezoid body
RP	raphe pallidus nucleus	TD	dorsal tegmental nucleus
RR	retrorubral region	TH	tyrosine hydroxylase
rs	rubrospinal tract	VIN	inferior vestibular nucleus
scp	superior cerebellar peduncle	VMN	medial vestibular nucleus
SG	substantia gelatinosus, subnucleus of the NTS	VS	superior vestibular nucleus
		vtb	ventral tegmental bundle
SM	medial nucleus of the NTS		

References

Agid, Y., Javoy, F. & Youdim, M. (1976). Monoamine oxidase and aldehyde dehydrogenase activity in the striatum of rats after 6-hydroxydopamine lesion of the nigrostriatal pathway. *British Journal of Pharmacology*, **48**, 175–7.

Arango, V., Ruggiero, D. A., Callaway, J. L., Anwar, M., Mann, J. J. & Reis, D. J. (1988). Catecholaminergic neurons in the ventrolateral medulla and nucleus of the solitary tract in the human. *Journal of Comparative Neurology*, **273**, 224–40.

Armstrong, D. M., Manley, L., Haycock, J. W. & Hersh, L. B. (1990). Co-localization of choline acetyltransferase and tyrosine hydroxylase within neurons of the dorsal motor nucleus of the vagus. *Journal of Chemical Neuroanatomy*, **3**, 133–140.

Arsenault, M. Y., Parent, P. & Descarries, L. (1988). Distribution and morphological characteristics of dopamine-immunoreactive neurons in the midbrain of the squirrel monkey *Saimiri sciureus*. *Journal of Comparative Neurology*, **267**, 489–506.

Astier, B., Kitahama, K., Denoroy, L., Jouvet, M. & Renaud, B. (1987a). Immunohistochemical evidence for the adrenergic medullary longitudinal bundle as a major pathway to the locus coeruleus. *Neuroscience Letters*, **74**, 214–6.

Astier, B., Kitahama, K., Denoroy, L., Jouvet, M. & Renaud, B. (1987b). Immunohistochemical evidence for the adrenergic medullary longitudinal bundle as a major pathway to the hypothalamus. *Neuroscience Letters*, **78**, 241–6.

Balthazart, J. & Ball, G. F. (1989). Effects of the noradrenergic neurotoxin DSP-4 on luteinizing hormone levels, catecholamine concentrations, α2-adrenergic receptor binding, and aromatase activity in the brain of the Japanese quail. *Brain Research*, **492**, 163–175.

Barnes, K. L., Chernicky, C. L., Block, C. H. & Ferrario, C. M. (1988). Distribution of catecholaminergic neuronal systems in the canine medulla oblongata and pons. *Journal of Comparative Neurology*, **274**, 127–41.

Battista, A., Fuxe, K., Goldstein, M. & Ogawa, M. (1972). Mapping of central monoamine neurons in the monkey. *Experientia*, **28**, 688–90.

Beckstead, R. M., Domesick, V. B. and Nauta, W. J. (1979). Efferent connections of the substantia nigra and ventral tegmental area in the rat. *Brain Research*, **175**, 191–217.

Bérod, A., Hartman, B. K., Keller, A., Joh, T. H. & Pujol, J. F. (1982). A new double labeling technique using tyrosine hydroxylase and dopamine-β-hydroxylase immunohistochemistry: evidence for dopaminergic cells lying in the pons of the beef brain. *Brain Research*, **240**, 235–43.

Björklund, A. & Lindvall, O. (1984). Dopamine-containing systems in the CNS. In *Handbook of Chemical Neuroanatomy, Vol.2: Classical Transmitters in the CNS, Part I*, ed. A. Björklund & T. Hökfelt. pp. 55–122. Amsterdam: Elsevier.

Blessing, W. W., Chalmers, J. P. & Howe, R. P. C. (1978). Distribution of catecholamine-containing cell bodies in the rabbit central nervous system. *Journal of Comparative Neurology*, **179**, 407–424.

Blessing, W. W., Frost, P. & Furness, J. B. (1980). Catecholamine cell groups of the cat medulla oblongata. *Brain Research*, **192**, 69–75.

Byrum, C. E. & Guyenet, P. G. (1987). Afferent and efferent connections of the A5 noradrenergic cell group in the rat. *Journal of Comparative Neurology*, **261**, 529–42.

Carlton, S. M., Honda, C. N., Willcockson, W. S., Lacrampe, M., Zhang, D., Denoroy, L., Chung, J. M. & Willis, W. D. (1991). Descending adrenergic input to the primate spinal cord and its possible role in modulation of spinothalamic cells. *Brain Research*, **543**, 77–90.

Carlton, S. M., Honda, C. N. & Denoroy, L. (1989). Distribution of phenylethanolamine N-methyltransferase cell bodies, axons, and terminals in monkey brainstem: an immunohistochemical mapping study. *Journal of Comparative Neurology*, **287**, 273–85.

Chan-Palay, V. & Asan, E. (1989a). Alterations in catecholamine neurons of the locus coeruleus in senile dementia of the Alzheimer type and in Parkinson's disease with and without dementia and depression. *Journal of Comparative Neurology*, **287**, 373–92.

Chan-Palay, V. & Asan, E. (1989b). Quantitation of catecholamine neurons in the locus coeruleus in human brains of normal young and older adults and in depression. *Journal of Comparative Neurology*, **287**, 357–72.

Chu, N. S. & Bloom, F. E. (1974). The catecholamine-containing neurons in the cat dorsolateral pontine tegmentum: distribution of the cell bodies and some axonal projections. *Brain Research*, **66**, 1–21.

Ciriello, J., Caverson, M. M. & Polosa, C. (1986). Function of the ventrolateral medulla in the control of the circulation. *Brain Research Review*, **11**, 359–91.

Clark, F. M. & Proudfit, H. K. (1991). The projection of noradrenergic neurons in the A7 catecholamine cell group to the spinal cord in the rat demonstrated by anterograde tracing combined with immunocytochemistry. *Brain Research*, **547**, 279–88.

Clement, H. W., Grote, C. & Wesemann, W. (1990). *In vivo* studies on the effect of monoamine oxidase inhibitors on dopamine and serotonin metabolism in rat brain areas. *Journal of Neural Transmission*, **32** (Suppl.), 85–8.

Colzi, A., d'Agostini, F., Kettler, R., Borroni, E. & Da Prada, M. (1990). Effects of selective and reversible MAO inhibitors on dopamine outflow in rat striatum: a microdyalysis study. *Journal of Neural Transmission*, **32** (Suppl.), 79–84.

Cornwell-Jones, C. A., Decker, M. W., Chang, J. W., Cole, B., Goltz. K. M., Tran, T. & McGaugh, J. L. (1989). Neonatal 6- hydroxydopa, but not DSP-4, elevates brainstem monoamines and impairs inhibitory avoidance learning in developing rats. *Brain Research*, **493**, 258–68.

Cumming, P., Von Krosigk, M., Reiner, P. B., McGeer, E. G. & Vincent, S. R. (1986). Absence of adrenaline neurons in the guinea pig brain. A combined immunohistochemical and high- performance liquid chromatography study. *Neuroscience Letters*, **63**, 125–30.

Dahlström, A. & Fuxe, K. (1964). Evidence for the existence of monoamine-containing neurons in the central nervous system. I. Demonstration of monoamines in the cell bodies of the brain stem neurons. *Acta Physiologica Scandinavica*, **62**, (Suppl. 232), 1–55.

Damarest, K., Smith, D. & Azzaro, A. (1980). The presence of the type A form of monoamine oxidase within nigrostriatal dopamine-containing neurons. *Journal of Pharmacology and Experimental Therapeutics*, **215**, 461–68.

Deutch, A. Y., Goldstein, M & Roth, R. H. (1986). The ascending projections of the dopaminergic neurons of the substantia nigra, zona reticulata: a combined retrograde tracer-immunohistochemical study. *Neuroscience Letters*, **71**, 257–63.

DiCarlo, V., Hubbard, J. E. & Pate, P. (1973). Fluorescence histochemistry of monoamine-containing cell bodies in the brain stem of the squirrel monkey (*Saimiri sciureus*). *Journal of Comparative Neurology*, **152**, 347–72.

Falck, B., Hillarp, N. A., Thieme, G. & Torp, A. (1962). Fluorescence of catecholamines and related compounds condensed with formaldehyde. *Journal of Histochemistry and Cytochemistry*, **10**, 348–54.

Fallon, J. H. & Moore, R. Y. (1978a). Catecholamine innervation of the basal forebrain. IV. Topography of the dopamine projection to the basal forebrain and neostriatum. *The Journal of Comparative Neurology*, **180**, 545–80.

Fallon, J. H. and Moore, R. Y. (1978b). Catecholamine innervation of basal forebrain. III. Olfactory bulb, anterior olfactory nuclei, olfactory tubercle and piriform cortex. *Journal of Comparative Neurology*, **180**, 533–44.

Felten, D. L. & Sladek, J. R. Jr (1983). Monoamine distribution in primate brain V. monoaminergic nuclei: anatomy, pathways and local organization. *Brain Research Bulletin*, **10**, 171–284.

Felten, D. L., Laties, A. M. & Carpenter, M. B. (1974). Monoamine containing cell bodies in the squirrel monkey brain. *American Journal of Anatomy*, **139**, 153–66.

Fort, P., Sakai, K., Luppi, P.H., Salvert, D. & Jouvet, M. (1989). Monoaminergic, peptidergic and cholinergic afferents to the cat facial nucleus as evidenced by a double immunostaining method with unconjugated cholera-toxin as a retrograde tracer. *Journal of Comparative Neurology*, **283**, 285–302.

Fritschy, J. M. & Grzanna. R. (1989). Immunohistochemical analysis of the neurotoxic effects of DSP-4 identifies two populations of noradrenergic axon terminals. *Neuroscience*, **30**, 181–97.

Fuxe, K. (1965). Evidence for the existence of monoamine-containing neurons in the central nervous system. IV. The distribution of monoamine terminals in the central nervous system. *Acta Physiologica Scandinavica*, **64** (Suppl. 247), 37–85.

Fuxe, K., Hamberger, B & Malmfors, T. (1966). Inhibition of amine uptake in tubero-infundibular and in catecholamine cell bodies of the area postrema. *Journal of Pharmacy and Pharmacology*, **18**, 543–44.

Garver, D. L. & Sladek, J. R Jr (1975). Monoamine distribution in primate brain. I. Catecholamine- containing perikarya in the brain stem of *Macaca speciosa*. *Journal of Comparative Neurology*, **159**, 289–304.

Geffard, M., Buijs, R. M., Séguéla, P., Poll, C. W. & LeMoal, M. (1984). First demonstration of highly specific and sensitive antibodies against dopamine. *Brain Research*, **82**, 161–5.

Gonzalo-Ruiz, A., Alonso, A., Sanz, J. M. and Llinás, R. R. (1992). A dopaminergic projection to the rat mammillary nuclei demonstrated by retrograde transport of Wheat Germ Agglutinin-Horseradish peroxidase and tyrosine hydroxylase immunohistochemistry. *Journal of Comparative Neurology*, **321**, 300–11.

Grzanna, R. & Molliver, M. E. (1980). The locus coeruleus in the rat, an immunohistochemical delineation. *Neuroscience*, **5**, 21–40.

Grzanna, R., Chee, W. K. & Akeyson, E. W. (1987). Noradrenergic projections to brainstem nuclei: Evidence for differential projections from noradrenergic subgroups. *Journal of Comparative Neurology*, **263**, 76–91.

Gwyn, D. G., Ritchie, T. C. & Coulter, J. D. (1985). The

central distribution of vagal catecholaminergic neurons which project into the abdomen in the rat. *Brain Research*, **358**, 139–44.

Halliday, G. M. & McLachlan, E. M. (1991a). A comparative analysis of neurons containing catecholamine- synthesizing enzymes in the ventrolateral medulla of rats, guinea- pigs and cats. *Neuroscience*, **43**, 531–50.

Halliday, G. M. & McLachlan, E. M. (1991b). Four groups of tyrosine hydroxylase immunoreactive neurons in the ventrolateral medulla of rats, guinea-pigs and cats identified on the basis of chemistry, topography and morphology. *Neuroscience*, **43**, 551–68.

Halliday, G. M. & Tork, I. (1986). Comparative anatomy of the ventromedial mesencephalic tegmentum in the rat, cat, monkey and human. *Journal of Comparative Neurology*, **252**, 423–45.

Halliday, G. M., Li, Y. W., Joh, T. H., Cotton, R. G. H., Howe, P. R. C., Geffen, L. B. & Blessing, W. W. (1988). Distribution of monoamine-synthesizing neurons in the human medulla oblongata. *Journal of Comparative Neurology*, **273**, 301–17.

Hazlett, J. C., Ho, R. H. & Martin, G. F. (1991). Organization of midbrain catecholamine-containing nuclei and their projections to the striatum in the north American opossum, *Didelphis virginiana*. *Journal of Comparative Neurology*, **306**, 585–601.

Hökfelt, T., Johansson, O & Goldstein, M. (1974). Immunohistochemical evidence for the existence of adrenaline neurons in the rat brain. *Brain Research*, **66**, 235–51.

Hökfelt, T., Johansson, O., Fuxe, K., Goldstein, M. & Park, D. (1976). Immunohistochemical studies on the localization and distribution of monoamine neuron system in the rat brain stem. I. Tyrosine hydroxylase in the meso- and diencephalon. *Medical Biology*, **54**, 427–453.

Hökfelt, T., Johansson O. & Goldstein M. (1984a). Central catecholamine neurons as revealed by immunohistochemistry with special reference to adrenaline neurons. In: *Handbook of Chemical Neuroanatomy, Vol. 2, Classical Transmitters In The CNS, Part I*, ed. A. Björklund & T. Hökfelt. pp. 157–276. Amsterdam: Elsevier.

Hökfelt, T., Martensson R., Björklund A., Kleinau S & Goldstein M. (1984b)., Distributional maps of tyrosine-hydroxylase- immunoreactive neurons in the rat brain. In *Handbook of Chemical Neuroanatomy, Vol. 2: Classical Transmitters in the CNS, Part 1*, ed. A. Björklund & T. Hökfelt, pp. 277–379. Amsterdam: Elsevier.

Hornykiewicz, O. & Kish, S. (1984). Neurochemical basis for dementia in Parkinson's disease. *Canadian Journal of Neurological Science*, **11**, 185–90.

Howe, R. P. C., Costa, M., Furness, J. B. & Chalmers, J. P. (1980). Simultaneous demonstration of phenylethanolamine N- methyltransferase immunofluorescent and catecholamine fluorescent nerve cell bodies in the rat medulla oblongata. *Neuroscience*, **5**, 2229–38.

Hubbard, J. E. & DiCarlo, V. (1973). Fluorescence histochemistry of monoamine-containing cell bodies in the brain stem of the squirrel monkey (*Saimiri scireus*). I. The locus coeruleus. *Journal of Comparative Neurology*, **147**, 553–66.

Ishikawa, M., Shimada, S. & Tanaka, C. (1975). Histochemical mapping of catecholamine neurons and fiber pathways in the pontine tegmentum of the dog. *Brain Research*, **86**, 1–16.

Iversen, L. L., Rossor, M. N., Reynolds, G. P., Hills, R., Roth, M., Mountjoy, C. Q., Footes, S. L., Morrison, J. H. & Bloom, F. E. (1983). Loss of pigmented dopamine-β-hydroxylase positive cells from locus coeruleus in senile dementia of Alzheimer's type. *Neuroscience Letters*, **39**, 95–100.

Iwamoto, G. A., Mitchell, J. H., Sadeq, M. & Kozlowski, G. P. (1989). Localization of tyrosine hydroxylase and phenylethanolamine N-methyltransferase immunoreactive cells in the medulla of the dog. *Neuroscience Letters*, **107**, 12–18.

Jacobowitz, D. & Kostrzewa, R. (1971). Selective action of 6- hydroxy-dopa on noradrenergic terminals: mapping of preterminal axons of the brain. *Life Science*, **10**, 1329–42.

Jacobowitz, D. M. & MacLean, P. D. (1977). A brain stem atlas of catecholaminergic and serotoninergic perikarya in the pygmy marmoset (*Cebuella pygmaea*). *Journal of Comparative Neurology*, **177**, 397–415.

Jaeger, C. B., Ruggiero, D. A., Albert, V. R., Park, D. H., Joh, T. H. & Reis, D. J. (1984). Aromatic L-amino acid decarboxylase in the rat brain: Immunocytochemical localization in neurons of the rat brain stem. *Neuroscience*, **11**, 691–713.

Jaeger, C. B., Teitelman, G., Joh, T. H., Albert, V. R., Park, D. H. & Reis, D. J. (1983). Some neurons of the rat central nervous system contain aromatic L-amino acid decarboxylase but not monoamines. *Science*, **219**, 1233–5.

Jiménez-Castellanos, J. & Graybiel, A. M. (1987). Subdivisions of the dopamine-containing A8–A9–A10 complex identified by their differential mesostriatal innervation of striosomes and extrastriosomal matrix. *Neuroscience*, **23**, 223–42.

Jones, B. E. & Moore, R. Y. (1974). Catecholamine-containing neurons of the nucleus locus coeruleus in the cat. *Journal of Comparative Neurology*, **157**, 43–52.

Jones, B. E. & Friedman, L. (1983). Atlas of catecholamine perikarya, varicosities and pathways in the brain stem of the cat. *Journal of Comparative Neurology*, **215**, 382–96.

Kalia, M. & Mesulam, M. M. (1980). Brain stem projections of sensory and motor components of the vagal complex in the cat. I. The cervical vagus and nodose ganglion. *Journal of Comparative Neurology*, **193**, 435–65.

Kalia, M., Fuxe, K. & Goldstein, M. (1985a). Rat medulla oblongata II. Dopaminergic, noradrenergic (A1 and A2) and adrenergic neurons, nerve fibers and presumptive terminal processes. *Journal of Comparative Neurology*, **233**, 308–22.

Kalia, M., Fuxe, K. & Goldstein, M. (1985b). Rat medulla oblongata. III. Adrenergic (C1 and C2) neurons, nerve fibers and presumptive terminal processes. *Journal of Comparative Neurology*, 233, 333–49.

Kalia, M., Fuxe, K., Goldstein, M., Harfstrand, A., Agnati, L. F. & Coyle, J.T. (1984). Evidence for the existence of putative dopamine, adrenaline and noradrenaline-containing vagal motor neurons in the brainstem of the rat. *Neuroscience Letters*, 50, 57–62.

Karasawa, N., Isomura, G., Yamada, K. & Nagatsu, I. (1991). Immunocytochemical localization of monoaminergic and non- aminergic neurons in the house-shrew (*Suncus Murinus*) brain. *Acta Histochemical Cytochemica*, 24, 465–75.

Kemper, C. M., O'Conner, D. T. & Westlund, K. N. (1987). Immunocytochemical localization of dopamine-β-hydroxylase in neurons of the human brain stem. *Neuroscience*, 23, 981–9.

Kitahama, K., Arai, R., Maeda, T. & Jouvet, M. (1986a). Demonstration of monoamine oxidase type B in serotonergic and type A in noradrenergic neurons in the cat dorsal pontine tegmentum by an improved histochemical technique. *Neuroscience Letters*, 71, 19–24.

Kitahama, K., Bérod, A., Denoyer, M. & Jouvet, M. (1987a). Visualization of tyrosine hydroxylase-immunoreactive neurons in the cat dorsal motor vagal cells after treatment with parachlorophenylalanine. *Neuroscience Letters*, 77, 155–60.

Kitahama, K., Buda, C., Sastre, J. P., Nagatsu, I., Raynaud, B., Jouvet, M. & Geffard, M. (1992). Dopaminergic neurons in the cat dorsal motor nucleus of the vagus, demonstrated by dopamine, AADC and TH immunohistochemistry. *Neuroscience Letters*, 146, 5–9.

Kitahama, K., Denney, R. M., Maeda, T. & Jouvet, M. (1991). Distribution of type B monoamine oxidase immunoreactivity in the cat brain with reference to enzyme histochemistry. *Neuroscience*, 44, 185–204.

Kitahama, K., Denoroy, L., Bérod, A & Jouvet, M. (1986b). Distribution of PNMT-immunoreactive neurons in the cat medulla oblongata. *Brain Research Bulletin*, 17, 197–208.

Kitahama, K., Denoroy, L., Goldstein, M., Jouvet, M. & Pearson, J. (1988b). Immunohistochemistry of tyrosine hydroxylase and phenylethanolamine N-methyltransferase in the human brain stem: description of adrenergic perikarya and characterization of longitudinal catecholaminergic pathways. *Neuroscience*, 25, 97–111.

Kitahama, K., Denoyer, M., Raynaud, B., Borri-Voltattorni, C., Weber, M. & Jouvet, M. (1990a). Aromatic L-amino acid decarboxylase immunohistochemistry in the cat lower brainstem and midbrain. *Journal of Comparative Neurology*, 302, 935–53.

Kitahama, K., Geffard, M., Okamura, H., Nagatsu, I., Mons, N & Jouvet, M. (1990b). Dopamine- and DOPA-immunoreactive neurons in the cat forebrain with reference to tyrosine-hydroxylase- immunohistochemistry. *Brain Research*, 518, 83–94.

Kitahama, K., Kimura, H., Maeda, T. & Jouvet, M. (1987b). Distribution of two types of monoamine oxidase-containing neurons in the cat medulla oblongata demonstrated by an improved histochemical method. *Neuroscience*, 20, 991–9.

Kitahama, K., Maeda, T., Denney, R. M. & Jouvet, M. (1993). Monoamine oxidase: distribution in the cat brain studied by enzyme- and immunohistochemistry: recent progress. *Progress in Neurobiology*, In press.

Kitahama, K., Mons, N., Okamura, H., Jouvet, M. & Geffard, M. (1988a). Endogenous L-DOPA, its immunoreactivity in neurons of midbrain and its projection fields in the cat. *Neuroscience Letters*, 95, 47–52.

Kitahama, K., Okamura, H., Goldstein, M., Nagatsu, I., Bérod, A. & Jouvet, M. (1989). A new group of tyrosine hydroxylase- immunoreactive neurons in the cat thalamus. *Brain Research*, 478, 156–60.

Kitahama, K., Pearson, J., Denoroy, L., Kopp, N., Ulrich, J., Maeda, T. & Jouvet, M. (1985). Adrenergic neurons in the human brain demonstrated by immunohistochemistry with antibodies to phenylethanolamine N-methyltransferase (PNMT): discovery of a new group in the nucleus tractus solitarius. *Neuroscience Letters*, 53, 303–8.

Köhler, C. & Goldstein, M. (1984). Golgi-like immunoperoxidase staining of dopamine neurons in the reticular formation of the rat brainstem using antibody to tyrosine-hydroxylase. *Journal of Comparative Neurology*, 223, 302–11.

Konradi, C., Svoma, E., Jellinger, K., Riederer, P., Denney, R. & Thibault, J. (1988). Topographic immunocytochemical mapping of monoamine oxidase-A, monoamine oxidase-B and tyrosine hydroxylase in human post mortem brain stem. *Neuroscience*, 26, 791–802.

Lambás-Señas, L., Chamba, G., Dennis, T., Renaud, B. & Scatton, B. (1985). Distribution of dopamine, noradrenaline and adrenaline in coronal sections of the rat lower brainstem. *Brain Research*, 347, 306–12.

Langston, J. W., Ballard, P., Tetrud, J. W. & Irwin, I. (1983). Chronic parkinsonism in human due to a product of meperidine- analog synthesis. *Science*, 219, 979–80.

Langston, J. W., Forno, L. S., Robert, C. S. & Irwin, I. (1984). Selective nigral toxicity after systemic administration of 1- methyl-4-phenyl-1,2,3,6-tetrahydropyridine (MPTP) in the squirrel monkey. *Brain Research*, 292, 390–4.

Léger, L., Charnay, Y., Chayvialle, J. A., Bérod, A., Dray, F., Pujol, J. F., Jouvet, M. & Dubois, P. M. (1983). Localization of substance P- and enkephalin-like immunoreactivity in relation to catecholamine- containing cell bodies in the cat dorsolateral pontine tegmentum: an immunofluorescence study. *Neuroscience*, 8, 525–46.

Levitt, P. & Moore, R. Y. (1979). Origin and organization of brainstem catecholamine innervation in the rat. *Journal of Comparative Neurology*, 186, 505–28.

Lindvall, O. & Björklund, A. (1974). The organization of the ascending catecholamine neuron systems in the rat brain as revealed by the glyoxylic acid fluorescence method. *Acta Physiologica Scandinavica*, 92 (Suppl.412), 1–48.

Luppi, P. H., Sakai, K., Fort, P., Salvert, D. & Jouvet, M.

(1988). The nuclei of origin of monoaminergic, peptidergic and cholinergic afferents to the cat nucleus reticularis. *Journal of Comparative Neurology*, **277**, 1–20.

Lyons, W. E. & Grzanna, R. (1988). Noradrenergic neurons with divergent projections to the motor trigeminal nucleus and the spinal cord: a double retrograde neuronal labeling study. *Neuroscience*, **26**, 681–93.

Lyons, W. E., Fritschy, J. M. & Grzanna, R. (1989). The noradrenergic neurotoxin DSP-4 eliminates the coeruleospinal projection but spares projections of the A5 and A7 groups to the ventral horn of the spinal cord. *Journal of Neuroscience*, **9**, 1481–9.

Maeda, T. & Shimizu, N. (1972). Projections ascendantes du locus coeruleus et d'autres neurones aminergiques pontiques au niveau du proencéphale du rat. *Brain Research*, **36**, 19–35.

Maeda, T., Pin, C., Salvert, D., Ligier, M. & Jouvet, M. (1973). Les neurons contenant des catécholamines du tegmentum pontique et leurs voies de projection chez le chat. *Brain Research*, **57**, 119–52.

Manier, M., Feuerstein, C. P, Mouchet, P., Mons, N., Geffard, M. & Thibault, J. (1990). Evidence for the existence of L-DOPA and dopamine- immunoreactive nerve cell bodies in the caudal part of the dorsal motor nucleus of the vagus nerve. *Journal of Chemical Neuroanatomy*, **3**, 193–205.

Manier, M., Mouchet, P. & Feuerstein, C. (1987). Immunohistochemical evidence for the coexistence of cholinergic and catecholaminergic phenotypes in neurons of the vagal motor nucleus in the adult rat. *Neuroscience Letters*, **80**, 141–5.

McLachlan, E. M., Anderson, C. R. & Sinclair, A. D. (1989). Are there bulbospinal catecholaminergic neurons in the guinea pig equivalent to the C1 cell group in the rat and rabbit? *Brain Research*, **481**, 274–85.

Mefford, I., Oke, A., Keller, R., Adams, R. N. & Jonsson, G. (1978). Epinephrine distribution in human brain. *Neuroscience Letters*, **9**, 227–31.

Mefford, I. N. (1987). Are there epinephrine neurons in rat brain? *Brain Research Review*, **12**, 383–95.

Meister, B., Hökfelt, T., Steinbusch, H. M. W., Skagerberg, G., Lindvall, O., Geffard, M., Joh, T. H., Cuello, A. C. & Goldstein, M. (1988). Do tyrosine hydroxylase-immunoreactive neurons in the ventral arcuate nucleus produce dopamine or only L-DOPA? *Journal of Chemical Neuroanatomy*, **1**, 59–64.

Miachon, S., Bérod, A., Léger, L., Chat, M., Hartman, B. K. & Pujol, J. F. (1984). Identification of catecholamine cell bodies in the pons and pons- mesencephalon junction of the cat brain, using tyrosine hydroxylase and dopamine-β-hydroxylase in immunohistochemistry. *Brain Research*, **305**, 369–74.

Mittendorf, A., Denoroy, L & Flügge, G. (1988). Anatomy of the adrenergic system in the medulla oblongata of the tree shrew: PNMT immunoreactive structures within the nucleus tractus solitarii. *Journal of Comparative Neurology*, **274**, 178–89.

Moll, G., Moll, R., Riederer, P., Gsell, W., Heinsen, H. & Denney, R. M. (1990). Immunofluorescence cytochemistry on the frozen sections of human substantia nigra for staining of monoamine oxidase A and monoamine oxidase B: a pilot study. *Journal of Neural Transmission*, **32**, 67–77.

Moore, R. Y & Bloom, F. E. (1979). Central catecholamine neurons systems: Anatomy and physiology of the norepinephrine and epinephrine systems. *Annual Review of Neuroscience*, **2**, 113–68.

Moore, R. Y & Card, J. P. (1984). Noradrenaline-containing neuron systems. In *Handbook of Chemical Neuroanatomy, Vol.2, Classical Transmitters in the CNS, Part I*, ed. A. Björklund & T. Hökfelt. pp. 123–56. Amsterdam: Elsevier.

Murray, H. M., Dominguez, W. F. & Martinez, J. E. (1982). Catecholamine neurons in the brain stem of the tree shrew (*Tupaia*). *Brain Research Bulletin*, **9**, 205–15.

Nagatsu, I., Sakai, M., Yoshida, M. & Nagatsu, T. (1988). Aromatic L-amino acid decarboxylase-immunoreactive neurons in and around the cerebrospinal fluid-contacting neurons of the central canal do not contain dopamine or serotonin in the mouse and rat spinal cord. *Brain Research*, **475**, 91–102.

Neff, N. H., Yang, H. Y. and Fuentes, J. A. (1974). The use of selective monoamine oxidase inhibitor drugs to modify amine metabolism in brain. In *Neuropharmacolgy of Monoamines and Their Regulatory Enzymes*, ed. Usdin, E. p.49–57, New York: Raven Press.

Nobin, A. & Björklund, A. (1973). Topography of the monoamine neuron systems in the human brain as revealed in fetuses. *Acta Physiologica Scandinavica*, **88** (Suppl. 388), 1–40.

Norgren, R. & Smith, G. P. (1988). Central distribution of subdiaphragmatic vagal branches in the rat. *Journal of Comparative Neurology*, **273**, 207–23.

Oades, R. D. & Halliday, G. M. (1987). Ventral tegmental (A10) system: neurobiology. 1. Anatomy and connectivity. *Brain Research Review*, **12**, 117–65.

Ochi, J. & Hosoya, Y. (1974). Fluorescence microscopic differentiation of monoamines in the hypothalamus and spinal cord of the lamprey, using a new filter system. *Histochemistry*, **40**, 263–6.

Ochi, J. & Shimizu, K. (1978). Occurrence of dopamine-containing neurons in the midbrain raphe nuclei of the rat. *Neuroscience Letters*, **8**, 317–20.

Okamura, H., Kitahama, K., Mons, N., Ibata, Y., Jouvet, M & Geffard, M. (1988*a*). L-DOPA-immunoreactive neurons in the rat hypothalamic tuberal region. *Neuroscience Letters*, **95**, 42–6.

Okamura, H., Kitahama, K., Nagatsu, I. & Geffard, M. (1988*b*). Comparative topography of dopamine- and tyrosine hydroxylase- immunoreactive neurons in the rat arcuate nucleus. *Neuroscience Letters*, **95**, 347–53.

Okamura, H., Kitahama, K., Raynaud, B., Nagatsu, I., Borri- Voltattorni, C. & Weber, M. (1988*c*). Aromatic L-amino acid decarboxylase (AADC)-immunoreactive cells in the tuberal region of the rat hypothalamus. *Biomedical Research*, **9**, 261–71.

Olson, L. & Fuxe, K. (1971). On the projections from the locus coeruleus noradrenaline neurons: the cerebellar innervation. *Brain Research*, **28**, 165–71.

Olson, L., Nyström, B. & Seiger, A. (1973). Monoamine fluorescence histochemistry of human postmortem brain. *Brain Research*, **63**, 231–47.

Olszewski, D. and Baxter, D. (1954). *Cytoarchitecture of the Human Brainstem*, Philadelphia: Lippincott.

Østergaard, K., Holm, I. E. & Zimmer, J. (1992). Tyrosine hydroxylase and acetylcholinesterase in the domestic pig mesencephalon: an immunocytochemical and histochemical study. *Journal of Comparative Neurology*, **322**, 149–66.

Palkovits, M., Mezey, E., Skirboll, L. R. & Hökfelt, T. (1992). Adrenergic projections from the lower brainstem to the hypothalamic paraventricular nucleus, the lateral hypothalamic area and the central nucleus of the amygdala in rats. *Journal of Chemical Neuroanatomy*, **5**, 407–15.

Palmer, A. M., Wilcock, G. K., Esiri, M. M., Francis, P. T. & Bowen, D. M. (1987). Monoaminergic innervation of the frontal and temporal lobes in Alzheimer's disease. *Brain Research*, **401**, 231–8.

Parent A., Poitras D. & Dubé L. (1984). Comparative anatomy of central monoaminergic systems. In *Handbook of Chemical Neuroanatomy, Vol.2: Classical Transmitters in the CNS, Part II*, ed. T. Hökfelt & A. Björklund. pp.409–39. Amsterdam: Elsevier.

Pearson, J., Goldstein, M., Markey, K. & Brandeis, L. (1983). Human brainstem catecholamine neuronal anatomy as indicated by immunocytochemistry with antibodies to tyrosine hydroxylase. *Neuroscience*, **8**, 3–32.

Pearson J., Halliday G., Sakamoto N. & Michel J. P. (1990). Catecholaminergic neurons. In *The Human Nervous System*, ed. G. Paxinos. pp. 1023–49. New York: Academic.

Pickel, V. M., Specht, L. A., Sumal, K. K., Joh, T. H., Reis, D. J. & Hervonen, A. (1980). Immunocytochemical localization of tyrosine hydroxylase in the human fetal nervous system. *Journal of Comparative Neurology*, **194**, 465–74.

Pin, C., Jones, B. & Jouvet, M. (1968). Topographie des neurones monoaminergique du tronc cérébral du chat: Etude par histofluorescence. *Comptes Rendus des séances de la Société de la Biologie*, **162**, 2136–41.

Poitras, D. & Parent, A. (1978). Atlas of the distribution of monoamine-containing nerve cell bodies in the brain stem of the cat. *Journal of Comparative Neurology*, **179**, 699–717.

Reiner, P. B. & Vincent, S. R. (1986). The distribution of tyrosine hydroxylase, dopamine-β-hydroxylase, and phenylethanolamine N-methyltransferase- immunoreactive neurons in the feline medulla oblongata. *Journal of Comparative Neurology*, **248**, 518–31.

Robert, O., Miachon, S., Kopp, N., Denoroy, L., Tommasi, M., Rollet, D. & Pujol, J. F. (1984). Immunohischemical study of the catecholaminergic systems in the lower brain stem of the human infant. *Human Neurobiology*, **3**, 229–34.

Ruggiero, D. A., Anwar, M. & Gootman, P. M. (1992). Presumptive adrenergic neurons containing phenylethanolamine N- methyltransferase immunoreactivity in the medulla oblongata of neonatal swine. *Brain Research*, **583**, 105–19.

Ruggiero, D. A., Gatti, P. J., Gillis, R. A., Norman, W. P., Anwar, M. & Reis, D. J. (1986). Adrenaline synthesizing neurons in the medulla of the cat. *Journal of Comparative Neurology*, **252**, 532–42.

Ruggiero, D. A., Ross, C. A., Anwar, M., Park, D. H., Joh, T. H. & Reis, D. J. (1985). Distribution of neurons containing phenylethanolamine N- methyltransferase in medulla and hypothalamus of rat. *Journal of Comparative Neurology*, **239**, 127–54.

Russel, G. V. (1955). The nucleus locus coeruleus (dorsolateralis tegmenti). *Texas Reports Biology and Medicine*, **13**, 939–88.

Saavedra, J. M., Brownstein, M. J. & Palkovits, M. (1976). Distribution of catechol-O-methyltransferase and monoamine oxidase in specific areas of the rat brain. *Brain Research*, **118**, 152–6.

Saavedra, J. M., Kvetnansky, R & Kopin, I. J. (1979). Adrenaline, noradrenaline and dopamine levels in specific brain stem areas of acutely immobilized rats. *Brain Research*, **160**, 271–80.

Saura Marti, J., Kettler, R., Da Prada, M. & Richards, J. G. (1990). Molecular neuroanatomy of MAO-A and MAO-B. *Journal of Neural Transmission*, **32**, 49–53.

Schalling, M., Friberg, K., Seroogy, K., Riederer, P., Bird, E., Schiffmann, S., Mailleux, P., Vanderhaeghen, J. J., Goldstein, M., Kitahama, K., Luppi, P. H., Jouvet, M. and Hökfelt, T. (1990). Analysis of expression of cholecystokinin in dopamine cells in the ventral mesencephalon of several species and humans with schizophrenia. *Proceedings of the National Academy of Sciences*, USA, **87**, 8427–31.

Schneider, J. S. & Markham, C. H. (1987). Immunohistochemical localization of monoamine oxidase-B in the cat brain: clues to understanding N-methyl-4-phenyl-1,2,3,6-tetrahydropyridine (MPTP) toxicity. *Experimental Neurology*, **97**, 465–81.

Schneider, J. S., Yuwiler, A. & Markham, C. H. (1986). Production of a Parkinson-like syndrome in the cat with N-methyl-4-phenyl- 1,2,3,6-tetrahydropyridine (MPTP): behavior, histology and biochemistry. *Experimental Neurology*, **91**, 293–307.

Seroogy, K. B., Dangaran, K., Lim, S., Haycock, J. W. & Fallon, J. H. (1989). Ventral mesencephalic neurons containing both cholecystokinin-and tyrosine hydroxylase-like immunoreactivities project to forebrain regions. *Journal of Comparative Neurology*, **279**, 397–414.

Shimada, S., Ishikawa, M. & Tanaka, C. (1976). Histochemical mapping of dopamine neurons and fiber pathways in dog mesencephalon. *Journal of Comparative Neurology*, **168**, 533–44.

Shimizu, N., Morikawa, N. & Okada, M. (1959). Histochemical studies of monoamine oxidase of the brain of rodents. *Zeitschrift für Zellforschung microskopische Anatomie*, **49**, 389–400.

Smeets, W. J. A. J. & Steinbusch, H. W. M. (1990). New insights into the reptilian catecholaminergic systems as revealed by antibodies against the neurotransmitters and their synthetic enzymes. *Journal of Chemical Neuroanatomy*, **3**, 25–43.

Smits, R. P. J. M., Steinbusch, H. W. M. & Mulder, A. H. (1990). Distribution of dopamine-immunoreactive cell bodies in the guinea-pig brain. *Journal of Chemical Neuroanatomy*, **3**, 101–23.

Stratford, T. R. & Wirtshafter, D. (1990). Ascending dopaminergic projections from the dorsal raphe nucleus in the rat. *Brain Research*, **511**, 173–6.

Sved, A. F. (1989). PNMT-containing catecholaminergic neurons are not necessarily adrenergic. *Brain Research*, **481**, 113–18.

Sved, A. F. (1990). Effects of monoamine oxidase inhibition on catecholamine levels: evidence for synthesis but not storage of epinephrine in rat spinal cord. *Brain Research*, **512**, 253–8.

Swanson, L. W. (1976). The locus coeruleus: a cytoarchitectonic, Golgi and immunohistochemical study in the albino rat. *Brain Research*, **110**, 39–56.

Swanson, L. W. (1982). The projections of the ventral tegmental area and adjacent regions: a combined fluorescent retrograde tracer and immunofluorescence study in the rat. *Brain Research Bulletin*, **9**, 321–53.

Swanson, L. W. & Hartman, B. K. (1975). The central adrenergic system. An immunofluorescence study of the location of cell bodies and their efferent connections in the rat utilizing dopamine-β-hydroxylase as a marker. *Journal of Comparative Neurology*, **163**, 467–506.

Takahashi, H., Nakashima, S., Ohama, E., Ikeda, S. & Ikuta, F. (1986). The subnucleus gelatinosus of the nucleus of tractus solitarii is a catecholaminergic nucleus: immunohistochemical and ultrastructural studies in the human fetal brain. *Biomedical Research*, **6**, 257–60.

Tanaka, C., Ishikawa, M. & Shimada, S. (1982). Histochemical mapping of catecholaminergic neurons and their ascending fiber pathways in rhesus monkey brain. *Brain Research Bulletin*, **9**, 255–70.

Tillet, Y. & Thibault, J. (1989). Catecholamine-containing neurons in the sheep brainstem and diencephalon: immunohistochemical study with tyrosine hydroxylase (TH) and dopamine-β-hydroxylase (DBH) antibodies. *Journal of Comparative Neurology*, **290**, 69–104.

Tillet, Y. (1988). Adrenergic neurons in sheep brain demonstrated by immunohistochemistry with antibodies to phenylethanolamine N-methyltransferase (PNMT) and dopamine-β-hydroxylase (DBH): absence of C1 cell group in the sheep brain. *Neuroscience Letters*, **95**, 107–12.

Touret, M., Valatx, J. L. & Jouvet, M. (1982). The locus coeruleus: a quantitative and genetic study in mice. *Brain Research*, **250**, 353–7.

Ungerstedt, U. (1971). Stereotaxic mapping of the monoamine pathways in the rat brain. *Acta Physiologica Scandinavica*, **82 (Suppl.367)**, 1–48.

Unguez, G. A. & Schneider, J. S. (1988). Dopaminergic dorsal raphe neurons in cats and monkeys are sensitive to the toxic effects of MPTP. *Neuroscience Letters*, **94**, 218–23.

Van Bockstaele, E. & Aston-Jones, G. (1992). Collateralized projections from neurons in the rostral medulla to the nucleus locus coeruleus, the nucleus of the solitary tract and the periaqueductal gray. *Neuroscience*, **49**, 653–68.

Van Bockstaele, E. J., Pieribone, V. A. & Astone-Jones, G. (1989). Diverse afferents converge on the nucleus paragigantocellularis in the rat ventrolateral medulla: Retrograde and anterograde tracing studies. *Journal of Comparative Neurology*, **290**, 561–84.

Vijayshankar, N. & Brody, H. (1979). A quantitative study of the pigmented neurons in the nuclei locus coeruleus and subcoeruleus in man as related to aging. *Journal of Neuropathology and Experimental Neurology*, **38**, 490–7.

Vincent, S. R. & Reiner, P. B. (1987). The immunohistochemical localization of choline acetyltransferase in the cat brain. *Brain Research Bulletin*, **18**, 371–415.

Vincent, S. R. & Hope, B. T. (1990). Tyrosine hydroxylase containing neurons lacking aromatic amino acid decarboxylase in the hamster brain. *Journal of Comparative Neurology*, **295**, 290–8.

Vincent, S. R. (1988). Distribution of tyrosine hydroxylase-dopamine-β-hydroxylase- and phenylethanolamine-N-methyl transferase- immunoreactive neurons in the brain of the hamster (*Mesocricetus auratus*). *Journal of Comparative Neurology*, **268**, 584–99.

Westlund, K. N. & Coulter, J. D. (1980). Descending projections of the locus coeruleus and subcoeruleus/medial parabrachial nuclei in monkey: axonal transport studies and dopamine-β-hydroxylase immunohistochemistry. *Brain Research Review*, **2**, 235–64.

Westlund, K. N., Denney, R. M., Kochersperger, L. M., Rose, R. M. & Abell, C. W. (1985). Distinct monoamine oxidase A and B populations in primate brain. *Science*, **230**, 181–3.

Westlund, K. N., Denney, R. M., Rose, R. M. & Abell, C. W. (1988). Localization of distinct monoamine oxidase A and monoamine oxidase B cell populations in human brainstem. *Neuroscience*, **25**, 439–56.

Wiklund, L., Léger, L. & Persson, M. (1981). Monoamine cell distribution in the cat brain stem. A fluorescence histochemical study with quantification of indolaminergic and locus coeruleus cell groups. *Journal of Comparative Neurology*, **203**, 613–47.

Williams, J. L. & Schneider, J. S. (1989). MPTP-induced ventral mesencephalic cell loss in the cat. *Neuroscience Letters*, **101**, 258–62.

9
Catecholaminergic neuronal systems in the diencephalon of mammals

Y. Tillet

Introduction

Since the pioneering work of Von Euler (1946) concerning the demonstration of the presence of catecholamine in nerve terminals, numerous studies have shown the role of catecholamines in the control of hormonal secretion (for review see Weiner & Ganong, 1978; Tuomisto & Männisto, 1985), or the regulation of autonomic functions (for review see Loewy & Spier, 1990). In addition to these physiological aspects, the formaldehyde induced fluorescence (FIF) method developed by Falck, Hillarp & Törp (1962) and immunohistochemistry of catecholamine synthesizing enzymes have been used to detect the presence of catecholamines in the brain of rat (Dahlström & Fuxe, 1964; Fuxe, 1965a, b; Hökfelt et al., 1984c).

Most of the immunohistochemical (and also FIF) studies concerning the distribution of central catecholamines were performed in rodents, mainly in the rat, as shown in Table 9.1. This species, however, cannot be considered as representative for all mammalian species. In the last ten years, therefore, numerous studies have been devoted to other species as well (Table 9.1).

In this chapter, our current knowledge of catecholaminergic innervation of the diencephalon of various mammalian species is summarized. The sections concern successively 1) the distribution of neuronal catecholaminergic groups (according to the Swedish nomenclature), 2) the distribution of catecholaminergic fibers 3) the morphological relationships with other neuronal systems, and 4) the functional implications of such relationships. The distribution of neuronal catecholaminergic groups will be illustrated on schematic drawings (Figs. 9.1 to 9.5) and on photographic plates (Figs. 9.6 to 9.12)

Catecholaminergic neuronal groups

In the diencephalon, catecholamine cell groups are usually designated as the A11–A15 cell groups (Fig. 9.1). In the present section, a survey is given of the distribution of catecholamine neurons in the diencephalon of different mammalian species. Interest is focussed on the different patterns of organization of these neurons since the 'general' distribution is now well known from the numerous studies of the rat brain by the Swedish school (for review see Björklund & Lindvall, 1984; Moore & Card, 1984; Hökfelt et al., 1984b, c).

Since comments on the different techniques are presented separately in chapter 1, only two general comments will be given here. First, within the diencephalon of mammals, TH immunohistochemistry gives a stronger immunolabeling signal than dopamine immunohistochemistry, and, secondly, dopamine immunolabeling becomes stronger after colchicine treatment (Kitahama et al., 1990). In the diencephalon, most catecholaminergic neurons are bipolar or multipolar, the dendritic trees are always stained and their perikaryal diameter ranges from 10 to 25 micrometers (Van Den Pol, Herbst & Powell 1984).

Caudal diencephalon

Group A10 (rostro-dorsal and ventral parts) In the caudal part of the diencephalon, a small gathering of THi neurons are observed on each side of the mam-

Table 9.1. *Extensive mapping of catecholamine containing structures in the diencephalon of mammalian species (immunohistochemical and FIF studies)*

Species	Methods	Antisera	References
*rat	IHC	TH	Hökfelt et al., 1976
rat	IHC	TH	Hökfelt et al., 1984c
rat	IHC	TH	Chan-Palay et al., 1984
rat	IHC	TH	Van Den Pol et al., 1984
rat	IHC	TH	Skagerberg et al., 1988
rat	IHC	PNMT	Hökfelt et al., 1984b
*mouse	IHC	TH	Ruggiero et al., 1984
*guinea pig	IHC	DA	Smits et al., 1990
*golden hamster	IHC	TH	Vincent, 1988
golden hamster	IHC	TH, AADC	Vincent & Hope, 1990
*house shrew	IHC	TH, AADC	Karasawa et al., 1991, 1992
*rabbit	FIF		Blessing et al., 1978
*cat	IHC, FIF	TH	Kitahama et al., 1987
cat	IHC	TH, DA	Kitahama et al., 1990
*sheep	IHC	TH, DBH	Tillet & Thibault, 1989
*monkey	FIF		Felten, 1976
monkey	FIF		Felten & Sladek, 1983
monkey	IHC	DBH	Ginsberg et al., 1993
*human	IHC	TH	Spencer et al., 1985
human	IHC	TH	Li et al., 1988
human	IHC	TH	Pearson et al., 1990a, b
human	FIF		Su et al., 1987

millary bodies (Fig. 9.4(h)–(i)) in the sheep (Tillet & Thibault, 1989) as well as in rats, where it is considered as a ventrorostral (vr) extension of the mesencephalic group A10 (Hökfelt et al., 1984c). In the sheep these neurons are bipolar and possess long, well stained dendritic trees.

In the dorsal part of the diencephalon, at the same level, some small neurons are scattered near the lateral habenula (Fig. 9.4(k)). They are mainly bipolar and weakly stained. This gathering is considered as the dorsorostral (dr) extension of the group A10 (Hökfelt et al., 1984b). This group has been described in sheep, rats and hamsters (Table 9.1) using anti-TH. In the house-shrew, some small weakly stained THi neurons were also observed in the lateral part of the habenula, but in this species they lacked AADCi (Karasawa et al., 1991, 1992). Neither A10vr nor A10dr groups were previously described by FIF method in the rat, indicating that the presence of dopamine or L-DOPA in these perikarya remained still questionable. In the guinea pig, using dopamine immunohistochemistry, the presence of group A10vr but not A10dr has been confirmed (Smits, Steinbusch & Mulder, 1990).

THi neurons observed around the mammillary bodies and in the supramammillary area as referenced A10vr in the rat send minor projections to the forebrain region. The most densely innervated area is the lateral septum (Sheppard, Mihailoff & German, 1988). These neurons are also considered as interneurons exhibiting local projections to the mammillary nucleus (Gonzalo-Ruiz et al., 1992).

Pineal gland THi neurons have been observed in the pineal gland of golden hamsters (Jin et al., 1988), but not in other species. These neurons are homogeneously distributed throughout the gland. Their immunostaining intensity and their number decrease during light illumination and increase during darkness. They do not appear to project outside the pineal gland (Shiotani et al., 1990).

Group A11 This group is constituted by catecholamine neurons along the caudal part of the third ventricle in the caudal diencephalon. It is observed in all species studied with anti-TH and FIF. The presence of dopamine in these cells was confirmed by means of antidopamine antibodies in the rat (Tison

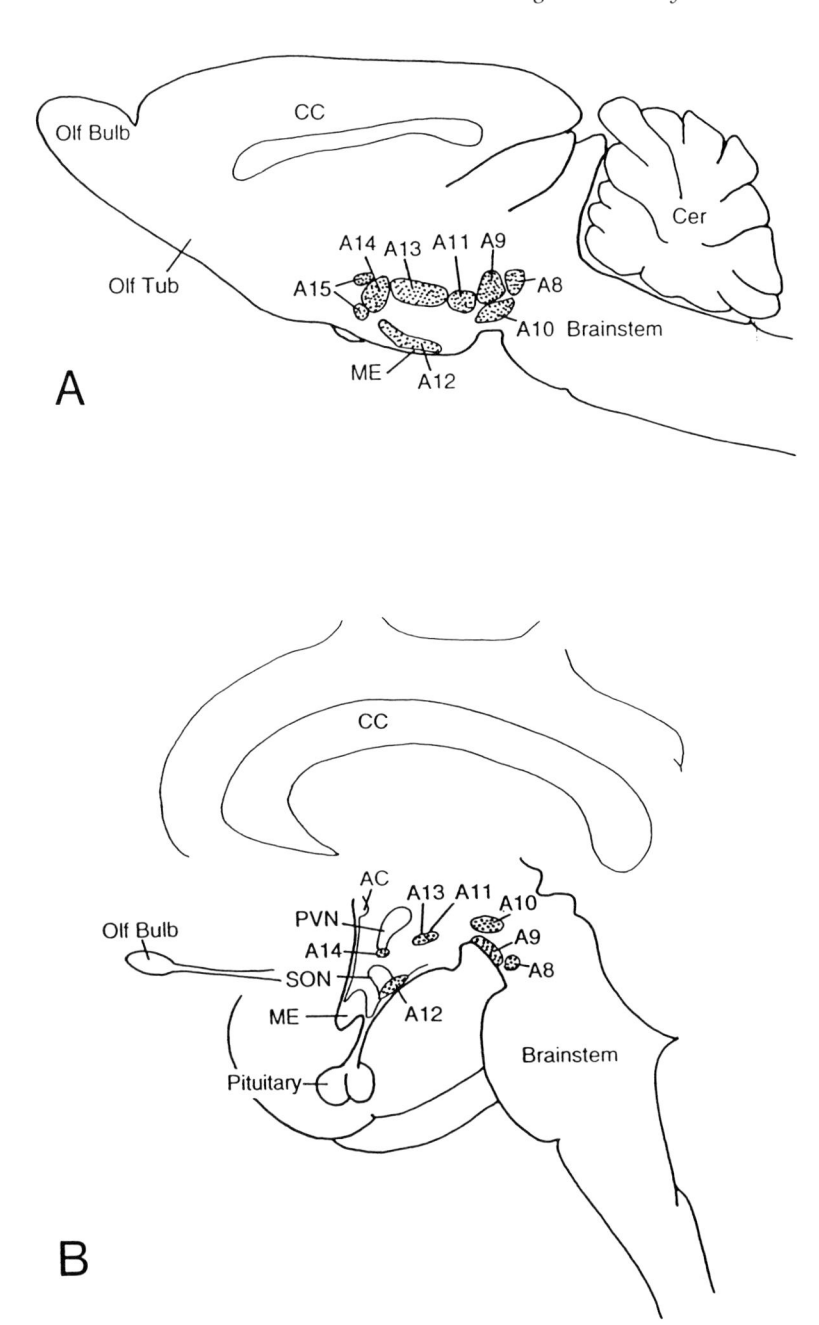

Fig. 9.1. Schematic drawings of parasagittal sections through the brain of rats (a) and primates (b) adapted from Ungerstedt (1971) and Nieuwenhuys (1985) respectively. Shaded areas represent the locations of catecholaminergic groups. In the primates, both PVN and SON contain THi neurons which may be related to A15.

et al., 1990), cat (Kitahama et al., 1990) (Figs. 9.3(e)–(g); 9.3(e)–(g)) and guinea pig (Smits, Steinbusch & Mulder, 1990). In rat and cat, the presence of L-DOPA is also identified. Species differences exist in the distribution of the cells within this group: whereas in the cat, THi neurons are more numerous around the ventral part of the ventricle (Kitahama et al., 1987), in rodents (Fig. 9.2(i)–(j)) they are equally distributed in the dorsal as well as ventral part (Hökfelt et al., 1984b). In sheep (Fig. 9.4(h)–(j)) (Tillet & Thibault, 1989) and human (Pearson et al., 1990a), this group contains only a small number of neurons, a very few are observed dorsally to the mammillary bodies. In all these species the neurons were in most cases multipolar.

Tract tracing methods have demonstrated that the neurons situated around the caudal part of the third ventricle have basically two major projection fields: (i) the spinal cord via a descending pathway by the dorsal longitudinal fasciculus of Schütz (Björklund & Skagerberg, 1979; Hökfelt, Philipson & Goldstein, 1979; Skagerberg et al., 1982; Skagerberg & Lindvall, 1985) and (ii) the periventricular area of the hypothalamus and preoptic area (Fuxe et al., 1985). Recent tracer studies demonstrated that some dopaminergic neurons of A11 group (of the subparafascicular nucleus) project also to several parts of the neocortex, and that a single neuron may project to the neocortex as well as to the spinal cord (Takada, Li & Hattori, 1988). The A11 group is also reported to project to the amygdala in several mammalian species (Takada, 1990; Veening, 1978; Otterson & Ben-Ari, 1979; Mehler, 1980). Furthermore, a dense dopaminergic innervation of the amygdala area is observed in the rat (Ungerstedt, 1971; Fallon, Koziell & Moore, 1978; Fuxe et al., 1974). However, such a dopaminergic innervation can vary among species since TH-immunoreactive (THi) neurons are species specifically demonstrated in the amygdala of hamster (Davis & Macrides, 1983; Vincent, 1988). A part of these neurons are actually dopaminergic as demonstrated in colchicine-treated hamster with anti-dopamine antibodies (Asmus, Kincaid & Newman, 1992).

Intermediate diencephalon

This area contains numerous catecholamine neurons which are mainly located in the ventral and dorsal part of the hypothalamus, and are referred to as the A12 and A13 groups, respectively.

Group A12 The highest concentration of neurons of the A12 group is found in the arcuate nucleus, or infundibular nucleus or tuberoinfundibular nucleus as denominated in rodents, sheep and human, respectively. In the mouse (Ruggiero et al., 1984), this area contains numerous THi neurons (Fig. 9.2(c)–(g)), and no clear-cut boundary is seen between the caudal part of the arcuate nucleus and group A11. In all species studied with FIF and/or immunohistochemistry, the neurons presented the same characteristics; they are round or oval and mainly bipolar with short processes. In the concerned areas, their density is important but their distribution in the mediobasal hypothalamus could be different according to the different studied species. In rat, mouse and hamster (Table 9.1), the perikarya are found close to the ventricle, on each side of the infundibulum whereas in sheep and in guinea pigs (Table 9.1) THi or dopamine-immunoreactive (DAi) neurons invade the dorsal part of the median eminence. In the cat (Figs. 9.3(c)–(f); 3(c)–(f)), a substantial number of neurons of the rostral part of the nucleus are situated in this area. In rodents, additional THi neurons are observed ventro-laterally to the arcuate nucleus (Hökfelt et al., 1984c) whereas such extension is not observed in cats and sheep. In the latter species the infundibular nucleus lies closer to the ventricle in its rostral part (Fig. 9.8(a)) than in its caudal part (Tillet & Thibault, 1989) (Fig. 9.4(e)–(g)). In the rostral part of the arcuate nucleus of the different species studied, THi neurons are also observed in the floor of the third ventricle. In human (Fig. 9.5(b)–(f)), the distribution appears slightly different, the THi perikarya of the A12 group are subdivided into two subgroups: one in the tuberal region while the second is found in the periventricular zone, in the floor and the ventral part of the wall of the third ventricle (Pearson et al., 1990a); this second subgroup could be related to the arcuate nucleus of the rodents.

Recently, it has been shown that the arcuate nucleus and the surrounding area can be divided into subgroups according to their catecholaminergic content. Since TH immunohistochemistry has demonstrated the presence of a higher number of catecholaminergic neurons than the FIF method did, for example, in rats (Hökfelt et al., 1984b) and cats (Kitahama et al., 1987), anti-dopamine and anti-L-DOPA have been used to check the catecholaminergic content of these THi perikarya. In the rat arcuate nucleus, dopamine immunoreactivity is identified only in neurons situated in the dorsomedial part of the nucleus, as studied by TH, AADC, and dopamine immunohistochemistry (Meister et al., 1988; Okamura et al., 1988b); some THi cells in the ventrolateral part contained only L-DOPA (Okamura et al., 1988a). The same distinction is supposed in human tuberoinfundibular nucleus since the ventrolateral THi cells fail to exhibit AADC immunoreactivity (Komori, Fujii & Nagatsu, 1991). Furthermore dopamine immunoreactivity is not observed in this area suggesting that these cells, like the corresponding ones in rats, contain L-DOPA. In the cat, the lateral part of the arcuate nucleus contains only a few THi and FIF labeled neurons (Kitahama et al., 1987). The division of the nucleus into dorsomedial and ventrolateral parts is not as clear as in rodents, and dopamine as well as L-DOPA are found in the dorsal part. In the sheep, a subdivision of infudibular nucleus is not observed, and THi perikarya are mainly observed close to the third ventricle, they contain also dopamine (Tillet, unpublished results). In the guinea pig homologous region where only dopamine immunoreactivity was studied, DAi neurons appear situated close to the ventricle, in the dorsal part of the arcuate nucleus (Smits et al., 1990) as in rats. The presence of L-DOPA is observed in the species where it was investigated (rat, cat) and suspected in other (human); in sheep and in guinea pig, it was not studied.

The catecholamine cells of group A12 constitute the tubero-infundibular system. This group was first suspected to project towards the median eminence and to the posterior lobe of the pituitary, on the basis of FIF and lesion studies (for review, see Fuxe et al., 1985). However, by means of more powerful tract tracing techniques, it has been demonstrated that the group A12 innervates only the external zone of the median eminence in rats and cats (Van Den Pol et al., 1984; Kawano & Daikoku, 1987; Yoshimoto et al., 1990).

Group A13 The cells of the group A13 lie in the dorsomedial hypothalamus and in the zona incerta. In rats, this group is characterized by a dense aggregate of THi neurons mainly situated in the zona incerta, but some neurons are also observed in the dorsomedial hypothalamic nucleus between the mammillothalamic fasciculus and the third ventricle. However, some variation has been observed between species with regard to numbers and arrangement. In mice (Fig. 9.2(d)–(i)) and rats this A13 cell group contains a large number of THi neurons. The number of neurons in the corresponding area in guinea pig is lower and the cells are more dispersed than in mice and rats. In the hamster (Vincent, 1988) and sheep (Fig. 9.4(e)–(g)) (Tillet & Thibault, 1989), THi neurons lie more laterally, mainly in the dorsomedial hypothalamic area but a dense cluster of neurons is not observed in an area corresponding to the rat zona incerta (Fig. 9.6(b)) whose cytoarchitecture is not well characterized in sheep. In this species the group A13 contains two morphologically different types of THi neurons: small bipolar perikarya intermingled with larger multipolar neurons, the largest being observed mainly in the dorsal part of the nucleus. In the cat (Figs. 9.3(c)–(e); 3(c)–(e); 9.6(a)), the distribution of the DAi and THi neurons are very similar to those described in the hamster. The neuronal density appears weaker but the neurons are more laterally extended than in the rodents. Since the A13 cell group merges with the A11 group caudally and their morphology are very similar, a clear separation is not possible (Chan-Palay et al., 1984; Hökfelt et al., 1984c). In the guinea pig, Smits, et al. (1990) considered, therefore, that A11 and A13 constitute only one group. In human, this group does not extend as laterally as in rodents and its rostral and caudal boundaries are not better evidenced than in other species (Spencer et al., 1985) (Fig. 9.5(f)).

In the rat, axons from the A13 group constitute the incerto-hypothalamic dopaminergic system (Björklund, Lindvall & Nobin, 1975) and project to the dorsal and anterior hypothalamus, and the medial preoptic area (Fuxe et al., 1985). Recent retrograde tracing studies, however, failed to confirm the projection to the medial preoptic area in the sheep (Tillet, Batailler & Thibault, 1993).

Ventral to the A13 group and to the fornix, several THi neurons were scattered throughout the medial diencephalon, up to the dorsal part of the arcuate nucleus, except in the ventromedial nucleus. These neurons are observed in all the studied species. The neuronal density appears lower in the mouse (Ruggiero et al., 1984) than in other rodents. The perikarya are fusiform or multipolar (Fig. 9.6(c)). In guinea pigs these neurons are DAi (Smits et al., 1990). These neurons have not been related to other hypothalamic catecholamine-containing groups.

Rostral diencephalon

Group A14 The rostral periventricular area of the third ventricle contains THi neurons which constitute the periventricular dopaminergic system and are referred to as the A14 group (Hökfelt et al., 1984c). In most mammalian species, the neurons of this group appear fusiform and bipolar with their long axis oriented parallel to the wall of the third ventricle (Fig. 9.7). The caudal part of this group is contiguous with the A13 group. The density of perikarya within this gathering varies among the different species. In rats, mice (Fig. 9.2(a)–(f)), and hamsters (Table 9.1),

MOUSE TH

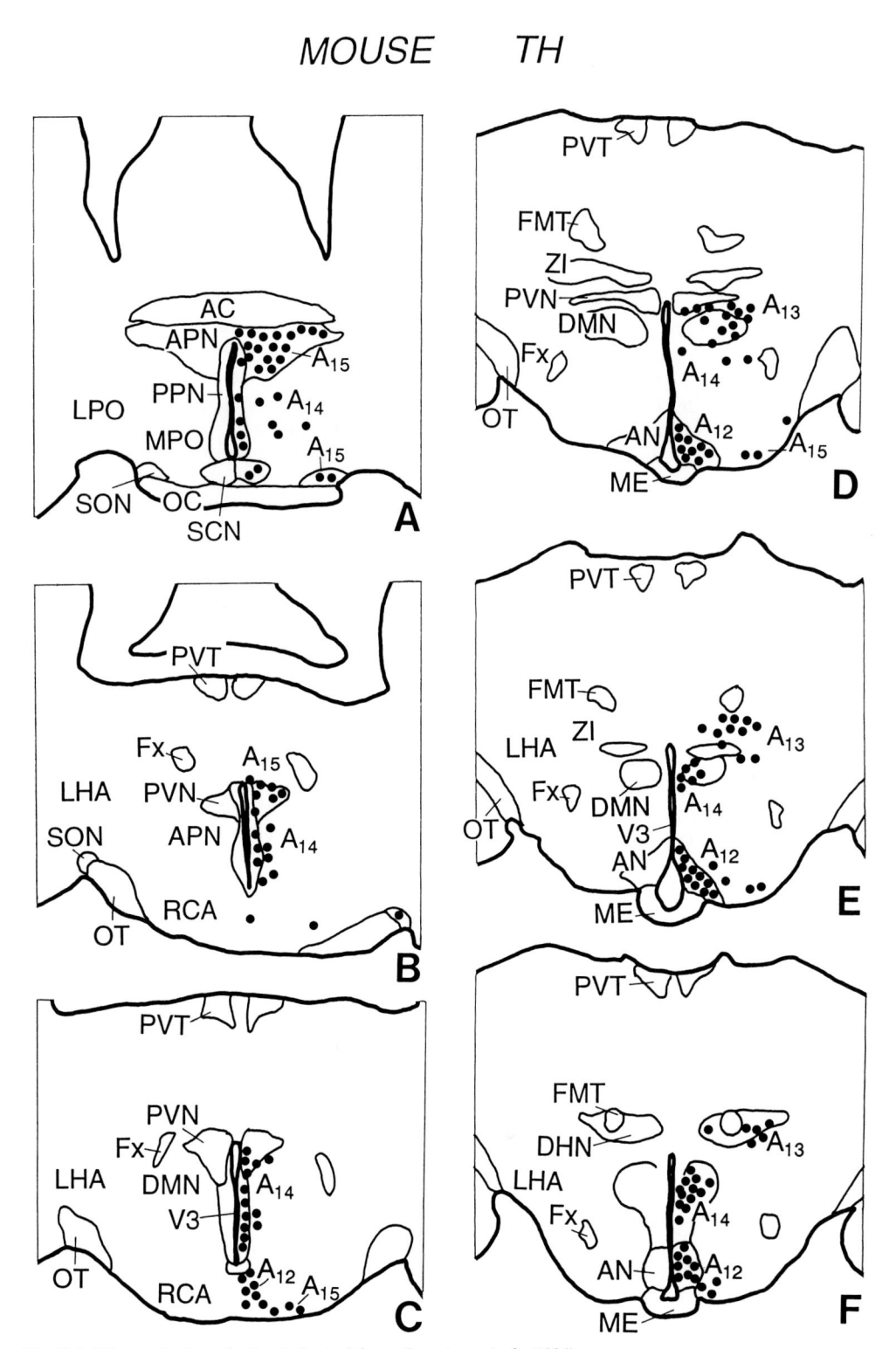

Fig. 9.2. Diencephalon of mice (adapted from Ruggiero et al., 1984).

Fig. 9.2–9.5. Schematic drawings of frontal sections through the diencephalon of several species, from rostral (a) to caudal (k) levels. Black dots represent THi neurons and open circles represent DAi neurons.

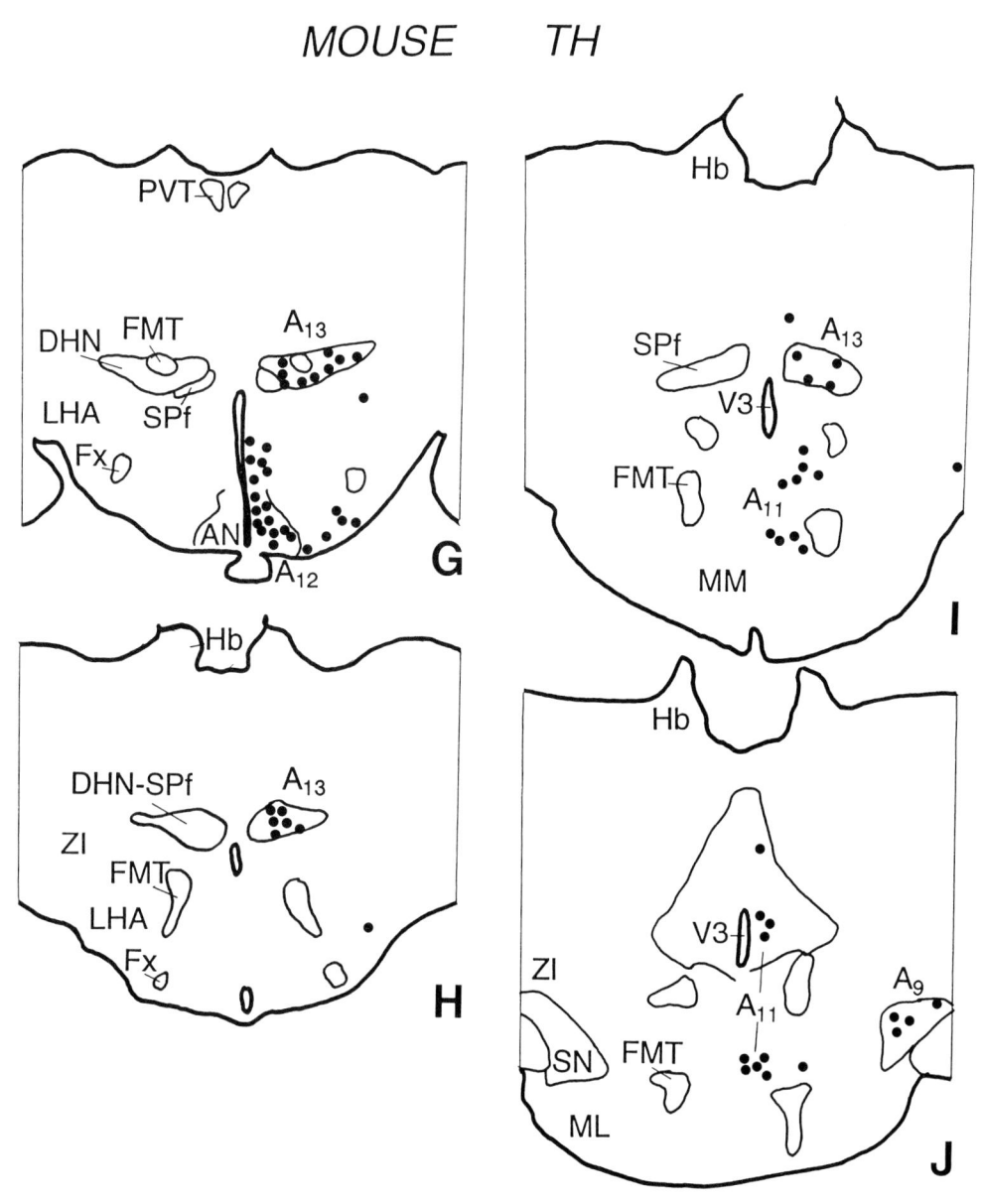

Catecholaminergic neuronal systems in diencephalon of mammals 213

Fig. 9.2. (cont.)

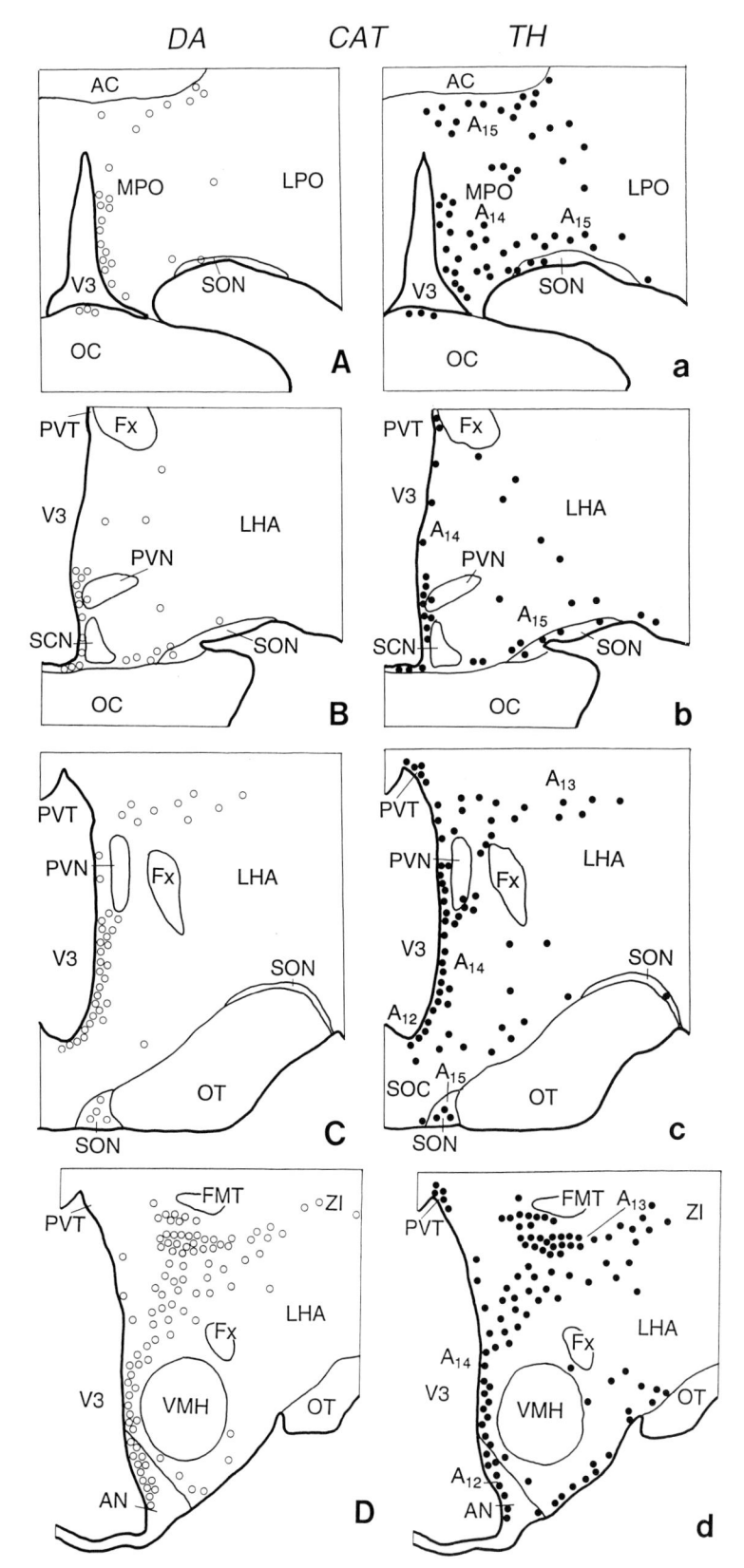

Fig. 9.3. Diencephalon of colchicine treated cats (adapted from Kitahama et al., 1990).

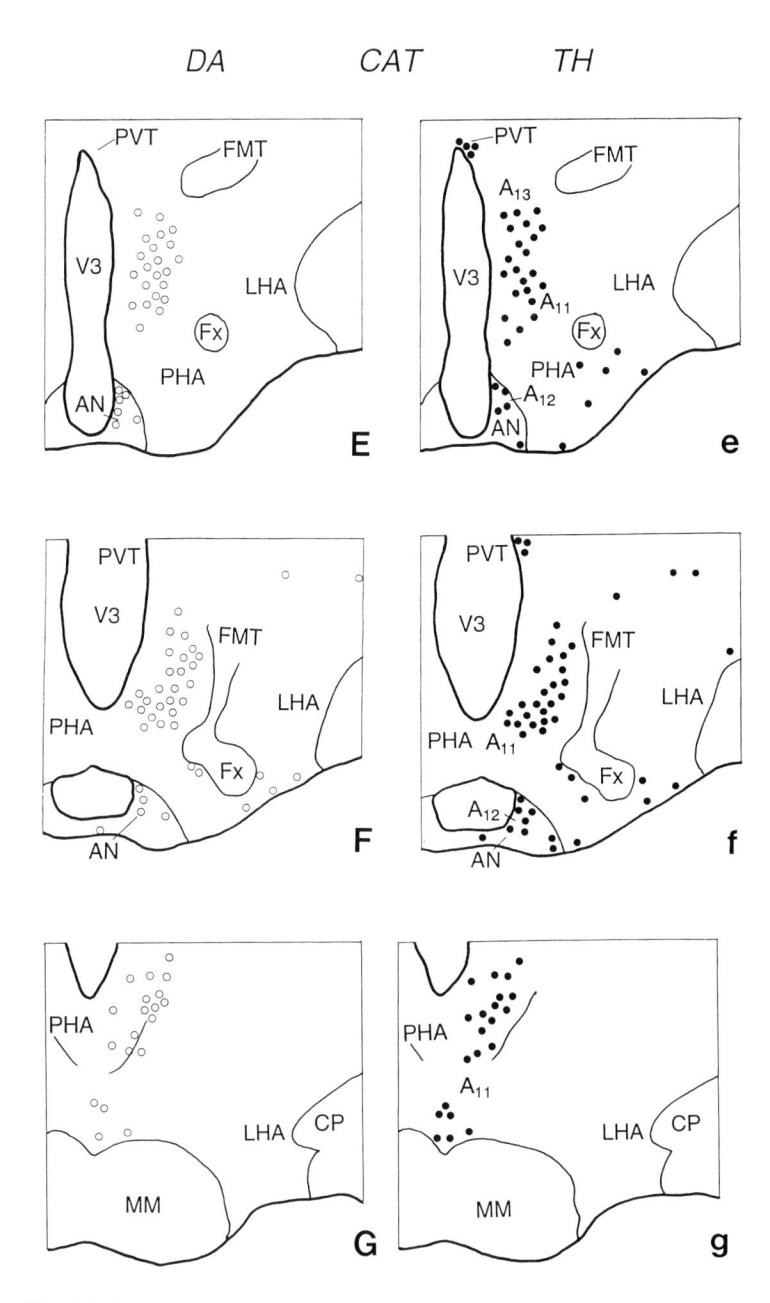

Catecholaminergic neuronal systems in diencephalon of mammals 215

Fig. 9.3. (cont.)

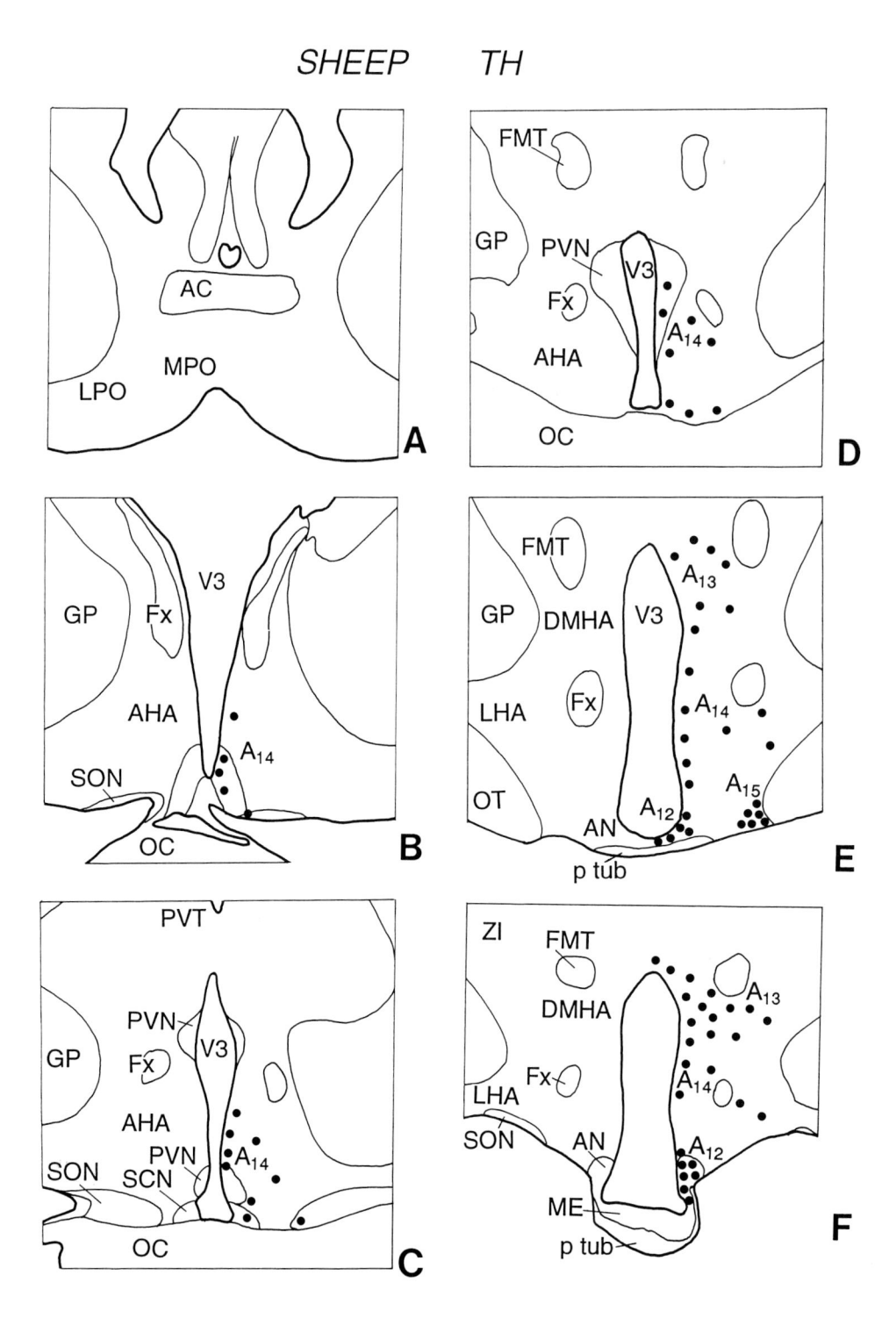

Fig. 9.4. Diencephalon of sheep (adapted from Tillet & Thibault, 1989).

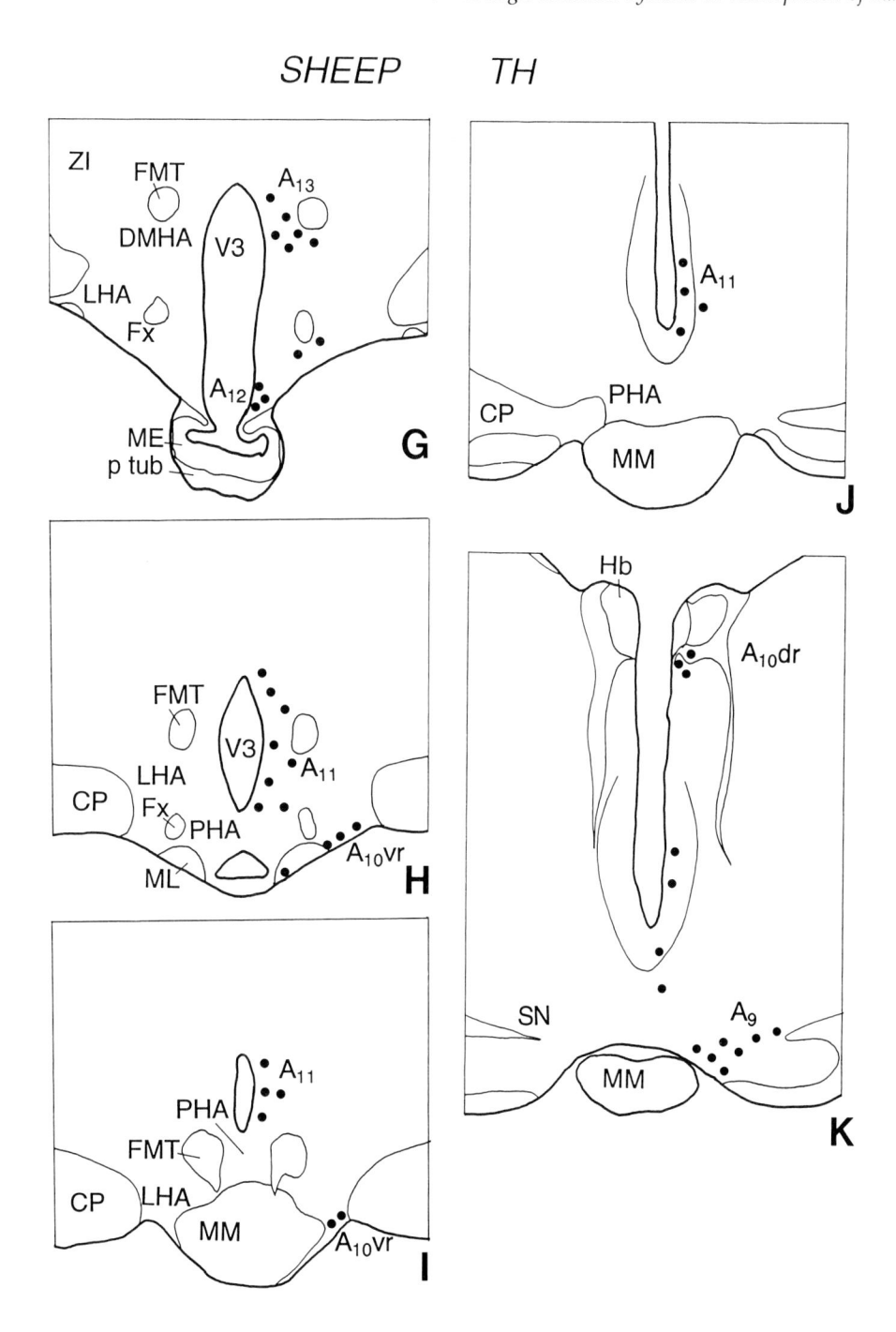

Fig. 9.4. (cont.)

HUMAN TH

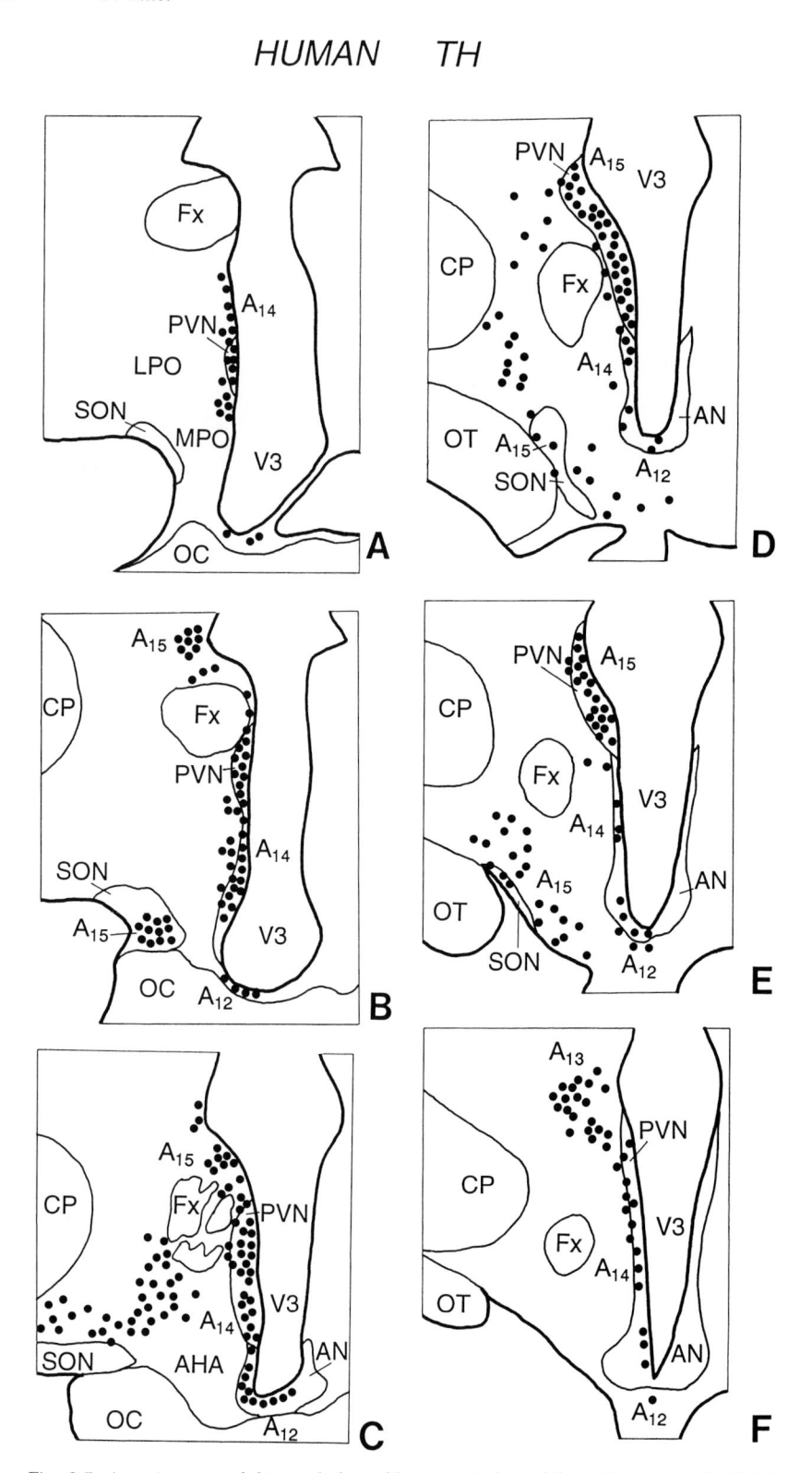

Fig. 9.5. Anterior part of diencephalon of humans (adapted from Spencer et al., 1985).

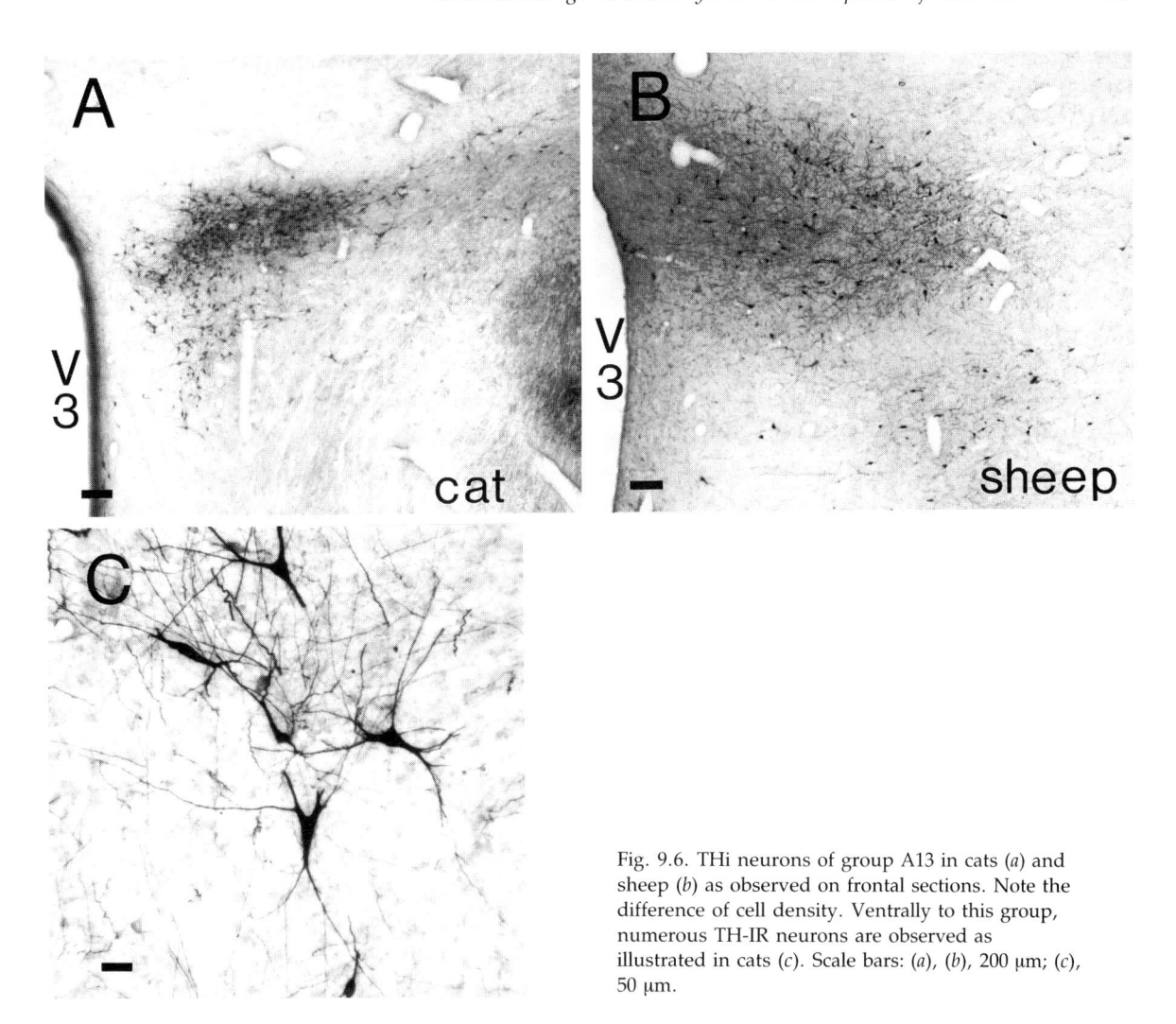

Fig. 9.6. THi neurons of group A13 in cats (a) and sheep (b) as observed on frontal sections. Note the difference of cell density. Ventrally to this group, numerous TH-IR neurons are observed as illustrated in cats (c). Scale bars: (a), (b), 200 μm; (c), 50 μm.

numerous THi neurons are not only observed in the dorsal and ventral parts of the wall of the third ventricle, but also in the lateral anterior hypothalamic area. In rats, the group A14 is subdivided into the dorsal (A14d) and ventral (A14v) parts (Hökfelt et al., 1984c). Contrary to other rodent species, the A14 group of guinea pigs contain less DAi neurons which are mainly located in the ventral part. According to these observations in different rodent species the presence of dopamine in all the neurons of this group remains questionable; these results might indicate a low level or an absence of dopamine in these neurons. Cats (Fig. 9.3(a)–(d); 3(a)–(d)) and sheep (Fig. 9.4(b)–(e)) present a similar pattern of distribution of THi neurons to those observed in guinea pigs: the perikarya are mainly found in the ventral part of the periventricular area. In addition, in the cat, some scattered THi neurons are seen in the lateral part. However, only those found close to third ventricle are also DAi (Kitahama et al., 1990) (Fig. 9.3(a), (a)). In the house shrew, the THi neurons lie close to the third ventricle, and the group A14 possesses a dorsal component (Karasawa et al., 1991), as in rats, but THi neurons are not found in the anterior hypothalamic area. In human, numerous THi neurons have been observed in the periventricular area (Figs. 9.5(a)–(f); 9.7(c)) (Spencer et al., 1985), but also in the lateral anterior hypothalamic area (Fig. 9.9(d)) (Pearson et al., 1990a).

In the lateral part of the A14 group a cluster of

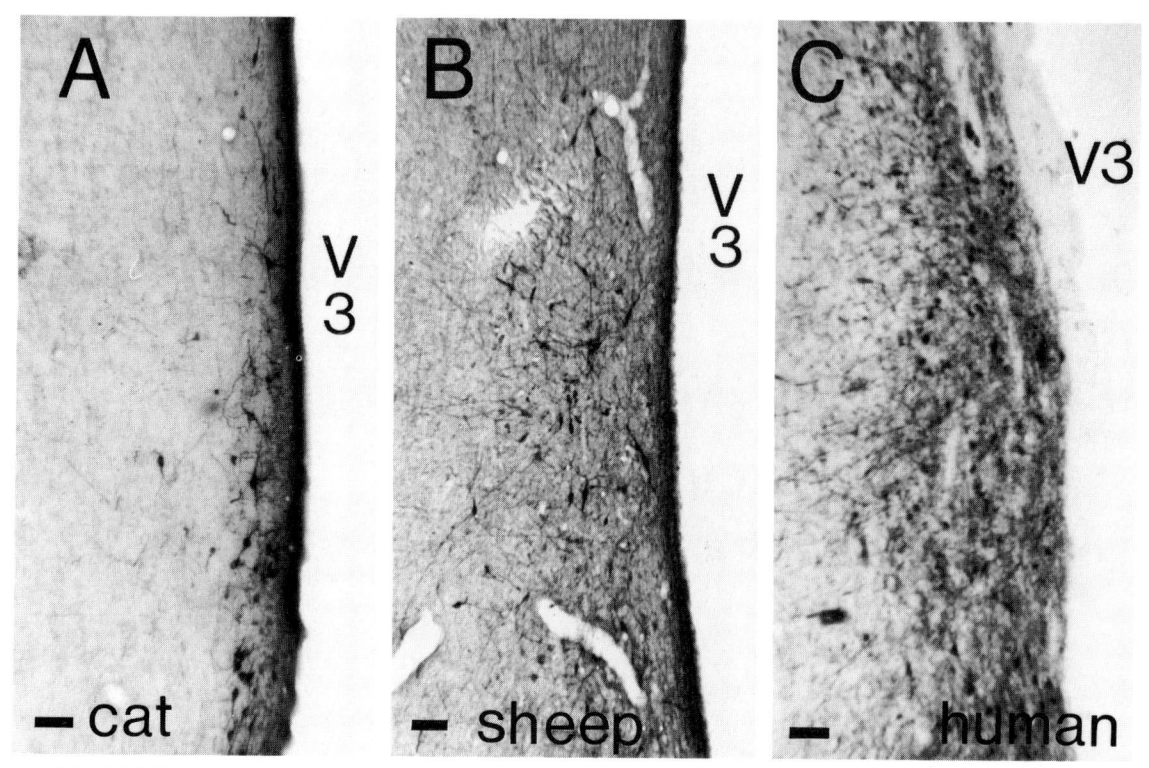

Fig. 9.7. THi neurons of the A14 group on the border of the third ventricle (frontal section) in cats (a), sheep (b), and humans (c). These neurons are mainly oriented parallel to the wall of the ventricle. Scale bars: (a), (b), (c), 100 µm.

THi neurons has been described in the medial preoptic area in rats and mice (Ruggiero et al., 1984). The medial preoptic nucleus, appears to be sexually dimorphic because in Nissl stained sections it contains a greater number of neurons in the male than in the female; the cytoarchitecture of this nucleus has been described in several rodents (Bleier, Byne & Siggelkow, 1982). Nevertheless, in the female rat, the number of THi neurons is higher than in the male (Simerly, Swanson & Gorski, 1985). The area corresponding to this structure appears to contain THi neurons in the hamster (Vincent, 1988), but lack DAi neurons in the guinea pig (Smits et al., 1990). In the cat, the anterior medial preoptic area contains TH and DAi neurons (Kitahama et al., 1990) although this area has not been characterized as sexually dimorphic. In sheep, the area related to the medial preoptic nucleus of rodents presents a few dispersed THi neurons, but a sexual dimorphism similar to those observed in rodents has not been evidenced in the cytoarchitecture of this area.

Group A15 The periventricular dopaminergic system contains another group which, in rat, is referred to as the A15 group (Hökfelt et al., 1984c). This cell group is composed of a dorsal and a ventral component. The dorsal subdivision lies rostrally to the ventral part of the bed nucleus of the stria terminalis and below the anterior commissure. The ventral subdivision lies primarily in the lateral retrochiasmatic area, but some neurons are also found in the supraoptic nucleus (SON) and above the optic chiasma. There is not a clear boundary between the rostral part of A14 and the caudal part of A15. The A15 group is well developed in the mouse (Fig. 9.2(a)–(d)), although its ventral part has been considered as a lateral extension of the arcuate nucleus by Ruggiero et al. (1984). In the hamster, this group is well evidenced with TH immunohistochemistry and numerous THi neurons are observed in the lateral preoptic area (Vincent, 1988). In the cat on the contrary, this group contains very few perikarya in its rostrodorsal part (Figs. 93 (a)–c; 3(a)–(c)), whereas in sheep THi cells are present only in the ventral subdivision (Fig. 94(e), 9.8(a)). In cats for example, the neuronal density could be increased after colchicine treatment (Kitahama et al., 1990). The true nature of this cell group is still questionable since in rodents (rat, guinea pig), the presence of dopamine immunoreactivity has not been observed (Mons, Tison & Geffard, 1990; Smits et al., 1990). In rats, some L-DOPAi neurons have been detected in this group (Mons et al., 1990) and in cats, some L-DOPA and DAi neurons were observed after colchicine treatment, but it has not been ascertained whether L-DOPA was the end product or a precursor of dopamine (Kitahama et al., 1990). In sheep, the presence of dopamine has been demonstrated by immunohistochemistry in the A15 group, as well as the presence of the enzymes involved in the synthesis of catecholamines (Tillet et al., 1990) (Fig. 9.8). In the house shrew, the exact product synthesized in the THi neurons observed in this nucleus has not been identified (Karasawa et al., 1991, 1992).

In human anterior diencephalon, the group A15 is specifically characterized by the presence of numerous THi neurons in the SON and paraventricular hypothalamic nucleus (PVN) (Figs. 9.5(a)–(e); 9.9(a)–(c)). Approximately 40% of the neurons of these groups were THi (Li et al., 1988; Panayotacopoulou et al., 1991). These neurons do not contain DBH immunoreactivity (Gaspar et al., 1985). The perikarya are large and possess the same morphological characteristics as the neurophysin-immunoreactive neurons. In other primates, numerous THi perikarya are also observed in the SON and PVN of cynomolgus macaques (Thind & Goldsmith, 1989). In the other species studied, only very few scattered THi neurons are observed in these nuclei under normal physiological conditions (Figs. 9.9(e), (f)), but their number increases dramatically when animals are subjected to hypersaline diet, as demonstrated in rats by immunohistochemistry (Kiss & Mezey, 1986), and in situ hybridization (Young III, Warden & Mezey, 1987). Under such conditions, the catecholamine end product of these neurons has not been clearly demonstrated although an increase in the synthesis of dopamine has been observed in dehydrated animals (Alper, Demarest & Moore, 1980). The presence of such a number of THi neurons in the SON and PVN of primates (including human) seems to be a characteristic of these species.

Both A14 and A15 periventricular groups have local projections i.e. in the diencephalon. It has been demonstrated in rat that A14 group innervates the neural and intermediate lobes of the pituitary (Kawano & Daikoku, 1987; Goudreau et al., 1992). In the cat, retrograde tract tracing methods have shown that the intermediate and neural lobe of the pituitary received THi fibers from both A14 and A15 groups (Luppi et al., 1986; Yoshimoto et al., 1990). In sheep, anterograde tracing studies from the retrochiasmatic area which contained the group A15 show similar results (Gayrard & Tillet, unpublished data). In the rats, other projection areas of these groups were found mainly in the diencephalon, i.e. the periventricular area, the preoptic area and the hypothalamus (Fuxe et al., 1985).

PNMT-like immunoreactive neurons

In rats treated with colchicine, the presence of PNMTi neurons has been demonstrated in various areas of the hypothalamus: around the fornix, in the lateral hypothalamic area near the dorsomedial hypothal-

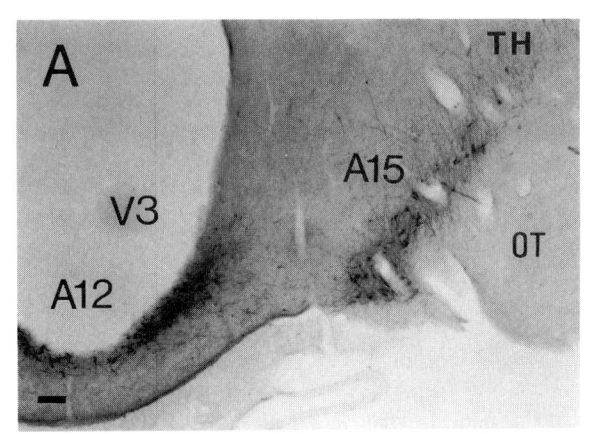

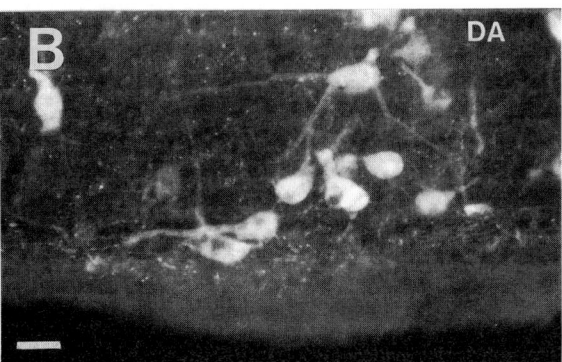

Fig. 9.8. Frontal sections through the mediobasal hypothalamus of sheep showing on (a), THi neurons of both A12 and A15 groups and on (b) DAi neurons of A15. Scale bars: (a), 200 μm; (b), 50 μm.

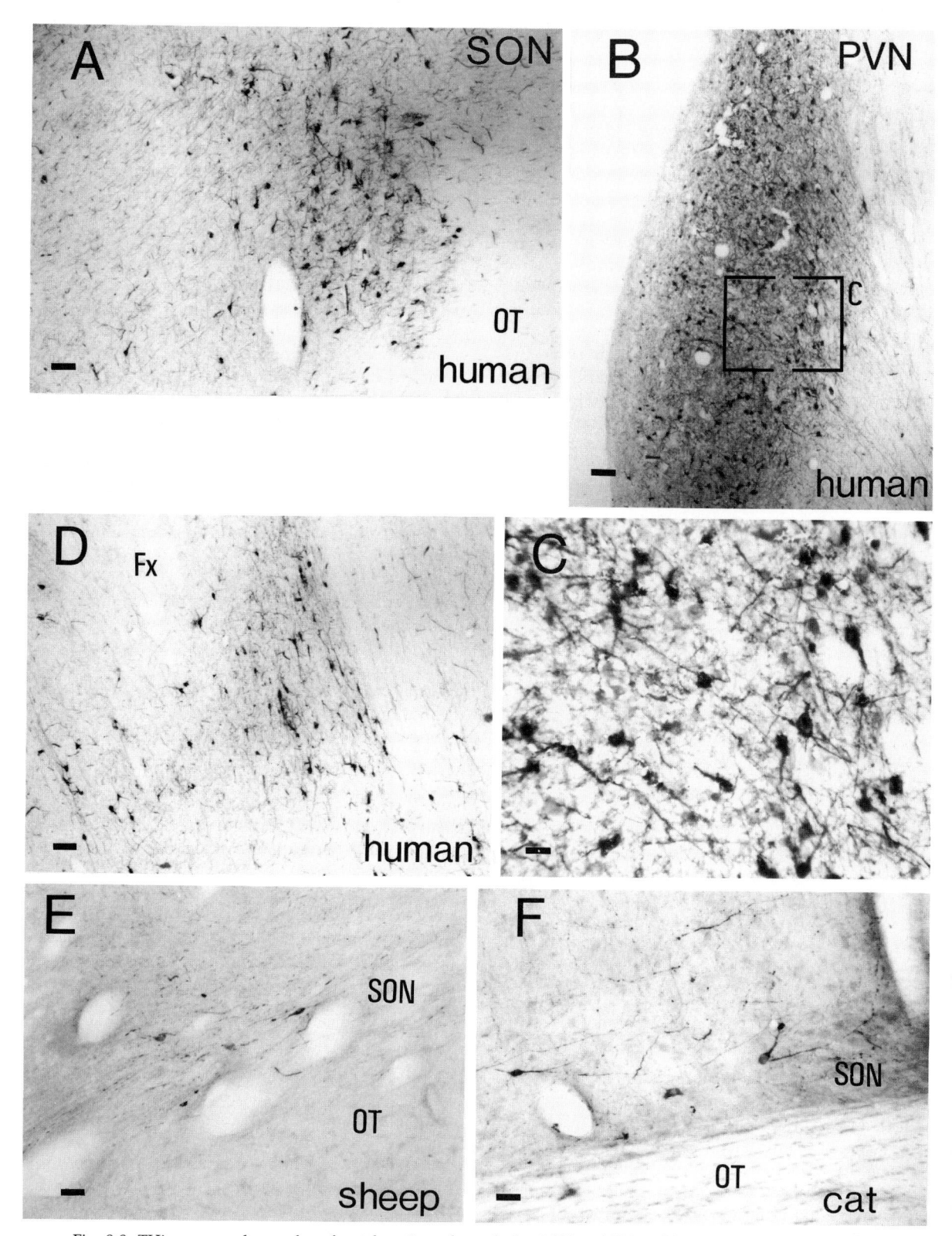

Fig. 9.9. THi neurons observed on frontal sections through the SON and PVN of humans (*a*), (*b*), (*c*), sheep (*e*) and cats (*f*). Note the numerous THi neurons of the SON (*a*) and PVN (*b*) of humans compared to the SON of sheep (*e*) and cats (*f*). In these latter species, only few cells are found in medial border of the nucleus. (*c*) higher magnification of (*b*). The lateral hypothalamic area between the SON and the PVN in humans also contains numerous THi perikarya, as shown in (*d*). Scale bars: (*a, b, d*), 200 μm; (*c*), (*e*), (*f*), 50 μm.

amic nucleus, in the dorsal hypothalamus in the medial and rostral zona incerta (Ross et al., 1984) and in the arcuate nucleus (Foster et al., 1985; Ruggiero et al., 1985). These neurons contain PNMT, but lack other enzymes of the catecholamine synthesis (TH, AADC, DBH). However, these neurons are not stained in untreated animals, and using different antisera, there is not good overlap between the different immunolabellings observed (Anderson & Howe, 1988). The discrepancies observed with different antisera in the labelling of rat hypothalamus has led the authors to hypothesize the existence of two different forms of PNMT in the medulla and the hypothalamus. Only antisera containing antibodies against common epitopes of both forms of PNMT would be able to recognize neurons in these two structures although the possibility of an artefactual staining cannot be excluded (Anderson & Howe, 1988).

Distribution of catecholamine-containing fibers in the diencephalon

In the different species, the distribution of catecholaminergic fibers has been less systematically and less accurately described than the distribution of perikarya. Moreover, the density of immunoreactive fibers is supposed to change in the same animal according to different physiological regulations or development. The different origins of the antisera used might also introduced variability in results. For all these reasons, comparisons of catecholaminergic innervations between different species have to be done with caution.

The distribution of dopamine-, noradrenaline-, and adrenaline-containing fibers was mainly studied using antisera against TH, DBH, and PNMT. Whereas the THi cell bodies in the diencephalon are generally considered to be dopaminergic (see above), the THi fibers may represent dopamine, noradrenaline as well as adrenaline fibers. At the microscopic level, proximal fibers corresponding to dendritic arborization appear without varicosities whereas distal fibers corresponding to axonal processes present numerous varicosities, DBHi varicosities seem to be larger in size than THi ones. Catecholaminergic fibers are mainly unmyelinated although some THi fibers surrounded with a thin sheath of myelin are demonstrated in the sheep hypothalamus by electron microscopy (Tillet & Thibault, 1993).

Within the diencephalon, catecholaminergic fibers were found in almost all the different nuclei, although the density of these fibers is not uniform. Dopaminergic fibers mainly originate within the diencephalon itself (Fuxe et al., 1985), whereas noradrenergic and adrenergic fibers arise from brainstem centers. The diencephalon is also traversed by the nigrostriatal bundles which run in proximity to the medial forebrain bundles. In rats and rabbits, the diencephalic noradrenergic fibers come mainly from the ventrolateral and dorsomedial medulla (Blessing et al., 1982; Day, Blessing & Willoughby, 1980; Sakumoto et al., 1978) but also from the locus coeruleus complex (Holets et al., 1988; Olson & Fuxe, 1972). In sheep, the preoptic area receives noradrenergic fibers equally from the locus coeruleus and the ventrolateral medulla (Tillet, Batailler & Thibault, 1993). Adrenergic innervation appears to originate from the caudal medulla (group C1, C2) in rat (Swanson et al., 1981; Sawchenko & Swanson, 1982, Palkovits et al., 1992) via the ascending medullary bundle (ventral tegmental bundle) (Astier et al., 1987). In fact, in other species such as guinea pigs or sheep, which have only a few PNMTi cells, the diencephalon does not seem to receive adrenergic inputs since PNMTi fibers have not been found (Tillet, 1988; Cumming et al., 1986). Moreover, in guinea pigs, adrenaline was not detected in brain tissue samples using HPLC assay for catecholamines (Cumming et al., 1986).

Caudal diencephalon

The *pineal gland*, which does not contain THi cells except for golden hamsters (see above), contains a dense plexus of TH/DBHi or FIF labeled fibers in all the species studied (Schröder, Fujisawa & Wollrath, 1987; Tillet & Thibault, 1989; Owman, 1965; Matsuura & Sano, 1983). They are mainly found around blood vessels and form small bundle-like arrangement (Fig. 9.10(a). These noradrenergic fibers do not appear to originate from the central nervous system but from the superior cervical ganglia of the peripheral nervous system which controls the secretion/synthesis of melatonin (Reiter, 1980). The level of noradrenaline turnover shows a 24-hour rhythm (Brownstein & Axelrod, 1974) although the pattern of aminergic innervation appears independent of the time of the day, at least in the rodent (Schröder, 1987).

The *mammillary bodies* receive a moderate density of THi and DBHi fibers as observed in rats (Chan-Palay et al., 1984; Ericson, Blomqvist & Köhler, 1989), sheep (Tillet & Thibault, 1989), cats (Cheung & Sladek, 1975). In monkeys, the mammillary complex contains little DBHi fibers (Ginsberg et al., 1993). In rats, the majority of the TH immunoreactive fibers appears to originate from the perimammillary area (Gonzalo-Ruiz et al., 1992). In sheep, the density is

224   Y. Tillet

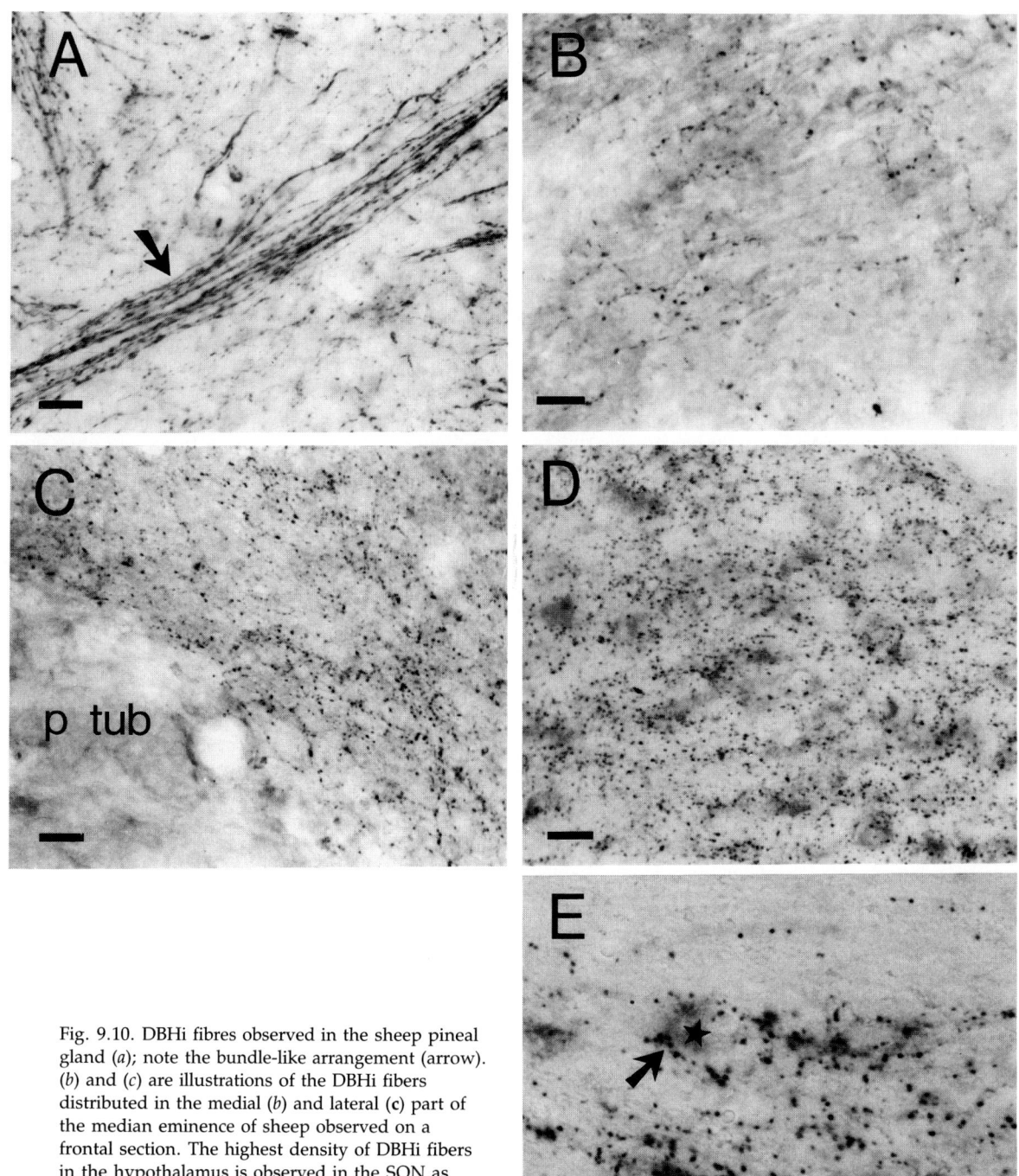

Fig. 9.10. DBHi fibres observed in the sheep pineal gland (*a*); note the bundle-like arrangement (arrow). (*b*) and (*c*) are illustrations of the DBHi fibers distributed in the medial (*b*) and lateral (*c*) part of the median eminence of sheep observed on a frontal section. The highest density of DBHi fibers in the hypothalamus is observed in the SON as illustrated in sheep (*d*) where close appositions (arrows) with immunonegative cells (stars) are observed (*e*). Scale bars: (*a*), (*c*), (*d*), 25 μm; (*b*), (*e*), 20 μm.

slightly lower than in rats. However, in both species immunoreactive fibers increase in number around the premammillary recess and in the supramammillary area. Very few PNMTi fibers have been described in the rat (Ericson, Blomqvist & Köhler, 1989).

Intermediate hypothalamus

Every hypothalamic region, in the different species studied contains THi and DBHi fibers, and no region could be found totally lacking catecholamine immunoreactive fibers; however, the density varies from region to region. Except in the medial forebrain bundles and surrounding areas, most of these fibers have numerous varicosities.

The *dorsal and lateral parts of the hypothalamus* receive moderate innervation from THi and DBHi fibers in rats (Swanson & Hartman, 1975; Chan-Palay *et al.*, 1984), humans (Pearson *et al.*, 1990*a*), monkeys (Ginsberg *et al.*, 1993) and sheep (Tillet & Thibault, 1989). The lowest density of TH/DBHi fibers is found in the ventromedial hypothalamic nucleus, in sheep and rats (Tillet & Thibault, 1989; Chan-Palay *et al.*, 1984). However, numerous fibers 'encapsulate' this nucleus and some of them cross its rostral part in rats (Chan-Palay *et al.*, 1984). In rats, the lateral hypothalamic area receives PNMTi fibers originating from the lower brainstem (Palkovits *et al.*, 1992).

The *median eminence* contains a high density of THi fibers mainly in the external layer of the palisade zone, where they are disposed parallel to each other. In all studied species this area contains the highest concentration of diencephalic THi fibers. Only rare DBHi fibers were observed there, but they are more numerous in sheep than in rats (Figs. 9.10(*b*), (*c*)). In rats as in cats, THi fibers of the median eminence are found to originate from A12 (Van Den Pol *et al.*, 1984; Yoshimoto *et al.* 1990). In rats, sparse PNMTi terminals are observed in the inner zone of the median eminence (Bosler, Beaudet & Denoroy, 1987; Hökfelt *et al.*, 1984*b*).

The *suprachiasmatic nucleus* shows one of the lowest THi fiber density in rats (Chan-Palay *et al.*, 1984) and sheep (Tillet & Thibault, 1989). They may originate locally from the suprachiasmatic nucleus (Van den Pol & Tsujimoto, 1985) or, as suggested by Kizer, Palkovits & Brownstein (1976) from the periventricular dopaminergic system of the midbrain. Conversely to the rat, the suprachiasmatic nucleus of the monkey receive a dense DBHi innervation (Ginsberg *et al.*, 1993).

Within the hypothalamus, one of the highest concentrations of catecholaminergic fibers is observed in the SON and PVN, in rats (Swanson & Hartman, 1975, Van Den Pol *et al.*, 1984; Chan-Palay *et al.*, 1984; Liposits, Phelix & Paull, 1986*a*), sheep (Tillet & Thibault, 1989), cats (Cheung & Sladek, 1975) and monkeys (Hoffman, Felten & Sladek, 1976; Ginsberg *et al.*, 1993). In rats and monkeys, THi, DBHi, and PNMTi fibers are abundant whereas in sheep only THi and DBHi fibers are found but the concentration of fibers restricted to these areas appear greater than in other species (Figs. 9.10(*d*)–(*e*)). The presence of a selective dopaminergic innervation of the PVN and SON nuclei has been demonstrated in rats after lesion of noradrenergic afferents at the rostral level of the mesencephalon (Lindvall, Björklund & Skagerberg, 1984; Liposits & Paull, 1989). This observation has been confirmed by direct observation of synaptic contacts between DAi or NAi fibers and neurons of these nuclei, as demonstrated by immunohistochemistry with antibodies raised against dopamine and noradrenaline (Buijs *et al.*, 1984; Decavel, Geffard & Calas, 1987). The noradrenergic fibers originate from the caudal medullary groups A1 and A2, via the medial forebrain bundles (Ungerstedt, 1971; Sawchenko & Swanson, 1982).

Anterior hypothalamus and preoptic area

In the anterior hypothalamus and preoptic area, the greatest density of DBHi fibers is observed in the medial part of these structures, around the preoptic recess of the third ventricle, in the organ vasculosum of the lamina terminalis, and in the bed nucleus of the stria terminalis close to the anterior commissure. The medial preoptic area appears more heavily innervated by DBHi fibers in sheep (Tillet & Thibault, 1989) and monkeys (Ginsberg *et al.*, 1993) than in rats. In these areas in sheep, the DBHi fibers are arranged in a basket-like structure, and appear to contact unlabeled perikarya (Figs. 9.11(*a*), (*b*)). These basket-like arrangement are commonly observed in rats (Swanson & Hartman, 1975), cats (Cheung & Sladek, 1975) and monkeys (Ginsberg *et al.*, 1993), and sheep (Tillet & Thibault, 1989). The lateral part of the rostral diencephalon contains a lower density of catecholaminergic fibers in rats than in sheep.

Thalamus

The thalamus contains less catecholaminergic fibers than the hypothalamus. In rats, sheep and monkeys the greatest density of fibers was observed in the PVT (Table 9.1). In rats this area also receives PNMTi fibers (Hökfelt *et al.*, 1984*b*; Astier *et al.*, 1987; Bosler *et al.*, 1987) but in sheep only DBH immunoreactive fibers are observed. In rats, THi fibers originate from the mesencephalic A8–A10 nuclei (Takada *et al.*,

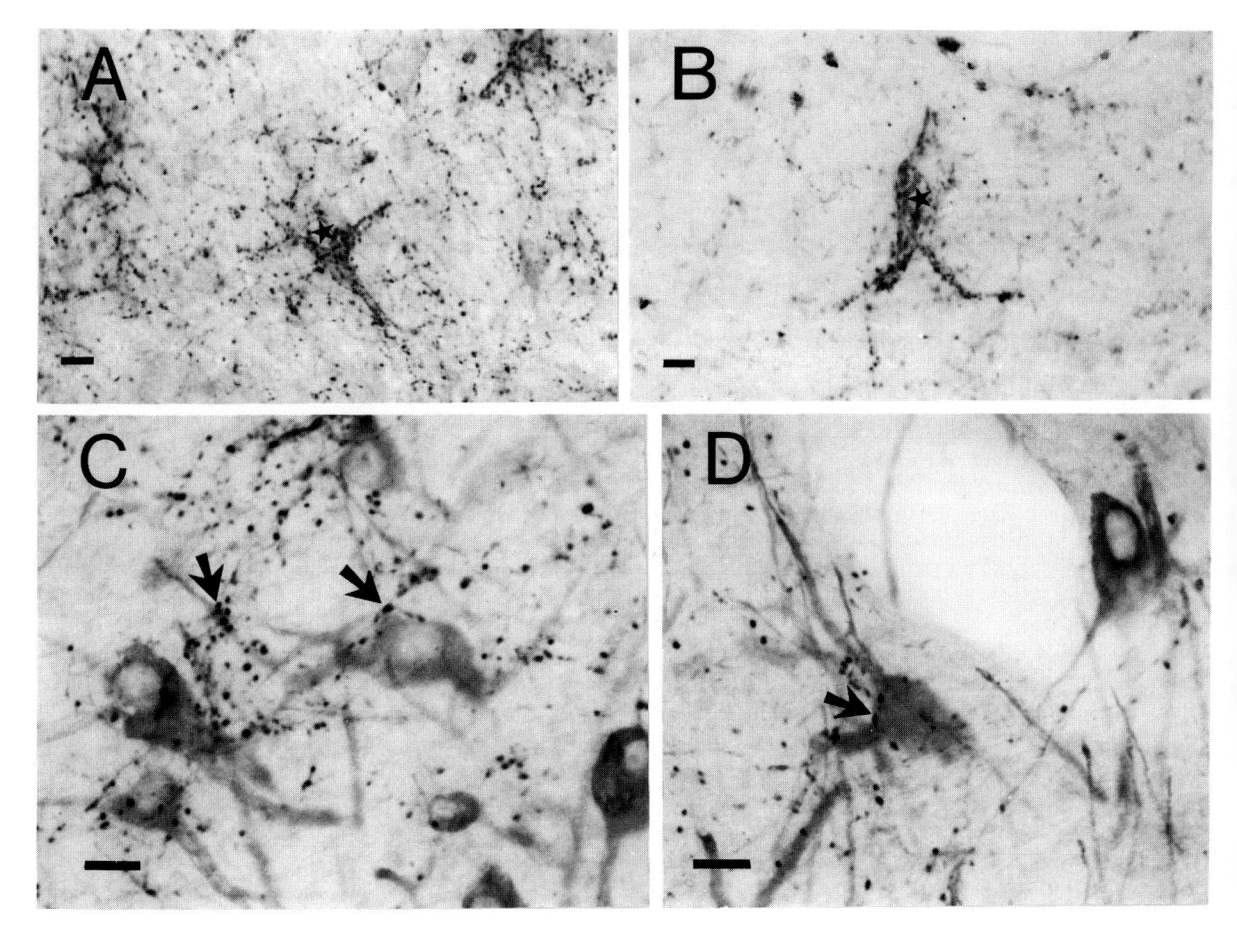

Fig. 9.11. DBHi fibres in the preoptic area and anterior hypothalamus of sheep. Basket-like structures of organized fibers near the bed nucleus of the stria terminalis (a) and in group A13 (b). These fibers appear to contact unlabeled perikarya (star). (c), (d) in group A15, double immunohistochemical labelings reveal that DBHi fibres (in black) are in close proximity with THi neurons (in grey). In some cases, synaptic contacts appear possible (arrows on (c) and (d)). Scale bars: (a), (b), 25 μm; (c), (d)), 20 μm.

1979). The majority of PNMTi fibers in the PVT originates from C1/C2 cell groups (Astier et al., 1987).

In the different mammalian species studied, the lateral and medial geniculate nuclei receive a dense network of THi/DBHi fibers. These fibers originate from the locus coeruleus (for review see Amaral & Sinnamon, 1977; Nieuwenhuys, 1985). In sheep (Tillet & Thibault, 1989) the density of DBHi fibers appears greater than in rats (Swanson & Hartman, 1975).

The thalamic reticular nucleus of the rat shows a dense plexus of DBHi fibers which originate from the locus coeruleus (Asanuma, 1992). Such a plexus of fibers has not been found in other mammalian species.

Innervation of catecholamine cell groups of the diencephalon

The noradrenergic innervation (demonstrated with DBH immunohistochemistry) of THi, dopamine cell groups appears different among the different species studied. In rats (Swanson & Hartman, 1975) the density of DBHi fibers is lower than in monkeys (Ginsberg et al., 1993) and sheep (Tillet & Thibault, 1989), where the infundibular nucleus (A12), the dorsomedial hypothalamus (A13), and the retrochiasmatic area (A15) receive numerous DBHi varicose fibers. In sheep, in these nuclei, these fibers appear to contact immunonegative perikarya or dendrites. Double labeling studies show close appositions of DBHi fibers on THi neurons (Figs. 9.11(c)(d)) and ultrastructural studies have demonstrated the presence of synaptic contacts between THi terminals and THi dendrites in the A15 cell group of the sheep (Tillet & Thibault, 1993). A similar observation (possible noradrenergic inputs onto dopaminergic neurons) has been also described in the A12 group of cats using formaldehyde-induced fluorescence (Cheung & Sladek, 1975).

In rats and cats, the nucleus A12 receives more DBHi fibers than the other diencephalic catecholaminergic nuclei (Cheung & Sladek, 1975). In rats and monkeys, the presence of synaptic contacts between THi fibers and THi perikarya indicating close relationships between supposed to be dopaminergic elements, has been confirmed at the electron microscopy level in the arcuate nucleus of rats (Leranth et al., 1985; Piotte et al., 1985) and in the ventral periventricular part of the hypothalamus of the monkey, (Thind & Goldsmith, 1986b) (see Table 9.4).

Interaction with other neuronal systems

Functional interactions between catecholaminergic systems and other neuronal groups can be appreciated by the morphological relationships between catecholaminergic neurons and other neurons characterized by their functions and/or their chemical contents. Colocalizations or demonstration of synaptic contacts involving catecholaminergic neurons illustrate these relationships.

In the last decade, numerous double immunohistochemical studies have demonstrated simultaneously the presence of neuropeptides or classical transmitters and catecholamines in the same neuron. Moreover, the presence of catecholaminergic afferents has been observed on numerous diencephalic neurons. From this point of view, one of the most extensively studied systems is the neuroendocrine system (for review see Hökfelt et al., 1986) and most of the colocalization concerns classical neurotransmitters (GABA, acetylcholine) and peptides (growth hormone-releasing hormone (GHRH), neurotensin, galanin . . .) (see Table 9.2).

However, studies of these relationships are mainly conducted in rodents and are more difficult in bigger animals such as primates or sheep, and they are quite impossible in human because of poor fixation. Consequently, the majority of the available data are derived from small species.

Monoaminergic systems (histamine, serotonin)

In the caudal diencephalon, the tuberomammillary area of mammals contains a dense gathering of histaminergic neurons (rat: Panula, Yang & Costa, 1984; cat: Lin et al., 1986; human: Panula et al., 1990). DBH/PNMTi fibers arising from the noradrenergic and adrenergic (C1/C3) groups of the caudal medulla synapse with unidentified neurons (Ericson et al., 1989). These two findings suggest possible interactions between noradrenaline/adrenaline containing neurons and histaminergic neurons, and consequently an influence of catecholamines on functions regulated by histamine. This might explain certain features of cardiovascular function in rats (Finch & Hichs, 1976), in which histamine as well as catecholamines are involved. Another example of such an interaction between catecholamine and histamine is the regulation of neuroendocrine and vegetative response (Robert & Calcutt, 1983; Schwartz, Garbarg & Pollard, 1986).

The coexistence of TH and serotonin has been demonstrated in fibers of the ventral periventricular area of baboons (Thind et al., 1987) but only in experimental animals with acute pituitary stalk section. In the pituitary gland of rats, TH and serotonin has been found in the same fibers (Saland et al., 1988). The presence of serotonin in neurons might result from uptake of the molecule. Therefore, the colocalization observed should be confirmed with the presence of tryptophan hydroxylase to insure that these serotoninergic fibers do synthesize their serotonin.

Other classical transmitters (GABA and acetylcholine)

The presence of *GABA* was demonstrated in the THi neurons of the arcuate nucleus of the rat (Everitt et al., 1984b) and in axon terminals of the median eminence in rats as well as cats, guinea pig, hare and rabbits (Schimchowitsch et al., 1991). Moreover, catecholamine fibers have also been shown to innervate GABAergic neurons in the mediobasal hypothalamus of monkey (Thind & Goldsmith, 1986a). These latter relationships have not been described in rodents.

The presence of choline acetyltransferase has also been described in a population of THi neurons of the rat arcuate nucleus (Tinner et al., 1989).

Peptidergic systems

Colocalization of catecholamines and peptides In rats, multiple peptides have been found to be colocalized in THi neurons of the arcuate nucleus (Everitt et al., 1986; Table 9.2). Concerning *neuropeptide Y* (NPY), numerous neurons are present in the arcuate nucleus (Nakagawa et al., 1985; Antonopoulos et al., 1989). Their co-existence with TH is controversial since

Table 9.2. *Colocalization of catecholamines and transmitters in perikarya of the diencephalon*

Amine[a]	Transmitter	Areas	Species	References
TH	GHRH	PVN	rat	Horvath et al., 1989
TH	GHRH	AN,TIN	rat	Meister et al., 1985, 1986
TH	GHRH	TIN	rat	Niimi et al., 1992
TH	GHRH	AN	rat	Okamura et al., 1985
TH	CCK	SupraMmb	rat	Seroogy et al., 1988
TH	PHI/VIP	SupraMmb	rat	Seroogy et al., 1988
TH	Gal	AN	rat	Melander et al., 1986
TH	NPY	PeVN,AHA	rat, mouse	Ciofi et al., 1991[c]
TH	NPY	AN,PVN[b]	rat, mouse	Ciofi et al., 1991[c]
TH	GABA/GAD	ME	rat, cat	Schimchowitsch et al., 1991
TH	GABA/GAD	ME	guinea pig	Schimchowitsch et al., 1991
TH	GABA/GAD	ME	hare	Schimchowitsch et al., 1991
TH	GABA/GAD	ME	rabbit	Schimchowitsch et al., 1991
TH	GABA	AN	rat	Everitt et al., 1984b
TH	SRIF	POA, PeVN	rat	Sakanaka et al., 1990
TH	SRIF	PeMmb	rat	Sakanaka et al., 1990
TH	ChAt	AN	rat	Tinner et al., 1989
TH	NT	AN, PeVN	rat	Ibata et al., 1983
TH	NT	AN	rat	Hökfelt et al., 1984a
TH	NT+GHRH	AN	rat	Everitt et al., 1986
TH	Gal+GHRH	AN	rat	Everitt et al., 1986
TH	GAD	AN	rat	Everitt et al., 1986
TH	ENK-8,DYN	AN	rat	Everitt et al., 1986
TH	Leu-ENK	AN	rat	Everitt et al., 1986
TH	OXT	PVN	rabbit	Schimchowitsch et al., 1983
TH	OXT	PVN	human	Li et al., 1988

[a] or aminergic markers like catecholaminergic synthesizing enzymes.
[b] periphery of the paraventricular nucleus.
[c] Colocalization observed with an anti Pre-Pro-NPY in lactating animals; Everitt et al., 1984a failed to observed NPY-TH colocalization.

Everitt et al. (1984a) failed to observe such colocalization, and Ciofi et al. (1991) described it only in lactating rodents. *Neurotensin* has been observed in THi neurons of the arcuate and periventricular nuclei of rats (Ibata et al., 1983; Hokfelt et al., 1984a). Numerous studies have demonstrated the co-existence of *GHRH* with dopamine in neurons of the rat mediobasal hypothalamus, particularly in the arcuate nucleus (Table 9.2). Dopamine was also observed colocalized with *somatostatin* (SRIF) in the preoptic periventricular region and around the ventromedial nucleus of the hypothalamus but not in the arcuate nucleus (Sakanaka, Magari & Inoue, 1990). In neurons of the PVN, TH is present together with oxytocin in human (Li et al., 1988) and rabbit (Schimchowitsch et al., 1983). In this latter species, only a small number of neurons contains both antigens. In human as in other mammalian species studied, TH is never observed in vasopressin containing neurons (Li et al., 1988). Other peptides like *opiates, peptide histidine-isoleucine* (PHI) or *vasoactive intestinal polypeptide* (VIP) are found within the diencephalic dopaminergic neurons (Table 9.2), but they represent only a small percentage of double labeled neurons. The physiological relationships between these peptides and dopamine are not understood.

Catecholaminergic inputs on peptidergic neurons
Numerous peptidergic neurons of the hypothalamus receive catecholaminergic afferents (Table 9.3) as it has been demonstrated by combined immunocytochemical and autoradiographic methods. The peptides involved in these relationships are mainly releasing or inhibiting factors of the hypothalamo-

Table 9.3. *Chemically specified intercellular relationships between catecholaminergic neurons and other neuronal systems in the diencephalon*

Substance involved	Areas	Catecholaminergic afferents			Species	References
		DA/TH	NA/DBH	Ad/PNMT		
ACTH	AN	+			rat	Kosawa & Nakaï, 1987
ACTH	AN	+			rat	Baker et al., 1986
LHRH	ME	+			rat	Ajika, 1979
LHRH	POA	+			rat	Chen et al., 1989
LHRH	AntHyp	+			rat	Hoffman et al., 1982
LHRH	Sept/Hyp	+			rat	Jennes et al., 1983
LHRH	ME	+			rat	Jennes et al., 1983
LHRH	ME	+			sheep	Kuljis & Advis, 1989
LHRH	POA	+	+		sheep	Tillet et al., 1989
GHRH	AN			+	rat	Liposits et al., 1989
GHRH	AN	+	+		rat	Sato et al., 1989[a]
SRIF	PeVN			+	rat	Liposits et al., 1990
NPY	AN	+			rat	Guy & Pelletier, 1988
POMC	PeVN		+		rat	Watson et al., 1980
Met-Enk	MDN	+	+		guinea pig	Mitchell et al., 1988[a]
TRH	PVN		+		rat	Liao et al., 1991
TRH	PVN		+		rat	Liposits et al., 1987
TRH, LHRH	ME	+			rat	Ajika, 1980
SRIF	ME	+			rat	Ajika, 1980
VP, OXT	PVN, SON	+	+		rat	Nakai et al., 1986[a] (review)
LHRH	POA, DBB	+	+		rat	Nakai et al., 1986[a] (review)
SRIF	PON, AN		+		rat	Nakai et al., 1986[a] (review)
TRH, CRF	PVN		+		rat	Nakai et al., 1986[a] (review)
ACTH	AN	+			rat	Nakai et al., 1986[a] (review)
β-Endo	AN	+			rat	Nakai et al., 1986[a] (review)
NT	PVN, AN	+			rat	Nakai et al., 1986[a] (review)
GABA	AN	+			monkey[b]	Thind & Goldsmith, 1986a
VP, OXT	PVN, SON		+[c]		monkey[b]	Sladek & Zimmerman, 1982
VP	PVN		+		rat	Silverman et al., 1983[a], 1985[a]
VP	PVN		+		rat	Nakada & Nakai, 1985[a]
VP	PVN		+		rat	Ochiai & Nakai, 1990
VP, OXT	PVN	+	+		rat	Hornby & Piekut, 1987
CRF	PVN	+			rat	Liposits et al., 1986b
CRF	PVN			+	rat	Liposits et al., 1986c, 1989
CRF	PVN	+			rat	Hornby & Piekut, 1989

[a] catecholaminergic afferents were demonstrated autoradiographically.
[b] *Macaca mulata*.
[c] catecholamines were revealed by FIF method.

hypophysial axis (GHRH, SRIF, CRF, TRH . . .) and endogenous opiates or other regulatory peptides (neurotensin, enkephalin, NPY . . .). Most of these studies have been done at the ultrastructural level in rats and guinea pigs (Table 9.3). In sheep, appositions between catecholaminergic afferents and luteinizing hormone-releasing hormone (LHRH) neurons have been observed in the medial preoptic area but the presence of synaptic contact between these elements needs confirmation at the electron microscopy level (Tillet, Caldani & Batailler, 1989).

The major catecholaminergic inputs on a peptidergic neuronal group might be on the vasopressin/oxytocin neurons of the SON and PVN and the corticotropin releasing factor (CRF) neurons of the PVN where the presence of synaptic contacts of noradrenergic, adrenergic and dopaminergic fibers has been demonstrated in rats (Buijs et al., 1984; Liposits, Phelix & Paull, 1986b; Liposits et al., 1986c; Decavel et al., 1987). The presence of synaptic contacts have also been demonstrated between PNMTi fibers and thyrotropin releasing hormone (TRH)-containing neurons (Shioda et al., 1986; Liposits et al., 1987). Both noradrenergic and adrenergic fibers originates from the A2/C2 groups (Ungerstedt, 1971; Sawchenko & Swanson, 1982) whereas dopamine originates from the ventral tegmental area (Iijima & Ogawa, 1981).

Catecholamines and steroids
Combined autoradiography of sex steroids and FIF method or immunohistochemistry for catecholaminergic synthesis enzymes have shown that dopaminergic and noradrenergic neurons are able to concentrate gonadal steroids in rats (Heritage, Grant & Stumpf, 1977; Sar, 1983). Similar findings have been observed in sheep using double immunohistochemical labeling (Batailler et al., 1992). In both cases, the steroid receptors are found in the arcuate nucleus where less than ten per cent of these dopaminergic neurons were doubly labeled.

Autoradiography with progesterone or its analogs, and immunohistochemical stainings for progesterone receptors have shown that a subpopulation of dopaminergic neurons of the arcuate nucleus in guinea pigs (Blaustein & Turcotte, 1989), rats (Sar, 1988), and monkeys (dorsal part of A12) (Kohama, Freesh & Bethea, 1992) are able to bind this steroid. In addition, these doubly labeled neurons are found in the A13 and A14 groups in rats and in the group A11 in the monkey. However, the percentage of double labeled neurons varies in these different species: it is low in guinea pigs, higher in rats (from 45% to 90%) and doubly labeled neurons are observed in the monkey only after estradiol and progestin treatment (Kohama et al., 1992).

In the ventrolateral hypothalamus of female guinea pig, DBHi fibers are observed in close association with estrogen receptor-immunoreactive cells (Tetel & Blaustein, 1991). This observation may support a direct action of noradrenaline on these cells.

Afferents to catecholaminergic neurons (Table 9.4)
Catecholaminergic neurons also receive synaptic contacts from peptidergic and other neuronal systems. Hypophysiotropic peptide-containing fibers like LHRH are observed in close proximity to dopamine neurons of the septo-hypothalamic area of rats (Jennes, Stumpf & Tappaz, 1983). In this species, the presence of CRF fibers on THi neurons has been suspected in the PVN (Liposits et al., 1986b, 1986c) and it has been confirmed in group A14 of the monkey (Thind & Goldsmith, 1989). Opiate peptide-containing fibers are also observed in close proximity to dopaminergic neurons of both A12 and A14 in rodents (Table 9.4), but these relationships are not observed in other species. NPY containing neurons, which receive dopaminergic afferents (Table 9.3), give synaptic inputs to dopaminergic neurons of the arcuate nucleus of rats (Guy & Pelletier, 1988). In addition to the colocalization of NPY and TH in the same neurons of the arcuate nucleus of lactating rodents (Ciofi et al., 1991), these observations point out the close relationships between this peptide and dopamine.

More classical transmitters (serotonin, GABA) have been observed synapsing with dopaminergic neurons. In rats, serotonin afferent fibers are observed not only on dopaminergic neurons of the arcuate nucleus (Kiss & Halasz, 1986), but also in the zona incerta and on dopaminergic fibers of the median eminence (Bosler, Joh & Beaudet, 1984).

GABA, in addition to its colocalization in TH-containing neurons (Table 9.2), is also observed in fibers synapsing with THi neurons in the ventral periventricular area of the third ventricle of monkeys (Thind & Goldsmith, 1986a) and in groups A12, A13, and A14 of rats (Van Den Pol, 1986).

Functional implications

Catecholaminergic neurons are involved in the control of many diencephalic functions, and most of these concern the control of pituitary hormone secretion (for review see Weiner & Gannong, 1978; Tuomisto & Männisto, 1985; Nakai et al., 1986; Weiner, Findell & Kordon, 1988).

Catecholamines and reproduction
The morphological relationships observed between catecholamines and the LHRH neuronal system point out the influence of catecholamine in the regulation of reproduction. Since the first observation of Sawyer, Markee & Hollinshead (1947), the role of catecholamines has been evidenced in rodents, primates (for review see Weiner et al., 1988) and sheep (for review see Thiéry & Martin, 1991). Their control of reproduction is conducted mainly through the

Table 9.4. *Identified transmitter containing afferents on dopaminergic neurons of the diencephalon*

Afferents neurones involved	Areas	Species	References
ACTH	AN	rat	Morel & Pelletier, 1986
ACTH	A12/A14	rat	Fitzsimmons et al., 1992[a]
DYN	A12/A14	rat	Fitzsimmons et al., 1992[a]
Met-Enk	A12/A14	rat	Fitzsimmons et al., 1992[a]
β-Endo	PeVN, ME[b]	rat	Horvath et al., 1992
NPY	AN	rat	Guy & Pelletier, 1988
CRF	A14	monkey[c]	Thind & Goldsmith, 1989
LHRH	Sept/Hyp	rat	Jennes et al., 1983
GAD	A12/A14	rat	Van Den Pol, 1986
GAD	DMH	rat	Van Den Pol, 1986
GAD	PeVN	monkey[c]	Thind & Goldsmith, 1986a
5HT	AN, ME, ZI	rat	Bosler et al., 1984
5HT	AN	rat	Kiss & Halasz, 1986
TH	AN	rat	Leranth et al., 1985
TH	AN	rat	Piotte et al., 1985
TH	PeVN	monkey	Thind & Goldsmith, 1986b
TH/DBH	A15	sheep	Tillet & Thibault, 1993

[a]afferents are more numerous in A12 than in A14.
[b]only in the lateral part of the ME.
[c]*Macaca fascicularis*.
[d]close proximity between DBH fibers and TH neurons were observed by light microscopy, but synapses between TH terminals and TH neurons were seen by electron microscopy.

LHRH/luteinizing hormone (LH) system and prolactin (PRL) secretion and these hormones are also controlled by sex steroids as initially demonstrated in sheep (Hammond, Hammond & Parkes, 1942).

Catecholamines and gonadal steroids. Double labeling has shown that catecholamines are able to concentrate sex steroids or contain steroid receptors (see above). Consequently, these steroids could modify the activity of catecholaminergic neurons. In A13 of steroid-treated rats, the turnover of dopamine is modified (Gunnett, Looklingland & Moore, 1986) and TH immunostaining is increased by gonadal steroids in A13 in both male and female rats (Sanghera et al., 1991). Moreover, changes in dopaminergic neuronal activity in the zona incerta, (A13) are correlated with the estrous cycle in female rats (MacKenzie, James & Wilson, 1988). Biochemical investigations have also demonstrated that testosterone inhibits the basal activity of tuberoinfundibular neurons in rats (Toney, Looklingland & Moore, 1991). At the ultrastructural level, morphological studies of neurons of the arcuate nucleus have shown a synaptic remodeling of cytoplasmic membrane after estradiol treatment or during the estrous cycle in rats (García-Segura, Baetens & Naftolin, 1986; Olmos et al., 1989). Thus, it appears that catecholamines mediate the role of steroids.

In addition, catecholamines may modulate the central effect of steroids. The concentration of functional estrogen receptors in the mediobasal hypothalamus is modulated by noradrenaline as demonstrated in female rat treated with DBH inhibitors (Blaustein, Brown & Swearengen, 1986). Morphological evidence for a direct effect of noradrenaline on steroid sensitive cells has been observed in the ventrolateral hypothalamus of the guinea pig (see above; Tetel & Blaustein, 1991).

Catecholamines and LHRH/LH. Catecholamines exert their regulatory effects on pulsatile LHRH and LH secretion, which is an important characteristic of their release (Caraty, Orgeur & Thiéry, 1982; Clarke & Cumming, 1982). However, the discrete role of catecholamines is still controversial, and it depends mainly on the physiological status of the animal (for review see Barraclough & Wise, 1982; Weiner et al.,

1988). To summarize, noradrenaline supresses pulsatile release of LH in ovariectomized female rats, but stimulates the LH secretion in estrogen treated animals. During preovulatory LH surge, noradrenaline and adrenaline stimulate the activity of LHRH neurons (Weiner et al., 1988). Dopamine inhibits pulsatile release of LH with or without steroids (Barraclough & Wise, 1982; Weiner et al., 1988). In human, an inhibitory role of dopamine is also suspected, but in monkeys, electrophysiological and pharmacological studies suggest a stimulatory role of both noradrenaline and dopamine on pulsatile release of LH. Noradrenaline involved in this regulation originates from the ventrolateral medulla (groups A1/A2) and dopamine appears to originate from the incerto-hypothalamic and periventricular dopaminergic systems (A13/A14) (MacKenzie et al., 1988).

In sheep, in addition to the role previously described for rodents and primates, dopamine is involved in the seasonal control of reproduction (Meyer & Goodman, 1986; for review see Thiéry & Martin, 1991), since it has been shown that the seasonal inhibition of pulsatile LH release induced by steroids is mediated by the dopaminergic neurons of the group A15 (Thiéry et al., 1989). The role of this dopaminergic group has not been examined in rats or monkeys, in which the reproduction is independent of the photoperiod.

Catecholamines and prolactin. In contrast to LH, PRL secretion is tonically inhibited by dopamine released from the tuberoinfundibular neurons (Fuxe, Hökfelt & Nilsson, 1969; Ben-Jonathan, 1985). In turn, dopaminergic neuron activity is influenced by PRL level (Gudelski & Porter, 1980). The role of noradrenaline is not so clear since noradrenaline can have stimulatory or inhibitory effects by acting through distinct intermediate neuronal systems (for review see Weiner et al., 1988).

The observations of GABA and acetylcholine as well as peptides in the dopaminergic neurons of the arcuate nucleus indicate possible interactions of these transmitters in control of LH or PRL secretion. This has been suspected in rats for GABA and for acetylcholine (Locatelli et al., 1985; Fuxe et al., 1988; for review see Weiner et al., 1988). Among peptides, neurotensin (Koenig et al., 1982; Tojo et al., 1986), NPY (Ciofi et al., 1991) and galanine (Sahu et al., 1987) have been found to participate to LH and PRL regulation.

Interactions between catecholamine–and opiate peptide-containing neurons

Dopaminergic neurons of the hypothalamus receive β-endorphin-containing afferents and send efferents towards these opiate-containing neurons in the hypothalamus (Tables 9.3 and 9.4). These morphological relationships point out physiological relationships between these neuronal systems. For instance it was shown in rats that dopamine decreases the level of β-endorphin content in the mediobasal hypothalamus (Locatelli et al., 1983). In a recent study, it has been demonstrated by *in situ* hybridization that dopamine might mediate the central action of sex steroids on the regulation of proopiomelanocortine (Tong & Pelletier, 1992). β-endorphin plays a major role in the preovulatory surge of LH (Kalra & Kalra, 1983) and this action is concomitant with enhanced output of dopamine from mediobasal hypothalamus and preoptic area (Kalra & Kalra, 1983). Moreover intracerebroventricular injection of β-endorphin induces a suppression of dopamine release, as well as modifications of dopamine turnover (Haskins et al., 1981; Lohse & Wuttke, 1981). Therefore, it appears that dopamine neurons mediate the effect of β-endorphin on gonadotropin secretion. Similarly dopamine neurons mediate the stimulatory effect of β-endorphin on PRL secretion (Bruni et al., 1977; Ferland et al., 1977) through an inhibition of dopamine release (Enjalbert et al., 1979).

Interaction between catecholaminergic neurons

The presence of synaptic contacts between THi terminals and THi perikarya in the arcuate nucleus of rat and in the ventral periventricular area of the monkey (Table 9.4) could constitute an anatomical support for a network between dopaminergic neurons. The role of such network might be the synchronization of neurons of the same group. (Leranth et al., 1985; Piotte et al., 1985; Piotte, Beaudet & Brawer, 1988). In sheep, whose THi neurons of A15 appear to receive DBH afferents (Tillet & Thibault, 1993), such contacts might support a regulatory role of noradrenaline on dopamine neurons. In this species, Goodman (1989) observed that dopamine and noradrenaline act 'in series' to inhibit LH pulsatility during anestrous. This author concluded that noradrenaline stimulates dopamine release which inhibits LH secretion, and moreover, this interaction takes place in the hypothalamus (Havern, Whisnant & Goodman, 1991). According to these observations, A15 might be the site of this interaction.

Interaction between catecholamine and serotonin

In rodents, serotoninergic inputs are observed on dopaminergic neurons involved in the control of pituitary hormone release and it has been hypothesized that the action of serotonin on pituitary hormones is mediated by dopamine neurons (Kordon et al., 1980). In female rats, functional relationships between dopamine and serotonin appears to exist since it was shown that the central administration of serotonin decreases TH catalytic activity and mRNA levels in the tuberoinfundibular dopaminergic neurons (Mathiasen, Arbogast & Vogt, 1992).

Interactions between catecholamines and GABA-containing neurons

In rat as in monkey hypothalamus, the presence of GABA containing terminals on dopaminergic neurons was observed (Table 9.4). GABA terminals inhibit dopaminergic neurons and the function of this innervation can be considered in the light of the well-known function of dopaminergic neurons themselves in the arcuate and periventricular nuclei. GABAergic fibers innervating dopaminergic neurons of the arcuate nucleus might be involved in the control of PRL secretion while those innervating dopaminergic neurons of the periventricular and dorsomedial hypothalamic area might control the LHRH secretion (Jennes et al., 1983). Numerous physiological data have demonstrated the influence of GABA in such regulation (for review see Weiner et al., 1988) as well as in the regulation of other pituitary hormone secretion (for review see Elias et al., 1982; Racagni et al., 1982)

Interactions between catecholamine- and neurotensin-containing neurons

Dopamine neurons are directly influenced by neurotensin which induces their activation (Gudelsky, Berry & Meltzer, 1989). Therefore, the regulatory effect of neurotensin in different functions appears, at least partly, mediated by dopaminergic neurons and might be explained by the known effect of the dopaminergic neurons themselves. Such a way of regulation has been demonstrated for neurotensin inhibition of PRL release which results in an activation of the inhibitory dopaminergic neurons for PRL secretion (Tojo et al., 1986; Koenig et al., 1982). The colocalization of neurotensin and dopamine in the same neuron (Table 9.2) indicates that neurotensin may be more generally considered as a modulator in dopaminergic tuberoinfundibular neurons (Hökfelt et al., 1984a), since neurotensin effects on anterior pituitary hormone secretion (LH, PRL, thyrotropin stimulating hormone (TSH)) are counteracted by TH inhibitors and dopamine receptor blocking agents (for review see Fuxe et al., 1984). Acting as a comodulator of dopamine secretion in dopaminergic tuberoinfundibular neurons, neurotensin might enhance postsynaptic dopaminergic receptor activity in the median eminence as hypothesized by Fuxe et al. (1984). It is also thought that neurotensin might stimulate dopamine neurons through its blocking effect on inhibitory dopaminergic somatodendritic autoreceptors as observed in the mesolimbic system (for review see Nemeroff & Cain, 1985; Shi & Bunney, 1992).

Interactions between catecholamines and NPY-containing neurons

NPY was observed uniquely in dopaminergic neurons of the hypothalamus of lactating rodents (Ciofi et al., 1991); this indicates that this peptide is involved in the control of PRL secretion. In tuberoinfundibular dopaminergic neurons, this peptide might play a role as a modulator of dopamine in the control of PRL release. The presence of NPY-containing afferents on THi neurons (Guy & Pelletier, 1988) shows that NPY could also directly modulate dopaminergic neuronal activity, as described for neurotensin and β-endorphin (see above). Such an effect is described after NPY injection in the lateral ventricle of rat, which induces a change in amine utilization in the tuberoinfundibular neurons (Harfstrand et al., 1987a, b). In rats, NPY and noradrenaline are involved together in the control of food intakes and appetite (Leibowitz, 1989) and reproduction (Crowley 1987; Allen, Crowley & Kalra, 1987). The mechanism of NPY action on catecholaminergic neurons is not well understood but Allen et al. (1987) have hypothesized that NPY action could depend on mechanism involving post synaptic adrenergic receptors.

Interaction between catecholamines and peptide-containing neurons of the PVN and SON

The observation of synaptic inputs of dopaminergic and/or noradrenergic terminals on vasopressin, oxytocin, and CRF neurons indicate clearly that these neurons are directly controlled by these catecholamines. An excitatory influence of dopamine has been demonstrated by a direct application of dopamine on these neurons (Moss, Urban & Cross, 1972; Mason, 1983). It was also shown that hypothalamic dopamine facilitates the reflex release of oxytocin induced by suckling and it also regulates the release of vasopressin in response to osmotic changes (Moos & Richard, 1982). In rats, noradrenaline is involved in the control of vasopressin release (Kühn, 1974; Sladek, Armstrong & Sladek, 1983), and adrenaline appears also to have a stimulatory effect on CRF/adrenocorticotropic hormone (ACTH) secretion (for review, see Al-Damluji, 1988). Using electrophysiological methods, it was demonstrated that stimulation of noradrenergic groups A1 or A2 (which innervate medial part of the PVN) facilitates firing of paraventricular tuberoinfundibular neurons (Day, Ferguson & Renaud, 1985). These observations support the hypothesis of a facilitatory role of these noradrenergic structures on ACTH secretion (Fuxe et al., 1970; Hedge, Van Ree & Versteeg, 1976).

In rodents, the number of THi neurons in the PVN and SON is low, but it can be increased by giving animals 2% NaCl to drink or by section of the ascending noradrenergic/adrenergic fibers from the brainstem, the section being performed 1 mm in front of the lambdoid suture (Kiss & Mezey, 1986). This increase of THi neurons is the result of a new synthesis of TH, since under the conditions of hyperosmotic fluid intake the level of TH mRNA is increased (Young III et al., 1987). Moreover, an increase in dopamine level is observed in dehydrated animals (Alper et al., 1980). It is therefore hypothesized that TH (and dopamine) is involved in the regulation of salt balance in the PVN and SON and that the signal generated by salt loading may be conveyed by noradrenergic/adrenergic fibers, at least in part, from the brainstem to the PVN and SON. Destruction of noradrenergic afferents by injection of 6-hydroxydopamine into the ventral noradrenergic bundle decreases the level of CRF in the PVN (Guillaume et al., 1987). The relationships between TH and osmoregulation is also demonstrated in Brattleboro rats. These animals (which lack vasopressin) possess more THi neurons than normal rats and a lower catecholaminergic innervation of the SON (Scholer & Sladek, 1981). After treatment with vasopressin, the number of THi neurons and catecholaminergic innervation return to control level (Kiss & Mezey, 1986). In man, the normal number of THi neurons is higher than in other species (see above). Since TH is colocalized with oxytocin (Li et al., 1988; Schimchowitsch et al., 1983), it may be hypothesized that dopamine may be released in the neural lobe of the pituitary and dopamine may modulate oxytocin release.

The noradrenergic/adrenergic fibers of the PVN and NSO originate from the nucleus of the solitary tract and the ventrolateral medulla (A1) which receive primary visceral afferents inputs (see Swanson et al., 1981, Sawchenko & Swanson, 1982). Therefore, the dense catecholaminergic innervation of both PVN and SON is involved in the integration of hypothalamo-neuroendocrine and autonomic response to visceral stimuli (for review see Swanson & Sawchenko, 1980; Sawchenko & Swanson, 1981, 1982). In fact, these catecholaminergic afferents constitute a part of the network for processing autonomic and neuroendocrine responses.

The PVN also contains TRH-containing neurons, which receive catecholaminergic innervation (Table 9.3). Catecholamines are involved in the control of this neuropeptide (Lamberton & Jackson, 1988); noradrenaline/adrenaline have a stimulatory effect on TRH system as shown by lesion studies (for review see Krulich, 1982). Conversely, dopamine might depress TSH secretion, such effects are described in rats and humans (Krulich, 1982).

Interactions between catecholamines and GHRH- and SRIF-containing neurons

The presence of PNMT containing fibers synapsing with SRIF- and GHRH-containing neurons (Table 9.3) constitutes a morphological proof for a direct action of adrenaline on the control of SRIF and GHRH release. Injection of noradrenaline and dopamine in the cerebral ventricle induces an increase of SRIF in the portal blood (Chihara, Arimura & Schally, 1979), however a decrease of growth hormone (GH) level is observed in animal treated with an inhibitor of adrenaline synthesis (Crowley, Terry & Johnson, 1982). GH secretion requires a precise action of GHRH and SRIF (Plotsky & Vale, 1985), with both under the control of catecholamines, mainly adrenaline. In the different species studied, it appears that stimuli which induce GH release act through central alpha-adrenergic receptors. Catecholamines, and particularly adrenaline, appear therefore to be in position to influence the ratio of GHRH and SRIF and consequently the control of GH (for review see Tuomisto & Männisto, 1985).

However, GHRH or SRIF co-exist with TH in the same cells and therefore, dopamine might influence the ratio of GHRH and SRIF. When dopamine is co-released with GHRH, its role might be to inhibit SRIF secretion and thus to potentiate the stimulatory effect of GHRH as hypothesized by Meister et al. (1986). In fact, intraventricular injection of dopamine induces GH release in rats (Vijayan, Krulich & McCann, 1978). In addition, the activity of dopaminergic cells of the arcuate nucleus, as demonstrated by voltammetry, is enhanced by intraventricular injection of GHRH in rats (Crespi et al., 1985). These close physiological relationships between dopamine and GHRH might indicate the existence of a positive 'feedback' of GHRH on its own secretion through a dopaminergic mediation.

Therefore, catecholamines may play an important role in the control of GH secretion by their actions through noradrenergic/adrenergic inputs to GHRH neurons and through the role of dopamine as companion transmitter in GHRH- and SRIF-containing neurons.

Conclusion

In all studied mammalian species, the diencephalon contains THi and dopaminergic neurons. In all mammalian species, THi neurons are disposed around the third ventricle where they form four horizontal columns, two dorsolateral and two ventrolateral to the ventricle. On each rostral and caudal part, THi neurons are found around the ventricle: both gatherings of the ipsilateral side meet each other. From this general pattern of distribution, species differences are observed. The rostrodorsal area around the third ventricle contains more numerous THi neurons in rodents than in sheep or primates, while the latter are characterized by the presence of numerous THi neurons in the SON and PVN. It appears, therefore, that the initial nomenclature used by Dahlström and Fuxe (1964) for describing the distribution of catecholamine cell groups in rats is not always applicable to the patterns observed in other mammalian species. For example, the lateral hypothalamus of sheep and humans contains several THi neurons observed below A13 in the ventral part of the diencephalon. To what group do these neurons belong? And the same question could be raised for the THi neurons observed in the pineal gland of hamsters. Moreover using more powerful tools such as antibodies against catecholamines themselves, some groups, previously described as homogenous appear now heterogenous such as the arcuate nucleus (A12), whose neurons contain dopamine and L-DOPA. Other THi neurons in the rostral and caudal parts of the third ventricle (corresponding to A14 dorsal part and A10 dorsorostral part, respectively) do not seem to synthesize dopamine since they are not labeled with antibodies against dopamine itself. The A15 ventral cell group is not DAi in rodents but it is in sheep.

Some diencephalic neurons appear to contain only one catecholamine synthesizing enzyme (TH or PNMT). The catecholaminergic end product of these neurons is unknown: either the enzyme is inactive or it synthesizes catecholamines after uptake of the necessary precursor. Therefore, it appears necessary to use antibodies against catecholamine synthesizing enzymes in conjunction with antibodies to neurotransmitters to distinguish actual catecholaminergic neurons from neurons that synthesize catecholamines only after uptake of the precursor or from neurons that stores catecholamines without synthesis.

The presence of the other transmitters (Table 9.2), mainly peptides, in THi neurons underlines also the heterogeneity of these groups. According to the companion transmitter observed, the role of the dopaminergic neuron might be different. This characteristic, and others such as the projection site and the functions in which these THi neurons are involved might be used to reconsider the classification of catecholaminergic structures in order to make it suitable for a greater number of species.

In the different mammalian species studied, all the diencephalic structures appear to receive catecholaminergic afferents. However, some species differences are noteworthy. For example, diencephalon in sheep and guinea pigs does not contain PNMTi fibers, although it does contain DBHi fibers as others species. In all species, PVN and SON and hypophysiotropic neurons receive a rather dense catecholaminergic innervation, which points out the important role of catecholamine in the regulation of endocrine functions and autonomic nervous system. However, the pattern of regulation might be different in different species. For example, in the control of osmoregulation, THi neurons in SON and PVN of rodents were less numerous than in primates but they are increased after dehydration. Therefore, the involvement of THi neurons in such regulation appears different in rodents and primates. Because of the wide distribution of catecholaminergic afferents in the diencephalon of the different studied species, it seems that all the information processed by this structure appears to be controlled by catecholamines and we could hypothesize the occurrence of a 'diencephalic sympathetic nervous system' equivalent to the peripheral sympathetic nervous system.

Acknowledgements

I am very greatful to Dr Jean Thibault for his generous gift of anti-TH and for his helpful discussions. I thank also Drs Jean Pelletier and Kunio Kitahama for their critical review of the manuscript, and Mr Alain Beguey for photographic assistance. In addition, Dr Kunio Kitahama kindly provided the immunolabeled sections of cat and human brains shown in this chapter.

Abbreviations

AADC	aromatic amino acid decarboxylase
AC	anterior commissure
ACTH	adrenocorticotropin hormone
Ad	adrenaline
AHA	anterior hypothalamic area
AN	arcuate nucleus
Ant Hyp	anterior hypothalamus
APN	anterior preoptic nucleus
β-Endo	β-endorphin
CC	corpus callosum
CCK	cholecystokinin
Cer	cerebellum
ChAt	acetylcholine
CP	cerebral peduncle
CRF	corticotropin releasing factor
DA	dopamine
DBB	diagonal band of Broca
DBH	dopamine-β-hyroxylase
DHN	dorsal hypothalamic nucleus
DMH	dorsomedial hypothalamic nucleus
DMHA	dorsomedial hypothalamic area
DYN	dynorphin
Enk	enkephalin
FIF	formaldehyde induced fluorescence
FMT	mammillothalamic tract
Fx	fornix
GABA	gamma aminobutyric acid
GAD	glutamic acid decarboxylase
Gal	galanin
GH	growth hormone
GHRH	growth hormone-releasing hormone
GP	globus pallidus
Hb	habenular nucleus
IHC	immunohistochemistry
Leu–Enk	leucin–enkephalin
LH	luteinizing hormone
LHA	lateral hypothalamic area
LHRH	luteinizing hormone-releasing hormone
LPO	lateral preoptic area
MDN	mediodorsal nucleus
ME	median eminence
Met–Enk	methionin–enkephalin
ML	lateral mammillary nucleus
MM	medial mammillary nucleus
MPO	medial preoptic area
NPY	neuropeptide Y
NA	noradrenaline
NT	neurotensin
OC	optic chiasm
Olf Bulb	olfactory bulb
Olf Tub	olfactory tubercle
OT	optic tract
OXT	oxytocin
PeMmb	perimammillary body area
PeVN	periventricular nucleus
PHA	posterior hypothalamic area
PHI	peptide histidine–isoleucine
PNMT	phenylethanolamine N-methyltransferase
POA	preoptic area
POMC	proopiomelanocortin
PON	preoptic nucleus
PPN	periventricular preoptic nucleus
PRL	prolactin
p tub	pars tuberalis
PVN	paraventricular hypothalamic nucleus
PVT	paraventricular thalamic nucleus
RCA	retrochiasmatic area
SCN	suprachiasmatic nucleus
Sept/Hyp	septo-hypothalamic area
SN	substantia nigra
SOC	supraoptic commissure
SON	supraoptic nucleus
SPf	subparafascicular nucleus
SRIF	somatostatin
SupraMmb	supramammillary body area
TH	tyrosine hydroxylase
TIN	tuberoinfundibular nucleus
TRH	thyrotropin releasing hormone
TSH	thyrotropin stimulating hormone
V3	third ventricle
VIP	vasoactive intestinal polypeptide
VMH	ventromedial hypothalamic nucleus
VP	vasopressin
ZI	zona incerta
5HT	serotonin

References

Ajika, K. (1979). Simultaneous localization of LHRH and catecholamines in rat hypothalamus. *Journal of Anatomy*, **128**, 331–47.

Ajika, K. (1980). Relationship between catecholaminergic neurons and hypothalamic hormone-containing neurons in the hypothalamus. In *Frontiers in Neuroendocrinology*, vol. 6 ed. L. Martini & W. F. Ganong, pp. 1–32. New York: Raven Press.

Al-Damluji, S. (1988). Adrenergic mechanisms in the control of corticotropin secretion. *Journal of Endocrinology*, **119**, 5–14.

Allen, L. G., Crowley, W. R. & Kalra, S. P. (1987). Interactions between neuropeptide Y and adrenergic systems in the stimulation of luteinizing hormone release in steroid-primed ovariectomized rats. *Endocrinology*, **121**, 1953–9.

Alper. R. H., Demarest, K. T. & Moore, K. E. (1980). Dehydration selectively increases dopamine synthesis in tuberohypophyseal dopaminergic neurons. *Neuroendocrinology*, **31**, 112–15.

Amaral, D. G. & Sinnamon, H. M. (1977). The locus coeruleus: Neurobiology of a central noradrenergic nucleus. *Progress in Neurobiology*, **9**, 147–96.

Anderson, C. R. & Howe, P. R. C. (1988). Is phenylethanolamine-N-methyltransferase (PNMT) contained in rat hypothalamic neurons? *Neuroscience Letters*, **93**, 164–9.

Antonopoulos, J., G. C. Papadopoulos, A. N. Karamanlidis, and H. Michaloudi (1989) Distribution of neuropeptides in the infundibular nucleus of the sheep. *Neuropeptides*, **14**, 121–8.

Asanuma, C. (1992). Noradrenergic innervation of the thalamic reticular nucleus: a light and electron microscopic immunohistochemical study in rats. *Journal of Comparative Neurology*, **319**, 299–311.

Asmus, S. E. Kincaid, A. E. & Newman. S. W. (1992). A species-specific population of tyrosine hydroxylase-immunoreactive neurons in the medial amygdaloid nucleus of the Syrian hamster. *Brain Research*, **575**, 199–207.

Astier, B., Kitahama, K., Denoroy, L., Jouvet, M. & Renaud, B. (1987). Immunohistochemical evidence for the adrenergic medullary longitudinal bundle as a major ascending pathway to the locus coeruleus. *Neuroscience Letters*, **74**, 132–8.

Baker, H., Ruggiero, D. A., Alden, S., Anwar, M. & Reis, D. J. (1986). Anatomical evidence for interactions between catecholamine- and adrenocorticotropin- containing neurons. *Neuroscience*, **17**, 469–84.

Barraclough, C. A. & Wise, P. M. (1982). The role of catecholamines in the regulation of pituitary luteinizing hormone and follicle stimulating hormone secretion. *Endocrine Research*, **3**, 91–119.

Batailler, M., Blache, D., Thibault, J. & Tillet, Y. (1992). Immunohistochemical colocalization of tyrosine hydroxylase and estradiol receptors in the sheep arcuate nucleus. *Neuroscience Letters*, **146**, 125–30.

Ben-Jonathan, N. (1985). Dopamine: a prolactin-inhibiting hormone. *Endocrine Reviews*, **6**, 564–89.

Björklund, A., & Lindvall, O. (1984). Dopamine-containing systems in the CNS. In *Handbook of Chemical Neuroanatomy. Vol. 2. Classical Neurotransmitters in the CNS, part I*, ed A. Björklund & T. Hökfelt, pp. 55–122. Amsterdam: Elsevier.

Björklund, A., Lindvall, O. & Nobin, A. (1975). Evidence of an incerto-hypothalamic dopamine neurone system in the rat. *Brain Research*, **89**, 29–42.

Björklund, A. & Skagerberg, G. (1979). Evidence for a major spinal cord projection from the diencephalic A11 dopamine cell group in the rat using transmitter-specific fluorescent retrograde tracing. *Brain Research*, **177**, 170–5.

Blaustein, J. D., Brown, T. J. & Swearengen, E. S. (1986). Dopamine-β-hydroxylase inhibitors modulate the concentration of functional estrogen receptors in female rat hypothalamus and pituitary gland. *Neuroendocrinology*, **43**, 150–8.

Blaustein, J. D. & Turcotte, J. C. (1989). A small population of tyrosine hydroxylase-immunoreactive neurons in the guinea-pig arcuate nucleus contains progestin receptor-immunoreactivity. *Journal of Neuroendocrinology*, **1**, 333–8.

Bleier, R., Byne, W. & Siggelkow, I. (1982). Cytoarchitectonic dimorphisms of the medial preoptic and anterior hypothalamic areas in guinea pig, rat, hamster, and mouse. *The Journal of Comparative Neurology*, **212**, 118–30.

Blessing, W. W., Chalmers, J. P. & Howe, P. R. C. (1978). Distribution of catecholamine-containing cell bodies in the rabbit central nervous system. *Journal of Comparative Neurology*, **179**, 407–24.

Blessing, W. W., Jaeger, C. B., Ruggiero, D. A. & Reis, D. J. (1982). Hypothalamic projection of medullary catecholamine neurons in the rabbit: combined catecholamine fluorescence and HRP transport study. *Brain Research Bulletin*, **9**, 279–86.

Bosler, O., Beaudet, A. & Denoroy, L. (1987). Electron-microscopic characterization of adrenergic axon terminals in the diencephalon of the rat. *Cell and Tissue Research*, **248**, 393–8.

Bosler, O., Joh, T. H. & Beaudet, A. (1984). Ultrastructural relationships between serotonin and dopamine neurons in the rat arcuate nucleus and medial zona incerta: a combined radioautographic and immunocytochemical study. *Neuroscience Letters*, **48**, 279–85.

Brownstein, M. & Axelrod, J. (1974). Pineal gland: 24-hour rhythm in norepinephrine turnover. *Science*, **184**, 163–5.

Bruni, J. F., Van Vugt, D., Marshall, S. & Meites, J. (1977). Effects of naloxone, morphine and methionine enkephalin on serum prolactin, luteinizing hormone, follicle stimulating hormone, thyroid stimulating hormone and growth hormone. *Life Science*, **21**, 461–6.

Buijs, R. M., Geffard, M., Pool, C. W. & Hoorneman, M. D. (1984). The dopaminergic innervation of the supraoptic and paraventricular nucleus. A light and electron microscopical study. *Brain Research*, **323**, 65–72.

Caraty, A., Orgeur, P. & Thiéry, J.-C. (1982). Mise en évidence d'une sécrétion pulsatile du LH-RH du sang porte hypophysaire chez la brebis par une technique originale de prélèvements multiples. *Comptes Rendus de l'Académie des Sciences (III)*, **295**, 103–6.

Chan-Palay, V., Zaborszky, L., Köhler, C., Goldstein, M. & Palay, S. L. (1984). Distribution of tyrosine-hydroxylase-immunoreactive neurons in the hypothalamus of rats. *Journal of Comparative Neurology*, **227**, 467–96.

Chen, W.-P., Witkin, J. W. & Silverman, A. -J. (1989). Gonadotropin releasing hormone (GnRH) neurons are directly innervated by catecholamine terminals. *Synapse*, **3**, 288–90.

Cheung, Y. & Sladek Jr, J. R. (1975). Catecholamine distribution in feline hypothalamus. *Journal of Comparative Neurology*, **164**, 339–60.

Chihara, K., Arimura, A. & Schally, A. V. (1979). Effect of intraventricular injection of dopamine, norepinephrine, acetylcholine, and 5-hydroxytryptamine on immunoreactive somatostatin release into the rat hypophyseal portal blood. *Endocrinology*, **104**, 1656–62.

Ciofi, P., Fallon, J. H., Croix, D., Polak, J. M. & Tramu, G. (1991). Expression of neuropeptide Y precursor-immunoreactivity in the hypothalamic dopaminergic tubero-infundibular system during lactation in rodents. *Endocrinology*, **128**, 823–34.

Clarke, I. J. & Cumming, J. T. (1982). The temporal relationship between gonadotropin-releasing hormone (GnRH) and luteinizing hormone (LH) in ovariectomized ewes. *Endocrinology*, **111**, 137–9.

Crespi, F., Paret, J., Keane, P. E., Morre, M., Coude, F. X. & Roncucci, R. (1985). Growth hormone-releasing factor modifies dopaminergic but not serotonergic activity in the arcuate nucleus of hypothalamus in the rat, as recorded *in vivo* by differential pulse voltammetry. *Brain Research*, **348**, 367–70.

Crowley, W. R. (1987). Control of luteinizing hormone secretion by ovarian hormone-monoamine-neuropeptide interactions. In *Integrative neuroendocrinology: Molecular, Cellular and Clinical Aspects* (1st International Congress of Neuroendocrinology, San Francisco, Calif. 1986), ed. S. M. Mc Cann & R. I. Weiner, pp. 54–69. Basel: Karger.

Crowley, W. R., Terry, L. C. & Johnson M. D. (1982). Evidence for the involvement of central epinephrine systems in the regulation of luteinizing hormone, prolactin, and growth hormone release in female rats. *Endocrinology*, **110**, 1102–7.

Cumming, P., Von Krosigk, M., Reiner, P. B., McGeer, E. G. & Vincent, S. R. (1986). Absence of adrenaline neurons in the guinea pig brain: a combined immunohistochemical and high-performance liquid chromatography study. *Neuroscience Letters*, **63**, 125–30.

Dahlström, A. & Fuxe, K. (1964). Evidence for the existence of monoamine containing neurons in the central nervous system. I. Demonstration of monoamines in cell bodies of brainstem neurons. *Acta Physiologica Scandinavica*, **62**, suppl. **232**, 1–55.

Davis, B. J. & Macrides, F. M. (1983). Tyrosine hydroxylase immunoreactive neurons and fibers in the olfactory system of the hamster. *Journal of Comparative Neurology*, **214**, 427–40.

Day, T. A., Blessing, W. W. & Willoughby, J. O. (1980). Noradrenergic and dopaminergic projections to the medial preoptic area of the rat. A combined horseradish peroxidase/catecholamine fluorescence study. *Brain Research*, **193**, 543–8.

Day, T. A., Ferguson, A. V. & Renaud, L. P. (1985). Noradrenergic afferents facilitate the activity of tuberoinfundibular neurons of the hypothalamic paraventricular nucleus. *Neuroendocrinology*, **41**, 17–22.

Decavel, C., Geffard, M. & Calas, A. (1987). Comparative study of dopamine- and noradrenaline-immunoreactive terminals in the paraventricular and supraoptic nuclei of the rat. *Neuroscience Letters*, **77**, 149–54.

Elias, A. N., Valenta, L. J., Szekeres, A. V. & Grossman, M. K. (1982). Regulatory role of gamma-aminobutyric acid in pituitary hormone secretion. *Psychoneuroendocrinology*, **7**, 15–30.

Enjalbert, A., Ruberg, M., Arancibia, S., Priam, M. & Kordon, C. (1979). Endogenous opiates block dopamine inhibition of prolactin secretion *in vitro*. *Nature (London)*, **280**, 595–7.

Ericson, H., Blomqvist, A. & Köhler, C. (1989). Brainstem afferents to the tuberomammillary nucleus in the rat brain with special reference to monoaminergic innervation. *Journal of Comparative Neurology*, **281**, 169–92.

Everitt, B. J., Hökfelt, T., Terenius, L., Tatemoto, K., Mutt, V. & Goldstein, M. (1984a). Differential co-existence of neuropeptide Y (NPY)-like immunoreactivity with catecholamines in the central nervous system of the rat. *Neuroscience*, **11**, 443–62.

Everitt, B. J., Hökfelt, T., Wu, J-Y. & Goldstein, M. (1984b). Coexistence of tyrosine hydroxylase-like and gamma-aminobutyric acid-like immunoreactivities in neurons of the arcuate nucleus. *Neuroendocrinology*, **39**, 189–91.

Everitt, B. J., Meister, B., Hökfelt, T., Melander, T., Terenius, L., Rökaeus, A., Theodorsson-Norheim, E., Dockray, G., Edwardson, J., Cuello, C., Elde, R., Goldstein, M., Hemmings, H., Ouimet, C., Walaas, J., Greengard, P., Vale, W., Weber, E., Wu, J-Y. & Chang, K-J. (1986). The hypothalamic arcuate nucleus-median eminence complex: immunohistochemistry of transmitters, peptides and DARPP-32 with special reference to coexistence in dopamine neurons. *Brain Research Reviews*, **11**, 97–155.

Falck, B., Hillarp, N. A. & Törp, A. (1962). Fluorescence of catecholamines and related compounds with formaldehyde. *Journal of Histochemistry and Cytochemistry*, **10**, 348–54.

Fallon, J. H., Koziell, D. A. & Moore, R. Y. (1978). Catecholaminergic innervation of the basal forebrain. II Amygdala, suprarhinal cortex and entorhinal cortex. *Journal of Comparative Neurology*, **180**, 509–32.

Felten, D. L. (1976). Catecholamine neurons in the squirrel monkey hypothalamus. *Journal of Neural Transmission*, **39**, 269–80.

Felten, D. L. & Sladek Jr, J. R. (1983). Monoamine distribution in primate brain V. Monoaminergic nuclei: anatomy, pathways and local organization. *Brain Research Bulletin*, **10**, 171–284.

Ferland, L., Fuxe, K., Eneroth, P., Gustafsson, J. A. & Skett, P. (1977). Effects of methionine-enkephalin on prolactin release and catecholamine levels and turnover in the median eminence. *European Journal of Pharmacology*, **43**, 89–90.

Finch, L. & Hichs, P. E. (1976). The cardiovascular effects

of intraventricularly administered histamine in the anaesthetised rat. *Naunyn-Schmiedebergs Archives of Pharmacology*, **293**, 151–7.

Fitzsimmons, M. D., Olschowka, J. A., Wiegand, S. J. & Hoffman, G. E. (1992). Interaction of opioid peptide-containing terminals with dopaminergic perikarya in the rat hypothalamus. *Brain Research*, **581**, 10–18.

Foster, G. A., Hökfelt, T., Coyle, J. T. & Goldstein, M. (1985). Immunohistochemical evidence for phenylethanolamine N-methyltransferase positive/tyrosine hydroxylase-negative neurons in the retina and the posterior hypothalamus of the rat. *Brain Research*, **330**, 183–8.

Fuxe, K. (1965a). Evidence for the existence of monoamine containing neurons in the central nervous system. III The monoamine nerve terminal. *Zeitschrift Für Zell Zelforschung Mikroskopische Anatomie*, **65**, 573–96.

Fuxe, K. (1965b). Evidence for the existence of monoamine containing neurons in the central nervous system. IV The distribution of monoamine nerve terminals in the central nervous system. *Acta Physiologica Scandinavica*, 64, suppl. **247**, 39–85.

Fuxe, K., Agnati, L. F., Andersson, K., Eneroth, P., Härfstrand, A., Goldstein, M. & Zoli, M. (1984). Studies on neurotensin-catecholamine interactions in the hypothalamus and in the forebrain of the male rat. *Neurochemistry International*, **6**, 737–50.

Fuxe, K., Agnati, L. F., Kalia, M., Goldstein, M., Anderson, K. & Harfstrand, A. (1985). Dopaminergic systems in the brain and pituitary. In *Basic and Clinical Aspects of Neurosciences: The Dopaminergic System*, ed. Sandoz advanced text, pp. 1–25. Berlin: Springer-Verlag.

Fuxe, K., Andersson, K., Eneroth, P., Härfstrand, A., Janson, A. M., von Euler, G., Agnati, L. F., Tinner, B., Köhler, C. & Hersh, L. B. (1988). Neuroendocrine and trophic actions of nicotine. In *The Pharmacology of Nicotine*, ed. M. J. Rand & K. Thurau, pp. 293–320. Oxford: IRL Press.

Fuxe, K., Corrodi, J., Hökfelt, T. & Jonsson, G. (1970). Central monoamine neurons and pituitary-adrenal activity. *Progress in Brain Research*, **32**, 236–422.

Fuxe, K., Hökfelt, T., Johansson, O., Jonsson, G., Lidbrink, P. & Ljungdahl, A. (1974). The origin of the dopamine nerve terminals in limbic and frontal cortex. Evidence for mesocortico dopamine neurons. *Brain Research*, **82**, 349–55.

Fuxe, K., Hökfelt, T. & Nilsson, O. (1969). Castration, sex hormones, and tubero-infundibular dopamine neurons, *Neuroendocrinology*, **5**, 107–20.

García-Segura, L. M., Baetens D. & Naftolin F. (1986). Synaptic remodelling in arcuate nucleus after injection of estradiol valerate in adult female rats. *Brain Research*, **366**, 131–6.

Gaspar, P., Berger, B., Alvarez, C., Vigny, A. & Henry, J. P. (1985). Catecholaminergic innervation of the septal area in man: immunocytochemical study using TH and DBH antibodies. *Journal of Comparative Neurology*, **241**, 12–33.

Ginsberg, S. D., Hof, P. R., Young, W. G. & Morrison, J. H. (1993). Noradrenergic innervation of the hypothalamus of rhesus monkeys: distribution of dopamine-β-hydroxylase immunoreactive fibers and quantitative analysis of varicosities in the paraventricular nucleus. *Journal of Comparative Neurology*, **327**, 597–611.

Gonzalo-Ruiz, A., Alonso, A., Sanz, J. M. & Llinas, R. R. (1992). A dopaminergic projection to the rat mammillary nuclei demonstrated by retrograde transport of wheat germ agglutinin-horseradish peroxidase and tyrosine hydroxylase immunohistochemistry. *Journal of Comparative Neurology*, **321**, 300–11.

Goodman, R. L. (1989). Functional organization of the catecholaminergic neuronal systems inhibiting luteinizing hormone secretion in anestrous ewes. *Neuroendocrinology*, **50**, 406–12.

Goudreau, J. L., Lindley, S. E., Looklingland, K. J. & Moore, K. E. (1992). Evidence that hypothalamic periventricular dopamine neurons innervate the intermediate lobe of the rat pituitary. *Neuroendorinology*, **56**, 100–5.

Gudelsky, G. A., Berry, S. A. & Meltzer, H. Y. (1989). Neurotensin activates tuberoinfundibular dopamine neurons and increases serum corticosterone concentrations in the rat. *Neuroendocrinology*, **49**, 604–9.

Gudelski, G. A. & Porter, J. C. (1980). Release of dopamine from tuberoinfundibular neurons into pituitary stalk blood after prolactin or haloperidol administration. *Endocrinology*, **106**, 526–9.

Guillaume, V., Conte-Devolx, B., Szafarczyk, A., Malaval, F., Pares-Herbute, N., Grino, M., Alonso, G., Assenmacher, I. & Oliver, C. (1987). The corticotropin-releasing factor release in rat hypophysial portal blood is mediated by brain catecholamines. *Neuroendocrinology*, **46**, 143–6.

Gunnett, J. W., Looklingland, K. J. & Moore, K. E. (1986). Comparison of the effects of castration and steroid replacement on incertohypothalamic dopaminergic neurons in male and female rats. *Neuroendocrinology*, **44**, 269–75.

Guy, J. & Pelletier, G. (1988). Neuronal interactions between neuropeptide Y (NPY) and catecholaminergic systems in the rat arcuate nucleus as shown by dual immunocytochemistry. *Peptides*, **9**, 567–70.

Hammond Jr, J., Hammond, J. & Parkes, A. S. (1942). Hormonal augmentation of fertility in sheep. I Induction of ovulation, superovulation and heat in sheep. *Journal of Agricultural Science*, **32**, 308–23.

Härfstrand, A., Eneroth, P., Agnati, L. & Fuxe, K. (1987a). Further studies on the effects of central administration of neuropeptide Y on neuroendocrine function in the male rat: relationship to hypothalamic catecholamines. *Regulatory Peptides*, **17**, 167–79.

Härfstrand, A., Fuxe, K., Agnati, L. F., Eneroth, P., Zini, I., Zoli, M., Andersson, K., von Euler, G., Terenius, L., Mutt, V. & Goldstein, M. (1987b). Studies on neuropeptide Y-catecholamine interactions in the hypothalamus and in the forebrain of the male rat. Relationship to neuroendocrine function. *Neurochemistry International*, **8**, 355–76.

Haskins, J. T., Gudelski, G. A., Moss, R. L. & Porter, J. C. (1981). Iontophoresis of morphine into the arcuate nucleus: effects of dopamine concentrations in hypophyseal portal plasma and serum prolactin concentrations. *Endocrinology*, **108**, 767–72.

Havern, R. L., Whisnant, C. S. & Goodman, R. L. (1991). Hypothalamic sites of catecholamine inhibition of luteinizing hormone in the anestrous. *Biology of Reproduction*, **44**, 476–82.

Hedge, G. A., Van Ree, J. M. & Versteeg, D. H. G. (1976). Correlation between hypothalamic catecholamine synthesis and ether stress-induced ACTH secretion. *Neuroendocrinology*, **21**, 521–4.

Heritage, A. S., Grant, L. D. & Stumpf, W. E. (1977). ^{3}H estradiol in catecholamine neurons of rat brain stem: combined localization by autoradiography and formaldehyde-induced fluorescence. *Journal of Comparative Neurology*, **176**, 607–30.

Hoffman, G. E., Felten, D. L. & Sladek Jr., J. R. (1976). Monoamine distribution in Primate brain. III. Catecholamine-containing varicosities in the hypothalamus of *Macaca mulatta*. *American Journal of Anatomy*, **147**, 501–14.

Hoffman, G. E., Wray, S., Goldstein, M. (1982). Relationship of catecholamines and LHRH: Light microscopic study. *Brain Research Bulletin*, **9**, 417–30.

Hökfelt, T., Everitt, E., Meister, B., Melander, T., Shalling, M., Johansson, O., Lundberg, J. M., Hulting, A.-L., Werner, S., Cuello, C., Hemming, H., Ouimet, C., Walaas, I., Greengard, P. & Goldstein, M. (1986). Neurons with multiple messengers with special reference to neuroendocrine systems. *Recent Progress in Hormone Research*, **42**, 1–70.

Hökfelt, T., Johansson, O., Fuxe, K., Goldstein, M. & Park, D. (1976). Immunohistochemical studies on the localization and distribution of monoamine neuron systems in the rat brain. I. Tyrosine hydroxylase in the mes- and diencephalon. *Medical Biology*, **54**, 427–53.

Hökfelt, T., Everitt, B. J., Theodorsson-Norheim, E. & Goldstein, M. (1984a). Occurrence of neurotensin-like immunoreactivity in subpopulations of hypothalamic, mesencephalic and medullary catecholamine neurons. *Journal of Comparative Neurology*, **222**, 543–59.

Hökfelt, T., Johansson, O. & Goldstein, M. (1984b). Central catecholamine neurons as revealed by immunohistochemistry with special reference to adrenaline neurons. In *Handbook of Chemical Neuroanatomy. Vol. 2. Classical Neurotransmitters in the CNS, part I*, ed A. Björklund & T. Hökfelt, pp. 157–276. Amsterdam: Elsevier.

Hökfelt, T., Martenson, R., Björklund, A., Kleinau, S. & Goldstein, M. (1984c). Distributional maps of tyrosine-hydroxylase-immunoreactive neurons in the rat brain. In *Handbook of Chemical Neuroanatomy. Vol. 2. Classical Neurotransmitters in the CNS, part I*, ed A. Björklund & T. Hökfelt, pp. 277–379. Amsterdam: Elsevier.

Hökfelt, T., Philipson, O. & Goldstein, M. (1979). Evidence for a dopaminergic pathway in the rat descending from the A11 cell group to the spinal cord. *Acta Physiologica Scandinavica*, **107**, 393–5.

Holets, V. R., Hökfelt, T., Rökaeus, A., Terenius, L. & Goldstein, M. (1988). Locus coeruleus neurons in the rat containing neuropeptide Y, tyrosine hydroxylase or galanin and their efferent projections to the spinal cord, cerebral cortex and hypothalamus. *Neuroscience*, **24**, 893–906.

Hornby, P. J. & Piekut, D. T. (1987). Catecholamine distribution and relationship to magnocellular neurons in the paraventricular nucleus of the rat. *Cell and Tissue Research*, **248**, 239–46.

Hornby, P. J. & Piekut, D. T. (1989). Opiocortin and catecholamine input to CRF-immunoreactive neurons in rat forebrain. *Peptides*, **10**, 1139–46.

Horvath, S., Mezey, E. & Palkovits, M. (1989). Partial coexistence of growth hormonereleasing hormone and tyrosine hydroxylase in paraventricular neurons in rats. *Peptides*, **10**, 791–5.

Horvath, T. L., Naftolin, F. & Leranth, C. (1992). β-endorphin innervation of dopamine neurons in the rat hypothalamus: a light and electron microscopic double immunostaining study. *Endocrinology*, **131**, 1547–55.

Ibata, Y., Fukui, K., Okamura, H., Kawakami, T., Tanaka, M., Obata, H. L., Tsuto, T., Terubayashi, H., Yanaihara, C. & Yanaihara, N. (1983). Coexistence of dopamine and neurotensin in hypothalamic arcuate and periventricular neurons. *Brain Research*, **269**, 177–9.

Iijima, K. & Ogawa, H. (1981). An HRP study on the distribution of all nuclei innervating the supraoptic nucleus in the rat brain. *Acta Histochemica*, **69**, 274–95.

Jennes, L., Stumpf, W. E. & Tappaz, M. L. (1983). Anatomical relationship of dopaminergic and gabaergic systems with the LHRH-systems in the septo-hypothalamic area. *Experimental Brain Research*, **50**, 91–9.

Jin, K. L., Shiotani, Y., Kawai, Y. & Kiyama, H. (1988). Immunohistochemical demonstration of tyrosine hydroxylase (TH)-positive but dopamine β-hydroxylase (DBH)-negative neuron-like cells in the pineal gland of golden hamsters. *Neuroscience Letters*, **93**, 28–31.

Kalra, S. P. & Kalra, P. S. (1983). Neural regulation of luteinizing hormone secretion in the rat. *Endocrine Reviews*, **4**, 311–51.

Karasawa, N., Isomura, G., Yamada, K. & Nagatsu, I. (1991). Immunocytochemical localization of monoaminergic and non-aminergic neurons in the house-shrew (*Suncus murinus*) brain. *Acta Histochemica Cytochemica*, **24**, 465–75.

Karasawa, N., Isomura, G., Yamada, K. & Nagatsu, I. (1992). Immunohistochemical localization of monoaminergic neurons in the house-shrew (*Suncus murinus*) brain. *Biogenic Amines*, **9**, 41–56.

Kawano, H. & Daikoku, S. (1987). Functional topography of the rat hypothalamic dopamine neuron systems: retrograde tracing and immunohistochemical study. *Journal of Comparative Neurology*, **265**, 242–53.

Kiss, J. & Halasz, B. (1986). Synaptic connections between serotoninergic axon terminals and tyrosine hydroxylase-immunoreactive neurons in the arcuate nucleus of the rat hypothalamus. A combination of electron microscopic autoradiography and immunocytochemistry. *Brain Research*, **364**, 284–94.

Kiss, J. Z. & Mezey, E. (1986). Tyrosine hydroxylase in magnocellular neurosecretory neurons. *Neuroendocrinology*, **43**, 519–25.

Kitahama, K., Geffard, M., Okamura, H., Nagatsu, I., Mons, N. & Jouvet, M. (1990). Dopamine- and DOPA-immunoreactive neurons in the cat forebrain with reference to tyrosine hydroxylase-immunohistochemistry. *Brain Research*, **518**, 83–94.

Kitahama, K., Luppi, P.-H., Berod, A., Goldstein, M. & Jouvet, M. (1987). Localization of tyrosine hydroxylase-immunoreactive neurons in the cat hypothalamus, with special reference to fluorescence histochemistry. *Journal of Comparative Neurology*, **262**, 578–93.

Kizer, J. S., Palkovits, M. & Brownstein, M. J. (1976). The projections of the A8, A9 and A10 dopaminergic cell bodies: evidence for a nigral-hypothalamic median eminence pathway. *Brain Research*, **108**, 363–70.

Koenig, J. I., Mayfield, M. A., McCann, S. M. & Krulich, L. (1982). On the prolactin-inhibitory effect of neurotensin. *Neuroendocrinology*, **35**, 277–81.

Kohama, S. G., Freesh, F. & Bethea, C. L. (1992). Immunocytochemical colocalization of hypothalamic progestin receptors and tyrosine hydroxylase in steroid-treated monkeys. *Endocrinology*, **131**, 509–17.

Komori, K., Fujii, T. & Nagatsu, I. (1991). Do some tyrosine hydroxylase-immunoreactive neurons in the human ventrolateral arcuate nucleus and globus pallidus produce only L-DOPA? *Neuroscience Letters*, **133**, 203–6.

Kordon, C., Enjalbert, A., Héry, M., Joseph-Bravo, P. I., Rotsztejn, W. & Ruberg, M. (1980). Role of neurotransmitters in the control of adenohypophyseal secretion. In *Handbook of the Hypothalamus*, vol.2, ed. P. J. Morgane & J. Panksepp, pp.253–306. New York: Dekker.

Kosawa, K. & Nakaï, Y. (1987). Electron-microscopic cytochemistry of the catecholaminergic innervation of ACTH-containing neurons in the rat hypothalamic arcuate nucleus. *Acta Anatomica*, **128**, 243–9.

Krulich, L. (1982). Neurotransmitter control of thyrotropin secretion. *Progress in Neuroendocrinology*, **35**, 139–47.

Kühn, E. R. (1974). Cholinergic and adrenergic release mechanisms of vasopressin in the male rat: a study with injections of neurotransmitters and blocking agents into the third ventricle. *Neuroendocrinology*, **16**, 255–64.

Kuljis, R. O. & Advis, J. P. (1989). Immunocytochemical and physiological evidence of a synapse between dopamine- and luteinizing hormone releasing hormone-containing neurons in the ewe median eminence. *Endocrinology*, **124**, 1579–81.

Lamberton, P. & Jackson, I. M. D. (1988). Neuroendocrine aspect of thyrotropin secretion. In *Clinical Endocrinology*, ed. R. Collu, G. M. Brown & G. R. Van Loon, pp. 287–309. Boston: Blackwell Scientific Publications.

Leibowitz, S. F. (1989). Hypothalamic neuropeptide Y and galanin: functional studies of coexistence with monoamines. In *Neuropeptide Y*, ed. V. Mutt et al., pp. 267–81. New York: Raven press.

Leranth, C., Sakamoto, H., Maclusky, N. J., Shanabrough, M. & Naftolin, F. (1985). Intrinsic tyrosine hydroxylase (TH) immunoreactive axons synapse with TH immunopositive neurons in the rat arcuate nucleus. *Brain Research*, **331**, 371–5.

Li, Y. W., Halliday, G. M., Joh, T. H., Geffen, L. B. & Blessing, W. W. (1988). Tyrosine hydroxylase-containing neurons in the supraoptic and paraventricular nuclei of the adult human. *Brain Research*, **461**, 75–86.

Liao, N., Bulant, M., Nicolas, P., Vaudry, H. & Pelletier, G. (1991). Anatomical interactions of proopiomelanocortin (POMC)-related peptides, neuropeptide Y (NPY) and dopamine β-hydroxylase (DβH) fibers and thyrotropin-releasing hormone (TRH) neurons in the paraventricular nucleus of rat hypothalamus. *Neuropeptides*, **18**, 63–7.

Lin, J. S., Luppi, P. H., Salvert, D., Sakai, K. & Jouvet, M. (1986). Histamine-containing neurons in the cat hypothalamus. *Comptes Rendus de l'Académie des Sciences (III)*, **303**, 371–6.

Lindvall, O., Björklund, A. & Skagerberg, G. (1984). Selective histochemical demonstration of dopamine terminal systems in rat di- and telencephalon: new evidence for dopaminergic innervation of hypothalamic neurosecretory nuclei. *Brain Research*, **306**, 19–30.

Liposits, Zs., Hrabovszky, E. & Paull, W. K. (1989). Catecholaminergic afferents to growth hormone-releasing hormone (GH-RH)-synthesizing neurons in the arcuate nucleus in the rat. *Biomedical Research*, **10** [suppl. 2], 163–72.

Liposits, Zs., Kallo, I., Barkovics-Kallo, M., Bohn, M. C. & Paull, W. K. (1990). Innervation of somatostastin synthesizing neurons by adrenergic, phenylethanolamine-N-methyltransferase (PNMT)-immunoreactive axons in the anterior periventricular nucleus of the rat hypothalamus. *Histochemistry*, **94**, 13–20

Liposits, Zs. & Paull, W. K. (1989). Association of dopaminergic fibers with corticotropin releasing hormone (CRH)-synthesizing neurons in the paraventricular nucleus of the rat hypothalamus. *Histochemistry*, **93**, 119–27.

Liposits, Zs., Paull, W. K., Wu, P., Jackson, I. M. D. & Lechan, R. M. (1987). Hypophysiotropic thyrotropin releasing hormone (TRH) synthesizing neurons. Ultrastructure, adrenergic innervation and putative transmitter action. *Histochemistry*, **88**, 1–10.

Liposits, Zs., Phelix, C. & Paull, W. K. (1986a). Electron microscopic analysis of tyrosine hydroxylase, dopamine-β-hydroxylase and phenylethanolamine-N-methyl-transferase immunoreactive innervation of the hypothalamic paraventricular nucleus in the rat. *Histochemistry*, **84**, 105–20.

Liposits, Zs., Phelix, C. & Paull, W. K. (1986b). Adrenergic innervation of corticotropin releasing factor

(CRF)-synthesizing neurons in the hypothalamic paraventricular nucleus of the rat. *Histochemistry*, **84**, 201–5.

Liposits, Zs., Sherman, D., Phelix, C. & Paull, W. K. (1986c). A combined light and electron microscopic immunocytochemical method for the simultaneous localization of multiple tissue antigens. Tyrosine hydroxylase immunoreactive innervation of corticotropin releasing factor synthesizing neurons in the paraventricular nucleus of rat. *Histochemistry*, **85**, 95–106.

Locatelli, V., Apud, J. A., Gudelski, G. A., Cocchi, D., Masotto, C., Casanueva, F., Racagni, G. & Muller, E. E. (1985). Prolactin in cerebrospinal fluid increases the synthesis and release of hypothalamic gamma-aminobutyric acid. *Journal of Endocrinology*, **106**, 323–8.

Locatelli, V., Petraglia, F., Penalva, A. & Panerai, A. E. (1983). Effect of dopaminergic drugs on hypothalamic and pituitary immunoreactive β-endorphin concentrations in the rat. *Life Science* **33**, 1711–17.

Loewy, A. D. & Spier, K. M. (eds.) (1990). *Central Regulation of Autonomic Functions*. New York: Oxford University Press.

Lohse, M. & Wuttke, W. (1981). Release and synthesis rates of catecholamines in hypothalamic, limbic and midbrain structures following intraventricular injection of β-endorphin in male rats. *Brain Research*, **229**, 389–402.

Luppi, P. H., Sakai, K., Salvert, D., Bérod, A. & Jouvet, M. (1986). Periventricular dopaminergic neurons terminating in the neuro-intermediate lobe of the cat hypophysis. *The Journal of Comparative Neurology*, **244**, 204–12.

MacKenzie, F. J., James, M. D. & Wilson, C. A. (1988). Changes in dopamine activity in the zona incerta (ZI) over the rat oestrous cycle and the effect of lesions of the ZI on the cyclicity: further evidence that the incertohypothalamic tract has a stimulatory role in the control of LH release. *Brain Research*, **444**, 75–83.

Mason, W. T. (1983). Excitation by dopamine of putative oxytocinergic neurones in the rat supraoptic nucleus *in vitro*: evidence of two classes of continuously firing neurons. *Brain Research*, **267**, 113–21.

Mathiasen, J. R., Arbogast, L. A. & Vogt, J. L. (1992). Central administration of serotonin decreases tyrosine hydroxylase catalytic activity and messenger ribonucleic acid signal levels in the hypothalamus of female rat. *Journal of Neuroendocrinology*, **4**, 631–9.

Matsuura, T. & Sano, Y. (1983). Distribution of monoamine-containing nerve fibers in the pineal organ of untreated and sympathectomized dogs. *Cell and Tissue Research*, **134**, 519–31.

Mehler, W. R. (1980). Subcortical afferent connections of the amygdala in the monkey. *Journal of Comparative Neurology*, **180**, 733–62.

Meister, B., Hökfelt, T., Steinbusch, H. W. M., Skagerberg, G., Lindvall, O., Geffard, M., Joh, T. H., Cuello, A. C. & Goldstein, M. (1988). Do tyrosine hydroxylase-immunoreactive neurons in the ventrolateral arcuate nucleus produce dopamine or only L-Dopa? *Journal of Chemical Neuroanatomy*, **1**, 59–64.

Meister, B., Hökfelt, T., Vale, W. W., Sawchenko, P. E., Swanson, L. & Goldstein, M. (1986). Coexistence of tyrosine hydroxylase and growth hormone-releasing factor in a subpopulation of tubero-infundibular neurons of the rat. *Neuroendocrinology*, **42**, 237–47.

Meister, B., Hökfelt, T., Vale, W. W. & Goldstein, M. (1985). Growth hormone releasing factor (GRF) and dopamine coexist in hypothalamic arcuate neurons. *Acta Physiologica Scandinavica*, **124**, 133–6.

Melander, T. Hökfelt, T., Rökaeus, A., Cuello, A. C., Oertel, W. H., Verhofstad A. & Goldstein, M. (1986). Coexistence of galanine-like immunoreactivity with catecholamines, 5-hydroxytryptamine, GABA and neuropeptides in the rat CNS. *The Journal of Neuroscience*, **6**, 3640–54.

Meyer, S. L. & Goodman, R. L. (1986). Separate neural systems mediate the steroid-dependent and steroid-independent suppression of tonic luteinizing hormone secretion in the anestrous ewe. *Biology of Reproduction*, **35**, 562–71.

Mitchell, V., Beauvillain J. C., Poulain, P. & Mazzuca, M. (1988). Catecholamine innervation of enkephalinergic neurons in guinea pig hypothalamus: Demonstration by an *in vitro* autoradiographic technique combined with a post-embedding immunogold method. *The Journal of Histochemistry and Cytochemistry*, **36**, 533–42.

Mons, N., Tison, F. & Geffard, M. (1990). Existence of L-DOPA immunoreactive neurons in the rat preoptic area and anterior hypothalamus *Neuroendocrinology*, **51**, 425–8.

Moore, R. Y., & Card, P. (1984). Noradrenaline-containing neuron systems. In *Handbook of Chemical Neuroanatomy. Vol. 2. Classical Neurotransmitters in the CNS, part 1*, ed A. Björklund & T. Hökfelt, pp. 123–156. Amsterdam: Elsevier.

Moos, F. & Richard, P. (1982). Excitatory effect of dopamine on oxytocin and vasopressin releases in the rat. *Brain Research*, **241**, 249–60.

Morel, G. & Pelletier, G. (1986). Endorphinic neurons are contacting the tuberoinfundibular dopaminergic neurons in the rat brain. *Peptides*, **7**, 1197–9.

Moss, R. L., Urban, I. & Cross, B. A. (1972). Microelectrophoresis of cholinergic and aminergic drugs on paraventricular neurons. *American Journal of Physiology*, **232**, 310–18.

Nakada, H. & Nakai, Y. (1985). Electron microscopic examination of the catecholaminergic innervation of the neurophysin- or vasopressin-containing neurons in the rat hypothalamus. *Brain Research*, **361**, 247–57.

Nakagawa, Y., Shiosaka, S., Emson, P. C. & Tohyama, M. (1985). Distribution of neuropeptide Y in the forebrain and diencephalon: an immunohistochemical analysis. *Brain Research*, **361**, 52–60.

Nakai, Y., Shioda, S., Ochiai, H. & Kozasa, K. (1986). Catecholamine–peptide interactions in the hypothalamus. *Current Topics in Neuroendocrinology*, **7**, 135–60.

Nemeroff, C. B. & Cain, S. T. (1985). Neurotensin-dopamine interactions in the CNS. *Trends in Pharmacological Sciences*, **6**, 201–5.

Nieuwenhuys, R. (1985). *Chemoarchitecture of the Brain*. Berlin, Heidelberg, New York, Tokyo: Springer-Verlag.

Niimi, M., Takahara, J., Sato, M. & Kawanishi, K. (1992). Identification of dopamine and growth hormone-releasing factor-containing neurons projecting to the median eminence of the rat by combined retrograde tracing and immunohistochemistry. *Neuroendocrinology*, **55**, 92–6.

Ochiai, H. & Nakai, Y. (1990). Ultrastructural demonstration of dopamine-β-hydroxylase immunoreactive nerve terminals on vasopressin neurons in the paraventricular nucleus of the rat by double-labeling immunocytochemistry. *Neuroscience Letters*, **120**, 87–90.

Okamura, H., Kitahama, K., Mons, N., Ibata, Y., Jouvet, M. & Geffard, M. (1988a). L-DOPA-immunoreactive neurons in the rat hypothalamic tuberal region. *Neuroscience Letters*, **95**, 42–6.

Okamura, H., Kitahama, K., Nagatsu, I. & Geffard, M. (1988b). Comparative topography of dopamine- and tyrosine hydroxylase-immunoreactive neurons in the rat arcuate nucleus. *Neuroscience Letters*, **95**, 347–53.

Okamura, H., Murakami, S., Chihara, K., Nagatsu, I. & Ibata, Y. (1985). Coexistence of growth hormone releasing factor-like and tyrosine hydroxylase-like immunoreactivities in neurons of the rat arcuate nucleus. *Neuroendocrinology*, **41**, 177–9.

Olmos, G., Naftolin, F., Perez, J., Tranque, P. A. & Garcia-Segura, L. M. (1989). Synaptic remodeling in the rat arcuate nucleus during the estrous cycle. *Neuroscience*, **32**, 663–7.

Olson, L., & Fuxe, K. (1972). Further mapping out of central noradrenaline neuron systems: projections of the 'subcoeruleus' area. *Brain Research*, **43**, 289–95.

Otterson, O. P. & Ben-Ari, Y. (1979). Afferent connections to the amygdaloid complex of the rat and cat. *Journal of Comparative Neurology*, **187**, 401–24.

Owman, Ch. (1965). Localization of neuronal and parenchymal monoamines under normal and experimental conditions in the mammalian pineal gland. *Progress in Brain Research*, **10**, 423–53.

Palkovits, M., Mezey, E., Skirboll, L. R. & Hökfelt, T. (1992). Adrenergic projections from the lower brainstem to the hypothalamic paraventricular nucleus, the lateral hypothalamic area and the central nucleus of the amygdala in rats. *Journal of Chemical Neuroanatomy*, **5**, 407–15.

Panayotacopoulou, M. T., Guntern, R., Bouras, C., Issidorides, M. R. & Constantinidis, J. (1991). Tyrosine hydroxylase-immunoreactive neurons in paraventricular and supraoptic nuclei of the human brain demonstrated by a method adapted to prolonged formalin fixation. *Journal of Neuroscience Methods*, **39**, 39–44.

Panula, P., Airaksinen, M. S., Pirvola, U. & Kotilainen, E. (1990). A histamine-containing neuronal system in human brain. *Neuroscience*, **34**, 127–32.

Panula, P., Yang, H. Y. T. & Costa, E. (1984). Histamine-containing neurons in the rat hypothalamus. *Proceedings of the National Academy of Science, USA*, **81**, 2572–6.

Pearson, J., Goldstein, M., Kitahama, K., Sakamoto, N. & Michel, J.-P. (1990a). Catecholaminergic neurons of the human central nervous system. In *An Introduction to Neurotransmission in Health and Desease*, ed. P. Riederer, N. Kopp & J. Pearson, pp. 22–36. Oxford: Oxford University Press.

Pearson, J., Halliday, G., Sakamoto, N. & Michel, J.-P. (1990b). Catecholaminergic neurons. In *The Human Nervous System*, ed. G. Paxinos, pp. 1023–49. San Diego CA: Academic Press.

Piotte, M., Beaudet, A. & Brawer, J. R. (1988). Light and electron microscopic study of tyrosine hydroxylase-immunoreactive neurons within the developing rat arcuate nucleus. *Brain Research*, **439**, 127–37.

Piotte, M., Beaudet, A., Joh, T. H. & Brawer, J. R. (1985). The fine structural organization of tyrosine hydroxylase immunoreactive neurons in the rat arcuate nucleus. *Journal of Comparative Neurology*, **239**, 44–53.

Plotsky, P. M. & Vale, W. (1985). Patterns of growth hormone-releasing factor and somatostatin secretion into the hypophysial-portal circulation of the rat. *Science*, **230**, 461–3.

Racagni, G., Apud, J. A., Cocchi, D., Locatelli, V. & Muller, E. E. (1982). Gabaergic control of anterior pituitary hormone secretion. *Life Science*, **31**, 823–38.

Reiter, R. J. (1980). The pineal and its hormones in the control of reproduction in mammals. *Endocrine Reviews*, **1**, 109–31.

Robert, F. & Calcutt, C. R. (1983). Histamine and hypothalamus. *Neuroscience*, **9**, 721–39.

Ross, C. A., Ruggiero, D. A., Meeley, M. P., Park; D. H., Joh, T. H. & Reis, D. J. (1984). A new group of neurons in hypothalamus containing phenylethanolamine N-methyltransferase (PNMT) but not tyrosine hydroxylase. *Brain Research*, **306**, 349–53.

Ruggiero, D. A., Baker, H., Joh, T. H. & Reis D. J. (1984). Distribution of catecholamine neurons in the hypothalamus and preoptic region of mouse. *Journal of Comparative Neurology*, **223**, 556–82.

Ruggiero, D. A., Ross, C. A., Anwar, M., Park, D. H., Joh, T. H. & Reis, D. J. (1985). Distribution of neurons containing phenylethanolamine N-methyltransferase in medulla and hypothalamus of rat. *Journal of Comparative Neurology*, **239**, 127–54.

Sahu, A., Crowley, W. R., Tatemoto, K., Balasubramaniam, A. & Kalra, S. P. (1987). Effects of neuropeptide Y, NPY analog (Norleucine[4]-NPY), galanin and neuropeptide K on LH release in ovariectomized (ovx) and ovx estrogen, progesterone-treated rats. *Peptides*, **8**, 921–6.

Sakanaka, M., Magari, S. & Inoue, N. (1990). Somatostatin co-localizes with tyrosine hydroxylase in the nerve cells of discrete hypothalamic region in rats. *Brain Research*, **516**, 313–17.

Sakumoto, T., Tohyama, M., Sato, K., Kimoto, V., Kinugasa, T., Tanizawa, O., Kurachi, K. & Shimizu N. (1978). Afferent fibre connections from the lower brain stem to hypothalamus studied by the horseradish peroxidase method with special reference to noradrenaline innervation. *Experimental Brain Research,* **31**, 81–94.

Saland, L. C., Wallace, J. A., Samora, A. & Guteriez, L. (1988). Co-localization of tyrosine hydroxylase (TH)-and serotonin (5HT)-immunoreactive innervation in the rat pituitary gland. *Neuroscience Letters,* **94**, 39–45.

Sanghera, M., Grady, S., Smith, W., Woodward D. J. & Porter, J. C. (1991). Incertohypothalamic A13 dopamine neurons: effect of gonadal steroids on tyrosine hydroxylase. *Neuroendocrinology,* **53**, 268–75.

Sar, M. (1983). Estradiol is concentrated in tyrosine hydroxylase-containing neurons of the hypothalamus. *Science,* **223**, 938–40.

Sar, M. (1988). Distribution of progestin-concentrating cells in rat brain: colocalization of [^{3}H]ORG.2058, a synthetic progestin, and antibodies to tyrosine hydroxylase in hypothalamus by combined autoradiography and immunocytochemistry. *Endocrinology,* **123**, 1110–18.

Sato, A., Shioda, S. & Nakai, Y. (1989). Catecholaminergic innervation of GRF-containing neurons in the rat hypothalamus revealed by electron-microscopic cytochemistry. *Cell and Tissue Research,* **258**, 31–4.

Sawchenko, P. E. & Swanson, L. W. (1981). Central noradrenergic pathways for the integration of hypothalamic neuroendocrine and autonomic response. *Science,* **214**, 685–7.

Sawchenko, P. E. & Swanson, L. W. (1982). The organization of noradrenergic pathways from the brainstem to the paraventricular and supraoptic nuclei in the rat. *Brain Research Reviews,* **4**, 275–325.

Sawyer, C. H., Markee, J. E. & Hollinshead, W. H. (1947). Inhibition of ovulation in the rabbit by the adrenergic blocking agent dibenamide. *Endocrinology,* **41**, 395–402.

Schimchowitsch, S., Stoekel, M. E., Vigny, A. & Porte, A. (1983). Oxytocinergic neurons with tyrosine hydroxylase-like immunoreactivity in the paraventricular nucleus of the rabbit hypothalamus. *Neuroscience Letters,* **43**, 55–9.

Schimchowitsch, S., Vuillez, P., Tappaz, M. L., Klein, M. J. & Stoeckel, M. L. (1991). Systematic presence of GABA-immunoreactivity in the tubero-infundibular and tubero-hypophyseal dopaminergic axonal systems: an ultrastructural immonogold study on several mammals. *Experimental Brain Research,* **83**, 575–86.

Scholer, J. & Sladek Jr, J. R. (1981). The supraoptic nucleus of the brattleboro rat has an altered afferent noradrenergic input. *Science,* **214**, 347–9.

Schröder, H. (1987). Aminergic innervation pattern of the rodent pineal gland: no apparent influence of time of day. *Acta Anatomica,* **129**, 22–6.

Schröder, H., Fujisawa, H. & Vollrath, L. (1987). Immunohistochemical characterization of rat pineal dopamine β-hydroxylase-containing structures: Use of a homologous monoclonal antibody. *Journal of Pineal Research,* **4**, 221–32.

Schwartz, J. C., Garbarg, M. & Pollard, H. (1986). Histaminergic transmission in the brain. In *Intrinsic Regulatory Systems of the Brain, Handbook of Physiology, Section 1, The Nervous System,* vol. 4, ed. F. E. Bloom, pp. 257–316. Bethesda, MD: American Physiological Society.

Seroogy, K., Tsuruoo, Y., Hökfelt T., Walsh, J., Fahrenkrug, J., Emson, P. C. & Goldstein, M. (1988). Further analysis of presence of peptides in dopamine neurons. Cholecystokinin, peptide histidine-isoleucine/vasoactive intestinal polypeptide and substance P in rat supramammillary region and mesencephalon. *Experimental Brain Research,* **72**, 523–34.

Sheppard, P. D., Mihailoff, G. A. & German, D. C. (1988). Anatomical and electrophysiological characterization of presumed dopamine-containing neurons within the supramammillary region of the rat. *Brain Research Bulletin,* **20**, 307–14.

Shi, W. X. & Bunney, B. S. (1992). Actions of neurotensin: a review of the electrophysiological studies. *Annals of the New York Academy of Sciences,* **668**, 129–45.

Shioda, S., Nakai, Y., Sato, A., Sunayama, S. & Shimoda, Y. (1986). Electron microscopic cytochemistry of the catecholaminergic innervation of TRH neurons in the rat hypothalamus. *Cell and Tissue Research,* **245**, 247–52.

Shiotani, Y., Kawai; Y., Jin, K. J., Kiyama, H. & Lin, L. P. (1990). Effects of constant light and darkness on the intrapineal neurons of golden hamsters, stained for tyrosine hydroxylase. A morphometric analysis. *Journal of Neural Transmission,* **82**, 231–7.

Silverman, A. J., Hou-Yu, A. & Oldfield, B. J. (1983). Ultrastructural identification of noradrenergic nerve terminals and vasopressin-containing neurons of the paraventricular nucleus in the same thin section. *Journal of Histochemistry and Cytochemistry,* **31**, 1151–6.

Silverman, A. J., Oldfield, B., Hou-Yu, A. & Zimmerman, E. A. (1985). The noradrenergic innervation of vasopressin neurons in the paraventricular nucleus of the hypothalamus: an ultrastructural study using radioautography and immunocytochemistry. *Brain Research,* **325**, 215–29.

Simerly, R. B., Swanson, L. W. & Gorski, R. A. (1985). The distribution of monoaminergic cells and fibers in a periventricular preoptic nucleus involved in the control of gonadotropin release: immunohistochemical evidence for a dopaminergic sexual dimorphism. *Brain Research,* **330**, 55–64.

Skagerberg, G., Björklund, A., Lindvall, O. & Schmidt, R. H. (1982). Origin and termination of the diencephalo-spinal dopaminergic system in the rat. *Brain Research Bulletin,* **9**, 237–44.

Skagerberg, G. & Lindvall, O. (1985). Organization of diencephalic dopamine neurones projecting to the spinal cord in the rat. *Brain Research,* **342**, 340–51.

Skagerberg, G., Meister, B., Hökfelt, T., Lindvall, O., Goldstein, M., Joh, T. & Cuello, A. C. (1988). Studies on dopamine-, tyrosine hydroxylase- and aromatic L-amino acid decarboxylase-containing cells in the rat diencephalon: comparison between formaldehyde-induced histofluorescence and immunofluorescence. *Neuroscience,* **24**, 605–20.

Sladek, C. D., Armstrong W. E. & Sladek Jr, J. R. (1983). Relationship between noradrenergic and osmotic control of vasopressin release. In *Structure and Function of Peptidergic and Aminergic Neurons*, ed. Y. Sano, Y. Ibata & E. A. Zimmerman, pp. 149–62. Tokyo: Japan Sci. Soc. Press; Utrecht: VNU Sci. Press BV.

Sladek, J. R. & Zimmerman, E. A. (1982). Simultaneous monoamine histofluorescence and neuropeptide immunohistochemistry: VI. Catecholaminergic innervation of vasopressin and oxytocin neurons in the rhesus monkey hypothalamus. *Brain Research Bulletin*, 9, 431–40.

Smits, R. P. J. M., Steinbusch, H. W. M. & Mulder, A. H. (1990). Distribution of dopamine-immunoreactive cell bodies in the guinea-pig brain. *Journal of Chemical Neuroanatomy*, 3, 101–23.

Spencer, S., Saper, C. B., Joh, T., Reis, D. J., Goldstein, M. & Raese, J. D. (1985). Distribution of catecholamine-containing neurons in the normal human hypothalamus. *Brain Research*, 328, 73–80.

Su, H.-S., Peng, Z.-C. & Li, Y.-W. (1987). Distribution of catecholamine-containing cell bodies in the human diencephalon. *Brain Research*, 409, 367–70.

Swanson, L. W. & Hartman, B. K. (1975). The central adrenergic system. An immunofluorescence study of the location of cell bodies and their efferent connections in the rat utilizing dopamine-beta-hydroxylase as a marker. *Journal of Comparative Neurology*, 163, 467–506.

Swanson, L. W. & Sawchenko, P. E. (1980). Paraventricular nucleus: a site for the integration of neuroendocrine and autonomic mechanisms. *Progress in Neuroendocrinology*, 31, 410–7.

Swanson, L. W., Sawchenko, P. E., Bérod, A., Hartman, B. K., Helle, K. B. & Vanorden, D. E. (1981). An immunohistochemical study of the organization of catecholaminergic cells and terminal fields in the paraventricular and supraoptic nuclei of the hypothalamus. *Journal of Comparative Neurology*, 196, 271–85.

Takada, M. (1990). The A11 catecholamine cell group: another origin of the dopaminergic innervation of the amygdala. *Neuroscience Letters*, 118, 132–5.

Takada, M., Campbell, K. J., Moriizumi, T. & Hattori, T. (1979). On the origin of the dopaminergic innervation of the paraventricular thalamic nucleus. *Neuroscience Letters*, 115, 33–6.

Takada, M., Li, Z. K. & Hattori, T. (1988). Single thalamic dopaminergic neurons project to both the neocortex and spinal cord. *Brain Research*, 455, 346–52.

Tetel, M. J. & Blaustein, J. D. (1991). Immunocytochemical evidence for noradrenergic regulation of estrogen receptors concentrations in the guinea pig hypothalamus. *Brain Research*, 565, 321–9.

Thiéry, J. C. & Martin, G. B. (1991). Neurohypophysiological control of the secretion of gonadotrophin-releasing hormone and luteinizing hormone in the sheep – a review. *Reproduction Fertility and Development*, 3, 137–73.

Thiéry, J -C., Martin, G. B., Tillet, Y., Caldani, M., Quentin, M., Jamain, C. & Ravault, J -P. (1989). Role of hypothalamic catecholamines in the regulation of luteinizing hormone and prolactin secretion in the ewe during seasonal anestrus. *Neuroendocrinology*, 49, 80–7.

Thind, K. K., Boggan, J. E., Song, T. & Goldsmith, P. C. (1987). Immunostaining reveals accumulation of serotonin and coexistence with tyrosine hydroxylase in hypothalamic neurons of acutely stalk-sectioned baboons. *Neuroendocrinology*, 45, 130–45.

Thind, K. K. & Goldsmith, P. C. (1986a). GABAergic and catecholaminergic synaptic interactions in the macaque hypothalamus: double label immunostaining with peroxidase-antiperoxidase and colloidal gold. *Brain Research*, 383, 215–27.

Thind, K. K. and Goldsmith, P. C. (1986b). Ultrastructural analysis of synapses involving tyrosine hydroxylase-containing neurons in the ventral periventricular hypothalamus of the macaque. *Brain Research*, 366, 37–52.

Thind, K. K. & Goldsmith, P. C. (1989). Corticotropin-releasing factor neurons innervate dopamine neurons in the periventricular hypothalamus of juvenile macaques. Synaptic evidence for a possible companion neurotransmitter. *Neuroendocrinology*, 50, 351–8.

Tillet, Y. (1988). Adrenergic neurons in sheep brain demonstrated by immunohistochemistry with antibodies to phenylethanolamine N-methyltransferase (PNMT) and dopamine-β-hydroxylase (DBH): absence of the C_1 cell group in the sheep brain. *Neuroscience Letters*, 95, 107–12.

Tillet, Y., Batailler, M., Krieger-Poullet, M. & Thibault, J. (1990). Presence of dopamine-immunoreactive cell bodies in the catecholaminergic group A15 of the sheep brain. *Histochemistry*, 93, 327–33.

Tillet, Y., Batailler, M. & Thibault, J. (1993). Neuronal projections to the medial preoptic area of the sheep, with special reference to monoaminergic afferents. Immunohistochemical and retrograde tract tracing studies. *Journal of Comparative Neurology*, 330, 195–220.

Tillet, Y., Caldani, M. & Batailler, M. (1989). Anatomical relationships of monoaminergic and neuropeptide Y-containing fibres with luteinizing hormone-releasing hormone systems in the preoptic area of the sheep brain: immunohistochemical studies. *Journal of Chemical Neuroanatomy*, 2, 319–26.

Tillet, Y. & Thibault, J. (1989). Catecholamine-containing neurons in the sheep brainstem and diencephalon: immunohistochemical study with tyrosine hydroxylase (TH) and dopamine-β-hydroxylase antibodies. *Journal of Comparative Neurology*, 290, 69–104.

Tillet, Y. & Thibault, J. (1993). Morphological relationships between tyrosine hydroxylase immunoreactive neurons and dopamine-beta-hydroxylase immunoreactive fibres in dopamine cell group A15 of the sheep. *Journal of Chemical Neuroanatomy*, 6, 69–78.

Tinner, B., Fuxe, K., Köhler, Ch., Hersh, L., Andersson, K., Jansson, A., Goldstein, M. & Agnati, L. F. (1989). Evidence for the existence of a population of arcuate neurons costoring choline acetyltransferase and tyrosine hydroxylase immunoreactivities in the male rat. *Neuroscience Letters*, **99**, 44–9.

Tison, F., Mons, N., Geffard, M. & Henry, P. (1990). Immunohistochemistry of endogenous L-DOPA in the rat posterior hypothalamus. *Histochemistry*, **93**, 655–60.

Tojo, K., Kata, Y., Kabayama, Y., Ohta, H., Inoue, T. & Imura, H. (1986). Further evidence that central neurotensin inhibits pituitary prolactin secretion by stimulating dopamine release from the hypothalamus. *Proceedings of the Society of Experimental Biology and Medicine*, **181**, 517–22.

Toney, T. W., Lookingland, K. J. & Moore, K. E. (1991). Role of testosterone in the regulation of tuberoinfundibular dopaminergic neurons in the male rat. *Neuroendocrinology*, **54**, 23–9.

Tong, Y. & Pelletier, G. (1992). Role of dopamine in the regulation of proopiomelanocortin (POMC) mRNA levels in the arcuate nucleus and pituitary gland of the female rat as studied by *in situ* hybridization. *Molecular Brain Research*, **15**, 27–32.

Tuomisto, J. & Männisto, P. (1985). Neurotransmitter regulation of anterior pituitary hormones. *Pharmacological Reviews*, **37**, 249–332.

Ungerstedt, U. (1971). Stereotaxic mapping of the monoamine pathways in the rat brain. *Acta Physiologica Scandinavica*, Suppl. 367, 1–48.

Van Den Pol, A. N. (1986). Tyrosine hydroxylase immunoreactive neurons throughout the hypothalamus receive glutamate decarboxylase immunoreactive synapses: a double pre-embedding immunocytochemical study with particulate silver and HRP. *Journal of Neuroscience*, **6**, 877–91.

Van Den Pol, A. N., Herbst, R. S. & Powell, J. F. (1984). Tyrosine hydroxylase-immunoreactive neurons of the hypothalamus: a light and electron microscopic study. *Neuroscience*, **13**, 1117–56.

Van Den Pol, A. N. & Tsujimoto, K. L. (1985). Neurotransmitters of the hypothalamic suprachiasmatic nucleus: immunocytochemical analysis of 25 antigens. *Neuroscience*, **15**, 1049–86.

Veening, J. G. (1978). Subcortical afferents of the amygdaloid complex in the rat: an HRP study. *Neuroscience Letters*, **8**, 197–202.

Vijayan, E., Krulich, L. & McCann, S. M. (1978). Catecholaminergic regulation of TSH and growth hormone release in ovariectomized and ovariectomized, steroid-primed rats. *Neuroendocrinology*, **26**, 174–85.

Vincent, S. R. (1988). Distribution of tyrosine hydroxylase-, dopamine-β-hydroxylase-, and phenylethanolamine-N-methyltransferase-immunoreactive neurons in the brain of the hamster (*Mesocricetus auratus*). *Journal of Comparative Neurology*, **268**, 584–99.

Vincent, S. R. & Hope, B. T. (1990). Tyrosine hydroxylase containing neurons lacking aromatic amino acid decarboxylase in the hamster brain. *Journal of Comparative Neurology*, **295**, 290–8.

Von Euler, U. S. (1946). A sympathomimetic ergone in adrenergic nerve fibers (sympathin) and its relation to adrenaline and noradrenaline. *Acta Physiologica Scandinavica*, **12**, 73–97.

Watson, S. J., Richard, C. W., Ciaranello, R. D. & Barchas, J. D. (1980). Interaction of opiates peptides and noradrenaline systems: light microscopic studies. *Peptides*, **1**, 23–30.

Weiner, R. I., Findell, P. R. & Kordon, C. (1988). Role of classic and peptide neuromediators in the neuroendocrine regulation of LH and Prolactin. In *The Physiology of Reproduction*, ed. E. Knobil, J Neill. *et al.*, pp. 1235–81. New York: Raven Press.

Weiner, R. I. & Ganong, W. F. (1978). Role of brain monoamines and histamine in regulation of anterior pituitary secretion. *Physiological Reviews*, **58**, 905–76.

Yoshimoto, Y., Sakai, K., Luppi, P. H., Fort, P., Salvert, D. & Jouvet, M. (1990). Catecholaminergic afferents to the cat median eminence as determined by double-labelling methods. *Neuroscience*, **36**, 491–501.

Young III, W. S., Warden, M. & Mezey, E. (1987). Tyrosine hydroxylase mRNA is increased by hyperosmotic stimuli in the paraventricular and supraoptic nuclei. *Neuroendocrinology*, **46**, 439–44.

10
Catecholaminergic innervation of the basal ganglia in mammals: anatomy and function

A. Reiner

Introduction

Since the work of Carlsson and his colleagues that led to their insightful conclusions on the role of dopamine depletion in the genesis of Parkinson's disease, dopamine has been known to play a major role in movement control via its influences on the basal ganglia (Carlsson, 1959; McGeer, Eccles & McGeer, 1978). In the intervening years, intensive research on this topic has generated a large amount of data on the localization of dopamine in the basal ganglia, its influence on single basal ganglia neurons, the pharmacology of its actions on the basal ganglia, and its global role in movement control. Although a comprehensive understanding of how dopamine influences movement has not yet been achieved, some broad understanding of its role has emerged. The present chapter will briefly review the major features of the anatomy and physiology of dopamine in the basal ganglia of mammals. Since the literature on this topic is massive, only some of it can be cited here. Note that the citations used here include many review papers, to which the reader is referred for more detail. In addition, we will review the small amount of data on the noradrenergic innervation of the basal ganglia.

General basal ganglia anatomy

The basal ganglia occupy the core and base of the mammalian cerebral hemispheres (Fig. 10.1) (Mehler, 1981; Reiner, Brauth & Karten, 1984; Parent, 1986; Graybiel, 1990). The basal ganglia can be considered to consist of two subdivisions, a somatic and a visceral/limbic (Heimer, Alheir & Zaborszky, 1985; Parent, 1986). The somatic subdivision, which is typically referred to as 'the basal ganglia' comprises a large field of relatively densely packed, medium-sized neurons and a smaller field of large, more sparsely packed neurons located ventromedial to the medium-sized cell field. The medium-sized cell field occupies much of the core of the telencephalon. Since this field is rendered striated in appearance by the many nerve bundles coursing through it, it has come to be called the striatum. Synonyms include the neostriatum and the caudate-putamen. The ventromedial large-sized cell field appears pale in cell stained material because it is cell sparse, and it has hence been termed the pallidum (or more fully the globus pallidus). Striatal and pallidal anatomy differ among mammals. I will focus on three mammalian groups (namely primates, carnivores and rodents), since members of these groups have received the most attention in studies of the mammalian basal ganglia. In primates, the striatum is divided into a dorsomedial region (termed the caudate nucleus) and a ventrolateral region (termed the putamen) by the fibers of the internal capsule passing between them (Fig. 10.2) (Parent, 1986; Graybiel, 1990). In contrast, in rodents and carnivores (rats and cats are the most well studied), corticofugal and corticopetal fibers do not form an internal capsule that divides the striatum (Fig. 10.1). Thus, a distinct caudate and putamen are not found in rodents and carnivores (Heimer et al., 1985; Parent, 1986). Although striatal neurons are generally uniform in their size and distribution, the use of various neurochemical markers has shown that the striatum consists of two major compartments, a patch (or striosomal) compartment and a matrix compartment, that can be recognized in all mammals (Graybiel, 1983; Parent, 1986; Gerfen, 1992). In primates, patch and matrix compartments can be distingu-

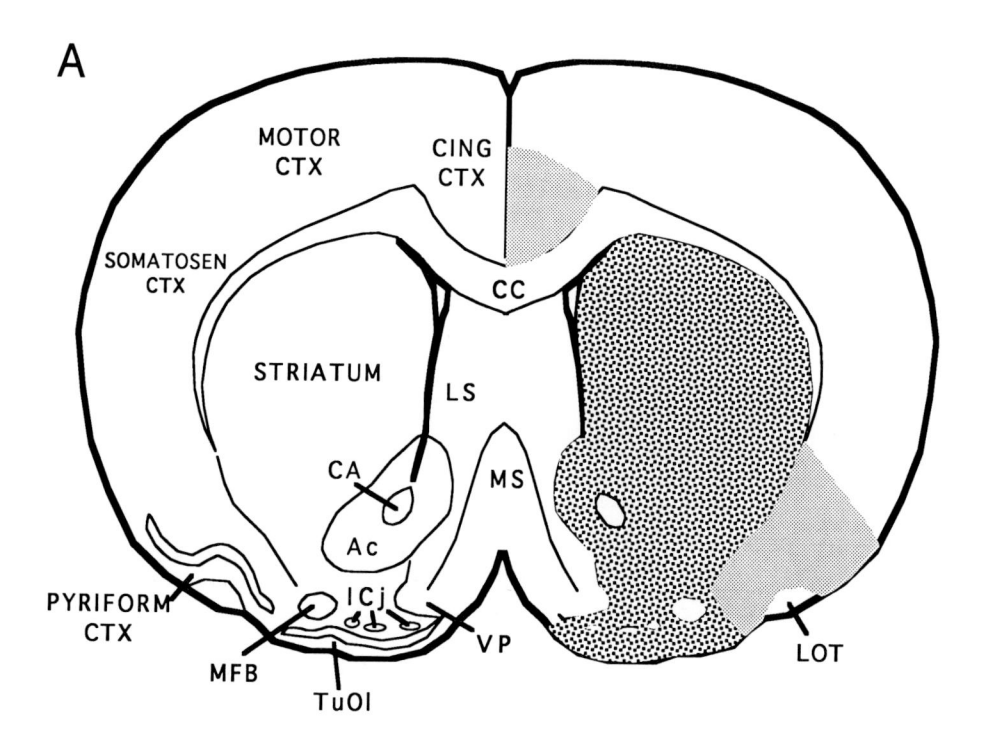

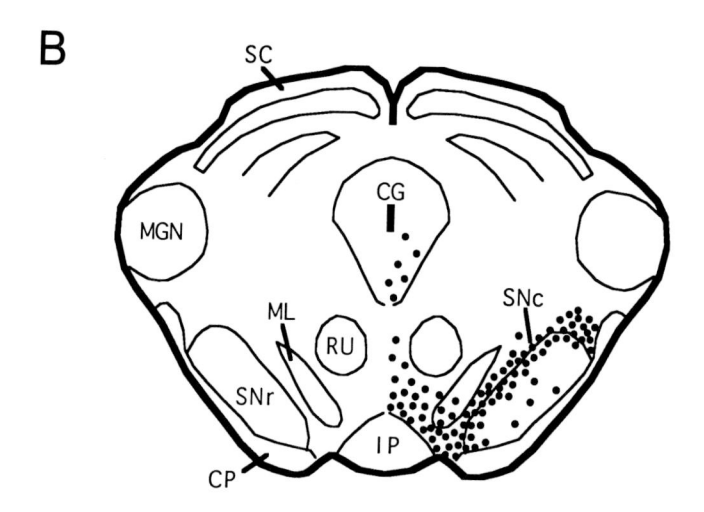

Fig. 10.1. Line drawing schematics illustrating the distribution of TH+ fibers in the rat telencephalon (a) and TH+ perikarya in the rat midbrain (b). Transverse sections of the brain are depicted. The shading density in (a) is proportional to the TH+ fiber abundance, while the perikaryal (dots) abundance in (b) is proportional to the TH+ perikaryal abundance. The distributions are based on those depicted in Hokfelt et al. (1984b). Abbreviations: Ac = nucleus accumbens; CA anterior commissure; CC = corpus callosum; CG = central gray; CING CTX = cingulate cortex; CP = cerebral peduncle; ICj = islets of Calleja; IP = interpeduncular nucleus; LOT = lateral olfactory tract; LS = lateral septum; MFB = medial forebrain bundle; MGN = medial geniculate nucleus; ML = medial lemniscus; MS = medial septum; PYRIFORM CTX = pyriform cortex; SOMATOSEN CTX = somatosensory cortex; SC = superior colliculus; SNc = substantia nigra, pars compacta; SNr = substantia nigra, pars reticulata; TuOl = olfactory tubercle; VP = ventral pallidum.

ished based on the greater cell packing density in the former (Goldman-Rakic, 1992). The patch compartment consists of an interconnected web of fingerlike extensions with irregular outlines that occupy about 15% of the striatum and are mainly located in the dorsomedial striatum (i.e. caudate). In any single plane of section, the interconnectedness of the patches is not apparent.

The pallidum too differs between primates on one hand and rats and cats on the other. In primates, the globus pallidus is divided into two segments, a lateral (GPL) and a medial (GPM) (Fig. 10.2). The GPL and GPM are sometimes also termed the external and internal segments of globus pallidus. In primates, these two segments are contiguous with one another and with the striatum (Mehler, 1981; Parent, 1986). The pallidum is also divided into two segments in carnivores and rodents: 1) a so-called globus pallidus (GP) that is contiguous with the striatum; and 2) an entopeduncular nucleus (EP) that is medial to but not contiguous with the globus pallidus (Parent, 1986; Heimer et al., 1985). Based on connectivity and neurotransmitter features of its input, the primate GPL is clearly homologous to the nonprimate GP and the primate GPM is considered homologous to the nonprimate EP (Haber & Nauta, 1983; Parent, 1986; Graybiel, 1990).

The second part of the basal ganglia is the visceral/limbic basal ganglia. While the somatic basal ganglia is thought to be involved in movement control, the visceral/limbic basal ganglia is thought to be involved in more affective and cognitive functions (Parent, 1986; Heimer et al., 1985). The visceral/limbic basal ganglia are sometimes called the ventral basal ganglia. This part of the basal ganglia also consists of striatal and pallidal parts, called the ventral striatum and ventral pallidum. The ventral striatum consists of diverse and scattered basal telencephalic cell groups, including the nucleus accumbens and the olfactory tubercle (Fig. 10.1). The central nucleus of the amygdala and the bed nucleus of the stria terminalis may also be included in the ventral striatum. The ventral pallidum includes the structure termed the ventral pallidum as well as deeper parts of the olfactory tubercle (Figs. 10.10).

General distribution of dopamine input to basal ganglia

As discussed further below, the basal ganglia receives a dopaminergic input from the A8–A10 dopaminergic neurons of the tegmentum. The general distribution of these terminals is very similar in all mammals studied, including diverse placental and marsupial species (Parent, 1986; Hazlett, Ho & Martin, 1991). Regardless of the limbic vs. somatic classification, all striatal parts of the basal ganglia are characterized by a rich dopaminergic innervation from the midbrain, while the pallidal parts are much poorer in such input (Figs. 10.1, 10.3) (Björklund & Lindvall, 1984; Hökfelt et al., 1984b; Voorn et al., 1986, 1988). The dopaminergic innervation of the striatum consists of a dense mat of terminals that label intensely for dopamine or tyrosine hydroxylase (TH) with immunolabeling methods or with formaldehyde induced fluorescence

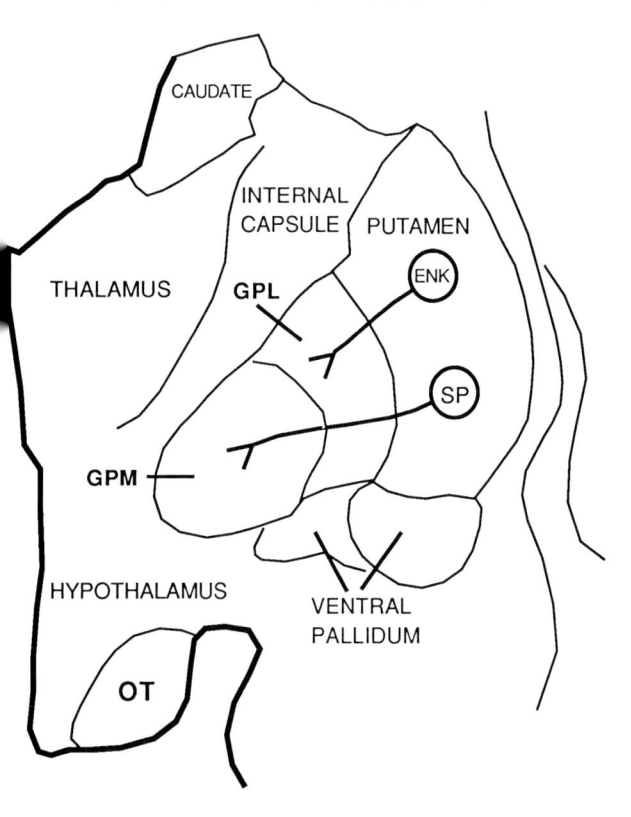

Fig. 10.2. Line drawing schematic illustrating the organization and some of the circuitry of the basal ganglia in primates. A transverse hemisection of the right side of the brain through the level of globus pallidus in rhesus monkey is depicted. The schematic indicates that GPL receives input from ENK+ striatal neurons while GPM receives input from SP+ striatal neurons. Note that the SP+ neurons also contain DYN and GABA, while the ENK+ neurons also contain GABA. See text for further details. Abbreviations: ENK = enkephalin; GPL = lateral segment of globus pallidus; GPM = medial segment of globus pallidus; OT = optic tract; SP = substance P.

250 A. Reiner

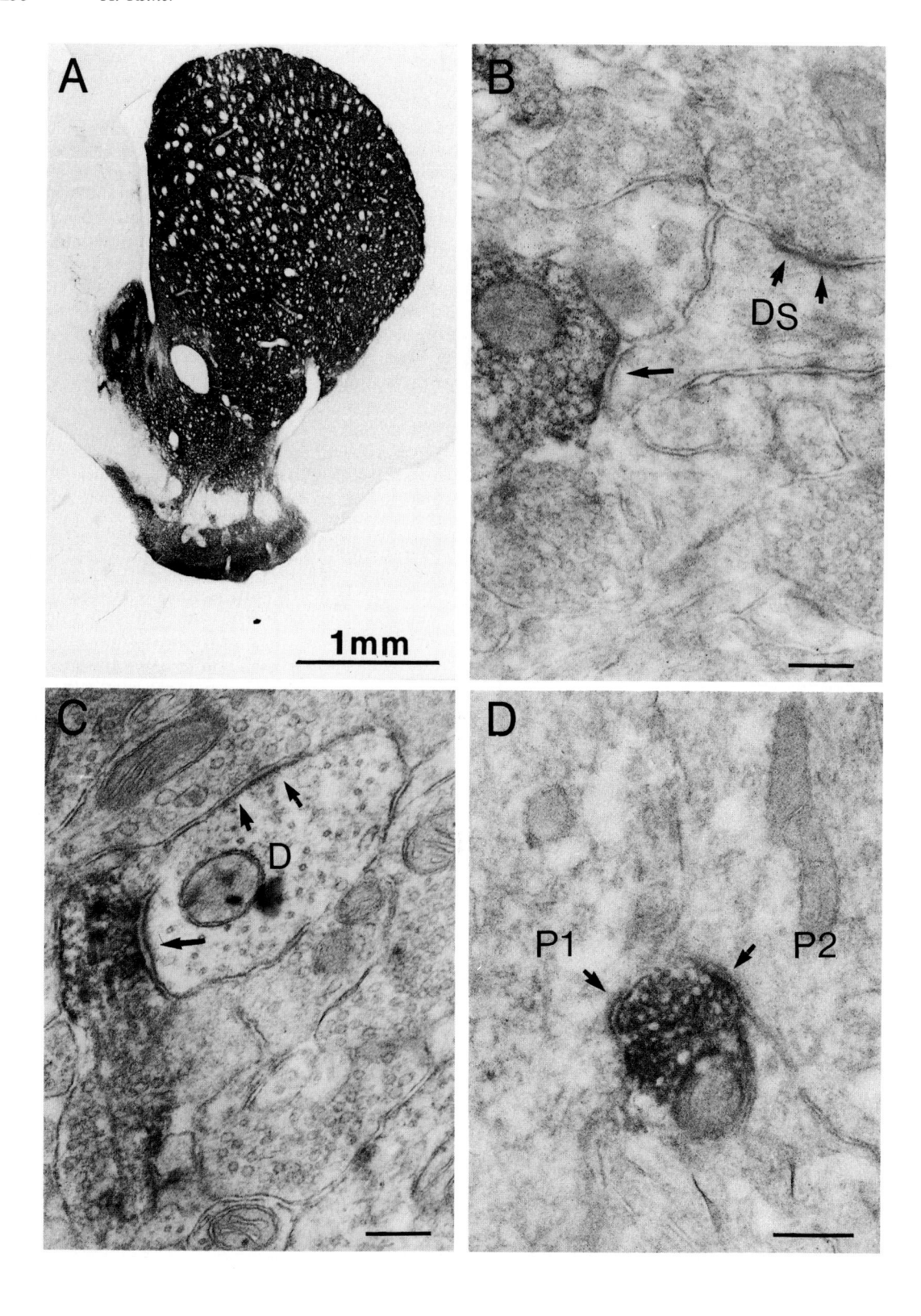

methods (Fig. 10.3). The striatal territories innervated include the dorsal (i.e. somatic) striatum, nucleus accumbens, the olfactory tubercle, the bed nucleus of the stria terminalis (BNST), and the central amygdala (Figs. 10.1, 10.3). Although not considered part of the striatum, the septum (which is a basal telencephalic cell group) contains dopaminergic (DA+) terminals that densely coat the perikarya and proximal dendrites of the lateral septal nucleus (Björklund & Lindvall, 1984; Hökfelt et al., 1984b; Voorn et al., 1986, 1988; Lavoie, Smith & Parent, 1989). Dopaminergic fibers are also conspicuous though much less abundant in the globus pallidus and the ventral pallidum. Although DA+ terminals in dorsal and ventral striatum are generally dense and highly abundant, there are some indications that there is heterogeneity in this innervation (Voorn et al., 1986; Graybiel, 1984; Graybiel, Hirsch & Agid, 1987; Lavoie et al., 1989). In particular, the dorsolateral rim of the dorsal striatum and some zones of the medial dorsal striatum that may correspond to the striosomal compartment appear richer in DA+ terminals (Fig. 10.3) (Voorn et al., 1986). The abundance of DA+ terminals in striosomes and in the dorsolateral rim appears particularly higher than in striatal matrix during late prenatal and early postnatal periods (Voorn et al., 1988; Graybiel, 1984).

Intrinsic organization and connectivity of basal ganglia

The striatum is the input part of the basal ganglia and has projections to pallidum and the tegmental dopaminergic cell fields (Figs. 10.2, 10.4) (Mehler, 1981; Reiner et al., 1984; Parent, 1986; Albin, Young & Penney, 1989; Graybiel, 1990). This is true in all mammals, as well as in birds and reptiles (Reiner et al., 1984; Parent, 1986), and it is true of both dorsal and ventral striatum (Alexander, DeLong & Strick, 1986; Parent, 1986; Alexander & Crutcher, 1990). It is important further to note that the striatal projections to the two parts of the GP (GPL and GPM) and to the two parts of the substantia nigra (the pars compacta and the pars reticulata) largely all arise from separate populations of neurons (Feger & Crossman, 1984; Parent, Bouchard & Smith, 1984; Beckstead & Cruz, 1986; Parent, 1986; Smith & Parent, 1986). Although the populations are separate, there may be some degree of collateral projection from one group to the target of another (Kawaguchi, Wilson & Emson, 1990). The striatal projections to both the GP and SN are topographically organized (Parent, 1986; Alexander & Crutcher, 1990; Selemon & Goldman-Rakic, 1990; Hedreen & DeLong, 1991).

Focussing on the dorsal striatum and the globus pallidus, the two different parts of the pallidum project to different targets. The GPL (or its non-primate homolog) projects to the subthalamic nucleus (and the tegmental DA cell field), while the GPM (or its nonprimate homologue) projects to a variety of thalamic nuclei (Fig. 10.6) (Haber et al., 1985; Parent, 1986; Alexander & Crutcher, 1990). The subthalamic nucleus (STN) projects back to the basal ganglia (particularly to the GPM), while the thalamic targets of the pallidum project to the striatum and/or to the cortex (Fig. 10.6). Of particular note, the GPM innervates thalamic nuclei that project to supplementary motor, premotor and motor cortex, thereby providing a route by which the basal ganglia can influence motor functions. The striatal input to the midbrain ends in either of two different parts of the substantia nigra (Figs. 10.5, 10.6). One input ends on

---

Caption to Fig. 10.3

Fig. 10.3. Photomicrograph illustrating the distribution of dopamine-containing terminals at a rostral transverse telencephalic level in rat (a), and EM photomicrographs showing the ultrastructural appearance of DA+ terminals in the rat striatum (b) – (d). Note that photomicrograph (a) shows DA+ fibers (as visualized using immunolabeling) to be very abundant in the dorsal striatum, nucleus accumbens and the olfactory tubercle. Note also, however, that both nucleus accumbens and the dorsal striatum contain patches that are richer in DA+ fibers than the surround and note that the dorsolateral rim of the striatum is also richer in DA+ fibers. Photomicrograph (b) shows an immunolabeled DA+ terminal making symmetric synaptic contact (arrow) with the neck of a dendritic spine (DS) that also receives an asymmetric synapse (pair of arrows) from an unlabeled round vesicle containing terminal (presumably a cortical or thalamic input). Photomicrograph (c) shows a DA+ terminal making symmetric synaptic contact (arrow) with a dendritic shaft (D) of a striatal neuron, which also receives an asymmetric input (pair of arrows) from a terminal containing round clear core and large dense core vesicles. Photomicrograph (d) shows a DA+ terminal seemingly synaptically contacting (arrows) two adjacent perikarya (P1 and P2). Photomicrograph (a) is from Voorn et al. (1988) and photomicrographs (b) – (d) are from Voorn and Buijs (1987). All photomicrographs are reproduced with the kind permission of P. Voorn and the publishers of these articles. Scale bars in (b) – (d) equals 200 nm.

252   A. Reiner

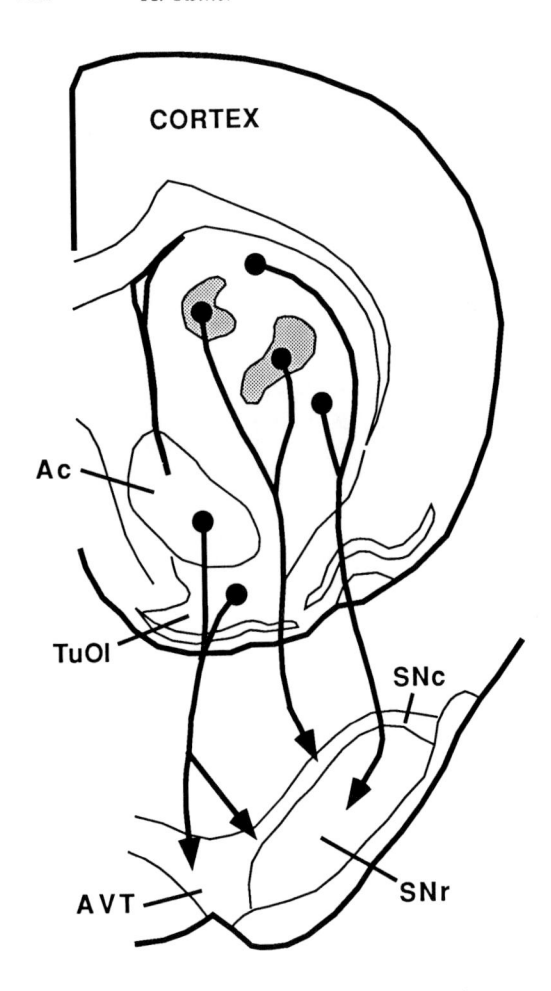

Fig. 10.4. Line drawing schematics illustrating the compartmental organization of projections from the dorsal and ventral striatum in rats to the AVT and nigral regions. Transverse hemisections of the right side of the brain are depicted. Note that striosomal neurons of the dorsal striatum (striosomes shown as shaded zones) project mainly to the dopaminergic neurons of the SNc, while matrix neurons of the dorsal striatum project mainly to the SNr. Neurons of the ventral striatum (nucleus accumbens and the olfactory tubercle) project mainly to the dopaminergic neurons of the AVT and medial SNc. The striatal neurons giving rise to these projections are mainly those containing SP, DYN and GABA. See text for further details. Abbreviations: Ac = nucleus accumbens; AVT = ventral tegmental area; SNc = substantia nigra, pars compacta; SNr = substantia nigra, pars reticulata; TuOl = olfactory tubercle.

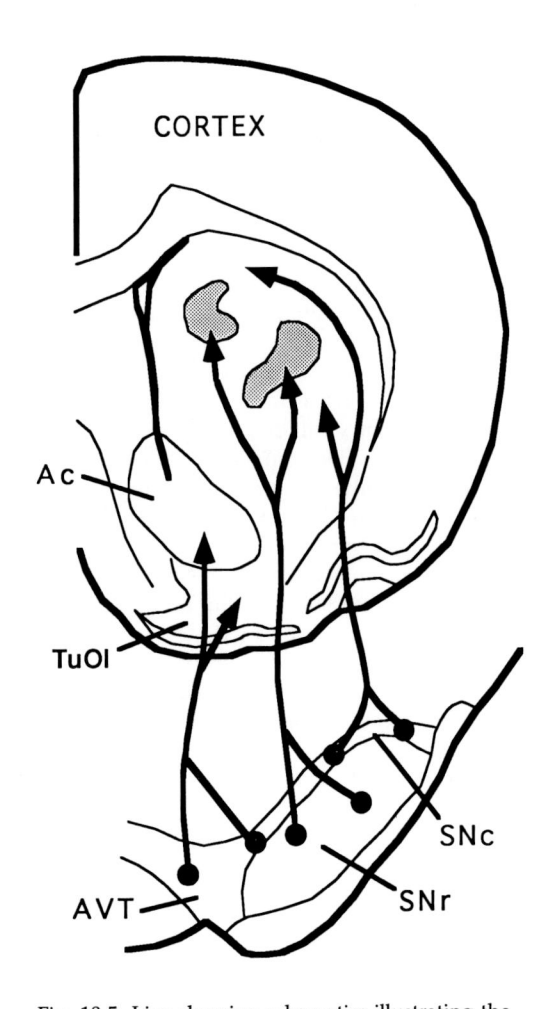

Fig. 10.5. Line drawing schematics illustrating the compartmental organization of projections from the AVT and nigral regions to the dorsal and ventral striatum in rats. Transverse hemisections of the right side of the brain are depicted. Note that striosomal neurons of the dorsal striatum (striosomes shown as shaded zones) receive input mainly from dopaminergic neurons of the SNr, while matrix neurons of the dorsal striatum receive their input mainly from dopaminergic neurons of the SNc. Neurons of the ventral striatum (nucleus accumbens and the olfactory tubercle) receive their input mainly from dopaminergic neurons of the AVT and medial SNc. See text for further details. Abbreviations: Ac=nucleus accumbens; AVT=ventral tegmental area; SNc=substantia nigra, pars compacta; SNr=substantia nigra, pars reticulata; TuOl=olfactory tubercle.

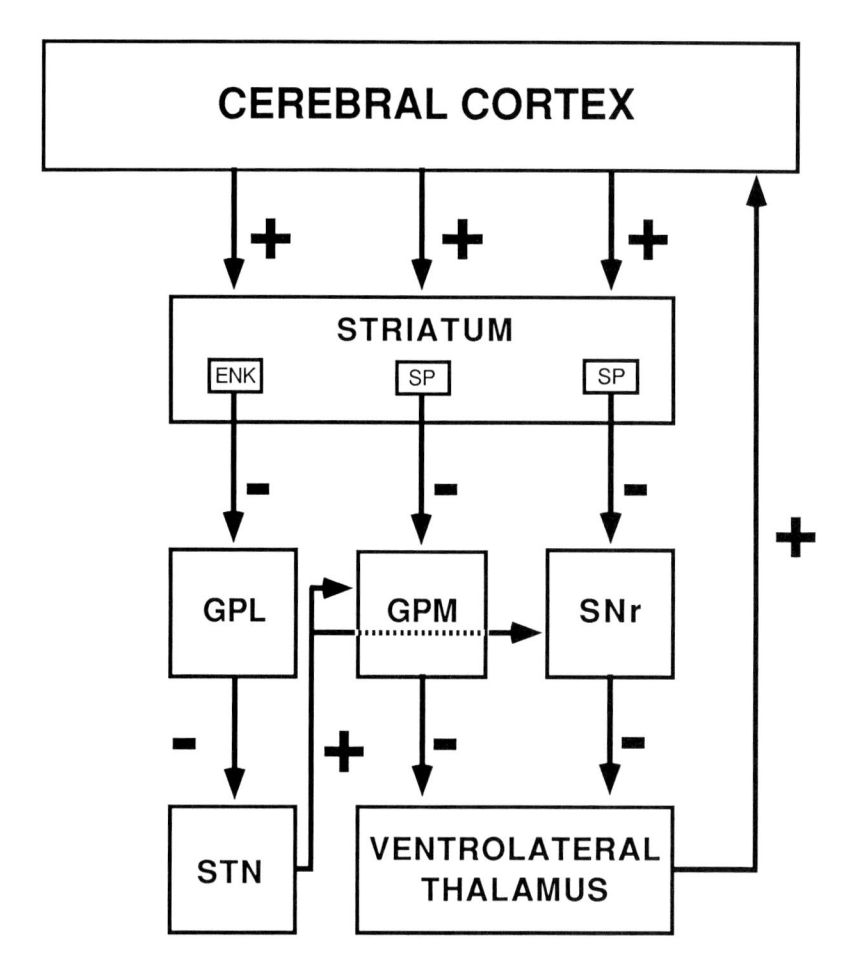

Figure 10.6. Simplified wiring diagram of the basal ganglia and its targets, emphasizing the functional organization of this circuitry as it relates to the motor functions of the basal ganglia. The pluses and minuses indicate whether the specific projections use an excitatory (plus) or inhibitory (minus) transmitter. Note that, while striatal and pallidal output is inhibitory, cortical and thalamic is excitatory. Similar diagrams have been presented previously by Albin et al. (1989) and Alexander and Crutcher (1990), among others. Note that the SP+ striatal neurons also contain DYN and GABA, while the ENK+ striatal neurons also contain GABA. See text for further details. Abbreviations: ENK=enkephalin; GPL=lateral segment of globus pallidus; GPM=medial segment of globus pallidus; SNc – substantia nigra, pars compacta; SNr=substantia nigra, pars reticulata; SP=substance P; STN= subthalamic nucleus.

pallidal-like neurons in the pars reticulata (SNr) (Alexander et al., 1986; Parent, 1986; Albin et al., 1989b; Alexander & Crutcher, 1990). These neurons project to the same general thalamic regions as the GPM (Fig. 10.6). The other striatonigral projection ends on dopamine neurons of the substantia nigra pars compacta (A9) and those of the retrorubral area (A8) (Fig. 10.4).

The visceral/limbic basal ganglia outputs are similar in organization (Haber et al., 1985; Heimer et al., 1985; Alexander et al., 1986; Parent, 1986; Alexander & Crutcher, 1990; Haber et al., 1990). The ventral striatum projects to midbrain dopamine neurons (A10 and medial A9) and to the ventral pallidum. The ventral pallidum in turn projects to various thalamic cell groups that project to frontal and prefrontal limbic cortices, as well as to the subthalamic nucleus. One major point of note regarding the ventral striatum output is that the projection of this region to the midbrain dopaminergic neurons is extensive, while that of the somatic striatum is more restricted (Berendse, Groenewegen & Lohman, 1992). Thus, although the dorsal striatum (which is more involved in movement control) is reciprocally connected with the midbrain dopamine neurons, the ventral striatum (which is more involved in affective and cognitive functions) is the major influence over the dopaminergic midbrain neurons innervating the dorsal striatum.

The striatum receives input from three major sites: the cortex, the intralaminar nuclei of the thalamus and the dopamine (DA) neurons of the midbrain (Mehler, 1981; Reiner et al., 1984; Parent, 1986; Albin et al., 1989b; Graybiel, 1990). It also receives serotonergic input from the raphe and is rich in cholinergic and GABAergic terminals from striatal neurons themselves (Reiner et al., 1984; Parent, 1986; Albin et al., 1989b; Graybiel, 1990). The ventral striatum receives similar inputs, except cortical inputs arise from 'limbic' cortex. In addition to striatal input, the pallidum receives input from few other regions. Of note, however, is the major input to the GPM from the subthalamic nucleus (Albin et al., 1989b). As noted above, the globus pallidus receives sparse DA+ input.

Organization of connections between striatum and dopaminergic midbrain neurons

Topography of DA input to striatum

Numerous pathway tracing studies or studies combining labeling of DA fibers with selective midbrain lesions in diverse mammalian species have shown a medial to lateral, an anterior to posterior and an inverted dorsal to ventral topography in the projection of midbrain DA neurons to the striatal territories of the basal telencephalon (Szabo, 1980; Björklund & Lindvall, 1984; Hazlett et al., 1991). The dopaminergic projection arises from the A10 (the ventral tegmental area, AVT), the A9 (the substantia nigra, SN) and the A8 (retrorubral) cell groups (Figs. 10.1, 10.5). Although a topography is present, individual dopaminergic neurons have widespread striatal projections and the overlap in projections among individual dopaminergic neurons is extensive (Björklund & Lindvall, 1984). It should be noted that a small projection to the striatum (less than 5% of the total projection) also arises from the nondopaminergic neurons of the AVT – SN – retrorubral regions (Guyenet & Crane, 1981; van der Kooy, Coscina & Hattori, 1981; Björklund & Lindvall, 1984). Superimposed on this general topography of the tegmentostriatal projection are some differences between dorsal and ventral striatum and between patch and matrix in the DA innervation from the midbrain (Fig. 10.5). Studies in diverse mammalian species have shown that the A10 and medial SN innervate the ventral striatum, while dopaminergic neurons in A8 and the dorsal tier of the SN pars compacta (SNc) innervate matrix of dorsal striatum, and dopaminergic neurons in the ventral tier of the SNc and in the SNr innervate patches (Gerfen, Herkenham & Thibault, 1987b; Jimenez-Castellanos & Graybiel, 1987; Langer & Graybiel, 1989). The dopaminergic fibers innervating the ventral striatum course within the medial forebrain bundle (MFB), while those innervating dorsal striatum travel partly in and partly dorsolateral to the MFB (Björklund & Lindvall, 1984).

Finally, it should be noted that the midbrain DA neurons also contain other neuroactive substances. For example, many DA neurons of A10 and A9 projecting to ventral striatum and cortex also contain cholecystokinin (CCK) in rats (Hökfelt et al., 1980a, b). Similarly, some midbrain DA neurons in rats contain neurotensin (Seroogy et al., 1989), while a population of DA neurons in the dorsal tier of the rat SNc contain the calcium binding protein calbindin (Gerfen, Bainbridge & Thibault, 1987a). The specific projection of the calbindin-containing SNc neurons to striatal matrix may account for the high levels of calbindin in the neuropil of the matrix (Gerfen et al., 1987a). As noted in the chapter by Berger & Gaspar, however, CCK and neurotensin appear to be largely absent from midbrain DA neurons in primates.

Topography of striatal input to A8–A10

The A8–A10 cell groups consist of dopaminergic neurons located in the ventral tegmentum, with the A10 (AVT) most rostral and medial, A9 more lateral and the A8 lateral and caudal (Figs. 10.1, 10.5). The A10 has a rostral extension that meets the hypothalamic A11 and caudal A10 neurons overlap the distribution of serotonergic raphe neurons (Hökfelt et al., 1984a). The A8–A10 are present in all mammals and in all cases distinct boundaries between them are ill-defined (Jacobowitz & MacLean, 1978; Murray, Dominguez & Martinez, 1982; Pearson et al., 1983; Felten & Sladek, 1983; Björklund & Lindvall, 1984; Hökfelt et al., 1984a, b; Haber & Groenewegen, 1990; Hazlett et al., 1991). The dendrites of the DA neurons in the A9 tend to be oriented in either the horizontal plane or in the vertical plane (Björklund & Lindvall, 1984). Further, mammals show some variation in the extent to which DA neurons are present in the SNr (Parent, 1986; Smeets, 1991). In primates, for example, the SNc has numerous fingerlike extensions of DA neurons that are found throughout the depth of SNr (Parent, 1986). The projection from the dorsal and ventral striatum to the A8–A10 appears to have the same general topographic order as the dopaminergic tegmento-striatal projection (Gerfen, 1985; Heimer et al., 1985). Several authors have further shown that different parts of the ventral striatum project differentially to A10, SNc and SNr, while the striosomes of the dorsal striatum project only to SNc and matrix only to SNr (Fig. 10.4) (Gerfen, 1985; Haber et al., 1990; Berendse, Groenewegen & Lohman, 1992). Among pallidal projections to the dopaminergic neurons, the GPL projects to the SN, while VP projects to the AVT and SN (Haber et al., 1985).

Organization of basal ganglia cell types

The striatum of all mammals consists of the same basic cell types. There are two categories of striatal neurons, spiny neurons that make up about 95% of striatal neurons, and aspiny neurons that make up the rest (Heimer et al., 1985; Gerfen, 1988). The aspiny types are all striatal interneurons and they include three major types: 1) cholinergic neurons; 2) neurons co-containing GABA, the neuropeptide LANT6, and the calcium binding protein parvalbumin; and 3) neurons cocontaining the neuropeptides somatostatin and neuropeptide Y (NPY) and the enzyme nitric oxide synthase (NOS) (Vincent et al., 1982; Reiner et al., 1984; Parent, 1986; Albin et al., 1990; Vincent & Kimura, 1992; Reiner & Anderson, 1993). The spiny neurons are all GABAergic, but about half of the spiny neurons also cocontain tachykinins (substance P and neurokinin A) and dynorphins (DYN), while the others contain enkephalins (ENK) (Anderson & Reiner, 1990; Graybiel, 1990; Reiner & Anderson, 1990; Gerfen, 1992). The spiny neurons are the striatal projection neurons and the neurochemically different types have different projections. For example, the GABA/ENK spiny neurons project mainly to the GPL (or its nonprimate homologue) (Fig. 10.2) (Parent, 1986; Reiner & Anderson, 1990). A small, and presumably separate, population of GABA/ENK neurons also projects to the midbrain dopamine neurons (Reiner & Anderson, 1990). In primates, separate populations of GABA/substance P/DYN neurons project to the GPM, the SNr and SNc (Figs. 10.2, 10.4) (Parent, 1986; Reiner et al., 1988). Similar populations are present in nonprimates, but they may have collateral projections to other striatal targets to a greater extent than is the case in primates (Kawaguchi et al., 1990). Although the spiny striatal neurons are relatively uniformly distributed in the striatum, there is some tendency for neuronal types with a common target to be clustered. For example, in cat and primates, striato-GPM neurons, striato-GPL neurons and striato-SNr neurons are clustered into separate interlacing domains within the striatal matrix (Jimenez-Castellanos & Graybiel, 1989; Selemon & Goldman-Rakic, 1990; Gimenez-Amaya & Graybiel, 1991). Similarly, the striatal neurons projecting to the pars compacta are clustered in the striosomes (Fig. 10.4) (Graybiel, 1983; Gerfen, 1992). Nonetheless, no striatal zone is occupied by exclusively ENK+ or substance P (SP)+ neurons (Gerfen 1992).

The cell types present in the ventral striatum are identical to those in the dorsal striatum. For example, ventral striatum also contains GABA+ projection neurons, SP+ projection neurons, DYN+ projection neurons, ENK+ projection neurons and the various interneurons (Heimer et al., 1985; Parent, 1986; Voorn, Gerfen & Groenewegen, 1989; Haber, Wolfe & Groenewegen, 1990; Reiner & Anderson, 1990; Martin et al., 1991). The patterns of neuropeptide and neurotransmitter colocalization for ventral striatum, however, have not been elucidated (Reiner & Anderson, 1990).

All pallidal neurons are GABAergic and they have also been shown to contain the neuropeptide LANT6 and the calcium binding protein parvalbumin in primate and rodent species (Reiner & Carraway, 1987; Reiner, 1987; Reiner & Anderson, 1993). This is also true of the pars reticulata of the substantia nigra (SNr). Since different segments of the somatic pallidum receive different striatal inputs, different parts of the pallidum are enriched in different neuropeptides (Fig. 10.2) (Haber & Elde, 1981; Haber & Nauta, 1983; Parent, 1986; Reiner et al., 1988; Reiner & Anderson, 1990). Thus, the GPM (or its non-primate homologue) is enriched in terminals containing SP and DYN, while the GPL (or its non-primate homologue) is enriched in ENK+ terminals. Both the SNr and SNc, as well as AVT and A8 are rich in SP+/DYN+ terminals. ENK+ terminals are present in AVT and SNc (Haber & Groenewegen, 1990; Reiner & Anderson, 1990; Sesack & Pickel, 1992).

Connections between midbrain dopamine neurons and striatal neurons at the ultrastructural and single cell level

Dopaminergic input to the striatum
Dopaminergic fibers in the striatum are thin and possess varicosities along their length (Gerfen et al., 1987; Gerfen, 1988). Individual midbrain DA neurons have

been estimated to give rise to up to 500 000 terminals in the striatum (Moore & Bloom, 1978). These varicosities form small terminals that range in size from 0.1–0.7μm and that make symmetric contacts with striatal neurons (Fig. 10.3) (Pickel et al., 1981; Arluison, Dietl & Thibault, 1984; Gerfen et al., 1987b). In contrast, nondopaminergic fibers and terminals from the A8–A10 region are 1.0–2.0μm (Gerfen et al., 1987). The dopaminergic terminals of the striatum (labeled with either anti-TH or anti-DA) reportedly make up about 20% of striatal terminals (Gerfen, 1988). Since spiny neurons are the predominant neuronal type in the basal ganglia, they are the predominant target of the dopaminergic terminals. Several studies have shown that this input is most common on the neck of dendritic spines (which receive cortical and thalamic input at their head) and somewhat less common on dendritic shafts (Bouyer et al., 1984; Freund, Powell & Smith, 1984). Dopaminergic terminals on cell bodies are rare. Multiple label studies have shown that SP+ striatonigral neurons and ENK+ (presumably striatopallidal) neurons all receive DA+ input, although the details of the possible differences between spiny neuron types in the somatodendritic distribution or abundance of the DA input has not been determined (Freund et al., 1984; Kubota, Inagaki & Kito, 1986a; Kubota et al., 1986b; Mendez et al., 1990). Nonetheless, the observation that the main types of striatal projection neurons receive DA+ input is consistent with electrophysiological studies showing that spiny striatal neurons invariably show monosynaptic responses to activation of their dopaminergic input (Kitai et al., 1976; Kitai, 1980). The DA+ input to striatal interneurons has not been well characterized. The somatostatin/NPY/NOS neurons appear to receive a sparse axosomatic and axodendritic DA+ input (Aoki & Pickel, 1988; Vuillet et al., 1989). Cholinergic neurons do not appear to receive direct DA+ input; rather, the well-documented dopaminergic effects on acetylcholine release may occur at the level of the terminals, since adjacent cholinergic and DA+ terminals apparently often contact common target structures (Gerfen, 1988).

Striatal input to midbrain dopaminergic neurons

Striatal terminals in the A8–A10 field contact both dopaminergic and non-dopaminergic neurons (Somogyi et al., 1982; Smith & Bolam, 1990). The non-dopaminergic neurons appear to be GABAergic (Albin et al., 1989b; Smith & Bolam, 1990). All of these striatal terminals contain GABA, but depending on the neuron of origin some also contain SP and DYN, while others contain ENK (Reiner & Anderson, 1990). Both the SP+ type and the ENK+ type striatal terminals have been shown to contact DA+ neurons (Chang, 1988; Mahalik, 1988; Bolam & Smith, 1990; Sesack & Pickel, 1992; Liang, Kozlowski & German, 1993). The striatal terminals typically make symmetric contacts on the dendrites of these neurons, and more rarely the perikarya. The nondopaminergic neurons of the SNr that receive striatal input in many cases project to the thalamic targets comparable to those of the GPM and to the superior colliculus (Alexander et al., 1986; Albin et al., 1989b). The striatorecipient SNr nondopaminergic neurons also appear to give rise to a local inhibitory input to tegmental DA neurons (Grace & Bunney, 1979, 1985a,b).

Receptors on striatal neurons and tegmental dopamine neurons

Since the work of Kebabian and his colleagues (Kebabian, Petzgold & Greengard, 1972; Kebabian & Calne, 1979), it has been known that the actions of dopamine are mediated by either of two types of receptors, a D1 type that stimulates cAMP synthesis and a D2 type that inhibits cAMP synthesis (Clark & White, 1987). Since cAMP serves to directly or indirectly affect various intracellular processes, including the opening or closing of ion channels, the two types of dopamine receptors produce their effects (in large part) by modulating cAMP levels (Clark & White, 1987). Since this early work defining two pharmacologically distinguishable types of dopamine receptors, it has become clear that D1 receptors activate adenylate cyclase via a stimulatory guanine nucleotide binding protein (called a G protein). Similarly, D2 receptors inhibit adenylate cyclase via an inhibitory G protein. Most recently, the genes for a number of different types of D1 and D2 receptors have been cloned and sequenced. Thus, the mammalian CNS contains several different types of D1 and D2 receptors, which are all fundamentally similar in their protein sequence. As a consequence of this progress in understanding dopamine receptors, studies of their distribution in the basal ganglia have progressed from ligand binding studies to localize D1 and D2 type receptor sites, to in situ hybridization studies for localizing which striatal neurons synthesize which specific type of dopamine receptor (as defined by amino acid sequence). Further, findings on intracellular messengers involved in transducing the dopamine signal has led to the ability to detect dopaminoceptive neurons based on their content of these messengers.

The first anatomical studies on dopamine receptors in the basal ganglia used ligand binding methods to describe the distribution of D1 type and D2 type receptors. Studies of the striatum have generally revealed that both D1 and D2 type receptors are distributed in a patchy fashion in the striatum (Joyce, Sapp & Marshall, 1986; Richfield, Young & Penney, 1987; Beckstead, Wooten & Trugman, 1988; Beckstead, 1984; Besson, Graybiel & Nastuk, 1988; Camps et al., 1989; Cortés et al., 1989; Berendse & Richfield, 1993). In general, D1 receptors appear more abundant than D2 receptors and some but not all of the patches in the dorsal striatum and ventral striatum align with histochemically definable compartments. Some authors have reported that some dopamine receptor patches of the dorsal striatum align specifically with the striosomal compartment, although the precise pattern and extent of correspondence may show some variability among mammalian species and during development (Joyce et al., 1986; Richfield et al., 1987; Besson et al., 1988; Camps et al., 1989; Cortés et al., 1989; Berendse & Richfield, 1993). Ligand binding studies (in some cases combined with selective lesions) have also revealed that D1 receptor levels are high on the terminals of striatal neurons projecting to the GPM (or its non-primate equivalent) and the SN (Altar & Hauser, 1987; Richfield et al., 1987; Beckstead et al., 1988; Besson et al., 1988; Cortés et al., 1989), while D2 receptors are high on the terminals of striatal neurons projecting to the GPL (or its nonprimate equivalent) (Richfield et al., 1987; Beckstead et al., 1988; Camps et al., 1989). This differential distribution of the two types of dopamine receptors on terminals in striatal target areas has led to the suggestion by many authors that D1 receptors are more abundant than D2 receptors on striatonigral and striato-GPM neurons (both of which contain SP), while D2 receptors are more abundant on striato-GPL neurons. To some extent, this notion has been supported by studies examining the identity of striatal neurons pharmacologically activated by selective D1 and D2 type agonists (Robertson, Vincent & Fibiger, 1990; 1992; Gerfen et al., 1990; Gerfen, 1992; Robertson, 1992; Zhang et al., 1992) and by studies examining the types of receptors lost from the striatum following selective destruction of striatonigral or striatopallidal projection neurons (Harrison, Wiley & Wooten, 1990, 1992).

The discovery using molecular cloning and sequencing methods of at least six different types of dopamine receptors complicates discussion of the localization of dopamine receptors on striatal neurons. The first two dopamine receptor genes discovered were termed D1 and D2 and it was clear that they coded for proteins having the pharmacology of D1 and D2 type receptors, respectively. It was briefly thought that they were *the* D1 and D2 receptors. Several additional dopamine receptor genes, however, were discovered soon thereafter (some coding for proteins with D1 type pharmacology and some for proteins with D2 type pharmacology), and they were named D3, D4 and D5 in sequence as they were discovered (Sibley, 1991; Van Tol et al., 1991; Sunahara et al., 1991; DeKeyser, 1993). The D2 gene was additionally found to give rise to two different splice variants (DeKeyser, 1993). The various dopamine receptors all were found to differ somewhat in their affinities for dopamine and its various agonists and antagonists. As a consequence of these discoveries, a terminological quagmire was created by the superimposition of six distinct types of dopamine receptors (as defined by amino acid sequence) upon the traditional division of dopamine receptors into two classes by pharmacological methods (Sibley, 1991). Thus, reference to a D1 receptor could now mean either the D1 identified by amino acid sequence or the D1 defined by pharmacology. This distinction is important because both the D1 and D5 receptors defined by molecular methods are D1 type receptors as defined by pharmacology, while D2 (short and long splice variants), D3 and D4 dopamine receptors are D2 type receptors as defined by pharmacology. Although the terminological problem has not been resolved with any consensus solution, investigators tend to use the terms 'D1 type' and 'D2 type' for the pharmacological level classification and terms such as 'D1' and 'D2' to refer to specific receptors as defined by amino acid sequence. Note that using this terminology it would be possible to find that a neuron contained, for example, D2 mRNA and D5 mRNA, but not D1 mRNA. Under these circumstances, this neuron would still be regarded as possessing both D1 type and D2 type receptors and, hence, be likely to show D1-type and D2-type pharmacology.

A number of studies have built on the molecular biology of dopamine receptors to determine the cellular localization of the protein or mRNA for specific dopamine receptors (Weiner, Levey & Brann, 1990; Bouthenet et al., 1991; Lemoine & Bloch, 1991; Meador-Woodruff et al., 1991b; Weiner et al., 1991). Prior to such studies it was thought that D1 type receptors were on striatal neurons, while D2 type receptors were located both presynaptically on dopaminergic and corticostriatal terminals (Clark & White, 1987; Maura, Giardi & Raiteri, 1988) and postsynaptically on neurons (Clark & White, 1987; Beckstead, 1984; DeKeyser, 1993). It is now clear that all six known dopamine receptor proteins are present postsynaptically in the striatum, while the presynaptic localization is more equivocal (DeKeyser, 1993; Surmeier et al., 1993). Because of the implication of the above noted studies that there is preferential localization of D1 and D2 type receptors on striatal projection neurons, a number of studies have combined *in situ* hybridization for mRNA for different dopamine receptors with methods for labeling specific types of striatal neurons, or have done double-labeling for dopamine receptor mRNA. These studies have shown that D1 mRNA appears more abundant in the perikarya of SP+ striatal neurons (Lemoine, Normand & Bloch, 1991; Gerfen, 1992), while D2 mRNA appears more abundant in the perikarya of ENK+ striatal neurons (Lemoine et al., 1990; Gerfen, 1992). Double-label studies, however, have shown that up to 40% of striatal neurons contain both D1 and D2 mRNA (Meador-Woodruff et al., 1991a). Several lines of evidence argue that this degree of colocalization is likely to be an underestimate: 1) pharmacological data examining the cellular locus of D1–D2 receptor synergy in the striatum (Walters et al., 1987; Weick & Walters, 1987; Bertorello et al., 1990; Piomelli et al., 1991); 2) electrophysiological data using D1 and D2 agonists and antagonists (Ohno, Sasa & Takaori, 1987; Surmeier et al., 1992, 1993); 3) single cell mRNA amplification data (Surmeier et al., 1992, 1993); 4) data on the localization of D1 and D2 receptor proteins (Ariano et al., 1992); 5) data on the localization of the D1 receptor coupled stimulatory G protein (Drinnan et al., 1991; Herve et al., 1993); and 6) data on the localization of the dopamine D1 receptor coupled phosphoprotein DARPP32 (Quimet et al., 1984, 1992; Anderson & Reiner, 1991). These various studies indicate that virtually all striatal projection neurons possess D1 receptors and D2 receptors. Single cell mRNA amplification data and anatomical localization of mRNA or protein further suggests that D3, D4 and D5 are also present in striatal neurons (Sibley, 1991; Van Tol et al., 1991; Sunahara et al., 1991; Surmeier et al., 1992, 1993). Thus, the extent of dopamine receptor colocalization appears to be extensive, at least for striatal projection neurons. In contrast, cholinergic neurons and somatostatin/NPY/NOS striatal interneurons appear to possess D2 receptors, but few or no D1 receptors (Anderson & Reiner, 1991; Dawson, Dawson & Wamsley 1990; Lemoine et al., 1991; Weiner et al., 1991; Stoof et al., 1992). Anatomical data thus supports the pharmacological data that dopamine produces its effects on acetylcholine release via D2 receptors (Stoof et al., 1992). Note that although colocalization of dopamine receptor types may be common for striatal projection neurons, there is still compelling evidence favoring the view that SP+ strianigral neurons preferentially possess more D1 type receptors on some portion of their surface, while ENK+ striatopallidal neurons preferentially possess D2 type receptors on some portion of their surface. The full details of the differential distribution of the different dopamine receptors on the soma, dendrites and axon terminals of individual types of striatal neurons, however, needs to be elucidated.

Finally, ligand binding studies have consistently shown that nigral dopamine neurons are rich in the D2 type dopamine receptor, but not in D1 type receptors (Palacios & Wamsley, 1984; Clark & White, 1987; Palacios et al., 1988; DeKeyser, 1993). Studies localizing dopamine receptor mRNA have specifically indicated that A8–A10 dopamine neurons synthesize D2 and D3 receptors, while A10 dopamine neurons (but not A8–A9 dopamine neurons) also synthesize D4 (Sokoloff et al., 1990; Bouthenet et al., 1991; Civelli et al., 1991; Lemoine & Bloch, 1991). The various D2 type receptors made by dopamine neurons appear to be autoreceptors by which dopamine neurons regulate their own state of activity (Clark & White, 1987). One distinctive feature of the localization of the D1–D5 proteins in the telencephalon and midbrain is the greater abundance of D3 and particularly D4 in the mesostriatal system (i.e. A10 neurons and their major targets, ventral striatum and prefrontal cortex) than in the nigrostriatal system (i.e. A8–A9 neurons and their major target, the dorsal striatum) (Bouthenet et al., 1991; Civelli et al., 1991; Van Tol et al., 1991). This differential distribution and the higher affinity of D3 receptors and particularly D4 receptors for atypical neuroleptics such as clozapine may explain why atypical neuroleptics do not produce movement disturbances – they have little effect on the movement related dopaminergic system, namely the nigrostriatal system (Sokoloff et al., 1990; Civelli et al., 1991; Van Tol et al., 1991).

Dopamine uptake mechanisms in the striatum

Dopamine released by terminals in the striatum is deactivated by energy-dependent uptake mechanisms, followed by enzymatic breakdown of the dopamine (McGeer et al., 1978). Studies of the uptake mechanism (called the transporter and now known to be a membrane protein) have shown that cocaine and other compounds such as mazindol and GBR 12935 are antagonists of dopamine uptake that can be used to study the localization of the uptake sites (Richfield, 1991). Such studies show the uptake sites to be high in the striatum, where they are located predominantly presynaptically (Richfield, 1991; Cline et al., 1992). A patchy striatal pattern of dopamine transporters has been reported in some species (Donnan et al., 1991). Cloning and sequencing studies have recently characterized the structure of a cocaine-sensitive dopamine transporter that is synthesized by dopaminergic midbrain neurons and presumably subsequently transported to terminals in the striatum (Kilty, Lorang & Amara, 1991; Shimada et al., 1991).

Functional considerations

Function of dopaminergic input to the striatum: cellular actions of dopamine

The disturbances in basal ganglia-mediated control of movement stemming from the loss of dopaminergic midbrain neurons in Parkinson's disease indicate that dopamine must have an important effect on basal ganglia function. Elucidating the cellular basis of this role requires determining the effects of dopamine on the activity of individual striatal neurons. Despite the seeming simplicity of this task, definitive conclusions on this topic have been very difficult to reach (Moore and Bloom, 1978). Some studies have reported the dopaminergic input to be excitatory, and some have reported it to be inhibitory (Moore & Bloom, 1978; Kitai, 1980; Grace & Bunney, 1985a; Akaoka, Saunier & Chouvet, 1987). While part of this past discrepancy may be attributable to technical problems (Moore & Bloom, 1978; Grace & Bunney, 1985a), it is now clear that much of it may be attributable to the fact that dopamine acts at multiple receptor sites. Further, with the recognition that dopamine receptors belong to the G-protein coupled family of receptors has come the realization that dopamine actions on neurons will not produce brisk changes in membrane potential but will instead modulate the overall responsivity of striatal neurons to their other inputs (meaning that the application of dopamine agonists may have no apparent effect on the activity of a quiescent neuron). It now seems clear that D1 agonists decrease the spontaneous or glutamate evoked discharge of striatal neurons, while D2 agonists increase the spontaneous or glutamate evoked discharge of striatal neurons (Uchimura, Higashi & Nishi, 1986; Hu & Wang, 1988; Surmeier et al., 1992). D2-type agonists selective for the D3 receptor may, however, also reduce evoked discharge (Hu & Wang, 1988; Surmeier et al., 1992). The effects of the different dopamine agonists on neuronal activity appear to be mediated via a G-protein mediated linkage of the D1, D2

and D3 receptors to voltage dependent sodium channels, with D1 and D3 agonists reducing currents through these channels and D2 agonists increasing them (Surmeier et al., 1992, 1993). The discovery of these receptor type specific effects of dopamine helps clarify why some prior studies had found dopamine to have both depolarizing and hyperpolarizing actions (Herrling & Hull, 1980; Akaoka et al., 1987). In addition to the effects of dopamine agonists on sodium ion channels, dopamine agonists also affect potassium ion channels in striatal neurons (Uchimura et al., 1986; Freedman & Weight, 1988; Kitai & Surmeier, 1993), though the implications of this for neuronal activity are less well understood.

A number of pharmacological, biochemical and anatomical studies have shown that the dopaminergic input to the striatum produces opposite effects on ENK neurons than on SP/DYN neurons, with the dopaminergic input apparently inhibiting ENK synthesis in the former (Tang, Costa & Schwartz, 1983; Normand et al., 1988; Reiner & Anderson, 1990) and increasing tachykinin synthesis in the latter (Nylander & Terenius, 1987; Reiner & Anderson, 1990; Zhang et al., 1992). This has raised the question of how dopamine produces these opposite effects on these two types of neurons and what the actions of dopamine really are on these neuronal types at the cellular level (since the above described effects may be indirectly mediated). One category of explanation that has been proposed is that differential abundance of the various types of dopamine receptors on the two types of neurons, coupled with the functional differences between the receptor types, may account for the differential responses to dopamine (Gerfen, 1992; Robertson, 1992). While at some level it is likely that this is true, the specific proposition that there is preferential abundance of inhibitory D2 receptors on ENK+ striatopallidal neurons and excitatory D1 receptors on SP+ striatonigral and striatopallidal neurons is clearly not correct since the above electrophysiological data unequivocally shows the dopamine action at the D1 site to be inhibitory and at the D2 site to be excitatory (Gerfen, 1992; Surmeier et al., 1992, 1993). Thus, the differential responses of ENK+ neurons and SP+ neurons to dopamine cannot be accounted for in a simple, direct way by a differential distribution of D1 type and D2 type receptors on SP+ and ENK+ striatal neurons.

One final interesting possible feature of the dopaminergic control of striatal neurons derives from the data suggesting that D2 receptors are located postsynaptic to tegmental dopaminergic input, while D1 receptors are located at extrajunctional sites

(Ohno, Sasa & Takaori, 1985, 1986; Akaoka et al., 1987; Bickford-Wimer et al., 1990). If this is true, it could have implications for understanding the organization and function of dopaminergic input to the striatum and for treating such disorders involving dopamine depletion as Parkinson's disease. It seems likely that with the increasing availability of antisera directed against specific dopamine receptor proteins this issue will be resolved (Ariano et al., 1992).

Function of striatal input to midbrain dopamine neurons: cellular and behavioral actions of GABA and neuropeptides

Dopamine neurons of the tegmentum have slow and irregular firing rates, with periods of bursting activity (Grace & Bunney, 1985a). Presumably this bursting activity is associated with release of dopamine from dopaminergic terminals within the striatum. The neurons also release dopamine locally within the tegmentum that serves to act on the D2 autoreceptors on dopaminergic neurons and suppress discharge (Björklund & Lindvall, 1984; Grace & Bunny, 1985a). The dopamine released locally in the tegmentum, however, appears not to act on the dopamine receptors present on the striatal terminals within the tegmentum (Ryan et al., 1989). The activity of dopamine neurons appears to increase in response to salient sensory stimuli or in response to stimuli associated with reward or reward itself, but typically not prior to or during movement (Strecker & Jacobs, 1985; Romo & Schultz, 1990; Schultz & Romo, 1990; Schultz, Apicella & Ljungberg, 1993). Thus, the dopaminergic neurons appear to play their role in movement by the modulation of striatal activity in relation to motivationally relevant stimuli. The dopamine neurons may receive this type of information from cortical, amygdaloid or ventral striatal afferents (Schultz & Romo, 1990). The dopaminergic neurons, however, are also under the direct (patch neurons) and indirect influence (matrix neurons) of dorsal striatal neurons, with the indirect influence being mediated by GABAergic local circuit neurons in the SNr (Olianas et al., 1978; Grace & Bunney, 1979, 1985b).

Studies on the postsynaptic actions of the various neurotransmitters and neuropeptides found in striatal projection neurons have largely focused on the individual roles of GABA, SP, DYN and ENK in the substantia nigra. For example, several studies show that SP and DYN have largely opposite effects on nigral neurons, with SP (as well as the related tachykinin neurokinin A) having excitatory effects on dopaminergic neurons (Innis, Andrade & Aghajanian, 1985; Kalivas, 1985; Reiner & Anderson, 1990) and increasing dopamine turnover in the striatum (Reiner & Anderson, 1990), and with DYN having inhibitory effects on nondopaminergic nigral neurons (Grace & Bunney, 1985b; Kalivas, 1985; Walker et al., 1987; Reiner & Anderson, 1990) and no stimulatory effect on dopamine turnover in the striatum (Reiner & Anderson, 1990). Despite the apparent fact that SP (and neurokinin A, NKA) act largely on different populations of nigral neurons than does DYN, when injected unilaterally into the nigra SP, NKA and DYN all produce similar behavioral effects, i.e. rotation away from the injected side (Reiner & Anderson, 1990). Since unilateral activation of nigral dopaminergic neurons is known to produce contralateral rotation (Ungerstedt et al., 1969; Ungerstedt & Arbuthnott, 1970), the effects of SP and NKA on both rotation and the release and turnover of dopamine by nigrostriatal terminals are consistent with the conclusion that SP and NKA activate nigral dopaminergic neurons. Such effects may be mediated by the direct synaptic input of tachykinin-containing striatal projection neurons onto tegmental dopaminergic neurons, or by synaptic input to nondopaminergic nigral neurons that have input to dopaminergic nigral neurons. The contralateral rotation that occurs following intranigral injections of DYN, however, is not mediated via nigral dopaminergic neurons (the effects occur even in the absence of nigral dopaminergic neurons), but instead appears to be mediated via nondopaminergic nigral neurons, either neurons projecting to the thalamus or superior colliculus or interneurons projecting to these projection neurons (Olianas et al., 1978; Parent, 1986; Reiner & Anderson, 1990). Thus, SP and DYN co-released from individual striatal terminals in the nigra may produce similar behaviors by acting on different target neurons. Along these lines, it would be of interest to know whether tachykinin receptors in the nigra are localized on different postsynaptic structures than are kappa opiate receptors.

ENK and its analogs also induce contralateral circling when injected unilaterally into the substantia nigra (Reiner & Anderson, 1990). This effect is inhibited by the depletion of dopaminergic nigral neurons, suggesting that the circling effect is mediated via the dopaminergic neurons. Consistent with this possibility, morphine is reported to exert an apparently indirect excitatory action on dopaminergic nigral neurons (Walker et al., 1987; Reiner & Anderson, 1990). Such effects may occur through inhibition of nigral nondopaminergic neurons that have inhibitory input to dopaminergic neurons or through presynaptic inhibition of striatal terminals that have inhibitory input to dopaminergic neurons.

The actions of GABA in the substantia nigra appear similar to those of DYN. Injections of GABA into the substantia nigra also yield contralateral rotation that is not dependent on the presence of dopaminergic nigrostriatal projection neurons (Reid et al., 1988; Reiner & Anderson, 1990). Further, electrophysiological studies have demonstrated that GABA has inhibitory effects on neurons in the substantia nigra. Unlike DYN, however, GABA inhibits both dopaminergic and nondopaminergic nigral neurons (Grace & Bunney, 1979, 1985b). Thus, like DYN, GABA released from striatonigral terminals may influence movement by inhibitory actions on nondopaminergic nigral neurons. Unlike DYN, GABA may secondarily influence movement through direct inhibitory effects on dopaminergic neurons in the substantia nigra (Grace & Bunney, 1979, 1985b).

Very little information is available on the actions of SP, DYN, ENK or GABA when applied in combination to tegmental neurons in striatal target areas. One study to date (Tan & Tsou, 1988) has examined the possible synergy between SP and DYN in the substantia nigra. Based on the results of the studies of the individual actions of SP and DYN in the substantia nigra, one might predict that SP and DYN infused together into the nigra would produce a potentiated response on the same behavior. The results of Tan & Tsou (1988) were consistent with this prediction, with SP and DYN injected together yielding a greater increase in dopamine turnover and contralateral circling than either SP or DYN alone. Nonetheless, very little is known about the functional significance of SP/DYN/GABA colocalization or ENK/GABA colocalization with respect to the actions of striatal terminals on tegmental dopaminergic neurons.

General scheme of basal ganglia functional organization

The basal ganglia are regarded as playing a role in the planning, initiation and execution of movement (Mardsen, 1980; DeLong et al., 1986; Albin et al., 1989b). It is thought that the diverse inputs that the striatum receives inform it of body position, motivational state, and goals, as well as the movement routines required to execute those goals (DeLong et al., 1986; Alexander & Crutcher, 1990; Apicella, Scarnati & Schultz, 1991; Gardiner & Kitai, 1992). The basal ganglia is then thought to integrate this diverse information and facilitate the behavior appropriate for the behavioral context. A variety of data from experimental studies in animals and from the clinical symptoms in human diseases of the basal ganglia have helped reveal the roles of specific parts of the basal ganglia circuitry in these functions (Fig. 10.6). In general, the tonically active GABAergic GPM and SNr inputs to the thalamus are thought to inhibit thalamic neurons capable of exciting cortical neurons directly responsible for movement control (Alexander et al., 1986; Alexander & Crutcher, 1990). The striatum is thought to release thalamic neurons from such inhibition when GABAergic striatopallidal projection neurons and striato-SNr projection neurons are excited in the appropriate fashion by their combination of inputs. As a consequence of such excitation, these striatal neurons inhibit their target pallidal and nigral neurons, thereby disinhibiting the thalamic target neurons of these pallidal and nigral neurons (Alexander et al., 1986; Albin et al., 1989b; Alexander & Crutcher, 1990).

Activation of a particular group of striatal neurons projecting to GPM or SNr is thought to be responsible for initiating the behavior inhibited by that part of the pallidum or nigra. Such specificity in the movements influenced by striatal activation appears to arise from the highly topographically ordered nature of the inputs and outputs of the basal ganglia (Parent, 1986; Tremblay & Filion, 1989). Loss or hypofunction of the striato-GPM and striato-SNr neurons (which contain SP and DYN) would consequently result in the clinical syndromes termed akinesia (inability to produce movement) and bradykinesia (slowed movement). Such symptoms occur in Parkinson's disease, in which the seemingly excitatory drive of the dopaminergic input to striatal SP+ neurons is lost (due to degeneration of nigral neurons), and in the late stages of Huntington's disease, in which SP+ striatal neurons themselves degenerate (Reiner et al., 1988; Albin et al., 1989b; DeLong, 1990).

If the striato-GPM and striato-SNr neurons promote movement, what is the role of the ENK+ striato-GPL neurons? The answer appears to be that this circuit is involved in suppressing unwanted movements (Reiner et al., 1988; Albin et al., 1989b, 1990). Ablation of the subthalamic nucleus (STN), which receives a prominent inhibitory GABAergic input from GPL and has a prominent excitatory (glutamatergic) input to GPM and SNr (Smith & Parent, 1989; Albin et al., 1989a), or injection of the GABA antagonist bicuculline into the GPL (which mimics the loss of inhibitory striato-GPL neurons) produces the excessive movements characteristic of chorea (Reiner et al., 1988; Albin et al., 1989b, 1990).

On this basis, it has been hypothesized that the striato-GPL-STN circuit might be involved in the suppression of unwanted motor programs and that selective dysfunction of striato-GPL projection neurons might occur in diseases (such as in early and middle stage Huntington's disease) that are characterized by excessive movements (i.e. hyperkinesia) (Reiner et al., 1988; Albin et al., 1989b, 1990). Such selective hypofunction of the ENK+ striato-GPL neurons would result in the disinhibition of GPL neurons projecting to the STN. The resulting increased inhibitory GPL output to the STN would cause a reduction of STN activity, thus mimicking the effects of STN ablation. This, in turn, would result in decreased excitatory input to the GPM and SNr from the STN, thereby decreasing activity of GPM and SNr neurons and releasing nondesired behaviors from inhibition, precipitating choreiform movements (Reiner et al., 1988; Albin et al., 1989b, 1990; DeLong, 1990). Some available data helps clarify how concomitant activation of striato-GPM/SNr neurons and striato-GPL neurons (as would presumably occur in a normal basal ganglia) could promote desired movement and inhibit unwanted movement. The data suggest that normal activation of a portion of the striato-GPM/SNr-thalamic pathway inhibits a focal group of GPM/SNr neurons, while the normal concomitant activation of the striato-GPL-STN-GPM pathway enhances the activity of the GPM/SNr neurons surrounding the focally inhibited neurons (Chevalier et al., 1985; Deniau & Chevalier, 1985; Tremblay & Filion, 1991). The consequence of this at the level of the thalamic target of GPM/SNr is that some specific behaviors are disinhibited (i.e. those controlled by the thalamic zone receiving input from the focally inhibited GPM/SNr neurons) while conflicting behaviors are inhibited (i.e. those controlled by the thalamic zone receiving input from the activated GPM/SNr zone surrounding the focally inhibited neurons). Within this conceptual framework, hypofunction of the striato-GPL neurons would yield an inability to suppress unwanted and conflicting movements during normal ongoing behavior.

Note that several authors have incorporated these ideas into models of basal ganglia function in which the SP+ and ENK+ striatopallidal pathways are assigned different roles (see Fig. 10.6) (Alexander et al., 1986; Albin et al., 1989b; Alexander & Crutcher, 1990). Note that the specific roles of the neuropeptides SP and ENK in influencing the GABAergic pallidal or nigral neurons is not well known; the neuropeptides in these circuits for now are mainly a way in which to distinguish the circuits. Finally, the ventral basal ganglia has been thought to play a role in sequencing and organizing affect and thought that is analogous to the role of the dorsal basal ganglia in ordering and sequencing movement (Alexander et al., 1986; Alexander & Crutcher, 1990).

Role of dopamine in the overall basal ganglia functions

Bilateral loss of the dopaminergic input to the striatum by experimental, accidental or naturally occurring loss of the tegmental dopamine neurons in all mammalian species studied (or administration of dopamine antagonists) leads to akinesia, catalepsy, a hunched posture, sensory inattention and behavioral unresponsiveness (due to an inability to coordinate movements and orient toward stimuli) (Björklund & Lindvall, 1984; Crossman, 1987; Albin et al., 1989b; Onn et al., 1990). Thus, dopamine released in the striatum is somehow permissive for movement. Conversely, excessive dopamine in the striatum (or dopamine agonists) promotes hyperkinetic movements. Thus, dopamine also appears to be important for suppressing unwanted movements. The recent studies showing that dopamine has opposite actions on SP+ striatonigral and striato-GPM neurons compared to ENK+ striato-GPL neurons helps explain the various symptoms that occur with paucity or excess of striatal dopamine. Since dopamine appears to facilitate the activity of SP+ neurons and suppress ENK+ neurons, one would expect dopamine's removal to diminish the activity of SP+ neurons and increase that of ENK+ neurons (Reiner & Anderson, 1990; Gerfen, 1992). One piece of evidence showing that this occurs is that experimentally induced depletion of striatal dopamine does lead to increased activity in SNr and GPM neurons (due to decreased inhibition from the now less active striatal SP+ neurons) and decreased GPL activity (due to increased inhibition from the now more active ENK+ striatal neurons) (Pan & Walters, 1988; Filion, Tremblay & Bedard, 1989; Filion & Tremblay, 1991; Yoshida, 1993). Based on the model of basal ganglia function presented above, this should lead to increased activity of the subthalamic neurons projecting to GPM/SNr and decreased activity of the thalamic target neurons of the GPM/SNr (due to both decreased inhibition of GPM/SNr neurons by striatum and increased excitation by the subthalamus), resulting in suppression of both desired and non-desired movements, which in fact appears to be the case (Yoshida, 1993). Chronic treatment with nonspecific dopamine agonists result in the opposite pattern of effects, with GPL activity increased due to suppression of ENK+ striato-GPL neurons, and GPM/SNr activity decreased due to diminished activity of SP+ striato-GPM/SNr neurons and increased GPL inhibition of subthalamic input to GPM/SNr. This would have the effect of promoting both desired and nondesired movements, i.e. hyperkinesis. Of further interest is that the loss of dopaminergic input to the striatum results in a loss of the topographic specificity of the influence of the striatum on the pallidum (Tremblay, Filion & Bedard, 1989). This could further account for the impairments in the ability to execute desired movements when striatal dopamine is depleted.

Human diseases involving dopamine and the basal ganglia

There are a number of human diseases involving the basal ganglia or its ventral counterpart in which dysfunction of the dopaminergic system is implicated. The most obvious of these is Parkinson's disease, in which dopaminergic neurons of the substantia nigra and retrorubral area are lost (Barbeau, 1986). Pathological studies in humans and experimental studies in animals indicate that loss of up to 90% of the nigral/retrorubral DA neurons is largely without symptoms, while loss of more is associated with the akinesia/bradykinesia of Parkinson's disease (Barbeau, 1986; Orr et al., 1986). The mechanism by which loss of DA might produce these symptoms was discussed above (Albin et al., 1989b). Supplying the remaining dopaminergic terminals with massive amounts of precursor for dopamine (i.e. L-DOPA) has proven an effective therapy for this disease, so long as there are surviving DA terminals in the striatum (Barbeau, 1986; Albin et al., 1989b). Dopamine is also implicated in such hyperkinetic disorders as Huntington's disease (HD) and Tourette Syndrome (TS). In the case of HD, the early stages are characterized by the occurrence of choreiform (unwanted movements) and are associated with the preferential loss of ENK+ striatopallidal neurons (Reiner et al., 1988). The early and relatively selective loss of ENK+ striato-GPL projections may account for the efficacy of dopamine antagonists in ameliorating the choreiform movements of HD and subthalamic lesions. Dopamine receptor antagonists appear to increase the activity of ENK+ striatopallidal neurons (Reiner & Anderson, 1990; Gerfen, 1992). Increases in the activity and neurotransmitter release by surviving striato-GPL neurons in HD patients could partially overcome the reduction in the number of ENK+ neurons projecting to GPL (Reiner et al., 1988; Albin et al., 1989b, 1990). Tourette syndrome is also a hyperkinetic syndrome, characterized by unwanted utterances and actions (Goetz, 1986). An excessive action of dopamine, perhaps in the striatum, appears implicated since dopamine antagonists are known to mitigate TS symptoms (Goetz, 1986). Numerous studies, however, have failed to detect any neuropathological changes in TS and there appear to be no alterations in dopamine levels or receptors in the striatum (Goetz, 1986). Recent studies do suggest that dopamine transporter mechanisms may be altered in TS (Factor, Sanchez-Ramos & Weiner, 1988; Singer, 1992). Finally, schizophrenia is a debilitating mood, thought and behavior disorder that affects a high percentage of the human population (Snyder et al., 1974). Dopamine is again implicated in schizophrenia in some unknown way since dopamine antagonists also eliminate or lessen its symptoms (Snyder et al., 1974). The abundance of D3 and D4 receptors in the mesolimbic dopamine neurons and their major target areas (the ventral striatum and prefrontal cortex) and the greater efficacy of D3/D4 specific antagonists (such as clozapine) in ameliorating schizophrenia suggest the mesolimbic system and ventral striatum (rather than the nigrostriatal system and dorsal striatum) as a site of disturbed neural function in schizophrenia. Whether there is some primary defect in dopamine transmission in this part of the brain (leading to exuberant dopamine action) or whether reducing DA transmission merely compensates for some other functional imbalance in mesolimbic circuitry is uncertain. Nonetheless, attributing the affective and cognitive disarray characteristic of schizophrenia to ventral striatal dysfunction is consistent with the hypothesized role of this region in organizing and sequencing affect and cognition.

Noradrenergic and adrenergic input to basal ganglia

Although the midbrain dopaminergic neurons and the dopaminergic innervation of the basal ganglia have received the most extensive interest, other catecholaminergic fibers also innervate the basal ganglia. Specifically, the striatum receives noradrenergic and adrenergic input from the locus coeruleus (A6) and the solitary tract region (A2/C2). Due to the massive dopaminergic input and the clinically demonstrable importance of this input, the noradrenergic and adrenergic input has largely been ignored. The small amount of available information on this innervation are discussed below.

Noradrenergic and adrenergic input topography and LM features

Noradrenergic (dopamine beta hydroxylase-containing, DBH+) fibers are found mainly medially in the striatum and in ventral striatal territories (Moore & Card, 1984). The locus coeruleus (LoC) of the isthmic central gray is the major noradrenergic cell group of the brain and is the major source of this DBH+ input, although the A2 cell group also appears to contribute. The LoC has extensive and widely collateralized projections throughout brain, including an extensive input to the telencephalon. The fibers of this projection ascend within the so-called dorsal bundle (which is located within the central tegmental bundle), as do ascending A2 fibers. These fibers give rise to a light innervation of ventral striatal areas such as nucleus accumbens and the olfactory tubercle, and moderate input to the central amygdaloid nucleus. The more medial and ventral parts of the bed nucleus of the stria terminalis (BNST) are densely innervated. The DBH+ input to the BNST is the densest among basal telencephalic regions. The central amygdaloid nucleus and the BNST also receive adrenergic fibers, as determined using immunolabeling for phenylethanolamine N-methyltransferase (PNMT) (Hökfelt, Johansson & Goldstein, 1984a). Fibers containing DBH are also present in the medial septal region (Gaspar et al., 1985). The DBH+ innervation of the dorsal striatum and globus pallidus is extremely sparse and much lighter than that to the ventral striatum. The ultrastructure of noradrenergic terminals in various regions outside the basal telencephalon consistently indicates them to form slightly asymmetric synapses, suggesting this may also be true in the basal telencephalon (Olschowka et al., 1981; Moore & Card, 1984).

Receptors, uptake mechanisms and functional role of noradrenergic and adrenergic input

Little detailed information is available on the distribution of noradrenergic/adrenergic receptors or uptake mechanisms in the basal ganglia. Receptors for noradrenaline and adrenaline are complex and include α-1 and α-2 adrenergic receptors, as well as β-1 and β-2 adrenergic receptors, as defined pharmacologically (Rogawski, 1985). Several of these have been cloned and sequenced, and they are of the G-protein coupled family of receptors (Dixon et al., 1986; Lomasney et al., 1991). In light of this and in light of the complexity of adrenergic receptor mechanisms, it is not surprising that the physiological effects of noradrenaline and adrenaline are complex and depend on the postsynaptic receptors present on the target cells (Moore & Bloom, 1979; Rogawski, 1985). In general, the actions are best described as modulatory. It seems likely that the effects of NA and adrenalin in the basal ganglia are mediated primarily via β-adrenergic receptors, since β-adrenergic receptors are highly abundant in the dorsal and ventral striatum while α-adrenergic receptors are sparse in the dorsal and ventral striatum (Palacios & Wamsley, 1984).

Acknowledgements

I would like to thank the following colleagues for their helpful discussions, which have influenced or led to many of the ideas presented in this manuscript: D. J. Surmeier, R. L. Albin, R. J. Nelson, W. J. A. J. Smeets, C. J. Wilson, K. D. Anderson and E. K. Richfield. I also thank Dr P. Voorn for allowing me to reproduce figures from some of his published papers in this review chapter.

References

Akaoka, H., Saunier, F. & Chouvet, G. (1987). Neuronal responses to dopamine in rat striatum: comparison between dopamine iontophoresis application and nigrostriatal pathway stimulation. *Biogenic Amines*, **4**, 407–12.

Albin, R. L., Aldridge, W., J., Young, A. B. & Gilman, S. (1989a). Feline subthalamic nucleus neurons contain glutamate-like but not GABA-like or glycine-like immunoreactivity. *Brain Research*, **491**, 185–8.

Young, A. B. (1990). Striatal and nigral neuron subpopulations in rigid Huntington's disease: implications for the functional anatomy of chorea and rigidity-akinesia. *Annals of Neurology*, **27**, 357–65.

Albin, R. L., Young, A. B. & Penney, J. B. (1989b). The functional anatomy of basal ganglia disorders. *Trends in Neurosciences*, **12**, 366–75.

Alexander, G. E., & Crutcher, M. D. (1990). Functional architecture of basal ganglia circuits: neural substrates of parallel processing. *Trends in Neurosciences*, **13**, 266–71.

Alexander, G. E., DeLong, M. R. & Strick, P. L. (1986). Parallel organization of functionally segregated circuits linking basal ganglia and cortex. *Annual Review of Neuroscience*, **9**, 357–81.

Altar, C. A. & Hauser, K. (1987). Topography of substantia nigra innervation by D1 receptor-containing striatonigral neurons. *Brain Research*, **410**, 1–11.

Anderson, K. D. & Reiner, A. (1990). The extensive co-occurrence of substance P and dynorphin in striatal projection neurons: an evolutionarily conserved feature of basal ganglia organization. *Journal of Comparative Neurology*, **295**, 339–69.

Anderson, K. D. & Reiner, A. (1991). Immunohistochemical localization of DARPP-32 in striatal projection neurons and striatal interneurons: implications for the localization of D1 dopamine receptors on different types of striatal neurons. *Brain Research*, **568**, 235–43.

Aoki, C. & Pickel, V. M. (1988). Neuropeptide Y-containing neurons in the rat striatum: ultrastructure and cellular relations with tyrosine hydroxylase-containing terminals and with astrocytes. *Brain Research*, **459**, 205–25.

Apicella, P., Scarnati, E. & Schultz, W. (1991). Tonically discharging neurons of monkey striatum respond to preparatory and rewarding stimuli. *Experimental Brain Research*, **84**, 672–5.

Ariano, M. A., Stromski, C. J., Smyk-Randall, E. M. & Sibley, D. R. (1992). D2 dopamine receptor localization on striatonigral neurons. *Neuroscience Letters*, **144**, 215–20.

Arluison, M., Dietl, M. & Thibault, J. (1984). Ultrastructural morphology of dopaminergic nerve terminals and synapses in the striatum of the rat using tyrosine hydroxylase immunocytochemistry: a topographical study. *Brain Research Bulletin*, **13**, 269–85.

Barbeau, A. (1986). Parkinson's disease: clinical features and etiopathology. In *Handbook of Clinical Neurology. Vol. 5 (49): Extrapyramidal Disorders.* ed. P. J. Vinken, G. W. Bruyn & H. L. Klawans. Elsevier Science Publishers, Amsterdam, Netherlands, pp. 87–152.

Beckstead, R. M. (1984). Association of dopamine D1 and D2 receptors with specific cellular elements in the basal ganglia of the cat: the uneven topography of dopamine receptors in the striatum is determined by intrinsic striatal cells, not nigrostriatal axons. *Neuroscience*, **27**(3), 851–863.

Beckstead, R. M. & Cruz, C. J. (1986). Striatal axons to the globus pallidus, entopeduncular nucleus and substantia nigra come mainly from separate cell populations in cat. *Neuroscience*, **19**, 147–58.

Beckstead, R. M., Wooten, G. F. & Trugman, J. M. (1988). Distribution of D1 and D2 dopamine receptors in the basal ganglia of the cat determined by quantitative autoradiography. *Journal of Comparative Neurology*, **268**, 131–45.

Berendse, H. W., Groenewegen, H. J. & Lohman, A. H. (1992). Compartmental distribution of ventral striatal neurons projecting to the mesencephalon in the rat. *Journal of Neuroscience*, **12**, 2079–103.

Berendse, H. W. & Richfield, E. K. (1993). Heterogeneous distribution of dopamine D1 and D2 receptors in the human ventral striatum. *Neuroscience Letters*, **150**, 75–9.

Bertorello, A. M., Hopfield, J. E., Aperia, A. & Greengard, P. (1990). Inhibition by dopamine of $(Na^+ + K^+)$ATPase activity in neostriatal neurons through D1 and D2 dopamine receptor synergism. *Nature*, **347**, 386–8.

Besson, M. J., Graybiel, A. M. & Nastuk, M. A. (1988). [^{3}H]SCH23390 Binding to D1 dopamine receptors in the basal ganglia of the cat and primate: delineation of striosomal compartments and pallidal and nigral subdivisions. *Neuroscience*, **26**, 101–19.

Bickford-Wimer, P., Kim, M., Boyajian, C., Cooper, D. M. F. & Freedman, R. (1990). Effects of pertussis toxin on caudate neuron electrophysiology: studies with dopamine D1 and D2 agonists. *Brain Research*, **533**, 263–7.

Björklund, A. & O. Lindvall (1984). Dopamine-containing systems in the CNS. In *Handbook of Chemical Neuroanatomy. Vol. 2: Classical Transmitters in the CNS. Part I*, ed. A. Björklund & T. Hökfelt, pp. 55–122. Amsterdam, the Netherlands: Elsevier Science Publishers.

Bolam, J. P. & Smith, Y. (1990). The GABA and substance P input to dopaminergic neurones in the substantia nigra of the rat. *Brain Research*, **529**, 57–78.

Bouthenet, M. L., Souil, E., Martres, M. P., Sokoloff, P., Giros, B. & Schwartz, J. C. (1991). Localization of dopamine D3 receptor mRNA in the rat brain using *in situ* hybridization histochemistry: comparison with D2 dopamine receptor mRNA. *Brain Research*, **564**, 203–19.

Bouyer, J. J., Park, D. H., Joh, T. H. & Pickel, V. M. (1984). Chemical and structural analysis of the relation between cortical inputs and tyrosine hydroxylase-containing terminals in rat neostriatum. *Brain Research*, **302**, 267–75.

Camps, M., Corets, R., Gueye, B., Probst, A. & Palacios, J. M. (1989). Dopamine receptors in human brain: Autoradiographic distribution of D2 sites. *Neuroscience*, **28**, 275–90.

Carlsson, A. (1959). Occurrence, distribution and physiological role of catecholamines in the nervous system. *Pharmacological Reviews*, **11**, 490–3.

Chang, H. T. (1988). Substance P-dopamine relationship in the rat substantia nigra: a light and electron microscopy study of double immunocytochemically labeled materials. *Brain Research*, **448**, 391–6.

Chevalier, G., Vacher, S., Deniau, J. M. & Desban, M. (1985). Disinhibition as a basic process in the expression of striatal functions. I. The striato-nigral influence on tecto-spinal/tecto-diencephalic neurons. *Brain Research*, **334**, 215–26.

Civelli, O., Bunzow, J. R., Grandy, D. K., Zhou, Q. Y. & Van Tol, H. H. M. (1991). Molecular biology of the dopamine receptors. *European Journal of Pharmacology – Molecular Biology Section*, **207**, 277–86.

Clark, D. & White, F. J. (1987). Review: D1 dopamine receptor – the search for a function: a critical evaluation of the D1/D2 dopamine receptor classification and its functional implications. *Synapse*, **1**, 347–88.

Cline, E. J., Scheffel, U., Boja, J. W., Mitchell, W. M., Carroll, F. I., Abraham, P., Lewin, A. H. & Kuhar, M. J. (1992). *In vivo* binding of [125I]RTI-55 to dopamine transporters: pharmacology and regional distribution with autoradiography. *Synapse*, **12**, 37–46.

Cortés, R., Gueye, B., Pazos, A., Probst, A. & Palacios, J. M. (1989). Dopamine receptors in human brain: autoradiographic distribution of D1 sites. *Neuroscience*, **28**, 263–73.

Crossman, A. R. (1987). Primate models of dyskinesia: the experimental approach to the study of basal ganglia-related involuntary movement disorders. *Neuroscience*, **21**, 1–40.

Dawson, V. L., Dawson, T. D. & Wamsley, J. K. (1990). Muscarinic and dopaminergic receptor subtypes on striatal cholinergic interneurons. *Brain Research Bulletin*, **25**, 903–12.

DeKeyser, J. (1993). Subtypes and localization of dopamine receptors in human brain. *Neurochemistry International*, **22**, 83–93.

Delong, M. R. (1990). Primate models of movement disorders of basal ganglia origin. *Trends in Neurosciences*, **13**, 281–5.

DeLong, M. R., Alexander, G. E., Mitchell, S. J. & Richardson, R. T. (1986). The contribution of basal ganglia to limb control. In *Progress in Brain Research, Vol. 64*, ed. H. J., Freund, U. Büttner, B. Cohen, & J. Noth, pp. 161–174. Amsterdam, the Netherlands: Elsevier Science Publisher.

Deniau, J. M. & Chevalier, G. (1985). Disinhibition as a basic process in the expression of striatal functions. II. The striato-nigral influence on thalamocortical cells of the ventromedial thalamic nucleus. *Brain Research*, **334**, 227–33.

Dixon, R. A. F., Kobilka, B. K., Strader, D. J., Benovic, J. L., Dohlman, H. G., Frielle, T., Bollanowski, M. A., Bennet, C. D., Rands, E., Diehl, R. E., Mumford, R. A., Slater, E. E., Sigal, I. S., Caron, M. G., Lefkowitz, R. J. & Strader, C. D. (1986). Cloning of the gene and cDNA for mammalian B-adrenergic receptor and homology with rhodopsin. *Nature*, **321**, 75–9.

Donnan, G. A., Kaczmarczyk, S. J., Paxinos, G., Chilco, P. J., Kalnins, R. M., Woodhouse, D. G. & Mendelsohn, F. A. O. (1991). Distribution of catecholamine uptake sites in human brain as determined by quantitative [3H]mazindol autoradiography. *Journal of Comparative Neurology*, **304**, 419–34.

Drinnan, S. L., Hope, B. T., Snutch, T. P. & Vinvent, S. R. (1991). Golf in the basal ganglia. *Molecular and Cellular Neurosciences*, **2**, 66–70.

Factor, S. A., Sanchez-Ramos, J. R. & Weiner, W. J. (1988). Cocaine and Tourette's syndrome. *Annals of Neurology*, **23**, 423–4.

Feger, J. & Crossman, A. R. (1984). Identification of different subpopulations of neostriatal neurones projecting to globus pallidus or substantia nigra in the monkey: a retrograde fluorescence double-labeling study. *Neuroscience Letters*, **49**, 7–12.

Felten, D. L. & Sladek, J. R. Jr. (1983). Monoamine distribution in primate brain V. Monoaminergic nuclei: anatomy, pathways and local organization. *Brain Research Bulletin*, **10**, 171–284.

Filion, M. & Tremblay, L. (1991). Abnormal spontaneous activity of globus pallidus in monkeys with MPTP-induced parkinsonism. *Brain Research*, **547**, 142–51.

Filion, M., Tremblay, L. & Bedard, P. J. (1989). Effects of dopamine agonists on the spontaneous activity of globus pallidus in monkeys with MPTP-induced parkinsonism. *Brain Research*, **547**, 152–61.

Freedman, J. E. & Weight, F. F. (1988). Single K+ channnels activated by D2 dopamine receptors in acutely dissociated neurons from rat corpus striatum. *Proceedings of the National Academy of Sciences*, USA, **85**, 3618–22.

Freund, T. F., Powell, J. F. & Smith, A. D. (1984). Tyrosine hydroxylase-immunoreactive boutons in synaptic contact with identified striatonigral neurons, with particular reference to dendritic spines. *Neuroscience*, **13(4)**, 1189–215.

Gardiner, T. W. & Kitai, S. T. (1992). Single-unit activity in the globus pallidus and neostriatum of the rat during performance of a trained head movement. *Experimental Brain Research*, **88**, 517–30.

Gaspar, P., Berger, B., Alvarez, C., Vigny, A. & Henry, J. P. (1985). Catecholaminergic innervation of the septal area in man: immunocytochemical study using TH and DBH antibodies. *Journal of Comparative Neurology*, **241**, 12–33.

Gerfen, C. R. (1985). The neostriatal mosaic: I. Compartmental organization of projections from the striatum to the substantia nigra in the rat. *Journal of Comparative Neurology*, **236**, 454–76.

Gerfen, C. R. (1988). Synaptic organization of the striatum. *Journal of Electron Microscopic Techniques*, **10**, 265–81.

Gerfen, C. R. (1992). The neostriatal mosaic: multiple levels of compartmental organization. *Trends in Neurosciences*, **15**, 133–9.

Gerfen, C. R., Engber, T. M., Mahan, L. C., Susel, Z., Chase, T. N., Monsma, F. J. Jr. & Sibley, D. R. (1990). D1 and D2 dopamine receptor-regulated gene expression of striatonigral and striatopallidal neurons. *Science*, **250**, 1429–32.

Gerfen, C. R., Bainbridge, K. G. & Thibault, J. (1987a). The neostriatal mosaic: III. Biochemical and developmental dissociation of patch-matrix mesostriatal systems. *Journal of Neuroscience*, **7**, 3935–44.

Gerfen, C. R., Herkenham, M. & Thibault, J. (1987b). The neostriatal mosaic: II. Patch- and matrix-directed mesostriatal dopaminergic and non-dopaminergic systems. *Journal of Neuroscience*, **7**, 3915–34.

Gimenez-Amaya, J. M. & Graybiel, A. M. (1991). Modular organization of projection neurons in the matrix compartment of the primate striatum. *Journal of Neuroscience*, **11**, 779–91.

Goetz, C. C. (1986) Tics: Gilles de la Tourette syndrome. In *Handbook of Clinical Neurology. Vol. 5 (49): Extrapyramidal Disorders*. ed. Vinken, P. J., Bruyn, G. W. & Klawans, H. L., pp. 627–639. Amsterdam, Netherlands: Elsevier Science Publisher.

Goldman-Rakic, P. S. (1992). Cytoarchitectonic heterogeneity of the primate neostriatum: subdivision into Island

and Matrix cellular compartments. *Journal of Comparative Neurology*, **205**, 398–413.

Grace, A. A. & Bunney, B. S. (1979). Paradoxical GABA excitation of nigral dopaminergic cells: Indirect mediation through reticulata inhibitory neurons. *European Journal of Pharmacology*, **59**, 211–18.

Grace, A. A. & Bunney, B. S. (1985a). Dopamine. In *Neurotransmitter Actions in the Vertebrate Nervous System*. ed. M. A. Rogawski & J. A. Barker. Plenum Publishing Co., New York, N Y, pp.285–319.

Grace, A. A. & Bunney, B. S. (1985b). Opposing effects of striatonigral pathways on midbrain dopamine cell activity. *Brain Research*, **333**: 271–84.

Graybiel, A. M. (1983). Compartmental organization of the mammalian striatum. In *Molecular and Cellular Interactions Underlying Higher Brain Functions. Progress in Brain Research*. ed. J. P. Changeux, J. Glowinski, M. Imbert & F. E. Bloom. pp.247–256 Publishers. Amsterdam, The Netherlands: Elsevier Science.

Graybiel, A. M. (1984) Correspondence between the dopamine islands and striosomes of the mammalian striatum. *Neuroscience*, **13**: 1157–7.

Graybiel, A. M. (1990). Neurotransmitters and neuromodulators in the basal ganglia. *Trends in Neurosciences*, **13**, 244–53.

Graybiel, A. M., Hirsh, E. C. & Agid, Y. A. (1987). Differences in tyrosine hydroxylase-like immunoreactivity characterize the mesostriatal innervation of striosomes and extrastriosomal matrix at maturity. *Proceedings of the National Academy of Sciences, USA, America* **84**, 303–7.

Guyenet, P. G. & Crane, J. K. (1981). Non-dopaminergic nigrostriatal pathway. *Brain Research*, **213**: 291–305.

Haber, S. N. & Elde, R. (1981). Correlation between met-enkephalin and substance P immunoreactivities in the primate globus pallidus. *Neuroscience*, **6**: 1291–7.

Haber, S. N. & Groenewegen, H. J. (1990). Interrelationship of the distribution of neuropeptides and tyrosine hydroxylase immunoreactivity in the human substantia nigra. *Journal of Comparative Neurology*, **290**: 53–68.

Haber, S. N., Groenewegen, H. J., Grove, E. A. & Nauta, W. J. H. (1985). Efferent connections of the ventral pallidum: Evidence of a dual striatopallidofugal pathway. *Journal of Comparative Neurology*, **235**: 322–35.

Haber, S. N., Lynd, E. Klein, C. & Groenewegen, H. J. (1990). Topographic organization of the ventral striatal efferent projections in the rhesus monkey: an anterograde tracing study. *Journal of Comparative Neurology*, **293**: 282–98.

Haber, S. N. & Nauta, W. J. H. (1983). Ramifications of the globus pallidus in the rat as indicated by patterns of immunohistochemistry. *Neuroscience*, **9**: 245–60.

Haber, S. N., Wolfe, D. P. & Groenewegen, H. J. (1990). The relationship between ventral striatal efferent fibers and the distribution of peptide-positive woolly fibers in the forebrain of the rhesus monkey. *Neuroscience*, **39**: 323–8.

Harrison, M. B., Wiley, R. G. & Wooten, G. F. (1990). Selective localization of striatal D1 receptors to striatonigral neurons. *Brain Research*, **528**: 317–22.

Harrison, M. B., Wiley, R. G. & Wooten, G. F. (1992). Changes in D2 but not D1 receptor binding in the striatum following a selective lesion of striatopallidal neurons. *Brain Research*, **590**: 305–10.

Hazlett, J. C., Ho, R. H. & Martin, G. F. (1991). Organization of midbrain catecholamine-containing nuclei and their projections to the striatum in the North American Opossum, *Didelphis virginiana*. *Journal of Comparative Neurology*, **306**: 585–601.

Hedreen, J. C. & DeLong, M. R. (1991). Organization of striatopallidal, striatonigral and nigrostriatal projections in the Macaque. *Journal of Comparative Neurology*, **304**: 569–95.

Heimer, L., Alhier, G. F. & Zaborszky, L. (1985). Basal ganglia. In *The Rat Nervous System* ed. G. Paxinos. Orlando: pp. 37–86. Academic Press.

Herrling, P. L. & Hull, C. D. (1980). Iontophoretically applied dopamine depolarizes and hyperpolarizes the membrane of cat caudate neurons. *Brain Research*; 441–62.

Herve, D., Levi-Strauss, M., Marey-Semper, I., Verney, C., Tassin, J. P., Glowinski, J. & Girault, J. A. (1993). Golf and Gs in rat basal ganglia: possible involvement of Golf in the coupling of dopamine D1 receptor with adenylate cyclase. *Journal of Neuroscience*, **13**; 2237–48.

Hökfelt, T., Johansson, O. & Goldstein, M. (1984a). Central catecholamine neurons as revealed by immunohistochemistry with special reference to adrenaline neurons. In *Handbook of Chemical Neuroanatomy. Volume 2 Classical Transmitters in the CNS. Part I*. ed. A. Björklund & T. Hökfelt, Elsevier Science Publishers, Amsterdam, the Netherlands, pp 157–276.

Hökfelt, T., Martensson, R., Björklund, A., Kleinau, S. & Goldstein, M. (1984b). Distributional maps of tyrosine hydroxylase-immunoreactive neurons in the rat brain. In *Handbook of Chemical Neuroanatomy. Volume 2 Classical Transmitters in the CNS. Part I*, ed. A. Björklund & T. Hökfelt, Amsterdam, the Netherlands: Elsevier Science Publishers. pp.227–379.

Hökfelt, T., Rehfeld, J. F. Skirboll, L. Ivemark, B., Goldstein, M. & Markey, K. (1980a). Evidence for coexistence of dopamine and CCK in meso-limbic neurones. *Nature*, **285**; 476–8.

Hökfelt, T., Skirboll, L., Rehfeld, J. F. Goldstein, M., Markey, K. & Dann, O. (1980b). A subpopulation of mesencephalic dopamine neurons projecting to limbic areas contains a cholecystokinin-like peptide: evidence from immunohistochemistry combined with retrograde tracing. *Neuroscience*, **5**; 2093–124.

Hu, X. T. & Wang R. Y. (1988). Comparison of effects of D1 and D2 dopamine receptor agonists on neurons in the rat caudate putamen: an electrophysiological study. *Journal of Neuroscience*, **8**; 4340–8.

Innis, R. B., Andrade, R. & Aghajanian, G. K. (1985). Substance K excites dopaminergic and non-dopaminergic neurons in rat substantia nigra. *Brain Research*, **335**, 381–3.

Jacobowitz, D. M. & Maclean, P. D. (1978). A brainstem atlas of catecholaminergic neurons and serotonergic perikarya in a pygmy primate (*Cebuella pygmaea*). *Journal of Comparative Neurology*, **177**, 397–415.

Jimenez-Castellanos, J. & Graybiel, A. M. (1987). Subdivisions of the dopamine-containing A8-A9-A10 complex identified by their differential mesostriatal innervation of striosomes and extrastriosomal matrix. *Neuroscience*, **23**, 223–42.

Jimenez-Castellanos, J. & Graybiel, A. M. (1989). Compartmental origins of striatal efferent projections in the cat. *Neuroscience*, **32**, 297–321.

Joyce, J. N., Sapp, D. W. & Marshall, J. F. (1986). Human striatal dopamine receptors are organized in compartments. *Proceedings of the National Academy of Sciences, USA*, **83**, 8002–6.

Kalivas, P. W. (1985). Interactions between neuropeptides and dopamine neurons in the ventromedial mesencephalon. *Neuroscience and Biobehavioral Reviews*, **9**, 573–87.

Kawaguchi, Y., Wilson, C. J. & Emson, P. C. (1990). Projection subtypes of rat neostriatal matrix cells revealed by intracellular injection of biocytin. *Journal of Neuroscience*, **10**, 3421–38.

Kebabian, J. W. & Calne, D. B. (1979). Multiple receptors for dopamine. *Nature*, **277**, 93–6.

Kebabian, J. W., Petzgold, G. L. & Greengard, P. (1972). Dopamine-sensitive adenylate cyclase in caudate nucleus of rat brain, and its similarity to the 'dopamine receptor'. *Proceedings of the National Academy of Sciences, USA*, **69**, 2145–9.

Kilty, J. E. Lorang, D. & Amara, S. G. (1991). Cloning and expression of a cocaine-sensitive rat dopamine transporter. *Science*, **154**, 578–9.

Kitai, S. T. (1980). Electrophysiology of the corpus striatum and brain stem integrating systems. In *Handbook of Physiology – The Nervous System II*. American Physiological Society, pp.997–1015.

Kitai, S. T., Kocsis, J. D., Preston, R. J. & Sugimori, M. (1976) Monosynaptic inputs to caudate neurons identified by intracellular injection of horseradish peroxidase. *Brain Research*, **109**, 601–6.

Kitai, S. T. & Surmeier, D. J. (1993). Cholinergic and dopaminergic modulation of potassium conductances in neostriatal neurons. In *Advances in Neurology. Vol. 60*. ed. H. Narabayashi, T. Nagatsu, N. Yanagisawa & Y. Mizuno. New York, NY: Raven Press, pp.40–51.

Kubota, Y., Inagaki, S. & Kito, S. (1986a). Innervation of substance P neurons by catecholaminergic terminals in the neostriatum. *Brain Research*, **375**, 163–7.

Kubota, Y., Inagaki, S., Kito, S., Takagi, H. & Smith, A. D. (1986b). Ultrastructural evidence of dopaminergic input to enkephalinergic neurons in rat neostriatum. *Brain Research*, **367**, 374–8.

Langer, L. F. & Graybiel, A. M. (1989). Distinct nigrostriatal projection systems innervates striosomes and matrix in primate striatum. *Brain Research*, **498**, 344–50.

Lavoie, B., Smith, Y. & Parent, A. (1989). Dopaminergic innervation of the basal ganglia in the squirrel monkey as revealed by tyrosine hydroxylase immunohistochemistry. *Journal of Comparative Neurology*, **289**, 36–52.

Lemoine, C. & Bloch, B. (1991). Rat striatal and mesencephalic neurons contain the long isoform of the D2 dopamine receptor mRNA. *Molecular Brain Research*, **10**, 283–9.

Lemoine, C., Normand, E., Guitteny, A. F., Fouque, B., Teoule, R. & Bloch, B. (1990). Dopamine receptor gene expression by enkephalin neurons in rat forebrain. *Proceedings of the National Academy of Sciences, USA*, **87**, 230–4.

Lemoine, C., Normand, E. & Bloch, B. (1991). Phenotypical characterization of the rat striatal neurons expressing the D1 dopamine receptor gene. *Proceedings of the National Academy of Sciences, USA*, **88**, 4205–9.

Liang, C. L., Kozlowski, G. P. & German, D. C. (1993). Leucine[5]-enkephalin afferents to midbrain dopaminergic neurons: Light and electron microscopic examination. *Journal of Comparative Neurology*, **332**, 269–81.

Lomasney, J. W., Cotecchia, S. Lefkowitz, R. J. & Caron, M. G. (1991). Molecular biology of α-adrenergic receptors: implications for receptor classification and for structure – function relationship. *Biochimica et Biophysica Acta*, **1095**, 127–39.

Mahalik, T. J. (1988). Direct demonstration of interactions between substance P immunoreactive terminals and tyrosine hydroxylase immunoreactive neurons in the substantia nigra of the rat: an ultrastructural study. *Synapse*, **2**, 508–15.

Marsden, C. D. (1980). The enigma of the basal ganglia and movement. *Trends in Neurosciences*, **3**, 1–4.

Martin, L. J., Hadfield, M. G., Dellovade, T. L. & Price, D. L. (1991). The striatal mosaic in primates: patterns of neuropeptide immunoreactivity differentiate the ventral striatum from the dorsal striatum. *Neuroscience*, **43**, 397–417.

Maura, G., Giardi, A. & Raiteri, M. (1988). Release-regulating D2 dopamine receptors are located on striatal glutametrgic nerve terminals. *Journal of Pharmacology and Experimental Therapeutics*, **247**, 680–4.

McGeer, P. L., Eccles, J. C. & McGeer, E. G. (1978). *Molecular Neurobiology of the Mammalian Brain*. New York, NY: Plenum Press.

Meador-Woodruff, J. H., Mansour, A., Healy, D. J., Kuehn, R., Zhou, Q. Y., Bunzow, J. R., Akil, H. O., Civelli, O. & Watson, S. J. (1991a). Comparison of distributions of D1 and D2 dopamine receptor mRNAs in rat brain. *Neuropsychopharmacology*, **5**, 231–42.

Meador-Woodruff, J. H., Mansour, A., Civelli, O. & Watson, S. J. (1991b). Distribution of D2 dopamine receptor mRNA in the primate brain. *Progress in Neuro-Psychopharmacology and Biological Pyschiatry*, **15**, 885–93.

Mehler, W. R. (1981). The basal ganglia – circa 1982. A review and commentary. In *Applied Neurophysiology. Vol. 44*. ed. P. L. Gildenberg. Basel, Switzerland: Karger, S. pp.261–290.

Mendez, I., Elisevich, K., Naus, C. C. G. & Flumerfelt, B. A. (1990). Neostriatal substance P and somatostatin neurons receive a direct nigral dopaminergic input: an ultrastructural double labeling immunocytochemical study. *Society for Neuroscience Abstracts*, **16**, 301.

Moore, R. Y. & Card, J. P. (1984). Noradrenalin-containing neurons. In *Volume 2 Classical Transmitters in the CNS. Part I*, ed. A. Björklund & T. Hökfelt, pp.123–156. Amsterdam, the Netherlands: Elsevier Science Publishers.

Moore, R. Y. & Bloom, F. E. (1978). Central catecholamine neuron systems: Anatomy and physiology of the dopamine systems. *Annual Review of Neuroscience*, 1, 129–69.

Moore, R. Y. & Bloom, F. E. (1979). Central catecholamine neuron systems: Anatomy and physiology of the norepinephrine and epinephrine systems. *Annual Review of Neuroscience*, 2, 113–68.

Murray, H. M., Dominguez, W. F. & Martinez, J. E. (1982). Catecholamine neurons in the brain stem of the tree shrew. *Brain Research Bulletin*, 9, 205–15.

Normand, E., Popovici, T., Onteniente, B., Fellmann, D., Piatier-Tonneau, D., Auffray, C. & Bloch, B. (1988). Dopaminergic neurons of the substantia nigra modulate preproenkephalin A gene expression in rat striatal neurons. *Brain Research*, 439, 39–46.

Nylander, I. & Terenius, L. H. (1987). Dopamine receptors mediate alterations in striatonigral dynorphin and substance P pathways. *Neuropharmacology*, 9, 1295–302.

Ohno, Y., Sasa M., & Takaori, S. (1985). Dopamine D-2 receptor-mediated excitation of caudate nucleus neurons from the substantia nigra. *Life Sciences*, 37, 1515–21.

Ohno, Y., Sasa, M. & Takaori, S. (1986). Excitation by dopamine D-2 receptor agonists, bromocriptine and Ly 171555, in caudate nucleus neurons activated by nigral stimulation. *Life Sciences*, 38, 1867–973.

Ohno, Y., Sasa, M. & Takaori, S. (1987). Coexistence of inhibitory dopamine D1 and excitatory D-2 receptors on the same caudate nucleus neurons. *Life Sciences*, 40, 1937–45.

Olianas, M. C., DeMontis, G. M., Maulas, G. & Tagliamonte, A. (1978). The striatal dopaminergic function is mediated by the inhibition of a nigral, non-dopaminergic neuronal system via a strio-nigral GABAergic pathway. *European Journal of Pharmacology*, 49, 233–41.

Olschowka, J. A., Molliver, M. E., Grzanna, R., Rice, F. L. & Coyle, J. T. (1981). Ultrastructural demonstration of noradrenergic synapses in the rat central nervous system by dopamine-hydroxylase immunocytochemistry. *Journal of Histochemistry and Cytochemistry*, 29, 271–80.

Onn, S. P., Balzer, J. R., Sidney, J. P., Stricker, E. M., Zigmond, M. J. & Berger, T. W. (1990). Lesions of the dopaminergic nigrostriatal system in neonatal rats: effects on the physiological activity of striatal neurons recorded during adulthood. *Brain Research*, 518, 274–8.

Orr, W. B., Gardiner, T. W., Stricker, E. M., Zigmond, M. J. & Berger, T. W. (1986). Short-term effects of dopamine-depleting brain lesions on spontaneous activity of striatal neurons: relation to local dopamine concentration and behavior. *Brain Research*, 376, 20–8.

Palacios, J. M. & Wamsley, J. K. (1984). Catecholamine receptors. In *Handbook of Chemical Neuroanatomy. Vol. 3: Classical Transmitters and Transmitter Receptors in the CNS. Part II*, ed. A. Björklund, T. Hökfelt & M. J. Kuhar, pp.325–351. Amsterdam, the Netherlands: Elsevier Science Publishers.

Palacios, J. M., Camps, M., Cortés, R. & Charuchinda, C. (1988). Characterization and distribution of brain dopamine receptors. In *Parkinson's Disease and Movement Disorders*. ed. J. Jankovbic & E. Tolosa. pp.27–36. Baltimore, MD: Urban & Schwarzenberg.

Pan, H. S. & Walters, J. R. (1988). Unilateral lesion of the nigrostriatal pathway decreases the firing rate and alters the firing pattern of globus pallidus neurons in the rat. *Synapse*, 2, 650–6.

Parent, A. (1986). *Comparative Neurobiology of the Basal Ganglia*. New York: John Wiley & Sons.

Parent, A., Bouchard, C. & Smith, Y. (1984). The striatopallidal and striatonigral projections: two distinct fiber systems in primate. *Brain Research*, 303, 385–90.

Pearson, J., Goldstein, M., Markey, K. & Brandeis, L. (1983). Human brainstem catecholamine neuronal anatomy as indicated by immunocytochemistry with antibodies to tyrosine hydroxylase. *Neuroscience*, 8, 3–32.

Pickel, V. M., Beckley, S. C., Joh, T. H. & Reis, D. J. (1981), Ultrastructural immunocytochemical localization of tyrosine hydroxylase in the neostriatum. *Brain Research*, 225, 373–85.

Piomelli, D., Pilon, C., Giros, B., Sokoloff, P., Martres, M. P. & Schwartz, J. C. (1991). Dopamine activation of the arachidonic acid cascade as a basis for D1/D2 receptor synergism. *Nature*, 353, 164–7.

Quimet, C. C., Miller, P. E., Hemmings, H. C., Jr., Walaas, S. I. & Greengard, P. (1984). DARPP-32, a dopamine and adenosine 3′:5′-monophosphate-regulated phosphoprotein enriched in dopamine-innervated brain regions. *Journal of Neuroscience*, 4, 111–24.

Quimet, C. C., Lamantia, A. S., Goldman-Rakic, P., Rakic, P. & Greengard, P. (1992). Immunocytochemical localization of DARPP-32, a dopamine and cyclic-AMP-regulated phosphoprotein in the primate brain. *Journal of Comparative Neurology*, 323, 209–18.

Reid, M., Herrera-Marschitz, M., Hökfelt, T., Terenius, L. & Ungerstedt, U. (1988). Differential modulation of striatal dopamine release by intranigral injection of gamma-aminobutyric acid (GABA), dynorphin A and substance P. *European Journal of Pharmacology*, 147, 411–20.

Reiner, A. (1987). A LANT6-like peptide that is distinct from Neuromedin N is present in striatal and pallidal neurons in the monkey basal ganglia. *Brain Research*, 422, 186–91.

Reiner, A., Albin, R. L., Anderson, K. D., D'Amato, C. J., Penney, J. B. & Young, A. B. (1988). Differential loss of substance P-containing striatopallidal and enkephalin-containing striatopallidal projections in Huntington's Disease. *Proceedings of the National Academy of Sciences* USA, 85, 5733–7.

Reiner, A. & Anderson, K. D. (1990). The patterns of neurotransmitter and neuropeptide co-occurrence among striatal projection neurons: conclusions based on recent findings. *Brain Research Reviews*, 15, 251–65.

Reiner, A. & Anderson, K. D. (1993). Co-occurrence of GABA, parvalbumin and the neurotensin-related hexapeptide LANT6 in pallidal, nigral and striatal neurons in pigeons and monkeys. *Brain Research*, in press.

Reiner, A., Brauth, S. E. & Karten, H. J. (1984). Evolution of the amniote basal ganglia. *Trends in Neurosciences*, 7, 320–5.

Reiner, A. & Carraway, R. E. (1987). Immunohistochemical and biochemical studies on Lys8-Asn9-Neurotensin8–13 (LANT6)-related peptides in the basal ganglia of pigeons, turtles and hamsters. *Journal of Comparative Neurology*, 257, 453–76.

Richfield, E. K. (1991). Quantitative autoradiography of the dopamine uptake complex in rat brain using [3H]GBR 12 935: binding characteristics. *Brain Research*, 540, 1–13.

Richfield, E. K., Young, A. B. & Penney, J. B. (1987). Comparative distribution of D-1 and D-2 receptors in the basal ganglia of turtles, pigeons, rats, cats, and monkeys. *Journal of Comparative Neurology*, 262, 446–63.

Robertson, H. A. (1992). Dopamine receptor interactions: some implications for the treatment of Parkinson's disease. *Trends in Neurosciences*, 15, 201–6.

Robertson, G. S., Vincent, S. R. & Fibiger, H. C. (1990). Striatonigral projection neurons contain D1 dopamine receptor-activated c-fos. *Brain Research*, 523, 288–90.

Robertson, G. S., Vincent, S. R. & Fibiger, H. C. (1992). D1 and D2 dopamine receptors differentially regulate c-fos expression in striatonigral and striatopallidal neurons. *Neuroscience*, 49, 285–96.

Rogawski, M. A. (1985). Norepinephrine. In *Neurotransmitter Actions in the Vertebrate Nervous System*. ed. M. A. Rogawski & J. A. Barker. pp. 242–284. New York, NY: Plenum Publishing Co.

Romo, R. & Schultz, W. (1990). Dopamine neurons of the monkey midbrain: contingencies of responses to active touch during self-initiated movements. *Journal of Neurophysiology*, 63, 592–606.

Ryan, L. J., Diana, M., Young, S. J. & Groves, P. M. (1989). Dopamine D1 heteroreceptors on striatonigral axons are not stimulated by endogenous dopamine either tonically or after amphetamine: evidence from terminal excitability. *Experimental Brain Research*, 77, 161–5.

Schultz, W., Apicella, P. & Ljungberg, T. (1993). Responses of monkey dopamine neurons to reward and conditioned stimuli during successive steps of learning a delayed response task. *Journal of Neuroscience*, 13, 900–13.

Schultz, W., & Romo, R. (1990). Dopamine neurons of the monkey midbrain: contingencies of responses to stimuli eliciting immediate behavioral reactions. *Journal of Neurophysiology*, 63, 607–23.

Selemon, L. D. & Goldman-Rakic, P. S. (1990). Topographic intermingling of striatonigral and striatopallidal neurons in the rhesus monkey. *Journal of Comparative Neurology*, 297, 359–76.

Seroogy, K. M., Dangaran, K., Lim, S., Haycock, J. W. & Fallon, J. H. (1989). Ventral mesencephalon neurons containing both cholecystokinin and tyrosine hydroxylase-like immunoreactivities project to forebrain regions. *Journal of Comparative Neurology*, 279, 397–414.

Sesack, S. R. & Pickel, V. M. (1992). Dual ultrastructural localization of enkephalin and tyrosine hydroxylase immunoreactivity in the rat ventral tegmental area: multiple substrates for opiate-dopamine interactions. *Journal of Neuroscience*, 12, 1335–50.

Shimada, S., Kitayama, S., Lin, C. L., Patel, A., Nanthakumar, E., Gregor, P., Kuhar, M. & Uhl, G. (1991). Cloning and expression of a cocaine-sensitive dopamine transporter complementary DNA. *Science*, 254, 576–8.

Sibley, D. R. (1991). Cloning of a D3 receptor subtype expands dopamine receptor family. *Trends in Pharmacological Sciences*, 12, 7–9.

Singer, H. S. (1992). Neurochemical analysis of postmortem cortical and striatal brain tissue in patients with Tourette syndrome. In *Advances in Neurology. Vol. 58*. ed. T. N. Chase, A. J. Friedhoff, & D. J. Cohen. pp. 135–144. New York, N. Y. Raven Press.

Smeets, W. J. A. J. (1991). Comparative aspects of the distribution of substance P and dopamine immunoreactivity in the substantia nigra of amniotes. *Brain, Behavior and Evolution*, 37, 179–88.

Smith, A. D. & Bolam, J. P. (1990). The neural network of the basal ganglia as revealed by the study of synaptic connections of identified neurons. *Trends in Neurosciences*, 13, 259–65.

Smith, Y. & Parent, A. (1986). Differential connections of caudate nucleus and putamen in the squirrel monkey (*Saimiri sciureus*). *Neuroscience*, 18, 347–71.

Smith, Y. & Parent, A. (1988). Neurons of the subthalamic nucleus in primates display glutamate but not GABA immunoreactivity. *Brain Research*, 453, 353–6.

Snyder, S. H., Barnerjee, S. P., Yamamura, H. I. & Greenberg, D. (1974). Drugs, neurotransmitters and schizophrenia. *Science*, 184, 1243–53.

Sokoloff, P., Giros, B. Martres, M. P., Bouthenet, M. L. & Schwartz, J. C. (1990). Molecular cloning and characterization of a novel dopamine receptor (D3) as a target for neuroleptics. *Nature*, 347, 146–51.

Somogyi, P., Priestley, J. V., Cuello, A. C., Smith, A. D. & Bolam, J. P. (1982). Synaptic connections of substance P-immunoreactive nerve terminals in the substantia nigra of the rat: a correlated light- and electron-microscopic study. *Cell and Tissue Research*, 223, 469–86.

Stoof, J. C., Drukarch, B., de Boer, P., Weterink, B. H. C. & Groenewegen, H. J. (1992). Regulation of the activity of striatal cholinergic neurons by dopamine. *Neuroscience*, 47, 755–70.

Strecker, R. E. & Jacobs, B. L. (1985). Substantia nigra dopaminergic unit activity in behaving cats: effects of arousal on spontaneous discharge and sensory evoked activity. *Brain Research*, 361, 339–50.

Stricker, E. M. & Zigmond, M. J. (1976). Recovery of function after damage to central catecholamine-containing neurons: a neurochemical model for the lateral hypothalamic syndrome. In *Progress in Psychobiology and*

Physiological Psychology. ed. J. M. Sprague & A. N. Epstein, pp. 121–188. N. Y: Academic Press.

Sunahara, R. K., Guan, H. C., O'Dowd, B. F., Seeman, P., Laurier, L. G., Ng, G., George, S. R., Torchia, J., Van Tol, H. M. M. & Niznik, H. B. (1991). Cloning of the gene for a human dopamine D$_5$ receptor with higher affinity for dopamine than D$_1$. *Nature*, **350**, 614–19.

Surmeier, D. J., Eberwine, J. Wilson, C. J. Cao, Y. Stefani, A. & Kitai, S. T. (1992). Dopamine receptor subtypes colocalize in rat striatonigral neurons. *Proceedings of the National Academy of Sciences, USA*, **89**, 10178–82.

Surmeier, D. J., Reiner A. Levine M. S. & Ariano, M. A. (1993). Are neostriatal dopamine receptors co-localized? *Trends in Neurosciences*, **16**, 299–305.

Szabo, J. (1980). Distribution of striatal afferents from the mesencephalon in the cat. *Brain Research*, **188**, 3–21.

Tan, D. P. & Tsou, K. (1988). Intranigral injection of dynorphin in combination with substance P on striatal dopamine metabolism in the rat. *Brain Research*, **443**, 310–14.

Tang, F., Costa, E. & Schwartz, J. P. (1983). Increase of proenkephalin mRNA and enkephalin content of rat striatum after daily injection of haloperidol for two to three weeks. *Proceedings of the National Academy of Sciences, USA*, **80**, 3841–4.

Tremblay, L. & Filion, M. (1989). Responses of pallidal neurons to striatal stimulation in intact waking monkeys. *Brain Research*, **498**, 1–16.

Tremblay, L. & Filion, M. (1991). Behavioral and neuronal effects of GABA agonist and antagonist injected locally in the globus pallidus of intact monkeys. *Society for Neuroscience Abstracts*, **16**, 428.

Tremblay, L., Filion, M. & Bedard, P. J. (1989). Responses of pallidal neurons to striatal stimulation in monkeys with MPTP-induced parkinsonism. *Brain Research*, **498**, 17–33.

Uchimura, N., Higashi, H. & Nishi, S. (1986). Hyperpolarizing and depolarizing actions of dopamine via D-1 and D-2 receptors on nucleus accumbens neurons, *Brain Research*, **375**, 368–72.

Ungerstedt, U. & Arbuthnott, G. W. (1970). Quantitative recording of rotational behavior in rats after 6-hydroxydopamine lesions of the nigrostriatal dopamine system. *Brain Research*, **24**, 485–93.

Ungerstedt, U., Butcher, L. L., Butcher, S. G., Anden, N. E. & Fuxe, K. (1969). Direct chemical stimulation of dopaminergic mechanisms in the neostriatum of the rat. *Brain Research*, **14**, 461–71.

van der Kooy, D., Coscina, D. V. & Hattori, T. (1981). Is there a non-dopaminergic nigrostriatal pathway? *Neuroscience*, **6**, 345–7.

Van Tol, H. H. M., Bunzow, J. R., Guan, H. C., Sunahara, R. K., Seeman, P., Niznik, H. B. & Civelli, O. (1991). Cloning of the gene for a human dopamine D4 receptor with high affinity for the antipsychotic clozapine. *Nature*, **350**, 610–14.

Vincent, S. R., Skirboll, L., Hökfelt, T., Johansson, O., Lundberg, J. M., Elde, R. P., Terenius, L. & Kimmel, J. (1982). Coexistence of somatostatin- and avian pancreatic polypeptide (APP)-like immunoreactivity in some forebrain neurons. *Neuroscience*, **7**, 439–46.

Vincent, S. R. and Kimura, H. (1992). Histochemical mapping of nitric oxide synthase in the rat brain. *Neuroscience*, **46**, 755–84.

Voorn, P. & Buijs, R. M. (1987). The ultrastructural demonstration of dopamine in the central nervous system. In *Monoaminergic Neurons: Light Microscopy and Ultrastructure*, ed. H. W. M. Steinbusch. International Brain Research Organization Handbook Series: Methods in the Neurosciences. Chichester: Wiley. pp. 241–264.

Voorn, P., Gerfen, C. R. & Groenewegen, H. J. (1989). Compartmental organization of the ventral striatum of the rat: immunohistochemical distribution of enkephalin, substance P, dopamine and calcium binding protein. *Journal of Comparative Neurology*, **289**, 189–201.

Voorn, P., Jorritsma-Byham, B. van Dijk, C. & Buijs, R. M. (1986). The dopaminergic innervation of the ventral striatum in the rat: A light- and electron-microscopical study with antibodies against dopamine. *Journal of Comparative Neurology*, **251**, 84–99.

Voorn, P., Kalsbeek, A. Jorritsma-Byham, B. & Groenewegen, H. J. (1988). The pre- and postnatal development of the dopaminergic cell groups in the ventral mesencephalon and the dopaminergic innervation of the striatum of the rat. *Neuroscience*, **25**, 857–87.

Vuillet, J., Kerkerian, L., Kachidian, P., Bosler, O. & Nieoullon, A. (1989). Ultrastructural correlates of functional relationships between nigral dopaminergic or cortical afferent fibers and neuropeptide Y-containing neurons in the rat striatum. *Neuroscience Letters*, **100**, 99–104.

Walker, J. M., Thompson, L. A., Frascella, J. & Friederich, M. W. (1987). Opposite effects of mu and kappa opiates on the firing-rate of dopamine cells in the substantia nigra of the rat. *European Journal of Pharmacology*, **134**, 53–9.

Walters, J. R., Bergstrom, D. A., Carlsson, J. H., Chase, T. N. & Braun, A. R. (1987). D1 dopamine receptor activation required for postsynaptic expression of D2 agonist effects. *Science*, **236**, 719–22.

Weick, B. G. & Walters, J. R. (1987). Effects of D1 and D2 dopamine receptor stimulation on the activity of substantia nigra pars reticulata neurons in 6-hydroxydopamine lesioned rats: D1/D2 coactivation induces potentiated responses. *Brain Research*, **405**, 234–46.

Weiner, D. M., Levey, A. I. & Brann, M. R. (1990). Expression of muscarinic acetylcholine and dopamine receptor mRNAs in rat basal ganglia. *Proceedings of the National Academy of Sciences, USA*, **87**, 7050–4.

Weiner, D. M., Levey, A. I. Sunahara, R. K., Noznik, H. B., O'Dowd, B. F., Seeman, P. & Brann, M. R. (1991). D1 and D2 dopamine receptor mRNA in rat brain. *Proceedings of the National Academy of Sciences USA*, **88**, 1859–63.

Yoshida, M. (1993). The neuronal mechanism underlying parkinsonism and dyskinesia, and differential roles of the putamen and caudate nucleus. In *Advances in Neurology.* Vol. 60. ed. H. Narabayashi, T. Nagatsu, N. Yanagisawa & Y. Mizuno. pp. 71–77. New York, NY: Raven Press.

Young, W. S. III, Bonner, T. I. & Brann, M. R. (1986). Mesencephalic dopamine neurons regulate the expression of neuropeptide mRNAs in the rat forebrain. *Proceedings of the National Academy of Sciences USA*, **83**, 9827–31.

Zhang, W. Q., Pennypacker, K. R., Ye, H., Merchenthaler, I. J., Grimes, L., Iadarola, M. J. & Hong, J. S. (1992). A 35 kDa Fos-related antigen is co-localized with substance P and dynorphin in striatal neurons. *Brain Research*, **577**, 312–17.

11
Telencephalic dopamine cells in monkeys, humans, and rats

M. Dubach

Introduction

The discovery of tyrosine hydroxylase-immunoreactive (TH-IR) neuronal cell bodies in the neostriatum (STR) in non-human primates (Dubach et al., 1986, 1987) adds another dimension to the problem of describing the role of dopamine (DA) in the basal ganglia, and suggests the importance of the comparative method in approaching this problem. A study of adult rats has detected a similar population of neurons, although in the rat they contain DA without detectable levels of TH (Voorn et al., 1988). DA neurons in the cortex have also been noted in monkeys and rats in these studies. The present contribution provides data on the variability and plasticity of telencephalic DA neurons in monkeys and compares the information available on their distribution and development in monkeys, humans, and rats.

Telencephalic dopamine cell distribution

Early reports of a few catecholamine cells in the adult monkey, human, and rat telencephalon were incorporated in studies of other topics (de la Torre, 1972; Hökfelt et al., 1977; Felten & Sladek, 1983; Hökfelt, Johansson & Goldstein, 1984a), but in the past ten years a number of investigations have focused specifically on such cells (Köhler et al., 1983; Dubach et al., 1986, 1987; Gaspar et al., 1987; Dubach, Schmidt & Bowden, 1988a; Trottier, Geffard & Evrard, 1989; Kuljis, Martin-Vasallo & Peress, 1989; Tashiro et al., 1989; Gouras et al., 1992).

Monkeys

The telencephalic DA neurons were originally detected as TH-IR cells in adult monkeys, mostly among and around the fibers of the olfactory tract near the ventral striatum (VS: nucleus accumbens and olfactory tubercle), numbering 4000–5000 cells per hemisphere in all three layers of the olfactory and piriform cortices (Köhler et al., 1983; Dubach et al., 1987); smaller numbers were noted nearby in the diagonal band of Broca. Similar cells are sparsely distributed more rostrally along the olfactory tract itself, in continuity with the dense population of DA neurons in the olfactory bulb (Dubach, unpublished observations); virtually no TH-IR neurons are located within the VS itself, which is clearly demarcated by its dense TH-IR neuropil, although many are located in white matter very near the border (Dubach et al., 1987). A caudal extension of this array of basal TH-IR neurons into the region of the basal nucleus of Meynert has also been described for monkeys as well as for humans (Gouras et al., 1992; see also Figure 11.3 below). In addition to these neurons near the VS, large numbers of additional TH-IR neurons also appear dorsally and rostrally in the putamen and lateral caudate in monkeys, especially at rostral levels of the STR from the anterior commissure forward (Dubach et al., 1987); peristriatal cells also occur outside striatal neuropil, in the corona radiata, external capsule, claustrum, and extreme capsule. At the most rostral level examined in this investigation, anterior to the rostral tip of the caudate nucleus, numerous TH-IR cell bodies were also observed in the deep layers of the prefrontal cortex and in adjoining white matter.

The numbers and distribution of telencephalic TH-IR cells were examined and compared in six long-tailed macaques to provide the data presented here, which have appeared previously in abstract from (Dubach et al., 1988a). Perfusions and immunohistochemical procedures have been described previously

(Dubach et al., 1987). Artefactual variation due to technique was minimized by treating corresponding sections from each of the 12 hemispheres simultaneously throughout the procedures, with 12 free-floating sections in each immunohistochemical container. The influence of the penetration of immunohistochemical reagents on the numbers of cells counted was tested by cross-sectioning a typical 40-micron section from a separate monkey after the same immunohistochemical treatment. This was accomplished by embedding the section in a supporting matrix of egg-yolk, cutting frozen sections on the microtome perpendicular to the original plane of section, and examining these cross-sections without further staining. While terminal labeling extended only a few microns into the tissue, labeled cell bodies appeared at all depths. Processes, which were often visible in the plane of section, apparently gave access even to most of the cells located in the middle depths; numbers of cells in each of three defined zones in the cross-section were roughly proportional to the size of each zone (Fig. 11.1).

In order to characterize the neurotransmitter identity of the telencephalic TH-IR cells, additional sections were stained for dopamine betahydroxylase and dopa decarboxylase immunoreactivity. No telencephalic cells were positive for dopamine betahydroxylase, although cells in the locus ceruleus were well labeled. This indicates that TH-IR cells do not synthesize norepinephrine or epinephrine. Many cells positive for dopa decarboxylase were detected in a distribution similar to that of the TH-IR cells, and in some cases the same cell was labeled by the two antibodies in adjacent sections. This supports the probability that the telencephalic TH-IR cells in monkeys are indeed DA cells.

Large numbers of TH-IR cells are associated with the STR and VS (Table 11.1; Figs. 11.2–11.4). For each subject and each of seven coronal intervals, the estimated total number of cells is listed in Table 11.1 and charted in Fig. 11.2. Cells were found within the dorsolateral caudate nucleus and the putamen, and dorsally a few of the cells lay outside the TH-IR neuropil in the associated white matter; a progressively larger proportion of the cells lay outside the striatal neuropil laterally near the putamen, ventrolaterally, and ventrally near the VS (first four levels: Fig. 11.3). Smaller numbers of cells occurred in the preoptic area, substantia innominata, septum, and prefrontal cortex. The estimated total number of telencephalic TH-IR cells for all regions averaged 32 910 ± 10 094 (mean and S.D.) per hemisphere, which can be compared with an estimated 81 680 ± 4042 mesencephalic DA neurons in the same species (German et al., 1988). These data indicate a high degree of individual variation of telencephalic TH-IR cells as compared with midbrain DA cells. Comparison of left and right sides from each monkey indicates that the variation among monkeys is much greater than variation between hemispheres within a given monkey. The mean

Penetration: Pooling by Zone

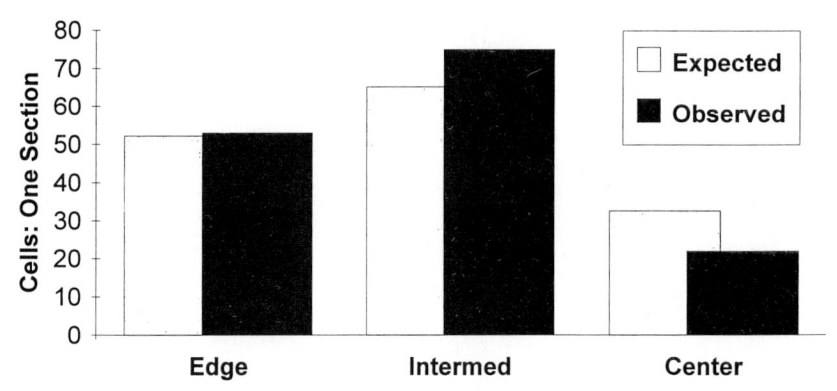

Fig. 11.1. A single section in the region of the middle head of the caudate was embedded in egg yolk and cross-sectioned to provide strips in which the depth of labeled cells within the original section could be measured. A 'center' zone was defined, bracketed by two 'intermediate' and two 'edge' zones in these cross-sectional strips. Given the total number of cells counted in all strips, the chart compares the number of cells located in each zone (observed) with the number of cells expected on the basis of the defined thickness of the zone (expected). A chi-square test indicated no significant difference between observed and expected values at $p = 0.05$.

Table 11.1. *Forebrain dopamine cells in six monkeys*

Monkey hemisphere:	155L	155R	157L	157R	207L	207R	026L	026R	030L	030R	088L	088R	Average	Std Dev
Coronal level:														
Frontal cortex	2 640	1 080	2 760	2 520	11 640	11 640	3 240	2 760	2 640	4 680	2 280	1 920	4 150	3 597
Anterior head of caudate	7 260	8 340	3 780	7 140	8 400	12 660	11 640	12 600	7 080	6 300	7 920	7 800	8 410	2 650
Middle head of caudate	5 880	5 880	9 000	5 700	10 320	11 700	10 920	9 780	7 320	8 460	5 100	5 100	7 930	2 402
Posterior head of caudate	6 360	6 000	6 060	5 160	10 080	11 160	7 680	6 420	12 600	8 220	9 120	7 140	8 000	2 309
Anterior body of caudate	1 260	1 320	1 020	900	3 120	2 700	3 000	1 380	360	1 200	720	1 260	1 520	908
Middle body of caudate	1 620	2 100	900	1 080	3 660	1 980	3 600	1 860	2 880	3 300	540	900	2 035	1 101
Posterior body of caudate	360	600	300	960	1 200	1 080	1 620	600	1 080	1 260	480	840	865	405
Total	25 380	25 320	23 820	23 460	48 420	52 920	41 700	35 400	33 960	33 420	26 160	24 960	32 910	10 094

Each column includes data for one hemisphere (L or R) from one animal (3-digit number). Labels on the left designate the anterior-posterior level of each coronal section. All coronal levels were sampled at 3-mm intervals except the first (level of the prefrontal cortex), which was 6 mm ahead of the second (level of the anterior head of caudate). The Abercrombie correction (Abercrombie, 1946) was made for the cell-count from each section; the result was multiplied to account for a full (usually 3-mm) series of coronal sections.

276   M. Dubach

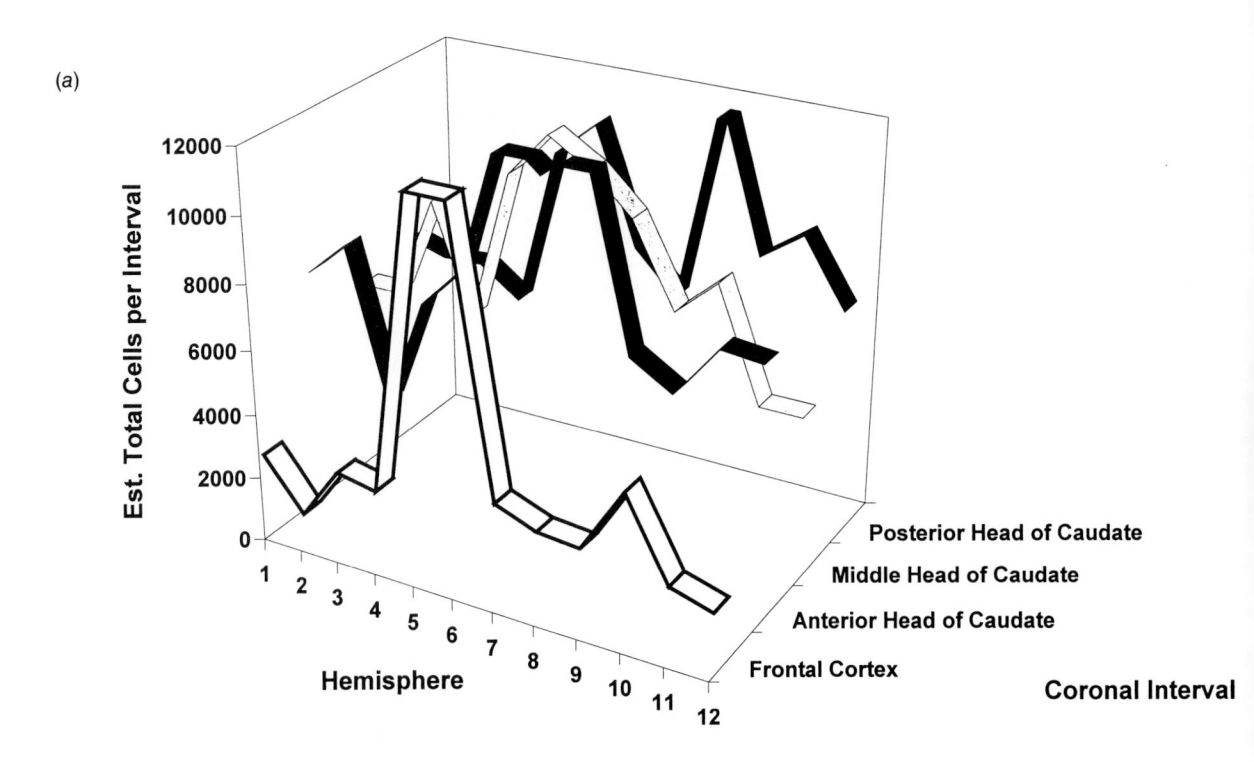

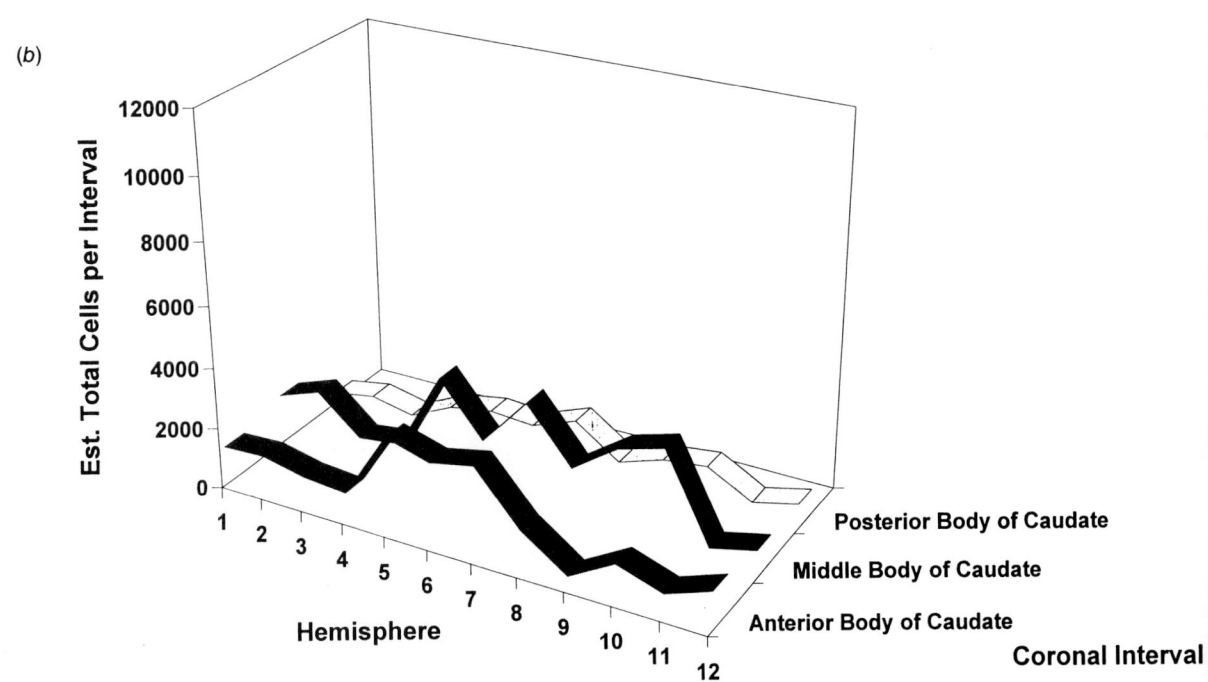

(a)

(b)

Fig. 11.2. Data from Table 11.1 are depicted row by row, each band representing one coronal level. Variation among subjects can be visualized along each band, representing one of the seven coronal segments for which cell numbers were estimated from counts in one section per hemisphere. Hemispheres are numbered in the order in which they appear in Table 11.1. Data from the level of the prefrontal cortex and three levels through the head of the caudate nucleus appear in Fig. 11.2(a); data from three levels through the body of the caudate nucleus appear in Fig. 11.2(b).

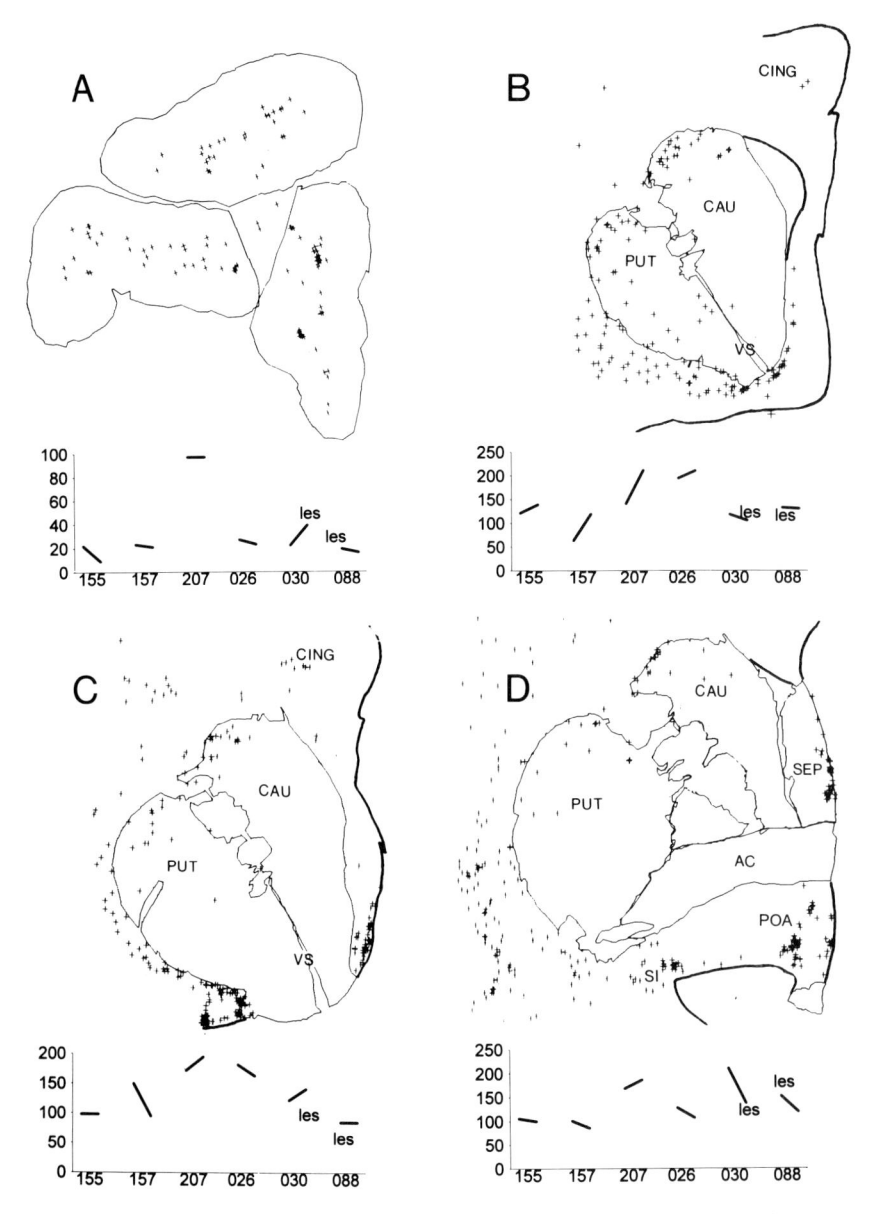

Fig. 11.3. TH-IR cell locations in sections from one hemisphere (207L) at the first four coronal levels. The chart beneath the distribution plot for each coronal level indicates for that level the number of cells counted (y-axis) in each hemisphere for each monkey (x-axis). Each line segment connects the number of cells in the left hemisphere (left) to the number in the right hemisphere (right). The two points labeled 'les' represent data from the two hemispheres lesioned by intranigral MPTP injections. A t-test of differences indicated no significant difference between the two hemispheres at p = 0.05 for any coronal level or for the totals of all coronal levels. (a). The first level contains only prefrontal cortex and subjacent white matter. (b),(c) The caudate, putamen, and ventral striatum are sharply defined in TH-IR sections as a continuous expanse of stained neuropil; the boundary of this neuropil is outlined in this computer-generated drawing. Note that the VS neuropil (nucleus accumbens/olfactory tubercle) stops short of the base of the brain at the second level (b), but extends fully to the base of the brain at the third level (c). (The double line dividing the VS dorsoventrally is an artifact of the data input procedure.) (d) At the fourth level, additional TH-IR cells appear in moderate numbers in the preoptic area, and in the substantia innominata, a region in which cholinergic cells of the basal nucleus of Meynert are also common. Cells in the infragranular layers of the cortex and subjacent white matter, ventrolateral to the putamen, were not included in the counts. Abbreviations for this figure: CAU, caudate nucleus; PUT, putamen; CING, cingulate gyrus; VS, ventral striatum; SEP, septum; AC, anterior commissure; POA, preoptic area; SI, substantia innominata.

number of TH-IR cells in each region (Fig. 11.4) is quite variable among individuals.

Unilateral DA lesions made in two monkeys (030R and 088L) by direct intranigral injections of 1-methyl-4-phenyl-1,2,3,6-tetrahydropyridine (Dubach et al., 1988b) had no consistent effect on the total number of TH-IR cells (Table 11.1). Some cells in the lesioned hemispheres, however, appeared to be hypertrophied (Fig. 11.5). One easily quantifiable feature of this change was an increase of the number of primary processes observed on each cell. Primary processes were counted for the lesioned and non-lesioned hemisphere at each of three coronal levels, and cells in the lesioned hemisphere had a higher average number of processes in each case (Fig. 11.6). Like other DA neurons (p. 289), the telencephalic TH-IR cells are thus capable of substantial phenotypic plasticity.

Humans

In humans, TH-IR cells occur in ventral locations from the hypothalamus to the olfactory tract. The paraventricular and supraoptic nuclei of the hypothalamus contain large numbers of TH-IR cells, including many of the magnocellular neurons (Spencer et al., 1985; Li et al., 1988; Pearson et al., 1990). Just anterior to this level, a few non-cholinergic magnocellular TH-IR neurons lie among the cholinergic neurons of the basal nucleus of Meynert, identifiable also by in situ hybridization with TH mRNA, in both monkeys and humans (Gouras et al., 1992). Larger numbers of both magnocellular and smaller TH-IR cells also occur more rostrally among the cholinergic cells of the diagonal band of Broca and the medial septal nucleus. The large TH-IR cells comprise 3–5% of all magnocellular neurons in these basal forebrain regions, and the small TH-IR cells are thought to be related to the extended amygdala (Gouras et al., 1992). Just anterior to the most rostral of these populations, separated from them medially by the extension of the VS to the base of the brain, TH-IR neurons in the basal forebrain are also associated with the olfactory tract and the anterior olfactory nucleus (Gaspar et al., 1985), in the anterior perforated substance ventral and lateral to the nucleus accumbens. Cells possibly corresponding to these and to the TH-IR cells near the VS in monkeys were observed in humans by fluorescence histochemistry 20 years ago (de la Torre, 1972).

Other investigators have described sparse populations of TH-IR neurons in the human cortex, somewhat larger and often with more primary dendrites than the monkey cells, but like them usually concentrated most heavily in the deep layers V and VI and in the subjacent white matter (Gaspar et al., 1987; Michel, Sakamoto & Pearson, 1988; Trottier et al., 1989; Kuljis et al., 1989). In situ hybridization with a mRNA probe for TH in humans has confirmed the

Structure: Estimated Total Cells

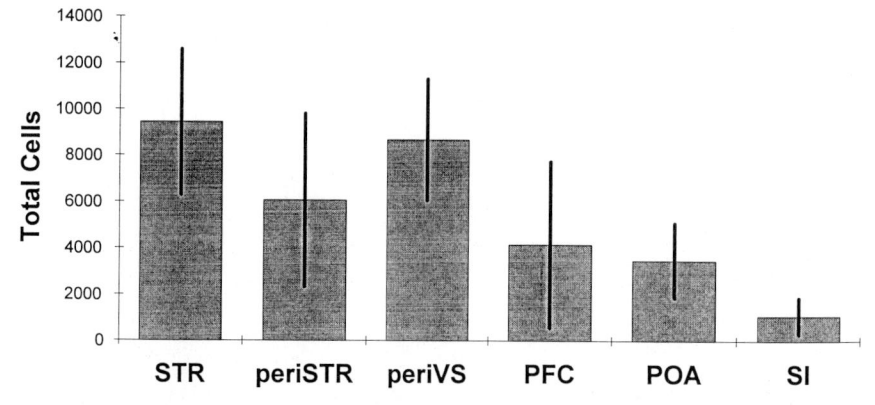

Fig. 11.4. Data from Table 11.1 are charted by anatomical area, summing across sections in which each area appears. The mean total number of cells and its standard deviation (y-axis) were estimated for each area on the basis of cell counts as in Table 11.1. Areas for which cells were counted (x-axis) included the caudate nucleus and putamen (STR), peristriatal white matter within 2 mm of striatal neuropil (periSTR), ventral forebrain ventral, medial and lateral to the VS, including cells in the septal area (periVS), the prefrontal cortex and subjacent white matter (PFC; counted in the first level only), the preoptic area (POA), and the substantia innominata (SI).

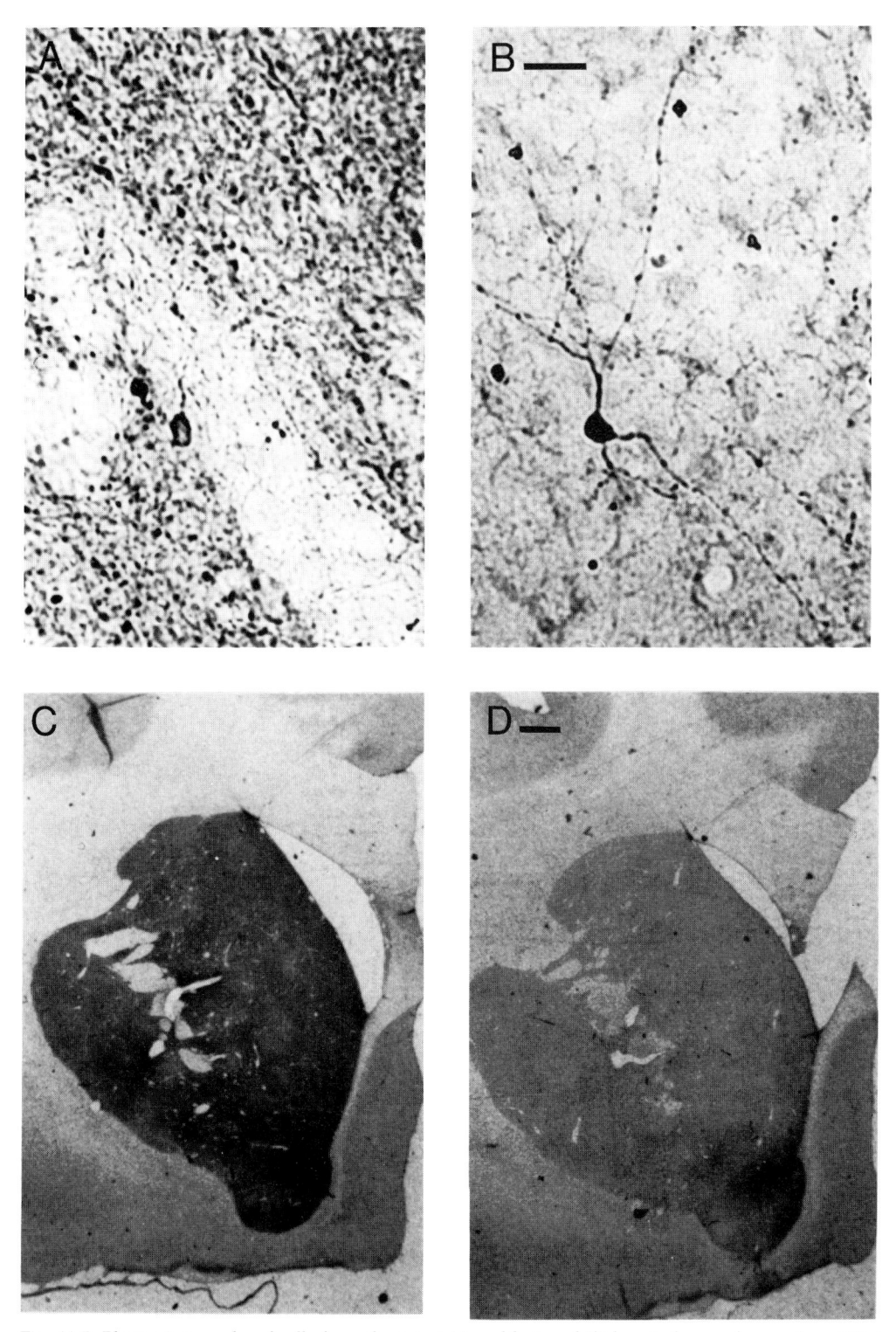

Fig. 11.5. Photomicrographs of cells from the intact (a) and lesioned (b) hemispheres of a monkey, 158 days after a lesion induced by intranigrally injected 1-methyl-4-phenyl-1,2,3,6-tetrahydropyridine, are paired with low-power photographs of intact (c) and lesioned (d) hemispheres from the same monkey. Scalebars (b),(d): 20 microns (a),(b); 1 millimeter (c),(d).

Lesion-induced Plasticity

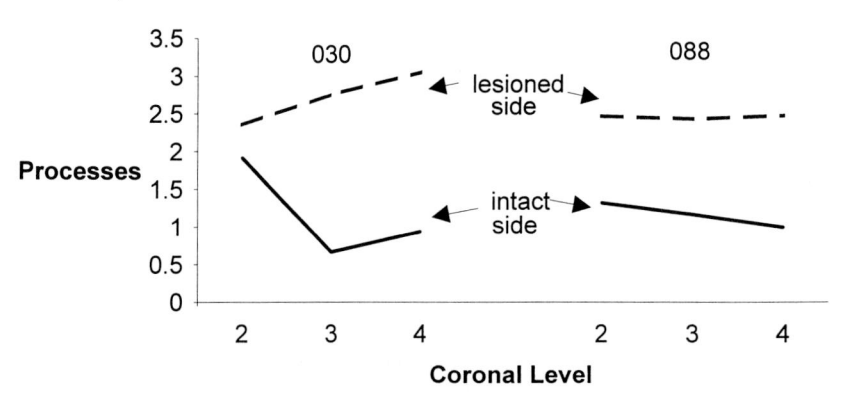

Fig. 11.6. The average number of primary processes associated with identified TH-LI neurons is charted for each of the three coronal levels that included the head of the caudate nucleus. The lesioned side in all cases contained cells with substantially more processes. A t-test of differences indicated a significant difference between lesioned and non-lesioned hemispheres at p = 0.01.

presence of the message as well as the enzyme (Kuljis et al., 1989). The cortical cells are more widespread in humans than in monkeys. As in monkeys, however, they are most common in frontal cortex. They are reported to number fewer than 0.1% of all cortical cells (Gaspar et al., 1987). The large size of the primate cortex, however, implies that though widely dispersed, this population of neurons is large enough to rival those of the midbrain in numbers (Michel et al., 1988; Dubach et al., 1988a; Pearson et al., 1990). The human cortical TH-IR cells reportedly do not contain the DA biosynthetic enzyme dopa decarboxylase (Gaspar et al., 1987; Kuljis et al., 1989), which has been detected in comparable cells in monkeys (Dubach et al., 1988a). Evidence has suggested that TH-IR cells are prominent among those affected by diffuse Lewy body disease (Kuljis et al., 1989), as was suggested by Gaspar et al. (1987) on the basis of the findings of Kosaka et al. (1984). The presence of cortical TH-IR neurons has also been indicated by the examination of biopsies taken at surgery from sites near or within tumor tissue (Hornung, Toerk & De Tribolet, 1989; Wakabayashi et al., 1989) or near focal pathologies in complex partial seizure patients (Zhu et al., 1990). These examinations have detected large numbers of TH-IR neurons, but adequate controls were not available to determine whether the disease states or superior fixation of small, fresh pieces of biopsy tissue enhanced expression or visualization of TH-IR.

Rats

Telencephalic DA cells have not been detectable in adult rats by TH-immunohistochemistry, but a population of DA-IR cell bodies similar in appearance and distribution has been noted in passing (Voorn et al., 1988). These cells were reported to appear in rats in the second postnatal week, in and near the STR and VS, and to remain throughout development to adulthood. These telencephalic DA-IR cells in rats, together with developmentally transient cortical TH-IR cells (p. 282), would complement well-known hypothalamic DA cells in rats, which include several discontinuous groups covering territory from the posterior hypothalamus to the periventricular region at the level of the magnocellular paraventricular and supraoptic nuclei, and even a few cells located in these nuclei themselves (Swanson et al., 1981; Chan-Palay et al., 1984; van den Pol, Herbst & Powell, 1984; Selemon & Sladek, 1986). Broadly comparable populations of presumptive DA neurons thus exist in monkeys, humans, and rats. Both cortical and striatal DA neurons are most often located near the gray/white border, either in gray matter or adjacent white matter.

Striatal dopamine cells: anatomical context

DA cells in rats and monkeys occur in close association with two specialized regions of the striatum, the marginal zone of the STR and the vicinity of the VS. The cells occur in and around the margin of the

STR where it contacts the surrounding white matter, in a dorsolateral region from the dorsolateral caudate to the lateral putamen, continuing into a ventral region adjacent to the VS. Only rarely do they occur deeper within the STR, or along the ventricular border, or dorsomedially under the corpus callosum. In the dorsolateral region, TH-IR cells are located primarily within the striatal neuropil and in smaller numbers in the adjacent white matter in monkeys (Dubach et al., 1987), and DA-IR cells are in a similar position in rats (Voorn et al., 1988). In the VS region, the cells occur in a denser, more concentrated distribution but do not lie within the neuropil; instead they appear in the adjacent white matter and the allocortex surrounding the olfactory tract in monkeys (Köhler et al., 1983; Dubach et al., 1987) and in a comparable position in rats (Voorn et al., 1988). Both the STR marginal zone and the VS not only differ from the rest of the striatum by virtue of the DA cells, but also deviate from the typical striosomal pattern of the majority of the STR. The dorsolateral marginal zone appears to be a single striosomal sheet covering most of the rostral dorsolateral STR, while the VS displays substantially more complex neurochemical variegation than the STR.

Neostriatum
Striosome and matrix compartments in the STR have been widely investigated (Graybiel & Ragsdale, 1983; Graybiel, 1990). Even in the earliest study of such patterns (Olson, Seiger & Fuxe, 1972), a dorsolateral specialization involving DA was apparent, but this 'marginal zone', comparable to smaller striosomes elsewhere, has received much less attention than the more typical central and medial patterning. Specialized marginal and dorsolateral patterns have been indicated by several methods, including DA immunohistochemistry (Voorn et al., 1986, 1988) and D2 receptor labeling (Joyce, Loeschen & Marshall, 1985; Altar et al., 1985) in intact rats, experimental manipulations of DA turnover (Olson et al., 1972; Hökfelt et al., 1984b), and measurements of DA uptake in DA-denervated tissue (Doucet, Descarries & Garcia, 1986). A dorsolateral band corresponding to the marginal zone also differs from other striosomes in its DA innervation from the midbrain, receiving input from a group of ventral DA neurons residing in isolated clusters in the ventral substantia nigra pars reticulata (SNpr) (Gerfen, Herkenham & Thibault, 1987b). Other features not directly involving DA but also marking a subcallosal and largely dorsolateral band include a light concentration of calcium-binding protein (CaBP) (Liu & Graybiel, 1992a) and a heavy concentration of mu opiate receptors (Atweh & Kuhar, 1977; Herkenham & Pert, 1980; Gerfen et al., 1987b; Sharif & Hughes, 1989). By these various measures the dorsolateral band or marginal zone appears to represent an unusually large and eccentrically positioned striosome wrapping rostrally around much of the dorsal and lateral STR.

Few investigations have provided any information on a dorsolateral specialization in the striatum of adult monkey species, although one report included indications of a low density of cholecystokinin receptors along the rostral dorsolateral margin of the caudate and putamen (Kritzer, Innis & Goldman-Rakic, 1990). In neonatal monkeys, the distributions of DA and cholinergic uptake sites, D2 dopamine receptors, and M1 muscarinic receptors include a marginal rim of high concentration, like internal striosomes at this age (Lowenstein et al., 1989). It may be that a distinct dorsolateral band in adult rats but not in adult monkeys represents a species difference. Although striosomal patterns in monkeys are generally similar to those in rats (Martin et al., 1991a), there are significant differences. Cell-dense islands occur in monkeys (Goldman-Rakic, 1982), in correspondence with substance P- and enkephalin-rich striosomes that are more sharply demarcated than in rats (Martin et al., 1991a). The average area of a striosome is also much larger in monkeys, although the total number of striosomes is about the same as in rats (Johnston et al., 1990). In monkeys, somatostatin processes occur in the striosomes as well as in the matrix (Martin et al., 1991a), while they are largely limited to the matrix in rats (Gerfen, 1984). TH is relatively less concentrated in striosomes than in matrix in adult monkeys (Graybiel, Hirsch & Agid, 1987; Martin et al., 1991a), and DA uptake sites and D2 receptors follow a similar adult pattern of distribution (Lowenstein et al., 1989).

Ventral striatum
Recent studies indicate that in rats the compartmentalization of much of the VS, particularly its peripheral portions, differs markedly from the predominant STR pattern (Berendse et al., 1992). The VS is particularly important to the striatal influence of DA, because it provides a direct projection to most of the mesencephalic DA-cell populations that innervate both the STR and the VS (Haber et al., 1985; Gerfen et al., 1987b; Berendse et al., 1992). In the olfactory tubercle region, moreover, a complex mixture of striatal, external pallidal, internal pallidal, amygdalar, and basal magnocellular nucleus neurons prevails (Heimer & Wilson, 1975; Heimer, 1978; Fibiger, 1982; Haber & Nauta, 1983; Haber et al., 1985; Schwaber et

al., 1987; Alheid & Heimer, 1988). The significance of these complex subdivisions in rats is not yet well understood, and very little information is available regarding compartmentalization of the nucleus accumbens and olfactory tubercle in monkeys, but the olfactory tubercle as a layered structure is much less prominent (Heimer, Van Hoesen & Rosene, 1977; Baron, Stephan & Frahm, 1987), and the olfactory bulbs are relatively much smaller. In and around this complex region, from which most midbrain DA cells receive their only striatal innervation, telencephalic DA cells are less densely concentrated in rats (Voorn et al., 1988) than in monkeys (Köhler et al., 1983; Dubach et al., 1987), and may represent a significant species difference affecting basal ganglia physiology.

Telencephalic dopamine cells: ontogeny in rats

The origin, final cell division, migration, and final differentiation of rat DA neurons have been widely studied from the perspective of the commonly recognized mesencephalic and diencephalic cell groups. A few of these investigations have touched upon the development of more rostral forebrain DA cells.

Somata containing TH are first detectable between E12 and E13. Even at this first appearance, they include cells anterior to the optic chiasm extending rostrally ventral to the developing striatum (Specht et al., 1981a). Within a day the prosencephalic cells comprise two discontinuous groups, (1) a few cells between the infundibulum and the POA, and (2) more rostrally a larger number of lightly stained cells ventral to the incipient striatum. Between E14 and E15, cells of the first group appear to be positioned as incipient hypothalamic groups A11–A14, and the lighter, more rostral cells are still ventral to the striatum. Axons and terminals reaching the ventrolateral STR first appear at this point. In spite of their position, these rostral cells have been designated the anterior outpost of the hypothalamic groups (Specht et al., 1981a). Evidence indicates that the final mitosis, the first detectable appearance of TH, and the completion of migration in these groups occur in that order. At E18, the same cell groups are visible in similar positions (Specht et al., 1981b). The most rostral cells are few in number, and continuous with cells that are more clearly a part of A14, extending into the lateral preoptic area and the lateral septum. In addition, however, a small number of strongly labeled cells have now appeared at the anterior pole of the frontal cortex. Even this late in development, striatal TH-IR terminals are visible only in the ventrolateral STR, together with fibers entering the STR along this margin from internal capsule bundles. At E21, striatal TH-IR terminals and ventrolateral axons from the internal capsule have extended their striatal territory, but they are still limited to the ventrolateral quadrant of the STR. The frontal cortex cells have disappeared (temporarily?), but TH-IR cells have appeared in the olfactory bulb (Specht et al., 1981b).

In neonatal rats, it has been proposed (Verney et al., 1982) that occasional cell bodies in the anteroventral frontal cortex at the level of the rhinal fissure, or in the caudal olfactory bulb, could correspond to those observed by Specht et al. (1981b). To this developmental scenario must be added the TH-IR cells in neocortex, observed by these authors at an AP level between the genu of the corpus callosum and the retrosplenial subiculum (Berger et al., 1985). Few such cells were visible between P8 and P12, but a marked increase from P14 to P18 brought their numbers to a maximum. This was followed by a gradual decline until the cells were undetectable at P30, although cortical TH-IR perikarya in adult rats have been reported (Kosaka, Hama & Nagatsu, 1987a). The first appearance of DA-IR cell bodies in and near the STR and VS also occurs in the second postnatal week (Voorn et al., 1988), coinciding approximately with the transient appearance of the large population of cortical TH-IR cells (Berger et al., 1985).

Several explanations have been suggested for the seemingly uneven development of telencephalic DA neurons. The adult cortical neurons do not appear to correspond to any of the prosencephalic neurons observed during prenatal development, but ventricular zone cells temporarily expressing TH at an earlier developmental date may again express TH at a later date (Berger et al., 1985). The sudden appearance of the cells in the olfactory bulb may indicate a migration of transiently TH-IR neurons from the frontal pole into the incipient olfactory bulb, or a late final mitosis of external granular neurons followed by the initiation of TH synthesis (Specht et al., 1981b). It seems reasonable to propose more generally (pp. 286–287) that cells variably producing TH and DA at different times during development are nonetheless consistently present along most of the ventral midline from the mesencephalon to the olfactory bulb.

Marginal zone ontogeny in rats

Given the relationship of striatal DA neurons to the margins of the STR, it is important also to consider the ontogeny of DA innervation, final cell division, and cellular differentiation in this region. The development of the marginal zone must be considered in the context of striosomal development. In normal perinatal development, striosomes rich in DA, TH, D1, mu opiate receptors, and substance P (Moon-Edley & Herkenham, 1984; Caboche, Rogard & Besson, 1991; Snyder-Keller, 1991) receive DA innervation earlier than the matrix (Gerfen, Baimbridge & Thibault, 1987a). TH-immunoreactive DA inputs to the matrix come a few days later, leading ultimately to an even distribution of detectable TH immunoreactivity (Liu & Graybiel, 1992a) and D1 receptors (Murrin & Zeng, 1989; Caboche et al., 1991). Only the dorsolateral region maintains a patchy DA distribution in adults (Voorn et al., 1986, 1988).

Prenatal marginal zone ontogeny

The lateral STR receives the earliest DA input prenatally, beginning with its marginal zone, from fibers that course rostrally from the SN under the entopeduncular nucleus to enter the STR from a ventrolateral direction (Voorn et al., 1988). The somata of origin of these fibers are among the earliest nigral DA cells to undergo the final mitotic division, and their axons invade the ventrolateral STR before their migration from the midline to the lateral SN has been completed (Voorn et al., 1988). Neurogenesis in the STR occurs earliest in the marginal zone (Marchand & Lajoie, 1986), but the first DA fibers arrive well before the period of most rapid final mitosis in this region (Voorn et al., 1988). DA innervation from the midbrain and neurogenesis in the STR are therefore probably contemporaneous. DA innervation proceeds in the rostral STR along a ventrolateral-to-dorsomedial gradient throughout prenatal development; more caudally, in a contrasting pattern, the DA innervation enters more ventromedially and neurogenesis also proceeds along a medial-to-lateral gradient (Marchand & Lajoie, 1986). The gradients of both neurogenesis (final mitosis) and cytodifferentiation (neuritogenesis) of striatal cells are very similar to these gradients of DA innervation throughout the STR and VS (Voorn et al., 1988). It is perhaps of more than passing interest that the course followed by DA fibers to reach the STR and VS in the rat (Voorn et al., 1988, Fig. 39), which marks the leading anatomical edge of these gradients, bears a striking resemblance to the course of the internal capsule in the monkey.

The marginal zone of extensive DA fiber ramification is thus the first striosomal specialization to appear in the STR. A thin rim of densely packed catecholamine-fluorescent varicosities in a 'pronounced marginal zone' was originally observed in fetal rats at age E18–19, after maternal treatment with a monoamine oxidase inhibitor; this dense pattern contrasted with sparse varicosities elsewhere, and appeared a day earlier than similarly detected striosomes deeper in the STR (Seiger & Olson, 1973). In non-treated rats one day older (E19–E20), DA-fluorescent striosomes and a 'lateral rim' of DA fluorescence and relatively dense mu opiate receptors at the border of the STR are detectable (Moon-Edley & Herkenham, 1984). The first islands detectable by DA immunohistochemistry appear in the same dorsolateral region (Voorn et al., 1988) and are in register with transiently CaBP-rich zones (Liu & Graybiel, 1992b). Also prenatally, a dorsolateral 'band' and the associated DA islands pass through a stage in which CaBP-positive cells and processes are present, but CaBP-IR quickly disappears from the dorsolateral zone (Liu & Graybiel, 1992b); a strongly TH-immunoreactive dorsolateral band is retained, however, and elsewhere in the STR the DA islands are also highly TH-immunoreactive perinatally (Liu & Graybiel, 1992a).

Postnatal marginal zone ontogeny

The lateral or dorsolateral band in early postnatal days is characterized by a high density of DA fluorescence, TH-IR, D1 receptors, mu opiate receptors, and AChE (Foster et al., 1987; Agnati et al., 1988; Murrin & Zeng, 1989). By the end of the third postnatal week DA fibers are evenly distributed throughout the STR except in the marginal zone, which retains a relatively higher density of DA fibers throughout development to adulthood (Voorn et al., 1988). During perinatal and later postnatal development, CaBP-immunoreactivity in the matrix also begins to appear in the caudoventromedial STR, expanding rostrodorsolaterally over a few days but leaving the dorsolateral portion CaBP-poor (Liu & Graybiel, 1992a).

Telencephalic dopamine cells: developmental significance

Distribution

As described above, the cortical DA cells in monkeys are located primarily in the deep cortical layers, which develop earlier than the superficial layers; the striatal DA cells are located at the early-developing periphery of the STR and VS. Neurogenesis and cytodifferentiation in the cortex follow the familiar pattern from inside to outside, beginning with the lowest layers; telencephalic DA cells are located in these layers and the subjacent white matter in the adult. In the striatum, these developmental processes are largely initiated at the outer margins, also near the locations of DA cells, as described above. It is therefore reasonable to propose, on the basis of their location at this early time in ontogeny, that these cells may play a role in development (Voorn et al., 1988), and specifically in the organization of early connections between the cortex and the striatum.

Plasticity

Unfortunately, however, striatal and cortical TH-IR/DA-IR neurons have not been consistently or continuously mapped in rats during perinatal development, and their own origin has not yet been clearly attributed to the ventromedial neuroepithelium from which other DA neurons originate. Rostral ventromedial cells that express TH at early points in development cannot be directly followed in embryos of intermediate ages to the locations of adult telencephalic TH-IR cells (Specht et al., 1981a,b; Verney et al., 1982; Berger et al., 1985; Verney et al., 1988).

A major difficulty in addressing this question may be the plasticity of these cells during development. DA neurons in many contexts are known to exhibit plasticity, and a survey of examples is instructive. (1) The TH phenotype can be delayed or impaired in certain types of cells. Expression of TH-immunoreactivity can be delayed until long after birth at some locations, such as the dopaminergic neurons of the dorsal root ganglia (Price & Mudge, 1983). TH-IR and TH mRNA expression in DA neurons in the olfactory bulb can be seriously affected by altering afferent input to the bulb (Baker et al., 1983; Kream et al., 1984; Stone et al., 1989). (2) TH-IR is known to be expressed temporarily in certain neurons, such as the neuroepithelial cells near the ventricle between E13 and E16 (Specht et al., 1978). Cells in the inferior colliculus express TH but not other catecholaminergic enzymes perinatally (Jaeger & Joh, 1983). TH expression, moreover, is known to be regulated, at least in adults, by dopamine itself and by cyclic AMP-dependent protein kinase and specific phosphatase inhibitors (Fujisawa & Okuno, 1989; Okuno & Fujisawa, 1991; Goldstein & Lieberman, 1992). (3) Monkeys deprived of social contact in the first year of life but surviving to old age exhibit several neurochemical alterations in the basal ganglia, including a reduction of TH-IR in the striatal matrix and 43% fewer TH-IR neurons in the substantia nigra pars compacta (SNpc) than socially reared control monkeys (Martin et al., 1991b). Normal numbers of Nissl-stained cells, however, are maintained in the SNpc of the socially deprived animals, suggesting that cells survive but change their phenotype. (4) At the other extreme, TH-IR neurons can become hypertrophied under certain conditions. Evidence for plasticity leading to the robust expression of TH has been observed in cultured (Lacovitti et al., 1987) and transplanted (Park, Joh & Ebner, 1986) fetal rat cortical neurons. Monkey TH-IR cells can enlarge and express exaggerated amounts of TH under extreme conditions – a partial lesion of the mesencephalic DA cell population can result in the hypertrophy of some of the remaining DA neurons in the midbrain, hypothalamus, and STR (Dubach et al., 1988a; Janson et al., 1991; see also Figs. 11.5 and 11.6).

Problems of unsteady DA expression may similarly attend the use of DA fluorescence and immunohistochemistry. (1) Hypothalamic cells that express TH-IR steadily during development express DA-IR on E13 and later on E17, but not on the intermediate day E15 (Voorn et al., 1988). (2) Levels of DA in the somata of DA neurons can be low enough to go undetected by some methods, as for example in the rat hypothalamus (Voorn et al., 1988), in the rat olfactory bulb (Seiger & Olson, 1973), or transiently in the rat amygdala (Verney et al., 1988). Apparently DA levels in telencephalic DA cells are so low that fluorescence methods have only rarely detected a few of the cells (de la Torre, 1972; Felten & Sladek, 1983). From these examples, it is clear that certain cells can alternately express or not express a dopaminergic phenotype. Investigation of the possible role of DA neurons in development may thus be hampered by plasticity that makes indicators of a DA phenotype difficult to detect continuously throughout development. In rats, the expression of TH-IR by striatal DA cells is developmentally transient at best. Investigators who have described telencephalic TH-IR cells in the rat in 'adults' have worked with young animals (Kosaka et al., 1987b: 5 weeks), or have seen only very small numbers of cells (Tashiro et al., 1989), or have only seen cortical TH-IR neurons after colchicine administration (Kosaka et al., 1987a). A population of telencephalic cells comparable to that seen in monkeys has been detected in rodents only by DA immunohistochemistry based on recently available antibodies, and its distribution has been only briefly described (Voorn et al., 1988).

Species differences in biochemistry

The lack of detectable TH in adult rat telencephalic DA cells might call into question their homology to comparable cells in primates. This discrepancy, however, may be related in some manner to major species differences in the biochemistry of DA. One relevant example is that TH itself is synthesized from four different mRNA sequences in humans and in monkeys, but only one in rats (Kobayashi et al., 1988; Nagatsu, 1991). Several other differences are associated with the high concentration of the DA deaminative metabolite 3,4-dihydroxyphenylacetic acid (DOPAC) relative to homovanillic acid in rats, but not in primates (Wilk, Watson & Glick, 1975; Bacopoulos et al, 1978; Rollema et al., 1989). DA deamination to form DOPAC in the rat striatum is mediated primarily by monoamine oxidase type A (MAO-A) (Fowler & Benedetti, 1983; Arbuthnott, Fairbrother & Butcher, 1990; Kumagae, Matsui & Iwat, 1991) and MAO-A constitutes a greater proportion of all MAO both inside and outside DA terminals in rat brains (Garrick & Murphy, 1980; Oreland et al., 1983). Most DA is metabolized in neuronal terminals (Stenstrom, Hardy & Oreland, 1987), but even in glial cells, deamination by MAO-A predominates (Kopin, 1987; Stenstrom et al. 1987). In primates, however, the majority of DA itself is metabolized extraneuronally by glial MAO-B (Garrick & Murphy, 1980; Stenstrom et al, 1987). The concentration of DOPAC in primates may also be diminished by the higher relative affinity of catechol-o-methyltransferase for acidic catechol substrates, thus favoring o-methylation of DOPAC over DA itself (White & Wu, 1975). Low levels of DOPAC in primates may signal a major reorganization of DA physiology, because DOPAC levels in rats may be related to distinctions among intraneuronal DA pools, modes of release, receptor types, and synaptic/extrasynaptic DA actions (Arbuthnott et al., 1990).

MAO also figures prominently in the susceptibility of primates, but not rats, to the DA neurotoxicity of peripherally administered 1-methyl-4-phenyl-1,2,3,6-tetrahydropyridine (MPTP). The mechanism by which MPTP in low doses causes nigral lesions in monkeys involves several steps (Trevor et al., 1987; Gerlach et al., 1991). The particular vulnerability of primates probably depends on the disposition of extraneuronal MAO-B, which can convert MPTP to its toxic metabolite MPP+ in both rats and monkeys (Fuller & Steranka, 1985; Glover, Gibb & Sandler, 1986a,b). MPTP in the brain is bound by MAO-B in monkeys, but by MAO-A in rats (Corsini et al., 1986; Del Zompo et al., 1986). In rats, MAO-B is present in high concentration only in the capillary endothelium, where it metabolizes MPTP to its neurotoxic but polar metabolite MPP+, which is slow to pass the blood-brain barrier into the brain (Kalaria & Harik, 1987; Riachi, Dietrich & Harik, 1990; Riachi & Harik, 1992). The monkey STR, on the other hand, is not protected by large amounts of endothelial MAO (Riachi & Harik, 1992). Furthermore, the nigrostriatal pathway in monkeys (but not rats) contains large quantities of MAO-B (Sandler et al., 1987; Willoughby, Glover & Sandler, 1988), and in monkeys (but not rats) MAO-B predominates in glial cells (Kopin, 1987; Stenstrom et al. 1987). MPTP neurotoxicity depends on the presence of intact glial cells in the SN and is prevented even when MPTP is directly injected intranigrally, as long as a gliotoxin is injected shortly before or along with MPTP (Takada, Li & Hattori, 1990). Long-term neurotoxicity may also depend on binding by neuromelanin in midbrain DA cells, which is not detectable in rats (Barden & Levine, 1983; Irwin & Langston, 1985), or on chronic storage in DA terminals in the striatum (Johannessen, 1991).

In sum, species differences include the identity of TH, levels of DOPAC and vascular MAO, the amount and distribution of MAO-B, the primary site of DA metabolism, COMT substrate affinities, and neuronal melanin content. In light of these major differences and in light of the evidence of DA-cell plasticity surveyed above, telencephalic DA cells are probably homologous populations in rats and primates, and their lack of detectable TH expression in rats is a species difference to be added to the list of biochemical and anatomical examples provided above and on pages 281–282.

Functional comments

Unfortunately, functional studies of telencephalic DA cells are not available; direct physiological recording will be difficult owing to the distribution and size of these cells. Their distribution, ontogeny, and neurochemistry, however, suggest a similarity to midbrain DA neurons, which are somewhat more accessible. Furthermore, their location in and near frontal cortex and striatum suggests that they may make local projections to structures that also receive DA innervation from the midbrain. It is therefore pertinent to comment on the response properties of mesencephalic DA neurons in monkeys. Midbrain DA neurons in anesthetized monkeys respond to noxious stimuli by either decreasing or (less commonly) increasing firing rates (Romo & Schultz, 1985; Schultz & Romo, 1987); each neuron responds in its own characteristic direction, regardless of the somatic location of the pain stimulus. Innocuous stimuli have no effect under anesthesia. In the waking monkey, during a spatial delayed alternation task, both a visual stimulus and a food reward altered firing rates in DA cells (65% and 52%, respectively), but during the delay (the mnemonic portion of the task), during which some STR and cortical cells specifically responded, no midbrain DA cells responded (Ljungberg, Apicella & Schultz, 1991). DA neurons thus appear to respond to attentional and motivational stimuli such as alerting stimuli, rewards, or pain, but do not modulate during motor or cognitive aspects of behavior (Ljungberg et al., 1991).

It is also instructive to compare the telencephalic DA neurons, as a widely distributed neuronal population, with basal forebrain cholinergic neurons, which have a similarly widespread distribution. These groups of cells have partially overlapping distributions, being intermingled in the septum, diagonal band, and substantia innominata. Both groups have early and very similar sets of cell birthdates, both in rats (Bayer, 1985) and probably in monkeys (Levitt & Rakic, 1982; Levitt, 1985; Kordower & Rakic, 1990), although only midbrain DA cells have been examined in studies of monkeys. Both groups are neurochemically related to cholinergic sympathetic neurons, which express a catecholaminergic phenotype during prenatal development (Cochard, Goldstein & Black, 1978; Teitelman et al., 1981; Landis & Keefe, 1983); a factor mediating this conversion has also been detected in the brain (Yamamori et al., 1989; Yamamori, 1991).

Response properties of basal forebrain cholinergic cells in monkeys are also similar to those of mesencephalic DA neurons. In a delayed response task, three-fourths of basal nucleus neurons responded, including two-thirds during the reward period (Richardson & DeLong, 1986). Most basal nucleus cells were activated in a go/no-go task that dissociated sensory and motor responses, but they displayed neither purely motor- nor purely sensory-related responses, and were considered to reflect either decision making or changes in arousal (Richardson & DeLong, 1990). Basal nucleus neurons can respond specifically to food, and can habituate after satiation to a particular food, without losing the capacity to respond to the sight or taste of other foods (Rolls et al., 1986). They can also display differential responses to either positively or negatively reinforcing tastes and their associated stimuli (Wilson & Rolls, 1990a,b). In tests that examined only movement parameters, some of these cholinergic cells were related to movement or load application, but others responded non-specifically (Mitchell et al., 1987). Basal forebrain cholinergic cells thus have response properties broadly comparable to those of midbrain DA neurons, in that both are related to attentional and reinforcement features rather than directly to motor or sensory aspects of behavior. While no functional information is available for telencephalic DA neurons, it is therefore reasonable to speculate that they also play a role within this arena.

Summary and conclusions

DA neurons have not previously been addressed in their entirety as a population extending from midbrain to frontal pole and olfactory bulb. The specialized rim of the dorsolateral striatum with which they are associated dorsally has rarely been the focus of anatomic investigation in rats, and never in monkeys. Neither the ontogeny of the forebrain DA cells, nor the compartmentalization of the VS which their ventral contingent surrounds has been investigated in monkeys. Much of the most pertinent information presented above has therefore been drawn from brief passages in disparate sources and assembled here for the first time. The following conclusions regarding the functional and developmental significance of the telencephalic DA cells are therefore speculative and primarily heuristic.

1. Numerous data from a number of investigators indicate, firstly, that TH and DA can be expressed variably and transiently by a substantial number of discontinuous cell groups near the ventral midline of the mesencephalon, diencephalon, and prosencephalon during prenatal development and, secondly, that DA neurons in the adult appear variably in a substantial number of discontinuous locales throughout these same areas. One interpretation of these data would suggest that an array of neurons that extends ventrally along the neural tube contains the progenitors of the classical adult DA cell groups from the retrorubral area to the olfactory bulb, also including the cells associated with ventral and dorsal striatum (anterior olfactory area and marginal zone of the STR) and the widespread cells of the deep layers of the cerebral cortex. The limited migration of the mesencephalic portion of this array has been compared topologically to an 'inverted fountain' (Hanaway, McConnell & Netsky, 1972). The striatal and cortical DA cells may migrate at an early point in development in a similar but more extensive pattern. This emergence would bring them, at an early point in telencephalic development, into position as two adjacent sheets between the incipient cortex and striatum, one in

the subplate zone just deep to the cortical plate (Kostovic & Rakic, 1990) and the other in an analogous position superficial to the striatal anlage. Given the circumstantial evidence for this possibility, an evaluation of monkey forebrain development and a re-evaluation of rat forebrain development to track the ontogeny of the telencephalic DA cells would be useful.

2. Concerning the functions of the DA neurons of the marginal zone and basal forebrain in the adult, considerations of anatomical position and relations provide the best available clues. Many of the STR neurons, and all of the VS neurons, are in the white matter just outside the DA terminal area. In this position, they may be beyond the reach of striatal afferents which condition the activity of mesencephalic DA neurons in the SNpc ventral tier and the SNpr (striosomes) and in the remainder of A9, A8, and A10 (VS). The cells are in an ideal position, on the other hand, for access by cortical afferents, which could specifically modulate an activity that is related to attentional and reinforcement features of stimuli. The cell bodies are very rarely in direct contact with one another, while brainstem DA neurons are tightly compacted. The telencephalic DA cells may therefore be capable of functioning more independently from one another than are midbrain DA cells. Their partial overlap with the distribution of basal forebrain cholinergic cells suggests the possibility of a similar origin, related roles in development, or functional synergies between telencephalic DA and cholinergic cells.

Acknowledgements

I wish to thank Joel Cummings and Renee Costello for technical assistance. This work was supported by NIH grant RR00166 to the Regional Primate Research Center at the University of Washington.

Abbreviations

CaBP	Calcium-binding protein
DA	Dopamine
DOPAC	3,4-dihydroxyphenylacetic acid
-IR	-immunoreactive
MAO	Monoamine oxidase
MPTP	1-methyl-4-phenyl-1,2,3,6-tetrahydropyridine
SNpc	Substantia nigra pars compacta
SNpr	Substantia nigra pars reticulata
STR	Neostriatum
TH	Tyrosine hydroxylase
VS	Ventral striatum

References

Abercrombie, M. (1946). Estimation of nuclear population from microtome sections. *Anatomical Record*, **94**, 239–47.

Agnati, L. F., Fuxe, K., Zoli, M., Ferraguti, F., Benfenati, F., Ouimet, C. C., Walaas, S. I., Hemmings, H. C. Jr., Goldstein, M. & Greengard, P. (1988). Morphometrical evidence for a complex organization of tyrosine hydroxylase-, enkephalin- and DARPP-32-like immunoreactive patches and their codistribution at three rostrocaudal levels in the rat neostriatum. *Neuroscience*, **27**, 785–97.

Alheid, G. F. & Heimer, L. (1988). New perspectives in basal forebrain organization of special relevance for neuropsychiatric disorders: the striatopallidal, amygdaloid, and corticopetal components of substantia innominata. *Neuroscience*, **27**, 1–39.

Altar, C. A., O'Neil, S. O., Walter, R. J. Jr. & Marshall, J. F. (1985). Brain dopamine and serotonin receptor sites revealed by digital subtraction autoradiography. *Science*, **228**, 597–600.

Arbuthnott, G. W., Fairbrother, I. S. & Butcher. S. P. (1990). Dopamine release and metabolism in the rat striatum: an analysis by *in vivo* brain microdialysis. *Pharmacology and Therapeutics*, **48**, 281–93.

Atweh, S. F. & Kuhar, M. J. (1977). Autoradiographic localization of opiate receptors in rat brain. III. The telencephalon. *Brain Research*, **134**, 393–405.

Bacopoulos, N. G., Bustos, G., Redmond, D. E., Baulu, J. & Roth, R. H. (1978). Regional sensitivity of primate brain dopaminergic neurons to haloperidol: alterations following chronic treatment. *Brain Research*, **157**, 396–401.

Baker, H., Kawano, T., Margolis, F. L. & Joh, T. H. (1983). Transneuronal regulation of tyrosine hydroxylase expression in olfactory bulb of mouse and rat. *Journal of Neuroscience*, **3**, 69–78.

Barden, H. & Levine, S. (1983). Histochemical observations on rodent brain melanin, *Brain Research Bulletin*, **10**, 847–51.

Baron, G., Stephan, H. & Frahm, H. D. (1987). Comparison of brain structure volumes in Insectivora and primates. VI. Paleocortical components, *Journal für Hirnforschung*, **28**, 463–77.

Bayer, S. (1985). Neurogenesis of the magnocellular basal telencephalic nuclei in the rat. *International Journal of Developmental Neuroscience*, **3**, 229–43.

Berendse, H. W., Groenewegen, H. J. & Lohman, A. H. M. (1992). Compartmental distribution of ventral striatal neurons projecting to the mesencephalon in the rat. *Journal of Neuroscience*, **12**, 2079–103.

Berger, B., Verney, C., Gaspar, P. & Febvret, A. (1985). Transient expression of tyrosine hydroxylase immunoreactivity in some neurons of the rat neocortex during postnatal development. *Developmental Brain Research*, **23**, 141–4.

Caboche, J., Rogard, M. & Besson, M. J. (1991). Comparative development of D1-dopamine and mu opiate receptors in normal and in 6-hydroxydopamine-lesioned neonatal rat striatum: dopaminergic fibers regulate mu but not D1 receptor distribution. *Developmental Brain Research*, **58**, 111–22.

Chan-Palay, V., Zaborszky, L., Köhler, C., Goldstein, M. & Palay, S. L. (1984). Distribution of tyrosine-hydroxylase-immunoreactive neurons in the hypothalamus of rats. *Journal of Comparative Neurology*, **227**, 467–96.

Cochard, P., Goldstein, M. & Black, B. (1978). Ontogenetic appearance and disappearance of tyrosine hydroxylase and catecholamines in the rat embryo. *Proceedings of the National Academy of Sciences USA*, **75**, 2986–90.

Corsini, G. U., Pintus, S., Bocchetta, A., Piccardi, M. P. & Del Zompo, M. (1986). Primate-rodent 3H-MPTP binding differences, and biotransformation of MPTP to a reactive intermediate *in vitro*. *Journal of Neural Transmission Supplement*, **2**, 55–60.

de la Torre, J. C. (1972). Catecholamines in the human diencephalon: A histochemical fluorescence study. *Acta Neuropathologica (Berlin)*, **21**, 165–8.

Del Zompo, M., Piccardi, M. P., Bernardi, F., Bonuccelli, U. & Corsini, G. U. (1986). Involvement of monoamine oxidase enzymes in the action of 1-methyl-4-phenyl-1,2,5,6-tetrahydropyridine, a selective neurotoxin, in the squirrel monkey: binding and biochemical studies. *Brain Research*, **378**, 320–4.

Doucet, G., Descarries, L. & Garcia, S. (1986). Quantification of the dopamine innervation in adult rat neostriatum. *Neuroscience*, **19**, 427–45.

Dubach, M., Schmidt, R. H. & Bowden, D. M. (1988a). Neostriatal and frontal cortical catecholamine cells in long-tailed macaques: individual variation and lesion-induced overdevelopment. *Society for Neuroscience Abstracts*, **14**, 1022.

Dubach, M., Schmidt, R. H., Martin, R., German, D. C. & Bowden, D. M. (1988b). Transplant improves hemiparkinsonian syndrome in nonhuman primate: intracerebral injection, rotometry, TH-immunohistochemistry. *Progress in Brain Research*, **78**, 491–6.

Dubach, M., Schmidt, R., Bowden, D. M., Kunkel, D., Martin, R., and German, D. (1986). Neurons containing tyrosine hydroxylase-like immunoreactivity in the caudate and putamen of a nonhuman primate. *Society for Neuroscience Abstracts*, **12**, 1327.

Dubach, M., Schmidt, R., Bowden, D. M., Kunkel, D., Martin, R. & German, D. (1987). Primate neostriatal neurons containing tyrosine hydroxylase: immunohistochemical evidence. *Neuroscience Letters*, **75**, 205–10.

Felten, D. L. & Sladek, J. R. Jr. (1983). Monoamine distribution in primate brain V. Monoaminergic nuclei: Anatomy, pathways and local organization. *Brain Research Bulletin*, **10**, 171–284.

Fibiger, H. C. (1982). The organization and some projections of cholinergic neurons of the mammalian forebrain. *Brain Research Reviews*, **4**, 327–88.

Foster, G. A., Schultzberg, M., Hökfelt, T., Goldstein, M. Hemmings, H. C. Jr., Ouimet, C. C., Walaas, S. I. & Greengard, P. (1987). Development of a dopamine- and cyclic adenosine 3':5'-monophosphate-regulated phosphoprotein (DARPP-32) in the prenatal rat central nervous system, and its relationship to the arrival of presumptive dopaminergic innervation. *Journal of Neuroscience*, **7**, 1994–2018.

Fowler, C. J. & Benedetti, M. S. (1983). The metabolism of dopamine by both forms of monoamine oxidase in the rat brain and its inhibition by cimoxatone. *Journal of Neurochemistry*, **40**, 1534–41.

Fujisawa, H. & Okuno, S. (1989). Regulation of the activity of tyrosine hydroxylase in the central nervous system. *Advances in Enzyme Regulation*, **28**, 93–110.

Fuller, R. W. & Steranka, L. R. (1985). Central and peripheral catecholamine depletion by 1-methyl-4-phenyl-tetrahydropyridine (MPTP) in rodents. *Life Sciences*, **36**, 243–7.

Garrick, N. A. & Murphy, D. L. (1980). Species differences in the deamination of dopamine and other substrates for monoamine oxidase in brain. *Psychopharmacology*, **72**, 27–33.

Gaspar, P., Berger, B., Alvarez, C., Vigny, A. & Henry, J. P. (1985). Catecholaminergic innervation of the septal area in man: immunocytochemical study using TH and DBH antibodies. *Journal of Comparative Neurology*, **241**, 12–33.

Gaspar, P., Berger, B., Febvret, A., Vigny, A., Krieger-Poulet, M. & Borri-Voltattorni, C. (1987). Tyrosine hydroxylase-immunoreactive neurons in the human cerebral cortex: a novel catecholaminergic group? *Neuroscience Letters*, **80**, 257–62.

Gerfen, C. R. (1984). The neostriatal mosaic: compartmentalization of corticostriatal input and striatonigral output systems. *Nature*, **311**, 461–4.

Gerfen, C. R., Baimbridge, K. G. & Thibault, J. (1987a). The neostriatal mosaic: III. Biochemical and developmental dissociation of patch-matrix mesostriatal systems. *Journal of Neuroscience*, **7**, 3935–44.

Gerfen, C. R., Herkenham, M. & Thibault, J. (1987b). The neostriatal mosaic: II. Patch- and matrix-directed mesostriatal dopaminergic and non-dopaminergic systems. *Journal of Neuroscience*, **7**, 3915–34.

Gerlach, M., Riederer, P., Przuntek, H. & Youdim, M. B.

(1991). MPTP mechanisms of neurotoxicity and their implications for Parkinson's disease. *European Journal of Pharmacology*, **208**, 273–86.

German, D. C., Dubach, M., Askari, S., Speciale, S. G. & Bowden, D. M. (1988). 1-methyl-4-phenyl-1,2,3,6-tetrahydropyridine-induced Parkinsonian syndrome in *Macaca fascicularis*: which midbrain dopaminergic neurons are lost? *Neuroscience*, **24**, 161–74.

Glover, V., Gibb, C. & Sandler, M. (1986a). Monoamine oxidase B (MAO-B) is the major catalyst for 1-methyl-4-phenyl-1,2,3,6-tetrahydropyridine (MPTP) oxidation in human brain and other tissues. *Neuroscience Letters*, **64**, 216–20.

Glover, V., Gibb, C. & Sandler, M. (1986b). The role of MAO in MPTP toxicity – a review. *Journal of Neural Transmission Supplement*, **20**, 65–76.

Goldman-Rakic, P. S. (1982). Cytoarchitectonic heterogeneity of the primate neostriatum: subdivision into Island and Matrix cellular compartments. *Journal of Comparative Neurology*, **205**, 398–413.

Goldstein, M. & Lieberman, A. (1992). The role of the regulatory enzymes of catecholamine synthesis in Parkinson's disease. *Neurology*, **42**, (Suppl. 4), 8–12.

Gouras, G. K., Rance, N. E., Young, W. S. III. & Koliatsos, V. E. (1992). Tyrosine-hydroxylase-containing neurons in the primate basal forebrain magnocellular complex. *Brain Research*, **584**, 287–93.

Graybiel, A. M. (1990). Neurotransmitters and neuromodulators in the basal ganglia. *Trends in Neurosciences*, **13**, 244–53.

Graybiel, A. M., Hirsch, E. C. & Agid, Y. A. (1987). Differences in tyrosine hydroxylase-like immunoreactivity characterize the mesostriatal innervation of striosomes and extrastriosomal matrix at maturity. *Proceedings of the National Academy of Sciences, USA*, **84**, 303–7.

Graybiel, A. M. & Ragsdale, C. W. Jr. (1983). Biochemical anatomy of the striatum. In *Chemical Neuroanatomy*, ed. P. C. Emson, pp. 427–504. New York: Raven Press.

Haber, S. N., Groenewegen, H. J., Grove, E. A. & Nauta, W. J. H. (1985). Efferent connections of the ventral pallidum: Evidence of a dual striatopallidofugal pathway. *Journal of Comparative Neurology*, **235**, 322–35.

Haber, S. N. & Nauta, W. J. H. (1983). Ramifications of the globus pallidus in the rat as indicated by patterns of immunohistochemistry. *Neuroscience*, **9**, 245–60.

Hanaway, J., McConnell, J. A. & Netsky, M. G. (1972). Histogenesis of the substantia nigra, ventral tegmental area of Tsai and interpeduncular nucleus: an autoradiographic study of the mesencephalon in the rat. *Journal of Comparative Neurology*, **142**, 59–74.

Heikkila, R. E., Manzino, L., Cabbat, F. S. & Duvoisin, R. C. (1984) Protection against the dopaminergic neurotoxicity of 1-methyl-4-phenyl-1,2,5,6-tetrahydropyridine by monoamine oxidase inhibitors. *Nature*, **311**, 467–69.

Heimer, L. (1978). The olfactory cortex and the ventral striatum. In *Limbic Mechanisms, the Continuing Evolution of the Limbic System Concept*, ed. K. E. Livingston & O. Hornykiewicz, pp. 95–187. New York: Plenum Press.

Heimer, L., Van Hoesen, G. W. & Rosene, D. L. (1977). The olfactory pathways and the anterior perforated substance in the primate brain. *International Journal of Neurology*, **12**, 42–52.

Heimer, L. & Wilson, R. D. (1975). The subcortical projection of the allocortex: similarities in the neural associations of the hippocampus, the piriform cortex, and the neocortex. In *Golgi Centennial Symposium: Perspectives in Neurobiology*, ed. M. Santini, pp.177–93. New York: Raven Press.

Herkenham, M. & Pert, C. B. (1980). *In vitro* autoradiography of opiate receptors in rat brain suggests loci of 'opiatergic' pathways. *Proceedings of the National Academy of Sciences, US. A.*, **77**, 5532–6.

Hirsch, E., Graybiel, A. M. & Agid, Y. A. (1988). Melanized dopaminergic neurons are differentially susceptible to degeneration in Parkinson's disease, *Nature*, **334**, 345–8.

Hökfelt, T., Johansson, O. & Goldstein, M. (1984a). Central catecholamine neurons as revealed by immunohistochemistry with special reference to adrenaline neurons. In *Handbook of Chemical Neuroanatomy, Volume 2: Classical Transmitters in the CNS, Part*, ed. A. Björklund & T. Hökfelt, pp. 157–276. Amsterdam: Elsevier.

Hökfelt, T., Johansson, O., Fuxe, K., Goldstein, M. & Park, D. (1977). Immunohistochemical studies on the localization and distribution of monoamine neuron systems in the rat brain II. Tyrosine hydroxylase in the telencephalon. *Medical Biology*, **55**, 21–40.

Hökfelt, T., Martensson, R., Björklund, A., Kleinau, S. & Goldstein, M. (1984b). Distributional maps of tyrosine-hydroxylase-immunoreactive neurons in the rat brain. In *Handbook of Chemical Neuroanatomy, Volume 2: Classical Transmitters in the CNS, Part I*, ed. A. Björklund & T. Hökfelt, pp.277–379. Amsterdam: Elsevier.

Hornung, J. P., Toerk, I. & De Tribolet, N. (1989). Morphology of tyrosine hydroxylase-immunoreactive neurons in the human cerebral cortex. *Experimental Brain Research*, **76**, 12–20.

Lacovitti, L., Lee, J., Joh, T. H. & Reis, D. J. (1987). Expression of tyrosine hydroxylase in neurons of cultured cerebral cortex: evidence for phenotypic plasticity in neurons of the CNS. *Journal of Neuroscience*, **7**, 1264–70.

Irwin, I. & Langston, J. W. (1985). Selective accumulation of MPP+ in the substantia nigra: a key to neurotoxicity? *Life Sciences*, **36**, 207–12.

Jaeger, C. B. & Joh, T. H. (1983). Transient expression of tyrosine hydroxylase in some neurons of the developing inferior colliculus of the rat. *Developmental Brain Research*, **11**, 128–32.

Janson, A. M., Fuxe, K., Goldstein, M. & Deutsch, A. Y. (1991). Hypertrophy of dopamine neurons in the primate following ventromedial mesencephalic tegmentum lesion. *Experimental Brain Research*, **87**, 232–8.

Johannessen, J. N. (1991). A model of chronic neurotoxicity: long-term retention of the neurotoxin 1-methyl-4-phenylpyridinium (MPP+) within catecholaminergic neurons. *Neurotoxicology*, **12**, 285–302.

Johnston, J. G., Gerfen, C. R., Haber, S. N. & van-der-

Kooy, D. (1990). Mechanisms of striatal pattern formation: conservation of mammalian compartmentalization. *Developmental Brain Research*, **57**, 93–102.

Jossan, S. S., Sakurai, E. & Oreland, L. (1989). MPTP toxicity in relation to age, dopamine uptake and MAO-B activity in two rodent species. *Pharmacology and Toxicology*, **64**, 314–18.

Joyce, J. N., Loeschen, S. K. & Marshall, J. F. (1985). Dopamine D-2 receptors in rat caudateputamen: the lateral to medial gradient does not correspond to dopaminergic innervation. *Brain Research*, **338**, 209–18.

Kalaria, R. N. and Harik, S. I. (1987). Blood–brain barrier monoamine oxidase: Enzyme characterization in cerebral microvessels and other tissues from six mammalian species, including human. *Journal of Neurochemistry*, **49**, 856–64.

Kobayashi, K., Kaneda, N., Ichinose, H., Kishi, F., Nakazawa, A., Kurosawa, Y., Fujita, K. & Nagatsu, T. (1988). Structure of the human tyrosine hydroxylase gene: alternative splicing from a single gene accounts for generation of four mRNA types. *Journal of Biochemistry*, **103**, 907–12.

Köhler, C., Everitt, B. J., Pearson, J. & Goldstein, M. (1983). Immunohistochemical evidence for a new group of catecholamine-containing neurons in the basal forebrain of the monkey. *Neuroscience Letters*, **37**, 161–66.

Kopin, I. J. (1987). MPTP: an industrial chemical and contaminant of illicit narcotics stimulates a new era in research on Parkinson's disease. *Environmental Health Perspectives*, **75**, 45–51.

Kordower, J. H. & Rakic, P. (1990). Neurogenesis of the magnocellular basal forebrain nuclei in the rhesus monkey. *Journal of Comparative Neurology*, **291**, 637–53.

Kosaka, K., Yoshimura, M., Ikeda, K. & Budka, H. (1984). Diffuse type of Lewy body disease: progressive dementia with abundant cortical Lewy bodies and senile changes of varying degree–a new disease? *Clinical Neuropathology*, **3**, 185–92.

Kosaka, T., Hama, K. & Nagatsu, I. (1987a). Tyrosine hydroxylase-immunoreactive intrinsic neurons in the rat cerebral cortex. *Experimental Brain Research*, **68**, 393–405.

Kosaka, T., Kosaka, K., Hataguchi, Y., Nagatsu, I., Wu, J. Y., Ottersen, O. P., Storm-Mathisen, J. & Hama, K. (1987b). Catecholaminergic neurons containing GABA-like and/or glutamic acid decarboxylase-like immunoreactivities in various brain regions of the rat. *Experimental Brain Research*, **66**, 191–210.

Kostovic, I. & Rakic, P. (1990). Developmental history of the transient subplate zone in the visual and somatosensory cortex of the macaque monkey and human brain. *Journal of Comparative Neurology*, **297**, 441–70.

Kream, R. M., Davis, B. J., Kawano, T., Margolis, F. L., & Macrides, F. (1984). Substance P and catecholaminergic expression in neurons of the hamster main olfactory bulb. *Journal of Comparative Neurology*, **222**, 140–54.

Kritzer, M. F., Innis, R. B. & Goldman-Rakic, P. S. (1990). Regional distribution of cholecystokinin binding sites in macaque basal ganglia determined by in vitro receptor autoradiography. *Neuroscience*, **38**, 81–92.

Kuljis, R. O., Martmn-Vasallo, P. & Peress, N. S. (1989). Lewy bodies in tyrosine hydroxylasesynthesizing neurons of the human cerebral cortex. *Neuroscience Letters*, **106**, 49–54.

Kumagae, Y., Matsui, Y. & Iwata, N. (1991) Deamination of norepinephrine, dopamine, and serotonin by type A monoamine oxidase in discrete regions of the rat brain and inhibition by RS-8359. *Japanese Journal of Pharmacology*, **55**, 121–8.

Landis, S. & Keefe, D. (1983). Evidence for neurotransmitter plasticity in vivo: developmental changes in properties of cholinergic sympathetic neurons. *Developmental Biology*, **98**, 349–72.

Levitt, P. (1985). Central monoamine neuron systems: their organization in the developing and mature primate brain and the genetic regulation of their terminal fields. In *Gilles de la Tourette Syndrome*, ed. A. J. Friedhoff & T. N. Chase, *Advances in Neurology*, **35**, 49–59. New York: Raven Press.

Levitt, P. & Rakic, P. (1982). The time of genesis, embryonic origin and differentiation of the brain stem monoamine neurons in the rhesus monkey. *Developmental Brain Research*, **4**, 35–57.

Li, Y. W., Halliday, G. M., Joh, T. H., Geffen, L. B. & Blessing, W. W. (1988). Tyrosine hydroxylase-containing neurons in the supraoptic and paraventricular nuclei of the adult human. *Brain Research*, **461**, 75–86.

Liu, F. C. & Graybiel, A. M. (1992a). Heterogeneous development of calbindin-D28K expression in the striatal matrix. *Journal of Comparative Neurology*, **320**, 304–22.

Liu, F. C. & Graybiel, A. M. (1992b). Transient calbindin-D28k-positive systems in the telencephalon: ganglionic eminence, developing striatum and cerebral cortex. *Journal of Neuroscience*, **12**, 674–90.

Ljungberg, T., Apicella, P. & Schultz, W. (1991). Responses of monkey midbrain dopamine neurons during delayed alternation performance. *Brain Research*, **567**, 337–41.

Lowenstein, P. R., Slesinger, P. A., Singer, H. S., Walker, L. C., Casanova, M. F., Raskin., L. S., Price, D. L. & Coyle, J. T. (1989). compartment-specific changes in the density of choline and dopamine uptake sites and muscarinic and dopaminergic receptors during the development of the baboon striatum: a quantitative receptor autoradiographic study. *Journal of Comparative Neurology*, **288**, 428–46.

Marchand, R. & Lajoie, L. (1986). Histogenesis of the striopallidal system in the rat. Neurogenesis of its neurons. *Neuroscience*, **17**, 573–90.

Martin, L. J., Hadfield, M. G., Dellovade, T. L., and Price, D. L. (1991a). The striatal mosaic in primates: patterns of neuropeptide immunoreactivity differentiate the ventral striatum from the dorsal striatum. *Neuroscience*, **43**, 397–417.

Martin, L. J., Spicer, D. M., Lewis, M. H., Gluck, J. P. & Cork, L. C. (1991b). Social deprivation of infant rhesus monkeys alters the chemoarchitecture of the brain: I. Subcortical regions. *Journal of Neuroscience*, **11**, 3344–58.

Michel, J.-P., Sakamoto, N. & Pearson, J. (1988). Catecholaminergic anatomy of the human forebrain. In *Progress in Catecholamine Research Part B: Central Aspects*, ed. M. Sandler, A. Dahlström & R. H. Belmaker, Neurology and Neurobiology 42B, pp. 175–78. New York: Alan R. Liss Inc.

Mitchell, S. J., Richardson, R. T., Baker, F. H. & DeLong, M. R. (1987). The primate nucleus basalis of Meynert: neuronal activity related to a visuomotor tracking task. *Experimental Brain Research*, **68**, 506–15.

Moon-Edley, S. & Herkenham, M. (1984). Comparative development of striatal opiate receptors and dopamine revealed by autoradiography and histofluorescence. *Brain Research*, **305**, 27–42.

Murrin, L. C. & Zeng, W. Y. (1989). Dopamine D1 receptor development in the rat striatum: early localization in striosomes. *Brain Research*, **480**, 170–7.

Nagatsu, T. (1991) Genes for human catecholamine-synthesizing enzymes. *Neuroscience Research*, **12**, 315–45.

Okuno, S. & Fujisawa, H. (1991). Conversion of tyrosine hydroxylase to stable and inactive form by the end products. *Journal of Neurochemistry*, **57**, 53–60.

Olson, L., Seiger, A. & Fuxe, K. (1972). Heterogeneity of striatal and limbic dopamine innervation: Highly fluorescent islands in developing and adult rats. *Brain Research*, **44**, 283–8.

Oreland, L., Arai, Y., Stenstrom, A. & Fowler, C. J. (1983). Monoamine oxidase activity and localization in the brain and the activity in relation to psychiatric disorders. *Modern Problems of Psychiatry*, **19**, 246–54.

Park, J. K., Joh, T. H. & Ebner, F. F. (1986). Tyrosine hydroxylase is expressed by neocortical neurons after transplantation. *Proceedings of the National Academy of Sciences*, USA, **83**, 7495–8.

Pearson, J., Halliday, G., Sakamoto, N. & Michell, J.-P. (1990). Catecholaminergic neurons. In *The Human Nervous System*, ed. G. Paxinos, pp. 1023–49. San Diego: Academic Press.

Price, J. & Mudge, A. W. (1983). A subpopulation of rat dorsal root ganglion neurones is catecholaminergic. *Nature*, 301, 241–3.

Riachi, N. J., Dietrich, W. D. & Harik, S. I. (1990). Effects of internal carotid administration of MPTP on rat brain and blood-brain barrier. *Brain Research*, **533**, 6–14.

Riachi, N. J. & Harik, S. I. (1992). Monoamine oxidases of the brains and livers of macaque and cercopithecus monkeys. *Experimental Neurology*, **115**, 212–17.

Richardson, R. T. & DeLong, M. R. (1986). Nucleus basalis of Meynert neuronal activity during a delayed response task in monkey. *Brain Research*, **399**, 364–8.

Richardson, R. T. & DeLong, M. R. (1990). Context-dependent responses of primate nucleus basalis neurons in a go/no-go task. *Journal of Neuroscience*, **10**, 2528–40.

Rollema, H., Alexander G. M., Grothusen, J. R., Matos, F. & Castagnoli, N. Jr. (1989). Comparison of the effects of intracerebrally administered MPP+(1-methyl-4-phenylpyridinium) in three species: microdialysis of dopamine and metabolites in mouse, rat and monkey striatum. *Neuroscience Letters*, **106**, 275–81.

Rolls, E. T., Murzi, E., Yaxley, S., Thorpe, S. J. & Simpson, S. J. (1986). Sensory-specific satiety: food-specific reduction in responsiveness of ventral forebrain neurons after feeding in the monkey. *Brain Research*, **368**, 79–86.

Romo, R. & Schultz, W. (1985). Prolonged changes in dopaminergic terminal excitability and short changes in dopaminergic neuron discharge rate after short peripheral stimulation in monkey. *Neuroscience Letters*, **62**, 335–40.

Sandler, M., Willoughby, J., Glover, V. & Gibb, C. (1987). Selegiline and the prophylaxis of Parkinson's disease. *Journal of Neural Transmission Supplement*, **25**, 35–43.

Schultz, W. & Romo, R. (1987). Responses of nigrostriatal dopamine neurons to high-intensity somatosensory stimulation in the anesthetized monkey. *Journal of Neurophysiology*, **57**, 201–17.

Schwaber, J. S., Rogers, W. T., Satoh, K. & Fibiger, H. C. (1987). Distribution and organization of cholinergic neurons in the rat forebrain demonstrated by computer-aided data acquisition and three-dimensional reconstruction. *Journal of Comparative Neurology*, **263**, 309–25.

Seiger, A. & Olson, L. (1973). Late prenatal ontogeny of central monoamine neurons in the rat: Fluorescence histochemical observations. *Zeitschrift für Anatomie und Entwicklungsgeschichte*, **140**, 281–318.

Selemon, L. D. & Sladek, J. R. Jr. (1986). Diencephalic catecholamine neurons (A-11, A-12, A-13, A-14) show divergent changes in the aged rat. *Journal of Comparative Neurology*, **254**, 113–24.

Sharif, N. A. & Hughes, J. (1989). Discrete mapping of brain Mu and delta opioid receptors using selective peptides: quantitative autoradiography, species differences and comparison with kappa receptors. *Peptides*, **10**, 499–522.

Snyder-Keller, A. M. (1991). Development of striatal compartmentalization following pre- or postnatal dopamine depletion. *Journal of Neuroscience*, **11**, 810–21.

Specht, L. A., Pickel, V. M., Joh, T. H. & Reis, D. J. (1978). Immunocytochemical localization of tyrosine hydroxylase in processes within the ventricular zone of prenatal rat brain. *Brain Research*, **156**, 315–21.

Specht, L. A., Pickel, V. M., Joh, T. H. & Reis, D. J. (1981a). Light-microscopic immunocytochemical localization of tyrosine hydroxylase in prenatal rat brain. I. Early ontogeny. *Journal of Comparative Neurology*, **199**, 233–53.

Specht, L. A., Pickel, V. M., Joh, T. H. & Reis, D. J. (1981b). Light-microscopic immunocytochemical localization of tyrosine hydroxylase in prenatal rat brain. II. *Journal of Comparative Neurology*, **199**, 255–76.

Spencer, S., Saper, C. B., Joh, T., Reis, D. J., Goldstein, M. & Raese, J. D. (1985). Distribution of catecholamine-containing neurons in the normal human hypothalamus. *Brain Research*, **328**, 73–80.

Stenstrom, A., Hardy, J. & Oreland, L. (1987). Intra- and extra-dopamine-synaptosomal localization of monoamine oxidase in striatal homogenates from four species. *Biochemical Pharmacology*, **36**, 2931–5.

Stone, D. M., Wessel, T., Paivarinta, H., Joh, T. H. & Baker, H. (1989). Transneuronally-mediated down regulation of olfactory bulb dopamine phenotype involves a loss of tyrosine hydroxylase messenger RNA. *Society for Neuroscience Abstracts*, **15**, 780.

Swanson, L. W., Sawchenko, P. E., Berod, A., Hartman, B. K., Helle, K. B. & Vanorden, D. E. (1981). An immunohistochemical study of the organization of catecholaminergic cells and terminal fields in the paraventricular and supraoptic nuclei of the hypothalamus. *Journal of Comparative Neurology*, **196**, 271–85.

Takada, M., Li, Z. K. & Hattori, T. (1990). Astroglial ablation prevents MPTP-induced nigrostriatal neuronal death. *Brain Research*, **509**, 55–61.

Tashiro, Y., Sugimoto, T., Hattori, T., Uemura, Y., Nagatsu, I., Kikuchi, H. & Mizuno, N. (1989). Tyrosine hydroxylase-like immunoreactive neurons in the striatum of the rat. *Neuroscience Letters*, **97**, 6–10.

Teitelman, G., Gershon, M. D., Rothman, T. P., Joh, T. H. & Reis, D. J. (1981). Proliferation and distribution of cells that transiently express a catecholaminergic phenotype during development in mice and rats. *Developmental Biology*, **86**, 348–55.

Trevor, A. J., Singer, T. P., Ramsay, R. R. & Castagnoli, N. Jr. (1987). Processing of MPTP by monoamine oxidases: implications for molecular toxicology. *Journal of Neural Transmission Supplement*, **23**, 73–89.

Trottier, S., Geffard, M. & Evrard, B. (1989). Co-localization of tyrosine hydroxylase and GABA immunoreactivities in human cortical neurons. *Neuroscience Letters*, **106**, 76–82.

van den Pol, A. N., Herbst, R. S. & Powell, J. F. (1984). Tyrosine hydroxylase-immunoreactive neurons of the hypothalamus: a light and electron microscopic study. *Neuroscience*, **13**, 1117–56.

Verney, C., Berger, B., Adrien, J., Vigny, A. & Gay, M. (1982). Development of the dopaminergic innervation of the rat cerebral cortex. A light microscopic immunocytochemical study using anti-tyrosine hydroxylase antibodies. *Developmental Brain Research*, **5**, 41–52.

Verney, C., Gaspar, P., Febvret, A. & Berger, B. (1988). Transient tyrosine hydroxylase-like immunoreactive neurons contain somatostatin and substance P in the developing amygdala and bed nucleus of the stria terminalis of the rat. *Developmental Brain Research*, **42**, 45–48.

Voorn, P., Jorritsma-Byham, B., van Dijk, C. & Buijs, R. M. (1986). The dopaminergic innervation of the ventral striatum in the rat: a light- and electron-microscopical study with antibodies against dopamine. *Journal of Comparative Neurology*, **251**, 84–99.

Voorn, P., Kalsbeek, A., Jorritsma-Byham, B. & Groenewegen, H. J. (1988). The pre- and postnatal development of the dopaminergic cell groups in the ventral mesencephalon and the dopaminergic innervation of the striatum of the rat. *Neuroscience*, **25**, 857–87.

Wakabayashi, K., Takahashi, H., Ikuta, F., Yamada, N., Watanabe, T. & Tanaka, R. (1989). [The occurrence of tyrosine hydroxylase-immunoreactive neurons in a parietal lobe ganglioglioma] *No To Shinkei*, **41**, 165–70.

White, H. L. & Wu, J. C. (1975). Properties of catechol O-methyltransferases from brain and liver of rat and human. *Biochemical Journal*, **145**, 135–43.

Wilk, S., Watson, E. & Glick, S. D. (1975). Dopamine metabolism in the tuberculum olfactorium. *European Journal of Pharmacology*, **30**, 117–20.

Willoughby, J., Glover, V. & Sandler, M. (1988). Histochemical localization of monoamine oxidase A and B in rat brain. *Journal of Neural Transmission*, **74**, 29–42.

Wilson, F. A. & Rolls, E. T. (1990a). Learning and memory is reflected in the responses of reinforcement-related neurons in the primate basal forebrain. *Journal of Neuroscience*, **10**, 1254–67.

Wilson, F. A. & Rolls, E. T. (1990b). Neuronal responses related to reinforcement in the primate basal forebrain. *Brain Research*, **509**, 213–31.

Yamamori, T. (1991). Localization of cholinergic differentiation factor/leukemia inhibitory factor mRNA in the rat brain and peripheral tissues. *Proceedings of the National Academy of Sciences USA*, **88**, 7298–302.

Yamamori, T., Fukada, K., Aebersold, R., Korsching, S., Fann, M. J. & Patterson, P. H. (1989). The cholinergic neuronal differentiation factor from heart cells is identical to leukemia inhibitory factor. *Science*, **246**, 1412–16.

Zhu, Z., Armstrong, D. L., Grossman, R. G. & Hamilton, W. J. (1990). Tyrosine hydroxylase-immunoreactive neurons in the temporal lobe in complex partial seizures. *Annals of Neurology*, **27**, 564–72.

12
Comparative anatomy of the catecholaminergic innervation of rat and primate cerebral cortex

B. Berger and P. Gaspar

Introduction

Within the cerebral cortex of mammals, monoaminergic projections constitute a particular category of cortical afferents characterized by widespread terminal fields that cross cytoarchitectonic and functional boundaries (for review see Björklund & Lindvall, 1984; Fallon & Loughlin, 1987). This extensive tangential extension of aminergic fibers contrasts with the limited tangential spread of the dominant cortical afferents, i.e. cortico-cortical or thalamic, that define cytoarchitectonic areas or cortical columns (Jones, 1981; Goldman-Rakic, 1988). Another characteristic of cortical aminergic fibers is that they originate in rather small nuclei in the brainstem and reach the cortical mantle after extensive divergence and collateralization. The distribution of the principal aminergic cell groups in the brainstem does not vary fundamentally among extant species (Parent, 1984), although there are quantitative differences in the number of aminergic cells (Swanson, 1976; Halliday & Törk, 1986, German et al., 1988, 1989; Baker et al., 1989; Chan-Palay & Asan, 1989) and possibly some modifications in the proportion of the aminergic cells that project to the cortex (Swanson, 1982; Schneiber & Törk, 1987; Gaspar, Stepniewska & Kaas, 1992). The most dramatic variation however, concerns the terminal arborization of aminergic neurons: this terminal network undergoes profound modifications, evolving in parallel with encephalization (Hofman, 1982), accompanying the development of new cortical areas and the increased differentiation in cortical layers (Killackey, 1990). This results in very different patterns of regional and laminar distributions of aminergic fibers between species, with changes being particularly marked for the dopaminergic mesocortical system.

The present account will focus on the cortical dopaminergic (DA) and noradrenergic (NA) afferents in species that are widely represented in biological research, rats and monkeys (macaques, squirrels, and owl monkeys). These observations will be related to those found in humans. Although systematic phylogenetic studies of aminergic innervation in the cortex of other mammals are still lacking, the observations available in these two mammalian orders (primates and rodents) indicate that the dopaminergic and noradrenergic innervations vary not only in their regional and laminar distributions but also in the organization of projection pathways and biochemical characters. The increased differentiation of these aminergic projections in primates relative to rodents may reflect an increased specificity of cortical targets in primates and the possibility for aminergic fibers to exert different modulatory effects.

General methodological remarks

The most widely used methods to reveal aminergic fibers rely on immunocytochemistry, since specific antibodies are available for the amines themselves and for the main enzymes of the catecholamines biosynthetic pathway, tyrosine hydroxylase (TH), dopamine-β-hydroxylase (DBH), and phenylethanol-amine-N-methyl-transferase (PNMT) that is only present in adrenergic neurons. In the cerebral cortex however, adrenergic innervation is not liable to interfere with visualization of DA and NA axons since a specific adrenergic innervation of the cerebral cortex has never been unambiguously demonstrated (see Hökfelt, Johansson & Goldstein, 1984a and discussion in Gaspar et al., 1989).

An important issue concerns the reliability of TH immunocytochemistry as a marker of DA fibers. Indeed, under specific circumstances, some TH antibodies appear to stain almost exclusively DA axons and terminals (Lewis et al., 1987; Gaspar et al., 1989; Noack & Lewis, 1989). However, depending upon the TH antibody employed (Noack & Lewis, 1989), the species analyzed and the developmental stage (Verney et al., 1982; Berger et al., 1985b), or even the region examined (Gaspar et al., 1989), a variable proportion of NA fibers can be labeled with TH antisera, in addition to the DA ones. Therefore, localization of TH-immunostaining should always be compared with other methods that selectively reveal NA innervation. Indeed, the other approach that allows specific visualization of DA or NA terminals relies on high affinity transport across the axonal membrane. This method can be made highly specific by loading sections, in vitro, with DA or NA (cold or tritiated) in the presence of selective uptake inhibitors, and it offers the advantage of completely filling the neuronal processes and visualizing unambiguously DA or NA terminals. Furthermore, it has been developed to allow the visualization of single labeled varicosities and thus permit their quantification (Descarries et al., 1987; Audet et al., 1988).

These different approaches used to visualize DA cell bodies and terminals rely in fact on distinct properties of the aminergic systems: synthesis of the transmitter (TH), endogenous stores of the amine, or high affinity transport. As such, they have sometimes provided conflicting results which in fact point to distinct metabolic properties. For instance, techniques that reveal endogenous DA, such as classical catecholamine fluorescence histochemistry or dopamine immunocytochemistry can give false-negative or weak labeling if the endogenous level is below the threshold of detection, for instance in the neonate (Schmidt et al., 1982) or in the anterior cingulate cortex of the rat (Van Eden et al., 1987). A promising method is the visualizing of dopaminergic terminals in the living human brain (Tedroff et al., 1988) with positron emission tomography, using appropriate radiotracers. This technique could offer considerable new perspectives when the sensitivity will be high enough to reveal the cortical DA terminal fields in addition to the striatal ones.

The cortical dopamine system

DA mesocortical system is, at the moment, the most striking and interesting example of phylogenetic modifications in the organization of ascending aminergic systems. Indeed, it has selectively restricted targets in rodents, while its terminal fields are considerably expanded and yet regionally specific in primates. Its biochemical constitution (colocalization with peptides) as well as its developmental pattern during ontogeny, differ in both mammalian orders. In spite of these differences, one may try to establish links between both sets of observations. This may appear easier if one considers that the DA mesocortical system is in fact heterogeneous, and probably comprizes functionally different subsystems.

The cortical dopamine system of rats

The existence of a cortical DA innervation independent of the NA system was first demonstrated in 1973 by Thierry et al., using neurotoxic lesions of the ascending noradrenergic bundle projecting to the cerebral cortex. The characteristics of the mesocortical DA systems, origin, pathways, collateralization patterns, distribution of terminal projections, ontogenetic development, have been extensively studied by several groups including ours and have been reported in several reviews (Björklund & Lindvall, 1984, Fallon & Loughlin 1987). The cortical DA innervation in rats is remarkable by its restricted distribution compared to the widespread cortical projections of the noradrenergic or serotonergic systems, but the striking and perhaps most significant feature is the heterogeneity of the mesocortical DA system, a notion that has emerged gradually over the last years. The following description is based on this point of view.

Distributional maps

Problems of cortical nomenclature in rats To facilitate interspecific comparisons, it is necessary to define cortical areas on the basis of multiple criteria comprising, in addition to cytoarchitectonics, connections and physiologically defined functions (Kaas, 1987). While such subdivisions are attempted in the primary visual and somatosensory systems, they are difficult to make in the 'association' or limbic cortical areas (precisely those areas where DA fibers predominate in rats) because the physiological correlates are ambiguous. Terminology based on cytoarchitectonics (Krettek & Price, 1977; Vogt & Peters, 1981) carries the risk of implying homologies between species while in fact they are not demonstrated. In this respect, the more 'neutral' nomenclature system proposed by Zilles (1985) which refers to stereotaxic coordinates is more pragmatic (see Fig. 12.1 for the correspondence of these nomenclatures).

A point of importance in this discussion is the medial frontal cortex of rats: an area receiving a dense

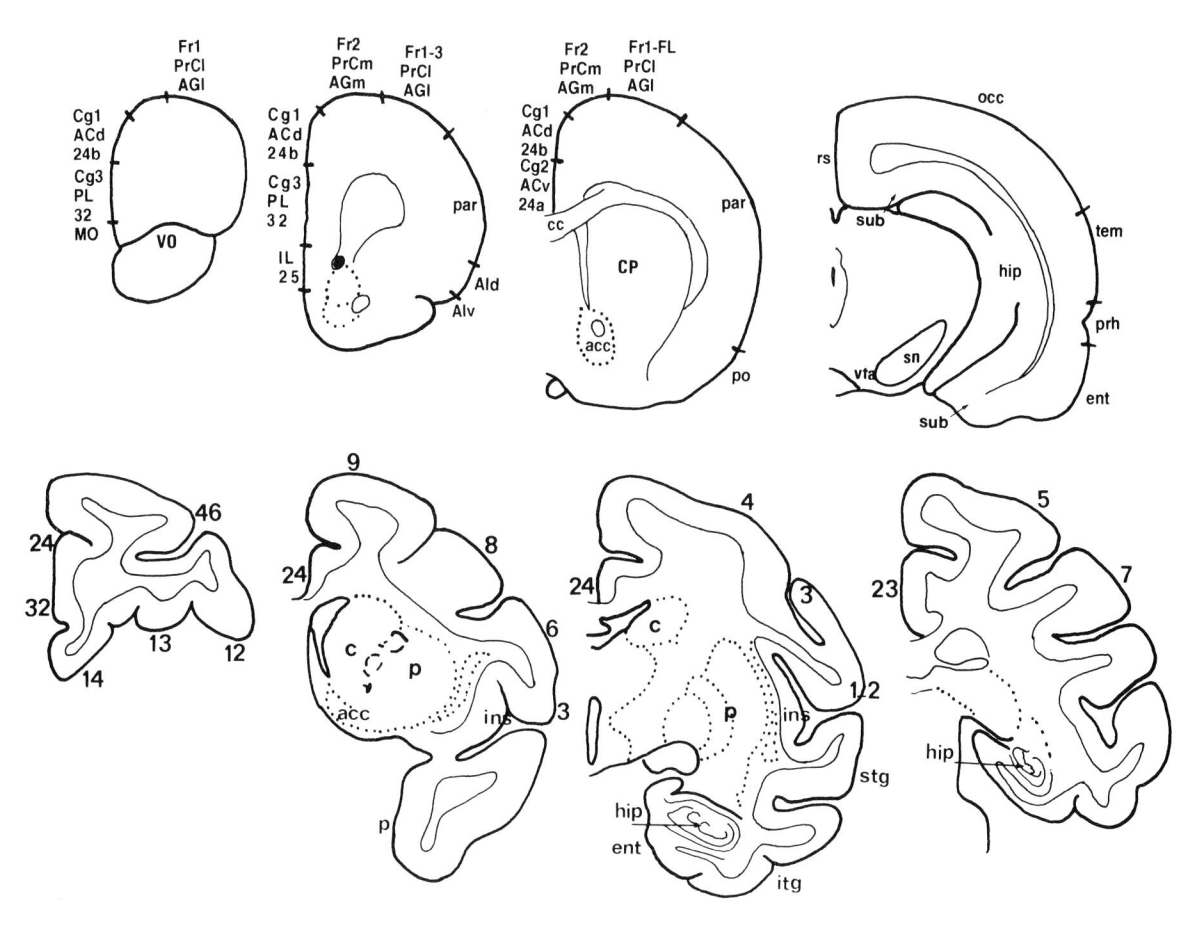

Fig. 12.1. Comparative, schematic rostrocaudal views in the coronal plane of the main cortical areas as defined by different authors in rat (top) and monkey (bottom), illustrated at different scales. Rat: 24a, b, 25 and 32 (Vogt & Peters, 1981); ACd, ACv, IL,MO, PL, PrCm, PrCl, VO (Krettek & Price, 1977); AGm, AGl (Donoghue & Wise, 1982); Ald,Alv, Cg1, 2, 3, Fr1, 2, 3, and FL (Zilles, 1985); Monkey: areas 8, 9, 12, 13, 14 and 46 are subdivisions of the prefrontal cortex; 1, 2, 3 are subdivisions of the primary somatosensory cortex; 4, is the primary motor cortex; 6, the premotor cortex; 5, 7 are somatosensory association areas; 23, posterior cingulate cortex; 24, anterior cingulate cortex; 32, prelimbic cortex.

dopaminergic input, and which is often taken, quite arbitrarily, to be homologous to the prefrontal cortex in primates (Reep, 1984; Markowitsch, 1988; Uylings & Van Eden, 1990). This equivalence was proposed by Leonard (1969) and Krettek & Price (1977), on the basis of connections with the mediodorsal (MD) nucleus of the thalamus: the connected areas in rats correspond to the rostral medial frontal wall, the orbital areas and the agranular insula (Fig. 12.1). However, this rostral medial frontal wall also encompasses areas such as the medial agranular field that is in fact a secondary motor area (Donoghue & Wise, 1982) and the pregenual extension (i.e. rostral to the genu of the corpus callosum) of the anterior cingulate cortex (Vogt & Peters 1981; Berger et al., 1985b). The corresponding areas in primates (premotor or cingulate) are not considered to belong to prefrontal cortex (Fig. 12.1). Furthermore, recent hodological studies in primates (Mufson & Mesulam, 1984; Goldman-Rakic and Porrino, 1985; Barbas, Haswell Henion & Dermon, 1991) indicate that the relation between the MD and the prefrontal cortex in primates is not exclusive since MD projects to other cortical areas, and the prefrontal cortex receives projections from other thalamic nuclei. It has been proposed that cortico-cortical connections would be more appropriate for defining the prefrontal cortex (Van Eden, Lamme & Uylings, 1992) but this suggestion merely transposes the problem, since the rat cerebral cortex lacks many of the association areas that characterize primate cortex (Pandya & Yeterian, 1985). In this chapter, we will present the cytoarchitectonic subdivisions with the corresponding terminology employed by different investigators, although this correspondence might not be absolute (Fig. 12.1).

Topography of the cortical DA terminal fields DA projections tend to be restricted to four main cortical terminal fields in rodents, with most of the primary sensorimotor and parieto-temporo-occipital association cortices being devoid of DA afferents, except in layer VI (Fig. 12.2). The major fields of DA terminals are: 1) the anteromedial frontal cortex (prelimbic cortex or area 32 or Cg3, and infralimbic cortex or area 25) with an extension into the medial and lateral orbital cortices; 2) the pregenual and supragenual anterior cingulate cortex (areas 24a, b, or ACd, ACv, or Cgl, Cg2); 3) the agranular insular cortex (AId, AIv) and perirhinal cortex; 4) the piriform (primary olfactory) and entorhinal cortex (Berger et al., 1974; 1976b, Lindvall et al., 1974, Lindvall, Björklund & Divac, 1978). With more sensitive techniques, we were able to demonstrate additional but smaller projections to the motor area, the medial agranular field (AGm/PrCm) and part of the lateral agranular field (AGl/PrCl); the retrosplenial (areas 29b–d) and visual cortices (mainly the secondary association areas) also receive a small contingent of DA fibers (Berger et al., 1985a; Phillipson, Kilpatrick & Jones, 1986; Van Eden et al., 1987; Papadopoulos, Parnavelas & Buijs, 1989a) as well as the ventral hippocampal formation (mainly in the subiculum-CA 1 field) (Verney et al., 1985).

Histochemical and immunocytochemical observations, confirmed by quantitative radioautography, showed a characteristic area-specific laminar distribution of these different cortical DA projections (Descarries et al., 1987). The highest density of dopamine innervation is in the superficial layers of the supragenual cingulate cortex, area 24a, particularly in layers II and III. A lower density, half that of 24a, was measured in the anteromedial frontal cortex where terminals predominate in layers V–VI, (being three times higher than in the upper layers). On the other hand, the agranular insula, the perirhinal, the piriform and the posterior entorhinal cortex contain a moderate dopamine innervation with varicosities in every layer. However, in the anterotheral part of the entorhinal cortex, DA afferents form cluster-like accumulations around cellullar islands in layers II–III. The parietal, temporal, primary visual, and lateral occipital neocortex contain the lowest density of terminals, mostly confined to layer VI.

Interestingly, there is some notion that the dopaminergic innervation may be lateralized. Biochemical measures showed an asymmetric distribution of the amine in the two hemispheres. Although the side with highest DA varied among studies, it may be correlated with handedness or preference to use left or right paw (Slopsema, Van der Gugten & de Bruin, 1982; Rosen et al., 1984; Barnéoud, Le Moal & Neveu, 1990, Le Moal & Simon, 1991).

Origin in the brainstem The cortical DA projections originate from a large band of about 40 000 neurons altogether, the A10, A9, A8 cell complex located in the ventral mesencephalon (Halliday & Törk, 1986; Oades and Halliday, 1987; German et al., 1989). No more than 2–3% of the projections are crossed (Swanson, 1982). A smaller number of DA mesocortical neurons are also found in the dorsal raphe nucleus (Yoshida et al., 1989; Stratford & Wirtshafter, 1990). However, it should be remembered that there are also non-DA projections arising from these areas; thus DA neurons were estimated to represent only 30% of the ventral tegmental area (VTA) neurons projecting to the frontal cortex (Swanson, 1982).

There is a broad mediolateral topography of the projections: those to the anteromedial field originate from the medial VTA, those to the cingulate cortex from the lateral VTA and medial substantia nigra (A9), those to the entorhinal cortex from the VTA and group A8 (Fallon & Loughlin, 1987; Deutch et al., 1988). Within this large continuum, a biochemical heterogeneity was demonstrated: immunocytochemical studies showed that distinct DA subpopulations were characterized by colocalization with calcium binding proteins, as calbindin D-28K (Gerfen, Bainbridge & Thibault, 1987), with neuropeptides such as cholecystokinin and neurotensin (Hökfelt et al., 1980, Hökfelt et al., 1984a) or with growth factors as basic fibroblast growth factor (Tooyama et al., 1992).

Distinct DA subpopulations contribute to the mesocortical DA innervation Further analysis of the morphological, developmental, biochemical and physiological characteristics of DA neurons projecting to the cerebral cortex indicates that the distinct area-specific density and laminar topography of the cortical projections is not random but reflects the participation of different DA subsets (Berger, Gaspar & Verney, 1991). We have proposed to distinguish two contingents on the following criteria (Table 12.1):

Table 12.1. *Heterogeneity of the mesocortical dopaminergic projections in rats*

Afferents to	the ant. medial cortex deep layers	the ant. cing. cortex superficial layers
Development	Prenatal	Postnatal
	E16-P0	P3-P21
Regional distribution	widespread	restricted
Laminar distribution	layers V–VI	layers I–II–III
Morphology	thin	thick
Coexistence with neurotensin	yes	no
Coexistence with cholecystokinin	yes	?
Origin	medial A10	lateral A10/medial A9
Collateralization	low	high
Endogenous DA content	high	low
Rate of DA utilization	low	high

Morphological aspect and developmental sequence A main difference that parallels the distinct laminar distribution and density of the cortical DA afferents concerns their morphological aspect and developmental sequence. The DA afferents that distribute predominantly to the deep cortical layers, V–VI or VI, are made up of thin and sinuous fibers with long smooth segments and sparsely distributed spindle-shaped varicosities. They develop largely prenatally, reaching the frontal cortex by the 16th day of embryonic life (Verney et al., 1982): this system can be fully visualized at birth, by the uptake of exogenous amine (Schmidt et al., 1982). However, the endogenous levels of TH and DA may lag behind, with a continuing postnatal maturation (Schmidt et al., 1982; Verney et al., 1982; Kalsbeek et al., 1988).

The second class of DA afferents that distributes to the superficial cortical layers I–III of the anterior cingulate cortex, medial agranular field, retrosplenial and visual cortices, reaches its target areas only during the first postnatal week. The highest density in layer III, characteristic of the adult age, is only observed from the third week on, and the density of DA input increases until P50. This superficial DA contingent is also characterized by a distinctive morphological aspect. Its fibers, which at early stages resemble those of the deep contingent, gradually differentiate into their adult form, becoming thicker and more densely varicose during the fourth postnatal week (Berger et al., 1985b; Kalsbeek et al., 1988).

Endogenous store and metabolic rate Other arguments support the hypothesis that these two classes of DA cortical afferents are provided by distinct neuronal populations differing in anatomy and function. DA terminals in the superficial layers are better visualized with TH immunocytochemistry (Berger et al., 1985b) or after *in vitro* loading (Descarries et al., 1987) than with DA immunocytochemistry (Van Eden et al., 1987; Kalsbeek et al., 1988). By contrast, all three methods similarly reveal DA fibers in the deep prelimbic cortex. This probably reflects the lower level of endogenous DA in the anterior cingulate and motor than in the prelimbic area (Tassin et al., 1978; Reader & Grondin, 1987) in contrast with the fourfold higher DA uptake over endogenous level, indicating a higher DA metabolic rate.

Ventral mesencephalic origin and collateralization pattern Another argument favoring the hypothesis of two different DA contingents is their origin from seemingly distinct neuronal subpopulations in the ventral mesencephalic DA groups (Björklund & Lindvall, 1984; Fallon & Loughlin, 1987). Combining various mesencephalic lesions with biochemical studies, Emson & Koob (1978) concluded that the medial substantia nigra projects only to the superficial layers of the anterior cingulate cortex, whereas the VTA projects to the deep layers of both the cingulate and anteromedial frontal cortex. This observation would obviously deserve further investigations with the sensitive tracing methods that are now available, but it is supported by other findings. DA neurons situated in the medial ventral tegmental area emit few collaterals as they project to their cortical (anteromedial, suprarhinal, perirhinal, piriform cortices) or subcortical targets (septum, and nucleus accumbens) (Deniau, Thierry & Feger, 1980; Swanson, 1982; Fallon & Loughlin, 1987; Takada & Hattori 1987); in contrast, DA neurons of the lateral VTA and medial substantia nigra appear to collateralize more

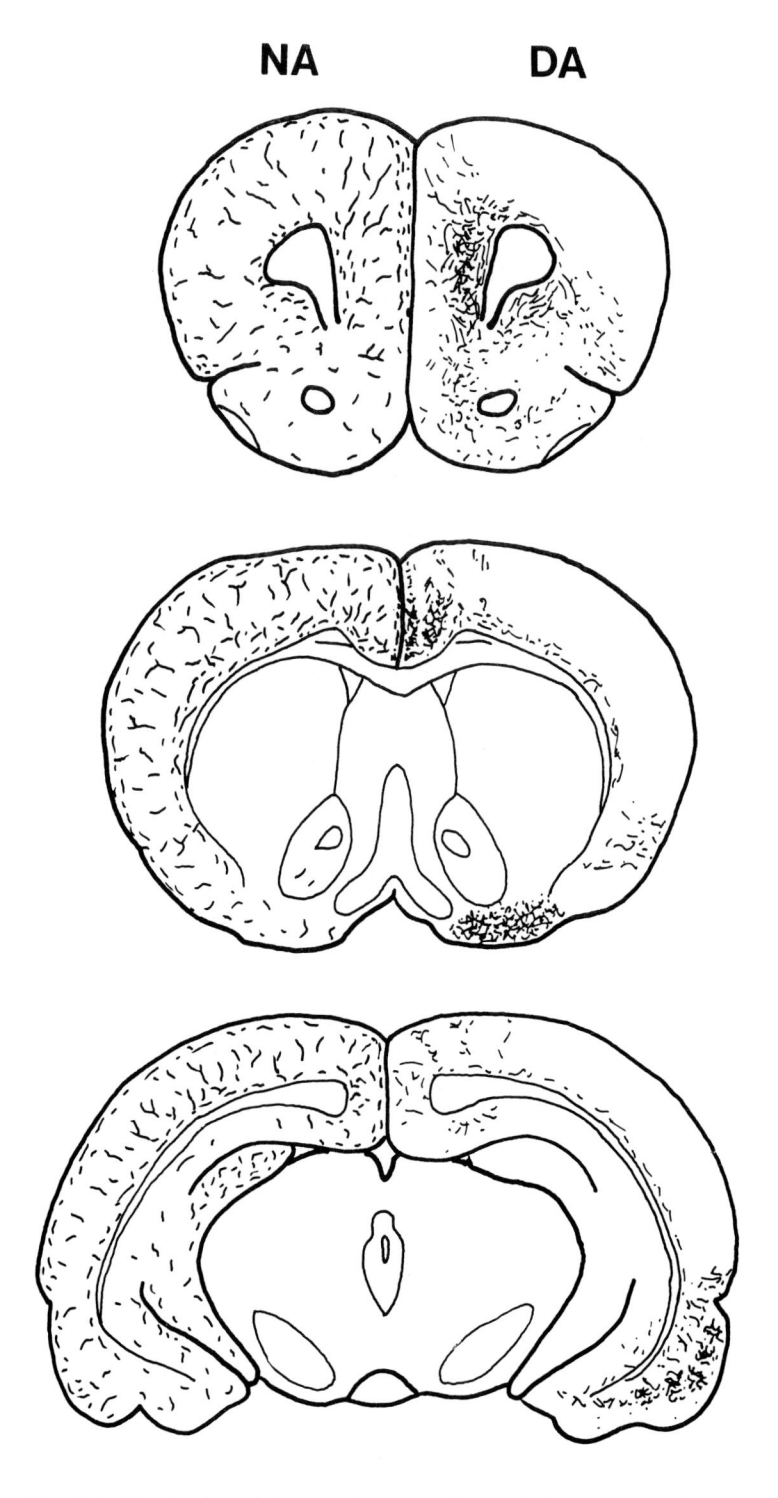

Fig. 12.2. Distribution of the noradrenergic (NA) and dopaminergic (DA) innervation in the rat cerebral cortex schematized on three coronal sections, corresponding to Fig 12.1.

towards their superficial cortical (anterior and posterior cingulate cortices) and their subcortical targets (striatum, nucleus accumbens and septum (Takada & Hattori, 1986; Fallon & Loughlin 1987). These differences between the two types of projections are paralleled by electrophysiological observations indicating that DA neurons projecting to the prefrontal cortex have a higher mean firing rate (9.3 spikes/second than those projecting to the cingulate cortex (5.9 spikes/second) (Chiodo, 1988).

Differential coexpression of neuropeptides and other chemical messengers. Different subsets of DA projections to the rat anteromedial frontal cortex, which originate in the ventral tegmental area and adjacent midline nuclei, contain either neurotensin (NT) or cholecystokinin or both peptides (Studler *et al.*, 1988; Seroogy *et al.*, 1988, 1989). In fact, the mixed NT/DA projections are not limited to the anteromedial frontal cortex but extend rostrocaudally to all the deep DA terminal fields (Febvret *et al.*, 1991). On the other hand, the DA subset which innervates the superficial anterior cingulate cortex and the other superficial targets of the premotor, retrosplenial and visual cortices lacks NT (Febvret *et al.*, 1991). The entorhinal cortex represents a particular configuration, innervated by DA, NT as well as mixed DA/NT subpopulations.

As will be discussed later, the lack of colocalization with neuropeptides seems to become dominant in primates and might characterize the whole DA cortical innervation in monkeys (Satoh & Matsumura, 1990; Oeth & Lewis, 1992) and humans (Gaspar, Berger, & Febvret, 1990).

A subset of DA neurons located in the ventral tegmental area and the dorsal tier of the substantia nigra pars compacta contains the calcium-binding protein, calbindin D-28K (Gerfen *et al.*, 1987). Whether these neurons also co-express neuropeptides and project to specific cortical targets is not yet known. The heterogeneity of the DA mesocortical projecting system is further indicated by the presence of basic fibroblast growth factor in a subpopulation of DA neurons in the VTA and medial substantia nigra, separate from the calbindin D-28K immunoreactive neurons, which seems to project predominantly to the prelimbic and infralimbic cortex but poorly to the anterior cingulate cortex (Tooyama *et al.*, 1992).

The cortical DA system in primates

An extensive DA innervation of the primate cerebral cortex had been suspected on biochemical bases (Brown, Crane & Goldman, 1979) and catecholamine fluorescence histochemistry (Levitt, Rakic & Goldman-Rakic, 1984); however, it was not demonstrated morphologically until 1986. The cortical DA innervation, considerably expanded in humans and monkeys relative to rodents (Fig. 12.3, 12.4), is characterized by three main features: a) a widespread distribution to all cortical areas from the frontal to the occipital poles with regionally specific variations in density; b) the highest density of terminals in the motor and anterior cingulate cortex resulting from a distinct laminar distribution in the agranular versus the granular cortex; c) a predominance of DA terminals in the molecular layer.

Distributional maps in macaque and squirrel monkeys As revealed with TH immunocytochemistry (Campbell *et al.*, 1987, Lewis *et al.*, 1987, 1988a) and with specific radioautographic detection of DA fibers after *in vitro* loading (Berger *et al.*, 1986, 1988), DA innervation reaches the entire cortical mantle (Fig. 12.3). The density of DA input follows a global rostrocaudal decreasing gradient such that the primary visual cortex receives the lowest amount of DA terminals, almost limited to layer I. However, there are regional differences in the distribution of dopaminergic afferents that follow cytoarchitectural trends. Maximal numbers of dopaminergic terminals are present in agranular cortices lacking a distinct layer IV: in all these agranular areas, dopaminergic fibers invade all layers. The primary motor and anterior cingulate cortices contain the densest innervation. In addition, a characteristic pattern is observed in layer III of the supplementary motor cortex (SMA). In upper layer III (IIIa), a dense uneven band of DA fibers form small islands approximately 400 μm wide, separated by similar areas of lower density. A similar configuration was observed in mesial area 4. However, this periodic organization in the motor cortices, observable with radioautographic detection of DA (Berger *et al.*, 1988) was not apparent with TH immunocytochemistry (Lewis *et al.*, 1987). This discrepancy may be due to the greater specificity of the autoradiographic method relative to TH immunodetection. Alternatively, the patterns we observed could reflect the distribution of axon terminals whereas TH labeling would demonstrate a greater number of axons 'de passage' with fewer varicosities per unit length.

In contrast to the agranular motor and cingulate cortex, the topography of DA afferents in granular cortical areas (association and somatosensory cortices) is characterized by the fact that layer IV, which receives the specific thalamocortical projec-

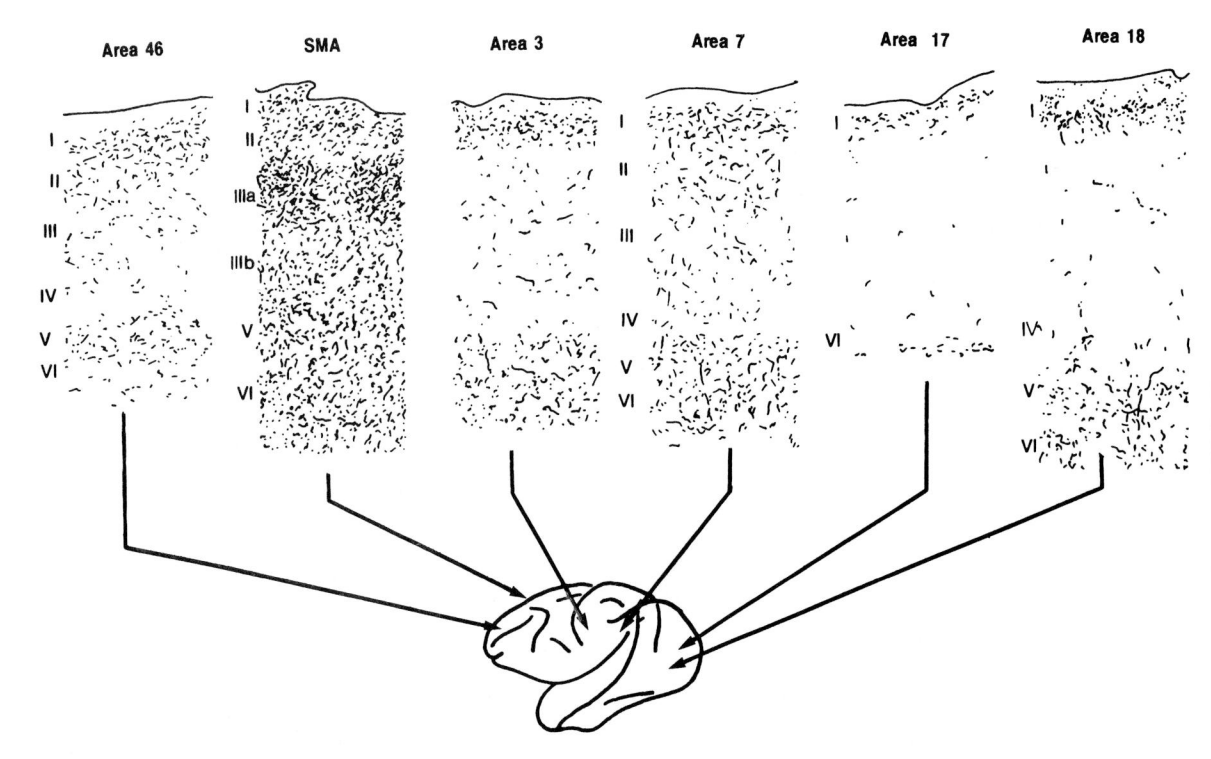

Fig. 12.3. Dopaminergic innervation of the cerebral cortex in Cynomolgus monkey. Accurate tracings of DA-containing fibers from photomontages of 50 μm thick sections processed by radioautography after specific uptake of 3HDA. Notice the dense widespread innervation of the supplementary motor cortex (SMA) with cluster-like arrangements in layer IIIa and the bilaminar pattern of distribution in the other, granular areas: 46, prefrontal cortex; 3, primary somatosensory cortex; 7, somatosensory association area; 17, 18, primary and secondary visual cortices. (Taken, with permission from Berger et al., 1988.)

tions almost completely lacks a DA innervation. This results in a bilaminar pattern of distribution with a particularly dense innervation of layers I and V–VI (47% of layer I) (Fig. 12.3). The lowest density is observed in layers III and IV. These differences in laminar distribution and density of terminals between the agranular and granular cortices are well illustrated in the prefrontal cortex where area 46 displays the lowest density of TH-labeled fibers, contrasting with the agranular areas 24 and 25 and the dysgranular medial area 9 (Lewis et al., 1988a). Variations in density can be related to other cytoarchitectonic features: primary somatosensory, visual, and auditory areas are less innervated than the corresponding association cortices (Campbell et al., 1987; Lewis et al., 1987; Berger et al., 1988); the dentate gyrus receives a dense DA innervation compared to Ammon's horn (Amaral & Campbell, 1986; Samson et al., 1990); in the rostral entorhinal cortex, characteristic accumulations of DA terminals accompany the cluster-like organization of layers II–III (Berger, Alvarez & Goldman-Rakic, 1993a; Akil & Lewis, 1993).

Whatever the amount of afferents in the other layers, the highest density of terminals is always found in layer I. In addition, granular cortices are further characterized by a predominance in the lower half of layer I particularly apparent in the depth of sulci.

Distributional maps in humans In the human postmortem brain, DA fibers were mapped by comparing TH and DBH- immunostaining (Gaspar et al., 1989) (Fig. 12.4). In fact, in our conditions, as in those of Lewis et al., (1987), TH appeared to label preferentially the DA fibers: only 10–50% of the DBH-IR axons were TH-IR. The distinctive morphology of the exclusively TH-LIR axons (i.e. very fine, convoluted with irregular varicosities) further aided the recogni-

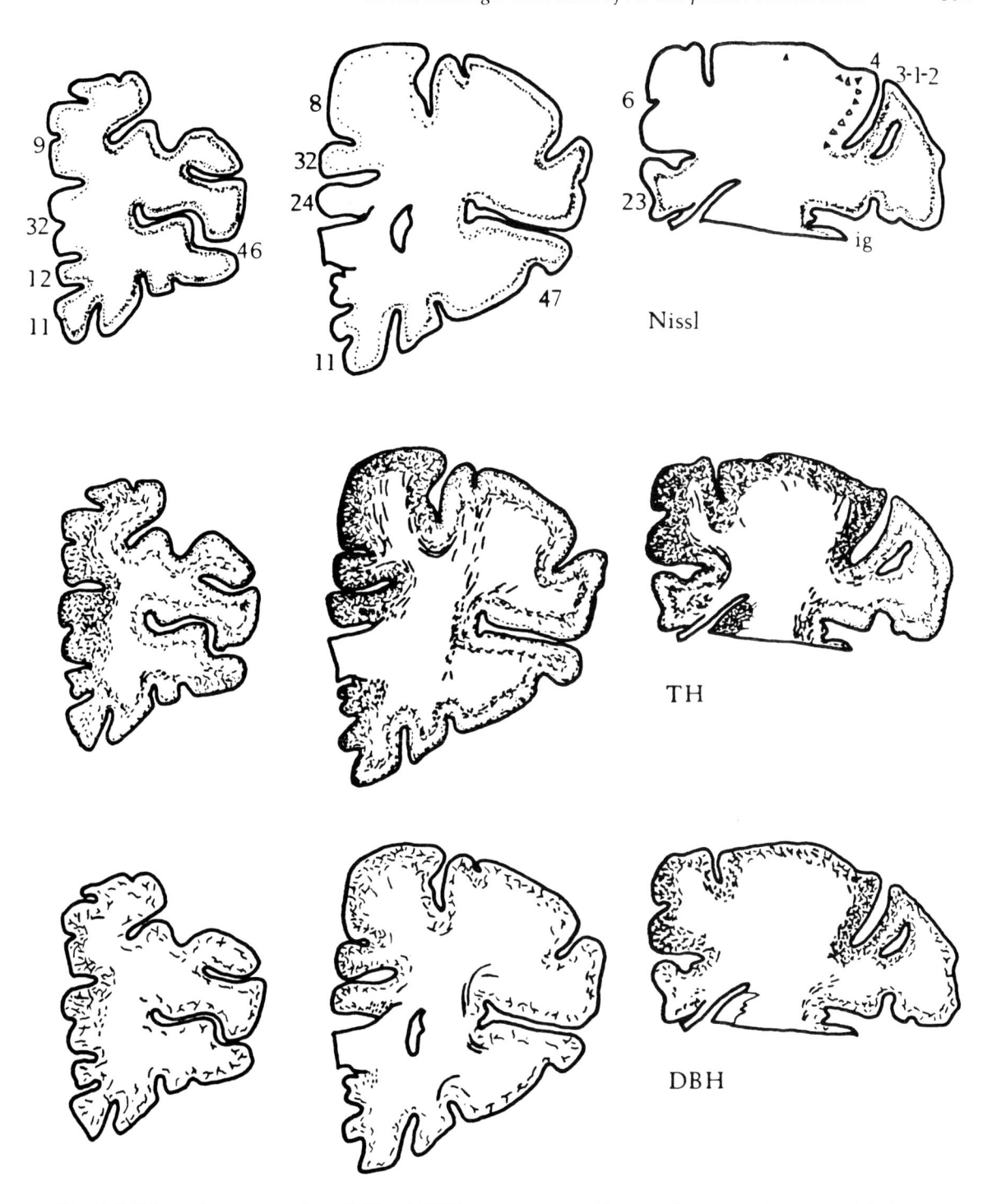

Fig. 12.4. Schematic representation of TH and DBH immunoreactive fibers on three coronal sections of the human frontal lobe. On the upper part of the figure, the localization of the main cytoarchitectonic areas is indicated, using Brodmann's numbering system (see legend of Fig. 12.1); on these schemes, granular layer IV is indicated with dots, dense in lateral prefrontal granular areas (46, 47, 11, 12), scattered in dysgranular areas (8, 32), or absent in agranular cortices (4, 6, and 24); the location of giant pyramidal cells (Betz cells) is figured with triangles.

tion of DA fibers. Such features were reminiscent of the DA axons visualized in human cortical biopsies (Berger, Escourolle & Moyne, 1976a).

Similarly to the observations made in monkeys, maximal numbers of TH-like immunoreactive (LIR) fibers were found in the agranular areas. Among these, the primary motor cortex and the anterior cingulate cortex exhibited the densest innervation followed by the insular cortex. In these agranular areas, the TH-LIR fibers were present in all layers but predominated in layer I and they occasionally formed small accumulations in layers II–III. The density of TH-LIR innervation was consistently lower in the granular cortices, such as the lateral prefrontal cortex, the primary somatosensory cortex (areas 3b-1), the posterior cingulate, or the occipital cortex, where a clearcut bilaminar pattern of distribution in deep layer I and layers V–VI was present. A more extensive analysis in the frontal human cortex, on large sections of the entire frontal lobe, revealed that there was an inverse relationship between the development of the granular layer and the density of TH-innervation (Fig. 12.4). As in monkeys, the density was lower in primary sensory cortices (somesthetic, auditory and particularly visual) than in association cortices. Diverging results have been reported regarding the DA innervation of the hippocampal formation (Gaspar et al., 1989; Torack & Morris, 1990) but no systematic study has yet been performed.

Characteristics of the mesocortical DA population in primates Because the widespread DA cortical innervation has been known more recently, we lack the extensive data gathered in rodents, for characterizing DA mesocortical populations in primates. However, the results already available point to some differences between these two mammalian orders.

Ventral mesencephalic origin Knowledge on the topographic organization of the mesocortical DA neurons in primates is still fragmentary. Only three studies, two in macaques and one in owl monkeys, have specifically examined mesencephalic afferents to the prefrontal, anterior cingulate, motor, and parietal cortices (Porrino & Goldman-Rakic, 1982; Lewis, Morrison & Goldstein, 1988b, Gaspar et al., 1992), while afferents to other cortical areas are still unexplored. Furthermore, in the earlier horseradish peroxidase study of Porrino & Goldman-Rakic (1982), both DA and non-DA neuron projections were included, complicating the analysis of results since as much as half of the mesencephalic neurons projecting to the cortex appear to be non-DA (Gaspar et al., 1992, Oeth & Lewis, 1992).

All three studies, however, made it clear that DA projections in monkeys originate from the entire mesencephalic DA complex that comprises about 150 000 neurons (Table 12.2); that is, not only the ventral tegmental area (VTA), but also the dorsal substantia nigra (SN) and the retrorubral A8 group (Fig. 12.5). In fact, the largest amount of DA neurons projecting to the motor and prefrontal areas were located laterally to the midline, in the n. parabrachialis pigmentosus, a lateral subgroup of the VTA, and in the dorsal diffuse part of the SN compacta (SNc); 10 to 20% of the DA mesocortical projections arose from the dorsal part of A8 (Gaspar et al., 1992).

In spite of a large topographic overlap of the mesocortical neurons projecting to the different cortical targets analyzed, there could be a crude mediolateral topography of the projections. Thus, DA projections to the supplementary motor area (close to the midline), originated from more medial groups in the VTA-SNc complex than those directed to lateral primary motor cortex, M1 (Gaspar et al., 1992). However Porrino & Goldman-Rakic (1982) reported an opposite topographic pattern, with projections to the anterior cingulate cortex (medial) arising more laterally in the SNc than projections to the lateral prefrontal cortex. This discrepancy could be due to the labeling of non-DA mesencephalic projections in the latter study; alternatively there may be a different organization in New and Old world monkeys.

Interestingly, the DA mesocortical projecting neurons were confined to the most dorsal tier of the three DA mesencephalic subdivisions in owl monkeys, at variance with observations in rats. In rats, DA mesocortical projections are distributed in both dorsal and ventral subnuclei of the VTA-SNc (e.g. n. paranigralis) and tend to be located in the most medial DA cell groups, except for the projections to the entorhinal cortex (Lindvall et al., 1978; Björklund & Lindvall, 1984; Fallon & Loughlin, 1987; Deutch et al., 1988). Although some of the observed differences may reflect differences in methodology or the fact that the examined areas were not homologous, the available data suggest that there is a dorsal shift of DA mesocortical neurons in primates relative to rats and cats (Schneiber & Törk 1987). The DA cortically projecting neurons have common histochemical characteristics in primates that set them apart from the other DA neurons in the mesencephalon; indeed, these neurons contain calbindin D-28K (Gaspar, Heizmann & Kaas, 1993) and are characterized by a distinctive noradrenergic and neurotensin innervation (Fig 12.5).

The question of a different origin of the deep and superficial innervations of the cerebral cortex has

Table 12.2. *Surface of the cerebral cortex and total number of neurons in the whole ventral mesencephalic group and the two-locus coerulens*

	Surface of the cerebral cortex (1)	DA neurons in VTA and S.Nigra (2, 3)	NA neurons in Locus coeruleus (4, 5, 6, 7, 8)
Rat	6.4 cm^2	40 000	3200
Macaque	250 cm^2	150 000	15 000
Human	2400 cm^2	450 000 to 1 100 000	40 000 to 60 000

(1) Hofman, 1985; (2) German *et al.*, 1989; (3) Halliday & Törk, 1986; (4) Swanson, 1982; (5) German *et al.*, 1988; (6) Baker *et al.*, 1989; (7) Chan-Palay & Asan, 1989; (8) Foote *et al.*, 1983.

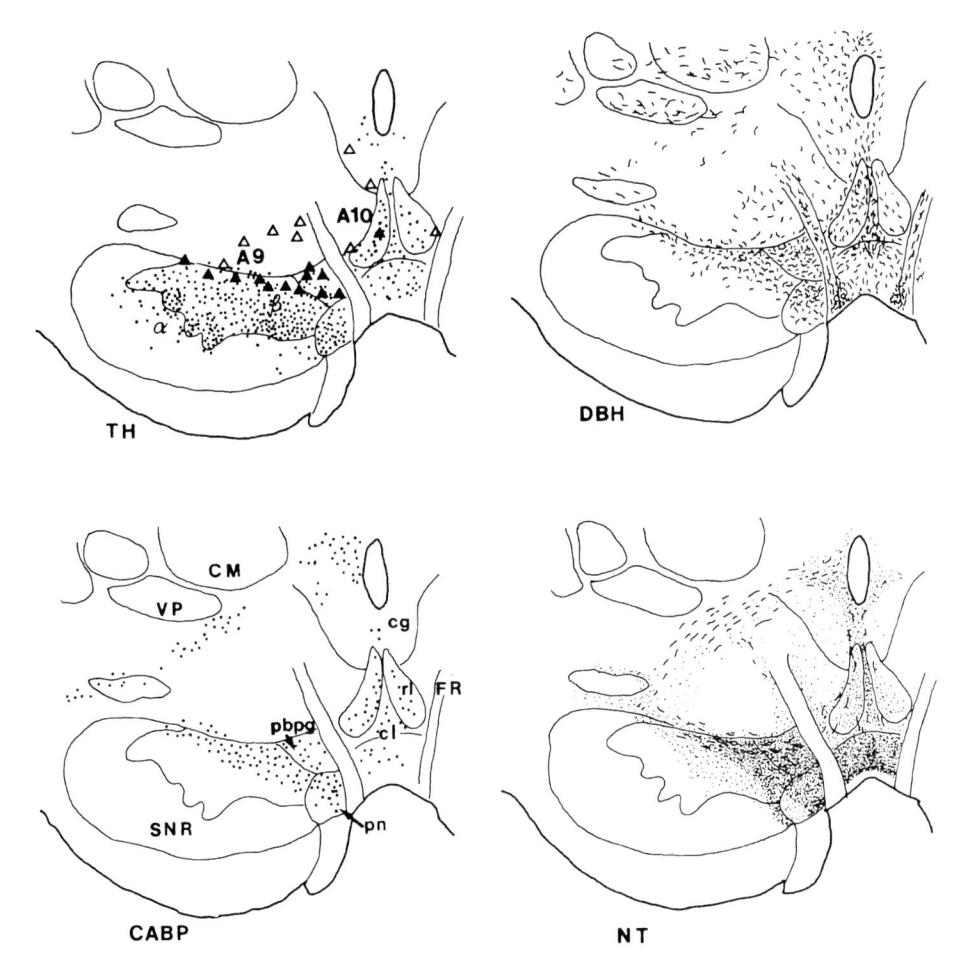

Fig. 12.5. Localization of cortically projecting neurons in the mesencephalon of owl monkeys, in relationship to chemoarchitectonic features of the region. TH: dots represent dopaminergic cell bodies in substantia nigra (A9) and ventral tegmental area (A10), triangles indicate neurons retrogradely labeled after injection of fluorochromes in the primary motor cortex or the lateral prefrontal cortex, filled triangles are for dopaminergic and open triangles for non-dopaminergic neurons. The topography of these cortically projecting neurons coincides with a distinct DA subgroup, characterized by calbindin D-28K (CaBP) immunoreactivity, and noradrenergic (DBH) and neurotensin positive (NT) innervation. (Modified from Gaspar *et al.*, 1992).

not yet been addressed, and it is not known whether projections to the deep or superficial cortical layers of the primate neocortex may differ on topographic or histochemical grounds. There is however an indication, albeit indirect, for the existence of such distinct subsets of DA projections: the differential loss of DA fibers in the superficial cortical layers of the frontal and motor cortices of Parkinsonian patients (Gaspar et al., 1991). This would be consistent with the well known massive cell loss in the SNc, contrasting with a less significant degeneration in the VTA (Hirsch, Graybiel & Agid, 1988; German et al., 1989). However, while the preferential laminar distribution of the DA depletion in Parkinson's disease might reflect a differential degeneration of two distinct dopamine populations in the mesencephalon, it could also represent a dying back process starting at the level of the most distal terminals.

Collateralization pattern Since the cortical surface has increased in primates (Table 12.2) in a much higher proportion (400 times larger in humans than in rats, Hofman, 1985) than the number of mesencephalic DA neurons (x 10 to 20 times, Halliday & Törk, 1986; German et al., 1989), one might expect an increase in the proportion of collateralized projections. In fact, this has been shown in owl monkeys (Gaspar et al., 1992). Collateralization within the frontal cortex did not seem to depend on whether the cortical areas were interconnected or not: approximately 30% of DA neurons projecting to the lateral prefrontal cortex sent collaterals to the primary motor cortex or to the supplementary motor cortex. Thus, individual DA axons may provide terminal arbors at cortical distances of at least 6 in mm either medio-lateral or rostrocaudal directions. Such collateralization patterns have not yet been explored in Old World monkeys. It is not known either, whether some DA neurons projecting to the cortex send collaterals to the striatum, as has been shown in rats (Takada & Hattori, 1986). However, the topographic distribution of mesostriatal projections in primates (Hedreen, 1991) suggests that they constitute a separate DA population.

Coexpression of neuropeptides In humans, neurotensin (NT) was not detected in TH-positive fibers in either the prefrontal cortex or in the other cortical DA terminal fields (Gaspar, Berger & Febvret, 1990). A similar segregation was suggested by the description of NT-like immunoreactivity in adult macaques (Satoh & Matsumura, 1990) and by our own data during fetal life in rhesus monkey (Berger et al., 1993, 1994).

In the ventral mesencephalon, several groups including ours, have failed to detect NT-LIR neurons with immunocytochemical methods, even after colchicine administration (Deutch et al., 1988; Gaspar et al., 1990; Bean et al., 1992). Recently however, Bean et al. (1992) demonstrated NT mRNA in about 3% of the DA neurons in the human mesencephalon. The positive neurons were located in the area of the n. parabrachialis pigmentosus and the retrorubral area, two regions which provide cortical DA afferents in both Old and New World monkeys (Porrino & Goldman-Rakic, 1982; Gaspar et al., 1992). These results raise more problems than answers: is the translation so low or the turnover so high that NT cannot be demonstrated in these neurons even after colchicine injections? Is the protein processed differently in different species leading to products that are not revealed by the available antibodies? Finally, do these neurons contribute to the NT innervation of the cerebral cortex?

Similarly, there appeared to be no mixed projections of DA and cholecystokinin (CCK) to the prefrontal cortex of adult macaques (Oeth & Lewis, 1992). A lingering question is whether this lack of colocalization reflects a permanent interspecific difference, as in hamster and guinea pig (Schalling et al., 1990) or merely results from transcriptional or posttranscriptional modifications. According to Palacios, Savasta & Mengod (1989), the amount of DA neurons expressing CCK mRNA in the ventral mesencephalon seemed to be lower in Old World monkeys than in squirrel monkeys, and lower in humans than in macaques. In humans, only a small number of CCK-positive neurons was present in the nucleus paranigralis but no signal was found in the substantia nigra. An intriguing result in this respect is the observation by Schalling et al. (1990), that in schizophrenic patients, there are numerous neurons containing CCK mRNA in the ventral mesencephalon, both VTA and SN. It may be hypothesized that nigral neurons in humans retain the capacity of expressing CCK mRNA, and that this unusual transcription may be induced either by the neuroleptic treatment or by the pathological state.

The dopaminoceptive population

The cortical dopaminoceptive population represents another domain where basic similarities exist between rats and primates but with marked differences also emerging.

Neuronal targets Several lines of evidence indicate that DA-containing varicosities form conventional synapses both in rats and primates. However, the proportion of synaptic contacts might well be target-dependent. Thus, Séguela, Watkins & Descarries (1988) using serial sections, found a synaptic incidence of 100% in the prefrontal cortex of rats and of only 50% in the insula. Symmetrical synaptic contacts were preferentially formed on dendritic shafts and spines of pyramidal cells (Van Eden et al., 1987, Séguela et al., 1988) but ultrastructural double labeling studies indicate that gabaergic neurons, in all likelihood interneurons, cannot be excluded as DA targets (Verney et al., 1990). Moreover, DA varicosities form triadic complexes that include, in addition to the symmetrical DA synapse, another bouton of unknown origin, forming an asymmetrical synapse, presumed to be excitatory, on the same dendritic spine or shaft and thus allowing for a direct modulation by dopamine of other afferents (Verney et al., 1990). Similar data have also been reported in monkey and human prefrontal cortex (Goldman-Rakic et al., 1989; Smiley et al., 1992). However it is not known whether there are variations of synaptic incidence within the numerous subdivisions of the prefrontal or other cortical areas.

Variety and topography of cortical DA receptors
Five types of DA receptors have presently been recognized either through pharmacological analyses using specific ligands or with molecular biological approaches. D1 and D2 are two widespread and predominant dopamine receptors (Civelli et al., 1991). Their regional and laminar densities in rat and monkey cerebral cortex is consistent with the regional and laminar dopaminergic innervation (Lidow et al., 1989; Richfield, Young & Penney, 1989; Sales et al., 1989; Goldman-Rakic, Lidow & Gallager, 1990). A subset of presynaptic D2 receptors, localized on the DA terminals themselves, regulates the release of DA (White & Wang, 1984). Two other receptors were localized in restricted DA terminal fields in rat, a subtype of receptor D1 (D1b or D5, Meador-Woodruff et al., 1992) and D3 (Bouthenet et al., 1991). A D4 receptor is cloned in humans, and relatively high levels of D4 mRNA were observed in the monkey frontal cortex (Van Tol et al., 1991). This localization might be of particular functional importance since the D4 receptor seems to be selectively involved in the antipsychotic action of the neuroleptic clozapine.

There is now increasing evidence that the dopaminoceptive neuronal population might include distinct sets of pyramidal cells and probably interneurons, differing in their receptor configuration and their laminar distribution. In rhesus monkey prefrontal cortex for instance, D1 receptors localized with radiolabeled specific ligands were present in highest relative concentration in superficial layers I–II–IIIa, whereas D2 receptors were distinguished by relatively high concentrations in the deep layer V (Goldman Rakic et al., 1990). An interdigitating laminar distribution of D1 and D2 receptors has also been demonstrated in the hippocampal region of rat and man (Kohler et al., 1991a, Kohler, Ericson & Radesater, 1991b).

In situ hybridization allows a further step in the characterization of the dopaminoceptive neuronal population at the cellular level. Two recent reports indicate that the DA receptor D1b or D5, which has been cloned in both rats and humans, displays a completely distinct pattern of cortical distribution in the two species: restricted to the hippocampus and the parafascicular nucleus of the thalamus in rats (Meador-Woodruff et al., 1992), it was observed in a large population of pyramidal cells in monkey and human motor cortex (Huntley et al., 1992). In addition, in the latter species, the three types of receptors (D1a and b and D2) might be co-expressed, particularly in the giant pyramidal Betz cells of the motor cortex. Co-expression of D1 and D2 receptors would offer a structural basis for a possible D1-D2 synergistic effect similar to that demonstrated recently in the rat prefrontal cortex (Retaux, Besson & Penit-Soria, 1991).

Other differences, both inter-areal and inter-specific in the neuronal population bearing D1 receptors, are suggested by the distribution of DARPP-32, a dopamine and cAMP-activated phosphoprotein enriched in dopaminoceptive neurons bearing D1 receptors (Hemmings et al., 1987; Schalling et al., 1990; Guennoun & Bloch, 1992; Gustafson et al., 1992). In monkey cortex, DARPP-32 labeled pyramidal cells are detected throughout layers V and VI, and their long ascending apical dendrites can be traced as far as layer I (Berger et al., 1990; Ouimet et al., 1992), indicating that DA might modulate primarily the vast array of cortico-subcortical output pyramidal neurons located in these layers. In contrast, DARPP-32-labeled neurons are not detected in layer V in rat, but in layer VI, mainly corresponding to the cortico-thalamic projections (Ouimet et al., 1984). Furthermore, if different classes of DA neurons provide the innervation of the superficial and deep cortical layers, a given pyramidal cell might be subject to different modulatory influences on its apical and basal dendrites; this type of complexity will add to that realized by the interdigitating distribution of various DA receptors.

Extensive *in situ* hybridization studies will be necessary for a systematic mapping of the individual dopaminoceptive neurons as a function of their types of DA receptors, neurotransmitters and connections. However, immunocytochemistry of DA receptors would be the method of choice for identifying the whole neuron including dendritic arborization and axonal projections (Huang et al., 1992). This should help to evaluate comparatively in rodents and primates the DA neuronal targets at different levels of the cortical circuitry.

A particular class of DOPA-ergic neurons in the cerebral cortex? There is another class of catecholaminergic neurons that we have first described in rat cerebral cortex some years ago (Berger et al., 1985c). These neurons express tyrosine hydroxylase (TH), but seem to lack the other catecholaminergic traits. For this reason, DOPA has been suggested as a possible neurotransmitter in these neurons. Similar neurons have now been observed in other species including human, by our and other groups (Gaspar et al. 1987; Hornung, Törk & Detribolet, 1989; Trottier, Geffard & Evrard, 1989; Satoh & Suzuki, 1990; Vincent & Hope, 1990), and the specificity of the immunocytochemical demonstration has been confirmed by *in situ* hybridization (Kuljis, Martin–Vasalo & Peress, 1989; Lewis et al., 1991). A prominent feature in both human and non human primates is their persistence in adulthood whereas in mice and rats, they are expressed mainly transitorily during the first month of postnatal life. This expression can be maintained in conditions of tissue culture (Iacovitti et al., 1987) or in grafts (Park, Joh & Ebner, 1986). In humans, they constitute less than 1% of the neuronal population but their number might increase in pathological conditions such as epileptic seizures (Zhu et al., 1990). They display a specific laminar and regional distribution pattern; concentrated in the infragranular layers, mainly layer VI, their density is lowest in primary cortical areas, and highest in association cortices such as the prefrontal cortex, and in limbic related areas (Gaspar et al. 1987, Lewis et al., 1991). These neurons display the morphology of interneurons and indeed contain GABA in addition to TH (Trottier et al., 1989). Although they seem very different from the classical DA neurons, they have been shown to contain Lewy bodies in some cases of Parkinson's disease, similarly to the neurons of the substantia nigra (Kuljis et al., 1989). This is an argument favoring the hypothesis that these neurons belong to a new class of catecholaminergic cells like other TH-IR neurons of the same type described in the striatum (Dubach et al., 1987) and in the basal forebrain (Kohler et al., 1983). However, the lack of other enzymes of the catecholaminergic chain still needs to be confirmed by transcriptional studies. Whatever the neuroactive substance released by these neurons, DOPA or dopamine, they could participate in the cortical dopaminergic pool, since DOPA might be either utilized as a neurotransmitter or serve as an additional source of precursor for monoaminergic afferents through uptake mechanisms. Alternatively, they might utilize another transmitter, unrelated to DA metabolism such as GABA or neuropeptides.

The cortical NA system

The cortical noradrenergic innervation is widespread to all cortical areas in both rodents and primates (for review see Lindvall & Björklund, 1978; Foote, Bloom & Aston-Jones, 1983; Fallon & Loughlin, 1987). However, although some differences in terminal arborization exist among the different cortical areas in rodents, the regional differentiation of NA innervation is much more striking in humans and monkeys. In contrast with other brain regions where NA innervation has multiple origins (locus coeruleus and medullary NA cell groups) and appears to be heterogeneous morphologically, the NA cortical system appears much more homogeneous, originating only in the locus coeruleus, with rather uniform fiber morphology, and similar vulnerability to neurotoxics such as DSP4 (Fritschy & Grzanna, 1989).

Distributional maps in rodents

NA innervation is widely distributed over all cortical areas and in the hippocampus (Fig. 12.2) (Blackstadt, Fuxe & Hökfelt, 1967; Fuxe, Hamberger & Hökfelt, 1968; Levitt & Moore, 1978; Lidov, Molliver & Zecevic, 1978a, Morrison et al., 1978). NA fibers reach all cortical layers forming a typical arborization pattern that is basically similar in all cortical areas (Fig. 12.2). Layer I contains the densest innervation with tangentially organized fibers oriented either sagittally or coronally, forming a grid-like pattern. Layers II–III contain the fewest fibers with a main radial orientation, while layers IV-V contain a rather dense terminal network. Long sagittally oriented fibers characterize layer VI. These fibers form intracortical NA pathways which, via local branches, innervate the cortex through which they pass. Indeed knife cuts throughout the width of the cortex, at lateral frontal levels, depleted NA in a 2 mm wide sector caudally to the site of transection, whereas transections involving only the upper cortical layers caused limited cortical NA depletion (Morrison et al., 1981). NA fibers traveling mediolaterally have also been described, some following peculiar oscillatory courses between layers II and IV (Papadopoulos, Parnavelas & Buijs, 1989b). Regional differences in the laminar distribution of NA fibers were observed with DBH-immunostaining, mainly regarding the extent of a terminal plexus in layer V: this plexus was well developed in primary somatosensory and primary motor areas, but was virtually absent in primary visual cortex, where fibers were mainly present in supragranular layers (Morrison et al., 1978). There are a few other examples where the pattern of NA innervation covaries with cytoarchitectonic boundaries. Thus, in the somatosensory cortex of the mouse, NA terminals appear to be concentrated inside the barrels while being sparser in the surround (Lidov, Rice & Molliver, 1978b). A specific laminar pattern of NA fibers was also described in the anterior cingulate cortex, where DBH positive fibers appeared concentrated in the lower part of layer I and in layers V–VI, a pattern complementary to that of DA fibers (Lewis et al., 1979).

However, regional differences in NA innervation were not confirmed with radioautographic analysis of terminals after specific uptake of tritiated NA (Audet et al., 1988). In this study, the density of NA terminals was similar in the anterior cingulate, primary somatosensory, medial frontal and piriform cortices. Overall density of NA terminals was estimated to be 1.2 million varicosities/mm^3 of tissue. A higher density of NA innervation in layer I was however confirmed, and quantified as being 1.5 to 2 times that of the other cortical layers.

Distributional maps in primates

In primates, both humans and monkeys, the distribution of NA is markedly uneven across the different cytoarchitectonic areas. This heterogeneity was already clear in biochemical analyses that indicated highest NA concentrations around the principal sulcus and in primary motor and sensory areas with decreasing gradients in both the rostral and caudal directions (Vogel, Orfei & Century, 1969; Björklund, Divac & Lindvall, 1978; Brown et al., 1979; Javoy-Agid et al., 1989). Similar indications were given by the histochemical analysis of Levitt et al. (1984), although the precise distinction between DA and NA fibers could not be made. More specific descriptions of NA innervation were made using DBH-immunostaining in squirrel monkeys (Morrison et al. 1982a,b; Morrison & Foote, 1986), cynomolgus monkeys (Morrison & Foote, 1986, Lewis & Morrison, 1989) and in humans (Powers et al., 1988, Gaspar et al., 1989). In these three species, the densest network of DBH-labeled fibers was found in the primary somatosensory (areas 3b,1,2) and motor cortices (area 4 and supplementary motor cortex). The density of fibers decreased caudally in the occipital cortex and rostrally in the frontal cortex, particularly in the lateral granular prefrontal cortex. The variations in the density of this terminal network are illustrated in a series of sections through the frontal cortex of the human brain (Fig. 12.4): density variations in the different cytoarchitectonic areas follow that of TH-immunostaining (putative DA innervation) although the layers innervated are not the same; a dissociation occurs however at the frontier between the primary motor and primary somatosensory cortex, where the number of TH fibers decreases whereas that of DBH fibers increases. The laminar distribution of NA fibers is different from that described in rodents: layer I tends to have the lowest density of innervation, few radially oriented fibers are present in layer II, while the bulk of NA terminal network is present in layers III to V, with a variable extension to layer IV, according to regions. The presence of long NA fibers in layer VI and the cortical-subcortical border (Morrison et al., 1982b; Gaspar et al., 1989), indicates that at least some axons travel intracortically; however these long fibers were oriented either in the sagittal or the frontal plane, suggesting that the contribution of the catecholaminergic fiber bundles in the cingulum and insula to NA innervation in the dorso-lateral cortex might be more important in gyrencephalic than in lissencephalic species.

Origin in the brainstem and collateralization pattern

In rodents, all the cortical NA innervation originates from the locus coeruleus, two symmetrical nuclei with about 1600 neurons each in rats (Swanson, 1976; Foote et al., 1983), situated in the pontine tegmentum, and 10% of the projections are crossed (Room, Postema & Korf, 1981). The pathway of the ascending NA axons has been described in great detail (Lindvall & Björklund, 1978): NA fibers ascend in the dorsal tegmental bundle in the mesencephalon and then course in the dorsomedial part of the medial forebrain bundle in the telencephalon, close to the mesocortical DA pathway.

In contrast with the mesocortical DA system,

the cortical NA system is highly collateralized. Combined injections of fluorescent tracers have shown that a given locus coeruleus neuron sends collaterals to sites as distant as the spinal cord, the cerebellum and the cerebral cortex (Nagai et al., 1981; Room et al., 1981). During their ascending course in the brain, NA axons provide numerous collaterals to subcortical structures, and within their intracortical trajectory, to the functionally different areas that they traverse. Indeed, Descarries and collaborators have calculated that a single locus coeruleus neuron might provide as much as 300 000 terminals (Séguela et al., 1990) inside the cortex. Using electrophysiological recordings, Sakaguchi & Nakamura (1987) found variable patterns of collateralization but no order between recording sites in the locus coeruleus and terminal fields in the cerebal cortex. However, more precise mapping, using 3D reconstructions of the whole nucleus, localized the origin of the cortical projections in the central portion of the nucleus in rats, and revealed within the locus coeruleus (LC) a fairly precise geometric distribution of neurons projecting to the different targets, with a crude topography such that more anterior cells project to anterior cortical regions (Loughlin, Foote & Bloom, 1986). They also reported that individual LC cells collateralized extensively in the anteroposterior dimension and less so in the medio-lateral extent of the cerebral cortex.

In primates, cortical NA projections also arise from the LC (Freedman, Foote & Bloom, 1975; Gatter & Powell, 1977). Each nucleus contains about 7500 neurons in macaques and 20 000 to 30 000 neurons in humans (German et al., 1988; Baker et al., 1989; Chan-Palay & Asan, 1989) (Table 12.2). While the projections arise mainly from the ipsilateral nucleus, 20 to 50% are crossed (Gatter & Powell, 1977). The fiber pathways have been studied with anterograde tracing studies (Bowden, German & Poynter, 1978): while two longitudinal fiber bundles are observable in the cingular bundle and in the external capsule below the insula, it is not completely clear how NA fibers travel intracortically, particularly in the highly convoluted cortex of humans. Using multiple injections of fluorescent tracers in the cerebral cortex of owl monkeys, we observed that a single neuron in the LC sent collateral projections to cortical areas located medially (supplementary motor cortex) and laterally (hand representation of the primary motor cortex) (Gaspar et al., 1992). Collateralization is found to be more important in primates, particularly in humans, than in rodents, since the volume of the cortex increases much more than the number of cells in the locus coeruleus (Table 12.2).

Heterogeneity within the NA coeruleo-cortical pathway?

Morphology and reactivity to degeneration. With catecholamine histofluorescence, Lidov et al. (1978a) had noted that fibers in the upper cortical layers appeared to contain more endogenous amine. No further evidence for a heterogeneity was shown however. NA fibers in the cerebral cortex of rodents and primates were described as having a rather uniform morphology with DBH immunostaining. Furthermore, using multiple injections of fluorescent dyes, Loughlin et al., (1986) indicated that individual LC neurons innervated both superficial and deep layers of a cortical region. Confirming these data, all the cortical NA terminals are damaged after administration of DSP4, a selective neurotoxic of NA fibers (Fritschy & Grzanna, 1989). This contrasts with other brain areas such as the septal area, where two types of NA fibers co-exist, differently affected by DSP4: one issued from the locus coeruleus, characterized by fine varicosities is DSP4 sensitive, while the other originating from noradrenergic neurons in the medulla (Lindvall & Stenevi, 1978) and characterized by large varicosities, is spared (Fritschy & Grzanna, 1989). In humans, most NA fibers have fine varicosities, but some NA fibers had large, swollen varicosities (Berger et al., 1976a; Powers et al., 1988; Gaspar et al., 1989); however, these may be degenerative changes, in aged patients. In Parkinson's disease, NA fibers were massively and uniformly reduced in all cortical layers and all areas, contrary to the laminar-specific depletion of DA fibers (Gaspar et al., 1991).

Colocalization with neuropeptides. Colocalization with peptides may define separate neural subpopulations within the locus coeruleus. In rats, galanin-like immunoreactivity was present in 82% of the neurons while neuropeptide-Y immunoreactivity was detected in only 23%. However, according to Holets et al. (1988), these neurons would represent a minority of the cortically projecting NA cells, 3% for galanin and 1% for NPY. These results which would suggest that only a very small population of locus ceruleus neurons project to the cerebral cortex await further confirmation. Using immunocytochemical methods and double-labeling, Forloni et al. (1987) stated that all NA cells in the locus coeruleus contained the acidic dipeptide N-acetyl-aspartylglutamate (NAAG), but this result has remained isolated. Recently, GABA was also reported to be co-localized with TH in the rat locus coeruleus (Iijima et al., 1992).

In humans, immunocytochemical studies indicated a high density of CRF, NPY, and neurotensin terminals in the locus coeruleus but no cell bodies were observed (Pammer, Gorcs & Palkovits, 1990). With immunocytochemistry to NPY in Cebus monkeys, we did not observe labeled neurons in the locus coeruleus (personal unpublished observations). However, since peptide levels can be very low in cell bodies and go undetected with immunocytochemistry, it is necessary to do *in situ* hybridization to look for expression of peptide mRNAs, before concluding on the lack of coexistence of NA with peptides in the locus coeruleus of primates.

The noradrenoceptive population

Neuronal targets The mode of termination of NA axons has remained for many years a controversial issue (for review see Foote *et al.*, 1983). Despite improved sensitivity and specificity in the techniques, there is a continuing argument between supporters of classical neurotransmission limited to specialized synaptic junctions (Zecevic & Molliver, 1978; Olschowska *et al.*, 1981; Papadopoulos *et al.*, 1989*b*) and supporters of NA release outside conventional synapses (Descarries *et al.*, 1977; Séguela *et al.*, 1990). These conflicting interpretations persist in the two most recent studies that used rather similar approaches, i.e. immunocytochemistry of NA and serial ultrathin sections. Indeed, Papadopoulos *et al.* (1989*b*) estimated that the vast majority (87%) of NA varicosities formed conventional synapses, whereas Séguela *et al.* (1990) estimated that these junctional complexes were rare (17–26%). This difference may arise from the difference in the number of serial sections examined, or from the stereological assumptions that were made to calculate the incidence of these synapses. Since all cortical layers of the frontoparietal and visual cortices were included in both studies, there is no reason to believe, as had been proposed (Foote *et al.*, 1983; Magistretti & Morrison, 1988), that the difference in the reported incidence of synapses results from a difference of synaptic incidence between NA fibers in the superficial or deep cortical layers. When synapses occur, they seem to be predominantly (Papadopoulos *et al.*, 1989*b*) or exclusively (Séguela *et al.*, 1990) of the symmetrical type. Cellular targets have been identified as interneurons or pyramidal neurons, mainly dendritic shafts or spines, and triadic arrangements with unlabeled varicosities have been described (Papadopoulos *et al.*, 1989*b*) as for the dopaminergic fibers (see above). Similar ultrastructural studies are presently lacking in primates.

Variety and topography of cortical adrenergic receptors The cortical targets of NA innervation may be defined by their differential content in adrenoreceptors of the alpha or beta type, or of their subtypes: α_{1A}, α_{1B}, α_2, β_1, β_2, most of which have been cloned (Harrison, Pearson & Lynch, 1991; Tota *et al.*, 1991). Cellular or behavioral responses to the activation of each receptor subtype is indeed different (see refs. in Schliebs & Gödicke, 1988; McCormick, 1992; Trovero *et al.*, 1992).

Alpha$_1$ receptors were densely represented in the cerebral cortex of the six mammalian species investigated by Palacios, Hoyer & Cortés (1987), but their distribution differed: in rodents, laminae IV and V were most enriched (but see Schliebs & Gödicke, 1988); in monkeys and humans, α_1 sites were distributed in a bilaminar fashion in layers I–III and VI. An α_1 receptor subtype, detected by binding with prazosin, may be of particular interest, as this receptor appears to be specifically responsible for interactions between the NA and the DA systems at cortical levels, and displays a strikingly specific regional distribution in motor cortical areas (Trovero *et al.*, 1992). Alpha$_2$ receptors were assumed to be located presynaptically, as they modulate NA release (see refs. in Cash *et al.*, 1986). However the laminar and regional distribution of α_2 binding sites in the cerebral cortex does not completely coincide with the pattern of NA innervation in human cortex as they are concentrated in layer I, with intermediate levels in layers II–III, and lowest in V–VI. In rat cortex they are densest in layers I and V (Pazos *et al.*, 1988; Schliebs & Gödicke, 1988), and the receptors are not modified by lesions of NA neurons (Cash *et al.*, 1986).

The β_1 and β_2 receptors constitute the other families of adrenergic receptors. β-adrenergic receptors seem to be present in a large population of neurons in the cerebral cortex (Wanaka *et al.*, 1989; Aoki, Joh & Pickel, 1987). Some ultrastructural studies on the distribution of β-adrenergic receptors, tend to support the notion that NA could act beyond identifiable synapses. Indeed, using an antibody against a fragment of the β_2 adrenoreceptor, immunoreactivity was almost exclusively associated with astrocytes, that were or were not in contact with catecholaminergic terminals (Aoki & Pickel, 1992). This glial localization of beta-adrenoreceptors, also suggested by biochemical studies (Stone & John, 1992) is intriguing. Among their possible functions, β-adrenergic receptor bearing astrocytes could mediate interactions between catecholaminergic and non-catecholaminergic neurons, or modulate axodendritic interactions by ensheathing synapses. Autoradiographic localizations of adrenorecepor subtypes have also been made in monkeys (Lidow et al., 1989; Palacios et al., 1987; Goldman-Rakic et al., 1990), and humans (Palacios et al., 1987; Pazos et al., 1988), indicating regional and laminar specific distributions of the α_1, α_2, and β receptors, that often differ from rodents.

Comparative developmental sequence of the DA and NA innervations in rats and primates

Two main developmental differences are oberved between rodents and primates. The major one, the most important from the functional point of view, is the protracted prenatal development of both the DA and NA systems in primates; the second is that a delay in the arrival of distinct DA contingents, comparable to that observed in rats has not been detected as yet in primates. The catecholaminergic fibers (DA and NA) reach the frontal cortex soon after the formation of the cortical plate but this corresponds to a much longer time of prenatal development in primates than in rats. In rats, as already mentioned, the precocious contingent of DA projections is detected at E16 and the NA fibers at E17, circa 5 days before birth (Levitt & Moore, 1978; Verney et al., 1984; Kalsbeek et al., 1988). In humans, catecholaminergic projections penetrate the anlage of the cerebral cortex at 11 weeks of gestation (Zecevic et al., 1991) and in rhesus monkey (gestation 23 weeks), between 7 and 8 weeks of fetal life (Berger et al., unpublished observations). Thereafter, limbic areas such as the cingulate cortex, the insula and the entorhinal cortex, receive a dense innervation earlier than other cortical areas but at midgestation, all cortical areas except the visual cortex are densely innervated (Berger, Verney & Goldman-Rakic, 1992). At that time, the typical difference between granular and agranular cortex in the pattern of laminar distribution of DA fibers is already clearly figured and the motor and anterior cingulate cortices appear the most densely innervated (Verney et al., 1993). This protracted prenatal development of the cortical catecholaminergic innervation sets the conditions for long lasting interactions with corticogenesis (Shatz, Chun & Luskin, 1988; Kostovic & Rakic, 1990), the precocious synaptogenesis (Rakic et al., 1986; Zecevic et al., 1989) and the early developing connections (Killackey & Chalupa, 1986; Goldman-Rakic, 1987a; Schwartz, Rakic & Goldman-Rakic, 1991). In this respect, the high number of DARPP-32 labeled neurons in the cortical subplate during the first half of gestation in rhesus monkeys suggests an early developement of the DA receptors, possibly of the D1 type (Berger et al., 1992). Finally the comparative results suggest that the development of catecholaminergic projections to the cerebral cortex might proceed faster in humans than in monkeys; we observed at midgestation, a higher amount of DA terminals in human motor cortex (Verney et al., 1993) than in monkeys (Berger et al., 1992).

Interactions between the catecholaminergic innervation and other neurotransmitters

Besides the synaptic or non-synaptic junction with the somatodendritic compartment of the dopaminoceptive and/or adrenoceptive cells, an interaction of the catecholaminergic afferents with the numerous other types of terminals in the cerebral cortex may be expected. DA afferents seem to interact with the other monoaminergic, noradrenergic, serotonergic and cholinergic systems. There are many indications for interactions between the NA and the DA systems. Some of them occur at cortical levels; for instance, DA is involved in the regulation of alpha and beta adrenergic receptors sensitivity (Hervé et al., 1990; Nowak, Zak & Superata, 1991); other interactions may occur in the mesencephalon, at the level of cortically projecting DA neurons, since these neurons receive a dense NA input (Gaspar et al., 1992). Recently some evidence has been provided that lesioning NA fibers could increase the destructive effect of DA neurotoxins such as MPTP (Mavridis et al., 1991).

Reciprocal interregulations of synthesis or release, have been demonstrated between cholinergic and DA or NA terminals in the cerebral cortex of rats and humans (Vizi, 1980; Beani et al., 1986, Jaffé & Hernandez, 1989; Mitchell et al., 1989; Yang & Mogenson, 1990, Vezina et al., 1992). Furthermore, cholinergic and adrenergic agonists facilitate synergistically the induction of long-term potentiation in cortical slices of rat brain (Bröcher, Artola & Singer, 1992). DA release is also enhanced by 5-HT3 receptor activation in the medial prefrontal cortex of freely moving rats (Chen et al., 1992). Another set of interactions occurs with neuropeptides. DA can stimulate somatostatin release (Thal et al., 1986) and may affect the neuronal transmission of cholecystokinin (Fukamauchi, Yoshikawa & Shibuya, 1992); conversely, somatostatin exerts a stimulatory effect on norepinephrine release (Tsujimoto & Tanaka, 1981). The co-expression and corelease of DA and neuropeptides in the same mesocortical DA neurons in rodents attracted much interest because of the role they might play in the pathophysiology of psychoses. However, much of the functional consequences of the corelease of NT and DA in rat prefrontal cortex (Bean & Roth, 1991) should be reconsidered in a different perspective in primates because of the apparent lack of colocalization. Interactions between NT and DA could nevertheless occur in the limbic cortical areas where both terminal plexuses are present (Gaspar et al., 1990). NT receptors may be present on cortical DA axons and have a presynaptic effect on DA terminals, on release-modulating sites for example, as shown in rodents (De Quidt & Emson, 1983; Seutin, Massotte & Dresse, 1989). Furthermore, the mesocortical DA neurons may be regulated at mesencephalic levels by the dense NT innervation that they receive (Fig. 12.5) both in rats and primates (Sadoul et al., 1984; Uhl et al., 1984, Szigethy & Beaudet, 1989; Gaspar et al., 1990, 1992). An interaction that has attracted much interest is that occurring between VIP containing neurons in the cerebral cortex and NA terminals. Indeed, VIP potentiates the effect of NA on cAMP production. The geometric organization of the two systems, VIP in radially oriented interneurons with a limited tangential extent and NA with a widespread tangential influence, led to the proposal that coactivation of the two systems could create metabolic hot-spots in cortical columns (Magistretti & Morrison, 1988).

Interactions with excitatory aminoacids are of particular importance since glutamate and aspartate appear to be the main transmitters of cortical projections. DA differentially affects the responses to excitatory aminoacids in human cortex (Cepeda et al., 1992). Conversely, glutamate exerts a tonic inhibitory regulation on DA transmission (Hata et al., 1990), and presynaptic NMDA receptors stimulate noradrenergic release (Fink, 1990). In vivo, physiological studies showed that activation of the mesoprefrontal DA system inhibits the spontaneous activity of cortical neurons and blocks the excitatory responses induced by electrical stimulation of the afferents from the MD nucleus of the thalamus or evoked by noxious peripheral stimuli (Mantz et al., 1988). This inhibition seems to involve both DA and Gabaergic components (Pirot et al., 1992), seemingly in agreement with the ultrastructural data (Verney et al., 1990). In contrast, stimulation of the locus coeruleus produced long-lasting post-stimulus inhibition of spontaneous activity that did not interfere with responses to tail-pinch or to thalamic stimulation (Mantz et al., 1988).

Functions of the cortical DA and NA system

The broad spectrum of physiological effects that depend on cortical dopaminergic and noradrenergic innervations are clearly beyond the scope of this review and the reader is referred to the excellent reviews of Foote et al., 1983; Foote & Morrison, 1987; Robbins et al., 1985; Chiodo, 1988; Le Moal & Simon, 1991; McCormick, 1992, for electrophysiological effects at the cellular level and behavioral studies. In fact, the functions of the DA and NA systems are as diverse as their targets; like the other aminergic groups, they play a regulatory role as opposed to the other sets of neurons that are involved in point to point transmission of the information.

Cortical DA system

The functions of DA have been intensively studied because of the major role of DA deficiency in Parkinson's disease and because DA dysfunction has been presumed in psychoses. Most physiological data on the functions of the cortical DA system have focused on the prefrontal cortex because of its role in higher integrative cortical functions. Dopamine was demonstrated to be necessary for cognitive functions both in rodents (Simon, Scatton & Le Moal, 1980) and monkeys (Brozoski et al., 1979), with D1 receptors being critically involved in mnemonic functions (Sawaguchi & Goldman-Rakic, 1991). Decreased cortical DA level has been implicated in the pathophysiology of schizophrenia (Weinberger, 1988) but primary alterations of DA neurons were never clearly demonstrated in psychoses.

The role of DA input to the motor areas in primates has not yet been analyzed because such an innervation was not suspected from anatomical studies in rodents. Since DA innervation predominates in the primary and secondary motor cortical areas, this amine could be important in the normal functioning of these areas in planning, initiation and sensory guidance of movements (Goldberg, 1985; Wise, 1985; Wiesendanger, 1986). Neurons sensitive to iontophoretically applied DA have been detected in all layers in the macaque motor cortex (Sawaguchi et al., 1986) and DA may have a modulatory role on neuronal activities involved in motor performances. In motor disorders such as Parkinsonian syndromes or tardive dyskinesias, DA depletion in the cortex could play a role in addition to the well-known nigrostriatal deficits. Indeed, a reduction of DA in the motor cortices was recently shown in Parkinson's disease (Gaspar et al., 1991) and alterations of movement-related activities are observed in the cortex of MPTP-treated monkeys (Doudet et al., 1990). Since single DA neurons can innervate both the motor and prefrontal cortex in owl monkeys (Gaspar et al., 1992), it is possible that these neurons are important in modulating simultaneously distant territories or in synchronizing their activities, thereby regulating sensory-motor integration. The inhibitory potency of DA on excitatory responses to various internal or external stimuli that is demonstrated in rodent prefrontal cortex (Mantz et al., 1988) could also hold true in the somatosensory and association areas in primates. Interestingly, DA agonist therapy has been shown to improve sensory neglect in human patients (Fleet et al., 1987). Besides its role in cognitive and motor functions, cortical DA has also been implicated in a broad set of functions related to the limbic system, underlying emotional states and motivational behaviors. The mesoprefrontal DA networks are activated in stressful situations (Lavielle et al., 1978; Kaneyuki et al., 1991); they are involved in self-stimulation behavior and reward and in self-administration of stimulant drugs (LeMoal & Simon, 1991; Koob, Maldonado & Stinus, 1992). On the other hand, the effects of the DA input are target-dependent. For instance, the DA projections to the anteromedial and rhinal part of the frontal cortex exert an inhibitory action on DA input in the nucleus accumbens, whereas mesoseptal DA projections have an opposite, facilitatory effect (Louilot et al., 1989). Lesions of DA terminals in the amygdala produce opposite changes in DA activity in prefrontal cortex and nucleus accumbens (Simon et al., 1988). In fact, a new conceptual framework tends to emerge, that of a functional interdependence as a general property of the DA mesencephalic groups and target areas, contributing to a coordinate modulation of different regions and integrative processes (Le Moal & Simon, 1991).

Cortical NA systems

The general notion is that the ascending NA systems are activated by sensory stimuli, particularly if they are novel or stressful, and that the result of this activation is an enhancement of the sensory signal, caused by the reduction of background (noise) activity (Foote & Morrison, 1987). The observations of Mantz et al. (1988) and of Sawaguchi et al. (1990) are interesting in this respect because they show the different effects of DA and NA on similar targets in the frontal cortex, and in the same physiological setting. In rats, as aforementioned, the *in vivo* electrical stimulation of the cortical noradrenergic pathway produces a depression of spontaneous cortical activity while enhancing the evoked excitatory responses, whereas the latter are completely suppressed by stimulation of the VTA. In contrast, iontophoretical application of NA and DA into the prefrontal cortex during a delayed response task in awake monkeys (Sawaguchi et al., 1990) produced a decrease and increase, respectively, of the neuronal activity during the different sequences of the task. At another behavioral level, adrenoceptor antagonists influence behaviour induced by DA agonists, and lesions of NA fibers reverse the locomotor deficits caused by a destruction of DA ascending neurons in rats (Taghzouti et al., 1988; Trovero et al., 1992).

Closely related to its role in attentional mechanisms, the NA coeruleo-cortical system may be of crucial importance in learning; the issues and behavioral studies related to this aspect have been reviewed by Robbins et al. (1985). In primates, administration of adrenergic agonists can improve DA-linked cognitive deficits in aged animals (Arnsten & Goldman-Rakic, 1990). This may be related to the aforementioned facilitatory effect of NA on the induction of long-lasting potentiation or depression in the hippocampus (Bröcher et al., 1992).

There have been many studies implicating NA in development and plasticity. The complexity of the available data makes it unclear as to what precisely are the effects of NA innervation. On the one hand, there are several examples where NA cortical innervation has been shown to promote plasticity in the neocortex: plasticity of ocular dominance columns in the visual cortex (Bear & Singer, 1986; Imamura & Kasamatsu, 1991), plastic rearrangements in the rat somatosensory cortex after partial sensory deprivation (Levin, Craik & Hand, 1988) and recovery of function after cortical damage (Boyeson, Callister & Cavazos, 1992; Kolb & Sutherland, 1992) are prevented or modified after NA depletion. On the other hand, NA could promote the normal regressive events that occur during cortical development, such as axonal pruning and synapse elimination (Blue & Parnavelas, 1982; Soto-Moyano et al., 1991). Indeed, *in vitro* studies suggest inhibitory effects of NA on neurite outgrowth (Lipton & Kater, 1989).

The role of NA dysfunctions in human diseases has been widely studied, more particularly in depression (for review see Gold, Goodwin & Chrousos, 1988) and in Parkinson's disease. In Parkinson's disease, it has been found that cortical NA innervation is severely depleted (Hornyckiewicz, 1982; Scatton et al., 1983; Gaspar et al., 1991), particularly in the motor areas, and that this denervation may be important in sensorimotor integration (Gaspar et al., 1991).

Indeed, the correction of this deficit seems to be important to correct postural and gait disorders in parkinsonian patients (Tohgi, Abe & Takahashi, 1993).

Phylogenetic trends and genetic differences of DA and NA neurons

Discussing phylogenetic trends in the catecholaminergic cortical projections with data available only in rats, monkeys and humans can only be a speculative enterprise. Clearly, more species should be analyzed both in the orders of primates and rodents, and in other orders. Given this restriction, one may retain four major characteristics of the catecholaminergic innervation, that are subject to modifications and adjustments in phylogeny: regional distribution within specific cortical areas, laminar distribution of aminergic fibers and the corollary distribution of target cells, colocalization with peptides, and extent of collateralization.

On a regional basis the DA cortical innervation is considerably expanded in primates relative to rodents and the NA innervation, already diffuse in rodents, becomes more differentiated in primates. These trends would be consistent with the notion that the number of cortical areas increases in higher primates, with an increasing functional and cytoarchitectonic differentiation between these areas. For example, primary motor cortex (M1) exists both in rats and primates, but a number of secondary motor areas and of prefrontal areas, identified in macaques, might be absent in rodents (Donoghue & Wise, 1982; Goldberg, 1985; Preuss & Goldman-Rakic, 1991; Uylings & Van Eden, 1990). Interestingly with respect to the DA innervation patterns that we observed, the supplementary motor area of primates was previously suggested to be phylogenetically related to the anterior cingulate cortex (Sanides, 1964, Goldberg 1985). We found DA innervation of the supplementary motor area of monkeys (Berger et al., 1988) to be organized very similarly to the DA input of the anterior cingulate cortex in rats, bringing further support to this notion. On the other hand, M1 in rats receives little DA fibers, whereas M1 in monkeys and humans has dense DA inputs. As regards the prefrontal cortex, DA innervation may be similarly distributed in the possibly homologous prelimbic and infralimbic areas of rodents, and in areas 32 and 25 of primates, whereas the granular lateral prefrontal cortex of primates (areas 9,10,46) would not exist in rodents. Thus, the regional distribution of both DA and NA in primates suggest that these cortical amines have different, possibly more specific targets in the cortex of primates than in rodents.

The laminar distribution of DA and NA also changes from rodents to primates, particularly in the upper cortical layers I-II. Indeed, NA terminal fibers are denser in layer I of rodents, but least dense in this layer in primates. Conversely, DA terminal fibers are scattered in layer I of rodents except in the anterior cingulate cortex, but densest in layers I-II in primates. One could suggest that there is a modification in the structure and integrative potential of this layer between rodents and primates. Modifications could affect a variety of extrinsic cortical and subcortical inputs in addition to the aminergic ones. In monkeys, the extended amygdalo-cortical projections, the 'non-specific' thalamo-cortical afferents (issued from the paracentral and midline thalamic nuclei) and the cortico-cortical connections of the feedback type (projections from higher order integrative cortical zones back to primary cortical fields), all terminate in layer I (Amaral & Price 1984; for review see Goldman-Rakic, 1987b). These three afferent systems increase in primates, in proportion with the development of association cortices. One might speculate that DA terminal innervation might increase in proportion with an increase of a particular class of target cells, e.g the cortico-cortically projecting neurons. However, subtle differences might be present in the laminar distribution of cortical aminergic afferents inside the primate order, between New and Old World monkeys for instance. Thus, the laminar complementarity between serotonin and NA innervation which is a striking feature in the squirrel monkey visual cortex is no longer apparent in cynomolgus monkey (Kosofsky et al., 1984).

We have attempted to establish a link between observations in rodents and primates in the perspective of a dual DA mesocortical system. In this view, the innervation of the deep cortical layers would be sustained in rats and primates, whereas the DA system innervating the superficial cortical layers, restricted to a few cortical regions in rats, would be considerably developed in primates. As discussed above however, the evidence for the existence of two mesocortical systems in primates is fragmentary. Interestingly, opossums, tree shrews, and possibly galagos, which are presumed to resemble ancestors of mammals and primates, respectively (Divac et al., 1978), might have an organization more similar to that of rodents, as no superficial DA system was found in these species. Some parallels may be drawn between the proposed dual mesocortical system and the organization of the DA mesostriatal system, comprising two main classes of afferents. The first to develop during embryonic life does not contain calbindin D-28K and projects to the striatal patches whereas the later developing contingent is colocalized with the calcium-binding protein and reaches the matrix compartment (Gerfen et al., 1987). A similar segregation of the corticostriatal projections to matrix and patches from cortical layers Va and Vb was shown (Gerfen 1992). It was proposed by this author that, as neocortical areas are added in primates, the 'limbic-related' striosomal system persists but the 'neocortical-related' matricial system becomes predominant. In this scheme, the superficial mesocortical components would be more related to neocortical development. As announced in the foreword, this scheme is at the moment purely speculative.

Collateralization, as already discussed, is more important in primates than in rodents since the cortical surface, and particularly the neocortical surface, increases at a much higher rate than DA mesencephalic neurons, or NA neurons in the locus coeruleus.

Colocalization of DA or NA with peptides seems to be subject to phylogenetic changes. Cholecystokinin and neurotensin are present in mesocortical DA projections and galanin and neuropeptide Y in coeruleocortical projections in rats, while these peptides could be absent or their expression differently regulated in the aminergic neurons of primates. These observations support the notion that extensive co-localization of neurotransmitters could be a primitive trait and that with evolution, subpopulations of neurons tend to specialize as far as neurotransmitter content is concerned (Hökfelt et al., 1987).

An interesting notion in phylogenetic analysis, is that of genetic variability within a given species, as this could be underlaid by genetic loci that would be the object of selection in evolution. There are a number of examples of such variability in mice with mutations affecting the number of neurons in the locus coeruleus and the ventral mesencephalon. The quaking mice for example are characterized by an increased number of locus coeruleus cells relative to wild type mice C57 BL6 (Maurin et al., 1985); interestingly, the 50% increase is precisely localized in the midportion of LC nucleus, suggesting that different loci may regulate the number of aminergic neurons within a single cell group. The C57 BL6 strain has 36% more neurons in the LC than the Balb/c strain (Berger et al., 1979) while in the latter, 50% more TH-containing neurons were observed in the mesencephalic DA A9-A10 groups than in the CBA/J strain (Ross et al., 1976). Variations between inbred rat strains have also been described with differences in the reactivity of DA mesolimbic projections to stress (LeMoal & Simon, 1991) or variations in the vulnerability to neurotoxins. It is not known whether such genetic variations in rats correspond or not to subtle differences in the anatomic organization of the system or to differences in the metabolic equipment.

Acknowledgements

We thank Chantal Alvarez and Aude Febvret for skilful technical assistance. Parts of the work presented here were realized in Yale University School of Medicine, Section of Neurobiology (Pr. P. S. Goldman-Rakic) and in Nashville, Vanderbilt University, (Pr. J. H. Kaas' laboratory). The support and technical assistance provided in these laboratories is acknowledged.

Abbreviations

acc	nucleus accumbens
AGm	medial agranular field
AGl	lateral agranular field
AId	dorsal agranular insula
AIv	ventral agranular insula
c	caudate nucleus
CaBP	calbindin D-28K
CCK	cholecystokinin
Cg1,2,3	cingulate cortex, areas 1,2,3
CP	caudatus putamen
DBH	dopamine-β-hydroxylase
ent	entorhinal cortex;
Fr1,2,3	frontal cortex, areas 1,2,3
FL	forelimb area
hip	hippocampus;
IL	infralimbic cortex
ins	insula
itg	inferior temporal gyrus
MO	medial ventral orbital area
NT	neurotensin
occ	occipital cortex
p	putamen
par	parietal cortex
PL	prelimbic cortex
po	piriform cortex
PrCm	medial precentral area
PrCl	lateral precentral area
prh	perirhinal cortex

rs   retrosplenial cortex
sn   substantia nigra
stg   superior temporal gyrus
sub   subiculum
tem   temporal cortex

TH   tyrosine hydroxylase
tp   temporal pole
VO   ventral orbital area
vta   ventral tegmental area.

References

Akil, M. & Lewis, D. A. (1993). The dopaminergic innervation of monkey entorhinal cortex. *Cerebral Cortex*, **3**, 533–50.

Amaral, D. & Campbell, M. J. (1986). Transmitter systems in the primate dentate gyrus. *Human Neurobiology*, **5**, 169–80.

Amaral, D. G. & Price, J. L. (1984). Amygdalo-cortical projections in the monkey (*Macaca fascicularis*). *Journal of Comparative Neurology*, **230**, 465–496.

Aoki, C. & Pickel, V. M. (1992). Ultrastructural relations between β-adrenergic receptors and catecholaminergic neurons. *Brain Research Bulletin*, **29**, 257–63.

Aoki, C., Joh, T. H. & Pickel, V. M. (1987). Ultrastructural localization of β-adrenergic receptor-like immunoreactivity in the cortex and neostriatum of rat brain. *Brain Research*, **437**, 264–82.

Arnsten, A. F. T. & Goldman-Rakic, P. S. (1990). Analysis of alpha-2 adrenergic agonist effects on the delayed nonmatch-to-sample performance of aged rhesus monkeys. *Neurobiology of Aging*, **11**, 583–90.

Audet, M. A., Doucet, G., Oleskevich, S. & Descarries, L. (1988). Quantified regional and laminar distribution of the noradrenaline innervation in the anterior half of the adult rat cerebral cortex. *Journal of Comparative Neurology*, **274**, 307–18.

Baker, K. G., Törk, I., Hornung, J. P. & Halasz, P. (1989). The human locus coeruleus complex: an immunohistochemical and three dimensional reconstruction study. *Experimental Brain Research*, **77**, 257–70.

Barbas, H., Haswell Henion, T. H. & Dermon, C. R. (1991). Diverse thalamic projections to the prefrontal cortex in the rhesus monkey. *Journal of Comparative Neurology*, **313**, 65–94.

Barnéoud, P., Le Moal, M. & Neveu, P.J. (1990). Asymmetric distribution of brain monoamines in left and right handed mice. *Brain Research*, **520**, 317–21.

Bean, A. J., Dagerlind, A., Hökfelt, T. & Dobner, P. R. (1992). Cloning of human neurotensin/neuromedin N genomic sequences and expression in the ventral mesencephalon of schizophrenics and age/sex matched controls. *Neuroscience*, **50**, 259–69.

Bean, A. J. & Roth, R. H. (1991). Extracellular dopamine and neurotensin in rat prefrontal cortex *in vivo*. Effects of median forebrain bundle stimulation frequency, stimulation pattern and dopamine autoreceptors. *Journal of Neuroscience*, **11**, 2694–702.

Beani, L., Tanganelli, S., Antonelli, T. & Bianchi, C. (1986). Noradrenergic modulation of cortical acetylcholine release is both direct and γ-aminobutyric acid-mediated. *Journal of Pharmacology and Experimental Therapeutics*, **236**, 230–6.

Bear, M.F. & Singer, W. (1986). Modulation of visual cortical plasticity by acetylcholine and noradrenaline. *Nature*, **320**, 172–76.

Berger, B., Alvarez, C. & Goldman-Rakic, P. S, (1994). Neurochemical development of the hippocampal region in the fetal rhesus monkey. II. Immunocytochemisty of peptides, calcium-binding proteins, Darpp-32 and monoamine innervation in the entorhinal cortex by the end of gestation. *Hippocampus*, in press.

Berger, B., Alvarez, C. & Goldman-Rakic, P. S. (1993). Neurochemical development of the hippocampal region in the fetal rhesus monkey. I. Early appearance of peptides, calcium binding proteins, DARPP-32 and the monoamine innervation in the entorhinal cortex during the first half of gestation (E47 to E90). *Hippocampus*, **3**, 279–306.

Berger, B., Escourolle, R. & Moyne, M. A. (1976a). Axones catécholaminergiques du cortex cérébral humain. Observation, en histofluorescence, de biopsies cérébrales dont 2 cas de maladie d'Alzheimer. *Revue Neurologique*, **132**, 183–94.

Berger, B., Febvret, A., Greengard, P. & Goldman-Rakic, P. S. (1990). DARPP-32, a phosphoprotein enriched in dopaminoceptive neurons bearing dopamine D1 receptors: distribution in the cerebral cortex of the newborn and adult rhesus monkey. *Journal of Comparative Neurology*, **299**, 327–48.

Berger, B., Gaspar, P. & Verney, C. (1991). Dopaminergic innervation of the cerebral cortex: unexpected differences between rodents and primates. *Trends in Neurosciences*, **14**, 21–7.

Berger, B., Hervé, D., Dolphin, A., Barthelemy, C., Gay, M. & Tassin, J. P. (1979). Genetically determined differences in noradrenergic input to the brain cortex: a histochemical and biochemical study in two in two inbred strains of mice. *Neuroscience*, **4**, 877–88.

Berger, B., Tassin, J. P., Blanc, G., Moyne, M. A. & Thierry, A. M. (1974). Histochemical confirmation for dopaminergic innervation of the rat cerebral cortex after destruction of the noradrenergic ascending pathways. *Brain Research*, **81**, 332–7.

Berger, B., Thierry, A M., Tassin, J. P. & Moyne, M. A. (1976b). Dopaminergic innervation of the rat prefrontal cortex: a fluorescence histochemical study. *Brain Research*, **106**, 133–45.

Berger, B., Trottier, S., Gaspar, P., Verney, C. & Alvarez, C. (1986). Major dopamine innervation of the cortical motor areas in the Cynomolgus monkey. A radioautographic study with comparative assessment of serotoninergic afferents. *Neuroscience Letters*, **72**, 121–7.

Berger, B., Trottier, S., Verney, C., Gaspar, P. & Alvarez, C. (1988). Regional and laminar distribution of the dopamine and serotonin innervation in macaque cerebral cortex. A radioautographic study. *Journal of Comparative Neurology*, **273**, 99–119.

Berger, B., Verney, C., Alvarez, C., Vigny, A. & Helle, K. B. (1985a). New dopaminergic terminal fields in the motor, visual (area 18b) and retrosplenial cortex in the young and adult rat. Immunocytochemical and catecholamine histochemical analyses. *Neuroscience*, **15**, 983–98.

Berger, B., Verney, C., Febvret, A., Vigny, A. & Helle, K. B. (1985b). Postnatal ontogenesis of the dopaminergic innervation in the rat anterior cingulate cortex (area 24). Immunocytochemical and catecholamine fluorescence histochemical analysis. *Developmental Brain Research*, **21**, 31–47.

Berger, B., Verney, C., Gaspar, P. & Febvret, A. (1985c). Transient expression of tyrosine hydroxylase immunoreactivity in some neurons of the rat neocortex during postnatal development. *Developmental Brain Research*, **23**, 141–4.

Berger, B., Verney, C. & Goldman-Rakic, P. S. (1992). Prenatal monoaminergic innervation of the cerebral cortex: differences between rodents and primates. In *Neurodevelopment, Aging and Cognition*, ed. I. Kostovic, S. Knezevic & G. Spilich, pp. 18–36. Boston: Birkhauser.

Björklund, A., Divac, I. & Lindvall, O. (1978). Regional distribution of catecholamines in monkey cerebral cortex, evidence for a dopaminergic innervation of the primate prefrontal cortex. *Neuroscience Letters*, **7**, 115–19.

Björklund, A. & Lindvall, O. (1984). Dopamine containing systems in the CNS. In *Handbook of Chemical Neuroanatomy Vol.2, Classical transmitters in the CNS, part 1*, ed. A. Björklund & T. Hökfelt, pp. 55–122. Elsevier.

Blackstad, T. W., Fuxe, K. & Hökfelt, T. (1967). Noradrenaline nerve terminals in the hippocampal region of the rat and the guinea pig. *Zeitschrift für Zellforschung*, **78**, 463–73.

Blue, M. E. & Parnavelas, J. G. (1982). The effect of neonatal 6-hydoxydopamine treatment on synaptogenesis in the visual cortex of the rat. *Journal of Comparative Neurology*, **205**, 199–205.

Bouthenet, M. L., Souil, E., Martres, M. P., Sokoloff, P., Giros, B. & Schwartz, J. C. (1991). Localization of dopamine D3 receptor mRNA in the rat brain using *in situ* hybridization histochemistry: comparison with dopamine D2 receptor mRNA. *Brain Research*, **564**, 203–19.

Bowden, D. M., German, D. C. & Poynter, W. D. (1978). An autoradiographic, semi-stereotaxic mapping of major projections from locus coeruleus and adjacent nuclei in *Macaca mulatta*. *Brain Research*, **145**, 257–76.

Boyeson, M. G., Callister, T. R. & Cavazos, J. E. (1992). Biochemical and behavioral effects of a sensorimotor cortex injury in rats pretreated with the noradrenergic neurotoxin DSP-4. *Behavioral Neuroscience*, **106**, 964–73.

Bröcher, S., Artola, A. & Singer, W. (1992). Agonists of cholinergic and noradrenergic receptors facilitate synergistically the induction of long-term potentiation in slices of rat visual cortex. *Brain Research*, **573**, 27–36.

Brown, R. M., Crane, A. M. & Goldman, P. S. (1979). Regional distribution of monoamines in the cerebral cortex and subcortical structures of the rhesus monkey: concentrations and *in vivo* synthesis rates. *Brain Research*, **168**, 133–50.

Brozoski, T. J., Brown, R. M., Rosvold, H. E. & Goldman, P. S. (1979). Cognitive deficit caused by regional depletion of dopamine in prefrontal cortex of rhesus monkey. *Science*, **205**, 929–32.

Campbell, M. J., Lewis, D. A., Foote, S. L. & Morrison, J. H. (1987). Distribution of choline acetyltransferase, serotonin, dopamine-β-hydroxylase, tyrosine hydroxylase immunoreactive fibers in monkey primary auditory cortex. *Journal of Comparative Neurology*, **261**, 209–21.

Cash, R., Raisman, R., Lanfumey, L., Ploska, A. & Agid, Y. (1986). Cellular localization of adrenergic receptors in rat and human brain. *Brain Research*, **370**, 127–35.

Cepeda, C., Radisavljevic, Z., Peacock, W., Levine, M. S. & Buchwald, N. A. (1992). Differential modulation by dopamine of responses evoked by excitatory aminoacids in human cortex. *Synapse*, **11**, 330–41.

Chan Palay, V. & Asan, E. (1989). Quantitation of catecholamine neurons in the locus coeruleus in human brains of normal young and older adults and in depression. *Journal of Comparative Neurology*, **287**, 357–72.

Chen, J. P., Paredes, W., Van Praag, H. M., Lowinson, J. H. & Gardner, E. L. (1992). Presynaptic dopamine release is enhanced by 5-HT3 receptor activation in medial prefrontal cortex of freely moving rats. *Synapse*, **10**, 264–6.

Chiodo, L. A. (1988). Dopamine-containing neurons in the mammalian central nervous system: electrophysiology and pharmacology. *Neuroscience Biobehavioral Review*, **12**, 49–91.

Civelli, O., Bunzow, J. R., Grandy, D. K., Zhou, Q. Y. & Van Tol, H. H. M. (1991). Molecular biology of the dopamine receptors. *European Journal of Pharmacology*, **207**, 277–86.

Deniau, J. M., Thierry, A. M. & Feger, J. (1980). Electrophysiological identification of mesencephalic ventromedial tegmental (VMT) neurons projecting to the frontal cortex, septum and nucleus accumbens. *Brain Research*, **189**, 315–26.

De Quidt, M. E. & Emson, P. C. (1983). Neurotensin facilitates dopamine release in vitro from rat striatal slices. *Brain Research*, **274**, 376–80.

Descarries, L., Lemay, B., Doucet, G. & Berger, B. (1987). Regional and laminar density of the dopamine innervation in adult rat cerebral cortex. *Neuroscience*, **21**, 807–24.

Descarries, L., Watkins, L. C. & Lapierre, Y. (1977). Noradrenergic axon terminals in the cerebral cortex of rat. III. Topometric ultrastructural analysis. *Brain Research*, **133**, 197–222.

Deutch, A. Y., Goldstein, M., Baldino, F. & Roth, R. H. (1988). Telencephalic projections of the A8 dopamine cell group. In Mesocorticolimbic dopamine system. *Annals of the New York Academy of Science*, **537**, 27–50.

Divac, I., Björklund, A., Lindvall, O. & Passingham, R. E. (1978). Converging projections from the mediodorsal thalamic nucleus and mesencephalic dopaminergic neurons to the neocortex in three species. *Journal of Comparative Neurology*, **180**, 59–72.

Donoghue, J. P. & Wise, S. P. (1982). The motor cortex of the rat: cytoarchitecture and microstimulation mapping. *Journal of Comparative Neurology*, **212**, 76–88.

Doudet, D. J., Gross, C., Arluison, M. & Bioulac, B. (1990). Modifications of precentral cortex discharge and EMG activity in monkeys with MPTP-induced lesions of dopaminergic nigral neurons. *Experimental Brain Research*, **80**, 177–88.

Dubach, M., Schmidt, R., Kunkel, D., Bowden, D. M., Marti, R. & German, D.C.(1987). Primate neostriatal neurons containing tyrosine hydroxylase: immunohistochemical evidence. *Neuroscience Letters*, **75**, 205–10.

Emson, P.C. & Koob, G. F. (1978). The origin and distribution of dopamine containing afferents to the rat frontal cortex. *Brain Research*, **142**, 249–68.

Fallon, J. H. & Loughlin, S. E. (1987). Monoamine innervation of cerebral cortex and a theory of the role of monoamines in cerebral cortex and basal ganglia. In *Cerebral Cortex*, vol. 6, ed. E. G. Jones & A. Peters, pp.41–127. New York: Plenum Press.

Febvret, A., Berger, B., Gaspar, P. & Verney, C. (1991). Further indication that distinct dopaminergic subsets project to the rat cerebral cortex: lack of colocalization with neurotensin in the superficial dopaminergic fields of the anterior cingulate, motor, retrosplenial and visual cortices. *Brain Research*, **547**, 37–52.

Fink, K. (1990). Presynaptic NMDA receptors stimulate noradrenergic release in the cerebral cortex. *European Journal of Pharmacology*, **185**, 115–18.

Fleet, W. S., Valenstein, E., Watson, R. T. & Heilman, K. M. (1987). Dopamine agonist therapy for neglect in humans. *Neurology*, **37**, 1765–70.

Foote, S. L., Bloom, F. E. & Aston-Jones, G. (1983). Nucleus locus ceruleus: new evidence of anatomical and physiological specificity. *Physiological Review*, **63**, 844–914.

Foote, S. L. & Morrison, J. H. (1987). Extrathalamic modulation of cortical function. *Annual Review of Neuroscience*, **10**, 67–96.

Forloni, G., Grzanna, R. Blakely, R. D. & Coyle, J. T. (1987). Co-localization of N-acetyl-aspartyl-glutamate in central cholinergic, noradrenergic, and serotonergic neurons. *Synapse*, **1**, 455–60.

Freedman, R., Foote, S. L. & Bloom, F. E. (1975). Histochemical characterization of a neocortical projection of the nucleus locus coeruleus in the squirrel monkey. *Journal of Comparative Neurology*, **164**, 209–32.

Fritschy, J. M. & Grzanna, R. (1989). Immunohistochemical analysis of the neurotoxic effects of DSP-4 identifies two populations of noradrenergic axon terminals. *Neuroscience*, **30**, 181–97.

Fukamauchi, F., Yoshikawa, T. & Shibuya, H. (1992). Presynaptic dopaminergic control mechanisms for CCK-8 like immunoreactivity in the rat medial frontal cortex. *Neuropeptides*, **23**, 55–60.

Fuxe, K., Hamberger, B. & Hökfelt, T. (1968). Distribution of noradrenaline nerve terminals in cortical areas of the rat. *Brain Research*, **8**, 125–31.

Gaspar, P., Berger, B. & Febvret, A. (1990). Neurotensin innervation of the human cerebral cortex: lack of colocalization with catecholamines. *Brain Research*, **530**, 181–95.

Gaspar, P., Berger, B., Febvret, A., Vigny, A. & Henry, J. P. (1989). Catecholamine innervation of the human cerebral cortex as revealed by comparative immunohistochemistry of tyrosine hydroxylase and dopamine-β-hydroxylase. *Journal of Comparative Neurology*, **279**, 249–71.

Gaspar, P., Berger, B., Febvret, A., Vigny, A., Krieger-Poulet, M. & Borri-Voltattorni, C. (1987). Tyrosine hydroxylase immunoreactive neurons in the human cerebral cortex: a novel catecholaminergic group? *Neuroscience Letters*, **80**, 257–62.

Gaspar, P., Duyckaerts, C., Alvarez, C., Javoy-Agid, F. & Berger, B. (1991). Alterations of dopaminergic and noradrenergic innervations in motor cortex in Parkinson's disease. *Annals of Neurology*, **30**, 365–74.

Gaspar, P., Heizmann, C. W. & Kaas, J. H. (1993). Calbindin D-28K in the dopaminergic mesocortical projection of a monkey (*Aotus trivirgatus*). *Brain Research*, **603**, 166–72.

Gaspar, P., Stepniewska, I. & Kaas, J. (1992). Topography and collateralization of the dopaminergic projections to motor and lateral prefrontal cortex in Owl monkeys. *Journal of Comparative Neurology*, **325**, 1–21.

Gatter, K. C. & Powell, T. P. S. (1977). The projection of the locus coeruleus upon the neocortex in the macaque monkey. *Neuroscience*, **2**, 441–5.

Gerfen, C. R. (1992). The neostriatal mosaic: multiple levels of compartmental organization. *Trends in Neurosciences*, **15**, 133–9.

Gerfen, C. R., Bainbridge, K. G. & Thibault, J. (1987). The neostriatal mosaic: III. Biochemical and developmental dissociation of patch-matrix mesostriatal systems. *Journal of Neuroscience*, **7**, 3935–44.

German, D. C., Manaye, K., Smith, W. K., Woodward, D. J. & Saper, C. B. (1989). Midbrain dopaminergic cell loss in Parkinson's disease: computer visualization. *Annals of Neurology*, **26**, 507–14.

German, D. C., Walker, B. S., Manaye, K., Smith, W. K., Woodward, D. J. & North A. J. (1988). The human locus coeruleus: computer reconstruction of cellular distribution. *Journal of Neuroscience*, **8**, 1776–88.

Gold, P. W., Goodwin, F. K. & Chrousos, G. P. (1988). Clinical and biochemical manifestations of depression. *New England Journal of Medicine*, **319**, 348–53.

Goldberg, G. (1985). Supplementary motor area structure

and function: review and hypotheses. *Behavioral Brain Science*, **8**, 567–87.

Goldman-Rakic, P. S. (1987a). Development of cortical circuitry and cognitive function. *Child Development*, **58**, 601–22.

Goldman-Rakic, P. S. (1987b). Circuitry of primate prefrontal cortex and regulation of behavior by representational memory. In *Handbook of Physiology*, vol. 5, ed. F. Plum & V. Mountcastle, pp. 373–417. Bethesda: MD. American Physiol. Soc.

Goldman-Rakic, P. S. (1988). Changing concepts of cortical connectivity: parallel distributed cortical networks. In *Neurobiology of Neocortex*, ed. P. Rakic & W. Singer, pp. 177–202. John Wiley & Sons.

Goldman-Rakic, P. S., Leranth, C., Williams, S. M., Mons, N. & Geffard, M. (1989). Dopamine synaptic complex with pyramidal neurons in primate cerebral cortex. *Proceedings of the National Academy of Science, USA*, **86**, 9015–19.

Goldman-Rakic, P. S., Lidow, M. S. & Gallager, D. W. (1990). Overlap of dopaminergic, adrenergic and serotoninergic receptors and complementarity of their subtypes in primate prefrontal cortex. *Journal of Neuroscience*, **10**, 2125–39.

Goldman-Rakic, P. S. & Porrino, L. J. (1985). The primate mediodorsal (MD) nucleus and its projections to the frontal lobe. *Journal of Comparative Neurology*, **242**, 535–60.

Guennoun, R. & Bloch, B. (1992). Ontogeny of D1 and DARPP-32 gene expression in the rat striatum: an in situ hybridization study. *Molecular Brain Research*, **12**, 131–9.

Gustafson, E. L., Ehrlich, M. E., Trivedi, P. & Greengard, P. (1992). Developmental regulation of phosphoprotein gene expression in the caudate-putamen of rat: an *in situ* hybridization study. *Neuroscience*, **51**, 65–75.

Halliday, G. M. & Törk, I. (1986). Comparative anatomy of the ventromedial mesencephalic tegmentum in the rat, cat, monkey and human. *Journal of Comparative Neurology*, **252**, 423–45.

Harrison, J. K., Pearson, W. R., and Lynch, K. R. (1991). Molecular characterization of α1- and α2-adrenoceptors. *Trends in Pharmacological Science*, **12**, 62–7.

Hata, N., Nishikawa, T., Umino, A. & Takahashi, K. (1990). Evidence for involvement of N-methyl-D-aspartate receptor in tonic inhibitory control of dopaminergic transmission in rat medial frontal cortex. *Neuroscience Letters*, **120**, 101–4.

Hedreen, J. C. (1991). Organization of striatopallidal, striatonigral, and nigrostriatal projections in the macaque. *Journal of Comparative Neurology*, **304**, 569–95.

Hemmings, H. C., Walaas, S. I., Ouimet, C. C. & Greengard, P. (1987). Dopaminergic regulation of protein phosphorylation in the striatum: DARPP 32. *Trends in Neurosciences*, **10**, 377–83.

Hervé, D., Trovero, F., Blanc, G., Vezina, P., Glowinski, J. & Tassin, J. P. (1990). Involvement of dopamine neurons in the regulation of β-adrenergic receptor sensitivity in rat prefrontal cortex. *Journal of Neurochemistry*, **54**, 1864–9.

Hirsch, E., Graybiel, A. M. & Agid, Y. (1988). Melanized dopaminergic neurons are differentially susceptible to degeneration in Parkinson's disease. *Nature*, **334**, 345–8.

Hofman, M. A. (1982). Encephalization in mammals in relation to the size of the cerebral cortex. *Brain Behavior and Evolution*, **20**, 84–96.

Hofman, M. A. (1985). Size and shape of the cerebral cortex in mammals. 1. The cortical surface. *Brain Behavior and Evolution*, **27**, 28–40.

Hökfelt, T., Everitt, B. J., Theodorsson-Norheim, E. & Goldstein, M. (1984b). Occurrence of neurotensin-like immunoreactivity in subpopulations of hypothalamic, mesencephalic and medullary catecholamine neurons. *Journal of Comparative Neurology*, **222**, 543–59.

Hökfelt, T., Johansson, O. & Goldstein, M. (1984a). Central catecholamine neurons as revealed by immunohistochemistry with special reference to adrenaline neurons. In *Handbook of Chemical Neuroanatomy*, vol. 2, part I, ed. A. Björklund & T. Hökfelt. pp. 157–176. Elsevier.

Hökfelt, T., Millhorn, D., Seroogy, K., Tsuruo, Y., Ceccatelli, S., Lindh, B., Meister, B., Melander, T., Schalling, M., Bartfai, T. & Terenius, L. (1987). Coexistence of peptides with classical neurotransmitters. *Experientia*, **43**, 768–80.

Hökfelt, T., Skirboll, L., Rehfeld, J. F., Goldstein, M., Markey, K. & Dann, O. (1980). A subpopulation of mesencephalic dopamine neurons projecting to limbic areas contains a cholecystokinin-like peptide: evidence from immunohistochemistry combined with retrograde tracing. *Neuroscience*, **5**, 2093–124.

Holets, V. R., Hökfelt, T., Rokaeus, A., Terenius, L. & Goldstein, M. (1988). Locus coeruleus neurons in the rat containing neuropeptide Y, tyrosine hydroxylase or galanin and their efferent projections to the spinal cord, cerebral cortex and hypothalamus. *Neuroscience*, **24**, 893–906.

Hornung, J. P., Törk, I. & Detribolet, N. (1989). Morphology of tyrosine hydroxylase immunoreactive neurons in the human cerebral cortex. *Experimental Brain Research*, **76**, 12–20.

Hornykiewicz, O. (1982). Brain neurotransmitter changes in Parkinson's disease. In *Movement Disorders*, ed C. D. Marsden & S. Fahn, pp 41–58. London: Butterworth

Huang, Q., Zhou, D., Chase, K., Gusella, J. F. & Aronin, N. (1992). Immunohistochemical localization of the D1 dopamine receptor in rat brain reveals its axonal transport, pre- and postsynaptic localization and prevalence in the basal ganglia, limbic system and thalamic reticular nucleus. *Proceedings of the National Academy of Sciences, USA*, **89**, 11988–92.

Huntley, G. W., Morrison, J. H., Prikhozhan, A. & Sealfon, S. C. (1992). Localization of multiple dopamine receptor subtype messenger RNAs in human and monkey motor cortex and striatum. *Molecular Brain Research*, **15**, 181–8.

Iacovitti, L., Lee, J., Joh, T. H. & Reis, D. J. (1987). Expression of tyrosine hydroxylase in neurons of cultured cerebral cortex: evidence for phenotypic plasticity in neurons of the CNS. *Journal of Neuroscience*, **74**, 1264–70.

Iijima, K., Sato, M., Kojima, N. & Ohtomo, K. (1992). Immunocytochemical and in situ hybridization evidence for the coexistence of gaba and tyrosine hydroxylase in the rat locus coeruleus. *Anatomical Record*, **234**, 593–604.

Imamura, K. & Kasamatsu, T. (1991). Ocular dominance plasticity restored by NA infusion to aplastic visual cortex of anesthetized and paralyzed kittens. *Experimental Brain Research*, **87**, 309–18.

Jaffé, E. H. & Hernandez, N. (1989). Release of (3H) dopamine from rat prefrontal cortex: modulation through presynaptic cholinergic heteroreceptors. *Neuroscience Letters*, **105**, 189–94.

Javoy-Agid, F., Scatton, B., Ruberg, M., L'Heureux, R., Cervera, P., Raisman, R., Maloteaux, J. M., Beck, H. & Agid, Y. (1989). Distribution of monoaminergic, cholinergic, and gabaergic markers in the human cerebral cortex. *Neuroscience*, **29**, 251–9.

Jones, E. G. (1981). Anatomy of cerebral cortex: columnar input-output organization. In *The Organization of the Cerebral Cortex*, ed. F.O. Schmitt, F. G. Worden, G. Adelman & S. G. Dennis, pp. 199–235. Cambridge, Mass: The MIT Press.

Kaas, J. H. (1987). The organization of neocortex in mammals: implications for theories of brain function. *Annual Review of Psychology*, **38**, 129–52.

Kalsbeek, A., Voorn, P., Buijs, R. M., Pool, C. W. & Uylings, H. B. M. (1988). Development of the dopaminergic innervation in the prefrontal cortex of the rat. *Journal of Comparative Neurology*, **269**, 58–72.

Kaneyuki, H., Yokoo, H., Tsuda, A., Yoshida, M., Mizuki, Y., Yamada, M. & Tanaka, M. (1991). Psychological stress increases dopamine turnover selectively in mesoprefrontal dopamine neurons of rats: reversal by diazepam. *Brain Research*, **557**, 154–61.

Killackey, H. P. (1990). Neocortical expansion: an attempt toward relating phylogeny and ontogeny. *Journal of Cognitive Neuroscience*, **2**, 1–17.

Killackey, H. P. & Chalupa, L. M. (1986). Ontogenetic change in the distribution of callosal projections neurons in the postcentral gyrus of the fetal rhesus monkey. *Journal of Comparative Neurology*, **244**, 331–48.

Köhler, C., Ericson, H., Hogberg, T., Halldin, C. & Chan-Palay, V. (1991a). Dopamine-D2 receptors in the rat, monkey and the postmortem human hippocampus. An autoradiographic study using the novel D2-selective ligand 1–125-NCQ-298. *Neuroscience Letters*, **125**, 12–14.

Köhler, C., Ericson, H. & Radesater, A. C. (1991b). Different laminar distributions of dopamine-D1 and dopamine-D2 receptors in the rat hippocampal region, *Neuroscience Letters*, **126**, 107–9.

Köhler, C., Everitt, B. J., Pearson, J. & Goldstein, M. (1983). Immunohistochemical evidence for a new group of catecholamine containing neurons in the basal forebrain of the monkey. *Neuroscience Letters*, **37**, 161–6.

Kolb, B. & Sutherland, R. J. (1992). Noradrenaline depletion blocks behavioral sparing and alters cortical morphogenesis after neonatal frontal cortex damage in rats. *Journal of Neuroscience*, **12**, 2321–30.

Koob, G. F., Maldonado, R. & Stinus, L. (1992). Neural substrates of opiate withdrawal. *Trends in Neurosciences*, **15**, 186–78.

Kosofsky, B. E., Molliver, M.E., Morrison, J. H. & Foote, S. L. (1984). The serotonin and norepinephrine innervation of primary visual cortex in the cynomolgus monkey (*Macaca fascicularis*). *Journal of Comparative Neurology*, **230**, 168–78.

Kostovic, I. & Rakic, P. (1990). Developmental history of the subplate transient zone in the visual and somatosensory cortex of the macaque monkey and human brain. *Journal of Comparative Neurology*, **297**, 441–70.

Krettek, J. E. & Price, J. L. (1977). The cortical projections of the mediodorsal nucleus and adjacent thalamic nuclei in the rat. *Journal of Comparative Neurology*, **17**, 157–92.

Kuljis, R. O., Martin-Vasalo, P. & Peress, N. S. (1989). Lewy bodies in tyrosine hydroxylase-synthesizing neurons of the human cerebral cortex. *Neuroscience Letters*, **106**, 49–54.

Lavielle, S., Tassin, J. P., Thierry, A. M., Blanc, G., Hervé, D., Barthelemy, C. & Glowinski, J. (1978). Blockade by benzodiazepines of the selective high increase in dopamine turnover induced by stress in mesocortical dopaminergic neurons of the rat. *Brain Research*, **168**, 585–94.

Le Moal, M. & Simon, H. (1991). Mesocorticolimbic dopaminergic network. Functional and regulatory roles. *Physiological Review*, **71**, 155–234.

Leonard, C. M. (1969). The prefrontal cortex of the rat. 1. Cortical projection of the mediodorsal nucleus. 2. Efferent connections. *Brain Research*, **12**, 321–43.

Levin, B. E., Craik, R. L. & Hand, P. J. (1988). The role of norepinephrine in adult rat somatosensory (SMI) cortical metabolism and plasticity. *Brain Research*, **443**, 261–71.

Levitt, P. & Moore, R. Y. (1978). Noradrenaline neuron innervation of the neocortex in the rat. *Brain Research*, **139**, 219–32.

Levitt, P., Rakic, P. & Goldman-Rakic, P. S. (1984). Region-specific distribution of catecholamine afferents in primate cerebral cortex: a fluorescence histochemical analysis. *Journal of Comparative Neurology*, **227**, 23–36.

Lewis, D. A., Campbell, M. J., Foote, S. L., Goldstein, M. & Morrison, J. H. (1987). The distribution of tyrosine hydroxylase immunoreactive fibers in primate neocortex is widespread but regionally specific. *Journal of Neuroscience*, **7**, 279–90.

Lewis, D. A., Foote, S. L., Goldstein, M. & Morrison, J. H. (1988a). The dopaminergic innervation of monkey prefrontal cortex: a tyrosine hydroxylase immunohistochemical study. *Brain Research*, **449**, 225–43.

Lewis, D. A., Melchitzky, D. S., Gioio, A., Solomon, Z. & Kaplan, B. B. (1991). Neuronal localization of tyrosine hydroxylase gene products in human neocortex. *Molecular and Cellular Neuroscience*, **2**, 228–34.

Lewis, M. S., Molliver, M. E., Morrison, J. H. & Lidov, H. G. W. (1979). Complementarity of dopaminergic and noradrenergic innervation in anterior cingulate cortex of the rat, *Brain Research*, **164**, 328–33.

Lewis, D. A. & Morrison, J. H. (1989). Noradrenergic innervation of monkey prefrontal cortex: a dopamine-β-hydroxylase immunohistochemical study. *Journal of Comparative Neurology*, **282**, 317–30.

Lewis, D. A., Morrison, J. H. & Goldstein, M. (1988b). Brainstem dopaminergic neurons project to monkey parietal cortex. *Neuroscience Letters*, **86**, 11–16.

Lidov, H. G. W., Molliver, M. E. & Zecevic, N. R. (1978a). Characterization of the monoaminergic innervation of immature rat neocortex: a histofluorescence analysis. *Journal of Comparative Neurology*, **181**, 663–80.

Lidov, H. G. W., Rice, F. L. & Molliver, M. E. (1978b). The organization of the catecholamine innervation of somatosensory cortex: the barrel field of the mouse. *Brain Research*, **153**, 577–84.

Lidow, M. S., Goldman-Rakic, P. S., Gallager, D. W., Geschwind, D. H. & Rakic, P. (1989). Distribution of major neurotransmitter receptors in the motor and somatosensory cortex of the rhesus monkey. *Neuroscience*, **32**, 609–27.

Lindvall, O. & Björklund, A. (1978). Organization of catecholamine neurons in the rat nervous system. In *Handbook of Psychopharmacology*, vol. 9, ed. L.L. Iversen, S.D. Iversen & S.H. Snyder. pp. 139–231. Plenum Publishing Corp.

Lindvall, O., Björklund, A. & Divac, I. (1978). Organization of catecholamine neurons projecting to the frontal cortex in the rat. *Brain Research*, **142**, 1–24.

Lindvall, O., Björklund, A., Moore, R. Y. & Stenevi, U. (1974). Mesencephalic dopamine neurons projecting to neocortex. *Brain Research*, **81**, 325–31.

Lindvall, O. & Stenevi, U. (1978). Dopamine and noradrenaline neurons projecting to the septal area in the rat. *Cell and Tissue Research*, **190**, 383–408.

Lipton, S. A. and Kater, S. B. (1989). Neurotransmitter regulation of neuronal outgrowth, plasticity and survival. *Trends in Neurosciences*, **12**, 265–70.

Loughlin, S. E., Foote, S. L. & Bloom, F. E. (1986). Efferent projections of locus coeruleus: topographic organization of cells of origin demonstrated by three dimensional reconstruction. *Neuroscience*, **18**, 291–306.

Louilot, A., Taghzouti, K., Simon, H. & Le Moal, M. (1989). Limbic system, basal ganglia and dopaminergic neurons. Executive and regulatory neurons and their role in the organization of behavior. *Brain Behavior and Evolution*, **33**, 157–61.

Magistretti, P. J. & Morrisson, J. H. (1988). Noradrenaline- and vasoactive intestinal peptide- containing neuronal systems in neocortex: functional convergence with contrasting morphology. *Neuroscience*, **24**, 367–78.

Mantz, J., Milla, C., Glowinski, J. & Thierry, A. M. (1988). Differential effects of ascending neurons containing dopamine and noradrenaline in the control of spontaneous activity and of evoked responses in the rat prefrontal cortex. *Neuroscience*, **27**, 517–26.

Markowitsch, H. J. (1988). Anatomical and functional organization of the primate prefrontal cortical system. In *Comparative Primate Biology*, vol 4, pp. 99–153. Alan R. Liss.

Maurin, Y., Berger, B., Le Saux, F., Gay, M. & Baumann, N. (1985). Increased number of locus ceruleus noradrenergic neurons in the convulsive mutant quaking mouse. *Neuroscience Letters*, **57**, 313–18.

McCormick, D. A. (1992). Neurotransmitter actions in the thalamus and cerebral cortex and their role in neuromodulation of thalamocortical activity. *Progress in Neurobiology*, **39**, 337–88.

Mavridis, M., Degryse, A. D., Lategan, A. J., Marien, M. R. & Colpaert, F. C. (1991). Effects of locus coeruleus lesions on parkinsonian signs, striatal dopamine and substantia nigra cell loss after 1-methyl-4-phenyl-1,2,3,6-tetrahydropyridine in monkeys: a possible role for the locus coeruleus in the progression of Parkinson's disease. *Neuroscience*, **41**, 507–23.

Meador-Woodruff, J. H., Mansour, A., Grandy, D. K., Damask, S. P., Civelli, O. & Watson, Jr. S. J. (1992). Distribution of D5 dopamine receptor mRNA in rat brain. *Neuroscience Letters*, **145**, 209–212.

Mitchell, S. N., Brazell, M. P., Joseph, M. H., Alavijeh, M. S. & Gray, J. A. (1989). Regionally specific effects of acute and chronic nicotine on rates of catecholamine and 5-hydroxytryptamine synthesis in rat brain. *European Journal of Pharmacology*, **167**, 311–22.

Morrison, J. H. & Foote, S. L. (1986). Noradrenergic and serotoninergic innervation of cortical, thalamic and tectal visual structures in Old and New World monkeys. *Journal of Comparative Neurology*, **243**, 117–38.

Morrison, J. H., Foote, S. L., Molliver, M. E., Bloom, F. E. & Lidov, H. G. W. (1982a). Noradrenergic and serotonergic fibers innervate complementary layers in monkey primary visual cortex. An immunohistochemical study. *Proceedings of the National Academy of Sciences, USA*, **79**, 2401–5.

Morrison, J. H., Foote, S. L., O'Connor, D. & Bloom, F. E. (1982b). Laminar, tangential and regional organization of the noradrenergic innervation of monkey cortex: dopamine-β-hydroxylase immunohistochemistry. *Brain Research Bulletin*, **9**, 309–19.

Morrison, J. H., Grzanna, R., Molliver, M. E. & Coyle, J. T. (1978). The distribution and orientation of noradrenergic fibers in neocortex of the rat: an immunofluorescence study. *Journal of Comparative Neurology*, **181**, 17–40.

Morrison, J. H., Molliver, M. E., Grzanna, R. & Coyle, J. T. (1981). The intra-cortical trajectory of the coeruleo-cortical projection in the rat: a tangentially organized cortical afferent. *Neuroscience*, **6**, 139–58.

Mufson, E. J. & Mesulam, M. M. (1984). Thalamic connections of the insula in the rhesus monkey and comments on the paralimbic connectivity of the medial pulvinar nucleus. *Journal of Comparative Neurology*, **227**, 109–20.

Nagai, T., Satoh, K., Imamoto, K. & Maeda, T. (1981). Fluorescent retrograde double labeling technique combined with wet histofluorescence used to demonstrate divergent axonal projections of central catecholamine neurons. *Cellular and Molecular Biology*, **27**, 403–12.

Noack, H. J. & Lewis, D. A. (1989). Antibodies directed against tyrosine hydroxylase differentially recognize noradrenergic axons in monkey neocortex. *Brain Research*, **500**, 313–24.

Nowak, G., Zak, J. & Superata, J. (1991). Role of dopaminergic neurons in denervation-induced α1-adrenergic up-regulation in the rat cerebral cortex. *Journal of Neurochemistry*, **56**, 914–16.

Oades, R. D. & Halliday, G. M. (1987). Ventral tegmental (A10) system: neurobiology. 1. Anatomy and connectivity. *Brain Research Review*, **12**, 117–66.

Oeth, K. M. & Lewis D. A. (1992). Cholecystokinin- and dopamine-containing mesencephalic neurons provide distinct projections to monkey prefrontal cortex. *Neuroscience Letters*, **145**, 87–92.

Olschowka, J. A., Molliver, M. E., Grzanna, R., Rice, F. L. & Coyle, J. T. (1981). Ultrastructural demonstration of noradrenergic synapses in the rat cental nervous system by dopamine-β-hydroxylase immunocytochemistry. *Journal of Histochemistry and Cytochemistry*, **29**, 271–80.

Ouimet, C. C., Lamantia, A. S., Goldman-Rakic, P., Rakic, P. & Greengard, P. (1992). Immunocytochemical localization of Darpp-32, a dopamine and cyclic-AMP-regulated phosphoprotein in the primate brain. *Journal of Comparative Neurology*, **323**, 209–18.

Ouimet, C. C., Miller, P. E., Hemmings, H. C., Walaas, S. J. & Greengard, P. (1984). DARPP-32, a dopamine and adenosine 3′ 5′ monophosphate regulated phosphoprotein enriched in dopamine-innervated brain regions. III. Immunocytochemical localization. *Journal of Neuroscience*, **4**, 111–24.

Palacios, J. M., Hoyer, D. & Cortés, R. (1987). α1-Adrenoceptors in the mammalian brain: similar pharmacology but different distribution in rodents and primates. *Brain Research*, **419**, 65–75.

Palacios, J. M., Savasta, M. & Mengod, G. (1989). Does cholecystokinin colocalize with dopamine in the human substantia nigra? *Brain Research*, **488**, 369–75.

Pammer, C., Gorcs, T. & Palkovits, M. (1990). Peptidergic innervation of the locus coeruleus cells in the human brain. *Brain Research*, **515**, 247–55.

Pandya, D. N. & Yeterian, E. H. (1985). Architecture and connections of cortical association areas. In *Cerebral Cortex*, vol.4, ed. A. Peters & E. G. Jones, pp. 3–61. New York: Plenum Press.

Papadopoulos, G. C., Parnavelas, J. C. & Buijs, R. M. (1989a). Light and electron microscopic immunocytochemical analysis of the dopamine innervation of the rat visual cortex. *Journal of Neurocytology*, **18**, 303–10.

Papadopoulos, G. C., Parnavelas, J. C. & Buijs, R. M. (1989b). Light and electron microscopic immunocytochemical analysis of the noradrenaline innervation of the rat visual cortex. *Journal of Neurocytology*, **18**, 1–10.

Parent, A. (1984). Functional anatomy and evolution of monoaminergic systems. *Annals of Zoology*, **24**, 783–90.

Park, J. K., Joh, T. H. & Ebner, F. F. (1986). Tyrosine hydroxylase is expressed by neocortical neurons after transplantation. *Proceedings of the National Academy of Sciences, USA*, **83**, 7495–8.

Pazos, A., Gonzalez, A. M., Pascual, J., Meana, J. J., Barturen, F. & Garcia-Sevilla, J. A. (1988). α2-Adrenoceptors in human forebrain: autoradiographic visualization and biomedical parameters using the agonist (3H) UK-14 304. *Brain Research*, **475**, 361–5.

Phillipson, O. T., Kilpatrick, I. C. & Jones M. W. (1986). Dopaminergic innervation of the primary visual cortex in the rat and some correlations with human cortex. *Brain Research Bulletin*, **18**, 621–33.

Pirot, S., Godbout, R., Mantz, J., Tassin, J. P., Glowinski, J. & Thierry, A. M. (1992). Inhibitory effects of ventral tegmental area stimulation on the activity of prefrontal cortical neurons: evidence for the involvement of both dopaminergic and gabaergic components. *Neuroscience*, **49**, 857–86.

Porrino, L. J. & Goldman-Rakic, P. S. (1982) Brainstem innervation of prefrontal and anterior cingulate cortex in the rhesus monkey revealed by retrograde transport of HRP. *Journal of Comparative Neurology*, **205**, 63–76.

Powers, R. E., Struble, R. G., Casanova, M. F., O'Connor, D. T., Kitt, C. A. & Price, D. L. (1988). Innervation of human hippocampus by noradrenergic systems: normal anatomy and structural abnormalities in aging and Alzeimer's disease. *Neuroscience*, **25**, 401–17.

Preuss, T. & Goldman-Rakic, P. S. (1991) Myelo- and cytoarchitecture of granular frontal cortex and surrounding regions in the Strepsirhine primate Galago and the anthropoid primate Macaca. *Journal of Comparative Neurology*, **310**, 429–74.

Rakic, P., Bourgeois, J. P., Eckenhoff, M. F., Zecevic, N. & Goldman-Rakic, P. S. (1986). Concurrent overproduction of synapses in diverse regions of the primate cerebral cortex. *Science*, **232**, 232–5.

Reader, T. A. & Grondin, L. (1987). Distribution of catecholamines, serotonin and their major metabolites in the rat cingulate, piriform, entorhinal, somatosensory and visual cortex: a biochemical survey using high-performance liquid chromatography. *Neurochemical Research*, **12**, 1087–97.

Reep, R. (1984). Relationship between prefrontal and limbic cortex: a comparative anatomical review. *Brain Behavior and Evolution*, **25**, 5–80.

Rétaux, S., Besson, M. J. & Penit-Soria, J. (1991). Synergism between D1 and D2 dopamine receptors in the inhibition of the evoked release of H3 Gaba in the rat prefrontal cortex. *Neuroscience*, **43**, 323–30.

Richfield, E. K., Young, A. B. & Penney, J. B. (1989). Comparative distributions of dopamine D1 and D2 receptors in the cerebral cortex of rats, cats and monkeys. *Journal of Comparative Neurology*, **286**, 409–26.

Robbins, T. W., Everitt, B. J., Cole, B. J., Archer, T. & Mohammed, A. (1985). Functional hypotheses of the coeruleocortical noradrenergic projection: a review of recent experimentation and theory. *Physiology and Psychology*, **13**, 127–50.

Room, P., Postema, F. & Korf, J. (1981). Divergent axon collaterals of rat locus coeruleus neurons: demonstration by a fluorescent double labeling technique. *Brain Research*, **221**, 219–30.

Rosen, G. D., Finklestein, S., Stoll, A. L., Yutzey, D. A. & Denenberg, V. H. (1984). Neurochemical asymmetries in the Albino rat's cortex, striatum and nucleus accumbens. *Life Science*, **34**, 1143–8.

Ross, R. A., Judd, A. B., Pickel, V. M., Joh, T. H. & Reis, D. J. (1976). Strain-dependent variations in number of midbrain dopaminergic neurones. *Nature*, **264**, 654–6.

Sadoul, J. L., Checler, F., Kitabgi, P., Rostene, W., Javoy-Agid, F. & Vincent, J. P. (1984). Loss of high affinity NT receptors in substantia nigra from Parkinsonian subjects. *Biochemical and Biophysical Research Communications*, **125**, 395–404.

Sakaguchi, T. & Nakamura, S. (1987). The mode of projections of single locus coeruleus neurons to the cerebral cortex in rats. *Neuroscience*, **20**, 221–30.

Sales, N., Martres, M. P., Bouthenet, M. L. & Schwartz, J. C. (1989). Ontogeny of dopaminergic D2 receptors in the rat nervous system: characterization and detailed autoradiographic mapping with 125I-iodosulpiride. *Neuroscience*, **28**, 673–700.

Samson, Y., Wu, J. J., Friedman, A. H. & Davis, J. N. (1990). Catecholaminergic innervation of the hippocampus in the Cynomolgus monkey. *Journal of Comparative Neurology*, **298**, 250–63.

Sanides, F. (1964). The cyto-myeloarchitecture of the human frontal lobe and its relation to phylogenetic differenciation of the cerebral cortex. *Journal für Hirnforschung*, **6**, 269–82.

Satoh, H. & Matsumura, H. (1990). Distribution of neurotensin-containing fibers in the frontal cortex of the macaque monkey. *Journal of Comparative Neurology*, **298**, 215–23.

Satoh, J. & Suzuki, K. (1990). Tyrosine hydroxylase-immunoreactive neurons in the mouse cerebral cortex during the postnatal period. *Developmental Brain Research*, **53**, 1–5.

Sawaguchi, T. & Goldman-Rakic, P. S. (1991). D1 dopamine receptors in prefrontal cortex: involvement in working memory. *Science*, **251**, 947–50.

Sawaguchi, T., Matsumura, M. & Kubota, K. (1986) Catecholamine sensitivities of motor cortical neurons of the monkey. *Neuroscience Letters*, **66**, 135–40.

Sawaguchi, T., Matsumura, M. & Kubota, K. (1990). Catecholaminergic effects on neuronal activity related to a delayed response task in monkey prefrontal cortex. *Journal of Neurophysiology*, **63**, 1385–400.

Scatton, B., Javoy-Agid, F., Rouquier, L., Dubois, B. & Agid, Y. (1983). Reduction of cortical dopamine, noradrenaline, serotonin and their metabolites in Parkinson's disease. *Brain Research*, **275**, 321–8.

Schalling, M., Friberg, K., Seroogy, K., Riederer, P., Bird, E., Schiffman, S. N., Mailleux, P., Vanderhaeghen, J. J., Kuga, S., Goldstein, M., Kitahama, K., Luppi, P. H., Jouvet, M. & Hökfelt, T. (1990). Analysis of expression of cholecystokinin in dopamine cells in the ventral mesencephalon of several species and in humans with schizophrenia. *Proceedings of the National Academy of Sciences, USA*, **87**, 8427–31.

Schliebs, R. & Gödicke, C. (1988). Laminar distribution of noradrenergic markers in rat visual cortex. *Neurochemistry International*, **13**, 481–6.

Schmidt, R. H., Björklund, A., Lindvall, O. & Loren, I. (1982). Prefrontal cortex: dense dopaminergic input in the newborn rat. *Developmental Brain Research*, **5**, 222–8.

Schneiber, T. & Törk I. (1987). Ventromedial mesencephalic tegmental (VMT) projections to ten functionally different cortical areas in the cat: topography and quantitative analysis. *Journal of Comparative Neurology*, **259**, 247–65.

Schwartz, M. L., Rakic, P. & Goldman-Rakic, P. S. (1991). Early phenotype expression of cortical neurons. Evidence that a subclass of migrating neurons have callosal axons. *Proceedings of the National Academy of Sciences, USA*, **88**, 1354–8.

Séguela, P., Watkins, K. C. & Descarries, L. (1988). Ultrastructural features of dopamine axon terminals in the anteromedial and the suprarhinal cortex of adult rat. *Brain Research*, **442**, 11–22.

Séguela, P., Watkins, K. C., Geffard, M. & Descarries, L. (1990). Noradrenaline axon terminals in adult rat neocortex: an immunocytochemical analysis in serial thin sections. *Neuroscience*, **35**, 249–64.

Seroogy, K., Ceccatelli, S., Schalling, M., Hökfelt, T., Frey, P., Walsh, J., Dockray, G., Brown, J., Buchan, A. & Goldstein, M. (1988). A subpopulation of dopaminergic neurons in rat ventral mesencephalon contains both NT and CCK. *Brain Research*, **455**, 88–98.

Seroogy, K. B., Dangaran, K., Lim, S., Haycock, J. W. & Fallon. J. H. (1989). Ventral mesencephalon neurons containing both cholecystokinin and tyrosine hydroxylase-like immunoreactivities project to forebrain regions. *Journal of Comparative Neurology*, **279**, 397–414.

Seutin, V., Massotte, L. & Dresse, A. (1989). Electrophysiological effects of neurotensin on dopaminergic neurons of the ventral tegmental area of the rat *in vitro*. *Neuropharmacology*, **28**, 949–54.

Shatz, C. J., Chun, J. J. M. & Luskin, M. B. (1988). The role of the subplate in the development of the mammalian telencephalon. In *Cerebral Cortex*, vol. 7, ed. A. Peters & E. G. Jones, pp. 35–58. New York: Plenum Press.

Simon, H., Scatton, B. & Le Moal, M. (1980). Dopaminergic A10 neurons are involved in cognitive functions. *Nature*, **286**, 150–1.

Simon, H., Taghzouti, K., Gozlan, H., Studler, J. M., Louilot, A., Hervé, D., Glowinski, J. & Tassin, J. P. (1988). Lesion of dopaminergic terminals in the amygdala produces enhanced locomotor responses to D-amphetamine and opposite changes in dopaminergic activity in prefrontal cortex and nucleus accumbens. *Brain Research*, **447**, 335–40.

Slopsema, J. S., Van der Gugten, J. & de Bruin, P. C. (1982). Regional concentrations of noradrenaline and dopamine in the frontal cortex of the rat: dopaminergic innervation of the prefrontal subareas and lateralization of prefrontal dopamine. *Brain Research*, **250**, 197–200.

Smiley, J. F., Williams, S. M., Szigeti, K. & Goldman-Rakic, P. S. (1992). Light and electron microscopic characterization of dopamine immunoreactive axons in human cerebral cortex. *Journal of Comparative Neurology*, **321**, 325–36.

Soto-Moyano, R., Hernandez, A., Pérez, H., Ruiz, S., Galleguillos, X. & Belmar, J. (1991). Yohimbine early in life alters functional properties of interhemispheric connections of rat visual cortex. *Brain Research Bulletin*, **26**, 259–63.

Stone, E. A. & John, S. M. (1992). Stress-induced increase of extracellular levels of cyclic AMP in rat cortex. *Brain Research*, **597**, 144–7.

Stratford, T. R. & Wirtshafter, D. (1990). Ascending dopaminergic projections from the dorsal raphe nucleus in the rat. *Brain Research*, **511**, 173–6.

Studler, J. M., Kitabgi, P., Tramu, G., Hervé, D., Glowinski, J. & Tassin, J. P. (1988). Extensive colocalisation of neurotensin with dopamine in rat mesocortico-frontal dopaminergic neurons. *Neuropeptides*, **11**, 95–100.

Swanson, L. W. (1976). The locus coeruleus: a cytoarchitectonic, golgi and immunohistochemical study in the albino rat. *Brain Research*, **110**, 39–56.

Swanson, L. W. (1982). The projections of the ventral tegmental area and adjacent regions: a combined fluorescent retrograde tracer and immunofluorescence study in the rat. *Brain Research Bulletin*, **9**, 321–53.

Szigethy, E. & Beaudet, A. (1989). Correspondence between high affinity 125 I-neurotensin binding sites and dopaminergic neurons in the rat substantia nigra and ventral tegmental area: a combined radioautographic and immunohistochemical light microscopic study. *Journal of Comparative Neurology*, **279**, 128–37.

Taghzouti, K., Simon, H., Hervé, D., Blanc, G., Studler, J. M., Glowinski, J., Le Moal, M. & Tassin, J. P. (1988). Behavioural deficits induced by an electrolytic lesion of the rat ventral mesencephalic tegmentum are corrected by a superimposed lesion of the dorsal noradrenergic system. *Brain Research*, **440**, 172–6.

Takada, M. & Hattori, T. (1986). Collateral projections from the substantia nigra to the cingulate cortex and striatum in the rat. *Brain Research*, **380**, 331–5.

Takada, M. & Hattori, T. (1987). Organization of ventral tegmental area cells projecting to the occipital cortex and forebrain in the rat. *Brain Research*, **418**, 27–33.

Tassin, J. P., Bockaert, J., Blanc, G., Stinus, L., Thierry, A. M., Lavielle, S., Prémont, J. & Glowinski, J. (1978). Topographical distribution of dopaminergic innervation and dopaminergic receptors of the anterior cerebral cortex of the rat. *Brain Research*, **154**, 241–51.

Tedroff, J., Aquilonius, S. M., Hartvig, P., Lundquist, H., Gee, A. G., Uhlin, J. & Langstrom, B. (1988). Monoamine re-uptake sites in the human brain evaluated *in vivo* by means of 11C-nomifensine and positron emission tomography: the effects of age and Parkinson's disease. *Acta Neurologica Scandinavica*, **77**, 192–201.

Thal, L. J., Laing, K., Horowitz, S. G. & Makman, M. H. (1986). Dopamine stimulates rat cortical somatostatin release. *Brain Research*, **372**, 205–9.

Thierry, A. M., Blanc, G., Sobel, A., Stinus, L. & Glowinski, J. (1973). Dopaminergic terminals in the rat cortex. *Science*, **182**, 499–501.

Tohgi, H., Abe, T., & Takahashi, S. (1993). The significance of norepinephrine deficiency in the pathogenesis of freezing phenomena and the effects of L-threo-DOPS. In: *Norepinephrine Deficiency*, ed: H. Narabayashi, and Y. Mizuno pp.77–88. Lancs, UK, New York,: The Parthenon Publishing Group.

Tota, M. R., Candelore M. L., Dixon, R. E., and Strader, C. D. (1991). Biophysical and genetic analysis of the ligand-binding site of the β-adrenoceptor. *Trends in Pharamacological Science*, **12**, 4–6.

Tooyama, I., Walker, D., Yamada, T., Hanai, K., Kimura, H., Mc Geer, E. G. & Mc Geer, P. L. (1992). High molecular weight basic fibroblast growth factor-like protein is localized to a subpopulation of mesencephalic dopaminergic neurons in rat brain. *Brain Research*, **593**, 274–80.

Torack, R. M. & Morris, J. C. (1990). Tyrosine hydroxylase-like (TH) immunoreactivity in human mesolimbic system. *Neuroscience Letters*, **116**, 75–80.

Trottier, S., Geffard, M. & Evrard, B. (1989). Co-localization of tyrosine hydroxylase and gaba immunoreactivities in human cortical neurons. *Neuroscience Letters*, **106**, 76–82.

Trovero, F., Blanc, G., Hervé, D., Vézina, P., Glowinski, J. & Tassin, J. P. (1992). Contribution of an α1-adrenergic receptor subtype to the expression of the 'ventral tegmental area syndrome'. *Neuroscience*, **47**, 69–76.

Tsujimoto, A. & Tanaka, S. (1981). Stimulatory effect of somatostatin on norepinephrine release from rat brain cortex slices. *Life Science*, **28**, 903–10.

Uhl, G. R., Whitehouse, P. J., Price, D. L., Tourtelotte, W. W. & Kuhar, M. J. (1984). Parkinson's disease: depletion of substantia nigra neurotensin receptors. *Brain Research*, **308**, 186–90.

Uylings, H. B. M. & Van Eden, C. G. (1990). Qualitative and quantitative comparison of the prefrontal cortex in rat and in primates, including humans. *Progress in Brain Research*, **85**, 31–62.

Van Eden, C. G., Hoorneman, E. M. D., Buijs, R. M., Matthissen, M. A. H., Geffard, M. & Uylings, H. B. M. (1987). Immunocytochemical localization of dopamine in the prefrontal cortex of the rat at the light and electron microscopic level. *Neuroscience*, **22**, 849–62.

Van Eden, C. G., Lamme, V. A. F. & Uylings, H. B. M. (1992). Heterotopic cortical afferents to the medial prefrontal cortex in the rat. A combined retrograde and anterograde tracer study. *European Journal of Neuroscience*, **4**, 77–97.

Van Tol, H. H. M., Bunzow, J. R., Guan, H. C., Sunahara, R. K., Seeman, P., Niznik, H. B. & Civelli, O. (1991). Cloning of the gene for a human dopamine D4 receptor with high affinity for the antipsychotic clozapine. *Nature*, **350**, 610–14.

Verney, C., Alvarez, C., Geffard, M. & Berger, B. (1990). Ultrastructural double labeling study of dopamine terminals and gaba-containing neurons in rat anteromedial cerebral cortex. *European Journal of Neuroscience*, **2**, 960–72.

Verney, C., Baulac, M., Berger, B., Alvarez, C., Vigny, A. & Helle, K. B. (1985). Morphological evidence for a dopaminergic field in the hippocampal formation of the young and adult rat. *Neuroscience*, **14**, 1039–52.

Verney, C., Berger, B., Adrien, J., Vigny, A. & Gay, M. (1982). Development of the dopaminergic innervation of the rat cerebral cortex. A light microscopic immunocytochemical study using antityrosine hydroxylase antibodies. *Developmental Brain Research*, **5**, 41–52.

Verney, C., Berger, B., Baulac, M., Helle, K. B. & Alvarez, C. (1984). Dopamine-β-hydroxylase-like immunoreactivity in the fetal cerebral cortex of the rat: noradrenergic ascending pathways and terminal fields. *International Journal of Developmental Neuroscience*, **2**, 491–503.

Verney, C., Milosevic, A., Alvarez, C. & Berger, B. (1993). Immunocytochemical evidence of well-developed dopaminergic and noradrenergic innervations in the frontal cerebral cortex of human fetuses at midgestation. *Journal of Comparative Neurology*, **336**, 331–44.

Verney, C., Zecevic, N., Nikolic, B., Alvarez, C. & Berger, B. (1991). Early evidence of catecholaminergic cell groups in 5 and 6 week-old human embryos using tyrosine hydroxylase and dopamine-β-hydroxylase immunocytochemistry. *Neuroscience Letters*, **131**, 121–4.

Vezina, P., Blanc, G., Glowinski, J. & Tassin, J. P. (1992). Nicotine and morphine differentially activate brain dopamine in prefrontocortical and subcortical terminal fields. Effects of acute and repeated injections. *Journal of Pharmacology and Experimental Therapeutics*, **261**, 484–90.

Vincent, S. R. & Hope, B. T. (1990). Tyrosine hydroxylase-containing neurons lacking aromatic aminoacid decarboxylase in the hamster brain. *Journal of Comparative Neurology*, **295**, 290–8.

Vizi, E. S. (1980). Modulation of cortical release of acetylcholine by noradrenaline released from nerves arising from rat locus coeruleus. *Neuroscience*, **5**, 2139–44.

Vogel, W. H., Orfei, V. & Century, B. (1969). Activities of enzymes involved in the formation and destruction of biogenic amines in various areas of human brain. *Journal of Pharmacology and Experimental Therapeutics*, **165**, 196–203.

Vogt, B. A. & Peters, A. (1981). Form and distribution of neurons in rat cingulate cortex: areas 32, 24, and 29. *Journal of Comparative Neurology*, **195**, 603–26.

Wanaka, A., Kiyama, H., Murakami, T., Matsumoto, M., Kamada, T., Malbon, C. C. & Tohyama, M. (1989). Immunocytochemical localization of β-adrenergic receptors in the rat brain. *Brain Research*, **485**, 125–40.

Weinberger, D. R. (1988). Schizophrenia and the frontal lobe. *Trends in Neurosciences*, **11**, 367–70.

White, F. J. & Wang, R. U. (1984). A10 Dopamine neurons: role of autoreceptors in determining firing rate and sensitivity to dopamine agonists. *Life Science*, **34**, 1161–70.

Wiesendanger, M. (1986). Recent developments in studies of the supplementary motor area of primates. *Review of Physiology and Biochemical Pharmacology*, **103**, 1–59.

Wise, S. P. (1985). The primate premotor cortex fifty years after Fulton. *Behavioral Brain Research*, **18**, 79–88.

Wise, R. A. & Rompre, P. P. (1989). Brain dopamine and reward. *Annual Review of Psychology*, **40**, 191–226.

Yang, C. R. & Mogenson, G. J. (1990). Dopaminergic modulation of cholinergic responses in rat medial prefrontal cortex. An electrophysiological study. *Brain Research*, **524**, 271–81.

Yoshida, M., Shirouzu, M., Tanaka, M., Semba, K. & Fibiger, H. C. (1989). Dopaminergic neurons in the nucleus raphe dorsalis innervate the prefrontal cortex in the rat: a combined retrograde tracing and immunohistochemical study using antidopamine serum. *Brain Research*, **496**, 373–6.

Zecevic, N., Bourgeois, J. P. & Rakic, P. (1989). Changes in synaptic density in motor cortex of rhesus monkey during fetal and postnatal life. *Developmental Brain Research*, **50**, 11–32.

Zecevic, N. R. & Molliver, M. E. (1978). The origin of monoaminergic innervation of immature rat neocortex: an ultrastructural analysis following lesions. *Brain Research*, **150**, 387–97.

Zecevic, N., Verney, C., Milosevic, A. & Berger, B. (1991). First description of the central catecholamine systems in 6–8 week-old human embryos. *Society for Neuroscience Abstracts*, **17**, 745.

Zhu, Z. Q., Armstrong, D. L., Grossman, R. G. & Hamilton, W. J. (1990). Tyrosine hydroxylase-immunoreactive neurons in the temporal lobe in complex partial seizures. *Annals of Neurology*, **27**, 564–72.

Zilles, K. (1985). *The Cortex of the Rat. A Stereotaxic Atlas.* Springer-Verlag Berlin.

Part II: Developmental aspects of catecholamine systems in the CNS of vertebrates

13

Development of central catecholaminergic neurons in teleosts

P. Ekström, T. Honkanen and B. Borg

Introduction

There is presently only one detailed study of the embryonic and early larval development of catecholaminergic neurons in a teleost (Ekström, Honkanen & Borg, 1992). This study deals with the three-spined stickleback (*Gasterosteus aculeatus* L.; Gasterosteiformes). The adult stickleback displays the same general distribution of catecholaminergic cell groups as other teleosts investigated to date (Ekström, Honkanen & Steinbusch, 1990), although minor interspecies differences do exist, especially with regard to the exact location of comparable cell groups and the distribution of fibers (Meek, Chapter 4). Therefore, the developmental sequence for the catecholaminergic cell groups and some of their projections in the stickleback may serve as a general example of the situation in teleosts.

Stickleback larvae hatch six days after fertilization when reared at 20 °C, i.e. at the age of 144 h or shortly thereafter. The stickleback embryos used in the present study hatched between 144 h and 168 h after fertilization. The hatched larvae remain in the nest, guarded by the male, for an additional four days, and begin their free-swimming life at the age of ten days. Four days after hatching the yolk sac is exhausted, the larvae begin active feeding, and their olfactory epithelium and central olfactory projections have attained adult features (Honkanen & Ekström, 1991). The visual system is probably functional by the same age, since retinal photoreceptors are well differentiated already one day after hatching (Ekström, Borg & van Veen, 1983).

Here we describe the early ontogenetic development of catecholaminergic neurons and their major axonal pathways, by use of specific antisera against dopamine (DA), tyrosine hydoxylase (TH) and dopamine β-hydroxylase (DBH). The time of appearance of catecholaminergic elements has been put in relation to the appearance of the early axonal scaffold of pathfinding fibers (Wilson et al., 1990), to the development of other neurotransmitter systems, and to the development of other known functional systems. The descriptions are based on the material used by Ekström et al. (1992), complemented by material subsequently prepared by the first author in the course of further studies of the early development of the central nervous system of the stickleback. This subsequent material has yielded more detailed information about the early differentiation of the brain, and has revealed some instances of earlier differentiation of structures than was reported by Ekström et al. (1992). This will be specifically commented upon.

The putative identification of different catecholamines has been discussed previously (Ekström et al., 1990, 1992). Briefly, DAir and THir/DAir neurons are considered as dopaminergic neurons if they are DBH-negative, whereas DBHir may be noradrenergic (locus coeruleus/subcoeruleus group) or either noradrenergic or adrenergic (caudal medullary group). The neurotransmitter identity of neurons that are THir but DA-negative and DBH-negative remains unknown (see Ekström et al., 1990, 1992).

Development of catecholaminergic neuronal cell groups and pathways

72 h embryo

At this stage, there is little or no cytoarchitectonic differentiation in the brain, as revealed on semi-thin plastic sections (Fig. 13.1). However, the main components of the early axonal scaffold, comparable to those present in the one-day old zebrafish embryo (Wilson et al., 1990), may be revealed already in the 60 h embryo by use of the monoclonal antibodies HNK-1 and 6-11B-1 (Ekström, unpublished observations). HNK-1 recognizes an epitope present on several glycoconjugates, including cell adhesion molecules, while 6-11B-1 recognizes acetylated α-tubulin, a component of microtubules (for references, see Wilson et al., 1990). These axonal pathways, together with prominent landmarks like the optic stalk and the optic recess, the primordia of the lateral recesses of the third ventricle, and the mesencephalic flexure make it possible to assess the extent of the major subdivisions of the brain. The rhombencephalon is undifferentiated in terms of cytoarchitectonic entities. Cranial nerve motor nuclei and reticular neurons cannot be labeled by acetylcholinesterase (AChE) histochemistry until in the 96 h embryo (Ekström, unpublished observations).

The first catecholaminergic neurons are detectable with immunocytochemistry in 72 h embryos. They are located in the diencephalon and in the rhombencephalon. The diencephalic neurons (Fig. 13.1(a)) are weakly TH immunoreactive (ir) and are located in the mantle layer, dorsal to the lateral extensions of the third ventricle and close to the ventral longitudinal tracts (Wilson et al., 1990). They will form part of the rostral subdivision of the so-called PVO- (paraventricular organ) accompanying cell group. At this stage, they do not possess neurites, nor express DA immunoreactivity.

In the rostral rhombencephalon, small numbers of faintly THir and DBHir neurons are located at the level of the future isthmus region (Fig. 13.1(b)). In some embryos, DBHir neurons with broad cytoplasmic processes can be discerned, indicating that neurite elongation has just commenced. These neurons constitute the earliest appearing cells of the noradrenergic locus coeruleus/subcoeruleus complex of the adult stickleback (Ekström et al., 1986).

96 h embryo

One day later, at 96 h, the development of the brain in general as well as the catecholamine systems has made considerable progress (Fig. 13.2). In the ventral preoptic area small THir/DAir neurons appear dorsal to the optic chiasm (Fig. 13.2(b)). The first immunoreactive neurons to appear in this area are invariably situated immediately adjacent to the optic tracts. Neurons in the preoptic area never exhibit DBH immunoreactivity and are, therefore, considered to be dopaminergic neurons.

The numbers of THir PVO-accompanying neurons have substantially increased. At this age they begin to show also DA immunoreactivity. As they never exhibit DBH immunoreactivity they are likely dopaminergic neurons. The neurons form a dense cluster immediately dorsal to the ventricular sulcus that marks the future location of the lateral recess of the third ventricle, or dorsal to the narrow lateral recess in those few specimens where it is already discernible (Fig. 13.2(d)). They do not show any contacts with the third ventricle, and thus clearly constitute

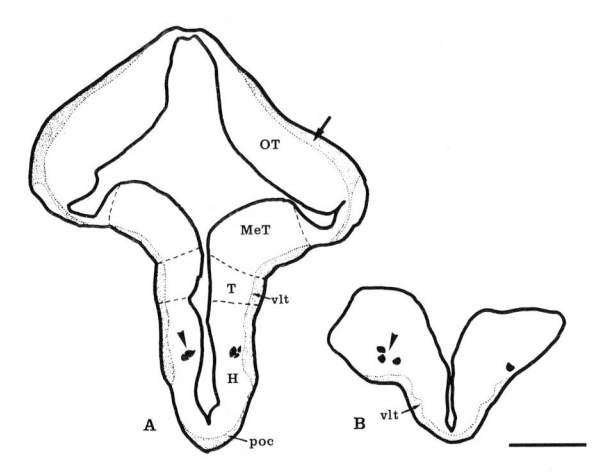

Fig. 13.1. Catecholaminergic neurons in the 72 h embryo. (a) THir neurons in the presumptive hypothalamus (arrowheads). (b) Noradrenergic neurons in the presumptive locus coeruleus (arrowheads). Stippled area shows the extent of the early axonal scaffold, as visualized by the monoclonal antibodies 6-11B-1 and HNK-1. The marginal zone of the optic tectum (arrow) is also labelled. Scale bar 100 μm. Camera lucida drawings. Adapted from Ekström et al. (1992).

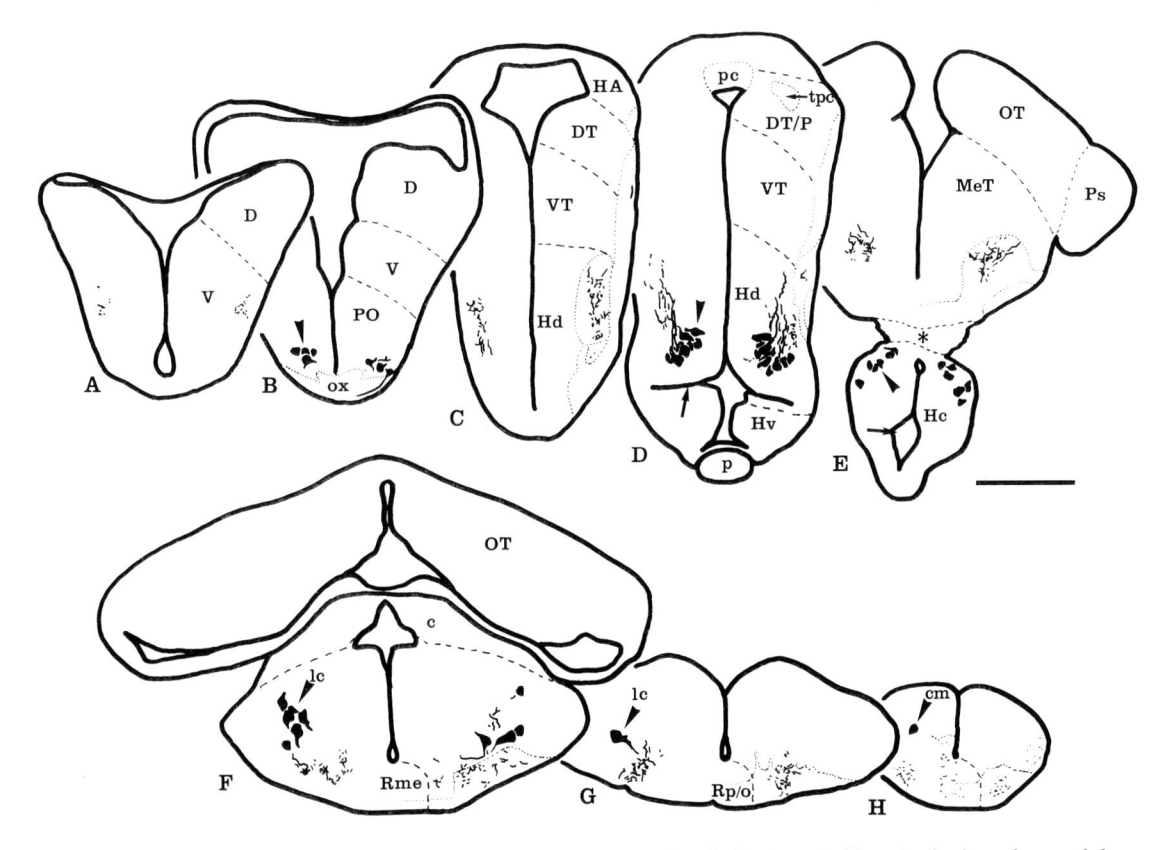

Fig. 13.2. Catecholaminergic neurons in the 96 h embryo. (*a*). Catecholaminergic fibres in the lateral part of the ventral telencephalon. (*b*). Dopaminergic neurons (*arrowhead*) in the ventral part of the preoptic nucleus. (*c*) Catecholaminergic axons lateral to the dorsal zone of the periventricular hypothalamus. (*d*) Dopaminergic PVO-accompanying neurons (*arrowhead*) in the rostral hypothalamus, dorsal to the lateral recesses of the third ventricle (*arrow*). (*e*). Dopaminergic neurons (*arrowhead*) in the caudal zone of the periventricular hypothalamus. *Asterisk* denotes the pial surface of the ventral flexure. (*f*, *g*). Noradrenergic neurons (*arrowheads*) in the locus coeruleus. (*h*). Adrenergic neurons (*arrowhead*) in the caudal brainstem. *Hatched lines* denote boundaries between cell masses, whereas *stippled lines* denote boundaries of some major axonal pathways. Scale bar 100 µm. Camera lucida drawings. Adapted from Ekström *et al.* (1992).

PVO-accompanying neurons. They emit dorsally directed axons, some of which turn laterally and then appear to join the marginal longitudinal tract. Caudal to the ventral flexure, the dopaminergic PVO-accompanying neurons occupy a more superficial position, immediately ventral to the mesencephalic tegmentum (Fig. 13.2(*e*)). At this age, the most caudally situated PVO-accompanying neurons appear to belong to the caudal hypothalamus as defined by Bergquist (1932), while the more rostrally situated are aligned along the border between the thalamus and hypothalamus, i.e. in an area similar to the zona incerta of mammals. However, it is important to note there are no landmarks defining the border between the posterior tuberculum and the hypothalamus. As will be discussed below, it appears that at later stages (at least part of) the caudally situated PVO-accompanying neurons belong to the posterior tuberculum. At this age the posterior recesses of the third ventricle have not yet evolved.

The numbers of noradrenergic neurons in the locus coeruleus/subcoeruleus complex (Fig. 13.2(*f*), (*g*)) have also increased. The neurons emit ventro-laterally directed axons that turn rostrally or caudally, forming the major ascending and descending catecholaminergic pathways. A faint dopamine immunoreactivity is now detected in some cell bodies.

By this age, DAir, THir and DBHir cell bodies without neurites appear in the caudal medulla (Fig. 13.2(*h*)). As at later stages they retain this immunoreactivity pattern, but do not express noradrenaline immunoreactivity (Ekström et al., 1986), they will be referred to as putatively adrenergic neurons (Ekström et al., 1990).

THir axons form a longitudinal tract that extends between the spinal cord and the telencephalon. In the brainstem it occupies a ventral, although not superficial, position (Fig. 13.2 (e)–(h)). The axons appear to join the dorsal aspect of the ventral longitudinal tract that courses along the ventral surface of the brainstem (Wilson et al., 1990). Further rostrally, the THir axons run immediately dorsal to the PVO-accompanying neurons and intermingle with their dorsally directed neurites (Fig. 13.2(d)). They then course lateral to the dorsal hypothalamus (Fig. 13.2 (c)) and the preoptic area, in close proximity to the optic tracts, and reach the lateral portion of the ventral telencephalic area (Fig. 13.2 (a)). The longitudinal pathway is formed by axons from the neurons in the locus coeruleus/subcoeruleus complex, and the PVO-accompanying neurons. Liquor-contacting neurons and neurons in the preoptic area do not contribute to the longitudinal catecholaminergic pathway at this stage.

DAir axons are present in all parts of the THir fiber system described above, albeit in smaller numbers. In addition, DAir axons occur in the postoptic commissure, and in the optic tracts *en route* for the rostral tectum.

Ascending DBHir axons from the locus coeruleus/subcoeruleus complex constitute a large portion of the longitudinal tracts up to the level of the mes-/diencephalic border. Farther rostrally, only very small numbers of axons (sometimes solitary axons with growth-cone like protrusions) can be detected in the areas that contain THir and/or DAir fibers. DBHir axons cannot be detected caudal to the locus coeruleus/subcoeruleus complex.

120 h embryo
At this stage, THir neurons appear for the first time in the telencephalon (Fig. 13.3). They are few in numbers, and are located in the dorsal part of the ventral telencephalon (Fig. 13.3(a)). Ascending THir axons appear in the telencephalon. The majority is confined to the ventral telencephalon, but some spread into the dorsal telencephalon and small numbers of axons reach far rostrally, even entering the developing olfactory epithelium.

The dopaminergic neurons in the preoptic area (Fig. 13.3(b), (c)) bear short THir neurites that extend ventrally and laterally towards the optic tracts. Numerous DAir/THir axons run in the optic tracts, in the caudal portion of the optic chiasm and in the postoptic commissure.

The dopaminergic PVO-accompanying neurons form a group that is located dorsal to the lateral recesses of the third ventricle and extends from the rostral point of emergence of the lateral recesses (Fig. 13.3(d)) into the caudal diencephalon (Fig. 13.3(e)). Thus, rostrally it constitutes a boundary between the hypothalamus and the ventral thalamus, further caudally it lies within the posterior tuberculum (cf. the subregio tuberculi posterioris of Bergquist (1932), which would encompass the dorsomedial PVO-accompanying neurons in Fig. 13.3(d)), and in its most caudal part, it lies again in the dorsal part of the caudal hypothalamus. However, as in the 96 h embryo, there is no clear demarcation of the posterior tuberculum.

DAir but TH-negative liquor-contacting neurons appear in the medial and dorsal walls of the lateral recess, between the PVO-accompanying neurons and the ventricular wall (Fig. 13.3(d)). The posterior recesses have now been formed and are, in a few specimens of this age, surrounded by DAir liquor-contacting neurons (Fig. 13.3(e)). This indicates that the delay between the first appearance of DAir liquor-contacting neurons in the rostral and caudal periventricular system may be shorter than concluded in our earlier report (Ekström et al., 1992).

In the pineal organ, THir unipolar neurons with short intrapineal neurites appear. They appear to persist in small numbers at least throughout larval development. DAir or DBHir neuronal elements were never observed in the pineal organ.

In the isthmus region, the numbers of noradrenergic neurons have reached adult levels. The neurons of the locus coeruleus/subcoeruleus complex occupy a relatively more dorsal position, and appear less densely clustered than at earlier stages (Fig. 13.3(f)), presumably because of the addition of non-catecholaminergic neurons that have left the mitotic cycle and migrated out of the ventricular zone. The most dorsally situated neurons lie close to the rostral margin of the developing cerebellum, and THir/DBHir neurons may be consistently observed within the cerebellar anlage at this age. Consistent with the increase in numbers of cell bodies, a larger number of axons contribute to the longitudinal catecholaminergic pathways (Fig 13.3(f)). The neurons in the locus coeruleus/subcoeruleus complex express a weak DA immunoreactivity.

The group of putative adrenergic neurons now extends between the entry of the vagus and the obex

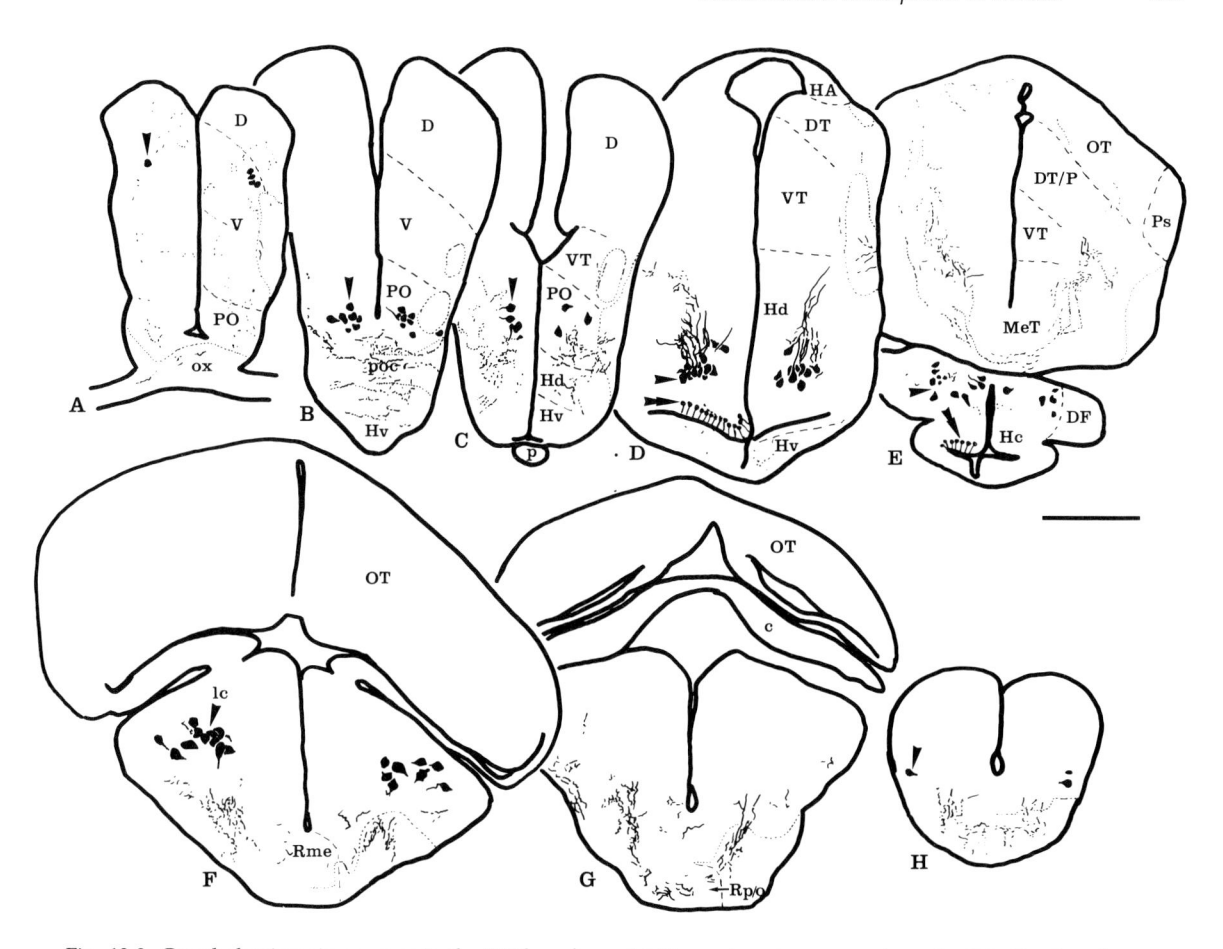

Fig. 13.3. Catecholaminergic neurons in the 120 h embryo. (*a*) Dopaminergic neurons (*arrowhead*) in the dorsal zone of the ventral telencephalon. (*b*) Dopaminergic neurons in the ventral portion of the preoptic nucleus (*arrowhead*), just dorsal to the postoptic commissure. (*c*). Dopaminergic neurons in the dorsal part of the preoptic nucleus (*arrowhead*). (*d*). Dopaminergic PVO-accompanying neurons (*arrowhead*) with dorsally directed axons, and dopaminergic liquor-contacting neurons (*double arrowhead*) lining the dorsal wall of the lateral recess. (*e*) Dopaminergic neurons (*arrowhead*) in the dorsal part of the caudal zone of the periventricular hypothalamus, and dopaminergic liquor-contacting neurons (*double arrowhead*) lining the dorsal wall of the posterior recess. (*f*). Noradrenergic neurons (*arrowhead*) in the locus coeruleus/subcoeruleus complex. (*g*). Catecholaminergic longitudinal axonal tract with axons decussating through the pallidus/obscurus raphe nucleus. (*h*). Adrenergic neurons (*arrowhead*) in the caudal medullary group. *Hatched lines* denote boundaries between cell masses, whereas *stippled lines* denote boundaries of some major axonal pathways. Scale bar 100 μm. Camera lucida drawings. Adapted from Ekström *et al.* (1992).

in the caudal medulla. They are endowed with short dendrites (especially visible in DBHir cells), indicating that by this age neurite growth has commenced in the adrenergic neurons of the lower brainstem. Also in the area postrema, weakly DAir and intensely THir neurons with neurites may be observed at this age. They never express DBH immunoreactivity, and are, therefore, considered to be dopaminergic neurons.

At this stage THir axons may be observed leaving the main longitudinal tracts. Some course in a dorsal direction into the mesencephalic tegmentum and the dorsal diencephalon, some even reaching the posterior commissure (Fig. 13.3(*e*)). Others leave in the ventral direction and enter the ventral hypothalamus (Fig. 13.3(*c*), (*d*). At the level of the optic chiasm, THir axons join the developing marginal optic tracts. THir fibers also decussate in the postoptic commissure (Fig. 13.3(b)) and the ventral tegmental commissural system (Fig. 13.3(e)–(h)). DAir fibers are present in all areas that contain THir fibers, but in consistently lower numbers.

The ascending DBHir fibers have increased in numbers. When reaching the level of the di-/mesencephalic border, they form two loosely arranged fiber bundles. One courses dorsally and contributes fibers to the ventrolateral aspect of the optic tectum. The other runs dorsal to the lateral recesses of the third ventricle, immediately dorsal to the PVO-accompanying neurons. The fibers continue rostralward through the preoptic region, where some fibers join the optic tracts, into the ventral telencephalon.

Descending DBHir fibers now reach the spinal cord. The adrenergic neurons in the caudal medulla do not seem to contribute any fibers to the longitudinal tracts.

144 h embryo

The THir neurons in the dorsal part of the ventral telencephalon still appear in only very small numbers (Fig. 13.4(a)). They extend short, laterally directed neurites. At this age, the neurons in this region express DA immunoreactivity. They retain a DAir/THir phenotype throughout life, and are considered dopaminergic neurons.

In the preoptic area (Fig. 13.4(b)), the organization of dopaminergic neurons has increased in complexity. The neurons adjacent to the optic tracts show two basically different orientations: superficially situated neurons emit medially directed axons that appear to be associated with the developing medial optic tracts, while neurons adjacent to the lateral forebrain bundle emit dorsolaterally directed axons that form a dense bundle dorsal to the lateral forebrain bundle. Part of the latter axons appear to join the marginal optic tract. Small numbers of THir neurons also appear further dorsally, in what presumably is the future preoptic nucleus. They emit laterally directed neurites that join the main longitudinal pathway.

Faintly THir immunoreactive neurons appear in the rostral part of the ventromedial thalamic nucleus (Fig. 13.4(b)). This group was not detected in the material previously described by Ekström et al. (1992) until in the four-day old larva (240 h after fertilization).

In the dorsal thalamic/pretectal region, at the level of the rostral margin of the posterior commissure, faintly THir bipolar neurons appear. They form one continuous cluster, with no indications of subdivision in two nuclei as in the 168 h larva. The neurons emit laterally directed neurites that spread over a small region in the pretectum and merge with the axons of the PVO-accompanying neurons (Fig. 13.4(c),(d)). They do not express DA immunoreactivity.

THir but DA- and DBH-negative cell bodies are still found in the pineal organ (Fig. 13.4(b)).

At this age, both the lateral and the posterior recesses of the third ventricle are well developed (Fig. 13.4.(d),(e)). The distribution of dopaminergic PVO-accompanying neurons around the third ventricle is essentially unchanged with respect to the 120 h stage (Fig. 13.4(c)–(e)), whereas the DAir liquor-contacting neurons surrounding the lateral and posterior recesses (Fig. 13.4(c)–(e) have increased in numbers. The PVO-accompanying neurons emit dorsally directed axons that reach into the pretectum (Fig. 13.4(c),(d)).

The noradrenergic neurons of the locus coeruleus/subcoeruleus complex (Fig. 13.4(f)) have gained a more distinct orientation (cf. Fig. 13.3(f)). Depending on the plane of section, the neurons in the caudal part of the nucleus sometimes appear to form two separate groups, as shown in Fig. 13.4(f)).

The adrenergic neurons of the caudal medulla (Fig. 13.4(g)) emit dorsolaterally directed axons. The dopaminergic neurons of the area postrema (Fig. 13.4(h)) show the same organization as in the 120 h embryo.

The distribution of THir axons is essentially similar to that observed in the 120 h embryo. However, it appears that DAir liquor-contacting neurons and dopaminergic neurons in the preoptic area now contribute to the previously established longitudinal pathways (not shown in the drawings). Moreover, the THir neurons in the dorsal thalamic/pretectal region give rise to fibers in the pretectum, and may contribute fibers to the rostral optic tectum (Fig. 13.4(c)). The distribution of DAir fibers overlaps with that of THir fibers, but DAir fibers are present in smaller numbers.

In addition to the previously described distribution, scattered DBHir fibers are now present in the superficial axon layer of the optic tectum, in the dorsal mesencephalic tegmentum, in the periventricular hypothalamus, in the inferior lobes, and in both the ventral and dorsal telencephalic areas. The adrenergic caudal medullary neurons now contribute axons to the longitudinal pathways.

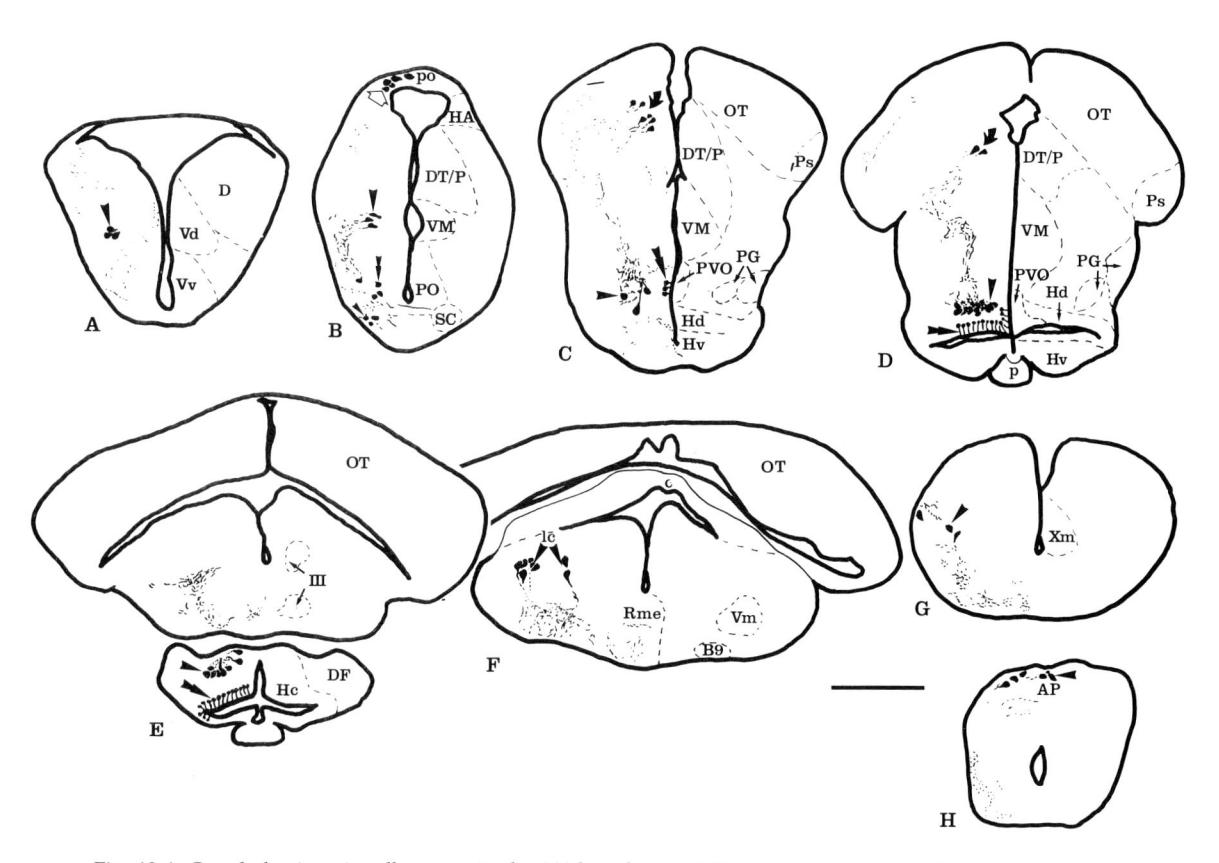

Fig. 13.4. Catecholaminergic cell groups in the 144 h embryo. (a) Dopaminergic neurons (*arrowhead*) in the dorsal nucleus of the ventral telencephalon. (b) THir neurons appear in the rostral part of the ventromedial thalamic nucleus. Dopaminergic neurons in the preoptic area have now separated in two groups, one in the suprachiasmatic nucleus (*small arrowhead*) and one in the preoptic nucleus (*small double arrowhead*). A population of transiently THir cells (*open arrow*) appears in the pineal organ. (c), (d) THir neurons (*curved arrow*) in the dorsal thalamic/pretectal area. Dopaminergic PVO-accompanying neurons (*arrowhead*) and liquor-contacting neurons in the paraventricular organ and the dorsal zone of the periventricular hypothalamus (*double arrowhead*). (e) Dopaminergic PVO-accompanying neurons (*arrowhead*) and liquor-contacting neurons in the caudal zone of the periventricular hypothalamus (*double arrowhead*). (f) Noradrenergic neurons in the locus coeruleus/subcoeruleus complex (*arrowhead*). (g) Adrenergic neurons in the caudal medullary group (*arrowhead*). (h) Dopaminergic neurons in the area postrema (*arrowhead*). *Hatched lines* denote cell group boundaries. Scale bar 100 μm. Camera lucida drawings.

One-day old larva (168 h after fertilization)

As shown in Figs. 13.5 and 13.6, the telencephalic dopaminergic neurons are still few in numbers, but their neurites are longer and appear to contribute to intratelencephalic projections. In general, they form one small circumscribed group, but in one specimen a tendency toward parcellation into a rostrodorsal (Fig. 13.6(a),(b)) and caudoventral (Fig. 13.6(d)) subdivision was observed. The presence of a rostral group of THir neurons may indicate that a dopaminergic phenotype has already been attained by some neurons adjacent to the future olfactory bulb, but that their levels of immunoreactive TH and dopamine are generally too low to be detected by immunocytochemistry. Otherwise, THir neurons are not observed in this region until in one month old juvenile sticklebacks.

The organization of the dopaminergic neurons in the preoptic area is similar to that seen in the 144 h embryo. The neurons are closely associated with the optic chiasm and the optic tracts. At the level of the optic chiasm, dopaminergic neurons in the ventral part of the preoptic nucleus are located directly adjacent to the marginal optic tract. These neurons

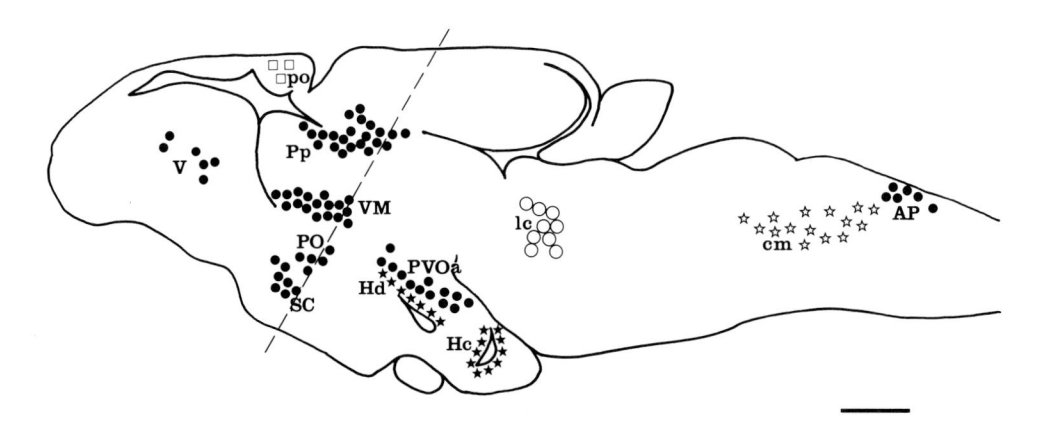

Fig. 13.5. Catecholaminergic cell groups in the newly hatched (168 h after fertilization) stickleback larva, as projected on a mid-sagittal plane. Rostral is to the left. *Black dots* denote dopaminergic neurons, *black stars* dopaminergic liquor-contacting neurons, *open circles* noradrenergic neurons, *open stars* putative adrenergic neurons, and *open squares* THir neurons. *PVOa*, PVO-accompanying. Hatched line shows plane of section of Fig.13.6 (k) Scale bar 100 μm.

correspond to the dopaminergic neurons of the suprachiasmatic nucleus in adults. They emit dorsally directed axons that course to the optic tectum, and caudally directed axons that appear to contribute to the innervation of the pituitary (see below). The THir neurons that are located along the dorsal margin of the preoptic nucleus now constitute a rather large group (Fig. 13.6(i),(j),(k)), and possess laterally directed neurites.

Numerous THir cells are present in the rostral and lateral parts of the pineal organ (Fig. 13.6(f),(g)). Their size and shape indicate that they are photoreceptor cells, but this remains to be explored at the electron microscopical level.

The THir neurons in the dorsal thalamic/pretectal region form two clusters, one dorsolateral and one ventromedial (Fig. 13.6(h),(i),(j),(k), that may be referred to as the periventricular pretectal

Caption to Fig. 13.6

Fig. 13.6. Catecholaminergic cell groups in the newly hatched (168 h after fertilization) stickleback larva. (a), (b) Rostral group of THir neurons (*arrowhead*) at the border between the olfactory bulb primordium and the dorsal and ventral telencephalon. (c) Level caudal to the olfactory bulb primordium (d) Caudal group of THir neurons (*arrowhead*). (e) Level of the anterior commissure and the rostralmost portions of the preoptic region, the thalamus and the habenula. (f), (g) THir cells in the pineal organ (*arrowhead*), but not in the parapineal organ. (h) The level of the posterior commissure. Dopaminergic neurons in the periventricular pretectal nucleus (*curved arrow*) and the ventromedial thalamic nucleus (*arrowhead*). (i) Dopaminergic neurons in the periventricular pretectal nucleus (*curved arrow*), the ventromedial thalamic nucleus (*arrowhead*) and the suprachiasmatic nucleus (*small arrowhead*). (j), (k) Dopaminergic neurons in the periventricular pretectal nucleus (*curved arrow*), the ventromedial thalamic nucleus (*arrowhead*), the suprachiasmatic nucleus (*small arrowhead*) and the preoptic nucleus (*double small arrowhead*). (l) Dopaminergic PVO-accompanying neurons (*arrowhead*). (m), (n), (o) Dopaminergic PVO-accompanying neurons (*arrowhead*) and liquor-contacting neurons in the paraventricular organ and the dorsal zone of the periventricular hypothalamus (*double arrowhead*). (p) Dopaminergic PVO-accompanying neurons (*arrow*) in the caudal zone of the periventricular hypothalamus and noradrenergic neurons in the locus coeruleus/subcoeruleus complex (*arrowhead*). (q) Liquor-contacting neurons in the caudal zone of the periventricular hypothalamus (*double arrowhead*) and noradrenergic neurons in the locus coeruleus/subcoeruleus complex (*arrowhead*). r. Level of the caudal part of the median raphe nucleus. (s) Level of the Mauthner cell bodies. (t) Level of the rostral abducens nucleus. (u) Level of the caudal abducens nucleus. (v) Level of the glossopharyngeal motor nucleus and the most rostral portion of the pallidus/obscurus raphe nucleus. (w), (x) Adrenergic neurons (*arrowhead*) in the caudal medullary group. (y) Dopaminergic neurons (*arrowhead*) in the area postrema. *Hatched lines* denote cell group boundaries. Scale bar 100 μm. 25 μm between consecutive sections from A to L and from N to P, 50 μm between sections L and M and between consecutive sections from P to Y. Camera lucida drawings.

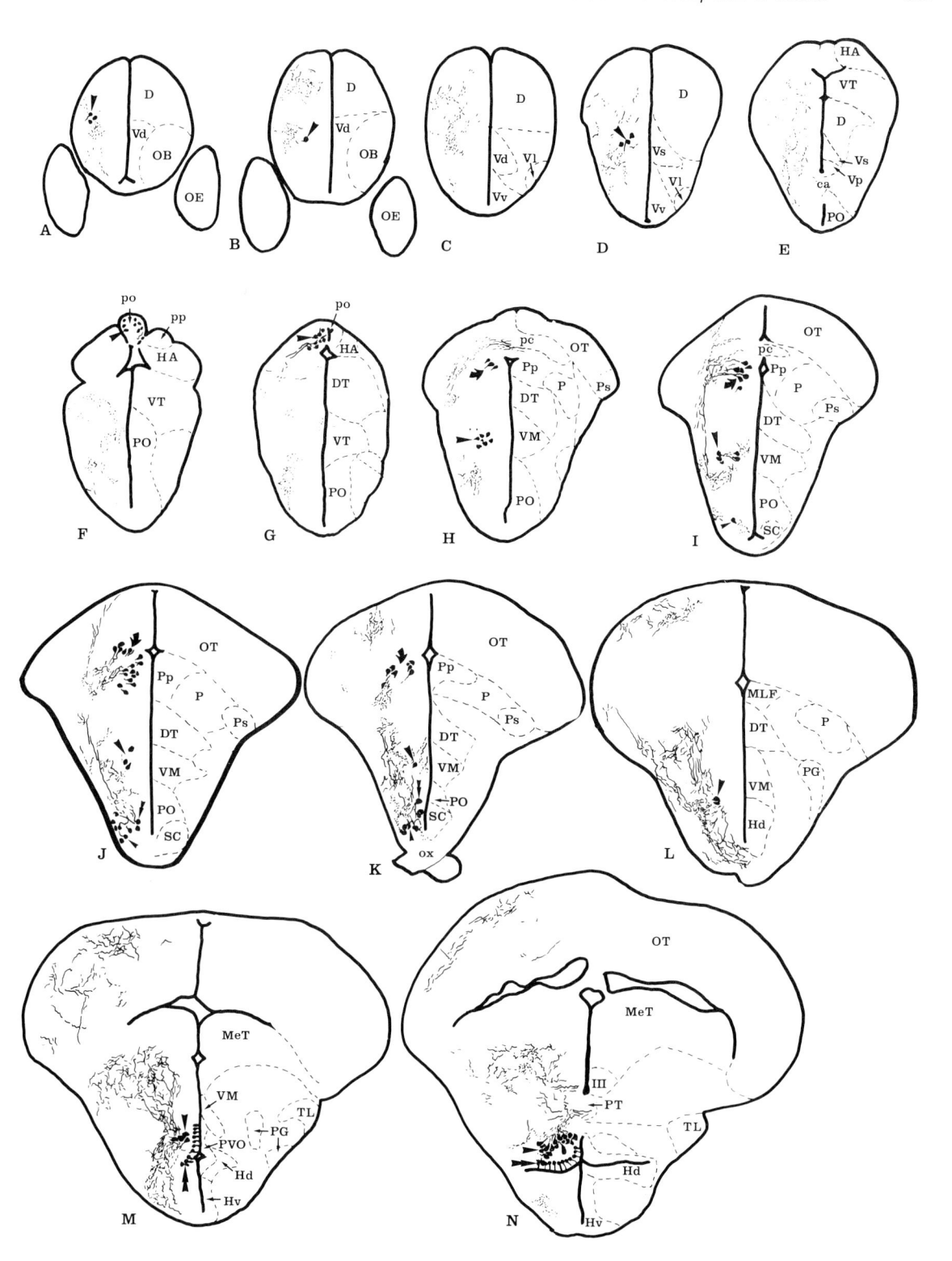

Caption on p. 332

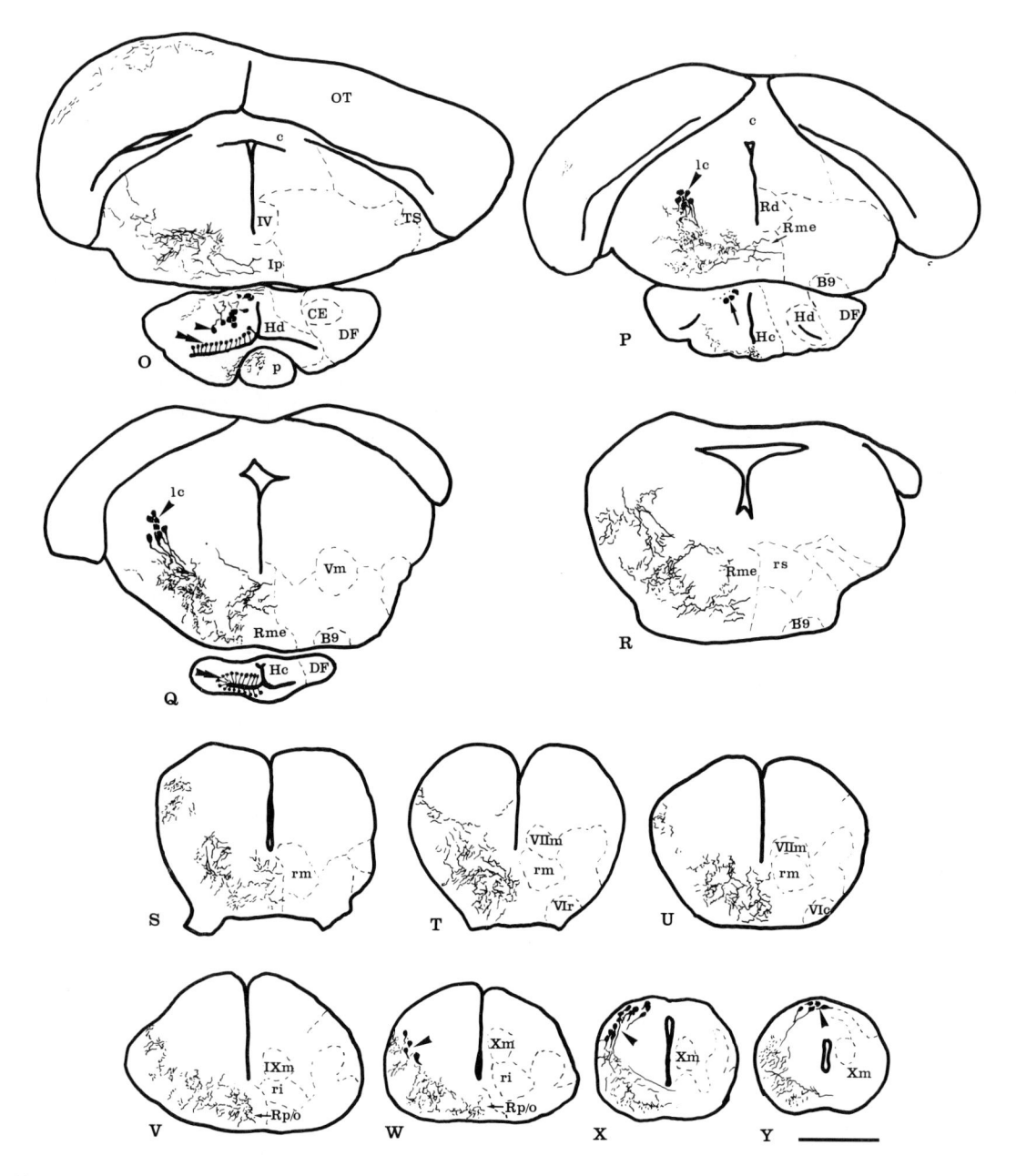

Fig. 13.6. (cont.)

dopaminergic groups. They emit axons that form two laterally directed bundles (one from each cluster; Fig. 13.6(j)). The axons encounter the dorsally directed axons emitted by the neurons in the preoptic region, and together they innervate the superficial tectal neuropil. Further caudally they intermingle with the axons of the PVO-accompanying neurons. DAir neurons can now be detected in the periventricular pretectal region. They occur in smaller numbers than THir neurons.

The THir neurons in the ventromedial thalamic nucleus (Fig. 13.6(h)–(k)) have increased in numbers.

Small numbers of DAir neurons may now be observed in the same location.

The dopaminergic PVO-accompanying neurons have increased in numbers, and emit intensely immunoreactive neurites in the dorsal direction (Fig. 13.6(*l*)–(*p*)). Since the 120 h stage, they form a continuous group that follows the border between the hypothalamus and ventral thalamus caudally through the posterior tuberculum, and into the caudal hypothalamus.

Part of the neurites emitted by PVO-accompanying neurons situated dorsal to the lateral recesses are very coarse, and may represent dendrites. In the rostral part of the nucleus, immediately rostral to the emergence of the lateral recesses, numerous axons are emitted in the ventral direction (Fig. 13.6(*l*),(*m*)). They encounter the caudally directed bundle of axons originating in the ventral preoptic nucleus. From this point of convergence a bundle of axons course caudally and terminate in the pituitary. However, it was not possible to determine the relative contribution to the pituitary innervation by the preoptic region and the PVO-accompanying neurons.

Surprisingly, weakly THir liquor-contacting neurons were observed lining the wall of the lateral recess in its most rostral and medial portion (level of Fig. 13.6(*m*). This is at variance with previous observations, where only DAir but TH-negative liquor-contacting neurons were observed at all developmental stages (Ekström et al., 1990, 1992). The significance of this finding is not clear, but it may be speculated that liquor-contacting neurons lining the third ventricle in the stickleback may express enough TH to be detected by immunocytochemistry during specific physiological conditions. DAir liquor-contacting neurons show basically the same distribution (Fig. 13.6(*m*)–(*q*) as in the 144 h embryo, but now surround the posterior recesses (Fig. 13.6(*q*).

The noradrenergic neurons of the locus coeruleus/subcoeruleus complex (Fig. 13.6(*p*),(*q*)) emit numerous intensely THir axons that join the main longitudinal catecholaminergic pathways. They exhibit a weak DA immunoreactivity.

In the caudal medulla, the adrenergic neurons occupy a position lateral to the inferior reticular nucleus and the vagal motor nucleus (Fig. 13.6(*w*),(*x*)). They form a group that extends between the level of the entry of the vagus nerve and the obex, where it is continuous with the dopaminergic neurons in the area postrema (Fig. 13.6(*y*)). Both groups of neurons emit ventrolaterally directed axons that collect in the lateral brainstem, from where they join the main descending fiber tracts. Only a small number of THir and DBHir fibers can be followed into the spinal cord. Thus, it appears that the majority of the projections from the catecholaminergic neurons in the caudal brainstem are of ascending nature.

In all, the distribution of THir fibers is still essentially similar to that observed already in 120 h and 144 h embryos. However, the numbers of immunoreactive axons have increased, particularly in the dorsoventral pathways of the diencephalon (Fig. 13.6(*i*)–(*m*)). Similarly, the numbers of DAir and DBHir fibers have increased. Still, several areas that receive profuse catecholaminergic innervation in the adult stickleback are only sparsely innervated or even devoid of DAir, THir and DBHir fibers. This is most evident regarding the innervation of the telencephalon (Fig. 13.6(*a*)–(*e*)), the ventral and caudal hypothalamus (Fig. 13.6(*n*)–(*q*)), and the optic tectum (Fig. 13.6(*i*)–(*r*)).

Four-day old larva (240 h after fertilization)
In general, the distribution of catecholaminergic neurons and fibres in the four-day old larva is very similar to that observed in the one-day old larva. Some further differentiation has occurred, though, and since this is the age when the larvae begin an active life this will be briefly described.

In the preoptic region, the dopaminergic neurons of the suprachiasmatic nucleus emit axons that form a caudally directed bundle which can be traced through the lateral tuberal nucleus into the pituitary.

The large dopaminergic PVO-accompanying neurons and the DAir liquor-contacting neurons emit dorsally and laterally directed bundles of axons. The dorsally directed axons merge with ascending axons and with axons from the THir/DAir neurons of the dorsal thalamic/pretectal cell groups, while the laterally directed axons turn ventrally and enter the inferior lobes of the hypothalamus.

The temporal appearance of catecholaminergic cell groups is summarized in Table 13.1 It should be noted that THir/DAir liquor-contacting neurons were not observed along the central canal of the spinal cord at any developmental stage. It is also noteworthy that THir/DAir neurons appear very late in the olfactory bulbs. In one month old juvenile sticklebacks, weakly THir neurons can be discerned in the dorsomedial part of the olfactory bulbs. At two months, the distribution of THir neurons is similar to that in adults, i.e. preferentially in the caudal part of the bulbs, close to the junction with the telencephalon. However, the number of THir neurons in the bulbs is still small. Unfortunately, we have as yet no data concerning

Table 13.1. *Appearance of catecholaminergic cell groups during embryonic and early posthatching development*

	72 h	96 h	120 h	144 h	168 h	240 h
Telencephalon: Vd			□	—	—	□
Preoptic area: nsc		□	—	—	—	□
npo			□	—	—	□
Hypothalamus/posterior tuberculum:						
PVO-acc. cells, rostral	▨	—	—	—	—	—
PVO-acc. cells, caudal		□	—	—	—	□
PVO (liquor-cont. cells)			■	■	■	■
nrp (liquor-cont. cells)				* ■	■	■
Pretectum: nPC				▨	□	□
PaC				▨	□	□
Thalamus: nVM				▨	□	□
Epithalamus: pineal organ			▨▨▨▨▨▨▨▨▨			
Brainstem:						
lc	▨	▰	▰	▰	▰	▰
Caudal medulla		▰	▰	▰	▰	▰
ap			□	—	—	□

▨▨▨▨▨	THir neurons
■■■■■	DAir neurons
□□□□□	DAir, THir neurons
▰▰▰▰▰	DAir, THir, DBHir neurons
▨▨▨▨▨	THir, DBHir neurons

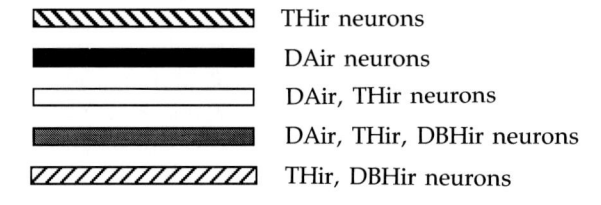

*Some specimens express ir cells earlier (see text)

the distribution of DAir neurons in one-month and two-month old sticklebacks.

Neurotransmitter identity of catecholaminergic neurons

In order to obtain a tentative identification of the transmitters used by the neurons visualized by DA, TH, and/or DBH immunoreactions in the stickleback embryo, we have used the following reasoning. Neurons expressing DA and TH immunoreactivity, but not noradrenaline (NA) immunoreactivity in the adult stickleback (Ekström et al., 1986), are dopaminergic. Neurons expressing DBH (and TH and DA to a variable extent), and NA immunoreactivity in the adult stickleback, are noradrenergic. Neurons expressing DBH, TH and DA, but not NA immunoreactivity in the adult, are adrenergic. However, we are well aware that this identification remains tentative, until conclusive experiments using antibodies against adrenaline and phenylethanolamine-N-methyltransferase have been performed.

Moreover, the dopaminergic nature of the DAir but (generally) TH-negative liquor-contacting neurons needs further study. It has been suggested by several authors that these neurons do not synthesize dopamine from tyrosine hydroxylase, but possess uptake mechanisms for dopamine and possibly L-DOPA (references in Ekström et al., 1992; Smeets & Reiner, Chapter 19). The picture is further complicated by the observation of NAir liquor-contacting neurons in some vertebrates (see Chapters 4–6). As weakly THir liquor-contacting neurons were observed in one-day old stickleback larvae, and strongly THir liquor-contacting neurons have been observed in goldfish (Hornby, Piekut & Demski, 1987) and in the Atlantic salmon (B. Holmqvist, B. Movérus, T. Östholm, P. Ekström, in preparation) some questions regarding the DAir liquor-contacting neurons may be raised. First, do DAir liquor-contacting neurons express TH and actually synthesize dopamine from L-tyrosine during specific physiological conditions? Secondly, do they express constitutionally low (undetectable by immunocytochemistry) levels of TH, and in addition possess an inordinately potent catecholamine uptake mechanism? Thirdly, are the NAir and DAir liquor-contacting neurons identical?

Another problem relates to the consistent observation of larger numbers of THir than DAir neuronal somata in the telencephalon, preoptic region and dorsal thalamic/pretectal region. There are two immediately obvious possible interpretations of this observation. First, dopamine content is below immunocytochemical detection levels in the cell bodies whereas TH is readily detected. This does not mean that these neurons do not utilize dopamine as transmitter at the presynaptic endings. Secondly, the neurons utilize L-DOPA rather than dopamine. In order to prove the latter possibility, the presence of L-DOPA and the absence of aromatic L-amino decarboxylase in these neurons should be specifically proven. Until this has been done, the first alternative offers the most parsimonious interpretation.

Absence of mesencephalic catecholaminergic neurons?

DAir, THir or DBHir neurons do not appear in the mesencephalon at any developmental stage. Thus, there is no trace of transiently expressed catecholaminergic neurons in the region corresponding to the ventral tegmental area and substantia nigra of adult elasmobranchs and amniotes (Smeets & Reiner, Chapter 19). However, dopaminergic neurons are present in the caudal diencephalon along the border to the mesencephalic tegmentum. Therefore, it may be pertinent to analyse the development of this region in greater detail.

Dopaminergic neurons are found in the periventricular hypothalamus and the posterior tuberculum. The largest numbers are of the liquor-contacting type, and they are predominantly located in the periventricular hypothalamus although significant numbers are found in the most rostral part of the posterior tuberculum, the paraventricular organ (PVO; according to the nomenclature of Braford and Northcutt (1983). However, numerous neurons do not exhibit contacts with the cerebrospinal fluid, and constitute the PVO-accompanying cell group. We use this term although this cell group extends far beyond the boundaries of the PVO proper. Rostrally, it is located along the border between the ventral thalamus and hypothalamus, and further caudally along the border between the diencephalon and mesencephalon (i.e. posterior tuberculum), from where it extends into the caudal hypothalamus. It is present already in 96 h embryos, although at this stage the posterior tubercular component is not evident.

In the 96 h embryo, the ventral flexure that marks the ventral border between the diencephalon and mesencephalon has become very sharp. The caudal portion of the posterior tuberculum and the caudal hypothalamus are folded back beneath the mesencephalic floor. All neurons in the diencephalon are gathered around the ventricles, forming a ventricular zone and a mantle layer that contains postmitotic neurons. There is only a thin cell-free marginal zone. In the caudal part of the hypothalamus the dorsally situated marginal zone of the caudal hypothalamus abuts a cell-free zone close to the pial surface rostral to the ventral flexure. This belongs to the most caudal portion of the posterior tuberculum according to Bergquist (1932). Thus, owing to the sharp folding at the ventral flexure the marginal zone of the caudal hypothalamus and posterior tuberculum adjoins that of the mesencephalon, and in frontal sections this region appears ventral to the mesencephalon (Fig. 13.2(e)).

When the first dopaminergic PVO-accompanying neurons appear ventral to the ventral flexure (i.e. in the 96 h embryo), they form bilateral groups located close to the marginal zone of the caudal hypothalamus (Fig. 13.2(e)). In frontal sections this appears to be exactly the same location as that of the first dopaminergic neurons of the ventral mesencephalon, i.e. rostral (ventral) to the ventral flexure (Specht et al., 1981). However, in sagittal sections it can be seen that the vast majority of the dopaminergic neurons in the stickleback embryo are clearly confined to the mantle zone rostral to the ventral flexure, i.e. the mantle zone of the caudal hypothalamus. Only later, from the 120 h stage, does it appear that an intermediate part of the group of PVO-accompanying cells are located within the posterior tuberculum. Whether they are generated within the posterior tuberculum, or have migrated from the hypothalamic groups, cannot be deduced from the present material.

Relationship with the development of sensory systems

Olfactory system
In the stickleback, the primary olfactory projections start to develop around 72 h and have reached the same targets seen in the adult stickleback 4 days after hatching (Honkanen and Ekström, 1990, 1991). However, it does not seem likely that catecholamines directly influence the early differentiation and pathway formation in the olfactory system. The olfactory pathways do not enter the forebrain until at hatching, and at this stage only small numbers of dopaminergic neurons and catecholaminergic fibers can be detected in the telencephalon. In the adult stickleback, the telencephalic areas with the highest densities of dopaminergic fibers, i.e. the lateroventral and laterodorsal parts of the dorsal telencephalon, coincide with primary olfactory projection areas. Moreover, dopaminergic neurons in the supra- and postcommissural nuclei of the ventral telencephalon also coincide with primary olfactory projection areas (Ekström et al., 1990; Honkanen & Ekström, 1990). Thus, it seems more likely that dopamine is involved in processing of olfactory information than in the development of the olfactory pathways.

It is not possible, in the absence of comparative data on other teleosts, to conclude whether the late appearance of dopaminergic neurons in the olfactory bulbs of the stickleback is a species-specific trait or a teleost-specific trait. It should be noted, though, that Ekström et al. (1990) reported relatively small numbers of DAir neurons in the adult olfactory bulbs, and THir neurons in a predominantly caudal location. Thus, the distribution of putatively dopaminergic neurons in the olfactory bulb of the stickleback may be atypical (cf. Chapter 4). However, it is also possible that, in this species, dopaminergic neurons of the olfactory bulbs possess constitutively low levels of both TH and DA.

Visual system
There are as yet no data concerning the development of the central visual pathways in the three-spined stickleback. However, cells in the ganglion cell layer of the central retina emit axons that enter the optic stalk already during day 3 after fertilization, and retinal photoreceptors (cones) are well developed by the time of hatching (Ekström et al., 1983). Thus it may be presumed that already at hatching stickleback larvae have a functional visual system, albeit not mature.

In the adult stickleback, dopaminergic neurons are located in the suprachiasmatic nucleus, and in the periventricular pretectal nucleus. The suprachiasmatic nucleus is the major hypothalamic retinorecipient nucleus, whereas the neuropil ventrolateral to the periventricular pretectal nucleus is the main pretectal retinorecipient area (Ekström, 1984). Moreover, large numbers of THir neurons are located in the ventromedial thalamic nucleus (Ekström et al., 1990), which also receives retinal innervation (Ekström, 1984). Thus, it appears that dopaminergic neurons in these areas are directly involved in visual processing, although it is presently not known whether they actually are postsynaptic to retinofugal axons.

It is noteworthy that in stickleback embryos and larvae, dopaminergic neurons both in the ventrolateral portion of the preoptic region (future suprachiasmatic nucleus) and the dorsal thalamic/pretectal region (future periventricular pretectal nucleus) give rise to axonal pathways to the optic tectum. These connections have not yet been demonstrated in adult sticklebacks, but appear to be present in the salmon (Holmqvist, Östholm & Ekström, 1993).

The photosensory pineal organ of the stickleback contains identifiable photoreceptor cells with outer segments already in the 72 h embryo (Ekström et al., 1983) and expresses several photoreceptor proteins and serotonin immunoreactivity at the same age or very shortly thereafter (van Veen et al., 1984; Östholm et al., 1988).

Relationship with the development of other neurotransmitter systems

Little information is available on the early development of other neurotransmitter systems in teleosts. Apart from a recent study of the ontogeny of substance P-immunoreactive elements in the goldfish brain (Vecino & Sharma, 1992) and one of the ontogeny of the serotoninergic system in a salmonid (Bolliet & Ali, 1992), all data available have been obtained in the three-spined stickleback.

Serotonin

The early development of the catecholaminergic neuronal system in the three-spined stickleback largely parallels that of the serotoninergic system (Ekström, Nyberg & van Veen, 1985). Serotoninergic neurons first appear in the hypothalamus and brainstem at approximately the same age (80 h embryo) as the first catecholaminergic neurons in the same regions. Serotoninergic neurons in the dorsal thalamus/pretectum appear later than the main cell groups of the hypothalamus and the brainstem. The subsequent differentiation of the hypothalamic cell groups and the raphe nuclei occurs within the same time span as that of the catecholaminergic cell groups. In the four-day old larva, i.e. when the larva starts active feeding from external sources, the organization of the serotoninergic cell groups is essentially the same as that seen in adult sticklebacks (Ekström & van Veen, 1984). However, there are a few marked differences between the development of the catecholaminergic system and the serotoninergic system.

There are several cell groups in the brainstem that express a serotoninergic phenotype during embryonic stages but lose it, or express it only weakly, during larval, juvenile and adult stages. These comprise the caudal raphe nuclei and the laterally migrated cell groups. Thus, the organization of the serotoninergic cell groups of the brainstem pass through an embryonic phase when they are very similar to that seen in mammalian embryos, whereas the catecholaminergic cell groups retain their 'teleostean' features throughout development.

Hypothalamic serotoninergic liquor-contacting cells appear earlier than their dopaminergic equivalents. They are actually the first differentiated serotoninergic cell type in the embryonic stickleback brain. In contrast, dopaminergic liquor-contacting neurons are only found after the dopaminergic PVO-accompanying cells have appeared.

Serotoninergic pathways appear to develop more rapidly than catecholaminergic pathways. This is most obvious for the ascending pathways to the telencephalon, which are well developed and encompass both ventral and dorsal telencephalon at the time of hatching.

In general, it appears that the serotoninergic system undergoes a faster differentiation during the embryonic phase, whereas the catecholaminergic system 'catches up' during the early larval phase. This may reflect a role for serotonin in the developing brain as a neurotrophic factor (Meier, Hertz & Schousboe, 1991). An almost identical developmental pattern of the serotoninergic system has recently been described in a salmonid (Bolliet & Ali, 1992).

Other neurotransmitters

In the goldfish embryo, the first substance P-immunoreactive neurons to appear are liquor-contacting neurons in the periventricular hypothalamus (Vecino & Sharma, 1992) at a developmental stage that appears close to that when the first THir and serotoninergic neurons appear in this region in the stickleback. However, there are no data regarding the development of other neurotransmitter systems in the goldfish, or regarding the development of the substance P system in the stickleback. Therefore, direct comparisons would be unfounded.

Functional aspects of the early catecholaminergic system

At the time when the stickleback larva begins its free-swimming life, i.e. four days after hatching, dopaminergic axons from neurons in the ventral preoptic area innervate the pituitary. Catecholamines exert pronounced hypophysiotrophic functions in adult teleosts. Dopamine stimulates the release of growth hormone (Wong, Chang & Peter, 1992). Dopamine inhibits, whereas noradrenaline stimulates secretion of gonadotropin (Peter et al., 1986; Chang, Van Goor & Acharya, 1991). Dopamine influences e.g. secretion of melanophore-stimulating hormone (Olivereau, Olivereau & Lambert, 1987), prolactin

(Olivereau et al., 1988; James & Wigham, 1984), thyrotropin and adrenocorticotropin (Olivereau et al., 1988). Little is known about temporal aspects of the development of pituitary control in teleosts, but it may be speculated that the appearance of dopaminergic fibers in the pituitary reflects a functional maturation of the hypothalamo-hypophyseal axis.

Dopaminergic neurons occur in or adjacent to primary visual centers (Ekström et al., 1990). In the salmon, it has been shown that neurons in these visual centers contribute to the dopaminergic innervation of the optic tectum (Holmqvist et al., 1993). The relatively late appearance of dopaminergic neurons in two of the corresponding areas in the stickleback, i.e. the dorsal thalamus/pretectum and the ventromedial thalamic nucleus, may reflect a later maturation of the retinal projections to these areas in comparison with the retinotectal projection (Presson & Fernald, 1986). However, it is notable that the dopaminergic neurons in the retinorecipient ventral preoptic area (suprachiasmatic nucleus), which encompass a hypophysiotrophic population, appear early during ontogenesis. This may reflect an early development of photic control of pituitary activity in the stickleback.

Dopamine and noradrenaline have pronounced effects on behavioural thermoregulation in the goldfish when injected into, or immediately rostral to the periventricular preoptic nucleus (Wollmuth, Crawshaw & Panayiotides – Djaferis, 1989). This region contains dopaminergic neurons in the goldfish as well as in the stickleback (Meek, Chapter 4). Whether it is a thermoregulatory centre in the stickleback is not known but, as mentioned above, this region contains dopaminergic hypophysiotrophic neurons that probably also receive retinal input.

Acknowledgements

The kind gifts of rabbit anti-dopamine (DA-9-5) by Dr H. Steinbusch, monoclonal antibodies 6-11B-1 by Dr G. Piperno and HNK-1 by Dr C. Stern are gratefully acknowledged. We also wish to thank Ms C. Rasmussen for excellent technical assistance. This study was supported by the Swedish Natural Science Research Council.

Abbreviations

AP	area postrema
B9	serotoninergic cell group B9
c	cerebellum
ca	anterior commissure
CE	central nucleus of the inferior lobe
cm	caudal medullary catecholaminergic (adrenergic) cell group
D	dorsal telencephalic area
DF	diffuse nucleus of the inferior lobe
DT	dorsal thalamus
Ip	interpeduncular nucleus
H	hypothalamus
Hc	caudal zone of periventricular hypothalamus
Hd	dorsal zone of periventricular hypothalamus
Hv	ventral zone of periventricular hypothalamus
HA	habenula
lc	locus coeruleus/subcoeruleus complex
MeT	mesencephalic tegmentum
MLF	nucleus of the medial longitudinal fascicle
OB	olfactory bulb primordium
OE	olfactory epithelium
ot	optic tract
ox	optic chiasm
p	pituitary
P	pretectum
pc	posterior commissure
PG	preglomerular nucleus
po	pineal organ
PO	preoptic area
poc	postoptic commissure
pp	parapineal organ
Pp	periventricular pretectal nucleus
Ps	superficial pretectal nucleus
PT	posterior tuberal nucleus
PVO	paraventricular organ
Rd	dorsal raphe nucleus
ri	inferior reticular nucleus
rm	medial reticular nucleus
Rme	median raphe nucleus
Rp/o	pallidus/obscurus raphe nuclei
rs	superior reticular nucleus
SC	suprachiasmatic nucleus
T	thalamus
TL	torus lateralis
TS	torus semicircularis
tpc	tract of the posterior commissure
V	ventral telencephalic area
Vd	dorsal nucleus of V
Vl	lateral nucleus of V
vlt	ventral longitudinal tract

VM	ventromedial thalamic nucleus	Vm	trigeminal motor nucleus
Vp	postcommissural nucleus of V	VIc	caudal abducens motor nucleus
Vs	supracommissural nucleus of V	VIr	rostral abducens motor nucleus
VT	ventral thalamus	VIIm	facial motor nucleus
Vv	ventral nucleus of V	VIc	caudal abducens motor nucleus
III	oculomotor nucleus	IXm	glossopharyngeal motor nucleus
IV	trochlear nucleus	Xm	vagal motor nucleus

References

Bergquist, H. (1932). Zur Morphologie des Zwischenhirns bei niederen Wirbeltieren. *Acta Zoologica (Stockholm)*, **13**, 57–303.

Bolliet, V. & Ali, M. A. (1992). Immunohistochemical study of the development of serotoninergic neurons in the brain of the brook trout *Salvelinus fontinalis*. *Brain, Behavior and Evolution*, **40**, 234–49.

Braford, M. R. & Northcutt, R. G. (1983). Organization of the diencephalon and pretectum of the ray-finned fishes. In *Fish Neurobiology, Vol. 2, Higher Brain Areas and Functions*, ed. R. E. Davis & R. G. Northcutt, pp. 117–63. Ann Arbor: The University of Michigan Press.

Chang, J. P., Van Goor, F. & Acharya, S. (1991). Influences of norepinephrine, and adrenergic agonists and antagonists on gonadotropin secretion from dispersed pituitary cells of goldfish, *Carassius auratus*. *Neuroendocrinology*, **54**, 202–10.

Ekström, P. (1984). Central neural connections of the pineal organ and retina in the teleost *Gasterosteus aculeatus* L. *Journal of Comparative Neurology*, **226**, 321–35.

Ekström, P. & van Veen, Th. (1984). Distribution of 5-hydroxytryptamine (serotonin) in the brain of the teleost *Gasterosteus aculeatus* L. *Journal of Comparative Neurology*, **226**, 307–20.

Ekström, P., Borg, B. & van Veen, Th. (1983). Ontogenetic development of the pineal organ, parapineal organ, and retina of the three-spined stickleback, *Gasterosteus aculeatus* L. (Teleostei). Development of photoreceptors. *Cell and Tissue Research*, **233**, 593–609.

Ekström, P., Nyberg, L. & van Veen, Th. (1985). Ontogenetic development of serotoninergic neurons in the brain of a teleost, the three-spined stickleback. An immunohistochemical analysis. *Developmental Brain Research*, **17**, 209–24.

Ekström, P., Reschke, M., Steinbusch, H. & van Veen, Th. (1986). Distribution of noradrenaline in the brain of the teleost *Gasterosteus aculeatus* L.: an immunohistochemical analysis. *Journal of Comparative Neurology*, **254**, 297–313.

Ekström, P., Honkanen, T. & Steinbusch, H. W. M. (1990). Distribution of dopamine-immunoreactive neuronal perikarya and fibres in the brain of a teleost, *Gasterosteus aculeatus*. Comparison with tyrosine hydroxylase- and dopamine β-hydroxylase-immunoreactive neurons. *Journal of Chemical Neuroanatomy*, **3**, 233–60.

Ekström, P., Honkanen, T. & Borg, B. (1992). Development of tyrosine hydroxylase-, dopamine- and dopamine β-hydroxylase-immunoreactive neurons in a teleost, the three-spined stickleback. *Journal of Chemical Neuroanatomy*, **5**, 481–501.

Holmqvist, B. I., Östholm, T. & Ekström, P. (1994). Neuroanatomical analysis of the visual and hypophysiotrophic systems in Atlantic salmon (*Salmo salar*) with emphasis on possible mediators of photoperiodic cues during parr-smolt transformation. *Proceedings of the 4th International Smolt Workshop, Aquaculture*, in press.

Honkanen, T. & Ekström, P. (1990). An immunocytochemical study of the olfactory projections in the three-spined stickleback, *Gasterosteus aculeatus*, L. *Journal of Comparative Neurology*, **292**, 65–72.

Honkanen, T. & Ekström, P. (1991). An immunocytochemical study of the development of the olfactory system in the three-spined stickleback (*Gasterosteus aculeatus* L., Teleostei). *Anatomy and Embryology*, **184**, 469–77.

Hornby, P. J., Piekut, D. T. & Demski, L. S. (1987). Localization of immunoreactive tyrosine hydroxylase in the goldfish brain. *Journal of Comparative Neurology*, **261**, 1–14.

James, V. A. & Wigham, T. (1984) Evidence for dopaminergic and serotonergic regulation of prolactin cell activity in the trout *Salmo gairdneri*. *General and Comparative Endocrinology*, **56**, 231–39.

Meier, E., Hertz, L. & Schousboe, A. (1991). Neurotransmitters as developmental signals. *Neurochemistry International*, **19**, 1–15.

Olivereau, M., Olivereau, J.-M. & Lambert, J.-F. (1987). In vivo effect of dopamine antagonists on melanocyte-stimulating hormone cells of the goldfish (*Carassius auratus* L.) pituitary. *General and Comparative Endocrinology*, **68**, 12–18.

Olivereau, M., Olivereau, J.-M. & Lambert, J.-F. (1988) Cytological responses of the pituitary (rostral pars distalis) and immunoreactive corticotropin-releasing factor (CRF) in the goldfish treated with dopamine antagonists. *General and Comparative Endocrinology*, **71**, 506–15.

Östholm, T., Ekström, P., Bruun, A. & van Veen, Th. (1988) Temporal disparity in pineal and retinal ontogeny. *Developmental Brain Research*, **42**, 1–13.

Peter, R. E., Chang, J. P., Nahorniak, C. S., Omeljaniuk, R. J., Sokolowska, M., Shih, S. H. & Billard, R. (1986) Interactions of catecholamines and GnRH in regulation of gonadotropin secretion in teleost fish. *Recent Progress in Hormone Research*, **42**, 513–48.

Presson, J. C. & Fernald, R. D. (1986) Development of the optic tract in the cichlid fish *Haplochromis burtoni*. *Developmental Brain Research*, **26**, 179–86.

Specht, L. A., Pickel, V. M., Joh, T. H. & Reis, D. J. (1981). Light-microscopic immunocytochemical localization of tyrosine hydroxylase in prenatal rat brain. I. Early ontogeny. *Journal of Comparative Neurology*, **199**, 233–53.

van Veen, Th., Ekström, P., Nyberg, L., Borg, B., Vigh-Teichmann, I. & Vigh, B. (1984). Serotonin and opsin immunoreactivities in the developing pineal organ of the three-spined stickleback, *Gasterosteus aculeatus* L. *Cell and Tissue Research*, **237**, 559–64.

Vecino, E. & Sharma, S. C. (1992). The development of substance P-like immunoreactivity in the goldfish brain. *Anatomy and Embryology*, **186**, 41–7.

Wilson, S. W., Ross, L. S., Parrett, T. & Easter, S. S. (1990). The development of a simple scaffold of axon tracts in the brain of the embryonic zebrafish, *Brachydanio rerio*. *Development*, **108**, 121–45.

Wollmuth, L. P., Crawshaw, L. I. & Panayiotides-Djaferis, H. (1989). The effects of dopamine on temperature regulation in goldfish. *Journal of Comparative Physiology*, **159B**, 83–9.

Wong, A. O. L., Chang, J. P. & Peter, R. E. (1992). Dopamine stimulates growth hormone release from the pituitary of goldfish, *Carassius auratus*, through the dopamine D1 receptors. *Endocrinology*, **130**, 1201–10.

14
Developmental aspects of catecholamine systems in the brain of anuran amphibians

A. González, O. Marín, R. Tuinhof and W. J. A. J. Smeets

Introduction

The anurans constitute the order of amphibians that includes frogs and toads. The members of this order undergo striking changes not only during the embryonic period but also during postembryonic development. The latter changes are generally referred to as metamorphosis. Usually, metamorphosis changes a larva, which is adapted to a specific environment, into a young adult with more or less different needs, imposed on it by its specific environment (Wald, 1981; Fritzsch, 1990). Whereas the external changes during the transformation from free-living tadpoles into froglets have been extensively analyzed and described, current knowledge of metamorphic changes in the CNS is limited.

A few studies by means of the formaldehyde induced fluorescence (FIF) technique have dealt with developmental aspects of the catecholamine systems of amphibians (Bartels, 1971; Terlou & Ploemacher, 1973; Notenboom, 1974; McKenna & Rosenbluth, 1975; Sims, 1977; Corio & Doerr-Schott, 1988). These studies were focused primarily on hypothalamohypophysial relationships and provide, therefore, no detailed information about the ontogeny of catecholaminergic cell groups and fibers. Another shortcoming of the latter studies is that, in general, only late larval stages were used.

The main purpose of this chapter is to get a better insight in the development of catecholamine systems in representatives of a class of vertebrates that marks a crucial point in the evolution, i.e. the transition from an aquatic to a terrestrial lifestyle. A better understanding of the development of these systems in amphibians may provide the answers to several fundamental questions, such as the origin of the midbrain dopaminergic cell groups of anamniotes and the nature of the dopamine containing CSF-contacting cells of the hypothalamic periventricular organ. We have studied, therefore, immunohistochemically the appearance and development of catecholaminergic cell groups and fiber systems in the brains of *Xenopus laevis* and *Rana ridibunda*. Antisera against the enzyme tyrosine hydroxylase (TH) were applied to reveal the overall distribution of catecholaminergic neuronal elements in the brain, whereas a dopamine (DA) antiserum was used to stain selectively dopamine containing cell bodies and fibers. Consequently, structures that are immunopositive for TH, but immunonegative for DA contain probably noradrenaline or adrenaline (Smeets & Steinbusch, 1990).

The South African clawed frog, *Xenopus laevis*, has been selected as core species, because the possibility of hormone induced breeding, the availability of an accurate timetable of development (Nieuwkoop & Faber, 1967), and the ease of maintenance of this species under laboratory conditions make it particularly suitable for the study of developmental aspects of neurotransmitter systems. An additional advantage is that neither the developing nor the adult brains of *Xenopus* contain much of the neuromelanin that obscures the immunoreactive cell bodies and fibers in other amphibians. The data obtained in *R. ridibunda*, an anuran that possess a brain with widely distributed neuromelanin during development (Mensah & Finger, 1975), are primarily used to assess the general features as well as the species differences in the ontogeny of anuran catecholamine systems.

Sixty *Xenopus laevis* tadpoles, ranging from stage 37 to stage 65, and 35 *Rana ridibunda* tadpoles, ranging from stage 30 to stage 50, were used. Staging was done according to Nieuwkoop & Faber (1967, *Xenopus*) and Manelli & Margaritora (1961, *Rana*). For comparison with other frogs, the staging by Taylor & Kollros (1946, *R. pipiens*) and Gona, Hauser & Uray (1982, *R. catesbeiana*) were used as references. Embryos and tadpoles were anaesthetized in a 0.3% solution of tricaine methanesulfonate (MS 222, Sandoz) and, subsequently, processed for TH- or DA-immunohistochemistry. The immunohistochemical procedures followed in the present study are essentially the same as those described in previous studies dealing with TH- and DA-immunoreactivities in amphibian brains (González & Smeets, 1991; González, Tuinhof & Smeets, 1993).

Notes on the development of anuran amphibians

The distribution of TH immunoreactive (THi) and DA immunoreactive (DAi) cell bodies and fibers has been investigated in animals of which the age varied from late embryonic up to juvenile stages. Since our observations involve tadpoles of two different anuran species (*Xenopus laevis*, *Rana ridibunda*), some notes on the development of these species seem to be in order.

In Fig. 14.1, a comparative staging of the development of *Xenopus* and *Rana* is presented. The end of the embryonic period extends over a rather long period of time beginning with the appearance of an operculum that covers the external gills and ending with the total resorption of the external gills. The larval period, marked by independent feeding, is generally subdivided into three sets of stages (Gona et al., 1982): 1) *premetamorphic stages*, in which the tadpole merely grows in size and the buds of the hindlimbs appear on the lateral side of the body; 2) *prometamorphic stages*, characterized by the progressive formation of the hindlimbs. This period ends when the length of the tail is at its maximum and the more drastic changes of the metamorphosis start; and 3) *metamorphic climax*, marking the period in which the transformation of the tailed larval form into the tailless, four-legged juvenile occurs.

In the following sections, we describe the development of catecholamine systems in anurans keeping these generalized sets of stages in mind. A detailed description of the catecholamine systems in the developing brain of *Xenopus* will be provided first, followed by a brief account of the major events during development in the brain of *Rana*.

Development of catecholamine systems in *xenopus*

The antibodies against TH and DA used in the present study revealed a pattern of immunostaining in the brain that was generally consistent among animals of the same stage. However, in some cases, variation in the intensity of immunostaining was observed among animals treated identically. This variation might be due either to variations in internal development between individuals that are staged exclusively on the basis of external morphological features or to small variations in the technical procedures.

Late embryonic stages

The earliest tadpoles of *Xenopus* that were analyzed correspond to developmental stages 38 to 42. All of them were individuals close to hatching and independent feeding. At these late embryonic stages, rather well developed catecholamine cell groups are already present (Fig. 14.2). Already in tadpoles of stage 38, distinct THi/DAi cell bodies are found in the spinal cord, ventral to the central canal (Table 14.1, cc). These cells extend throughout the length of the spinal cord but do not reach caudal rhombencephalic levels at this early stage. At caudal spinal cord levels, the number of cells is larger and the cells are more packed together. The THi/DAi cells in the spinal cord are characterized by having a short, thick process that protrudes into the central canal and thin processes that course laterorostrally (Figs. 14.2, 14.3).

At stage 39, another group of large, THi/DAi cell bodies appears at caudal diencephalic levels, i.e. in the posterior tubercle (TP) dorsal to the infundibulum. Slightly later, at stages 40/41, three other groups of CA containing cells are recognized in the diencephalon, viz. the nucleus of the periventricular organ (NPv), its accompanying cells (aNPv), and the suprachiasmatic nucleus (SC) (Fig. 14.2). A remarkable feature of the DAi, liquor (CSF)-contacting cells of the periventricular organ is that they never show immunoreactivity with TH antibodies. The accompanying cells of the periventricular organ, on the contrary, do not contact the CSF but are immunopositive for both TH and DA antibodies (Fig. 14.2(*f*)). The THi/DAi cell bodies of the suprachiasmatic nucleus constitute a small population of cells dorsal and rostral to the chiasmatic ridge (Figs. 14.2(*e*), 14.4, 14.5). Although these cells lie very close to the wall of the preoptic recess, a distinct relation with the ependymal layer cannot be recognized.

At about the same time, two new groups of THi

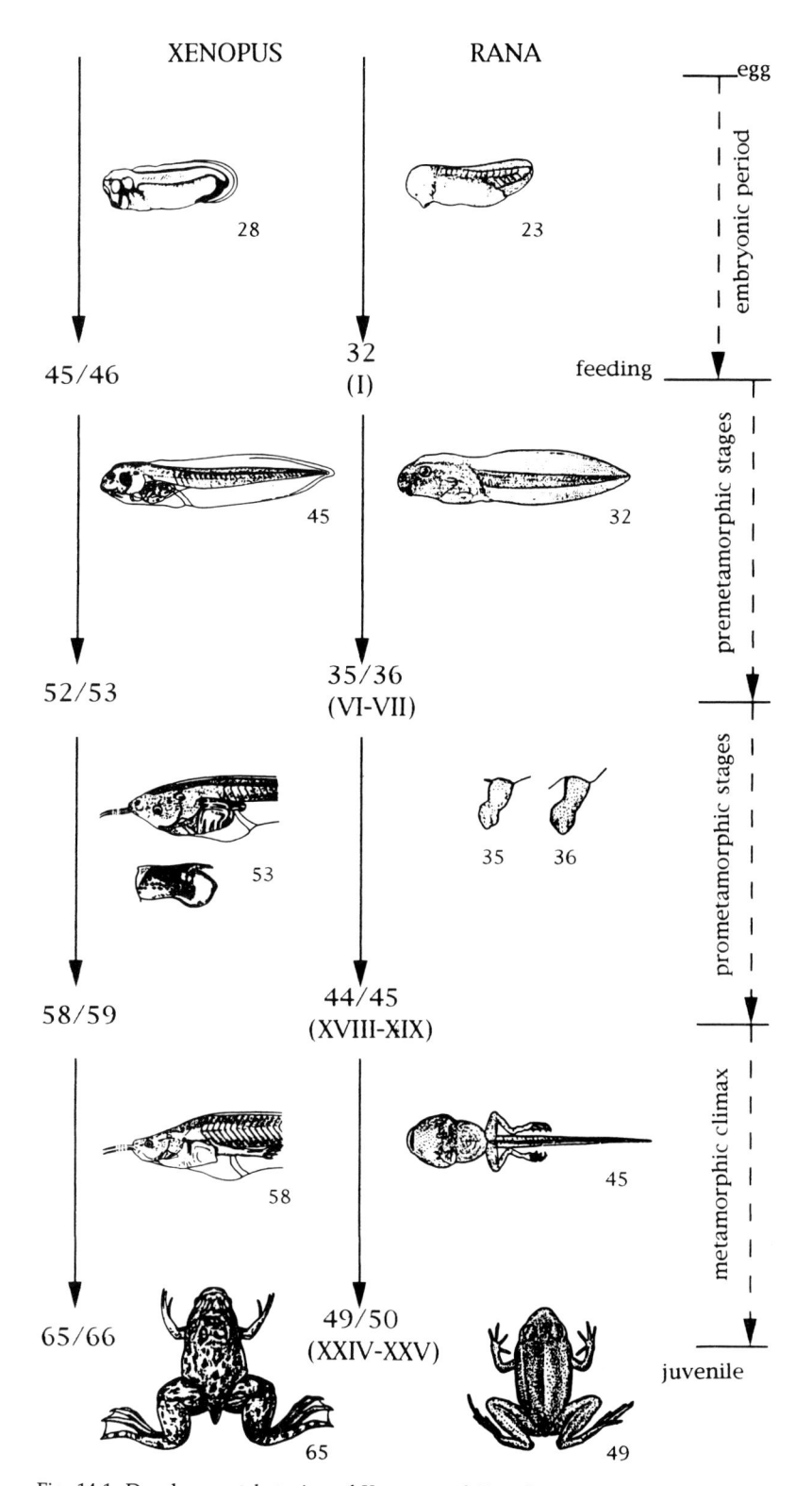

Fig. 14.1. Developmental staging of *Xenopus* and *Rana* from egg to juvenile. The periods of development, as indicated on the right side, have been adapted for *Xenopus* (left column) after Nieuwkoop & Faber (1967), for *Rana esculenta* (Arabic numbers in central column) after Manelli & Margaritora (1961), and for *Rana pipiens* (Roman numbers in central column) after Taylor & Kollros (1946).

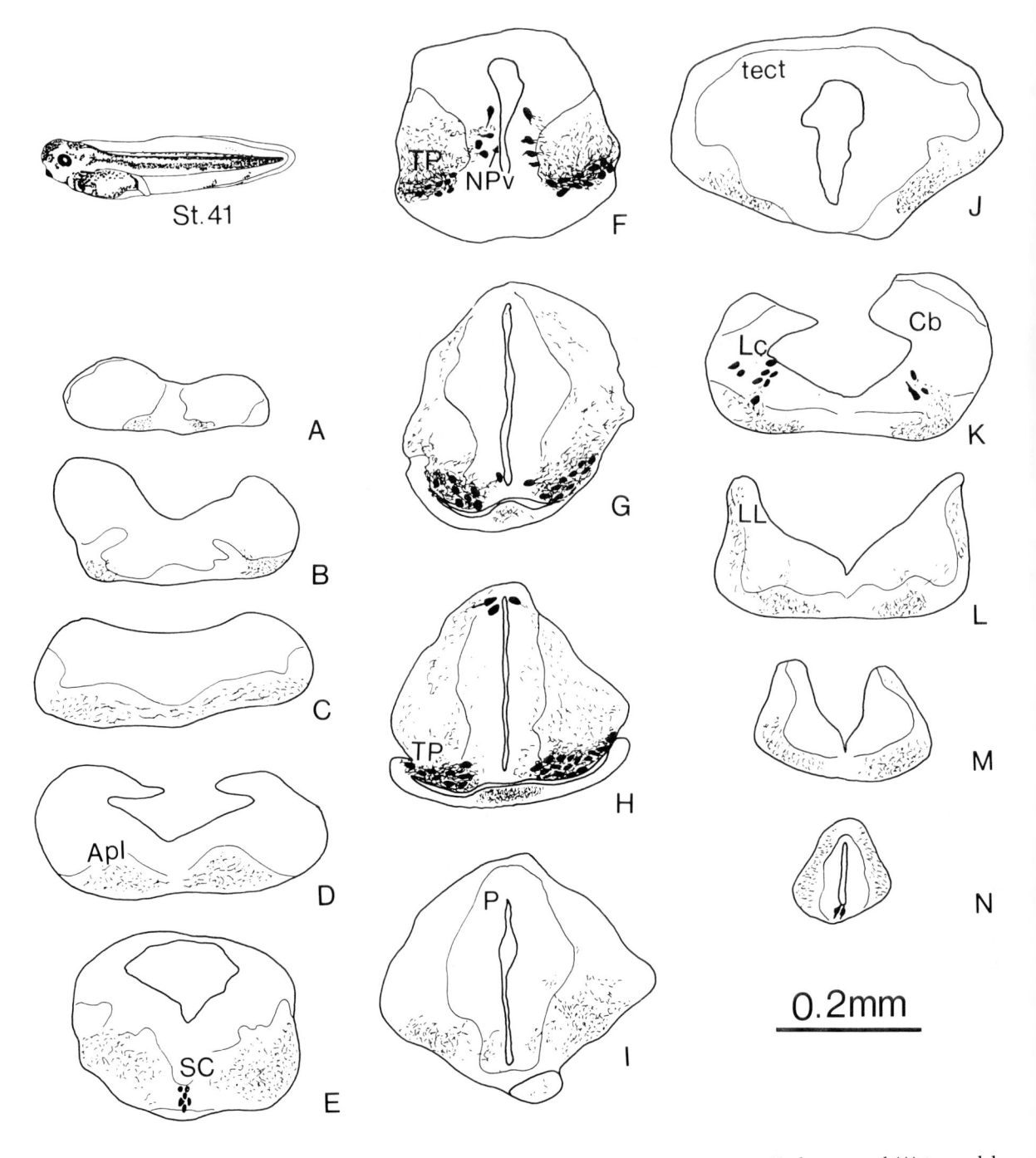

Fig. 14.2. Diagrams of transverse sections through the brain of *Xenopus laevis* at stage 41, from rostral (A) to caudal (N), showing the position of THi cell bodies (large dots) and fibers (small dots, wavy lines).

Table 14.1. Timetable of appearance of CA cell groups in the CNS of *Xenopus laevis* from embryonic stage 38 through climax stage 61. Horizontal bars indicate the presence of cell groups containing only THi cell bodies (small dots), THi/DAi cell bodies (solid black) or cells that are exclusively DAi (large dots).

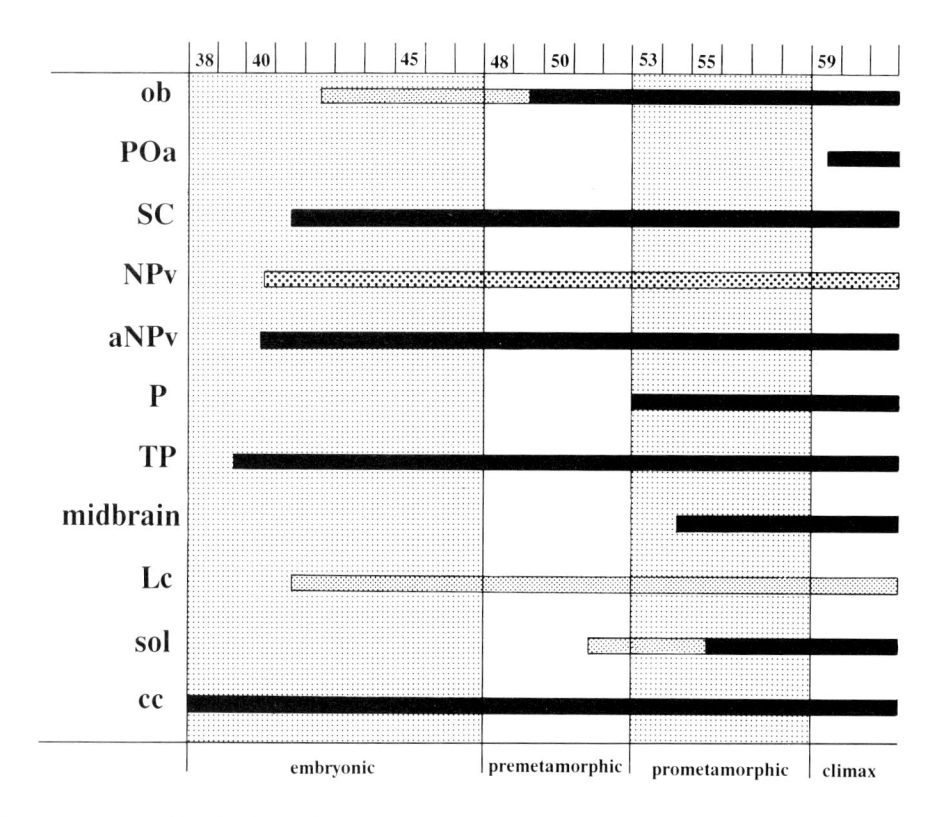

cells appear outside the hypothalamus. One group of weakly immunoreactive cells lies in the immature olfactory bulb (stage 42, Fig. 14.5), the other group, also containing weakly stained cells, is found in the isthmic region at the level of the developing cerebellum (stage 41, Figs. 14.2(k), 14.7). Some cells of the latter group lie close to the ventricle, whereas others are located at the border between the cell zone and the fiber zone. Weakly stained processes of the THi cells extend ventrolaterally into the fiber zone. In the case of one tadpole at stage 41, but not others at this stage, a few THi cell bodies were found in the dorsal part of the diencephalon, close to the midline (Figs. 14.2(h), 14.6). However, in subsequent developmental stages, neither THi nor DAi cell bodies were observed in a corresponding position.

At stage 41, several fiber systems are observed that are immunoreactive with TH and DA antisera. THi/DAi fibers can already be traced to the hypophysis, the basal forebrain, the ventral and ventrolateral tegmental parts of the midbrain and rhombencephalon, and the marginal zone of the spinal cord (Figs. 14.2, 14.6–14.8). At stages 40/41, some fibers leave the main stream of fibers in the ventrolateral rhombencephalic tegmentum. They course to the octavolateral area, where they terminate primarily in the area that receives lateral line input (Figs. 14.2(l), 14.8).

Several features of the distribution pattern of immunoreactivity in larvae just before premetamorphosis (stages 43–45) deserve comments. First the number of THi/DAi cell bodies in the posterior tubercle has dramatically increased, particularly in its mediocaudal part. A second feature to be mentioned is the appearance of THi/DAi cell bodies in the retina (Figs. 14.9, 14.10). The cells, which are few in number, lie in the innermost cell row of the inner nuclear layer. They have processes which arborize profusely within the inner plexiform layer but do not constitute distinct laminae. Dopaminergic cells in the retina are consistently observed in later stages of development, but the pattern, as seen in adults, is not gained before late prometamorphic stages (Figs. 14.11, 14.12).

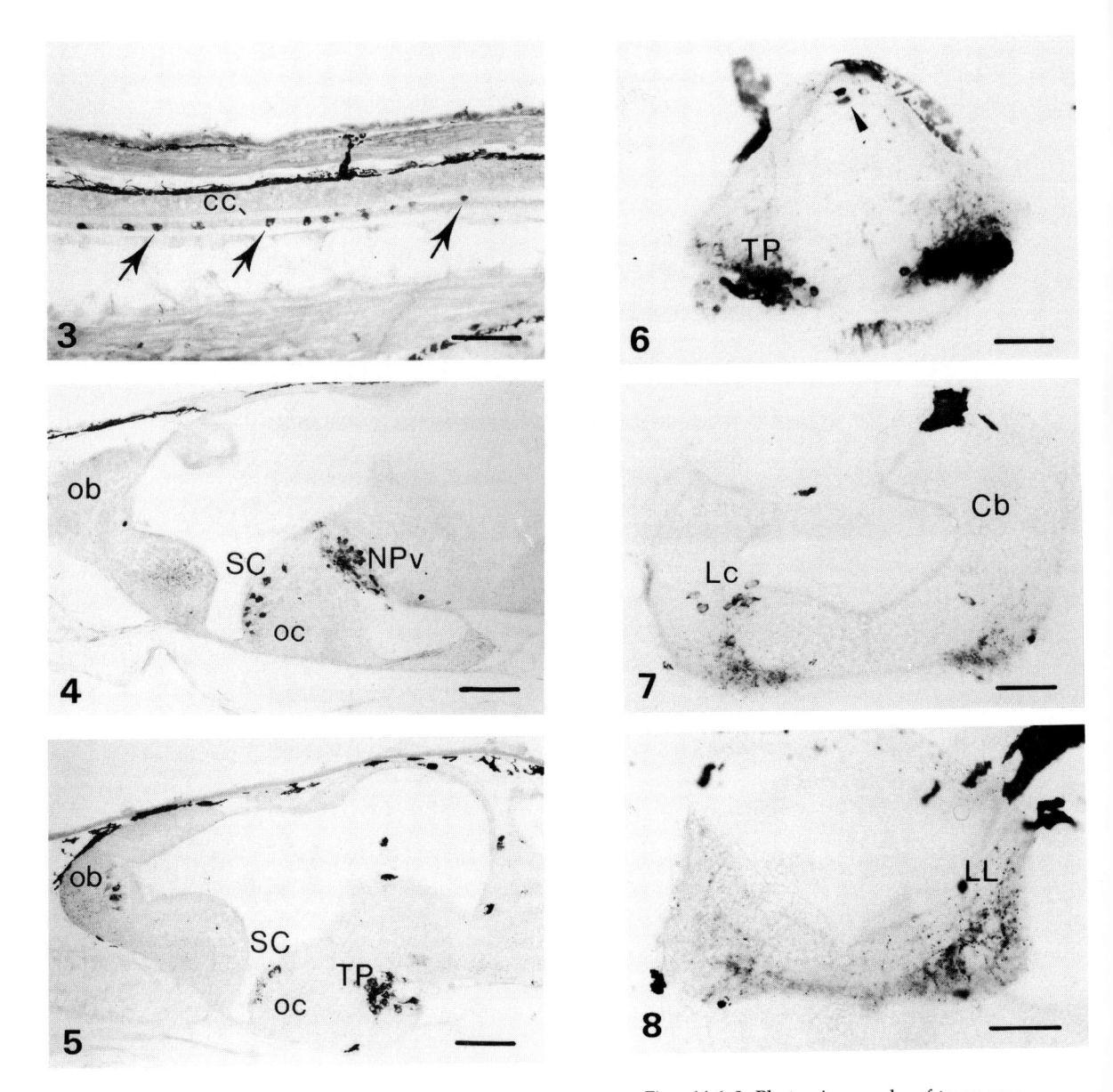

Figs. 14.3–14.5 Photomicrographs of sagittal sections of *Xenopus laevis* showing DAi cell bodies (arrows) ventral to the central canal (Fig.14.3, stage 40), in the nucleus of the periventricular organ and the suprachiasmatic nucleus (Fig.14.4, stage 42), and THi cells in the olfactory bulb, suprachiasmatic nucleus and posterior tubercle (Fig.14.5, stage 42). The section shown in Fig.14.5 is slightly more lateral to that of Fig.14.4. Note that cell bodies in the olfactory bulb stain with TH but not with DA antibodies. Bar = 200 μm.

Figs. 14.6–8. Photomicrographs of transverse sections through the brain of *Xenopus laevis* (stage 41) stained with the TH antiserum. Fig.14.6 shows THi cells in the posterior tubercle (TP) and in the dorsal part of the diencephalon (arrow head). at a level comparable to that of Fig.14.2(*H*). The other two figures show the longitudinal tracts in the brainstem at the level of the locus coeruleus (Fig. 14.7) and the caudal rhombencephalon (Fig. 14.8). Bar = 50 μm.

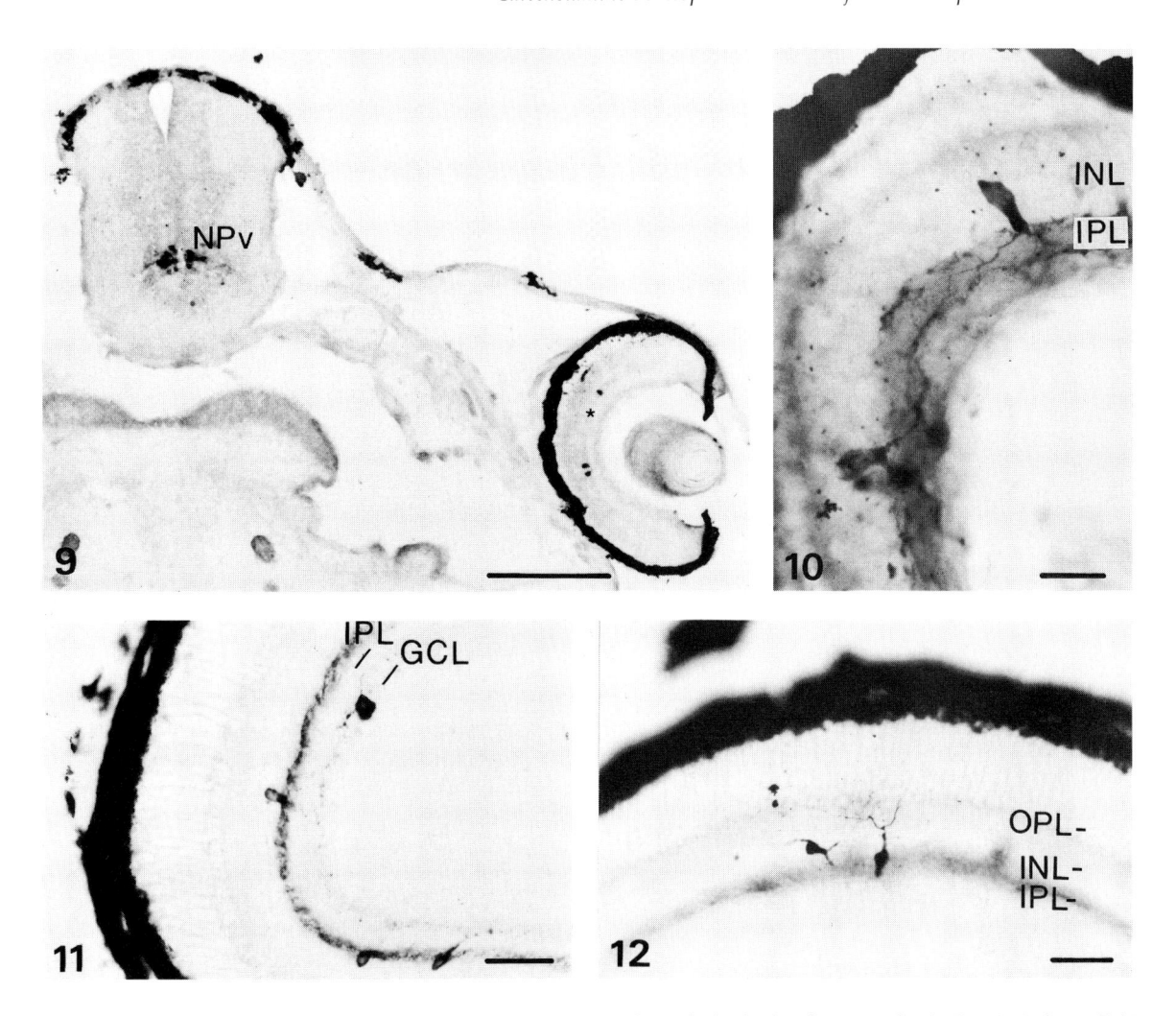

Figs. 14.9–14.12. Photomicrographs of transverse sections through the body of *Xenopus laevis*. Fig. 14.9 shows DAi cell bodies in the nucleus of the periventricular organ (NPv) and in the retina (asterisk) at stage 43. Fig. 14.10 is a higher magnification of the retina, at the same stage, showing the branching of the THi cells in the internal plexiform layer, whereas Fig. 14.11 and 14.12 illustrate the THi amacrine and displaced amacrine cells (Fig. 14.11) and interplexiform cells (Fig. 14.12). Bar = 200 μm (Fig. 14.9), 25 μm (Fig. 14.10), and 45 μm (Figs. 14.11, 14.12).

Premetamorphic stages

The development of CA systems in the brain of *Xenopus* during premetamorphic stages is characterized by a progressive maturation of THi/DAi cell bodies and fibers. From the beginning of this period, the immunoreactive cell bodies in the spinal cord display a more regularly spaced distribution and they are now more numerous at rostral spinal cord levels. In the isthmic region, the THi cells are clearly separated from the ventricle. Their main processes are directed to the ventrolateral tegmentum, whereas other, thinner processes extend dorsally into the lateral aspect of the cerebellum.

A new group of weakly stained, THi cells is, beginning with stage 51, observed in the proximity of the solitary tract. In the hypothalamus, the cell groups gain individuality. The THi/DAi cell bodies in the suprachiasmatic nucleus can easily be subdivided into two subgroups: one in the midline beneath the ventricular tip, the other in the lateral part of the nucleus. The medial subdivision contains cells that send short processes to the ventricular surface and are, therefore, considered to be liquor contacting. The THi/DAi accompanying cells of the periventricular organ have long processes that arborize extensively in the lateral parts of the diencephalon. Also the cell group in the posterior tubercle has matured considerably. Not only the number of cells has increased, but also their location has undergone changes. The cell group extends into the infundibular region (Fig. 14.13). Moreover, the cells that are located more medially, reach the midline in the diencephalic-mesencephalic transition area. A distinct, separate midbrain THi/DAi cell group is not observed yet.

At late premetamorphic stages (stages 52/53), THi/DAi cells are also present in the posterior thalamic nucleus. The cells are located near the ventricle and have processes that reach the lateral thalamus, the pretectal area, and even the midbrain tectum. Another change is noted in the olfactory bulb, where the CA cells not only show a increasing laminar organization, but also display, starting with stage 49, DA immunoreactivity.

During the premetamorphic stages, a steady increase of the number of THi/DAi cells in the retina occurs. Compared to previous developmental stages, the cells still lie exclusively in the inner nuclear layer, but their processes are now laminarly arranged within the scleral and vitreal sublaminae of the inner plexiform layer.

With respect to the development of CA fiber systems during the premetamorphic stages, it should be noted that considerably denser THi/DAi plexuses are found in the lateral line area (Fig. 14.14), the cervical and thoracic spinal cord segments (Fig. 14.15), and, starting from stage 51, in the nucleus accumbens and the striatum. Moreover, during the same period, immunoreactive fibers invade new targets, such as the torus semicircularis, tectum and thalamus.

Prometamorphic stages

The prometamorphosis covers a rather long period of time during which the previously described CA cell groups and fiber plexuses mature primarily by increasing in number (Figs. 14.16–14.22). The intensity of TH immunoreactivity in the brains of prometamorphic larvae largely resembles that observed in the adult brain. In the hypothalamus, long processes of cells located in the suprachiasmatic nucleus can be traced to the median eminence and, further caudally, to the intermediate lobe of the hypophysis (Fig. 14.17). From stage 54 on, a separate group of THi/DAi cells is found in the midbrain tegmentum (Fig. 14.17). In the olfactory bulb, the THi cells display a laminar organization with large, strongly immunoreactive

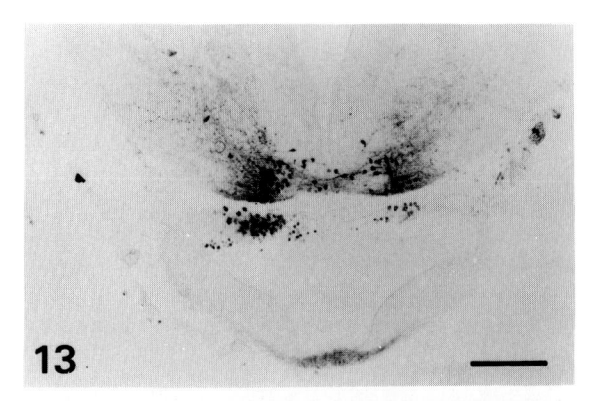

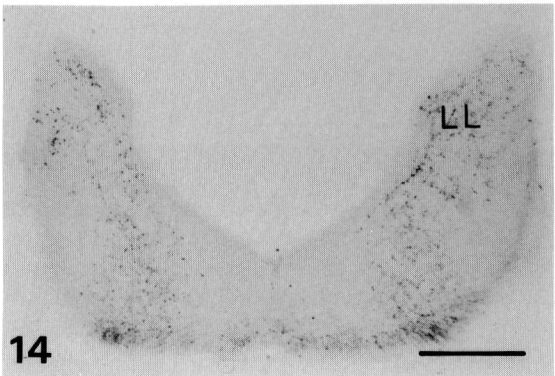

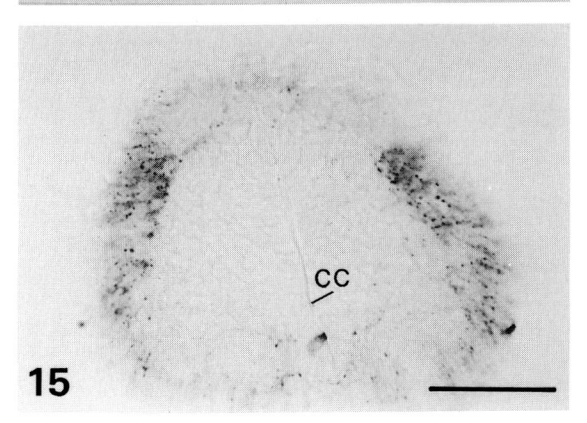

Figs. 14.13–14.15. Photomicrographs of transverse sections through the brain of *Xenopus laevis* during premetamorphosis (stages 49–50) showing THi cells in the di-mesencephalic transition zone (Fig. 14.13), and the distribution of THi fibers in the rhombencephalon (Fig. 14.14) and spinal cord (Fig. 14.15). Bar 200 μm (Fig. 14.13), 100 μm (Figs. 14.14, 14.15).

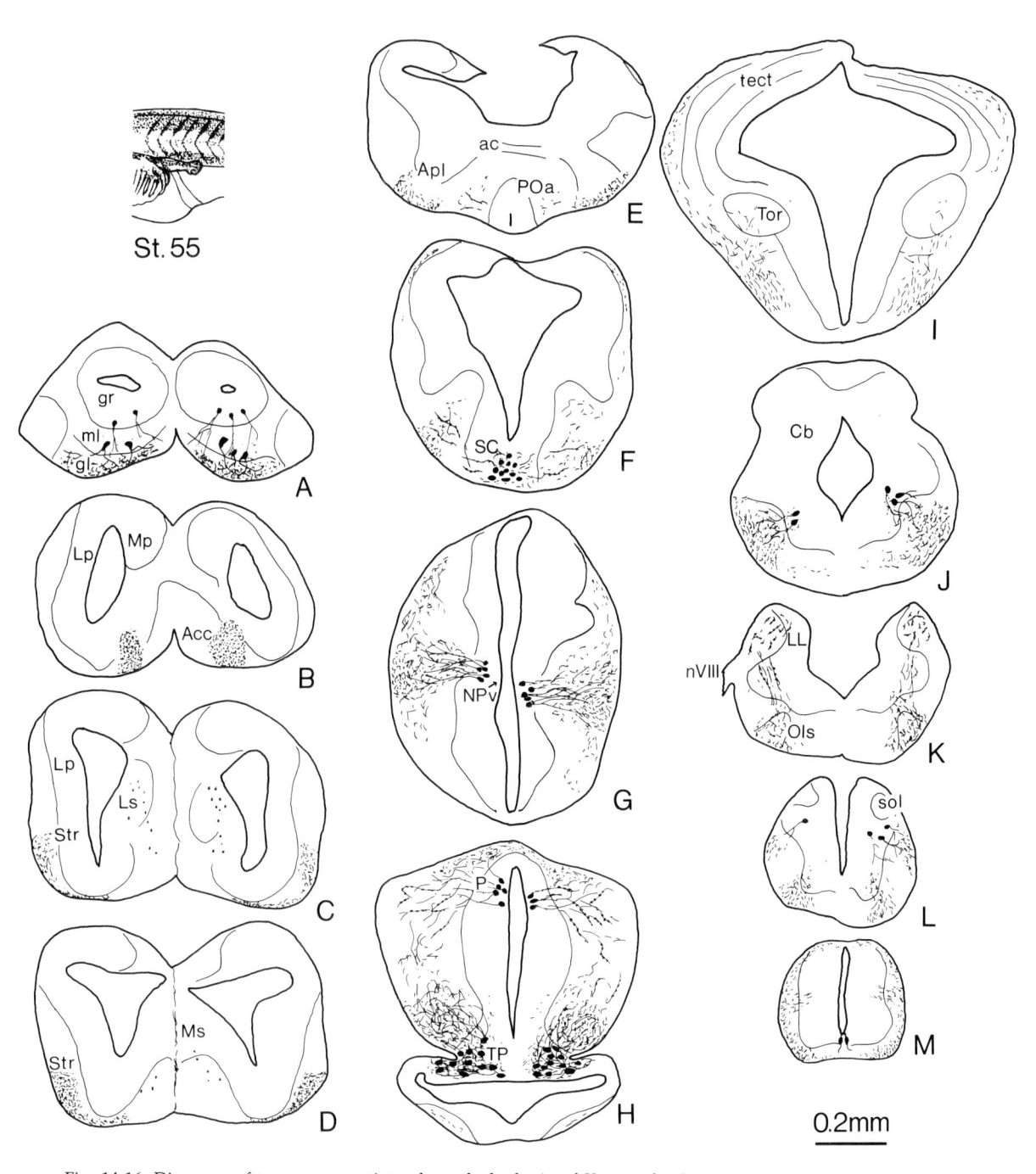

Fig. 14.16. Diagrams of transverse sections through the brain of *Xenopus laevis* at prometamorphic stage 55 showing, from rostral (A) to caudal (M), the distribution of THi cell bodies (large dots) and fibers (small dots, wavy lines).

cells located ventrally, around the glomeruli, and small, weakly THi cells in the granular layer (Figs. 14.16, 14.19, 14.20). In the posterior thalamic nucleus, the THi/DAi cells have long processes which extend into the lateral diencephalon where they intermingle with fibers arising from the hypothalamic CA cell groups (Figs. 14.16, 14.22). The isthmic group, i.e. the locus coeruleus, has expanded rostrocaudally and lies dorsal to well-developed tracts that course longitudinally throughout the brainstem (Figs. 14.16, 14.17). TH immunostaining of the few cells around the solitary tract is still weak during prometamorphic stages. A difference with previous stages, however, is that the cells are also weakly immunoreactive for DA antibodies from stage 55 on.

The distribution of THi/DAi fibers in brains of prometamorphic larvae is almost identical to that observed in adult *Xenopus*. Well-developed plexuses of immunoreactive fibers occur in the rostral nucleus accumbens, striatum, septum, brainstem ventrolateral tegmentum, lateral line area and spinal cord (Figs. 14.16–14.19).

As the prometamorphosis proceeds, the dopaminergic neuronal elements in the retina reach a high degree of organization. Thus, at stage 57, numerous THi/DAi cells are found in the inner nuclear layer. These cells show the typical morphology of amacrine cells with an extensive arborization of their processes into the vitreal and, in particular, scleral laminae of the inner plexiform layer. In addition, immunoreactive cells are present in the ganglion cell layer (*displaced amacrine cells*) with processes also branching within the inner plexiform layer (Fig. 14.11). Moreover, THi/DAi *interplexiform cells* are found with processes extending not only to the inner plexiform layer, but also to the outer plexiform layer (Fig. 14.12).

Metamorphic climax

At the time the juvenile stages are reached, a basic pattern of CA organization is observed that is similar to that present in the adult brain (see Chapter 5). The brain grows in size and complexity and the CA systems achieve their final development. The major events that occur during these stages concern the maturation of the innervation of certain brain structures, such as the olfactory bulb, thalamus, midbrain tectum and torus semicircularis. Moreover, two new groups of THi/DAi cell bodies appear during the metamorphic climax. One group, recognizable at stage 59, consists of CSF-contacting cells along the preoptic recess. The other group is a caudal continuation of the CA group around the solitary tract (now

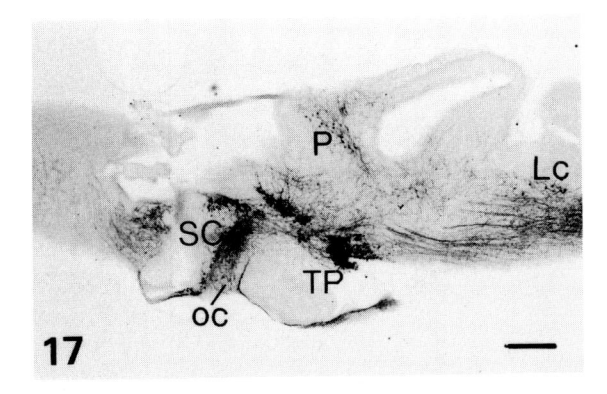

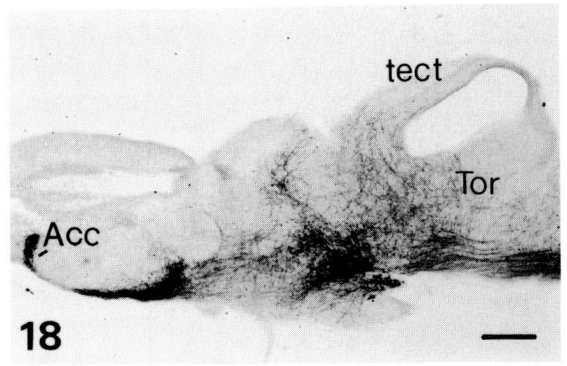

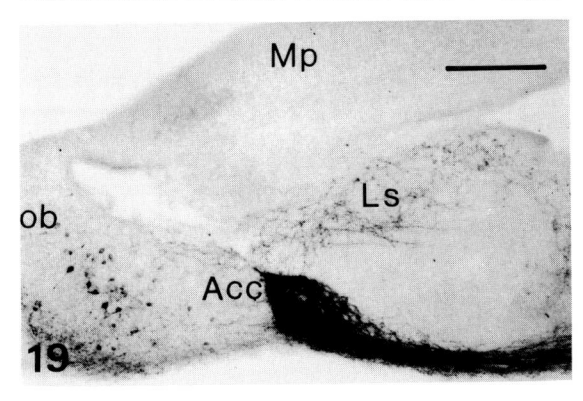

Figs. 14.17–14.19. Photomicrographs of sagittal sections through the brain of *Xenopus laevis* at prometamorphic stage 55 showing THi cell bodies and fibers in the forebrain and rostral brainstem. Fig. 14.17 is taken of a section that lies slightly lateral to that shown in Fig. 14.18. Fig. 14.19 is a higher magnification of the rostral telencephalon showing the dense THi innervation of the nucleus accumbens and the immunoreactive cell bodies in the olfactory bulb. Bar = 200 μm.

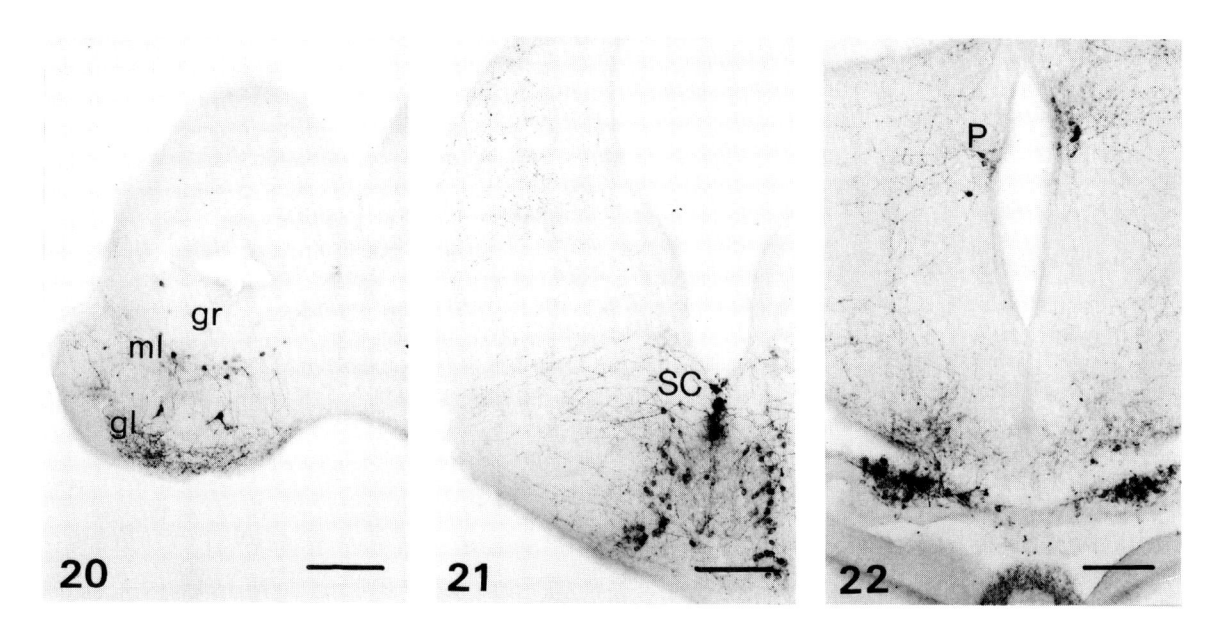

Figs. 14.20–14.22. Photomicrographs of transverse sections of *Xenopus laevis* (stages 55–59) showing THi cell bodies and fibers in the olfactory bulb (Fig. 14.20), the suprachiasmatic region (Fig. 14.21), and the caudal part of the diencephalon (Fig. 14.22). Bar = 100 μm.

well stained with TH and DA antibodies) into the presumptive area postrema which is located dorsal to the central canal, immediately caudal to the obex. The cells in the area postrema are not found until the end of the metamorphosis.

Development of CA systems in the brain of *Rana ridibunda*

The development of CA systems was also studied with TH antibodies in tadpoles of *Rana ridibunda* ranging from late embryonic up to juvenile stages. Through the embryonic period, the CA systems develop progressively and at stage 31, just before the beginning of the larval period, six distinct THi cell groups are observed. As shown in Fig. 14.23, these groups are found in the olfactory bulb (Fig. 14.24), suprachiasmatic nucleus (Fig. 14.25), posterior tubercle up to the diencephalomesencephalic transition zone (Fig. 14.26), isthmic region, solitary tract/area postrema (Fig. 14.27), and ventral to the central canal of the spinal cord. In particular, the early appearance of numerous cell bodies at caudal brainstem levels was noted.

The distribution of THi fibers in the brain of tadpoles at late embryonic stages is rather limited (Fig. 14.23). Dense plexuses of immunoreactive fibers are found in the glomeruli of the olfactory bulb (Figs. 14.23(a), 14.24), the ventral and ventrolateral portions of the diencephalon (Figs. 14.23(d),(e), 14.25), the ventrolateral tegmentum throughout the brainstem (Figs. 14.23(f)–(j), 14.27), and the lateral line area (Fig. 14.27). A weak to moderate plexus is observed in striatal and amygdalar regions of the telencephalon (Fig. 14.23(b),(c)).

During the premetamorphic period (stages 32–39), no new groups of THi cell bodies or fiber plexuses appear but the maturation of CA systems proceeds by increase of the number of cells and fibers. Particularly, the innervation of the nucleus accumbens, septum, thalamus, tectum and torus semicircularis (Fig. 14.29) is well developed. In this period, the CA innervation of the lateral line area reaches its maximum (Fig. 14.30).

In the subsequent prometamorphic period (stages 39–44/45), the distribution of THi cell bodies and fibers shows more and more resemblance with the pattern observed in adults (Fig. 14.28, see also Chapter 5). Additional THi cell groups are now observed in the posterior thalamic nucleus, midbrain tegmentum, and around the midline at caudal rhombencephalic levels. In the infundibulum, a group of cells is present which are separated from but extend processes to the ventricular surface suggesting that

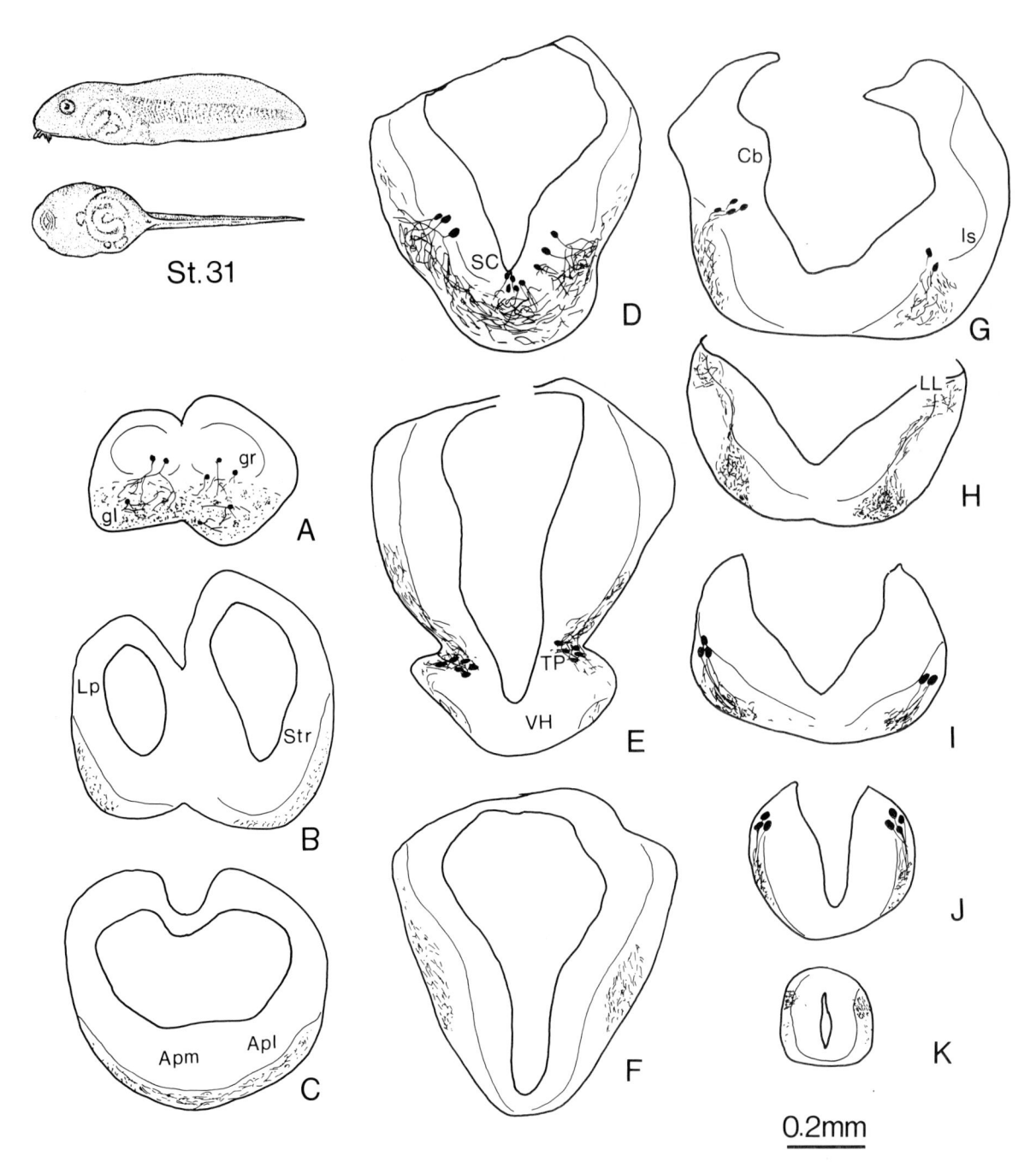

Fig. 14.23. Diagrams of transverse sections through the brain of *Rana ridibunda* (stage 31) showing, from rostral (A) to caudal (K), the distribution of THi cell bodies (large dots) and fibers (small dots, wavy lines).

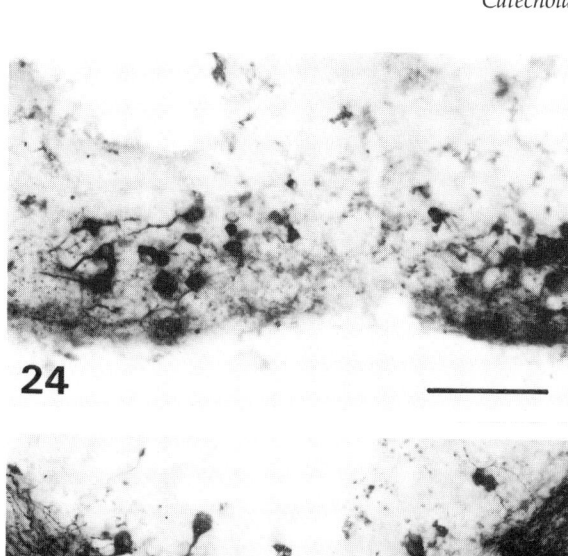

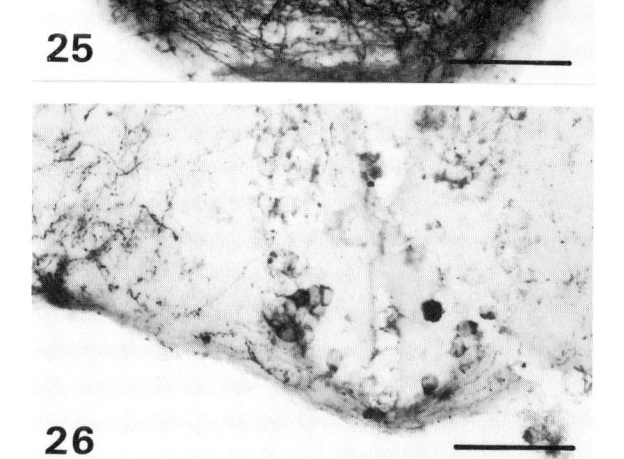

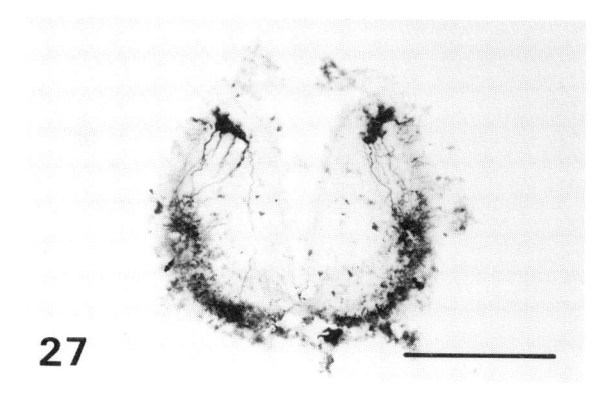

they contact the CSF (Fig. 14.31). However, cells in the nucleus of the periventricular organ never express immunoreactivity for TH. In the isthmic region, the THi cells in the locus coeruleus stain intensely and possess long, thin processes that enter the cerebellum (Fig. 14.32).

Notable features of the development of CA systems during the metamorphic climax are the appearance of THi cell bodies in the anterior preoptic area along the ventricular recess and the regression of the CA innervation of the rhombencephalic alar plate. The latter phenomenon parallels the progressive loss of the lateral line system that takes place in this anuran species during metamorphosis.

Concluding remarks

Comparative aspects of development of CA systems in the brain of amphibians

From the previous account, it is clear that the catecholamine systems in the two anuran species studied develop early during embryonic stages and that they show a largely similar temporal sequence of appearance of CA neuronal elements. There are, however, a few differences between the species that deserve comment. First, whereas in tadpoles of *Xenopus* THi cell bodies adjacent to the solitary tract are not seen until midpremetamorphosis, in *Rana* numerous THi cell bodies are already observed in a corresponding position at the end of the embryonic period. Secondly, an earlier appearance of THi cell bodies is also noted in the suprachiasmatic region of *Rana*. Moreover, cells in the olfactory bulb of *Rana* are strongly THi early in development, whereas in *Xenopus* these cells are only weakly immunoreactive and do not show DA immunoreactivity until premetamorphic stages. Since no data about DA immunoreactivity are available for *Rana*, it is currently unclear whether DA is already present in immunohistochemically detectable amounts at these early developmental stages.

Caption to Figs. 14.24–27

Figs. 14.24–14.27. Photomicrographs of transverse sections though the brain of *Rana ridibunda* tadpoles (stages 30–31) showing THi cell bodies and fibers in the olfactory bulb (Fig. 24), the suprachiasmatic nucleus (Fig. 14.25), the ventral part of the di-mesencephalic transition zone (Fig. 14.26), and in the caudal brainstem, close to the obex (Fig. 14.27). Bar = 100 μm.

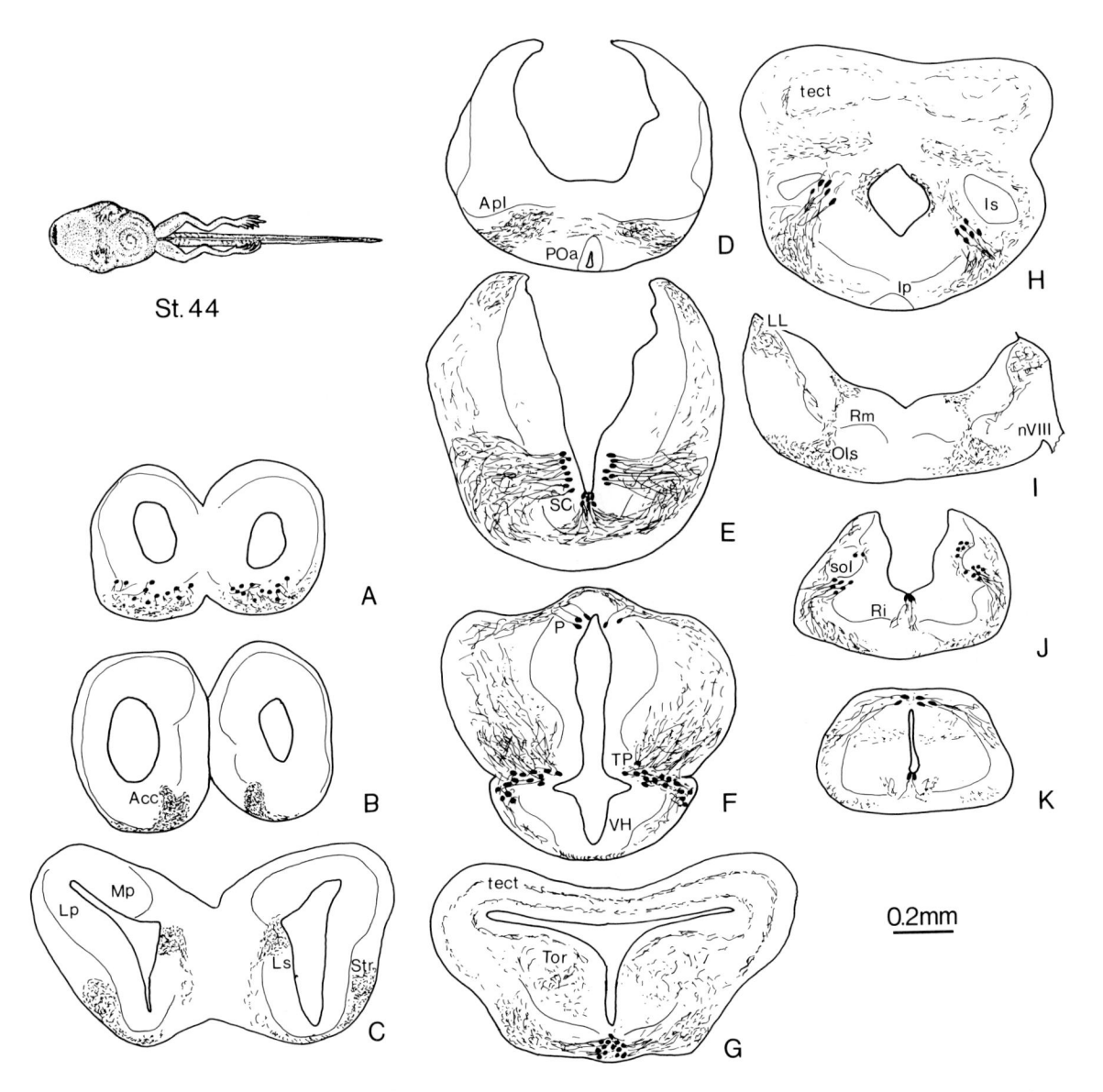

Fig. 14.28. Diagrams of transverse sections through the brain of *Rana ridibunda* at the end of the prometamorphosis (stage 44) showing, from rostral (A) to caudal (K) the distribution of THi cell bodies (large dots) and fibers (small dots, wavy lines).

When our results are compared with those of previous studies by means of the formaldehyde-induced fluorescence technique (Bartels, 1971; Terlou & Ploemacher, 1973; Notenboom, 1974; McKenna & Rosenbluth, 1975; Corio & Doerr-Schott, 1988), it becomes clear that TH immunohistochemistry enables an earlier detection of CA cell groups and fiber systems in amphibians. We found that these systems already start to organize at stages 39–41 (*Xenopus*) and 31–32 (*Rana*) which coincide with the moment that embryos start feeding, constantly swimming (without tactile stimulation), and using their gill structures, visual and lateral line systems. A recent study of the development of THi cell groups and fibers in the diencephalon and mesencephalon of *Rana catesbeiana* during premetamorphic, prometamorphic and climax stages (Carr, Norris & Samora, 1990) revealed a similar sequence of appearance of TH cell groups in these brain divisions as reported by us. However, whereas we found a distinct THi/DAi hypothalamohypophysial tract at early prometamorphic stages, Carr et al. were unable to identify immunoreactivity in that tract until the beginning of the climax period.

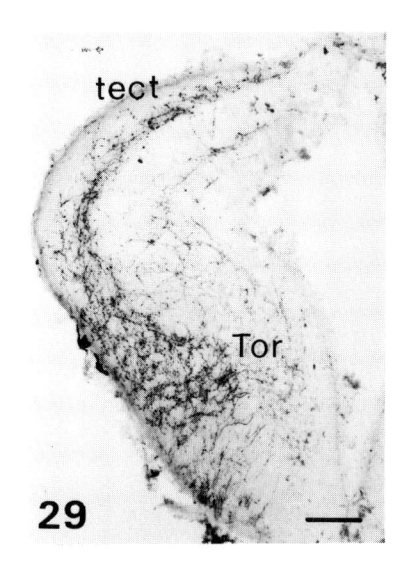

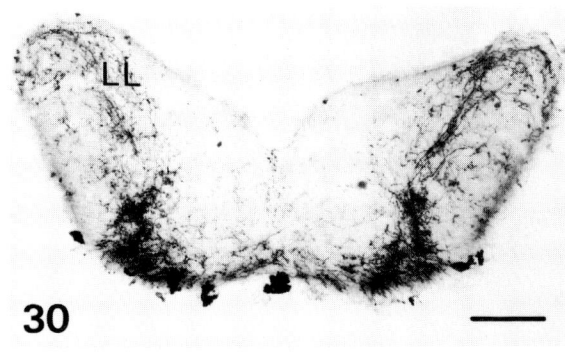

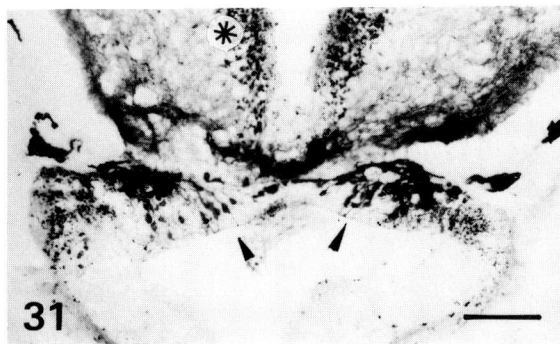

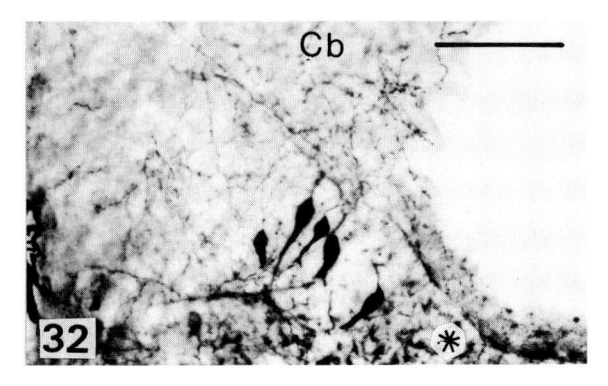

Temporal correlation between the development of CA systems and other developmental aspects of the CNS of anurans

There are relatively few studies that deal with changes in the CNS of anurans during development. The data are scarce and very diverse depending upon the point of interest of the authors. Nevertheless, in this section we attempt to correlate the data in the literature about developmental aspects of the CNS of anurans with events that take place, at the same time, in the development of the CA systems.

Spinal cord A remarkable finding of the present study is the presence of numerous immunoreactive fibers in the marginal zone of the spinal cord of the two anuran species studied. Even though the cell population ventral to the central canal of *Xenopus* has been shown to be immunoreactive for TH (Chen & Heathcote, 1992), it seems unlikely that the longitudinal THi fibers arise exclusively from these cells. On the basis of the results of retrograde tracing studies, candidates of supraspinal CA input to the anuran spinal cord are the nucleus of the solitary tract, the locus coeruleus, the posterior tubercle or the preoptic nucleus (ten Donkelaar et al., 1981; Tóth, Csank & Lázár, 1985). However, the appearance of the longitudinal THi fibers in the spinal cord preceeds that of THi cells in the areas just mentioned, except for those in the posterior tubercle. Moreover, developmental tract tracing studies have revealed that hypothalamo-spinal projections in *Xenopus* are not established until stage 57 (ten Donkelaar & de Boer-van Huizen, 1982),

Caption to Figs. 14.29–32

Figs. 14.29–14.32. Photomicrographs of transverse sections through the brain of *Rana ridibunda* during premetamorphosis (Figs. 14.29, 14.30) and prometamorphosis (Figs. 14.31, 14.32) showing THi fibers in the midbrain (Fig. 14.29) and rhombencephalon (Fig. 14.30), and THi cell bodies in the infundibulum (Fig. 14.31, arrowheads point to CSF-contacting cells) and the locus coeruleus (Fig. 14.32). Asterisks indicate neuromelanin. Bar = 100 μm.

whereas projections arising from the locus coeruleus and the solitary tract nucleus are not present until stage 43 and 58/60, respectively (van Mier & ten Donkelaar, 1984; Norlander, Baden & Ryba, 1985). Double labeling experiments with tract tracing techniques and immunohistochemistry are necessary for a better understanding of the origin of CA fibers to the spinal cord. A similar discrepancy may exist in ranid frogs (cf. Forehand & Farel, 1982; this chapter). The early appearance of THi fibers in the marginal zone of the spinal cord resembles the previously described serotoninergic supraspinal input (Soller, 1977; van Mier et al., 1986) and suggests an involvement of both monoaminergic systems in early locomotor behavior.

Rhombencephalic lateral line area. In most anuran amphibians, e.g. *Rana*, the neuromasts, afferents and second order cells of the brainstem that are related to the lateral line system, are lost during metamorphosis (Wahnschaffe, Bartsch & Fritzsch, 1987). However, several anurans, including *Xenopus*, retain a functional lateral line system throughout life (Fritzsch, Nikundiwe & Will, 1984). The appearance of THi fibers within the rhombencephalic lateral line area at early premetamorphic stages preceeds slightly the moment of which the lateral line nuclei can be recognized (Jacobi & Rubinson, 1983: *Rana catesbeiana*). The density of THi fibers increases during the prometamorphosis in both *Rana* and *Xenopus*. Whereas in the latter species a distinct plexus of THi fibers is present in the alar plate also during the climax, in *Rana* the innervation of the lateral line area decreases in parallel with the loss of lateral line neurons and octaval cell proliferation (Kollros, 1981; Wahnschaffe et al., 1987).

Mesencephalon and diencephalon. The development of the catecholaminergic innervation of the midbrain tectum parallels that of tectal growth following a sequence from rostral to caudomedial (Straznicky & Gaze, 1972; Lewis & Straznicky, 1979). THi/DAi fibers start to innervate lateral and superficial layers of the developing tectum and at the end of the premetamorphosis, when the pattern of retinotectal projections is established (Fawcett & Gaze, 1982), a laminar pattern of CA fibers is achieved at the same time.

The generation of diencephalic cells starts in the caudoventrolateral portion of diencephalon and continues in rostrodorsomedial direction (Clairambault, 1976; Tay & Straznicky, 1982). In the hypothalamus, the pre- and postchiasmatic parts display similar cytological characteristics but the prechiasmatic region does not differentiate until the end of the metamorphosis (Clairambault, 1976). The appearance of hypothalamic THi cell bodies follows the same wave of development, beginning with the lateral aspect of the posterior tubercle followed by cell groups that lie more medially such as the accompanying cells of the periventricular organ and the nucleus suprachiasmaticus. In line with this pattern is also the relatively late appearance of THi/DAi cells in the anterior preoptic area. Probably the most remarkable finding of the present study is that the THi/DAi cells in the midbrain of anurans seem to originate from the caudal hypothalamic cell group. It is unknown yet, whether the midbrain TH/DA cells migrate out of the hypothalamus or constitute a continuous field of CA cells that spans the hypothalamus and the midbrain, with the hypothalamic ones maturing first. The time of appearance and the rostrocaudal distribution of the cells favors the first suggestion. This finding may have consequences for our concepts of evolution of mesotelencephalic DA projections in vertebrates. It would imply that, despite the similarity in adult brains, the midbrain TH/DA cell groups of anamniotes and amniotes have different sites of origin which may reflect fundamental differences in basal ganglia organization.

Telencephalon. As in the other brain regions, the ingrowth of THi fibers into the telencephalon parallels the development of the hemispheres (cf. Clairambault, 1976). The first fibers innervating the telencephalon are found in amygdalar and striatal areas followed by the nucleus accumbens and the lateral septum. Curiously, in *Rana* the fibers form initially a distinct neuropil external to the striatal cell plate, but in the metamorphic climax, the THi/DAi fibers lie within the cellular plate, as is the case in the adult brain. In *Xenopus*, on the contrary, the THi/DAi fibers stay external to the striatal cell layer.

Retina. Our study confirms the presence of THi amacrine cells in the retina of amphibians as reported previously (Watt et al., 1988; Zhu & Straznicky, 1990, 1991). It also shows that already at early stage 39 these cells are present in the inner nuclear layer of the retina of *Xenopus* and that they contain DA. Moreover, our results reveal THi/DAi putative displaced amacrine cells in the ganglion cell layer and interplexiform cells. Similar observations have been made in a urodele, the tiger salamander *Ambystoma tigrinum* by Yang et al., (1991). The latter authors found that about 10% of the THi cells lie in the ganglion cell layer, whereas one percent of the THi cells of the inner nuclear layer possess interplexiform processes.

Acknowledgements

The authors are much indebted to Mrs B. Jorritsma-Byham and Ms R.G.P. Wismans for technical assistance, and Mr D. de Jong for preparing the photomicrographs. The study was financially supported by NATO Collaborative Research Grant CRG 910970.

Abbreviations

ac	anterior commissure
Acc	nucleus accumbens
aNPv	accompanying cells of the periventricular organ
Apl	amygdala, pars lateralis
Apm	amygdala, pars medialis
Cb	cerebellum
cc	central canal
GCL	ganglion cell layer of the retina
gl	glomerular layer of the olfactory bulb
gr	granular layer of the olfactory bulb
INL	inner nuclear layer of the retina
Ip	nucleus interpeduncularis
IPL	inner plexiform layer of the retina
Is	nucleus isthmi
Lc	locus coeruleus
LL	lateral line nucleus
Lp	lateral pallium
Ls	lateral septum
ml	mitral cell layer of the olfactory bulb
nVIII	octaval nerve
Mp	medial pallium
Ms	medial septum
NPv	nucleus of the periventricular organ
ob	olfactory bulb
oc	optic chiasm
Ols	superior olivary nucleus
OPL	outer plexiform layer of the retina
P	posterior thalamic nucleus
POa	anterior preoptic area
Ri	nucleus reticularis inferior
Rm	nucleus reticularis medius
SC	nucleus suprachiasmaticus
sol	solitary tract
Str	striatum
tect	tectum mesencephali
Tor	torus semicircularis
TP	tuberculum posterius
VH	ventral hypothalamic nucleus

References

Bartels, W. (1971). Die Ontogenese der amin-haltigen Neuronen-systeme im Gehirn von *Rana temporaria*. *Zeitschrift für Zellforschung und mikroskopische Anatomie*, **116**, 94–118.

Carr, J. A., Norris, D. O. & Samora, A. (1990). Organization of tyrosine hydroxylase-immunoreactive neurons in the di- and mesencephalon of the American bullfrog (*Rana catesbeiana*) during metamorphosis. *Cell and Tissue Research*, **263**, 155–63.

Chen, A. & Heathcote, R. D. (1992). Development of an interneuronal pattern within the spinal cord. *Abstracts Society for Neuroscience*, **18**, 621.

Clairambault, P. (1976). Development of the prosencephalon. In *Frog Neurobiology*, ed. R. Llinas & W. Precht, pp.924–44, Heidelberg, Springer-Verlag.

Corio, M. & Doerr-Schott, J. (1988). The monoaminergic system in the diencephalon of the newt tadpole, *Triturus alpestris* (Mert). A histofluorescence study. *Journal für Hirnforschung*, **29**, 377–84.

Fawcett, J. W. & Gaze, R. M. (1982). The retinotectal fibre pathways from normal and compound eyes in *Xenopus*. *Journal of Embryology and Experimental Morphology*, **72**, 19–37.

Forehand, C. J. & Farel, P. B. (1982). Spinal cord development in anuran larvae: II. Ascending and descending pathways. *Journal of Comparative Neurology*, **209**, 395–408.

Fritzsch, B. (1990). The evolution of metamorphosis in amphibians. *Journal of Neurobiology*, **21**, 1011–21.

Fritzsch, B., Nikundiwe, A. M. & Will, U. (1984). Projection patterns of lateral-line afferents in anurans: a comparative HRP study. *Journal of Comparative Neurology*, **229**, 451–69.

Gona, A. G., Hauser, K. F. & Uray, N. J. (1982). Ultrastructural studies on Purkinje cell maturation in the cerebellum of the frog tadpole during spontaneous and thyroxine-induced metamorphosis. *Brain, Behavior and Evolution*, **20**, 156–71.

González, A. & Smeets, W. J. A. J. (1991). Comparative analysis of dopamine and tyrosine hydroxylase immunoreactivities in the brain of two amphibians, the anuran *Rana ridibunda* and the urodele *Pleurodeles waltlii*. *Journal of Comparative Neurology*, **303**, 457–77.

González, A., Tuinof, R. & Smeets, W. J. A. J. (1993). Distribution of tyrosine hydroxylase and dopamine immunmoreactivities in the brain of the South African clawed frog *Xenopus laevis*. *Anatomy and Embryology*, **187**, 193–201.

Jacobi, J. & Rubinson, K. (1983). The acoustic and lateral line nuclei are distinct in the premetamorphic frog, *Rana catesbeiana*. *Journal of Comparative Neurology*, **216**, 152–61.

Kollros, J. J. (1981). Transitions in the nervous system during amphibian metamorphosis. In *Metamorphosis: A Problem in Developmental Biology*, ed. L. I. Gilbert & E. Frieden, pp. 445–59, New York: Plenum Press.

Lewis, S. & Straznicky, C. (1979). The time of origin of the mesencephalic trigeminal neurons in *Xenopus*. *Journal of Comparative Neurology*, **183**, 633–45.

McKenna, O. C. & Rosenbluth, J. (1975). Ontogenetic studies of a catecholamine-containing nucleus of the toad hypothalamus in relation to metamorphosis. *Experimental Neurology*, **46**, 496–505.

Manelli, H. & Margaritora, F. (1961). Tavole cronologiche dello sviluppo di *Rana esculenta*. *Rendiconti dell' Accademia Nazionale dei XL*, Vol. XII, 1–15.

Mensah, P. & Finger, T. (1975). Neuromelanin: a source of possible error in HRP material. *Brain Research*, **98**, 183–8.

Nieuwkoop, P. D. & Faber, J. (1967). *Normal table of Xenopus laevis (Daudin)*, Amsterdam, North-Holland.

Norlander, R. H., Baden, S. T. & Ryba, T. M. J. (1985). Development of early brainstem projections to the tail spinal cord of *Xenopus*. *Journal of Comparative Neurology*, **231**, 519–29.

Notenboom, C. D. (1974). The relation between the caudo-dorsal region of the preoptic nucleus and the pars nervosa of the pituitary gland in *Xenopus laevis* tadpoles. An investigation based on hypothalamic lesions. *Cell and Tissue Research*, **149**, 457–71.

Sims, T. J. (1977). The development of monoamine-containing neurons in the brain and spinal cord of the salamander *Ambystoma mexicanum*. *Journal Comparative Neurology*, **172**, 319–36.

Smeets, W. J. A. J. & Steinbusch, H. W. M. (1990). New insights into the the reptilian catecholaminergic systems as revealed by antibodies against the neurotransmitters and their synthetic enzymes. *Journal of Chemical Neuroanatomy*, **3**, 25–43.

Soller, R. W. (1977). Monoaminergic inputs to frog motoneurons: An anatomical study using fluorescence histochemical and silver degeneration techniques. *Brain Research*, **122**, 445–58.

Straznicky, C. & Gaze, R. M. (1972). The development of the tectum in *Xenopus laevis*: an autoradiographic study. *Journal of Embryology and Experimental Morphology*, **28**, 87–115.

Tay, D. & Straznicky, C. (1982). The development of the diencephalon in *Xenopus*. An autoradiographic study. *Anatomy and Embryology*, **163**, 371–88.

Taylor, A. C. & Kollros, J. J. (1946). Stages in the normal development of *Rana pipiens* larvae. *Anatomical Record*, **94**, 7–23.

ten Donkelaar, H. J., de Boer-van Huizen, R., Schouten, F. T. M. & Eggen, S. J. H. (1981). Cells of origin of descending pathways to the spinal cord in the clawed toad (*Xenopus laevis*). *Neuroscience*, **6**, 2297–312.

ten Donkelaar, H. J., de Boer-van Huizen, R., Schouten, F. T. M. & Eggen, S. J. H. (1981). Cells of origin of descending pathways to the spinal cord in the clawed toad (*Xenopus laevis*). *Neuroscience*, **6**, 2297–312.

Terlou, M. & Ploemacher, R. E. (1973). The distribution of monoamines in the tel-, di- and mesencephalon of *Xenopus laevis* tadpoles, with special reference to the hypothalamo-hypophysial system. *Zeitschrift für Zellforschung und mikroskopische Anatomie*, **137**, 521–40.

Tóth, P., Csank, G. & Lázár, G. (1985). Morphology of the cells of origin of descending pathways to the spinal cord in *Rana esculenta*. A tracing study using cobaltic–lysine complex. *Journal für Hirnforschung*, **26**, 335–83.

van Mier, P., Joosten, H. W. J., van Rheden, R. & ten Donkelaar, H. J. (1986). The development of serotoninergic raphespinal projections in *Xenopus laevis*. *International Journal of Developmental Neuroscience*, **4**, 465–75.

van Mier, P. & ten Donkelaar, H. J. (1984). Early development of descending pathways from the brain stem to the spinal cord in *Xenopus laevis*. *Anatomy and Embryology*, **170**, 295–306.

Wahnschaffe, U., Bartsch, U. & Fritzsch, B. (1987). Metamorphic changes within the lateral-line system of Anura. *Anatomy and Embryology*, **175**, 431–42.

Wald, G. (1981). Metamorphosis: an overview. In *Metamorphosis: A Problem in Developmental Biology*, ed. L. T. Gilbert & E. Frieden, pp. 1–42, New York, Plenum Press.

Watt, C. B., Yang, S-Z, Lam, D. M. K. & Wu, S. M. (1988). Localization of tyrosine-hydroxylase-like immunoreactive amacrine cells in the larval tiger salamander retina. *Journal of Comparative Neurology*, **272**, 114–26.

Yang, C-Y, Lukasiewicz, P., Maguire, G., Werblin, F. S. & Yazulla, S. (1991). Amacrine cells in the tiger salamander retina: morphology, physiology, and neurotransmitter identification. *Journal of Comparative Neurology*, **312**, 19–32.

Zhu, B. & Straznicky, C. (1990). Dendritic morphology and retinal distribution of tyrosine hydroxylase-like immunoreactive amacrine cells in *Bufo marinus*. *Anatomy and Embryology*, **181**, 365–71.

Zhu, B. & Straznicky, C. (1991). Morphology and retinal distribution of tyrosine hydroxylase-like immunoreactive amacrine cells in the retina of developing *Xenopus laevis*. *Anatomy and Embryology*, **184**, 33–45.

15
Ontogenesis of catecholamine systems in the brain of the lizard *Gallotia galloti*

L. Medina, L. Puelles and W.J.A.J. Smeets

Introduction

The adult brains of most vertebrates have several features in common with respect to the distribution of catecholamine (CA) containing cell groups. For example in all vertebrates CA cell bodies are observed in the olfactory bulbs, hypothalamus, midbrain tegmentum, isthmic region and caudal brainstem (see Chapter 20 for review). Developmental studies of brains of birds and mammals have shown that CA cell bodies in many of these locations are already present at early embryonic stages (Tennyson, Mytilineou & Barrett, 1973; Lauder & Bloom, 1974; Specht et al., 1981a; Yurkewicz et al., 1981; Guglielmone & Panzica, 1985; Verney et al., 1991). Evidence has been obtained that the cell groups do not arise at once: CA cell bodies appear first in the hypothalamus and the isthmic region and, subsequently, in the midbrain tegmentum and other brain regions. A similar sequence of appearance of CA cell groups is observed also in the developing brain of the teleost, *Gasterosteus aculeatus* (Ekström, Honkanen & Borg, 1992). However, in the latter species, CA cells in the midbrain tegmentum were observed neither in the developing nor in the adult brain.

In order to better understand the evolution and ontogeny of the CA systems of vertebrates, in the present study we have analyzed the development of the CA systems of a reptile, the lizard *Gallotia galloti*. This study has been carried out using antibodies against the enzyme tyrosine hydroxylase (TH) as well as against dopamine (DA), which enable us to differentiate between the dopamine system and the noradrenaline/adrenaline system.

Material and methods

Twenty four embryos of the lizard *Gallotia galloti* (Reptilia, Lacertidae), from developmental stage 32 (S32) to hatching, were used in the present study (see Table 15.1). Embryos were removed from the egg, immersed in Ringer solution, and carefully staged according to Ramos Steffens (1980) and Dufaure & Hubert (1961). The embryonic brains were fixed by immersion in 5% glutaraldehyde, 1% $Na_2S_2O_5$ in 0.1 M, pH 7.4 phosphate buffer. After 15 to 30 minutes (depending on the size of the brain) the brains were immersed in phosphate buffered 30% sucrose containing 1% $Na_2S_2O_5$. Subsequently, the brains were cut on a cryostat at 50 µm in a sagittal, corrected horizontal (parallel to the optic tract), or frontal plane. Sections were serially collected in 0.05 M, pH 7.6 Tris-buffered saline (TBS) containing 1% $Na_2S_2O_5$. One series of sections was processed for tyrosine hydroxylase (TH) immunohistochemistry, whereas the other series was processed for dopamine (DA) immunohistochemistry.

In addition, seven adult specimens of *Gallotia galloti* were deeply anesthetized and perfused transcardially with saline solution followed by a phosphate buffered solution of either 4% paraformaldehyde ($n=4$) or 5% glutaraldehyde and 1% $Na_2S_2O_5$ ($n=3$). The brains were cryoprotected and sectioned on a cryostat (sagittal, corrected horizontal and transverse sections, 40 µm thick). Sections were then processed for either TH or DA immunohistochemistry. The steps in these immunohistochemical procedures are largely the same as previously described for other reptiles (Smeets & Steinbusch, 1990). In brief, for DA immunohistochemistry, sections were incubated in:

Table 15.1. *Number and development stages of embryos of* Gallotia galloti *used for the present study*

Developmental stage	Number of embryos
S32	1
S33	2
S34	1
S35	4
S36	3
S37	3
S38	2
S39	2
S40	2
Hatching	4

1 Rabbit antidopamine antiserum (generously provided by Dr R.M. Buijs, Netherlands Institute for Brain Research, Amsterdam), diluted 1 : 3,000 for 16 hours.
2 Swine antirabbit antiserum (Nordic), diluted 1 : 50 for 1 hour.
3 Rabbit peroxidase antiperoxidase complex (Dakopatts), diluted 1 : 800 for 1 hour.

For TH immunohistochemistry, sections were incubated in:
1 Mouse antityrosine hydroxylase antiserum (Incstar), diluted 1 : 2000 for 16 hours.
2 Goat antimouse antiserum (Nordic), diluted 1 : 100 for 1 hour.
3 Mouse peroxidase antiperoxidase complex (Dakopatts), diluted 1 : 500 for 1 hour.

Control sections were processed omitting the primary antiserum. No specific labeling of somata or fibers was found in these sections. Additional series of embryonic brains stained with cresyl violet or acetylcholinesterase were available for facilitating the interpretation of the location of immunoreactive elements.

Development of catecholamine systems in *Gallotia galloti*.

Adult brain

The distribution of TH immunoreactive (THi) and dopamine immunoreactive (DAi) cell bodies and fibers in the adult brain of the lizard *Gallotia galloti* is largely the same as that reported for another lizard species, i.e. *Gekko gecko* (Smeets, Hoogland & Voorn, 1986; see also Chapter 6). Only minor differences were observed between the two species in the dopaminergic innervation of the superficial tectal layers and in the retinorecipient thalamic nuclei (Medina & Smeets, 1992).

Embryonic development of the brain

Lacertid lizards are oviparous animals, whose embryonic development lasts for approximately 45 days at an average temperature of 25 °C. On the basis of morphological features of the embryo, this period has been divided in 40 developmental stages for the lizard *Lacerta muralis* (Dufaure & Hubert, 1961), and correspondents of the last 19 developmental stages of *L. muralis* (i.e. S22–S40) have been described for *Gallotia galloti* (Ramos Steffens, 1980). When the female lays eggs, the embryo is already at a developmental stage that ranges approximately from stage 24 to stage 28. At the moment of egg laying, the embryo (mainly the head and upper part of the body) is bent over itself, and the trunk is open on its ventral side (Ramos Steffens, 1980). Rudiments of the heart and the intestine are present in the embryo, and limb buds start to develop at stage 28 (Ramos Steffens, 1980). The number of somites ranges between 17 pairs (S24) and 36 pairs (S28), and there are one to four branchial arches. In the head the neuropore is closed, and eye and ear vesicles are observed. The neural tube is still at an early proliferative period, with most cells accumulated in a very thick ventricular zone, although its major divisions of prosencephalon, mesencephalon, and rhomben-cephalon are already recognised. The amniotic membrane totally covers the embryo, and a periembryonic vascular system is present (Ramos Steffens, 1980).

The first weakly TH immunoreactive (THi) cell bodies are found at stage 32 (S32) and, therefore, our survey on the development of reptilian catecholamine systems starts at this stage. In this section we first briefly describe the main events that happen in the body and head of *G. galloti* at each developmental stage as described by Ramos Steffens (1980), followed by a description of the results on TH and DA immunohistochemistry in the embryonic brain at the corresponding stage. The results on the development of TH and DA immunohistochemistry in the brain of *G. galloti* are summarized in Fig. 15.1, in which the THi and DAi perikarya and fibers are depicted in schematics of sagittal sections of the brain at each developmental stage. In addition, on the right side of each sagittal section in Fig. 15.1, a drawing of the embryo of *G. galloti* at the corresponding developmental stage is shown.

Developmental Stages 32 and 33 (S32, S33) At S32, the trunk of the embryo remains open through a narrow umbilicus, which connects the embryo to the

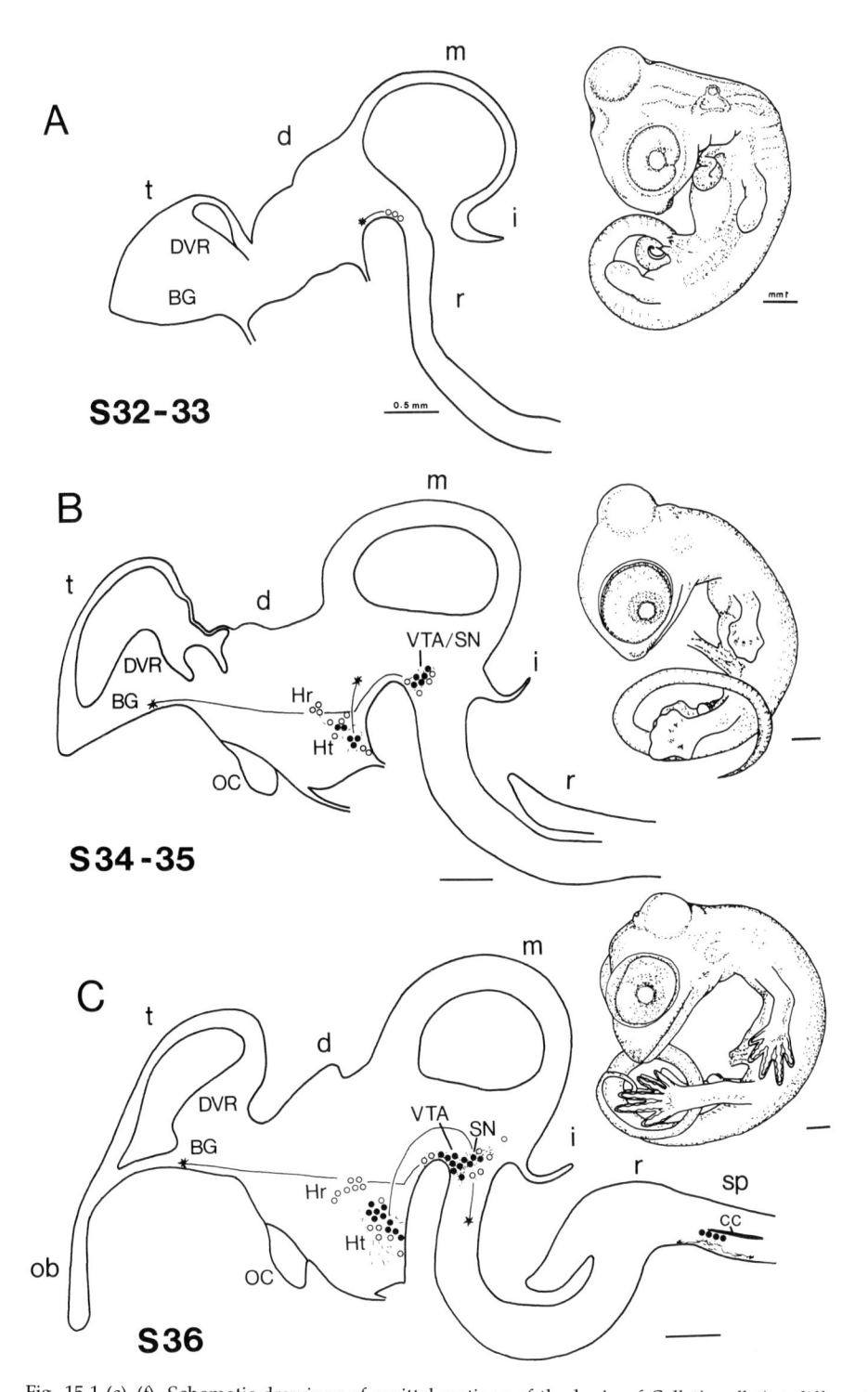

Fig. 15.1 (a)–(f). Schematic drawings of sagittal sections of the brain of *Gallotia galloti* at different developmental stages (S32-S40), representing the CA groups observed at each stage in the embryonic brain. Empty circles represent cell bodies that are THi, but DA negative. Filled circles represent cell bodies that are both THi and DAi. Triangles represent cell bodies that are DAi but TH negative. The distribution of immunoreactive fibers at each stage is indicated by fine lines and dots, and the developing CA fiber bundles are represented by a single line that ends in an asterisk, which represents a growth cone. Drawings of the whole embryo of *Gallotia galloti* at different developmental stages (taken from Ramos Steffens, 1980) are presented on the right side of the brains. Bar: brain = 0.5 mm; embryo = 1 mm.

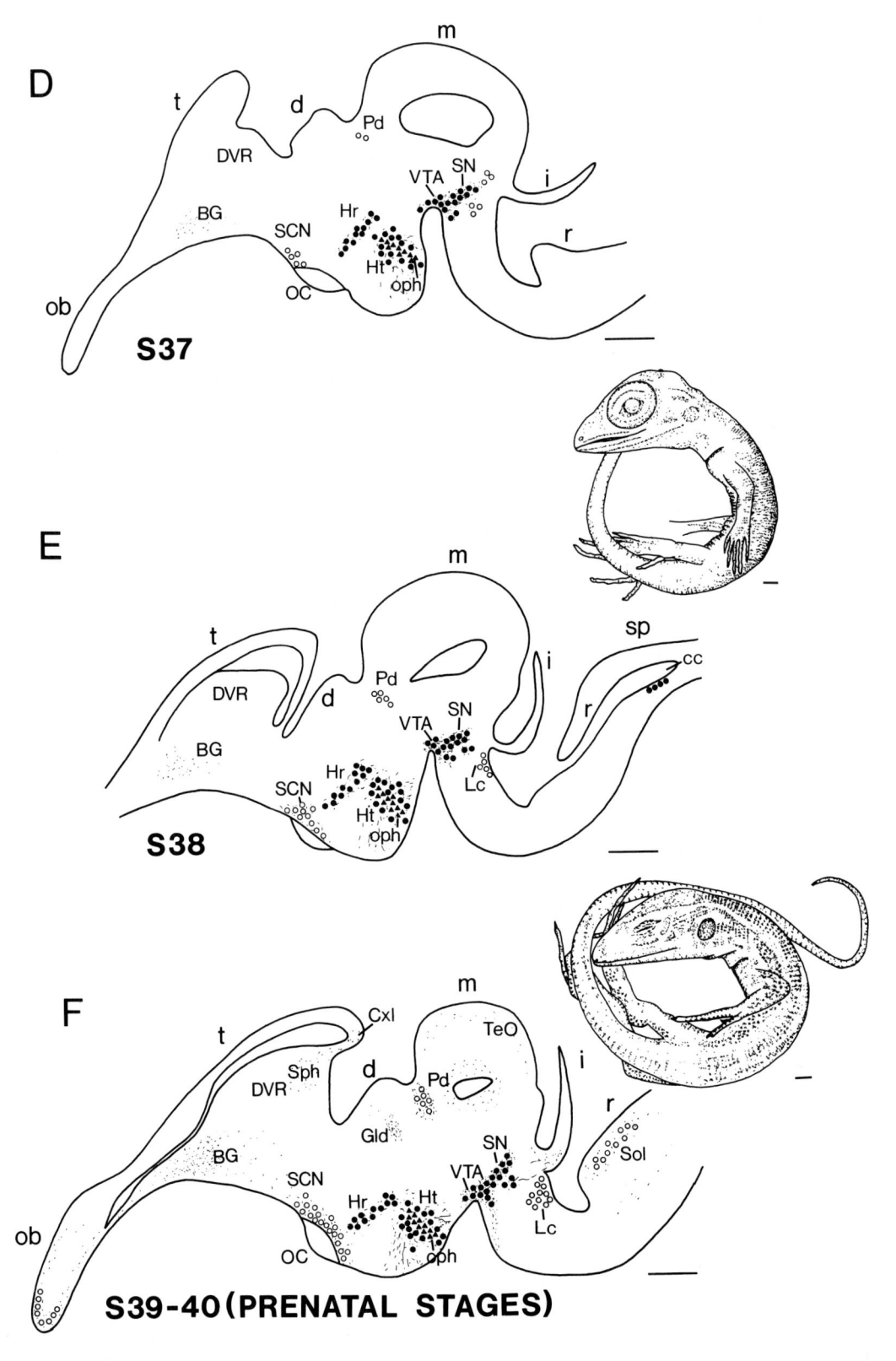

Fig. 15.1 (cont.)

yolk sac and embryonic membranes (Fig. 15.1(a). Rudiments of the liver and of the excretory and reproductive systems are already present. Four limbs are visible and each of them presents a flattening at the end which will become feet at later stages. The branchial arches are in regression and early jaw prominences are formed around the stomodeum. The olfactory vesicle is formed, and pigmentation has started in the eyes. In the brain, the prosencephalon is clearly divided into telencephalon and diencephalon. In addition to the high number of cells still located in the ventricular zone, a number of neuroblasts and immature neurons are observed in the mantle of the brain. In some regions, it is possible to distinguish clusters of cells in the mantle which represent the beginnings of nuclei such as the pretectal geniculate nucleus in the pretectum (Medina, 1991), or the ventral geniculate nucleus in the ventral thalamus (Trujillo, 1982). In the basal telencephalon, cell bodies located in the mantle begin to form the basal ganglia, but it is not possible to distinguish between striatum and pallidum at this early stage (Yanes et al., 1989). The boundary between the early basal ganglia and the dorsal ventricular ridge (DVR) is also unclear at S32 (Yanes et al., 1989).

At S32, a few, weakly THi cell bodies are found in the midbrain tegmentum, dorsal to the cephalic flexure and close to the midline (Fig. 15.1(a)). One stage later, at S33, both the number of THi cell bodies in this position and the intensity of their immunoreactivity have increased substantially (Figs. 15.1(a), 15.2(a,b)). The cell bodies display some immature features, such as the absence of dendrites and the presence of a long, thick process which is directed either rostralwards, lateralwards, or dorsalwards (Figs. 15.2(a), (b)). A few THi neuroblasts show a long process directed caudalwards. A small, THi fiber bundle arises in this midbrain cell group and courses rostralwards following the longitudinal axis of the brain (Fig. 15.2(a)). This fiber bundle reaches as far as the basal region of the caudal diencephalon.

At late S33, the THi perikarya of the midbrain tegmentum are even more numerous and strongly immunoreactive than observed at early S33 (Fig. 15.2(b)). Some THi cell bodies are located at more lateral positions in the tegmentum. The ascending THi fibers arising in the CA cell group of the midbrain tegmentum can be traced as rostral as the hypothalamus. Additionally, a few slightly THi perikarya are observed at tuberomammillary hypothalamic levels, where they are located close to the midline. These cell bodies are either round without any process or monopolar with a single, very thick process.

Developmental stage 34 (S34) At S34, toes start to differentiate in the limbs of the embryo, and all branchial arches have disappeared. In the head, jaw prominences have merged, and a rudimentary palate appears. The tongue is already observed in the mouth. In the brain, more neuroblasts and immature neurons are located in the mantle, which has become thicker than in previous stages. This is specially noticeable in the rhombencephalic and mesencephalic tegmentum, the pretectum, the ventral thalamus, and the tuberomammillary hypothalamus.

At this developmental stage, the THi cell group in the midbrain tegmentum largely resembles the corresponding group as observed at late S33. In a sagittal section, most THi perikarya are located in the caudal part of the mesencephalic tegmentum, where a sharp boundary defines the limit between the mesencephalic and isthmic tegmental regions (Fig. 15.3). Only a few THi perikarya are found caudal to this limit or crossing it (Fig. 15.3). Other THi cell bodies are displaced rostralwards from this main mesencephalic group, and lie between and parallel to the ascending THi fibers (Fig. 15.3). At this stage, the growing THi axons reach the caudal telencephalon.

At S34, the THi cell group in the tuberomammillary hypothalamus has increased considerably both in the number of cells and their immunoreactivity (Fig. 15.3). Some of these perikarya show the beginning of dendritic arborization, which gives them a more mature appearance. Others are still monopolar neuroblasts. A number of THi perikarya are located close to the infundibular region (Fig. 15.3).

Developmental stage 35 (S35) At S35, individual toes connected by a membrane are distinguished in the limbs of the embryo. The reproductive system is more developed, but still not sexually differentiated. In the head, a rudimentary eardrum is present. Vascularization starts in the brain at S35 (Trujillo, 1982; Yanes et al., 1989). The mantle of the brain is considerably thicker than in previous stages, and many nuclei are already distinguishable, such as the oculomotor nuclei, the ventral tegmental area and the substantia nigra in the mesencephalon; the ventral geniculate nucleus, the ventrolateral thalamic nucleus and the area triangularis in the ventral thalamus (Trujillo, 1982); and the periventricular and ventromedial nuclei in the hypothalamus. In the telencephalon, the limit between the striatum and DVR becomes clear. Synaptogenesis starts in the thalamus at S35 (Trujillo, 1982).

Compared to previous stages, the THi cell group of the midbrain tegmentum has undergone

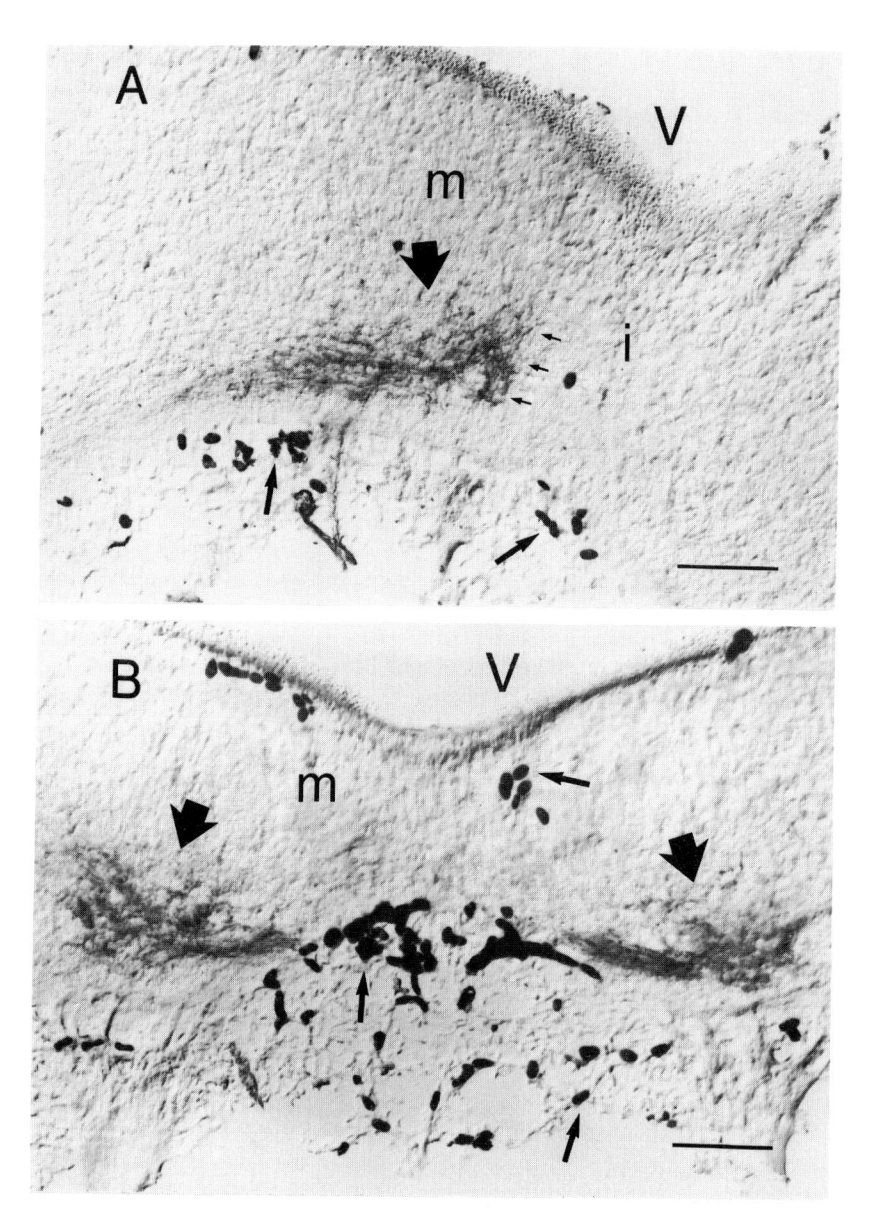

Fig. 15.2 (a) Photomicrograph of a sagittal section of the brain of *Gallotia* at S33, showing THi perikarya located in the midbrain tegmentum (large arrow), just above the cephalic flexure. The ventricle is observed in the dorsal part of the photograph (V). Note that THi perikarya are located adjacent to the limit between midbrain (m) and isthmus (i) (limit is indicated with small arrows), and from here they spread rostralwards in the midbrain. The immunoreactive cells observed in the mesencephalon have a fuzzy appearance and are lightly labeled. Very intensely labeled blood cells are also observed in the photomicrograph (medium-sized arrows). (b) Photomicrograph of a corrected horizontal section (parallel to the optic tract) of the brain at late S33, showing THi cell bodies in the midbrain tegmentum, at both sides of the midline (large arrows). The ventricle is observed in the dorsal part of the photograph (V). Very intensely labeled blood cells are also observed in the photomicrograph (medium-sized arrows). Bar: A, B = 100 μm.

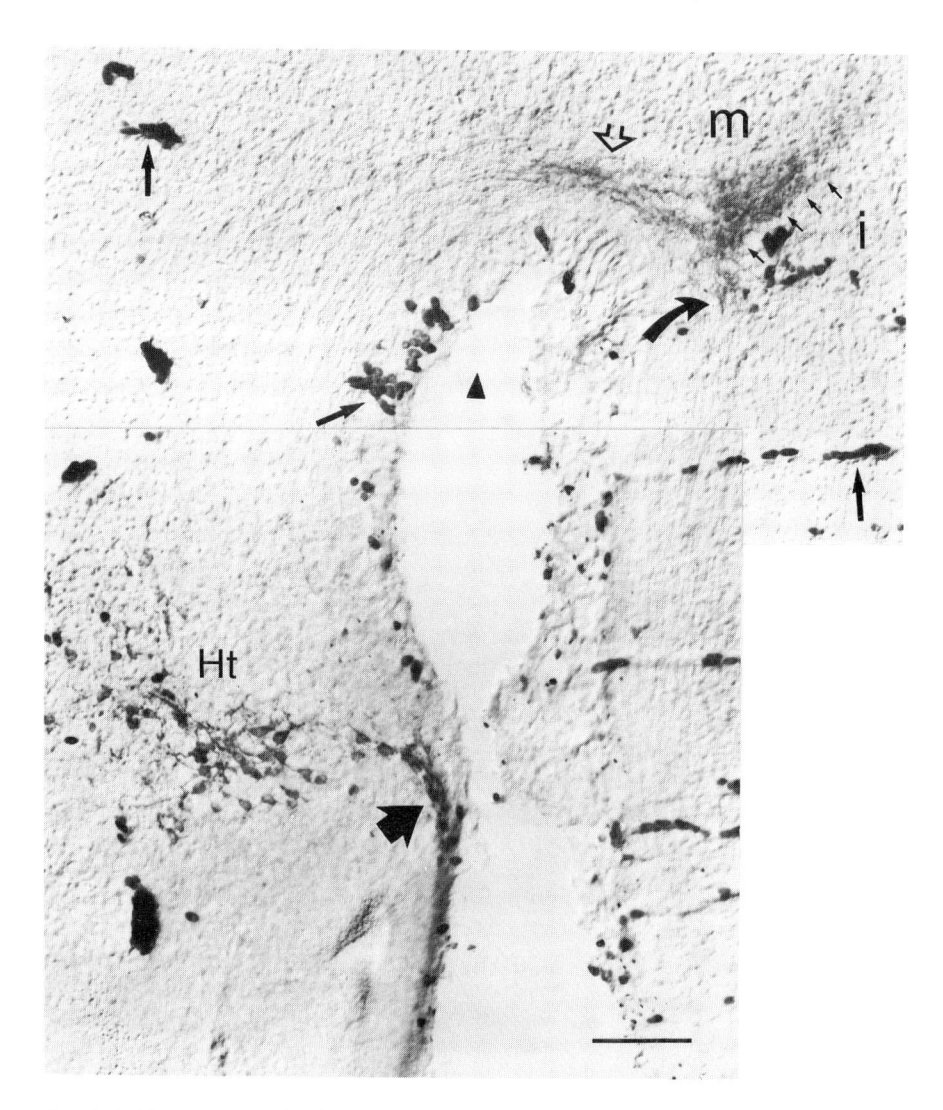

Fig. 15.3. Photomicrograph of a sagittal section of the brain at S34, showing THi perikarya located in the midbrain tegmentum (m) and the tuberomammillary hypothalamus (Ht). Note the sharp difference between the immunoreactive cells observed in the hypothalamus, which are clearly outlined and intensely labeled, and the immunoreactive cells observed in the mesencephalon, which have a fuzzy appearance and are lightly labeled. Most of the THi perikarya of the midbrain are located adjacent to the limit between mesencephalon (m) and isthmus (i) (limit indicated with small arrows). A few THi cell bodies are displaced caudal to the limit, in the isthmic tegmentum (curved arrow). A fiber bundle originates in the THi cell group of the midbrain and courses rostralwards (empty arrow). In the hypothalamus, some THi cell bodies are located close to the infundibular region (large arrow). Note the high degree of curvature of the cephalic flexure (arrowhead). This high curvature causes the angle of the longitudinal (rostro-caudal) axis of the brain to change continuously. The reader can get a better idea of the orientation of this photograph by comparing it with the schematics of sagittal sections of the embryonic brain shown in Fig. 15.1. Very intensely labeled blood cells are also observed in the photomicrograph (medium-sized arrows). Bar = 100 μm.

dramatic changes. These are: 1) a remarkable increase in the number of immunoreactive cells; 2) the cells show a stronger TH immunoreactivity; and 3) the presence of first and second order dendrites, which gives the cells a more mature appearance (Figs. 15.1(b), 15.4). The majority of the cell bodies located in the midbrain tegmentum are immunopositive with the DA antiserum. Another notable change is that the midbrain cell group is now clearly subdivided in a lateral and a medial cell cluster (Fig. 15.4). The cells in the lateral group most likely give rise to the future substantia nigra (SN). Most of the cells in the medial group will likely constitute the ventral tegmental area (VTA) dopaminergic cell group. Some THi cells in the medial group, however, show an immature appearance, and lie either very close to the midline with a process crossing it (Fig. 15.4(b)), or in the transition zone between the midbrain and isthmus (Fig. 15.4(b)).

With respect to the hypothalamus, the THi perikarya of the tuberomammillary hypothalamic cell group (Ht) are now located more laterally. Most of these cell bodies lie perpendicular to the ventricle and are monopolar or bipolar neuroblasts. At this stage, numerous cells of the tuberomammillary hypothalamic group are also immunopositive with the DA antiserum. A bundle of immunoreactive axons originates in the latter group and can be traced caudally, following the longitudinal axis of the brain (Fig. 15.1(b)). At S35, another THi cell group appears at rostrodorsal hypothalamic levels (Hr), consisting of a small number of cells which lie close to the midline and have an immature appearance (Fig. 15.1(b)).

Developmental stage 36 (S36) At S36, the reproductive system of the embryo starts to show sexual dimorphism, and the abdominal skin shows scales. In the brain, several new nuclei which were not clearly observed in previous stages are outlined, such as the locus coeruleus, the magnocellular and parvocellular isthmic nuclei, the central nucleus of the torus semicircularis, the nucleus rotundus of the dorsal thalamus (Trujillo, 1982), and the periventricular organ in the hypothalamus. In the telencephalon, it is possible to distinguish the nucleus accumbens.

At this stage, strongly THi / DAi perikarya are observed, for the first time, ventral to the central canal of the spinal cord (Figs. 15.1(c), 15.5). These cell bodies contact the ventricle by means of a short, thick process. A bundle of THi axons arises in these spinal cell bodies and courses rostral- or caudalwards.

Compared to the previous stage, there are several features of the midbrain THi / DAi cell groups at S36 that deserve comment. First, the subdivision into a medial cell group (VTA) and a lateral cell group (SN) has become more distinct. Moreover, a few THi cell bodies, are found in a more dorsal position, close to the periaqueductal gray. Some other THi cells are, at this stage, observed in the isthmic tegmentum, adjacent to the midline (Fig. 15.1(c)). These cell bodies constitute a caudal part of the VTA. Other THi cell bodies are observed rostral to the main group of CA cells of the midbrain. These cells are located in the basal region of the caudal diencephalon, and constitute the rostral part of the VTA (Fig. 15.1(c)).

In the hypothalamus, the THi / DAi cell bodies in the tuberomammillary region lie lateral to the periventricular hypothalamic organ (oph), in which neither THi nor DAi cell bodies are observed yet. Most of the immunoreactive cell bodies of the tuberomammillary hypothalamic group (Ht; Fig. 15.1(c)) are located in the future lateral hypothalamic area, whereas a few cells lie in a ventrocaudal part of the periventricular hypothalamic nucleus. In general, the tuberomammillary cells have a triangular or bipolar shape and possess already second order dendrites. The number of THi perikarya observed in the rostrodorsal periventricular hypothalamic group (Hr) at S36 has increased considerably (Fig. 15.1(c)). The majority of these rostrodorsal cells consist of monopolar neuroblasts that lie parallel to the ventricle, extending from dorsocaudal to rostroventral levels of the periventricular hypothalamus (Fig. 15.1(c)). A number of THi cell bodies located in the dorsocaudalmost portion of this hypothalamic group are arranged radial to the ventricle, and some of them occupy more lateral positions.

Developmental Stage 37 (S37) At S37, the sexual dimorphism of the reproductive system is more evident than at S36. In the limbs of the embryo, the membrane between toes has withdrawn, and claws start to develop (Ramos Steffens, 1980). The eyelids are partially closed. Most of the grisea in the brain are outlined, except for the superficial layers of the tectum and the cortex, which are still forming.

With respect to the pattern of immunoreactivity, there are several events at stage 37 that are noteworthy. First, the appearance of three new catecholamine cell groups, viz. the periventricular hypothalamic organ, the pretectal posterodorsal nucleus and the suprachiasmatic nucleus (Fig. 15.1(d)). Whereas, in general, developing catecholamine cells first display TH immunoreactivity, the cells in the hypothalamic periventricular organ are strongly immunoreactive for DA but never show TH immuno-

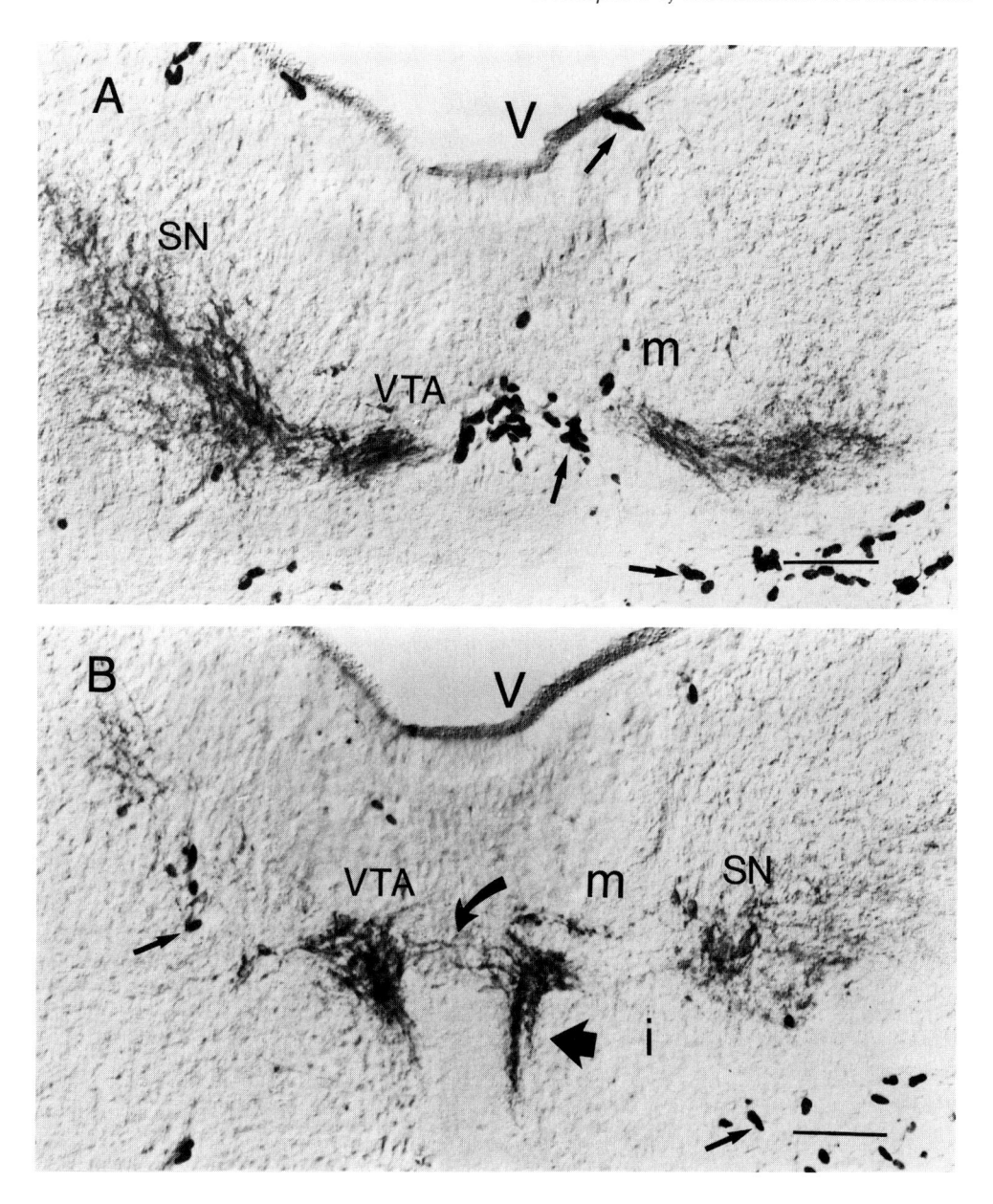

Fig. 15.4 (a), (b). Photomicrographs of transverse sections of the brain at S35, showing the THi cell cell bodies at rostral (a) and caudal levels (b) of the midbrain tegmentum, at both sides of the midline. Note that two different subgroups can be distinguished in the midbrain (m): one located in a medial position (future VTA); and another located in a lateral position (future SN). At caudal levels (b), some THi neuroblasts of the medial group are displaced into the isthmus (i) (large arrow), whereas some other THi cell bodies are located closer to the midline and have a process crossing it (curved arrow). The ventricle is observed in the dorsal part of the photograph (V). Very intensely labeled blood cells are also observed in the photomicrographs (small arrows). Bar: A, B = 100 μm.

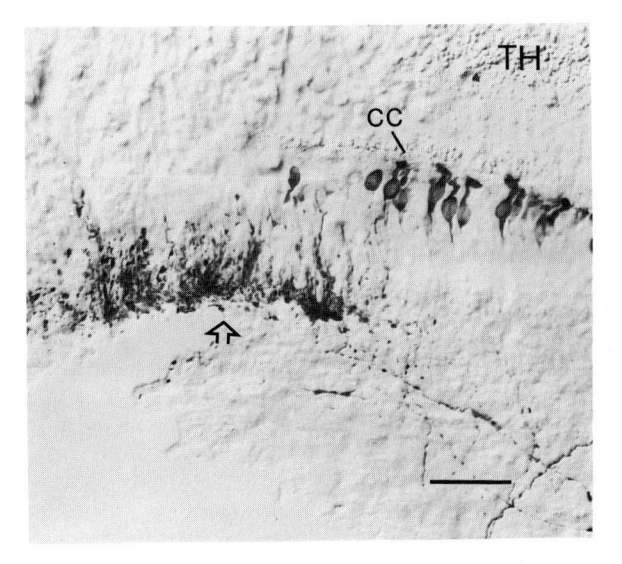

Fig. 15.5. Photomicrograph of a sagittal section of the brain at S36, showing THi cell bodies located in the spinal cord, ventral to the central canal (cc). Axons originated in these immunoreactive cells decussate in a commissure ventral to them (arrow). Bar = 50 μm.

reactivity either in the developing or in the adult lizard brain (Fig. 15.6). All DAi cells in the latter organ seem to contact the cerebrospinal fluid of the ventricle. The other two new groups of catecholaminergic cell bodies, i.e. the pretectal posterodorsal nucleus and the suprachiasmatic nucleus, contain a few weakly THi perikarya (Fig. 15.1(d)). The cell bodies of the posterodorsal nucleus and the suprachiasmatic nucleus are DA immunonegative.

The CA cell bodies of the midbrain/isthmic group and the hypothalamus show a higher degree of maturation than at S36. This is specially noticeable for the cell bodies of the rostrodorsal periventricular hypothalamic group (Hr), which show first and second order dendritic processes and are also DAi. In the midbrain, a few THi cell bodies are observed in the periaqueductal gray, close to the ventricle. The cell bodies of the periaqueductal gray are DA immunonegative.

Immunoreactive fibers are observed in the regions where the CA cell bodies are located (Fig. 15.1(d)). Further, lightly stained fibers are observed in the basal ganglia of the telencephalon, although varicosities cannot be distinguished.

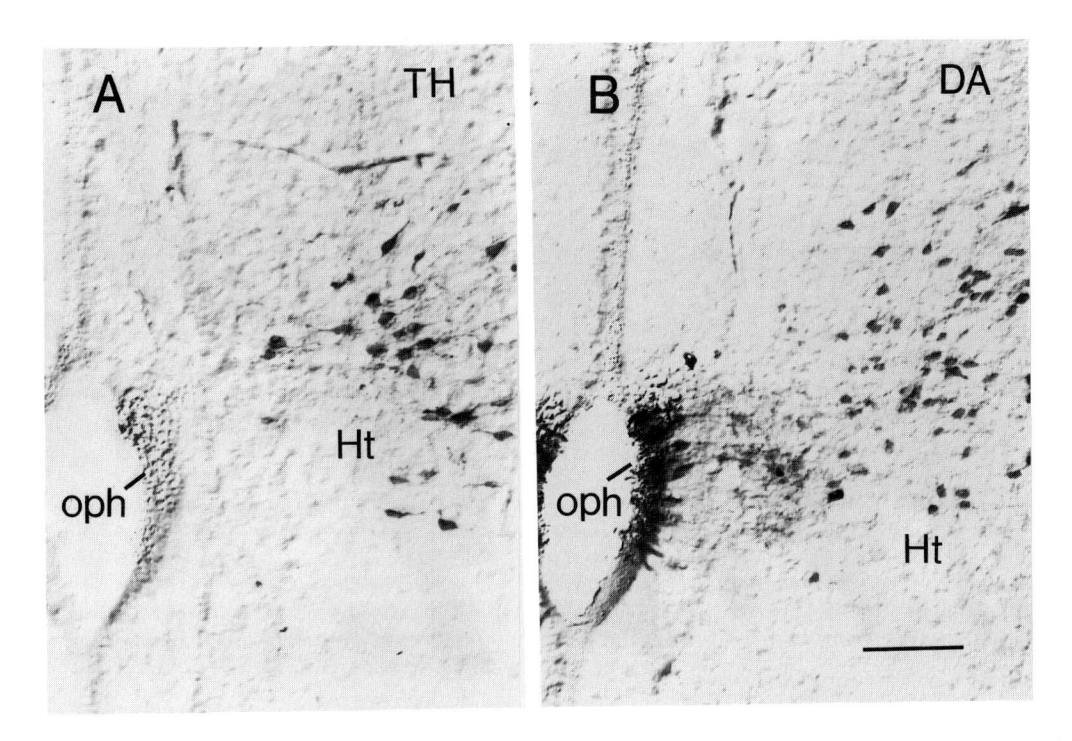

Fig. 15.6. (a), (b). Photomicrographs of corrected horizontal (parallel to the optic tract), adjacent sections of the brain at S37 stained for TH (a) or DA (b), showing immunoreactive perikarya in the tuberomammillary hypothalamic cell group (Ht), and the periventricular organ (oph). Bar = 100 μm.

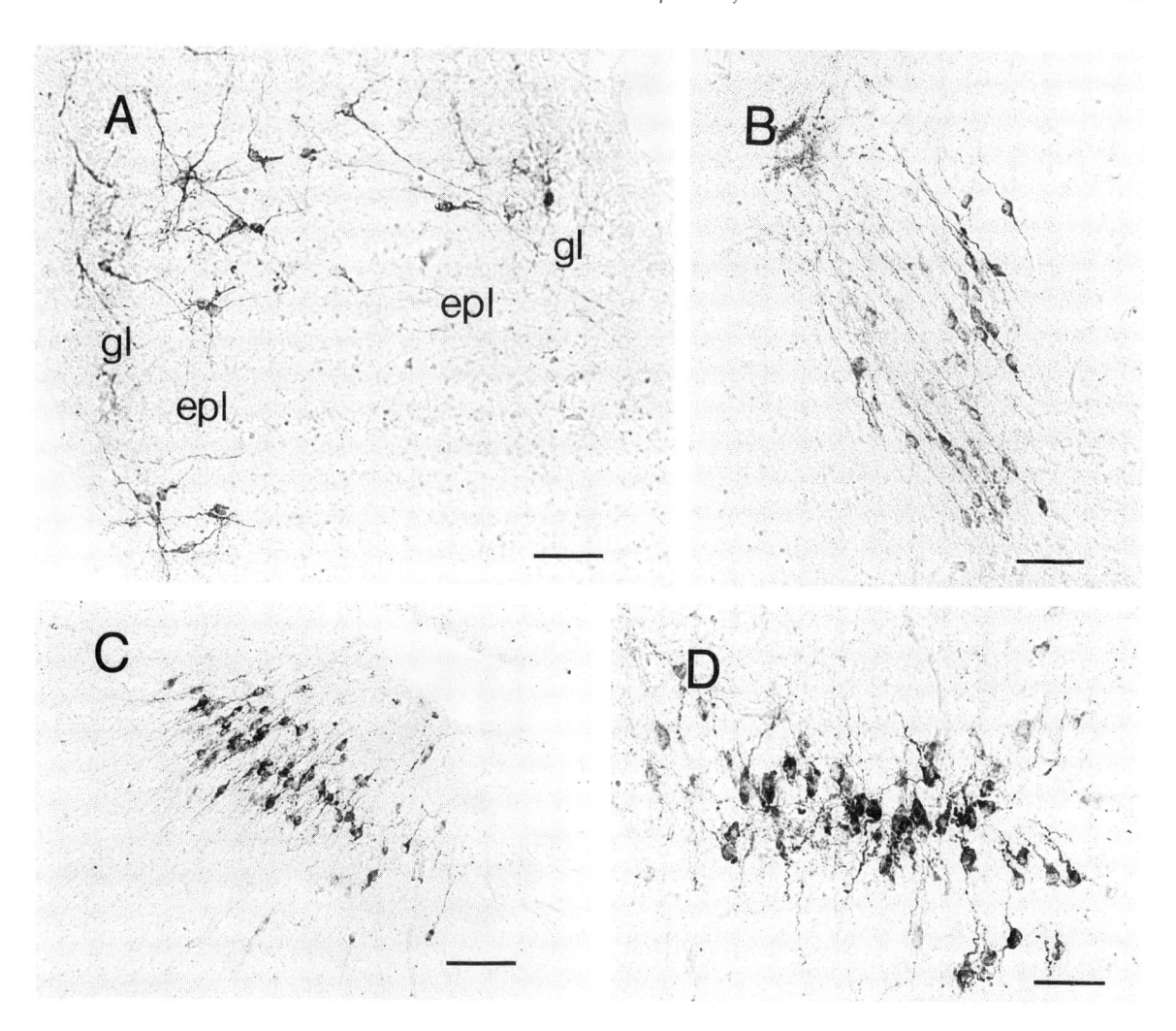

Fig. 15.7 (a). Photomicrograph of a horizontal section through the olfactory bulb of *Gallotia galloti* at S39, showing THi cell bodies in the glomerular layer (gl) and the external plexiform layer (epl). In this photograph, rostral is to the top. (b), (c), (d) Photomicrographs of sagittal sections of the brain at S39, showing THi cell bodies in the pretectal posterodorsal nucleus (b), locus coeruleus (c), and nucleus of the solitary tract (d). In these photographs, rostral is to the left, and dorsal is to the top. Bar: A, B, D = 50 μm; C = 100 μm.

Developmental stage 38 (S38) At S38, scales and a light, diffuse pigmentation start to differentiate in the skin of the dorsal part of the embryo. In the brain, most cell groups are formed and many neurons are undergoing cytodifferentiation, such as dendritic and axonal branching, and synaptogenesis. Synaptogenesis starts in the striatum at S38 (Yanes et al., 1989), although in rostral areas like the thalamus this process started at S35 and goes on through subsequent stages (Trujillo, 1982).

At S38, weakly THi cell bodies are observed for the first time in the locus coeruleus (Lc) (Fig. 15.1(e)). The number of THi cells and the immunoreactivity in the pretectal posterodorsal nucleus and the suprachiasmatic nucleus is considerably increased (Fig. 15.1). Still, the cells in the latter two nuclei (Pd, SCN) do not express DA immunoreactivity.

Strongly immunoreactive plexuses of fibers and varicosities are observed in the VTA, SN and hypothalamus. A moderate staining is observed in the basal ganglia of the telencephalon, but varicosities cannot be recognized yet.

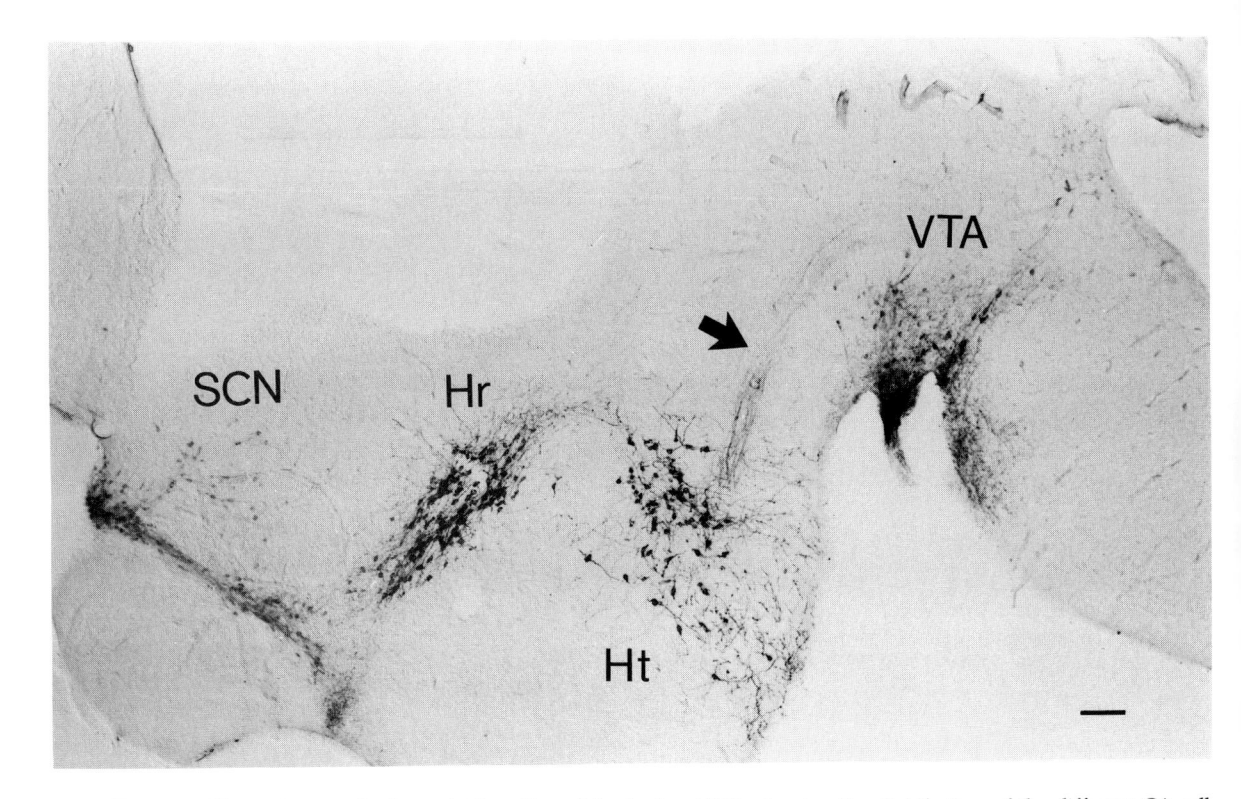

Fig. 15.8. Photomicrograph of a sagittal section of the brain at S40, showing the distribution of the different CA cell groups in the midbrain (VTA), tuberomammillary hypothalamus (Ht), rostrodorsal periventricular hypothalamus (Hr), and suprachiasmatic nucleus (SCN). Note the immunoreactive fiber bundle that arises in the tuberomammillary hypothalamus and courses caudalwards (arrow). Bar = 100 μm.

Prenatal developmental stages (S39, S40) During these prenatal stages, the gonads of females become internalized, and scales and pigmentation differentiate in the skin of the head and body, so that the embryo acquires an adult-like aspect. During these stages, the lamination of the superficial tectal layers of the brain achieves the adult state (Medina, 1991). The neurons of the brain undergo the final steps of their cytodifferentiation (formation of the dendritic arbor, acquisition of dendritic and somatic spines, completion of synaptogenesis), and the brain acquires an adult-like appearance.

At S39, THi cell bodies are observed for the first time in the olfactory bulb and the caudal brainstem (nucleus of the solitary tract and ventrolateral tegmentum) (Figs. 15.1(f), 15.7, 15.8).

During these stages, THi fibers become apparent throughout the brain, and varicosities are recognized at S40. The number of CA fibers and varicosities increases dramatically at late S40, giving the brain an adult-like CA innervation pattern just prior to hatching. Moderate to dense plexuses of fibers and varicosities are observed in the nucleus accumbens, striatum, olfactory tubercle, septum, dorsal ventricular ridge, nucleus sphericus, posterior lateral cortex, dorsal geniculate nucleus, midbrain tectum, as well as in all the regions that contain CA cell bodies.

Spatio-temporal sequence of appearance of CA cell bodies in the embryonic brain of *Gallotia*

Relation to cytodifferentiation

The sequence of appearance of CA cell bodies is represented in Fig. 15.9. Three different groups of CA cell bodies can be distinguished in the embryonic brain of *Gallotia* on the basis of their degree of cytodifferentiation at the moment when they first become immunoreactive:

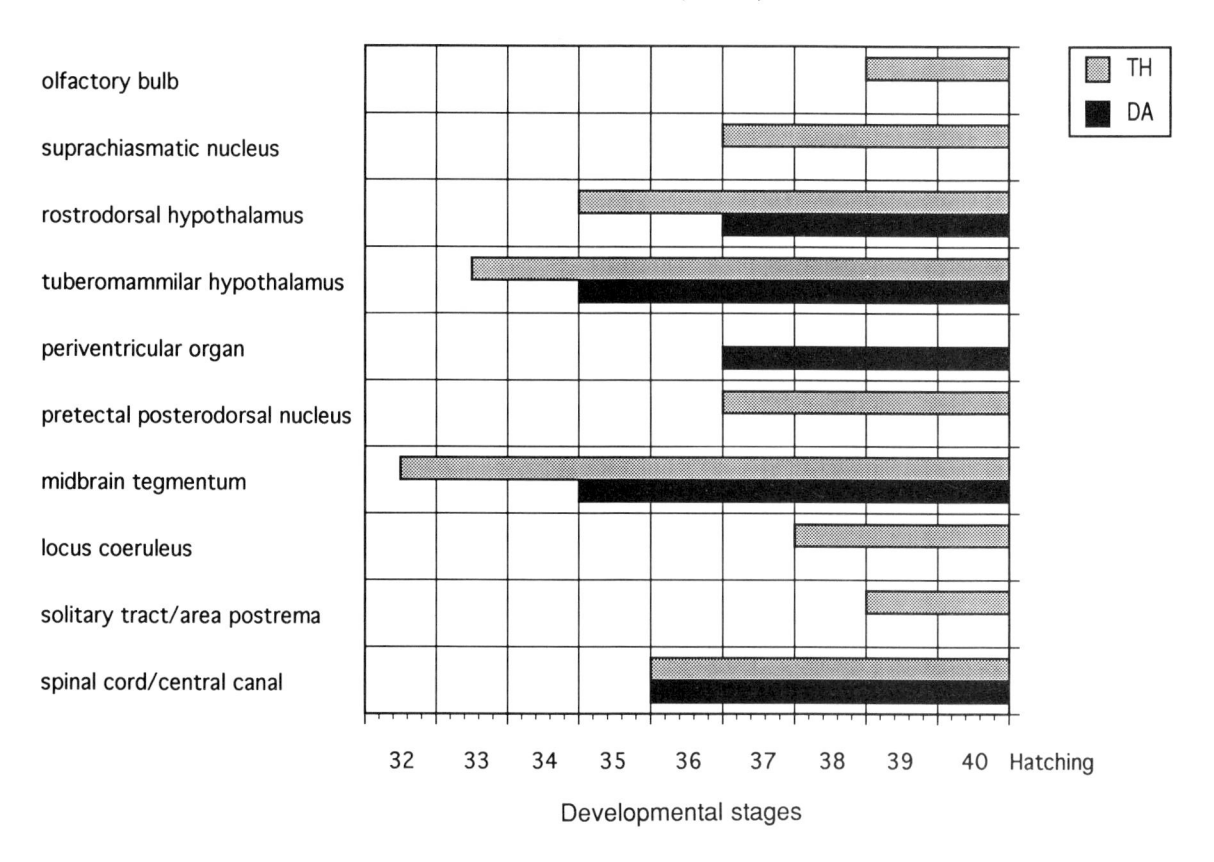

Fig. 15.9. Schematic showing the sequence of appearance of THi (hatched bar) and DAi (solid bar) perikarya in the different structures of the brain of *Gallotia galloti* during ontogeny (developmental stages 32 to hatching).

Cell bodies that become THi at a very early moment of their cytodifferentiation When these cell bodies become immunoreactive, they are located close to the ventricle, have a round, monopolar or bipolar shape (with thick varicose processes), and are dopamine immunonegative. The CA cell bodies of the midbrain tegmentum, the tuberomammillary hypothalamus and the rostrodorsal periventricular hypothalamus can be included in this category. At subsequent stages, CA cell bodies in these groups are gradually located in more lateral positions in the mantle, and show signs of maturation (become DAi and acquire second and third order dendrites). This suggests that these cell bodies first become THi prior to their migration into the mantle and maturation. However, other kind of experiments are needed to confirm this suggestion.

Cell bodies that become THi at a late moment of their cytodifferentiation In general, when these CA cell bodies become immunoreactive, they are located further away from the ventricle, in the position that they occupy in the adult brain, and possess two or more dendritic processes. This suggests that these cell bodies become immunoreactive after they have finished their migration into the mantle, and have started maturing. The CA cell bodies of the olfactory bulb, preoptic region, pretectal posterodorsal nucleus, locus coeruleus, and nucleus of the solitary tract are included in this category.

Cell bodies that remain attached to the ventricle during and after their cytodifferentiation This is the case of the CA perikarya of the periventricular hypothalamic organ (only DAi) and spinal cord (both THi and DAi). These cell bodies stay attached to the ventricle during all development and do not migrate. Since these cell bodies do not develop a dendritic arbor, it is difficult to know at which moment of their cytodifferentiation they are when they become immunoreactive. The observation of THi/DAi growing axons arising in the CA cell bodies of the spinal cord at S36 indicates that these spinal neurons start to express TH and DA at a very early moment of their cytodifferentiation.

A segmental approach as a help in understanding CA cell movements during development

During development, the neural tube becomes subdivided into transverse segmental domains called neuromeres, which represent regions of increased cell proliferation (Bergquist and Källén, 1954; Vaage, 1969; Puelles, Amat & Martínez-de-la-Torre, 1987; Lumsden & Keynes, 1989; Noden, 1991). To summarize, the following segments have been described in the brain of diverse vertebrates: 7–8 rhombencephalic segments or rhombomeres (rh1–rh8); a large isthmocerebellar segment (sometimes referred to as rh1); 1 mesencephalic segment or mesomere; and 6 prosencephalic segments (p1–p6) (for details, see Puelles & Medina, Chapter 17). In support of the existence of these segments, several homeobox genes (which produce important regulatory factors affecting the expression of other genes) show a segmented expression pattern during development of the neural tube that coincides with the morphologically-defined segmental boundaries (Keynes & Lumsden, 1990; Noden, 1991; Price et al., 1991; Bulfone et al., 1993). Several lines of evidence indicate that interneuromeric boundaries may act as barriers to longitudinal cell movement (Noden, 1991). Although this may be a general rule for the development of the brain, many exceptions to it are found during development, one example being the facial motoneurons in mammals and, probably also in reptiles. In these animals, the facial motoneurons are born in rhombomeres 4–5, and migrate into the more caudal rhombomeres 6–7 (Medina et al., 1993). In the context of the development of CA systems, a number of studies including this one have shown the expression of CA by cell bodies at a very early time in their cytodifferentiation, probably before the completion of their migration (references and discussion above). In the lizard *Gallotia*, these early CA cell bodies are located in the midbrain tegmentum, the tuberomammillary hypothalamus (basal part of p4) and the rostrodorsal periventricular hypothalamus (alar parts of p4–p5). As discussed above, there is some morphological evidence that CA cell bodies of these groups in the lizard *Gallotia* may migrate from a periventricular into a more lateral position. This migration may be over a great distance for the CA cells of the midbrain tegmentum that constitute the lateral parts of the substantia nigra, as has also been proposed for mammals (Marchand & Poirier, 1983). In contrast, this migration should be very short for the CA cells of the rostrodorsal periventricular hypothalamus which constitute part of the periventricular hypothalamic nucleus, and for the CA cells of the midbrain tegmentum which constitute the VTA. In addition to these standard migrations into the mantle within the same segment, our material has provided morphological evidence of immature CA cells of the midbrain tegmentum that may be migrating into the isthmic tegmentum (see Fig. 15.4(*b*)), and morphological evidence for such a migration across the limit between mesencephalon and isthmus has been found also in chick embryos (Puelles & Medina, 1994). In our material, we have observed THi/DAi cell bodies located in the basal region of p1 at late prenatal developmental stages, as well as in adults. These neurons are caudally continuous with the main group of CA cell bodies of the midbrain tegmentum, and they constitute a rostral part of the VTA in reptiles, as in birds and mammals (Puelles & Medina, 1994). The origin of the CA neurons of the basal region of p1 is unclear, and no clear evidence for a migration from the midbrain into p1 has been obtained in reptiles or birds (Puelles & Medina, 1994). With respect to the rostroventral periventricular hypothalamic cell group, it is interesting to note that at S36 the dorsocaudalmost cells of the group are arranged radial to the ventricle, in contrast to the rest of them which are arranged parallel to it. Using neuromeric terms, the dorsocaudal cells of this group are probably located in the alar part of p3 and may correspond to the zona incerta of mammals (Hökfelt et al., 1984), whereas the rest of the CA cells of this group are probably located in the alar parts of p4 and p5 and correspond to 'rostral' parts of A14 of mammals (Hökfelt et al., 1984). Our results provide some morphological evidence of some CA neuroblasts of p4–p5 that may migrate rostralwards during development. No evidence for such a migration has been obtained in mammals or birds (see Puelles & Medina, 1994), and from a segmental viewpoint, the probability for an intersegmental migration is very low, although not impossible (see discussion above).

Comparison to the spatio-temporal sequence of other vertebrates

As in *Gallotia*, catecholamines have been detected in embryonic brains of mammals and birds in cell bodies that are at a very early moment in their cytodifferentiation (Specht et al., 1981a; Guglielmone & Panzica, 1984, 1985; Di Porzio et al., 1990; Shults et al., 1990; Verney et al., 1991; Puelles & Medina, 1994). However, in mammals and birds these early CA cell bodies are located not only in the hypothalamus and midbrain tegmentum, but also in the locus coeruleus and medulla (Specht et al., 1981a; Guglielmone & Panzica, 1985; Di Porzio et al., 1990; Shults et al.; 1990; Verney et al., 1991; Puelles & Medina, 1994). A more detailed study in the chick embryo has revealed that during development of the brain, CA cell bodies are first detected in the tuberomammillary hypothalamus (5.5–6 days of incubation), and later they are detected in the substantia nigra, locus coeruleus and medulla (8–9 days of incubation) (Puelles & Medina, 1994). This sequence of appearance is different from that observed in Gallotia, in which THi cell bodies are observed first in the midbrain tegmentum (S32–33), shortly thereafter in the tuberomammillary hypothalamus (late S33–S34), and much later in the locus coeruleus and medulla (S38–S39). The observation that the locus coeruleus of Gallotia is cytologically apparent by S36, but its neurons become THi only at S38 indicates that these cells have a long waiting period between their origin and their first expression of the enzyme TH.

As in Gallotia, in the chicken embryo dopamine is detected in the cell bodies of the paraventricular organ several days later than in other hypothalamic areas (Guglielmone & Panzica, 1984; Puelles & Medina, 1994). An interesting finding of the present study is that the cell bodies of the hypothalamic periventricular organ never express TH immunoreactivity during any developmental stage, and they become strongly DAi at S37 without any apparent presence of TH, an enzyme that is needed for the synthesis of dopamine. In adult specimens of all non-mammalian vertebrates studied so far, cell bodies of the periventricular organ are DAi, but non-THi (Ekström, Honkanen & Steinbusch, 1990; Smeets & González, 1990; Smeets & Steinbusch, 1990; González & Smeets, 1991). To explain this, Smeets, Kidjan & Jonker (1991) have proposed that the cell bodies of the periventricular organ do not synthesize dopamine, but instead accumulate it from the CSF. If this is the case, it is interesting to note that just one stage before the first detection of DA immunoreactivity in the cell bodies of the periventricular organ (S37), CSF-contacting cell bodies located ventral to the central canal of the spinal cord become strongly THi and DAi (S36). This raises the possibility that dopamine may be released into the ventricle by CSF-contacting cell bodies of the spinal cord, and accumulated from the CSF by the cell bodies of the periventricular organ of the hypothalamus.

In the lizard brain, the last CA cell group that becomes THi during development is that in the olfactory bulb (S39). This finding agrees with the data on the development of CA systems in rats, where TH is detected in the CA cell bodies of the olfactory bulb also at a very late, prenatal stage (Specht et al., 81b).

Transient expression of CA cell bodies in the brain of Gallotia

CA cell bodies are not present in the central gray of the adult brain of Gallotia. This indicates that the THi perikarya observed in the central gray of the embryonic brain of Gallotia from S37 until prenatal stages represent transient CA neurons. The presence of transient CA neurons in the developing brain and body has been reported many times in mammals (for example, Cochard, Goldstein & Black, 1978, 1979; Teitelman et al., 1979, 1981; Specht et al., 1981b; Jaeger & Joh, 1983; Jonakait et al., 1984; Berger et al., 1985; Verney et al., 1988). In fact, the transient expression of a catecholaminergic phenotype is a widespread phenomenon characteristic of a variety of cells of diverse embryonic origin. Therefore, it is now thought that this transient expression is not just a phenomenon related to cells that die after aberrant development, but may represent a more complex phenomenon of cell bodies that transiently express detectable levels of TH and retain this ability during the entire life (Jonakait et al., 1984). From a phylogenetic point of view, our finding of transient CA cell bodies in the central gray of Gallotia is very interesting. Although CA cell bodies are not observed in the central gray of some adult reptiles such as the lizards Gallotia and Gekko, or the turtles and snakes (Smeets et al., 1986; Smeets, Jonker & Hoogland, 1987; Smeets, 1988; see also Chapter 6), THi neurons are present in the central gray of other adult reptiles, such as the lizards Varanus and Anolis (Wolters, Ten Donkelaar & Verhofstad, 1984; see also Chapter 6) and crocodiles (L. Medina, unpublished observations). Dopaminergic neurons are present in the central gray of several birds (chicks and pigeons: Guglielmone & Panzica, 1984; Puelles & Medina, 1994; Reiner et al., 1994) and rats (aqueduct of Sylvius -dorsal A10, Hökfelt et al., 1984), and it is probable that primitive reptiles also contained dopaminergic cells in such a location.

Development of CA fiber tracts and innervation

In the present study we observed growing axons containing DA and/or TH immunoreactivity in the embryonic brain of the lizard *Gallotia*. Catecholamines and TH have previously been detected in growing axons of other vertebrates (Tennyson *et al.*, 1973; Specht *et al.*, 1981a; Voorn *et al.*, 1988; Pindzola, Ho & Martin, 1990; Reisert *et al.*, 1990). In fact, catecholamines and TH are not only present in the growing axonal shaft, but also in the axonal growth cones (present results; Specht *et al.*, 1981a), indicating that they may play a role in finding the migration pathway or may influence the development of the target. Supporting this idea, recent experiments *in vitro* have provided evidence that stimulation of D2 dopamine receptor results in both neurite branching and outgrowth (Todd, 1992), whereas stimulation of D1 dopamine receptor inhibits growth cone motility and neurite outgrowth (Lankford, DeMello & Klein, 1987, 1988), suggesting that catecholamines may play an important role as morphogens during development (Lauder & Krebs, 1984; Lankford *et al.*, 1988; Fiszman *et al.*, 1991; Todd, 1992).

In the present study we could observe the outgrowth of three different fiber bundles that were THi (one of them also DAi): (1) ascending THi fibers arising from the CA cell group of the midbrain/isthmic tegmentum; (2) descending THi fibers from the CA cell group of the tuberomammillary hypothalamus; (3) ascending and descending THi/DAi fiber bundles arising in the CA cell bodies of the spinal cord. These CA fiber bundles were observed in the embryonic brain from S33–S36, and after growing, they constitute or contribute to some important CA pathways of the brain, such as the nigrostriatal pathway which provides an important CA innervation to the basal ganglia (see below). The CA cell groups that become positive at a late moment of their differentiation, such as the posterodorsal pretectal nucleus, the locus coeruleus, and the nucleus of the solitary tract, also contribute to some of the CA fiber tracts and innervation that are present in the adult brain (such as the ascending noradrenergic fiber bundle arising in the locus coeruleus, which provides an important noradrenergic innervation to the telencephalon). However, since cell bodies in these groups become immunoreactive after they have undergone a considerable part of their cytodifferentiation, by the time they are first positive their main connections are probably already established.

Development of the nigrostriatal CA pathway The THi fiber bundle originated in the CA cell group of the midbrain/isthmic tegmentum starts growing at S32–33, and reaches the caudal telencephalon at S34. However, it seems that these CA fibers do not innervate the lateral striatum until S37–S38, and an adult-like innervation of the striatum and nucleus accumbens (with clear varicosities) is observed at prenatal stage S40, just prior to hatching. These results agree with those of Yanes *et al.* (1989) on the development of the striatum in the lizard *Gallotia galloti*, which indicate that striatal neurons (born from S31 to S35) mature between S38–S40, and that the synaptogenesis (mainly axo-dendritic synapses) starts at S38. Yanes *et al.* (1989) also noted that axo-dendritic synapses acquire an adult-like morphology by S40, coinciding with the moment that we observed THi varicosities in the striatum. The early arrival of the THi axons in the striatum suggests that they may have an influence on the maturation of the striatal neurons. This idea has also been suggested in mammals, where the CA innervation of the rostral striatum follows a lateroventral to dorsomedial gradient which coincides with the maturation gradient of the striatal neurons (Tennyson *et al.*, 1973; Specht *et al.*, 1981b; Voorn *et al.*, 1988). The molecular mechanisms by which dopaminergic axons may influence the development of the striatum are not known yet, but in any case they must be mediated by specific receptors. In line with this, there is recent evidence that during the development of the rat striatum, dopamine receptors are detected soon after the arrival of THi fibers (Foster *et al.*, 1987; De Vries, Mulder & Schoffelmeer, 1992), and that the postsynaptic D1 dopamine receptors become functionally active much earlier than the postsynaptic D2 dopamine receptors (De Vries *et al.*, 1992). This differential developmental profile of the D1 and D2 dopamine receptors may play an important role in the development of the striatum, considering that stimulation of D1 or D2 receptor results in different effects on neurite outgrowth.

Acknowledgements

We thank the members of the UDI of Cell Biology of the University of La Laguna (Tenerife, Spain), especially C. Díaz, E. Martí, C. M. Trujillo, A. Lancha, E. Del Castillo and J. Milán, for their help in obtaining the lizard embryos that we used for this study. We also thank E. J. Karle for her valuable help with the illustrations for this chapter, Dr R. M. Buijs (Netherlands Institute for Brain Research, Amsterdam, The Netherlands) for generously providing the antiserum against dopamine, and Dr A. Reiner (Dept. Anat. & Neurobiol., University Tennessee,

Memphis, USA) for partially supporting this work (NIH Grant NS-19620) and for his helpful suggestions for improving this chapter. This work was also partially supported by the Spanish DGICYT PB87-0688-C01-C02 and PB90-0296-C01-C02 to L. P., and by postdoctoral fellowships from the Spanish Ministerio de Educacion y Ciencia to L. M.

Abbreviations

BG	basal ganglia
cc	central canal
Cxl	lateral cortex
d	diencephalon
DA	dopamine
DVR	dorsal ventricular ridge
epl	external plexiform layer
gl	glomerular layer
Ht	tuberomammillary hypothalamus
Hr	rostrodorsal periventricular hypothalamus
i	isthmus
Lc	locus coeruleus
m	mesencephalon
ob	olfactory bulb
OC	optic chiasm
oph	periventricular hypothalamic organ
Pd	posterodorsal nucleus
r	rhombencephalon
SCN	suprachiasmatic nucleus
SN	substantia nigra
Sol	nucleus of the solitary tract
sp	spinal cord
Sph	nucleus sphericus
t	telencephalon
TeO	optic tectum
TH	tyrosine hydroxylase
VTA	ventral tegmental area

References

Berger, B., Verney, C., Gaspar, P. & Febvret, A. (1985). Transient expression of tyrosine-hydroxylase immunoreactivity in some neurons of the rat neocortex during postnatal development. *Developmental Brain Research*, 23, 141–4.

Bulfone, A., Puelles, L., Porteus, M. H., Frohman, M. A., Martin, G. R. & Rubenstein, J. L. R. (1993). Spatially restricted expression of Dlx–1, Dlx–2 (Tes–1), Gbx–2 and Wnt–3 in the embryonic day 12.5 mouse forebrain defines potential transverse and longitudinal segmental boundaries. *Journal of Neuroscience*, 13, 3155–72.

Bergquist, H., & Källén, B. (1954). Notes on the early histogenesis and morphogenesis of the central nervous system in vertebrates. *Journal of Comparative Neurology*, 100, 627–60.

Cochard, P., Goldstein, M. & Black, I. B. (1978). Ontogenetic appearance and disappearance of tyrosine hydroxylase and catecholamines in the rat embryo. *Proceedings of the National Academy of Sciences, USA*, 75, 2986–90.

Cochard, P., Goldstein, M. & Black, I. B. (1979). Initial development of the noradrenergic phenotype in autonomic neuroblasts of the rat embryo in vivo. *Developmental Biology*, 71, 100–14.

De Vries, T. J., Mulder, A. H. & Schoffelmeer, A. N. M. (1992). Differential ontogeny of functional dopamine and muscarinic receptors mediating presynaptic inhibition of neurotransmitter release and postsynaptic regulation of adenylate cyclase activity in rat striatum. *Developmental Brain Research*, 66, 91–6.

Di Porzio, U., Zuddas, A., Cosenza-Murphy, D. B. & Barker, J. L. (1990). Early appearance of tyrosine hydroxylase immunoreactive cells in the mesencephalon of mouse embryos. *International Journal of Development and Neuroscience*, 8, 523–32.

Dufaure, J. P., & Hubert, J. (1961). Table de développement du lézard vivipare (Lacerta vivipara Jacquin). *Archives d'Anatomie Microscopique et de Morphologie Expérimentale*, 50, 309–27.

Ekström, P., Honkanen, T. & Steinbusch, H. W. M. (1990). Distribution of dopamine-immunoreactive neuronal perikarya and fibers in the brain of a teleost, Gasterosteus aculeatus L. Comparison with tyrosine hydroxylase- and dopamine-β-hydroxylase-immunoreactive neurons. *Journal of Chemical Neuroanatomy*, 3, 233–60. 233–60.

Ekström, P., Honkanen, T. & Borg, B. (1992). Development of catecholaminergic neurons in the brain of a teleost fish, the three-spined stickleback (Gasterosteus aculeatus). In *Abstracts of the 7th International Catecholamine Symposium*, pp.82, Amsterdam.

Fiszman, M. L., Zuddas, A., Masana, M. I., Barker, J. L. & diPorzio, U. (1991). Dopamine synthesis precedes dopamine uptake in embryonic rat mesencephalic neurons. *Journal of Neurochemistry*, 56, 392–9.

Foster, G. A., Schultzberg, M., Hökfelt, T., Goldstein, M., Hemmings, H. C., Ouimet, C. C., Walaas, S. I. & Greengard, P. (1987). Development of a dopamine- and cyclic adenosine 3′: 5′-monophosphate-regulated phosphoprotein (DARPP-32) in the prenatal rat central nervous system, and its relationship to the arrival of presumptive dopaminergic innervation. *Journal of Neuroscience*, **7**, 1994–18.

González, A. & Smeets, W. J. A. J. (1991). Comparative analysis of dopamine and tyrosine hydroxylase immunoreactivities in the brain of two amphibians, the anuran *Rana ridibunda* and the urodele *Pleurodeles waltlii*. *Journal of Comparative Neurology*, **303**, 457–77.

Guglielmone, R. & Panzica, G. C. (1984). Typology, distribution and development of the catecholamine-containing neurons in the chicken brain. *Cell and Tissue Research*, **237**, 67–79.

Guglielmone, R. & Panzica, G. C. (1985). Early appearance of catecholaminergic neurons in the central nervous system of precocial and altricial avian species. A fluorescence-histochemical study. *Cell and Tissue Research*, **240**, 381–4.

Hökfelt, T., Martensson, R., Björklund, A., Kleinau, S. & Goldstein, M. (1984). Distributional maps of tyrosine-hydroxylase-immunoreactive neurons in the rat brain. In *Handbook of Chemical Neuroanatomy. Vol. 2: Classical Transmitters in the CNS, Part I.*, ed. A. Björklund & T. Hökfelt, pp. 277–379. Amsterdam, Elsevier.

Jaeger, C. B., & Joh, T. J. (1983). Transient expression of tyrosine hydroxylase in some neurons of the developing inferior colliculus of the rat. *Developmental Brain Research*, **11**, 128–32.

Jonakait, G. M., Markey, K. A., Goldstein, M. & Black, I. B. (1984). Transient expression of selected catecholaminergic traits in cranial sensory and dorsal root ganglia of the embryonic rat. *Developmental Biology*, **101**, 51–60.

Keynes, R. & Lumsden, A. (1990). Segmentation and the origin of regional diversity in the vertebrate central nervous system. *Neuron*, **2**, 1–19.

Lankford, K., DeMello, F. G. & Klein, W. L. (1987). A transient embryonic dopamine receptor inhibits growth cone motility and neurite outgrowth in a subset of avian retina neurons. *Neuroscience Letters*, **75**, 169–74.

Lankford, K., DeMello, F. G. & Klein, W. L. (1988). D1-type dopamine receptors inhibit growth cone motility in cultured retina neurons: evidence that neurotransmitters act as morphogenic growth regulators in the developing central nervous system. *Proceedings of the National Academy of Sciences, USA*, **85**, 4567–71.

Lauder, J. M. & Bloom, F. E. (1974). Ontogeny of monoamine neurons in the locus coeruleus, raphe nuclei and substantia nigra of the rat. *Journal of Comparative Neurology*, **155**, 469–82.

Lauder, J. M. & Krebs, H. (1984). Humeral influences on brain development. *Advances in Cell Neuroscience*, **5**, 3–51.

Lumsden, A., & Keynes, R. (1989). Segmental patterns of neuronal development in the chick hindbrain. *Nature*, **337**, 424–8.

Marchand, R. & Poirier, L. J. (1983). Isthmic origin of neurons of the rat substantia nigra. *Neuroscience*, **9**, 373–81.

Medina, L. (1991). *Estudio Ontogenético e Inmunohistoquímico de los Centros Visuales Primarios de Reptiles*. PhD Dissertation, Universidad de La Laguna, Spain.

Medina, L. & Smeets, W. J. A. J. (1992). Cholinergic, monoaminergic and peptidergic innervation of the primary visual centers in the brain of the lizards *Gekko gecko* and *Gallotia galloti*. *Brain, Behavior and Evolution*, **40**, 157–81.

Medina, L., Smeets, W. J. A. J., Hoogland, P. V. & Puelles, L. (1993). Distribution of choline acetyltransferase immunoreactivity in the brain of the lizard *Gallotia galloti*. *Journal of Comparative Neurology*, **331**, 261–85.

Noden, D. M. (1991). Vertebrate craniofacial development: the relation between ontogenetic process and morphological outcome. *Brain, Behavior and Evolution*, **38**, 190–225.

Pindzola, R. R., Ho, R. H. & Martin, G. F. (1990). Development of catecholaminergic projections to the spinal cord in the North American opossum, *Didelphis virginiana*. *Journal of Comparative Neurology*, **294**, 399–417.

Price, M., Lemaistre, M., Pischetola, M., Di Lauro, R., & Duboule, D. (1991). A mouse gene related to Distal-less shows a restricted expression in the developing forebrain. *Nature*, **351**, 748–51.

Puelles, L., Amat, J. A. & Martínez-de-la-Torre, M. (1987). Segment-related, mosaic neurogenetic pattern in the forebrain and mesencephalon of early chick embryos: I. Topography of AChE-positive neuroblast up to stage HH18. *Journal of Comparative Neurology*, **266**, 247–68.

Puelles, L. & Medina, L. (1994). Development of neurons expressing tyrosine hydroxylase and dopamine in the chicken brain. In *Phylogeny and Development of Catecholamine Systems in the CNS of Vertebrates*, ed. W. J. A. J. Smeets & A. Reiner. Cambridge, Cambridge Univ. Press.

Ramos Steffens, A. (1980). *Tabla de desarrollo embrionario de Lacerta galloti galloti (periodo de organogénesis), y aspectos de su reproducción*. Universidad de La Laguna, Spain.

Reiner, A., Karle, E. J., Anderson, K. D. & Medina, L. (1994) Catecholaminergic perikarya and fibers in the avian nervous system. In *Phylogeny and Development of Catecholamine Systems in the CNS of Vertebrates*, ed. W. J. A. J. Smeets & A. Reiner, pp. 381–404. Cambridge, Cambridge Univ. Press.

Reisert, I., Schuster, R., Zienecker, R. & Pilgrim, C. (1990) Prenatal development of mesencephalic and diencephalic dopaminergic systems in the male and female rat. *Developmental Brain Research*, **53**, 222–9.

Smeets, W. J. A. J. (1988). Distribution of dopamine immunoreactivity in the forebrain and midbrain of the snake *Python regius*: a study with antibodies against dopamine. *Journal of Comparative Neurology*, **271**, 115–29.

Smeets, W. J. A. J. & González, A. (1990). Are putative dopamine-accumulating cell bodies in the hypothalamic periventricular organ a primitive brain character of non-mammalian vertebrates? *Neuroscience Letters*, **114**, 248–52.

Smeets, W. J. A. J., Hoogland, P. V. & Voorn, P. (1986).

The distribution of dopamine immunoreactivity in the forebrain and midbrain of the lizard *Gekko gecko*: an immunohistochemical study with antibodies against dopamine. *Journal of Comparative Neurology*, **253**, 46–60.

Smeets, W. J. A. J., Jonker, A. J. & Hoogland, P. V. (1987). Distribution of dopamine in the forebrain and midbrain of the red-eared turtle, *Pseudemys scripta elegans*, reinvestigated using antibodies against dopamine. *Brain, Behavior and Evolution*, **30**, 121–42.

Smeets, W. J. A. J., Kidjan, M. N. & Jonker, A. J. (1991). a-MPT does not affect dopamine levels in the periventricular organ of lizards. *NeuroReport*, **2**, 369–72.

Smeets, W. J. A. J. & Steinbusch, H. W. M. (1990). New insights into the reptilian catecholaminergic systems as revealed by antibodies against the neurotransmitters and their synthetic enzymes. *Journal of Chemical Neuroanatomy*, **3**, 25–43.

Shults, C. W., Hashimoto, R., Brady, R. M. & Gage, F. H. (1990). Dopaminergic cells align along radial glia in the developing mesencephalon of the rat. *Neuroscience*, **38**, 427–36.

Specht, L. A., Pickel, V. M., Joh, T. H. & Reis, D. J. (1981a). Light-microscopic immunocytochemical localization of tyrosine hydroxylase in prenatal rat brain. I. Early ontogeny. *Journal of Comparative Neurology*, **199**, 233–53.

Specht, L. A., Pickel, V. M., Joh, T. H. & Reis, D. J. (1981b). Light-microscopic immunocytochemical localization of tyrosine hydroxylase in prenatal rat brain. II. Late ontogeny. *Journal of Comparative Neurology*, **199**, 255–76.

Teitelman, G., Baker, H., Joh, T. H. & Reis, D. J. (1979). Appearance of catecholamine-synthesizing enzymes during development of rat sympathetic nervous system: possible role of tissue environment. *Proceedings of the National Academy of Sciences, USA*, **76**, 509–13.

Teitelman, G., Gershon, M. D., Rothman, T. P., Joh, T. H. & Reis, D. J. (1981). Proliferation and distribution of cells that transiently express a catecholaminergic phenotype during development in mice and rats. *Developmental Biology*, **86**, 348–55.

Tennyson, V. M., Mytilineou, C. & Barrett, R. E. (1973). Fluorescence and electron microscopic studies of the early development of the substantia nigra and area ventralis tegmenti in the fetal rabbit. *Journal of Comparative Neurology*, **149**, 233–58.

Todd, R. D. (1992). Neural development is regulated by classical neurotransmitters: dopamine D2 receptor stimulation enhances neurite outgrowth. *Biological Psychiatry*, **31**, 794–807.

Trujillo, C. M. (1982). Ontogénesis de los Núcleos Talámicos en Gallotia galloti (Reptil, Lacertidae): Estudio Estructural y Ultrastructural. PhD Dissertation, Universidad de La Laguna, Spain.

Vaage, S. (1969). Segmentation of the primitive neural tube in chick embryos. A morphological, histochemical and autoradiographical investigation. *Advances in Anatomy, Embryology and Cell Biology*, **41**, 1–88.

Verney, C., Gaspar, P., Febvret, A. & Berger, B. (1988). Transient tyrosine hydroxylase-like immunoreactive neurons contain somatostatin and substance P in the developing amygdala and bed nucleus of the stria terminalis of the rat. *Developmental Brain Research*, **42**, 45–58.

Verney, C., Zecevic, N., Nikolic, B., Alvarez, C. & Berger, B. (1991). Early evidence of catecholaminergic cell groups in 5- and 6-week-old human embryos using tyrosine hydroxylase and dopamine-β-hydroxylase immunohistochemistry. *Neuroscience Letters*, **131**, 121–4.

Voorn, P., Kalsbeek, A., Jorritsma-Byham, B. & Groenewegen, H. J. (1988). The pre-and postnatal development of the dopaminergic cell groups in the ventral mesencephalon and the dopaminergic innervation of the striatum of the rat. *Neuroscience*, **25**, 857–87.

Wallace, J. A., Mondragon, R. M., Allgood, P. C., Hoffman, T. J. & Maez, R. R. (1987). Two populations of tyrosine hydroxylase-positive cells occur in the spinal cord of the chick embryo and hatchling. *Neuroscience Letters*, **83**, 253–8.

Wolters, J. G., Ten Donkelaar, H. J. & Verhofstad, A. A. J. (1984). Distribution of catecholamines in the brain stem and spinal cord of the lizard *Varanus exanthematicus*: an immunohistochemical study based on the used of antibodies to tyrosine hydroxylase. *Neuroscience*, **13**, 469–93.

Yanes, C., Perez-Batista, M. A., Martin-Trujillo, J. M., Monzon, M. & Rodriguez, A. (1989). Development of the ventral striatum in the lizard *Gallotia galloti*. *Journal of Anatomy*, **164**, 93–100.

Yurkewicz, L., Marchi, M., Lauder, J. M. & Giacobini, E. (1981). Development and aging of noradrenergic cell bodies and axon terminals in the chicken. *Journal of Neuroscience Research*, **6**, 621–41.

16
Development of neurons expressing tyrosine hydroxylase and dopamine in the chicken brain: a comparative segmental analysis

L. Puelles and L. Medina

Introduction

A few studies have addressed the early development of catecholaminergic (CA) cell populations in the avian brain (Yurkewicz et al., 1981; Guglielmone & Panzica, 1984, 1985; Wallace et al., 1987), as well as the corresponding mature topographic distribution (Fuxe & Ljunggren, 1965; Sharp & Follett, 1968; Ikeda & Gotoh, 1971, 1974; Dubé & Parent, 1981; Kiss & Péczely, 1987; Smeets, 1991; Reiner et al., this volume). Compared with data available for mammals, the earliest expression of catecholaminergic traits in avian brains apparently occurs in a slightly different spatial pattern and relatively later in development. There is also the implicit suggestion that some avian CA cell groups do not follow the migration routes found in mammals. For example, the respective nigral populations acquire very different final positions (near the subthalamus in mammals, but near the isthmus in birds and reptiles). This chapter aims to re-examine these spatiotemporal patterns within the context of a segmental (neuromeric) conception of the developing vertebrate brain (Rendahl, 1924; Bergquist & Källén, 1954; Vaage, 1969, 1973; Keyser, 1972; Puelles et al., 1987a, 1991; Lumsden & Keynes, 1989). Readers may wish to consult other papers from this laboratory developing this viewpoint (Puelles & Martínez-de-la-Torre, 1987; Puelles et al., 1987b, 1991, 1992; Martínez et al., 1992; Medina et al., 1993; Bulfone et al., 1993a,b; Puelles & Rubenstein, 1993).

The stereotyped distribution of CA populations has apparently never been examined from the point of view of morphologic position within longitudinal and transversal parcellations of the brain. In most contributions, only passing reference is made, if at all, to the issues of (1) alar or basal origin of the young CA neurons, (2) segmental origin of the diverse cell groups and/or precise relationships to brain vesicle boundaries and (3) relationship of CA-containing axonal tracts to longitudinal or transverse landmarks of the brain wall.

These issues hold nowadays some importance in view of the increasing amount of marker molecules and early developmental genes that are being mapped to specific sectors or cell populations of the neural tube. Developments in this field may soon provide common ground for hypotheses on the hitherto unassailable issue of how specific neuronal populations select similar neurotransmitter phenotypes in different parts of the brain. It is a prerequisite for this that the relative position and chronological unfolding of the various matrix areas for given cell categories are known with sufficient precision. Similarly, more details are needed on the corresponding differentiation timetables as well as on the eventual migratory phenomena which may disperse or displace given cell categories from their postmitotic positions.

This sort of analysis is performed here of necessity in a preliminary, non-exhaustive way; we explore in some detail the possibilities offered by this novel approach for developmental understanding as well as for comparative inquiries. We focus our main interest in forebrain CA neuronal patterns.

Material and methods

The embryonic material analysed comprises two different series of chick embryos fixed at various stages of development.

Series 1 The embryos were perfused and postfixed

for 1–5 hours with 5% glutaraldehyde, 1% Na$_2$S$_2$O$_5$ in 0.1 M, pH 7.4 phosphate buffer. The brains then were dissected out and cryoprotected in buffered 30% sucrose solution containing 1% Na$_2$S$_2$O$_5$. Sagittal frozen sections, 50 μm thick, were collected in parallel series in 0.05 M, pH 7.6 Tris-buffered saline with 1% Na$_2$S$_2$O$_5$. One series of sections was processed for tyrosine hydroxylase immunohistochemistry (INCSTAR monoclonal antibody) and the other was processed for dopamine immunohistochemistry (rabbit antiserum generously provided by Dr Buijs, Netherlands Institute for Brain Research, Amsterdam), as described in Medina et al. (this book). Technical difficulties impeded extension of this series below 8 days of incubation (stage 34).

Series 2 This group of embryos was aimed to cover the developmental period before stage 34. Chick embryos between stages 26 and 35 (5 to 9 days of incubation) were immersed overnight in 4% paraformaldehyde in 0.1M, pH 7.4 phosphate buffer. The heads were then quickly dehydrated, cleared in butanol and embedded in Paraplast. Sagittal, horizontal or transverse sections were obtained which were subsequently tested for tyrosine hydroxylase immunoreactivity (as before) and counterstained with cresyl violet (See Fig. 16.1(a)–(d)).

Morphological background

The brains of chick embryos (like those of other vertebrates) may be subdivided cytoarchitectonically into *segmental brain domains*. This term means transverse architectural subdivisions of the neuraxis which are known (or postulated) to derive embryologically from simple, ringlike neuroepithelial regions called *neuromeres*. These are supposed to be metameric, that is, to have some serially repeated morphological pattern, notwithstanding their different specific identities and fates. The commonest structural pattern seems to be a subdivision into floor, basal, alar and roof parts of each neuromere (Puelles & Rubenstein, 1993). Depending on the brain part referred to, the segments may be called: myelomeres, rhombomeres, mesomeres or prosomeres. The classical reference works are the review of Bergquist and Källen (1954) and the monograph by Vaage (1969). These should be complemented or corrected with the diverse modifications introduced 1) for relationships with cranial nerves and branchial arches (Lumsden & Keynes, 1989; Noden, 1991), 2) for the peculiar isthmocerebellar region (Vaage, 1973; Puelles and Martinez-de-la-Torre 1987; Martinez & Alvarado-Mallart, 1989;

Hallonet, Teillet & Le Douarin, 1990) and 3) for the prosencephalon (Puelles et al., 1987a,b, 1991, 1992; Puelles & Rubenstein, 1993; Bulfone et al., 1993a,b).

Seven or eight rhombomeres are recognized in the rhombencephalon (Vaage, 1969). These neuromeres are observable in the early embryo (up to 4–5 days of incubation), but subsequent thickening of the brain wall disguises the boundaries. A detailed mapping of the early interrhombomeric boundaries onto the mature rhombencephalic cytoarchitecture is not available yet (but see Hallonet et al., 1990; Tan & LeDouarin, 1991). However, the fixed pattern of brainstem motor nuclei and nerve roots (Lumsden & Keynes, 1989) allows an estimation of the segmental 'level' within this region.

The mesencephalon is classically described as having two neuromeres (Vaage, 1969). However, recent experimental data demonstrated that the caudal mesomere, together with a part of the first prorhombomere (RhA), is involved in the formation of the isthmocerebellar region, rather than forming any of the conventional 'mesencephalic' centers (Martinez and Alvarado-Mallart, 1989; Alvarado-Mallart, Martinez & Lance-Jones, 1990; Hallonet et al., 1990). The caudally directed neuroepithelial flux that mediates this phenomenon occurs between 2.5 and 5 days of incubation (Martinez-de-la-Torre, 1991). Our present description starts at 5.5–6 days, when a single *secondary mesencephalic vesicle* remains after the prospective isthmocerebellar neuroepithelium has been segregated caudalwards. This secondary midbrain contains the substantia nigra and other CA populations, whereas the locus coeruleus and subcoeruleus CA cell groups originate in the isthmocerebellar domain.

Three prosencephalic segments (prosomeres) are clearly observable caudal to a plane extending from the velum transversum (dorsal telencephalo-diencephalic boundary) to the transverse retromammillary furrow in the floor of the brain. They are the well-known *synencephalon, posterior parencephalon* and *anterior parencephalon* segments (Rendahl, 1924; schema modified by Puelles et al., 1987a, 1991). For simplicity, these may be named p1-p3 prosomeres, respectively, according to Bulfone et al. (1993a), Puelles & Rubenstein (1993) and Fernandez et al. (submitted). The numbering of prosomeres proceeds caudorostrally in accordance with their temporal order of appearance (Trujillo & Alvarado-Mallart, 1991). Their prospective differentiated domains in the developing and mature diencephalon have been traced (Fig. 16.1; see also Puelles et al., 1987a, 1991). The large alar plate fields of these prosomeres differentiate variously as the pretectum (p1), the dorsal thalamus/epithalamus (p2) and the ventral thalamus (p3) (PT; DT; VT in Fig. 16.1*d*). The subjacent basal fields encompass the basal synencephalic (p1), tuberculum posterior (p2) and retromammillary (p3) tegmental regions, respectively (bs; tp; rm in Fig. 16.1). This basal region may be further subdivided longitudinally into a basal zone, that differentiates precociously, and a paramedian zone, adjacent to the floor region, that shows retarded neurogenesis (Puelles *et al.*, 1987*a*).

The forebrain macrodomain (proneuromere) extending rostrad to these first three prosomeres combines the whole hypothalamus, basally, with a much enlarged alar plate region that contains the supraoptic, preoptic, peduncular and commissural regions, as well as the evaginated optic and telencephalic vesicles. This *secondary prosencephalon* (Puelles *et al.*, 1987*a*), may be also empirically subdivided in transverse regions by reference to the rostral ending of the longitudinal alar/basal boundary behind the optic stalks (Fig. 16.1*d*; see Bulfone *et al.*, 1993*a,b*; Puelles & Rubenstein, 1993). The evidence for the development of these regions from early prosomeres is based on a descriptive approach (p4–p6 prosomeres of Fernandez *et al.*, submitted) and on gene mapping data (Bulfone *et al.*, 1993*a,b*; Puelles & Rubenstein, 1993). While the evidence for a morphologic and genetic transverse subdivision is strong enough to warrant tentative employment of the p4–p6 schema, we must await various kinds of complementary developmental data (Martínez *et al.*, 1992; Figdor & Stern, 1993) to definitely conclude on the actual *number* of boundaries and segments. The mammillary and tuberoinfundibular hypothalamic regions (whose mutual boundary marks the postulated p4/p5 interprosomeric limit) jointly represent the enlarged rostral end of the paramedian longitudinal zone (Puelles *et al.*, 1987*a*), whereas the basal plate proper ends rostrally at the retrochiasmatic commissural region (rch; Fig. 16.1). The topographic arrangement of all these various transverse domains and longitudinal subdivisions of the forebrain is illustrated in Fig. 16.1, as observed just at the appearance of earliest CA traits. In the following description, the relative radial depth of CA cells within the parcellations shown in Fig. 16.1 shall be indicated by the self-explanatory terms: ventricular, periventricular or subventricular, intermediate, superficial and subpial.

Development of TH/DA cells and fiber tracts

Early stages (5–7 days of incubation)

Young neurons expressing detectable tyrosine hydroxylase first appear at 5.5–6 days of incubation. They are located in the mammillary area and tuberomammillary boundary region of the longitudinal paramedian plate (p4 segment; Fig. 16.1(*d*)). All these neurons have a fusiform shape and their axonal processes tend to be oriented caudally; they have a small, scarcely branched dendritic process at the opposite cellular pole. Immunoreactive cells in the mammillary area occupy generally a subventricular position, deep to a dense aggregate of unstained neurons which later on becomes the subthalamic nucleus, or nucleus ansae lenticularis anterior (mm, st; Figs. 16.1(*c*); 16.2(*a*),(*b*)). However, some of the CA cells lie superficially at the boundary with the retromammillary area. THi cells at the tuberomammillary boundary zone mostly occupy intermediate or superficial positions and reach dorsally the fibers of the ventral supraoptic commissure at the alar/basal boundary (tm; Figs. 16.1(*d*), 16.2(*a*),(*b*)).

The same cell populations are THi at 7–7.5 days of incubation, displaying an increased number of differentiated cells, with slightly larger cell sizes and more developed cell processes. The immunoreactive axons of these cells form a distinct, slightly dispersed tract that follows the mammillotegmental tract at least as far back as the isthmomesencephalic boundary.

At 7.5 days, a separate group of faintly stained small fusiform cells was detected at the ventralmost part of the alar plate of the anterior parencephalon (p3 segment; Fig. 16.1).

Intermediate stages (8–10 days of incubation)

At 8 days of incubation several other CA cell groups are THi, besides those observed previously (Fig. 16.3(*a*)). A fraction of all these THi young neurons are also DAi.

Secondary prosencephalon Mammillary and tuberomammillary CA cells [p4] are rather well developed, showing triangular profiles and longer, sparsely branched dendrites. Those in the mammillary subventricular region surround the developing hypothalamic periventricular organ and partly accumulate superficially near the mammillary midline (Fig. 16.2(*c*)). THi cells in the tuberomammillary subpial region are densely aggregated caudoventrally, but appear less compactly arranged than in earlier

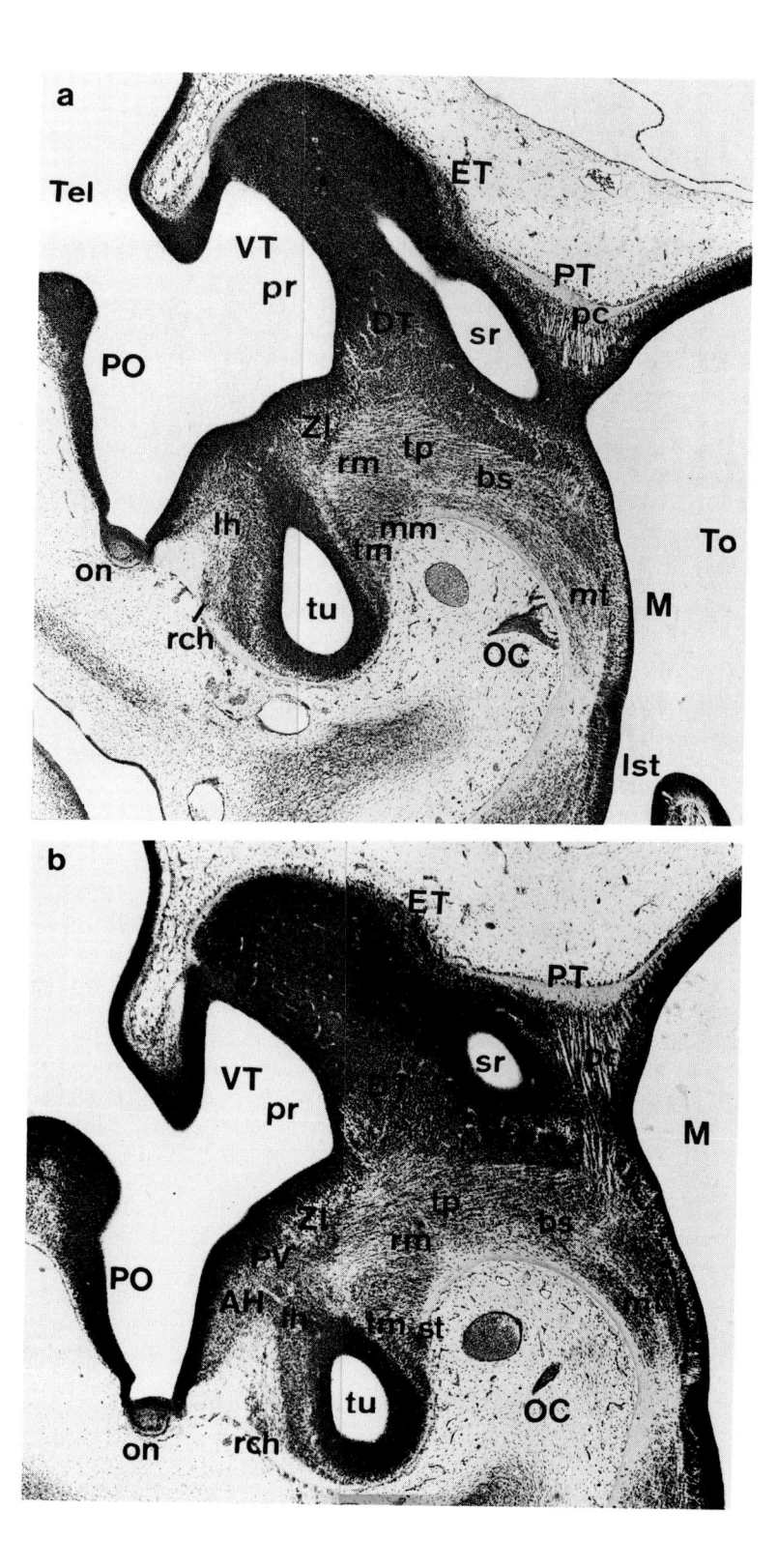

Fig. 16.1.

embryos in the neighborhood of the ventral supraoptic commissural tract. These cells invade a large extent of the caudal lateral hypothalamus (see tm, csv, lh in Fig. 16.1(c),(d)). Only scarce cells and axons in this THi complex were DAi.

Moreover, an important new group of fusiform THi cells that are oriented longitudinally was found at subventricular levels in an alar region of the secondary prosencephalon. This was estimated to correspond to the prospective supraoptic and paraventricular zones of the anterior hypothalamus, according to our observations at later stages. More dispersed, immature-looking THi cells appear in the anterior peduncular region and anterior preoptic zone, extending up to the anterior commissure behind the course of the septomesencephalic tract (Figs. 16.2(c); 16.3(a)).

Many immunoreactive fibers extend between the vicinity of the tuberomammillary complex and the basal telencephalon, as an ascending component of the medial forebrain bundle. It seems that most of these fibers originate more caudally (within the mesencephalon or rhombencephalon); they extend longitudinally within the basal plate up to the p4 segment and then bend sharply dorsalward into the telencephalic peduncle after they pass the p3/p4 boundary. Stained axonal growth cones are visible at their tips as they invade the paleostriatal anlage at 8 days.

Contrarily, cells in the mammillo-tuberomammillary complex mostly seem to send their axons backwards into the brainstem tegmentum as a descending component of the CA pathway (Fig. 16.2(c)).

Diencephalon Numerous THi young neurons are present at the anterior parencephalic basal and alar plates (p3). Cells in the basal and paramedian plates are oriented transversely within the retromammillary area and extend up to the level of the ventral supraoptic commissural tract. Alar cells appear dispersed at a restricted periventricular zone estimated to correspond to the prospective zona incerta (ZI; Figs. 16.2(c); 16.3(a)); this is placed rostroventrally, just below the ventral end of the pseudosulcus parencephalicus, the triangular (in sagittal sections) ventricular recess that represents a remnant of the primitive p3 cavity or parencephalic recess (see ZI, pr in Fig.16.1(a),(b). This cell group seems to be partly continuous rostrally with the cell population found in the supraoptic-paraventricular area. It contained also a number of DAi neurons. Only scarce THi neurons were found in the diencephalon caudal to the retromammillary area, appearing as small separate groups

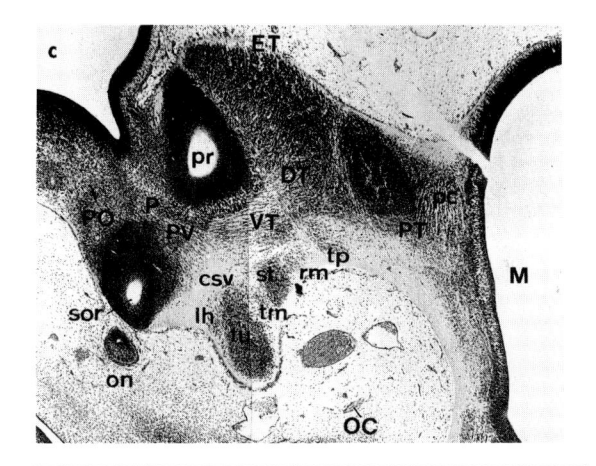

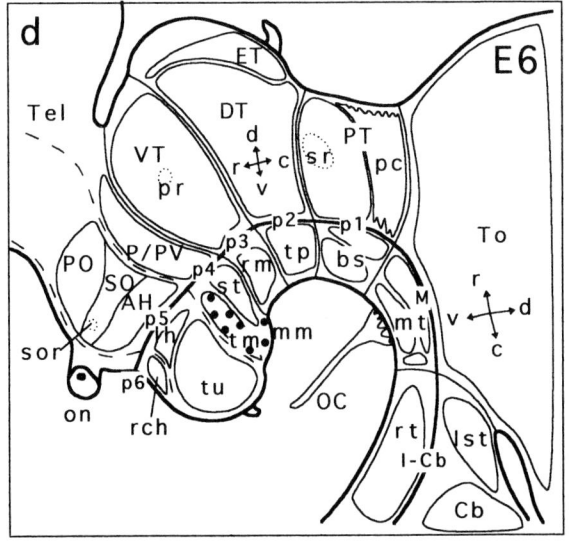

Fig. 16.1. Three sagittal sections through the forebrain of a 5.5 day-old chick embryo. (a) is medialmost and (c) is lateralmost. The drawing in (d) is a graphic reconstruction of this region, illustrating, jointly with the photographs, the morphologic landmarks employed in a segmental analysis and the location of the earliest THi cells (black circles). Several characteristic structural regions and areas, separated by cell-poor gaps in the sections, are identified as well. The identification of prosomeres (p1-p6) and other segmental domains (M, I-Cb) is placed along the curved line that represents the approximate cytoarchitectonic boundary between alar and basal longitudinal zones, that is, the length axis of the brain. The bending of this line at the cephalic flexure introduces a change in the apparent position of the topological dorsoventral and rostrocaudal directions, as indicated by the crossed small arrows within the DT and the To regions. The less evident postulated boundaries between the p4, p5 and p6 domains were traced with dashed lines.

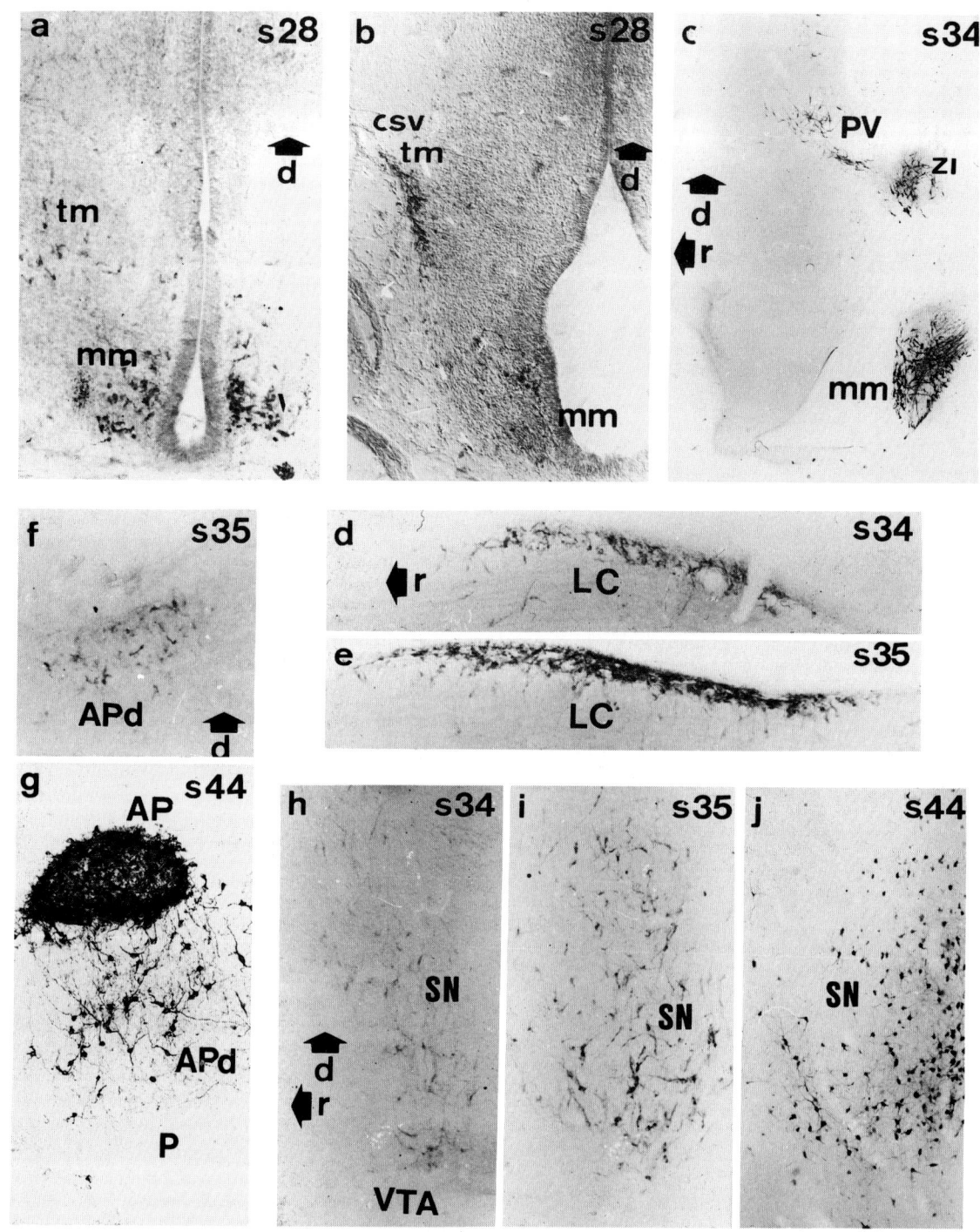

Fig. 16.2. Photographs of developing CA neuronal populations in chick embryos of various ages (Hamburger – Hamilton stage marked at upper right corner). Arrows indicate orientation (d=dorsal; r=rostral). (a) transverse section of mammillary cell group at E6; (b) transverse section of intermediate and subpial tuberomammillary cells at E6; (c) sagittal view of mammillary, zona incerta and paraventricular cell groups at E8; (d), (e) sagittal sections showing the embryonic locus coeruleus at E8 and E9; (f), (g) sagittal sections through the area praetectalis diffusa when it first appears at E9 (f) and in its mature state, together with AP, at E18 (g); (h), (i), (j) sagittal sections through the maturing substantia nigra at E8, E9 and E18.

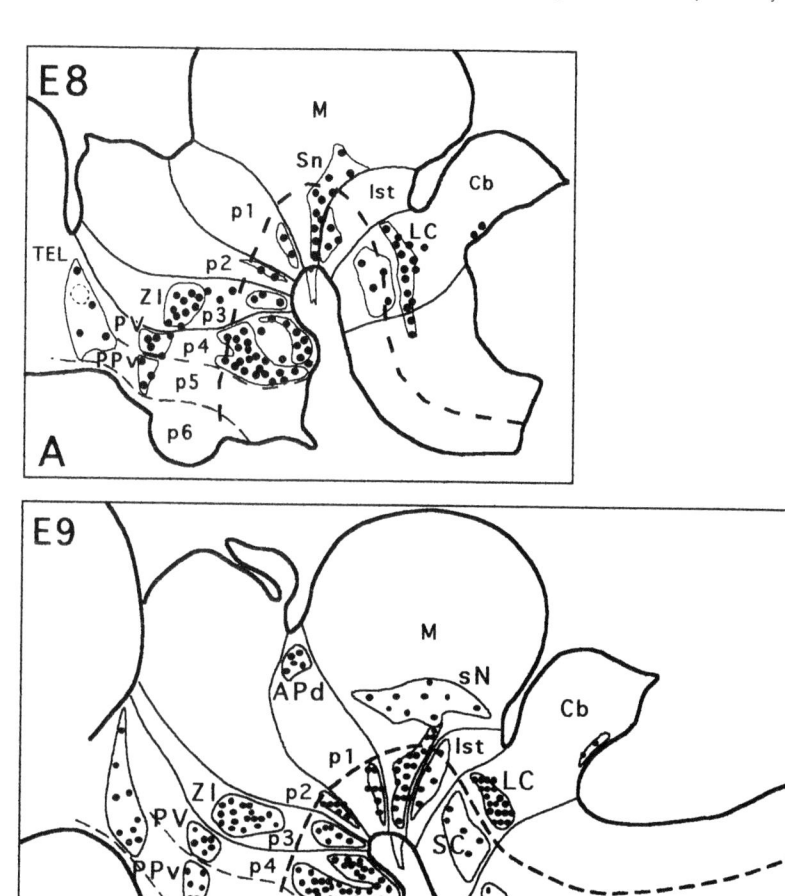

Fig. 16.3. Graphic reconstruction of the distribution of chicken CA cell bodies relative to the segmental Bauplan at 8 and 9 days of incubation (E8, E9). The axial reference, the alar/basal boundary, is shown as a thick dashed line. Clearcut segmental boundaries are traced as continuous thin lines and postulated boundaries as dashed thin lines.

within the basal regions of the posterior parencephalon (p2) and synencephalon (p1) (Fig. 16.3(a)).

Mesencephalon Palely THi and DAi small neurons first appear at 8 days in the mesencephalon, dispersed from ventromedial to dorsolateral in the caudal tegmentum, adjacent to the isthmomesencephalic boundary (Figs. 16.2(h); 3(a)). The ventralmost few cells occupy the paramedian plate, possibly constituting the anlage of the ventral tegmental area, whereas the remaining cells largely represent the primordium of the substantia nigra in the basal plate plus a few supranigral alar cells. These dorsalmost cells lie within the periventricular gray. The axons of all these cells converge ventromedially on the ascending longitudinal tract that bends around the cephalic flexure and proceeds across the tuberomammillary stratum externum (p4) into the telencephalic peduncle.

Isthmocerebellar domain The rostralmost basal and paramedian isthmus contains at 8 days a dispersed group of THi/DAi cells that seems continuous with the ventral part of the substantia nigra and the ventral tegmental area across the isthmomesencephalic boundary. We conceive this population as a retronigral primordium, or, alternatively, as a caudal extension of the ventral tegmental area. These cells are partly mixed with the rostralmost raphe cell population, dorsal to the interpeduncular nucleus (Fig. 16.3(a)). More caudally, the dorsolateral tegmentum shows disperse THi (but not DAi) cells of the prospective n.subcoeruleus and more compact and numerous, palestaining THi (DA-negative) cells of the locus coeruleus within the dorsolateral periventricular stratum (Fig. 16.2(d)). This primordium lies near the superior cerebellar peduncle, along which a few THi neurons were found; some of them lie subventricularly inside the cerebellum (Fig. 16.3(a)). The longitudinal CA axonal tract passes ventral to the locus coeruleus and receives axons from all these neurons.

At 9 and 10 days of incubation all these CA cell groups and tracts are much better differentiated (Figs. 16.2(e); 16.2(i); 16.3(b)), showing larger cell bodies, longer dendrites and more intense immunoreactivity. In addition, a faintly stained, slightly dispersed new cell group appeared dorsally in the alar plate of the synencephalon (p1), in the primordium of the nucleus areae praetectalis diffusa, or nucleus pretectalis medialis (Fig. 16.2(f)). These cells send their thin axons dorsally and bilaterally (through the posterior commissure) to a small subpial neuropile, which is the anlage of the nucleus areae praetectalis (Fig. 16.2(g)). An increased THi cell group was seen basally in the synencephalon, apposed to the *rostral* boundary of the mesencephalic tegmentum; it is thus separated from the caudally placed substantia nigra by a space lacking THi cells that contains the growing nucleus ruber. This cell group fuses ventrally with the ventral tegmental area (Fig. 16.3(b)).

Nigral cells in the caudal mesencephalon are continuous with a dispersed population of larger cells distributed caudorostrally within the mesencephalic dorsolateral periventricular gray, called by us the supranigral group. This stratum also shows a sparse plexus of subventricular axons and terminals (Fig. 16.3(b)).

The periventricular locus coeruleus shows greater immunoreactivity of its neurons, some of which seem to segregate ventrally into the subcoeruleus cell group (Figs. 16.2(e); 16.3(b)). A few isolated THi neurons appear as well subventricularly within the cerebellum (Fig. 16.3(b)).

Another new faintly stained and rather compact cell group was detected in a caudal region of the dorsolateral medulla which represents the CA population later found at the nucleus of the solitary tract and area postrema (Fig. 16.3(b)).

Up to the 10 day stage, the hypothalamic periventricular organ lacks TH or DA immunoreactivity.

Late embryonic stages – mature pattern
Distribution of TH and DA was also studied in advanced embryos at 18 days of incubation. These showed a fully mature pattern of CA cell groups, as compared with material from 1 month-old hatchlings kindly placed at our disposal by M. Martinez-de-la-Torre. We will report for each brain region the final positions of the diverse CA cell groups already found at 10 days of incubation, adding the few late-appearing cell groups whose exact moment of appearance between 10 and 18 days is not known. All these observations were extensively cross-correlated with alternate sections stained with cresyl-violet or acetylcholinesterase, using a camera lucida device. Since in most cases the TH and DA antibodies stained similar numbers and types of cells and fibers, we will only state expressly the cases in which differential staining occurred. We do not know whether individual cells express both markers, since we did not do double immunostaining, though this seems probable in many cases.

Secondary prosencephalon The precocious and voluminous THi/DAi population occupying deep and superficial regions of the mammillary area (including the tuberomammillary cell group) contains large-and medium-sized multipolar or triangular neurons and a rich plexus of immunoreactive processes (Fig. 16.4(c)). The mammillary recess is continuous dorsally with the hypothalamic periventricular organ, whose liquor-contacting cells stain strongly now with the anti-DA antibody (but not with the anti-TH one). A number of periventricular THi/DAi neurons lie near the organ, constituting a nucleus of the periventricular organ.

Various CA cell groups were also found in the alar plate of the secondary prosencephalon at a distance from the foregoing paramedian and basal plate formations. These alar cells lie roughly at cross-sectional planes across the preoptic and supraoptic regions (Fig. 16.4(c)). The longitudinal band of fusiform cells detected at earlier stages decomposes into two more or less distinct cell groups that surround shell-like the system of magnocellular basophilic neurons of the paraventricular nucleus and the n.preopticus periventricularis (PV and PPv, Fig. 16.4(c)). Many of these neurons extend dendrites to the periventricular stratum and their extremely thin axons can be followed through the preoptico-infundibular tract into the eminentia media. At the boundary between

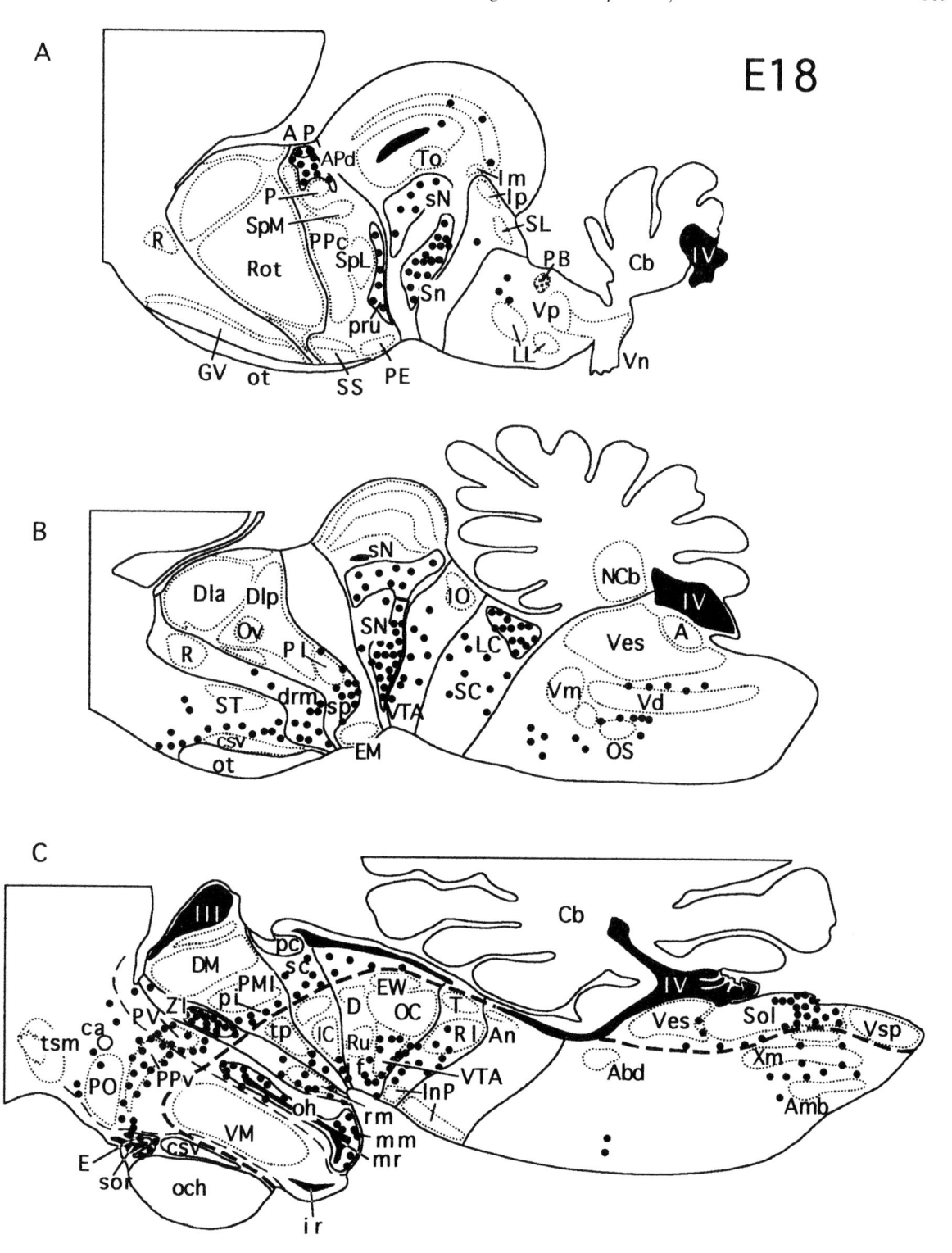

Fig. 16.4. Mapping of CA (TH/DAi) cells (black dots) in the brain of an E18 chick embryo, represented by three spaced parasagittal sections going from lateral (*a*) to medial (*c*). The segmental boundaries are traced as thin continuous or discontinuous (postulated limits) lines. The alar/basal longitudinal limit (thick dashes) is shown only in C for simplicity. These limits and other landmarks (dotted lines) were identified in cresyl-violet-stained adjacent sections with a camera lucida. The mapping was also correlated with immunoreacted and counterstained hatchling material.

the organum vasculosum laminae terminalis and the optic chiasm a slightly separate third group of small CA neurons appears which partly fuses across the midline with its contralateral counterpart. This may be described as an epichiasmatic CA cell group (E; Fig. 16.4(c)). The limits between these three alar aggregates coincide with the postulated borders between the p6, p5 and p4 prosomeres (Figs. 16.1(d); 16.3; 16.4(c)).

In addition to these localized alar cell groups there appear much sparser small immunoreactive neurons distributed variously (generally as isolated cells) within the lateral and anterior preoptic area, anterior peduncular area, caudalmost septum and nucleus of the stria medullaris (Figs. 16.4(b), (c)). These are mostly non-reacting with the anti-DA antibody. Other DA-negative groups of small THi neurons are found deep to the optic tract, surrounding externally and internally the descending tract of the ventral supraoptic commissure as it passes across the stratum cellulare externum of the hypothalamus (Fig 16.4(b)). These cells apparently represent those found previously at the early superficial tuberomammilar population.

At 18 days of incubation THi (DA negative) cells are also present in the deep layers of the olfactory bulb. The moment of appearance of immunoreactive cells in the retina was not investigated.

Diencephalon The anterior parencephalon (p3 segment) contains several CA cell populations differing in dorsoventral position. There is a moderate number of medium-sized neurons at the paramedian retromammillary area (rm; Fig. 16.4(c)). More dorsally and laterally a group of large cells extends dorsalward just caudal to the nucleus of the OPH, but occupying a slightly less deep position, that is, lying at intermediate depth. This may be identified as a dorsal retromammillary cell group (drm; Fig. 16.4(b)). Finally, the alar plate of p3, or 'ventral thalamus area', shows a large number of medium-sized neurons in the periventricular zona incerta (ZI; Fig. 16.4(c)); these cells are more separated from the ventricular lining than those more rostral ones that form a shell around the paraventricular nucleus.

The posterior parencephalon or p2 segment also contains three different cell groups. A few cells are present within the paramedian plate in the posterior tuberculum area (tp; Fig. 16.4(c)). More dorsally, a larger cell group is found caudoventrally relative to the nucleus posterointermedius or n.ansae lenticularis posterior of the dorsal thalamus. This group may be named n.sub-posterointermedius (sp; Fig. 16.4(b)). Finally, the periventricular stratum just ventral to n.paramedianus intermedius and the tail of n.subrotundus, caudal to the zona incerta, shows a small number of CA neurons that we named postincertal nucleus (pi; Fig. 16.4(c)).

The synencephalon (p1 segment) has a paramedian CA population that seems to form a rostral part of the ventral tegmental area, or nucleus interfascicularis (if; Fig. 16.4(c)). A number of dispersed neurons are found more laterally in the caudal synencephalic basal plate, rostrally adjacent to the nucleus ruber of the mesencephalon, forming a prerubral CA cell group (pr; Fig. 16.4(a)). The alar plate contains superficially the highly developed complex of small and slightly dispersed THi/DAi neurons of the n. areae praetectalis diffusa, or n. pretectalis medialis, together with similar cells aggregated inside the densely immunoreactive neuropile of the n. areae praetectalis (APd; AP; Figs. (16.2(g), 16.4(a)). Axons from this pretectal complex can be traced across the posterior commissure to the contralateral side. Additionally, there appear dispersed large neurons occupying a large extent of the periventricular stratum ventral to the posterior commissure – the n. subcommissuralis (sc; Fig. 16.4(c)).

Mesencephalon This territory displays the densely populated ventral tegmental area in the paramedian region; this appears divided by the oculomotor nerve root into a small medial part and a larger lateral part. Its cells are small or medium-sized (VTA; Fig. 16.4(c)). The caudolateral tegmentum is extensively colonized by the less dense and slightly larger cells of the substantia nigra, whose size increases towards the brain surface (SN; Figs. 16.2(j); 16.4(a),(b)). The alar supranigral cell group is similar to the synencephalic periventricular population; it displays a number of rather large, sparsely ramified neurons distributed through the periventricular gray. Caudally some of them even surround the aqueduct and approach the dorsomedian brain surface rostral to the isthmo-optic nucleus (sN; Figs. 16.4(a),(b),(c)). A few dispersed multipolar neurons show pale TH immunoreactivity (no DA) inside the tectum, in the stratum griseum centrale, or, less commonly, in the periventricular stratum (Fig. 16.4(a)). Some such cells are still present one month after hatching.

Isthmus and rhombencephalic domains Rostralmost isthmic structures appear to have been colonized by a number of CA cells escaped from the large caudal mesencephalic CA complex. A caudal isthmic part of the ventral tegmental area may be distinguished in relation to the rostral end of the interpeduncular nucleus, which is wholly isthmic according to us (Figs. 16.4(*b*),(*c*)). Similarly, the nucleus raphe linearis and the tegmentum lateral to it also contain a number of smaller CA neurons, forming a retronigral area (Figs. 16.4(*b*),(*c*)).

Just caudal to this there is the locus coeruleus/subcoeruleus complex, extending respectively across periventricular and intermediate levels of the alar plate, dorsal to the nuclei of the lateral lemniscus (Figs. 16.4(*a*),(*b*)). The dorsal parabrachial nucleus shows a round patch of dense immunoreactive terminals superficially (stippling in PB; Fig. 16.4(*a*)).

The rh2 and rh3 segments apparently each have a group of subtrigeminal CA medium-sized cells, placed ventro-lateral to the rostral and caudal motor nuclei of the trigeminal nerve, respectively (Fig. 16.4(*b*)). A similar but more disperse cell group occupies the neighborhood of the various motor subnuclei of the facial nerve (rh4 segment). These rh4 cells do not react with the anti-DA antibody.

Deep to and surrounding caudally the superior olive (which possibly lies in the rh5 segment) there is a distinct population of small CA neurons (OS; Fig. 16.4(*b*)). The caudalmost rhombomeric domains (rh6–rh8) contain two different types of CA neurons. Dispersed medium-sized cells surround dorsomedially the nucleus ambiguus (Amb; Fig. 16.4(*c*)). On the other hand, a large population of very small and small THi/DAi neurons was found densely aggregated in the area postrema and the dorsocaudal sector of the nucleus of the solitary tract (Sol; Fig. 16.4(*c*)). Similar but much less numerous cells lie interstitially in a thin shell around the lateral, ventral and medial boundaries of this nucleus. This extension can be followed rostrally along the solitary tract on the dorsal aspect of the descending trigeminal column, at least as far as the rh4 segment (Vd; Figs. 16.4(*b*),(*c*)). At the level of n. laminaris other small cells form a thin sheet that extends from the ventricular lining down to the supratrigeminal strip, intercalated between n. laminaris and nucleus vestibularis medialis (not shown).

Starting at the rostral spinal cord, three different new cell groups were visible exclusively with anti-TH antibody: a ventral strip of ependymal cells, accompanied by sparse free small neurons and a number of dorsolateral small- and medium-sized cells occupying the nucleus of the dorsolateral funiculus and other dorsomedian parts of the dorsal horn basis (not shown).

Segmental localization pattern

The present approach introduces some novelty as regards description and localization of CA cell groups. Stepping on previous studies on reptilian, avian and mammalian brains (Puelles and Martínez-de-la-Torre, 1987; Puelles *et al.*, 1987*a*,*b*, 1991, 1992; Medina *et al.*, 1993; Bulfone *et al.*, 1993*a*,*b*; Puelles & Rubenstein, 1993; Fernández *et al.*, submitted), we employed a more detailed structural scaffolding than is usual in the literature: the transverse boundaries of diverse segmental domains and the common longitudinal subdivisions of the neural wall at each one of them (paramedian, basal and alar). This embodies a conception of the basal/alar boundary as a general reference landmark that is bent around the cephalic flexure and crosses the midline somewhere in the postoptic region (Puelles *et al.*, 1987*a*, 1991; Puelles & Rubenstein, 1993; Ross, Parrett & Easter, 1992). Differences introduced by this approach with respect to conventional use are probably readily apparent to the reader of this book (Figs. 16.5, 16.6); we expect that they may clarify some previously obscure points.

One important example appears in the boundary of the mesencephalon with the diencephalon. The earliest and most influential version of this limit was given by His 100 years ago (His, 1893). For convenience, he proposed a line connecting the posterior commissure with the mammillary bodies. However, embryologists studying the growth of the early brain vesicles soon concluded that a more consistent limit could be traced in all vertebrates. This limit is a curved plane passing just behind the dorsoventrally-coursing fibers of the posterior commissure (Palmgren, 1921; Vaage, 1969; Puelles *et al.*, 1987). Nowadays there is gene-mapping and various other experimental evidence supporting the same conclusion (Figdor & Stern, 1993; for review see Puelles & Rubenstein, 1993).

Reference to our Figure 1 shows that the conventional commissuro-mammillary line disregards the cytoarchitectural parcellations observable in sagittal sections (Fig. 16.1). Unfortunately, explicit or implicit assumption of this boundary is found in most descriptions of the development of CA cells (see Fig. 1 of Specht *et al.*, 1981), leading to potential confusion in the understanding of important CA cell groups like the ventral tegmental area and the substantia nigra. We would thus like to make the point that whatever the apparent usefulness of His' boundary in adults, it should not be taken seriously at the developmental level (we also think that a non-dogmatic conception of adult anatomy should constantly adapt to the best

392 L. Puelles and L. Medina

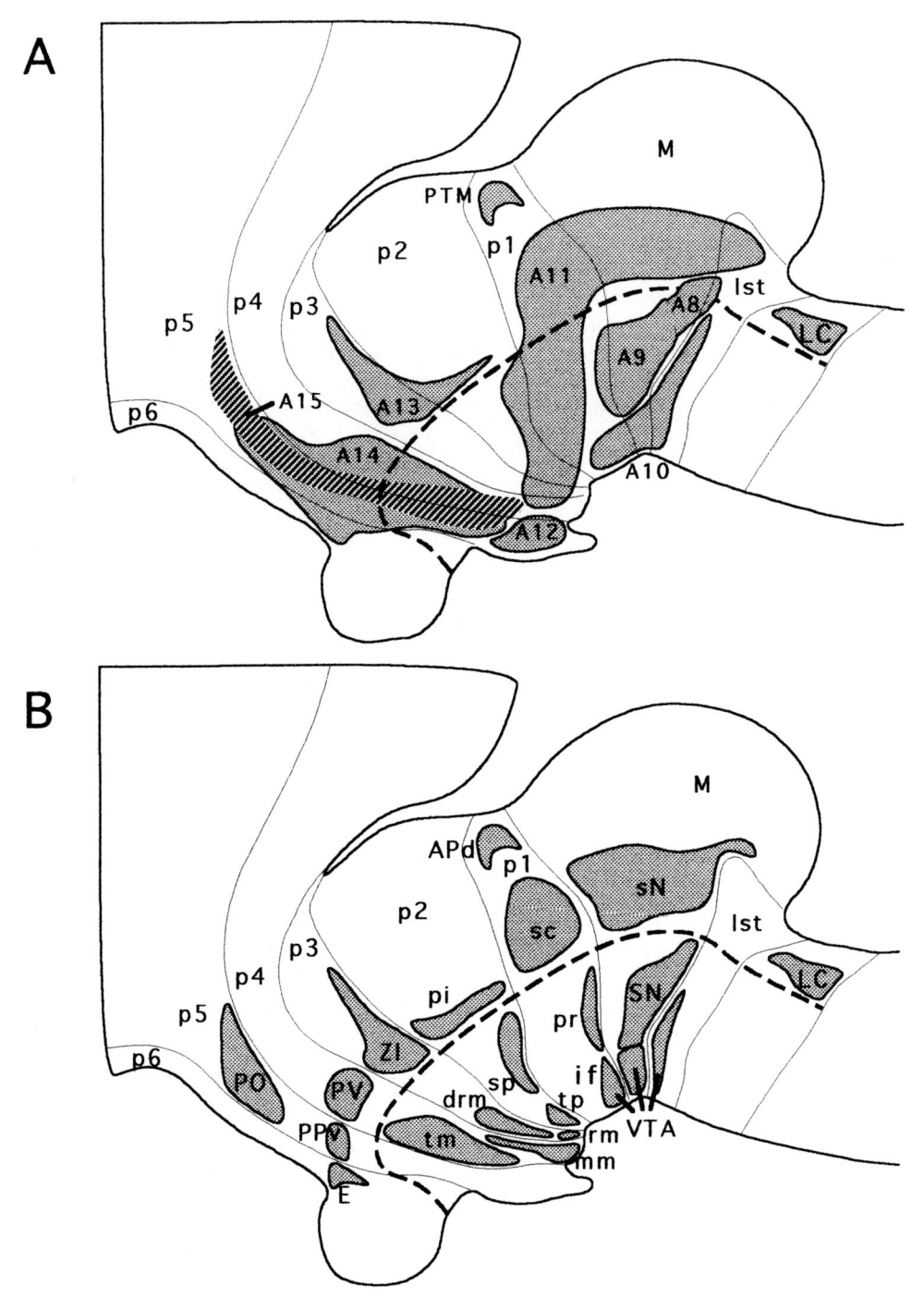

Fig. 16.5. Schema of avian forebrain and midbrain CA populations (shaded or striped). (a) as mapped in pigeons by Reiner et al. (this book) using the 'mammalian' alphanumeric terms and a conventional anatomical schema, but shown here projected upon our segmental and alar/basal boundaries; (b) as presently mapped in the chick by us, in accordance with segmental parcellations. The pigeon A12 cell group does not appear in the chick. All other populations are comparable.

available developmental knowledge). Our approach allocates various sectors of the extensive retromammillary/prerubral tegmentum to different diencephalic segmental domains, thus allowing fine analysis of histogenetic patterns at this locus.

A more precise classification of cell populations (particularly those in the forebrain) according to their longitudinal and dorsoventral position suggests in this case the convenience of some regrouping with regard to generally used schemata, as, for example, the system developed by Dahlström and Fuxe (1964), Fuxe, Hökfelt & Ungerstedt (1969), Björklund & Nobin (1973) and Hökfelt et al. (1984) (comparison in our Figs. 16.5 and 16.6). One important reason for introducing that nomenclature was that many CA populations seemed to cross conventional neuroanatomic boundaries. We suggest now that the set of segmental plus longitudinal boundaries resolves that problem; the limits of most CA cell groups agree with the morphological parcellations of the segmental schema. This approach highlights instead small differences in cell size, typology, position and density, correlated with variable TH/DA immunochemical profiles and heterochronic patterns of appearance (or of marker expression) that were not hitherto apparent. We hold that our modifications, even if they introduce greater apparent complexity, are justified by the increased developmental and comparative orderliness conveyed by the segmental viewpoint. This kind of order is advantageous in the search of patterning mechanisms, insofar as segments and longitudinal subdivisions of the brain seem to have an ancestral basis encoded by early gene expression patterns (Bulfone et al., 1993a,b; Puelles & Rubenstein, 1993).

There are two overall forebrain patterns that stand out. First, expression of CA traits is a histogenetic characteristic that is repeated in most (if not in all) neuromeric domains, that is, most segments seem to produce one or more CA cell populations. The causal basis of this differentiation trait would thus seem to be generally distributed. Notwithstanding this general statement, note that each segmental domain displays a peculiar pattern of CA differentiation; this reveals the existence of specific developmental aspects presumably derived from the differential genetic specification of each segment and histogenetic area of the neural tube.

A second overall pattern is found in the dorsoventral distribution of CA cell populations. Separate sets of CA cell groups seem to originate either basally or dorsally in the diverse segmental domains (Figs. 16.3 and 16.4). Basal cell groups are relatively precocious, whereas alar cell groups appear slightly later in development. Basal populations start appearing within the paramedian longitudinal zone and divide thereafter into groups of smaller cells remaining in situ (like the mammillary and retromammillary cells, or the ventral tegmental area population) and groups of medium- to large-sized cells that spread into the adjacent basal tegmentum (hypothalamic stratum externum cells, tuberomammillary cells, n.of the paraventricular hypothalamic organ, retromammillary nucleus, n.subposterointermedius, synencephalic tegmental group, substantia nigra, etc.). It is not yet clear whether these two sorts of basal CA neurons originate in the same paramedian matrix area or in different areas that are adjacent.

Alar CA populations can be also clearly divided into two separate sets. One of them comprises those cell groups that develop within ventral* (*= this term refers to position relative to the bent longitudinal axis; see Fig. 16.1) periventricular regions of the respective alar plate domains: epichiasmatic nucleus (p6), cells within or around PPv and PV (p5, p4), zona incerta (p3), postincertal cells ventral to PMI (p2), subcommissural large neurons of the synencephalon (p1), supranigral large neurons (mesencephalon), locus coeruleus (isthmocerebellar domain) (Fig. 16.4(c)). The other, less extensive set is represented by small cells found away from the periventricular stratum near the dorsalmost parts of the alar domains. The most notable example is the large population of the area praetectalis diffusa (p1), but the dispersed cells found separately at the telencephalic peduncle (anterior preoptic region, anterior peduncular area, septum and n.of the stria medullaris) also fall within this definition (Figs. 16.4(a),(c)). Note that the telencephalic vesicle is 'hyper-alar' in our schema (Fig. 16.1), so that CA cells in the olfactory bulb constitute a third, hyperdorsal set of forebrain alar CA formations, together with those developing in the retina. The postulated p5 and p6 prosomeres have alar cell groups but lack paramedian and basal CA cells in birds and non-mammals in general, although these appear in mammals (Fig. 16.5).

The rhombencephalic pattern is slightly different in the rostral isthmocerebellar domain than in the caudal part derived from rhombomeres. The caudal isthmus contains the large locus coeruleus population (A6), possibly the most numerous population developing among the set of CA centers developing periventricularly within the ventralmost part of the alar plate (see above). The intracerebellar cell group (A4) may have migrated tangentially from the locus coeru-

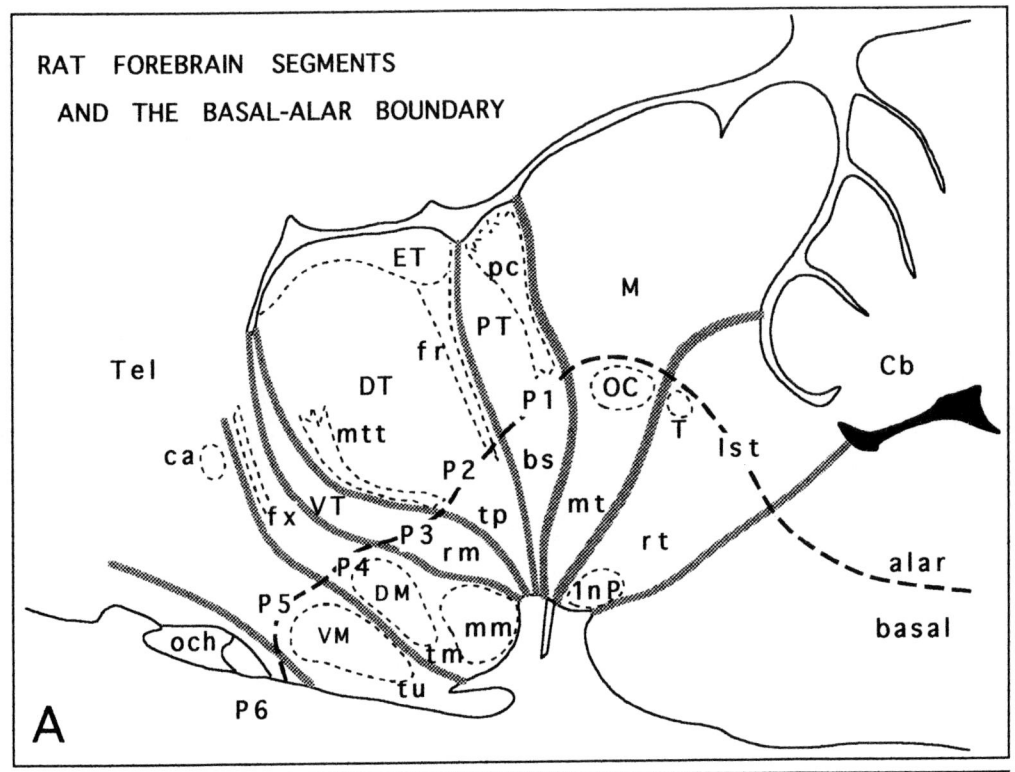

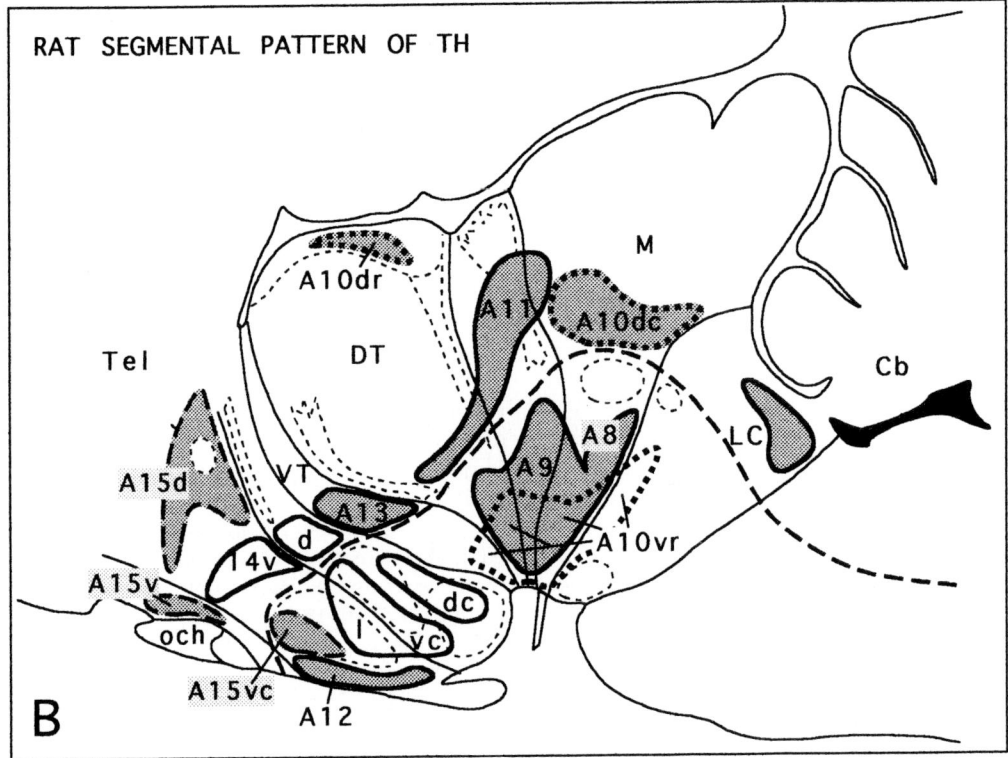

Fig. 16.6.

leus primordium. The subcoeruleus dispersed cell population apparently migrates ventrolaterally from the same locus. This displacement probably is in fact a radial movement towards the brain surface along bent glial guidelines. Paramedian and basal CA cells of the isthmus are only found close to the isthmomesencephalic limit. As found also in lizard embryos (Medina et al., this book), it is our impression that these caudal 'ventral tegmental' or 'retronigral' cells have migrated backwards from the large mesencephalic source.

Positions of CA cell groups more caudally in the rhombencephalon are stereotyped (Dahlström & Fuxe, 1964). They comprise a set of alar periventricular small cells (A2/C2), like those related to n.laminaris and n.vestibularis medialis, or the more numerous cells within n.solitarius, and other cells that migrate radially to a dorsolateral tegmental locus (as indicated by occasional cells stopped midway along the migration route). The latter include the pretrigeminal (A7), prefacial (A5), periolivary (A3 ?), periambiguus (C1/A1) and small supratrigeminal cells. Given that the branchiomotor nuclei (trigeminal, facial, ambiguus) occupy their definitive locations after variously complex migratory processes, it is an open question whether the CA cell groups placed near them lie in the alar plate or in the basal plate. An example of a CA cell group developing at the dorsalmost part of the alar plate is found as well in the population of the area postrema.

Finally, the spinal cord shows clear examples of basal (both paramedian and intermediate tegmental) CA cell populations, together with other alar CA cells.

Temporal patterns and the problem of migrations

There were no previous immunocytochemical data on the earliest appearance of CA traits in the chick forebrain. We observed the earliest, immature-looking THi young neurons in the mammillary area (posterior hypothalamus; basal p4) at 5.5 days of incubation. This is several days earlier than was suggested by means of the histofluorescent method (9 days i.o., according to Guglielmone & Panzica, 1984). The precociousness of these cells is corroborated by the fact that they always lead the other cell groups in their somatic and dendritic maturation. Preliminary analysis of autoradiographic data (Puelles, unpublished observations) indicates that these earliest CA cells may be born still earlier, perhaps as early as at 2 days for the tuberomammillary cells. Caution is thus needed regarding the interpretation of the position of all studied THi populations, since some migratory movements might have occurred before they express the marker. In fact, tuberomammillary CA cells are first detected in a subpial position.

Superior sensitivity of the immunoreaction is evident as well in our data for 7–9 days of incubation; during these three days most definitive THi cell groups start to express the marker and many cells also begin to express DA-immunoreactivity. Though there is heterochrony between the diverse segments, the basal populations generally mature in advance of the respective alar ones, a common pattern in segmental domains (Puelles et al., 1987a). The fact that the substantia nigra and locus coeruleus become first visible at 8 days i.o. does not exclude earlier cell birthdates for their neurons. This is supported by the

Fig. 16.6. Comparison of the pattern of TH in the rat given in the literature with the rat segmental schema. (a) Structural elements taken from various sagittal plates of the Paxinos and Watson rat atlas were employed to produce a schematic representation of our segmental Bauplan as applied to the rat forebrain. Significant reference landmarks comparable to those found in the chick were drawn with dashed lines and the basal – alar boundary (thick dashes) and the segmental limits (gray mesh) are emphasized. (b) This drawing contains the approximate distribution of mammalian CA cell groups relative to segmental domains in 16.6(a), identified according to the literature cited in the text, in order to compare with the patterns in reptiles and birds (Fig. 16.5(b)). On representing these CA populations (all shaded except the four 'parts' of A14), relevant grisea (particularly the substantia nigra) were taken slightly modified from the Paxinos and Watson atlas. Inadequacy of the conventional alphanumeric classification is suggested by the tendency of these cell groups to cross longitudinal or transverse boundaries (A14, A11, A9), or, in some cases, to be widely and discontinuously dispersed, in addition (A15, A10). Note that the A14 complex, composed of v, d, l, dc and vc parts (unshaded continuous outlines), was 'trimmed' somewhat, excluding the sparsest cells. In actual preparations the different parts may seem less separate than they appear here. The A9 and A8 cell groups are shown as forming a plurisegmental complex, similarly to the A10vr subpopulation (dashed unfilled outline).

fact that cell bodies, ascending tract and axonal growth cones in the basal telencephalon become immunoreactive simultaneously. The earliest THi nigral and supranigral cells, though showing small size, undifferentiated fusiform shapes and weak immunoreactivity, already occupy all the dorsoventral range of positions of the mature pattern at 8 days (Figs. 16.2 (h)–(j)). They must be interpreted therefore as being essentially postmigratory. The belief in a paramedian source and subsequent tangential migration for nigral cells (Hanaway, McConnel & Netsky, 1971; Marchand & Poirier, 1983) is partially supported by the fact that the early fusiform THi cells are not disposed radially with respect to the nearest ventricular zone and lie instead parallel to the pia. Additionally, some cells seem to cross the isthmomesencephalic boundary into the rostralmost isthmic raphe and tegmental regions.

Similarly inconclusive data were obtained with respect to evidence for large-scale intra- or intersegmental migrations in the forebrain, since most cell groups already occupy the whole range of their mature distribution when they first appear (compare Figs. 16.3 (a), (b) and 16.4 (c)). The fact that similarly placed mature CA neuronal populations are found in amphibians and fishes (chapters in this book) – animals which normally lack widespread neuronal migrations –, inclines us to be cautious about the existence of forebrain tangential CA cell migratory processes in birds or reptiles (Medina et al., this book). This conclusion implies that most CA cell groups lie near to their respective ventricular matrix areas. A possible exception is the apparent displacement of tuberomammillary cells into the neighborhood of the ventral supraoptic commissural tract (or lateral hypothalamus), which may be a case of a paramedian-tegmental migration that is lacking in anamniotes, unless it simply results from a passive dispersal of the cells by invading fiber tracts.

The comparative reasoning may be argued in favour of the postulated mediolateral nigral migration, since the anamniote mesencephalic CA cell populations tend to be restricted to the basal paramedian region, from which they presumably migrate dorsolaterally in amniotes.

These considerations suggest that limited migrations may occur within the basal plate, both intrasegmentally (from a sector in the paramedian zone into the respective basal tegmental zone) and intersegmentally (e.g., across the isthmic boundary), but probably not between the basal and alar domains of the brain. Migrations within the alar plate are generally restricted to radial movement from a deep periventricular to a more superficial locus. Such is the case of the sauropsidan pretectal cells, whose homologues in anamniotes occupy periventricular positions (González & Smeets, 1991 and this book). The issue of CA cell migrations in the avian brain needs to be reexamined using markers that identify prospective CA populations at their earliest postmitotic stages of development. It would be useful to have neuronal birth data for these various cell groups.

Comparison with other Tetrapods

Amphibians

Comparable immunohistochemical data for amphibian embryos are restricted to those on *Rana catesbeiana* of Carr, Norris & Samora. (1991), which complement descriptions of the adult pattern in various anuran and urodele species (see Gonzalez & Smeets, 1991, also for review of other reports; this book). These data may be easily translated into our segmental schema. The amphibian preoptic/anterior suprachiasmatic CA cells may correspond to our avian PV/PPv cell groups. It is interesting that they also partly colocalize with magnocellular cells of this region (Carr, Norris & Samora, 1991). The amphibian suprachiasmatic nucleus also comprises truly epichiasmatic cells like those found in the chick (Carr et al., 1991). A dorsoventrally elongated group of cells in the basal 'ventral thalamus' and 'ventral part' of the tuberculum posterior may include cells homologous to those in the zona incerta and the retromammillary areas of the chick (Carr et al., 1991). Catecholamine populations in the pretectal ('posterior thalamic') and mammillary ('dorsal infundibulum') regions are also readily comparable. Periventricular ventral-alar cell groups in p1, p2 and mesencephalon, as well as a migrated substantia nigra, seem to be lacking in amphibia.

Reptiles

Comparison with our own developmental data for the lizard *Gallotia galloti* also suggests close topographic similarity (Medina et al., this book; Fig. 15.1). The lizard 'rostrodorsal periventricular hypothalamic' cell group mainly encompasses the PV/PPv complex of the chick, while its dorsocaudal part possibly represents a cell group homologous to the avian ZI. The lizard suprachiasmatic cell group comprises cells disposed similarly as the avian epichiasmatic ones. Sparse cells found dorsally in the preoptic/peduncular region may correspond to the dispersed peduncular CA neurons found in chicks. The 'ventral tegmental area' in the lizard clearly contains synencephalic and isthmic portions, in addition to the main mesencephalic part of VTA.

There are some important differences in the spatiotemporal sequence in which various groups express the TH and DA markers. Earliest CA cells in the lizard appear in the *caudal mesencephalon*, though followed shortly thereafter by those in the mammillary region. Interestingly, given the marked precocity of the latter in the chick, the mammillary cells in the lizard have a more differentiated appearance (larger somata with more and longer dendritic processes) than the nigral cells that became immunoreactive before them. This suggests that the difference may not lie in a different pattern of proliferation, but in a variation in the postmitotic waiting period before the neurons express detectable levels of TH.

A similar difference obtains in the case of the lizard locus coeruleus; this appears several (five) stages later than the substantia nigra, but its neurons look relatively more mature from the beginning (see Fig. 15.7 (c) in Medina *et al.*, this book). On the whole, it may be concluded that the avian nigral complex has a retarded expression of CA traits relative to the schedule seen in *Gallotia*, whereas the avian locus coeruleus possibly accelerates slightly in comparison to the lizard one.

Mammals

There are several descriptions of the development of CA systems in mammals, including recent ones employing immunocytochemical techniques (Specht *et al.*, 1981a,b; Foster *et al.*, 1985; Daikoku *et al.*, 1986; Jaeger, 1986; Voorn *et al.*, 1988; Ugrumov *et al.*, 1989; Reisert *et al.*, 1990; Kalsbeek, Voorn & Buijs, 1992; Jaeger & Teitelman, 1992; Foster, 1992), *in situ* hybridization with TH-antisense RNA probes (Burgunder & Young, 1990) and ontogeny of autoradiographically mapped dopamine receptors (Sales *et al.*, 1989). The early appearance of these various traits in the rat and other mammals can be helpful for determining the general validity of our segmental mapping, insofar as segmental domains have also been described in the mammalian diencephalon and mesencephalon (Bergquist, 1954; Ströer, 1956; Coggeshall, 1964; Keyser, 1972; Gribnau & Geijsberts, 1985; Lakke *et al.*, 1988; Bulfone *et al.*, 1993a; Puelles & Rubenstein, 1993; Fernandez *et al.*, submitted).

None of the authors studying CA system development or adult topography considered doing a segmental morphological analysis. Published illustrations are mostly based on transverse sections and these provide in fact the least helpful material for our purposes, since most segmental boundaries are cut obliquely (Fig. 16.6(a)). Correlation with our approach is nevertheless possible when topographic reference was made to clearcut segmental landmarks, such as various tracts that course transversely close to a segmental boundary (Fig. 16.6(a)): the posterior commissure (M/p1 limit), the retroflex tract (p1/p2), the zona limitans tract and the mammillothalamic tract (p2/p3), the lateral and medial forebrain bundles (p3/p4) or the fornix tract (p4/p5) (Coggeshall, 1964; Keyser, 1972; Gribnau & Geijsberts, 1984; Fernandez *et al.*, submitted). Note that most of these can be identified in all vertebrates (Ross *et al.*, 1992; Figdor & Stern, 1993).

There are many indications that the overall pattern of CA cells in the brain is roughly similar in birds and mammals (Parent, Poitras & Dubé, 1984; Kiss & Péczely, 1987; Reiner *et al.*, this volume). A shortcut to dealing here with the ample literature is provided by those few reports that refer sagittal mappings of CA cell groups to some of the axonal landmarks cited above (*rat*: Björklund & Nobin, 1973; Björklund, Lindvall & Nobin, 1975; Chan-Palay *et al.*, 1984; *rabbit*: Blessing, Chalmers & Howe, 1978; *bat*: Yoshida *et al.*, 1982). We have used all these data, together with the detailed TH mappings and nomenclature of Hökfelt *et al.* (1984), to generate our Fig. 16.6(b). This shows our provisional interpretation of the presumed segmental relationships of the CA cell groups identified in mammals. Note that this figure simply represents how the concepts used in the literature relate to the segmental schema and not our agreement with the resulting pattern (we would like to split some populations and group others, – see below). The schema is nevertheless useful for comparing with the avian pattern (Fig. 16.5) and for centering our subsequent discussion of some developmental data.

The distribution and boundaries of CA cell groups roughly correspond with the postulated segmental domains and there also appears a rough subdivision into typologically distinct alar and basal sets. Major differences include the well-populated A12 cell group in the tuberal region of p5 and the A10dr cell group in p2 (both non-existent in non-mammals), the absence of a dorsal pretectal cell group and the apparent scarcity of CA cells in the mammillary region of p4 (note that the hypothalamic periventricular organ is absent in mammals).

Some of the groups identified initially with the fluorescent methods (A15, A14, A11, A10, A9) were subsequently expanded to encompass additional cells found immunohistochemically (Hökfelt *et al.*, 1984). The anatomical heterogeneity of these additions was tentatively resolved by *ad hoc* definition of various 'parts' (for example: dorsal, ventral, lateral, caudal parts of A14; dorsorostral, dorsocaudal, ventrorostral parts of A10; dorsal, ventral and caudal parts of A15). These stand out in our schema as being stretched, some of them rather forcedly, across one or several transverse or longitudinal boundaries.

An alternative resolution of the problem of added CA populations implies giving to each distinct group its own name or number. The developmental approach implicit in our segmental analysis clearly favors this sort of solution, since otherwise we shall eventually come into difficulties at the level of histogenetic and genetic causal explanation. This urges the convenience of 'splitting' these subdivided groups at least into distinct alar or basal groups and ideally into segmental units as well. Our argument holds as long as convincing evidence for *migrations* explaining the apparent pluritopic distribution is not brought forward. For example, the fact that the alar A11 cell group clearly crosses the intersegmental plane defined by the tractus retroflexus in all vertebrates (as shown by Björklund & Nobin, 1973; Björklund *et al.*, 1975; Chan-Palay *et al.*, 1984) suggests its cells are born in two different segments (if there is no migration) and should accordingly be split into a small group in p2 and a larger group in p1. These would be topographically homologous to the avian postincertal and subcommissural periventricular populations defined by us, respectively (Fig. 16.4(*c*)).

The nigral A9 complex also seems stretched across two or more segments and should ideally be split (though this seems difficult for practical reasons, the substantia nigra being so well established as a single anatomical entity). We would like to point out that the 'laterodorsal' A9 part of Hökfelt *et al.* (1984) is placed so close to the medial geniculate body that it must lie either in the dorsal thalamus (p2; possibly corresponding to the n. subposterointermedius in the chick – Figs. 16.4(*b*), 16.5(*b*)) or in the pretectal segment (p1), which separates the dorsal thalamus from the mesencephalon (compare prerubral group in Figs. 16.4(*a*); 16.5(*b*)). The 'caudal' A9 part and the A8 cell group jointly are topographically coincident with the population identified as 'substantia nigra' in sauropsids (Smeets, 1991; Medina *et al.*, this volume; present report). Fig. 16.5(*b*) and 16.6(*b*) also explain that A8 is strictly dorsal to caudal A9, although conventional 'cross-sections' give the impression that it lies caudal to it, due to oblique sectioning of the isthmo-mesencephalic boundary. It is probably this appearance of a changing position which led Dahlström & Fuxe (1965) to name them differently.

In any case, this 'caudal' A9/A8 population is the only one which is clearly mesencephalic and which theoretically should retain the name 'S.nigra'.

The A10 complex seems most unnatural in its 'extended' formulation (Hökfelt *et al.*, 1984; Reiner *et al.*, this volume – see Figs. 16.5(*a*); 16.6(*a*)), since it apparently contains mixed entities that not only stretch across several intersegmental boundaries within the prerubral and retrorubral basal tegmentum (the ventrorostral part), but also 'jump' across the basal/alar boundary of the mesencephalon and diencephalon (dorsocaudal/dorsorostral parts). In fact, the 'dorsocaudal' part of A10 coincides with our supranigral population in the chick and the 'dorsorostral' part (absent in sauropsids) falls within the dorsalmost, epithalamic domain of p2! (Fig. 16.6(*b*)). The characteristic of being 'periventricular' is not distinctive enough to validate such heterogeneous 'lumping'. The term A10 should thus at least be restricted to the paramedian basal populations, which nevertheless comprise distinct retromammillary, posterior tuberculum, basal synencephalic, mesencephalic and isthmic portions that should ideally receive each a separate name or number. We think that one unfortunate consequence of the conventional rostral limit of the mesencephalon established by His (1893) (and by the *Nomina Anatomica* thereafter) is a tendency to subsume diverse tegmental structures at the apex of the cephalic flexure into single anatomical concepts (another example: the nucleus ruber and its two 'parts' extending up to the retroflex tract).

With this background in mind, we shall comment on some relevant embryonic evidence on the development of the substantia nigra and ventral tegmental area. The pioneering immunohistochemical study of Specht *et al.* (1981*a,b*) in the rat discovered the earliest THi cells appearing simultaneously in the 'ventral prosencephalon' (preoptic and postoptic cell groups), in the rostrolateral and caudodorsal rhombencephalon and in the 'ventral mesencephalon' at E12.5. Cells in the 'ventral mesencephalon' were stated to appear 'along the rostral portion of the cephalic flexure'. The Figures of Specht *et al.* (1981*a*) clearly show that these cells were outside the mesencephalon (as delimited by us) and actually extended through the paramedian longitudinal zone from p1 to p4 (that is, from the basal synencephalon to the mammillary region). It was only one day later (E13.5) that this cell column extended backwards within the mesencephalon (their Fig. 7), reaching periventricular strata by E14.5 (their Fig. 12). A similar relative precocity of the p1-p4 prosomeres relative to the mesencephalon was suggested by Voorn *et al.* (1988), who found earliest DAi cells in the *ventral diencephalon* at E13 and only one day later in the mesencephalon. Reisert *et al.* (1990) also mention early THi cells in the 'premammillary' and posterior preoptic regions, apart of those in the 'ventral mesencephalon'. Additionally, expression of the TH gene (Burgunder & Young, 1990) seems to begin in the rostral part of the cephalic flexure at E12, extending backwards into the caudal mesencephalon. All these data correlate with a rostrocaudal gradient of cell birthdays in the substantia nigra/VTA complex (Altman & Bayer, 1981; Marchand & Poirier, 1983). This population is produced between E12 and E15, suggesting that the cells begin to express the TH gene and protein shortly after mitosis. We conclude that the sequence of paramedian groups developing CA traits is similar in the chick and rat embryos, though rats seem to have accelerated drastically the differentiation of the enzymatic chain for catecholamine synthesis and other specific markers (Sales *et al.*, 1989). Similar acceleration may be deduced for the locus coeruleus and the prosencephalic cell groups.

As regards the postulated caudal origin of the CA cells of the substantia nigra and ventral tegmental area in a 'medial cell stream' (MSC) arising from the fovea isthmi (Marchand & Poirier, 1983), we would like to point out, first, that the fovea seems to lie just behind the stream as seen in sagittal sections (their Figs. 3(*d*), (*e*); secondly, that the 'MSC' marked by Marchand and Poirier (1983) in their Fig. 2(*f*) obviously lies in the isthmus, but is not the MSC and probably corresponds to the VLS (ventrolateral stream), given the laterality of the sagittal section; thirdly, that the MSC clearly extends as a distinct periventricular formation as far rostrally as the retromammillary area in horizontal sections (their Figs. 3(*a*),(*b*),(*c*),(*f*),(*g*) and 4a–d; see also Keyser, 1972, his Fig. 60 and text). Since we trace the lineal boundary between isthmus and mesencephalon just *in front of* the fovea isthmi (Palmgren, 1921; Vaage, 1969, 1973; see Fig. 16.1), we are inclined to interpret the MSC as a mesencephalic and diencephalic, but not isthmic, structure. The lateral 'wings' of the MSC contain the nigral cells that presumably migrate dorsolaterally; available evidence does not disprove that these cells are produced everywhere along the length of the MSC continuum. This raises in turn the possibility that *there is no caudorostral streaming of nigral cells* at all in the rat, but that each segmental sector of the paramedian proliferative zone produces its own part of the substantia nigra and of the ventral tegmental area. This hypothesis is supported by the data cited above: the earliest *rostral* nigral neurons are born at E12 and *immediately* begin to express the TH protein in the *rostral* part of the paramedian continuum. If they had to migrate rostrally from the caudal mesencephalon (or the isthmus), the earliest THi cells would appear caudally and move progressively forwards. There is simply not enough time for such a migration. The caudal A9 and A8 cells are produced last and also mature slightly later (Specht *et al.*, 1981*a*). There are no important mediolateral differences in the times of cell generation (Marchand & Poirier, 1983).

The possible existence of plurisegmental 'wings' that form the complete A8/A9 complex and a plurisegmental medial continuum that forms the paramedian A10 complex represents a pattern that may be compared with that obtained in chick embryos, where the number of CA cells is not as high as in mammals and where clearly separate 'ventral tegmental' and 'tegmental' cell groups appear in a rostrocaudal segmental sequence extended from the caudal ventral diencephalon to the caudal mesencephalon (Fig. 16.5(*b*)). This interpretation conveys that the whole set of sauropsidan retromammillary, posterior tuberculum and synencephalic tegmental and paramedian CA cell groups may as a group be homologous to rostral and intermediate parts of the mammalian A9 and A10 complexes, though remaining largely separate one from another and from the 'substantia nigra', or caudal A8/A9 complex, whereas homologous segmental components appear fused together in the mammalian brain due to increased cell number. The conundrum of why mammals have a substantia nigra extending into the subthalamus whereas sauropsids have a caudal one adjacent to the isthmus is tentatively resolved by the deduction that the term 'substantia nigra' is semantically ambiguous in its present usage and means a large plurisegmental complex in mammals but only one single segmental component in sauropsids and other vertebrates. Sauropsids and anamniotes also have less massive CA cell populations extending to the subthalamus (p4). The alternative solution found in the literature assumes a massive rostralward nigral migration in mammals for which there is scarce evidence (Keyser, 1972; Marchand & Poirier, 1983).

The overall picture accordingly indicates a considerable degree of developmental and topographic homology of CA neuronal populations in mammals, birds and other tetrapods and suggests the need to explore cell migrations more thoroughly. The comparative analysis of the whole tegmental region is set under a different light, consistent with an emerging segmental and genetic theory of the brain (Puelles & Rubenstein, 1993).

Acknowledgements

The authors wish to thank W. J. A. J. Smeets for providing the initial incentive to perform this study. We also thank Dr R. M. Buijs (Netherlands Institute for Brain Research, Amsterdam) for generously supplying the antiserum against dopamine. Margaret Martínez-de-la-Torre kindly allowed us to examine her preparations of chicken immunostained for TH. F. Marín helped with photographic work. The expert technical assistance of Ma. C. Morga is gratefully acknowledged. Work supported by Spanish DGICYT grants PB87–0688–C01 and PB90–0296–C01 to L. P.

Note added in proof:
We recently reacted additional early chick embryos with a different TH-antibody reported to be more sensitive and found the earliest THi cells at 3.5 days of incubation (stage 22). These results will be reported separately.

Abbreviations

A	nucleus angularis
Abd	nucleus abducens
AH	anterior hypothalamus
Amb	nucleus ambiguus
An	nucleus angularis
AP	nucleus areae praetectalis/area praetectalis diffusa
APd	area praetectalis diffusa
bs	basal synencephalic area
c	caudal
ca	commissura anterior
Cb	cerebellum
csv	commissura supraoptica ventralis
D	nucleus of Darkschewitsch
d	dorsal
dc	dorsocaudal
drm	dorsal retromammillary cell group
DLa	nucleus dorsolateralis anterior
DLp	nucleus dorsolateralis posterior
DM	nucleus dorsomedialis
dr	dorsorostral
DT	dorsal thalamus
E	epichiasmatic dopaminergic population
EM	nucleus ectomammillaris
ET	epithalamus
EW	nucleus of Edinger-Westphal
fr	fasciculus retroflexus
fx	fornix
GV	nucleus geniculatus ventralis
III	third ventricle
IC	nucleus interstitialis of Cajal
I-Cb	isthmocerebellar segmental domain
if	interfascicular cell group
Im	nucleus isthmi pars magnocellularis
InP	interpeduncular nucleus
IO	nucleus isthmo-opticus
Ip	nucleus isthmi pars parvocellularis
ir	infundibular recess
Ist	isthmic region
IV	fourth ventricle
LC	locus coeruleus
lh	lateral hypothalamic area
LL	lateral lemniscal nuclei
M	mesencephalic segmental domain
mm	mammillary area
mr	mammillary recess
mt	mesencephalic tegmental area
mtt	mammillothalamic tract
NCb	cerebellar nucleus
OC	oculomotor nucleus and root
och	optic chiasm
oh	organum periventriculare hypothalami
on	optic nerve
OS	superior olive
ot	optic tract
Ov	nucleus ovoidalis
P	pretectal nucleus
p1-p6	prosomeres (prosencephalic segmental domains)
PB	dorsal parabrachial nucleus
pc	posterior commissure
PE	nucleus pretectalis externus
PI	nucleus posterointermedius (or n. ansae lenticularis posterior)
pi	post-incertal cell group
PMI	nucleus posteromedianus intermedius
PO	preoptic region
PPc	nucleus principalis praecommissuralis
PPv	preoptic periventricular nucleus
P/PV	peduncular/paraventricular region
pr	parencephalic recess
pru	pre-rubral cell group
PT	pretectal region
PV	paraventricular nucleus
R	reticular nucleus
r	rostral

rch	retrochiasmatic area	T	trochlear nucleus
Rl	nucleus raphe linearis	Tel	telencephalon
rm	retromammillary area	tm	tuberomammillary area
Rot	nucleus rotundus	To	tectum opticum
rt	rhombencephalic tegmental area	tp	tuberculum posterior area
Ru	nucleus ruber	tsm	tractus septomesencephalicus
SC	locus subcoeruleus	tu	tuberal area
sc	subcommissural cell group	v	ventral
SL	semilunar isthmic nucleus	vc	ventrocaudal
Sn	substantia nigra *sensu* Smeets (1991)	Vd	descending trigeminal column
sN	supranigral cell group	Ves	vestibular complex
sp	sub-posterointermedius cell group	VM	ventromedial hypothalamic nucleus
SO	supraoptic region	Vm	trigeminal motor nucleus
Sol	nucleus solitarius	Vn	trigeminal root
sor	supraoptic recess	Vp	principal sensory nucleus of the trigeminus
SPl	nucleus spiriformis lateralis		
SPm	nucleus spiriformis medialis	vr	ventrorostral
sr	synencephalic recess	Vsp	spinal trigeminal nucleus
SS	nucleus superficialis synencephali	VT	ventral thalamus
ST	subthalamic nucleus (or nucleus ansae lenticularis anterior)	VTA	ventral tegmental area
		Xm	motor nucleus of the vagus
st	subthalamic nucleus primordium	ZI	zona incerta

References

Altman, J. & Bayer, S. A. (1981). Development of the brain stem in the rat. V. Thymidine-radiographic study of the time of origin of neurons in the midbrain tegmentum. *Journal of Comparative Neurology*, **198**, 677–716.

Alvarado-Mallart, R. M., Martínez, S. & Lance-Jones, C. C. (1990). Pluripotentiality of the 2-day-old avian germinative neuroepithelium. *Developmental Biology*, **139**, 75–88.

Bergquist, H. (1954). Ontogenesis of diencephalic nuclei in vertebrates. A comparative study. *Lunds Universitets Arsskrift*, Avd.2, **65**, 1–34.

Bergquist, H. & Källen, B. (1954). Notes on the early histogenesis and morphogenesis of the central nervous system in vertebrates. *Journal of Comparative Neurology*, **100**, 627–59.

Björklund, A. & Nobin, A. (1973). Fluorescence histochemical and microspectrofluorometric mapping of dopamine and noradrenaline cell groups in the rat diencephalon. *Brain Research*, **51**, 193–205.

Björklund, A., Lindvall, O. & Nobin, A. (1975). Evidence of an incerto-hypothalamic dopamine neurone system in the rat. *Brain Research*, **89**, 29–42.

Blessing, W. W., Chalmers, J. P. & Howe, P. R. C. (1978). Distribution of catecholamine-containing cell bodies in the rabbit central nervous system. *Journal of Comparative Neurology*, **179**, 407–24.

Bulfone, A., Puelles, L., Porteus, M. H., Frohman, M. A., Martin, G. R. & Rubenstein, J. L. R. (1993a). Spatially restricted expression of *Dlx-1*, *Dlx-2* (*Tes-1*), *Gbx-2* and *Wnt-3* in the embryonic day 12.5 mouse forebrain defines potential transverse and longitudinal segmental boundaries. *The Journal of Neuroscience*, **13**, 3155–72.

Bulfone, A., Kim, H-J., Puelles, L., Porteus, M. H., Grippo, J. F. & Rubenstein, J. L. R. (1993b). The mouse *Dlx-2(Tes-1)* gene is expressed in spatially restricted domains of the forebrain, face and limbs in midgestation mouse embryos. *Mechanisms of Development*, **40**, 129–40.

Burgunder, J-M. & Young, W. S., III (1990). Ontogeny of tyrosine hydroxylase and cholecystokinin gene expression in the rat mesencephalon. *Developmental Brain Research*, **52**, 85–93.

Carr, J. A., Norris, D. O. & Samora, A. (1991). Organization of tyrosine hydroxylase-immunoreactive neurons in the di- and mesencephalon of the American bullfrog (*Rana catesbeiana*) during metamorphosis. *Cell and Tissue Research*, **263**, 155–63.

Chan-Palay, V. Záborzky, L., Köhler, C., Goldstein, M. & Palay, S. (1984). Distribution of tyrosine-hydroxylase-immunoreactive neurons in the hypothalamus of rats. *Journal of Comparative Neurology*, **227**, 467–96.

Coggeshall, R. E. (1964). A study of diencephalon development in the albino rat. *Journal of Comparative Neurology*, **122**, 241–70.

Dahlström, A. & Fuxe, K. (1964). Evidence for the existence of monoamine-containing neurons in the central nervous system. I. Demonstration of monoamines in the cell bodies of brain stemm neurons. *Acta Physiologica Scandinavica*, suppl.62, **232**, 1–55.

Daikoku, S., Kawano, H., Okamura, Y., Tokuzen, M. &

Nagatsu, I. (1986). Ontogenesis of immunoreactive tyrosine hydroxylase – containing neurons in rat hypothalamus. *Developmental Brain Research*, **28**, 85–98.

Dubé, L. & Parent, A. (1981). The monoamine-containing neurons in avian brain: I. A study of the brain stem of the chicken (*Gallus domesticus*) by means of fluorescence and acetylcholinesterase histochemistry. *Journal of Comparative Neurology*, **196**, 695–708.

Figdor, M. C. & Stern, C. D. (1993). Segmental organization of the embryonic diencephalon. *Nature*, **363**, 630–4.

Foster, G. A. (1992). Phenylethanolamine N-methyltransferase – the adrenaline-synthesizing enzyme. In *Handbook of Chemical Neuroanatomy. Vol.10. Ontogeny of Transmitters and Peptides in the CNS*, eds A. Björklund, T. Hökfelt and M. Tohyama, pp. 133–56, Amsterdam: Elsevier.

Foster, G. A., Schultzberg, M., Goldstein, M. & Hökfelt, T. (1985). Ontogeny of phenylethanolamine N-methyltransferase – and tyrosine hydroxylase-like immunoreactivity in presumptive adrenaline neurons of the foetal rat central nervous system. *Journal of Comparative Neurology*, **236**, 348–81.

Fuxe, K. & Ljunggren, L. (1965). Cellular localization of monoamines in the upper brain stem of the pigeon. *Journal of Comparative Neurology*, **148**, 61–90.

Fuxe, K., Hökfelt, T. & Ungerstedt, U. (1969). Distribution of monoamines in the mammalian central nervous system by histochemical studies. In *Metabolism of Amines in the brain*, ed G. Hooper, pp. 10–22, London: Macmillan.

González, A. & Smeets, W. J. A. J. (1991). Comparative analysis of dopamine and tyrosine hydroxylase immunoreactivities in the brain of two amphibians, the anuran *Rana ridibunda* and the urodele *Pleurodeles waltlii*. *Journal of Comparative Neurology*, **303**, 457–77.

González, A. & Smeets, W. J. A. J. (1994). Catecholamine systems in the CNS of amphibians. In *Phylogeny and development of catecholamine systems in the CNS of vertebrates*, eds W. J. A. J. Smeets & A. Reiner, pp. 77–102 Cambridge, Cambridge University Press.

Gribnau, A. A. M. & Geijsberts, L. G. M. (1985). Morphogenesis of the brain in staged Rhesus monkey embryos. *Advances in Anatomy, Embryology and Cell Biology*, **91**, 1–69.

Guglielmone, R. & Panzica, G. C. (1984). Typology, distribution and development of the catecholamine-containing neurons in the chicken brain. *Cell and Tissue Research*, **237**, 67–79.

Guglielmone, R. & Panzica, G. C. (1985). Early appearance of catecholaminergic neurons in the central nervous system of precocial and altricial avian species. A fluorescence-histochemical study. *Cell and Tissue Research*, **240**, 381–84.

Hallonet, M. R., Teillet, M. A. & LeDouarin, N. M. (1990). A new approach to the development of the cerebellum provided by the quail chick marker system. *Development*, **108**, 19–31.

Hanaway, J., McConnel, J. A. & Netsky, M. G. (1971). Histogenesis of the substantia nigra, ventral tegmental area of Tsai and interpeduncular nucleus: an autoradiographic study of the mesencephalon in the rat. *Journal of Comparative Neurology*, **142**, 59–74.

His, W. (1893). Vorschläge zur Einteilung des Gehirn. *Archiv für Anatomie und Entwicklungsgeschichte (Leipzig)*, **17**, 172–9.

Hökfelt, T., Martensson, R., Björklund, A., Kleinau, S. & Goldstein, M. (1984). Distributional maps of tyrosine-hydroxylase-immunoreactive neurons in the rat brain. In *Handbook of Chemical Neuroanatomy. Vol.2. Classical Transmitters in the CNS. Part.I*, eds A. Björklund and T. Hökfelt, pp. 277–379, Amsterdam: Elsevier.

Ikeda, H. & Gotoh, J. (1971). Distribution of monoamine-containing cells in the central nervous system of the chicken. *Japanese Journal of Pharmacology*, **21**, 763–84.

Ikeda, H. & Gotoh, J. (1974). Distribution of monoamine-containing terminals and fibers in the central nervous system of the chicken. *Japanese Journal of Pharmacology*, **24**, 831–41.

Jaeger, C. B. (1986). Aromatic L-amino acid decarboxylase in the rat brain: immunocytochemical localization during prenatal development. *Neuroscience*, **18**, 121–50.

Jaeger, C. B. & Teitelman, G. (1992). Immunocytochemical distribution of aromatic L-amino acid decarboxylase (AADC) in rat. In *Handbook of Chemical Neuroanatomy. Vol.10. Ontogeny of Transmitters and Peptides in the CNS*, ed. A. Björklund, T. Hökfelt and M. Tohyama, pp. 113–32, Amsterdam: Elsevier.

Kalsbeek, A., Voorn, P. & Buijs, R. M. (1992). Development of dopamine-containing systems in the CNS. In *Handbook of Chemical Neuroanatomy. Vol.10. Ontogeny of Transmitters and Peptides in the CNS*, eds A. Björklund, T. Hökfelt and M. Tohyama, pp. 63–112, Amsterdam: Elsevier.

Keyser, A. (1972). The development of the diencephalon of the Chinese hamster. An investigation of the validity of the criteria of subdivision of the brain. *Acta Anatomica*, suppl.59, **83**, 1–178.

Kiss, J. Z. & Péczely, P. (1987). Distribution of tyrosine-hydroxylase (TH)-immunoreactive neurons in the diencephalon of the pigeon (*Columba livia domestica*). *Journal of Comparative Neurology*, **257**, 333–46.

Lakke, E. A. J. F., van der Veeken, J. G. P. M., Mentink, M. M. T. & Marani, E. (1988). A SEM study on the development of the ventricular surface morphology in the diencephalon of the rat. *Anatomy and Embryology*, **179**, 73–80.

Lumsden, A & Keynes, R. (1989). Segmental patterns of neuronal development in the chick hindbrain. *Nature*, **337**, 424–8.

Marchand, R. & Poirier, L. J. (1983). Isthmic origin of neurons of the rat substantia nigra. *Neuroscience*, **9**, 373–81.

Martínez, S. & Alvarado-Mallart, R. M. (1989). Rostral cerebellum originates from the caudal portion of the so-called 'mesencephalic' vesicle: a study using chick/quail transplants. *European Journal of Neuroscience*, **1**, 549–60.

Martínez, S., Geijo, E., Sánchez-Vives, M. V., Puelles, L.

and Gallego, R. (1992). Reduced junctional permeability at interrhombomeric boundaries. *Development*, **116**, 1069–76.

Martínez-de-la-Torre, M. (1991). Marcaje experimental del flujo neuroepitelial en el límite rombomesencefálico. *Abstracts IV Meeting of the Spanish Society of Neuroscience* (Alicante), **16**,12-P.

Medina, L., Smeets, W. J. A. J., Hoogland, P. V. & Puelles, L. (1993). Distribution of choline acetyltransferase immunoreactivity in the brain of the lizard *Gallotia galloti*. *Journal of Comparative Neurology*, **331**, 261–85.

Medina, L., Puelles, L. & Smeets, W. J. A. J. (1994). Ontogenesis of catecholamine systems in the brain of the lizard *Gallotia galloti*. In *Phylogeny and Development of Catecholamine Systems in the CNS of Vertebrates*, eds. W. J. A. J. Smeets and A. Reiner, pp. 361–79 Cambridge: Cambridge University Press.

Noden, D. M. (1991). Vertebrate craniofacial development: the relation between ontogenetic process and morphological outcome. *Brain, Behavior and Evolution*, **38**, 190–225.

Palmgren, A. (1921). Embryological and morphological studies on the midbrain and cerebellum of vertebrates. *Acta Zoologica (Stockholm)*, **2**, 1–94.

Parent, A., Poitras, D. & Dubé, L. (1984). Comparative anatomy of central monoaminergic systems. In *Handbook of Chemical Neuroanatomy, Vol.2. Classical Transmitters in the CNS, Part I*, ed A. Björklund & T. Hökfelt, pp. 409–39. Amsterdam: Elsevier.

Puelles, L. & Martínez-de-la-Torre, M. (1987). Autoradiographic and Golgi study on the early development of n.isthmi principalis and adjacent grisea in the chick embryo: a tridimensional viewpoint. *Anatomy and Embryology*, **176**, 19–34.

Puelles, L., Amat, J. A. & Martinez-de-la-Torre, M. (1987a). Segment-related, mosaic neurogenetic pattern in the forebrain and mesencephalon of early chick embryos: I. Topography of AChE-positive neuroblasts up to stage HH18. *Journal of Comparative Neurology*, **266**, 247–68.

Puelles, L., Domenech–Ratto, G. & Martínez-de-la-Torre, M. (1987b). Location of the rostral end of the longitudinal brain axis: Review of an old topic in the light of marking experiments on the closing rostral neuropore. *Journal of Morphology*, **194**, 163–71.

Puelles, L., Guillén, M. & Martínez-de-la-Torre, M. (1991). Observations on the fate of nucleus superficialis magnocellularis of Rendahl in the avian diencephalon, bearing on the organization and nomenclature of neighboring retinorecipient nuclei. *Anatomy and Embryology*, **183**, 221–3.

Puelles, L. & Rubenstein, J. L. R. (1993). Expression patterns of homeobox and other putative regulatory genes in the embryonic mouse forebrain suggest a neuromeric organization. *Trends in Neuroscience*, **16**, 472–9.

Puelles, L., Sánchez, M. P., Spreafico, R. & Fairén, A. (1992). Prenatal development of calbindin immunoreactivity in the dorsal thalamus of the rat. *Neuroscience*, **46**, 135–47.

Reiner, A., Karle., E. J., Anderson, K. D. & Medina, L. (1994). Catecholaminergic perikarya and fibers in the avian nervous system. In *Phylogeny and Development of Catecholamine Systems in the CNS of Vertebrates*, eds. W. J. A. J. Smeets and A. Reiner, pp. 135–181 Cambridge: Cambridge University Press.

Reisert, I., Schuster, R., Zienecker, R. & Pilgrim, C. (1990). Prenatal development of mesencephalic and diencephalic dopaminergic systems in the male and female rat. *Developmental Brain Research*, **53**, 222–9.

Rendahl, H. (1924). Embryologische und morphologische Studien über das Zwischenhirn beim Huhn. *Acta Zoologica (Stockholm)*, **5**, 241–344.

Ross, L. S., Parrett, T. & Easter, S. S., Jr. (1992). Axonogenesis and morphogenesis in the embryonic zebrafish brain. *Journal of Neuroscience*, **12**, 467–82.

Sales, N., Martres, M. P., Bouthenet, M. L. & Schwartz, J. C. (1989) Ontogeny of dopaminergic D-2 receptors in the rat nervous system: characterization and detailed autoradiographic mapping with [^{125}I] iodosulpiride. *Neuroscience*, **28**, 673–700.

Sharp, P. J. & Follett, B. K. (1968). The distribution of monoamines in the hypothalamus of the japanese quail, *Coturnix coturnix japonica*. *Zeitschrift für Zellforschung*, **90**, 245–62.

Smeets, W. J. A. J. (1991). Comparative aspects of the distribution of substance P and dopamine immunoreactivity in the substantia nigra of amniotes. *Brain Behavior and Evolution*, **37**, 179–88.

Specht, L. A., Pickel, V. M., Joh, T. H. & Reis, D. J. (1981a). Light-microscopic immunocytochemical localization of tyrosine hydroxylase in prenatal rat brain. I. Early ontogeny. *Journal of Comparative Neurology*, **199**, 233–53.

Specht, L. A., Pickel, V. M., Joh, T. H. & Reis, D. J. (1981b). Light-microscopic immunocytochemical localization of tyrosine hydroxylase in prenatal rat brain. II. Late ontogeny. *Journal of Comparative Neurology*, **199**, 255–76.

Ströer, W. F. H. (1956). Studies on the diencephalon. I. The embryology of the diencephalon of the rat. *Journal of Comparative Neurology*, **105**, 1–24.

Tan, K. & LeDouarin, N. M. (1991). Development of the nuclei and cell migration in the medulla oblongata. *Anatomy and Embryology*, **183**, 321–43.

Trujillo, C. M. & Alvarado-Mallart, R. M. (1991). Estudio de la cartografía de areas diencefálicas por medio de embriones quimera pollo/codorniz. *Abstracts IV Meeting of the Spanish Society of Neuroscience*, (Alicante), **23**. 1–P.

Ugrumov, M. V., Taxi, J., Tixier-Vidal, A., Thibault, J. & Mitskevich, M. S. (1989). Ontogenesis of tyrosine-hydroxylase-immunoreactive structures in the rat hypothalamus. An atlas of neuronal cell bodies. *Neuroscience*, **29**, 139–56.

Vaage, S. (1969). Segmentation of the primitive neural tube in chick embryos. A morphological, histochemical and autoradiographical investigation. *Advances in Anatomy, Embryology and Cell Biology*, **41**, 1–88.

Vaage, S. (1973). The histogenesis of the isthmic nuclei in

chick embryos (*Gallus domesticus*). *Zeitschrift für Anatomie und Entwicklungsgeschichte*, **142**, 283–314.

Voorn, P., Kalsbeek, A., Jorritsma-Byham, B. & Groenewegen, H. J. (1988). The pre- and postnatal development of the dopaminergic cell groups in the ventral mesencephalon and the dopaminergic innervation of the striatum of the rat. *Neuroscience*, **25**, 857–87.

Wallace, J. A., Mondragon, R. M., Allgood, P. C., Hoffman, T. J. & Maez, R. R. (1987). Two populations of tyrosine hydroxylase-positive cells occur in the spinal cord of the chick embryo and hatchling. *Neuroscience Letters*, **83**, 253–8.

Yoshida, M., Nagatsu, I., Kondo, Y., Karasawa, N., Spatz, M. & Nagatsu, T. (1982). Immunohistochemical localization of catecholamine-synthetizing enzymes and serotonin in the bat brain. *Acta Histochemica Cytochemica*, **15**, 116–28.

Yurkewicz, L., Marchi, M., Lauder, J. M. & Giacobini, E. (1981). Development and aging of noradrenergic cell bodies and axon terminals in the chicken. *Journal of Neuroscience Research*, **6**, 621–41.

17
Ontogeny of catecholaminergic neurons in the central nervous system of mammalian species: general aspects

G. A. Foster

Introduction

The ontogeny of catecholamine-containing neurones in the mammalian central nervous system has been studied extensively over the past few decades. Initial investigations made use of the Falck–Hillarp histofluorescence method, which permitted visualization of monoaminergic neurones in fixed tissue. Although variations in the basic Falck–Hillarp methodology have led to improvements in sensitivity and specificity, the advent of immunohistochemistry has yielded a still greater ability to detect putative catecholaminergic neurones at earlier stages of gestational life. In addition, it has also been possible to discriminate more easily between dopamine, noradrenaline and adrenaline cells. However, the great disadvantage of immunohistochemistry, at least until recently, has been that antibodies have been available against only the marker enzymes of catecholamine synthesis but not against the transmitters themselves. As a result, some doubt exists as to whether, and at what rate, the demonstrated enzyme is actually utilized and, if so, whether the reaction product is subsequently processed to form a biologically active neurotransmitter. For instance, there are many examples in both the developing and in the mature animal of the expression of just one catecholaminergic enzyme (usually tyrosine hydroxylase (TH), but sometimes L-aromatic acid decarboxylase (AADC) or phenylethanolamine N-methyltransferase (PNMT)). Theoretically, such neurones are unlikely to be synthesizing a catecholaminergic neurotransmitter. Perhaps coincidentally, the expression of this single enzyme has often turned out to be only transient.

More recently, antibodies have been raised which react specifically with one or other of the various catecholamines. Most notable in ontogenetic studies has been the use of antibodies against dopamine, the results of which have largely resolved the problems described above in defining the transmitter phenotype of developing dopaminergic neurones.

Finally, the messenger RNA coding for tyrosine hydroxylase has been visualized in sections of the developing rat brain by *in situ* hybridization histochemistry. Perhaps the greatest promise of this methodology lies in its ability largely to sidestep interpretative issues inherent in the immunohistochemical embryological studies (such as ill-defined specificity of the antibodies, or altered post-translational processing of the synthetic enzyme).

The aim of the present chapter is to provide a series of maps in which the development of the catecholaminergic cell bodies and nerve terminals is illustrated. Since the bulk of investigations have been in the rat, for the sake of brevity, this is the only species that will be presented pictorially. In the text, however, the ontogeny of catecholamines and their synthetic enzymes in both this and other species will be described, although for further detail the reader is referred to the papers listed in Table 17.1. Information on the development of catecholamine receptors and other post-synaptic markers will also be provided. The few discrepancies in the maturation of catecholaminergic neurones between the various mammalian species will be discussed briefly, but the disparate appearance of one or more parameters of catecholamine expression in the same neurone, both temporally and topographically, will be dealt with at greater length. Finally, the ascription of function to a given prenatal catecholaminergic cell group is largely restricted to assuming that it performs the foetal

Table 17.1. *List of citations describing the ontogeny of catecholamines in the central nervous system of various species*

Rat:	Olson & Seiger (1972), Seiger & Olson (1973), Specht et al. (1981a,b), Teitelman et al. (1983), Foster et al. (1985a), Bohn et al. (1986), Jaeger (1986), Voorn et al. (1988), Ugrumov et al. (1989a,b) and Kalsbeek et al. (1992).
Mouse:	Björklund, Enemar & Falck (1968), Golden (1972, 1973), Foster et al. (1988), Wulle & Schnitzer (1989), DiPorzio et al. (1990) and Satoh & Suzuki (1990).
New World bats:	Studholme, Yazulla & Phillips (1987).
Rabbit:	Barrett, Cote & Tennyson (1971), Tennyson et al. (1973), Fung et al. (1982) and Parkinson & Rando (1984).
Guinea-pig:	Maeda & Astic (1972), Parkinson et al. (1985) and Kalaria & Prince (1988).
Opossum:	Martin, Ho & Hazlett (1989) and Pindzola et al. (1990).
Cat:	Connor & Neff (1970), Loup & Cadilhac (1970).
Hamster/gerbil:	Mitrofanis and Finlay (1990).
Norwegian lemming:	Hissa & Pyhälä (1971).
Swine:	Ruggiero, Anwar & Gootman (1992).
Human:	Olson & Ungerstedt (1970), Hyyppä (1972), Nobin & Björklund (1973), Olson et al. (1973), Pearson, Brandeis & Goldstein (1980), Pickel et al. (1980) and Verney et al. (1991).

equivalent of that in the mature animal. It is perhaps more valuable, therefore, to discuss some of the mechanisms governing the expression of the catecholamine phenotype in the central nervous system, and this is the subject of the last section in the chapter.

Distributional maps of catecholaminergic neurones in the developing rat brain

Abbreviations are listed at the end of the chapter. Maps indicate the disposition of TH-immunoreactive (THir), AADCir and PNMTir neuronal somata (dots) and processes in the foetal rat at days 13 (Fig. 17.1), 15 (Fig. 17.2), 17 (Fig. 17.3), 19 (Fig. 17.4) and 21 (Fig. 17.5) of gestation. Although areas of overlap of the three antigens will be apparent, no attempt has been made to identify those neurones definitively identified as containing more than one antigen; such information does exist, however, for TH- and PNMT-immunoreactivity (PNMTi) (Foster et al., 1985a). Information has been drawn from Specht et al. (1981a,b), Foster et al. (1985a) for TH, Teitelman et al. (1983), Jaeger (1986) for AADC and Foster et al. (1985a) for PNMT. Note that in the maps relating to the ontogeny of AADC, the AADCir, non-catecholaminergic cell groups have not generally been knowingly included, except for where they may overlap catecholaminergic cell groups. Note also that although analysis has been carried out (Jaeger, 1986), the disposition of cells at E21, and of AADC-containing fibres has not been documented, and is therefore not drawn in the Figures here. For a complete identification of all the structures delineated in the maps, see Foster (1994).

Dopamine

Mesencephalon The first parameter of CA expression to be recognized during mammalian CNS ontogeny is TH, which appears in the ventricular layer of the ventral mesencephalon (ie. the future substantia nigra) of the mouse at days 8.5–9.0 of gestation (Di Porzio et al., 1990). The equivalent embryonic stage in the rat is about day 10.5 (Rugh, 1968), but it is not until days 11.5–12.5 at the earliest that THir cells are visualized in the rat mesencephalon (Rothman et al., 1980; Specht et al., 1981a: Foster et al. 1985a). However, AADCi is demonstrable in the rat embryo along the whole notochord and, more particularly, in the ventral mesencephalon already at day 10.5 (Teitelman et al., 1983). The AADC-positive cells are located in the ependymal layer of the neural tube, and have been shown by autoradiographic studies still to be replicating (Teitelman et al., 1983). It seems, therefore, that partial differentiation to a catecholaminergic phenotype can be specified before the final division of a cell. Secondly, although these putative dopaminergic neurones start to exhibit differentiation at about the same time in the two species, it may be manifest via expression of different neurochemical parameters.

The mRNA coding for TH is first apparent at day 13 of gestation (Burgunder & Young, 1990), a little later than THi is detectable. Histofluorescent cells (Olson & Seiger, 1972) and DAi (Voorn et al., 1988) are visualized for the first time in rat mesencephalic cells at day 13–13.5 of gestation, or slightly later (Maeda & Dresse, 1968; Cadilhac & Pons, 1976), but by this time the neurones have migrated out of the ependymal layer and are therefore likely to have ceased replication. Fluorescence histochemistry was also used to demonstrate cells in the ventral mesencephalon of the mouse and rabbit at E13–14, (Golden, 1972; 1973; Tennyson, Mytilineou & Barrett, 1973), rather later in the mouse than has been found for the first appearance of THi (Foster et al., 1988; Di Porzio et al., 1990). In human brain, histofluorescent cells were discovered at E7–9 weeks (Olson, Boréus & Seiger, 1973), while THi was first demonstrable at E5.5–6 weeks (Verney et al., 1991). By embryonic days 14 and 15 in the rat, the THir, AADCir, DAir, green fluorescent cells of the three presumptive adult mesencephalic dopaminergic neurones groups (A8–A10) have become aggregated into distinct, but not separate, entities. Over the next 2 days, the dopamine neurones of the substantia nigra start to become organized into the pars compacta and reticulata regions (Figs. 17.1, 17.7). This spatial development continues during days 18–21, the A9 group of the substantia nigra becoming better delineated from the ventral tegmental cells, the A10 group, by the day of birth. From then to P7, the primordial cell groups of the mesencephalon take on the appearance of the

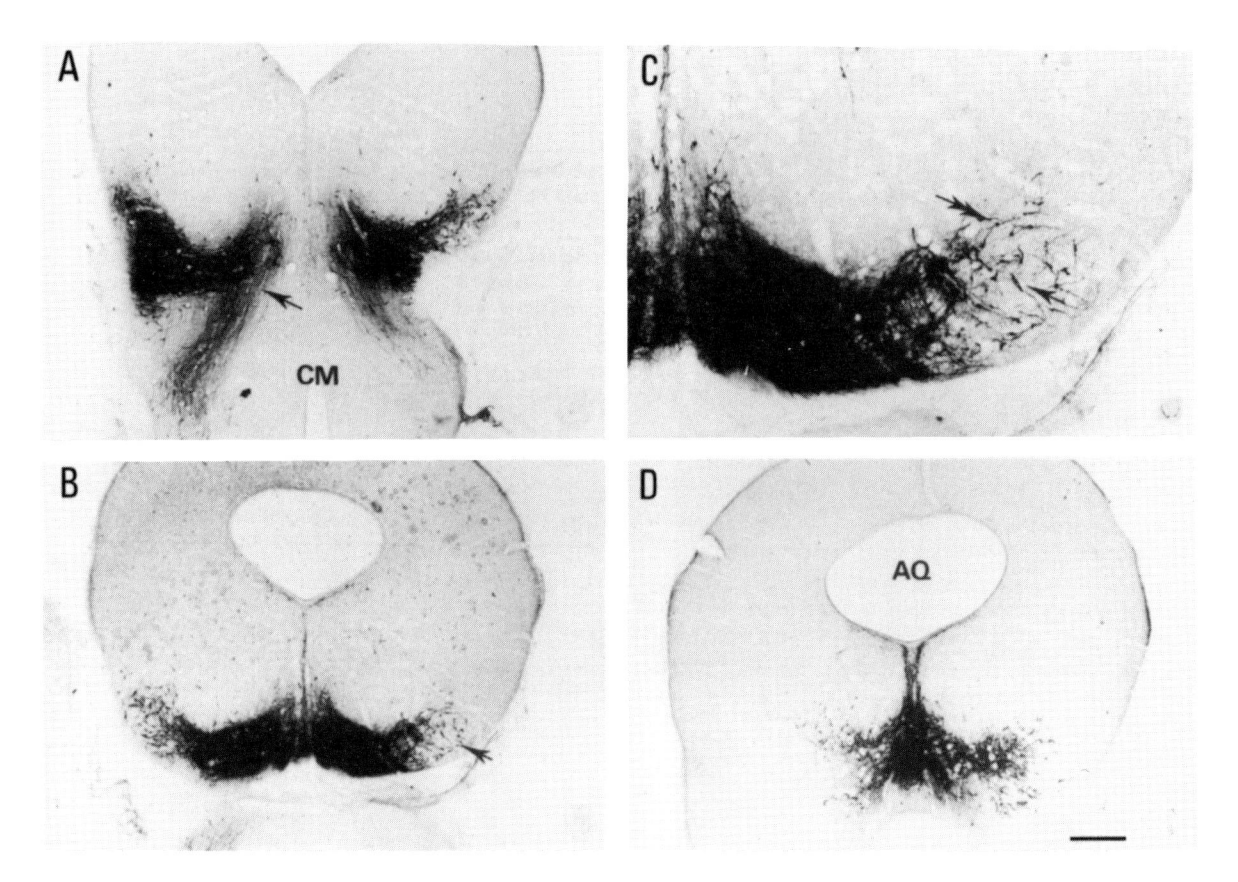

Fig. 17.1. Light micrographs of transverse sections through the E17 rat mesencephalon after immunoperoxidase staining with antibodies against DA. Rostrally, the two dopaminergic cell groups are separated, in lateral positions (a), whereas more caudally they begin to merge (b), higher magnification in (c). Note the dorsomedial/ventrolateral (single-headed arrows) and dorsolateral/ventromedial (double-headed arrows) orientation of some dendrites. At more caudal levels still, there exists a single cell group in the midline, extending dorsally as far as the Sylvian aqueduct (AQ). CM = mammillary body. Scale bar indicates 250 μm in (a), (b) and d, and 62.6 μm in (c). Modified with permission, from Kalsbeek et al. (1992).

408 G. A. Foster

Atlas TH

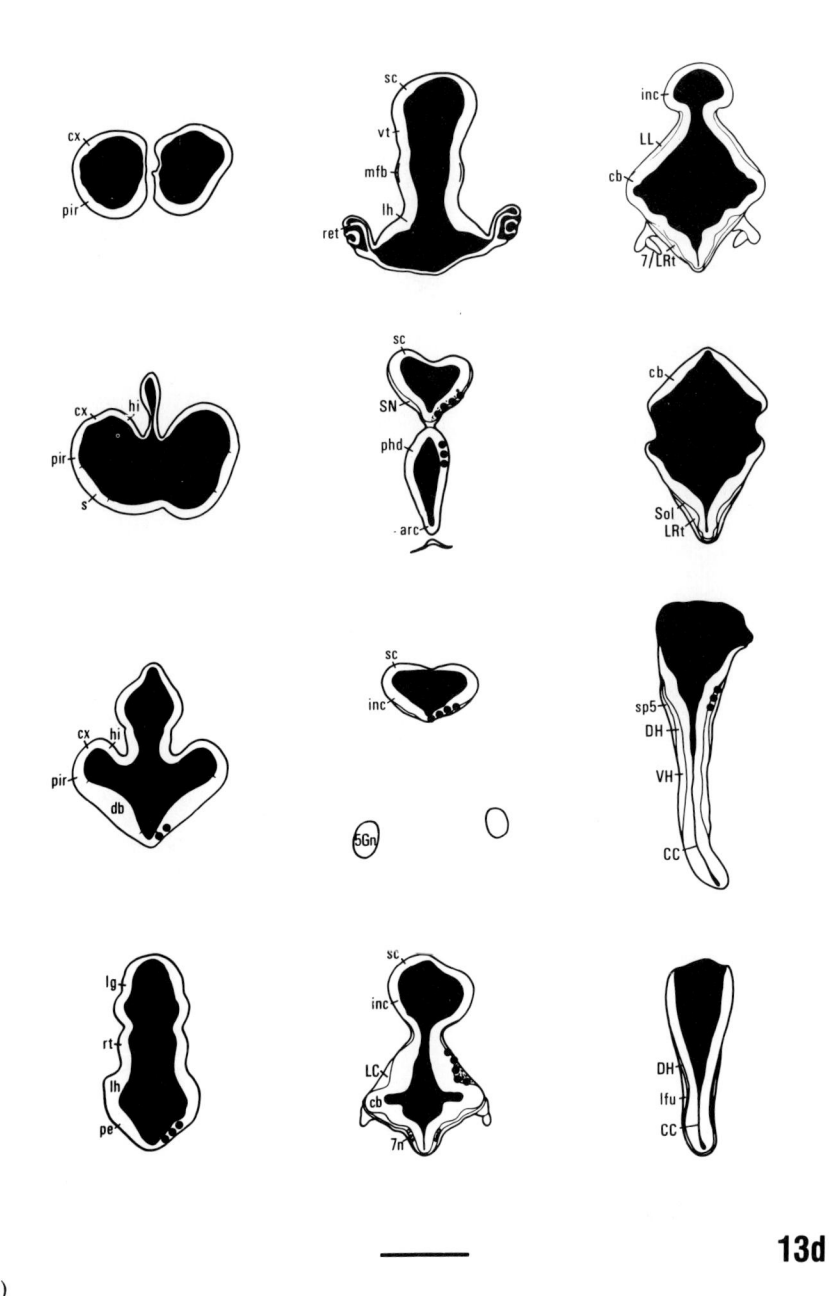

Fig. 17.2(a)

13d

Figs. 17.2–17.6. Maps of the development of THi, AADCi and PNMTi in the rat brain at E13, E15, E17, E19 and E21 of gestation. Brains were sectioned at 12 levels, starting most rostrally in the top left of each Figure, and moving caudally down each column. The bottom right of each figure is therefore the most caudal part of the brain. The sections were taken at approximately equivalent levels at each age. The plane of sectioning was frontal at the level of the mesencephalon, and was parallel throughout the brain. Note, however, that the variations in the flexing of the brain during ontogeny has resulted in some sections appearing almost horizontal. Dots approximate to the number of immunoreactive cells per 15–30 μm section. Fibers are also shown. Scale bar indicates 1mm.

Ontogeny of catecholaminergic neurons in CNS of mammals

AADC PNMT

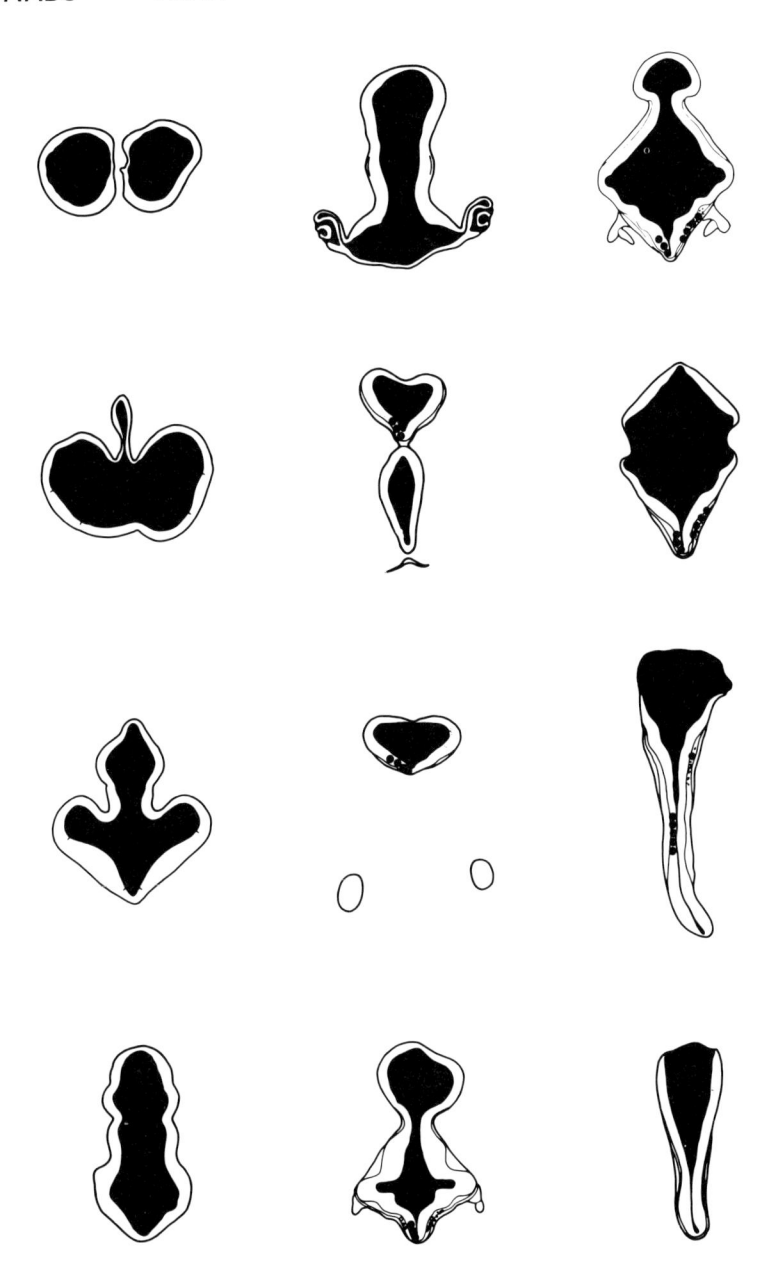

13d

Fig. 17.2(b)

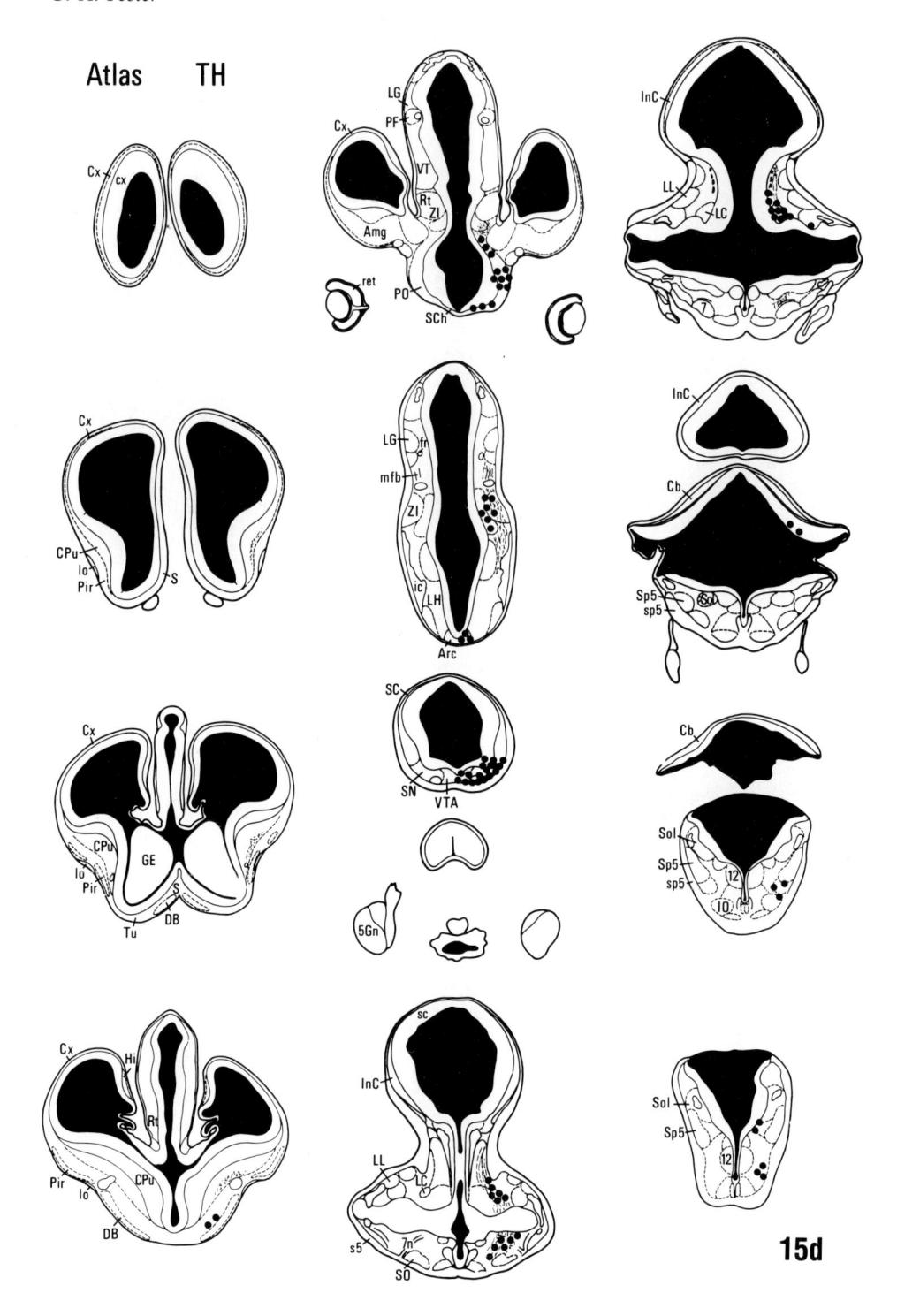

Fig. 17.3(a)

Ontogeny of catecholaminergic neurons in CNS of mammals 411

AADC   PNMT

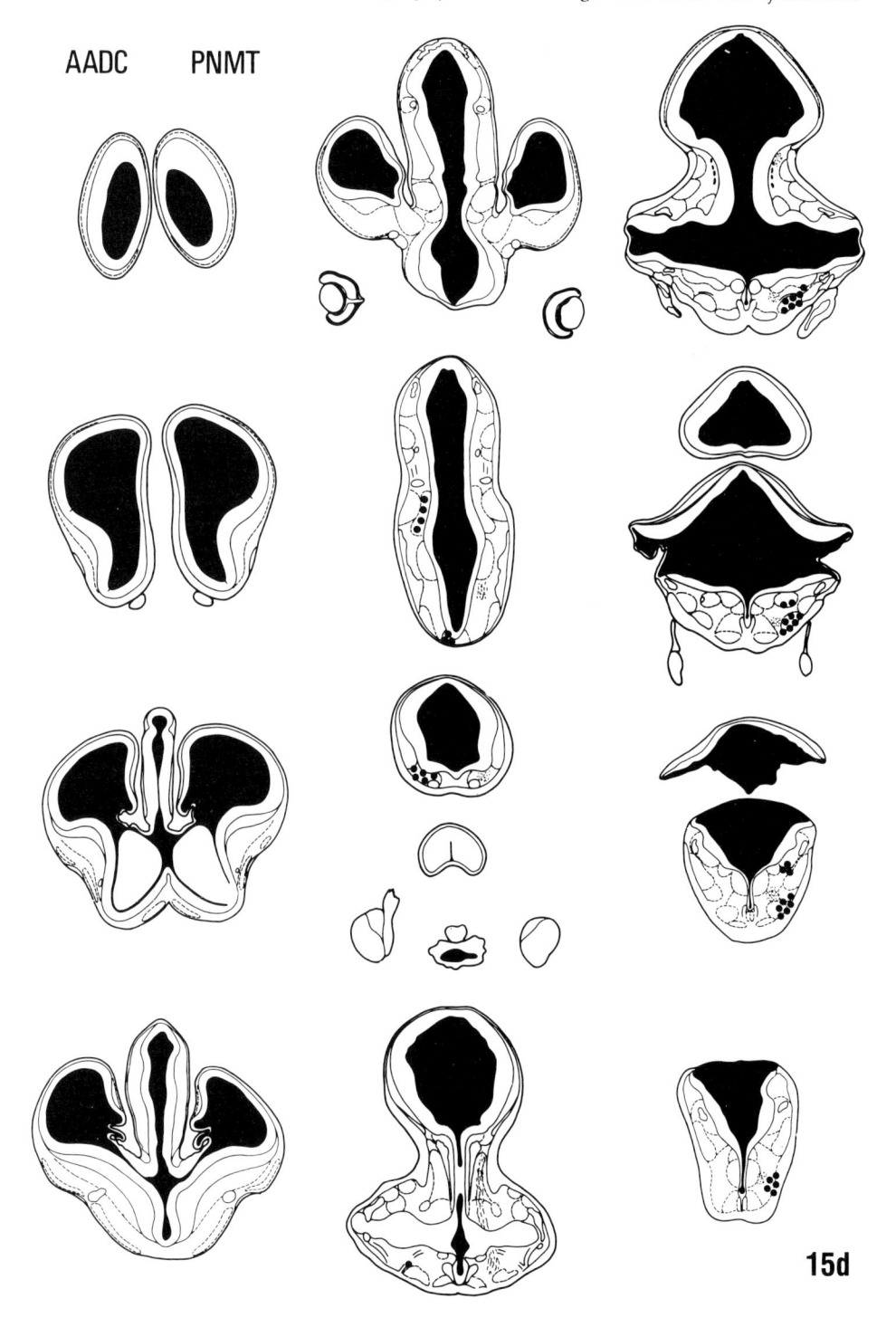

15d

Fig. 17.3(*b*)

412   G. A. Foster

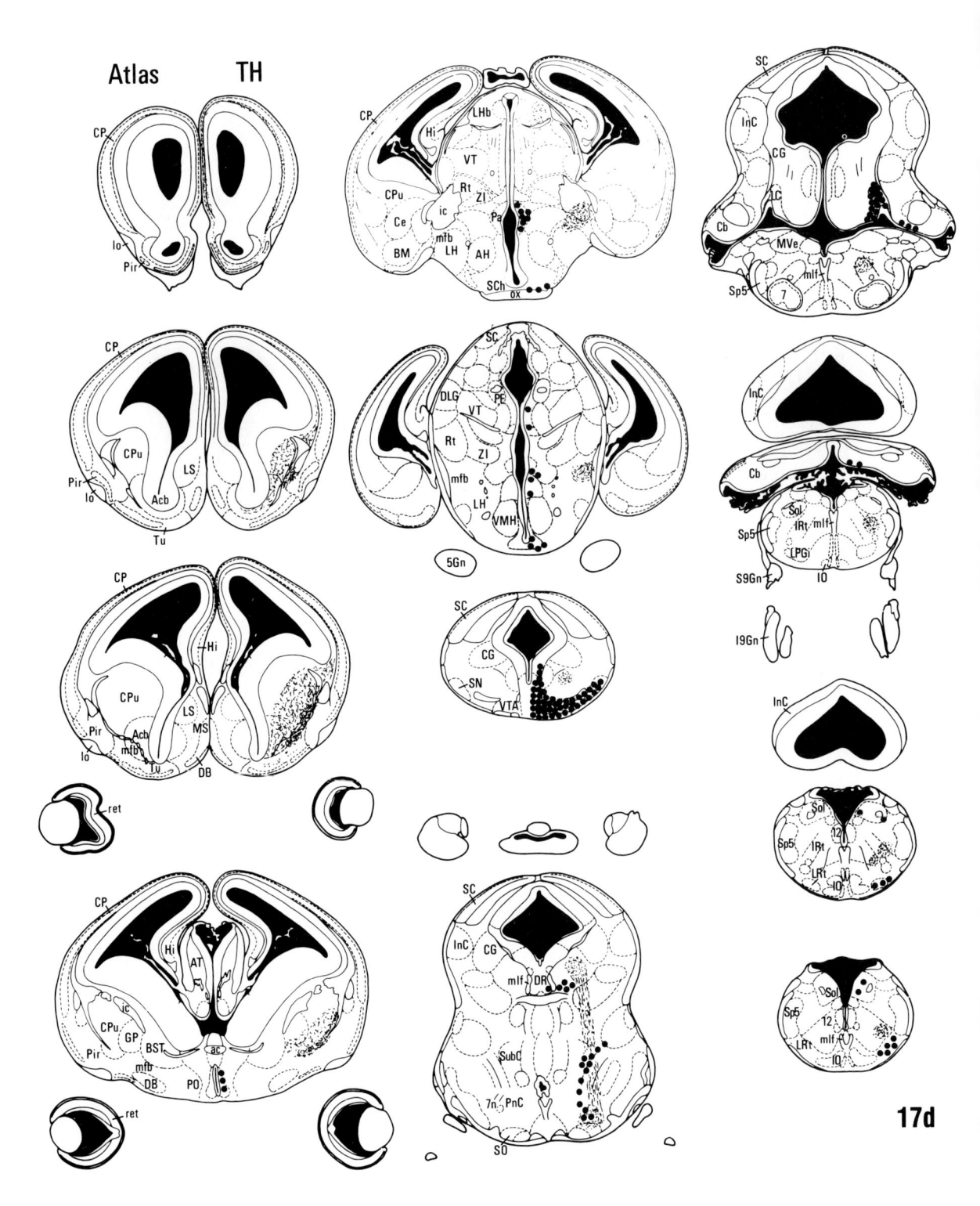

Fig. 17.4(a)

Ontogeny of catecholaminergic neurons in CNS of mammals 413

AADC   PNMT

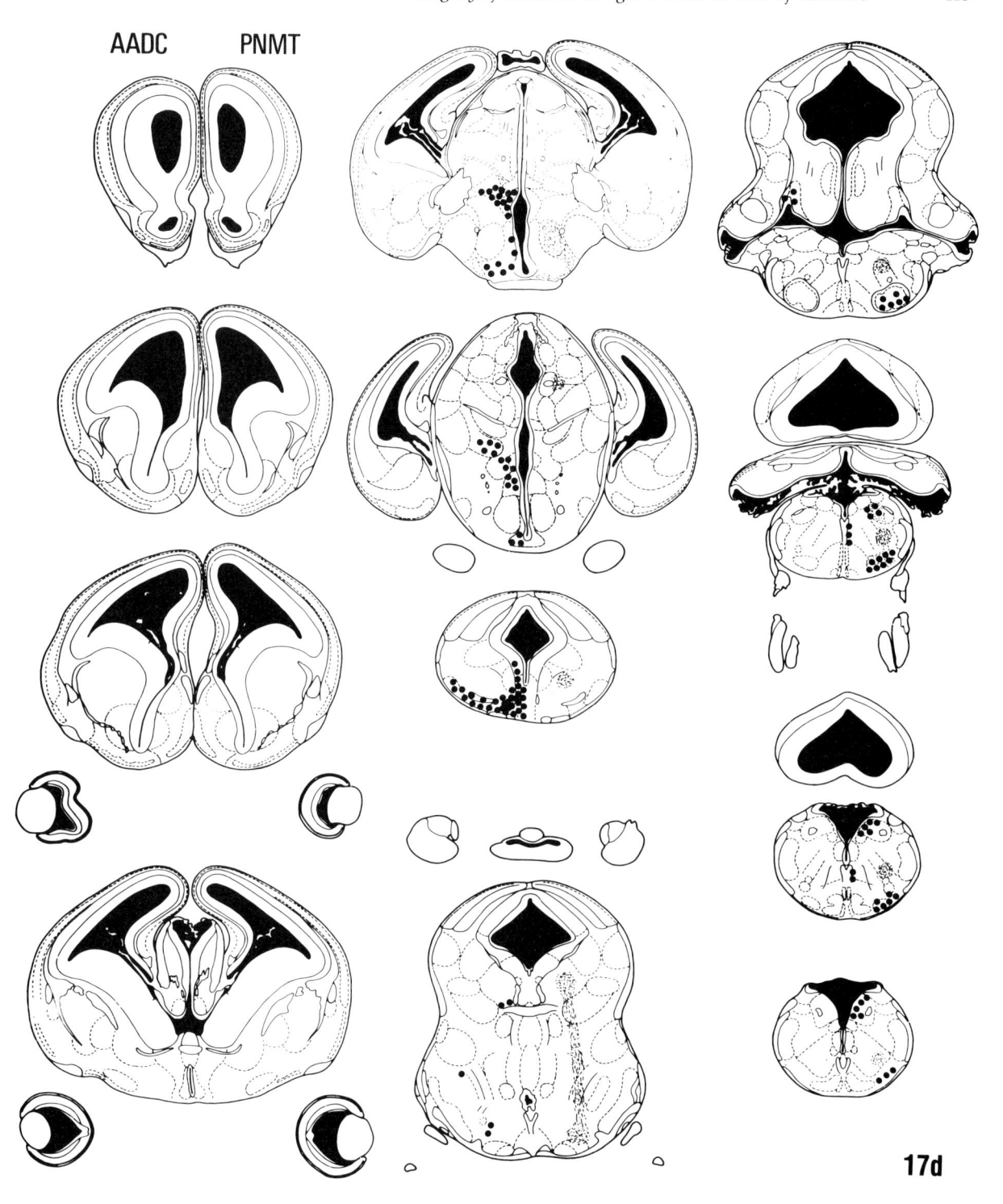

Fig. 17.4(*b*)

17d

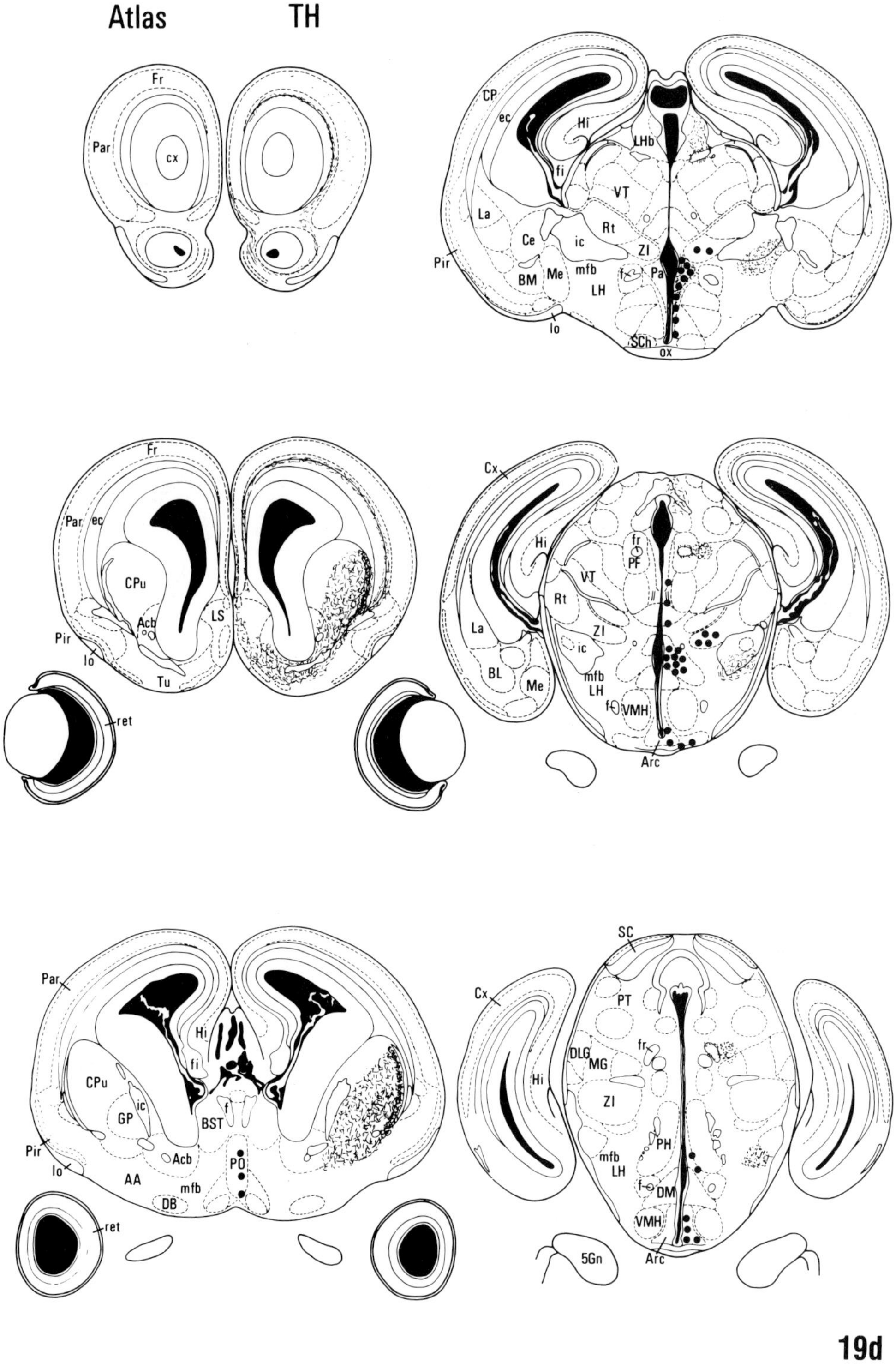
Atlas
TH
Fr
Par
cx
CP
ec
Hi
LHb
fi
VT
La
Rt
Ce
ic
ZI
Pir
BM
Me
mfb
f
Pa
LH
lo
SCh
ox
Fr
Par
ec
CPu
Acb
LS
Pir
lo
Tu
ret
Cx
Hi
fr
PF
VT
Rt
La
ZI
ic
BL
Me
mfb
LH
f
VMH
Arc
Par
Hi
fi
CPu
ic
GP
f
BST
Pir
lo
Acb
PO
AA
mfb
DB
ret
SC
Cx
PT
DLG
MG
fr
Hi
ZI
PH
mfb
LH
f
DM
VMH
5Gn
Arc
19d

Fig. 17.5(*a*)

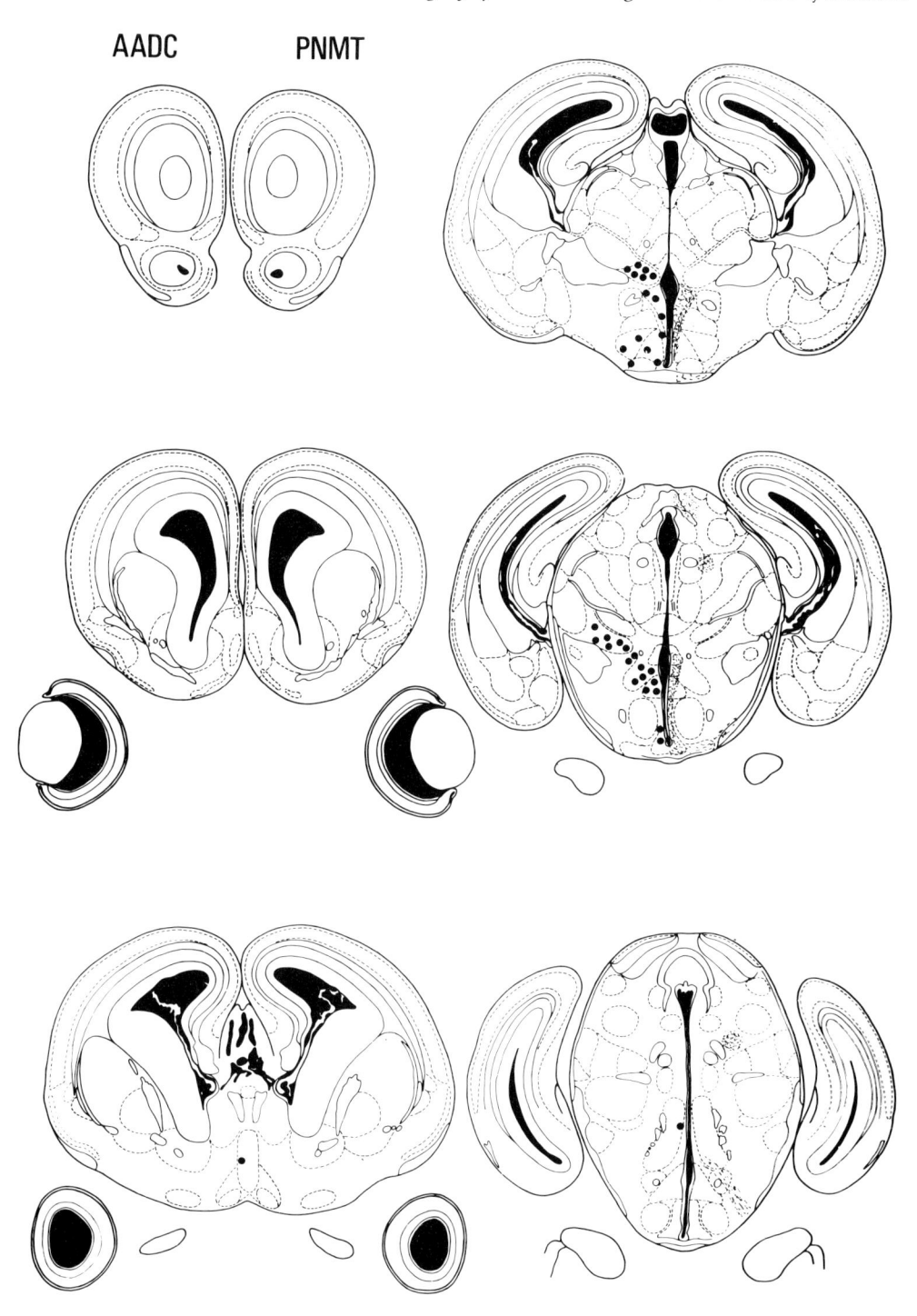

Ontogeny of catecholaminergic neurons in CNS of mammals 415

AADC   PNMT

19d

Fig. 17.5(*b*)

Fig. 17.5(c)

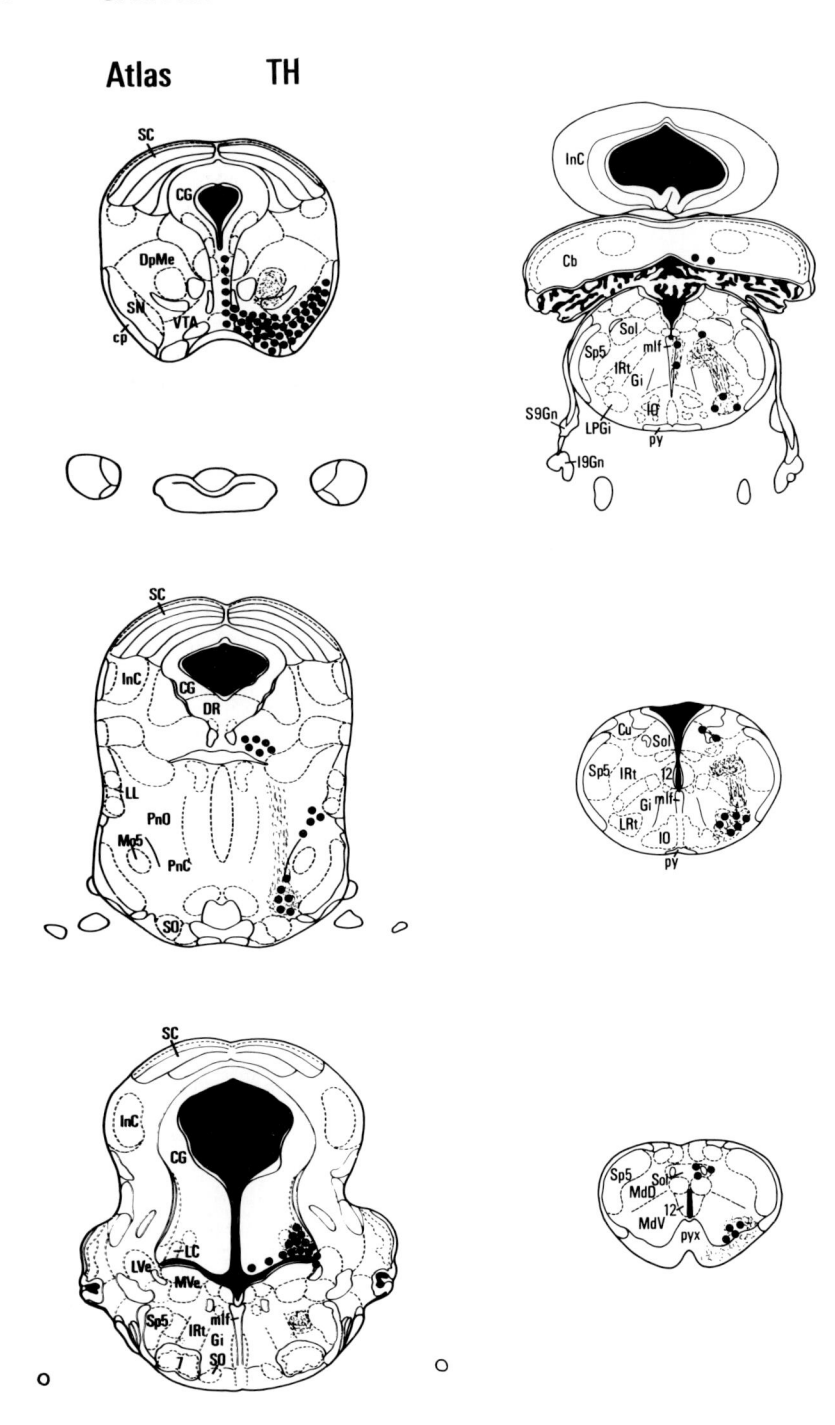

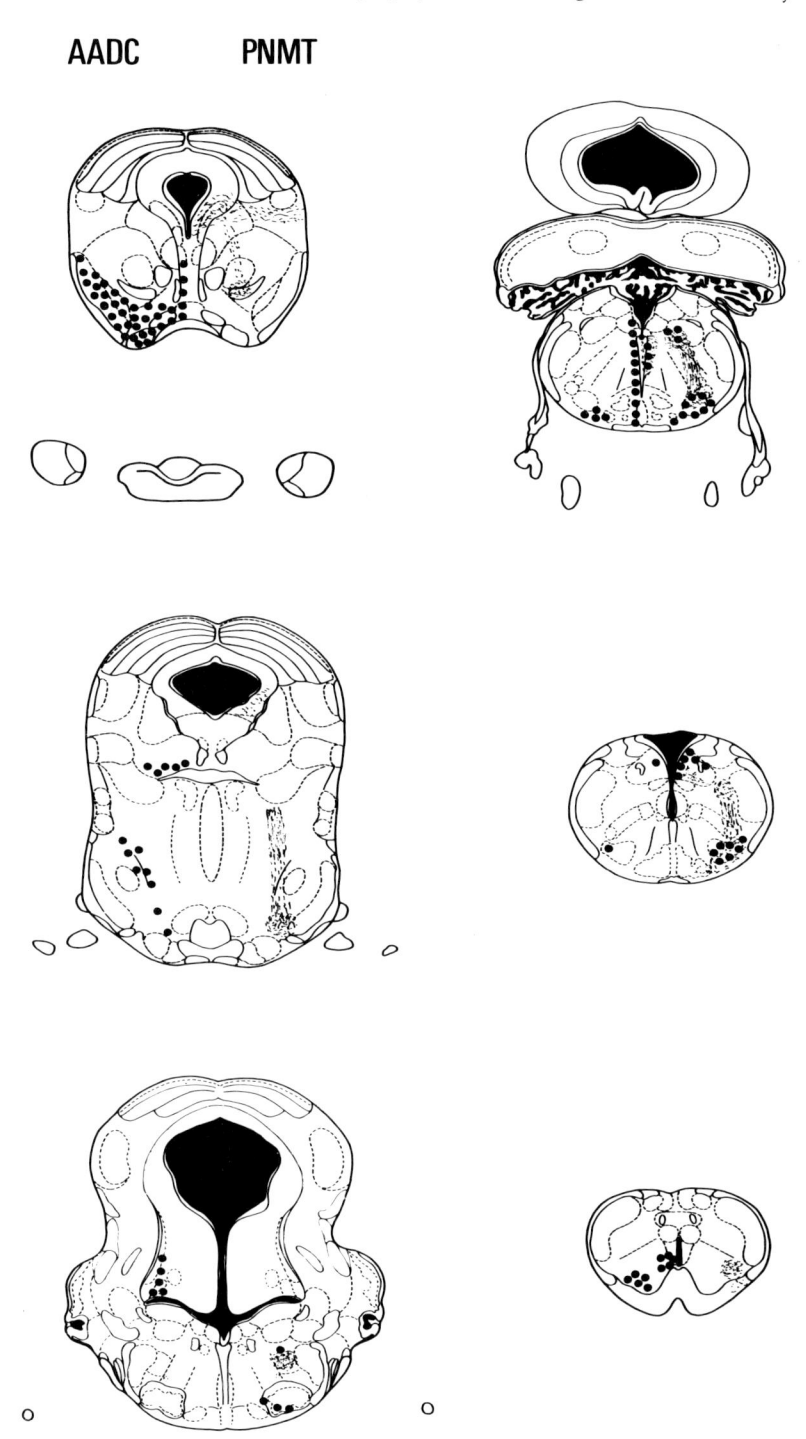

Ontogeny of catecholaminergic neurons in CNS of mammals 417

AADC PNMT

19d

Fig. 17.5(d)

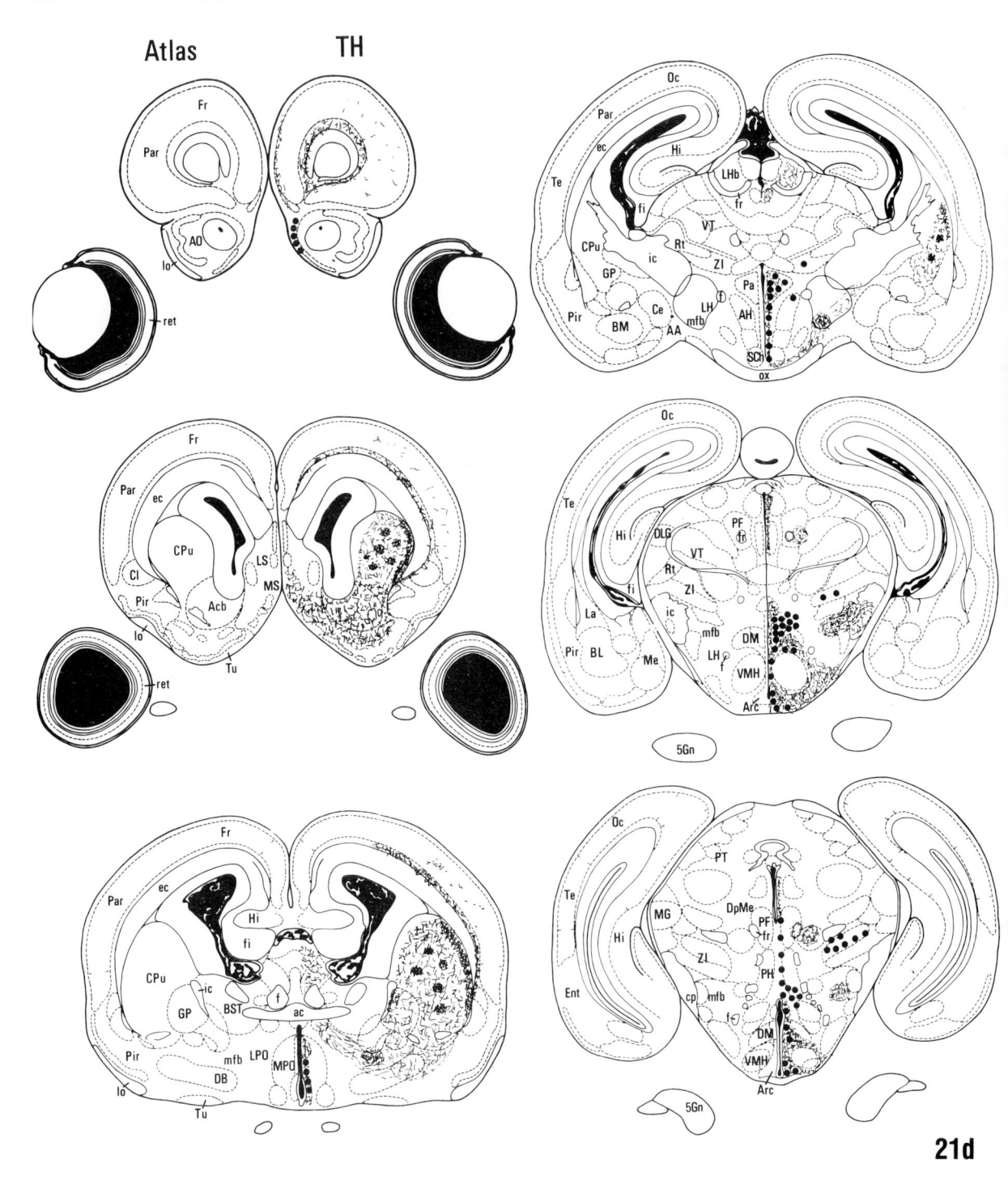

Fig. 17.6(a)

Ontogeny of catecholaminergic neurons in CNS of mammals 419

AADC PNMT

21d

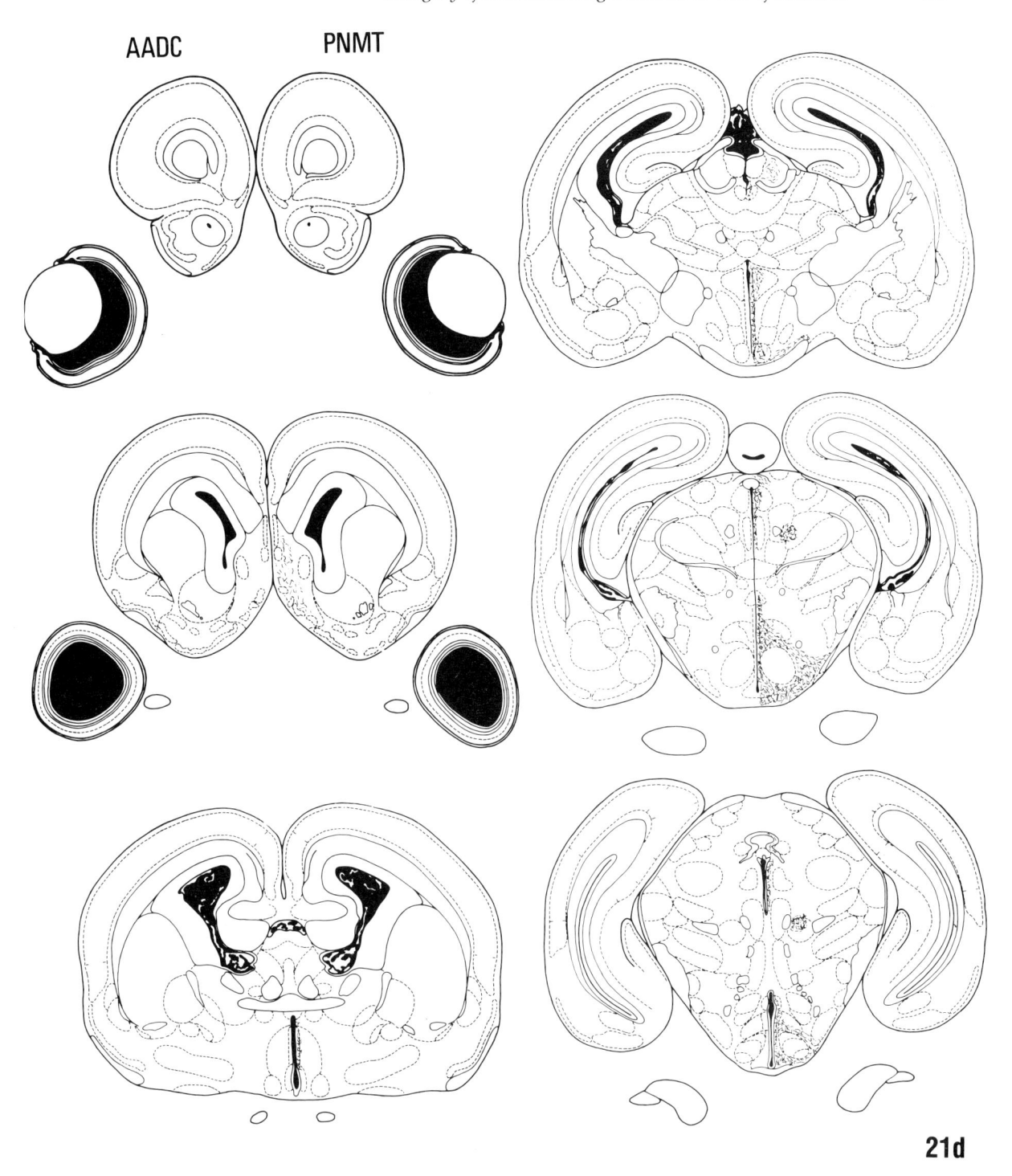

Fig. 17.6(*b*)

Fig. 17.6(c)

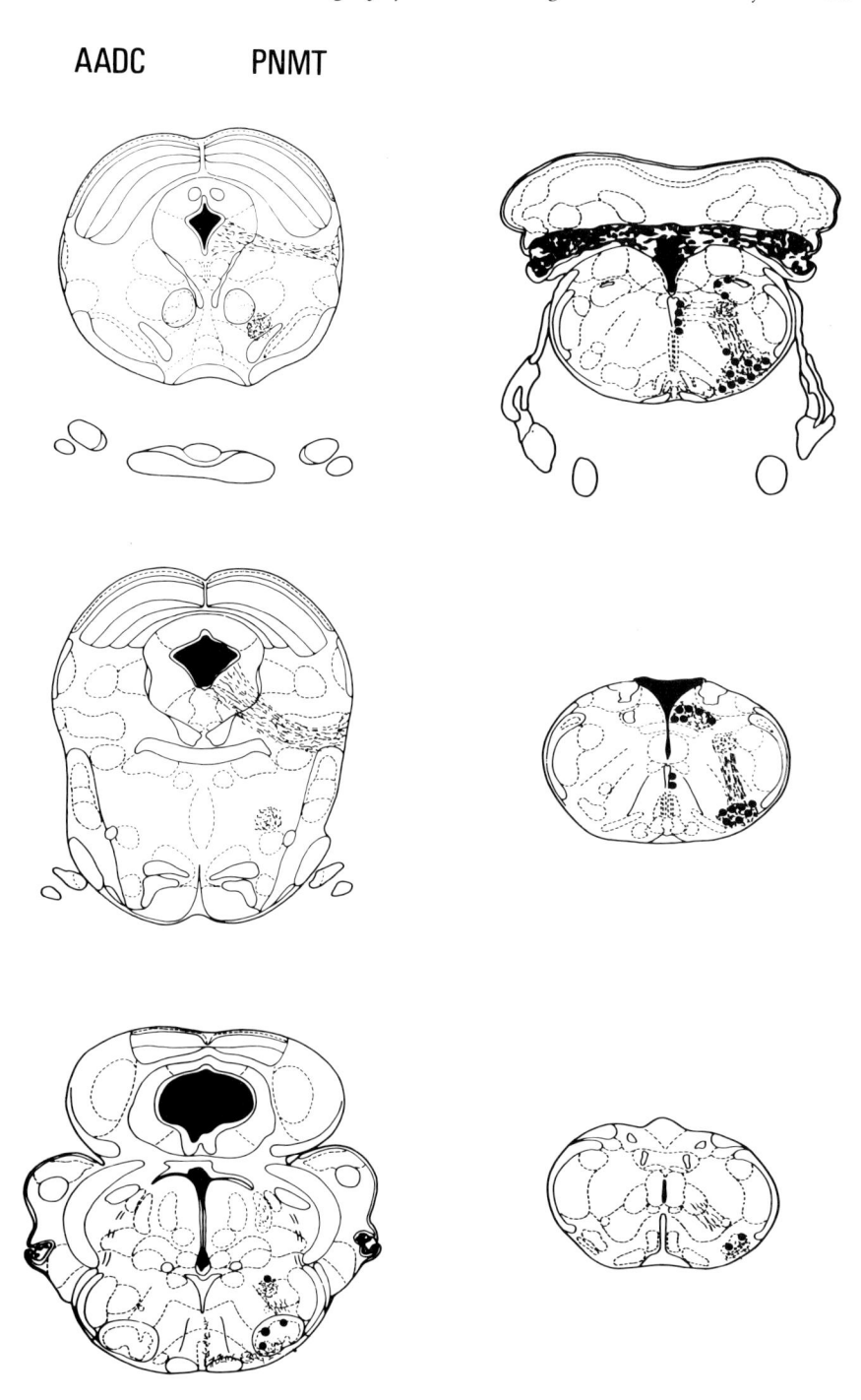

Ontogeny of catecholaminergic neurons in CNS of mammals 421

AADC   PNMT

21d

Fig. 17.6(d)

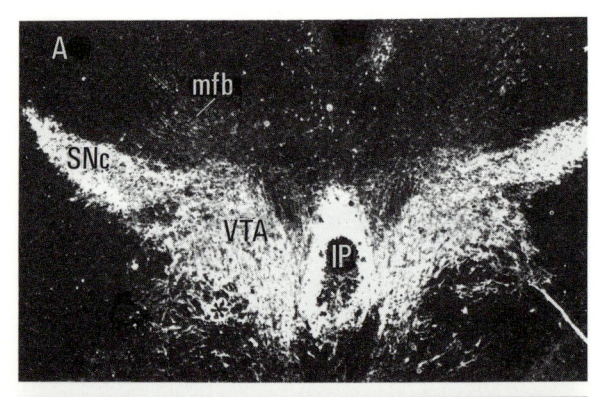

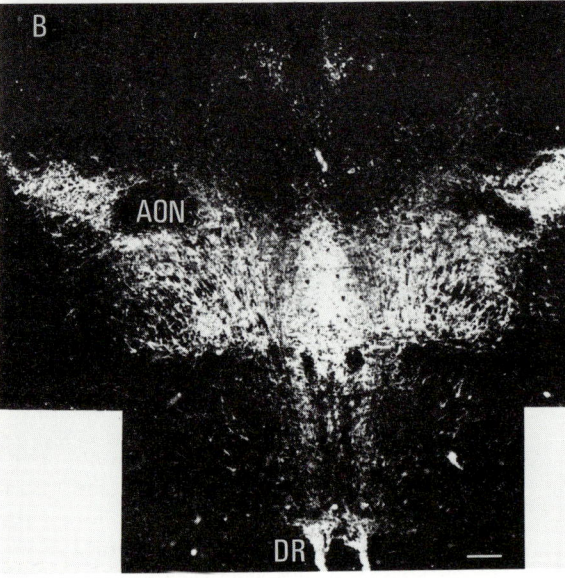

Fig. 17.7. Darkfield micrographs of horizontal sections of E17.5 rat ventral mesencephalon after immunoperoxidase labelling with AADC antibodies. The substantia nigra, pars compacta (SNc) is continuous with the ventral tegmental area (VTA) in ventral regions (*a*), but is separate more dorsally (*b*). AON = medial nucleus of the accessory optic system, DR = dorsal raphé nucleus, IP = interpeduncular nucleus, mfb = medial forebrain bundle. Scale bar indicates 100 μm. Modified with permission, from Jaeger, (1986).

ically and neurochemically. The analysis by histofluorescence (Olson & Seiger, 1972; Seiger & Olson, 1973), and by DA- (Voorn *et al.*, 1988) and TH-immunolabeling (Specht *et al.*, 1981a,b; Foster *et al.*, 1987*a*) has demonstrated that fibres enter the ventrolateral region of the striatum at about E14–15 in the rat and in the mouse (Golden, 1973; Foster *et al.*, 1988). The rabbit striatum exhibits such catecholaminergic fibers at E19 (Tennyson *et al.*, 1973) and the guinea-pig at E38 (Maeda & Astic, 1972). Some processes extend into the neurogenetic subventricular zone of the ganglionic eminence. As development proceeds, the fibers increase in number and density and expand dorsomedially towards the lateral ventricle, and medially around its ventral tip into the nucleus accumbens. A lateral expansion also takes place. At about this time (E17) in the rat, TH activity and release and re-uptake of DA are first detectable (Nomura, Naitoh & Segawa, 1976; Nomura, Yotsumoto & Segawa, 1981; Yotsumoto & Nomura, 1981). At about E21 and the day of birth in the both the rat and the mouse (Foster *et al.*, 1987*a*; 1988; Voorn *et al.*, 1988), the THir fibers of the striatum became reorganized, changing from a relatively homogeneous distribution to one containing a number of fiber-dense patches surrounded by fiber-sparse zones (Fig. 17.8). In the rat, these patches in the medial striatal areas persist only until post-natal week 2, whereas those in the lateral striatum remain in the adult.

The biochemical levels of dopamine in the striatum tend to peak later in development (P60) (Keller, Bartholini & Pletscher, 1973; Giorgi *et al.*, 1987) than the degree and type of fiber innervation as measured histofluorimetrically or by TH-immunohistochemistry (by P21) (Voorn *et al.*, 1988). Nonetheless, there is still a reasonable correlation between the various species of the maturation of the dopaminergic projection to the striatum (Connor & Neff, 1970; Loizou & Salt, 1970; Tennyson *et al.*, 1972; Nomura *et al.*, 1976; Crawford, Connor & Doller, 1984; Kalaria & Prince, 1988).

Some DA fibers in the ventrolateral part of the striatum continue during maturation to extend into the cortex. These mesocortical fibres are first found, specifically, using DA antibodies, at E15 entering the intermediate zone (Voorn *et al.*, 1988). Two to three days later, the DAir processes extend as far rostrally as the olfactory bulb, and extend dorsally into the ventral, then dorsal parts of the frontal cortex. The dopamine innervation of the amygdala begins at E18, and one day later invasion of the septum is also apparent. At E19–21, dopamine fibers become demonstrable in progressively more caudal parts of the

adult nuclei, and by P14 the morphology and orientation of their dendritic arbors are very similar to those of the mature animal.

The innervation of forebrain regions by catecholaminergic cells of the mesencephalon of the rat and other species has been described both histochem-

The ontogeny of pre- and post-synaptic markers for dopamine innervation (such as dopamine- and cyclic adenosine 3':5' -monophosphate-regulated phosphoprotein (DARPP-32), and some DA-receptor subtypes) have also been analyzed. DARPP-32i first becomes manifest in the rat caudate putamen at E14, and in the mouse caudate at E12, its appearance in both species preceding that of the TH innervation of the same regions (Foster et al., 1987a 1988). In addition, the reorganization of the dopaminergic fiber networks into patches occurs only after the same patches have been formed by DARPP-32-positive cell bodies (Foster et al., 1987a) (Fig. 17.8). The analysis of D_1 dopamine receptor ontogeny in the rat by quantitative autoradiography was carried out only postnatally; up to 50% of the adult level was already present at birth (Murrin & Zeng, 1990), particularly in caudal regions of the forebrain, whereas in the frontal cortex, D_1 receptor number only began to increase at P10. The mRNA for the D_1 dopamine receptor (444 amino acid form) is present in whole brain at E14, whereas the 415 amino acid form does not appear until E17 (Mack, O'Malley & Todd, 1991). Both forms only reach their adult levels of expression some time after P15, both in whole brain (Mack et al., 1991) and in striatum (Chen & Weiss, 1991).

Hypothalamus Tyrosine-hydroxylase immunoreactive cells appear at about E12.5–E13.5 in the rostral and lateral hypothalamus, and may constitute the anlage of the A14 and A15 groups (Specht et al., 1981a,b; Daikoku et al., 1986; Ugrumov et al., 1989a). Further caudal are found the primordial TH-positive cells of the A11–A13 groups. By E13–E15, two separate clusters become distinguishable caudally: the A12 group, and what are probably A14 and A15v cells in the basal hypothalamus, lateral to the median eminence (see Hökfelt et al., 1984), and the presumptive A11 and A13 groups in the dorsal hypothalamus in the periventricular region. Hypothalamic TH activity was detected first at E13 (Friedman et al., 1988, 1989), rising 12-fold prenatally and a further 4-fold over the first month of post-natal life (Breese & Traylor, 1972; Coyle & Axelrod, 1972). The amount of mRNA coding for TH rose by the same amount over this entire period (Kedsierski & Porter, 1990). AADCi was first demonstrable in the cells of the hypothalamus (groups A12 – A15) at E15 (Jaeger, 1986). However, with the exception of the A12 arcuate group, there was very little spatial overlap of AADC positivity with TH positivity; in the arcuate there was overlap, but it is not clear if the AADCir neurones were indeed

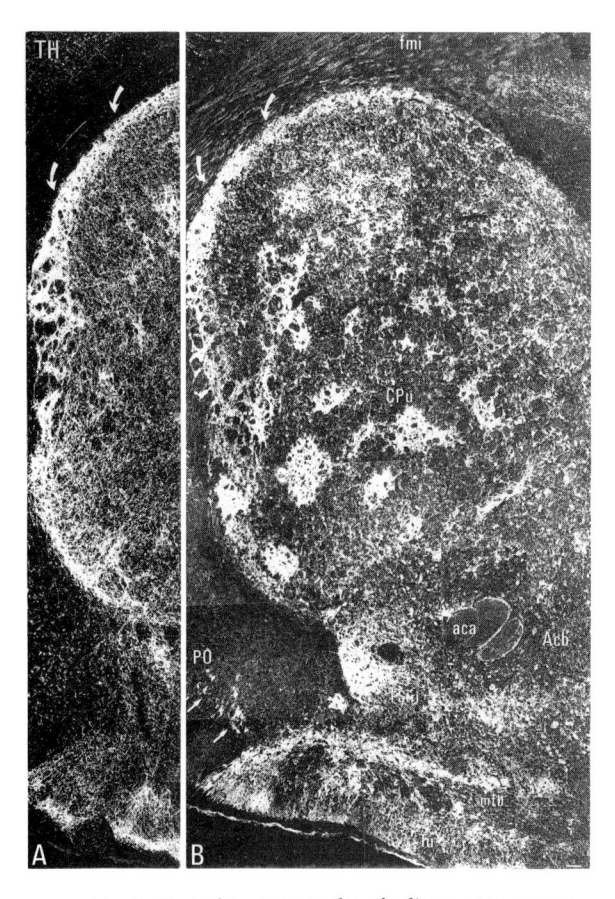

Fig. 17.8. Light micrographs of adjacent transverse sections of P1 rat caudate putamen (CPu) after immunofluorescent labeling with antibodies against TH (A) or DARPP-32 (B). DARPP-32ir cells are found throughout the caudate, but are more densely packed in some clusters within the caudate and around its dorsolateral rim (curved arrows). These clusters coincide with denser than average networks of THir fibers. Note also the DARPP-32-containing cells in the primary olfactory cortex (PO), nucleus accumbens (Acb), fundus striati (FStr) and olfactory tubercle (Tu). aca = anterior part of the anterior commissure, fmi = forceps minor of the corpus callosum, mfb = medial forebrain bundle. Scale bar indicates 50 μm. Modified with permission, from Foster et al. (1987a).

cortex, and also now innervate regions of the basal cortical plate itself. Post-natally, more superficial layers of the cortex become innervated, and the projection is essentially complete by P10–12, with some subsequent localized increases in fiber density in, for example, the prefrontal cortex (Kalsbeek, 1988).

the same as those expressing THi. BY E16 – E19, though, AADC- and TH-positive cells now exhibited substantial overlap in the primordial A11, A13, A14 and A15d groups.

Coincidental with the appearance of TH-positive and AADC-positive neurones in the same hypothalamic regions, DAir cells first become demonstrable in the rostral hypothalamus (Kalsbeek, Voorn & Buijs, 1992). They were only weakly immunoreactive at E15, and remained so despite having dramatically increased in number by E17 and E18 in the periventricular (A14) and arcuate nuclei (A12) and in the A11/A13 caudal groups. Tissue processed for fluorescence histochemistry exhibited fluorescent cells for the first time at E18 (Seiger & Olson, 1973) or later (Hyyppä, 1972; Loizou, 1971) in these regions also. From this age until birth, the cells became more dispersed, to adopt a disposition similar to that of the adult.

Tyrosine hydroxylase-containing axon bundles were found in the hypothalamus at E14 (Reisert et al., 1990), but these later disappeared. By contrast, the first DAir fibers entered the hypothalamus at E21, but were still rare as late as P4 (Kalsbeek et al., 1992). In thalamic regions, however, the DA innervation, presumably from cells in the ventral tegmental area via the fasciculus retroflexus (Skagerberg, Lindvall & Björklund, 1984), was complete by this age (Kalsbeek et al., 1992). At P9, a more dense network of DA-containing fibres had developed, particularly in the paraventricular nucleus and by P14, in the dorsomedial hypothalamus also. Adult levels of innervation were reached by about P3 weeks.

For a fuller account of the development of catecholamines in the diencephalon, see Chapter 18.

Caudal brainstem Although the bulk of the catecholamine neurones in the caudal brainstem are either adrenergic or noradrenergic, there is now considerable evidence that some of them may synthesize DA as an end-product. A coordinated analysis of the development of the full complement of the catecholamine synthetic enzymes in the brainstem has not been pursued. It is impossible, therefore, to determine with the available information whether the appearance of DAir cells in the locus coeruleus at E16, or in the A1–A3 cell groups at E18 (Kalsbeek et al., 1992), represents the development of dopaminergic neurones proper, rather than of NA/A neurones utilizing DA as an intermediate. Nevertheless, it has been proposed that the darker staining neurones in the A1 and A2 cells possibly reflects their synthesis of dopamine as a neurotransmitter.

Dopamine-containing fibers probably originating in the A11 cells of the hypothalamus begin to innervate the brainstem and spinal cord from about E15 onwards (Kalsbeek et al., 1992). Some fibers terminate in the A1–A3 cell groups post-natally, although it is not clear whether these are collaterals of the major diencephalospinal DA projection, or a minor input only to the brainstem.

Olfactory bulb Tyrosine hydroxylase-immunoreactivity appears in the glomerular layer of the rat olfactory bulb at E20 (Specht et al., 1981b; Matsutani, Senba & Tohyama, 1988; McLean & Shipley, 1988), and weak DAir cells are first seen shortly afterwards at E21 (Kalsbeek et al., 1992). These correspond to the A16 cell group. They increase dramatically in number over the next two weeks, post-natally. Neurones of the external glomerular layer, at least in the mouse, undergo their final division perinatally, so it is just about possible that these dopaminergic neurones arise from that region. Alternatively, it has been suggested that they have migrated from the frontal cortex, where they are briefly present at E18 (Specht et al., 1981b). It should be noted that the large numbers of dopamine neurones occurring in the olfactory bulb of various non-mammalian species (e.g. turtle (Halasz et al., 1982)) constitute a different population of cells from that occurring in rats and mice and, accordingly, their embryonic origin is also likely to be different.

Retina Tyrosine hydroxylase-immunoreactive cells of the rat retina (A17 group) are first demonstrable at P2 – 3 in the inner nuclear layer (Nguyen-Legros, Vigny & Gay, 1983; Foster et al., 1985c). Some cells first appear as small and unipolar, are most numerous at P15, but then decline and disappear entirely by adulthood. It has also been proposed that they may instead become transformed into large stellate cells by P10. A second type of DAir neurons appears at P5. These cells have larger perikarya, and are also situated in the inner nuclear layer but at its border with the inner plexiform layer (Foster et al., 1985c). These persist in the adult retina. Similar cells are found in the mouse retina at P6 (Wulle & Schnitzer, 1989). In the rabbit, TH activity and dopamine are found only post-natally (Parkinson & Rando, 1984), although some aspects of the dopaminergic phenotype (such as dopamine uptake) can be detected prenatally (Fung, Kong & Lam, 1982). In the hamster and gerbil, TH-positive cells were first discovered only at P8 and P6, respectively, in the cytoblast layer (Mitrofanis & Finlay, 1990). Most of these later migrated to the inner part of the inner nuclear layer. As might be expected in a precocial species, the development of dopaminergic neurones in the guinea-pig appears to occur earlier, in pre-natal life (Parkinson et al., 1985). The ontogeny of TH-positive, presumptive dopamine neurones in the human retina has also been described, with the first appearance of such cells occurring at the 6th post-ovulatory week (Versaux-Botteri et al., 1992). Both in relation to the time of birth and, more pertinently, with respect to eye maturation, the development of THi in the human retina occurs much earlier than in the rat.

Noradrenaline

Rostral rhombencephalon Noradrenergic cells of the A6 group of the rat locus coeruleus, and of the A4 and A5 groups in the pons are first seen by tyrosine hydroxylase immunohistochemistry at E12 (Specht et al., 1981a; König, Wilkie & Lauder, 1988). The neurones in these regions undergo their final division at E10–E13, with a peak at E12 (Lauder & Bloom, 1974). At E13.5, the A6 cells have increased in number, and become closely clustered. Occasional THir processes extend into the ventricular layer. Ventrally, in the pons proper, TH-positive cells of the A5 group are clearly delineated at this age, lined up along the ventral and lateral surface of the rhombencephalon. They are linked by a continuous collection of cells (the so-called arc (Olson & Seiger, 1972; Seiger & Olson, 1973) running up to the subcoeruleus (A6v). The same cell groups (A4–A6) are delineated by fluorescence histochemistry at E14 (Olson & Seiger, 1972), shortly after they demonstrable by TH-immunohistochemistry. By E17 the borders of the A5 cells and the arc can be seen histofluorimetrically (Seiger & Olson, 1973), while at E18 the A4 cells in addition are identifiable as a few intensely fluorescent neurones in the ventral cerebellum, continuous with the A6 neurones. Also extending from the A6 group, but in a rostral and medial direction are the neurones constituting the A7 group. These cells were first observed at E14.5 with antibodies against TH (Specht et al., 1981a), and by histofluorescence at E17 (Seiger & Olson, 1973).

AADC-immunoreactivity was detectable in neurones of the locus coeruleus only at E16 (Jaeger, 1986), some 3½ days after THi was first found. Its appearance coincides with the first demonstration of DAi (Kalsbeek et al., 1992) in these cells, but is later by two days than the first visualization of histofluorescent cells in this region. The first detection of dopamine β-hydroxylase activity in rat brain also occurred at about this time (Coyle & Axelrod, 1972).

Caudal rhombencephalon A few TH-positive cells are seen in the caudal medulla oblongata at E12 – 12.5 (Specht et al., 1981a; König et al., 1988), just rostral to the cervical flexure. These are likely to constitute the future noradrenergic A1 and A2 groups. One day later, the cells can be divided into ventrolateral (A1) and dorsomedial (A2) parts, and by E14.5, they have aggregated into distinct nuclei at the caudalmost medulla. The A1 cells more rostrally and, more medially, the A3 cells are continuous in the ventrolateral medulla, but are not easily delineated at this stage. Using Falck–Hillarp histochemistry, the A1/A3 cells are first demonstrable at E15, and by E16 are clearly separated from the now detectable A2 cells (Seiger & Olson, 1973). All three cell groups were largely mature by E18, the A3 cells being characterized by their weaker fluorescence and more scattered distribution. The A1–A3 cells groups were first visible at a similar time in the opossum, at birth (which is 12–13 days post-conception), and fibres from some of these cells have innervated as far caudally as the lumbosacral spinal cord by P5 (Pindzola, Ho & Martin, 1990). However, in the mouse, only rarely were histofluorescent neurones seen prior to birth (Golden, 1973).

Dopamine β-hydroxylase-immunoreactive cells can also be visualized in the medulla oblongata at E12.5–13 (Bohn, Goldstein & Black, 1986; Heap & Foster, 1994). However, at this stage, the bulk of the immunoreactive cells in the ventrolateral medulla, at least, belong to the adrenergic C1 group more rostrally, rather than the caudally located A1 noradrenergic group (Heap & Foster, 1994).

Some AADC-containing cells are found in the medulla oblongata at E15 (Jaeger, 1986), but they also do not overlap with either the A1 or A2 groups. At E16, however, some AADCir neurones are found amongst the TH-positive cells of the A1 group, but others are situated lateral to, but not within, the A2 cells until about E19, at which time the two cell types show considerable overlap. Since intervening times were not reported, these observations are not inconsistent with the first histofluorimetric demonstration of the A1/A2 cells at E15–16. It has been suggested (Jaeger, 1986) that the AADC-positive cells which are not catecholaminergic or indoleaminergic may in fact belong to a population of APUD neurones – amine precursor uptake and decarboxylating cells (Pearse, 1969).

Adrenaline

Phenylethanolamine N-methyltransferase (PNMT)-containing neurones are first discovered in the rat brain at E13 (Foster et al., 1985a) (Fig. 17.9). They are situated in the ventrolateral medulla (the C1 group) and dorsomedially (the C2 group) in the primordial solitary tract nucleus. Both groups overlap or are just rostral to, respectively, the A1 and A2 noradrenergic cell groups, and their numbers are similar to those found in the adult. Cells possessed developing processes which extended in a coronal plane towards a bundle of longitudinally oriented PNMTir fibers. The third group of PNMT-positive cells became apparent at day 16 of gestation, and was located close to the midline at about the same rostrocaudal level as the C2 cells. Comparison of adjacent sections, or elution and restaining of the same section, revealed that THi did not appear in many of the PNMT-positive cells until several days later ie. E17–E21 (Fig. 17.10). However, DβHi was expressed in some presumptive adrenergic cells of the ventrolateral medulla oblongata as early as E12.5–13 (Bohn et al., 1986; Heap & Foster, 1994). In addition, three days later, at E16, was the first evidence of AADCi amongst the PNMTir cells of the ventrolateral medulla oblongata and the nucleus of the solitary tract (Jaeger, 1986). Again, though, the degree of colocalization within individual neurones of AADC with PNMT and/or TH is not known.

The first PNMTir fibers in the hypothalamus, principally in its lateral region, were apparent at E14 in the rat (Foster et al., 1985a); descending varicose fibers in the lateral funiculus of the cervical spinal cord first appeared at this age as well. Short, transversely oriented fibers coursing between the dorsomedial and ventrolateral PNMT-containing cells clusters of the medulla oblongata were initially visualized at E16; elsewhere a few fine fibers could now be demonstrated in the periaqueductal central gray, just ventral to the Sylvian aqueduct, in the retrochiasmatic nucleus and in the dorsal thalamus. A dense bundle of axons running longitudinally in the brain stem was also present, and at rostral levels this gave rise at E18 to a branching network of fibers fanning out over the whole of the ventral and lateral surface of the Sylvian aqueduct. By the day of birth, virtually all the other structures containing PNMTir processes in the adult now displayed at least part of their complement.

Disparate appearance of catecholamine parameters

Much of the apparently discrepant expression of various catecholaminergic parameters during ontogeny is found in ephemerally demonstrable populations of neurones. For example, neurones containing PNMTi are found in the floor of the embryonic pons at day E14 of gestation only to disappear 1 day later (Bohn et al., 1986). Similarly, in the post-natal rat, PNMTir cells were found in the bed nucleus of the stria terminalis and the amygdala at P7, but could no longer be immunostained at P35 (Mezey, 1989). However, some THir cells could still be visualized in these regions in the adult (Verney et al., 1988). Of the two types of TH-containing neurone demonstrable in the developing rat retina, only one persists to adulthood

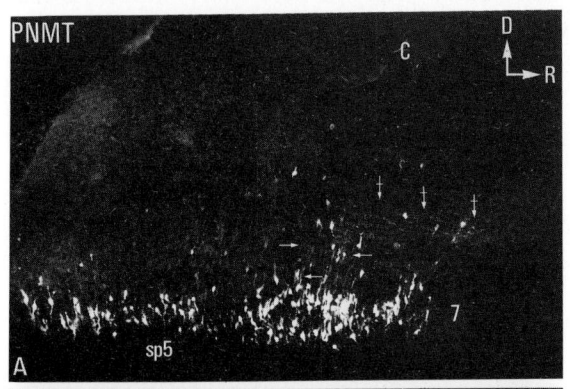

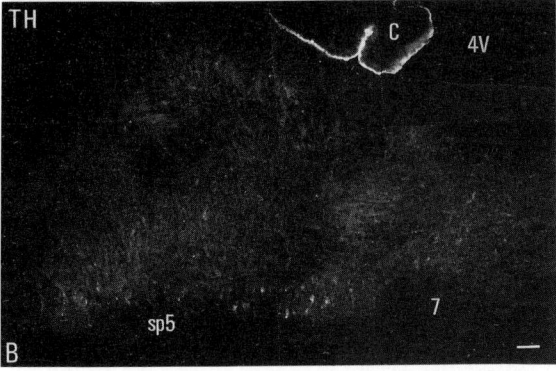

Fig. 17.9. Light micrographs of adjacent parasagittal sections of E13 rat medulla oblongata after immunofluorescent labelling with antibodies against PNMT (a) or TH (b). Although many PNMTir cells are demonstrable at this age, virtually none of them also contains THi. In addition, neither the short dorsoventral PNMT fibres (arrows) nor the long, rostrocaudally oriented processes (crossed arrows) display THi. C = cerebellum, D = dorsal direction, R = rostral direction. Scale bar indicates 50 μm. Reproduced with permission from Foster et al. (1985a).

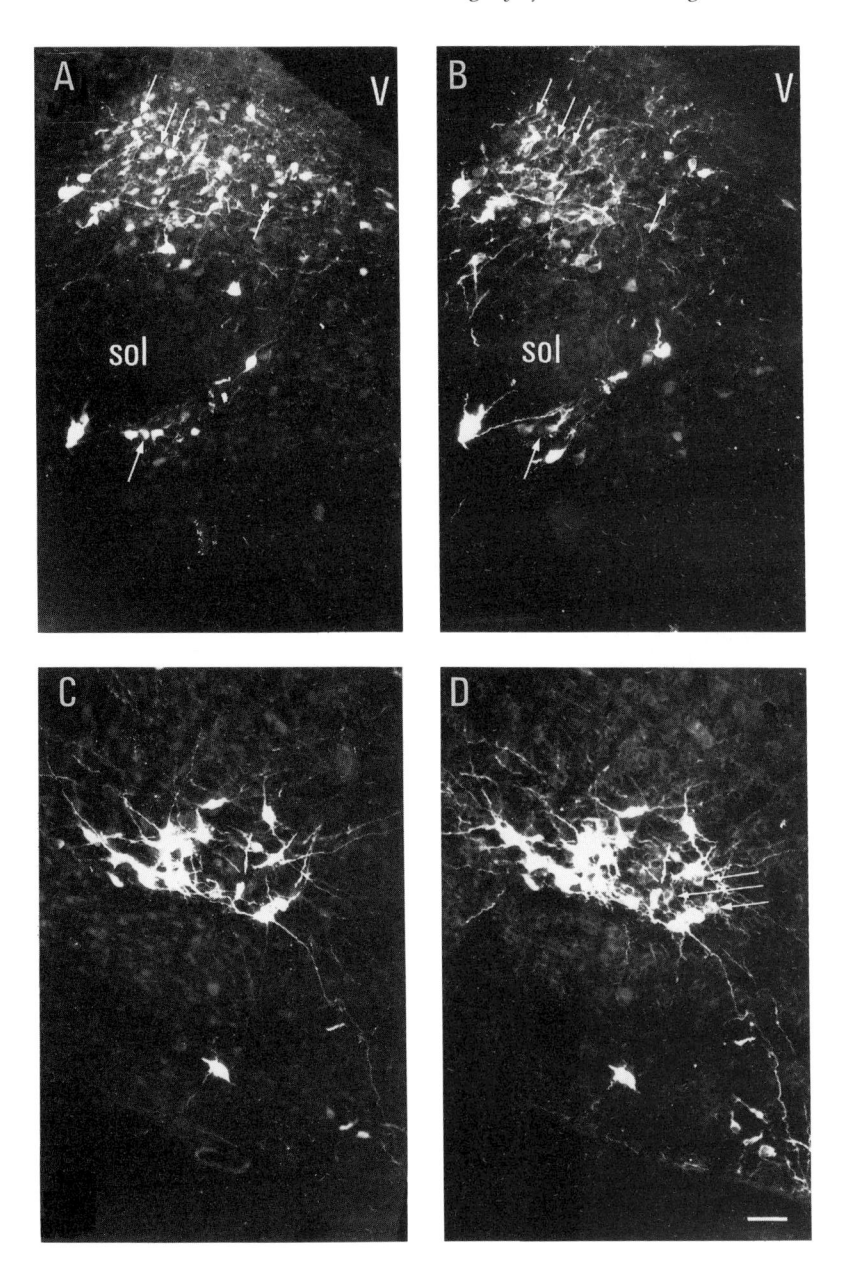

Fig. 17.10. Light micrographs of the same transverse section of P1 rat medulla oblongata after immunofluorescent labeling with antibodies against PNMT (a), (c) or, after elution and restaining, against TH (b), (d). Many cells in the dorsal vagal complex (a), (b) are now both PNMTir and THir: however, there are PNMTir cells in this region which are still apparently devoid of THi (arrows). In the ventrolateral medulla, most PNMT-containing neurones are by now also TH-positive (c), (d). Arrows indicate probable cells from the A1 noradrenaline group. Scale bar indicates 50 μm. Modified with permission from Foster et al. (1984).

(Foster et al., 1985c); the other is first seen at P2, is maximal at about P15, but declines completely by adulthood. TH-containing cells have been described in the olfactory placode and ventral spinal cord of the E11 rat, but are reported to disappear within 48–72h of their first manifestation (Foster et al., 1985b; Heap & Foster, 1994). No AADCi, DβHi, NAi or DAi was demonstrable in these cells while they were TH-positive. TH-immunoreactive cells are also found briefly in the rat frontal cortex at E18, and disappear at E21 (Specht et al., 1981b), while a later transient expression of such cells has also been described (Berger et al., 1985). At P3, TH-positive cells can be seen in in layers V and VI of the mouse cortex (Satoh & Suzuki, 1990). The most intense labeling occurs at P13, but then declines completely by P60. Since these cells contain no DBH-LI, they are unlikely to be producing NA. Neurones ephemerally expressing THi are evident between P3 and P21 in the rat inferior colliculus (Jaeger & Joh, 1983). Finally, in the mouse anterior olfactory nucleus, a transitory population of TH-positive neurones has been described (Nagatsu et al., 1990). In all these examples, no catecholamine, and no other catecholaminergic function, have yet been discovered. Interestingly, it has proved possible to induce the expression of THi in rat cortical neurones either by transplanting them to the adult rat cortex (Park, Joh & Ebner, 1986), or by growing them in culture (Iacovitti et al., 1987). One factor active in promoting the expression of TH in such cells is muscle-derived differentiation factor, which has now been partially purified (Iacovitti & Du, 1992). This factor also affects the expression of TH in neurones which normally synthesize it, such as dopaminergic cells from the mesencephalon (Iacovitti et al., 1992). Other substances can also affect TH expression and or dopamine concentrations, such as gonadal steroids (Crowley, O'Donohue & Jacobowitz, 1978; Leret & Fraile, 1985; Reisert et al., 1987; Simerly, 1989) and dopaminergic agonists (De Vitry et al., 1991). Moreover, many trophic factors, including FGF (Ferrari et al., 1989), EGF (Hadjiconstantinou et al., 1991), brain derived neurotrophic factor (Hyman et al., 1991) and the ganglioside GM1 (Hadjiconstantinou & Neff, 1988) exert a protective action against the effects of lesioning TH-containing mesencephalic neurones, part of which might be via up-regulation of a dopaminergic phenotype. It is possible, then, that many more neurones than those normally expressing TH are also receptive to factors like MDF. Coincidental, but temporary receptivity of such cells combined with the presence of a given factor will result in the fleeting expression of TH.

Not all neurones expressing a single catecholaminergic trait, however, are transient in nature. For example, cells in the ventrolateral arcuate hypothalamic nucleus which express THi, but not AADCi, persist into adulthood (Meister et al., 1988; Okamura et al., 1988). Similarly, PNMTir neurones containing no demonstrable THi or DβHi are found in both the developing and fully mature rat dorsal (Ross et al., 1984) and posterior hypothalamus and retina (Foster et al., 1985d). In contrast to the epigenetic regulation of TH expression, it has been suggested that the control of PNMT expression in the CNS may briefly involve dexamethasone during maturation (Bohn et al., 1986). However, unlike in the adrenal gland (Bohn, Goldstein & Black, 1981), it appears not to be sensitive to other glucocorticoids, either in vivo or in vitro (Bohn et al., 1987).

In the final category of discrepant expression of catecholaminergic traits fall those neurones which are destined to possess the full complement of synthetic enzymes by adulthood, but which at certain times during embryogenesis contain only one. Perhaps the most analyzed neurones in this group are the adrenergic neurones of the medulla oblongata. Although PNMTi is found in nearly all cells of the C1 and C2 groups early in gestation ie. at E13, elution of the PNMT antibodies followed by restaining with TH antiserum showed that most of these cells contained no detectable THi (Foster et al., 1985a). Only the most caudal neurones of the ventrolateral C1 group displayed THi, these also overlapping with the positive A1 cells. At later stages of development THi began to appear in progressively more rostrally located cells of the C1 group. Although C2 cells also started to express THi, there was no obvious spatial pattern as to which. The caudal to rostral wave of appearance of THi in the PNMTir cells of the ventrolateral medulla continued over the remaining days of gestation, such that by about the day of birth virtually all neurones of this C1 group now contained both antigens. Similarly, the C3 group, which at its first visualization at E16 exhibited only PNMTi, by birth also displayed THi in all cells. However, of the C2 cells expressing PNMTi in and around the solitary tract nucleus at birth, only 50–60% contained THi, the double-stained neurones existing predominantly in the rostral part of the cell group.

Neurochemical analysis of the development of PNMT and TH activity in the dorsal and ventral parts of the medulla oblongata during gestation (Foster et al., 1987b) confirmed the immunohistochemical findings. In addition, the concentrations of the catecholamines themselves, and their principal metabolites, were measured. Despite the almost complete lack of TH-LI and enzymatic activity in both the C1 and C2 groups of the E14 medulla oblongata, the steady-state levels of adrenaline were similar to those of the adult. By contrast, levels of noradrenaline closely mirrored the activity of TH, being some 15–30 times less than in the adult. It is possible, therefore, that the synthesis of adrenaline, at least in the embryonic brainstem, does not utilize exactly the same pathway as that proposed for its production in the adrenal gland (Blaschko, 1939).

In summary, the development of catecholamine neurones has been described by measurement of a wide variety of parameters. Although the bulk of the histochemical and neurochemical data are mutually supportive, some anomalous observations still remain to be investigated further.

Abbreviations

5Gn	trigeminal ganglion
7	facial nucleus
7n	facial nerve or its root
12	hypoglossal nucleus
AA	anterior amygdaloid area
ac	anterior commissure
Acb	accumbens nucleus
AH	anterior hypothalamic nucleus
Amg	amygdaloid complex
AO	anterior olfactory nucleus
AP	area postrema
Arc	arcuate hypothalamic nucleus
arc	arcuate nucleus neuroepithelium
AT	anterior thalamus
BL	basolateral amygdaloid nucleus
BM	basomedial amygdaloid nucleus
Cb	cerebellum
cb	cerebellar neuroepithelium
CC	central canal
Ce	central amygdaloid nucleus
CG	central (periaqueductal) gray
CP	cortical plate
cp	cerebral peduncle, basal part
CPu	caudate putamen
Cu	cuneate nucleus
Cx	cerebral cortex
cx	cortical neuroepithelium
DB	nuclei of the diagonal band
db	diagonal band neuroepithelium
DH	dorsal hypothalamus
DLG	dorsal lateral geniculate nucleus
DM	dorsomedial hypothalamic nucleus
DpMe	deep mesencephalic nucleus
DR	dorsal raphé nucleus
ec	external capsule
f	fornix
fi	fimbria of the hippocampus
Fr	frontal cortex
fr	fasciculus retroflexus
GE	ganglionic eminence
Gi	gigantocellular reticular nucleus
GP	globus pallidus
Gr	gracile nucleus
Hi	hippocampal formation
hi	hippocampal neuroepithelium
I9Gn	inferior glossopharyngeal ganglion
ic	internal capsule
IMLF	interstitial nucleus of the medial longitudinal fasciculus
InC	inferior colliculus
inc	inferior collicular neuroepithelium
IO	inferior olive
IRt	intermediate reticular zone
La	lateral amygdaloid nucleus
LC	locus coeruleus
lfu	lateral funiculus
LG	lateral geniculate nucleus
lg	lateral geniculate neuroepithelium
LH	lateral hypothalamic area
lh	lateral hypothalamic neuroepithelium
LHb	lateral habenular nucleus
LL	nuclei of the lateral lemniscus
lo	lateral olfactory tract
LPGi	lateral paragigantocellular nucleus
LPO	lateral preoptic nucleus
LRt	lateral reticular nucleus
LS	lateral septal nucleus
LVe	lateral vestibular nucleus
MdD	medullary reticular nucleus, dorsal part
MdV	medullary reticular nucleus, ventral part
Me	medial amygdaloid nucleus
mfb	medial forebrain bundle
MG	medial geniculate nucleus
ml	medial lemniscus
mlf	medial longitudinal fasciculus
Mo5	motor trigeminal nucleus
MPO	medial preoptic nucleus
MS	medial septal nucleus
MVe	medial vestibular nucleus
Oc	occipital cortex
ox	optic chiasm
Pa	paraventricular hypothalamic nucleus

Par	parietal cortex	s5	sensory root of the trigeminal nerve
PCRt	parvocellular reticular nucleus	S9Gn	superior glossopharyngeal ganglion
pe	periventricular hypothalamic neuroepithelium	SC	superior colliculus
		sc	superior collicular neuroepithelium
PF	parafascicular thalamic nucleus	SCh	suprachiasmatic nucleus
PH	posterior hypothalamic area	SN	substantia nigra
phd	posterior hypothalamic dorsal neuroepithelium	SO	superior olive
		Sol	nucleus of the solitary tract
Pir	piriform cortex	Sp5	spinal trigeminal nucleus
pir	piriform cortical neuroepithelium	sp5	spinal trigeminal tract
Pn	pontine nuclei	SubC	subcoeruleus nucleus
PnC	pontine reticular nucleus, caudal part	Te	temporal cortex
PnO	pontine reticular nucleus, oral part	Tu	olfactory tubercle
		VCA	ventral cochlear nucleus, anterior part
PO	preoptic nuclei	VH	ventral hypothalamic nucleus
PT	pretectal region	VMH	ventromedial hypothalamic nucleus
py	pyramidal tract	VT	ventral thalamus
pyx	pyramidal decussation	vt	ventral thalamic neuroepithelium
R	red nucleus	VTA	ventral tegmental area
ret	retinal neuroepithelium	xscp	decussation of the superior cerebellar peduncle
Rt	reticular thalamic area		
rt	reticular thalamic neuroepithelium	ZI	zona incerta
S	septum		
s	septal neuroepithelium		

References

Barrett, R. E., Cote, L. & Tennyson, V. M. (1971). The histofluorescence and fine structure of the developing neostriatum of the rabbit correlated with biochemical levels of dopamine. *Anatomical Record*, **169**, 273.

Berger, B., Verney, C., Gaspar, P. & Febrvet, A. (1985). Transient expression of tyrosine hydroxylase immunoreactivity in some neurons of the rat neocortex during postnatal development. *Developmental Brain Research*, **23**, 141–4.

Björklund, A., Enemar, A. & Falck, B. (1968). Monoamines in the hypothalamo-hypophyseal system of the mouse with special reference to the ontogenetic aspects. *Zeitschrift für Zellforschung*, **89**, 590–607.

Blaschko, H. (1939). The specific action of L-dopa decarboxylase. *Journal of Physiology*, **96**, 50-1P.

Bohn, M. C., Goldstein, M. & Black, I. B. (1981). Role of glucocorticoids in expression of the adrenergic phenotype in rat embryonic adrenal gland. *Developmental Biology*, **82**, 1–10.

Bohn, M. C., Goldstein, M. & Black, I. B. (1986). Expression and development of phenylethanolamine N-methyltransferase (PNMT) in rat brainstem: studies with glucocorticoids. *Developmental Biology*, **114**, 180–93.

Bohn, M. C. Dreyfus, C. F. Friedman, W. J. & Markey, K. A. (1987). Glucocorticoid effects on phenylethanolamine N-methyltransferase (PNMT) in explants of embryonic rat medulla oblongata. *Developmental Brain Research*, **37**, 257–66.

Breese, G. R. & Traylor, T. D. (1972). Developmental characteristics of brain catecholamines and tyrosine hydroxylase in the rat: effects of 6-hydroxydopamine. *British Journal of Pharmacology*, **44**, 210–22.

Burgunder, J. M. & Young, W. S. (1990). Ontogeny of tyrosine hydroxylase and cholecystokinin gene expression in the rat mesencephalon. *Developmental Brain Research*, **52**, 85–93.

Cadilhac, J. & Pons, F. (1976). Le développement prénatal des neurones à monoamines chez le rat. *Comptes Rendus des Seances de Societe de Biologie*, **170**, 25–30.

Chen, J. F. & Weiss, B. (1991). Ontogenetic expression of D_2 dopamine receptor mRNA in rat corpus striatum. *Developmental Brain Research*, **63**, 95–104.

Connor, J. D. & Neff, N. H. (1970). Dopamine concentrations in the caudate nucleus of the developing cat. *Life Sciences*, **9**, 1165–8.

Coyle, J. T. & Axelrod, J. (1972). Tyrosine hydroxylase in rat brain: developmental characteristics. *Journal of Neurochemistry*, **19**, 1117–23.

Crawford, I. L., Connor, J. D. & Doller, H. J. (1984). Monoamine concentrations in brain regions of the developing rabbit. *International Journal of Developmental Neuroscience*, **2**, 415–19.

Crowley, W. R., O'Donohue, T. L. & Jacobowitz, D. M. (1978). Sex differences in catecholamine content in discrete brain nuclei of the rat: effects of neonatal castration or testosterone treatment. *Acta Endocrinologica*, **89**, 20–8.

Daikoku, S., Kawano, H., Okamura, Y., Tokuzen, M. & Nagatsu, I. (1986). Ontogenesis of immunoreactive tyrosine hydroxylase-containing neurons in the rat hypothalamus. *Developmental Brain Research*, **28**, 85–98.

De Vitry, F., Hillion, J., Catelon, J., Thibault, J., Benoliel,

J. J. & Hamon, M. (1991). Dopamine increases the expression of tyrosine hydroxylase and aromatic acid decarboxylase in primary cultures of fetal neurons. *Developmental Brain Research*, **59**, 123–31.

Di Porzio, U., Zuddas, A., Cosenza-Murphy, D. B. & Barker, J. L. (1990). Early appearance of tyrosine hydroxylase immunoreactive cells in the mesencephalon of mouse embryos. *International Journal of Developmental Neuroscience*, **8**, 523–32.

Ferrari, G., Minozzi, M. -C., Toffano, G., Leon, A. & Skaper, S. D. (1989). Basic fibroblast growth factor promotes the survival and development of mesencephalic neurons in culture. *Developmental Biology*, **133**, 140–7.

Foster, G. A. (1994). *An Atlas of the Prenatal Rat Brain and Spinal Cord*. Oxford, Oxford University Press.

Foster, G. A., Schultzberg, M., Goldstein, M. & Hökfelt, T. (1984). Delayed appearance of tyrosine hydroxylase-like immunoreactivity in PNMT-immunoreactive neurones of the foetal rat brain. *Acta Physiologica Scandinavica*, **122**, 429–32.

Foster, G. A., Schultzberg, M., Goldstein, M. & Hökfelt, T. (1985a). Ontogeny of phenylethanolamine N-methyltransferase- and tyrosine hydroxylase-like immunoreactivity in presumptive adrenaline neurones of the foetal rat central nervous system. *Journal of Comparative Neurology*, **236**, 348–81.

Foster, G. A., Schultzberg, M., Dahl, D., Goldstein, M. & Verhofstad, A. A. J. (1985b). Ephemeral existence of a single catecholamine synthetic enzyme in the olfactory placode and spinal cord of the embryonic rat. *International Journal of Developmental Neuroscience*, **3**, 597–608.

Foster, G. A., Schultzberg, M., Goldstein, M. & Hökfelt, T. (1985c). Differential ontogeny of three putative catecholamine cell types in the postnatal rat retina. *Developmental Brain Research*, **22**, 187–96.

Foster, G. A., Hökfelt, T., Coyle, J. T. & Goldstein, M. (1985d). Immunohistochemical evidence for phenylethanolamine-N-methyltransferase-positive/tyrosine hydroxylase-negative neurones in the retina and posterior hypothalamus of the rat. *Brain Research*, **330**, 183–8.

Foster, G. A., Schultzberg, M., Hökfelt, T., Goldstein, M., Hemmings, H. C. Jr., Ouimet, C. C., Walaas, S. I. & Greengard, P. (1987a). Development of a dopamine- and cyclic adenosine 3':5'-monophosphate-regulated phosphoprotein (DARPP-32) in the prenatal rat central nervous system, and its relationship to the arrival of presumptive dopaminergic innervation. *Journal of Neuroscience*, **7**, 1994–2018.

Foster, G. A., Sundström, E., Helmer-Matyjek, E., Goldstein, M. & Hökfelt, T. (1987b). Abundance in the embryonic brain stem of adrenaline during the absence of detectable tyrosine hydroxylase activity. *Journal of Neurochemistry*, **48**, 202–7.

Foster, G. A., Schultzberg, M., Hökfelt, T., Goldstein, M., Hemmings, H. C. Jr., Ouimet, C. C., Walaas, S. I. & Greengard, P. (1988). Ontogeny of the dopamine- and cyclic adenosine 3':5'-monophosphate-regulated phosphoprotein (DARPP-32) in the pre- and post-natal mouse central nervous system. *International Journal of Developmental Neuroscience*, **6**, 367–86.

Friedman, W. J., Dreyfus, C. F., McEwen, B. S. & Black, I. B. (1988). Presynaptic transmitters and depolarizing influences regulate development of the substantia nigra in culture. *Journal of Neuroscience*, **8**, 3616–23.

Friedman, W. J., Dreyfus, C. F., McEwen, B. S. & Black, I. B. (1989). Developmental regulation of tyrosine hydroxylase in the mediobasal hypothalamus. *Developmental Brain Research*, **48**, 177–85.

Fung, S. C., Kong, Y. C. & Lam, D. M. K. (1982). Prenatal development of GABAergic, glycinergic, and dopaminergic neurons in the rabbit retina. *Journal of Neuroscience*, **2**, 1623–32.

Giorgi, O., De Montis, G., Porceddu, M. L., Mele, S., Calderini, G., Toffano, G. & Biggio, G. (1987). Developmental and age-related changes in D1-dopamine receptors and dopamine content in the rat striatum. *Developmental Brain Research*, **35**, 283–90.

Golden, G. S. (1972). Embryologic development of a nigrostriatal projection in the mouse. *Brain Research*, **44**, 278–82.

Golden, G. S. (1973). Prenatal development of the biogenic amine system of the mouse brain. *Developmental Biology*, **33**, 300–11.

Hadjiconstantinou, M. & Neff, N. H. (1988). Treatment with GM1 ganglioside restores striatal dopamine in the 1-methyl-n-phenyl-1, 2, 3, 6-tetrahydropyridine-treated mouse. *Journal of Neurochemistry*, **51**, 1190–6.

Hadjiconstantinou, M., Fitkin, J. G., Dalia, A. & Neff, N. H. (1991). Epidermal growth factor enhances striatal dopaminergic parameters in the 1-methyl-4-phenyl-1, 2, 3, 6-tetrahydropyridine-treated mouse. *Journal of Neurochemistry*, **57**, 479–82.

Halasz, N., Nowycky, M., Hökfelt, T, Shepherd, G. M., Markey, K. & Goldstein, M. (1982). Dopaminergic periglomerular cells in the turtle olfactory bulb. *Brain Research Bulletin*, **5**, 425–36.

Heap, M. & Foster, G. A. (1994). Asynchronous expression of catecholamine synthetic enzymes in rat brainstem neurones during prenatal development. *Journal of Anatomy*, (in press).

Hissa, R. & Pyhälä, L. (1971). Ontogenetic changes of monoamines in the diencephalon of the Norwegian lemming. *Annales Zoologici Fennici*, **8**, 367–73.

Hökfelt, T., Johansson, O., Fuxe, K. Goldstein, M. & Park, D. (1976). Immunohistochemical studies on the localization and distribution of monoamine neuron systems in the rat brain. I. Tyrosine hydroxylase in the mes- and diencephalon. *Medicine and Biology*, **54**, 427–53.

Hökfelt, T., Mårtensson, R., Björklund, A. & Kleinau, S. & Eddstein, M. (1984). Distributional maps of tyrosine – hydroxylase – immunoreactive neurons in the rat brain. In *Handbook of Chemical Neuroanatomy, Vol. 2: Classical Transmitters in the CNS, part I*, eds. A. Björklund & T. Hökfelt, pp. 277–379. Amsterdam: Elsevier.

Hyman, C., Hofer, M., Barde, Y. A., Juhasz, M., Yancopoulos, G. D., Squinto, S. P. & Lindsay, R. M. (1991).

BDNF is a neurotrophic factor for dopaminergic neurones of the substantia nigra. *Nature*, **350**, 230–5.

Hyyppä, M. (1972). Hypothalamic monoamines in human fetuses. *Neuroendocrinology*, **9**, 257–66.

Iacovitti, L., Lee, J., Joh, T. J. & Reis, D. (1987). Expression of tyrosine hydroxylase in neurons of cultured of cerebral cortex: evidence for phenotype plasticity in neurons of the CNS. *Journal of Neuroscience*, **7**, 1264–70.

Iacovitti, L. & Du, X. (1992). Biochemical isolation and characterization of muscle derived differentiation factor: a novel factor (S) which induces expression of the TH gene in non-catecholamine neurons in culture. *American Society for Neuroscience Abstract*, **18**, 392.

Iacovitti, L., Evinger, M. J. & Stull, M. D. (1992) Muscle-derived differentiation factor increases expression of the TH gene and enzyme activity in cultures dopamine neurons from the rat midbrain. *Molecular Brain Research*, **16**, 215–22.

Jaeger, C. (1986). Aromatic L-amino acid decarboxylase in the rat brain: immunocytochemical localization during prenatal development. *Neuroscience*, **18**, 121–50.

Jaeger, C. & Joh, T. J. (1983). Transient expression of tyrosine hydroxylase of some neurons of the developing inferior colliculus of the rat. *Developmental Brain Research*, **11**, 128–32.

Kalaria, R. N. & Prince, A. K. (1988). Neurochemical development of the striatum in a precocial (guinea pig) and an altricial (rat) species. *International Journal of Developmental Neuroscience*, **6**, 161–6.

Kalsbeek, A., Voorn, P., Buijs, R. M. & Uylings H. B. M. (1988). Development of the dopaminergic innervation in the prefrontal cortex of the rat. *Journal of Comparative Neurology*, **269**, 58–72.

Kalsbeek, A., Voorn, P., & Buijs, R. M. (1992). Development of dopamine-containing systems in the CNS. In *Handbook of Chemical Neuroanatomy, Vol. 10: The Ontogeny of Transmitters and Peptides in the CNS*, eds. A. Björklund, T. Hökfelt & M. Tohyama, pp 63–112. Amsterdam: Elsevier.

Kedzierski, W. & Porter, J. C. (1990). Quantitative study of tyrosine hydroxylase mRNA in catecholaminergic neurons and adrenals during development and aging. *Molecular Brain Research*, **7**, 45–51.

Keller, H. H., Bartholini, G. & Pletscher, A. (1973). Spontaneous and drug-induced changes of cerebral dopamine turnover during postnatal development of rats. *Brain Research*, **64**, 371–8.

König, N., Wilkie, M. B. & Lauder, J. M. (1988). Tyrosine hydroxylase and serotonin containing cells in embryonic rat rhombencephalon: a whole-mount immunocytochemical study. *Journal of Neuroscience Research*, **20**, 212–23.

Lauder, J. M. & Bloom, F. E. (1974). Ontogeny of monoamine neurons in the locus coeruleus, raphe nuclei and substantia nigra of the rat. *Journal of Comparative Neurology*, **155**, 469–82.

Leret, M. L. & Fraile, A. (1985). Effect of gonadectomy on brain catecholamines during the postnatal period. *Comparative Biochemistry and Physiology*, **81**, 405–9.

Loizou, L. A. (1971). The postnatal development of monoamine-containing structures in the hypothalamo-hypophyseal system of the albino rat. *Zeitschrift für Zellforschung*, **114**, 234–52.

Loizou, L. A. & Salt, P. (1970). Regional changes in monoamines of the rat brain during postnatal development. *Brain Research*, **40**, 395–418.

Loup, M. & Cadilhac, J. (1970) Le développement des neurones à monoamines du cerveau chez le chaton. *Comptes Rendus des Séances de Société de Biologie Paris*, **164**, 1582–7.

Mack, K. J., O'Malley K. L. & Todd, R. D. (1991). Differential expression of dopaminergic D2 receptor messenger RNAs during development. *Developmental Brain Research*, **59**, 249–51.

Maeda, K. & Astic, L. (1972). Developpement des neurones monoaminergiques centraux chez le foetus de Cobaye. *Comptes Rendus des Séances de Société de Biologie Paris*, **166**, 1014–7.

Maeda, K. & Dresse, A. (1968). Possibilités d'étude du trajet des fibres cérébrales monoaminergiques chez le rat nouveau-né. *Comptes Rendus des Séances de Société de Biologie Paris*, **166**, 1626–9.

Martin, G. F., Ho, R. H. & Hazlett, J. C. (1989). The early development of major projections to the dorsal striatum in the North American opossum. *Developmental Brain Research*, **7**, 9–12.

Matsutani, S., Senba, E. & Tohyama, M. (1988). Neuropeptide- and neurotransmitter-related immunoreactivities in the developing rat olfactory bulb. *Journal of Comparative Neurology*, **272**, 331–42.

McLean, J. H. & Shipley, M. T. (1988). Post-mitotic, postmigrational expression of tyrosine hydroxylase in olfactory bulb dopaminergic neurons. *Journal of Neuroscience*, **8**, 3658–69.

Meister, B., Hökfelt, T., Steinbusch, H. W. M., Skagerberg, G., Lindvall, O., Geffard, M., Joh, T. J., Cuello, A. C. & Gulden, M. (1988). Do tyrosine hydroxylase-immunoreactive neurons in the ventrolateral arcuate nucleus produce dopamine or only L-DOPA? *Journal of Chemical Neuroanatomy*, **1**, 59–64.

Mezey, E. (1989). Phenylethanolamine N-methyltransferase-containing neurons in the limbic system of the young rat. *Proceedings of the National Academy of Sciences, USA*, **86**, 347–51.

Mitrofanis, J. & Finlay, B. L. (1990). Developmental changes in the distribution of retinal catecholaminergic neurones in hamsters and gerbils. *Journal of Comparative Neurology*, **292**, 480–94.

Murrin, L. C. & Zeng, W. (1990). Ontogeny of dopamine D1 receptors in rat forebrain: a quantitative autoradiographic study. *Developmental Brain Research*, **57**, 7–13.

Nagatsu, I., Komori, K., Takeuchi, T., Sakai, M., Yamada, K. & Karasawa, N. (1990). Transient tyrosine hydroxylase immunoreactive neurons in the region of the anterior olfactory nucleus of pre- and postnatal mice do not contain dopamine. *Brain Research*, **511**, 55–62.

Nguyen-Legros, J., Vigny, A. & Gay, M. (1983). Postnatal development of TH-like immunoreactivity in the rat retina. *Experimental Eye Research*, **37**, 23–32.

Nobin, A. & Björklund, A. (1973). Topography of the monoamine neuron systems in the human brain as revealed in fetuses. *Acta Physiologica Scandinavica Supplement*, **388**, 1–4C.

Nomura Y., Naitoh, F. & Segawa, T. (1976). Regional changes in monamine content and uptake of the rat brain during postnatal development. *Brain Research*, **101**, 305–15.

Nomura, Y., Yotsumoto, I. & Segawa, T. (1981) Ontogenetic development of high potassium- and acetylcholine-induced release of dopamine from striatal slices of the rat. *Developmental Brain Research*, **1**, 171–7.

Okamura, H., Kitahama, K., Mons, N., Ibata, Y., Jouvet, M. & Geffard, M. (1988). L-DOPA-immunoreactive neurons in the rat hypothalamic tuboreal region. *Neuroscience Letters*, **95**, 42–6.

Olson L. & Ungerstedt, U. (1970). A simple high capacity freeze-drier for histochemical use. *Histochemie*, **22**, 8–19.

Olson, L. & Seiger, A. (1972). Early prenatal ontogeny of central monoamine neurons in the rat: fluorescence histochemical observation. *Zeitschrift für Anatomie und Entwicklung Geschichte*, **137**, 301–16.

Olson, L., Boréus, L. O. & Seiger, A. (1973). Histochemical demonstration and mapping of 5-hydroxytryptamine and catecholamine-containing neurons in human fetal brain. *Zeitschrift für Anatomie und Entwicklung Geschichte*, **139**, 259–82.

Park, J. K., Joh, T. H. & Ebner, F. F. (1986). Tyrosine hydroxylase is expressed by neocortical neurons after transplantation. *Proceedings of the National Academy of Science 5, USA*, **83**, 7495–8.

Parkinson, D. & Rando, R. R. (1984). Ontogenesis of dopaminergic neurons in the postnatal rabbit retina: pre- and post-synaptic elements. *Developmental Brain Research*, **13**, 207–17.

Parkinson, D., Spira, A., Wyse, J. P. & Patten, M. (1985). The ontogenesis of the dopaminergic cell in the pre- and postnatal guinea pig retina. *International Journal of Developmental Neuroscience*, **3**, 157–67.

Pearse, A. G. E. (1969). The cytochemistry and ultrastructure of polypeptide hormone-producing cells of the APUD series and the embryologic, physiologic and pathologic implications of the concept. *Journal of Histochemistry and Cytochemistry*, **17**, 303–13.

Pearson, J., Brandeis, L. & Goldstein, M. (1980). Appearance of tyrosine hydroxylase immunoreactivity in the human embryo. *Developmental Neuroscience*, **3**, 140–50.

Pickel, V. M., Specht, L. A., Sumal, K. K., Joh, T. J., Reis, D. J. & Hervonen, A. (1980). Immunocytochemical localization of tyrosine hydroxylase in the human fetal nervous system. *Journal of Comparative Neurology*, **194**, 465–74.

Pindzola, R. R., Ho,. R. H. & Martin, G. F. (1990). Development of catecholaminergic projections to the spinal cord in the North American opossum, *Didelphus virginiana*. *Journal of Comparative Neurology*, **294**, 399–417.

Reisert, I., Han, V., Lieth, E., Toran-Allerand, D., Pilgrim, C. & Lauder, J. (1987). Sex steroids promote neurite growth in mesencephalic tyrosine hydroxylase immunoreactive neurons in vitro. *International Journal of Developmental Neuroscience*, **5**, 91–8.

Reisert, I., Schuster, R., Zienecker, R. & Pilgrim, C. (1990). Prenatal development of mesencephalic and diencephalic dopaminergic systems in the male and female rat. *Developmental Brain Research*, **53**, 222–9.

Ross, C. A., Ruggiero, D. A., Park, D. H., Joh, T. J. & Reis, D. J. (1984). A new group of neurons in the hypothalamus containing phenylethanolamine N-methyltransferase (PNMT) but not tyrosine hydroxylase. *Brain Research*, **306**, 349–52.

Rothman, T. P., Specht, L., Gershon, M. D., Joh, T. J., Teitelman, G., Pickel, V. M. & Reis D. J. (1980). Catecholamine biosynthetic enzymes are expressed in replicating cells of the peripheral but not the central nervous system. *Proceedings of the National Academy of Science*, **77**, 6221–5.

Ruggiero, D. A., Anwar, M. & Gootman, P. M. (1992). Presumptive adrenergic neurons containing phenylethanolamine N-methyltransferase immunoreactivity in the medulla oblongata of neonatal swine. *Brain Research*, **583**, 105–19.

Rugh, R. (1968). *The Mouse: Its Reproduction and Development*. Minneapolis: Burgess.

Satoh, J. & Suzuki, K. (1990). Tyrosine hydroxylase-immunoreactive neurons in the mouse cerebral cortex during the postnatal period. *Developmental Brain Research*, **53**, 1–5.

Seiger, A. & Olson, L. (1973). Late prenatal ontogeny of central monoamine neurons in the rat: fluorescence histochemical observations. *Zeitschrift Anatomie und Entwicklung Geschichte*, **140**, 281–318.

Simerly, R. B. (1989). Hormonal control of the development and regulation of tyrosine hydroxylase expression within a sexually dimorphic population of dopaminergic cells in the hypothalamus. *Molecular Brain Research*, **6**, 297–310.

Skagerberg, G., Lindvall, O. & Björklund, A. (1984). Origin, course and termination of the mesohabenular dopamine pathway in the rat. *Brain Research*, **307**, 99–108.

Specht, L. A., Pickel, V. M., Joh, T. J. & Reis, D. J. (1981a). Light-microscopic immunocytochemical localization of tyrosine hydroxylase in prenatal rat brain. I. Early ontogeny. *Journal of Comparative Neurology*, **199**, 233–53.

Specht, L. A., Pickel, V. M., Joh, T. J. & Reis, D. J. (1981b). Light-microscopic immunocytochemical localization of tyrosine hydroxylase in prenatal rat brain. I. Late ontogeny. *Journal of Comparative Neurology*, **199**, 255–76.

Studholme, K. M., Yazulla, S. & Phillips, M. (1987). Interspecific comparisons of immunohistochemical localization of retinal transmitters in four species of bats. *Brain Behaviour Zoology*, **30**, 163–73.

Teitelman, G., Jaeger, C. B., Albert, V., Joh, T. J. & Reis, D. J. (1983). Expression of amino acid decarboxylase in proliferating cells of the neural tube and notochord of developing rat embryo. *Journal of Neuroscience*, **3**, 1379–88.

Tennyson, V. M., Barrett, R. E., Cohen, G., Cote, L., Heikkila, R. & Mytilineou, C. (1972). The developing histochemistry, electronmicroscopy, endogenous dopamine levels, and 3[H] dopamine uptake. *Brain Research*, **46**, 251–85.

Tennyson, V. M., Mytinileou, C. & Barrett, R. E. (1973). Fluorescence and electron microscopic studies of the early development of the substantia nigra and area ventralis tegmenti in the fetal rabbit. *Journal of Comparative Neurology*, **149**, 233–58.

Ugrumov, M. V., Tixier-Vidal, A., Taxi, J., Thibault, J. & Mitskevich, M. S. (1989a). Ontogenesis of tyrosine hydroxylase-immunopositive structures in the rat hypothalamus. An atlas of neuronal cell bodies. *Neuroscience*, **29**, 135–56.

Ugrumov, M. V., Tixier-Vidal, A., Taxi, J., Thibault, J. & Mitskevich, M. .S. (1989b). Ontogenesis of tyrosine hydroxylase-immunopositive structures in the rat hypothalamus. Fiber pathways and terminal fields. *Neuroscience*, **29**, 157–66.

Verney, C., Gaspar, P., Febrvet, A. & Berger, B. (1988). Transient tyrosine hydroxylase-like immunoreactive neurons contain somatostatin and substance P in the developing amygdala and bed nucleus of the stria terminalis of the rat. *Developmental Brain Research*, **42**, 45–58.

Verney, C., Zecevic, N., Nikolic, B., Alvarez, C. & Berger, B. (1991). Early evidence of catecholaminergic cell groups in 5- and 6-week-old human embryos using tyrosine hydroxylase and dopamine-β-hydroxylase immunocytochemistry. *Neuroscience Letters*, **131**, 121–4.

Versaux-Botteri, C., Verney, C., Zecevic, N. & Nguyen-Legros, J. (1992). Early appearance of tyrosine hydroxylase immunoreacivity in the retina of human embryos. *Developmental Brain Research*, **69**, 283–7.

Voorn, P., Kalsbeek, A., Jorritsma-Byham, B. & Groenewegen, H. J. (1988). The pre- and postnatal development of the dopaminergic cell groups in the ventral mesencephalon and the dopaminergic innervation of the striatum of the rat. *Neuroscience*, **25**, 857–77.

Wulle, I. & Schnitze, J. (1989). Distribution and morphology of tyrosine hydroxylase-immunoreactive neurons in the developing mouse retina. *Developmental Brain Research*, **48**, 59–72.

Yotsumoto, I. & Nomura, Y. (1981). Ontogenesis of the dopamine uptake into P2 fractions and slices of the rat brain. *Japanese Journal of Pharmacology*, **31**, 296–300.

18
Hypothalamic catecholaminergic systems in ontogenesis: development and functional significance

M. V. Ugrumov

Introduction

Over recent years, numerous data have been accumulated on the important functional significance of the hypothalamic catecholamines (CAs) adrenaline, noradrenaline (NA) and dopamine (DA), for neuroendocrine regulations (Weiner, Findell & Kordon, 1988; Collu, 1989). Catecholamines are either synthesized in the hypothalamus, or transferred via axons from mesencephalon, pons and medulla oblongata (Hökfelt et al., 1984). In adults, CAs provide regulating reversible action on the targets at different hierarchical levels of the neuroendocrine system. They play a role as: 1) neurotransmitters controlling the functional activity of the hypothalamic neurosecretory neurons; 2) neuromodulators, e.g. regulating the release of neurohormones from the adjacent neurosecretory axons in the median eminence; and 3) neurohormones transferred via the hypothalamo-hypophysial portal circulation to the target adenohypophysial cells (Weiner et al., 1988; Collu, 1989).

In ontogenesis, CAs influence the development, initially, of the whole embryo and, later, of the central nervous system including the hypothalamus long before the onset of the neuroendocrine functions (Lauder, 1983; Buznikov, 1991). In this case, CAs provide irreversible organizing effects on the targets over critical periods (Lauder, 1983; Swaab & Mirmiran, 1986; Jarzab et al., 1990b). In turn, the differentiation of CA neurons and expression of their phenotype are not strictly predetermined genetically, being under neurohormonal control (Lauder, 1983; De Vitry et al., 1991).

Thus, the purpose of this review is to summarize the data on: 1) the development of the hypothalamic CA system; 2) hormonal regulation of the CA system development; and 3) the functional significance of the hypothalamic CA system in the developing organism.

Architectonics of the hypothalamic catecholamine system

Cell bodies

A general review of the development of CA neurons in the brain of mammals is presented in the previous chapter. Here, we confine ourselves to a more detailed analysis of the development of the hypothalamic CA cell groups and their functional significance.

Whole hypothalamus until E18. Rare tyrosine hydroxylase immunopositive (THi) and fluorescent cells were first observed in the marginal zone of the rat hypothalamus at the 12–13th embryonic day* (E) (Fig. 18.1) (Schlumpf et al., 1980; Specht et al., 1981; Ugrumov et al., 1989a). One day later, the fluorescent neurons are most abundant in small accumulations in the posterior hypothalamus around the 3rd ventricle (Schlumpf et al., 1980). At E15–E16, the THi and aromatic L-amino acid decarboxylase immunopositive (AADCi) neurons are present in large, bilateral dorsomedial accumulations in the caudal hypothalamus (Fig. 18.2) (Daikoku et al., 1986; Ugrumov et al., 1989a). Although there is overlap in their distribution, the number of AADCi neurons and the area of their distribution exceed those of THi neurons (Fig. 18.2(d)) (Jaeger, 1986). These data suggest the existence of neurons expressing AADC, but lacking TH

---
* The day of conception is considered as the 1st embryonic day (E1).

(Jaeger, 1986). At E15–E16, as earlier, small accumulations of fluorescent neurons occupy the dorsomedial position around the 3rd ventricle (Fig. 18.2(d)) (Schlumpf et al., 1980; Borisova et al., 1991). First at E15, rare DA immunopositive (DAi) cells appear in the rostral hypothalamic region around the 3rd ventricle (Fig. 18.2(c)) (Kalsbeek, Voorn & Buijs, 1992).

At E17 additional THi and AADCi cells are detected at the level of the preoptic area. However, AADCi cells are mainly localized lateral to the less numerous THi cells (Jaeger, 1986). From E18 until puberty, CA neurons become widely distributed throughout the hypothalamus concentrating in certain neurosecretory nuclei (Figs. 18.3, 18.4, 18.5) (Ugrumov et al., 1989a; Borisova et al., 1991), as in adults (Hökfelt et al., 1984).

Septo-preoptic area: E18-puberty. At E18, rare THi neurons are seen in the organum vasculosum of the lamina terminalis and the diagonal band (Fig. 18.3(b)). More caudally, rare THi neurons in small clusters are scattered in the medial and lateral preoptic area, as well as along the lateroventral surface of the hypothalamus and above the optic chiasma (Fig. 18.3(c)). At E20, THi neurons become evident in accumulations around the 3rd ventricle. The majority of THi neurons are localized between the optic nerves and chiasma ventrally, and the anterior commissure dorsally (Fig. 18.4(b),(c)). According to the generally accepted classification by Hökfelt et al. (1984), these neurons belong to the A15 group.

By E20, the number of AADCi neurons in the septo-preoptic area increases significantly compared to E15. The highest concentration of AADCi neurons lies interior to the anterior commissure (corresponding to A15 group). Moreover, AADCi cells are widely distributed along the lateroventral surface of the hypothalamus (Fig. 18.4(b),(c)) (Jaeger, 1986). From E18 onward, occasionally fluorescent cells were observed scattered around the 3rd ventricle and in the medial preoptic area (Figs. 18.3(c), 18.4(c)), as well as in the rostral portion of the paraventricular nucleus (Figs. 18.4(c), 18.5(c)) (Borisova et al., 1991).

Anterior hypothalamus and retrochiasmatic region: E18-puberty. At E18, a number of THi neurons is

Fig. 18.1. E12/13

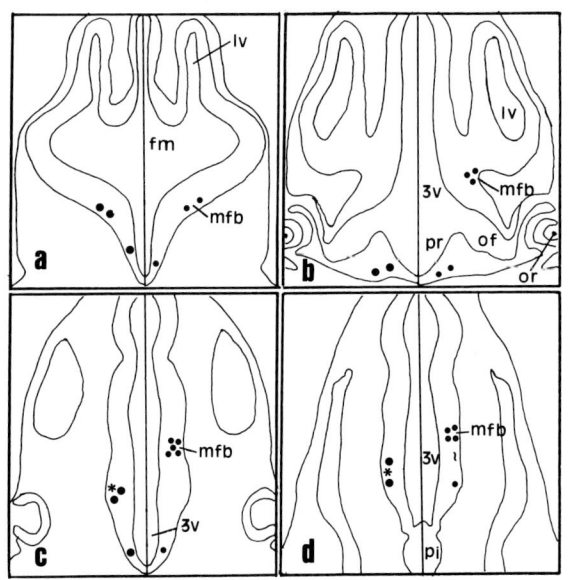

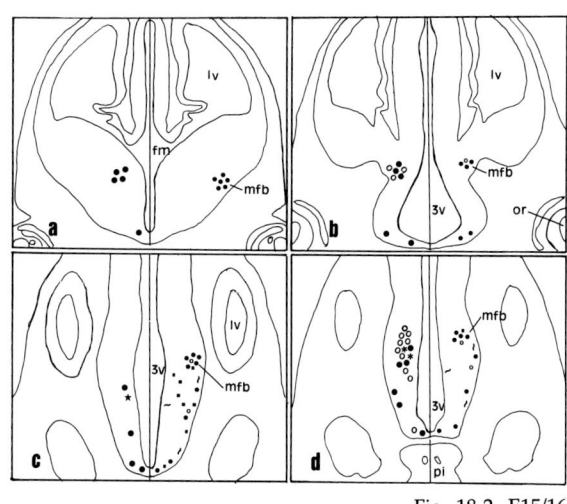

Fig. 18.2. E15/16

Figs. 18.1–18.5. Schematic representations of catecholamine cell bodies (left sides) and fibers (right sides) in the rat hypothalamus at the 12–13th (Fig. 18.1), 15–16th (Fig. 18.2), 18–19th (Fig. 18.3), and 20–21st (Fig. 18.4) fetal days, as well as in postnatal rats (Fig. 18.5). The atlas summarizes the most principal immunocytochemical and histochemical data on the distribution of catecholamines and their synthetic enzymes in the developing hypothalamus (modified after: Schlumpf et al., 1980; Foster et al., 1985; Jaeger, 1986; Ugrumov et al., 1989a, b; Borisova et al., 1991; Kalsbeek et al., 1992). The frontal sections are placed in alphabetical order from rostral to caudal hypothalamic levels. Large dots, THi cells; small dots, THi fibers; large circles, AADCi cells; small circles, AADCi fibers; squares, PNMTi fibers; asterisks and wavy strings, fluorescent cell bodies and fibers, respectively; large stars, DAi cell bodies; small stars, DAi fibers.

Fig. 18.3. E18/19.

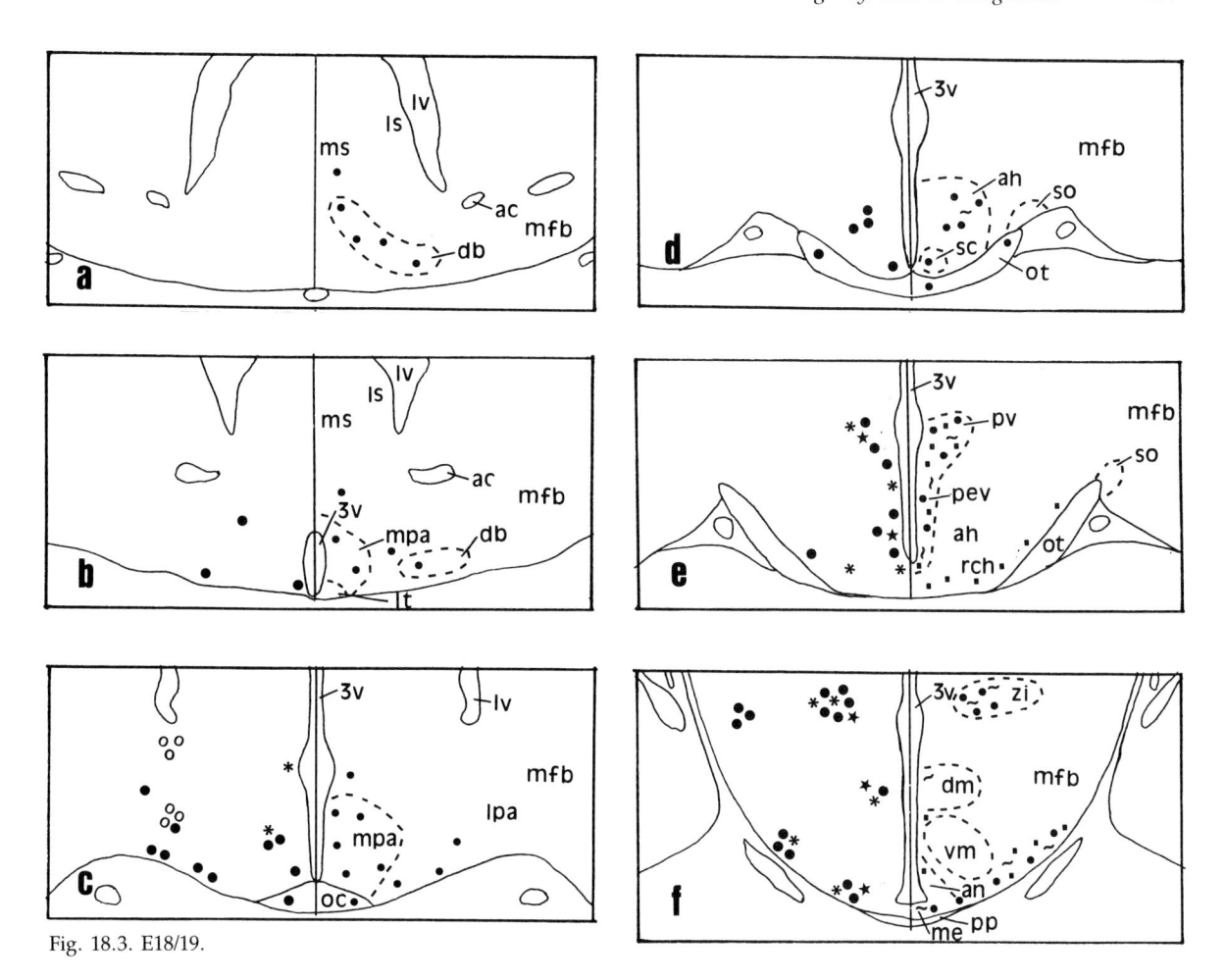

found in the anterior hypothalamic periventricular (Pev) and paraventricular (pv) nuclei (Fig. 18.3(d),(e)) (corresponding to A14 group). Rare neurons are also observed within the suprachiasmatic nucleus (sc), optic chiasm and optic tracts (Fig. 18.3(d),(e)). From E20 onwards, the major accumulation of THi neurons (A14 cell group) extends along the 3rd ventricle in ventrodorsal direction occupying the periventricular nucleus and partly overlapping the paraventricular nucleus, anterior hypothalamic nucleus, suprachiasmatic and ventral retrochiasmatic region (Fig. 18.4(d),(e), 18.5(d),(e)), (Ugrumov et al., 1989a). In neonates, the periventricular accumulation of THi neurons spreads towards the supraoptic nucleus.

At E20, AADCi neurons are located in the paraventricular nucleus, anterior hypothalamic nucleus and periventricular nucleus (corresponding to A14 cell group), as well as in the close vicinity of the supraoptic nucleus (Fig. 18.4(d),(e)). At E18, the fluorescent neurons are mainly localized in the paraventricular nucleus and periventricular nucleus (Fig. 18.3(d), (e)) (A14 cell group). Occasionally, neurons are scattered in the anterior hypothalamic nucleus and retrochiasmatic region (Fig. 18.3(d),(e)) (Borisova et al., 1991). The number of fluorescent neurons and the area of their distribution become maximal at E20, though the intensity of fluorescence increases continuously until P21* (Fig. 18.4(d),(e), 18.5(d),(e)) (Loizou, 1971; Borisova et al., 1991). The histofluorescent data are in agreement with the observations of relatively numerous DAi cells within the paraventricular nucleus and periventricular nucleus

---

* The day of parturition is considered as the 1st postnatal day (P1).

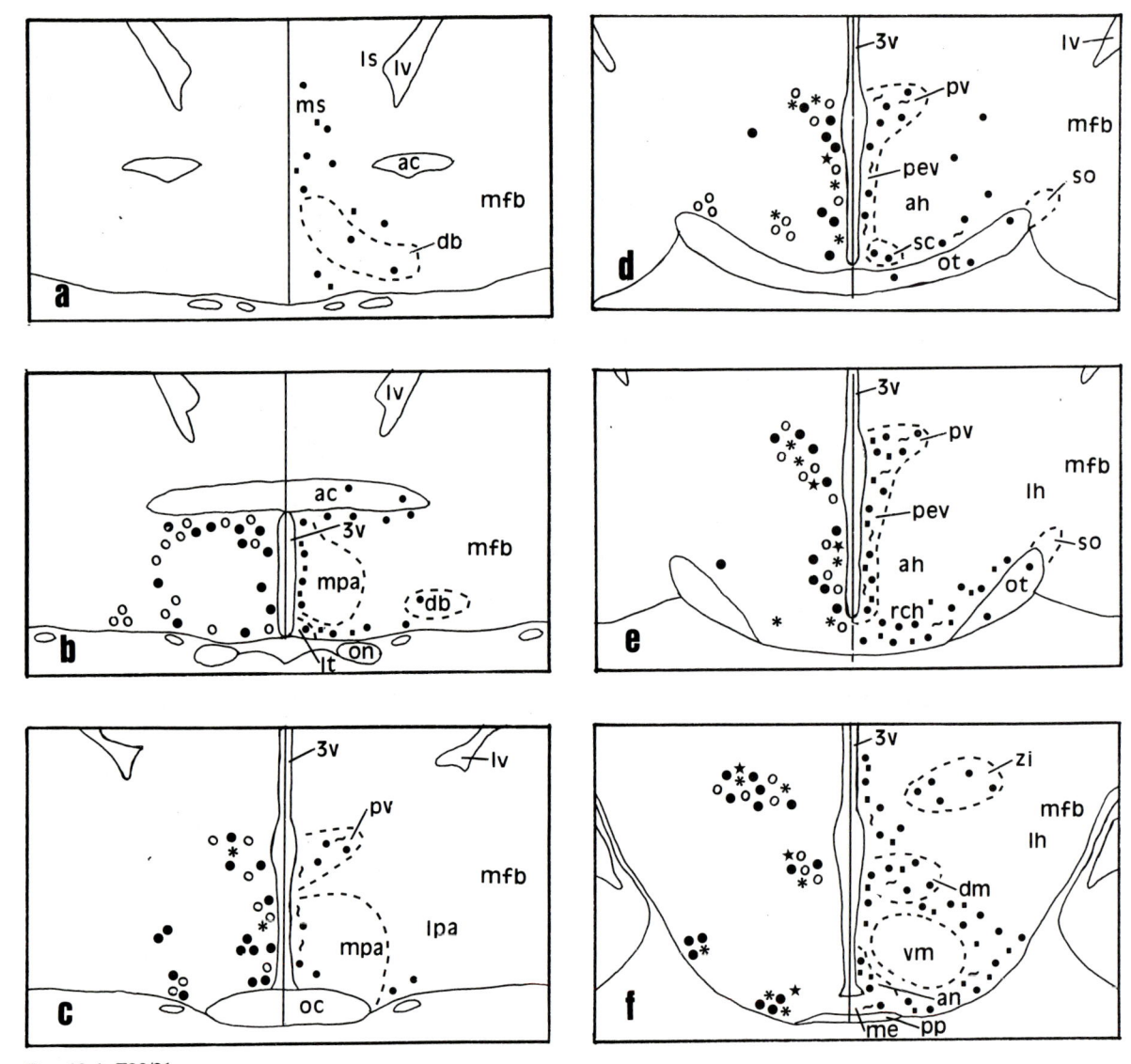

Fig. 18.4. E20/21.

(corresponding to A14 cell group) from E17–E18 (Fig. 18.3e) onwards (Kalsbeek et al., 1992).

Infundibular and postinfundibular regions: E18-puberty. At E18, a considerable number of THi neurons are concentrated at infundibular and postinfundibular levels. Their major, dorsal clusters occupy the zona incerta (Fig. 18.3(f)) (A13, A11 cell groups). With age, THi neurons of this accumulation increase progressively in number (Fig. 18.4(f)). In neonates, this cluster extends to a massive periventricular accumulation of THi neurons, on the one hand, and to the dorsomedial nucleus and posterior hypothalamic nucleus, on the other (Fig. 18.5(f)).

Less numerous THi neurons gave rise to bilateral ventral accumulations in the mediobasal hypothalamus immediately lateral to the median eminence. These clusters are observed only in perinatal rats (E18–P3) (Figs. 18.3(f), 18.4(f)). From P3 onwards, only few THi neurons remain at the ventral surface of the mediobasal hypothalamus while similar neuronal accumulations appear concomitantly in the arcuate nucleus (Fig. 18.5(f)) (corresponding to A12 cell group). According to Piotte et al. (1988), THi neurons

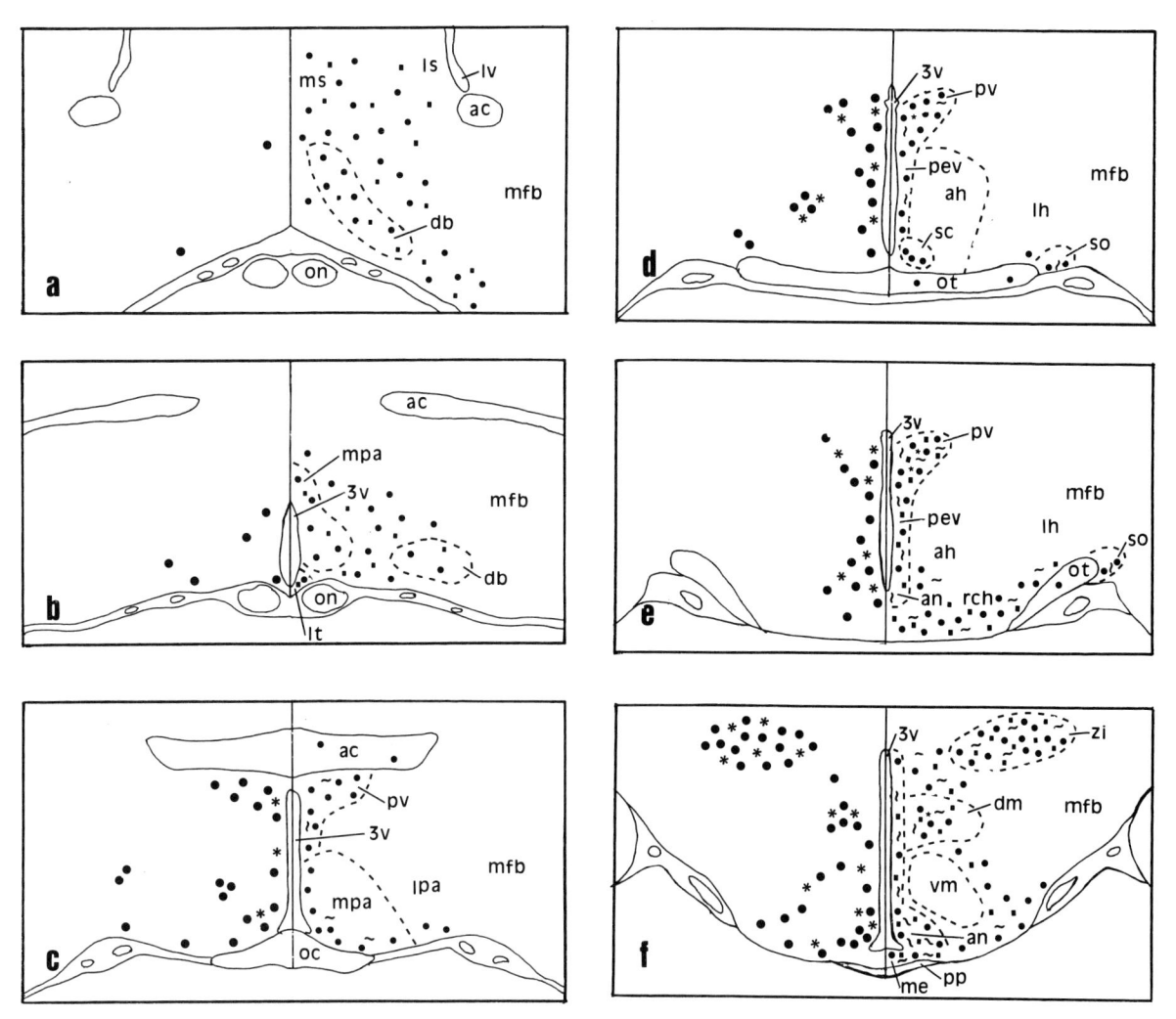

Fig. 18.5. Postnatal.

occupy their definitive position in the rat arcuate nucleus not earlier than at P15. It is still uncertain whether this is a result of retarded neuronal migration or of delayed expression of TH synthesis.

Similarly to THi cells, the major wedge-shaped cluster of AADCi neurons is related to the zona incerta at E20 (Fig. 18.4(f)) (A13, A11 cell groups). This accumulation extends ventrally to the dorsomedial nucleus and periventricular nucleus (Fig. 18.4(e), (f)) (Jaeger, 1986). Phenylethanolamine N-methyltransferase immunopositive (PNMTi) neurons are never observed in the hypothalamus in perinatal rats (Foster et al., 1985).

From E18 until puberty, the most prominent accumulations of fluorescent cell bodies are observed in the zona incerta and dorsomedial nucleus (corresponding to A13, A11 cell groups), dorsally, as well as in the ventrolateral portion of the arcuate nucleus and under the medial forebrain bundle (mfb), ventrally (A12 cell group). Occasionally, fluorescent neurons are found scattered around the ventromedial nucleus but never within the nucleus (Figs 18.3(f), 18.4(f), 18.5(f)) (Loizou, 1971; Borisova et al., 1991).

From E17–E18 onwards, the DAi cells are also detected in the zona incerta and the dorsomedial nucleus (corresponding to A13 and partly A11 cell groups), as well as in the mediobasal hypothalamus lateral to the median eminence (Figs 18.3(f), 18.4(f)) (Kalsbeek et al., 1992).

Thus, CA neurons appear long before birth.

Over perinatal period they occupy their definitive positions concentrating mainly in the zona incerta, periventricular nucleus and arcuate nucleus.

Nerve fibers

Whole hypothalamus until E18. THi fibers, apparently belonging to the neurons of the mesencephalon and pons (Specht *et al.*, 1981), first enter the hypothalamus via the mfb as early as at E13 (Fig. 18.1) (Ugrumov *et al.*, 1989b). The concentration of THi fibers in the mfb increases progressively. Some fibers leave the bundle projecting to the diagonal band, septum, anterior commissure, optic chiasm and tracts, paraventricular nucleus, as well as to the median eminence, zona incerta, dorsomedial nucleus and other target-regions (Figs. 18.1–18.5) (Ugrumov *et al.*, 1989b). Moreover, a prominent number of THi fibers arise from the hypothalamic neurons. It is obvious that the places of high concentrations of THi neurons become occupied by extensive networks of THi fibers (Figs. 18.1–18.5). However, similar networks are also observed in the hypothalamic regions being poor in THi neurons (Ugrumov *et al.*, 1989b).

In addition to THi fibers, AADCi and PNMTi fibers enter the marginal hypothalamic region via the mfb from E16 (Fig. 18.2) and E14, respectively. Two days later, these fibers at first sprout to the medial retrochiasmatic region (Foster *et al.*, 1985; Jaeger, 1986).

At E12 and E14, fluorescent fibers are occasionally observed in the caudodorsal hypothalamus (Fig. 18.1(*d*), close to mammillotegmental tract (Schlumpf *et al.*, 1980). Two days later, the fibers are regularly observed within the marginal and periventricular zones (Fig. 18.2(*c*),(*d*)). Some fibers are directed towards the ventricular lumen (Borisova *et al.*, 1991). From E18 onwards, CA fibers are widely distributed in the hypothalamus (Figs. 18.3–18.5).

Septo-preoptic area: E18-puberty. THi fibers are scattered in the diagonal band and septum from E18 (Fig. 18.3(*a*),(*b*)). Over the subsequent perinatal period, their concentration increases progressively (Figs. 18.4(*a*),(*b*); 18.5(*a*),(*b*)) (Ugrumov *et al.*, 1989b). At E21, PNMTi fibers become evident in the medial septum and the organum vasculosum of the lamina terminalis (Fig. 18.4(*a*),(*b*)). By birth, they extend to the lateral septum and the diagonal band (Fig. 18.5(*a*),(*b*)) (Foster *et al.*, 1985).

Anterior hypothalamus and retrochiasmatic region: E18-puberty. At E18, THi fibers first appear in the suprachiasmatic nucleus (Fig. 18.3(*d*)) and their number remains insignificant until birth (Fig. 18.4(*d*)). Over the first nine days of postnatal life, the concentration of THi fibers increases abruptly (Fig. 18.5(*d*)) (Ugrumov *et al.*, 1989b) whereas they have practically disappeared by adulthood (Van den Pol & Tsujimoto, 1985). These data suggest either the provisional character of THi fibers or the transient expression of TH synthesis in the corresponding neurons. Although a few THi fibers project to the paraventricular nucleus even in fetuses (Figs. 18.3(*e*), 18.4(*d*),(*e*)), their concentration becomes considerable only after birth (Fig. 18.5(*d*),(*e*)) (Ugrumov *et al.*, 1989b).

PNMTi fibers are first observed in the anterior hypothalamus and the retrochiasmatic region at E19 (Foster *et al.*, 1985). They are present in the paraventricular nucleus and periventricular nucleus extending laterally to the mfb (Fig. 18.3(*e*)). Over the subsequent perinatal period, the concentration of PNMT fibers increases significantly, particularly in the paraventricular nucleus (Figs. 18.4(*e*), 18.5(*e*)) (Foster *et al.*, 1985).

From E18 onwards, the most numerous fluorescent fibers are concentrated in the paraventricular and periventricular nuclei (Figs. 18.3(*e*), 18.4(*d*),(*e*), 18.5(*d*),(*e*)). In the paraventricular nucleus, fluorescent fibers either surround so-called magnocellular neurons or project to the 3rd ventricle (Khachaturian & Sladek, 1980; Borisova *et al.*, 1991). In the periventricular nucleus, the fibers are mainly oriented in parallel to the 3rd ventricle (Borisova *et al.*, 1991). Fluorescent fibers are also visible in the supraoptic nucleus from E22 (Hyyppä, 1969). The number of the fibers and the intensity of their fluorescence increase permanently during subsequent pre- and postnatal life (Figs. 18.4(*d*),(*e*), 18.5(*d*),(*e*)) (Hyyppä, 1969; Borisova *et al.*, 1991).

Infundibular and postinfundibular regions: E18-puberty. THi fibers reach the median eminence and terminate there at E18 (Fig. 18.3(*f*)). Their number increases continuously (Fig. 18.5(*f*)), reaching maximum around puberty (Ugrumov, 1991, 1992). PNMTi fibers are first observed at infundibular and postinfundibular levels at E19 (Fig. 18.3(*f*)) (Foster *et al.*, 1985). At the infundibular level, they are distributed either around the 3rd ventricle or along the ventral surface of the hypothalamus extending from the arcuate nucleus to the mfb (Fig. 18.3(*f*)). At the postinfundibular level, the accumulation of PNMTi fibers overlaps the zona incerta. Over the subsequent perinatal period, the PNMTi fibers occupy not only the zona incerta, but also the dorsomedial nucleus, arcuate nucleus and the median eminence (Figs. 18.4(f), 18.5(f)) (Foster et al., 1985).

From E18 onwards, the fluorescent fibers are mainly accumulated in the same regions as the cell bodies, i.e., in the zona incerta, periventricular nucleus and the mediobasal hypothalamus. Although occasional fibers are seen in the rat median eminence as early as at E18 (Smith & Simpson, 1970; Borisova et al., 1991), their concentration in the external zone become appreciable only several days after birth (Fig. 18.5(f)) (Ibata et al., 1982), reaching adult levels around puberty (Hyyppä, 1969; Smith & Simpson, 1970; Loizou, 1971). In addition to the external zone, fluorescent fibers are regularly observed in the internal zone and hypophysial stalk, projecting towards the posterior lobe and adenohypophysis (Smith & Simpson, 1970; Loizou, 1971). Fluorescent fibers first enter the neurointermediate lobe at P2, and their concentration rise permanently until P14 (Davis et al., 1984).

In the available literature, there are only few data on the innervation of the hypothalamus by DAi fibers in ontogenesis. Thus, DAi fibers were first observed in the rat hypothalamus at E21. The network of DAi fibers by P14 became as widely distributed as in adults, but the density of the network even at P21 remained less extensive than in adulthood (Kalsbeek et al., 1992).

Thus, from E12 until puberty the hypothalamus becomes innervated by CA fibers belonging either to the hypothalamic neurons or arriving via the mfb from outside the hypothalamus. Over this period, CA fibers innervate such target regions as the diagonal band, septum, SCN, paraventricular nucleus, etc. and sprout to the median eminence and the 3rd ventricle.

Morpho-functional characteristics of hypothalamic catecholamine neurons

Genesis and morphology

It is generally accepted that differentiation of neurons begins after the last mitotic division of the ventricular (matrix) cell precursors. According to long-survival [³H]thymidine autoradiography combined with TH immunocytochemistry, three populations of hypothalamic THi neurons are distinguished: the earliest population in the zona incerta, the middle population in the periventricular nucleus (Borisova et al., unpublished observations), and the latest one in the arcuate nucleus (Fig. 18.6) (Borisova et al., 1993).

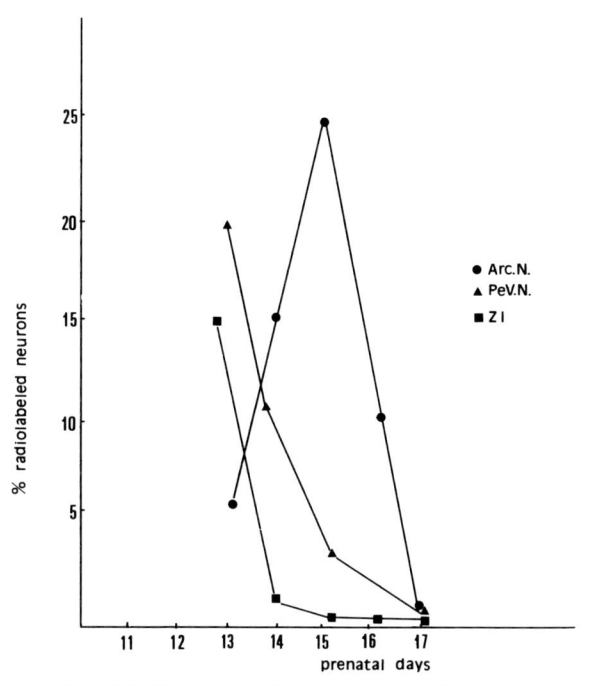

Fig. 18.6. Percentage of tyrosine hydroxylase and radioactively labeled cells being born on fetal days 13, 14, 15, and 16 in the zona incerta (ZI), periventricular (PeV.N.), and arcuate (Arc.N) nuclei of rats. The curve represents the percentage of the cells forming on each day of gestation.

Light microscopy. THi neurons undergo striking morphological modifications in the course of differentiation. The first small THi neurons, observed at E13, possess one or two short unbranched processes terminating with growth cones. Two days later, multipolar neurons appear, additionally. From E18 onwards, THi cell bodies increase in size, and their processes in length (Ugrumov et al., 1989a). After birth, three neuronal types might be distinguished (Ugrumov et al., 1989a). The first one consists of small uni- and bipolar neurons with short and narrow unbranched processes. These neurons are localized ventrally, mainly in the mediobasal hypothalamus including the arcuate nucleus. The second type, composed of large multipolar neurons with long ramified processes, is distributed dorsally, in the zona incerta, dorsomedial nucleus and paraventricular nucleus (Ugrumov et al., 1989a). A third group of rather large bipolar neurons of the third type occupies the periventricular nucleus.

The populations mentioned above appear to differ not only in their morphology and the time of origin, but also in the delay between the neuron origin and expression of specific synthesis. Thus, the neurons of the dorsal clusters (second type), express TH and CA synthesis just after their origin. Conversely, the neurons of the arcuate nucleus (first type) seem to be characterized by a lag period, at least for several days, between the onset of TH and CA synthesis (Borisova et al., 1993).

Electron microscopy. THi neurons of the suprachiasmatic nucleus (Ugrumov et al., 1994) and arcuate nucleus (Piotte et al., 1988) belonging to the first type (see above) have been studied at the electron microscopic (EM) level. The neurons of the suprachiasmatic and arcuate nuclei are oval in shape and small in size both in fetuses and postnatal rats. They possess relatively large nuclei and scanty cytoplasm, as well as poorly developed Golgi complex and granular endoplasmic reticulum (Fig. 18.7(a)) (Ugrumov et al., in press). Over the perinatal period, their size slightly increases, their nuclei become indented, dense core vesicles appear in the area of the moderately developed Golgi complex, showing the onset of the secretory activity (Fig. 18.7(a)) (Ugrumov et al., in press). THi neurons of the arcuate nucleus, being studied only in postnatal rats, are similar in appearance to those of the suprachiasmatic nucleus. Piotte and co-authors (1988) emphasized that these cells remained unchanged from P2 until puberty, suggesting the onset of their functioning in newborns.

Expression of biochemical phenotype

Tyrosine hydroxylase. TH activity was evaluated biochemically only in developing mice. This activity was detected in the mouse hypothalamus as early as at E13. Then, it increased 12-fold by the end of fetal life and 4-fold postnatally, until puberty (Friedman et al., 1989). These data correlate well with the continuous increase of THi cell bodies and fibers in number over the fetal and neonatal periods in rats (2.1. Cell bodies).

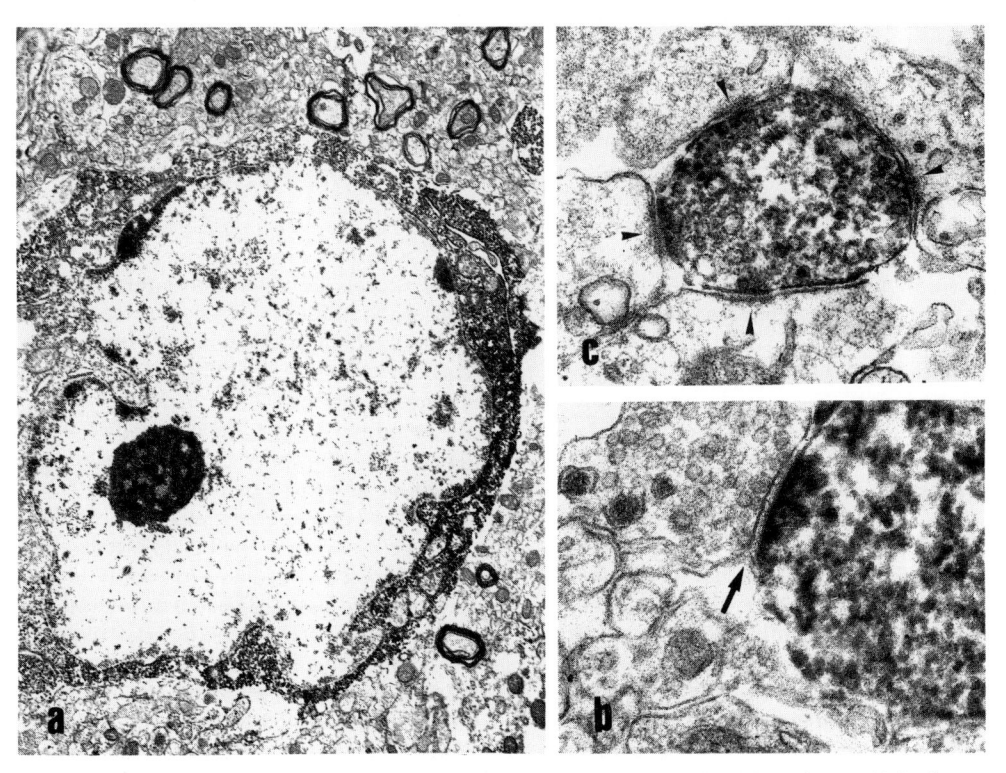

Fig. 18.7. Tyrosine hydroxylase immunopositive (THi) cell body (a), dendrite (b) and axon (c) in the suprachiasmatic nucleus of rats at the 2nd (c) and 21st (a), (b) postnatal days. Arrow, asymmetric synapse between THi dendrite and immunonegative axon; arrowhead, symmetric synapse between THi axon and immunonegative dendrite.

Aromatic L-amino acid decarboxylase. According to immunocytochemical results mentioned above, AADC is first expressed in the hypothalamus as early as at E16. Over subsequent prenatal period, the number of AADC-containing neurons increases progressively and is followed by their wide distribution throughout the hypothalamus (Jaeger, 1986).

Phenylethanolamine N-methyltransferase. In the only available immunocytochemical study, no PNMT-containing neurons were observed in the developing hypothalamus of rats (Foster et al., 1985). Still, in rats PNMT is recognized in the nerve fibers reaching the hypothalamus at least from E13 (Foster et al., 1985).

It should be mentioned that the discrepancy in the earliest appearance among the enzymes for the CA synthesis might be explained by different sensitivities of the applied techniques. Moreover, according to Teitelman's et al. (1983), AADC is expressed prior to the last mitotic division of the ventricular cell precursors, while the TH synthesis begins only after arrival of differentiating neurons to their final location.

Catecholamines. Histofluorescent techniques have showed the first CA-containing neurons in the hypothalamus as early as at E12 (Schlumpf et al., 1980). Their number increased continuously until the end of fetal life. The fluorescence of individual neurons also rose progressively up to E20, followed by its drop around P9 and subsequent increase by P21 (Borisova et al., 1991).

We failed to find in the available literature the biochemical data on the DA content in the hypothalamus of young fetuses. According to our high performance liquid chromatography data, at E21 the DA concentration is as high as at P11 (Trembleau et al., 1991). On the contrary, the concentration of NA increases fourfold over the perinatal period, exceeding considerabaly the DA level. Adrenaline remains at the lowest level compared with other CAs both in fetuses and neonates (Trembleau et al., 1991). Postnatally, the content of NA in the hypothalamus increases continuously up to age 1.5–2 months (Weiner & Ganong, 1972; Nomura, Naiton & Segawa, 1976). Over the same period, the DA level in diencephalon practically does not change (Nomura et al., 1976).

From the comparison of the immunocytochemical data, it follows that there is a delay of 3–4 days between the first appearance of THi and DAi cells in the developing hypothalamus (see section on cell bodies). Thus, a few weakly stained DAi cells are first observed in the hypothalamus at E15. Over subsequent 3–5 days, their number increases progressively, but, still, the cells remain weakly immunostained (Kalsbeek et al., 1992).

Catecholamines are partly transferred via axons from the hypothalamus to the neurointermediate lobe, being detectable there from P2 (Davis et al., 1984; Eberle, 1988). The CA content (ng/lobe) and concentration (ng/mg protein) increase continuously in the neurointermediate lobe until P14. From this time, the concentration of CAs decreased until the end of the 2nd month, whereas their absolute content is not changed. From birth until adulthood, the concentration and content of DA exceeds those of NA 10 times (Davis et al., 1984).

Uptake and release of catecholamines. The specific uptake of intraventricularly injected [^{3}H]DA by hypothalamic CA neuronal elements is first detected with autoradiography in fetal rats at E18 (Ugrumov, 1992). Further biochemical 'isotopic' study *in vitro* showed the simultaneous onset of the uptake and K$^+$-stimulated Ca^{2+}-dependent release of CAs even earlier, at E16 (Fig. 18.8) (Borisova et al., 1991). The uptake increases 2.6 times over subsequent two days and, then, doubles between E20 and P9, reaching the adult level. The K$^+$-stimulated release of CAs increases considerably by E17, remaining at the same level in older fetuses and neonates. It doubles between P9 and P45 (Fig. 18.8) (Borisova et al., 1991).

Genetic sexual dimorphism

According to recent data, sexual dimorphism of DA neurons is manifested initially in the diencephalon without any influence of sexual steroids (SS) (Reisert, Engele & Pilgrim, 1989). Thus, the DA neurons taken into tissue culture from rat fetuses at E14, i.e. before the onset of androgen secretion, show more intense outgrowth of processes and higher uptake of [3H]DA in females, than in males. The neurons, taken into tissue culture three days later, show only the increased [^{3}H]dopamine uptake. Moreover, the hypothalamic THi neurons of fetal males exceed in size those of fetal females both *in vivo* and in tissue culture. These differences become less prominent in newborns and disappear in adults (Kolbinger et al., 1991). Kolbinger et al. (1991) hold the opinion that the initial sexual dimorphism in the CA neurons might be a result of the earlier origin of these neurons in females than in males (Reisert et al., 1989).

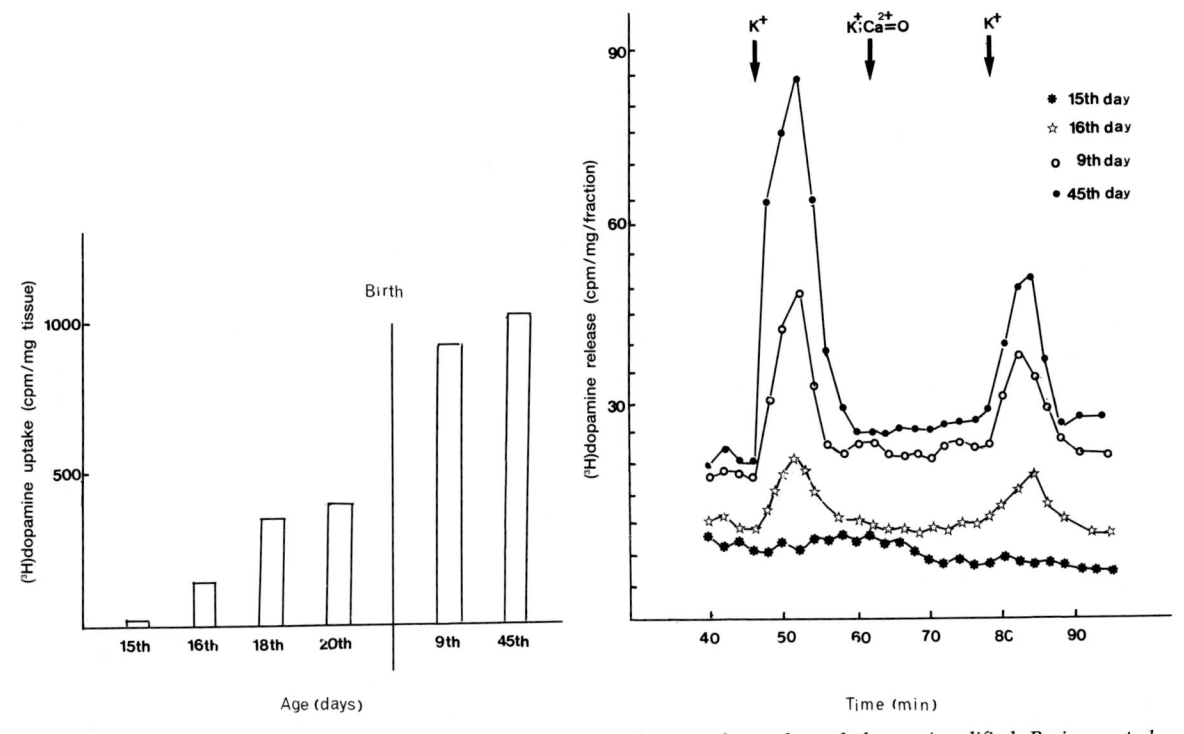

Fig. 18.8. Specific [^{3}H]dopamine uptake and K-stimulated release in the rat hypothalamus (modified, Borisova et al., 1991).

Afferent innervation
Currently, only two EM immunocytochemical studies have been devoted to the innervation of the hypothalamic CA neurons in ontogenesis (Piotte et al., 1988; Ugrumov et al., 1994).

Suprachiasmatic nucleus. Immunonegative axons possessing synaptic and dense core vesicles make specialized contacts with both THi cell bodies and dendrites in the sprachiasmatic nucleus of fetuses and postnatal rats (Ugrumov et al., 1994). Initially at E22, these contacts look like the immature synapses (presynapses) (Fig. 18.9(a)) showing small accumulation of synaptic vesicles at the presynaptic membrane, as well as slight thickening of pre- and postsynaptic membranes. After birth, the specialized neuronal contacts gain the ultrastructural characteristics, first, of symmetric (Gray-type II) and, then, also of asymmetric (Gray-type I) synapses (Figs. 18.7(b), 18.9(c)) (Ugrumov et al., in press). At P21, the innervation of THi neurons is characterized by high diversity that is manifested, e.g. by the appearance of 'crest' synapses with the participation of THi postsynaptic spines. Moreover, single THi cell bodies and dendrites become innervated by several immunonegative axons giving rise to different types of synapses: symmetric and asymmetric, with one or several active zones (Fig. 18.9) (Ugrumov et al., 1994). Although the nature of immunonegative axons remains uncertain, it is necessary to take into account that the innervation of THi neurons coincides with the establishment of the retinohypothalamic (Klein, Moore & Reppert, 1991) and serotoninergic (Ugrumov et al., 1994) inputs to the suprachiasmatic nucleus.

Arcuate nucleus. From P2 onwards, the THi neurons of the arcuate nucleus are innervated by immunonegative axons (Piotte et al., 1988). In newborns, these axons contact synaptically only with THi perikarya, while by puberty synapses with THi dendrites predominate (Piotte et al., 1988).

Efferent innervation
Only a few papers are available on the topographic relations of CA fibers with their targets in the developing hypothalamus.

Catecholaminergic systems in ontogenesis 445

innervated by THi fibers triple over the same period (Wray & Hoffman, 1986).

Suprachiasmatic nucleus. According to EM data, the suprachiasmatic nucleus of rats is innervated by THi axons at least from E22 (Ugrumov et al., 1994). In fetuses, the THi axons make presynapses with immunonegative dendrites (Fig. 18.9(a)) while in neonates, first (P2), symmetric (Gray-type II) (Figs 18.7(c), 18.9(a)) and, then (P9), asymmetric synapses (Gray-type I) appear. The number of synapses and their diversity increase continuously over postnatal life (Fig. 18.9(i)) (Ugrumov et al., 1994). For instance, single THi axons give rise to synapses with several immunonegative dendrites, as described for adults (Klein et al., 1991).

When evaluating the functional significance of THi axons in the developing suprachiasmatic nucleus, it is necessary to take into account that: 1) there is an overlapping in the distribution of the peptidergic neurons–pacemakers of the circadian rhythms (Klein et al., 1991) with the terminal field of THi axons (see above); 2) the onset of the functioning of these neurons (Klein et al., 1991) coincides with the innervation of the suprachiasmatic nucleus by THi axons; and 3) DA provides specific action on certain neurons in fetal suprachiasmatic nucleus via D1 receptors (Weaver, Rivkees & Reppert, 1992).

Arcuate nucleus. From P2 until puberty, THi axons innervate immunonegative neurons in the rat arcuate nucleus giving rise mainly to axodendritic synapses (Piotte et al., 1988). Less numerous axosomatic synapses are also observed but only in newborns (P2). According to Piotte and coauthors (1988), the axosomatic synapses are replaced by the axodendritic ones in the course of dendritic development.

Median eminence. Double labeling technique showed, first, at the light microscopic level (Ibata et al., 1981, 1982) but later also at the EM level (Ugrumov, 1991), that the axoaxonal contacts between LHRH and CA fibers are established in the rat median eminence over the second postnatal week, while the DA 'innervation' of somatostatin axons is delayed for a week (Ibata et al., 1982). The formation of the axoaxonal contacts is supposed to be followed by the onset of the DA regulation of the adenohypophysiotropic neuropeptide release (Ugrumov, 1991), as it occurs in adults (Kuljis & Advis, 1989).

3rd ventricle. THi fibers spread to the periventricular region as early as at E15, while at E18 and E20 they

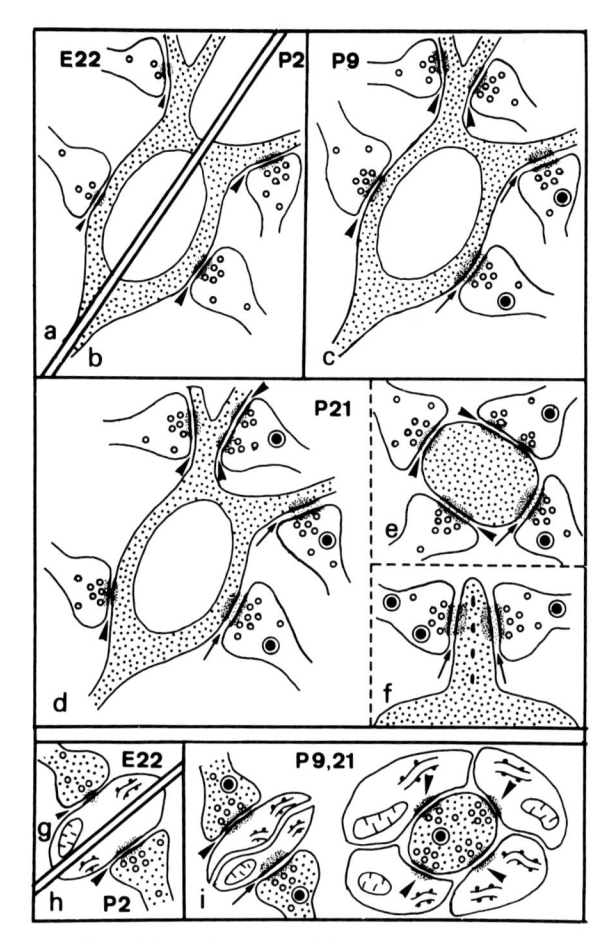

Fig. 18.9. Schematic representation of synaptogenesis in the suprachiasmatic nucleus in rats at the 22nd embryonic day (E22; (a), (g) as well as at the 2nd (P2; (b), (h), 9th (P9; (c), (i) and 21st (P21; (d)–(f), (i) postnatal days. Presynapses and synapses are made either by tyrosine hydroxylase immunopositive (THi) cell bodies and fibers (dots) with immunonegative axons (a)–(f) or by THi axons (dots) and immunonegative dendrites (g)–(i). Small arrowhead, presynapse; large arrowhead, symmetric synapse; arrow, asymmetric synapse.

Septo-preoptic area. Double immunostaining at the light microscopic level shows that from P2 some THi fibers terminate on the luteinizing hormone-releasing hormone (LHRH) neurons in the septo-preoptic region. The number of smooth LHRH neurons innervated by THi fibers remains at a constant level from P2 until P90. Conversely, the frequency of the LHRH neurons possessing spine-like processes and have reached the ventricular lumen, suggesting the CA release into the cerebrospinal fluid. Following this time, the frequency of axoventricular contacts drop being practically absent after birth (Ugrumov et al., 1989b). The cerebrospinal fluid, therefore, is considered as a pathway for CA transfer through the brain in fetuses several days before birth.

Neurohormonal control of the catecholamine system

Hypothalamo-hypophysial (neuro)hormonal factors

According to recent data, the chronic treatment of tissue cultures of the rat embryonic hypothalamus (E15–E16) with DA or apomorphine (DA agonist) increases significantly the number of cells expressing TH and AADC. De Vitry et al. (1991) held the opinion that CAs trigger the differentiation of embryonic cell precursors towards CA neurons by acting via autoreceptors.

Some adenohypophysial factors are also suggested to control the development of the hypothalamic CA system. Thus, coculturing of embryonic hypothalamus with the intermediate lobe resulted in a transient increase of the [^{3}H]DA specific uptake by the hypothalamic tissue. This effect, observed after 6 days of tissue culture, disappears by the 12th day (Charli et al., 1993).

More clear data have been obtained on the imprinting effect of α-melanotropin in the establishment of the feedback regulation of the DA neurons in the arcuate nucleus of neonatal rats (Lichtensteiger & Schlumpf, 1986). This feedback is established between P4 and P8, which coincides with a melanotropin peak in plasma (Davis et al., 1984). The passive immunization with antiserum against melanotropin over this period make the DA neurons non-sensitive to this hormone in adults (Lichtensteiger & Schlumpf, 1986).

Hormones of peripheric endocrine glands

Androgens. Sexual differences in the content of CAs, mainly of NA, are observed in the hypothalamus of rats from P10–P12 onwards: the NA level in females exceeds that in males. These data first suggested the possible role of SS in the development of the CA system (Wilson & Agrawal, 1979; Siddiqui & Gilmore, 1988). This was further confirmed with different models, e.g. neonatally castrated male rats and neonatally androgenized females. The NA content in androgenized females dropped by adulthood to the level observed in intact adult males (Siddiqui & Gilmore, 1988). Conversely, the NA level in neonatally castrated adult males became as high as in intact adult females (Siddiqui & Gilmore, 1988).

In contrast to NA, no clear sexual difference is observed in the hypothalamic DA content in neonates (Wilson & Agrawal, 1979). Neonatal castration of males does not modify the DA content until P120, while at P180 its level highly exceeds that of intact males and neonatally androgenized females (Siddiqui & Gilmore, 1988).

Androgen dependent differences in the CA content were observed not only in the whole hypothalamus, but also in its local regions, e.g. in the arcuate nucleus and median eminence (Crowley O'Donohue & Jacobowitz, 1978; Demarest et al., 1981). Further morphological studies provided additional information on the mechanisms of androgen action on the developing hypothalamic CA system. The DA neuron in the anteroventral periventricular nucleus, being three times as abundant in adult females as in males is a good model to examine this issue (Simerly et al., 1985). The neonatal castration of male rats increases while androgenization of females decreases the number of THi and TH mRNA-containing neurons in the anteroventral periventricular nucleus (Simerly et al., 1985; Simerly, 1989). This means that the differentiation of hypothalamic CA neurons is under androgen control over the critical neonatal period. Although the mechanism of androgen action remains uncertain, these hormones probably either promote the death of DA cells or limit their migration to the anteroventral periventricular nucleus (Simerly et al., 1985; Simerly, 1989).

From the physiological point of view, the DA neurons of the anteroventral periventricular nucleus seem to be involved in the mediation of the androgen effect on phasic gonadotropin secretion. The failure to show phasic luteinizing hormone secretion in response to estrogen and progesterone in adult females after either neonatal androgenization (with subsequent reduction of DA neuron population in the anteroventral periventricular nucleus) or microsurgical lesion of this hypothalamic region favour this hypothesis (Simerly et al., 1985).

Estrogens. The functional significance of estrogens in the development of the hypothalamic CA system is not as clear as that of androgens. The attempts to evaluate their role in the control of TH activity both *in vivo* and *in vitro* did not give any positive results (Friedman et al., 1989).

Thyroid hormones. The role of thyroid hormones in the differentiation of CA neurons has been proved in hypothalamic tissue culture of fetal mice (E16). Triiodothyronine results in an increased size but not number of DA neurons. Moreover, it stimulates growth and arborization of DA processes associated with a rising of [^{3}H]DA specific uptake (Puymirat et al., 1983).

In an *in vivo* study, the pretreatment of rats at P8–P10 with thyroxine resulted in increased levels of NA and DA at P16–P21. By P45, this was followed by decreased NA and DA contents in pretreated animals to their levels in the control animals (Lengvari, Branch & Taylor, 1980). These data point to the transient stimulating effect of thyroid hormones on the brain development.

Hypothyroidism produced by methimazole, applied either over intrauterine development or for three postnatal weeks, did not modify hypothalamic CA metabolism in adulthood (Harris et al., 1987).

Insulin. Hyperinsulinism, produced by daily injections of insulin from P1 to P15, caused a significant increase in the hypothalamic concentrations of NA and DA (Stahl et al., 1986). A relatively high level of NA remained at least until the 5th month of age.

Maternal hormones

In addition to hormones of the developing organism, maternal hormones also contribute to the maturation of the hypothalamic CA system. Thus, maternal prolactin, penetrating to the pups in the course of suckling, provides a long-term effect on DA neurons of the arcuate nucleus (Shyr, Crowley & Grosvenor, 1986). The depletion of maternal prolactin in pups over the neonatal critical period (P2–P5) by daily injections of bromocriptine to lactating females resulted in a depression of DA turnover and in elevation of the plasma prolactin level in adulthood (Shyr et al., 1986).

Thus, hormones control the development of the hypothalamic CA system, modifying neuronal differentiation and metabolism, networks of nerve fibers, synaptic organization, neuroendocrine regulation and behavior. The list of (neuro)hormonal factors being potent in this respect certainly is not limited to those mentioned above.

Functional significance of catecholamines

Neuropeptide gene expression

Over recent years, the attention of neurobiologists has been attracted to the regulatory role of neurohormonal factors including CAs, in the differentiation of the target neurons and in expression of their specific phenotype (De Vitry et al., 1991; Schilling et al., 1991). In order to evaluate the role of CAs in the differentiation of target peptidergic neurons, neuropeptide gene expression has been studied in fetuses and in neonatal rats following chronic pharmacological depletion of CAs (Ugrumov et al., unpublished observations). The vasopressin- and oxytocinergic neurons of the supraoptic nucleus, highly innervated by CA fibers in adults (Sladek & Armstrong, 1987), were considered as a good model to examine this issue. The chronic depletion of CAs did not modify the contents of vasopressin and oxytocin mRNAs in fetuses (E21) whereas in neonates (P9) this resulted in the significant augmentation of both mRNAs levels. These data suggest that CAs inhibit gene expression of vasopressin and oxytocin postnatally, which coincides with synaptogenesis in this region (Boer, 1987). Apparently, CA action on neuropeptide gene expression in neonates is via the CA synaptic input, as it has been demonstrated in adults (Baldino et al., 1988).

Sexual differentiation of the hypothalamus

The role of CAs in 'sexual differentiation' of the hypothalamus is widely discussed over the last decade. Catecholamines have been suggested to mediate the organizing effects of SS during the critical neonatal period, inducing permanent sexual differences in: 1) morphology of the hypothalamic target regions; 2) gonadotropin secretion patterns; and 3) sexual behavior (Table 18.1). The mechanisms of SS and monoamines interaction still remain far from complete understanding.

Morphologically, the dominant role of CAs in sexual differences was proved by the enlargement of 'the sexually dimorphic nucleus of the preoptic area' in adult male and female rats following neonatal (P1–P5) stimulation of β2-adrenergic agonist, salbutamol (Jarzab et al., 1990a). It is noteworthy that the pre- and neonatal treatment of males even with high concentrations of androgens or estrogens failed to increase the volume of this region (Döhler et al., 1984). According to the authors' suggestion, the effect of salbutamol was accounted for either by its promotion of neuron migration to the sexually dimorphic nucleus of the preoptic area, or by preventing neuron death in this region (Jarzab et al., 1990a).

Biochemically, SS controls the number of α- and β-receptors (Jarzab et al., 1990b) as well as influences the metabolism of NA (Reznikov & Nosenko, 1983). In turn, NA influences the concentration of SS

Table 18.1. *Effects of neonatally applied sexual steroids and drugs related to either the physiological action or metabolism of catecholamines (Ugrumov, 1992)*

Experimental model	Physiological test	Drug	Mechanism of action	Effect
Neonatal androgenization of females (Raum & Swerdloff, 1981)	Persistent estrus in adulthood	a) phenoxybenzamine b) phentolamine c) propranolol d) a & c e) b & c f) α-methyl-p-tyrosine	αAN αAN βAN Inhibitor of CA synthesis	↓ ↓ — — — —
Neonatal androgenization of female & gonadectomy in adulthood (Jarzab et al., 1990b)	Female sexual behavior	a) clonidine b) prazosin c) yohimbine	α_2A α_1AN α_2AN	— — —
Intraventricular injection of [³H]testosterone to female at P4 (Raum et al., 1984)	Nuclear accumulation of [³H]estradiol in hypothalamus	a) phenoxybenzamine b) isoproterenol c) isosuprine d) hydroxybenzylpindolol e) a & d f) b & d g) c & d	αAN βA βA βA	↓ ↓ ↓ — ↑ ↑ ↑
Neonatal treatment of females with drugs & gonadectomy in adulthood (Jarzab et al., 1989, 1990b)	LH response to estradiol and progesterone in adulthood	a) clonidine b) prazosin c) yohimbine d) salbutamol e) alprenolol f) isoprenaline	α_2A α_1AN α_2AN β_2A βAN βA	— ↑ ↑ ↓ ↓ ↓
Neonatal treatment of females with drugs & gonadectomy in adulthood (Jarzab et al., 1990b)	female sexual behavior	a) clonidine b) prazosine c) yohimbine d) salbutamol e) isoprenaline	α_2A α_1AN α_2AN β_2A βA	↓ — — ↑ ↑
Neonatal treatment of males with drugs. Castration and estradiol and progressive in adulthood (Jarzab et al., 1990b).	female sexual behavior	a) isoprenaline b) salbutamol c) alprenolol	α_2A β_2A βAN	↑ — —

α – α-receptor; A – agonist; AN – antagonist; β – β-receptor; ↑ – increased; ↓ – decreased; – no effect.

receptors (Blaustein & Turcotte, 1987). Moreover, NA stimulated β-receptors are supposed to prevent androgenization of the hypothalamus via an inhibition either of testosterone aromatization or of estradiol uptake by the neuronal nuclei. The last mechanism is closely related to the SS modulation of the genome (Raum & Swerdloff, 1981; Raum, Marcano & Swerdloff, 1984). Moreover, stimulation of β-receptors intensified female sexual behavior, but it inhibited female type gonadotropin secretion. On the contrary, the blockade of α-receptors increased the luteinizing hormone-surge release, while decreasing female sexual behavior. Activation of α2-receptors suppressed female sexual behavior, though it had no effect on luteinizing hormone release (Table 18.1) (Jarzab et al., 1989).

Physiologically, interfering with the action of CAs and SS on gonadotropin secretion is of particular interest. Thus, gonadotropin secretion is under DA inhibitory control in female rats, but not in males, until P20. The physiological response of both females and males was completely reversed after their neonatal androgenization and castration, respectively (Lacau de Mengido, Becu-Villalobos & Libertum, 1987). These data indirectly support the hypothesis that the DA neurons in the anteroventral periventricular nucleus of the preoptic area mediate the SS effects on gonadotropin secretion (Simerly et al., 1985, Simerly, 1989).

Thus, CAs are believed to mediate the organizing effects of SS on the hypothalamus over the critical period of ontogenesis inducing the permanent sexual differences in morphology of the hypothalamic target regions (e.g. sexually dimorphic nucleus of the preoptic area), the gonadotropin secretion patterns, and sexual behavior.

Adenohypophysial hormone secretion

Among the hypothalamic monoamines, only DA plays the role of the adenohypophysiotropic neurohormone, inhibiting α-melanocyte-stimulating hormone and prolactin secretions. Although DA receptors and α-melanocyte-stimulating hormone are present in intermediate lobe even in fetuses (Eberle, 1988), the DA receptor antagonist, *cis*-flupenthixol, provokes melanocyte-stimulating hormone release only from P8. These data show the establishment of DA inhibitory control over melanocyte-stimulating hormone secretion at the beginning of the second postnatal week.

The same strategy, as for α-melanocyte-stimulating hormone, has been used to evaluate the onset of DA inhibitory control over prolactin secretion. The inhibitor of the adenohypophysial DA receptors, pimozide, becomes effective first at P3 in increasing prolactin plasma level, thus, showing the establishment of DA control (Ojeda & McCann, 1974).

Conclusions

1. The development of the hypothalamic CA system is manifested, first, in the appearance of CA neurons in the hypothalamus and brainstem long before birth (E12–E13). Further, neuron differentiation results in an expression of the key-enzymes and CA synthesis, the onset of specific uptake and K^+-stimulated release of CAs, growing of CA axons to the hypophysial portal circulation, 3rd ventricle and the target neurons, followed by the establishment of specialized contacts: axovascular, axoventricular and synapses.
2. The development of the CA system is under hormonal control. Some hormones (e.g. thyroid hormones) provide a transient stimulating effect on neuronal differentiation, while others (e.g. SS) have a programming irreversible action realized during the critical perinatal period. In the last case, the hormones modify neuron differentiation and metabolism, networks of fibers, synaptic organization, and, finally, neuroendocrine regulations and behavior.
3. The hypothalamic CAs that appeared early in ontogenesis are characterized by a wide spectrum of physiological actions. First, they control the genesis and further differentiation of the target neurons. Later, over the critical perinatal period, they mediate the hormonal programing influence, modifying the architectonics, metabolism, receptors, etc. of the hypothalamic target regions and, finally, the adenohypophysial hormone secretion and behavior.

Acknowledgements

This work has been partly supported by the grants from IBRO (1989), Ministère de la Recherche et de la Technologie, France (1991) and NATO No CRG 920758 (1992).

Abbreviations

AADC	aromatic L-amino acid decarboxylase
ac	anterior commissure
an	arcuate nucleus
CA(s)	catecholamine(s)
DA	dopamine
db	diagonal band
dm	dorsomedial nucleus
E	embryonic day
EM	electron microscopic
fm	foramen Monroi

lh	lateral hypothalamus	ot	optic tract
LHRH	luteinizing hormone releasing hormone	P	postnatal day
lpa	lateral preoptic area	pev	periventricular nucleus
ls	lateral septum	pi	pituitary
lt	lamina terminalis	pp	primary portal plexus
lv	lateral ventricle	pv	paraventricular nucleus
me	median eminence	PNMT	phenylethanolamine N-methyltransferase
mfb	medial forebrain bundle		
mpa	medial preoptic area	rch	retrochiasmatic region
ms	medial septum	sc	suprachiasmatic nucleus
NA	noradrenaline	so	supraoptic nucleus
oc	optic chiasma	SS	sexual steroids
of	optic fissure	3v	third ventricle
on	optic nerve	vm	ventromedial nucleus
or	optic rudiment	zi	zona incerta

References

Baldino, F., O'Kane, T. M., Fitzpatrick-McElligott, S. & Wolfson, B. (1988). Coordinate hormonal and synaptic regulation of vasopressin messenger RNA. *Science*, **241**, 978–81.

Blaustein, J. D. & Turcotte, J. (1987). Further evidence of noradrenergic regulation of rat hypothalamic estrogen receptor concentration: possible non-functional increase and functional decrease. *Brain Research*, **436**, 253–64.

Boer, G. J. (1987). Development of vasopressin systems and their functions. In *Vasopressin; Principles & Properties*, ed. D. M. Gash & G. J. Boer, pp. 117–74. New York: Plenum Press.

Borisova, N. A., Sapronova, A. Y., Proshlyakova, E. V. & Ugrumov, M. V. (1991). Ontogenesis of the hypothalamic catecholaminergic system in rats. Synthesis, uptake and release of catecholamines. *Neuroscience*, **43**, 223–9.

Borisova, N. A., Ugrumov, M. V., Balan, I. S. & Thibault, J. (1993). Development of the tuberoinfundibular system in rats. Birthdates of the tyrosine hydroxylase immunopositive neurons. *Developmental Brain Research*, **73**, 173–6.

Buznikov, G. A. (1991). *Neurotransmitters in Embryogenesis*. London: Harwood Acad. Publ.

Charli, J. L., Faivre-Bauman, A., Loudes, C. & Kordon, C. (1993). Co-culture of rat melanotrophs with hypothalamic cells enhances differentiation of dopaminergic neurons. *Molecular and Cellular Research*, **4**, 55–63.

Collu, R. (1989). Neuroendocrine control of pituitary hormone secretion. In *Pediatric Endocrinology*. ed. R. Collu, J. R. Ducharme & H. J. Guyda, pp. 1–36. New York: Raven Press.

Crowley, W. R., O'Donohue, T. L. & Jacobowitz, D. M. (1978). Sex differences in catecholamine content in discrete brain nuclei of the rat: effects of neonatal castration on testosterone treatment. *Acta Endocrinologica (Copenhagen)*, **89**, 20–8.

Daikoku, S., Kawano, H., Okamura, Y., Tokuzen, M. & Nagatsu, I. (1986). Ontogenesis of immunoreactive tyrosine hydroxylase-containing neurons in rat hypothalamus. *Developmental Brain Research*, **28**, 85–98.

Davis, M. D., Lichtensteiger, W., Schlumpf, M. & Bruinink, A. (1984). Early postnatal development of pituitary intermediate lobe control in the rat by dopamine neurons. *Neuroendocrinology*, **39**, 1–12.

Demarest, K. T., McKay, D. W., Riegle, G. D. & Moore, K. E. (1981). Sexual differences in tuberoinfundibular nerve activity induced by neonatal androgen exposure. *Neuroendocrinology*, **32**, 108–13.

De Vitry, F., Hillion, J., Catelon, J., Thibault, J., Beneliel, J. J. & Hamon, M. (1991). Dopamine increases the expression of tyrosine hydroxylase and aromatic amino acid decarboxylase in primary cultures of fetal neurons. *Developmental Brain Research*, **59**, 123–31.

Döhler, K. -D., Coquelin, A., Davis, F., Hines, M., Shryne, J. E. & Gorski, R. A. (1984). Pre- and postnatal influence of testosterone propionate and diethylstilbestrol on differentiation of the sexually dimorphic nucleus of the preoptic area in male and female rats. *Brain Research*, **302**, 291–5.

Eberle, A. N. (1988). *The Melanotropins. Chemistry, Physiology and Mechanisms of Action*. Basel: Karger.

Foster, G. A., Schultzberg, M., Goldstein, M. & Hökfelt, T. (1985). Ontogeny of phenylethanolamine N-methyltransferase and tyrosine hydroxylase-like immunoreactivity in presumptive adrenaline neurons of the foetal rat central nervous system. *Journal of Comparative Neurology*, **236**, 348–81.

Friedman, W. J., Dreyfus, C. F., McEwen, B. S. & Black, I. B. (1989). Developmental regulation of tyrosine hydroxylase in the mediobasal hypothalamus. *Developmental Brain Research*, **48**, 177–85.

Harris, P. E., Lewis, B. M., Dieguez, C., Hall, R. & Scanlon,

M. F. (1987). The effects of neonatal hypothyroidism on brain catecholamine turnover in adult rats: assessment by a steady-state method. *Clinical Science*, **72**, 621–7.

Hökfelt, T., Martensson, R., Björklund, A., Kleinau, S. & Goldstein, M. (1984). Distribution maps of tyrosine-hydroxylase-immunoreactive neurons in the rat brain. In *Handbook of Chemical Neuroanatomy. vol. 2. Classical Neurotransmitters in the CNS*. Part 1, ed. A. Björklund & T. Hökfelt, pp. 277–379. Amsterdam: Elsevier.

Hyyppä, M. (1969). A histochemical study of the primary catecholamines in the hypothalamic neurons of the rat in relation to the ontogenetic and sexual differentiation. *Zeitschrift für Zellforschung und Mikroskopische Anatomie*, **98**, 550–60.

Ibata, Y., Fukui, K., Obata, H. L., Tanaka, M., Hisa, Y. & Sano, Y. (1982). Postnatal ontogeny of catecholamine and somatostatin neuron systems in the median eminence of the rat as revealed by a colocalization technique. *Brain Research Bulletin*, **9**, 407–15.

Ibata, Y., Tani, N., Obata, H. L., Tanaka, M., Kubo, S., Fukui, K., Fujimoto, M., Knoshita, H., Watanabe, K. & Sano, Y. (1981). Correlative ontogenetic development of catecholamine- and LHRH-containing nerve endings in the median eminence of the rat. *Cell and Tissue Research*, **216**, 31–8.

Jaeger, C. B. (1986). Aromatic L-amino acid decarboxylase in the rat brain: immunocytochemical localization during prenatal development. *Neuroscience*, **18**, 121–50.

Jarzab, B., Gubala, E., Achtelik, W., Lindner, G., Pogorzelska, E. & Döhler, K.-D. (1989). Postnatal treatment of rats with beta-adrenergic agonists or antagonists influences differentiation of sexual brain functions. *Experimental and Clinical Endocrinology*, **94**, 61–71.

Jarzab, B., Kaminski, M., Gubala, E., Achtelik, W., Wagiel, J. & Döhler, K. D. (1990a). Postnatal treatment of rats with the B₂-adrenergic agonist salbutamol influences the volume of the sexually dimorphic nucleus in the preoptic area. *Brain Research*, **516**, 257–62.

Jarzab, B., Kokocińska, D., Kaminski, M., Gubala, E., Achtelik, W., Wagiel, J. & Döhler, K. D. (1990b). Influence of neurotransmitters on sexual differentiation of the brain: relationship between the volume of the SDA-POA and functional characteristics. In *Hormones, Brain and Behavior in Vertebrates. I. Sexual Differentiation, Neurochemical Aspects, Neurotransmitters and Neuropeptides. Comparative Physiology*. vol. 8, ed. J. Balthazart, pp. 41–50. Basel: Karger.

Kalsbeek, A., Voorn, P. & Buijs, R. M. (1992). Development of dopamine-containing systems in the CNS. In *Handbook of Chemical Neuroanatomy. Vol. 10. Ontogeny of Transmitters and Peptides in the CNS*, ed. A. Björklund & T. Hökfelt, pp. 63–112. Amsterdam: Elsevier.

Khachaturian, H. & Sladek, J. R. (1980). Simultaneous monoamine histofluorescence and neuropeptide immunocytochemistry: III. Ontogeny of catecholamine varicosities and neurophysin neurons in the rat supraoptic and paraventricular nuclei. *Peptides*, **1**, 77–95.

Klein, D. C., Moore, R. Y. & Reppert S. M. (eds) (1991). *Suprachiasmatic Nucleus. The Mind's Clock*. New York: Oxford University Press.

Kolbinger, W., Trepel, M., Beyer, C., Pilgrim, C. & Reisert, I. (1991). The influence of genetic sex on sexual differentiation of diencephalic dopaminergic neurons *in vitro* and *in vivo*. *Brain Research*, **544**, 349–52.

Kuljis, R. O. & Advis, J. P. (1989). Immunocytochemical and physiological evidence of synapse between dopamine- and luteinizing hormone releasing hormone-containing neurons in the ewe median eminence. *Endocrinology*, **124**, 1579–81.

Lacau de Mengido J., Becu-Villalobos D. & Libertun C. (1987). Sexual differences in the dopaminergic control of luteinizing hormone secretion in the developing rat. *Developmental Brain Research*, **35**, 91–5.

Lauder, J. M. (1983). Hormonal and humoral influences on brain development. *Psychoneuroendocrinology*, **8**, 121–55.

Lengvari, I., Branch, B. J. & Taylor, A. N. (1980). Effects of perinatal tyroxine and/or corticosterone treatment on the ontogenesis of hypothalamic and mesencephalic norepinephrine and dopamine content. *Developmental Neuroscience*, **3**, 59–65.

Lichtensteiger, W. & Schlumpf, M. (1986). Permanent alteration of peptide feedback on dopamine neurons after injection of α-melanotropin antiserum at a critical period of postnatal development. *Brain Research*, **368**, 205–10.

Loizou, L. A. (1971). The postnatal development of monoamine-containing structures in the hypothalamo-hypophyseal system of the albino rat. *Zeitschrift für Zellforschung und mikroskopische Anatomie*, **114**, 234–52.

Meister, B., Hökfelt, T., Steinbusch, H. W. M., Skagerberg, G., Lindvall, O., Geffard, M., Joh, T. H., Cuello, A. C. & Goldstein, M. (1988). Do tyrosine hydroxylase-immunoreactive neurons in the ventralateral arcuate nucleus produce dopamine or only L-Dopa? *Journal of Chemical Neuroanatomy*, **1**, 59–64.

Nomura, Y., Naiton, F. & Segawa, T. (1976). Regional changes in monoamine content and uptake of the rat brain during postnatal development. *Brain Research*, **101**, 305–15.

Ojeda, S. R. & McCann, S. M. (1974). Development of dopaminergic and estrogenic control of prolactin release in the female rat. *Endocrinology*, **95**, 1499–505.

Piotte M., Beaudet A., Joh T. H. & Brawer J. R. (1988). Light and electron microscopic study of tyrosine hydroxylase-immunoreactive neurons within the developing rat arcuate nucleus. *Brain Research*, **439**, 127–37.

Puymirat, J., Barret, A., Picart, R., Vigny, A., Loudes, C., Faivre-Bauman, A. & Tixier-Vidal, A. (1983). Triiodothyronine enhances the morphological migration of dopaminergic neurons from fetal mouse hypothalamus cultured in serum-free medium. *Neuroscience*, **10**, 801–10.

Raum, W. J., Marcano, M. & Swerdloff, R. S. (1984). Nuclear accumulation of estradiol derived from the aromatization of testosterone is inhibited by hypothalamic beta-receptor stimulation in the neonatal female rat. *Biology of Reproduction*, 30, 388–96.

Raum, W. J. & Swerdloff, R. S. (1981). The role of hypothalamic adrenergic receptors in preventing testosterone-induced androgenization in the female rat brain. *Endocrinology*, 109, 273–8.

Reisert, I., Engele, J. & Pilgrim, Ch. (1989). Early sexual differentiation of diencephalic dopaminergic neurons of the rat *in vitro*. *Cell and Tissue Research*, 255, 411–17.

Reznikov, A. G. & Nosenko, N. D. (1983). It is possible that noradrenaline is the biogenic amine responsible for androgen-dependent sexual brain differentiation. *Experimental and Clinical Endocrinology*, 81, 91–3.

Schilling, K., Schmale, H., Oeding, P. & Pilgrim, Ch. (1991). Regulation of vasopressin expression in cultured diencephalic neurons by glucocorticoids. *Neuroendocrinology*, 53, 528–35.

Schlumpf, M., Lichtensteiger, W., Shoemaker, W. J. & Bloom, F. E. (1980). Fetal monoamine systems: early stages and cortical projections. In *Biogenic Amines in Development*, ed. H. Parvez & S. Parvez, pp. 567–90. Amsterdam: Elsevier.

Shyr, S. W., Crowley, W. R. & Grosvenor, C. E. (1986). Effect of neonatal prolactin deficiency on prepubertal tuberoinfundibular and tuberohypophysial dopaminergic neuronal activity. *Endocrinology*, 119, 1217–21.

Siddiqui, A. & Gilmore, D. P. (1988). Regional differences in the catecholamine content of the rat brain: effects of neonatal castration and androgenization. *Acta Endocrinologica (Copenhagen)*, 118, 483–94.

Simerly, R. B. (1989). Hormonal control of the development and regulation of tyrosine hydroxylase expression within a sexually dimorphic population of dopaminergic cells in the hypothalamus. *Molecular Brain Research*, 6, 297–310.

Simerly, R. B., Swanson, L. W., Handa, R. J. & Gorski, R. A. (1985). Influence of perinatal androgen on the sexually dimorphic distribution of tyrosine hydroxylase-immunoreactive cells and fibers in the anteroventral periventricular nucleus of the rat. *Neuroendocrinology*, 40, 501–10.

Sladek, C. D. & Armstrong, W. E. (1987). Effect of neurotransmitters and neuropeptides on vasopressin release. In *Vasopressin*, ed. D. M. Gash, & G. J. Boer, pp. 275–333. New York: Plenum Publication.

Smith, G. G. & Simpson, R. W. (1970). Monoamine fluorescence in the median eminence of foetal, neonatal and adult rats. *Zeitschrift für Zellforschung und mikroskopische Anatomie*, 104, 541–56.

Specht, L. A., Pickel, V. M., Joh, T. H. & Reis, D. J. (1981). Light microscopic immunocytochemical localization of tyrosine hydroxylase in prenatal rat brain. I. Early ontogeny. *Journal of Comparative Neurology*, 199, 233–53.

Stahl, F., Poppe, I., Gotz, F., Docke, F. & Dörner, G. (1986). Environment-dependent changes of neurotransmitter levels in the developing brain of rats. In *Monogr. Neural Sci.*, vol. 12, ed. M. M. Cohen, pp. 185–90. Basel: Karger.

Swaab, D. F. & Mirmiran, M. (1986). Functional teratogenic effects of chemicals on the developing brain. In *Monographs of Neural Science.*, vol. 12, ed. M. M. Cohen, pp. 45–57. Basel: Karger.

Teitelman, G., Jaeger, C. B., Albert, V., Joh, T. H. & Reis, D. J. (1983). Expression of amino acid decarboxylase in proliferating cells of the neural tube and notochord of developing rat embryo. *Journal of Neuroscience*, 3, 1379–88.

Trembleau, A., Ugrumov, M. V., Bernabe, J. & Calas, A. (1991). Development of the hypothalamic catecholaminergic system and its pharmacological depletion during ontogenesis. *Society for Neurosciences Abstracts*, Vol. 17, part 2, p. 1039.

Ugrumov, M. V. (1991). Developing hypothalamus in differentiation of neurosecretory neurons and in establishment of pathways for neurohormone transport. *International Review of Cytology*, 129, 207–67.

Ugrumov, M. V. (1992). Development of the hypothalamic monoaminergic system in ontogenesis. Morphofunctional aspects. *Zoological Science*, 9, 17–36.

Ugrumov, M. V., Taxi, J., Tixier-Vidal, A., Thibault, J. & Mitskevich, M. S. (1989a). Ontogenesis of tyrosine hydroxylase-immunopositive structures in the rat hypothalamus. An atlas of neuronal cell bodies. *Neuroscience*, 29, 135–56.

Ugrumov, M. V., Tixier-Vidal, A., Taxi, J., Thibault, J. & Mitskevich, M. S. (1989b). Ontogenesis of tyrosine hydroxylase-immunopositive structures in the rat hypothalamus. Fiber pathways and terminal fields. *Neuroscience*, 29, 157–66.

Ugrumov, M. V., Popov, A. P., Vladimirov, S. V., Kasmambetova, S. & Thibault, J. (1994) Development of the suprachiasmatic nucleus during ontogenesis. Tyrosine hydroxylase immunopositive cell bodies and fibers, *Neuroscience*, 58, 151–160.

Van den Pol, A. N. & Tsujimoto, K. L. (1985). Neurotransmitters of the hypothalamic suprachiasmatic nucleus: immunocytochemical analysis of 25 neuronal antigens. *Neuroscience*, 15, 1049–86.

Weaver, D. R., Rivkees, S. A. & Reppert, S. M. (1992). D_1-dopamine receptors activate *c-fos* expression in the fetal suprachiasmatic nuclei. *Proceedings of the National Academy of Sciences, USA*, 89, 9201–4.

Weiner, R. I. & Ganong, W. F. (1972). Norepinephrine concentration in the hypothalamus, amygdala, hippocampus, and cerebral cortex during postnatal development and vaginal opening. *Neuroendocrinology*, 9, 65–71.

Weiner, R. I., Findell, P. R. & Kordon C. (1988). Role of classic and peptide neuromediators in the neuroendocrine regulation of LH and prolactin. In *The Physiology of Reproduction*, ed. E. Knobil, J. Neill *et al.*, pp. 1235–81. New York: Raven Press.

Wilson, W. E. & Agrawal, A. K. (1979). Brain regional levels of neurotransmitter amines as neurochemical correlates of sex-specific ontogenesis in the rat. *Developmental Neuroscience*, 2, 195–200.

Wray, S. & Hoffman, G. (1986). Catecholamine innervation of LH–RH neurons: a developmental study. *Brain Research*, 399, 327–31.

19
Sexual differentiation of central catecholamine systems

I. Reisert, E. Küppers and C. Pilgrim

Introduction

Sexually dimorphic patterns of behavior and neuroendocrine control are found not only in mammals but also in other vertebrate phyla as well as in invertebrates. This concerns predominantly but not exclusively brain functions related to reproduction (Beatty, 1979; Kelley, 1988; Hutchison, 1991). Reproductive behavior, for instance, involves sex-specific, highly complex and stereotyped motor patterns during sexual orientation, courtship, copulation as well as parental care. Sex-specific organization of neuroendocrine systems is characterized by sex differences in patterns of release of gonadotropins (Dyer, 1984), prolactin (Arey, Averill & Freeman, 1989) and growth hormone (Jansson, Edén & Isaksson, 1985) from the pituitary. Results of morphological, pharmacological and physiological investigations point to the existence of sexually dimorphic neural circuits that form the somatic basis of sex-specific brain functions (for review see Dörner et al., 1991). Catecholamine (CA) systems are of particular interest in this regard because they are known to influence sexual behavior as well as pituitary functions (for reviews see Everitt, 1977; Kordon, 1988; Numan, 1988; Barclay & Cheng, 1992). In fact, there are numerous reports on sex differences in morphology and function of dopamine (DA) and noradrenaline (NA) systems of the brain. Most of these results have been acquired in the rat and concern distribution of transmitter-specific nerve cells and fibers, levels and turnover of CA, activities of synthesizing and metabolizing enzymes, receptors as well as CA-related behaviors. For a survey of sexual dimorphisms of adult CA systems, the reader is referred to a recent review article (Reisert & Pil-

grim, 1991). Relevant data may also be found in Table 19.1.

Sex-specific development of CA systems *in vivo*

Sexual dimorphisms of the mammalian brain are generally ascribed to organizational effects of androgens during ontogenesis. There is ample evidence for the existence of a developmental 'critical' or 'sensitive' period for sexual differentiation of the brain the timing of which may vary from species to species (MacLusky & Naftolin, 1981; Ford & D'Occhio, 1989; Hutchison, 1991). In the rat, the beginning of the critical period coincides with the peak of testosterone (T) secretion from the fetal testis at embryonic day (ED) 18 (Weisz & Ward, 1980). After entering the brain, T may be aromatized to 17β-estradiol (E), the steroid thought to be responsible for the establishment of a male brain (Hutchison, 1991). It is believed that E locally converted from T induces defeminization and masculinization of certain neural circuits as a prerequisite for sex-specific regulation of reproductive behavior and endocrine systems. According to this concept, development of a female brain is the 'default pathway', i.e., a passive event not requiring the presence of gonadal hormones. Evidence for an androgen-linked mechanism of sexual differentiation of the brain is provided by the effects of castration or hormonal treatment during development. Such experiments, when carried out in the critical period, may result in a sex-reversion of reproductive behavior and brain morphology (Gorski, 1988; Hutchison, 1991; Pilgrim & Reisert, 1992). It may be advantageous to distinguish this 'organizational' phase of steroid action, during which sex-specific neural cir-

Table 19.1. *Sex differences of catecholaminergic systems in the rat and the influence of manipulations of the hormonal environment*

Brain area	Age	Sex differences (prior to experimental manipulations)		Results of experimental manipulations	References
locus coeruleus	ED21-PD13	decrease in numbers of neurons	♂>♀	♀TP (ED18)=♂C	Tobet & Fox, 1989
locus coeruleus	PD90	numbers of neurons	♀>♂	♀TP (P0)≈♂C ♂gonex (P0)≈♀C	Guillamón, de Blas & Segovia, 1988
hypothalamus (anterior periventricular nucleus)	PD67	numbers of TH-mRNA- and TH-containing cells	♀>♂	♀TP (PD5)≈♂C ♂gonex (P0)≈♀C	Simerly, 1989
hypothalamus (anterior periventricular nucleus)	PD130–144	numbers of cells and density of fibers IR for TH	♀>♂	♀TP (ED16-PD10)≈♂C	Simerly, Swanson & Gorski, 1985; Simerly et al., 1985
whole brain	PD2	NA levels	♂>♀	♀TP (PD5)≈♂C at PD10 only	Hardin, 1973
hypothalamus	PD12	NA levels	♀>♂	♂gonex (PD1)>♀C	Wilson & Agrawal, 1979
hypothalamus (arcuate and suprachiasmatic nucleus)	PD66	NA levels	♂>♀	♀TP (PD4) and ♂gonex (PD1) – no effect	Crowley, O'Donohue & Jacobowitz, 1978
striatum	PD60	NA levels	♀>♂	♀gonex and ♂gonex (P0) – no effect	Leret et al., 1987
midbrain	PD12	Da levels	♂>♀	♂DES (PD2,4+6)≈♀C	Wilson & Agrawal, 1979
cortex (anterior)	PD10	Da levels	♀>♂	♀TP (P0)=♂C	Stewart, Kühnemann & Rajabi, 1991
	PD21	DA and DOPAC levels	♂>♀	♀TP (P0)=♂C	
hypothalamus (arcuate nucleus)	PD66	DA levels	♂>♀	♀TP (PD4)≈♂C ♂gonex (P0)≈♀C	Crowley, O'Donohue & Jacobowitz, 1978
hypothalamus (median eminence)	PD100	DOPA synthesis	♀>♂	♂gonex (PD1)≈♀C ♀TP (PD4)=♂C	Demarest et al., 1981
hypothalamus	PD12	MAO activity	♂>♀	♀TP (P0)=♂C ♂gonex (P0)=♀C	Gaziri & Ladosky, 1973
hypothalamus	PD30	D_2 receptor levels	♂>♀	♂ADT (PD2)=♀C ♀TP (PD3)=♂C	Herdon & Wilson, 1985

ADT – aromatase inhibitor androst-4-ene-3,4,17-trione. C – control – no manipulations of hormonal environment. DA – dopamine. DES – diethylstilbestrol. DOPA – dihydroxyphenylalanine. DOPAC – dihydroxyphenylacetic acid. ED – embryonic day. gonex – gonadectomy. IR – immunoreactive. MAO – monoamine oxydase. NA – noradrenaline. PD – postnatal day. P0 – day of birth. TH – tyrosine hydroxylase. TP – testosterone proprionate.

cuits are laid down, from 'activational' effects of gonadal steroids in the adult although there is considerable overlap in time and also in modes of cellular responses elicited by sex hormones (Arnold & Breedlove, 1985).

Essentially, two arguments speak for the assumption that the above concept may also apply to the sexual differentiation of CA systems. First, sex differences in organization of CA systems and in functional indices of CA transmission begin to appear during the perinatal critical period (Hardin, 1973; Gaziri & Ladosky, 1973; Vaccari, 1980; Flügge, Wuttke & Fuchs, 1986; Lacau de Mengido, Becú-Villalobos & Libertun, 1987; Tobet & Fox, 1989; Reisert et al., 1990; Kolbinger et al., 1991; Ovtscharoff et al., 1992). However, there are exceptions as to the time period when sexual dimorphisms may first develop or may be further modified (Becker & Ramirez, 1981; Negro-Vilar et al., 1984; Segovia, Guillamón & de Blas, 1990; Arbogast & Voogt, 1991). Secondly, experimental manipulation of the hormonal environment, such as gonadectomy and/or administration of sex steroids, during development leads to persisting changes in the sex-specific properties of CA neurons. Table 19.1 gives a synopsis of a number of such experiments carried out on rat embryos or pups. It should be noted that some details of the investigators' experimental design and results were omitted from Table 19.1 for the sake of brevity. The purpose is to give an impression of the plasticity of sex differences in morphology and function of CA systems. It may be seen, for instance, that androgen treatment of females during the critical period may result in masculinization of CA neuron numbers, CA levels or metabolism as well as receptor densities. Conversely, gonadectomy of male pups may prevent masculinization of CA neuron numbers as well as DA levels or metabolism. Again, not all experimental results seem to fit the androgen hypothesis: sex differences in NA levels, e.g. do not always change after gonadectomy or androgen treatment or may even be augmented by gonadectomy of female pups. Such exceptions from the rule may be taken as hints that the view of the process of sexual differentiation outlined above is too simplistic and that other factors and mechanisms must be invoked to fully explain this phenomenon. In fact, there are a number of observations indicating that the regulation of sexual differentiation of the brain is complex. Some investigators have obtained results that suggest that development of a female brain is not entirely intrinsic and determined by the mere absence of androgen but requires the presence of estrogen (Döhler et al., 1984).

This appears to hold specifically for the maturation of sex-specific regulation of prolactin secretion by DA (Watanobe & Takebe, 1987). Male differentiation of certain CNS regions may be under direct control of androgens effective without conversion to estrogen (Goldstein & Sengelaub, 1990). Various experimental results support the idea that sexual differentiation is determined by an interaction of gonadal steroids with other agents or mechanisms, such as adrenal steroids (Kincl, 1990), neurotransmitters, growth factors (Toran-Allerand, Ellis & Pfenninger, 1988; Yanase et al., 1988) or proto-oncogene-mediated signal transduction (Whorf & Tobet, 1992). It thus appears doubtful that sex-specific development of CA systems can be regarded as an epigenetic phenomenon solely dependent on the absence or presence of androgens.

Sex-specific development of CA systems *in vitro*

If sexual differentiation of the brain is controlled by more than a single factor, the investigation of this process requires a careful analysis of the development of each neuronal phenotype in order to determine whether and how its development is affected by gender. Since cause–effect relationships are often difficult to analyze in the brain developing *in vivo* because of the impracticability of distinguishing direct from indirect effects of experimental manipulations, an alternative may be seen in the employment of brain cell cultures. Several investigators have shown that *in vitro* models are suited to study the differentiation of CA neurons (Ahnert-Hilger et al., 1986) and to test various agents for developmental effects (Friedman et al., 1988; Sklair & Segal, 1990; Engele & Bohn, 1991). In order to distinguish effects of gonadal hormones from other sex-linked factors, a new experimental approach was adopted by our group. Instead of raising cultures from pooled male and female rat embryos, we prepared gender-specific dissociated cell cultures after having separated the genders by inspection of the gonads (for details see Reisert, Engele & Pilgrim, 1989; Engele, Pilgrim & Reisert, 1989b). The cells were obtained from the main subdivisions of the ED 14 embryonic rat brainstem. These cultures contain considerable numbers of tyrosine hydroxylase-immunoreactive (TH-IR) neurons which develop DA or NA properties according to the location in the brainstem from which they were taken (for a discussion of the nature of TH-IR cells see Ahnert-Hilger et al., 1986; Engele et al., 1989a; Küppers, Pilgrim & Reisert, 1991). It is important to note that the serum used to supplement the

Table 19.2. *Differentiation of CA neurons in gender-specific cultures of embryonic rat brain stem and effects of gonadal steroids*

Brain area	Days in vitro	Sex differences in the absence of sex steroids		Results of long-term treatment with sex steroids	References
diencephalon	6	size of TH-IR cell bodies	♂>♀	♂ E,T and ♀ E,T – no effect	Kolbinger et al., 1991
mesencephalon	3–13	numbers of TH-IR neurons	♀>♂	♂ E,T and ♀ E,T – no effect	Beyer, Pilgrim & Reisert, 1991
rhombencephalon	3,6	lengths of TH-IR neurites	3 DIV ♂=♀	♀ E,T – increase; ♂ E,T – no effect	Küppers, Pilgrim & Reisert, 1991
			6 DIV ♂≈♀	♀ E,T – no effect; ♂ E,T – increase	
diencephalon	6	uptake of (^{3}H)DA/ culture well	♀>♂	♂ E,T and ♀ E,T – no effect	Reisert, Engele & Pilgrim, 1989
mesencephalon	6	uptake of (^{3}H)DA/ TH-IR neuron	♂>♀	♀ E=♂; ♀ T and ♂ E,T – no effect	Engele, Pilgrim & Reisert, 1989b
rhombencephalon	3,6	uptake (^{3}H)NA/ TH-IR neuron	3 Div ♂≈♀	♂ E,T and ♀ E,T – no effect	Unpublished observations
			6 DIV ♂>♀	♂ E,T and ♀ E,T – no effect	
diencephalon	10,13	DA levels/TH-IR neuron	♀>♂	♂ E,T and ♀ E,T – decrease	Beyer, Pilgrim & Reisert, 1991
rhombencephalon	3,6	NA levels/TH-IR neuron	3 DIV ♂>♀	♂ E,T and ♀ E,T – no effect	Unpublished observations
			6 DIV ♂≈♀	♀ E,T > ♀ E,T – no effect	
diencephalon	10	DOPA synthesis/ culture well	♀ > ♂	♂ E,T and ♀ E,T – decrease	Beyer et al., 1992
rhombencephalon	3,6	DOPA synthesis/ TH-IR neuron	3 DIV ♂ > ♀	♀ E=♂; ♂ E – no effect	Unpublished observations
			6 DIV ♂≈♀	♂ E > ♂; ♀ E – no effect	

DA – dopamine. DIV – days *in vitro*. DOPA – dihydroxyphenylalanine. E – 17β-estradiol. IR – immunoreactive. NA – noradrenaline. T – testosterone. TH – tyrosine hydroxylase.

culture medium was from castrated horses and did not contain measurable amounts of T or E. Other sex steroids, such as androstenedione and androstenetriol sulfate were present in concentrations of at least two orders of magnitude below physiological levels.

Surprisingly, region-specific morphological and functional sex differences of CA neurons were observed in control cultures raised in the absence of sex steroids (Table 19.2). Female mesencephalic cultures were found to contain about 25% higher numbers of DA cells than male cultures. This was different with diencephalic DA neurons and rhombencephalic NA neurons which did not exhibit statistically significant sex differences in cell numbers. It is presently not clear whether the mesencephalic sex differences are due to differential proliferation of committed precursor cells prior to removal of the tissue, to differential survival of DA cells *in vitro*, or to a sex-specific schedule in the expression of TH. Morphometric measurements of soma sizes yielded a hormone-independent sex difference in diencephalic but not in mesencephalic cultures. Male DA neurons were about 30% larger than female neurons. A conclusive functional parameter to assess the maturation of neurons is their capacity for transmitter (re)uptake (Tennyson et al., 1972; Coyle, 1977; Azmitia & Whitaker-Azmitia, 1987). Thus specific uptake of radiolabeled CA was assessed by subtracting non-specific uptake measured in the presence of an appropriate inhibitor from the total uptake. In the absence of measurable amounts of sex steroids, uptake of (^{3}H)DA was twice as high in female than in male diencephalon. The reverse was true for the mesencephalon where male cells took up 1.5 times more (^{3}H)DA than female cells. A sex difference in uptake of (^{3}H)NA was seen in rhombencephalic cultures cultured for 6 days in vitro (DIV) but not yet in 3 DIV cultures. Investigations of developmental profiles of endogenous DA and metabolites by HPLC, too, yielded hormone-independent sex differences. Higher levels of DA were measured in female than in male control cultures of both diencephalon and mesencephalon. However, when DA levels were related to numbers of TH-IR cells, the sex differences disappeared in mesencephalic but not in diencephalic cultures. Dihydroxyphenylacetic acid (DOPAC) levels and vesicular storage capacity did not exhibit sex differences. Sex differences in NA levels per TH-IR neuron were present in 3 DIV but no longer in 6 DIV rhombencephalic cultures. TH activity was assessed by measuring accumulation of dihydroxyphenylalanine (DOPA) in the presence of an inhibitor of aromatic amino acid decarboxylase. Diencephalic cultures raised from females produced markedly more DOPA than those raised from males. A temporary sex difference in DOPA synthesis was also seen in rhombencephalic cultures but not in mesencephalic cultures (for ref. see Table 19.2).

In view of the androgen theory of sexual differentiation and the well-documented effects of sex steroids on CA systems developing in vivo (see pp. 453–455), part of the cultures were treated daily with E or T in physiological concentrations (10^{-12} or 10^{-9} M, respectively) from the beginning up to 13 DIV. The results are also listed in Table 19.2. A number of parameters, such as cell numbers, cell sizes, NA uptake and NA levels did not respond to steroid treatment. The most impressive effects were present regarding levels of endogenous DA and DOPA formation in diencephalic cultures. Both E and T decreased these functions in a dose-dependent manner without altering the sex differences. Down-regulation of DOPA synthesis, irrespective of the gender, would be in line with results of most investigations carried out in vivo. Long-term treatment with estrogen decreases TH activity in the hypothalamus of adult male and female rats (Luine, McEwen & Black, 1977; Arita & Kimura, 1987). This is at least partly due to down-regulation of the expression of mRNA coding for the TH protein (Blum, McEwen & Roberts, 1987; Morrell et al., 1989; Simerly, 1989). No influence of E on TH activity in embryonic mouse hypothalamic explant cultures was detected by Friedman et al. (1989) although the effect on TH activity in the adult was confirmed. Androgens, too, inhibit CA synthesis in the limbic forebrain and striatum (Engel et al., 1979; Abreu et al., 1988). On the other hand, it was reported that three daily injections of E increased DOPA accumulation in the median eminence (Demarest & Moore, 1980) and castration increased TH-IR cell numbers in the hypothalamus of male rats (Brawer, Bertley & Beaudet, 1986). It could well be that the discrepancies are accounted for by the complex connectivity developing in the brain in vivo. Inputs from other steroid-sensitive neurons may mask the direct effects of sex steroids on DA neurons. At present we are not able to decide whether the steroid-induced decrease in DOPA synthesis in our cultures is due to a genomic effect mediated by nuclear receptors. Such receptors have been described in hypothalamic explant cultures (Toran-Allerand, Gerlach & McEwen, 1980). An alternative to genomic effects would be direct inhibition of TH by metabolites of E such as 2-OH-estradiol which competes with the pteridine cofactor of the enzyme (Lloyd & Weisz, 1978).

The above hormone effects cannot be taken as evidence for an involvement of gonadal hormones in the generation of the sexual dimorphisms detected in our cultures. More interesting in this respect may be those hormone effects listed in Table 19.2 which were not consistently observed under all circumstances. For instance, NA neurons did not respond to E or T treatment with an enhanced outgrowth of neurites except in 3 DIV females and in 6 DIV males. Treatment with E increased DA uptake in female but not in male mesencephalic cultures. Apparently we are faced with an unexpectedly complex situation. Effects of sex steroids may be restricted to specific CA phenotypes, specific neuronal properties, specific time windows and to one gender. Although the idea that the exact timing of the critical period of sexual differentiation may differ for a given CNS structure or function has been entertained before (MacLusky & Naftolin, 1981; Rhees, Shryne & Gorski, 1990), the present results appear to take us one step further. We have to consider the possibility that such time windows are not identical for males and females. If such a gender-specific schedule of steroid sensitivity really exists, it is hardly conceivable how it could be brought about by sex differences in hormonal environment.

The salient point of the *in vitro* investigations on CA neurons developing in gender-specific cultures is that sexual dimorphisms of CA neurons may develop independently of the action of gonadal steroids. This is in contradiction to the generally accepted theory which holds that sexual differentiation of the brain is caused by the organizational effect of gonadal steroids present during a critical period of brain development (cf. pp. 453–455). The brain tissue used to raise the above cultures was removed at ED 14, i.e. well before sex differences in hormonal environment of the embryo are known to develop (Weisz & Ward, 1980; Baum, Woutersen & Slob, 1991). Even the classical targets of steroids during development, the genital ducts, do not exhibit sex differences prior to ED 15.5 (Eusterschulte, Reisert & Pilgrim, 1992). In order to cope with the remote possibility that sex steroid-dependent determinative events occur *in utero* before the brain tissue is taken into culture, additional cultures were prepared from embryos whose mothers had been treated with the estrogen antagonist tamoxifen or the androgen antagonist cyproterone acetate. These antisteroids competitively inhibit effects of the gonadal hormones on target cells by interference with the uptake process, receptor binding, translocation to the nucleus, and/or DNA binding of the ligand-receptor complex. In spite of this pretreatment, DOPA synthesis rates were again higher in diencephalic cultures raised from females than from males (Beyer et al., 1992). In conclusion, epigenetic control by the hormonal environment cannot be the only mechanism responsible for generation of sexual dimorphisms described above. Other mechanisms, such as cell-autonomous realization of a sex-specific genetic program, must be invoked to explain sexual differentiation of CA systems.

Although it is obvious that most of the above results could not have been acquired if it were not for the availability of gender-specific neural cultures, there is no doubt that their relevance to normal processes of development *in vivo* remains to be established. Still there are several clues indicating that the observations made *in vitro* are specific and are not culture artifacts. The existence, albeit transient, of sex differences in DA cell numbers was confirmed in the mesencephalon of ED 17 intact rat embryos (Reisert et al., 1990). Also, the sex differences in soma sizes of diencephalic DA neurons are qualitatively and quantitatively similar *in vitro* and *in vivo* (Kolbinger et al., 1991). Predictions made on the basis of the *in vitro* findings have turned out to be true. Recently as yet unknown sex differences in the organization of the prenatal rat striatum have been detected *in vivo* (Ovtscharoff et al., 1992). It is conceivable that some of the mechanisms of sexual differentiation have been obscured to date by the complex situation of the developing brain *in vivo*. It may therefore be appropriate at this point to cite a phrase from the book of Olsen (1992): 'The overpowering effects of testicular hormones on sexual behavior differentiation make it easy to ignore genetic influences'.

Sex-specific vulnerability of CA systems during development

In Table 19.1, we have cited evidence that manipulation of the hormonal environment during ontogenesis may affect sexual differentiation of CA systems. Here we want to draw attention to the fact that this process may also be disturbed by interfering with CA transmission itself, predominantly during the same period of development which has been found to be critical for organizational effects of sex hormones. Table 19.3, similarly to Table 19.1, gives a synopsis of experiments in which rat embryos or pups were treated with various drugs known to affect the availability of CA transmitters or the occupancy of CA receptors. Alternatively, ligands were used that act on GABAergic or nicotinic receptors known to be localized to CA neurons. It is important to note that the effects of this kind of drug treatment are sex-specific. Interference with CA transmission during development may even result in sex reversal of CA levels or in masculinization of CA neuron numbers in later life. These neuroanatomical or neurochemical results correlate with results of behavioral studies. Prenatal nicotine treatment has adverse effects on motor development of male but not of female rat pups (Schlumpf et al., 1988) and abolishes the sex difference in saccharin preference of the adult offspring (Lichtensteiger & Schlumpf, 1985). Although it can presently not be excluded that some of the above drugs act by generating imbalances of gonadal hormones, the possibility of direct teratogenic effects due to a sex-specific vulnerability of developing CA neurons must be considered. There is circumstantial evidence that such a phenomenon may also be found in humans. Neurologic and affective disorders that have been related to alterations of CA functions including chorea minor, tardive dyskinesia, infantile hyperkinetic syndrome, Parkinson's disease, depression, schizophrenia, and anorexia nervosa show a sex-specific prevalence (for review see Pilgrim & Reisert, 1992). Some or all of these disorders may

Table 19.3. *Sex-specific developmental effects of manipulations of catecholaminergic transmission in the rat*

Brain area	Age	Sex differences (prior to experimental manipulation)		Results of experimental manipulation	References
locus coeruleus	PD16	volume of LC, numbers of neurons	♀>♂	♀=♂ following diazepam treatment of P0 females	Segovia et al., 1991
medial preoptic area	PD10	NA levels	♀>♂	♂>♀ following muscimol treatment of ED16–PD9 females	Flügge, Wuttke & Fuchs, 1984
forebrain	PD10	NA levels	♀>♂	♂>♀ following nicotine treatment of ED12–19 males	Lichtensteiger et al., 1988
hypothalamus	PD30	NA levels	♀=♂	♀>♂ following reserpine treatment of PD4 females	Hyyppä & Rinne, 1971
limbic system	P0	NA levels	♀>♂	♂>♀ following DA treatment of P0 males and females	González & Leret, 1992
forebrain	PD15	DA levels	♀>♂	♂>♀ following nicotine treatment of ED12–19 males	Lichtensteiger et al., 1988
limbic system	PD80	DA levels	♀=♂	♀>♂ following DA treatment of P0 males and females	González & Leret, 1992
striatum	PD14	D_2 receptor binding sites	♀=♂	♀>♂ following nicotine treatment of ED1–21 males	Fung & Lau, 1989
	PD12,20	serum LH and PRL levels	♀=♂	♀>♂ following haloperidol treatment of PD12 and PD20 females	Lacau de Mengido, Becú-Villalobos & Libertun, 1987

DA – dopamine. ED – embryonic day. LC – locus coeruleus. LH – luteinizing hormone. NA – noradrenaline. PD – postnatal day. P0 – day of birth. PRL – prolactin.

indeed have a neurodevelopmental etiology. As discussed on pp. 455–458, CA neurons develop morphological and functional sexual dimorphisms which appear to be determined not so much by the hormonal environment as by primary somatic effects of sex-specific genes. It is conceivable that such genes confer upon these neurons a sex-specific vulnerability to temporary imbalances of neuroactive substances in the internal milieu or to other adverse conditions in their environment. This would then result in a sex-specific predisposition towards manifestation of a disease.

Acknowledgement

Support by the Deutsche Forschungsgemeinschaft is gratefully acknowledged (Re 413/2–4, Pi 68/9–1).

References

Abreu, P., Hernandez, G., Calzadilla, C. H. & Alonso, R. (1988). Reproductive hormones control striatal tyrosine hydroxylase activity in the male rat. *Neuroscience Letters*, **95**, 213–17.

Ahnert-Hilger, G., Engele, J., Reisert, I. & Pilgrim, C. (1986). Different developmental schedules of dopaminergic and noradrenergic neurons in dissociation culture of fetal rat midbrain and hindbrain. *Neuroscience*, **17**, 157–65.

Arbogast, L. A. & Voogt, J. L. (1991). Ontogeny of tyrosine hydroxylase mRNA signal levels in central dopaminergic neurons: development of a gender difference in the arcuate nuclei. *Developmental Brain Research*, **63**, 151–61.

Arey, B. J., Averill, R. L. W. & Freeman, M. E. (1989). A sex-specific endogenous stimulatory rhythm regulating prolactin secretion. *Endocrinology*, **124**, 119–23.

Arita, J. & Kimura, F. (1987). Direct inhibitory effect of long term estradiol treatment on dopamine synthesis in tuberoinfundibular dopaminergic neurons: *in vitro* studies using hypothalamic slices. *Endocrinology*, **121**, 692–8.

Arnold, A. P. & Breedlove, S. M. (1985). Organizational and activational effects of sex steroids on brain and behavior: a reanalysis. *Hormones and Behavior*, **19**, 469–98.

Azmitia, E. C. & Whitaker-Azmitia, P. M. (1987). Target cell stimulation of dissociated serotonergic neurons in culture. *Neuroscience*, **20**, 47–63.

Barclay, S. R. & Cheng, M. -F. (1992). Role of catecholamines in the courtship behavior of male ring doves. *Pharmacology, Biochemistry and Behavior*, **41**, 739–47.

Baum, M. J., Woutersen, P. J. A. & Slob, A. K. (1991). Sex difference in whole-body androgen content in rats on fetal days 18 and 19 without evidence that androgen passes from males to females. *Biology of Reproduction*, **44**, 747–51.

Beatty, W. W. (1979). Gonadal hormones and sex differences in nonreproductive behaviors in rodents: organizational and activational influences. *Hormones and Behavior*, **12**, 112–63.

Becker, J. B. & Ramirez, V. D. (1981). Experimental studies on the development of sex differences in the release of dopamine from striatal tissue fragments *in vitro*. *Neuroendocrinology*, **32**, 168–73.

Beyer, C., Eusterschulte, B., Pilgrim, C. & Reisert, I. (1992). Sex steroids do not alter sex differences in tyrosine hydroxylase activity of dopaminergic neurons *in vitro*. *Cell and Tissue Research*, **270**, 547–52.

Beyer, C., Pilgrim, C. & Reisert, I. (1991). Dopamine content and metabolism in mesencephalic and diencephalic cell cultures: sex differences and effects of sex steroids. *Journal of Neuroscience*, **11**, 1325–33.

Blum, M., McEwen, B. S. & Roberts, J. L. (1987). Transcriptional analysis of tyrosine hydroxylase gene expression in the tuberoinfundibular dopaminergic neurons of the rat arcuate nucleus after estrogen treatment. *Journal of Biological Chemistry*, **262**, 817–21.

Brawer, J., Bertley, J. & Beaudet, A. (1986). Testosterone inhibition of tyrosine hydroxylase expression in the hypothalamic arcuate nucleus. *Neuroscience Letters*, **67**, 313–18.

Coyle, J. T. (1977). Biochemical aspects of neurotransmission in the developing brain. *International Review of Neurobiology*, **20**, 65–103.

Crowley, W. R., O'Donohue, T. L. & Jacobowitz, D. M. (1978). Sex differences in catecholamine content in discrete brain nuclei of the rat: effects of neonatal castration or testosterone treatment. *Acta Endocrinologica (Copenh)*, **89**, 20–8.

Demarest, K. T., McKay, D. W., Riegle, G. D. & Moore, K. E. (1981). Sexual differences in tuberoinfundibular dopamine nerve activity induced by neonatal androgen exposure. *Neuroendocrinology*, **32**, 108–13.

Demarest, K. T. & Moore, K. E. (1980). Accumulation of L-dopa in the median eminence: an index of tuberoinfundibular dopaminergic nerve activity. *Endocrinology*, **106**, 463–8.

Döhler, K. D., Hancke, J. L., Srivastava, S. S., Hofmann, C., Shryne, J. E. & Gorski, R. A. (1984). Participation of estrogens in female sexual differentiation of the brain: neuroanatomical, neuroendocrine and behavioral evidence. *Progress in Brain Research*, **61**, 99–117.

Dörner, G., Poppe, I., Stahl, F., Kölzsch, J. & Uebelhack, R. (1991). Gene- and environment-dependent neuroendocrine etiogenesis of homosexuality and transsexualism. *Experimental and Clinical Endocrinology*, **98**, 141–50.

Dyer, R. G. (1984). Sexual differentiation of the forebrain – relationship to gonadotrophin secretion. *Progress in Brain Research*, **61**, 223–36.

Engel, J., Ahlenius, S., Almgren, O. & Carlsson, A. (1979). Effects of gonadectomy and hormone replacement on brain monoamine synthesis in male rats. *Pharmacology, Biochemistry and Behavior*, **10**, 149–54.

Engele, J. & Bohn, M. C. (1991). The neurotrophic effects of fibroblast growth factors on dopaminergic neurons *in vitro* are mediated by mesencephalic glia. *Journal of Neuroscience*, **11**, 3070–8.

Engele, J., Pilgrim, C., Kirsch, M. & Reisert, I. (1989a). Different developmental potentials of diencephalic and mesencephalic dopaminergic neurons *n vitro*. *Brain Research*, **483**, 98–109.

Engele, J., Pilgrim, C. & Reisert, I. (1989b). Sexual differentiation of mesencephalic neurons *in vitro*: effects of sex and gonadal hormones. *International Journal of Developmental Neuroscience*, **7**, 603–11.

Eusterschulte, B., Reisert, I. & Pilgrim, C. (1992). Absence of sex differences in size of the genital ducts of the rat prior to embryonic day 15.5–16.0. *Tissue and Cell*, **24**, 483–9.

Everitt, B. J. (1977). Cerebral monoamines and sexual behavior. In *Handbook of Sexology*, ed. J. Money & H. Musaph, pp. 429–48. Amsterdam: Elsevier/North Holland.

Flügge, G., Wuttke, W., & Fuchs, E. (1984). Perinatal application of a GABA agonist disturbs normal sexual differentiation. In *Fetal Neuroendocrinology*, ed. F. Ellendorff, P. D. Gluckman & N. Parvizi, pp. 161–3. Ithaka, NY: Perinatology Press.

Flügge, G., Wuttke, W. & Fuchs, E. (1986). Postnatal development and sexual differentiation of central nervous catecholaminergic systems. *Biogenic Amines*, 3, 249–55.

Ford, J. J. & D'Occhio, M. J. (1989). Differentiation of sexual behavior in cattle, sheep and swine. *Journal of Animal Science*, 67, 1816–23.

Friedman, W. J., Dreyfus, C. F., McEwen, B. & Black, I. B. (1988). Presynaptic transmitters and depolarizing influences regulate development of the substantia nigra in culture. *Journal of Neuroscience*, 8, 3616–23.

Friedman, W. J., Dreyfus, C. F., McEwen, B. S. & Black, I. B. (1989). Developmental regulation of tyrosine hydroxylase in the mediobasal hypothalamus. *Developmental Brain Research*, 48, 177–85.

Fung, Y. K. & Lau, Y. -S. (1989). Effects of prenatal nicotine exposure on rat striatal dopaminergic and nicotinic systems. *Pharmacolology, Biochemistry and Behavior*, 33, 1–6.

Gaziri, L. C. J. & Ladosky, W. (1973). Monoamine oxidase variation during sexual differentiation. *Neuroendocrinology*, 12, 249–56.

Goldstein, L. A. & Sengelaub, D. R. (1990). Hormonal control of neuron number in sexually dimorphic spinal nuclei of the rat: IV. Masculinization of the spinal nucleus of the bulbocavernosus with testosterone metabolites. *Journal of Neurobiology*, 21, 719–30.

González, M. I. & Leret, M. L. (1992). Neonatal catecholaminergic influence on behaviour and sexual hormones. *Physiology and Behavior*, 51, 527–31.

Gorski, R. A. (1988). Structural sex differences in the brain: their origin and significance. In *Neural Control of Reproductive Function*, ed. J. M. Lakoski, J. R. Perez-Polo, D. K. Rassin, C. R. Gustavson & C. S. Watson, pp. 33–44. New York: Alan R. Liss, Inc.

Guillamón, A., de Blas, M. R. & Segovia, S. (1988). Effects of sex steroids on the development of the locus coeruleus in the rat. *Developmental Brain Research*, 40, 306–10.

Hardin, C. M. (1973). Sex differences and the effects of testosterone injections on biogenic amine levels of neonatal rat brain. *Brain Research*, 62, 286–90.

Herdon, H. J. & Wilson, C. A. (1985). Changes in hypothalamic dopamine D-2 receptors during sexual maturation in male and female rats. *Brain Research*, 343, 151–3.

Hutchison, J. B. (1991). Hormonal control of behaviour: steroid action in the brain. *Current Opinion in Neurobiology*, 1, 562–70.

Hyyppä, M. & Rinne, U. K. (1971). Hypothalamic monoamines after the neonatal androgenization, castration or reserpine treatment of the rat. *Acta Endocrinologica (Copenh)*, 66, 317–24.

Jansson, J.-O., Edén, S. & Isaksson, O. (1985). Sexual dimorphism in the control of growth hormone secretion. *Endocrine Reviews*, 6, 128–50.

Kelley, D. B. (1988). Sexually dimorphic behaviors. *Annual Review of Neuroscience*, 11, 225–51.

Kincl, F. A. (1990). *Hormone Toxicity in the Newborn*. Berlin, Heidelberg, New York: Springer.

Kolbinger, W., Trepel, M., Beyer, C., Pilgrim, C. & Reisert, I. (1991). The influence of genetic sex on sexual differentiation of diencephalic dopaminergic neurons *in vitro* and *in vivo*. *Brain Research*, 544, 349–52.

Kordon, C. (1988). Hypothalamo-hypophyseal mechanisms involved in the regulation of hormones and behaviour. In *New Concepts in Depression*, ed. M. Briley & G. Fillion, pp. 235–46. Basingstoke, England: Macmillan Press.

Küppers, E., Pilgrim, C. & Reisert, I. (1991). Sex-specific schedule in steroid response of rhombencephalic catecholaminergic neurons *in vitro*. *International Journal of Developmental Neuroscience*, 9, 537–44.

Lacau de Mengido, I., Becú-Villalobos, D. & Libertun, C. (1987). Sexual differences in the dopaminergic control of luteinizing hormone secretion in the developing rat. *Developmental Brain Research*, 35, 91–5.

Leret, M. L., González, M. I., Tranque, P. & Fraile, A. (1987). Influence of sexual differentiation on striatal and limbic catecholamines. *Comparative Biochemistry and Physiology*, 86C, 299–303.

Lichtensteiger, W., Ribary, U., Schlumpf, M., Odermatt, B., & Widmer, H. R. (1988). Prenatal adverse effects of nicotine on the developing brain. *Progress in Brain Research*, 73, 137–57.

Lichtensteiger, W. & Schlumpf, M. (1985). Prenatal nicotine affects fetal testosterone and sexual dimorphism of saccharin preference. *Pharmacology, Biochemistry and Behavior*, 23, 439–44.

Lloyd, T. & Weisz, J. (1978). Direct inhibition of tyrosine hydroxylase activity by catechol estrogens. *Journal of Biological Chemistry*, 253, 4841–3.

Luine, V. N., McEwen, B. S. & Black, I. B. (1977). Effect of 17β-estradiol on hypothalamic tyrosine hydroxylase activity. *Brain Research*, 120, 188–92.

MacLusky, N. J. & Naftolin, F. (1981). Sexual differentiation of the central nervous system. *Science*, 211, 1294–303.

Morrell, J. I., Rosenthal, M. F., McCabe, J. T., Harrington, C. A., Chikaraishi, D. M. & Pfaff, D. W. (1989). Tyrosine hydroxylase mRNA in the neurons of the tuberoinfundibular region and zona incerta examined after gonadal steroid hormone treatment. *Molecular Endocrinology*, 3, 1426–33.

Negro-Vilar, A., Tesone, M., Johnston, C. A., DePaolo, L., & Justo, S. N. (1984). Sex differences in regulation of gonadotropin secretion: Involvement of central monoaminergic and peptidergic systems and brain steroid receptors. In *Sexual Differentiation: Basic and Clinical Aspects*, ed. M. Serio, M. Motta & M. Zanisi, pp. 107–18. New York: Raven Press.

Numan, M. (1988). Maternal behavior. In *The Physiology of Reproduction, Vol. 2*, ed. E. Knobil & J. Neill, pp. 1569–645. New York: Raven Press.

Olsen, K. L. (1992). Genetic influences on sexual behavior differentiation. In *Sexual differentiation. Vol. 11, Handbook* of *Behavioral Neurobiology*, ed. A. A. Gerall, H. Moltz & I. L. Ward, pp. 1–40. New York: Plenum Publ.Co.

Ovtscharoff, W., Eusterschulte, B., Zienecker, R., Reisert, I. & Pilgrim, C. (1992). Sex differences in densities of dopaminergic fibers and GABAergic neurons in the prenatal rat striatum. *Journal of Comparative Neurology*, **323**, 299–304.

Pilgrim, C. & Reisert, I. (1992). Differences between male and female brains – developmental mechanisms and implications. *Hormone and Metabolic Research*, **24**, 353–9.

Reisert, I., Engele, J. & Pilgrim, C. (1989). Early sexual differentiation of diencephalic dopaminergic neurons of the rat *in vitro*. *Cell and Tissue Research*, **255**, 411–17.

Reisert, I. & Pilgrim, C. (1991). Sexual differentiation of monoaminergic neurons – genetic or epigenetic? *Trends in Neurosciences*, **14**, 468–73.

Reisert, I., Schuster, R., Zienecker, R. & Pilgrim, C. (1990). Prenatal development of mesencephalic and diencephalic dopaminergic systems in the male and female rat. *Developmental Brain Research*, **53**, 222–9.

Rhees, R. W., Shryne, J. E. & Gorski, R. A. (1990). Onset of the hormone-sensitive perinatal period for sexual differentiation of the sexually dimorphic nucleus of the preoptic area in female rats. *Journal of Neurobiology*, **21**, 781–6.

Schlumpf, M., Gähwiler, M., Ribary, U. & Lichtensteiger, W. (1988). A new device for monitoring early motor development: prenatal nicotine-induced changes. *Pharmacology, Biochemistry and Behavior*, **30**, 199–203.

Segovia, S., Guillamón, A. & de Blas, M. R. (1990). Age-related changes in cell population of the locus coeruleus in the rat. *Medical Science Research*, **18**, 29–30.

Segovia, S., Pérez-Laso, C., Rodríguez-Zafra, M., Calés, J. M., del Abril, A., de Blas, M. R., Collado, P., Valencia, A. & Guillamón, A. (1991). Early postnatal diazepam exposure alters sex differences in the rat brain. *Brain Research Bulletin*, **26**, 899–907.

Simerly, R. B. (1989). Hormonal control of the development and regulation of tyrosine hydroxylase expression within a sexually dimorphic population of dopaminergic cells in the hypothalamus. *Molecular Brain Research*, **6**, 297–310.

Simerly, R. B., Swanson, L. W. & Gorski, R. A. (1985). The distribution of monoaminergic cells and fibers in a periventricular preoptic nucleus involved in the control of gonadotropin release: immunocytochemical evidence for a dopaminergic sexual dimorphism. *Brain Research*, **330**, 55–64.

Simerly, R. B., Swanson, L. W., Handa, R. J. & Gorski, R. A. (1985). Influence of perinatal androgen on the sexually dimorphic distribution of tyrosine hydroxylase-immunoreactive cells and fibers in the anteroventral periventricular nucleus of the rat. *Neuroendocrinology*, **40**, 501–10.

Sklair, L. & Segal, M. (1990). Target cell stimulation and inhibition of norepinephrine uptake in dissociated rat locus coeruleus cultures. *Developmental Brain Research*, **52**, 191–9.

Stewart, J., Kühnemann, S. & Rajabi, H. (1991). Neonatal exposure to gonadal hormones affects the development of monoamine systems in rat cortex. *Journal of Neuroendocrinology*, **3**, 85–93.

Tennyson, V. M., Barrett, R. E., Cohen, G., Cote, L., Heikkila, R. & Mytilineou, C. (1972). The developing neostriatum of the rabbit: Correlation of fluorescence histochemistry, electron microscopy, endogenous dopamine levels, and (3H)dopamine uptake. *Brain Research*, **46**, 251–85.

Tobet, S. A. & Fox, T. O. (1989). Androgen regulation of an antigen expressed in regions of developing brainstem monoaminergic cell groups. *Developmental Brain Research*, **46**, 253–61.

Toran-Allerand, C. D., Ellis, L. & Pfenninger, K. H. (1988). Estrogen and insulin synergism in neurite growth enhancemnet *in vitro*: mediation of steroid effects by interactions with growth factors? *Developmental Brain Research*, **41**, 87–100.

Toran-Allerand, C. D., Gerlach, J. L. & McEwen, B. S. (1980). Autoradiographic localization of [3H]estradiol related to steroid responsiveness in cultures of the newborn mouse hypothalamus and preoptic area. *Brain Research*, **184**, 517–22.

Vaccari, A. (1980). Sexual differentiation of monoamine neurotransmitters. In *Biogenic Amines in Development*, ed. H. Parvez & S. Parvez, pp. 327–52. Amsterdam: Elsevier/North-Holland Biomedical Press.

Watanobe, H. & Takebe, K. (1987). Involvement of postnatal gonads in the maturation of dopaminergic regulation of prolactin secretion in female rats. *Endocrinology*, **120**, 2212–19.

Weisz, J. & Ward, I. L. (1980). Plasma testosterone and progesterone titers of pregnant rats, their male and female fetuses, and neonatal offspring. *Endocrinology*, **106**, 306–16.

Whorf, R. C. & Tobet, S. A. (1992). Expression of the Raf-1 protein in rat brain during development and its hormonal regulation in hypothalamus. *Journal of Neurobiology*, **23**, 103–19.

Wilson, W. E. & Agrawal, A. K. (1979). Brain regional levels of neurotransmitter amines as neurochemical correlates of sex-specific ontogenesis in the rat. *Developmental Neuroscience*, **2**, 195–200.

Yanase, M., Honmura, A., Akaishi, T. & Sakuma, Y. (1988). Nerve growth factor-mediated sexual differentiation of the rat hypothalamus. *Neuroscience Research*, **6**, 181–5.

Part III: Catecholamines in the CNS of vertebrates: current concepts of evolution and functional significance

20

Catecholamines in the CNS of vertebrates: current concepts of evolution and functional significance

W. J. A. J. Smeets and A. Reiner

Introduction

In the foregoing chapters, a huge amount of information has been presented on catecholamines in both developing and adult brains of representatives for each class of vertebrates. As mentioned in the preface, each chapter contributor was asked to minimize out-class comparisons to avoid, chapter after chapter, extensive and repetitious comparisons of catecholamine distributions among classes (and particularly to mammals). In the present, final chapter, however, all such information is brought together, and an effort is made to draw out the key evolutionary implications and the basic principles of organization and function of brain catecholamine systems. Before making general statements or drawing conclusions, it seems appropriate, however, to comment on current concepts of evolutionary processes and on both quantitative and qualitative aspects of immunohistochemical methods.

Evolutionary considerations

Most of our knowledge with respect to the changes in brains and the processes that have produced these changes is derived from the patterns of variation exhibited by the brains of living vertebrates. Except for some information on the relative size and external shape of brains from endocranial casts (Jerison, 1990), no data are available on brain organization in extinct vertebrates because brains, like most other soft structures, do not fossilize. Despite this limitation, the variation in brain organization among living vertebrates can often be used in combination with knowledge of the evolutionary relationships among these groups to infer many of the major changes in brain structure and function that have occurred during the course of vertebrate evolution. The notion that evolution is not a linear process is now generally accepted. Vertebrate phylogeny is characterized by several groups of organisms evolving in parallel lines away from a common ancestral population for as long as 500 million years. Thus, the living descendants of those ancestral groups comprise a mosaic of traits, some of which have been retained with little or no change (primitive) and others of which have arisen or changed greatly over time (derived). Moreover, many of the derived traits have occurred in more than one group indepedently and at different times.

Reconstruction of the course of evolutionary change in brain organization, although not an easy task, can be achieved if adequate and thorough data are available for the structures or systems of interest and if cladistic principles are applied (as outlined by Northcutt, 1984). This holds also for the organization of catecholamine systems in the CNS of vertebrates. In 1984, André Parent and his colleagues summarized the then existing literature on that topic in Volume 2 of the *Handbook of Chemical Neuroanatomy Series* edited by Björklund and Hökfelt. From that review it became clear that, compared to mammals, little information was available on the structure and organization of catecholamine systems in nonmammalian vertebrates. The then available data for non-mammalian vertebrates were exclusively based on results obtained by the relatively insensitive formaldehyde induced fluorescence (FIF) technique. During the last decade, however, we have witnessed a dramatic increase in the number of immunohistochemical studies dealing with catecholamines in the CNS of non-mammalia using antibodies against catecholamines or their synthesizing enzymes, as evidenced by the chapters in this book. Moreover, in that same volume, the reviews on catecholamine systems of mammals dealt primarily with the brains of rats. More recent investigations, however, in other mammalian species including primates and human, have revealed notable differences in the organization of these systems among mammals. A timely review of comparative aspects of catecholamine systems in the brain of vertebrates (including mammals), therefore, seemed to be warranted in order to characterize the evolution of catecholamine systems. We believe that the analysis of interspecies variation that such an exercise would necessarily entail would also address the issue of the degree to which it is legitimate to generalize the results obtained in one species to another. This question is particularly important in view of the general practice in research on mammals to readily generalize from rats to humans. As we will see, in many cases, catecholamine systems in reptiles and birds more closely resemble those in humans than do those in rats.

Evaluation of the techniques used in catecholamine research

As discussed in Chapter 1, immunohistochemical labeling methods have become the method of choice for studying the localization of catecholamines in the nervous system. The availability of antisera against specific enzymes involved in catecholamine synthesis and against specific catecholamines has made it possible to use immunohistochemical methods to determine which neurons contain specific catecholamine enzymes or specific catecholamines themselves. This provides a strategy for determining more precisely than previously with FIF methods the types of catecholamines produced as a biologically active end product by the various catecholaminergic neurons of the brain. For example, an investigator can determine that a neuron is dopaminergic if it contains TH and AADC, but not DBH or PNMT. If these cells also label for dopamine, but not for noradrenaline or adrenaline, then it is be doubly clear that the cells in question are dopaminergic. Similarly, a noradrenergic neuron is one that contains TH, AADC and DBH, as well as noradrenaline. Such a neuron might also label for dopamine, since in this case dopamine would serve as a precursor for noradrenaline. Application of this strategy within a comparative context is extremely valuable, since it makes it possible to compare specific CA cell groups across vertebrate species in terms of the precise catecholamine they contain and in terms of how that CA may have changed during evolution. Such a comparative approach, however, is not without pitfalls. The absence of immunolabeling for a particular CA enzyme may either mean that the enzyme is absent or that the antiserum used does not crossreact avidly with the evolutionary variant of the enzyme found in the species under study. For example, the general absence of immunolabeling for PNMT in the A2 region in anamniotes does not necessarily mean that there are no adrenergic neurons present in this region. To some extent, the use of immunolabeling for the catecholamines themselves can be used as a way to cross-check the information provided by immunolabeling for enzymes. For example, one could use anti-adrenaline antisera to determine if adrenergic neurons were indeed absent from the A2 region in anamniotes. Unfortunately, antibodies against adrenaline reveal immunolabeling in the adrenal medulla, but not in the CNS.

Even the approach with antisera directed against catecholamines themselves must be used with caution, however, since the relative abundance of different catecholamines in neurons could more reflect their accumulation as a precursor pool (due to slow conversion to the next CA) than as a final endproduct. Finally, all immunolabeling studies need to recognize that absence of immunolabeling for a particular substance does not prove absence of that substance. The levels of that substance may be low but functionally significant. Thus, in applying a comparative immunolabeling strategy, the possibilities that antisera sensitivity could vary or that enzyme or catecholamine levels could be variable (thereby leading to false negatives) need to be considered. Nonetheless, even with these caveats, clear conclusions can be reached about the evolution of many CA cell groups. A greater impediment, in fact, is the relative paucity of data for CAs in many groups of vertebrates. These problems are acute for groups

Table 20.1. *Distribution of the main catecholaminergic cell groups in the CNS of vertebrates. For explanation of symbols, see Table 20.2*

	Caudal rhomb	Rostral rhomb	Midbrain	Diencephalon	Olfactory bulb	Retina
Cyclostomes	+	+	−	+	+	+
Chondrichthyans	+	+	±	+	+	+
Osteichthyans	+	+	±	+	+	+
Amphibians	+	+	+	+	+	+
Reptiles	+	+	+	+	+	+
Birds	+	+	+	+	+	+
Mammals	+	+	+	+	+	+

such as agnathans, but are even present for amniote groups if the goal is to make a systematic cladistic analysis of the distribution and evolution of specific traits. Even with all of the above noted limitations, however, the available data make possible many insights and new conclusions on the evolution of CA systems among vertebrates as discussed below.

Comparative analysis of catecholamine systems of vertebrates using the A1–A17/C1–C3 nomenclature

In this section, the catecholaminergic cell groups in the CNS of vertebrates will be surveyed using the nomenclature as originally proposed by Dahlström and Fuxe (1964). In their classic paper, these authors described the distribution of catecholamine containing cell bodies in the brain of rats as demonstrated by means of the formaldehyde induced fluorescence technique. Dahlström and Fuxe recognized 12 groups of catecholaminergic neurons which they labeled A1 to A12, from caudal to rostral. Hökfelt and his colleagues (1984a,b) more recently re-analyzed the original classification of catecholamine cell groups in the brain of rats with immunohistochemical methods using antibodies against the biosynthetic enzymes TH, DBH and PNMT. These immunohistochemical studies have not only largely confirmed the presence of the catecholamine cell groups as previously demonstrated with the fluorescence techniques, but also identified additional CA cell groups. Moreover, with antisera against the different enzymes it became possible to discriminate between the different catecholamines.

Generally, six main groups of catecholamine cells are recognized in the brain of vertebrates: (1) a caudal rhombencephalic group (A1–A3/C1–C3); (2) a rostral rhombencephalic group (A4–A7); (3) a mesencephalic group (A8–A10); (4) a diencephalic group (A11–A15); (5) an olfactory bulb group (A16); and a retinal group (A17). As shown in Table 20.1, the main groups of catecholamine cell bodies are present in the CNS of all vertebrates studied so far. However, the midbrain is the area that, speaking in general terms, shows the greatest variation in the distribution of CA cell bodies among vertebrates. These differences may be partly due to the immunohistochemical techniques, but they certainly also reflect true differences in the organization of CA systems between the different classes of vertebrates. THi/DAi or fluorescent cells have never been observed in the midbrain of cyclostomes, holocephalians (the sistergroup of elasmobranchs), and actinopterygian fish, a finding that will be discussed later in more detail (see pp. 468–477).

Whereas most catecholamine cell groups are found in roughly corresponding locations in the brain of vertebrates, a closer examination of their position as well as their actual catecholamine content reveal notable differences not only between different classes of vertebrates, but also within a single class. In the following sections, we have attempted to analyse the distribution of CA cell groups in the brains of vertebrates studied so far in a comparative way using the classification as originally proposed by Dahlström & Fuxe (1964) but updated and extended by Hökfelt et al. (1984b) (see Table 20.2).

Caudal rhombencephalon: A1–A3/C1–C3

Usually, two distinct CA cell groups are recognized in the caudal rhombencephalon of vertebrates: a ventrolateral tegmental group and a dorsomedial group in the nucleus tractus solitarii/area postrema complex. Within the ventrolateral tegmental group, noradrenergic (A1) and adrenergic (C1) cell populations are distinguished. Similarly, the dorsomedial group

Table 20.2. Comparative analysis of catecholamine cell bodies in the brain of vertebrates using the A1–A17/C1–C3 nomenclature

	A1	A2	A3	A4	A5	A6	A7	A8	A9	A10	A11	A12	A13	A14	A15	A16	A17	C1	C2	C3
Cyclostomes	?	?	–	–	–	+	?	–	–	–	?	?	?	?	?	+	+	?	?	–
Chondrichthyans	+	+	–	–	±	+	?	–	±	±	?	?	?	?	?	+	+	?	?	–
Osteichthyans	+	+	–	–	–	+	?	–	–	±	?	?	?	?	?	+	+	+	+	–
Amphibians	+	+	–	–	–	+	?	+	?	+	+	+	?	+	+	+	+	?	?	–
Reptiles	+	+	–	–	±	+	±	+	+	+	+	+	?	+	+	+	+	+	±	–
Birds	+	+	–	±	+	+	?	+	+	+	+	+	+	+	+	+	+	+	+	–
Mammals	+	+	±	+	+	+	+	+	+	+	+	+	?	+	+	+	+	+	±	±

+ present in all species studies so far.
± present in some species, but not in others.
? present, but not recognizable as a separate entity.
– not found.

consists of noradrenergic (A2) and adrenergic (C2) cell populations. As shown in Table 20.2, A1 and A2 cell groups are present in the caudal brainstems of all vertebrates studied. Even in cyclostomes, THi cells are found at caudal rhombencephalic levels, yet it is currently unclear how these cell bodies located ventral to the motor nucleus of the vagal nerve in the lamprey brain relate to the A1/A2 cell groups of other vertebrates. An A3 cell group was originally described for rat brains as consisting of a few small, very weakly to weakly fluorescent cells within the dorsal accessory inferior olive (Dahlström & Fuxe, 1964), but could not be confirmed by subsequent immunohistochemical studies (Armstrong et al., 1982; Hökfelt et al., 1984b). In fact, it may be questioned whether an A3 group exists, since studies of other mammalian species as well as reptiles and birds have never revealed catecholaminergic cell bodies in a corresponding position. The same holds for the C3 adrenergic cell group. In rats, the cells constituting this group lie along the midline within and dorsal to the medial longitudinal fasciculus. However, in other mammalian species, with the possible exception of hamsters (Vincent, 1988), as well as in reptiles and birds, such a cell group is not evident. Thus, on the basis of TH immunohistochemistry, it may be concluded that A1–A2/C1–C2 cell groups are general, primitive traits of catecholamine systems in the CNS of vertebrates, whereas A3 and C3 cell groups are most likely derived features of the rodent brain.

A closer examination of the distribution of catecholamine cell bodies in the A1 and A2 groups reveal considerable differences in location and numbers of cells between species, even within a single class. Both in mammals (Kitahama et al., Chapter 8) and reptiles (Smeets, Chapter 6), these cell groups are represented by compactly aggregated cells in some species (rodents, lizards), whereas in others (primates, carnivores, turtles) the cells show a more dispersed distribution. Studies in mammals and reptiles have also raised questions about the actual catecholamine involved in the caudal rhombencephalic cell groups. Previously, it was thought that the A1/A2 cell groups consisted of noradrenergic cell bodies, but studies by means of antibodies raised against the various catecholaminergic biosynthetic enzymes (e.g. Hökfelt et al., 1984a; Jaeger et al., 1984) have provided evidence that some cells in the A2 group may have dopamine or L-DOPA as their endproduct. Recently, this notion got further support by immunohistochemical studies with antibodies against dopamine in bony fish (Meek, Joosten & Steinbusch, 1989), amphibians (González & Smeets, 1991; González, Tuinhof & Smeets, 1993) and reptiles (Smeets & Steinbusch, 1990), which demonstrated strongly immunoreactive cell bodies in the dorsomedial group. Although double labeling studies have not been performed yet in these species, the number and distribution of THi, DAi, NAi, DBHi or PNMTi cell bodies suggest that cells containing dopamine, noradrenaline or adrenaline intermingle in the A2 group. Dopaminergic cells probably also intermingle with other catecholaminergic cell groups in the ventrolateral tegmentum (A1/C1 groups), as shown for reptiles.

Studies with PNMT antisera have revealed also considerable differences in the distribution of putative adrenergic cell bodies belonging to the C1 and C2 groups. Among mammals, for example, sheep lack a C1 group (Tillet, 1988), but the most discrepant pattern is found in guinea pigs. Neither antibodies against PNMT nor high-performance liquid chromatography have demonstrated the presence of the enzyme or its synthetic product, i.e. adrenaline, in the brain of this species (Fuller & Hemrick-Luecke, 1983; Cumming et al., 1986). Similar discrepancies are observed in reptiles. Whereas in turtles PNMTi cell bodies are present in locations that correspond to those of the C1 and C2 of mammals, in lizards only a C1 group seems to exist (Smeets & Jonker, 1990). The meaning of these findings and their possible bearing on our concepts of catecholamine systems will be discussed below (pp. 473–476).

Rostral rhombencephalon: A4–A7
In the rat pons, noradrenergic cells are classified into four groups, i.e. A4, A5, A6 and A7. Of these cell groups, the A6 (i.e. locus coeruleus) is the most prominent one. As mentioned previously (Kitahama et al., Chapter 8), the A6 has been subdivided into dorsal (locus coeruleus), ventral (locus subcoeruleus), and lateral (parabrachial, Kolliker-Fuse nuclei) subgroups. In all non-mammalian species studied, a THi cell group, comparable to the A6 group of mammals, has been recognized in the rostral rhombencephalon (Table 20.2). Studies with DBH- and NA antisera have further fostered the notion that the presence of the A6 cell group is a primitive or general trait of the noradrenaline system in vertebrates. There are, however, considerable differences among species, even within a single class, as demonstrated for reptiles (Smeets, Chapter 6) and mammals (Kitahama et al., Chapter 8), with respect to the packing density and number of cells that constitute the A6 group.

In rats, the A4 group consists of cells that form a dorsolateral continuation of the A6 complex and lie in the lateral part of the roof of the fourth ventricle, closely related to the ependymal cell lining. A similar cell group is present in most mammalian species studied, but not in sheep. In non-mammalians, an A4 group is generally not recognized (Table 20.2). Even in birds, the presence of a comparable cell group has not been unequivocally demonstrated (Reiner et al., Chapter 7).

The remaining two rostral rhombencephalic groups, i.e. A5 and A7, have not always been recognized as separate cell groups in the brain of mammals (Kitahama et al., Chapter 8). The same holds for other classes of vertebrates (Table 20.2). It is, therefore, not surprising that many authors have proposed that the A4–A7 cell groups be grouped together under one label, namely locus coeruleus.

Midbrain: A8–A10

Originally, the midbrain DA histofluoroscent cells of rats were classified into three groups, i.e. A8 (retrorubral), A9 (substantia nigra) and A10 (ventral tegmental area), on the basis of their localization (Dahlström & Fuxe, 1964). Later, immunohistochemical studies led to a further division of the A9 cell group into a ventral (A9v) and a lateral (A9l) subgroup, and of the A10 group into dorsorostral (A10dr), dorsocaudal (A10dc), ventrorostral (A10vr) and caudal (A10c) subgroups (Hökfelt et al., 1984b). The recent, updated classification is adaptable to most mammalian species, but hardly to the human midbrain (Kitahama et al., Chapter 8). Apart from the more or less distinct A8–A10 cell groups, the midbrain of the various mammalian species contains many diffusely arranged THi/DAi cell bodies throughout the tegmentum. In primates, including humans, these latter cells are quite numerous. Among non-mammalians, distinct midbrain DA cell groups are present in reptiles and birds. In both classes of vertebrates, a tripartite division of these cells on the basis of position is also possible. Although THi/DAi cell bodies are also present in the midbrain of amphibians, two differences are noted when compared to amniotes. Firstly, the size of the amphibian midbrain group is limited in rostrocaudal direction, extending only from the di-mesencephalic transition zone to the exit of the oculomotor nerve. Secondly, further subdivision into A8–A10 groups is not possible since the THi/DAi neurons in amphibians are only found along the midline and, therefore, appear to correspond only to the A10.

Until recently, it was thought that midbrain DA cell groups did not occur in fish. This notion was corroborated by immunohistochemical studies that failed to stain cells in the midbrain of various actinopterygian species (Meek, Chapter 4). A study of the lungfish Protopterus, however, revealed the existence of a midbrain THi cell group in this member of the Sarcopterygii, the sister group of the Actinopterygii (Reiner & Northcutt, 1987). Thus, ray-finned bony fish, but not such lobe-finned bony fish as lungfishes, seem to lack a midbrain DA cell group. This is in sharp contrast with cartilaginous fishes where all elasmobranchs (sharks, skates and rays) studied so far have been shown to possess a midbrain DA cell group (Stuesse et al., Chapter 3) comparable to that observed in amphibians in its rostrocaudal extent. In elasmobranchs, however, two groups of DA cells, one located along the midline and another group lateral to the red nucleus, are usually recognized. Remarkably, such groups were not observed in the sister group of elasmobranchs, the Holocephali which are generally considered as more primitive than elasmobranchs. In cyclostomes, immunohistochemical studies have not revealed a midbrain CA cell group (Pierre et al., Chapter 2), although the presence of distinct CA cell groups in other parts of the brain could be clearly established. The possible evolutionary significance of the apparent variation in the mesencephalic A8–A10 cell group among vertebrates will be discussed on pp. 473–476.

Diencephalon (A11–A15)

In all vertebrates studied, numerous THi/DAi cell bodies are found in the diencephalon. In mammals, in particular rats, they have been classified into five groups, labeled A11 through A15. From the present survey by Tillet (Chapter 9), it becomes clear that in all mammalian species studied, THi cell bodies constitute two rostrocaudally oriented columns on each side of the brain, i.e. a dorsolateral and a ventrolateral group, interconnected by CA cell bodies that lie along the ventricular wall. On the basis of their position in transverse sections, the CA cell bodies had been divided originally into four groups (for review, see Björklund & Lindvall, 1984): A11 (caudal diencephalic cell group located in the periventricular gray matter of thalamus, hypothalamus and rostral midbrain), A12 (tuberal cell group located in the arcuate nucleus and in the adjacent part of the periventricular nucleus), A13 (dorsal hypothalamic cell group within the zona incerta), and A14 (rostral periventricular cell group). More recent studies (see Hökfelt et al., 1984b; Tillet, Chapter 9) revealed more THi cell bodies leading not only to the defining of a new cell group (A15), but also to a further subdivision of diencephalic CA cell groups. Instead of adding new numbers, Hökfelt et al. (1984b) preferred to use letters which indicate the location of those cell groups. Ventral and dorsal parts have been recognized in the A12 (A12d, A12v), the A14 (A14d, A14v) and the A15 (A15d, A15v) cell group. An additional lateral extension of the A14 (A14l) was also found. The A13 cell group consists not only of CA cells within the zona incerta, but also of cells extending caudally along the ventromedial aspects of the mammillothalamic tract (A13c) and rostrally into the internal capsule (A13l). Studies in a variety of mammalian species have shown that there are notable differences in the extent, number and location of the diencephalic CA cell groups among the different species (Tillet, Chapter 9). For example, in humans and primates, numerous THi cells belonging to the A15 group are located within the supraoptic nucleus and the paraventricular hypothalamic nucleus, whereas only a few scattered THi cells are observed in the other mammalian species studied (rodents, cats, cows, sheep).

The general pattern of interconnected dorsolateral and ventrolateral CA cell groups is also observed in birds (Chapter 7), reptiles (Chapter 6), and, to a lesser extent, in anamniotes (Chapters 2–5). For example, on the basis of position, the THi/DAi cell bodies in the suprachiasmatic nucleus of amphibians and in the rostroventral part of the periventricular hypothalamic nucleus of reptiles resemble the A12 group of birds and mammals. The THi/DAi cells that accompany the nucleus of the amphibian periventricular organ and the DAi cell bodies in the caudodorsal part of the periventricular hypothalamic nucleus of reptiles may correspond to the A14 group of birds and mammals. Similarly, the CA cell group observed in the posterior tubercle of anamniotes partly resembles the A11 group. However, a solid comparison of the various cell groups between different classes of vertebrates is hampered by the lack of developmental and connectional data. Moreover, as noted by Puelles and Medina (Chapter 16), a classification merely based on the anatomical position in transverse sections is potentionally misleading (see also pp. 473–476).

Olfactory bulb (A16)

The most rostral dopamine cells in the brain of vertebrates are found in the olfactory bulb and designated as the A16 group. In the main olfactory bulb of rats, the majority of these cells are located in the glomerular layer, but some are found in the external plexiform layer. A similar pattern is observed in other mammalian species, although differences in number of cells occur. For example, whereas in rats, cats, guinea pigs and rabbits, the number of dopamine cells in the external plexiform layer is low, in the Syrian hamster a large population of THi cells is observed in the corresponding layer (Baker, 1986). Although not much attention has been paid to it, in most mammalian species studied additional THi cell bodies were found in the deep layers of the main olfactory bulb including the mitral cell layer, the internal plexiform layer and the granule cell layer (Davis & Macrides, 1983; Baker, 1986; Tillet, Thibault & Dubois, 1987).

Dopamine cells are a general feature of the vertebrate olfactory bulb as can be concluded from Table 20.2. The exact location of these cell bodies in the olfactory bulbs of different vertebrates, however, deserves some comment. By means of fluorescent techniques and by immnunohistochemical studies with DA antibodies, the majority of olfactory bulb DA cells were recognized in a periglomerular position in reptiles (Halasz et al., 1982; Smeets, 1988a,b; Smeets & Steinbusch, 1990), birds (Reiner et al., Chapter 7) and mammals (Halasz, Ljungdahl & Hökfelt, 1978; Hökfelt et al., 1984a,b). Also in elasmobranchs, a periglomerular position of DAi cell bodies prevails (Meredith & Smeets, 1987), whereas such neurons are present, but less numerous, in amphibians (González & Smeets, 1991; González et al., 1993). In elasmobranchs, amphibians and amniotes, a smaller number of cells has consistently been found in the external plexiform layer by means of antibodies against DA or with the FIF technique. With DA immunohistochemistry, a different pattern is, however, observed in lampreys and bony fishes, where the majority of immunoreactive cells are found in the deep cellular layers of the olfactory bulb (Pierre et al., Chapter 2; Roberts, Meredith & Maslam, 1989; Ekström, Honkanen & Steinbusch, 1990).

Studies with TH antibodies in elasmobranch fish (Northcutt et al., 1988), lungfish (Reiner & Northcutt, 1987), teleost (Alonso et al., 1989), amphibians (González & Smeets, 1991; González et al., 1993) and reptiles (Halasz et al., 1982; Smeets, 1988a,b; Smeets & Steinbusch, 1990) have demonstrated that immunoreactive cell bodies are additionally present in the internal granular layer of the olfactory bulb. Since the latter cells do not contain dopamine in immunohistochemical detectable amounts, at least in amphibians and reptiles, it has been suggested that these cells contain DOPA as the endproduct of the catecholamine biosynthetic pathway (Smeets & Steinbusch, 1990).

Even though it must be kept in mind that TH antibodies may reveal cell bodies that do not contain DA, it is obvious that there are considerable differences between species. It may be speculated that a predominantly periglomerular location of DAi neurons in the olfactory bulb is a shared derived character of amniotes, whereas a location in deeper bulbar layers is the primitive condition. If so, the distribution of DA cell bodies in the olfactory bulbs of amphibians represents a transitional stage, whereas that of elasmobranchs might be considered as an independently derived condition.

Retina (A17)

Many neuroactive substances, e.g. GABA, glycine, glutamate and catecholamines have been recognized as neurotransmitters or neuromodulators in the vertebrate retina (for reviews, see Brecha, 1983; Ehinger, 1983; Massey & Redburn, 1987). Dopamine is found in high concentrations in retina and is also considered a major retinal neurotransmitter. Noradrenaline and adrenaline have also been detected in retinas of teleost and several mammalian species (see e.g. Hadjiconstantinou, Cohen & Neff, 1983; Osborne & Nesselhut, 1983; Hadjiconstantinou et al., 1984; Jaffe, Urbina & Drujan, 1991) although these catecholamines are present in much smaller quantities than dopamine. The presence of dopamine cells in the retina of vertebrates has been demonstrated by several techniques, such as the FIF-technique, immunohistochemistry and high-affinity uptake techniques (see e.g. Brecha, 1983; Massey & Redburn, 1987; Mariani & Hokoc, 1988; Nguyen-Legros, 1988; Tauchi, Madigan & Masland, 1990; Crooks & Kolb, 1992; Chapters 2–7, this volume). It appears that, in the retinas of all species studied so far, the majority of DA-containing cell bodies are found at the inner border of the inner nuclear layer among amacrine cells, where they constitute only a small proportion of the cells in this region. In addition, some cells are displaced to either the inner plexiform layer or the ganglion cell layer. Generally, two types of dopaminergic amacrine cells are recognized. Type 1 CA cells are strongly TH immunoreactive, are found exclusively within the inner nuclear layer and have dendrites that ramify in a dense plexus in stratum 1 of the inner plexiform layer. Type 2 CA cells are less immunoreactive, lie in the inner nuclear layer, the inner plexiform layer or the ganglion cell layer and emit dendrites that are stratified in stratum 3 and at the junction of layers 4 and 5 of the inner plexiform layer (Tauchi, Madigan & Masland, 1990). A subpopulation of type 1 CA cells, i.e. the interplexiform cells, possess additional dendrites that pass through the inner nuclear layer to reach the outer plexiform layer (Savy et al., 1989; Larabi & Nguyen-Legros, 1991; Schütte, 1991).

From the available data one gets the impression that the distribution of retinal THi cells in all vertebrates is quite similar, but the presence of THi interplexiform cells and cells in the inner plexiform layer and ganglion cell layer has not been unequivocally demonstrated in all species studied (see, e.g. Ehinger, 1976; Witkovsky, Eldred & Karten, 1984; Studholme, Yazulla & Phillips, 1987). The reported differences may be partly due to staining problems (non-detectable levels of TH or DA content), but true variations between even closely related species cannot be excluded (see, e.g. Studholme et al., 1987).

Non-classified catecholamine cell groups in the CNS of vertebrates

Apart from the CA cell groups mentioned in the previous section, there are numerous catecholaminergic neurons in the brain of vertebrates that were not observed or identified as a group in previous studies. A reason for this neglect could be that most studies, particularly those in non-mammalian vertebrates, were carried out with the less sensitive fluorescent techniques. Moreover, some cell groups were found in one species but not in others, thus making conclusions hazardous. However, as witnessed in the previous chapters, enough data have now been accumulated to separate corn from wheat. In the next paragraphs, CA cell groups in the following brain structures, from caudal to rostral, will be discussed (see Table 20.3): spinal cord, pretectum and habenular region, periventricular organ (OPH), cortex and basal forebrain.

Spinal cord

An interesting result of catecholamine research during the last decade is the finding of a catecholamine, putatively dopamine, cell group in the spinal cord of non-mammalian vertebrates. CSF-contacting dopaminergic cells, ventral to the central canal, have been reported in lampreys (Pierre et al., Chapter 2), cartilaginous fish (Stuesse et al., Chapter 3), amphibians (González & Smeets, Chapter 5), reptiles (Smeets, Chapter 6), and birds (Reiner et al., Chapter 7). In bony fishes, such cells are demonstrated in the chondrostean fish, *Acipenser ruthenus* (Kotrschal, Krautgartner & Adam, 1985) and the holostean fish, *Lepisosteus osseus* (Parent & Northcutt, 1982), both of which may be considered as models of the primitive actinopterygian stock. In advanced actinopterygians,

Table 20.3. *Comparative analysis of non-classified catecholamine cell groups in the CNS of vertebrates*

	Spinal cord	Pretectum	Habenular region	OPH	Cortex	Basal forebrain
Cyclostomes	+	−	−	+	−	+
Chondrichthyans	+	−	+	+	+	+
Osteichthyans	+	+	−	+	−	+
Amphibians	+	+	−	+	−	−
Reptiles	+	+	−	+	±	−
Birds	+	+	−	+	−	−
Mammals	±	−	+	−	+	+

i.e. teleost fish, CSF-contacting cells in the spinal cord are not consistently found. For example, they are present in the European eel (Roberts, Meredith & Maslam, 1989) but not in other teleosts, such as goldfish (Hornby, Piekut & Demski, 1987), mormyrid fish (Meek, Joosten & Steinbusch, 1989), gymnotid fish (Sas, Maler & Tinner, 1990) and perciform fish (Batten et al., 1993).

Similarly conflicting results have been reported for mammalian species. Earlier studies by means of fluorescence histochemical and immunohistochemical studies led to the conclusion that all catecholamine fibers in the spinal cord have a supraspinal origin. However, later studies with more sensitive modifications of the FIF technique in combination with colchicine treatment (Singhaniyom, Wreford & Güldner, 1983) or by means of TH immunohistochemistry (Dietl et al., 1985; Mouchet et al., 1986) revealed CA containing cells in the spinal cord of rats. In contrast to the CA cells observed in the spinal cord of non-mammalian vertebrates, the cells in the rat brain are located in the dorsal half of the spinal cord and do not possess processes that contact the CSF. Furthermore, a recent study with DA and NA antibodies failed to demonstrate the presence of catecholamine perikarya in the spinal cord of rats (Mouchet, Manier & Feuerstein, 1992).

An important finding which has received unfortunately little attention has been reported by Pindzola, Ho & Martin (1990) for the North American opossum, *Didelphis virginiana*. Investigating the development of catecholaminergic projections to the spinal cord in pouch-young animals ranging from postnatal day (PD) 1 to PD105, these authors observed THi cell bodies in the spinal cord from PD1 to PD90. At PD1, these cells are sparse and found mainly in the ventral horn at cervical levels. However, their number increases with age becoming particularly numerous between PD32 and PD90; and also, the distribution of the cells changes, extending, at these later stages, from cervical to lumbosacral levels. Moreover, THi cells are found at these stages not only in the ventral and dorsal horn, but also around the central canal. Some of the cells located in the ependymal zone have processes that extend into the canal itself, thus resembling the THi/DAi cell bodies in the spinal cord of non-mammalian species. After PD90, the number of THi cells decreases, but in adult animals an occasional immunoreactive cell is found around the central canal at cervical levels (Pindzola et al., 1990). In this respect, it is also important to note that Jaeger et al. (1984) found an extensive group of AADCi cells, labeled by them as the D1 group, in the spinal cord of rats. The cells are found throughout the spinal cord but their density varies, being highest at cervical levels. A characteristic feature of the D1 cells is that they possess a dendritic process that extends into the central canal. These remarkable, seemingly conflicting results concerning CA cell bodies in the spinal cord of vertebrates and their possible bearing on our concepts of evolution of the spinal cord CA system will be discussed further on pp. 473–476.

Pretectum and habenular region

A pretectal CA cell group was first detected in reptiles, i.e. the monitor lizard, *Varanus exanthematicus*, by Wolters, Ten Donkelaar & Verhofstad (1984) with antibodies to TH. Subsequent studies with DA antisera have demonstrated that a dopaminergic pretectal posterodorsal cell group is a general feature of reptiles (Smeets, Hoogland & Voorn, 1986; Smeets, 1988b; Smeets & Steinbusch, 1990). Immunohistochemical studies with TH- and/or DA- antibodies have revealed comparable cell groups in bony fishes, amphibians and birds (Chapters 4, 5 and 7). Although it is difficult to compare the pretectal cell group of different vertebrates, it should be noted that they invariably lie close to the posterior commissure. Such a dopaminergic cell group seems to be absent in the pretectum of cartilaginous fish and mammals. However, in these two classes of vertebrates, DAi or THi cell bodies are present in the habenular region (Hökfelt et al., 1984b; Meredith & Smeets, 1987) (Table 20.3). Remarkably, studies with AADC antibodies in rats (Jaeger et al., 1984) labeled not only cells in the habenular region (D6 cell group) but also numerous cells scattered dorsally and ventrally to the posterior commissure. This latter cell group, labeled the D5 group, resembles in its position the pretectal CA cell group of nonmammalian species, suggesting that a pretectal group should be considered a primitive trait of CA systems in the CNS of vertebrates (see also pp. 473–476).

Hypothalamic periventricular organ

An intriguing feature of catecholamine systems in the CNS of vertebrates is the presence of hypothalamic cells that are fluorescent with the FIF technique and immunoreactive to antibodies directed against catecholamines, but that do not stain with antibodies against their synthetic enzymes. Such cells occur in the hypothalamic periventricular (or paraventricular) organ (OPH) of nonmammalian vertebrates (for review, see Smeets & Steinbusch, 1990; Chapters 2–7; Table 20.3). It is now firmly established that cells in the OPH contain dopamine and, at least in some teleost fish (Meek, Joosten & Hafmans, 1993), amphibians (González & Smeets, 1993) and reptiles (Smeets & Steinbusch, 1989, 1990), noradrenaline. The most straightforward explanation for the apparent lack of TH- and/or DBH- immunoreactivity in the presumed catecholamine cells of the OPH is that the cells lack the enzymes involved in the biosynthesis of the monoamines. This would imply that the DA- and NA-immunoreactivity in these cells is the result of a local uptake mechanism. It has been suggested that the cells could accumulate DA and NA from the CSF of the third ventricle (Smeets & González, 1990; Smeets & Steinbusch, 1990), since catecholamine cells that lack TH- and/or DBH-immunoreactivity, are in direct contact with the ventricle. Uptake of catecholamines from the CSF by these neurons would be in line with the proposed receptive nature of the CSF-contacting processes of the cells in the OPH (Vigh-Teichman & Vigh, 1974). In agreement with this suggestion is the finding that, after intraventricular injection of 3H-dopamine in the brain of frogs, radioactive labeling appears in the intraventricular processes of the CA cells in the infundibular recess (Nakai, Ochiai & Shioda, 1977). Furthermore, a recent study in the lizard *Gekko gecko* has shown that intraventricular injections of α-methylparatyrosine, a DA synthesis inhibitor, have no effect on the DA immunoreactivity in the CSF-contacting cells of the OPH, but cause a dramatic decrease in immunoreactivity in other DA cell groups (Smeets, Kidjan & Jonker, 1991). These findings further corroborate the suggestion that cells in the OPH accumulate their catecholamines from an external source.

Other explanations for the observed discrepancy are a low concentration of TH and/or DBH or a different form of TH and DBH, not recognized by the TH/DBH antibodies used in studies of nonmammalian CA systems. However, the possibility of histochemically non-detectable concentrations of TH and DBH combined with intense staining for DA and NA seems to be highly unlikely. Moreover, it should be noted that THi cells are never observed in the OPH of developing brains of amphibians (González et al., Chapter 14), reptiles (Medina, Puelles & Smeets, Chapter 15) and birds (Puelles & Medina, Chapter 16), giving further support to their accumulating nature.

Mammals do not possess DAi and/or NAi CSF-contacting cells. The presence of CA-accumulating perikarya in the periventricular organ of nonmammalian vertebrates is, therefore, considered to be a primitive trait of CA systems in the vertebrate brain that has been lost in mammals.

Cortex and basal forebrain

From the foregoing chapters, it is evident that catecholaminergic neurons are also present in the telencephalon proper of many vertebrate species. Whereas fluorescent studies failed to detect such neurons, about a decade ago, Specht et al. (1981) and Berger et al. (1985) demonstrated the existence of neurons that transiently expressed TH immunoreactivity in the developing, but not adult, rat brain. Subsequent studies (for references, see Dubach, Chapter 11 and Berger & Gaspar, Chapter 12) revealed that TH immunoreactive cells are present in adult primates, including human. In non-mammalian vertebrates, THi cell bodies were also identified in both developing and adult brains. As shown in Table 20.3, the exact location of these cells varies among vertebrate classes. For example, THi cells are observed in the basal forebrain of cyclostomes, cartilaginous fish and bony fish. On the contrary, cortical THi cell bodies are found only in cartilaginous fish and some reptiles. Until now, THi cell bodies were not reported for the telencephalon proper of amphibians and birds.

Since FIF studies never showed cells in cortical

and basal forebrain areas, the results obtained by TH immuno-histochemistry were reported with caution. However, more recent studies with antibodies against dopamine have reinforced the notion that catecholaminergic, most likely dopaminergic, neurons are present in the telencephalon of vertebrates. The sparse data available indicate that they appear relatively late in development (Verney et al., 1991; Ekstrom et al., Chapter 13).

Finally, there is now evidence that some Purkinje cells in the mammalian cerebellum contain TH (Hess & Wilson, 1991; Berthie et al., 1993). Transient THi cells have been demonstrated also in the cerebellum of birds (Puelles & Medina, Chapter 16). There is much debate whether such neurons are to be considered as true catecholaminergic neurons (see e.g. Seil et al., 1992), a point that will be further discussed on p. 475.

Comparative analysis of catecholamine systems of vertebrates using a segmental approach

Numerous recent studies have shown that the brain of vertebrates develops in a segmental fashion (reviewed in Chapter 16 by Puelles & Medina), with the segments being prominent as bulges that are transiently present early in development. The hindbrain at an early stage of development consists of 7–8 rhombomeres, the midbrain of one mesomere and the bulk of the diencephalon of 3 or more prosomeres (Bergquist & Källén, 1954; Vaage, 1969. 1973; Keyser, 1972; Puelles et al., 1987; Lumsden & Keynes, 1989; Noden, 1991; Figdor & Stern, 1993). The segmental development of more rostroventral parts of the diencephalon and the telencephalon has not yet been fully characterized. Each identified brainstem segment possesses a regional and serial identity, with specific cell groups arising within specific segments. The segmental pattern of development leaves its imprint on the adult brain and governs the location and topographic relationships of the various brainstem cell groups to each other. In the adult brain, these segments can best be viewed in the sagittal or horizontal plane. The organization and development of the catecholaminergic cell groups of the brain can be readily analyzed with respect to the segmental plan of brain organization. The boundaries of the different segments in the adult brainstem can be identified in many cases using such landmarks as cranial nerves, cranial nerve nuclei and various fiber bundles (Vaage, 1969; Lumsden & Keynes, 1989; Noden, 1991; Puelles et al., 1987). Note that the conventional transverse plane employed for nearly all vertebrate species is not parallel to the plane of division between segments. Rather, the conventional transverse plane tends to include cell groups of different segmental origin. Thus, the boundaries between segments do not conform necessarily to the conventionally accepted boundaries for the major subdivisions of the brain. For example, the pretectum and rostral parts of the tegmentum arise from diencephalic prosomeres. Because, the conformation of the adult brain varies markedly among vertebrate species, exclusive use of the transverse plane for analyzing cell group distribution may obscure the common segmental origins and adult locations of some specific cell groups. In contrast, examination of the segmental organization of the catecholamineric cell groups of the brain of different species provides a clear means for characterizing the location of each cell group and for characterizing the topographic relationships of specific groups to one another. A segmental approach for analyzing catecholaminergic cell group organization is also valuable because it helps clarify the developmental origin of specific cell groups, and because it provides a very concrete framework within which to view evolutionary variation in brain organization. Further, because of the developmental underpinnings of segmental analysis, interspecies differences can be related to potential developmental differences. In the discussion below in which we compare catecholaminergic system organization among vertebrates, we will show the advantages of a segmental outlook for some brainstem cell groups. As we will show, a segmental approach can be particularly revealing for clarifying the evolution of the dopaminergic cell groups of the midbrain.

Current concepts of evolution of catecholamine systems in the CNS of vertebrates

As mentioned in the preface, in the final part of this book we wish to point out some major evolutionary implications. In their review on central monoaminergic systems in vertebrates, Parent and his colleagues (1984) noted the existence of striking differences between vertebrates. Two major evolutionary trends were reported by the latter authors: 1) In fishes and in amphibians, the largest number of catecholamine cells occurs in the hypothalamus, whereas in amniotes the most prominent population of CA cell bodies is found in the midbrain tegmentum; 2) From fishes to mammals there is a significant increase in the number of midbrain CA cells, which coincides with a decrease in the the number of hypothalamic CA cells. Parent *et al.* (1984) were well aware that living vertebrates could no longer be considered as representatives of one linear series of ever increasing complexity, the so-called *scala naturae*. Nevertheless, they managed to present a histogram of number of CA cells in the diencephalon and midbrain of different species which would perfectly advocate such a view. The same authors also noted that the hypertrophy of the midbrain CA cell groups coincided with the appearance of the first true terrestrial vertebrates, i.e. the reptiles. Given the dramatic increase in available data on CA systems in vertebrates (not only in number of species, but also by means of more sensitive techniques), the statements on the evolution of CA systems of vertebrates need serious revision.

Our survey of CA cell groups in the CNS of vertebrates (pp. 465–473) has shown that there are many similarities, but also differences between vertebrate classes and even within a single class. It is, therefore, not easy to recognize clearcut trends in the evolution of catecholamine systems of vertebrates. Yet, several hypotheses on the changes catecholamine systems have undergone during evolution, may be brought forward. First, there seems to be no solid basis for the statement that the number of midbrain DA cells hypertrophies at the cost of the number of DA cells in the diencephalon. Thus, while it is true that midbrain cell groups are well developed in reptiles, and, in particular, birds and mammals, the hypothalamus of both anamniotes and amniotes contains numerous DA cell bodies and it seems, therefore, not justified to relate the number of CA cell bodies in the midbrain to that in the hypothalamus. Moreover, distinct DA cell groups are now known to be present in the midbrain tegmentum of elasmobranchs, lungfishes and amphibians, making the correlation between the development of a midbrain DA cell group and the transition from an aquatic to a terrestrial lifestyle less likely.

The other notion of Parent *et al.* (1984), i.e. a reduction of hypothalamic CSF-contacting CA cell bodies, is supported by more recent immunohistochemical studies. However, as mentioned on p. 472, the CA cells in the periventricular organ contain dopamine and/or noradrenaline, which they do not produce themselves but accumulate from the ventricle. Such cells are apparently absent in mammals, suggesting that they are to considered as a primitive feature of CA systems of vertebrates which has been lost in mammals. It does appear to be the case, however, that CSF-contacting CA neurons are abundant throughout the hypothalamus of agnathans, ray-finned fish and amphibians, while in amniotes they are limited to the OPH.

Our comparative analysis of the A1–A17/C1–C3 has revealed several other important findings which may have implications for the evolution of CA systems in vertebrates. For example, studies of the A1–A3/C1–C3 cell groups with antibodies against TH did not suggest the existence of major differences between vertebrates. The THi cells in the caudal brainstem of reptiles, birds and mammals are always located in a radially oriented zone that extends from the region containing the nucleus of the solitary tract and the dorsal motor nucleus of the vagus (A2, C2) to the ventrolateral medullary region (A1, C1). However, the fact that in some species PNMTi cell bodies are partially or totally lacking is now well established. Ontogenetic studies in rats have shown that at embryonic day 12 a single group of THi cell bodies is present at caudal brainstem levels (for ref., see Foster, Chapter 17). Later in development, this cell mass segregates into a dorsal group and a ventral group. Differences in location of PNMTi cells in adult specimens of the various species studied may, therefore, reflect differences in the degree of migration of the future adrenergic cells within the radial column. If so, the brains of the lizard, *Gekko gecko*, and of sheep may represent two extremes: in *Gekko* all putative adrenergic perikarya have migrated to the ventrolateral medulla, whereas in sheep the cells have stayed within the dorsomedial medulla.

In support of this notion is the finding that the innervation pattern of the single PNMTi cell group in *Gekko* matches that of the three PNMTi cell groups in rats. Although variation in migration of PNMTi cells may be an explanation for the reported differences, it should be noted that Foster *et al.* (1985) found that PNMT in fetal rat brains appears in an 'explosive' manner on embryonic day 13 in cells that later become the C1 and C2 group. The same authors also reported that TH is found considerably later in these cells. Several explanations for the existence of PNMT-positive/TH-negative neurons have been offered (Foster *et al.*, 1985; Foster, Chapter 17). One of these suggests a phylogenetic older route involving octopamine synthesis which is later replaced by a route involving L-dopa. The functional significance of the delayed appearance of TH immunoreactivity in the PNMTi cells is still unknown.

Another explanation for the observed differences in the location of PNMTi cell bodies in the various species studied is that some cells have lost the capacity to produce PNMT or contain PNMT in a non-detectable quantity. The apparent lack of PNMT immunoreactivity in the brain of guinea pigs combined with non-detectable levels of adrenaline points in that direction for this species.

In line with this latter explanation is the observation of several non-classified CA cell groups such as those in the spinal cord, the pretectum and habenular region, and in the forebrain. Whereas previous studies were mostly focused on comparing the CA cell groups in non-mammalians with those in mammals using the A1–A17/C1–C3 nomenclature, the present survey has yielded some results that may lead to a profound revision of our understanding of the evolution of catecholamine systems. Generally, it is thought that CA systems in non-mammalians are less complex, both in numbers of cells and numbers of cell groups. Evolution of these systems would involve an increase in numbers of CA cell bodies as well as a further differentiation of these cell groups. To a certain extent that notion might be correct. For example, the brains of anamniotes contain notably fewer CA cells than those in amniotes and many CA cell groups are not recognizable as a separate entity (Table 20.2). However, it should be pointed out that the subdivisions in the CA cell groups, as reported for mammals, do not necessarily correlate with differences in function. In this respect, it is worth mentioning that a segmental approach provides a different view on the organization of CA cell groups in vertebrates and, consequently, on the evolution of CA systems. For example, a drastic reorganization of hypothalamic cell groups might be expected, when it is taken for granted that CA cells migrate intrasegmentally rather than intersegmentally (see Puelles & Medina, Chapter 16).

As already extensively discussed in Chapter 16 by Puelles and Medina, the segmental approach has a serious impact on our concepts of the organization, development and evolution of the A8–A10 CA cell groups. The A10 complex seems the most unnatural group, since it not only stretches across several intersegmental boundaries, it also crosses the alar/basal boundary of mesencephalic and diencephalic segments. It has been suggested by Puelles and Medina that the paramedian proliferation zone of each of the caudal one or two diencephalic segments and the mesencephalic segment produces its own part of the ventral tegmental area. A similar, multisegmental origin may hold for the substantia nigra. If so, then the restricted rostrocaudal extent of the CA cells in the midbrain of elasmobranchs and amphibians may be explained by assuming that only the basal parts of those segments contributing to the rostral midbrain of adults (the caudal two diencephalic segments, termed the posterior parencephalon and the synencephalon) develop neurons that express TH/DA. This would imply that the midbrain DA cell groups of anamniotes are only partly comparable to the A8–A10 group of amniotes. Further, the relationship of the CA neurons of the posterior tubercle region in anamniotes is partly clarified by the segmental approach to brain organization. The posterior tubercle region of the adult anamniote brain has been regarded as hypothalamic by some authors and mesencephalic by others. The segmental approach indicates that the posterior tubercle region in anamniotes is the adult derivative of the basal part of the posterior parencephalon (the intermediate diencephalic segment). This same embryonic region is generally regarded as a rostral (prerubral) part of the midbrain in amniotes and no distinct posterior tubercle region is evident. This line of reasoning suggests that the CA neurons of the posterior tubercle of anamniotes may in fact be homologous to what are considered rostral tegmental neurons in amniotes. The posterior tubercle region may also include parts of the amniote A11.

At the outset of this chapter, our goal was to seek answers to such questions as the extent to which catecholamine systems in different vertebrates are comparable to each other and the nature of the changes these systems have undergone during evolution. The overview presented here shows that while we certainly have gained a clearer insight into these issues, additional fundamental questions are raised that need to be addressed in the near future. One of these questions concerns the significance of the presence or absence of a particular CA cell group. It is tempting to consider negative findings with fluorescent-, immunohistochemical-, or *in situ* hybridization techniques in certain vertebrates as evidence for the absence of such a cell group in those vertebrates. Although that might be true, there is now compelling evidence that several circumstances may influence the expression of catecholamine phenotypes. For example, olfactory peduncle lesions induce TH expression in rat forebrain neurons that receive olfactory nerve input (Guthrie & Leon, 1989). Another example is that, in rats, glia from the substantia nigra but not glia from the neostriatum is capable of inducing development or promoting survival of dopamine cells in the striatum (Beyer et al., 1991). Furthermore, it is well known that gonadal steroids may influence catecholamine systems (for review, see Reisert & Pilgrim, 1991; Reisert et al., Chapter 19). Although such findings may dampen the enthusiasm of investigators who wish to draw clearcut conclusions on the phylogeny and development of CA systems in the CNS of vertebrates, they underline the importance of fundamental research on the genetic and epigenetic factors involved in CA expression. It is hoped that such studies are pursued in the near future, since it is likely that they will further clarify the evolution and development of brain CA systems.

Functional significance of catecholamine systems in the CNS of vertebrates

The chapters in the present book have focused largely on the morphology and development of brain CA systems in the diverse classes of vertebrates. We chose this emphasis because, in fact, until recently there was little data on the anatomy of CA systems for many vertebrate groups, and (as a consequence) even less on their functions. The goal of anatomical studies, however, is to explicate function. In the present case, the attempt to bring together the current anatomical data for the diverse vertebrate groups is intended to help provide the framework for more detailed future studies on the functions of these systems. Even with the current state of knowledge, there are some general points that can be made about the comparative aspects of CA systems, both in terms of what is known about function and what is implied by the anatomy as a direction of future study.

In our presentation in this final chapter, we have dealt with CA neurons on a system by system basis, namely those of the retina, olfactory bulb, hypothalamus, tegmentum, isthmic region (locus coeruleus), and caudal medulla. For the retina, olfactory bulb, rostral rhombencephalic and caudal rhombencephalic CA systems, the available anatomical data shows a high degree of conservatism in the presence and general location of these cell groups. Although there are certainly likely to be major differences for these across species in connections, the precise CAs utilized and the precise functions, the anatomical conservatism is suggestive of a broadly conserved role. In contrast, hypothalamic and midbrain CA cell groups are more variable in their presence and precise location across vertebrate species, suggestive of some greater divergence in role among vertebrate species.

Among the conserved CA cell groups, retinal CA cell and olfactory CA cell are likely to share similar functions across vertebrates, since: 1) retinal CA neurons are typically DA amacrine cells, and more sporadically DA interplexiform cells; and 2) olfactory DA cells are typically found in a periglomerular position, and more sporadically in a deep position. Dopaminergic amacrine cells clearly play a role in the transmission of information from photoreceptors (via bipolar and other amacrine cells) to ganglion cells, while the interplexiform DA cells appear to influence outer retinal function. In some species, the interplexiform cells play quite specialized roles in outer retinal function, e.g. RPE pigment retinomotor movements in fish (Kolbinger et al., 1990). The periglomerular dopamine cells of the olfactory bulb in all species presumably modulate the olfactory nerve input to mitral cells, thus underlining the notion that basic olfactory bulb processing mechanisms may be very ancient and conserved.

The NA cell group of the rostral rhombencephalon (i.e. the locus coeruleus cluster) may be the most conservative cell group of the brain, being present in all vertebrate species. It would be interesting to explore the functional significance of this ubiquity. The locus coeruleus in mammals and birds has been shown to have widespread projections throughout the brain and spinal cord and to exert a modulatory influence over these diverse regions. Despite our current inability to more precisely define this modulatory influence, it must clearly be a critical feature of brain organization and function since the locus coeruleus seems to be part of the vertebrate brain from the outset.

The functional role of the caudal rhombencephalic cells groups are somewhat more clearcut: they play a role in linking visceral sensory pathways of the brain with the sympathetic nervous system. In birds and mammals, these CA neurons are known to have ascending projections to parabrachial and hindbrain sites and descending projections to sympathetic preganglionic spinal cord neurons (whereby they influence vascular tone in particular). It seems likely that these connections and this involvement in sympathetic functions are also present in reptiles and anamniotes. A fruitful topic for study would be investigation of the changes in anatomical and functional organization of the caudal rhombencephalic CA system that accompany evolutionary variation in the organization of the sympathetic nervous system and the nature of its role in control of autonomic functions and the vasculature.

Hypothalamic CA systems will require further study in order to elucidate their evolution. Among amniotes many of the same cell groups are recognizable, but among anamniotes the hypothalamic CA system (particular in bony fish and agnathans) consists of numerous neurons throughout the hypothalamus, the majority of which contact the CSF. The functional significance of the loss of CSF contacting hypothalamic neurons in mammals and their widespread presence in nonterrestrial vertebrates (including the paraventricular organ of amphibians, reptiles and birds) is uncertain. There is good evidence that some portion of the hypothalamic CA system or the CA input to some portion of the hypothalamus plays a sexually dimorphic role in reproductive behavior. Other functions of hypothalamic CA systems, such as in feeding, require further study. As noted above, the midbrain CA system shows evolutionary variation between amniotes and anamniotes. In amniotes, it consists of a large field of tegmental DA neurons that innervate various telencephalic targets, most notably the striatal part of the basal ganglia. The dopaminergic input to the striatum in birds and mammals plays a key role in the ability of the basal ganglia to play its role in movement control. Loss of this input (as in human Parkinson's disease) or pharmacological blockade of this input slows movement and impairs its initiation. Conversely, pharmacological augmentation of dopaminergic transmission in birds and mammals promotes excessive movements, particularly stereotyped movements. Similar results are observed in reptiles (Andersen, Baestrup & Randrup, 1975). Despite the fact that amphibians only appear to have DA neurons present in their rostral midbrain, these neurons do innervate the amphibian basal ganglia and play a similar role in movement control as in amniotes (Barbeau et al., 1985). In light of the uncertainty about the nature of the DA projection from the posterior tubercle to the presumed striatum in anamniotes, it would be interesting to explore the effect of dopamine on movement in various anamniote species. In any case, the data presented in this book strongly support the view that the dopaminergic tegmental projection to the striatum is very similar in organization and function among amniotes. Further study of this system in reptiles and birds could thus be profitable for furthering understanding of the overall functions of this system.

Finally, we have noted several CA neuronal systems that are prominent in nonmammals but absent or lacking some of the CA enzymes in mammals. Prominent among these are the pretectal/habenular CA neurons and the spinal cord CA neurons (which contact the CSF of the central canal) that are evident in diverse non-mammals. Nothing is known of the functions of these cell groups. Exploration of their function would be of interest for explaining why they have been lost as mature CA neurons in mammals. As noted above, the neurons of these cell groups may be retained but they may have lost the ability to produce all of the enzymes necessary for CA synthesis.

Concluding remarks

An edited book containing contributions by multiple authors is often a collection of papers, each dealing with a particular aspect of the topic under consideration. Usually, each chapter stands on its own and can be read separately. The present book, although also multi authored, differs because the chapter contributors were asked to write their chapters following some restricting guidelines that called for largely deferring discussion of global issues until the final chapter. Nevertheless, each chapter has its own value as it shows the state of knowledge on brain catecholamine systems for each vertebrate class. The overall value of the comparative approach for studying CA systems is provided in this final chapter. A careful scrutiny of the distribution of cell bodies that contain catecholamines and their synthesizing enzymes has revealed some major evolutionary trends. For example, evolution of CA systems has not necessarily involved an increase in number and further differentiation of pre-existing CA cell groups. Loss of (or reduced) capacity to produce catecholamines by neurons in some cell groups also contributes to the differences observed in CA systems in vertebrates. There are, however, still numerous gaps in our knowledge that should be addressed in the near future.

As we have seen, TH immunohistochemistry has been successfully applied to the brains of representatives of each major vertebrate class. The same holds for DA immunohistochemistry, although it must be admitted that data obtained with DA antibodies are still sparse or absent for cyclostomes, cartilaginous fish, non-teleostean actinopterygian fish and birds. Except for rats, DA antisera have not been used extensively in other mammalian species. A remarkable gap is present in our knowledge of noradrenaline and adrenaline in the CNS of vertebrates. Whereas specific antibodies against noradrenaline are available but not applied widely yet, data of adrenergic cell bodies and fibers are exclusively based on PNMT immunohistochemistry. Future studies with antibodies against NA and PNMT (or against adrenaline itself) in a variety of vertebrates may certainly reveal important information not only on these CA systems but also on the relationships among different noradrenergic and adrenergic cell groups.

Another gap concerns knowledge of connections of the various CA cell groups. Whereas in mammals, the number of studies combining tract-tracing techniques with immunohistochemistry increases gradually, such studies are almost lacking for non-mammalian vertebrates. Yet, a detailed knowledge of the connections of CA cells is needed to get more insight into the basic organization of CA systems. Such information may, for example, answer whether subdivions such as those made for rhombencephalic and diencephalic CA cell groups in mammals have a functional significance. If so, similar studies in non-mammalian species may reveal whether such cell groups are present but not parcellated (Ebbesson, 1984), or even absent in non-mammalian species.

Another point that deserves attention is the development of catecholamine systems. Whereas substantial information is available for rats, for other mammalian species and for non-mammalians data are sparse or completely lacking. In fact, chapters 14-16 are the first reports in which detailed timetables of the development of cell bodies and fibers are presented on catecholamine systems in amphibians, reptiles and birds. We are well aware, however, that the conclusions from these studies have to be yet considered as preliminary. Since immunodetection of catecholamines or their synthetic enzymes depends on the amount of these substances in neurons and their processes, the *in situ* hybridization techniques may be a welcome additional tool for investigating the development of CA systems. The latter technique may be particularly helpful in determining the delay between cell birth and the first expression of CA synthesis. Nevertheless, there is presently ample evidence that some cells already express CA immunoreactivity shortly after neurogenesis, whereas others become immunoreactive much later in their development. This is important because a better understanding of migratory processes requires early detection of CA cells. It also makes it possible to test the hypotheses on CA cell group development and identity offered by the segmental approach.

A serious shortcoming in our knowledge of catecholamine systems of vertebrates is the paucity of information on receptors in non-mammalian vertebrates. Such information is desperately needed if we are to understand better the evolution of the structure and function of CA systems in the CNS of vertebrates. Equally needed are data on colocalization of catecholamines with other neurotransmitter systems. In the various chapters of this book, these matters have already been addressed to some extent, but certainly not exhaustively, largely owing to the scarcity of these kinds of data in non-mammals.

From the foregoing survey it should be clear that while great progress has been made in understanding the comparative organization and evolution of brain CA systems, many intriguing problems pertaining to the phylogeny and development of CA systems in the CNS of vertebrates still await solution. It has also become clear that a comparative approach is very illuminating, if not indispensable, for a clearer understanding of the basic principles and variation in the structural organization of these systems. This book will have served its purpose if it fosters interest in these fundamental questions and if it kindles within some of its readers enthusiasm to work on these issues.

References

Alonso, J. R., Covenas, R., Lara, J., Arevalo, R., de Leon, M. & Aijon, J. (1989). Tyrosine hydroxylase immunoreactivity in a subpopulation of granule cells in the olfactory bulb of teleost fish. *Brain, Behavior and Evolution*, **34**, 318–24.

Andersen, H, Baestrup, C. & Randrup, A. (1975). Apomorphine-induced stereotyped biting in the tortoise in relation to dopaminergic mechanisms. *Brain, Behavior and Evolution*, **11**, 365–73.

Armstrong, D. M., Ross, C. A., Pickel, V. P., Joh, T. H. & Reis, D. J. (1982). Distribution of dopamine, noradrenaline and adrenaline-containing cell bodies in the rat medulla oblongata: demonstration by immunocytochemical localization of catecholamine biosynthetic enzymes. *Journal of Comparative Neurology*, **212**, 173–87.

Baker, H. (1986). Species differences in the distribution of substance P and tyrosine hydroxylase immunoreactivity in the olfactory bulb. *Journal of Comparative Neurology*, **252**, 206–26.

Barbeau, A., Dallaire, L., Buu, N. T., Veilleux, F., Boyer, H., de Lanney, L. E., Irwin, I., Langston, E. B. & Langston, J. W. (1985). New amphibian models for study of 1-methyl-4-phenyl-1,2,3,6-tetrahydropyridine (MPTP). *Life Sciences*, **36**, 1125–34.

Batten, T. F. C., Berry, P. A., Maqbool, A., Moons, L. & Vandesande, F. (1993). Immunolocalization of catecholamine enzymes, serotonin, dopamine and L-Dopa in the brain of *Dicentrarchus labrax* (Teleostei). *Brain Research Bulletin*, **31**, 233–52.

Berger, B., Verney, C., Gaspar, P. & Febvret, A. (1985). Transient expression of tyrosine-hydroxylase immunoreactivity in some neurons of the rat neocortex during postnatal development. *Developmental Brain Research*, **23**, 141–4.

Bergquist, H. & Källén, B. (1954). Notes on the early histogenesis and morphogenesis of the central nervous system in vertebrates. *Journal of Comparative Neurology*, **100**, 627–60.

Berthie, B., Axelrad, H., Verney, C. & Marc, M. E. (1993). A small population of Purkinje cells in the posterior vermis is specifically labeled by a tyrosine hydroxylase antibody. In *Serotonin, the Cerebellum, and Ataxia*, ed. P. Truillas & K. Fuxe, pp. 121–127, New York: Raven Press.

Beyer, C., Pilgrim, C., Reisert, I. & Misgeld, U. (1991). Cells from embryonic rat striatum cocultured with mesencephalic glia express dopaminergic phenotypes. *Neuroscience Letters*, **128**, 1–3.

Björklund, A., & Lindvall, O. (1984). Dopamine-containing systems in the CNS. In *Handbook of Chemical Neuroanatomy. Vol.2, Classical Transmitters in the CNS, part I*, ed. A. Björklund & T. Hökfelt, pp. 55–122. Amsterdam: Elsevier.

Brecha, N. (1983). Retinal neurotransmitters: histochemical and biochemical studies. In *Chemical Neuroanatomy*, ed. P. C. Emson, pp. 85–129.

Crooks, J. & Kolb, H. (1992). Localization of GABA, glycine, glutamate and tyrosine hydroxylase in the human retina. *Journal of Comparative Neurology*, **315**, 287–302.

Cumming, P., von Krosigk, M., Reiner, P. B., McGeer, E. G. & Vincent, S. R. (1986). Absence of adrenaline neurons in the guinea pig brain: a combined immunohistochemical and high-performance liquid chromatography study. *Neuroscience Letters*, **63**, 125–30.

Dahlström, A. & Fuxe, K. (1964). Evidence for the existence of monoamine-containing neurons in the central nervous system. I.Demonstration of monoamines in the cell bodies of brain stem neurons. *Acta Physiologica Scandinavica*, **62**, Suppl. 232, 1–155.

Davis, B. J. & Macrides, F. (1983). Tyrosine hydroxylase immunoreactive neurons and fibers in the olfactory system of the hamster. *Journal of Comparative Neurology*, **214**, 427–40.

Dietl, M., Arluison, M., Mouchet, P., Feuerstein, C., Manier, M. & Thibault, J. (1985) Immunohistochemical demonstration of catecholaminergic cell bodies in the spinal cord of the rat. *Histochemistry*, **82**, 385–9.

Ebbesson, S. O. E. (1984). Evolution and ontogeny of neural circuits. *The Behavioral and Brain Sciences*, **7**, 321–66.

Ehinger, B. (1976). Biogenic monoamines as transmitters in the retina. In *Transmitters in the Visual Process*, ed. S. L. Bonting, pp.145–63. Oxford: Pergamon Press.

Ehinger, B. (1983). Neurotransmitter systems in the retina. *Retina*, **2**, 305–21.

Ekström, P., Honkanen, T. & Borg, B. (1992). Development of tyrosine hydroxylase-, dopamine- and dopamine-β-hydroxylase-immunoreactive neurons in a teleost, the three-spined stickleback. *Journal of Chemical Neuroanatomy*, **5**, 481–501.

Ekström, P., Honkanen, T., & Steinbusch, H. W. M. (1990). Distribution of dopamine-immunoreactive neuronal perikarya and fibers in the brain of a teleost, *Gasterosteus aculeatus*. Comparison with TH- and DBH-IR neurons. *Journal of chemical Neuroanatomy*, **3**, 233–60.

Figdor, M. C. & Stern, C. D. (1993). Segmental organization of embryonic diencephalon. *Nature*, **363**, 630–4.

Foster, G. A., Schultzberg, M., Goldstein, M. & Hökfelt, T. (1985). Ontogeny of phenylethanolamine N-methyltransferase- and tyrosine hydroxylase-like immunoreactivity in presumptive adrenaline neurones of the foetal rat central nervous system. *Journal of Comparative Neurology*, **236**, 348–81.

Fuller, R. W. & Hemrick-Luecke, S. K. (1983). Species differences in epinephrine concentration and norepinephrine N-methyltransferase activity in hypothalamus and brain stem. *Comparative Biochemistry and Physiology*, **74**C, 47–9

González, A. & Smeets, W. J. A. J. (1991). Comparative analysis of dopamine and tyrosine hydroxylase immunoreactivities in the brain of two amphibians, the anuran *Rana ridibunda* and the urodele *Pleurodeles waltlii*. *Journal of Comparative Neurology*, **303**, 457–77.

González, A. & Smeets, W. J. A. J. (1993) Noradrenaline in the brain of the South African clawed frog *Xenopus laevis*: A study with antibodies against noradrenaline and dopamine-β-hydroxylase. *Journal of Comparative Neurology*, **331**, 363–74.

González, A., Tuinhof, R. & Smeets, W. J. A. J. (1993) Distribution of tyrosine hydroxylase- and dopamine-immunoreactivities in the brain of the South African clawed frog *Xenopus laevis*. *Anatomy and Embryology*, **187**, 193–201.

Guthrie, K. M. & Leon, M. (1989). Induction of tyrosine hydroxylase expression in rat forebrain neurons. *Brain Research*, **497**, 117–31.

Hadjiconstantinou, M., Cohen, J. & Neff, N. H. (1983). Epinephrine: A potential neurotransmitter in retina. *Journal of Neurochemistry*, **41**, 1440–4.

Hadjiconstantinou, M., Mariani, A. P., Panula, P., Joh, T. H. & Neff, N. H. (1984). Immunohistochemical evidence for epinephrine-containing retinal amacrine cells. *Neuroscience*, **13**, 547–51.

Halasz, N., Ljungdahl, A. & Hökfelt, T. (1978). Transmitter histochemistry of the rat olfactory bulb. II. Fluorescence histochemical, autoradiographic and electron microscopic localization of monoamines. *Brain Research*, **154**, 253–71.

Halasz, N., Nowycky, M., Hökfelt, T., Shepherd, G. M., Markey, K. & Goldstein, M. (1982). Dopaminergic periglomerular cells in the turtle olfactory bulb. *Brain Research Bulletin*, **9**, 383–9.

Hess, E. J. & Wilson, M. C. (1991). Tottering and leaner mutations perturb transient developmental expression of tyrosine hydroxylase in embryologically distinct Purkinje cells. *Neuron*, **6**, 123–32.

Hökfelt, T., Johansson, O. & Goldstein, M. (1984a). Central catecholamine neurons as revealed by immunohistochemistry with special reference to adrenaline neurons. In *Handbook of Chemical Neuroanatomy, Vol.2, Classical Transmitters in the CNS*, Part I, ed. A. Björklund & T. Hökfelt, pp. 157–276. Amsterdam: Elsevier.

Hökfelt, T., Martensson, R., Björklund, A., Kleinau, S. & Goldstein, M. (1984b). Distributional maps of tyrosine-hydroxylase-immunoreactive neurons in the rat brain. In *Handbook of Chemical Neuroanatomy, Vol.2, Classical Neurotransmitters in the CNS, Part I*, pp. 277–379.

Hornby, P. J., Piekut, D. T. & Demski, L. S. (1987). Localization of immunoreactive tyrosine hydroxylase in the goldfish brain. *Journal of Comparative Neurology*, **261**, 1–15.

Jaeger, C. B., Ruggiero, D. A., Albert, V. R., Park, D. H., Joh, T. H. & Reis, D. J. (1984). Aromatic L-amino acid decarboxylase in the rat brain: immunocytochemical localization in neurons of the brain stem. *Neuroscience*, **11**, 619–713.

Jaffe, E. H., Urbina, M. & Drujan, B. D. (1991). Possible neurotransmitter role of noradrenaline in the teleost retina. *Journal of Neuroscience Research*, **29**, 190–5.

Jerison, H. J. (1990). Fossil evidence on the evolution of the neocortex. In *Cerebral Cortex, Vol. 8A, Comparative Structure and Evolution of Cerebral Cortex, Part I*, ed. E. G. Jones & A. Peters, pp. 285–309. New York Plenum Press.

Keyser, A. (1972). The development of the diencephalon of the Chinese hamster. An investigation of the validity of the criteria of subdivision of the brain. *Acta Anatomica*, **83**, (Suppl. 59), 1–178.

Kolbinger, W., Kohler, K., Oetting, H. & Weiler, R. (1990). Endogeneous dopamine and cyclic events in the fish retina, I: HPLC assay of total content, release and metabolic turnover during different light/dark cycles. *Vision Neuroscience*, **5**, 143–9.

Kotrschal, K., Krautgartner, W-D. & Adam, H. (1985). Distribution of aminergic neurons in the brain of the sterlet, *Acipenser ruthenus* (Chondrostei, Actinopterygii). *Journal fur Hirnforschung*, **26**, 65–72.

Larabi, Y. & Nguyen-Legros, J. (1991). Morphology, density and distribution of tyrosine hydroxylase immunoreactive cells in the retina of the gerbil *Meriones shawi*. Relationships with horizontal cells. *Journal fur Hirnforschung*, **32**, 387–95.

Lumsden, A. & Keynes, R. (1989). Segmental patterns of neuronal development in the chick hindbrain. *Nature*, **337**, 424–8.

Mariani, A. & Hokoc, J. N. (1988). Two types of tyrosine hydroxylase-immunoreactive amacrine cell in the rhesus monkey retina. *Journal of Comparative Neurology*, **276**, 81–91.

Massey, S. C. & Redburn, D. A. (1987). Transmitter circuits in the vertebrate retina. *Progress in Neurobiology*, **28**, 55–96.

Meek, J., Joosten, H. W. J. & Hafmans, T. G. M. (1993). Distribution of noradrenaline-immunoreactivity in the brain of the mormyrid teleost *Gnathonemus petersii*. *Journal of Comparative Neurology*, **328**, 145–60.

Meek, J., Joosten, H. W. J. & Steinbusch, H. W. M. (1989). The distribution of dopamine-immunoreactivity in the brain of the mormyrid teleost *Gnathonemus petersii*. *Journal of Comparative Neurology*, **281**, 362–83.

Meredith, G. E. & Smeets, W. J. A. J. (1987). Immunocytochemical analysis of the dopamine system in the forebrain and midbrain of *Raja radiata*: evidence for a substantia nigra and ventral tegmental area in cartilaginous fish. *Journal of Comparative Neurology*, **265**, 530–48.

Mouchet, P., Manier, M., Dietl, M., Feuerstein, C., Berod, A., Arluison, M., Denoroys, L. & Thibault, J. (1986) Immunohistochemical study of catecholaminergic cell bodies in the rat spinal cord. *Brain Research Bulletin*, **16**, 341–53.

Mouchet, C., Manier, M. & Feuerstein, C. (1992). Immunohistochemical study of the catecholaminergic innervation of the spinal cord of the rat using specific antibodies against dopamine and noradrenaline. *Journal of Chemical Neuroanatomy*, **5**, 427–40.

Nakai, Y., Ochiai, H. & Shioda, S. (1977) Cytological evidence for different types of cerebrospinal fluid-contacting subependymal cells in the preoptic and infundibular recesses of the frog. *Cell and Tissue Research*, **176**, 317–334.

Nguyen-Legros, J. (1988). Morphology and distribution of catecholamine neurons in mammalian retina. *Progress in Retinal Research*, **7**, 113–47.

Noden, D. M. (1991). Vertebrate craniofacial development: the relation between ontogenetic process and morphological outcome. *Brain, Behavior and Evolution*, **38**, 190–225.

Northcutt, R. G. (1984). Evolution of the vertebrate nervous system: Patterns and processes. *American Zoologist*, **24**, 701–16.

Northcutt, R. G., Reiner, A. & Karten, H. (1988). Immunohistochemical study of the telencephalon of the spiny dogfish, *Squalus acanthias*. *Journal of Comparative Neurology*, **277**, 250–67.

Osborne, N. N. & Nesselhut, T. (1983). Adrenaline occurrence in the bovine retina. *Neuroscience Letters*, **39**, 33–6.

Parent, A. & Northcutt, R. G. (1982). The monoamine-containing neurons in the brain of the garfish, *Lepisosteus osseus*. *Brain Research Bulletin*, **9**, 189–204.

Parent, A., Poitras, D. & Dubé, L. (1984). Comparative anatomy of central monoaminergic systems. In *Handbook of Chemical Neuroanatomy, Vol. 2, Classical Transmitters in the CNS, Part I*, ed. A. Björklund & T. Hökfelt, pp. 409–39. Amsterdam: Elsevier.

Pindzola, R. R., Ho, R. H. & Martin, G. E. (1990). Development of catecholaminergic projections to the spinal cord in the North American opossum, *Didelphis virginiana*. *Journal of Comparative Neurology*, **294**, 399–417.

Puelles, L., Amat, J. A. & Martnez-de-la-Torre, M. (1987). Segment-related, mosaic neurogenetic pattern in the forebrain and mesencephalon of early chick embryos: I. Topography of AChE-positive neuroblast up to stage HH18. *Journal of Comparative Neurology*, **266**, 247–68.

Reiner, A. & Northcutt, R. G. (1987). An immunohistochemical study of the telencephalon of the African lungfish, *Protopterus annectens*. *Journal of Comparative Neurology*, **256**, 463–81.

Reisert, I. & Pilgrim, C. (1991). Sexual differentiation of monoaminergic neurons-genetic or epigenetic? *Trends in Neurosciences*, **14**, 468–73.

Rexed, B. (1954). A cytoarchitectonic atlas of the spinal cord in the cat. *Journal of Comparative Neurology*, **100**, 297–379.

Roberts, B. L., Meredith, G. E. & Maslam, S. (1989). Immunocytochemical analysis of the dopamine system in the brain and spinal cord of the European eel, *Anguilla anguilla*. *Anatomy and Embryology*, **180**, 401–12.

Sas, E., Maler, L. & Tinner, B. (1990). Catecholaminergic systems in the brain of a gymnotiform teleost fish: an immunohistochemical study. *Journal of Comparative Neurology*, **292**, 127–62.

Savy, C., Yelnik, J., Martin-Martinelli, E., Karpouzas, I. & Nguyen-Legros, T. d. T. (1989). Distribution and spatial geometry of dopamine interplexiform cells in the rat retina: I. Developing retina. *Journal of Comparative Neurology*, **289**, 99–110.

Schutte, M. (1991). [125]SCH 23982, a new tool for rapid visualization of dopaminergic neurons in lower vertebrate retinas. *Neuroscience Letters*, **121**, 29–33.

Seil, F. J., Johnson, M. L., Nishi, R. & Nilaver, G. (1992) Tyrosine hydroxylase expression in non-catecholaminergic cells in cerebellar cultures. *Brain Research*, **569**, 164–8.

Singhaniyom, W., Wreford, N. G. M. & Güldner, F. H. (1983). Asymmetric distribution of catecholamine-containing neuronal perikarya in the upper cervical spinal cord of rat. *Neuroscience Letters*, **41**, 91–7.

Smeets, W. J. A. J. (1988a). The monoaminergic systems in the forebrain and midbrain of reptiles investigated with specific antibodies against serotonin, dopamine and noradrenaline. In *The Forebrain of Reptiles: Current Concepts of Structure and Function*. ed. W. K. Schwerdtfeger & W. J. A. J. Smeets, pp.97–109. Basel: Karger.

Smeets, W. J. A. J. (1988b). Distribution of dopamine immunoreactivity in the forebrain and midbrain of the snake *Python regius*: a study with antibodies against dopamine. *Journal of Comparative Neurology*, **271**, 115–29.

Smeets, W. J. A. J. & González, A. (1990) Are putative dopamine-accumulating cell bodies in the hypothalamic periventricular organ a primitive brain character of non-mammalian vertebrates? *Neuroscience Letters*, **114**, 248–52.

Smeets, W. J. A. J., Hoogland, P. V. & Voorn, P. (1986). The distribution of dopamine immunoreactivity in the forebrain and midbrain of the lizard *Gekko gecko*. An immunohistochemical study with antibodies against dopamine. *Journal of Comparative Neurology*, **253**, 46–60.

Smeets, W. J. A. J. & Jonker, A. J. (1990). Distribution of phenylethanolamine-N-methyltransferase-immunoreactive perikarya and fibers in the brain of the lizard *Gekko gecko*. *Brain, Behavior and Evolution*, **36**, 59–72.

Smeets, W. J. A. J., Kidjan, M. N. & Jonker, A. J. (1991). α-MPT does not affect dopamine levels in the periventricular organ of lizards. *NeuroReport*, **2**, 369–72.

Smeets, W. J. A. J. & Steinbusch, H. W. M. (1989). Distribution of noradrenaline immunoreactivity in the forebrain and midbrain of the lizard *Gekko gecko*. *Journal of Comparative Neurology*, **285**, 453–66.

Smeets, W. J. A. J. & Steinbusch, H. W. M. (1990). New insights into the reptilian catecholaminergic systems as revealed by antibodies against the neurotransmitters and their synthetic enzymes. *Journal of Chemical Neuroanatomy*, **3**, 25–43.

Specht, L. A., Pickel, V. M., Joh, T. J. & Reis, D. J. (1981). Light-microscopic immunocytochemical localization of tyrosine hydroxylase in prenatal rat brain. I. Early ontogeny. *Journal of Comparative Neurology*, **199**, 233–53.

Studholme, K. M., Yazulla, S. & Phillips, C. J. (1987). Interspecific comparisons of immunohistochemical localization of retinal neurotransmitters in four species of bats. *Brain, Behavior and Evolution*, **30**, 160–73.

Tauchi, M., Madigan, N. K. & Masland, R. H. (1990). Shapes and distributions of the catecholamine-accumulating neurons in the rabbit retina. *Journal of Comparative Neurology*, **293**, 178–89.

Tillet, Y. (1988). Adrenergic neurons in sheep brain demonstrated by immunohistochemistry with antibodies to phenylethanolamine-N-methyltransferase (PNMT) and dopamine-β-hydroxylase (DBH): absence of the C1 cell group in the sheep brain. *Neuroscience Letters*, **95**, 107–12.

Tillet, Y., Thibault, J. & Dubois, M. P. (1987). Immunocytochemical demonstration of catecholamine and serotonin neurons in the sheep olfactory bulb. *Neuroscience*, **20**, 1011–22.

Vaage, S. (1969). The segmentation of the primitive neural tube in chick embryos (*Gallus domesticus*). A morphological, histochemical and autoradiographical investigation. *Advances in Anatomy, Embryology and Cell Biology*, **41**, 1–88.

Vaage, S. (1973) The histogenesis of the isthmic nuclei in chick embryos (*Gallus domesticus*). I. A morphological study. *Zeitschrift für Anatomie und Entwicklungs-Geschichte*, **142**, 283–314.

Verney, C., Zecevic, N., Nicolic, B., Alvarez, C. & Berger, B. (1991). Early evidence of catecholaminergic cell groups in 5-and 6-week-old human embryos using tyrosine hydroxylase and dopamine-β-hydroxylase immunocytochemistry. *Neuroscience Letters*, **131**, 121–4.

Vigh-Teichman, I. & Vigh, B. (1974). The infundibular cerebrospinal fluid contacting neurons. *Advances in Anatomy, Embryology and Cell Biology*, **50**, 1–90.

Vincent, S. R. (1988). Distributions of tyrosine hydroxylase-, dopamine-β-hydroxylase-, and phenylethanolamine-N-methyltransferase-immunoreactive neurons in the brain of the hamster (*Mesocricetus auratus*). *Journal of Comparative Neurology*, **268**, 584–99.

Witkovsky, P., Eldred, W. D. & Karten, H. J. (1984). Catecholamine- and indoleamine-containing neurons in the turtle retina. *Journal of Comparative Neurology*, **228**, 217–25.

Wolters, J. G., ten Donkelaar, H. J. & Verhofstad, A. A. J. (1984). Distribution of catecholamines in the brain stem and spinal cord of the lizard *Varanus exanthematicus*: an immunohistochemical study based on the use of antibodies to tyrosine hydroxylase. *Neuroscience*, **13**, 469–93.

Index

Page numbers in *italics* refer to figures and tables. Abbreviations: CA = catecholamine; DA = dopamine; NA = noradrenaline; TH = tyrosine hydroxylase.

A1-17 system, 5, 465–70
N-acetyl-aspartylglutamate (NAAG), 308
acetylcholine, 125–6, 166, 258
acetylcholinesterase (AChE), 16
Acipencer ruthenus, 71
actinopterygian brains, 54, 71–2
adenohypophysial hormone secretion, 449
adrenaline, 1, *2*, 37, 443
  amphibians, 100
  reptile brain, 118–19
  retina, 470
  teleosts, 70
adrenergic input into basal ganglia, 264
β-adrenergic receptors, 309–10
adrenergic receptors, 171, 264, 309–10
adrenocorticotropin, 340
afferent innervation, hypothalamus, 444
Alzheimer's disease, 190
amacrine cells of birds, 162, 164
Amblystoma tigrinum, 358–9
aminergic neurons, 293–4
amphibians, *38*, 77, 396
  adrenaline concentration, 100
  brain structure innervation, 97–8
  CSF contacting cells, 99
  DA/NA fiber sites of origin, 98–9
  DAi cells, 85
  dopamine system, 85, 87–9, *90*, 91
  locus coeruleus cells, 96
  NAi cell bodies, 91–2
  noradrenaline system, 94–6
  olfactory bulb, 85, 99
  PNMTi cell bodies, 96
  PNMTi fibers, 96–7
androgens, 447, 453, 455, 456, 457, 458

Anolis carolinensis, 111, *114*, 115
anurans, 77, 78, 355–7
  brain development, 343–4, 355–7
  cell bodies, 78–82
  CNS development, 357–9
  DAi fibers, 82, 84
  development, 344, *345*
  lateral line system, 358
  midbrain DA cell groups, 99, 343
  retina development, 358–9
  spinal cord development, 357–8
  THi fibers, 82, *83*, 84
apomorphine, 168, 169, 446
arcuate nucleus, rat, 442, 444–6
arginine-vasotocin, 127
aromatic L-amino acid decarboxylase (AADC), 3, 4, 5, 189, 197, 443
  rat brain neuron development, 407–20
axoaxonal contact formation, 445

basal forebrain, CA cell groups, 472–3
basal ganglia, 247, *248*, 249, *252*, *253*, 255
  adrenergic input, 263–4
  adrenergic receptors, 264
  connectivity, 251, *252*, 253–4
  dopamine input, 249, *250*, 251
  dopamine receptors, 256–8
  dopamine role, 259, 262–3
  functional organization, 261–2
  human disease involvement, 263
  intrinsic organization, 251, 253–4
  noradrenergic input, 263–4
  noradrenergic receptors, 264
  thalamic neuron inhibition, 261
basal nucleus, monkey, 286
bed nucleus of stria terminalis (BNST), 251, 264

behavior
  initiation in striatal neurons, 261
  reproductive in birds, 164–5
  sexual dimorphism, 453
bichir, Senegal, 71
birds, *38*, 135
  adrenergic receptors, 171
  amacrine cells, 162, 164
  basal ganglia function, 168–9
  diencephalon, 165
  hypothalamus, 164–5
  mesencephalon, 165–6, *167*, 168–70
  olfactory bulb, 164
  pretectum, 165
  reproductive behaviour, 164–5
  rhombencephalon, 170–2
bombesin, 41
bony fish *see* teleosts
brain
  fish, 39–40, 54, 72
  mammalian, *192*, 193, *194*, 195–7, 398, 399
  rat, *394*, 407–20, 421, *422*, 423–6, *427*, 428–9
  reptile, 109–11, 114–16, 122–4
  vertebrate, 1, 97–8, 463–4
calbindin, 254, 299, 314
calcitonin gene-related peptide, 41
calcium-binding protein (CaBP), 281
cartilaginous fish, 21, 22, 23, 32–7, 39–41, 49
castration in rats, 446
catechol-*o*-methyl transferase (COMT), 3
catecholamine, 1, 2, 3, 4, 5, 13
  cartilaginous fishes, 23, 40–1
  CSF-contacting cell bodies, 474
  drugs affecting transmitter availability, 458, *459*
  innervation, 310–11

mammalian innervation, 376
peptide interaction, 13–15, 40–1
receptors, 3, 458, *459*, 478
research techniques, 464–5
transmission, 455, *459*
catecholamine neurons
  cell cultures, 455–6
  distribution maps, 406, *407–20*, 421–6
  GABAergic receptors, 458
  homology, 399
  mesencephalic in rat brain, 406, 421–3
  nicotinic receptors, 458
  ontogeny in mammalian CNS, 405–6
  sex-specific vulnerability, 458–9
catecholamine systems, 473–8
  analysis, 465, *466*, 467–70, 473
  sex-specific development, 453, *454*, 455–9
catfish, 49, *51*
caudal brainstem, rat, 424
caudal rhombencephalon, 425, 465, 467, 476
caudate-putamen *see* striatum
cell cultures, 455–6, 457
central nervous system, vertebrates, 1, 463–78
  basal forebrain CA cell groups, 472–3
  CA cell groups, 465–9, 470–3, 474, 475, 476
  CA system comparative analysis, 465, *466*, 467–70
  evolution of catecholamine systems, 473–6
  evolutionary considerations, 463–4
  segmental approach to catecholamine systems, 473
cerebral cortex, 296, *298*, 299, 472–3
  human, 278, 280, 300, *301*, 302
  primate, 295–6, 299–300, *301*, 304, 307, 308
  rat, 294–6, *298*, 306–8
cerebral cortex, mammalian, 293–4, *303*, 304–6, 311–12
  adrenergic receptors, 309–10
  DA afferents, 297
  DA neuron collaterals, 297, 299
  DA projections, 310
  noradrenergic system, 306–10
  noradrenoceptive population, 309–10
cerebrospinal fluid, CA transfer, 446
chick embryo brain, 375, 381–3, 393, 395–6
  commissuro-mammillary line, *384–5*, 391
  diencephalon, 385, *386*, 387, *389*, 390
  early stage development, 383, *384–5*, *386*
  intermediate stage, 383, 385, *386*, 387–8

isthmocerebellar domain, *386*, 387–8
isthmus domain, *389*, 390–1
late stage, 388, *389*, 390–1
mesencephalon, 382, *386*, 387
rhombencephalon, 382, *389*, 390–1, 393, 395
secondary prosencephalon, 383, 385, *386*, 388, *389*, 390
segmental localization pattern, 391, *392*, 393, 395
TH/DA cell and fiber development, 383, *384*, 385, *386*, 387–8, *389*, 390–1
Chinemys reevesii, 125
cholecystokinin, 14, 126, 127, 304
  DA colocalization, 314
  DA effects on release, 310
  DA neuron projections, 299
  midbrain, 186, 254
  receptors in monkey striatum, 281
choline acetyltransferase (ChAT), 125, 227
chondrichthyes *see* cartilaginous fish
clozapine, 258, 263
coelacanth, 53
column of Terni, 172
corticogenesis, 310
corticotropin releasing factor (CRF), 229, 230, 308
crocodilians, 103
cyclostomes, 7
cyproterone acetate, 458

diencephalon, 8–11, 165, 326, 468–9
  anuran development, 358
  cell culture DA neurons, 456–7
  chick, 385, *386*, 387, *389*, 390, 391
  pigeon, 136, *139–41*, 146–7, *153*, *156*, 157–9, *158*
diencephalon, mammalian, 207–8, *209*, 210–11, *212–18*, 219–21, *222*, 223, 225–7
  afferents to catecholaminergic neurons, 230
  choline acetyltransferase, 227
  DBHi fibers, 223, *224*, 225, 226, 235
  dopaminergic fibers, 223, 235
  functional implications of catecholaminergic neurons, 230–5
  neuronal system interactions, 227–30
  paraventricular nucleus, 233–4, 235
  peptidergic system interactions, 227–30
  pineal gland, 208, 223, *224*
  supraoptic nucleus, 233–4, 235
  thalamus, 225–6
  THi neurons, 210, 211, *212–14*, 219–20, 221, 235
  transmitter colocalization, *228*
3,4-dihydroxyphenylacetic acid (DOPAC), 189, 285
dihydroxyphenylalanine *see* DOPA

L-DOPA, 15, 189
DOPA, 2, 3, 5, 457
DOPA-ergic neurons, 306
dopamine, 1, 2, 4, 5, 16, 262, 263
  acetylcholine release, 258
  acquiring cells, 65, *66*
  agonists, 259
  antibodies against, 78
  basal ganglia, 168–9, 249, *250*, 251
  cells, 36, *275*, *276*, 282, 283–7
  cellular actions, 259–60
  colocalization with peptides, 314
  cortical system, 294–7, *298*, 299–300, *301*, 302, *303*, 304–6
  deamination, 285
  expression in hypothalamus, 443
  fiber distribution in amphibians, 78
  function in birds, 165
  gonadotropin secretion control, 449
  lamprey, 7, 12, 16
  LH regulation, 232
  mammalian midbrain, 189
  mesencephalic complex, 302, *303*, 304
  mesocortical, 294, 302, *303*, 304
  mesostriatal system, 281, 313–14
  neurotransmitter function, 63, *64*, 65
  opiate peptide-containing neuron interactions, 232
  origins in cerebral cortex, 297
  Parkinson's disease, 247, 260
  pigeon mesencephalon, *142*, 147–8, *161*
  prenatal development, 286
  pressor responses, 1
  rat embryonic hypothalamus, 446
  reactivity in teleosts, 58, *59*, 60
  release during development, 375
  reptile olfactory system, 122
  reptiles, 111, *112–13*, 114
  retina, 470
  striatal cells, 1, 280–2
  terminals, 296, 297, 299, 300
  transporter, 258–9
  unsteady expression, 284
  uptake mechanisms in striatum, 258–9
  varicosities forming synapses, 305
  vertebrate distribution, 37
  visualization methods, 294
dopamine β-hydroxylase (DBH), 69, 190, *191*, 193
  antibodies against, 77, 91, 92, 94
  antiserum, 92, 96
dopamine ß-hydroxylase (DBH)
  fibers in pigeon
    diencephalon, *151*, *156*, 159
    pretectum/mesencephalon, 160
    rhombencephalon, *143–5*, *152*, *156*, 160, 162, *163*
    spinal cord, *145*, *152*, 162
    telencephalon, 155, *156*, 157
dopamine fibers, 98–9, 313
  human cortex, 300, *301*, 302

dopamine fibers—*cont.*
  striatum, 255–6
  TH immunocytochemistry, 293–4
dopamine immunoreactivity (DAi), 99
  amphibians, 78–82, *83*, 84, 85
  fibers, 61–3, *67*, 82, *83*, 84
  neurons, 60, 61, 65, *67*
  paraventricular nucleus (PVO) cells, 60, 65
  teleost brain, 60, *62*, 68
dopamine innervation, 310, 311, 312
  cortical density, 296, *298*
  DA subpopulations in mesocortex, 296–7, *299*
  ontogeny in rat telencephalon, 283
  primate cerebral cortex, 299, *300*, *301*
  rat cerebral cortex, 294, *298*
dopamine neurons, 260–1
  A8-A10 cell groups, 254
  collaterals, 297, *299*
  dendrites, 254
  diencephalic cell culture, 456–7, *458*
  distribution maps in developing rat brain, 406, 421–5
  function of marginal zone/basal forebrain, 287
  genetic differences, 313–14
  mesocortical, 296, 302
  midbrain, 183, 254–5
  monkey, 285–6
  network, 232
  neurokinase effects, 260
  phylogenetic trends, 313–14
  rat telencephalon, 282
  sexual dimorphism, 443
  tegmental, 256–8
  in telencephalon, 273
  teleost, 65, *66*, 72
  terminals, 287, 294
dopamine receptors, 126, 256–8
  bird, 168, 170
  development in rat, 376
  location, 259–60
  mammalian cortex, 305–6
  mammalian neostriatum, 281
dopamine system
  amphibian, 78, 85, 87–9, *90*, 91
  function of cortical, 311–12
  lamprey, 16
  reptiles, 104, *105–6*, 107, *108*, 109–11, *112–13*, 114–16
  sex differences, 453
dopaminoceptive neurones, 305
dorsal bundle, 264
dorsomedial nucleus, rat, 441
dynorphin, 166, 255, 260, 261

eel, European, 49, *51*
efferent innervation, hypothalamus, 444–6
elasmobranchs, 21, *38*
β-endorphin, 14, 232

enkephalin, 41, 259, 260–1, 262
  birds, 166, 168–9
  Huntington's disease, 263
  striatal neurons, 255
Eptatretus burgeri, 16
17β-estradiol, 453, 457
estrogen, 231, 446, 447, 457, 458

Falck–Hillarp histofluorescence method, 405
cis-flupenthixol, 449
forebrain, *394*, *396*, 422
formaldehyde-induced fluorescence (FIF), 3–4, 464
  amphibians, 77
  bony fish, 49, 71, 72
  mammalian brain, 183, 207
  reptiles, 103

G protein, 256, 259
GABA, 255, 256, 261, 306
  bird mesencephalon, 166
  mammalian diencephalon, 230
  transmission blocking, 169
GABA-containing neurons, 233
galanin, 314
Gallotia galloti, 361–2, *369*, 372–6, 396–7
  CA system development, 362, *363–4*, 365, *366–7*, 368, *369*, 370–2
  DAi cell bodies, 362, *363–4*, 373–4
  embryonic brain development, 362, *363–4*, 365, *366–7*, 368, *369*, 370–2
  nigostriatal CA pathway development, 376
  stage 32 and 33, 362, *363*, 365, *366*
  stage 35, *363*, 365, *366*, *369*
  stage 36, *363*, 368, *370*
  stage 37, *364*, 368, 370
  stage 38, *371*
  stage 39 and 40, *364*, *371*, 372
  THi cell bodies, 362, *363–4*, 365, *366*, *367*, 368, *369*, 370–2, 373–4
  THi fibers, 365, 376
  THi/DAi cell groups, 368
gar fish, 71–2
Gasterosteus aculeatus L., 325
Gekko gecko, 104, *105–6*, 107, *108*, 109–22
  cells staining with antibodies against TH, 120
  ChAT immunoreactive cell bodies, 125
  DA immunoreactive cell bodies/fibers, 121
  DAi fibers, 123–4
  dopamine containing fibers, 109, 114, 115
  dopamine-acetylcholine interaction, 126
  NAi cells, 114
  NAi fibers, 116, 123–4
  NAi/DBHi cells, *117*, 118

NAi/DBHi fibers, 116, 118
PNMTi cell bodies, *117*, 118–19
PNMTi fibers, 119
serotonin fiber overlap, 125
striatum, 124
globus pallidus, 249, 251
glyoxylic acid localization technique, 4
goldfish, 49, *51*
gonadotropin hormone-releasing hormone positive (GNRHi) cells, 41
gonadotropin secretion, 339, 449
growth hormone releasing hormone (GHRH), 228, 234–5
Gymnophiona, 77

habenular region, 471–2
hagfish, 7, 16
haloperidol, 168
His boundary, 391, 398
histamine, 13, 125
holocephalians, 21, *28–30*, 30
5-HT3 receptor activation, 310
5-HT *see* serotonin
5-HTP, 197
human diseases, 263
Huntington's disease, 263
hypothalamic catecholaminergic systems, 435–41
  afferent innervation, 444
  androgen control, 446–7
  biochemical phenotype expression, 442–3
  efferent innervation, 444–6
  estrogen control, 446
  infundibular region, 438–40, 440–1
  insulin control, 447
  maternal hormones, 447
  nerve fibers, *437–9*, 440–1
  neurohormonal control, 446–9
  neurons, 441–6
  neuropeptide gene expression, 447
  sexual dimorphism, 443
  thyroid hormone control, 447
hypothalamic periventricular organ, 472
hypothalamo-hypophysial hormonal factors, 446
hypothalamus
  adenohypophysial hormone secretion, 449, 476–7
  birds, 164–5
  rat, 423–4, 426, *427*
  sexual differentiation, 447–9
immunohistochemistry, 2, 3, 4–5
  bony fish, 72
  mammalian brain, 183, 207
incerto-hypothalamic DA+ projection system, 165
indoleamines, 4
insulin, 447

lacertids, 362
Lampetra sp, 7

lamprey, 7, 8, 13–15, 16
  basal ganglia, 15
  brain, *9–10*, 12
  brainstem, 15–16
  diencephalon, 8–11
  motor system, 15–16
  noradrenergic system, 17
  olfactory system, 15
  rhombencephalon, 12
  spinal cord, *9–10*, *11*, 12, 15–16
  tyrosine hydroxylase (TH) distribution map, 7–12
  visual system, 15
LANT6, 255
lateral line, 53–4, 350, 352, 353, 358
Lepisosteus osseus, 71–2
Lewy bodies, 306
lizards, 111, 114, 115, 118
  adrenergic system, 119
  olfactory bulb, 375
lungfish, 53, 72
luteinizing hormone (LH), 165, 231–2
luteinizing hormone releasing hormone (LHRH), 14, 16, 229, 230, 231–2

mammals
  brainstem CA cell groups, *38*
  CA cell development comparison with chick, *392*, *394*, 397–400
matricial system, neocortical, 314
medial cell stream, 399
median eminence, rat, 445, 446
medulla oblongata, mammalian, *192*, *194*, 195–7
alpha-melanocyte-stimulating hormone, 449
melanophore-stimulating hormone, 339
alpha-melanotropin, 446
mesencephalon, 397, 398–9, 406, 421–3
  anuran development, 358
  birds, 165–6, *167*, 168–70
  chick, *386*, 387, 391
  dopamine receptors, 170
  limit *see* His boundary
  pallidal input in birds, 169–70
  pigeon, *141–3*, *156*, *158*, 159–60, *161*
  striatal neurons, 166
  teleost, 337
mesocortical system, dopaminergic, 293
mesomeres, 382
mesotocin, 127
met-enkephalin, 14
1-methyl-4-phenyl-1,2,3,6-tetrahydropyridine (MPTP), 285
alpha-methylparatyrosine (alpha-MPT), 121–2
midbrain, 468
  mammalian, 183, *184–5*, 186, *187*, 188–9
monoamine oxidase (MAO), 3

cartilaginous fish, 21, 23, 30
distribution in lamprey, 7
holocephalians, 30
mammalian midbrain, 189
spinal cord, 197
type A (MAO-A), 285
type B (MAO-B), 285
mormyrid fish, 49, *51*
myelomeres, 382

neo-striatum *see* striatum
neural circuits, sexually dimorphic, 453
neuroendocrine systems, 453
neurohypophyseal peptides, 127
neurokinase, 260
neurokinin A, 255
neuroleptics, atypical, 258
neuromeres, 382
neuropeptide gene expression, 447
neuropeptide Y, 126, 127, 233
  bird mesencephalon, 166
  cartilaginous fish, 41
  DA colocalization, 314
  locus coeruleus, 308
  mammalian diencephalon, 227, 229, 230
  reptiles, 126, 127
  striatal neurons, 255
neurotensin, 14
  coexpression in primate cortex, 304
  DA colocalization, 314
  DA neuron projections, 299
  interactions with DA, 311
  locus coeruleus, 308
  mammalian diencephalon, 228
  midbrain DA neurons, 254
neurotensin-containing neurons, 233
nitric oxide synthase, 255
noradrenaline, 1, 2, 3
  acquiring cells, 70
  amphibians, 78
  antibodies against, 78
  axons, 308, 309
  cell colocalization with neuropetides, 308–9
  coeruleo-cortical pathway, 308–9
  colocalization with peptides, 314
  distribution in vertebrates, 37
  dopaminergic neurons, 232
  fibers, 69–70, 78, 98–9, 116–18, 264, 313
  hypothalamus, 429, 443
  innervation, *298*, 306–8, 310–11
  lamprey, 7, 12
  LH regulation, 232
  localization, 4, 5
  NAi cell bodies, 69, 91–2, 99
  receptors in basal ganglia, 264
  reptiles, 116–18, 122
  retina, 470
  sex steroid effects, 447–8
noradrenaline neurons, 313–14
  distribution maps in rat brain, 425–6

rhombencephalic cell culture, 456, 457
sex steroids, 457
teleosts, 73
terminals, 294
noradrenaline system, 312–13
  amphibians, 94–6
  cortical, 306–10
  lamprey, 17
  reptiles, *105–7*
  sex differences, 453
noradrenoceptive population, cortical, 309–10

olfactory bulb, 469–70, 476
  amphibian, 85
  birds, 164
  dopamine cells, 164, 469, 470
  lizard, 375
  monkey, 282
  pigeon, 136, *137*, 150
  rat, 424
  reptile, 122
olfactory system, stickleback, 326, 338
olfactory tract, monkey, 273
ontogenesis, 435
opiate peptide-containing neurons, 232
mu opiate receptors, 281
osteichtyes, 49
ovipary, 362
oxytocin, 228, 447

pallidum, mammalian, 249
pancreatic polypeptide, 14
paraventricular nucleus, rat, 441
paraventricular organ, teleost, 337
Parkinsonian syndromes, 311
Parkinson's disease, 1, 2
  dopamine depletion, 247, 260, 311
  dopaminergic neuron loss, 259
  L-DOPA effects, 263
  Lewy bodies, 306
  locus coeruleus cell loss, 190
  NA dysfunction, 312–13
parvalbumin, 166, 255
peptidergic neurons, 228–30, 233–4
peptides, 13–15, 126–7
  reptiles, 126–7
  transmitters in mammalian diencephalon, 235
peripheral nervous system neurotransmission, 1
periventricular pretectal nucleus, stickleback, 338
phenylethanolamine-N-methyltransferase (PNMT), 3, 4, 5
  antibodies, 77
  biochemical phenotype expression in hypothalamus, 443
  immunoreactive cells, 70–1
  immunoreactivity (PNMTi), 96–7
  system in reptiles, *105–7*
photoreceptor cells, stickleback, 339

pigeon, 150, *151*, 153–5, *156*, 157–60, 162
  adrenergic cell groups, 149–50
  catecholamine distribution, 135–6
  DBH+ fibers in spinal cord, *145, 152*, 162
  diencephalon, 136, *139–41*, 146–7, *153, 156*, 157–9
  displaced ganglion cells (DGC) in retina, 164
  hypothalamic cell groups, 136, *139–41*, 146–7
  mesencephalon, *141–3, 142*, 147–8, *156, 158*, 159–60, *161*
  noradrenergic cell groups, 149
  olfactory bulb, 136, *137*, 150
  pretectum, *141–3*, 147, *156, 158*, 159–60, *161*
  retina, 136, 150, 162, 164
  rhombencephalon, *143–5*, 148–50, *152, 156, 158*, 160, 162, *163*
  spinal cord, *145*, 150, *152*, 162
  telencephalon, *137–42*, 150, *151*, 153–5, *156*, 157
  TH+ fibers in rhombencephalon, *143–5, 152, 156*, 160, 162, *163*
  TH+ fibers in spinal cord, *145, 152*, 162
pimozide, 449
pineal organ, 208, 339
pituitary hormone, 230
Pleurodeles waltlii, 78
  conditions of noradrenaline system, 94, 95
  DAi cell bodies, 81–2, *83–4*
  DAi fibers, 84–5
  DAi/THi cell bodies, 87
  DAi/THi fibers, *86*, 87, 88, 89
  DBH antibodies, 94
  locus coeruleus cells, 96
  NAi cell bodies, 92
  NAi fibers, 94, *95*
  NAi/DBHi fiber distribution, 95
  PNMTi cell bodies, *95*, 96
  PNMTi fibers, *97*
  striatum, 85
  THi cell bodies, 81–2, *83–4*, 85
  THi fibers, 84–5
Polypterus senegalus, 71
pons, mammalian, *184*, 189–90, *191–2*, 193
potassium ion channels, 259
preoptic area, rat, 447
pretectum, 471–2
  birds, 165
  pigeon, *141–3, 156, 158*, 159–60, *161*
primary motor cortex (M1), 313
progesterone, 230
prolactin, 232, 340, 447, 449, 455
proopiomelanocortin, 232
prosomeres, 382, 383
Protopterus, 72
Pseudemys scripta elegans, 119, 120, 121

CSF-contacting cells, 111, 121
DA containing cell bodies, 111
DA containing fibers, 114, 115, 116
DAi/THi cell bodies, 111, *112*
NAi cell body distribution, 118
NAi/DBHi cells/fibers, *117*, 118
PNMTi cell bodies, *117*, 119
Purkinje cells, 473
Python regius, 111, *113*, 114

Rana catesbeiana, 94–5, 96, 356–7
Rana ridibunda, 78, 343–4, 353, *354*, 355
  DAi fibers, 96
  DAi/THi cell bodies, 87, *90*
  DAi/THi fibers, 87–8, 89, *90*, 91
  DBHi fibers, 96
  development, 345
  locus coeruleus cells, 96
  premetamorphosis, *357*
  prometamorphosis, *356, 357*
  THi cell bodies, *354, 355, 357*
  THi fibers, 353, *354, 355, 357*
reproduction, CA neurone regulation, 230–2
reproductive behaviour, birds, 164–5
reptiles, 119–24, 125–6
  adrenaline distribution, 118–19, 128
  adrenergic cell bodies, 118–19
  CA cell development comparison with chick, 396–7
  CA system development, 361
  ChAT immunoreactivity, 126
  CNS, 103–4
  CSF-contacting cells, 111
  DAi/THi fibers, 114
  dopamine system, 104, *105–6*, 107, *108*, 109–11, *112–13*, 114–16
  major brainstem catecholaminergic cell groups, *38*
  neurohypophyseal peptide interactions, 127
  neurotransmitter systems, 124–7
  noradrenaline distribution, 116–18, 128
  noradrenaline system, *105–7*, 118
  peptide interaction with catecholamines, 126–7
  periventricular organ, 122
  PNMT system, *105–7*
  sensory system, 122–4
  spinal cord, 124
  striatum, 124
  visual system, 122–3
reserpine, 1
retina, 470, 476
  adrenaline, 470
  development in anurans, 358–9
  dopamine, 470
  dopaminergic neurons, 424–5
  neurotransmitters, 470
  noradrenaline, 470
  pigeon, 136, 150, 162, 164
  reptiles, 123
  THi cells, 164, 470

Xenopus laevis, 347, *348, 349*, 350
rhombencephalon, 465–8, 476
  bird, 170–2
  cell culture NA neurons, 456, 457
  lamprey, 12
  pigeon, *143–5*, 148–50, *152, 156, 158*, 160, 162, *163*
rhombomeres, 382
rostral rhombencephalon, 425, 467–8, 476

salbutamol, 447
sarcopterygian brains, 54, 72
schizophrenia, 263
septo-preoptic area, rat, 445
serotinergic system, stickleback, 339
serotonin, 13, 16, 125, 230, 233
sex steroids, 230, 231, 447–9
sex-specific development, 455–8
sexual differentiation, 453–9
  androgen theory, 457
  androgen-linked mechanism, 453
  catecholamines in, 447–9
  sexual dimorphism, 453
shark, 21, 22, 23
  monoamine oxidase (MOA), 30
  myelencephalon, 30–2
  spinal cord, 23, *24–7*, 30
  telencephalon, 23
Siren lacertina, 78, 96
  DAi cell bodies, 88
  DAi/THi cell bodies, 87
  DAi/THi fibers, 88, 89, *90*
  THi cell bodies, 87, *90*, 95
snakes, 111
sodium ion channels, 259
somatostatin, 14, 234–5
  bird mesencephalon, 166
  cartilaginous fish, 41
  mammalian diencephalon, 228
  processes in neostriatum, 281
  release stimulation by DA, 310
  striatal neurons, 255
song production, 165
spinal cord, 470–1
  development in anurans, 357–8
  pigeon, *145*, 150, *152*, 162
  reptile DAi/THi fibers, 111
sterlet, 71
stickleback, 49, *51*, 325
  72 h embryo, 326
  96 h embryo, 326–8, 337–8
  120 h embryo, 328–30, 338
  144 h embryo, 330, *331*
  adrenergic neurons, 328–9, *334*, 335
  adrenocorticotropin, 340
  CA cell group development, 326–32, *333, 334–5, 336*, 337
  CA neuron development, 325–6
  DAir fibers, 330, *331*–2, *334, 335*, 337, 338
  DAir/THir axons, 328, 330, 335
  DBHir neurons, 326, 330, 335
  diencephalon neurons, 326, 328

L-DOPA uptake by DAir neurons, 337
dopamine uptake by DAir neurons, 337
dopaminergic neurons, 325–6, 337, 338
four-day old larva, 335, 337
functional aspects of catecholaminergic system, 339–40
gonadotropin secretion, 339
liquor-contacting neurons, 328, 330, 335, 337, 339
melanophore-stimulating hormone, 339
mesencephalon, 337–8
neurotransmitter identity, 337
noradrenergic neurons, 326, 327, 328, *331, 334*, 335, 337
olfactory bulb THir neurons, 55
olfactory projections, 338
olfactory system, 326, 338
one-day old larva, 331–2, *333, 334–5*
optic tectum, 340
paraventricular organ-accompanying neurons, 326, 327, 328, 330, 335, 337, 338
periventricular pretectal nucleus, 338
photoreceptor cells, 339
pineal organ, 339
prolactin, 340
rhombencephalon neurons, 326
sensory system development, 338–9
serotinergic system, 339
substance P, 339
suprachiasmatic nucleus, 338, 340
thermoregulation, 340
THir cells, 332, *333*
THir neurons, 326, 328–32, *333, 334–7*
thyrotropin, 340
visual system, 338–9, 340
striatal neurons, behavior initiation, 261
striato-GPL neurons, 261–2
striatopallidal pathway, 262
striatum, 247, 254–5
dopamine fibers in rat, 422–3
dopamine uptake mechanisms, 258–9
dopaminergic innervation, *250*, 251
dopaminergic input, 254, 255–60
dopaminergic terminals, 251, 256
input to midbrain dopamine neurons, 256
mammalian dopamine cells, 280–2
neurons, 247, 249, *252*, 255, 256
receptors on neurons, 256–8
rim of dorsolateral, 286
teleost, 65, 67–9
terminals, 256
TH-IR cells, 281

TH-IR expression by DA cells, 284
ventral, 253, 254
striosomal system, limbic-related, 314
striosomes, 281, 283
substance P, 14, 255
bird mesencephalon, 166
cartilaginous fish, 41
catecholamine interactions in reptiles, 126–7
dopamine effects on neurons, 259, 260, 261, 262
GABAergic neuron influence, 262
stickleback, 339
substantia nigra, 2, 251, *252*, 253
mammalian brain, 398, 399
rat dopamine neuron development, 421
suprachiasmatic nucleus, 442, 444, 445
stickleback, 338, 340
supraoptic nucleus, rat, 447
synaptogenesis, precocious, 310

tamoxifen, 458
tardive dyskinesia, 311
telencephalon
anuran development, 358
lamprey, 8
mammalian, 273–4, *275–7*, 278, 280, 283–7
pigeon, *137–42*, 150, *151*, 153–5, *156*, 157
rat, 280, 282–3, 284
telencephalon, monkey, 274, *275, 276–7*, 278, *279*, 280
DA neurons, 273–4, 278, 286
TH-IR neurons, 273
teleosts, 49, 58, 325–6
adrenaline, 70
brain organization, 53–4
brainstem CA cell groups, *38*
catecholamines in brain, *52*, 54–5, *56–7*, 58, *59*, 60–3, *64*, 65, *66*, 67–71
DAir fibers, 61–3, 67
DAir neurons, 60, 61, 65, *67*
DBH immunoreactive cell bodies, 69
DBH immunoreactive fibers, 69
dopamine acquiring cells, 65
dopamine immunoreactivity, 58, *59*, 60, *62*, 68
dopamine innervation, 63
dopamine neurotransmitter function, 63, *64*, 65
dopaminergic neurons, 65, *66*, 72
formaldehyde-induced fluorescence (FIF) studies, 71, 72
immunohistochemical methods, 72
lateral line specialization, 53–4
lobe-finned, 49–50
mesencephalon THir neurons, 58
NA acquiring cells, 70
NA immunoreactive cell bodies, 69
noradrenaline, 69–70

noradrenergic neurons, 69–70, 73
olfactory bulb, 55
phylogenetic relations, 49, *50–2*, 53
PNMT-immunoreactive cells, 70–1
sensory specializations, 53
striatum, 65, 67–9
telencephalon, 53, 55
THir neurons, 55, *56–7*, 58
testosterone, 231, 453, 457
thalamus, 254, 261
thermoregulation of stickleback, 340
third ventricle, efferent innervation, 445–6
thyroid hormones, 447
thyrotropin, 340
thyrotropin releasing hormone (TRH), 229, 230
Tourette syndrome, 263
turtles, 103
tyrosine, 2, 3
tyrosine hydroxylase, 4
antibodies in amphibians, 77
antibodies in reptiles, 103–4
distribution map in lamprey, 7–12
dopaminergic innervation to striatum, 249
expression by cortical neurons, 306
expression during prenatal development, 286
fibers in column of Terni, 172
fibers in pigeon rhombencephalon, *143–5, 152, 156,* 160, *162, 163*
fibers in pigeon spinal cord, *145, 152,* 162
gene expression, 399
hypothalamus, 442
hypothalamus of pigeon, 136, 146–7
lamprey antibodies, 7
mammalian diencephalon, 227–8
mesencephalon of pigeon, *142,* 147–8, *161*
messenger RNA, 405, 421, 423, 446
monkey neostriatum, 281
neuron sexual dimorphism, 443
olfactory bulb of pigeon, 136, 150
pretectum of pigeon, 147, *158*
retina of pigeon, 136, 150
rhombencephalon of pigeon, *143–5,* 148–50, *158*
spinal cord of pigeon, *145,* 150
teleost brain, 54–5, *56–7,* 58, *59,* 60
tyrosine hydroxylase immunoreactivity (THi), 7, 8, 10–11, 12
cartilaginous fish brain, 23, 32, *33, 34, 35, 36–7*
cartilaginous fish motor system, 39
cell bodies in Anura, 78–82
fibers in amphibians, 82, *83,* 84
fibers in teleosts, 58
holocephalians, *28–30*
human cortex, 278, 280
lamprey, 15
multipolar cells, 31

tyrosine hydroxylase
    immunoreactivity—*cont.*
  neuronal cell bodies, 273
  neurons, 55, *56, 57*
  noradrenergic cells, 12
  sarcopterygians, 72
  shark brain, *24–7,* 31, *33*
  spiny dogfish, 31
  telencephalon, 273–4, *277, 278, 279–80*
  teleost hypothalamic cells, 55, 58

urodeles, 77, 78, 84–5, 99
  cell bodies, 81–2, *83–4*

vasopressin, 233–4, 447
vasotocin, 127
ventral striatum, 273, 281–2

Xenopus laevis, 78, 343–4, *346,* 347, *348,* 349–50, *351,* 352–3
  brain structure innervation, 97–8
  cell bodies, 78–82
  DAi cells, 78–82, *348, 349*
  DAi/THi cell bodies, 87
  DAi/THi fibers, *86,* 87, 89, 91
  DBHi fibers, *93*
  development, *345*
  immunoreactive cell bodies, *79–81*
  late embryonic stage, 344, *345–6,* 347, *348–9*
  lateral line, 350, 352
  liquor contacting cells, 344
  locus coeruleus cells, 96
  metamorphic climax, 352–3
  NAi cell bodies, 91
  NAi fibers, 92–4
  NAi/DBHi fiber distrubition, 95
  noradrenaline system, 94
  perikarya, 78, 82
  premetamorphic stages, 349–50
  prometamorphic stages, 350, *351,* 352, *353*
  retina, 347, *348, 349,* 350
  spinal cord development, 357–8
  THi cell bodies, *346,* 347, *351,* 355
  THi cells, 79–82, 349, *350,* 352
  THi/DAi cell bodies, 344, 347, *348,* 349
  THi/DAi cells, 350, 352
  THi/DAi fibers, 347, *348,* 352

zona incerta, rat, 441